Invertebrate Pathology

Invertebrate Pathology

EDITED BY

Andrew F. Rowley

Department of Biosciences, Faculty of Science and Engineering, Swansea University, Wales, UK

Christopher J. Coates

Department of Biosciences, Faculty of Science and Engineering, Swansea University, Wales, UK

Miranda M.A. Whitten

Institute of Life Science, Swansea University Medical School, Wales, UK

OXFORD
UNIVERSITY PRESS

OXFORD

UNIVERSITY PRESS

Great Clarendon Street, Oxford, OX2 6DP,
United Kingdom

Oxford University Press is a department of the University of Oxford.
It furthers the University's objective of excellence in research, scholarship,
and education by publishing worldwide. Oxford is a registered trade mark of
Oxford University Press in the UK and in certain other countries

Published in the United States of America by Oxford University Press
198 Madison Avenue, New York, NY 10016, United States of America

British Library Cataloguing in Publication Data
Data available

Library of Congress Control Number: 2021938123

ISBN 978–0–19–885375–6

DOI:10.1093/oso/9780198853756.001.0001

Printed and bound by
CPI Group (UK) Ltd, Croydon, CR0 4YY

For Brenda, Louise, and Dylan,
Susan and Lonán,
Jef and Jan.

Preface

Our goal for undertaking this book was to produce a taxonomically rich one-stop-shop for aspiring invertebrate pathologists. Concurrent molecular and bioinformatics developments over the last decade have catalysed a renaissance in invertebrate pathology. High-throughput sequencing, handheld diagnostic kits and the move to disruptive technologies has done so much to shed light on what imperils ecologically and economically favoured invertebrates.

Just like invertebrate-pathogen relationships, this book has evolved over time into two distinct sections. It is a fool's errand to try to cover the innate immune responses of all invertebrates and the weird and wonderful eccentricities of some taxa—as such—we decided upon a different approach in Part I (Chapters 1 and 2). We wrote these chapters from the point of view of the host—how do you defend against a seemingly endless diversity of biotic and abiotic insults? They achieve this by building physical barriers, reinforcing them when necessary, incorporating a surveillance system (cellular immunity) and deploying chemical weaponry (humoral immunity) when compromised. Some of these are fundamental processes, such as phagocytosis, that have survived throughout evolution, while others such as the prophenoloxidase system, have been lost. The concluding chapter in this section, Chapter 3, gives a brief overview on approaches to disease diagnosis. It is necessarily brief and focusses on approaches rather than the practicalities of each method.

For Part II, the objective is simple. We wish the reader to be able to select any chapter and gain a broad appreciation of the main diseases and pathologic manifestations of that invertebrate assemblage. To achieve this, we assembled leading experts in their field to provide succinct overviews particularly aimed at readers new to the field. We thank them for their perseverance during the gestation period of this volume.

Andrew F. Rowley, Chistopher J. Coates and Miranda M.A. Whitten
Swansea, UK, March 2021

Contents

List of Contributors

Heba Abdelgaffar Department of Entomology and Plant Pathology, University of Tennessee, USA

David Bass Centre for Environment, Fisheries and Aquaculture Science, UK

Kelly S. Bateman 2OIE Collaborating Centre for Emerging Aquatic Animal Diseases, CEFAS, UK

S. Anne Boettger Department of Biology, West Chester University, USA

David G. Bourne College of Science and Engineering, James Cook University and Australian Institute of Marine Science, Australia

Maria Byrne School of Life and Environmental Sciences, University of Sydney, Australia

Christopher J. Coates Department of Biosciences, Faculty of Science and Engineering, Swansea University, Wales, UK

Sarah C. Culloty School of Biological, Earth and Environmental Sciences, University College Cork, Ireland

Arun K. Dhar Aquaculture Pathology Laboratory, School of Animal and Comparative Biomedical Sciences, The University of Arizona, USA

Roberto Cruz-Flores Aquaculture Pathology Laboratory, School of Animal and Comparative Biomedical Sciences, The University of Arizona, USA

Keryn B. Gedan Department of Biological Sciences, George Washington University, USA

Andreas Heyland Department of Integrative Biology, University of Guelph, Canada

Juan Luis Jurat-Fuentes Department of Entomology and Plant Pathology, University of Tennessee, USA

Trevor Jackson AgResearch Ltd, Lincoln Research Centre, New Zealand

Shin-Ichi Kitamura Centre for Marine Environmental Studies, Ehime University, Japan

Diana L. Lipscomb Department of Biological Sciences, George Washington University, USA

Matt Longshaw Q2 Solutions, Central Laboratories, Scotland, UK

Heidi M. Luter Australian Institute of Marine Science, Australia

Sharon A. Lynch School of Biological, Earth and Environmental Sciences, University College Cork, Ireland

Audrey J. Majeske Department of Biological Sciences, Oakland University, USA

Jenny Makkonen Department of Environmental and Biological Sciences, University of Eastern Finland and Biological Safety Solutions Ltd./Oy, Finland

Shelagh K. Malham School of Ocean Sciences, Bangor University, Wales, UK

Remziye Nalçacioğlu Insect Virology, Karadeniz Technical University, Turkey

Almudena Ortiz-Urquiza Department of Biosciences, Faculty of Science and Engineering, Swansea University, Wales, UK

Cathie A. Page Australian Institute of Marine Science, Australia

Delphine Panziera Biointeractions and Plant Health, Wageningen University and Research, The Netherlands

Jirka Manuel Petersen Laboratory of Virology, Wageningen University and Research, The Netherlands

Jonathan P. Rast Department of Medical Biophysics and Department of Immunology,

University of Toronto, Canada; Department of Pathology and Laboratory Medicine, Emory University and Emory Vaccine Center, USA

Vera I.D. Ros Laboratory of Virology, Wageningen University and Research, The Netherlands

Andrew F. Rowley Department of Biosciences, Faculty of Science and Engineering, Swansea University, Wales, UK

Eugene Ryabov Bee Research Laboratory, USDA, USA

Nicholas W. Schuh Department of Integrative Biology, University of Guelph and Department of Medical Biophysics, University of Toronto, Canada

Jeffrey D. Shields Department of Aquatic Health Sciences, Virginia Institute of Marine Science, William & Mary, USA

Andy Shinn Benchmark R&D (Thailand) Ltd, Chonburi, Thailand and Centre for Sustainable Tropical Fisheries and Aquaculture, James Cook University, Australia

Hillary A. Smith College of Science and Engineering, James Cook University, Australia

L. Courtney Smith Department of Biological Sciences, George Washington University, USA

Linsheng Song Laoning Key Laboratory of Marine Animal Immunology, Dalian Ocean University, PRC China

Jacqueline L. Stroud Soil Systems Research Group, SRUC, Scotland, UK

Ghada Tafesh-Edwards Department of Biological Sciences, George Washington University, USA

Monique M. van Oers Laboratory of Virology, Wageningen University and Research, The Netherlands

Lingling Wang Laoning Key Laboratory of Marine Animal Immunology, Dalian Ocean University, PRC China

Nicole S. Webster Australian Institute of Marine Science, Townsville, Queensland and Australian Centre for Ecogenomics, University of Queensland, Australia

Miranda M.A. Whitten Medical School, Swansea University, Wales, UK

Zhuang Xue Laoning Key Laboratory of Marine Animal Immunology, Dalian Ocean University, PRC China

Zichao Yu Laoning Key Laboratory of Marine Animal Immunology, Dalian Ocean University, PRC China

PART I

Host Defences and Approaches to Disease Detection

CHAPTER 1

Host defences of invertebrates to pathogens and parasites

Christopher J. Coates, Andrew F. Rowley, L. Courtney Smith, and Miranda M.A. Whitten

1.1 Introduction

If I have seen further it is by standing on ye shoulders of giants,

Isaac Newton (1676)

Elie (Ilya) Metchnikoff (1845–1916) had a fabled moment of inspiration while the rest of his family went to the local circus when on holiday by the coast in Messina (Gordon 2016). During this time, he took the transparent bipinnaria larvae of a sea star and impaled them with rose prickles from his garden. The next day he observed cells (presumably coelomocytes) migrating to and encapsulating these foreign bodies. This and other experiments using a range of invertebrate models including the transparent water flea, *Daphnia*, resulted in his theory that phagocytosis is a key component of the cellular immune systems of all animals including humans (Metchnikoff 1893). So, not only was the science of cellular immunology firmly established but the use of invertebrates to study such events was also forged. Over 70 years later, Hans Boman et al. (1991) developed another invertebrate model, the larval stage of the cecropia moth, *Hyalophora cecropia*, to decipher the nature and diversity of haemolymph-borne humoral antimicrobial systems. These caterpillars, with their large volumes of haemolymph, made ideal animals to study the structure and biosynthesis of antimicrobial proteins

and peptides using the protein purification techniques available at that time. Boman and co-workers discovered nine factors (P1–9) and P9 consisted of two small molecules later named cecropins a and b that were the first animal-produced antimicrobial peptides (AMPs) to be identified (Steiner et al. 1981). From these observations arose the discovery of the myriad of AMPs in all animals regardless of their phylogenetic position. Finally, Jules Hoffmann and co-workers pioneered studies on Toll and their signalling pathways in *Drosophila* (e.g. Lemaitre et al. 1996) that were later identified as conserved in 'higher' animals. Both Metchnikoff and Hoffmann were awarded Nobel prizes in Physiology or Medicine in 1908 and 2011, respectively. These and other key researchers including Evelyn Bachère (Montpellier, France), Christopher J. Bayne (Oregon, USA), Edwin L. Cooper (UCLA, USA), Kenneth Söderhäll (Upsala, Sweden), Norman A. Ratcliffe (Swansea, UK), Timothy Yoshino (Wisconsin, USA), Gerardo Vasta (Baltimore, USA) and the late Valerie J. Smith (St. Andrews, UK) have provided the foundations of invertebrate immunology.

It is important to appreciate that the interest of most researchers who study invertebrate immunity is multifaceted. Some seek to trace the phylogeny of immunity, others are interested in the ecological consequences of the immune system (now referred to as 'ecological immunity') and others seek to develop model animals that may provide a better understanding of host pathogen interactions in humans. Several invertebrates have become prominent in this latter goal including the nematode,

Christopher J. Coates et al., *Host defences of invertebrates to pathogens and parasites*. In: *Invertebrate Pathology*. Edited by Andrew F. Rowley, Christopher J. Coates and Miranda M.A. Whitten, Oxford University Press. © Oxford University Press (2022). DOI: 10.1093/oso/9780198853756.003.0001

Caenorhabditis elegans, the fruit fly, *Drosophila* and the wax moth, *Galleria mellonella*. For instance, in *C. elegans*, with their short life cycle and a multitude of genomic tools, it has become a popular model to study innate immunity and human diseases including Alzheimer's and Parkinson's (Apfield and Alper 2018), while in *G. mellonella* their larvae are increasingly used to examine evolutionarily conserved host defences to human bacterial and fungal pathogens (e.g. Champion et al. 2016; Jemel et al. 2020; Lim et al. 2018) as well as identifying the immuno-toxicological responses to environmental factors such as toxins in harmful algal blooms (e.g. Coates et al. 2019; Chapter 2). One of the strategic goals of these approaches is to replace whole mammal models because of their high cost and ethical implications.

Textbooks on immunology rarely spend more than a few pages describing the principles of invertebrate immunology. Coming away from these often gives the reader the impression that this immune system is simple and non-specific in nature yet with over 1.3 million extant species of invertebrates, there is inevitably a great diversity of strategies to defend against disease. Furthermore, invertebrates have unique mechanisms of defence including the prophenoloxidase system only found in arthropods, molluscs, echinoderms and tunicates (e.g. Lu et al. 2014; Parrinello et al. 2015) and some forms of 'acquired' immunity where the host has an apparent specific heightened response on second exposure to a parasite or pathogen (e.g. Armitage et al. 2017; Pradeu and Du Pasquier 2018).

This chapter aims to provide a concise overview of the invertebrate defence mechanisms to parasites and pathogens, and how such agents can overcome and/or circumvent these defences to cause disease. Because there is extensive literature, the reader is directed to timely reviews for further detail as shown in the Recommended Further Reading section at the end of this chapter.

1.2 Social distancing—behavioural avoidance of infection, and other mechanisms to avoid infection of hosts

It is important for social invertebrates, such as bees, ants and wasps to have mechanisms to avoid contact with other infected conspecifics. Such behaviours have been widely studied in these insects and often referred to as 'social immunity' (Wilson-Rich et al. 2009; Cremer et al. 2018). Social immunity can be accomplished by a number of mechanisms including nest hygiene, which in some cases may disinfect and kill infected conspecifics (e.g. Pull et al. 2018), and by changes in the cuticular hydrocarbon content following disease that honeybees employ to distinguish between healthy and diseased workers (e.g. Salvy et al. 2001). When honeybee (*Apis mellifera*) colonies are subject to *Varroa* mite infestation they increase foraging activities for resin collection (Pusceddu et al. 2019). This resin is believed to reduce the microbial load in the colony—a phenomenon referred to as 'self-medication'. Bumblebees also avoid flowers containing parasites—an example of prophylactic avoidance behaviour (see Section 13.2). Finally, healthy Caribbean spiny lobsters, *Panulirus argus*, avoid sharing a 'den' with lobsters infected with the *Panulirus argus* virus 1 (Behringer et al. 2006) and although the mechanism of this avoidance behaviour is currently unknown, it is likely to be pheromonal in nature. Non-social invertebrates also show a number of activities to reduce the chance of infection. For example, female fruit flies, *Drosophila melanogaster*, threatened by parasitoid wasps oviposit into food with a high ethanol content that helps their progeny avoid parasitism (Kacsoh et al. 2013).

1.3 Outer coverings—chemical and physical defences

1.3.1 Integument

Many parasites and pathogens need to invade their hosts to achieve the appropriate environment to allow them to survive, replicate and in some cases complete their lifecycle. Hence, they employ both active and passive mechanisms to gain entry to the tissues of their hosts—using several routes to achieve this (Figure 1.1). Hosts have formidable barriers against invasion. The structure and complexity of the integument is highly variable within the invertebrates. In many species it consists of a single layer of epithelial cells sometimes with an outer thin cuticle. In such cases, there may be additional barriers against parasite invasion such as mucus. For instance, in the tube dwelling hemichordates,

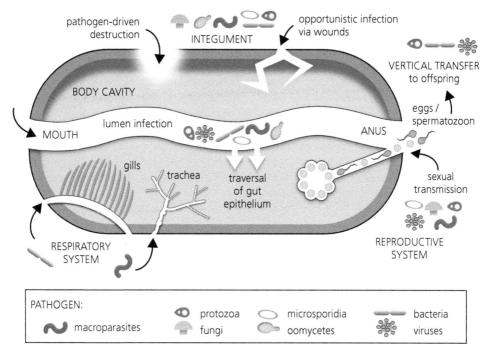

Figure 1.1 Major routes of invasion of parasites and pathogens in invertebrates with body cavities. The various pathogen and parasite types most commonly associated with each invasion route are indicated by the symbols. Oral ingestion (per os) and entry via the alimentary canal is a common strategy used by many types of pathogens and parasites. Occasionally, entry is via the anus. Infections may be localised to the gut lumen or result in systemic infection when the gut epithelium is traversed. The excreta of such hosts represent a very common vehicle for horizontal transmission of pathogens and parasites to fresh hosts. Opportunistic infections can also occur via pre-existing gaps or damage to the integument, and invertebrates are particularly vulnerable during moulting. The integument can also be breached chemically (enzymatically) or mechanically by representatives of most types of parasites and pathogens, particularly by fungi and macroparasites. Natural openings such as the gills or tracheal system are also exploited. Parasites and pathogens can also infect the reproductive tract and be transmitted sexually. Furthermore, representatives of the viruses, bacteria and protozoa are transmitted vertically to the host's offspring via eggs, spermatozoa, or reproductive tract secretions.

Saccoglossus spp., mucus is produced by glandular cells within the epithelium. It has a strong smell of iodine and has potent antibacterial activity associated with low molecular weight, heat stable molecules, possibility either halogenated indoles and/or bromophenols (King 1986; Millar and Ratcliffe 1987a). Agglutinins found in the mucus of these animals may also have a defensive function (Millar and Ratcliffe 1987b). Other antimicrobial factors in invertebrate mucus include lysozyme-like activity in some sabellid (polychaete) worms (Giangrande et al. 2014). Finally, heat sensitive proteins in the epidermal mucus layer of some molluscs are important in determining the outcome of interaction with parasitic trematodes. When mucus from both compatible and incompatible hosts is evaluated for protection against miracidial invasion by the trematode *Fascioloides magna*, mucus from resistant hosts rapidly kills the miracidia while that from susceptible hosts has no such activity (Coyne et al. 2015).

Arthropods and many nematodes are covered by a toughened outer covering, the cuticle. This is secreted by the epidermal cells and consists of the procuticle (exo- and endo-cuticle) mainly composed of chitin and protein with an outermost epicuticle that has waxes and lipids as its main components (Figure 1.2). In insects, this is composed of chitin and chitoproteins and is often flexible to permit some growth, while in crustaceans, such as crabs, the cuticle is hardened with the addition of calcium salts. The intact cuticle can only be penetrated directly by a few pathogens, mainly fungi and oomycetes. These produce chitinases and proteases to aid in their penetration through the cuticle. In many cases, there is a host reaction to the presence of carbohydrates in the surface coat of the hyphae that results in the activation of the host's prophenoloxidase system and the formation of melanin and toxic quinones (Figure 1.2; see Section 1.6.4). The cuticle of nematodes is generally thinner than that of arthropods (for example *ca.* 0.5 µm in diameter for *C. elegans*; Taffoni and Pujol 2015) and is predominantly formed from collagens, cuticulins, glycoproteins, and lipids. Mechanical injury to the cuticle results

in an increase in antimicrobial peptide gene expression in the underlying epidermis (Taffoni and Pujol 2015).

Urochordates (Phylum Chordata, Subphylum Urochordata) such as ascidians, are covered by a living tunic consisting of an extracellular matrix composed largely of tunicin, a cellulose-like polymer, plus a variety of host cells and symbiotic bacteria including cyanobacteria (Figure 1.3). The rigidity of the tunic is highly variable; for instance, in *Styela clava* it is leathery and extremely robust

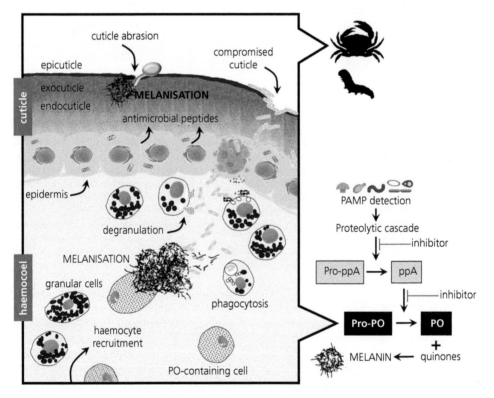

Figure 1.2 The arthropod cuticle—structure and defence mechanisms. The arthropod cuticle is a multifunctional suit of armour that surrounds and protects the underlying viscera, as well as providing the body with shape and stability. The outermost layer is the epicuticle, which is reinforced with proteins, melanin, and calcium salts. It is often waxy, as is the case for insect larvae. Should a pathogen breach the cuticle—depicted by the degradative (green) oomycete—melanin is synthesised to help seal the wound and coat the pathogen, thereby impeding invasion into the haemocoel. If the cuticle is compromised, then opportunists—depicted by blue bacteria—will gradually overwhelm the weakened tissues and make their way into the haemolymph. Circulating haemocytes that encounter the bacteria will respond in several ways depending on the cell sub-type present: (1) phagocytosis, (2) degranulation (factors target microbes, as well as recruit other haemocytes), (3) phenoloxidase (PO) release, (4) wound repair/haemolymph gelation. Inset, a stylised prophenoloxidase (proPO) activation cascade is depicted. Detection of microbial ligands (PAMPs) activates a serine-protease cascade in insects and decapods. A series of pro-phenoloxidase activating proteins (pro-ppA) are converted into their active forms (ppA), which ultimately convert the inactive proPO to PO. In the presence of mono- and di-phenols, melanin precursors (quinones) are generated rapidly. The quinone derivatives undergo cyclisation to form melanin.

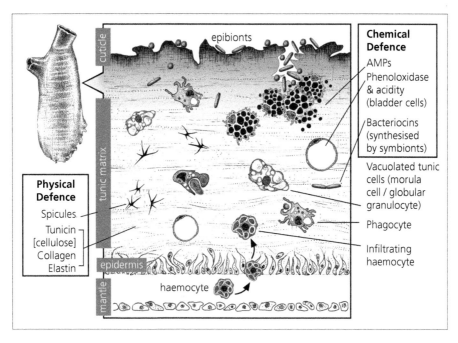

Figure 1.3 The tunic of urochordates—structure and defence mechanisms. The tunic combines physical, chemical, and cell-mediated defences, with contributions from the resident microbiota. The outer cuticle is a thin but tough layer of fibres in a watery matrix, and it may harbour harmless epibionts on its surface. Beneath is the tunic matrix or 'ground substance', a thick amorphous layer composed of tunicin (cellulose), collagen and elastin fibres secreted by epidermal cells at its base. In some species, defensive calcareous spicules are present. Scattered throughout the matrix is a variety of cell types that vary between species; only those with defensive roles are illustrated. Haemocytes infiltrate the matrix from the mantle, and thus an infected or damaged tunic appears 'inflamed'. Globular granular cells (morula cells) are highly characteristic and release several immune factors including phenoloxidase (PO). Clusters of granulocytes plug zones of damage (there is no coagulation) and release fibrogranular contents including AMPs. Other infiltrating cells are phagocytic and clear microbes and debris. The vacuoles of bladder cells occupy the majority of the cell volume and contain a defensive acidic material as well as PO. Antimicrobial factors such as bacteriocins are furthermore produced by resident symbionts as part of the natural tunic microbiota.

while in *Ciona intestinalis* the tunic is soft and transparent. In terms of structure, the outermost layer is thin but tough and referred to as the cuticle, while immediately underneath is the main tunicin-based matrix that in some species contains calcareous spicules that probably have a defensive role against predation (Hirose 2009). The cell types in the main tunic include phagocytic and non-phagocytic haemocytes, and tunic bladder cells that are highly vacuolated and contain an acidic material that may have general defensive properties (Figure 1.3). Other cell types include tunic net cells (myocytes), pigment cells, luminocytes, and phycocytes (Hirose 2009) that are probably not involved in defence against disease agents. Finally, the tunic contains a bacterial microbiome that as well as having biosynthetic ability of product generation

utilised by the host, also produces metabolites for chemical defence against predation and to give protection from invading pathogens (Chen et al. 2018; Dror et al. 2019).

1.3.2 Alimentary canal

The alimentary canal or gut presents a balancing act for all invertebrates because of the requirement to eliminate potential pathogens or disable toxins, while maintaining a benign commensal or beneficial mutualist/symbiotic microbiota. The gut has several vulnerabilities: it can be breached mechanically by microbes and parasites leading to systemic infection, it can suffer an imbalance in the diversity, location or population load of its microbiota (a dysbiosis), and toxins can cross intact gut barriers to affect the tissues (as discussed in Chapter 2). Although there is huge variety in gut anatomy and diet across phyla, several defences are core to all invertebrates, some of which are simply the consequence of processing food. Many ingested microbes do not survive the grinding activity in specialised foregut structures; other microbes are inactivated by digestive enzymes or unfavourable pH conditions (e.g. the highly alkaline midgut of many caterpillars), and others succumb to extreme redox potentials in specific gut regions.

Most invertebrates possess a three-part gut that consists of a foregut, midgut and hindgut (e.g. Emery et al. 2019). The gut barrier is formed by an epithelium composed mostly of enterocytes but also a small number of specialised cells such as endocrine cells and a pool of replenishing stem cells that maintain the integrity of the epithelium. Irrespective of whether the gut is complete, or blind ending (as in cnidarians and ctenophores for example), a common feature is a chitinous barrier on the luminal side that prevents microbes from directly contacting the gut epithelium. In arthropods and annelids, the foregut and hindgut are lined by a chitinous cuticle, and in the midgut, enterocytes continually secrete a delaminating peritrophic matrix or envelope (PM) composed of chitin fibrils embedded in a glycoprotein mesh. This mesh has a pore size of 20–300 nm, which is small enough to trap most microbes within the endoperitrophic space. There may also be a mucus layer between the epithelium and the PM. Other invertebrates have structures comparable to the PM, for example in *C. intestinalis* endostyle cells secrete a 'mucus net' composed of chitin, cellulose and mucin (Nakashima et al. 2018). Both the PM and the mucus net contain toxin-binding proteins (e.g. Kuraishi et al. 2011) and opsonic factors (Dishaw et al. 2016).

While effector molecules are commonly found in the milieu of the gut lumen, those molecules involved in non-self-recognition tend to be localised to the epithelium of the gut barrier. Pattern recognition receptors (Section 1.6.1) are concentrated on the epithelium, which trigger a range of signalling pathways upon detecting a potential pathogen. These pathways culminate in synthesis by enterocytes of microbicidal effectors such as antimicrobial peptides (AMPs; (see Section 1.6.2) and reactive oxygen species that diffuse into the lumen along with microbicidal lysozyme and antiviral factors (e.g. serine proteases and lipases). Antioxidant and stress-response factors are concomitantly triggered to protect host tissues from the deleterious effects of these species. Phenoloxidase activity resulting in melanisation (Section 1.6.4) is sometimes evident in the gut, though its source (local synthesis *vs.* the haemolymph) is seldom elucidated. It is thought that gut immunity is permanently activated at a low basal level (Krams et al. 2017) and when a pathogen is detected the response is up-regulated. Under certain specific conditions, with sufficient provocation, it may escalate to a systemic immune response (e.g. Zaidman-Rémy et al. 2006). Basal immunity is important because even the relationship a host has with its 'benign' and 'beneficial' microbiota is precarious. For example, in the bactivorous nematode, *C. elegans*, there is a strong inverse correlation between the load of commensal gut bacteria and worm longevity (Portal-Celhay et al. 2012); in other words as immune competence diminishes during ageing, the indigenous bacteria reproduce to dangerous levels. In honeybees, the commensal microbiota induces AMP release both locally in the gut and systemically into the haemolymph (Kwong et al. 2017), which also indicates that there is crosstalk between the rest of the body and the gut. Some invertebrates use diapause or moulting as an opportunity to reduce parasite and pathogen burdens, such as the sea cucumber *Parastichopus californicus*, whose gut atrophies along with many other organs in autumn. These organ remnants, together with any infecting pathogens and parasites, are shed via the cloaca at the start of the winter diapause (Fankboner 2003). Finally, anorexia, defensive diarrhoea, and vomiting are highly conserved behavioural responses to infection that usually result in improved invertebrate survival, despite the obvious drawbacks.

Although the vast majority of microbes and parasites pass through the gut without posing a direct threat to an invertebrate, they can nevertheless represent dangerous competition to the beneficial microbes, many of which help to maintain the health of their host (reviewed by Daffonchio et al. 2016). It is therefore not surprising that 'defensive symbionts' evolved strategies to eliminate competition in their gut niche and prevent dysbiosis, which indirectly facilitate host immunity in two key ways: (i) by priming innate immunity via pattern recognition receptor mediated mechanisms (Sansone et al. 2015), and (ii) by contributing their own antimicrobial factors to the gut lumen. Although not fully understood, the native microbiome of the bumblebee is protective against infection from the protozoan pathogen *Crithidia bombi* (Mockler et al. 2018). Secretion of mundticin, a type of bacteriocin antimicrobial peptide, by the symbiont *Enterococcus mundtii* in the gut of caterpillars selectively inhibits opportunistic pathogens such as *Enterococcus faecalis* but not commensal bacteria (Shao et al. 2017). Bacteriocins have been found widely in gut bacteria from insects and molluscs in particular, and such factors have useful commercial and medical applications.

1.3.3 Other routes of infection

Figure 1.1 illustrates that other routes of infection can occur, especially where there are openings through the integument. These include sensory pits, openings of the respiratory systems in insects and genital apertures. A recent discovery has highlighted the potential of openings termed nephropores of the antennal gland in crustaceans as a site of invasion by pathogenic viruses and bacteria (De Gryse et al. 2020). The antennal gland is part of an excretory system that exits in the head region of many crustaceans. Previous studies have reported infections within this system in crabs (e.g. Thrupp et al. 2013) suggesting that the mikrocytid causative agents gained entry and exit via the nephropores.

1.4 Wound healing and haemostasis

Invertebrates have two key mechanisms of haemostasis; (i) cellular aggregation without extensive plasma gelation as in annelids, molluscs, and urochordates (Figure 1.4), and (ii) cellular aggregation with extensive plasma gelation as in insects and crustaceans, and other arthropods. Echinoderms exhibit a mixture of both mechanisms These effectively help to close wounds to avoid infectious agents gaining access to the

internal tissues and to reduce fluid loss. This process of haemostasis has been extensively studied in arthropods (see Cerenius and Söderhäll 2011 and Eleftherianos and Revenis 2011 for reviews) and echinoderms (Bertheussen and Seljelid 1978; Hillier and Vacquier 2003; Cerenius and Söderhäll 2011). The first stage in this process in insects is the formation of a soft clot when granulocytes (granule-containing haemocytes) degranulate resulting in the formation of a sticky gel. This entraps potentially invading microbes at the site of wounding and plugs the opening. The next phase results in the formation of a hard clot when the prophenoloxidase pathway and transglutaminase are activated following granule exocytosis. At a later stage, phagocytic plasmatocytes attach to the clot and assist in the regeneration of the epidermal layer of the damaged cuticle. Essentially, this is the same series of events that occur during nodule formation and encapsulation of parasites and pathogens by insects (see Section 1.5.3).

The mechanism of haemostasis has also been investigated in sea urchins. Although these animals show the extensive plasma gelation reaction, under certain circumstances the coelomocytes can aggregate independently of protein clotting to form syncytia-like structures in response to wounding. This reaction involves at least two factors: (i) amassin (Hillier and Vacquier 2003) and (ii) arylsulphatase (D'Andrea-Winslow et al. 2012). Amassin is a 75 kDa plasma protein that aggregates through extensive disulfide-bond formation that traps cells in the protein clot. Its key importance in clotting has been demonstrated by the addition of a reducing agent that breaks the disulfide bonds, dissociates the amassin proteins, and releases the cells. Antiserum to amassin can block the formation of the protein clot and the aggregation of sea urchin coelomocytes (Hillier and Vacquier 2003). A recent proteomic study of coelomic fluid of the sea star, *Asterias rubens* failed to find any evidence of the presence of amassin suggesting that coagulation in sea urchins and sea stars may be distinct processes (Shabelnikov et al. 2019). Finally, when arylsulphatase, which is localised to aggregated sea urchin coelomocytes is inhibited, this results in the loss of clotting but its role in clot formation remains unclear.

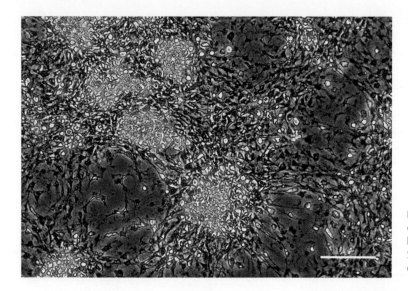

Figure 1.4 Cellular aggregation by circulating coelomocytes following bleeding in the starfish, *Asterias rubens.* Scale bar = 100 μm. Micrograph courtesy of Dr K. Hamilton.

1.5 Internal defences—cellular events

1.5.1 The cells and tissues of invertebrate immune systems

Most coelomate invertebrates contain freely circulating coelomocytes and/or haemocytes—some of which have defensive functions. There is an enormous diversity in such cells in terms of their morphology (Table 1.1; Figures 1.5 and 1.6) and function and this has resulted in a great deal of confusion in the literature brought about by incorrect or ambiguous assignment of their allocated names (see Box 1.1). Based only on the functional role of these cells, they can be classified as progenitors, phagocytic, haemostatic, nutritive, flagellated or pigmented cells (Table 1.1; Ratcliffe et al. 1985). Some invertebrates may only have a single type of haemocyte/coelomocyte in circulation (e.g. the gastropod *Crepidula fornicata*; Figure 1.5a) while others may have a larger number of distinct cell types. For example, the coelomic fluid of sea stars (class Asteroidea), including *Asterias rubens* consists of *ca.* 95% phagocytic amoebocytes, while in sea urchins (class Echinoidea) there is a multitude of cells types including a variety of phagocytes, coloured and colourless spherule cells, and vibratile cells with single flagellae (Figure 1.6; Smith et al. 2018). Similarly, haemocyte types in arthropods show great diversity of form and function. In lepidopterous insects, as well as phagocytic

plasmatocytes, there are also granular cells (= granulocytes and cystocytes/coagulocytes), spherule cells, and oenocytoids found in the haemolymph (Table 1.1; Figure 1.5b), while in crustaceans there is less diversity of haemocyte types with hyaline cells, semi-granular cells, and granular cells (Figure 1.5c). A pertinent example of immune cell heterogeneity in terms of morphology and function comes from sea urchins or echinoids more broadly (see Box 1.2). Not only is there great diversity in the numbers and types of cells in the body cavities of invertebrates but this complexity is amplified by the combination of the various sites of haemopoiesis and the developmental interrelationships of each cell type, which is unclear for many invertebrates.

Invertebrates also have fixed phagocytic and pinocytotic cells strategically located to deal with a variety of infectious and non-infectious agents found on circulation. For example, pinocytotic cells called nephrocytes are found in the central stem of crustacean gills. They appear to remove soluble products following the breakdown of microbial agents (e.g. Smith and Ratcliffe 1981). In several insect orders, collections of haemocytes are positioned within or adjacent to the incurrent valves of the heart (ostia), which is efficient for detecting and removing microbes due to the fast haemolymph flow (e.g. Sigle and Hillyer 2016). There are descriptions of 'phagocytic organs' in a variety of invertebrates including insects,

Table 1.1 Haemocyte and coelomocyte classification based on functional roles (Based on Ratcliffe et al. 1985)

Group	Progenitors	Phagocytic cells	Haemostatic cells (and mechanisms of haemostasis)	Nutritive cells	Pigmented cells	Miscellaneous cells
Annelids	Lymphocyte-like cells	Amoebocytes	Amoebocytes (cellular aggregation)	Eleocytes, chloragogen cells	Erythrocytes	
Molluscs	Leucoblast	Amoebocytes (granulocytes and hyalinocytes)	Amoebocytes (cellular aggregation)			
Insects	Prohaemocytes	Plasmatocytes, granular cells, lamellocytes and thrombocytoids	Granular cells, cystocytes, crystal cells (plasma gelation) Thrombocytoids in dipterans (cellular aggregation)	Spherule cells (?)		Oenocytoids
Crustaceans	Stem cells	Hyaline cells, semi-granular cells, granular cells	Semi-granular cells (mainly plasma gelation)		Cyanocytes (fixed)	
Echinoderms	Progenitors	Bladder amoebocytes	Amoebocytes (cellular aggregation)	Spherule cells?	Spherule cells	Vibratile cells, crystal cells
Urochordates	Stem cells, haemoblasts	Amoebocytes (granular and hyaline)	Amoebocytes (cellular aggregation)	Morula cells, compartment cells?	Pigment cells	

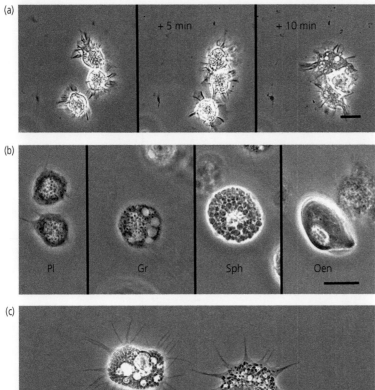

Figure 1.5 Examples of the morphological diversity of invertebrate haemocytes. (a) Amoeboid haemocytes of the gastropod, *Crepidula fornicata* viewed over a time series. (b) Haemocyte types in the lepidopteran, *Galleria mellonella* comprising plasmatocytes (Pl), granular cells (Gr), spherule cells (Sph) and oenocytoids (Oen). (c) Haemocyte types in the crustacean, *Carcinus maenas* comprising agranular hyaline cells (Hy), semi-granular cells (S-G) and granular cells (Gr). Scale bars = 10 μm.

echinoderms, and urochordates and some of these are haemopoietic in nature.

1.5.2 Phagocytosis

Phagocytosis by free and fixed cells of invertebrate immune systems has been widely studied both in terms of the mechanisms of recognition and intracellular killing. All invertebrates with a coelom and/or a haemocoel have professional phagocytes either freely circulating or fixed on or

within a particular tissue. The mode of intracellular killing of internalised foreign cells show distinct similarities with mammalian professional phagocytes in terms of mechanisms and molecules (e.g. Browne et al. 2013). These can be subdivided into oxygen-dependent and oxygen-independent mechanisms Oxygen-dependent products include various oxygen radicals and hydrogen peroxide (H_2O_2) generated by the respiratory burst (also known as the oxidative burst), and reactive nitrogen species such as nitric oxide (NO), while lysozyme and

---

Box 1.1 Nomenclature of freely circulating immune related cells in invertebrates—controversy and a lack of standardisation

There are numerous schemes employed to categorise these cells, from which a number of classification schemes can be identified including:

1. Classification based on morphology and behaviour of live cells (e.g. granular cells, spherule cells of insects, amoebocytes of numerous taxa.)
2. Classification based on proposed similarity to mammalian leucocytes (e.g. granulocytes, macrophages, lymphocytes, progenitor cells.)
3. Classification based on differential staining properties of fixed cells (e.g. basophilic granulocytes, eosinophilic granulocytes.)

Clearly, each of these schemes has pitfalls. In particular, the use of terms given for mammalian leucocytes is ill-advised as there is no evidence of a phylogenetic link between mammalian and invertebrate immune cells. Although the general concept of a 'macrophage-like' cell can describe an ancient lineage of phagocytic cells that appears to feature throughout the animal kingdom, the term is too broad to facilitate a useful classification in invertebrates, and there exist mammalian cell types within the same lineage that do not bear the name 'macrophage' or macrophage-like cell. Moreover, invertebrates do not have true lymphocytes, yet the term is still occasionally used. Despite attempts by some to propose a unified classification based on function alone (e.g. Ratcliffe et al. 1985; Table 1.1) these have not become established in the literature and a wide variety of names still exist even for the same species of invertebrate.

---

antimicrobial peptides are good examples of products of oxygen-independent pathways.

1.5.3 Nodule formation (nodulation) and encapsulation

The defence reactions of nodule formation and encapsulation function in invertebrates with body cavities, namely the haemocoel or coelom, which include the majority of haemocytes or coelomocytes, respectively. The process has been carefully studied in several arthropods including lepidopterans (e.g. *G. mellonella*) and crustaceans (e.g. the shore crab, *Carcinus maenas*) particularly with respect to the timescale and structural development. Nodules form when bacteria or other microbes gain entry into these cavities via a two-stage process. The initial phase is a rapid degranulation of sensitive cells such as granular cells or cystocytes (coagulocytes) in insects and semi-granular cells in crustaceans and is triggered by the presence of microbes and their products (Figure 1.7). Degranulation of these cells leads to the activation of the prophenoloxidase system resulting in melanin deposition around the microbes (Ratcliffe and Gagen 1976,1977; Figures 1.7 and 1.8a). The second phase of nodulation occurs several hours later when phagocytic cells such as

plasmatocytes and lamellocytes in insects, and hyaline and semi-granular cells in crustaceans, leave circulation, attach, and spread over the core of each nodule. Mature nodules, which take 24–48 hours to fully form, consist of many layers of enveloping cells that effectively wall-off the initiating microbes (Figure 1.8a; Ratcliffe and Gagen 1976). The process of encapsulation pertains to parasites and pathogens that are too large to be encased in a nodule or cannot be phagocytosed. The process is fundamentally the same as nodule formation in terms of cell types involved and mature capsules wall-off parasites and damaged tissues (Figure 1.8b).

1.5.4 Extracellular chromatin traps (ETosis)

Immune cell death is a fundamental part of invertebrate innate immunity (i.e. necrobiology)—whether it is apoptosis after degranulation, lysis to promote haemolymph gelation, or cellular self-sacrifice whilst ensheathing pathogens (see Section 1.5.3). The most striking example of this is the controlled release of nuclear-derived chromatin that functions as a net to entrap and kill pathogens, called extracellular traps (ETosis; first described as NETosis (neutrophils) by Brinkmann et al. 2004).

Box 1.2   Diversity among the defensive coelomocytes of echinoderms

Coelomocytes are cells in the coelomic fluid of echinoderms that mediate the immune response by recognising, phagocytosing, and neutralising pathogens. There are a variety of coelomocytes in sea urchins, sea stars, and sea cucumbers (Smith 1981; Ramírez-Gómez et al. 2010), however little is known about these cells in other echinoderms (reviewed by e.g., Smith et al. 2018). Coelomocytes are characterised based on morphology, with phagocytes, spherule cells (also called spherulocytes, amoebocytes, morula cells, and granulocytes), and vibratile cells receiving the most research focus (Figure 1.6). Haemocytes, progenitor cells (which are stem cell-like, also called lymphocytes based on their

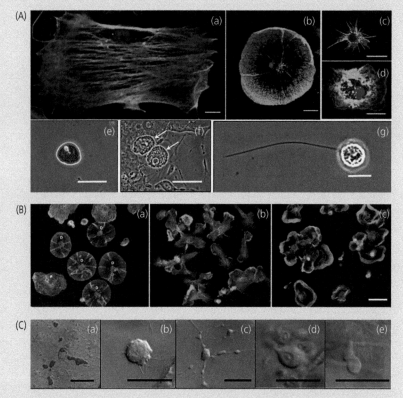

Figure 1.6   Immune cells in echinoderms. A. Coelomocytes in the adult sea urchin, *Strongylocentrotus purpuratus*, are (a) polygonal phagocyte, (b) discoidal phagocyte, (c) small phagocyte, (d) medium phagocyte, (e) red spherule cell and (f) colourless spherule cells from the sea urchin, *Paracentrotus lividus*, that are identical to those in *S purpuratus*, and (g) vibratile cell a-d. Immunocytochemistry of the phagocytes in A(a-d) shows the structure of the actin cytoskeleton (green), nuclear DNA (blue), and expression of the SpTransformer proteins (red).
B. Large phagocytes spread on glass show similar cytoskeletal structure from (a) the sea urchin, *Strongylocentrotus droebachiensis*, (b) the sea star, *Asterial forbesi*, and (c) the sea cucumber, *Sclerodactyla briareus*. Immunocytochemistry shows the actin cytoskeleton (green), nuclear DNA (blue), and tubulin (red).
C. Larval immune cells in the sea urchin, *Strongylocentrotus purpuratus*. (a) A pigment cell in stellate morphology in the ectoderm of a larva with echinochrome containing vesicles in the extended filopodia. Several types of immune cells are present in the larval blastocoel: (b) globular cell, (c) filopodial cell, (d) an ovoid cell, and (e) an amoeboid cell. Scale bars in Aa-d, g are 10 μm. Scale bars in Ae, f, B, and C are 20 μm. Source: the images in Aa, Ac and Ad are published with permission from © Golconda et al. (2019) *Frontiers in Immunology* 10: 870. doi: 10.3389/fimmu.2019.00870. Source: The image in Ab was originally published in Majeske et al. (2014). *Journal of Immunology* 195: 5678–5688. Copyright © [2014] The American Association of Immunologists, Inc. The images in Ag and B are published from Smith et al. (2018) with permission from Springer Nature; *Advances in Comparative Immunology*, E.L. Cooper, (ed.) © 2018. The images in C are published with permission from Ho et al. (2016) under the terms of the Creative Commons Attribution Non-Commercial License CC BY-NC.

Box 1.2 *Continued*

morphology), and crystal cells (reviewed by Smith et al. 2018) are also encountered, albeit in fewer echinoderm species. Phagocytes are the most abundant cell type in coelomic fluid, although their numbers vary with species, the individual animal, and its immunological status (Figures 1.6.Aa–d; B), and the cells can be further subdivided based on morphologies when spread on glass (Figure 1.6.Aa, b; B). Spherule cells are spherical when in suspension and become amoeboid when spread on glass. They possess large vacuoles (spherules), the contents of which determine cell colour (Figure 1.6.Ae, f). The striking red spherule cells from sea urchins contain the pigment echinochrome A and other naphthoquinones (Coates et al. 2018; Hira et al. 2020), while these chemicals are absent from the rarer colourless spherule cells. Vibratile cells are also spherical with large vesicles, and they are motile based on the activity of a single flagellum (Figure 1.6.Ag) (Pinsino et al. 2007).

The immune cells of planktonic echinoderm larvae are located in the ectoderm and the blastocoel, and studies with sea urchin larvae identified many shared characteristics between adult coelomocytes and larval immune cells (reviewed in Smith et al. 2018). For example, the larval pigment cells are similar to adult red spherule cells because both contain echinochrome, and both cells migrate to sites of infection or injury (Figure 1.6.Ca) (e.g., Coffaro and Hinegardner 1977; Ho et al. 2016). A variety of additional immune cell types are found in sea urchin larvae including globular cells (Figure 1.6.Cb), phagocytic filopodial cells (Figure 1.6.Cc), motile phagocytic ovoid cells (Figure 1.6.Cd), and amoeboid cells which are similar to the colourless spherule cells (Figure 1.6.Ce). These interact with other cells in the blastocoel during immune responses.

Coelomocytes in adult sea urchins have a broad range of immune functions. These activities include, but are not limited to, phagocytosis (Smith and Davidson 1994), syncytia formation (Majeske et al. 2013), secretion of clotting factors (e.g., Hillier and Vacquier 2003), and the promotion of cellular clot formation (e.g., Edds 1977; Henson and Schatten 1983). Coelomocytes also secrete or exocytose a range of antimicrobial molecules including echinochrome A, antimicrobial peptides, and antibacterial proteins and they can produce lysozyme, reactive oxygen species, prophenoloxidase, and hydrogen peroxide (Smith et al. 2018 and references therein. Phagocytosis of pathogens is a key activity of the cellular immune response in sea cucumbers and sea urchins (Eliseikina and Magarlamov 2002; Ramírez-Gómez et al. 2010), and it can be augmented by several factors including LPS-binding β-integrin (Wang et al. 2018), and a complement component homologue with opsonic activity (e.g. Clow et al. 2004). Nevertheless, the phagocytic capacity of coelomocytes and the associated production of reactive oxygen species (ROS) can be significantly reduced in echinoderms afflicted with some diseases (see Section 18.4.1).

Immune cells in larval echinoderms (Figure 1.6.C) act similarly to those in adults by responding quickly to pathogen invasion and by releasing a number of immune proteins with antimicrobial activity (e.g., Ho et al. 2016). Thus, in both phases of the life history of sea urchins, their effective immune system functions through the activities of the immune cells.

In the presence of LPS, bacteria, or the pharmacological agent phorbol 12-myrisate 13-acetate (PMA), invertebrate immune cells isolated from insects, crustaceans, molluscs, cnidarians, and annelids discharge chromatin meshes embedded with histone-derived antimicrobial cryptides, proteases, and homologues of myeloperoxidase to produce biocidal hypochlorous acid (Altincicek et al. 2008; Homa 2018; Ng et al. 2013; Poirier et al. 2014; Robb et al. 2014). The traps are fibrous in structure, with distinct globular domains. In shore crabs, there is strong evidence that encapsulation of the bacterium *Listonella* (*Vibrio*) *anguillarum* is initiated by ETosis in the gill lamellae, whereby the chromatin scaffold promotes haemocyte attachment and recruitment (Robb et al. 2014).

1.6 Internal defences—recognition and effector molecules

1.6.1 Signalling pathways

There is a wide variety of pattern recognition receptors (PRRs) on invertebrate phagocytes and other immune cells that recognise and bind to pathogen-associated molecular patterns (PAMPs) (also termed microbe-associated molecular patterns-MAMPs), including Toll receptors that were first

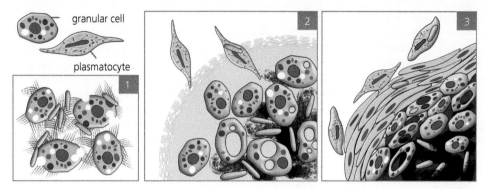

Figure 1.7 Cellular activity in nodule formation as seen in insects. (1) On random contact with bacteria, granular cells (pink) discharge their contents to entrap bacteria (green) in flocculent material forming a loose matrix. Clusters of cells and bacteria begin to form. (2) Triggering of the proPO activation cascade results in melanin deposition as the matrix and cell clusters become more compact. Large numbers of plasmatocytes (blue), some of which contain intracellular bacteria, are attracted to the early nodule. (3) A mature nodule is formed by 12–24hr, consisting of an outer region of newly attached cells, a middle region of extremely flattened cells, and an inner core containing melanised inclusions and trapped bacteria.

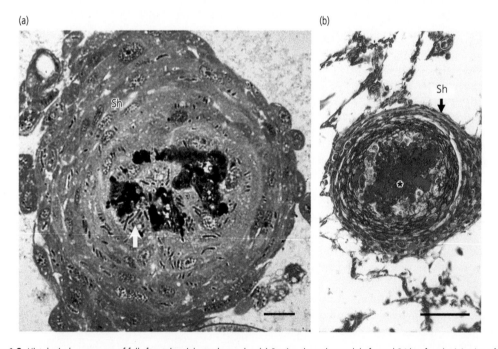

Figure 1.8 Histological appearance of fully formed nodules and capsules. (a) Section through a nodule formed 24 hr after the injection of *Bacillus cereus* into the lepidopteran, *Pieris brassicae*. Bacteria (arrow) are found in the inner melanised core surrounded by a multicellular sheath (Sh) of plasmatocytes to effectively wall-off the bacteria. Scale bar = 10 μm. (b) Encapsulation of necrotic material in the crab, *Carcinus maenas*. Note multicellular sheath (Sh) surrounding the melanised core (*) of necrotic material. Scale bar =100 μm.

discovered in insects (reviewed by Lemaitre and Hoffman 2007), peptidoglycan recognition proteins (Wang et al. 2019), and lectins among others (see Section 1.6.3). Membrane-bound PRRs may also indirectly recognise targets that are first coated in soluble, circulating PRR molecules that act as opsonins. Commonly examined PAMPs/MAMPs include LPS, lipoteichoic acid, peptidoglycan, and non-methylated DNA of bacteria, β-1,3-glucans in fungi and oomycetes, various glycans on the external surfaces of parasites, and viral DNA or RNA. The PRRs initiate a variety of conserved signalling pathways. In insects, the immune deficiency (Imd) and dual oxidase (Duox), c-Jun N-terminal kinase (JNK; stress response), Janus kinase/signal transducer and activator of transcription (Jak/STAT), the epidermal growth factor receptor (EGFR; homeostatic), and Notch (developmental) pathways are all triggered in response to pathogens either in contact with the gut epithelium or the cells within the haemocoel/coelom. Equivalent pathways have also been discovered in other invertebrates including the model nematode, *C. elegans* (Chávez et al. 2009; Ha et al. 2009; Portal-Celhay et al. 2012). Because these signalling pathways have already been exhaustively reviewed elsewhere (e.g. Hetru and Hoffmann 2009; Nie et al. 2018; Bang 2019) they will not be covered further in this chapter. Instead, the reader is referred to these and other reviews for more detailed information.

1.6.2 Antimicrobial peptides

Ubiquitously synthesised throughout nature, antimicrobial peptides (AMPs) are amphipathic and usually have a net positive charge (cationic), and typically contain fewer than 100 amino acid residues (~ 10 kDa or less). They exert direct pathogen killing activities as well as functions to detoxify microbial toxins, to modulate the host immune system, and they can act synergistically (e.g. Zanchi et al. 2017). Antibacterial, antiviral, antifungal, antiparasitic, symbiostatic, and even insecticidal, and antitumour activities have been ascribed to various AMPs. Their activities tend to be broad-spectrum, targeted to broad classes of pathogens rather than to a particular pathogen species. Hence, a single AMP can often kill a wide

variety of pathogens. Although there is huge diversity in the amino acid sequences and specificities of AMPs depending on the host species, AMPs form three structural groups and the mode of action falls broadly into two categories: (i) those AMPs that target and disrupt the pathogen surface—often but not always, a component of the cell wall (carbohydrates), membrane (lipid), or receptors (proteins); or (ii) those AMPs that interact with intracellular targets such as proteins, DNA, and RNA (reviewed in Wojda et al. 2020). The cationic and amphipathic structure promotes interactions with negatively charged moieties, especially in microbial membrane lipids and nucleic acids. Permeation and disruption of the membrane lipids may be the most common mechanism of invertebrate AMPs, with the caveat that the exact mode of action is rarely elucidated in detail. It is assumed that eukaryotic cells avoid self-targeting by self AMPs based on a more positively charged cell membrane. By contrast, infected, damaged, or neoplastic host cells exhibit changes in the plasma membrane such as exposure of negatively charged lipids that would be bound by AMPs.

Antimicrobial peptides are synthesised mainly by the fat body (insects), hepatopancreas (crustaceans), or equivalent tissues in other invertebrates, and are secreted into the haemolymph/coelomic fluid where they constitute the main humoral antipathogen defence factors. Tightly controlled local synthesis also occurs in tissues such as the gut, ganglia, respiratory system, reproductive tract, in haemocytes, and in venom glands. Synthesis can be constitutive or enhanced in response to infection. Commonly, an AMP gene is either intron-less, or has few introns, which is a hallmark of an early-response gene. This, of course, is beneficial because the synthesis of an immune factor must outpace the doubling time of the target pathogen. Multiple copies of AMP genes are often present, for example *D. melanogaster* has four antibacterial Attacin and two Diptericin genes (Hanson and Lemaitre 2020). The expression of AMPs is—in insects at least—regulated by the Imd NF-κB and Toll signalling pathways (Section 1.6.1). Less is currently known about the regulation of AMP expression in other invertebrates. Extensive lists of AMPs in various invertebrate phyla have been compiled recently for shrimp (Destoumieux-Garzón et al. 2016), insects (Wu et al. 2018), echinoderms (Li et al. 2015), oysters (Wang et al. 2018), and tunicates (Casertano et al. 2020).

1.6.3 Lectins/agglutinins

Lectins, sometimes referred to as agglutinins or haemagglutinins, are proteins that bind specific sugar residues as part of complex carbohydrates. Because of their fundamental roles, they are widespread in nature in microbes, plants, and animals. Lectins have been widely described in both acoelomate and coelomate invertebrates ranging from sponges (Gardères et al. 2015) to cephalochordates (see review by Gao and Zhang 2018). There are several main groups of lectins but in invertebrates most is known about C-type lectins and galectins (S-type lectins). C-type lectins are so-called because of their structure and functional requirement for calcium for binding based on their characteristic carbohydrate binding domains. Some C-type lectins may not require calcium for binding despite their original classification. Most have a single carbohydrate-recognition domain but some have two or multiple domains that give these molecules additional functional significance such as crosslinking or agglutinating target molecules. The numbers of different C-type lectins in single species can be highly variable. For instance, in insects this can range from four in the pea aphid, *Acyrthosiphon pisum* to forty in the yellow fever mosquito, *Aedes aegypti* (see review by Xia et al. 2019), whereas in the purple sea urchin, there are about 300 lectin coding exons in the genome (Hibino et al. 2006).

Galectins are a family of beta-galactoside-binding molecules that are identified by a unique sequence motif in the carbohydrate binding domain. They are evolutionarily conserved, and, like C-type lectins, they are widely distributed from protists to mammals (Vasta 2020; Vasta and Wang 2020). Invertebrate galectins often have multiple carbohydrate binding domains (Vasta et al. 2015; Table 1.2) and are expressed in a range of tissues including gills and haemocytes/coelomocytes and are frequently upregulated following microbial challenge (e.g. Zhang et al. 2020). Both C-type lectins and galectins are thought to be key recognition molecules in

invertebrates, and, in some instances, they may have direct killing ability of various microbes (e.g. Su et al. 2020). In the case of galectins, it is thought that they can cross-link glycans on the surface of haemocytes (i.e. 'self' molecules) with 'non-self', carbohydrate-containing molecules on the surface of viruses, bacteria, fungi, and parasites that enhances phagocytosis and clearance from circulation (Vasta and Wang 2020).

Ficolins are one of the several initiating proteins of the complement lectin pathway that function through carbohydrate binding. They contain two domains, a C-terminal fibrinogen-like and a collagen domain. Ficolin homologues have been found in advanced deuterostomate invertebrates including urochordates (Franchi and Ballarin 2017), cephalochordates (Huang et al. 2011), and echinoderms (Hibino et al. 2006) and there may be a functional lectin pathway in these invertebrates (Li and Xu 2015). However, the end products generated and their importance in the overall immune response is still unclear, and they may act to amplify opsonisation rather than direct microbial killing.

1.6.4 The phenoloxidase activating system

Melanisation of damaged tissues and pathogenic intruders is a key modality of invertebrate allostasis. In crustaceans and insects, the biogenesis of melanin is invariably a result of the prophenoloxidase (proPO) activation cascade (reviewed by Cerenius and Söderhäll 2004; Whitten and Coates 2017; Figure 1.2). Briefly, microbial ligands (e.g. LPS, β-glucans) are recognised by receptors embedded along the haemocyte surface, and their activation triggers an intracellular signalling cascade that ultimately leads to the extra-cellularisation of phenoloxidase (PO) via degranulation. PO activation is coordinated by serine proteinases and protease inhibitors regulate the cascade to avoid collateral damage. The generated melanotic polymers are microbiostatic—binding to pathogen surfaces, retarding growth and movement throughout the haemocoel. Such polymers are also deposited at sites of mechanical injury and clot formation to act as a sealant. The toxic by-products and unstable quinone intermediates of melanin synthesis assist

Table 1.2 Functions of galectins in selected invertebrates

Source (and nomenclature)	No of carbohydrate binding domains	Agglutinating activity	Microbial and carbohydrate binding activity	Microbial killing activity	Other immune activities	Reference(s)
Insects						
Silkworm, *Bombyx mori* (*Bm*Galectin-4)	2 (CRD1 and 2 low sequence similarity)	+ (G+ and G-bacteria)	LPS, lipoteichoic acid, peptidoglycan, laminarin + (G + and G- bacteria)	–*	Expressed by haemocytes, highly expressed in fertilised eggs. *E. coli* and *Staph. aureus* induces reduction in expression of Bm Galetin-4 by haemocytes	Rao et al. 2016
Drosophila melanogaster (Dmgal)	2				Haemocyte and neural tissue expressed	Pace et al. 2002
Crustaceans						
Kuruma shrimp, *Marsupenaeus japonicus* (*Mj*Gal)		–	LPS, lipoteichoic acid + (G+ and G- bacteria)		Promotes *in vivo* clearance of *Listonella* (*Vibrio*) *anguillarum* and phagocytosis. RNAi silencing of *Mj*Gal results in reduced bacterial clearance	Shi et al. 2014
Shrimp, *Litopenaeus vannamei* (*Lv*Gal)	1	+ (G+ and G- bacteria)	+ (mainly against a G- bacterium, *L. anguillarum*)	–*	Promotes *in vivo* phagocytosis of bacteria	Hou et al. 2015
Molluscs						
Eastern oyster, *Crassostrea virginica* (*Cv*Gal1)	4				Facilitates recognition of bacteria and the pathogen, *Perkinsus marinus*	Feng et al. 2013; Tasumi and Vasta 2007; Vasta and Wang 2020

continued

Table 1.2 *Continued*

Source (and nomenclature)	No of carbohy-drate binding domains	Agglutinating activity	Microbial and carbohydrate binding activity	Microbial killing activity	Other immune activities	Reference(s)
Pacific oyster, *Crassostrea gigas* (*Cg*Gal 2 and 3)	2	+ (*Cg*Gal-3 only)	LPS, glucan, peptidoglycan (*Cg*Gal-3) + (G- bacteria and fungi—*Cg*Gal-2) + (G+ and G- bacteria and fungi—*Cg* Gal-3)		Both galectins promote *in vitro* encapsulation of agarose beads	Huang et al. 2018
Bay scallop, *Argopecten irradians* (*Ai*Gal1/2)	4	+ (G+ and G- bacteria)			Promotes encapsulation of agarose beads	Song et al. 2011
Razor clam, *Solen grandis* (*Sg*Gal-1)		+ (G+, G- and fungi)	Peptidoglycan, beta-glucan		Promotes *in vitro* encapsulation of agarose and phagocytosis of *E. coli*	Zhao et al. 2018
Echinoderms						
Sea cucumber, *Apostichopus japonicus* (*Aj*Gal-1)	3	+ (G+ and G-bacteria and fungi)	LPS, peptidoglycan, mannose	+ (G+ and G-bacteria) but not fungus tested (*Pichia pastoris*)		Zhang et al. 2020

* N.B. authors employed a relatively insensitive method to assess this, so results may not be correct.

in killing Gram-positive and Gram-negative bacteria, yeast, and oomycetes (Cerenius et al. 2010; Coates and Talbot 2018; Quinn et al. 2020). Several enzymes with PO-like activity have been characterised among invertebrates. Chelicerates and some molluscs do not seem to possess a true PO but instead rely on laccase or haemocyanin-derived PO activities (Coates and Costa-Paiva 2020). Tyrosinase and haemocyanin in the presence of certain allosteric factors (Nillus et al. 2008) catalyze the hydroxylation of mono-phenols into diphenols. Tyrosinase, catecholoxidase, laccase, and haemocyanin-derived PO are all capable of oxidising diphenols into quinones, which undergo further enzyme-dependent/independent reactions to form indole-based eumelanin in the haemolymph. Alternatively, melanin is directed to the cuticle (or insect gut) to reinforce these multi-layered tissue barriers.

1.6.5 Lysozymes

Lysozymes (E.C. 3.2.1.17) are a large family of hydrolytic enzymes that hydrolyse sugar residues found in microbial peptidoglycan. Hence, Gram-positive bacteria with peptidoglycan exposed on their surfaces are rapidly damaged by exposure to these enzymes. Invertebrates have three main types of lysozymes, namely c (chicken egg or conventional)-type, g (goose egg)-type and i (invertebrate)-type (Van Herreweghe and Michiels 2012). The i-type lysozymes have been described in a wide range of invertebrates including sponges, molluscs, annelids, nematodes, arthropods and hemichordates (Van Herreweghe and Michiels 2012) while g-type lysozymes appear to be restricted to molluscs and urochordates (Di Falco et al. 2017), and c-type to arthropods and molluscs (Van Herreweghe and Michiels 2012). Some invertebrates have multiple isoforms of i-, c- and g-type lysozymes with differing sites of synthesis in the body, perhaps inferring some functional diversity. For instance, Pacific white shrimp, *Litopenaeus* (*Penaeus*) *vannamei* have three lysozyme genes *lyz*-c, *lyz*-i1 and *lyz*-i2, and the resulting products differ in their amino acid sequences and their lytic activity against different Gram-positive and Gram-negative bacteria (Peregrino-Uriarte et al. 2012; Chen et al. 2016). Of interest is the observation

that some invertebrate lysozymes have activity not just against Gram-positive bacteria but also Gram-negative forms in which peptidoglycan may be shielded including various species of vibrios (e.g. Chen et al. 2016). It is widely accepted that lysozymes are important in both immune defence and digestion. There are numerous reports that lysozyme gene expression in invertebrates can be up- and occasionally down-regulated following various immune challenges (e.g. Chen et al. 2016; Xie et al. 2019).

1.6.6 Lipoproteins

Lipoproteins are lipid/protein complexes in invertebrates that function to transport lipids among tissues through the aqueous medium of the haemolymph/coelomic fluid. Many lipoproteins are highly conserved throughout evolution, and some are also remarkably multifunctional immune factors. The ability to act as PRRs by binding common PAMPs such as microbial β-1,3-glucan is particularly notable (e.g. Duvic and Söderhäll 1990; Duvic and Brehélin 1998). Lipophorin, which is a type of lipoprotein, is present in insect haemolymph and is a complex of three apolipophorins: apoLp-I, apoLp-II, and apoLp-III. There are three categories of lipoproteins in crustaceans, of which high-density lipoprotein/β-glucan binding protein (HDL-BGBP; may appear as βGBP in older literature) is the most immunologically relevant. The description of lipophorins by Wynant et al. (2014) as 'general scavengers for pathogens' is extremely apt. Insect apoLp-III binds to microbial cell wall components including LPS, lipoteichoic acid, β-1,3-glucan, and even dsRNA (a viral PAMP). The PAMP binding activity results in detoxification of LPS, bacterial and fungal cell clumping, and activation of AMP synthesis. Insect apoLp-III is taken up by granular haemocytes and appears to induce degranulation, and furthermore it can influence directly haemocyte adhesion properties, and enhance encapsulation/nodulation (Whitten et al. 2004 and references therein). Specific receptors for binding β-glucan-associated HDL-BGBP complexes have been identified on insect and shrimp haemocytes, and both apoLp-III and HDL-BGBP regulate PO activity (Section 1.6.4) via β-glucan binding. In tarantulas (Chelicerata), where phenoloxidase enzymes are lacking, high-density lipoprotein HDL-2 converts the oxygen-carrying molecule haemocyanin into a catecholoxidase with PO-like immune activity (Schenk et al. 2015). Lipoproteins also contribute to haemolymph clotting in chelicerates, insects, and crustaceans (Duvic and Brehélin 1998; reviewed by Hoegar and Schenk 2020).

1.7 Does the invertebrate immune system have specificity and memory?

One of the hallmarks of the adaptive immune system of jawed vertebrates is acquired immunity based on somatic recombination and clonal proliferation of lymphocytes following exposure to various antigens and pathogens. This is the basis of vaccination whereupon secondary exposure to the same agent there is a heightened and antigen-specific reaction. Despite various experiments pointing to the possibility of some form of specific (acquired) immunity in invertebrates in the 1960s (see Pradeu and Du Pasquier 2018 for a critical review), by two decades later it became clear that invertebrates do not have the 'machinery' for recombination of antibody-like genes, nor do they undergo true clonal amplification of lymphocytes or other immune cells. Despite an apparent lack of a molecular mechanism to explain how upon a second exposure to a pathogen an invertebrate can elicit a specific immune response, a number of researchers found evidence of such an event, particularly in arthropods. They christened this phenomenon by various terms including 'immune priming', 'specific immune priming', 'trans-generational immune priming', 'innate immune memory' and 'strain-specific immunity' (e.g. Little and Kraaijeveld 2004; Sadd et al. 2005; Rowley and Powell 2007; Tetreau et al. 2019; Lafont et al. 2020). While immune priming without any specificity can be easily explained by short-term elevation in the numbers of haemocytes/coelomocytes and/or transcriptional changes in antimicrobial peptides and other killing factors, specific immune priming when the host defence is elevated only occurs following challenge with the same foreign agent, cannot be explained by such

phenomena. For instance, while AMPs may have selectivity in activity at a wide level, for example Gram-negative *vs.* Gram-positive bacteria, they do not differentiate in their mode of action between closely related species. Because some of these initial reports had no realistic mechanistic explanation of their observations, this led to scepticism of the validity of such descriptions (e.g. Hauton and Smith 2007).

Watson et al. (2005) found a potential molecular mechanism for some of these observations in insects based on a surprising candidate, the Down syndrome cellular adhesion molecule (Dscam) that functions in the ontogeny of the nervous system in *Drosophila* (Schmucker et al. 2000). The gene responsible for Dscam (*Dscam1*) is composed of about 100 exons, of which some are alternatively spliced to code for large numbers of isoforms These isoforms are thought to have variable binding capacity to various PAMPs (i.e. they act as PRRs). Watson et al. (2005) and other groups working with other model species were able to show that specific changes in levels of phagocytosis mediated by Dscam isoforms provided a potential explanation for the apparent specificity of the phenomenon of 'specific immune priming' in these arthropods (see Box 1.3). Hence, 'specific immune priming' may be based on directed alternative splicing to generate antigen-specific variants of Dscam to act as PRRs and/or opsonins in the interaction between professional phagocytes and specific microbes.

Trans-generational (usually maternal) immune priming in which there is a level of specificity of action is a more difficult phenomenon to explain through mechanistic models, and while various suggestions have been made (e.g. epigenetic mechanisms) no consensus exists for invertebrates. Simple transfer of immunity from parent (mother) to offspring, without any specificity, can be accomplished by, for example, the transfer of antimicrobial peptides from mother to eggs during oogenesis (e.g. Esteves et al. 2009) while reports of quasi-specificity of trans-generational changes in immune status (e.g. Little et al. 2003) are currently not fully explained and like other aspects of this field are still controversial. Indeed, when the model host arthropod, *Daphnia magna* was evaluated in response to various strains of the model pathogen,

---

Box 1.3 Down syndrome adhesion molecule—a generator of antigen binding diversity in arthropods?

It is now increasingly recognised that although invertebrates lack functional immunoglobulin and true lymphocytes, some groups are capable of some form of specific immunity. One of the molecules that function in this novel form of acquired immunity is the Down syndrome cellular adhesion molecule (Dscam) of some arthropods (Watson et al. 2005; Armitage et al. 2017; Ng and Kurtz 2020). The hypervariable gene form (*Dscam-hv* = *Dscam1*) has only been found in crustaceans (various species of shrimp, crabs and water fleas) and in insects (e.g. bees, flour beetles, fruit flies, mosquitoes, and silkworms), while in chelicerates (ticks) and myriapods (centipedes) it has limited alternative splicing activity (Armitage et al. 2017). Alternative splicing of the variable exons of *Dscam1* can result in large numbers of variant protein forms. For example, in *D. melanogaster* there is an estimated 38,016 isoforms of Dscam (Ng and Kurtz 2020) while in the Chinese mitten crab, *Eriocheir sinensis* there are 30,600 potential isoforms (Li et al. 2018). The largest number of potential isoforms of Dscam is in shrimp, *Litopenaeus vannamei* with 116,288 isoforms (Chou et al. 2009). The Dscam protein contains 10 immunoglobulin (Ig1–10) and 6 fibronectin type III (FNIII-1–6) domains characteristic of the immunoglobulin superfamily proteins, together with a transmembrane domain. The Dscam protein may occur as a membrane-bound form putatively acting as a hypervariable pattern-recognition receptor (PRR) on various cell types including phagocytes, or as a tail-less secreted form, in some crustaceans (Chou et al. 2009; Wang et al. 2013) perhaps with opsonic activity. Experimentally induced infections usually result in upregulation of *Dscam-hv* transcription with variable time courses depending on the host species. The resulting isoforms are thought to selectively bind to the original foreign agent used in the priming regime (Meijers et al. 2007; Wojtowicz et al. 2007, Sawaya et al. 2008; Li et al. 2018). It is therefore the interaction between these isoforms with a range of microbes and parasites plus the host phagocytes, that likely results in the enhanced levels of pathogen-specific phagocytosis in insects and crustaceans following the initial infection (e.g. Watson et al 2005; Pope et al. 2011).

Whether the universality of Dscam-mediated specific changes in disease resistance is a character of pancrustaceans (insects and crustaceans) is not known. It has been argued that the estimated variable number of *Dscam1* gene isoforms may not provide the same level of variation in Dscam for binding pathogens as is achieved by vertebrate immunoglobulin (Armitage et al. 2017; Ng and Kurtz 2020). Furthermore, not all analyses of various microbes and parasites have shown specific elevation of the immune system following a second homologous challenge. For instance, when ants (*Formica selyi*) are initially exposed to a sublethal priming dose of a natural fungal pathogen (*Beauveria bassiana*) and later challenged with a lethal dose of the same pathogen, they showed no sign of heightened resistance (Reber and Chapuisat 2012). There was also no evidence of any 'specific immune priming' using combinations of homologous and heterologous fungi with initial exposure followed by later challenge. Similar approaches with juvenile shrimp, *Litopenaeus vannamei*, showed that priming with heat-killed *Vibrio harveyi* resulted in an increase in the percentage of phagocytic haemocytes towards this vibrio (Pope et al. 2011). More specifically, when monolayers of shrimp haemocytes were overlaid with a mixture and *V. harveyi* and *Bacillus subtilis*, the bacterial uptake of *V. harveyi* was significantly increased while that of *B. subtilis* was unaffected by priming. However, if shrimp were injected with *B. subtilis* and their haemocytes incubated with the same mix of bacteria seven days later, no increase in uptake of *B. subtilis* was observed. Overall a cautious approach is required in considering the importance and universality of immune priming for all invertebrates and their parasites and pathogens (Armitage et al. 2017; Ng and Kurtz 2020).

Pasteuria ramosa, heightened trans-generational protection was demonstrated following challenge but did not differ based on the strain of *P. ramosa* (Ben-Ami et al. 2020). For further discussion of the potential role of trans-generational immune priming in ecology and host fitness see Tetreau et al. (2019) and Robinson and Green (2020).

1.8 Interaction between viruses and the invertebrate immune system

Unlike microbial intruders, viruses are incapable of autonomous replication, relying on hijacking host intracellular machinery and plundering the cytoplasmic resources. Ostensibly, the detection of viral particles inside invertebrate haemocytes initiates pro-apoptotic pathways and to a lesser extent autophagy (Lamiable et al. 2016) to thwart a virus from using the cell as a factory for replication. However, once progeny have been assembled, some viruses induce apoptosis to avoid pro-inflammatory processes. Many viruses, particularly those with compact RNA-based genomes, deploy multifunctional proteins to disarm pro-apoptotic proteins such as caspases and p53 (Everett and McFadden 1999; Liu et al. 2009). The immune-signalling cascades like JakSTAT play a role in coordinating cellular responses to viral presence in the haemolymph, and at least in mosquitoes and fruit flies, the Domeless receptor complexes with viral nucleic acids to activate the transcription of immune-mediators. Among the invertebrates, a plethora of immune factors have been linked to viral inhibition and destruction, including C-type lectins (Xu et al. 2014), haemocyanin (Coates and Nairn 2014) and hemolin (Terenius 2008).

The RNA interference (RNAi) system provides protection against viruses by targeting RNA and inhibiting the translation of mRNA (Li et al. 2002; Robalino et al. 2005; Liu et al. 2009; Huang and Zhang 2012; Hauton 2017). This gene regulatory mechanism works by degrading viral nucleic acids, and in doing so, restricts replication within cells and dissemination throughout the haemocoel. Viral nucleic acids—usually long dsRNAs—are intercepted by a sensor helicase called Dicer that belongs to the RNAase III family and cuts the invading dsRNA into double stranded sequences of 21 nucleotides that are processed into small interfering (si) RNAs. In siRNA form, they are complexed by Argnonaute-2, a member of a conserved family of RNA silencing proteins, plus several other proteins to form the RNA inducing silencing complex (RISC). Once bound to this effector protein complex, the siRNAs are used as templates to track mRNAs or single stranded viral genomes for cleavage and destruction. In addition to RNAi, antiviral defence in invertebrates has also been linked to other small RNA pathways modulated by Piwi-interacting RNA (piRNA) and microRNAs (miRNA) (Nayak et al. 2013). Unsurprisingly, viruses can circumvent host defences by targeting aspects of the RNAi system, which is usually referred to as viral

suppressors of RNA silencing (Samuel et al. 2018). For example, in mosquitoes, Cymbium Ringspot Virus binds siRNAs so they cannot be incorporated into RISC for degradation (Lakatos et al. 2004). Information on RNAi has been attained mostly from insects (reviewed by Palmer et al. 2018), and to a lesser extent crustaceans and bivalves (Huang et al. 2017; Green and Speck 2018), yet such evidence suggests that these defences are conserved and highly effective among the broader invertebrate groups described in this volume.

1.9 Interaction between bacteria and the invertebrate immune system

Pathogenic bacteria have many methods to avoid or circumvent the cells of the invertebrate immune system in their susceptible hosts. Bacteria can hinder phagocytosis, which is a key defence strategy of most invertebrates. They can: (i) inhibit uptake by cells using various mechanisms such as anti-phagocytic capsules and slime, (ii) stimulate adhesion and uptake to allow invasion of professional phagocytes, replicate intracellularly, and thereby hide from the immune system (e.g. *Vibrio splendidus* in *Crassostrea gigas* phagocytes; Duperthuy et al. 2011), (iii) inhibit lysosomal fusion to block formation of phagolysosomes and avoid exposure to antibacterial factors, (iv) interfere with antimicrobial activities of phagocytes such as inhibiting reactive oxygen species production, (v) cause phagocytes to undergo apoptosis, and (vi) secrete lytic factors such as cytotoxins and phospholipases that kill phagocytes. Vibrios that infect bivalve molluscs employ several mechanisms to block immune responsiveness (Table 1.3). Cytotoxic activity by various species of vibrios has been widely reported, especially in these molluscs. For example, a zinc metalloprotease in extracellular products of virulent strains of *V. aestuarianus* causes changes to the morphology of adherent haemocytes and instigates a significant reduction in their adhesive and phagocytic properties (Labreuche et al. 2010). *Vibrio tasmaniensis* toxicity to oyster haemocytes is dependent on an intracellular mechanism following their ingestion, while in the same host, virulent strains of *V. crassostreae* employ a contact-dependent mechanism that involves an exported protein coded by the

Table 1.3 Examples of mechanisms utilised by vibrios to interfere with haemocyte mediated phagocytosis in bivalve molluscs

Mode of action	Examples	References
Inhibition of uptake	*V. aestuarianus* inhibits phagocytic activity of Pacific oyster, *Crassostrea gigas* haemocytes	Labreuche et al. 2006
Attachment to and invasion of phagocytes	Outer membrane proteins on *V. splendidus* in *C. gigas* phagocytes. Acts as an adhesin. Allows vibrios to invade and colonise haemocytes	Duperthuy et al. 2011
Cytotoxic activity against phagocytes	Zinc metalloprotease secreted by *V. aestuarianus* causes changes in adhesive, phagocytic and morphology of oyster (*C. gigas*) haemocytes	Labreuche et al. 2010
Intracellular toxicity	*V. tasmaniensis* only kills oyster haemocytes following its ingestion	Rubio et al. 2019
Contact cytotoxicity	*V. crassostreae* kills oyster haemocytes by contact dependent mechanism	Rubio et al. 2019
Inhibition of intracellular killing	*V. splendidus* in oyster haemocytes affects phagolysosome formation. Molecular mechanism of action unclear	Duperthuy et al. 2011
Inhibition of intracellular killing and haemocyte lysis	*V. splendidus* reduces lysosomal membrane stability within phagocytes of *Mytilus galloprovincialis* causing cell destruction	Balbi et al. 2013

conserved virulence gene r5.7 (Rubio et al. 2019). *Vibrio splendidus* strains that cause infections in the Pacific oyster, *C. gigas* have a protein in the outer membrane that facilitates invasion into the haemocytes (Duperthuy et al. 2011). This protein also contributes to resistance against antimicrobial peptides and is considered a key virulence factor in this strain of *V. splendidus* (Duperthuy et al. 2010).

Members of the Bacillaceae such as *Bacillus thuringiensis* and *B. cereus* are pathogens of many insects. As well as the various toxins and phospholipases produced by these bacteria, the ability to form resistant endospores is of importance in overcoming the cellular defences of insects. Pathogenic (*B. cereus*) and non-pathogenic (*E. coli*) bacteria within nodules of *G. mellonella* are both subsumed into nodules by 4 hours post-injection (Walters and Ratcliffe 1983). However, by 18 hours the viable population of *B. cereus* in the nodules increases by 1,270% of the initial injection and only 3% of the non-pathogenic *E. coli* survive by 18 hours. This suggests that the endospores and vegetative bacteria survive in the nodules and break out later. As various toxins produced by Bacillaceae have haemolytic and cytolytic

activity and also cause haemocyte lysis in various insects including *G. mellonella* (e.g. Salamitou et al. 2000), these are likely to be important virulence factors in facilitating bacterial survival within nodules.

1.10 Interaction between fungi and the invertebrate immune system

Pathogenic fungi display several infectious morphotypes, with many being dimorphic. Spore-like propagules tend to adhere to the exoskeleton whereupon they develop hyphal-like structures (or penetration pegs) that force passage across the cuticle/tissue barriers and into the haemocoel. Few pathogens are equipped to breach the mineralised exoskeletons—the crustacean carapace, the insect integument, and bivalve shells, with the notable exceptions of fungi and oomycetes (Söderhäll and Unestam 1975; Leger et al. 1986). Of course, ecdysis assists in shedding fungi that have yet to penetrate the cuticle, however the new soft cuticle can make them vulnerable to many disease-causing agents. Host surfaces can be hostile platforms to dissuade fungal establishment, with many insects producing antimicrobial epicuticular waxes, fatty acids, and proteases (Ortiz-Urquiza and Keyhani 2013). In turn, entomopathogenic fungi deploy protease inhibitors, chitinolytic factors, detoxifying enzymes, and symbiotic bacteria (Butt et al. 2016). Horseshoe crabs secrete a cytolytic surfactant from hypodermal glands lining the outer cuticle, which keeps the carapace clean of epibionts and fungi (Harrington et al. 2008). The barrier epithelium beneath the endocuticle can detect tissue damage and pathogen presence, and in response, release signalling molecules to communicate with the circulating haemocytes and antimicrobial peptides to antagonise the fungus (Imler and Bulet 2005). Epithelial layers with access to or contact with the environment, including the alimentary canal, reproductive tract, gills, and Malpighian tubules, all encounter microbes and represent an important line of defence among invertebrates (Rosenstiel et al. 2009) and vertebrates alike. Toll-like receptor signalling mediated by epithelia is an ancient form of immunity—present in animals lacking a mesoderm (e.g. sponges; Wiens et al. 2007) and can be triggered by the fungal β-glucans. The immediate response to fungal presence in the haemolymph is to deploy melanin at the site of entry, and recruit haemocytes to form palisades in a similar manner to bacterial nodules/capsules. Should the fungus escape, to avoid alerting the patrolling haemocytes, some species will coat themselves in host carbohydrates making it difficult to detect the immunogenic β-glucans. Fungal proliferation is rapid and results in yeast-like progeny in the few observations of mycosis from infected crabs (Davies et al. 2020), and blastospores or hyphal bodies in mycosis in insects (Butt et al. 2016). Pre-formed soluble factors within the haemolymph/coelomic fluid tend to have biostatic properties, such as apolipophorins, whereas inducible factors tend to have biocidal properties such as the antifungal peptides drosomycin and metchnikowin from fruit flies (Tassanakajon et al. 2015). Immobilising or tagging fungi with opsonins, lectins, and β-glucan binding proteins makes them targets for haemocytes to ingest or encapsulate (as for bacteria, see Section 1.9). To avoid such fates, fungi produce secondary metabolites, cyclic peptides (e.g. destruxins; Vey et al. 2002), detoxicants and proteases to compromise haemocyte and organelle (lysosomal,

mitochondrial) membrane integrity, and to neutralise harmful oxidising and nitrosative radicals.

1.11 Interaction between parasites and the invertebrate immune system

Parasites have a number of approaches to either avoid or circumvent the activities of the host immune system. They may colonise regions of the host in which few leucocytes are present or where these cells cannot gain access, as in the case of insect parasitoids that lay their eggs in nervous tissue (see Chapter 13). Some parasites implement mimicry of host tissues, such that the immune system is 'unaware' of their presence. This may involve acquisition of host antigens with which they masquerade as 'self'. Parasites may also employ active inhibition of the host immune response, for example by reducing the production of circulating leucocytes thereby inhibiting the cellular and humoral response. Leucocyte reduction can leave the host vulnerable to secondary infections. Parasites may use the host response to proliferate within phagocytes such as *Perkinsus marinus* and *Haplosporidium nelsoni* that cause dermo and MSX diseases respectively, in oysters and other bivalves (see Chapter 8). Other examples of parasites avoiding the host immune response include: (i) the blood fluke, *Schistosoma mansoni*, in their intermediate snail host, *Biomphalaria glabrata*, (ii) the protistan parasite, *Perkinsus marinus*, in oysters and other bivalves, and (iii) parasitic dinoflagellates, *Hematodinium* sp., in crabs.

1.11.1 Dermo disease

Dermo disease of eastern oysters, *Crassostrea virginica*, is caused by *Perkinsus marinus*, a protist parasite belonging to the proposed phylum Perkinsozoa (see Chapter 8 for more details on the host range and pathology of this disease). These parasites invade host tissues, and once phagocytosed by the haemocytes in circulation they multiply after overcoming the intracellular processes that kill other parasites and pathogens. Hence, this parasite uses the defence reaction of phagocytosis to provide an intracellular environment that is outside of the reach of humoral and cellular immune responsiveness. The activities of galectins (see Section 1.6.2) in recognition and subsequent uptake of *P. marinus* by phagocytic haemocytes has been carefully evaluated to understand the glycans on the surface of both the parasites and the haemocytes and the linkage formed between them by the oyster galectins including CvGal1 and CvGal2 (see Table 1.2; Vasta 2020; Vasta and Wang 2020; Vasta et al. 2020).

1.11.2 Schistosomes and the snail, *Biomphalaria glabrata*

Human schistosomiasis is a devastating disease that affects over 200 million people worldwide (Pila et al. 2017) that is caused by infection from the trematode, *Schistosoma mansoni* and related species. The infection occurs when free swimming cercariae penetrate human skin and approximately 6 to 8 weeks later, adult worms in the portal vein in the abdomen mate and produce eggs that leave the host in the faeces and urine, and hatch in the environment to produce motile miracidia. The miracidia of *S. mansoni* tend to penetrate the foot of freshwater snails, including *Biomphalaria glabrata*, which act as the obligate intermediate host. Once inside the molluscan host, these miracidia develop through various stages of sporocysts that can migrate within the tissues. They are eventually shed from the infected snail to the environment as cercariae in search of the definitive host.

Not all snail strains of *B. glabrata* are susceptible to parasitisation by *S. mansoni*. There are compatible/susceptible and incompatible/resistant strains and employing these combinations experimentally has enabled a mechanistic understanding of why some snails are susceptible and others are not. Three fundamental observations have been made using this approach over the last few decades. First, resistant snail strains recognise *S. mansoni* sporocysts as non-self, and their haemocytes migrate to and encapsulate the parasites, whereas susceptible strains neither recognise nor react to parasites that develop fully within their tissues (see reviews by Coustau et al. 2015 and Pila et al. 2017). Encapsulation of sporocysts results in their destruction by virtue of oxidative killing methods including the generation of H_2O_2 (Hahn et al. 2001a) and NO (Hahn et al. 2001b). Hence, encapsulation is a key cellular defence against schistosomes such as *S. mansoni*. Second, passive transfer of resistance is accomplished by injecting cell-free haemolymph from resistant to susceptible snail strains, which implies that humoral factors are important in determining the ability of the host to deal with *S. mansoni* effectively (Granath and Yoshino 1984). Finally, resistance (immunity?) to *S. mansoni* can be achieved by challenging snails with either inactivated or single doses of parasites. The secondary challenge appears to elicit a humoral rather than a cellular response (Pinaud et al. 2016), leaving snails resistant. This implies that a selective activation of the snail immune system can be mediated by generating soluble factors that have some level of specificity or selectivity in their mode of action.

A great deal of attention has been placed on determining the nature of the snail host's protective humoral factors. Of the several candidate molecules identified, the most striking is a group of lectin-like molecules with fibrinogen-related domains (which are therefore called fibrinogen-related proteins or FREPs). These proteins have the capacity for somatic diversification by alternative splicing similar to that observed for Dscam in arthropods (see Box 1.3) suggesting that they may be involved in elevated resistance to trematodes following primary challenge (reviewed in Adema and Loker 2015; Pila et al. 2017). There are several forms of FREPs in *B. glabrata* (*Bg*FREPs). For instance, *Bg*FREP3 can bind bacteria, fungi, and *S. mansoni* (Hanington et al. 2010; Pinaud et al. 2016). *Bg*FREP3 binds mucins associated with different strains of *S. mansoni* leading to their recognition and encapsulation by the haemocytes. Hence, compatibility between snail and parasite may depend on the highly polymorphic nature of both *S. mansoni* surface molecules (i.e. mucins) and host (multiple FREP isoforms). Knockdown of *Bg*FREP3 in snails results in a modest (*ca.* 30%) reduction in resistance of snails to *S. mansoni* (Hanington et al. 2012) implying that other factors are likely involved in the detection and destruction of this parasite. Candidates for these include the cytolytic molecules biomphalysin (Galinier et al. 2013) and glabralysin (Lassalle et al. 2020), granulins (Hambrook et al. 2019), and thioester-containing proteins (*Bg*TEPs). It is feasible that multiple types of anti-pathogen molecules function together as for *Bg*TEP1, *Bg*FREP3, and *Bg*FREP2 that may provide a comprehensive explanation of resistance of some strains of *B. glabrata* to *S. mansoni* (Li et al. 2020), although this may prove to be an oversimplification of what is most likely a complex host-parasite relationship.

1.11.3 *Hematodinium* infections of crabs

Dinoflagellates belonging to *Hematodinium* spp. are important protist parasites of over 40 species of crustaceans (see Chapter 17). The parasite multiplies without interference in the tissues of the host and while the timescale of the period from infection to death may differ widely depending on the species of host, the end point is usually death. During the infection period, the parasites proliferate in the haemolymph while the total haemocyte concentration gradually declines. Whether this reduction in haemocytes is a direct consequence of the infection (e.g. by parasite induced suppression of haematopoiesis) is unclear. However, a consequence of increasing metabolic exhaustion from active immune responsiveness, may be that the host redirects resources away from haemopoiesis. A key observation for understanding this process is that the haemocytes do not recognise and interact with the increasingly large numbers of parasites and there is no evidence of phagocytosis, encapsulation, or nodule formation to encase the parasites. For instance, in infected crabs where nodules and capsules are occasionally present, they rarely have any entrapped trophonts of *Hematodinium* spp. implying that the trigger for nodule formation is unrelated to the detection of *Hematodinium*. There are data showing that these parasites reduce immune gene expression in infected crabs (e.g. Li et al. 2015) but the process of bacterial clearance in edible crabs, *Cancer pagurus* is unaffected by early infection with *Hematodinium* (Rowley et al. 2015). It may be that some form of molecular mimicry is involved in this parasite interaction with the host (Rowley et al. 2015). Hence, it appears that these parasites use specific mechanisms to avoid a host response rather than general immune suppression.

1.12 Future directions

1.12.1 The invertebrate microbiome

There is increasing evidence that the microbiota associated with a given host (the microbiome) has a complex and not fully explained interaction with the immune system, in addition to the general physiology of the host. While this has been explored in mammals, invertebrates may provide simpler models to study such phenomena (Douglas 2019). For example, studies using *Hydra* spp. have revealed that the profile of AMPs produced by several host species can select for commensal bacteria (Franzenburg et al. 2013) likely by providing competition and hence protection from colonisation by pathogenic bacteria. Furthermore, such commensal and symbiont associations are important in host-parasite co-evolution (Gerardo et al. 2020).

1.12.2 The invertebrate immune system in a changing world

Because invertebrates are largely ectotherms, they are strongly influenced by changes in temperature including the extremes of cold and hot climes. There is an increasing volume of literature showing immune suppression caused by recent higher summer temperatures across both aquatic and terrestrial invertebrates but some of the most striking examples come from aquatic animals including corals and farmed shellfish. In this latter case, outbreaks of diseases resulting from heatwaves have been widely reported (see Chapter 8; review by Burge et al. 2014 and Henroth and Baden 2018). Additional stressors resulting from these events include hypoxia, alterations of salinity, higher levels of toxic harmful algal blooms, together with increased microbial growth and virulence—all resulting in additional pressure on the immune system. The literature is replete with reports of changes in immune potential, and hence disease resistance, in bivalve molluscs of commercial importance subject to temperature, pH, and other anthropogenic stressors. A review of this is beyond the scope of the present account but some reports are worth elaborating as illustrations of this field. The Pacific oyster, *Crassostrea gigas* is an important aquaculture species in many countries. Mass mortalities have been reported in summer months as a result of infections by vibrios and viruses such as Oyster herpes virus (see Chapter 8). *C. gigas* spawn in summer months and because this event is energetically costly these individuals show depressed function of their immune systems that leaves them susceptible to disease after spawning.

Thermal stress is highest in oysters post-spawning and the rates of clearance and phagocytosis are significantly reduced in animals under the combination of both temperature stress and post-spawning, compared with pre-spawning animals (Wendling and Wegner 2013). Increases in summer temperatures resulting from local heatwaves hence place aquaculture production of bivalves, such as *C. gigas* and the blue mussel *Mytilus edulis*, at risk in a warming world.

1.13 Summary

- The invertebrate immune system consists of a wide variety of cells and molecules and is innate in nature.
- Immune cells in invertebrates do not include lymphocytes or cells capable of clonal proliferation or production of antigen-binding immunoglobulin.
- Nearly all invertebrates have circulating immune surveillance cells, termed haemocytes or coelomocytes, which have defensive and haemostatic functions and that mediate the invertebrate immune responses.
- Nodule formation and encapsulation are effective at sequestering and killing invading microbes and macroparasites.
- Invertebrates have novel immune molecules that are not retained through vertebrate evolution including FREPs, the phenoloxidase activating system, a wide variety of lectins that are species-specific, and a significant amplification of many PRR genes in some invertebrate genomes.
- The invertebrate immune system appears to have some form of immunological memory, best termed 'innate immune memory' based on alternative splicing of genes such as *Dscam-1* but its impact in host—parasite/pathogen interaction outside of the arthropods is not fully established.
- Pathogens and parasites have a multitude of mechanisms to either circumvent or avoid the host immune system, and these play out and expand in the context of the host-pathogen evolutionary arms race. While the interaction between invertebrate host and pathogen has been explored in detail in some models (e.g. *Perkinsus marinus* in oysters and vibrio infections in bivalve molluscs), in others it is not understood (e.g. *Hematodinium* spp. in crustaceans).
- Several invertebrate model species have been developed to facilitate detailed study of immune cells and molecules including *Drosophila* and *Caenorhadbitis*. These invertebrate models are increasingly important in fundamental studies in both human diseases, and behavioural and ecological immunology. Investigations of the invertebrate microbiome and its interaction with host defences is of increasing importance, particularly as invertebrates confront rapidly changing climates and environments.

Recommended further reading

Social distancing—behavioural avoidance of infection and other mechanisms to avoid contact between host and parasite/pathogen

Cremer, S., Pull, C.D., and Fürst, M.A. 2018. Social immunity: emergence and evolution of colony-level disease protection. *Annual Review of Entomology* 63:105–123 (Excellent overview of concepts of social immunity in insects)

Chemical and physical barriers to infection

Ortiz-Urquiza, A. and Keyhani, N.O. 2013. Action on the surface: Entomopathogenic fungi versus the insect cuticle. *Insects* 4: 357–374

Hirose, E. 2009. Ascidian tunic cells: Morphology and functional diversity of free cells outside the epidermis. *Invertebrate Biology* 128: 83–96

Wound healing and haemostasis

Eleftherianos, I. and Revenis, C. 2011. Role and importance of phenoloxidase in insect hemostasis. *Journal of Innate Immunity* 3:28–33

Cerenius, L. and Söderhäll, K. 2011. Coagulation in invertebrates. *Journal of Innate Immunity* 3: 3–8 (Review of blood clotting mechanisms in invertebrates with an emphasis on arthropods)

Phagocytosis

Browne, N., Heelan, M., and Kavanagh, K. 2013. An analysis of the structural and functional similarities of insect hemocytes and mammalian phagocytes. *Virulence* 4: 597–603

Nazario-Toole, A.E. and Wu, L.P. 2017. Phagocytosis in insect immunity. In *Advances in Insect Physiology* 52, P. Ligoxygakis. (ed.) pp. 35–82. London: Academic Press (Review of receptors and signalling pathways involved in phagocytosis)

Nodule formation and encapsulation

Dubovskiy, I.M., Kryukova, N.A., Glupov, V.V., and Ratcliffe, N.A. 2016. Encapsulation and nodulation in insects. *Invertebrate Survival Journal* 13: 229–246

Signalling pathways

Bang, S. 2019. JAK/STAT signaling in insect innate immunity. *Entomological Research* 49: 339–353 (Recent review on signalling pathways in different insect groups)

Hetru, C. and Hoffmann, J.A. 2009. NF- kB in the immune response of Drosophila. *Cold Spring Harbour Perspectives in Biology* 1: a000232

Antimicrobial peptides

Hanson, M.A. and Lemaitre, B. 2020. New insights on Drosophila antimicrobial peptide function in host defense and beyond. *Current Opinion in Immunology* 62: 22–30

Mylonakis, E., Podsiadlowski, L., Muhammed, M., and Vilcinskas, A. 2016 Diversity, evolution, and medical applications of insect antimicrobial peptides. *Philosophical Transactions of the Royal Society B* 371: 20150290

Smith, V.J. and Dyrynda, E.A. 2015. Antimicrobial proteins: From old proteins, new tricks. *Molecular Immunology* 68: 383–398

Lectins

Vasta G.R. 2020. Galectins in host–pathogen interactions: Structural, functional, and evolutionary aspects. In *Lectin in Host Defense Against Microbial Infections*. Hsieh .SL. (ed.) *Advances in Experimental Medicine and Biology* vol. 1204. pp. 169–196. Singapore: Springer (Wide ranging review of galectins in invertebrates and vertebrates)

Vasta, G.R. and Wang, J.X. 2020. Galectin-mediated immune recognition: Opsonic roles with contrasting outcomes in selected shrimp and bivalve mollusk species. *Developmental & Comparative Immunology* 110: 103721

The prophenoloxidase activating system

Cerenius, L. and Söderhäll, K. 2004. The prophenoloxidase-activating system in invertebrates. *Immunological Reviews* 198: 116–126

Whitten, M.M. and Coates, C.J. 2017. Re-evaluation of insect melanogenesis research: Views from the dark side. *Pigment Cell & Melanoma Research* 30: 386–401

Lysozymes

Van Herreweghe, J.M. and Michiels, C.W. 2012. Invertebrate lysozymes: Diversity and distribution, molecular mechanism, and in vivo *function. Journal of Biosciences* 37: 327–348 (Comprehensive review of invertebrate lysozymes)

Does the invertebrate immune system have specificity and memory?

Armitage, S.A.O., Kurtz, J., Brites, D., Dong, Y., Du Pasquier, L., and Wang, H.-C. 2017. Dscam1 in pancrustacean immunity: Current status and a look to the future. *Frontiers in Immunology* 8: 662

Ng, T.H. and Kurtz, J. 2020. Dscam in immunity: A question of diversity in insects and crustaceans. *Developmental & Comparative Immunology* 106: 103559

Pradeu, T. and Du Pasquier, L. 2018. Immunological memory: What's in a name? *Immunological Reviews* 28: 3.7–20. (Wide-ranging review taking a critical analysis of literature)

Tetreau, G., Dhinaut, J., Gourbal, B., and Moret, Y. 2019. Trans-generational immune priming in invertebrates: Current knowledge and future prospects. *Frontiers in Immunology* 10: 1938 (Comprehensive review of literature and pitfalls in experimental design and interpretation)

Interaction between viruses and the invertebrate immune system

Green, T.J., Raftos, D., Speck, P., and Montagnani, C. 2015. Antiviral immunity in marine molluscs. *Journal of General Virology* 96: 2471–2482

Liu, H., Söderhäll, K., and Jiravanichpaisal, P. 2009. Antiviral immunity in crustaceans. *Fish & Shellfish Immunology* 27: 79–88

Wang, P.H., Weng, S. P., and He, J.G. 2015. Nucleic acid-induced antiviral immunity in invertebrates: An evolutionary perspective. *Developmental & Comparative Immunology* 48: 291–296

Interaction between bacteria and the invertebrate immune system

McAnulty, S.J. and Nyholm, S.V. 2017. The role of hemocytes in the Hawaiian bobtail squid, *Euprymna scolopes*: A model organism for studying beneficial host–microbe interactions. *Frontiers in Microbiology* 7: 2013 (Overview of the interaction between *Vibrio fischeri* and its squid host)

Destoumieux-Garzón, D., Canesi, L., Oyanedel, D., Travers, M.-A., Charrière, G.M., Pruzzo, C., and Vezzulli, L. 2020. Vibrio–bivalve interactions in health and disease. *Environmental Microbiology* 22: 4323–4341 (Review of strategies employed by vibrios to circumvent to immune system in bivalve molluscs)

Interaction between fungi and the invertebrate immune system

Butt, T.M., Coates, C.J., Dubovskiy, I.M., and Ratcliffe, N.A. 2016. Entomopathogenic fungi: new insights into host–pathogen interactions. *Advances in Genetics* 94: 307–364

Interaction between parasites and the invertebrate immune system

Castillo, J.C., Reynolds, S.E., and Eleftherianos, I. 2011. Insect immune responses to nematode parasites. *Trends in Parasitology* 27: 537–547 (Review of interaction between nematodes and insect defences)

Castillo, M.G., Humphries, J.E., Mourão, M.M., Marquez, J., Gonzalez, A., and Montelongo, C.E. 2020. *Biomphalaria glabrata* immunity: post-genome advances. *Developmental & Comparative Immunology* 104: 103557

Pila, E.A., Li, H., Hambrook, J.R., Wu, X., and Hanington, P.C. 2017. Schistosomiasis from a snail's perspective: Advances in snail immunity. *Trends in Parasitology* 33: 845-857 (Comprehensive overview of snail immunity to trematodes)

Rowley, A.F., Smith, A.L., and Davies, C.E. 2015. How does the dinoflagellate parasite *Hematodinium* outsmart the immune system of its crustacean hosts? *PLoS Pathogens* 11: e1004724 (Brief opinion piece on this key parasite of decapods)

Vasta, G.R. and Wang, J.X. 2020. Galectin-mediated immune recognition: Opsonic roles with contrasting outcomes in selected shrimp and bivalve mollusk species. *Developmental & Comparative Immunology* 110: 103721 (Recent review of the roles of galectins in immune recognition)

Acknowledgements

AFR and CJC were part-funded by the European Regional Development fund through the Ireland Wales Cooperation Programme, BLUEFISH, and the BBSRC/NERC ARCH UK Aquaculture Initiative (BB/P017215/1).

References

Adema, C.M. and Loker, E.S. 2015. Digenean-gastropod host associations inform on aspects of specific immunity in snails. *Developmental & Comparative Immunology* 48: 275–283

Altincicek, B., Stötzel, S., Wygrecka, M., Preissner, K.T., and Vilcinskas, A. 2008. Host-derived extracellular nucleic acids enhance innate immune responses, induce coagulation, and prolong survival upon infection in insects. *The Journal of Immunology* 181: 2705–2712

Apfield, J. and Alper, S. 2018. What can we learn about human disease from the nematode *C. elegans*? In *Disease Gene Identification Methods and Protocols*. J.K. DiStefano (ed.) pp. 53–75. Totowa, N.J., USA: Humana Press

Armitage, S.A.O., Kurtz, J., Brites, D., Dong, Y., Du Pasquier, L., and Wang, H.-C. 2017. *Dscam1* in pancrustacean immunity: Current status and a look to the future. *Frontiers in Immunology* 8: 662

Balbi, T., Fabbri, R., Cortyese, K. et al. 2013. Interactions between *Mytilus galloprovincialis* hemocytes and the bivalve pathogens *Vibrio aestuarianus* 01/032 and *Vibrio splendidus* LGP32. *Fish & Shellfish Immunology* 35: 1906–1915

Bang, S. 2019. JAK/STAT signaling in insect innate immunity. *Entomological Research* 49: 339–353

Behringer, D.C., Butler, M.J., and Shields, J.D. 2006. Avoidance of disease by social lobsters. *Nature* 441: 421

Ben-Ami, F., Orlic, C., and Regoes, R.R. 2020. Disentangling non-specific and specific transgenerational immune priming components in host-parasite interactions. *Proceedings of the Royal Society B* 287: 20192386

Bertheussen, K. and Seljelid, R. 1978. Echinoid phagocytes *in vitro*. *Experimental Cell Research* 111: 401–412

Boman, H.G., Faye, I., Gudmundsson, G.H., Lee, J.Y., and Lidholm, D.A. 1991. Cell-free immunity in *Cecropia*—a model system for antibacterial proteins. *European Journal of Biochemistry* 201: 23–31

Brinkmann, V., Reichard, U., Goosmann, C. et al. 2004. Neutrophil extracellular traps kill bacteria. *Science* 303: 1532–1535

Browne, N., Heelan, M., and Kavanagh, K. 2013. An analysis of the structural and functional similarities of insect hemocytes and mammalian phagocytes. *Virulence* 4: 597–603

Burge, C.A., Eakin, C.M., Friedman, C.S. et al. 2014. Climate change influences on marine infectious diseases: Implications for management and society. *Annual Review of Marine Science* 6: 249–277

Butt, T.M., Coates, C.J., Dubovskiy, I.M., and Ratcliffe, N.A. 2016. Entomopathogenic fungi: New insights into host–pathogen interactions. *Advances in Genetics* 94: 307–364

Casertano, M., Menna, M., and Imperatore, C. 2020. The ascidian-derived metabolites with antimicrobial properties. *Antibiotics* 9: 510

Cerenius, L., Babu, R., Söderhäll, K., and Jiravanichpaisal, P. 2010. *In vitro* effects on bacterial growth of phenoloxidase reaction products. *Journal of Invertebrate Pathology* 103: 21–23

Cerenius, L. and Söderhäll, K. 2004. The prophenoloxidase-activating system in invertebrates. *Immunological Reviews* 198: 116–126

Cerenius, L. and Söderhäll, K. 2011. Coagulation in invertebrates. *Journal of Innate Immunity* 3: 3–8.

Champion, O.L., Wagley, S., and Titball, R.W. 2016. *Galleria mellonella* as a model host for microbiological and toxin research. *Virulence* 7: 840–845

Chávez, V., Mohri-Shiomi, A., and Garsin, D.A. 2009. Ce-Duox1/BLI-3 generates reactive oxygen species as a protective innate immune mechanism in *Caenorhabditis elegans*. *Infection and immunity* 77: 4983–4989

Chen, L., Hu, J.-S., Xu, J.-L., Shao, C.-L., and Wang, G.-Y. 2018. Biological and chemical diversity of ascidian-associated microorganisms. *Marine Drugs* 16: 362

Chen, T., Ren, C., Wang, Y. et al. 2016. Molecular cloning, inducible expression and antibacterial analysis of a novel i-type lysozyme (lyz-i2) in Pacific white shrimp, *Litopenaeus vannamei*. *Fish & Shellfish Immunology* 54: 197–203

Chou, P.H., Chang, H.S., Chen, I.T. et al. 2009. The putative invertebrate adaptive immune protein *Litopenaeus vannamei* Dscam (LvDscam) is the first reported Dscam to lack a transmembrane domain and cytoplasmic tail. *Developmental & Comparative Immunology* 33: 1258–1267

Clow, L.A., Raftos, D.A., Gross, P.S., and Smith, L.C. 2004. The sea urchin homologue, SpC3, functions as an opsonin. *Journal of Experimental Biology* 207: 2147–2155

Coates, C.J. and Costa-Paiva, E.M. 2020. Multifunctional roles of hemocyanins. In *Vertebrate and Invertebrate Respiratory Proteins, Lipoproteins and other Body Fluid Proteins*. U. Hoeger. and J.R. Harris (eds.) pp. 233–250. Springer, Cham.

Coates, C.J., Lim, J., Harman, K., Rowley, A.F., Griffiths, D.J., Emery, H., and Layton, W. 2019. The insect, *Galleria mellonella*, is a compatible model for evaluating the toxicology of okadaic acid. *Cell Biology and Toxicology* 35: 10.1007

Coates, C.J., McCulloch, C., Betts J., and Whalley, T. 2018. Echinochrome A release by red spherule cells is an iron-withholding strategy of sea urchin innate immunity. *Journal of Innate Immunology* 10: 119–130

Coates, C.J. and Nairn, J. 2014. Diverse immune functions of hemocyanins. *Developmental & Comparative Immunology* 45: 43–55

Coates, C.J. and Talbot, J. 2018. Hemocyanin-derived phenoloxidase reaction products display anti-infective properties. *Developmental & Comparative Immunology* 86: 7–51

Coffaro, K.A. and Hinegardner, R.T. 1977. Immune response in the sea urchin *Lytechinus pictus*. *Science* 197: 1389–1390

Coustau, C., Gourbal, B., Duval, D., Yoshino, T.P., Adema, C.M., and Mitta, G. 2015. Advances in gastropod immunity from the study of the interaction between the snail *Biomphalaria glabrata* and its parasites: A review of research progress over the last decade. *Fish & Shellfish Immunology* 46: 5–16

Coyne, K., Laursen, J.R., and Yoshino, T.P. 2015. *In vitro* effects of mucus from the mantle of compatible (*Lymnaea elodes*) and incompatible (*Helisoma trivolvis*) snail hosts on *Fascioloides magna* miracidia. *Journal of Parasitology* 101: 351–357

Cremer, S., Pull, C.D., and Fürst, M.A. 2018. Social immunity: Emergence and evolution of colony-level disease protection. *Annual Review of Entomology* 63: 105–123

D'Andrea-Winslow, L., Radke, D., Utecht, T., Kaneko, T., and Akasaka, K. 2012. Sea urchin coelomocyte arylsulfatase: A modulator of the echinoderm clotting pathway. *Integrative Zoology* 7: 61–73

Daffonchio, D., Alma, A., Favia, G., Sacchi, L., and Bandi, C. 2016. Invertebrate gut associations. In *Manual of Environmental Microbiology* 4th edn. M.V. Yates, C.H. Nakatsu, R.V. Miller, and S.D. Pillai (eds.) Washington: ASM Press

Davies, C.E., Malkin, S.H., Thomas, J.E., Batista, F.M., Rowley, A.F., and Coates, C.J. 2020. Mycosis is a disease state encountered rarely in shore crabs, *Carcinus maenas. Pathogens* 9: 462

De Gryse, G.M.A., Van Khuong, T., Descamps, B. et al. 2020. The shrimp nephrocomplex serves as a major portal of pathogen entry and is involved in the molting process. *Proceedings of the National Academy of Sciences USA* 117: 28374–28383

Destoumieux-Garzón, D., Rosa, R.D., Schmitt, P. et al. 2016. Antimicrobial peptides in marine invertebrate health and disease. *Philosophical Transactions of the Royal Society London B, Biological Sciences* 371: 20150300

Di Falco, F., Cammarata, M., and Vizzini, A. 2017. Molecular characterisation, evolution and expression analysis of g-type lysozymes in *Ciona intestinalis. Developmental & Comparative Immunology* 67: 457–463

Dishaw, L.J., Leigh, B., Cannon, J.P. et al. 2016. Gut immunity in a protochordate involves a secreted immunoglobulin-type mediator binding host chitin and bacteria. *Nature Communications* 7: 10617

Douglas, A.E. 2019. Simple animal models for microbiome research. *Nature Reviews Microbiology* 17: 764–775

Dror, H., Novak, L., Evans, J.S. López-Legentil, S., and Skenkar, N. 2019. Core and dynamic microbial communities of two invasive ascidians: Can host–symbiont dynamics plasticity affect invasion capacity? *Microbial Ecology* 78: 170–184

Duperthuy, M., Binesse, J., Le Roux, F. et al. 2010. The major outer membrane protein OmpU of *Vibrio splendidus* contributes to host antimicrobial peptide resistance and is required for virulence in the oyster *Crassostrea gigas. Environmental Microbiology* 12: 951–963

Duperthuy, M., Schmitt, P., Garzón, E. et al. 2011. Use of OmpU porins for attachment and invasion of *Crassostrea gigas* immune cells by the oyster pathogen *Vibrio splendidus. Proceedings of the National Academy of Sciences of the USA* 108: 2993–2998

Duvic, B. and Brehélin, M. 1998. Two major proteins from locust plasma are involved in coagulation and

are specifically precipitated by laminarin, a β-1,3-glucan. *Insect Biochemistry and Molecular Biology* 28: 959–967

Duvic, B. and Söderhäll, K. 1990. Purification and characterization of a beta-1,3-glucan binding protein from plasma of the crayfish *Pacifastacus leniusculus. The Journal of Biological Chemistry* 265, 9327–9332.

Edds, K.T. 1977. Dynamic aspects of filopodial formation by reorganization of microfilaments. *Journal of Cell Biology* 73(2): 479–491

Eleftherianos, I. and Revenis, C. 2011. Role and importance of phenoloxidase in insect hemostasis. *Journal of Innate Immunity* 3: 28–33

Eliseikina, M.G. and Magarlamov, T.Y. 2002. Coelomocyte morphology in the holothurians *Apostichopus japonicus* (Aspidochirota: Stichopodidae) and *Cucumaria japonica* (Dendrochirota: Cucumariidae). *Russian Journal of Marine Biology* 28: 197–202

Emery, H., Johnston, R., Rowley, A.F., Coates, C.J. 2019. Indomethacin-induced gut damage in a surrogate insect model, *Galleria mellonella. Archives of Toxicology* 93(8): 2347–2360.

Esteves, E., Fogaca, A.C., Maldonado, R. et al. 2009. Antimicrobial activity in the tick *Rhipicephalus* (*Boophilus*) *microplus* eggs: Cellular localization and temporal expression of microplusin during oogenesis and embryogenesis. *Developmental & Comparative Immunology* 33: 913–919

Everett, H. and McFadden, G. 1999. Apoptosis: An innate immune response to virus infection. *Trends in Microbiology* 7: 160–165

Fankboner, P.V. 2003. Digestive system of invertebrates. *eLS*. American Cancer Society.

Feng, C., Ghosh, A., Amin, M.H. et al. 2013. The galectin CvGal1 from the eastern oyster (*Crassostrea virginica*) binds to blood Group A oligosaccharides on the hemocyte surface. *The Journal of Biological Chemistry* 288: 24394–24409

Franchi, N. and Ballarin, L. 2017. Morula cells as key hemocytes of the lectin pathway of complement activation in the colonial tunicate *Botryllus schlosseri. Fish & Shellfish Immunology* 63: 157–164

Franzenburg, S., Walter, J., Künzel, S. et al. 2013. Distinct antimicrobial peptide expression determines host species-specific bacterial associations. *Proceedings of the National Academy of Sciences USA* 110: E3730–E3738

Galinier, R., Portela, J., Mone, Y. et al. 2013. Biomphalysin, a new β pore-forming toxin involved in *Biomphalaria glabrata* immune defense against *Schistosoma mansoni. PLoS Pathogens* 9: e1003216

Gao, Z. and Zhang, S. 2018. Cephalochordata: *Branchiostoma*. In *Advances in Comparative Immunology*. E.L.

Cooper (ed.) pp. 593–635. Cham, Switzerland: Springer International Publishing AG

Gardères, J., Bourguet-Kondracki, M.-L., Hamer, B., Batel, R., Schröder, H.C., and Müller, W.E.G. 2015. Porifera lectins: Diversity, physiological roles and biotechnological potential. *Marine Drugs* 13: 5059–5101

Gerardo, N.M., Hoang, K.L., and Stoy, K.S. 2020. Evolution of animal immunity in the light of beneficial symbioses. *Philosophical Transactions of the Royal Society B Biological Sciences* 375: 20190601

Giangrande, A., Licciano, M., Schiosi, R., Musco, L., and Stabili, L. 2014. Chemical and structural defensive external strategies in six sabellid worms (Annelida). *Marine Ecology* 35: 36–45

Gordon, S. 2016. Elie Metchnikoff, the man and the myth. *Journal of Innate Immunity* 8: 223–227

Granath, W.O. and Yoshino, T.P. 1984. *Schistosoma mansoni* passive transfer of resistance by serum in the vector snail, *Biomphalaria glabrata*. *Experimental Parasitology* 58: 188–193

Green, T.J. and Speck, P. 2018. Antiviral defense and innate immune memory in the oyster. *Viruses* 10: 133

Ha, E.-M., Lee, K.-A., Park, S.-H. et al. 2009. Regulation of DUOX by the Galphaq-phospholipase Cbeta-Ca^{2+} pathway in *Drosophila* gut immunity. *Developmental Cell* 16: 386–397

Hahn, U.K., Bender, R.C., and Bayne, C.J. 2001a. Killing of *Schistosoma mansoni* sporocysts by hemocytes from resistant *Biomphalaria glabrata*: Role of reactive oxygen species. *Journal of Parasitology* 87: 292–299

Hahn, U.K., Bender, R.C. and Bayne, C.J. 2001b. Involvement of nitric oxide in killing of *Schistosoma mansoni* sporocysts by hemocytes from resistant *Biomphalaria glabrata*. *Journal of Parasitology* 87: 778–785

Hambrook J.R., Gharamah A.A., Pila E.A., Hussein S., and Hanington P.C. 2019. *Biomphalaria glabrata* granulin increases resistance to *Schistosoma mansoni* infection in several *Biomphalaria* species and induces the production of reactive oxygen species by haemocytes. *Genes* (Basel) 11: 38

Hanington, P.C., Forys, M.A., Dragoo, J.W., Zhang, S.-M., Adema, C.M., and Loker, E.S. 2010. Role for a somatically diversified lectin in resistance of an invertebrate to parasite infection. *Proceedings of the National Academy of Sciences USA* 107: 21087–21092

Hanington, P.C., Forys, M.A., and Loker, E.S. 2012. A somatically diversified defense factor, FREP3, is a determinant of snail resistance to schistosome infection. *PLoS Neglected Tropical Diseases* 6: e1591

Hanson, M.A. and Lemaitre, B. 2020. New insights on *Drosophila* antimicrobial peptide function in host

defense and beyond. *Current Opinion in Immunology* 62: 22–30

Harrington, J.M., Leippe, M., and Armstrong, P.B. 2008. Epithelial immunity in a marine invertebrate: A cytolytic activity from a cuticular secretion of the American horseshoe crab, *Limulus polyphemus*. *Marine Biology* 153: 1165–1171

Hauton, C. 2017. Recent progress toward the identification of anti-viral immune mechanisms in decapod crustaceans. *Journal of Invertebrate Pathology* 147: 111–117

Hauton, C. and Smith, V.J. 2007. Adaptive immunity in invertebrates: A straw house without a mechanistic foundation. *Bioessays* 29: 1138–1146

Henroth, B.E. and Baden, S.P. 2018. Alteration of host-pathogen interactions in the wake of climate change—increasing risk for shellfish associated infections? *Environmental Research* 161: 425–438

Henson, J.H. and Schatten, G. 1983. Calcium regulation of the actin-mediated cytoskeletal transformation of sea urchin coelomocytes. *Cell Motility* 3(5–6): 525–534

Hetru, C. and Hoffmann, J.A. 2009. NF- kB in the immune response of *Drosophila*. *Cold Spring Harbour Perspectives in Biology* 1: a000232

Hibino, T., Loza-Coll, M., Messier, C. et al. 2006. The immune gene repertoire encoded in the purple sea urchin genome. *Developmental Biology* 300: 349–365

Hillier, B.J. and Vacquier, V.D. 2003. Amassin, an olfactomedin protein, mediates the massive intercellular adhesion of sea urchin coelomocytes. *Journal of Cell Biology* 160: 597–604

Hira, J., Wolfson, D., Andersen, A.J.C., Haug, T., and Stensvåg, K. 2020. Autofluorescence mediated red spherulocyte sorting provides insights into the source of spinochromes in sea urchins. *Scientific Reports* 10(1): 1149

Hirose, E. 2009. Ascidian tunic cells: Morphology and functional diversity of free cells outside the epidermis. *Invertebrate Biology* 128: 83–96

Ho, E.C.H., Buckley, K.M., Schrankel, C.S., et al. 2016. Perturbation of gut bacteria induces a coordinated cellular immune response in the purple sea urchin larva. *Immunology and Cell Biology* 94: 861–874

Hoeger, U. and Schenk, S. 2020. Crustacean hemolymph lipoproteins. In *Vertebrate and Invertebrate Respiratory Proteins, Lipoproteins and Other Body Fluid Proteins. Subcellular Biochemistry* Vol 94. U. Hoeger and J. Harris (eds.) pp. 35–61. Springer International Publishing AG

Homa, J. 2018. Earthworm coelomocyte extracellular traps: Structural and functional similarities with neutrophil NETs. *Cell & Tissue Research* 371: 407–414

Hou, F., Liu, Y., He, S. et al. 2015. A galectin from shrimp *Litopenaeus vannamei* is involved in immune recognition and bacteria phagocytosis. *Fish & Shellfish Immunology* 44: 584–591

Huang, B., Zhang, L., Du, Y., Xu, F., Li, L., and Zhang, G. 2017. Characterization of the mollusc RIG-I/MAVS pathway reveals an archaic antiviral signalling framework in invertebrates. *Scientific Reports* 7: 1–13

Huang, H., Huang, S., Yu, Y. et al. 2011. Functional characterization of a ficolin-mediated complement pathway in amphioxus. *The Journal of Biological Chemistry* 286: 36739–36748

Huang, M., Zhou, T., Wu, Y. et al. 2018. Characterisation and functional comparison of single-CRD and multidomain galectins CgGal-2 and CgGal-3 from oyster *Crassostrea gigas*. *Fish & Shellfish Immunology* 78: 238–247

Huang, T. and Zhang, X. 2012. Functional analysis of a crustacean microRNA in host-virus interactions. *Journal of Virology* 86: 12997–13004

Imler, J.L. and Bulet, P. 2005. Antimicrobial peptides in *Drosophila*: Structures, activities and gene regulation. In *Mechanisms of Epithelial Defense* 86. D. Kabelitz and J.M. Schroeder (eds.) pp. 1–21. New York: Karger Publishers

Jemel, S., Guillot, J., Kallel, K., Botterel, F., and Dannaoui, E. 2020. *Galleria mellonella* for the evaluation of antifungal efficacy against medically important fungi, a narrative review. *Microorganisms* 8: E390

Kacsoh, B. Z. Lynch, Z.R., Mortimer, N.T., and Schenke, T.A. 2013. Fruit flies medicate offspring after seeing parasites. *Science* 339: 947–950

King, G.M. 1986. Inhibition of microbial activity in marine sediments by a bromophenol from a hemichordate. *Nature* 323: 257–259

Krams, I.A., Kecko, S., Jõers, P. et al. 2017. Microbiome symbionts and diet diversity incur costs on the immune system of insect larvae. *The Journal of Experimental Biology* 220: 4204–4212

Kuraishi, T., Binggeli, O., Opotsa, O., Buchon, N., and Lemaitre, B. 2011. Genetic evidence for a protective role of the peritrophic matrix against intestinal bacterial infection in *Drosophila melanogaster*. *Proceedings of the National Academy of Sciences of the United States of America* 108: 15966–15971

Kwong, W.K., Mancenido, A.L. and Moran, N.A. 2017. Immune system stimulation by the native gut microbiota of honey bees. *Royal Society Open Science* 4: 170003

Labreuche, Y., Lambert, C., Soudant, P., Boulo, V., Huvet, A. and Nicolas, J.-L. 2006. Cellular and molecular hemocyte responses of the Pacific oyster, *Crassostrea gigas*, following bacterial infection with *Vibrio aestuarianus* strain 01/32. *Microbes and Infection* 8: 2715–2724

Labreuche, Y., Le Roux, F., Henry, J. et al. 2010. *Vibrio aestuarianus* zinc metalloprotease causes lethality in the Pacific oyster *Crassostrea gigas* and impairs the host cellular immune defenses. *Fish & Shellfish Immunology* 29: 753–758

Lafont, M., Vergnes, A., Vidal-Dupiol, J. et al. 2020. A sustained immune response supports long-term antiviral immune priming in the Pacific oyster, *Crassostrea gigas*. *Mbio* 11. e02777–19

Lakatos, L., Szittya, G., Silhavy, D., and Burgyán, J. 2004. Molecular mechanism of RNA silencing suppression mediated by p19 protein of tombusviruses. *The EMBO Journal* 23: 876–884

Lamiable, O., Arnold, J., De Faria, I.J.D.S. et al. 2016. Analysis of the contribution of hemocytes and autophagy to *Drosophila* antiviral immunity. *Journal of Virology* 90: 5415–5426

Lassalle, D., Tetreau, G., Pinaud, S. et al. 2020. Glabralysins, potential new β-pore-forming toxin family members from the schistosomiasis vector snail *Biomphalaria glabrata*. *Genes* 11: 65

Leger, R.S., Cooper, R.M., and Charnley, A.K. 1986. Cuticle-degrading enzymes of entomopathogenic fungi: Cuticle degradation *in vitro* by enzymes from entomopathogens. *Journal of Invertebrate Pathology* 47: 167–177

Lemaitre, B., and Hoffmann, J. 2007. The host defense of *Drosophila melanogaster*. *Annual Review of Immunology* 25: 697–743

Lemaitre, B., Nicolas, E., Michaut, L., Reichhart, J.M., and Hoffmann, J.A. 1996. The dorsoventral regulatory gene cassette spätzle/Toll/cactus controls the potent antifungal response in *Drosophila* adults. *Cell* 86: 973–983

Li, C., Blencke, H.-M., Haug, T. and Stensvåg, K. 2015. Antimicrobial peptides in echinoderm host defense. *Developmental & Comparative Immunology* 49: 190–197

Li, H., Hambrook, J.R., Pila, E.A. et al. 2020. Coordination of humoral immune factors dictates compatibility between *Schistosoma mansoni* and *Biomphalaria glabrata*. *eLife* 9: e51708

Li, H., Li, W. X., and Ding, S.W. 2002. Induction and suppression of RNA silencing by an animal virus. *Science* 296: 1319–1321

Li, M., Li, C., Wang, J., and Song. S. 2015. Immune response and gene expression in hemocytes of *Portunus trituberculatus* inoculated with the parasitic dinoflagellate *Hematodinium*. *Molecular Immunology* 65: 113–122

Li, R. and Xu, A. 2015. The complement system of amphioxus. In *Amphioxus Immunity: Tracing the Origins of Human Immunity*. A.-L Xu (ed.) pp. 141–152. London: Academic Press

Li, X.-J., Yang, L., Li, D., Zhu, Y.-T., Wang, Q., and Li, W.-W. 2018. Pathogen- specific binding soluble Down syndrome cell adhesion molecule (Dscam) regulates phagocytosis via membrane-bound Dscam in crab. *Frontiers in Immunology* 9: 801

Lim, J., Coates, C.J., Seoane, P.I. et al. 2018. Characterizing the mechanisms of nonopsonic uptake of cryptococci by macrophages. *The Journal of Immunology* 200: 3539–3546

Little, T.J. and Kraaijeveld, A.R. 2004. Ecological and evolutionary implications of immunological priming in invertebrates. *Trends in Ecology & Evolution* 19: 58–60

Little, T.J., O'Connor, B., Colegrave, N., Watt, K., and Read, A.F. 2003. Maternal transfer of strain-specific immunity in an invertebrate. *Current Biology* 13: 489–492

Liu, H., Söderhäll, K. and Jiravanichpaisal, P. 2009. Antiviral immunity in crustaceans. *Fish & Shellfish Immunology* 27: 79–88

Lu, A., Zhang, Q., Zhang, J. et al. 2014. Insect prophenoloxidase: The view beyond immunity. *Frontiers in Physiology* 5: 252

Majeske, A.J., Bayne, C.J., and Smith, L.C. 2013. Aggregation of sea urchin phagocytes is augmented *in vitro* by lipopolysaccharide. *PLoS ONE* 8: e61419

Meijers, R., Puettmann-Holgado, R., Skiniotis, G. et al. 2007. Structural basis of Dscam isoform specificity. *Nature* 449: 487–491

Metchnikoff, E. 1983. *Lectures on the Comparative Pathology of Inflammation Delivered at the Pasteur Institute 1891* (translated from the French by F.A. Starling and E.H. Starling) London: Kegan Paul

Millar D.A. and Ratcliffe, N.A. 1987a. The antibacterial activity of the hemichordate *Saccoglossus horsti* (Enteropneusta). *Journal of Invertebrate Pathology* 50: 191–200

Millar, D.A. and Ratcliffe, N.A. 1987b. Activity and preliminary characterisation of a hemagglutinin from the hemichordate *Saccoglossus ruber*. *Developmental & Comparative Immunology* 11: 309–320

Mockler, B.K. Kwong, W.K., Moran, N.A., and Koch, H. 2018. Microbiome structure influences infection by the parasite *Crithidia bombi* in bumble bees. *Applied and Environmental Microbiology* 84: e02335–17

Nakashima, K., Kimura, S., Ogawa, Y. et al. 2018. Chitin-based barrier immunity and its loss predated mucus-colonization by indigenous gut microbiota. *Nature Communications* 9: 3402

Nayak, A., Tassetto, M., Kunitomi, M., and Andino, R. 2013. RNA interference-mediated intrinsic antiviral immunity in invertebrates. In *Intrinsic Immunity*. B.R. Cullen (ed.) pp. 183–200. Berlin, Heidelberg: Springer

Ng, T.H., Chang, S.H., Wu, M.H., and Wang, H.C. 2013. Shrimp hemocytes release extracellular traps that kill bacteria. *Developmental & Comparative Immunology* 41: 644–651

Ng, T.H. and Kurtz, J. 2020. Dscam in immunity: A question of diversity in insects and crustaceans. *Developmental & Comparative Immunology* 106: 103559

Nie, L., Cai, S.-Y., Shao, J.-Z., and Chen, J. 2018. Toll-like receptors, associated roles, and signaling networks in non-mammals. *Frontiers in Immunology* 9: 1523

Nillius, D., Jaenicke, E., and Decker, H. 2008. Switch between tyrosinase and catecholoxidase activity of scorpion hemocyanin by allosteric effectors. *FEBS Letters* 582: 749–754

Ortiz-Urquiza, A. and Keyhani, N.O. 2013. Action on the surface: Entomopathogenic fungi versus the insect cuticle. *Insects* 4: 357–374

Pace, K.E., Lebestky, T., Hummel, T., Arnoux, P., Kwan, K., and Baum, L.G. 2002. Characterization of a novel *Drosophila melanogaster* galectin. Expression in developing immune, neural, and muscle tissues. *The Journal of Biological Chemistry* 277: 13091–13098

Palmer, W.H., Varghese, F.S., and Van Rij, R.P. 2018. Natural variation in resistance to virus infection in dipteran insects. *Viruses* 10: 118

Parrinello, D., Sanfratello, M.A., Vizzini, A., and Cammarata, M. 2015. The expression of an immune-related phenoloxidase gene is modulated in *Ciona intestinalis* ovary, test cells, embryos and larva. *Journal of Experimental Zoology Part B: Molecular and Developmental Evolution* 324: 141–151

Peregrino-Uriarte, A.B., Muhlia-Almazan, A.T., Arvizu-Flores, A.A. et al. 2012. Shrimp invertebrate lysozyme i-lyz: Gene structure, molecular model and response of c and i lysozymes to lipopolysaccharide (LPS). *Fish & Shellfish Immunology* 32: 230–236

Pila, E.A., Li, H., Hambrook, J.R., Wu, X., and Hanington, P.C. 2017. Schistosomiasis from a snail's perspective: Advances in snail immunity. *Trends in Parasitology* 33: 845–857

Pinaud, S., Portela, J., Duval, D. et al. 2016. A shift from cellular to humoral responses contributes to innate immune memory in the vector snail *Biomphalaria glabrata*. *PLoS Pathogens* 12: e1005361

Pinsino, A., Thorndyke, M.C., and Matranga, V. 2007. Coelomocytes and posttraumatic response in the common sea star *Asterias rubens*. *Cell Stress Chaperones* 12: 331–341

Poirier, A.C., Schmitt, P., Rosa, R.D. et al. 2014. Antimicrobial histones and DNA traps in invertebrate immunity evidences in *Crassostrea gigas*. *The Journal of Biological Chemistry* 289: 24821–24831

Pope, E.C., Powell, A., Roberts, E.C., Shields, R.J., Wardle, R., and Rowley, A.F. 2011. Enhanced cellular immunity in shrimp (*Litopenaeus vannamei*) after 'vaccination'. *PLoS One* 6: e20960

Portal-Celhay, C., Bradley, E.R., and Blaser, M.J. 2012. Control of intestinal bacterial proliferation in regulation of lifespan in *Caenorhabditis elegans*. *BMC Microbiology* 12: 49

Pradeu, T. and Du Pasquier, L. 2018. Immunological memory: What's in a name? *Immunological Reviews* 28: 3.7–20

Pull, C.D., Ugelvig, L.V., Weisenhofer, F. et al. 2018. Destructive disinfection of infected brood prevents systemic disease spread in ant colonies. *eLife* 7: e32073

Pusceddu, M., Piluzza, G., Theodorou, P. et al. 2019. Resin foraging dynamics in *Varroa destructor*-infested hives: A case of medication of kin? *Insect Science* 26: 297–310

Quinn, E.A., Malkin, S.H., Rowley, A.F., and Coates, C.J. 2020. Laccase and catecholoxidase activities contribute to innate immunity in slipper limpets, *Crepidula fornicata*. *Developmental & Comparative Immunology* 110: 103724

Ramírez-Gómez, F., Aponte-Rivera, F., Méndez-Castaner, L., and García-Arrarás, J.E. 2010. Changes in holothurian coelomocyte populations following immune stimulation with different molecular patterns. *Fish & Shellfish Immunology* 29: 175–185

Rao, X.-J., Wu, P., Shahzad, T. et al. 2016. Characterization of a dual-CRD galectin in the silkworm *Bombyx mori*. *Developmental & Comparative Immunology* 60: 149–159

Ratcliffe, N.A. and Gagen, S.J. 1976. Cellular defense reactions of insect hemocytes *in vivo*: Nodule formation and development in *Galleria mellonella* and *Pieris brassicae* larvae. *Journal of Invertebrate Pathology* 28: 373–382

Ratcliffe, N.A. and Gagen, S.J. 1977. Studies on the *in vivo* cellular reactions of insects: an ultrastructural analysis of nodule formation in *Galleria mellonella*. *Tissue & Cell* 9: 73–85

Ratcliffe, N.A., Rowley, A.F., Fitzgerald, S.W., and Rhodes, C.P. 1985. Invertebrate immunity: Basic concepts and recent advances. *International Review of Cytology* 97: 184–350

Reber, A. and Chapuisat, M. 2012. No evidence for immune priming in ants exposed to a fungal pathogen. *PLoS ONE* 7: e35372

Robalino, J., Bartlett, T., Shepard, E. et al. 2005. Double-stranded RNA induces sequence-specific antiviral silencing in addition to nonspecific immunity in a marine shrimp: Convergence of RNA interference and innate immunity in the invertebrate antiviral response? *Journal of Virology* 79: 13561–13571

Robb, C.T., Dyrynda, E.A., Gray, R.D., Rossi, A.G., and Smith, V.J. 2014. Invertebrate extracellular phagocyte traps show that chromatin is an ancient defence weapon. *Nature Communications* 5: 1–11

Robinson, A.N. and Green, T.J. 2020. Fitness costs associated with maternal immune priming in the oyster. *Fish & Shellfish Immunology* 103: 32–36

Rosenstiel, P., Philipp, E.E., Schreiber, S., and Bosch, T.C. 2009. Evolution and function of innate immune receptors–insights from marine invertebrates. *Journal of Innate Immunity* 1: 291–300

Rowley, A.F. and Powell, A. 2007. Invertebrate immune systems–specific, quasi-specific, or nonspecific? *Journal of Immunology* 179: 7209–7214

Rowley, A.F., Smith, A.L., and Davies, C.E. 2015. How does the dinoflagellate parasite *Hematodinium* outsmart the immune system of its crustacean hosts? *PLoS Pathogens* 11: e1004724

Rubio, T., Oyanedel, D., Labreuche, Y. et al. 2019. Species-specific mechanisms of cytotoxicity toward immune cells determine the successful outcome of *Vibrio* infections. *Proceedings of the National Academy of Science U.S.A.* 116: 14238–14247

Sadd, B.M., Kleinlogel, Y., Schmid-Hempel, R., and Schmid-Hempel, P. 2005. Trans-generational immune priming in a social insect. *Biology Letters* 1: 386–388

Salamitou, S., Ramisse, F., Brehélin, M. et al. 2000. The *plcR* regulon is involved in the opportunistic properties of *Bacillus thuringiensis* and *Bacillus cereus* in mice and insects. *Microbiology UK* 146: 2825–2832

Salvy, M., Martin, C., Bagneres, A.G. et al. 2001. Modifications of the cuticular hydrocarbon profile of *Apis mellifera* worker bees in the presence of the ectoparasitic mite *Varroa jacobsoni* in brood cells. *Parasitology* 122: 145–159

Samuel, G.H., Adelman, Z.N., and Myles, K.M. 2018. Antiviral immunity and virus-mediated antagonism in disease vector mosquitoes. *Trends in Microbiology* 26: 447–461

Sansone, C.L., Cohen, J., Yasunaga, A. et al. 2015. Microbiota-dependent priming of antiviral intestinal immunity in *Drosophila*. *Cell Host & Microbe* 18: 571–581

Sawaya, M.R., Wojtowicz, W.M., Andre, I. et al. 2008. A double S shape provides the structural basis forth extraordinary binding specificity of Dscam isoforms. *Cell* 134: 1007–1018

Schenk, S., Schmidt, J., Hoeger, U., and Decker, H. 2015. Lipoprotein-induced phenoloxidase-activity in tarantula hemocyanin. *Biochimica et Biophysica Acta—Proteins and Proteomics*. 1854: 939–949

Schmucker, D., Clemens, J.C., Shu, H. et al. 2000. *Drosophila* Dscam is an axon guidance receptor exhibiting extraordinary molecular diversity. *Cell* 101: 671–684

Shabelnikov, S.V., Bobkov, D.E., Sharlaimova, N.S., and Petukhova, O.A. 2019. Injury affects coelomic fluid proteome of the common starfish, *Asterias rubens. Journal of Experimental Biology* 222: jeb198556

Shao, Y., Chen, B., Sun, C., Ishida, K., Hertweck, C., and Boland, W. 2017. Symbiont-derived antimicrobials contribute to the control of the lepidopteran gut microbiota. *Cell Chemical Biology* 24: 66–75

Shi, X.Z., Wang, L., Xu, S. et al. 2014. A galectin from the kuruma shrimp (*Marsupenaeus japonicus*) functions as an opsonin and promotes bacterial clearance from hemolymph. *PLoS One* 9: e91794

Sigle, L. and Hillyer, J. 2016. Mosquito hemocytes preferentially aggregate and phagocytose pathogens in the periostial regions of the heart that experience the most hemolymph flow. *Developmental and Comparative Immunology* 55: 90–101

Smith, L.C., Arizza, V., Barela, M.A. et al. 2018. Echinodermata: The complex immune system of echinoderms. In *Advances in Comparative Immunology*. E.L. Cooper (ed.) pp. 409–509. Cham, Switzerland: Springer International Publishing AG

Smith, L.C. and Davidson, E.H. 1994. The echinoderm immune system. Characters shared with vertebrate immune systems and characters arising later in deuterostome phylogeny. *Annals of the New York Academy of Sciences* 712: 213–226

Smith, V.J. 1981. The echinoderms. In *Invertebrate Blood Cells*. N.A. Ratcliffe and A.F. Rowley (eds.) pp. 513–562. New York: Academic Press.

Smith, V.J. and Ratcliffe, N.A. 1981. Pathological changes in the nephrocytes of the shore crab, *Carcinus maenas*, following injection of bacteria. *Journal of Invertebrate Pathology* 38: 113–121

Söderháll, K. and Unestam, T. 1975. Properties of extracellular enzymes from *Aphanomyces astaci* and their relevance in the penetration process of crayfish cuticle. *Physiologia Plantarum* 35: 140–146

Song, X., Zhang, H., Wang, L. et al. 2011. A galectin with quadruple-domain from Bay scallop *Argopecten irradians* is involved in innate immune response. *Developmental & Comparative Immunology* 35: 592–603

Steiner, H., Hultmark, D., Engstrom, A., Bennich, H., and Boman, H.G. 1981. Sequence and specificity of two antibacterial proteins involved in insect immunity. *Nature* 292: 246–248

Su, Y., Liu, Y., Gao, F., and Cui, Z. 2020. A novel C-type lectin with a YPD motif from *Portunus trituberculatus* (PtCLec1) mediating pathogen recognition and opsonization. *Developmental & Comparative Immunology* 106: 103609

Taffoni, C. and Pujol, N. 2015. Mechanisms of innate immunity in *C. elegans* epidermis. *Tissue Barriers* 3: e1078432

Tassanakajon, A., Somboonwiwat, K., and Amparyup, P. 2015. Sequence diversity and evolution of antimicrobial peptides in invertebrates. *Developmental & Comparative Immunology* 48: 324–341

Tasumi, S. and Vasta, G.R. 2007. A galectin of unique domain organization from hemocytes of the eastern oyster (*Crassostrea virginica*) is a receptor for the protistan parasite *Perkinsus marinus. Journal of Immunology* 179: 3086–3098

Terenius, O. 2008. Hemolin—a lepidopteran anti-viral defense factor? *Developmental & Comparative Immunology* 32: 311–316

Tetreau, G., Dhinaut, J., Gourbal, B., and Moret, Y. 2019. Trans-generational immune priming in invertebrates: Current knowledge and future prospects. *Frontiers in Immunology* 10: 1938

Thrupp, T.J., Lynch, S.A., Wootton, E.C. et al. 2013. Infection of juvenile edible crabs, *Cancer pagurus* by a haplosporidian-like parasite. *Journal of Invertebrate Pathology* 114: 92–99

Van Herreweghe, J.M. and Michiels, C.W. 2012. Invertebrate lysozymes: Diversity and distribution, molecular mechanism and *in vivo* function. *Journal of Biosciences* 37: 327–348

Vasta G.R. (2020) Galectins in host–pathogen interactions: Structural, functional and evolutionary aspects. In *Lectin in Host Defense Against Microbial Infections. Advances in Experimental Medicine and Biology*, vol. 1204. S.L. Hsieh (ed.) pp. 169–196. Singapore: Springer

Vasta, G.R., Feng, C., Bianchet, M.A., Bachvaroff, T.R., and Tasumi, S. 2015. Structural, functional, and evolutionary aspects of galectins in aquatic mollusks: From a sweet tooth to the Trojan horse. *Fish & Shellfish Immunology* 46: 94–106

Vasta, G.R., Feng, C., Tasumi, S. et al. 2020. Biochemical characterization of oyster and clam galectins: Selective recognition of carbohydrate ligands on host hemocytes and *Perkinsus* parasites. *Frontiers in Chemistry* 8: 98

Vasta, G.R. and Wang, J.X. 2020. Galectin-mediated immune recognition: Opsonic roles with contrasting outcomes in selected shrimp and bivalve mollusk species. *Developmental & Comparative Immunology* 110: 103721

Vey, A., Matha, V., and Dumas, C. 2002. Effects of the peptide mycotoxin destruxin E on insect haemocytes and on dynamics and efficiency of the multicellular immune reaction. *Journal of Invertebrate Pathology* 80: 177–187

Walters, J.B. and Ratcliffe, N.A. 1983. Studies on the *in vivo* cellular reactions of insects: Fate of pathogenic and non-pathogenic bacteria in *Galleria mellonella* nodules. *Journal of Insect Physiology* 29: 417–424

Wang, J., Wang, L., Gao, Y. et al. 2013. A tailless Dscam from *Eriocheir sinensis* diversified by alternative splicing. *Fish & Shellfish Immunology* 35: 249–261

Wang, L., Song, X., and Song, L. 2018. The oyster immunity. *Developmental & Comparative Immunology* 80: 99–118

Wang, Q., Ren, M., Liu, X., Xia, H., and Chen, K. 2019. Peptidoglycan recognition proteins in insect immunity. *Molecular Immunology* 106: 69–76

Wang, Z., Lv, Z., Li, C., Shao, Y., Zhang, W., and Zhao, X. 2018. An invertebrate β-integrin mediates coelomocyte phagocytosis via activation of septin2 and 7 but not septin10. *International Journal of Biological Macromolecules* 113: 1167–1181

Watson, F.L., Püttmann-Holgado. R., Thomas, F. et al. 2005. Extensive diversity of Ig-superfamily proteins in the immune system of insects. *Science* 309: 1874–1878

Wendling, C.C and Wegner, K.M. 2013. Relative contribution of reproductive investment, thermal stress and *Vibrio* infection to summer mortality phenomena in Pacific oysters. *Aquaculture* 412–413: 88–96

Whitten, M.M. and Coates, C.J. 2017. Re-evaluation of insect melanogenesis research: Views from the dark side. *Pigment Cell & Melanoma Research* 30: 386–401

Whitten, M.M.A., Tew, I.F., Lee, B.L., and Ratcliffe, N.A. 2004. A novel role for an insect apolipoprotein (apolipophorin III) in beta-1,3-glucan pattern recognition and cellular encapsulation reactions. *Journal of Immunology* 172: 2177–2185

Wiens, M., Korzhev, M., Perović-Ottstadt, S. et al. 2007. Toll-like receptors are part of the innate immune defense system of sponges (Demospongiae: Porifera). *Molecular Biology and Evolution* 24: 792–804

Wilson-Rich, N., Spivak, M., Fefferman, N.H., and Starks, P.T. 2009. Genetic, individual, and group facilitation of disease resistance in insect societies. *Annual Review of Entomology* 54: 405–423

Wojda, I., Cytryńska, M., Zdybicka-Barabas, A., and Kordaczuk, J. 2020. Insect defense proteins and peptides. In *Vertebrate and Invertebrate Respiratory Proteins, Lipoproteins and other Body Fluid Proteins, Subcellular Biochemistry* 94. U. Hoeger and J.R. Harris (eds.) pp. 81–121. Switzerland AG: Springer Nature

Wojtowicz, W.M., Wu, W., Andre, I., Qian, B., Baker, D., and Zipursky, S.L. 2007. A vast repertoire of Dscam binding specificities arises from modular interactions of variable Ig domains. *Cell* 130: 1134–1145

Wu, Q., Patočka, J., and Kuča, K. 2018. Insect antimicrobial peptides, a mini review. *Toxins* 10: 461

Wynant, N., Duressa, T.F., Santos, D. et al. 2014. Lipophorins can adhere to dsRNA, bacteria and fungi present in the hemolymph of the desert locust: A role as general scavenger for pathogens in the open body cavity. *Journal of Insect Physiology* 64: 7–13

Xie, J.-W., Cheng, C.-H., Ma, H.-L. et al. 2019. Molecular characterization, expression and antimicrobial activities of a c-type lysozyme from the mud crab, *Scylla paramamosain*, *Developmental & Comparative Immunology* 98: 54–64

Xu, Y.H., Bi, W.J., Wang, X.W., Zhao, Y.R., Zhao, X.F. and Wang, J.X. 2014. Two novel C-type lectins with a low-density lipoprotein receptor class A domain have antiviral function in the shrimp *Marsupenaeus japonicus*. *Developmental & Comparative Immunology* 42: 323–332

Zaidman-Rémy, A., Hervé, M., Poidevin, M. et al. 2006. The *Drosophila* amidase PGRP-LB modulates the immune response to bacterial infection. *Immunity* 24: 463–473

Zanchi, C., Johnston, P.R., and Rolff, J. 2017. Evolution of defence cocktails: Antimicrobial peptide combinations reduce mortality and persistent infection. *Molecular Ecology* 26: 5334–534

Zhang, C., Xue, Z., Yu, Z. et al. 2020. A tandem repeat galectin-1 from *Apostichopus japonicus* with broad PAMP recognition pattern and antibacterial activity. *Fish & Shellfish Immunology* 99: 167–175

Zhao, T., Wei, X., Yang, J., Wang, S., and Zhang, Y. 2018. Galactoside-binding lectin in *Solen grandis* as a pattern recognition receptor mediating opsonization. *Fish & Shellfish Immunology* 8: 183–189

CHAPTER 2

Host defences of invertebrates to non-communicable diseases

Christopher J. Coates

2.1 Introduction

Most often, disease is considered the outcome of a collision between a pathogenic agent and a susceptible host, yet it was Hippocrates of Kos (460–370 BC) who described disease as being a 'disharmony' of the body (or imbalance of the four humours). Cellular and tissue degradation are pathognomonic of bacterial septicaemia and mycosis—but the aetiology of non-communicable diseases (NCDs) (or stressors) is difficult to outline and has received much less coverage. NCDs are not transmissible between hosts. Secondary and opportunistic infections linked to a primary abiotic disruptor can be misdiagnosed as the causative agent. Moreover, stress from fluctuating environmental parameters and pollutants, and injury from predators/conspecifics can compromise the health of an invertebrate, paving the way for opportunistic infections (Sparks 1972). In Chapter 1, the reader is made aware of the broad defences designed to thwart micro- and macro-parasites, yet terrestrial and aquatic invertebrates must also contend with a myriad of xenobiotic contaminants, such as hydrocarbons, heavy metals, pesticides, plastics, pharmaceuticals, and toxins. These contaminants can be found in the environments they reside in (air, water, and soil), the foods they eat due to bioaccumulation, and across trophic levels due to biomagnification.

Herein, invertebrate defences toward 'sterile' diseases and stress are discussed, with a particular emphasis placed on the 'chemical defensome', which is made-up of REDOX enzymes, detoxicants, antioxidants, ATP-dependent efflux transporters, and the factors that regulate and maintain the detoxification machinery.

2.2 Stress avoidance—defensive behaviours in response to physical threats and injury

As stated in Chapter 1, (Section 1.2), social invertebrates will avoid and ostracise infected conspecifics, groom and self-medicate with antimicrobials to prevent the contraction and spread of disease. Examples exist of invertebrates displaying defensive behaviours that avert non-communicable conditions from stress, injury, and intoxication, but are distinct from those strategies used to avoid infection. Perhaps the most obvious behavioural defence is **autotomy** or self-amputation among arthropods. For example, orb-weaver spiders (*Argiope* spp.) will undergo limb autotomy when envenomed by predatory hymenopterans and hemipterans (e.g. the ambush bug *Phymata fasciata*; Eisner and Camazine 1983). In laboratory trials, > 65% and ~ 80% of spiders injected with honeybee and wasp venom, respectively, discarded the inoculated limb. Autotomy is common among marine invertebrates, e.g. crab claws/legs and starfish arms, when they are stressed (warding off predators) or wounded (reviewed by Lindsay 2010). An extreme example

Christopher J. Coates, *Host defences of invertebrates to non-communicable diseases*. In: *Invertebrate Pathology*. Edited by Andrew F. Rowley, Christopher J. Coates and Miranda M.A. Whitten, Oxford University Press. © Oxford University Press (2022). DOI: 10.1093/oso/9780198853756.003.0002 is **evisceration** in sea cucumbers (Holothuroidea), which involves the propulsion of internal organs such as the respiratory tree, haemal vessels, tentacles, and gut through the anus—a process that is accomplished by gross muscle contraction, connective tissue alterations, and visceral tearing. Sea cucumbers such as *Eupentacta quinquesemita* eviscerate when targeted by fish or crabs in an attempt to dissuade their attacker, and in some cases, immobilise them (Byrne 2001). The autotomised tissues will regenerate fully after about 2 to 5 weeks depending on the species.

Other defences to predators—displayed almost exclusively by insects—are **auto-haemorrhaging** and **autothysis** (altruistic suicide). Armoured ground crickets (*Acanthoplus discoidalis*) project acrid smelling haemolymph up to 30 mm distance from connective tissues in the legs (trochanter and coxa), and the pronotum where fluid seeps out (Bateman and Fleming 2009 and references cited therein). At least five insect orders are capable of reflex bleeding/squirting, and when haemolymph is mixed with toxic, sticky, and corrosive substances, it is an effective chemical weapon (Moore and Williams 1990; Bateman and Fleming 2009). In soldiers of the termite *Globitermes sulphureus*, contraction of muscles around a frontal gland leads to dehiscence and rupturing of tissue through the integument, causing a tar-like substance to spill out onto their enemy (Bordereau et al. 1997). Older termite workers of the species *Neocapritermes taracua* mix copper-containing crystals with a salivary gland extract in order to explode onto their targets and protect the colony from an acute threat (Šobotník et al. 2012). Additional self-destructive behaviours include suicidal death grips and adhesive self-sacrifice (Shorter and Rueppell 2012).

Altogether, the previous examples represent costly responses to non-pathogenic assault—losing a limb or masses of solid/liquid tissues is risky as the actor is likely to be left vulnerable, at least in the immediate aftermath, to other stressors.

2.3 Injury repair and sterile inflammation

In humans, NCDs are synonymous with autoimmune disorders, cancers, nutritional imbalance, and environmental diseases caused by the pollution of soil, water, and air (reviewed by Fuller et al. 2018). For invertebrates, 'stress' and idiopathic conditions are likely consequences of acute and chronic exposure to environmental agitators (Figure 2.1; Figure 2.2). Terrestrial and aquatic invertebrates co-exist with their would-be microbial antagonists, where the number of infected hosts are in the minority because of how effective innate immune defences are at repelling and controlling infectious agents (see Chapter 1). Invertebrates must detect and repair injuries rapidly to remove effete and dead cells, and to keep those areas aseptic—a collective process known as inflammation (Box 2.1). The inflammatory programmes of invertebrates are initiated regardless of whether the damage is biological, chemical, or physical in origin.

Traumatic (mechanical) injuries from blunt force, predation, conspecific aggression, overcrowding, environmental 'stress' (e.g. sedimentation; Box 2.2), and mishandling in managed settings can range from superficial epidermal and shell discolouration, fracturing and disfigurement of structural features, focal lesions, and localised tissue loss, to transcuticle and subdermal destruction. The vital organs of most known invertebrates are situated within the body cavity (haemocoel) and are in direct contact with the blood (haemolymph, coelomic fluid) in an open circulatory system; therefore, a wound must be sealed to curtail lymph/fluid loss and pathogen entry. From molecular to cellular components, clotting and wound healing have been resolved in a diverse range of invertebrates, e.g. shrimp, insects, corals (Rowley and Ratcliffe 1978; Palmer et al. 2011; Perdomo-Morales et al. 2019). Although most of our understanding of clotting comes from studies on arthropods (e.g. Schmid et al. 2019), the process follows a conserved pattern of events across taxa. An initial Phase 1 soft (protein) clot is formed with the aid of haemostatic and proteolytic factors, fibrillar polymers, and cross-linking proteins (usually a transglutaminase) to temporarily seal the breach and entrap microbes in the process. In fact, the coagulin-based clot in horseshoe crabs (*Limulus polyphemus*) is so effective at immobilising bacteria, the flagella of *Vibrio alginolyticus* stop moving and there is a lack of Brownian motion. Isakova and Armstrong (2003) refereed to this as imprisonment on death row, where the bacteria will be killed by serum antimicrobials within hours. Phase 2, or the second hardening phase, involves reinforcement

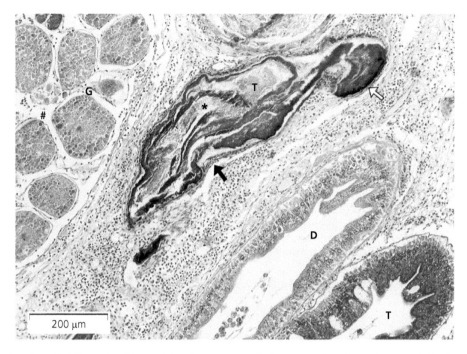

Figure 2.1 Putative sterile inflammation of damaged tissue in a crustacean, *Carcinus maenas*. There is gross haemocyte infiltration and reactivity within the inter-tubular spaces of the hepatopancreas. The asterisk (*) denotes tissue degeneration and loss of cellular structure. The black arrow points to haemocyte layers walling-off the damaged tissue and the white arrow points to melanised debris. The hashtag (#) denotes haemocytes present within the gonadal tissue (but there is no sign of tissue trauma). D, digestive tissue; G, gonadal tissue; T, tubule. This shore crab was examined as part of a yearlong disease survey conducted by Davies et al. (2019). The micrograph (H & E staining) was captured from a female crab (collected in December 2017, Mumbles Pier (Swansea), Wales UK) with no measurable signs of infection. Microscopic inspection of haemolymph did not reveal any visible bacterial/fungal cells. Plating haemolymph on TSA (+2% NaCl) medium did not lead to microbial colony growth. Using tissue histology and PCR-based methods, there was no evidence of microsporidians, haplosporidians, *Hematodinium* spp., *Sacculina carcini*, trematodes, paramyxids, mikrocytids, *Vibrio* spp. or fungal species. The aetiology is of unknown origin.

of the clot through haemocyte adhesion driven by agents like peroxinectin and insoluble melanin scaffolds generated through phenoloxidase-mediated cascades (Theopold et al. 2004). The resulting 'scab' acts as a temporary extracellular matrix for the re-establishment of an epithelial barrier. See Chapter 1, Section 1.4 for details on blood gelation in other invertebrates.

In addition to physical assault, harm caused by chemicals (xenobiotics) can induce sterile inflammation—host tissue repair and immune activation in the absence of a pathogen (Figure 2.2). If the stressor persists or the internal damage is not resolved, it can cause morbidity and death. The body must be equipped to detect so-called damage-associated molecular patterns (DAMPs), which are inextricably linked to haemostasis, and occur via receptor-ligand interactions in a similar manner to the detection of pathogen-associated

molecular patterns (PAMPs; Chapter 1, Section 1.6). DAMPs are derived from 'self' tissues, whereas PAMPs are non-self, exogenous signals of microbial surface ligands. DAMPs are conserved among invertebrates and vertebrates; examples include oxidised phospholipids and lipoproteins, extra-cellularised inner membrane phospholipids such as phosphatidylserine, liberated mitochondrial components, and actin (Coates et al. 2013, 2018; Srinivasan et al. 2016). Seong and Matzinger (2004) proposed that the leaking and uncontrolled release of usually intracellular macromolecules acts as an alarm/danger signal due to their hydrophobicity, thereby representing an ancient immune-surveillance system. Several signalling cascades involving human toll-like receptors (TLRs) and dendritic cell-specific sensor of tissue damage receptor (DNGR-1) detect DAMPs and trigger repair responses—this process appears similar in

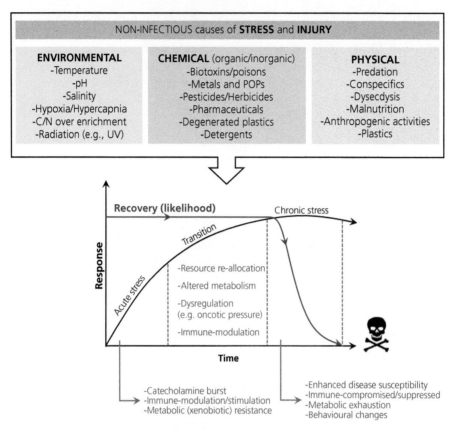

Figure 2.2 Stress-associated outcomes in invertebrates. A single, acute stress event tends to have a transient, immune-stimulatory impact on an invertebrate. Prolonged exposure (subacute) to stress or several sources of stress (environmental, physical, chemical), leads to a systemic response, in which there is a metabolic shift to maintain homeostasis (allostasis) and repair putative damage. Chronic stress, e.g. from pollutant exposure over several weeks to months, can have serious detrimental impacts on an invertebrate, e.g. enhanced disease susceptibility or tissue damage to such an extent that repair is no longer possible. C/N, carbon/nitrogen; POPs, persistent organic pollutants. Nardocci et al. (2014) inspired the lower panel of this figure.

invertebrates. Administration of actin (DAMP) to *Drosophila melanogaster* activated a signalling pathway involved in dead cell clearance and wound repair; expression of tyrosine kinases in the fat body led to the production of cytokines (Upd), which act as ligands for the JakSTAT-coupled receptor Domeless that switches on the transcription of STAT-responsive genes (Srinivasan et al. 2016).

2.4 Anti-tumour defences in invertebrates

Disease and damage can manifest in many forms independent of pathogens or have unknown aetiologies (Table 2.1)—tumourigenesis being one of them. Tumour establishment or neoplasms have been reported in invertebrates as early as the 1800s (reviewed by Scharrer and Lochhead 1950), yet knowledge remains scant with respect to how these ectopic masses arise and what anti-tumour defences are deployed. Spontaneous genetic abnormalities, hereditary predisposition, and the failure of molecular controls (i.e. evasion of apoptosis) must be contributing factors to neoplasia, but links to pollutant exposure (e.g. carcinogens) in the field are disparate, as are viral sources of oncogenesis, and both lack sufficient experimental validation. Key features that define vertebrate tumour formation are not directly applicable to invertebrates, such as angiogenesis (Robert 2010), and there has been

---

Box 2.1 Inflammation in invertebrates

In humans, there are five signs of acute inflammation; heat (*calor*), pain (*dolor*), redness (*rubor*), swelling (*tumour*), and loss of function (*functio laesa*), which are not all shared by invertebrates. Nevertheless, inflammation can be described as the local, protective tissue response to injury and microbial intrusion by the coordinated efforts of blood/lymph vessels, immune cells, and chemical mediators—processes that do occur in spineless animals (Sparks 1972). In the following table, the classical signs of inflammation are compared/contrasted; **loss of function** is apparent across all metazoans and can be attributed to tissues/organs being damaged beyond repair from the initial stimulus, or as a consequence of prolonged swelling, cellular infiltration, and (hyper)reactivity (i.e. collateral damage, mobility inhibition).

Sign	Mammal	Invertebrate
Heat and **Redness**	**YES.** Heat and redness are visible reactions to tissue trauma caused by vascular dilation and permeability, and increased blood flow (containing erythrocytes) from core/interior regions.	**NO.** Cold blooded (ectothermic). **NO.** Invertebrate haemolymph (functional equivalent to blood) tends to be blueish in colour due to copper-based haemocyanin (e.g. crustaceans, chelicerates, gastropods), pink/purple in colour due to iron-based haemerythrin (e.g. annelids, sipunculids), or yellowish-green caused by pigments (e.g. insects). Deoxygenated haemolymph is colourless (clear to opaque). In some cases, it is red due to iron-based (usually extracellular) haemoglobin (e.g. blood clams *Tegillarca* species; Coates and Decker 2017).
Swelling	**YES.** Fluid retention and pooling of blood influxes from the circulatory system.	**YES.** Oedema does occur in certain tissues (e.g. gills; see Figure 9.6g), but in contrast to the vertebrate circulatory system, the invertebrate organs bathe in the haemolymph/coelomic fluid within a body cavity. Immune cells (haemocytes, coelomocytes) are drawn to sites of damage/infection, which can also lead to swelling. Interestingly, Yan and Hillyer (2020) demonstrated that phagocytic haemocytes aggregate at key sites of haemolymph flow (heart valves) during infection—providing compelling evidence for the integration of the immune and circulatory systems in insects
Pain	**YES.** Release of bradykinin and histamine to stimulate nerves, as well as regional swelling, tissue distortion, and exudate congestion putting pressure on nerves.	**YES** and **NO.** Nociceptors are widespread, yet invertebrate pain and sentience remain controversial topics. Evidence does exist in favour for so-called advanced invertebrates, i.e. cephalopods and decapods, through physiological changes, (e.g. crustacean hyperglycaemic hormone), avoidance learning, and protective 'motor' reactions (Elwood 2011 and 2019).

limited effort to identify derivation of abnormal cells using immunohistochemistry.

Some features like tissue invasion can be difficult to confirm in an open circulatory system, whereas sustained growth and anaplasia are clearer under microscopic examinations. Of course, just like vertebrates, there are preventative measures in place, such as analogues of p53 tumour suppressors involved in DNA damage repair and cell cycle control (including apoptosis; Muttray and

Table 2.1 Examples of non-infectious and sterile inflammatory conditions in invertebrates

Disease/Syndrome/Neoplasia	Host	Pathologic features	References
Carcinoma-like neoplasm	King crab (*Paralithodes camtschatica*)	Tissue mass present on the ventral side of the anterior hindgut. Epithelial cells were large and pleomorphic, with prominent hypertrophied and amorphous nuclei (anaplastic). The extent of tissue invasion ranged from diffuse to excessive. There was clear evidence of haemocyte reactivity—phagocytic granular cells engulfed individual abnormal cells, and some 'free' sections of the tumour were seemingly encapsulated.	Sparks and Morado (1987)
Dehydration, Emaciation	Arachnids, Insects, Entognatha	Manifesting externally as shrinkage or cavitation of the abdominal segments, and internally, the digestive gland/tissues are usually contracted with increased levels of vacuolation and granulation of pyramidal cells.	Newton and Smolowitz (2018)
Distress syndrome	Freshwater, pulmonate gastropods (e.g. *Biomphalaria glabrata*)	Caused by copper toxicity. Features include cardiac arrhythmia, muscular spasms, immobility, swelling of tentacles, and cephalopedal mass, epidermal sloughing, and loss of foot adherence to a substrate.	Cheng and Sullivan (1973); LaDouceur et al. (2016)
Eversion syndrome	Jellyfish (scyphomedusae)	The bell everts and becomes concave, and thereby exposes the gastric cavity. This idiopathic disease is characterised by swelling of striated muscle, necrosis, and mesogleal deterioration in the form of fibre loss and thinning.	Freeman et al. (2009)
Hamartoma (or papilloma)	-Lobster (*Homaraus americanus*) -Shrimp (*Penaeus aztecus*)	Excessive proliferation of exophytic tumour-like tissue protruding through the carapace (abdominal somites). The gross architecture is rugose/corrugated with a hyper-pigmented (blue) apex. The tumours are fibrous and incorporate some irregular cuticle structures and epithelioid cells displaying hypertrophy (vacuolation), and hyperchromasia when assessed histologically.	Shields and Small (2013); Sparks and Lightner (1973)
Hemic (disseminated) neoplasia	Bivalves Fruit flies Shrimp	Large, spherical, anaplastic cells that are up to four-times larger than haemocytes (with distinct nucleoli) present in connective tissue, lymph vessels, muscle, visceral mass, and mantle. The transformed cells are known to invade gonadal tissue, causing atrophy, and degeneration that leads to gametogenesis arrest. Invading cells also cause displacement, compression, and necrosis in connective and gill tissues. Eventually, the tissues become deregulated to such an extent that morbidity and mortality are inevitable.	Farley (1969a/b); Sparks (1972); Gateff (1978); Shrestha and Gateff (1986); Lightner and Brock (1987); reviewed by Tascedda and Ottaviani (2014)
Leukemic disease (Hematopoietic neoplasia)	Soft-shell clam (*Mya arenaria*)	Circulating cells within the haemolymph with evidence that the untransformed cell is a connective tissue cell.	Smolowitz et al. (1989)
Moult death syndrome (MDS)	Decapods (larval)	Dysecdysis leading to missing/deformed limbs and *in exitus*. Internally, multi-focal calcium deposits are visible along the exuvium, e.g. claws. MDS is considered a result of nutrient and diet deficiencies, notably lipids. Juvenile *H. americanus* fed on casein-based diets suffered high levels of death due to an inability to complete ecdysis. Addition of soybean-derived phospholipids (i.e. choline) and glycerides were effective at reducing the occurrence of MDS syndrome in shrimp *Penaeus japonicas*.	Bowser and Rosemark (1981); Coutteau et al., (1997); Wang et al. (2016)
Necrotic dermatitis syndrome (or bell rot)	Jellyfish species	Associated with blunt trauma (hitting tank walls, conspecifics) and inadequate housing conditions (environmental stress) in aquaria. The bell (umbrella) becomes eroded and ulcerative lesions can develop. Histologically, nuclear aberrations, epidermal necrosis, cytolysis, and hyperplasia in umbrellar layers are common, as well as immune cell (amoebocyte) infiltration.	LaDouceur et al. (2013)

Box 2.2 Sedimentation-driven damage in Cnidaria

Siltation and the excessive displacement of sediment due to agricultural run-off, drilling, and dredging can cause sedimentation stress in cnidarians, notably corals and inhabitants of coastal reef systems. Particulate size, as well as organic and nutrient contents are key contributing factors to sedimentation stress (Weber et al. 2006). Prolonged exposure can reduce photosynthetic outputs of corals (Riegl and Branch 1995; Philipp and Fabricius 2003), and cause tissue-wide damage. Exposure of star coral colonies (*Montastraea cavernosa*) to >200 mg cm^{-2} sediment for 4 weeks in laboratory settings led to drastic deteriorations in histopathological condition, from swelling and polyp retraction within

1 week, to increased granularity of gastrodermal cells and the calicoblastic epithelium, as well as tissue atrophy, necrosis, and debris accumulation along lower and middle polyp regions (Vargas-Angel et al. 2006). Such pathologic features were also observed in field-collected corals from different sites, with evidence of hypertrophic mucocytes, amoebocyte accumulation, and reduced numbers of zooxanthellae (Vargas-Angel et al. 2007). Ostensibly, injured tissues were infiltrated by amoebocytes, in what has been confirmed as a conserved cellular response to damage/wounding among scleractinian corals and metazoans (Palmer et al. 2011).

Vassilenko 2018). Interestingly, > 50% of the known proteins associated with disease and uncontrolled cell proliferation in humans have homologues in fruit flies (Mirzoyan et al. 2019).

Mollusc and insect tumours appear to be the most frequently reported and are likely encountered due to the intensive study and commercial importance of their members, yet tumour occurrence in their crustacean counterparts is comparatively rare (Peters et al. 2012). With that said, there is clear evidence from two decapod studies on the role of haemocytes in circumscribing tumour spread. Circulating granular haemocytes of the red king crab (*Paralithodes camtschaticus*) targeted and phagocytosed neoplastic cells (Sparks and Morado 1987). Tissue micrographs depicted vast numbers of co-located haemocytes containing eosinophilic granules trying to encapsulate the primary neoplasm situated at the anterior of the hindgut (Table 2.1). More recently, Shields and Small (2013) described a so-called hamartoma protruding from the abdominal dorsum of an intermoult American lobster (*Homarus americanus*) with distinct blue colouration and multi-focal melanisation (black in appearance; Figure 2.3a). The authors postulated that the aberrant growth originated from deregulated wound repair linked to dysecdysis (cuticle rupture). This interpretation may also explain the gross blue and black pigmentation that is likely

from the incorporation of haemolymph containing oxy-haemocyanin (blue) and derived phenoloxidase activities (black eumelanin). Necropsy and histologic examination of the tumour revealed pervasive granular haemocyte infiltration (Figure 2.3b). Haemocyte-mediated defences that are traditionally associated with pathogen clearance, namely encapsulation, nodulation, and melanisation, form part of the anti-tumour defences. In insects, haemocytes can distinguish between healthy and neoplastic cells, isolating the latter and forming melanin-enriched aggregates (capsules) to occlude neoplastic tissue areas (Nappi 1984; Vogt 2008). Melanotic tumour formation in drosophilids are heritable (Sparrow 1978)—linked to mutation in genes called *wizard* and *dappled* (Rodriquez et al. 1996). Hyperactivation of the JakSTAT pathway—associated with PAMP and DAMP responses—can provoke epithelial tumours (Amoyel et al. 2014).

The most reported invertebrate tumours are hemic (or disseminated) neoplasia caused by transformed cells in the haemolymph, i.e. bivalve leukaemia (although other invertebrates have liquid tumours). The transformed cells that cause this disease are genetically distinct from host haemocytes and are transmissible among closely related taxa (Carballal et al. 2015; Metzger et al. 2016; Yonemitsu et al. 2019; see Chapter 8).

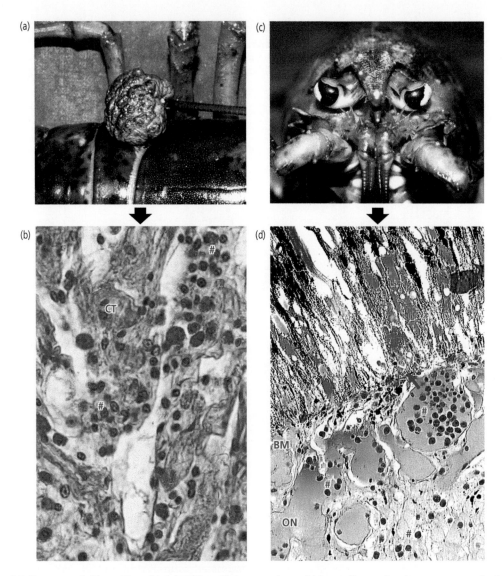

Figure 2.3 Non-communicable conditions of American lobster, *Homarus americanus*. (a) Tumour-like growth, or hamartoma, protruding from an abdominal somite of a lobster caught off Maine (USA). Note, the distinct blue and black colouration. (b) Tissue micrograph of the stroma within the hamartoma. The stroma is replete with infiltrating haemocytes (#), and irregular (fibrous and spongy) connective tissue (CT) scaffolds. Some cells have shrunken (pyknotic) nuclei. (c) Lobster with idiopathic blindness in both eyes. (d) Example tissue micrograph from an affected region of the eye showing haemocytes responding to damage. Haemocytes (#) accumulated in the haemal sinuses of the optic nerve (ON) area, alongside the clumping of pigment (arrow), basement membrane (BM). Source: Prof Jeff Shields (Virginia Institute of Marine Science, USA) kindly provided the original images for (a-d).

2.5 Cell stress responses and metabolic resistance

The evolutionarily conserved gene and protein networks that exercise the various phases of metabolic resistance described below are referred to collectively as the 'chemical defensome'. Goldstone et al. (2006) and Goldstone (2008) surveyed the genomes of the Pacific purple sea urchin (*Strongylocentrotus purpuratus*) and starlet sea anemone

(*Nematostella vectensis*) to reveal ~400 and 266 genes, respectively, associated with protection against chemical stress. Moreover, the chemical defensome keeps the invertebrate body safe from natural agents such as heavy metals, macro-eukaryotic and microbe-derived products, and those that are manufactured, e.g. hydrocarbons, herbicides, plastics, nanoparticles, and drugs. With few exceptions, there is general conformity among higher-order gene groups responsible for sensing and signalling xenobiotic presence—usually gustatory and olfactory receptors, e.g. aryl hydrocarbon receptor and nuclear factor modulate the CYP450 and multioxidase system (Goldstone 2008, and references therein).

2.5.1 Invertebrate defences to changeable environments

Environmental instability or disruption—usually in the form of anthropogenic-derived pollutants and climate shocks—drives episodic outbreaks of disease, and mass die-off events. There has been much scrutiny of the putative effects of climate change scenarios on the health of invertebrates (biodiversity, fecundity, disease outcomes), which is beyond the scope of this chapter, but these topics are embedded throughout the taxonomic chapters of this volume.

Acute stress events lead to immune-stimulation, but their gross negative impact is most often transient, and homeostasis can be restored (phenotypic plasticity). Sub-acute, prolonged, and chronic exposure to a stressful stimulus or a combination of stressors is debilitating, broadly immunosuppressive, and a drain on metabolic resources (Figure 2.2; e.g. Goulson et al. 2015; Johnson et al. 2016). Of course, the challenges of coping with stress for an invertebrate are influenced by spatial and temporal factors, the magnitude of the insult, as well as the acquisition and allocation of resources to fuel physiological reprogramming (see Adamo (2012) for neuroendocrine mediators). Adverse aquatic conditions in the form of pH and salinity fluctuations, C/N overloading, temperature extremes, oxygen depletion (hypoxia, hypercapnia), and excess radiation (UV/ionising) can induce a stress state and the accumulation of noxious by-products such as reactive oxygen/nitrogen radicals derived from impaired

macromolecules (Coates and Söderhäll 2021; Kett et al. 2020). The latter three variables also burden terrestrial animals.

When faced with unfavourable salinity levels for example, aquatic invertebrates trigger osmotic control mechanisms to modulate the free amino acid pool and ion composition of intra- and extra-cellular fluids regardless of whether they are osmoconformers or osmoregulators (reviewed by Hauton 2016). Acclimation and the continued control of cell volume through ion regulation are essential for proper cell functioning (e.g. enzyme-ligand interactions). In decapods, high concentrations of ion transporters and membrane channels (sodium, potassium, chloride, calcium) are found along the gills and digestive canals in addition to a multitude of carbonic anhydrases for acid-base maintenance and pH control (e.g. Towle et al. 2011)—all of which show marked increases in activity and synthesis when cells are stressed.

The **cellular stress response** (CSR) is an ancient, conserved, and ubiquitous defence mechanism of pathophysiological maintenance (Kültz 2005). Regardless of the source of stress (e.g. UV, temperature, pollutant), the sensing of derived macromolecular damage (protein, DNA, lipid) triggers a series of generalised responses in all invertebrates. Variability in the response to a stressor arises as a function of the tissue type, species, or population being investigated. **Heat shock proteins** (HSPs) were originally characterised in fruit flies (*D. melanogaster*) in response to unfavourable temperatures (Tissiéres et al. 1974). Since then, their roles as chaperones have been elucidated, in addition to the discovery of extended family members, ATP-dependent HSP 60s, 70s, and 90s, and small ATP-independent HSP20s. HSPs refold denatured proteins, discourage protein aggregation, and if proteins are damaged beyond repair, they chaperone the deformed macromolecules through degradation pathways (polyubiquitination and proteasomal hydrolysis). Roles of HSPs extend beyond thermal stress (cold and hot) to include other abiotic factors like salinity (e.g. sea cucumbers; Meng et al. 2011) and anoxia (e.g. insects; King and MacRae 2015), stress hardening and cross tolerance (Sejerkilde et al. 2003). In humans, some HSPs are immune-modulators, and recent investigations have aligned them with invertebrate innate immunity and pathogen resistance, e.g. shrimp HSP70 and *Vibrio parahaemolyticus* (Junprung et al. 2019).

Responding to suboxia and anoxia represents another conserved approach of the CSR. Oxygen is vital for ATP generation, tissue alimentation, wound healing, and immune defences, e.g. NADPH-dependent respiratory burst, melanin biogenesis (Coates and Decker 2017). **Hypoxia inducible factors** (HIFs) accumulate in cells when oxygen is in short supply (ischemia), but not during normoxia as they are regulated by prolyl hydroxylase domain enzymes (PHDs; Piontkivska et al. 2011; Giannetto et al. 2015; Wang et al. 2015). PHDs are distributed widely among metazoans and are considered oxygen sensors. When oxygen supply is sufficient, they hydroxylate subunits of HIF to restrict them to the cytoplasm. Inadequate oxygen tension or hypercapnia disrupt PHD suppression of HIF, thereby freeing the subunits to enter the nucleus, switch-on the expression of hypoxia response elements (e.g. anaerobic glycolysis) and proteins that work towards oxygen conservation and delivery to tissues, e.g. enhanced production of oxygen transporters in crustaceans and tracheal remodelling in insects (Gorr et al. 2004, 2006).

2.5.2 Coping with oxidative stress

Routine mitochondrial respiration generates potentially harmful oxygen-containing by-products known as reactive oxygen species (ROS) or radicals (e.g. hydroxyl radical $^{\bullet}$OH). A series of common antioxidant proteins (Table 2.2) and pathways disarm these. Intra- and extracellular levels of ROS and nitrosative radicals increase when pathogens, environmental factors, and pollutants antagonise cells, e.g. phagocyte-mediated killing of microbes, annelids exposed to silver nanoparticles in soil (Gomes et al. 2015), salinity-induced oxidative stress in coastal invertebrates (Rivera-Ingraham and Lignot 2017). Excessive production and release of ROS is detrimental, as these unstable molecules react with macromolecular structures:

(1) Striking cell and organelle membranes, causing lipid peroxidation and leakiness;

(2) Breaking DNA strands and forming base adducts (e.g. 8-hydroxy-2′-deoxyguanosine);

(3) Amino acid oxidation (e.g. tryptophan), enzyme inactivation, protein misfolding, and structural damage.

If cell structures cannot be maintained or recycled (autophagy), then cell death (apoptosis) is often the end-point of radical accumulation. Mitogen-activated protein kinase (MAPK) pathways, oxidoreductases, dehydrogenases, and hydrolases manage REDOX sensing, communication, and regulation. (Hatanaka et al. 2009). Constitutive and inducible REDOX enzymes work in concert to control ROS (see Table 2.2)—superoxide dismutase uses $2H^+$ to convert superoxide anions ($2x\ O_2^{\bullet-}$) into hydrogen peroxide (H_2O_2) and dioxygen (O_2), and subsequently catalase converts hydrogen peroxide ($2x\ H_2O_2$) into water ($2x\ H_2O$) and dioxygen. Glutathione peroxidase can perform the second reaction, and alternatively, uses glutathione to reduce lipid hydroperoxides (R-OOH) to alcohols (ROH). The tripeptide glutathione is a key antioxidant found in animal, plant, and microbial cells. Glutathione scavenges ROS directly, and crucially assists antioxidant enzymes by acting as a reducing equivalent. Glutathione is oxidised to glutathione disulphide during detoxification events, and importantly, it is recycled into its reduced form by the enzyme glutathione reductase. Such antioxidant machinery has been reported across the invertebrate spectrum in response to radicals generated through stress exposure, e.g. metals, toxins (annelids, Adeel et al. 2019; crustaceans, Bell and Smith 1995; echinoids, Goldstone et al. 2006; insects, Dubovskiy et al. 2011).

2.5.3 Xenobiotic metabolism and disposition

Invertebrates differ greatly in terms of physiologies and life history, and are exposed to a battery of xenobiotics, e.g. polychlorinated biphenyls (PCBs), pesticides, and pharmaceuticals. Despite the vast chemical diversity encountered, most xenobiotics are hydrophobic (lipophilic, nonpolar, partially ionised at physiological pH)—a property that promotes their passage across tissue barriers (gills, integument, digestive, and reproductive tracts) and into cells. The main subcellular sites of metabolic resistance are the endoplasmic reticulum, the cytosol, mitochondria, and lysosomes, with the

Table 2.2 Common protein components of the 'chemical defensome'

	Antioxidant and xenobiotic metabolism activities	References[#]
ATP binding cassette transporters (e.g. P-glycoproteins)	**Phase III** Export bio-transformed xenobiotics out of the cell. Expansion of these efflux transporters—gene amplifications and upregulation—leads to multi-xenobiotic resistance (e.g. pesticides, pharmaceuticals).	Venn et al. 2009; Li et al. 2019
Catalase [EC 1.11.1.6]	Converts hydrogen peroxide to water and oxygen	Adeel et al. 2019; Bell and Smith 1995
Cytochrome P450 monooxygenases [EC 1.14.14.1]	**Phase I** A superfamily of enzymes that incorporate a hydroxyl (OH) group to a substrate/target (i.e. mono-oxygenase activity)	Snyder 2000; Van Leeuwen and Dermauw 2016; Li et al. 2007
Esterases (non-specific) e.g. Carboxyl/cholinesterase [EC 3.1.1.1]	**Phase I** A diverse family of hydrolases—using water to split esters into an alcohol and an acid. E.g. Carboxyl/cholinesterases play a key role in dietary detoxification processes in insects.	Van Leeuwen and Dermauw 2016; Haites et al. 1972
Glutathione reductase [EC 1.8.1.7]	Reduces glutathione disulphide (GSSG) to the sulfhydryl form (GSH), which is used to prevent oxidative stress.	Gamble et al. 1995; Regoli and Principato 1995
Glutathione-S-transferase [EC 2.5.1.18]	**Phase II** Conjugates the reduced (sulfhydryl) form of glutathione to xenobiotics (to make them more water soluble, i.e. detoxification)	Saint-Denis et al. 1998; Van Leeuwen and Dermauw 2016
Peroxidases, e.g. Glutathione peroxidase [EC 1.11.1.9]	Large family of proteins. Protection from oxidative damage; reduce steroid/lipid hydroperoxides to alcohols, and reduce hydrogen peroxide to water	Bell and Smith 1994, Gamble et al. 1995; Adeel et al. 2019
Peroxiredoxins (broad family of enzymes) [EC 1.11.1.24]	These enzymes are ubiquitous antioxidants using REDOX-active (peroxidatic) cysteine residues to scavenge peroxide and peroxynitrite. The active site cysteine attacks the peroxide substrate and is oxidised to sulfenic acid.	Abbas et al. 2019; Dai et al. 2018; Tu et al. 2018; Zhang and Lu 2015
Superoxide dismutase (SOD) [EC 1.5.1.1]	Catalyses the disproportionation (partitioning) of superoxide radicals into dioxygen and hydrogen peroxide. Cu- and Zn-conjugated SODs are found widespread in the cytoplasm of eukaryotic cells, whereas Mn-conjugated SODs are located within mitochondria.	Adeel et al. 2019; Bell and Smith 1994. Richier et al. 2003

Examples of literature form terrestrial and aquatic invertebrates. It is not an exhaustive list.

hepatopancreas and fat body representing the main detoxification organs in arthropods (Dallinger 1993; Ahearn et al. 2004; Li et al. 2019). Gills, gonads, and lymph/fluid cells are additional sites of detoxification in arthropods, molluscs, and echinoderms (Snyder 2000).

The processing of xenobiotics can be categorised broadly into three phases:

- Modification or functionalisation (I). Location, endoplasmic reticulum (microsomes).

Some toxicants and metals are repelled immediately from cells without the need for Phases II and I, as ATP-binding cassette (ABC) transporters (i.e. efflux pumps) can block entry. These transporters may not recognise anthropogenic-derived toxicants, especially if a complex mix of xenobiotics saturate binding sites—hence, they circumvent the plasma membrane and make their way inside (Epel et al. 2008). From microbes to mammals, cytochrome P450 monooxygenase enzymes (or mixed function oxygenases) represent a superfamily of proteins, with members acting as one of the major first-line defences used against xenobiotics (Snyder 2000). Invertebrates like *C. elegans* and *D. melanogaster* have dozens of CYP genes. The most studied is CYP1A1—induced by dioxins such as PAHs and PCBs. CYP450-mediated reactions include hydroxylation, epoxidation, oxidative deamination, S/N/O-de-alkylation, and dehalogenation. Depending on the substrate, some products of CYP450s may be more reactive and damaging to the host. Phase-I detoxicants also reduce unsaturated bonds to saturated, hydrolyse amides, and ester bonds (amidases and esterases, respectively; Table 2.2).

- Conjugation (II). Location, cytoplasm.

The outputs of Phase-I events yield active products with 'polar handles' such as amino (NH_2), carboxyl (-COOH), and hydroxyl (OH) functional groups. Conjugation with endogenous, charged molecules—glycine, sulphate, glucuronic acid—increase hydrophilicity, neutralise active metabolic intermediates, and facilitates elimination (reviewed by Chen 2020). Broad specificity transferases act on any hydrophobic compound with electrophilic or nucleophilic moieties. A key Phase-II detoxicant is the bisubstrate enzyme glutathione S-transferase (GST). These function downstream from CYP450s, which introduce electrophilic functional groups, thereby promoting glutathione transfer. GST-mediated glutathione transfer to a xenobiotic is intended to neutralise such electrophilic sites, increasing the molecular weight, and rendering the target hydrophilic (water-soluble). Biotransformation can result in activation (rather than deactivation), reactive metabolites, and radicals that may be more toxic, mutagenic, or carcinogenic when compared to the parent compound, e.g. bivalve metabolism of dinoflagellate derived toxins, the azaspiracids, can result in dozens of congeners (Jauffrais et al. 2013).

- Excretion (III)

Outputs from Phase-II activities can require further processing, but usually they are removed from cells through active transport. The conjugated xenobiotic can no longer move freely across membranes, and in the case of glutathione, the anionic group acts as an affinity tag for ABC (efflux) transporters (Table 2.2). Subfamily members of the ABC transporters, such as P-glycoproteins, can provide resistance to xenobiotics (e.g. reef corals exposed to pharmaceuticals, Venn et al. 2009), as can the major facilitator family of proteins (e.g. mites, Van Leeuwen and Dermauw 2016).

2.5.4 Metals—a balance between regulation and detoxification

Excessive amounts of bioavailable, essential metals such as copper (micronutrients) are toxic to aquatic invertebrates, some semi-aquatic and terrestrial soil dwellers, but toxicosis and pathology vary due to the broad physiologies, life, and developmental stages of extant taxa (Grosell et al. 2007). Metals can make their way across the integument, gills, and gut where they are intercepted by cellular detoxification and regulatory pathways, isolated in mitochondria and lysosomes, undergo efflux into the lymph (*via* basolateral route), or complexed with antioxidants like metallothioneins (MT)s. The metal load of an invertebrate needs to be carefully managed as some, such as copper and iron, are incorporated into the catalytic sites of proteins (i.e. apoenzymes) for respiration, central metabolism, and innate immunity (Ahearn et al. 2004).

Histologically, the appearance of gross lesions is not specific to copper toxicity but is a common aberration of pollutant exposure, notably heavy 'metal overload': vacuolation, sloughing, and necrosis of epithelial cells in multiple organs, immune cell infiltration of connective tissue, and digestive gland atrophy. Such pathologic features are the result of metal-induced oxidative stress, DNA damage, altered ion regulation, reduced ability to uptake/carry oxygen and metabolic derangement (Sparks, 1972; Newton and Smolowitz 2018). Interestingly, high levels of manganese have been linked to ocular lesions and idiopathic blindness in American lobsters (Maniscalco and Shields 2006; Ochs et al. 2020; Figure 2.3c/d). Mussels (*Mytilus edulis*) transplanted to areas of high copper pollution led to reductions in the density/size of lysosomes and neutral lipids, atrophy of the digestive epithelium, and an increase in the presence of lipofuscin (pigmented granules containing high levels of metals and oxidised unsaturated fatty acids from organelles; Zorita et al. 2006). Lipofuscin can complex inorganic cations, e.g. copper, zinc, iron, magnesium, and manganese, with excess metal exposure correlating positively with lipofuscin levels in marine molluscs (Zorita et al. 2006; see Figure 2.4). In fact, lipofuscin levels and other signatures of oxidative stress—lipidosis, dysfunctional autophagy, protein carbonyls, lysosomal membrane instability—in the digestive glands of mussels (*Mytilus galloprovincialis*) and periwinkles (*Littorina littorea*) increased in a dose-dependent manner when incubated for 7 days in contaminated waters; 0.1–10 µM copper and several xenobiotics (e.g. chlorpyrifos, malathion, phenanthrene; Shaw et al. 2019).

MTs are a prevalent family of cysteine rich proteins that are responsible for binding, transporting, storing, and detoxifying metals. Expression is induced by mono- and divalent metals, physiological (iron, copper, zinc) and exogenous (cadmium,

mercury). Protein isoforms can differ substantially with respect to their tissue location (e.g. intra- versus extra-cellular), expression patterns, pH and temperature optima, but their functions are highly conserved across metazoans. In the crab, *Callinectes sapidus*, there are three isoforms with differential metal binding preferences, MT-I (cadmium, zinc, and copper), MT-II (cadmium and zinc), and MT-III (copper only; Brouwer et al. 2002). Other macromolecules such as ferritins, oxygen-transport proteins (Box 2.3), and phytochelatin synthases (PCS) work concurrently with MTs to curtail oxidative stress (Salinas-Clarot et al. 2011; Bundy et al. 2014). The free thiol groups of the MT cysteines, which represent >30% of the amino acid content, capture superoxide and hydroxyl radicals, and in the process, oxidise cysteine to cystine (which liberates any bound metals). PCS produce phytochelatin, which is a polymer of glutathione, and acts as a potent metal chelator. Moreover, cellular defences against copper toxicity in adult oysters and clams can cause green shell discolouration, as haemocytes sequester copper and form clusters in the mantle and gills (Arnold et al. 2010; Newton and Smolowitz 2018). There is also evidence that crustacean haemocytes act as a sink for zinc (reviewed by Ahearn et al. 2004).

Enhanced resistance to chemicals does evolve through the overexpression of detoxification enzymes and amplification of the associated genes (transposon mediated). There are numerous cases of insects demonstrating resistance to organophosphates and pyrethroids via CYP450 and GST selection (reviewed extensively by Li et al. 2007). Interestingly, the biological machinery used to develop insecticide resistance is also used for counteracting allelochemicals (e.g. esterases). Interrogating the genomes of chelicerate mites revealed GST, CYP450, ABC transporters, and cholinesterase expansion and radiation events responsible for their recalcitrance to xenobiotics (Van Leeuwen and Dermauw 2016). Exogenous metals and toxins can have profound effects on invertebrate immune-functioning—but this prodigious topic is beyond the scope of this chapter (see Galloway and Depledge 2001; Coates and Söderhäll 2021).

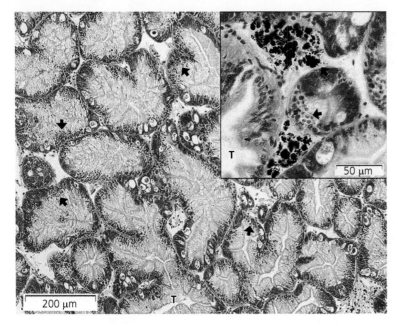

Figure 2.4 Lipofuscin granule presence in the digestive gland of a gastropod, *Crepidula fornicata*. Pigmented materials reminiscent of lipofuscin and eumelanin are distributed among all visible tubules (T), deposited at the basal surface. The black arrows point to some examples of pigment accumulation. Inset: region of higher magnification illustrating pigment accumulation in the inter- and intra-tubular spaces. This slipper limpet was examined as part of an initial assessment of phenoloxidase enzyme activities of tissue carried out by Quinn et al. (2020). Source: the histology micrograph (H & E staining) was captured from a seemingly healthy limpet collected in June 2019, Hazelbeach (Pembrokeshire), Wales UK. In the absence of infection, the accumulation of lysosomal lipofuscin granules in the digestive gland may be linked directly to metal exposure, notably copper.

Box 2.3 Oxygen transport proteins are linked intimately to defence, metal cycling, and detoxification.

The two major invertebrate oxygen transport proteins (OTPs), haemocyanin and haemerythrin are multifunctional and involved mainly in respiration and innate immunity (Coates and Decker 2017). Both haemocyanin and hemerythrin are metallated proteins containing dicopper and diiron active sites, respectively. Mechanisms must exist to direct metals (essential and non-essential) down metabolic pathways for proteostasis and/or detoxification, yet intriguingly, invertebrate OTPs also appear to have a role in regulating metal contaminants of the haemolymph.

Monomeric forms of haemerythrin act as detoxicants by binding and storing cadmium. Levels of haemerythrin mRNAs across several tissue types, such as muscle, increase dramatically when annelids (e.g. *Hediste diversicolor)* are exposed to heavy metals. Haemocyanins also bind metals, and in fact, horseshoe crab haemocyanin contains multiple binding sites for a range of transition metals, which can also act as allosteric factors. In the garden snail *Cornu aspersum*, two metallothioneins exist, one for copper regulation and another for cadmium detoxification, with the former linked to haemocyanin synthesis (Pedrini-Martha et al. 2020). Several recent studies have reported on the subunit isoform diversity that make up haemocyanin oligomers (hexameric in crustaceans), and posited that such heterogeneity may be responsible for the broad functional repertoire of this protein beyond oxygen transport/storage (Wang et al. 2019; Yao et al. 2019; Mendoza-Porras et al, 2020; Coates and Costa-Paiva 2020).

2.6 Summary

- Disentangling the individual effects of any stressor is difficult, as invertebrates *in situ* contend with a cocktail of xenobiotic disruptors and environmental flux.
- Tissue damage caused by non-communicable chemical, physical, or environmental factors are repaired in the same manner as damage caused by pathogens.

- DAMPs in the form of compromised proteins, nucleic acids, and fats represent a universal trigger for activating the cell stress response.
- Defence against xenobiotics is a recurrent process among invertebrates. Xenobiotic → crosses cell membrane → Phase I (oxidation, reduction, hydroxylation; biotransformation) → Phase II (conjugated with endogenous compounds; bioinactivation) → Phase III (exported by transporters).
- Prolonged exposure to xenobiotics causes immune suppression resulting in potential changes to infectious disease incidence and severity.
- Aberrant tissue growths (neoplastic, anaplastic, ectopic) are circumscribed by immune cell-mediated defences in invertebrates (ensheathment, melanisation).

Recommended further reading

Stroud, J. Diseases of Annelids, Chapter 7, this volume (Detailed examples of toxicosis and pollutant-induced histopathological changes.)

Coates, C.J. and Söderhäll, K. 2021. The stress–immunity axis in shellfish. *Journal of Invertebrate Pathology* 186: e107492

Canesi, L. and Corsi, I. 2016. Effects of nanomaterials on marine invertebrates. *Science of the total Environment* 565: 933–940

Dar, M.I., Green, I.D., and Khan, F.A. 2019. Trace metal contamination: Transfer and fate in food chains of terrestrial invertebrates. *Food Webs* 20: e00116

Galloway, T.S. and Depledge, M.H. 2001. Immunotoxicity in invertebrates: Measurement and ecotoxicological relevance. *Ecotoxicology* 10(1): 5–23

Goldstone, J.V., Hamdoun, A., Cole, B.J., Howard-Ashby, M., Nebert, D.W., Scally, M., Dean, M. et al. 2006. The chemical defensome: Environmental sensing and response genes in the *Strongylocentrotus purpuratus* genome. *Developmental Biology* 300: 366–384

James, R.R. and Xu, J. 2012. Mechanisms by which pesticides affect insect immunity. *Journal of Invertebrate Pathology* 109(2): 175–182

Li, X., Schuler, M.A., and Berenbaum, M.R. 2007. Molecular mechanisms of metabolic resistance to synthetic and natural xenobiotics. *Annual Review of Entomology* 52: 231–253

Newton, A.L. and Smolowitz, R. 2018. Invertebrates. In *Pathology of Wildlife and Zoo Animals* (pp. 1019–1052). Academic Press. https://www.sciencedirect.com/book/9780128053065/pathology-of-wildlife-and-zoo-animals#book-description

Perez, D.G. and Fontanetti, C.S. 2011. Hemocitical responses to environmental stress in invertebrates: A review. *Environmental Monitoring and Assessment* 177(1–4): 437–447

Acknowledgements

Operations were part funded by the European Regional Development Fund through the Ireland Wales Cooperation programme BlueFish. I would like to thank Susan and Lonán Coates for their support throughout the writing of this text.

References

Abbas, M.N., Kausar, S., and Cui, H. 2019. The biological role of peroxiredoxins in innate immune responses of aquatic invertebrates. *Fish & Shellfish Immunology* 89: 91–97

Adamo, S.A. 2012. The effects of the stress response on immune function in invertebrates: An evolutionary perspective on an ancient connection. *Hormones and Behavior* 62: 324–330

Adeel, M., Ma, C., Ullah,C. et al. 2019. Exposure to nickel oxide nanoparticles insinuates physiological, ultrastructural and oxidative damage: A life cycle study on *Eisenia fetida*. *Environmental Pollution* 254: 113032

Ahearn, G.A., Mandal, P.K., and Mandal, A. 2004. Mechanisms of heavy-metal sequestration and detoxification in crustaceans: A review. *Journal of Comparative Physiology B* 174: 439–452

Amoyel, M., Anderson, A.M., and Bach, E.A. 2014. JAK/STAT pathway dysregulation in tumors: A *Drosophila* perspective. *Seminars in Cell and Developmental Biology* 28: 96–103

Arnold, W.R., Cotsifas, J.S., Ogle, R.S., DePalma, S.G., and Smith, D.S. 2010. A comparison of the copper sensitivity of six invertebrate species in ambient salt water of varying dissolved organic matter concentrations. *Environmental Toxicology and Chemistry* 29: 311–319

Bateman, P.W. and Fleming, P.A. 2009. There will be blood: Autohaemorrhage behaviour as part of the defence repertoire of an insect. *Journal of Zoology* 278: 342–348

Bell, K.L. and Smith, V.J. 1995. Occurrence and distribution of antioxidant enzymes in the haemolymph of the shore crab *Carcinus maenas*. *Marine Biology* 123: 829–836

Bordereau, C., Robert, A., Van Tuyen, V., and Peppuy, A. 1997. Suicidal defensive behaviour by frontal gland dehiscence in *Globitermes sulphureus* Haviland soldiers (Isoptera). *Insectes Sociaux* 44: 289–297

Bowser, P.R. and Rosemark, R. 1981. Mortalities of cultured lobsters, *Homarus*, associated with a molt death syndrome. *Aquaculture* 23: 11–18

Brouwer, M., Syring, R., and Brouwer, T.H. 2002. Role of a copper-specific metallothionein of the blue crab, *Callinectes sapidus*, in copper metabolism associated with degradation and synthesis of hemocyanin. *Journal of Inorganic Biochemistry* 88: 228–239

Bundy, J.G., Kille, P., Liebeke, M., and Spurgeon, D.J. 2014. Metallothioneins may not be enough—the role of phytochelatins in invertebrate metal detoxification. *Environmental Science and Technology* 48: 885–886

Byrne, M. 2001. The morphology of autotomy structures in the sea cucumber *Eupentacta quinquesemita* before and during evisceration. *Journal of Experimental Biology* 204: 849–863

Carballal, M.J., Barber, B.J., Iglesias, D., and Villalba, A. 2015. Neoplastic diseases of marine bivalves. *Journal of Invertebrate Pathology* 131: 83–106

Chen, G. 2020. Xenobiotic metabolism and disposition. In *An Introduction to Interdisciplinary Toxicology*. Carey Pope and Jing Liu (eds.) pp. 31–42. Academic Press. https://books.google.co.uk/books?hl=en&lr=&id=Sx

XRDwAAQBAJ&oi=fnd&pg=PR13&dq=An+Introduction+to+Interdisciplinary+Toxicology+From+Molecules+to+Man&ots=l97CDakStO&sig=fm6w4frxnHaDNaypbRrvR9Waids#v=onepage&q&f=false

Cheng, T.C. and Sullivan, J.T. 1973. The effect of copper on the heart-rate of *Biomphalaria glabrata* (Mollusca: Pulmonata). *Comparative and General Pharmacology* 4: 37–41

Coates, C.J. and Costa-Paiva, E.M. 2020. Multifunctional roles of hemocyanins. In *Vertebrate and Invertebrate Respiratory Proteins, Lipoproteins and other Body Fluid Proteins*. Ulrich Hoeger and J. Robin Harris (eds.) pp. 233–250. Cham, New York: Springer

Coates, C.J. and Decker, H. 2017. Immunological properties of oxygen-transport proteins: Hemoglobin, hemocyanin and hemerythrin. *Cellular and Molecular Life Sciences* 74: 293–317

Coates, C.J. and Söderhäll, K. 2021. The stress–immunity axis in shellfish. *Journal of Invertebrate Pathology* 186: e107492.

Coates, C.J., McCulloch, C., Betts, J., and Whalley, T. 2018. Echinochrome A release by red spherule cells is an iron-withholding strategy of sea urchin innate immunity. *Journal of Innate Immunity* 10: 119–130

Coates, C. J., Whalley, T., Wyman, M., and Nairn, J. 2013. A putative link between phagocytosis-induced apoptosis and hemocyanin-derived phenoloxidase activation. *Apoptosis* 18: 1319–1331

Coutteau, P., Geurden, I., Camara, M.R., Bergot, P., and Sorgeloos, P. 1997. Review on the dietary effects of phospholipids in fish and crustacean larviculture. *Aquaculture* 155: 149–164

Dai, L.S., Yu, X.M., Abbas, M.N., Li, C.S., Chu, S.H., Kausar, S., and Wang, T. T. 2018. Essential role of the peroxiredoxin 4 in *Procambarus clarkii* antioxidant defense and immune responses. *Fish & Shellfish Immunology* 75: 216–222

Dallinger, R. 1993. Strategies of metal detoxification in terrestrial invertebrates. *Ecotoxicology of Metals in Invertebrates* 245: 290

Davies, C.E., Batista, F.M., Malkin, S.H., Thomas, J.E., Bryan, C.C., Crocombe, P., Coates, C.J., and Rowley, A.F. 2019. Spatial and temporal disease dynamics of the parasite *Hematodinium* sp. in shore crabs, Carcinus maenas. *Parasites and Vectors* 12: 472

Dubovskiy, I.M., Grizanova, E.V., Ershova, N.S., Rantala, M.J., and Glupov, V.V. 2011. The effects of dietary nickel on the detoxification enzymes, innate immunity and resistance to the fungus *Beauveria bassiana* in the larvae of the greater wax moth *Galleria mellonella*. *Chemosphere* 85: 92–96

Eisner, T. and Camazine, S. 1983. Spider leg autotomy induced by prey venom injection: An adaptive response to 'pain'?. *Proceedings of the National Academy of Sciences USA* 80: 3382–3385

Elwood, R.W. 2011. Pain and suffering in invertebrates? *ILar Journal* 52: 175–184

Elwood, R.W. 2019. Discrimination between nociceptive reflexes and more complex responses consistent with pain in crustaceans. *Philosophical Transactions of the Royal Society B* 374: 20190368

Epel, D., Luckenbach, T., Stevenson, C.N., MacManus-Spencer, L.A., Hamdoun, A., and Smital, A.T. 2008. Efflux transporters: Newly appreciated roles in protection against pollutants. *Environmental Science and Technology* 42: 3914–3920

Farley, C.A. 1969a. Probable neoplastic disease of the haematopoietic system in oysters *Crassostrea virginica* and *Crassostrea gigas*. *Monographs of the National Cancer Institute* 31: 541–566

Farley, C.A. 1969b. Sarcomatoid proliferative disease in a wild population of blue mussels (*Mytilus edulis*). *Journal of the National Cancer Institute* 43: 509–516

Freeman, K.S., Lewbart, G.A., Robarge, W.P., Harms, C.A., Law, J.M., and Stoskopf, M.K. 2009. Characterization of eversion syndrome in captive Scyphomedusa jellyfish. *American Journal of Veterinary Research* 70: 1087–1093

Fuller, R., Rahona, E., Fisher, S., Caravanos, J., Webb, D., Kass, D. et al. 2018. Pollution and non-communicable disease: Time to end the neglect. *The Lancet Planetary Health* 2: e96–e98

Galloway, T.S. and Depledge, M.H. 2001. Immunotoxicity in invertebrates: Measurement and ecotoxicological relevance. *Ecotoxicology* 10: 5–23

Gamble, S.C., Goldfarb, P.S., Porte, C., and Livingstone, D.R. 1995. Glutathione peroxidase and other antioxidant enzyme function in marine invertebrates (*Mytilus edulis*, *Pecten maximus*, *Carcinus maenas* and *Asterias rubens*). *Marine Environmental Research* 39: 191–195

Gateff, E. 1978. Malignant neoplasms of genetic origin in *Drosophila melanogaster*. *Science* 200: 1448–1459

Giannetto, A., Maisano, M., Cappello, T., Oliva, S., Parrino, V., Natalotto, A. et al. 2015. Hypoxia-inducible factor α and Hif-prolyl hydroxylase characterization and gene expression in short-time air-exposed *Mytilus galloprovincialis*. *Marine Biotechnology* 17: 768–781

Goldstone, J.V. 2008. Environmental sensing and response genes in cnidaria: The chemical defensome in the sea anemone *Nematostella vectensis*. *Cell Biology and Toxicology* 24: 483–502

Goldstone, J.V., Hamdoun, A., Cole, B.J., Howard-Ashby, M., Nebert, D.W., Scally, M., Dean, M. et al. 2006. The chemical defensome: Environmental sensing and response genes in the *Strongylocentrotus purpuratus* genome. *Developmental Biology* 300: 366–384

Gomes, S.I., Hansen, D., Scott-Fordsmand, J.J., and Amorim, M.J. 2015. Effects of silver nanoparticles to soil invertebrates: Oxidative stress biomarkers in *Eisenia fetida*. *Environmental Pollution* 199: 49–55

Gorr, T.A., Cahn, J.D., Yamagata, H., and Bunn, H.F. 2004. Hypoxia-induced synthesis of hemoglobin in the crustacean Daphnia magna is hypoxia-inducible factor-dependent. *The Journal of Biological Chemistry* 279: 36038–36047

Gorr, T.A., Gassmann, M., and Wappner, P. 2006. Sensing and responding to hypoxia via HIF in model invertebrates. *Journal of Insect Physiology* 52: 349–364

Goulson, D., Nicholls, E., Botías, C., and Rotheray, E.L. 2015. Bee declines driven by combined stress from parasites, pesticides, and lack of flowers. *Science* 347: 1255957

Grosell, M., Blanchard, J., Brix, K.V., and Gerdes, R. 2007. Physiology is pivotal for interactions between salinity and acute copper toxicity to fish and invertebrates. *Aquatic Toxicology* 84: 162–172

Haites, N., Don, M., and Masters, C.J. 1972. Heterogeneity and molecular weight inter-relationships of the esterase isoenzymes of several invertebrate species. *Comparative Biochemistry and Physiology* 2: 303–322

Hatanaka, R., Sekine, Y., Hayakawa, T., Takeda, K., and Ichijo, H. 2009. Signaling pathways in invertebrate immune and stress response. *Invertebrate Survival Journal* 6: 32–43

Hauton, C. 2016. Effects of salinity as a stressor to aquatic invertebrates. In *Stressors in the Marine Environment: Physiological and ecological responses; societal implications*. Solan, M. and Whiteley, N. (eds.) pp. 3–24. Oxford: Oxford University Press

Isakova, V. and Armstrong, P.B. 2003. Imprisonment in a death-row cell: The fates of microbes entrapped in the *Limulus* blood clot. *The Biological Bulletin Woods Hole* 205: 203–204

Jauffrais, T., Kilcoyne, J., Herrenknecht, C., Truquet, P., Séchet, V., Miles, C.O., and Hess, P. 2013. Dissolved azaspiracids are absorbed and metabolized by blue mussels (*Mytilus edulis*). *Toxicon* 65: 81–89

Johnson, L., Coates, C.J., Albalat, A., Todd, K., and Neil, D. 2016. Temperature-dependent morbidity of 'nicked' edible crab *Cancer pagurus*. *Fisheries Research* 175: 127–131

Jin, M., Liao, C., Chakrabarty, S., Zheng, W., Wu, K., and Xiao, Y. 2019. Transcriptional response of ATP-binding cassette (ABC) transporters to insecticides in the cotton bollworm *Helicoverpa armigera*. *Pesticide Biochemistry and Physiology* 154, 46–59

Junprung, W., Supungul, P., and Tassanakajon, A. 2019. *Litopenaeus vannamei* heat shock protein 70 (LvH-SP70) enhances resistance to a strain of *Vibrio parahaemolyticus*, which can cause acute hepatopancreatic necrosis disease (AHPND), by activating shrimp immunity. *Developmental & Comparative Immunology* 90: 138–146

Kett, G.F., Culloty, S.C., Lynch, S.A., and Jansen, M.A. 2020. Solar UV radiation modulates animal health and pathogen prevalence in coastal habitats knowledge gaps and implications for bivalve aquaculture. *Marine Ecology Progress Series* 653: 217–231

King, A.M. and MacRae, T.H. 2015. Insect heat shock proteins during stress and diapause. *Annual Review of Entomology* 60: 59–75

Kültz, D. 2005. Molecular and evolutionary basis of the cellular stress response. *Annual Review of Physiology* 67: 225–257

LaDouceur, E.E.B., Garner, M.M., Wynne, J., Fish, S., and Adams, L. 2013. Ulcerative umbrellar lesions in captive moon jelly (*Aurelia aurita*) medusae. *Veterinary Pathology* 50: 434–442

LaDouceur, E.E.B., Wynne, J., Garner, M.M., Nyaoke, A., and Keel, M.K. 2016. Lesions of copper toxicosis in captive marine invertebrates with comparisons to normal histology. *Veterinary Pathology* 53: 648–658

Li, S., Yu, X., and Feng, Q. 2019. Fat body biology in the last decade. *Annual Review of Entomology* 64: 315–333

Li, X., Schuler, M.A., and Berenbaum, M.R. 2007. Molecular mechanisms of metabolic resistance to synthetic and natural xenobiotics. *Annual Review of Entomology* 52: 231–253

Lightner, D.V. and Brock, J.A. 1987. A lymphoma-like neoplasm arising from hematopoietic tissue in the white shrimp, *Penaeus vannamei* Boone (Crustacea: Decapoda). *Journal of Invertebrate Pathology* 49: 188–193

Lindsay, S.M. 2010. Frequency of injury and the ecology of regeneration in marine benthic invertebrates. *Integrative and Comparative Biology* 50: 479–493

Nardocci, G., Navarro, C., Cortés, P.P., Imarai, M., Montoya, M., Valenzuela, B. et al. 2014. Neuroendocrine mechanisms for immune system regulation during stress in fish. *Fish & Shellfish Immunology* 40: 531–538

Newton, A.L. and Smolowitz, R. 2018. Invertebrates. In *Pathology of Wildlife and Zoo Animals*. Karen A. Terio, Denise Mcaloose and Judith G. St. Leger (eds.) pp. 1019–1052. Academic Press. https://books.google.co.uk/books?hl=en&lr=&id=rUwADQAAQBAJ&oi=fnd&pg=PP1&dq=Pathology+of+Wildlife+and+Zoo+Animals&ots=ew9P1oOcba&sig=X2XkNuX4zFSD8X_1ScF_tMDIjzU#v=onepage&q=Pathology%20of%20Wildlife%20and%20Zoo%20Animals&f=false

Maniscalco, A.M. and Shields, J.D. 2006. Histopathology of idiopathic lesions in the eyes of Homarus americanus from Long Island Sound. *Journal of Invertebrate Pathology* 9: 88–97.

Mendoza-Porras, O., Kamath, S., Harris, J.O., Colgrave, M.L., Huerlimann, R., Lopata, A.L., and Wade, N.M. 2020. Resolving hemocyanin isoform complexity in haemolymph of black tiger shrimp *Penaeus* monodon-implications in aquaculture, medicine and food safety. *Journal of Proteomics* 218: 103689

Meng, X.L., Dong, Y.W., Dong, S.L., Yu, S.S., and Zhou, X. 2011. Mortality of the sea cucumber, *Apostichopus japonicus* Selenka, exposed to acute salinity decrease and related physiological responses: Osmoregulation and heat shock protein expression. *Aquaculture* 316: 88–92

Metzger, M.J., Villalba, A., Carballal, M.J., Iglesias, D., Sherry, J., Reinisch, C. et al. 2016. Widespread transmission of independent cancer lineages within multiple bivalve species. *Nature* 534: 705–709

Mirzoyan, Z., Sollazzo, M., Allocca, M., Valenza, A.M., Grifoni, D., and Bellosta, P. 2019. *Drosophila melanogaster*: A model organism to study cancer. *Frontiers in Genetics* 10: 51

Moore, K.A. and Williams, D.D. 1990. Novel strategies in the complex defense repertoire of a stonefly (*Pteronarcys dorsata*) nymph. *Oikos* 57: 49–56

Muttray, A.F. and Vassilenko, K. 2018. Mollusca: Disseminated neoplasia in bivalves and the p53 protein family. In *Advances in Comparative Immunology*, Edwin L. Cooper (ed.) pp. 953–979. New York: Springer, Cham. https://link.springer.com/book/10.1007/978-3-319-76768-0

Nappi, A.J. 1984. Hemocyte reactions and early cellular changes during melanotic tumor formation in *Drosophila melanogaster*. *Journal of Invertebrate Pathology* 43: 395–406

Ochs, A.T., Shields, J.D., Rice, G.W., and Unger, M.A. 2020. Acute and long-term manganese exposure and subsequent accumulation in relation to idiopathic blindness in the American lobster. *Homarus americanus. Aquatic Toxicology* 219: 105379

Palmer, C.V., Traylor-Knowles, N.G., Willis, B.L., and Bythell, J.C. 2011. Corals use similar immune cells and wound-healing processes as those of higher organisms. *PLoS One* 6: e23992

Pedrini-Martha, V., Schnegg, R., Schäfer, G.G., Lieb, B., Salvenmoser, W., and Dallinger, R. 2020. Responsiveness of metallothionein and hemocyanin genes to cadmium and copper exposure in the garden snail *Cornu aspersum. Journal of Experimental Zoology Part A: Ecological and Integrative Physiology* DOI: 10.1002/jez. 2425

Perdomo-Morales, R., Montero-Alejo, V., and Perera, E. 2019. The clotting system in decapod crustaceans: History, current knowledge and what we need to know beyond the models. *Fish & Shellfish Immunology* 84: 204–212

Peters, E., Smolowitz, R., and Reynolds, T. 2012. Neoplasia. In *Invertebrate Medicine*. Lewbart, G.A. (ed.) pp. 448–525. Chichester: John Wiley & Sons

Philipp, E. and Fabricius, K. 2003. Photophysiological stress in scleractinian corals in response to short-term sedimentation. *Journal of Experimental Marine Biology and Ecology* 287: 57–78

Piontkivska, H., Chung, J.S., Ivanina, A.V., Sokolov, E.P., Techa, S., and Sokolova, I.M. 2011. Molecular characterization and mRNA expression of two key enzymes of hypoxia-sensing pathways in eastern oysters *Crassostrea virginica* (Gmelin): Hypoxia-inducible factor α (HIF-α) and HIF-prolyl hydroxylase (PHD). *Comparative Biochemistry and Physiology Part D: Genomics and Proteomics* 6: 103–114

Quinn, E.A., Malkin, S.H., Rowley, A.F., and Coates, C.J. 2020. Laccase and catecholoxidase activities contribute to innate immunity in slipper limpets, *Crepidula fornicata*. *Developmental & Comparative Immunology* 110: e103724

Regoli, F. and Principato, G. 1995. Glutathione, glutathione-dependent and antioxidant enzymes in mussel, *Mytilus galloprovincialis*, exposed to metals under field and laboratory conditions: Implications for the use of biochemical biomarkers. *Aquatic Toxicology* 31: 143–164

Richier, S., Merle, P.L., Furla, P., Pigozzi, D., Sola, F., and Allemand, D. 2003. Characterization of superoxide dismutases in anoxia-and hyperoxia-tolerant symbiotic cnidarians. *Biochimica et Biophysica Acta-General Subjects* 1621: 84–91

Riegl, B. and Branch, G.M. 1995. Effects of sediment on the energy budgets of four scleractinian (Bourne 1900) and five alcyonacean (Lamouroux 1816) corals. *Journal of Experimental Marine Biology and Ecology* 186: 259–275

Rivera-Ingraham, G.A. and Lignot, J.H. 2017. Osmoregulation, bioenergetics and oxidative stress in coastal marine invertebrates: Raising the questions for future research. *Journal of Experimental Biology* 220: 1749–1760

Rodriguez, A., Zhou, Z., Tang, M.L., Meller, S., Chen, J., Bellen, H., and Kimbrell, D.A. (1996). Identification of immune system and response genes, and novel mutations causing melanotic tumor formation in Drosophila melanogaster. Genetics, 143(2), 929–940

Robert, J. 2010. Comparative study of tumorigenesis and tumor immunity in invertebrates and nonmammalian vertebrates. *Developmental & Comparative Immunology* 34: 915–925

Rowley, A.F. and Ratcliffe, N.A. 1978. A histological study of wound healing and hemocyte function in the wax-moth *Galleria mellonella*. *Journal of Morphology* 157: 181–199

Saint-Denis, M., Labrot, F., Narbonne, J.F., and Ribera, D. (1998). Glutathione, glutathione-related enzymes, and catalase activities in the earthworm *Eisenia fetida andrei*. *Archives of Environmental Contamination and Toxicology* 35: 602–614

Salinas-Clarot, K., Gutiérrez, A.P., Núñez-Acuña, G., and Gallardo-Escárate, C. 2011. Molecular characterization and gene expression of ferritin in red abalone (*Haliotis rufescens*). *Fish & Shellfish Immunology* 30: 430–433

Scharrer, B. and Lochhead, M.S. 1950. Tumors in the invertebrates: A review. *Cancer Research* 10: 403–419

Schmid, M.R., Dziedziech, A., Arefin, B., Kienzle, T., Wang, Z., Akhter, M., . . . and Theopold, U. 2019. Insect hemolymph coagulation: Kinetics of classically and non-classically secreted clotting factors. *Insect Biochemistry and Molecular Biology* 109: 63–71

Seong, S.Y. and Matzinger, P. 2004. Hydrophobicity: An ancient damage-associated molecular pattern that initiates innate immune responses. *Nature Reviews Immunology* 4: 469–478

Sejerkilde, M., Sørensen, J.G., and Loeschcke, V. 2003. Effects of cold-and heat hardening on thermal resistance in *Drosophila melanogaster* . *Journal of Insect Physiology* 49: 719–726

Shaw, J.P., Moore, M.N., Readman, J.W., Mou, Z., Langston, W.J., Lowe, D.M. et al. (2019). Oxidative stress, lysosomal damage and dysfunctional autophagy in molluscan hepatopancreas (digestive gland) induced by chemical contaminants. *Marine Environmental Research* 152: e104825.

Shields, J.D. and Small, H.J. 2013. An unusual cuticular tumor-like growth on the abdomen of a lobster *Homarus americanus. Journal of Invertebrate Pathology* 114: 245–249

Shorter, J.R. and Rueppell, O. 2012. A review on self-destructive defense behaviors in social insects. *Insectes Sociaux* 59: 1–10

Shrestha, R. and Gateff, E. 1986. Ultrastructure and cytochemistry of the tumorous blood cells in the mutant lethal (3) malignant blood neoplasm of *Drosophila melanogaster. Journal of Invertebrate Pathology* 48: 1–12

Smolowitz, R.M., Miosky, D., and Reinisch, C.L. 1989. Ontogeny of leukemic cells of the soft-shell clam. *Journal of Invertebrate Pathology* 53: 41–51

Snyder, M.J. 2000. Cytochrome P450 enzymes in aquatic invertebrates: Recent advances and future directions. *Aquatic Toxicology* 48: 529–547

Šobotník, J., Bourguignon, T., Hanus, R., Demianová, Z., Pyteliková, J., Mareš, M. et al. 2012. Explosive backpacks in old termite workers. *Science* 337: 436–436

Sparks, A.K. 1972. *Invertebrate Pathology. Noncommunicable Diseases*. London and New York: Academic Press (Elsevier).

Sparrow, J.C. 1978. Melanotic 'tumours'. In *The Genetics and Biology of Drosophila*. M. Ashburner and T.R.F. Wright (eds.) Vol. 2, pp 277–313. London/New York: Academic Press

Sparks, A.K. and Lightner, D.V. 1973. A tumorlike papilliform growth in the brown shrimp (*Penaeus aztecus*). *Journal of Invertebrate Pathology* 22: 203–212

Sparks, A.K. and Morado, J.F. 1987. A putative carcinoma-like neoplasm in the hindgut of a red king carb, *Paralithodes camtschatica. Journal of Invertebrate Pathology* 50: 45–52

Srinivasan, N., Gordon, O., Ahrens, S., Franz, A., Deddouche, S., Chakravarty, P. et al. 2016. Actin is an evolutionarily-conserved damage-associated molecular pattern that signals tissue injury in *Drosophila melanogaster*. *eLife* 5: e19662

Tascedda, F. and Ottaviani, E. 2014. Tumors in invertebrates. *Invertebrate Survival Journal* 11: 197–203

Theopold, U., Schmidt, O., Söderhäll, K., and Dushay, M.S. 2004. Coagulation in arthropods: Defence, wound closure and healing. *Trends in Immunology* 25: 289–294

Tissiéres, A., Mitchell, H.K., and Tracy, U.M. 1974. Protein synthesis in salivary glands of *Drosophila melanogaster*: Relation to chromosome puffs. *Journal of Molecular Biology* 84: 389–398

Towle, D.W., Henry, R.P., and Terwilliger, N.B. 2011. Microarray-detected changes in gene expression in gills of green crabs (*Carcinus maenas*) upon dilution of environmental salinity. *Comparative Biochemistry and Physiology Part D: Genomics and Proteomics* 6: 115–125

Tu, D.D., Zhou, Y.L., Gu, W.B., Zhu, Q.H., Xu, B.P., Zhou, Z.K. et al. 2018. Identification and characterization of six peroxiredoxin transcripts from mud crab *Scylla paramamosain*: The first evidence of peroxiredoxin gene family in crustacean and their expression profiles under biotic and abiotic stresses. *Molecular Immunology* 93: 223–235

Van Leeuwen, T. and Dermauw, W. 2016. The molecular evolution of xenobiotic metabolism and resistance in chelicerate mites. *Annual Review of Entomology* 61: 475–498

Vargas-Ángel, B., Peters, E.C., Kramarsky-Winter, E., Gilliam, D.S., and Dodge, R.E. 2007. Cellular reactions to sedimentation and temperature stress in the Caribbean coral *Montastraea cavernosa. Journal of Invertebrate Pathology* 95: 140–145

Vargas-Ángel, B., Riegl, B., Gilliam, D.S., and Dodge, R.E. 2006. An experimental histopathological rating scale of sedimentation stress in the Caribbean coral *Montastraea cavernosa. Proceedings of the 10th International Coral Reef Symposium*, 1168–1173.

Venn, A.A., Quinn, J., Jones, R., and Bodnar, A. 2009. P-glycoprotein (multi-xenobiotic resistance) and heat shock protein gene expression in the reef Marine Biotechnolog *Montastraea franksi* in response to environmental toxicants. *Aquatic Toxicology* 93: 188–195

Vogt, G. 2008. How to minimize formation and growth of tumours: Potential benefits of decapod crustaceans for cancer research. *International Journal of Cancer* 123: 2727–2734

Wang, J.T., Han, T., Li, X.Y., Hu, S.X., Jiang, Y.D., and Wang, C.L. 2016. Effects of dietary phosphatidylcholine (PC) levels on the growth, molt performance and fatty acid composition of juvenile swimming crab, *Portunus trituberculatus. Animal Feed Science and Technology* 216: 225–233

Wang, J., Janech, M.G., and Burnett, K.G. 2019. Protein-level evidence of novel β-type hemocyanin and heterogeneous subunit usage in the Pacific whiteleg shrimp *Litopenaeus vannamei. Frontiers in Marine Science* 6: 687

Wang, L., Cui, S., Ma, L., Kong, L., and Geng, X. 2015. Current advances in the novel functions of hypoxia-inducible factor and prolyl hydroxylase in invertebrates. *Insect Molecular Biology* 24: 634–648

Weber, M., Lott, C., and Fabricius, K.E. 2006. Sedimentation stress in a scleractinian coral exposed to terrestrial and marine sediments with contrasting physical, organic and geochemical properties. *Journal of Experimental Marine Biology and Ecology* 336: 18–32

Yan, Y. and Hillyer, J.F. 2020. The immune and circulatory systems are functionally integrated across insect evolution. *Science Advances* 6: eabb3164

Yao, T., Zhao, M.M., He, J., Han, T., Peng, W., Zhang, H. et al. 2019. Gene expression and phenoloxidase activities of hemocyanin isoforms in response to pathogen infections in abalone *Haliotis diversicolor. International Journal of Biological Macromolecules* 129: 538–551

Yonemitsu, M.A., Giersch, R.M., Polo-Prieto, M., Hammel, M., Simon, A., Cremonte, F. et al. 2019. A single clonal lineage of transmissible cancer identified in two marine mussel species in South America and Europe. *eLife* 8: e47788

Zhang, Y. and Lu, Z. 2015. Peroxiredoxin 1 protects the pea aphid *Acyrthosiphon pisum* from oxidative stress induced by *Micrococcus luteus* infection. *Journal of Invertebrate Pathology* 127: 115–121

Zorita, I., Ortiz-Zarragoitia, M., Soto, M., and Cajaraville, M.P. (2006). Biomarkers in mussels from a copper site gradient (Visnes, Norway): an integrated biochemical, histochemical and histological study. Aquatic toxicology, 78, S109–S116

CHAPTER 3

Diagnostic approaches in invertebrate pathology

David Bass, Andrew F. Rowley, and Christopher J. Coates

3.1 Introduction

The study of diseases of invertebrates dates back to ancient times. Aristotle in his *Historia Animalium* used his observational ability to describe a disease of bees; presumably those used by the ancient Greeks to produce honey. Observation, be it with the naked eye or with the assistance of microscopes, still remains a fundamental part of any disease diagnosis even over 2000 years after Aristotle died in 322 BC. The last 50 years have seen a renaissance in the study of invertebrate diseases aided by advances in microscopy, microbiology, biochemistry and the advent of molecular biology. Most recently, the development of so-called next generation sequencing (more informatively referred to as high throughput sequencing), has resulted in fundamental changes in our understanding of disease states including their polymicrobial aetiology and pathobiomes (Bass et al. 2019).

There are many excellent guides to the methods available in the pathologists' toolkit (see Recommended Further Reading section) and so the role of this brief chapter is not to provide protocols for these. Instead, our aim is to (1) provide an overview of the techniques available and (2) review recent developments in high throughput sequencing (HTS) and how this is influencing our understanding of disease processes.

3.2 Sampling regimes

Surveys/screens require careful planning so that temporal patterns of disease can be accurately established in invertebrate populations. Of great importance is the need to ensure an appropriately powered experiment (i.e., to have adequate sample sizes) and to accurately record biometric data (size, sex, moult stage, physical damage, missing limbs, discolouration, epibionts, and commensals) of affected invertebrates—such as crustaceans—that are likely to have bearing on disease incidence and severity. Consulting tables in a standard statistical text, for example Simon and Schill (1984), can guide the number of samples required to detect an infected host at different levels of confidence in a defined population size. Unlike exploratory disease surveys, where selecting a robust sample size is challenging and resources may be limited, having a particular target or disease-causing agent of interest can help to formulate *a priori* predictions (i.e. a systematic approach). For example, the parasitic dinoflagellate *Hematodinium* sp. was found to infect 10–15% of juvenile brown crabs (*Cancer pagurus*) at the nadir of the annual disease cycle within the waters surrounding Mumbles Head, Swansea (Wales, UK; Smith et al. 2015). The prevalence datum was subsequently used to calculate a robust sample size for estimating prevalence of the same

David Bass, Andrew F. Rowley, and Christopher J. Coates, *Diagnostic approaches in invertebrate pathology*. In: *Invertebrate Pathology*. Edited by Andrew F. Rowley, Christopher J. Coates and Miranda M.A. Whitten, Oxford University Press. © Oxford University Press (2022). DOI: 10.1093/oso/9780198853756.003.0003 pathogen at the same site for another decapod, the common shore crab *Carcinus maenas* (Davies et al. 2019). They selected an alpha (statistical significance) level of 0.05, a desired power of $\geq 80\%$, and a two-tailed test as this allows for testing of both lower and higher rate(s) of infection—yielding, a minimum of 48 animals per site per time-point (monthly). Determining an acceptable sample size will not only ensure the veracity of the study but also will avoid over-sampling and reduce the amount of time and resources invested for the same outcome.

Sampling in remote areas away from laboratory facilities can provide additional challenges of sample storage and processing, as does the frequency (biweekly, monthly) and intensity (sample size) of survey work. Transport of marine invertebrates in newspaper soaked in chilled seawater to the laboratory for necropsy may be necessary. Temporary housing of aquatic invertebrates in aquaria is to be avoided, if at all possible, because of potential changes in infected animals. If unavoidable, then keep aquatic animals in isolated (closed circulation) tanks where the water can be screened later; for example, some infected animals will shed parasites into the water after the 'stress' of translocation.

Figure 3.1 provides a schematic overview of the stages in disease diagnosis. The following text is organised to explain this figure and provide some examples of the approaches taken.

3.3 Stage 1—Observations of changes in external anatomy and behaviour

While pathologists working with invertebrates do not have the historical basis of pathology of Celsus and others including the cardinal signs of inflammation (calor, dolor, tumor, rubor, and functio laesa; see Chapter 2, Table 2.1) to guide them, observation of affected animals remains the initial stage in any disease diagnosis. For example, current nomenclature of coral disease is based on direct observation (e.g. black band disease; see Chapter 4). Similarly, insects and arachnids infected by fungi may have extensive fruiting bodies emerging from their tissues including those produced by the fungal families Cordycipitaceae and Ophiocordycipitaceae

(Figure 3.2a,b). Some insects may show rectal prolapses following viral infections and cadavers of infected insects may be odiferous (Lacey and Solter 2012). Of course, disease does not always manifest as external signatures or behavioural changes, many invertebrates are asymptomatic hosts. Rowley et al. (2020) tell a cautionary tale regarding the ectoparasite *Sacculina carcini* and the common shore crab *Carcinus maenas*. Quantitation of external root structures or scarring from such tissue invasion on the ventral reproductive opening of the crab can grossly underestimate the prevalence of this parasite within a population—histological examination of 221 animals identified the presence of *S. carcini* in $\sim 24\%$, as opposed to $\sim 6\%$ and $\sim 2\%$ from externa and scars, respectively (Rowley et al. 2020).

Careful recording of abnormalities in the integument of shelled animals and any unusual epibionts should be made. The presence of parasites on gill surfaces in crustaceans and molluscs also needs to be recorded. Rigorous logging of all environmental parameters such as air or water temperature, salinity, and oxygen tension are key parts of this initial disease survey, especially if disease outbreaks may be associated with changes in environmental conditions.

Changes in behaviour are sometimes useful instruments in the invertebrate pathologist's toolkit. Examples include limp lobster disease where limbs sag and are unresponsive to mechanical stimulation (Tall et al. 2003), and loss of burying behaviour in cockles following parasitisation affecting the hinge or foot musculature (Longshaw and Malham 2013). In the case of changes in burrowing behaviour in cockles, both bacteria (Blanchet et al. 2003) and a variety of digenean parasites (Longshaw and Malham 2013) can result in cockles appearing on the surface when exposed by the tide rather than burrowed. Hence, changes in the burying behaviour alone cannot inform the observer of the causative agent in such conditions without further investigation. Pathogens can have distinctive and important behavioural affects on their hosts. For example, the normally gregarious juvenile spiny lobsters *Panulirus argus* become solitary prior to infectiousness when playing host to the virus Panulirus argus virus 1 (PAV1), and when heavily infected they become

Stage 0
-Survey type (targeted *vs.* exploratory)
-Sample site selection
-Sample size selection
-Diagnostic tests and sample handling/storage considerations

Stage 1 External Observations

• Host anatomy
-Physical deformities, such as, surface discolouration, pitting, dysecdysis, missing limbs, deformed wings,
-Presence of epibionts, ectoparasites, and commensals

• Host behaviour
-sagging limbs, anorexia or lethargy
-tail flips (shrimp, langoustines)
-limp spines on sea urchin test
-change in gregarious invertebrates (e.g., insects)
-burying behaviour (e.g., cockles)

• Environmental parameters
-pH (soil, water)
-temperature (air, water)
-salinity, oxygen tension, and dissolved nutrients (water)

Stage 2 Internal Observations
• Solid and liquid tissues
-discolouration of the haemolymph/coelomic fluid
-solid tissue 'squash' for parasite spillage
-total and differential haemocyte counts (fixed lymph)
-view raw haemolymph under phase contrast

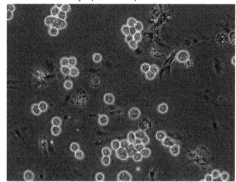

Decapod haemocytes flatten when in contact with a surface, and reveal the presence of refractile *Hematodinium*

Stage 3 Isolation and identification of putative disease-causing agent
-Dilution series of tissue fluid, and plating on general versus specific media (TSA and TCBS, respectively) for bacterial colony forming unit enumeration (i.e. microbial load)
-use of liquid medium or cell lines for fungal, viral and protist propagation
-of limited use when presented with poly-microbial aetiology

Stage 4 Histology approaches
• Traditional paraffin wax embedding
-H & E staining
• *In situ* hybridisation
• Immunohistology
• Ideal for visualising pathogens *in situ*, host response(s), and tissue damage simultaneously

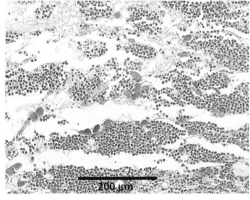

Complete effacement of decapod connective tissue by the prolific replication of the parasite *Hematodinium*

Stage 5 Electron microscopy
• Scanning and transmission
-a key tool for distinguishing morphology

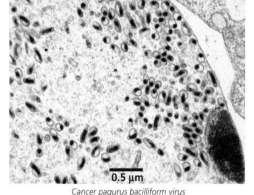

Cancer pagurus bacilliform virus

Stage 6 Nucleic acids assessment
• PCR-based identification and quantification
-genomic and complementary DNAs (gDNA, cDNA)
• High through-put sequencing
-Key tool for poly-microbial assessment

Figure 3.1 Overview of the stages and approaches recommended for accurate disease diagnosis (and pathologic assessment). Source: Dr Kelly Bateman (CEFAS, UK) provided the original micrograph of *Cancer pagurus* bacilliform virus. TSA - tryptone soy agar; TCBS - thiosulphate citrate bile salts sucrose agar.

(a)

(b)

Figure 3.2 *Cordyceps locustiphila* infection of an orthopteran insect Romaleidae; (a) found in Rio Claro, Antioquia (Colombia). Typical club-shaped stromata and prominent necks of perithecia are visible. *Cordyceps sp. indet* infection of an orthopteran insect (b) found in Iquitos, Upper Amazon (Peru). The buried perithecial necks are emerging. Exposure to light leads to the violent discharge of such perithecia containing the sexual ascospores. Source: Harry C. Evans (Centre for Agriculture and Bioscience International, Surrey UK) kindly provided the original images.

sedentary and stop feeding before death (Behringer et al. 2011).

3.4 Stage 2—Observation of tissues

Simple observation of squashes of tissues and body fluids (haemolymph and coelomic fluid) using light microscopy with vital stains or by phase contrast are an invaluable second stage in any assessment of a diseased invertebrate. There are many publications showing how to aseptically recover haemolymph and coelomic fluid from a wide range of invertebrates for such approaches. Infected haemolymph or coelomic fluid of invertebrates with sepsis-type infections will be cloudy-milky in appearance, although this alone will not identify the type of causative agents involved. Similarly, dramatic reduction in haemocyte or coelomocyte numbers in circulation will leave such fluids unusually clear. Changes in differential and total haemocyte/coelomocyte counts may also be useful to measure using haemocytometer counts and Cytospin preparations but a note of caution is necessary, as these are influenced by a wide range of factors (e.g. environmental temperature, nutritional status, age, moult stage, tidal cycle for intertidal animals) that are independent of any disease condition (Coates and Söderhäll 2021).

3.5 Stage 3—Isolation and cultivation of putative pathogens and parasites

A central tenet of Koch's postulates is the ability to isolate and grow in pure culture the causative agents of microbial diseases. While this is a realistic approach with many heterotrophic bacteria, protists, and fungi (but not all), viruses and obligate intracellular bacteria are more challenging because of their requirement for live cells in which to multiply. Insect cell lines were originally developed in the 1950s for the production of baculoviruses as control agents of pest insects and to date over 900 different cell lines have been developed from a variety of cell types in lepidopterans and dipterans (Arif and Pavlik 2013; Drugmand et al. 2012). Many viruses that infect insects can be grown in these cell lines that only require simple culture media. However, there are no immortal cell lines for other invertebrates, including molluscs and crustaceans, and this is a barrier to virus research in these animals, so their propagation has to be made in whole animals.

The pathologist has a wide variety of commercially available defined soluble and solid media for bacterial and fungal isolation from invertebrate tissues. Tryptone soy agar and broth with and without sterile blood products are useful general growth media for many bacteria that can be adjusted with NaCl (2%) for invertebrate tissues isosmotic with seawater. While most of the selective media available were originally developed for the cultivation and identification of human bacterial and fungal pathogens, some can be useful in invertebrate pathology. For instance, in the case of vibrios, the selective medium, thiosulphate-citrate-bile salts-sucrose agar was originally developed for the growth and identification of the human pathogens, *Vibrio cholerae* and *V. parahaemolyticus* in environmental samples. It has been used for such a purpose with invertebrates, although it does not permit the growth of all species of vibrios that are pathogens of these aquatic animals and so care must be taken with its use to determine total numbers of vibrios in tissues.

3.6 Stage 4—Histology, immunohistology, and *in situ* hybridisation

Histological examination of affected tissues is a key approach to invertebrate pathology because, not only does it allow for the visualisation of parasites and pathogens *in situ* but also the damage caused by these and the associated host responses. It is particularly important where pathogens have been identified in tissues by their molecular signals only (see Section 3.8) to demonstrate that such animals are actually infected and that these agents are multiplying in the tissues. Detailed descriptions of the methods used in general histology are outside the scope of this chapter and these are widely available to the reader from numerous sources (e.g. Maynard et al. 2014). What needs to be stressed here is the importance of the quality of the initial fixation in allowing accurate interpretation, as incomplete fixation can result in post-mortem artefacts that result in cellular damage that is unrelated to the actual infection studied. Most fixatives for light microscope histology include formaldehyde in a final concentration range of 5–10%. Commonly used fixatives include formol-saline (*ca.* 10% formaldehyde in an appropriate neutral buffered saline). For marine tissues, Davidson's seawater fixative (a mixture of formaldehyde, ethanol, acetic acid and seawater) for 12–24 hours, works well. If the samples are to be subsequently used for *in situ* hybridisation (ISH; see the following section), they should be transferred within 48 hours of initial fixation into 70% ethanol. Bouin's solution, a mixture of formaldehyde, picric acid and acetic acid is a powerful fixative but the user needs to be aware that once dry, picric acid can become spontaneously explosive! Unlike Davidson's, samples can be stored for many months in Bouin's prior to any processing back in the laboratory.

The most commonly used staining methods for invertebrate pathology are haematoxylin and eosin (H&E) and Giemsa (a mixture of methylene blue, eosin and Azure B) but a large number of additional methods can be used including Gram stain for bacteria, and Grocott–Gömöri's methenamine silver and periodic acid Schiff stains for fungi. Masson's trichrome consists of three stains, usually haematoxylin, a red dye such as Biebrich scarlet and aniline blue. The latter of these, stains connective tissue a bright blue colour and may allow for better differentiation of any abnormal tissues in some invertebrates.

'Traditional' histology is based on embedding and sectioning material in paraffin wax such as Histoplast. While it is possible to cut sections of *ca.* 5 μm thickness using this support medium, thinner sections are only possible using resins such as methacrylate and epoxy resins that were designed for electron microscopy. Sections cut for light microscope histology using these resins can be *ca.* 0.5 μm thick and this results in improved resolution. These so-called 'thick plastic' sections can be quickly stained with dyes such as mixtures of toluidine blue and methylene blue (see Figure 15. 2e in Chapter 15 as an example of this form of histology). Thick plastic histology is not widely used by invertebrate pathologists, yet it has much to offer in its improved resolution and fewer processing artefacts.

Light microscope histopathology has been strengthened by the development of immunohistochemistry and ISH to label and identify pathogens and parasites in histological sections. This latter method can be used both for histological sections of frozen and formaldehyde-fixed material, as well as whole mounts (Nielsen 2014). Within invertebrate pathology, ISH has become widely employed to detect and identify infectious agents using different probes and fluorescent and non-fluorescent labels. Immunohistopathology makes use of the specific interaction between the antigen (usually the infectious agent) and antibodies (monoclonal or polyclonal) raised against the agent under investigation. Because antibodies to many pathogens of invertebrates are not usually available commercially, this method relies on investigators raising antibody to their chosen infectious agent. Therefore, ISH is fast becoming the method of choice rather than immunocytochemistry. ISH fulfils the role of a 'bridge' between molecular detection of pathogens (including specific diagnostic assays) and their visualisation, localisation, and quantification in host tissues, providing evidence of infection (or

not), which cannot be inferred from molecular evidence alone.

3.7 Stage 5—Electron microscopy

Both transmission and scanning electron microscopy are widely used within invertebrate pathology to supplement routine light microscopy largely because of their improved resolution. Tissues are normally fixed in ice-cold glutaraldehyde followed by osmium tetroxide. Like light microscopy, this stage in processing is critically important because rapid penetration of the fixative speeds up the cross-linking of biological molecules within cells prior to autolysis setting in. Rapid penetration will strongly influence the validity of this method and reduce the incidence of artefacts caused by post-mortem autolysis of cells. Tissues must be fixed immediately—if necessary, in the field—to avoid these effects during transportation and storage. Although glutaraldehyde penetrates biological material quickly, the block size of tissue must be kept as small as possible (1–2 mm^3) to ensure rapid fixation. Tissues fixed in the field using ice-cold glutaraldehyde can be stored in this for > 12 hours and held in a suitable buffered wash before post-fixation in osmium tetroxide back in the laboratory. Both fixatives should be handled with care avoiding contact with liquid or vapour forms of these toxic chemicals and with appropriate waste disposal. Protocols for dehydration, embedding, resin polymerisation, ultramicrotomy and staining are widely available in the literature, and are not covered in this brief overview.

Most pathogens and parasites of invertebrates have been examined using these approaches and, in some cases, morphological features have been employed as a taxonomic tool. For instance, members of the Order Haplosporida that are known primarily as parasites of mainly molluscs and crustaceans, have been extensively studied using electron microscopy and their taxonomy defined by a series of ultrastructural features including those within spore stages (e.g. Hine et al. 2009). However, such approaches have limitations and studies often use a combination of histology, electron microscopy and gene sequences (e.g. Azevedo et al. 2006, Urrutia et al. 2019) to provide evidence of their taxonomic and phylogenetic relationships.

3.8 Stage 6—Nucleic acid approaches to disease diagnosis, advances, and pitfalls

While some pathogens are relatively well characterised and are diagnosed reliably using molecular and/or pathology-based assays, others are far less tractable. These fall into three main categories:

(1) Previously unknown pathogens of unknown phylogenetic/taxonomic affiliation, and therefore without established diagnostic assays, often highly genetically divergent.
(2) Pathogen 'clusters': pathogens with very closely related genotypes but distinct pathological characteristics (e.g. virulence, host relationships), which are difficult to separate using short diagnostic marker genes, and for which more specific markers and/or longer gene regions are necessary for reliable diagnostics.
(3) Syndromic conditions with unknown and/or multiple pathogenic agents involving pathogens from all domains of life (bacteria, viruses, eukaryotes), and possibly multiple aetiologies (e.g. pathobiotic systems; Bass et al 2019). The aetiology of most of these diseases remains unknown.

Additionally, pathogen dispersal and infectivity in the aquatic environment differs significantly from terrestrial systems. Pathogens are easily transported to susceptible hosts by moving water and/or alternative hosts/vectors and are often present as latent or low infections. Diagnostic methods powerful enough to detect such infections, which are particularly important with respect to movement of animals and dense culture associated with aquaculture activities, are therefore essential.

The challenges presented by these scenarios can be addressed in the following ways:

(1) Increasing the detection sensitivity of existing broadly targeted diagnostic sequencing methods to detect and identify low-level infections, common in aquatic systems, which are subject to pathogenetic switches from a multitude of environmental cues. This can be achieved by a range of standard polymerase chain reaction (PCR) techniques, such as maximising taxonomic specificity of an assay to exclude amplification of non-target sequences, minimising the length of the target PCR region (amplicon) and increasing PCR cycle number. Increasingly, techniques are being developed for more selective metagenomic sequencing, for example the ReadUntil facility of Oxford Nanopore, which targets predetermined sequence types (Loose et al. 2016) is currently being evaluated for use in aquatic animal health settings.

We caution, however, that using end-point PCR directed at gDNA only to diagnose/screen a host for a pathogen is not a sufficient measure of pathologic condition and is vulnerable to both Type I (false positive) and Type II (false negative) results.

(2) Developing more broadly inclusive sequencing approaches that are not dominated by sequences from the host, either via PCR or shotgun sequencing is the way forward. This can be achieved by reducing host representation in nucleic acid sampling, for which a range of techniques exist, for example: selective fractionation of non-host cells/spores (e.g. Watts et al. 2017), laser dissection microscopy (e.g. Lutz et al. 2016), and differential DNA extraction techniques (Heravi et al. 2020). Alternatively/additionally, molecular approaches can be designed to avoid sequencing of the host, for example by the use of primers specifically targeting certain pathogen lineages (Bass et al. 2015) but this is often not viable in the many cases where the pathogen is unknown and cannot easily be assigned even to a broad taxonomic group, or multiple pathogens are implicated. In the former case, metagenomic sequencing can be employed to sample all of the DNA sequences in a sample (e.g. infected tissue),

following which bioinformatic and phylogenetic analyses can be used to genetically identify and characterise the pathogen (e.g. Hartikainen et al. 2014).

Where multiple symbionts/pathogens are implicated in a pathobiotic system (i.e. reduced health status as a result of interactions between symbionts, the host, and the environment), a battery of HTS approaches can be applied to identify which symbionts are present, and the pattern of their distribution and activity in relation to the pathological signs observed. These approaches can involve/combine taxon/group-specific PCR assays, metabarcoding targeting a marker gene of interest (e.g. the bacterial 16S ribosomal ribonucleic acid (rRNA) gene), and transciptomic sequencing. In 'multi-omic' studies, combinations of HTS techniques such as these, ideally informed by microscopy-based techniques, can provide a powerful multi-dimensional insight into disease aetiology and pathogenesis (e.g. Broberg et al. 2018). Other pathobiotic systems and analyses are reviewed in Bass et al. (2019).

While the technological and analytical challenges of multi-omic datasets are demonstrably being addressed, interpretation of symbiont/pathogen data at the taxonomic level is often overlooked. It is easy to produce taxonomic outputs from HTS data but they need to be understood in context and require further interpretation. Data type and reference database are key. For example, the annotation of metagenomic/transcriptomic data using databases dominated by genomic references will generate outputs biased towards organisms for which (partial) genomes are available. This is more of a problem for eukaryotes and viruses, which are more incompletely represented in genomic databases than bacteria. Even taxonomic outputs based on 16S and 18S amplicon sequences are potentially misleading: (i) the choice of reference database can have a strong influence on the taxonomic profile produced; (ii) a large proportion of the sequences amplified from many sample types will be significantly < 100% match to reference sequences and therefore represent different species, genera, or higher taxonomic

level discrepancies; and (iii) marker regions differ in the taxonomic resolution they offer, which is an important consideration when interpreting 'user friendly' pie and stacked bar charts, and taxon lists. For example, even if separated at the highest resolution (e.g. using amplicon sequence variants (ASVs) in high resolution sample inference from Illumina amplicon data (DADA2)), 16S and 18S generally cannot resolve species-level differences as the corresponding gene regions do not evolve quickly enough to do so. An added complication for the taxonomic assignment of very closely related sequence representations (ASVs, operational taxonomic units (OTU)s) is that the taxonomic signal can be obscured by other sources of sequence variation (base misincorporation during PCR, sequencing errors, polymorphisms between multiple copies of the same gene within a genome) discussed further below.

(3) Accessing more genetic information (i.e. larger proportions of genomes) rather than short-targeted marker gene approaches allow. Pathogen 'clusters' as defined previously are increasingly apparent and challenge both accurate diagnostics (and mitigating actions based on these) and policy/trade implications which rely on high-resolution detection of a tightly defined entity. Examples are growing in all groups of pathogens: viruses (e.g. the fish viruses koi herpesvirus disease (CyHV-3)/koi herpes virus (KHV), and viral haemorrhagic septicaemia (VHS)), bacteria (closely related strains within many pathogenic genera e.g. *Vibrio, Aeromonas, Yersinia*) (Bayliss et al. 2017), and eukaryotes (e.g. *Marteilia* spp.—Kerr et al. 2018, *Gyrodactylus salaris*—Stentiford et al. 2014). In such cases, the usually short diagnostic gene regions (often of a relatively conserved and easily accessible gene such as the eukaryotic 18S rRNA), which were designed based on pathogens of interest before the genetic diversity of lineages related to them was apparent, do not have sufficient taxonomic/phylogenetic signal to distinguish one pathogenic lineage from another. Even when this can be achieved at the level of sequence signatures (consistent differences in single nucleotide polymorphism

(SNP)s or short motifs between lineages) the phylogenetic signal to sequence polymorphism/error noise ratio may be too low to robustly separate the taxa, thereby leading to taxonomic uncertainty and lack of resolution. An example of this is the case of oyster pathogens, *Marteilia refringens* and *M. pararefringens*, which are separable on the basis of five sequence positions in internal transcribed spacer regions (ITS1s) rDNA but isolates of each species do not form monophyletic clades when only short gene regions (ITS1 and intergenic spacer (IGS) rDNA) are used for phylogenetic analysis (Kerr et al. 2018). Only when longer regions of the rRNA gene array are used is the phylogenetic signal relatively strong enough to recover strongly supported, mutually exclusive clades for each species.

3.9 Summary

- A systematic approach to disease diagnosis is of importance in the development of invertebrate pathology.
- The pathologist working on invertebrates has a number of methods available ranging from traditional microscopy, microbial cultivation and cell culture through to high throughput sequencing.
- Combinations of molecular techniques and traditional histopathology are important in disease diagnosis.
- Calculating accurate sample size will not only ensure the veracity of the study but also will avoid over-sampling and promote resource efficiency.
- Molecular sequence-based methods of pathogen detection and diagnostics need to take into account the high levels of genetic/lineage diversity of pathogens in animals and other environmental samples in order to be sufficiently discriminating.
- Disease should be considered as the result of interactions between all host symbionts (even when a primary pathogen is identified), the host, and the environment (the pathobiome concept).
- HTS and multi-omic approaches to disease aetiology and pathogenesis provide new insight into animal diseases.
- Taxonomic outputs of HTS/multi-omic analyses require additional interpretation in order to reliably reflect pathogens and symbionts in a system.

Suggested further reading

General

Frasca, S., Gast, R.J., Bogomolni, A.L., and Szczepanek, S.M. 2020. Diagnosing marine diseases. In: *Marine Disease Ecology*. D.C. Behringer, B.R. Silliman, and K.D. Lafferty (eds.). pp. 213–231. Oxford: Oxford University Press.
Lacey, L. (ed.) 2012. *Manual of Techniques in Invertebrate Pathology* 2nd edn. pp. 504. London: Academic Press. (Overview of methods in insect pathology with limited coverage of molluscs).
Howard, D.W., Lewis, E.J., Keller, B.J., and Smith, C.S. 2004. *Histological Techniques for Marine Bivalve Mollusks and Crustaceans* 2nd edn, p. 218. NOAA Technical Memorandum NOS NCCOS 5.
Shields, J.D. 2017. Collection techniques for the analyses of pathogens in crustaceans. *Journal of Crustacean Biology* 37: 753–763.

Histology

Maynard, R., Downes, N., and Finney, B. 2014. *Histological Techniques: An Introduction for Beginners in Toxicology*. p. 334. London: Royal Society of Chemistry.

Suvarna, K.S., Layton, C., and Bancroft, J.D. 2019. *Bancroft's Theory and Practice of Histological Techniques*, 8th edn, p. 584. Oxford: Elsevier. (Practical-based overview of histology but with a strong emphasis of human histopathology.)

In situ hybridisation

Nielsen, B.S. 2014. *In situ Hybridisation Protocols*, 4th edn, p. 275. N.Y: Humana Press.. (Comprehensive review of methods including full practical details.)

High throughput sequencing

Bass. D., Stentiford, G.D., Littlewood, T.D., and Hartikainen, H. 2015. Diverse applications of environmental DNA methods in parasitology. *Trends in Parasitology* 31: 499–513.

Bass, D., Stentiford, G.D., Wang, H.-C., Kostella, B., and Tyler, C.R. 2019. The pathobiome in animal and plant diseases. *Trends in Ecology and Evolution* 34: P996–P1008.

Acknowledgements

DB was supported by funding from the UK Department of Environment, Food and Rural Affairs (Defra) under contracts FC1214 and FC1215. AFR and CJC were supported by funding from EU Ireland Wales Interreg programme (Bluefish) and UKRI (ARCH-UK-Shellfish Aquaculture Network). We thank Dr Christopher Cunningham (Swansea University) for providing comments on a draft version of the text and Drs Kelly Bateman (Cefas, UK) for the electron micrograph in Figure 3.1 and Harry C. Evans (Centre for Agriculture and Bioscience International, Surrey UK) for the photographs in Figure 3.2.

References

Arif, B. and Pavlik, L. 2013. Insect cell culture: Virus replication and applications to biotechnology. *Journal of Invertebrate Pathology* 112: S138–S141

Azevedo, C., Balseiro, P., Casal, G., Gestal, C., Aranguren, R., Stokes, N.A., Carnegie, R.B., Novoa, B., Burreson, E.M., and Figueras, A. 2006. Ultrastructural and molecular characterization of *Haplosporidium montforti* n. sp., parasite of the European abalone *Haliotis tuberculata*. *Journal of Invertebrate Pathology* 92: 23–32

Bass, D., Stentiford, G.D., Wang, H.-C., Kostella, B., and Tyler, C.R. 2019. The pathobiome in animal and plant diseases. *Trends in Ecology and Evolution* 34: P996–P1008

Bass, D., Stentiford, G.D., Littlewood, T.D., and Hartikainen, H. 2015. Diverse applications of environmental DNA methods in parasitology. *Trends in Parasitology* 31: 499–513

Bayliss, S.C., Verner-Jeffreys, D.W., Bartie, K.L., Aanensen, D.M., Sheppard, S.K., Adams, A., and Feil, E.J. 2017. The promise of whole genome pathogen sequencing for the molecular epidemiology of emerging aquaculture pathogens. *Frontiers in Microbiology* 8: 121

Behringer, D.C., Butler, M.J. IV, Shields, J.D., and Moss, J. 2011. Review of Panulirus argus virus 1- a decade after its discovery. *Diseases of Aquatic Organisms* 94: 153–160

Blanchet H., Raymond N., De Montaudouin X., Capdepuy M., and Bachelet G. 2003. Effects of digenean trematodes and heterotrophic bacteria on mortality and burying capability of the common cockle *Cerastoderma edule* (L.). *Journal of Experimental Marine Biology and Ecology* 293: 89–105

Broberg, M., Doonan, J., Mundt, F., Denman, S., and McDonald, J.E. 2018. Integrated multi-omic analysis of host-microbiota interactions in acute oak decline. *Microbiome* 6: 21

Callahan, B.J., McMurdie, P.J., and Holmes, S. P. 2017. Exact sequence variants should replace operational taxonomic units in marker-gene data analysis. *ISME Journal* 11: 2639–2643

Coates, C.J. and Söderhäll, K. 2021. The stress-immunity axis in shellfish. *Journal of Invertebrate Pathology* 186: e107492.

Davies, C.E., Batista, F. M., Malkin, S.H., Thomas, J.E., Bryan, C.C., Crocombe, P., Coates, C.J., and Rowley, A.F. 2019. Spatial and temporal disease dynamics of the parasite *Hematodinium sp.* in shore crabs, *Carcinus maenas*. *Parasites & Vectors* 12: e472.

Drugmand, J.-C., Schneider, Y.-J., and Agathos, S.N. 2012. Insect cells as factories for biomanufacturing. *Biotechnology Advances* 30: 1140–1157

Hartikainen H., Stentiford G.D., Bateman K.S., Berney C., Feist S.W., Longshaw M., Okamura B., Stone D., Ward G., Wood C., and Bass D. 2014. Mikrocytids are a broadly distributed and divergent radiation of parasites in aquatic invertebrates. *Current Biology* 24: 807–812

Heravi, F.S., Zakrzewski, M., Vickery, K., and Hu, H. 2020. Host DNA depletion efficiency of microbiome DNA enrichment methods in infected tissue samples. *Journal of Microbiological Methods* 170: 105856

Hine, P.M., Carnegie, R.B., Burreson, E.M., and Engelsma, M.Y. 2009. Inter-relationships of haplosporidians deduced from ultrastructural studies. *Diseases of Aquatic Organisms* 83: 247–256

Kerr, R., Ward, G.M., Stentiford, G.D., Alfjorden, A., Mortensen, S., Bignell, J.P., Feist, S.W., Villalba, A., Carballal, M.J., Cao, A., Arzul, I., Ryder, D., and Bass, D. 2018. *Marteilia refringens* and *Marteilia pararefringens* sp. nov. are distinct parasites of bivalves and have different European distributions. *Parasitology* 145: 1483–1492

Lacey, L.A. and Solter, L.F. 2012. Initial handling and diagnosis of diseased invertebrates. In *Manual of Techniques in Invertebrate Pathology*, L.A. Lacey (ed.) pp. 1–14., London: Academic Press.

Longshaw, M. and Malham, S.K. 2013. A review of the infectious agents, parasites, pathogens and commensals of European cockles (*Cerastoderma edule* and *C. glaucum*). *Journal of the Marine Biological Association of the United Kingdom* 93: 227–247

Loose, M., Malla, S., and Stout, M. 2016. Real-time selective sequencing using nanopore technology. *Nature Methods* 13: 751–754

Lutz, H.L., Marra, N.J., Grewe, F., Carlson, J.S., Palinauskas, V., Valkiūnas, G., and Stanhope, M.J. 2016. Laser capture microdissection microscopy and genome sequencing of the avian malaria parasite, *Plasmodium relictum*. *Parasitology Research* 115: 4503–4510

Nielsen, B.S. 2014. *In situ* Hybridisation Protocols, 4ᵗʰ edn., p. 275. N.Y: Humana Press.

Rowley, A.F., Davies, C.E., Malkin, S.H., Bryan, C.C., Thomas, J.E., Batista, F.M., Coates, C.J. 2020. Prevalence and histopathology of the parasitic barnacle, *Sacculina carcini* in shore crabs, *Carcinus maenas*. *Journal of Invertebrate Pathology* 171: e107338

Simon R.C., and Schill, W.B. 1984. Tables of sample size requirements for detection of fish infected by pathogens: Three confidence levels for different infection prevalence and various populations sizes. *Journal of Fish Diseases* 7: 515–520

Smith, A.L., Hirschle, L., Vogan, C.L., and Rowley, A.F. 2015. Parasitization of juvenile edible crabs (*Cancer pagurus*) by the dinoflagellate, *Hematodinium sp.*: Pathobiology, seasonality and its potential effects on commercial fisheries. *Parasitology* 142: 428–438

Stentiford, G.D., Feist, S.W., Stone, D.M., Peeler, E.L., and Bass, D. 2014. Policy, phylogeny and the parasite. *Trends in Parasitology* 30: 274–281

Tall, B. D., Fall, S., Pereira, M. R., Ramos-Valle, M., Curtis, S. K., Kothary, M.H., Chu, D.M., Monday, S.R., Kornegay, L., Donkar, T., Prince, D., Thunberg, R. L., Shangraw, K. A., Hanes, D.E., Khambaty, F. M., Lampel, K. A., Bier, J.W., and Bayer, R.C. 2003. Characterization of *Vibrio fluvialis*-like strains implicated in limp lobster disease. *Applied and Environmental Microbiology* 69: 7435–7446

Urrutia, A., Bass, D., Ward, G., Ross, S., Bojko, J., Marigomez, I., and Feist, S. 2019. Ultrastructure, phylogeny and histopathology of two novel haplosporidians parasitising amphipods, and importance of crustaceans as hosts. *Diseases of Aquatic Organisms* 136: 87–103

Watts, E.A., Dhara, A., and Sinai, A.P. 2017. Purification of *Toxoplasma gondii* tissue cysts using Percoll gradients. *Current Protocols in Microbiology* 45: 20C.2.1–20C.2.19

PART II

The Diseases

Diseases of Acoelomate and Coelomate Protostomes

CHAPTER 4

Diseases of scleractinian corals

David G. Bourne, Hillary A. Smith, and Cathie A. Page

4.1 Introduction

There is a large diversity of coral species, with over 1,300 extant species in the order Scleractinia (phylum Cnidaria, class Anthozoa; World Register of Marine Species (WoRMS)). The Scleractinia (stony corals) represent the ecological framework builders of modern coral reefs, providing habitat and refuge for countless other marine species that inhabit coral reef ecosystems As ecosystem engineers, corals provide important goods and services that are both ecologically and economically important, including sustaining high biodiversity, protecting coastlines, supporting fisheries, and contributing an estimated US$ 36 billion to the global economy directly from tourism-based industries (Spalding et al. 2017).

4.1.1 Coral biology, ecology, and microbiology

Corals build colonies by modular additions of single polyps to achieve complex morphologies. Within the Scleractinia, polyps are monomorphic (with the exception of the genus *Acropora*), and all perform the same functions, including deposition of calcium carbonate, feeding, and reproduction. Modular colonial organisation of corals allows for the unique capability of sharing resources among polyps through the connective tissue (coenosarc), responding to environmental cues across a colony, and allowing a coral animal to experience partial mortality. Within the Scleractinia, symbiosis with unicellular photosynthetic dinoflagellates belonging to the family Symbiodinaceae is a common characteristic (LaJeunesse et al. 2018) and underpins the coral animal's success in oligotrophic waters.

Symbiodiniaceae cells are situated within the coral gastrodermal layer and are particularly abundant in gastrodermal cells adjacent the free body wall epidermis. Photosynthesis by these symbionts contributes to coral metabolism by translocating fixed carbon to coral animal cells, providing a significant proportion of their nutritional needs. Additionally, algal photosynthesis requires sunlight, and therefore this symbiosis plays an important role in determining the depth distribution of coral species.

An array of other microbial communities has been demonstrated to be associated with corals, including bacteria, Archaea, fungi, and viruses (reviewed in Bourne et al. 2016b). Collectively, this consortium of host and microbial cells has been termed the coral holobiont, with the stability of the microbiome linked to whole organism health. In addition to the coral host's immune defences (i.e. antioxidant production and melanisation processes; see Chapter 1 for further details), coral-associated microbial communities are likely involved in coral immunity either directly through production of antimicrobial compounds (Ritchie 2006) or indirectly through niche exclusion of opportunistic or pathogenic organisms (Rosenberg et al. 2007; Shnit-Orland and Kushmaro 2009). While some microbial taxa are directly linked to disease onset (Sweet and Bulling 2017), disturbance of the coral host's normal microbial community composition (i.e. dysbiosis) can induce disease or disease-like signs (Egan and Gardiner 2016; Sweet and Bulling 2017).

The coral holobiont is well studied, and in particular the delicate symbiosis between the coral

David G. Bourne, Hillary A. Smith, and Cathie A. Page, *Diseases of scleractinian corals*. In: *Invertebrate Pathology*. Edited by Andrew F. Rowley, Christopher J. Coates and Miranda M.A. Whitten, Oxford University Press. © Oxford University Press (2022). DOI: 10.1093/oso/9780198853756.003.0004 animal and Symbiodiniaceae, with environmental stress (particularly temperature increases) resulting in a breakdown of this partnership and ejection of Symbiodiniaceae cells from the coral tissues (termed bleaching). Climate driven ocean warming is the biggest threat to coral reefs globally (Hughes et al. 2017a), pushing corals and their relationship with Symbiodiniaceae above their thermal threshold, leading to sub-optimal coral health and increased incidence of bleaching and disease (Hoegh-Guldberg et al. 2017). Severe mass bleaching events can result in mass coral mortality, and subsequent extensive impacts on reef ecosystems, with these impacts predicted to be exacerbated under future climate scenarios (Hughes et al. 2017b). Synergy between heat stress, bleaching, and disease has been repeatedly demonstrated, leading to further concerning reduction in coral cover globally (Brodnicke et al. 2019). Combined with localised impacts including eutrophication, overfishing, and coastal development, reefs globally are in a precarious predicament.

4.1.2 Coral disease background

Black band disease (BBD) was identified in the early 1970's in the Caribbean through the work of Antonius (1973). BBD is often described as the first coral disease to be identified, though descriptions of growth anomalies (GA) on corals were reported a decade earlier (Squires 1965). Since those early studies of rare occurrences of diseased colonies (Antonius 1973), coral disease has increased in terms of prevalence, diversity of disease types, and impacts on coral communities and reefs. Initial studies of coral disease focussed on Caribbean corals, in part due to the detection of novel diseases and significant disease outbreaks in this region (Gladfelter 1982; McClanahan and Muthiga 1998; Williams et al. 1999). Disease mediated declines in *Acropora palmata* and *Acropora cervicornis* have been so dramatic throughout their Caribbean range that these species were nominated for inclusion in the United States' endangered species register (Diaz-Soltera 1999). In part due to disease outbreaks, Caribbean acroporid corals are now rare on reefs on which they had previously been the dominant coral taxa for at least 95,000 years (Pandolfi and Jackson 2006).

Until more recently, coral disease was thought to occur only rarely and have little impact on the function of Indo-Pacific corals and coral reefs (reviewed in Sutherland et al. 2004). Significant disease outbreaks have, however, now been recorded from numerous Indo-Pacific regions (Antonius 1999; Willis et al. 2004; Aeby 2005; Williams et al. 2011; Richards and Newman 2019). It is unclear whether the recent emergence of novel coral diseases and outbreaks on Indo-Pacific reefs represent a real increase in coral disease, or an increase in disease-focussed research in this region. Nonetheless, disease is now ubiquitous though generally low in prevalence outside of outbreak conditions on reefs throughout the Indo-Pacific (Willis et al. 2004; Page et al. 2009; Onton et al. 2011; Aeby et al. 2011).

Despite over 50 years of research which has characterised an extensive array of coral syndromes, the causative agents of most have remained largely unidentified (Montilla et al. 2019). Currently, there are over 40 different coral disease syndromes reported in the literature that detail compromised coral health, though many have poor macroscopic and diagnostic descriptions with little epizootic and aetiologic information. In addition, many reported diseases lack wide distributions, being reported only from few localised reefs. As a result, many of the names associated with coral disease syndromes (i.e. WP Type I and Type II, white plague-like, WB type I, type II, type III etc.) are imprecise and lack clear definitions that discriminate the individual syndromes (Bruckner 2016b).

4.2 Principal diseases

There are nine major scleractinian diseases reported from the Caribbean, including growth anomalies (GA), black band disease (BBD), white plague (WP), white band disease (WBD), white pox (WPX), yellow band disease (YBD), dark spot syndrome (DSD), Caribbean ciliate infestations, and Caribbean stony coral tissue loss disease (SCTLD), and six in the Indo-Pacific Region including black band disease (BBD), white syndromes (WSs), brown band disease (BrB), skeletal eroding band (SEB), ulcerative white syndrome (UWS), and growth anomalies (GA). These diseases reported for corals are summarised in Table 4.1. While several other diseases have been detailed, these main diseases have a broad geographical footprint and repeated observations. Coral bleaching, representing the loss of the endosymbiotic dinoflagellates (Symbiodinaceae) is

Disease	Disease causing agent(s)	Host range	Geographical range	Pathology	Key references
Growth Anomalies (GAs)	Unknown	Wide host range, including *Astreopora*, *Fungia*, *Pavona*, *Platygyra*, *Montipora*, *Acropora*, *Madrepora*, *Montastraea*, *Diplora*, *Pocillopora*, *Sideras-trea*, *Mycetophyllia*, *Leptastrea* and *Porites*	Global; documented from all major coral reefs systems	Acroporids display distinct smooth to rugose, vari-ably sized skeletal masses, having reduced or absent calices, polyps overgrown by pale to colourless tissues. Excessive malformed accretion is observed in the coenosteum regions. Abnormal development of gas-trovascular cavities is also observed. Hyperplasia of the basal body wall is observed in histological analyses along with reduced mesenterial filaments, no tenta-cles and reduced or lack of Symbiodinaceae in the gastrodermal cells. Differences in GAs and healthy tissues of other species is less well described. *Porites compressa* GAs present increased thickness of basal body wall, reduced Sym-biodinaceae cells, and increased size and granularity of the calicodermis	Loya et al. 1984; Peters et al. 1986; Hanahan and Weinberg 2030; Kaczmarsky 2036; Kaczmarsky and Richardson 2007; Work et al. 2008; Work et al. 2015; Preston and Richards 2020
Black band disease (BBD)	Polymicrobial disease mat, Cyanobacteria (*Rose-ofilum reptotaenium*) dominates the micro-bial mat in addition to a diverse community of sul-phate reducing, sulphate oxidising, and heterotroph-ic bacteria. Synergistic activities of microbial con-sortium promote lesion progression	Wide host range; at least 42 Caribbean and 57 Indo-Pacific taxa	Global; documented from all major coral reefs systems	Darkly pigmented microbial band or mat that pro-gresses across coral colonies at rates up to 2 cm day⁻¹. Clear demarcation between the band (which can be black, brown, or red depending on cyanobac-terial pigmentation), live coral tissues and denuded coral skeleton. Band can be mms to cms wide and mms thick. Anoxic and sulphidic conditions within and at the base of the microbial mat promote necrosis in underlying coral tissues. Cyanotoxins are implicated in tissue necrosis in some case studies. Cyanobacte-ria have a potential metabolic adaption to the high sulphide levels within the mat. BBD typically progress-es from a peripheral or central point within a coral colony, radiating outwards in a circular or semi-circular pattern on massive or encrusting colonies. On branch-ing colonies, the band typically originates basally, progressing up coral branches	Rützler and Santavy 1983; Rutzler et al. 1983; Kuta and Richardson 1996; Bruckner et al. 1997; Richardson 1997; Kuta and Richardson 2002; Richardson 2004; Sutherland et al. 2004; Casamatta et al. 2012; Sato et al. 2016; Meyer et al. 2017; Sato et al. 2017
White Plague (Inclusive of WPs type I, type II and type III)	WP-type I: Unknown	Initially *Mycetophyllia ferox* and *Siderastrea sidera*, though has been reported from 30 Caribbean coral species	Northern Florida Keys including Carysfort Reef and Key Largo Reef	Lesion present as a distinct boundary between appar-ently healthy tissue and recently denuded white skeleton. Lesions originate at variable locations on a colony and expand at a rate of ~ 3 mm day⁻¹	Dustan 1977; Bythell et al. 2004; Sutherland et al. 2004; Bruckner 2016; Williams et al. 2020
	WP-type II: *Aurantimonas coralicida* originally iden-tified and satisfied Koch's postulates, though later studies showed colonies were no longer susceptible	First identified in *Dichocoenia stokesi* colonies in the Flori-da Keys, though since reported from 43 coral taxa	Florida Keys and wider Caribbean region	Gross lesion morphology similar to WP-1 though lesion progression is faster and always originates basally, progressing towards the colony apex. Bleached tissue may be observed at the tissue lesion interface. Maximum progression rate was recorded as 2 cm day⁻¹	Richardson et al. 1998; Denner et al. 2003; Bythell et al. 2004; Lesser et al. 2007; Bruckner 2016

continued

Table 4.1 *Continued*

Disease	Disease causing agent(s)	Host range	Geographical range	Pathology	Key references
	WP-type III: Unknown	*Montastraea* and *Colpophyllia*	Florida Keys	Similar to WP Type II, though lesion progression rates 5–10 times faster (exceeding 10 cm day⁻¹) with lesions originating in colony centre expanding outwards across a broad front	Bythell et al. 2004; Bruckner 2016
	WP-like (Red Sea): *Thalassomonas loyana* sp.	*Favia* and *Goniastrea*	Red Sea	A distinct boundary between apparently healthy coral tissue and freshly exposed coral skeleton, with no build-up of microorganisms or necrotic tissues at this interface	Barash et al. 2005; Thompson et al. 2006; Bruckner 2016
	WP-like (USVI): Unknown, though *Roseovarius crassostreae* identified in diseased tissues	*Montastraea*	US Virgin Islands	See description of WP Type II	Pantos et al. 2003; Bruckner, 2016
WBD	WB type I; Unknown. Some case studies show rod-shaped bacteria, bacterial aggregates, and *Bacillus* sp., *Lactobacillus* sp. and a *Rickettsiales*-like organism (RLO) at lesion interface, though causation not confirmed through Koch's postulates	*Acropora palmata* and *Acropora cervicornis*	Caribbean-wide	Diffuse linear or annular lesion originating basally with distinct band of tissue loss that reveals intact, bare coral skeleton that is well differentiated from the discoloured and algae overgrown distal skeleton. Lesions originate basally and progress to the colony apex along furcated branches	Gladfelter 1982; Green and Bruckner 2000; Bythell et al. 2004; Bruckner 2009; Gignoux-Wolfsohn et al. 2020
	WB type-II: Unknown. *Vibrio harveyi*/*Vibrio carchariae*	*Acropora cervicornis*	Bahamas, Puerto Rico, Florida	Original differentiation from WBD type I is lesion boundary preceded by a band of bleached tissue (of variable width 2–20 cm). The advancing lesion may progress past the bleaching region. Lesions often originate basally and progress upward but can originate at colony apex and progress to colony base. Faster lesion progression rate than WBD type 1 (i.e. up to 10 cm day⁻¹)	Ritchie and Smith, 1998; Bythell et al., 2004, Bruckner, 2016
WPX	Acroporid serratiosis causative agent: *Serratia marcescens*	*Acropora palmata*	Florida	Focal and multifocal irregular circular, oblong or pyriform shaped lesions devoid of coral tissue exposing bare white skeleton. Lesions are basal to apically distributed and surrounded by a front of necrotic pigmented living tissue. Most lesions are small but may be > 80 cm². Lesions can enlarge at an average rate of 2.5 cm² day⁻¹ and coalesce to cause extensive tissue loss and whole colony mortality. Colony mortality and disease severity was greater for outbreaks from 1994–2004 compared to 2008–2014	Patterson et al. 2002; Sutherland and Ritchie 2004; Sutherland et al. 2010; Lesser and Jarett 2014; Joyner et al. 2015; Sutherland et al. 2015a, 2016b

	Causative agent	Host	Geographic location	Gross lesion description	References
	Other WPX syndromes: Unknown	*Acropora palmata*	USVI, Mexico, Puerto Rico, Bahamas	Similar gross lesion description to Acroporid serratiosis WPX, though no causative agent identified and *S. marcescens* not associated with lesions	Bruckner and Bruckner 1997; Rodriguez-Martinez et al. 2001; Weil 2C04; Muller et al. 2C08; Sutherland et al. 2016
White syndromes Indo-Pacific (WSs)	Multiple identified biological agents including vibrios (*V. corallilyticus*, *V. harveyi*), ciliates, viruses, parasites, and helminths, as well as cellular apoptosis. Vibrios or other microbes (Rhodobacteraceae) may contribute, or opportunistically proliferate in compromised tissues	Wide host range, but most prevalent amongst acroporid and pocilloporid species and in particular the tabulate *A. hyacinthus* and *A. cytherea*	Indo-Pacific wide; Great Barrier Reef, Western Australian reefs, Palmyra, Red Sea, Western Indian Ocean, Hawaii	Diffuse linear (or annular) band or irregular patch comprised of recently exposed coral skeleton adjacent to healthy tissues. Tissue loss can be peripheral, basal, or central and affects polyps and coenosarc. Often originating from a small lesion front and manifests as a band front across the entire colony. Lesions have a linear shape and diffuse border resulting in a continuous pattern of white exposed skeleton. Progression rate is moderate (subacute) to rapid (acute) Host factors including population density contribute to disease outbreaks and often outbreaks are correlated with warm seawater anomalies	Willis et al. 2004; Work and Rameyer 2C05; Ainsworth et al. 2007a; Ainsworth et al. 2007b; Bruno et al. 2007b; Sussman et al. 2007; Heron et al. 2008; Sweet and Bythell 2012; Bourne et al. 2015; Pollock et al. 2017; Smith et al. 2020
	Acute *Montipora* WS (aMWS): *Vibrio corallilyticus* strain OCN008 and *Pseudoalteromonas piratica* strain OCN003 Chronic *Montipora* WS (cMWS): *Vibrio owensii* strain OCN002	*Montipora capitata*	Hawaii	Disease manifests in acute and chronic forms based on gross lesions. aMWS presents as locally extensive or diffuse lesions resulting in rapid tissue loss, characterised by > 5 cm of recently denuded white skeleton and resulting in high rates of colony mortality. cMWS presents as multifocal lesions found year-round, with lesions progressing slowly and < 5 cm of recently denuded white skeleton at lesion boundary	Aeby et al. 2010; Ushijima et al. 2012; Work et al. 2012; Ushijima et al. 2014; Beurmann et al. 2017
UWS; Indo-Pacific	Unknown	Multiple *Porites* sp., *Echinopora*, *Goniastrea*, *Heliopora*, *Dispsastrea* (previously *Favia*), *Montipora*	Indo-Pacific, including Philippines, East Africa, Indonesia, Guam, Palau, Great Barrier Reef, India, Maldives	Focal or multi-focal tissue loss that manifests typically in a central location but forms anywhere on the coral colony. May start as white tissue discolouration leading to full tissue-thickness ulceration which affects polyps and coenosarc. The multifocal lesions often coalesce and appear circular to oblong in shape. Lesions may have a smooth or diffuse border. Lesion size is small (~ 3–5 mm diameter) and can have varying rates of progression from chronic (not progressing) to subacute	Raymundo et al. 2003; Raymundo et al. 2005; Haapkyla et al. 2007; Myers and Raymundo 2009; Page et al. 2009; Montano et al. 2016

continued

Table 4.1 Continued

Disease	Disease causing agent(s)	Host range	Geographical range	Pathology	Key references
DSS	Unknown	*Montastraea annularis*, *Siderastrea siderea*, and *Stephanocoenia intersepta*	Caribbean	Brown, purple or black irregular multifocal spots of varying sizes resulting in coral tissue loss and exposure of bare skeleton that is secondarily colonised by algae and other fouling organisms Lesion can expand by 3 mm month^{-1}. Symbiodinaceae populations associated with lesions appear swollen and darkly pigmented.	Cervino et al. 2001; Garzón-Ferreira et al. 2001; Gil-Agudelo et al. 2004; Weil 2004; Borger 2005; Gil-Agudelo et al. 2007; Renegar et al. 2008; Porter et al. 2011; Sweet et al. 2013
Caribbean Yellow Band Disease (CYBD)	Unknown; *Vibrio* species; including *Vibrio alginolyticus* associated with disease lesions, though Koch's postulates not satisfied	*Montastraea annularis* complex and sometimes, though rarely, reported in *M cavernosa* and other favids	Caribbean reefs (diseases with similar reported gross signs in Atlantic and Pacific reefs though potentially disease with different aetiology)	Lesions initially characterised by pale to yellow coral tissue that presents as focal, multifocal, linear, or annular bands that radiate over time with multifocal lesions often coalescing. Lesions appear on colonies in random, lateral, basal, or apical areas. As the lesion expands, the inner areas suffer tissue necrosis and mortality resulting in colonisation of the skeleton by algae. Histopathology indicates degeneration of tissues and cells in the lesion interface. *Vibrio* sp. infection of Symbiodinaceae cells have been reported in early stages of lesion onset which results in loss of pigments within the dinoflagellate cells. Cells also appear depleted of organelle structures and chloroplasts are reduced, fragmented, and lose cellular integrity	Cervino et al. 2001; Cervino et al. 2004; Bruckner and Bruckner 2006; Cervino et al. 2008; Cunning et al. 2009; Bruckner and Riegl 2016
Skeletal Eroding Band (SEB) and Caribbean Ciliate Infection (CCI)	Ciliate infestation *Halofolliculina corallasia*	Wide host range, 82 species reported from the Indo-Pacific and Red Sea and 26 species from the Caribbean	Global (Indo-Pacific, Red Sea and Caribbean)	Lesions are characterised by a dark linear or annular band of variable width (1–10 cm thick) that separates apparently healthy tissue from recently exposed skeleton. The dark band is comprised of high densities of the *Halofolliculina corallasia* ciliate. At low ciliate densities the band can appear speckled and/or light green in colour. Coral skeleton denuded of tissue can appear eroded as a result of boring and potential chemical activity of the ciliates. Bands often originate basally or peripherally and progress at variable rates across colonies or upward to colony apex	Antonius 1999; Cróquer et al. 2006; Page and Willis 2008; Rodriguez et al. 2009; Page et al. 2016; Montano et al. 2020

Disease	Cause	Host range	Distribution	Description	References
Brown Band Disease (BBD)	Ciliate infestation *Porpostoma guamense*	Wide host range	Global (Indo-Pacific, Red Sea and Caribbean)	Lesion is characterised by a brown linear or annular band of variable width (typically < 5 cm wide) separating apparently healthy tissue from denuded skeleton. A narrow zone of bleached tissue may be present between the brown band and live coral tissue. The brown band is comprised of a high density of the protozoan scuticociliates (Class Oligohymenophorea; genus *Porpostoma*). Lesions and bands often originate basally or peripherally and progress rapidly across a colony or apically	Bourne et al. 2008; Page et al. 2009; Qiu et al. 2010; Looban et al. 2011; Sweet and Bythell 2012; Weil et al. 2C12; Nicolet et al. 2C13; Sweet et al. 2C14; Randall et al. 2C15
SCTLD	Unknown	Multiple coral genera including *Agaricia, Colpophyllia, Dendrogyra, Dichocoenia, Diploria, Eusmilia, Meandrina, Montastraea, Madracis Mycetophyllia, Orbicella, Porites, Pseudodiploria, Siderastrea, Solenastrea, Stephanocoenia*	Florida Reef Tract, Caribbean (Jamaica, Quintana Roo (Mexico), St. Maarten, St. Thomas and St. John (USVI), Dominican Republic, Turks & Caicos Islands, Belize, St. Eustatius (Netherlands), Puerto Rico, and Grand Bahama)	Focal or multi-focal diffuse area of tissue loss resulting in denuded coral skeleton. Tissues bordering the lesions are occasionally bleached, though normally indistinct bands of pallor are observed with progress to normal pigmentation away from the denuded skeleton. Tissue loss is distributed basally, peripherally, or both and can be acute (3–4 cm day^{-1}) to subacute. Histopathology indicates signs of degradation, fragmentation, and swelling of the gastrodermal basal body wall which progresses to the oral surface and manifests as necrosis	Precht et al. 2016; Walton et al. 2018; Aeby et al. 2019; Alvarez-Filip et al. 2019; Meyer et al. 2019; Weil et al. 2019; Muller et al. 2020; Rosales et al. 2020

Figure 4.1 Images of common coral diseases. (a) Acroporid from on the Great Barrier Reef displaying growth anomaly. (b) *Montipora* sp. on the Great Barrier Reef displaying black band disease. (c) *Acropora cervicornis* from Florida with white band disease. (d) *Acropora hyacythus* from the Great Barrier Reef displaying white syndrome. (e) *Siderastrea* sp. from the Caribbean with dark spot disease. (f) *Orbicella* sp. from Belize with yellow band disease. (g) *Acropora muricata* from the Indo-Pacific region displaying skeletal eroding band (h) *Acropora muricata* from the Great Barrier Reef displaying brown band disease. (i) *Montastrea* sp. from the Caribbean with stony coral tissue loss disease. Source: images from the Great Barrier Reef and Indo-Pacific supplied by Cathie Page, images from the Caribbean supplied by Greta Aeby.

by definition also a disease (i.e. impairment of normal function). Thermally induced bleaching is a physiological response to environmental stress and is investigated extensively elsewhere, hence is not presented in this chapter. Comparatively less research has been undertaken to understand diseases of octocorals compared to those affecting scleractinians, with the exception of Aspergillosis infection of gorgonians. Diseases such as BBD and GA have also been documented within octocorals (Weil et al. 2015; Kim 2016), however, this chapter specifically focusses on diseases of scleractinian corals. The diseases are presented in rough chronological order of when they were first reported in the literature (see Sutherland et al. 2004), though all white diseases are grouped for consistency.

4.2.1 Growth anomalies (GAs)

GAs were the first (Squires 1965), and still are one of the most commonly reported syndromes globally (Work et al. 2015). These present as exuberant distinct tissue and skeletal growths on coral colonies, having grown at a faster rate than surrounding areas (see Figure 4.1a). Characteristics include aberrant corallite and coenosarc shapes, often as a result of rapid skeletal accretion (Peters et al. 1986; Work et al. 2008a). They have been reported from a large number of coral genera, with prevalence ranging from low (< 5%) up to 70% in some reef regions (Aeby et al. 2011; Work et al. 2015). They are most commonly reported for corals in the families Acroporidae and Poritidae (Aeby et al. 2011). The classical appearance of GAs in acroporids is described by smooth to rugose, variably sized skeletal masses with reduced or absent calices. The tissue overlaying the abnormal growth is often pale or translucent due to the mass of aragonite skeleton underlying the tissue and reduced densities of Symbiodiniaceae. Plating acroporids often have GAs on the upper surfaces, though in branching colonies they can be found randomly (Work et al. 2008b). Across all acroporids, the abnormal development of the gastrovascular canals and excessive accretion of the coenosteum (i.e. skeletal material secreted by the coenosarc) is consistently observed (Preston and Richards 2020). Visually different GAs have been described for *Porites* spp., including a skeletal

growth resulting in a raised appearance demarcated by paler tissues, while a second type displays very rough surfaces with pale or white patches and occasional pink pigmentation (Kaczmarsky 2006).

The aetiology of GAs is unknown, with previous studies proposing environmental factors, genetic predisposition, and infectious agents including bacteria, fungi, and viruses (Loya et al. 1984; Peters et al. 1986; Domart-Coulon et al. 2006; Kaczmarsky 2006; Kaczmarsky and Richardson 2007; Aeby et al. 2011). Ageing and senescence leading to inadequate cell repair have also been proposed (Hanahan and Weinberg 2000; Work et al. 2015). Elevated activity of the enzyme phenoloxidase in GAs and the healthy tissues of GA-affected colonies, compared to colonies having no GAs, suggests an immune response in affected colonies (Palmer and Baird 2018). The impacts of GAs on corals appear to differ depending on species impacted and may reflect the previously described investment in immunity. In acroporids, GAs impact the host negatively as they can progress and spread, therefore reducing fitness through reduced growth and reproductive output. GAs have been categorised as neoplasia conditions (cancerous) and some aspects of the lesions support this. However, diagnosis of neoplasia in invertebrates is challenging. The nature of GAs indicates aberrant cell division, though further work is required to appropriately classify these lesions and Work et al. (2015) argue the classification of GAs as true cancers in other coral genera to be premature.

4.2.2 Black band disease (BBD)

BBD represents arguably the most comprehensively understood coral disease. First reported in the 1970s, it has a global distribution that infects a wide range of scleractinian taxa (Richardson 2004; Page and Willis 2006; Sato et al. 2016; Spalding et al. 2017). BBD is often reported to infect dominant coral taxa on reefs, irrespective of morphology, with at least 42 Caribbean and 57 Indo-Pacific taxa being susceptible. The prevalence of the disease is generally low (< 1%), though localised outbreaks have been reported that can impact coral population demographics (Edmunds 1991; Kuta and Richardson 1996; Green and Bruckner 2000; Sato et al. 2009). The lesion presents as a darkly pigmented band that migrates across corals (see Figure 4.1b), killing the underlying tissues and leaving behind dead and denuded coral skeleton that is quickly overgrown by algae and other fouling organisms. Progression rates are, on average 3 mm day^{-1} though can range from 1 mm day^{-1} up to 2 cm day^{-1} in some case studies (Richardson 2004), with higher progression rates resulting in whole colony mortality in days to weeks. Water-borne transmission of BBD is supported by spatiotemporal patterns in the spread of the disease to other colonies often following the direction of predominant current and waves (Bruckner et al. 1997; Zverlov et al. 2005). Gastropods, polychaetes, and reef fishes may also act as vectors, or may facilitate transmission through physical injury to the coral host (Aeby and Santavy 2006; Chong-Seng et al. 2011; Nicolet et al. 2018).

BBD is characterised as a polymicrobial disease, whereby a consortium of microorganisms forms a microbial mat that creates and maintains anoxic and sulfidic biochemical conditions that are toxic to the underlying coral tissues. The microbial communities associated with BBD lesions have been well characterised and are remarkably consistent across global case studies (reviewed in Sato et al. 2016). Filamentous cyanobacteria affiliated to *Roseofilum reptotaenium* (Casamatta et al. 2012) (Figures 4.2 a,b), represent the largest biomass in the microbial mat, though highly diverse communities of heterotrophic bacteria (Sekar et al. 2008; Sato et al. 2010; Meyer et al. 2017; Sato et al. 2017), a phylogenetically novel lineage of Archaea (Sato et al. 2013), plus a high abundance of sulphur-oxidising and sulphate-reducing bacteria have been characterised from BBD lesions (Bourne et al. 2011; Bourne et al. 2013). Histopathology of BBD demonstrates the incursion of cyanobacterial filaments into the tissues at the lesion interface, with cyanobacterial toxins implicated in contributing to tissue necrosis in some studies (Richardson et al. 2007). The integrated metabolic pathways that support the microbial consortium in the lesion have been characterised through metagenomic and meta-transcriptomic approaches, with cyanobacterial-derived photosynthetically fixed CO_2 enhancing productivity and promoting sulphide production within the lesion (Meyer et al. 2017; Sato et al. 2017). A putative sulphide tolerance metabolic pathway exists in the dominant *Roseofilum reptotaenium* cyanobacterium that explains its success and persistence in the lesion (Sato et al. 2017).

4.2.3 White syndromes (WSs)

Within the literature, there is a historical legacy of different coral diseases being reported that lack clear and unambiguous descriptions of the gross lesions. This has resulted in descriptions of coral diseases that are open to subjective interpretation, leading to a proliferation of coral disease names that display similar visible signs of tissue loss. This is particularly evident for many of the 'white' diseases reported in the Western Atlantic, for which names such as WP, WB, and WPX are frequently reported. Despite varying names, all of these 'white' diseases display a tissue loss pattern characterised by a distinct separation of apparently healthy tissue and freshly exposed white coral skeleton and the absence of visible causative agents at the tissue-skeleton interface. Distinction between diseases have been made based on taxa affected, rates of tissue loss, slight variations in pattern of appearance, and lesion progression, locations of the lesions, and if a zone of bleached tissue is present between healthy and necrotic coral tissues (Bruckner 2016b). Without the standardised and systematic framework for describing coral lesions (Work and Aeby 2006), and with little definitive identification of the underlying causation of the diseases, these various disease names have proliferated in the literature and are often used interchangeably, making it very difficult to compare diseases between different case studies. As a result of the limited understanding of the epizoology and aetiology of many white disease case studies, and to avoid confusion, here we discuss the pathology of individual reported case studies using the original disease nomenclature allocated to disease signs.

4.2.3.1 White plague (WP)

WP was first described from the Florida Reef Tract in the mid-1970s and characterised by macroscopic signs of an advancing distinct margin between apparently healthy coral tissue and newly exposed white skeleton, often resulting in whole colony mortality (Dustan 1977; Bythell et al. 2004). A subsequent outbreak of WP was reported in 1995 on the same reefs, though it was named WP type II and distinguished from the first reported outbreak (WP type I) based on the more rapid rate of lesion progression, higher prevalence and greater number of susceptible coral species (Richardson 1998). Work and Aeby (2006) provide a definitive gross description of WP-Type II as diffuse, peripherally and basally situated large, irregular, and distinct areas of tissue loss revealing intact bare white skeleton. An additional type of WP, WP type III was reported in 1999 on reefs in Florida and characterised by very rapid rates of lesion progression (Richardson et al. 2001). WP type II and type III have been reported to have spread throughout the wider Caribbean (Nugues 2002; Borger 2003; Croquer et al. 2003) and other reef systems across the Indo-Pacific and Red Sea (Barash et al. 2005). Confusion in descriptions of gross lesion characteristics combined with poor understanding of the underlying aetiologies have resulted in more recent naming of lesions displaying these signs as WP-like disease (Pratte and Richardson 2016), similar to the grouping of white syndromes of Indo-Pacific corals (see Section 4.2.3.4), until aetiologies of distinct diseases can be established.

The aetiologic diagnosis of WP-like diseases are reported for some case studies. The bacterium *Aurantimonas coralicidia* was identified as the causative agent of WP type II based on isolation of this strain from diseased colonies of *Dichocoenia stokesi* followed by inoculation of healthy colonies that subsequently displayed the same disease signs (Denner et al. 2003). However, later studies highlighted that apparently healthy colonies are not susceptible to the original or subsequently isolated bacterial strains (Lesser et al. 2007). Another bacterial agent, *Thalassomonas loyana* (Barash et al. 2005; Thompson et al. 2006), has been implicated in disease causation for WP in Red Sea corals (*Favia favus*), with inoculation of the bacterium supplemented with seawater filtrate containing an unknown factor from the aquarium water of diseased colonies resulting in disease onset on apparently healthy

colonies (Barash et al. 2005). Microbial community profiling of WP-like diseased tissues of *Montastrea annularis* colonies from the US Virgin Islands found a high abundance of sequences related to *Roseovarius crassostreae*, a previously identified bacterium involved in juvenile oyster disease, which was suggested to have a role in WP-like disease causation (Pantos et al. 2003). In all these cases the aetiological agent was not unequivocally confirmed and therefore definitive causation is still lacking. It is possible that different pathogens or bacterial consortia produce similar disease phenotypes in different coral species across broad geographic regions (Sweet and Bulling 2017).

4.2.3.2 White band disease (WBD)

The general distinction between WBD and WP diseases is that WBD is restricted to Caribbean acroporid species (see Figure 4.1c), while WP occurs on massive and encrusting non-*Acropora* growth forms, though this distinction can be argued to be historically arbitrary (Bythell et al. 2004). Similar to WP, two forms of WBD have been reported in the literature. White band disease (type I) was first reported in the 1970s in the US Virgin Islands (Gladfelter et al. 1977; Gladfelter 1982). Over the ensuing decades, the disease spread throughout the Caribbean basin (now reported from more than 27 countries), severely impacting the populations of two of the dominant framework corals of Caribbean reefs, *Acropora palmata* and *A. cervicornis* (Green and Bruckner 2000; Bruckner et al. 2002). At some sites, the abundance and cover of these corals declined by > 90%, leading to a designation of these two species as critically endangered (The International Union for Conservation of Nature (IUCN) Red List of Threatened species; Hogarth 2006). WBD type II was first identified in San Salvador, Bahamas, and contrasts to WB type I as it affects only *A. cervicornis*. The gross lesions of both type I and type II are described by a distinct band of white exposed skeleton that separates live tissue and algal colonised dead skeleton behind the lesion front (Bruckner 2009). The two types are distinguished based on the rates of tissue loss, which is faster in WBD type II (up to 10 cm day^{-1}) compared to WBD type I (average 5.5 mm day^{-1}). Type II can be further distinguished by the presence of a bleached tissue region at the interface of apparently normal tissue and denuded skeleton (Ritchie and Smith 1998). Like WP, there is a lack of clear distinctions between type I and type II, plus additional poorly characterised syndromes of acroporids in the Caribbean are reported (i.e. shut-down reaction; Antonius 1977; rapid tissue loss; Williams and Miller 2005). Hence, the general term WBD is now used to describe any visible white band that results from tissue death and loss that is not related to bleaching, predation or other band diseases (e.g. BBD or ciliate infections). The unifying characteristics of WBD describe a diffuse, basally situated lesion with distinct areas of tissue loss that reveal intact, bare skeleton that is well differentiated from the distal skeleton, which is discoloured and overgrown with algae (Work and Aeby 2006).

Identification of the causative agents of WBD have remained elusive. Transmission studies have variably indicated an infectious biotic agent, with direct contact, biological vectors, and the water column as possible modes of spread of the disease (Vollmer and Kline 2008; Gignoux-Wolfsohn et al. 2012; Certner et al. 2017). Histopathology of the disease lesion demonstrates atrophy, necrosis, and lysing of the surface and basal body wall and polyp structures at the tissue-lesion margin (Gignoux-Wolfsohn et al. 2020). Early studies also identified bacterial aggregates in higher abundance in tissue regions adjacent to WBD type I lesions, though bacterial aggregates are also generally found in healthy coral tissues (Peters 1984; Santavy and Peters 1997; Work and Aeby 2014). Earlier studies of WBD-type II identified a higher abundance of *Vibrio* spp. (most closely related to *V. carchariae*; synonym *V. harveyi*) in diseased tissue relative to healthy tissues, though causation was not demonstrated (Ritchie and Smith 1998; Gil-Agudelo et al. 2006). Other bacterial agents are associated with WBD diseased tissues including *Bacillus* sp., *Lactobacillus* sp. and a *Rickettsiales*-like organism (Casas et al. 2004; Sweet et al. 2014), though the latter are consistently observed in both visibly healthy and diseased tissues of WBD infected colonies (Gignoux-Wolfsohn et al. 2020). Recent studies have suggested that shifts in the microbial consortia associated with the host coral are linked to gross lesion signs and compromised host health (Sweet and Bulling 2017). Hence, rather than

identifying a single biotic causative agent, it is proposed that dysbiosis of the associated microbiome results in a transition toward a pathobiome that is characteristic of the disease.

4.2.3.3 White pox (WPX)

The first report of WPX disease was in 1996 on reefs off Key West in Florida (Patterson et al. 2002), though in following years the disease was also reported more widely across the Caribbean (Porter et al. 2001; Rodriguez-Martinez et al. 2001; Patterson et al. 2002; Sutherland and Ritchie 2004). The disease affects *Acropora palmata* corals and manifests as variable sized, irregular white patches devoid of coral tissues that can develop simultaneously on different regions of the same colony. Lesions are characterised as focal to multifocal, basal to apically distributed, small to medium in size, and varying shapes from circular, oblong or pyriform with extensive tissue loss leaving intact bare white skeleton (Work and Aeby 2006). The lesions progress at an average of 2.5 cm day^{-1}, with progression greatest at elevated temperatures (Patterson et al. 2002). Lesions vary in size from generally a few cm in area to > 80 cm^2 and can coalesce, resulting in whole colony mortality (Sutherland et al. 2004). Additional case studies have reported syndromes with similar gross lesion characteristics at sites across the Caribbean, with various names including white-patch disease, patchy necrosis, and necrotic patches (Bruckner and Bruckner 1997; Rodriguez-Martinez et al. 2001), though these are now generally grouped under WPX since no comparative pathology studies were conducted.

Early reports of WPX highlighted it was highly contagious, spreading to neighbouring colonies and rapidly being transmitted between reefs in the Florida Keys National Marine Park Sanctuary. It is estimated that 88% of remaining *A. palmata* corals were lost in this region between 1996 to 2002 (Porter et al. 2001; Patterson et al. 2002; Sutherland et al. 2004). In other regions such as the United States Virgin Islands, Puerto Rico, and Mexico, mortality rates were variable and patterns in disease outbreaks followed seasonal drivers (Bruckner 2016a). More recent studies assessing WPX outbreaks highlight that disease severity and colony mortality was high between 1994–2004 and low in the period between 2008–2014, suggesting that changes in host, causative agent, and environmental dynamics influence the disease aetiology and complicate diagnosis (Sutherland et al. 2016a).

The causative agent of the disease was originally identified and confirmed by Koch's postulates as the bacterium *Serratia marcescens* and was linked to possible human sewage pollution (Patterson et al. 2002; Sutherland et al. 2010). Subsequent studies failed to recover *S. marcescens* from coral with WPX disease signs and questioned its role as the aetiological agent of the disease (Lesser and Jarett 2014; Joyner et al. 2015). WPX is still characterised on gross disease signs, though one form of WPX is termed acroporid serratiosis and is diagnosed if the classic lesion signs on the *A. palmata* host co-occur with the presence of the established pathogen *S. marcescens* (Sutherland et al. 2016a). However, WPX does not always co-occur with *S. marcescens*, and hence the same gross signs may be caused by more than one aetiological agent (Sutherland et al. 2016a).

Histopathology of apparently healthy tissue (1 to 10 cm away from the gross lesion margin) demonstrate tissue and cellular degeneration characterised by rounding of gland cells in mesenterial filaments and epidermis of the pharynx, potentially indicating involvement of bacterial derived toxins (for acroporid serratiosis; Sutherland 2015). Atrophy and necrosis in coenenchyme tissues and disruption of the epithelia of the gastrovascular cavity were also observed in apparently healthy tissues. Bacteria were not directly evident in diseased tissues, though rickettsia-like bacteria have been observed (Sutherland et al. 2016b). Histopathology of the apparently healthy tissues and disease lesion tissues of WPX colonies was similar, indicating a potential systemic, whole-colony response, which may partially explain the patchy distribution of lesions (Sutherland et al. 2016b).

4.2.3.4 White syndromes (WSs) of Indo-Pacific corals

Willis et al. (2004) highlighted the difficulties of assigning causative agents and causal relationships to diseases that display similar signs. Therefore, it was recommended to group diseases in the Indo-Pacific for which the underlying disease aetiology is unknown, but with macroscopic signs of irregular white bands or patches as a consequence of tissue loss, collectively as White Syndromes (WSs). Similar to the Caribbean, there are likely to be many diseases with varying underlying causative agents that result in gross WS signs in Indo-Pacific corals. However, this grouping prevents the emergence of multiple disease names which have few distinguishing diagnostic features (Willis et al. 2004). Once the underlying aetiologies are determined, then more definitive descriptions using established frameworks can be made (Work and Aeby 2006; Work et al. 2008b; Beeden et al. 2012). Until then, this grouping prevents the erection of distinct disease names that can result in confusion for researchers and managers.

WS gross lesions are characterised by a diffuse linear (or annular) band or irregular patch comprised of recently exposed coral skeleton adjacent to healthy tissues (see Figure 4.1d). In some cases, a region of bleached tissue may precede the advancing lesion, though more commonly there is no transition zone between healthy coral tissues and freshly denuded coral skeleton (Bourne et al. 2015; Bourne et al. 2016a). WSs were first recorded from the Great Barrier Reef in 1998/1999, affecting 17 species or growth forms of corals, though disease signs are particularly prevalent in the fast-growing tabular and branching species of *Acropora* (Willis et al. 2004). It is noted, however, that earlier reports of lesions consistent with WSs were observed in the Philippines (Antonius 1985). Multiple localised reports of tissue loss consistent with WS signs were quickly reported from regions across the Indo-Pacific. The signs were reportedly affecting tabulate acroporids from the North-Western Hawaiian Islands (Aeby 2005), Marshall Islands (Jacobson et al. 2006), Pilbara region reefs, and Christmas and Cocos Islands, North Western Australia (Hobbs and Frisch 2010; Page and Stoddart 2010) and the US remote Pacific Islands (Vargas-Angel 2009; Aeby et al. 2011). WSs were also reported in other coral genera across this region including *Pachyseris*, *Montipora*, *Pocillopora*, *Goniastrea*, and *Platygyra* (Sussman et al. 2008; Page et al. 2009; Vargas-Angel 2009; Aeby et al. 2011).

Consistent with descriptive frameworks, additional information has been introduced for many WSs, including *Montipora* white syndrome, which displays both acute and chronic lesion progression rates in Hawaii (Aeby et al. 2010; Work et al. 2012). A number of diseases have been reported with signs consistent with WSs in massive *Porites* corals, and alternative names have been suggested, including *Porites* white patch syndrome (Séré et al. 2012), *Porites* tissue loss (Williams et al. 2010), and *Porites* bleaching with tissue loss (Lawrence et al. 2004; Sudek et al. 2012). However, no underlying aetiology is provided to support alternative naming, and therefore it is recommended that they should be categorised under WSs, and more specifically *Porites* white syndrome.

The challenges associated with assigning causation to WSs have been detailed previously in Bourne et al. (2015). At the gross level, there are generally no visual signs of microbial communities at the lesion interface. In some case studies, however, ciliates have been observed at the lesion boundary, and hence ciliates have been suggested to be one causative agent (Sweet and Bythell 2012), noting that visible ciliate bands are characteristic of BrB (presented in more detail in Section 4.2.7). Not surprisingly, considering that this grouping of diseases likely encompasses a range of causative agents, evidence for disease causation at the cellular level has been varied. Histologically, a range of cellular patterns has been observed for WS lesion tissues. Work and Rameyer (2005) observed cell necrosis associated with filamentous algae, fungi, and ciliates in histological sections from a range of corals including tabular acroporids, *Montipora*, *Porites*, and *Echinopora* genera. On the other hand, other case studies displayed no microbial cells associated with diseased tissue, but instead observed tissue necrosis characterised by fragmentation, dissolution, or swelling of cell nuclei (Work and Rameyer 2005). Ainsworth et al. (2007b) investigated WS-infected *Acropora* corals from Heron Island and did not detect bacteria in histological sections, nor through florescence *in situ* hybridisation (FISH) microscopy at the lesion border or in healthy tissue up to 3 cm away from the lesion (Ainsworth et al. 2007b). Similarly, investigations of *Acropora* WSs from three disperse geographic locations, Lizard Island, Western Australia,

and Palmyra Atoll, observed no signs of necrosis, fragmentation, tissue swelling, ciliates, or cyanobacteria in the healthy tissues immediately preceding the disease lesion (Pollock et al. 2017; Smith et al. 2020). Helminths and fungi were observed in a minority of samples, though these organisms were also observed in healthy tissues with no gross WS signs. In contrast, high levels of necrosis and tissue fragmentation characterised the lesion front of samples subjected to histological investigations (Pollock et al. 2017). FISH also detected no bacteria in healthy tissues (less than 1 cm ahead of WS lesion fronts), though bacteria were detected in all lesion tissues both at and immediately behind the WS lesion front (Pollock et al. 2017; Smith et al. 2020). In other coral species (*Hydnophora* sp. and *Stylophora pistillata*), extensive tissue necrosis was partnered with extensive bacterial proliferation penetrating the healthy tissue layers (Ainsworth et al. 2007a), further highlighting likely differing underlying aetiologies for Indo-Pacific WSs. Studies on WS of the dominant Hawaiian coral species *Montipora capitata* demonstrated that at the cellular level, different microorganisms were associated with a rapidly progressing acute phase (i.e. ciliates) compared to a slow progressing chronic phase (i.e. helminths and chimeric parasites) (Aeby et al. 2010; Work et al. 2012). The variety of observations at a cellular level highlights the complexity of disease causation for WSs, which have a number of potential multiple causes, as well as varying host responses (Work et al. 2012; Aeby et al. 2016).

A number of studies have detected an increase in Rhodobacteraceae-affiliated sequences in compromised WS tissues (Sunagawa et al. 2009; Cárdenas et al. 2012; Roder et al. 2014; Pollock et al. 2017). This bacterial family is emerging as a potential indicator of compromised tissue health, though its direct role in disease onset or progression is unknown (Smith et al. 2020). A number of other studies have identified *Vibrio* affiliated bacteria as potential causative agents in WS diseases. Ben-Haim et al. (2003) isolated *Vibrio coralliilyticus* from necrotic *Pocillopora damicornis* tissues sampled from reefs in the Red Sea. Sussman et al. (2008), satisfying Henle-Koch's postulates, similarly reported *V. coralliilyticus* strains associated with WS lesions from multiple disease outbreaks in the Great

Barrier Reef (*Montipora aequituberculata*), Palau (*Pachyseris* speciosa), and the Marshall Islands (tabulate acroporids). *V. coralliilyticus* possessed high metalloprotease activity, which bleached and lysed the Symbiodiniaceae cells within the coral gastrodermal tissues and resulted in cleavage of coral connective tissues and paracellular perturbations (Ben-Haim et al. 2003; Sussman et al. 2008; Sussman et al. 2009). *Vibrio harveyi* was also implicated in WSs in Indonesian coral studies (Luna et al. 2010). Two pathogenic vibrios have also been implicated in inducing disease signs consistent with *Montipora* WS in Hawaii. *Vibrio owensii* (strain OCN002) was identified as a bacterial pathogen linked to the chronic phase characterised by diffuse tissue loss (chronic *Montipora* WS) and *V. coralliilyticus* (strain OCN088) was identified as inducing acute *Montipora* WS characterised by faster progressive tissue loss (Ushijima et al. 2012; Ushijima et al. 2014). However, vibrios were not detected in high abundance using FISH and 16S rRNA gene amplicon sequencing in WSs across three Indo-Pacific sites (Smith et al. 2020). These disparate patterns highlight the complexity of the involvement of this group in WS causation, suggesting that it may have a role in some, but not all, WSs. More recently, a bacterium, *Pseudoalteromonas piratica* (strain OCN003) was reported as another aetiological agent of acute *Montipora* WS, inducing a switch from the chronic to the acute form of the disease and suggested that similar disease signs (acute *Montipora* WS on *M. capitata*) is caused by multiple bacterial pathogens (Beurmann et al. 2017). The variety of microbial taxa identified in WS tissues highlights the possibility that dysbiosis caused by many detrimental taxa may lead to disease signs, rather than a single bacterial agent. Novel studies have applied microscale tracking combining microfluidics with stable isotopes to view high resolution interactions of putative bacterial pathogens and host coral tissues (Shapiro et al. 2016; Gibbin et al. 2018). In these studies, the penetration and dispersal of the coral pathogen *V. coralliilyticus*, inoculated onto individual *Pocillopora damicornis* polyps were visualised, with most pathogen cells located in the oral epidermis (Gibbin et al. 2018).

In addition to microbial-mediated pathways, programmed cell death, and virus-like particles have been implicated in some WS case studies. Programmed cell death can be instigated by biotic or abiotic factors in other organisms and can be an effective method for removing pathogens from a host. Ainsworth et al. (2007b) reported apoptotic cell nuclei within epithelial and gastrodermal tissues at the lesion borders, with transmission electron microscopy images supporting this. However, Pollock et al. (2017), using the same approach, did not detect programmed cell death in WS lesion tissues for corals sampled at Lizard Island on the Great Barrier Reef. Patten et al. (2008) observed virus like particles in WS lesions, though there was no correlation of such particles with diseased versus healthy tissues (Patten et al. 2008). In contrast, Pollock et al. (2014) reported 65% higher virus like particle numbers in acroporid tissues displaying WS at Lizard Island.

4.2.3.5 Ulcerative white syndromes (UWS)

Another disease consistent with WSs for Indo-Pacific corals is reported as ulcerative white spots (UWS). First reported from *Porites* sp. in the Philippines, the name originally erected was *Porites* ulcerative white spot disease (Raymundo et al. 2003). Reports of similar lesions on other coral genera including *Montipora*, massive morphologies, and the octocoral *Heliopora* resulted in the more general nomenclature to accommodate the increasing range of hosts displaying characteristic lesions for which the underlying aetiology remains unknown. UWS lesions are distinct from WSs by displaying multifocal patterns of bleached tissue (3 to 5 mm in diameter) which manifest as small circular or oblong areas that can either regress or progress to full tissue-thickness ulcerations, resulting in patches of bare white coral skeleton. The small lesions can occasionally coalesce, resulting in whole colony mortality (Raymundo et al. 2003). Reports from reefs in the Philippines highlight high prevalence during outbreaks (72% on some reefs; Raymundo et al. 2005), though generally levels are low but widely distributed across other reef areas including the Wakatobi Marine National Park, Indonesia (Haapkyla et al. 2007), Zanzibar and Kenya (Harvell et al. 2007), Guam (Myers and Raymundo 2009), Palau (Page et al. 2009), Maldives (Montano et al. 2016), and the Great Barrier Reef (Willis et al.

2004). Little is known of causative agents of the lesions, though *Vibrio* affiliated bacteria have been implicated through isolation and inoculation studies attempting to satisfy Henle-Koch's postulates, though no direct characterisation of microorganisms within active lesions was conducted (Arboleda and Reichardt 2010).

4.2.4 Dark spot syndrome (DSD)

Dark spot syndrome (or dark spot disease; DSD) was first reported in association with a bleaching event in the 1990s on Colombian Reefs (Solano et al. 1993; Gil-Agudelo et al. 2004; Borger 2005). Subsequently, it has been widely reported across the entire Caribbean region. It has been found predominantly affecting three coral species (*Montastraea annularis*, *Siderastrea siderea*, and *Stephanocoenia intersepta*) (Gil-Agudelo et al. 2004; Weil 2004), though further studies identified possible wider species susceptibility (Garzón-Ferreira et al. 2001). On Floridian reefs, the disease was highly prevalent, infecting more than 70% of corals surveyed, highlighting its potential impact on coral populations (Porter et al. 2011).

The disease lesion manifests as dark spots (brown, purple or black), that subsequently result in tissue loss, exposing skeleton that is colonised by algae or other fouling invertebrates (see Figure 4.1e); (Cervino et al. 2001; Borger 2005). The aetiology of DSD is unknown, though Borger (2005) suggested that lesions may represent a stress response in the coral. This response is proposed to cause disruption to the endosymbiotic dinoflagellates, with the Symbiodiniaceae cells appearing pigmented and swollen, many with disruption of their organelles (Renegar et al., 2008). Both the Symbiodiniaceae populations primarily, and the host tissues secondarily, have been suggested to be compromised through an unknown aetiological agent (Cervino et al. 2001). Histological analysis identified fungi associated with disease tissues, similar to *Aspergillus sydowii*, the pathogen linked to aspergillosis in sea fans (Renegar et al. 2008; Sweet et al. 2013). Studies have also profiled microbial communities associated with infected tissues and linked putative pathogens *Vibrio carchariae*, a cyanobacterium *Oscillatoria* (associated with BBD), and a fungal plant pathogen, *Rhytisma acerinum*, as

potential causative agents (Gil-Agudelo et al. 2007; Sweet et al. 2013). These studies only used molecular profiling and no studies have demonstrated direct causation or microbial pathogens directly linked with diseased tissues.

4.2.5 Yellow-Band disease (YBD)

YBD has been reported extensively throughout the Caribbean, however diseases with similar gross descriptions have been found on corals from other regions including Pacific and Arabian Gulf reefs (Bruckner and Riegl 2016). These yellow diseases have been termed Pacific YBD and Arabian YBD to avoid confusion with the originally described YBD affecting Caribbean corals. Most studies, including the most thoroughly described disease lesions, have emerged from the Caribbean and are focused upon here.

First reported to be affecting *Montastrea annularis* and *M. faveolata* colonies in 1994 on Florida reefs (Reeves 1994), Caribbean yellow-band disease is characterised by a band or patch of pale yellow tissue that radiates outwards from the central lesion (see Figure 4.1f; Cervino et al. 2001; Bruckner and Riegl 2016). Central areas of the lesion may suffer tissue mortality, with the lesion boundary forming annular bands. Multiple lesions can form on a single colony and may coalesce, resulting in extensive partial colony mortality. Typical reported rates of lesion progression are < 1 cm per month, though lesions can be persistent and therefore have extensive accumulated impacts on coral colonies. There are also seasonal differences in lesion progression, with warmer conditions promoting the disease prevalence and rates of progression (Cervino et al. 2001; Bruckner and Bruckner 2006). The disease can have a high prevalence in the *M. annularis* complex (> 90%) and therefore has contributed to significant mortality of these corals on Caribbean reefs in the 2000s (Bruckner and Bruckner 2006; Bruckner and Hill 2009), with longer term impacts on reproductive output also impacting coral populations (Weil et al. 2009). Original outbreaks of Caribbean yellow-band disease occurred following coral bleaching events, and both temperature and light have been correlated with the prevalence and virulence of the disease (Bruckner and Bruckner 2006; Harvell et al. 2009). Experimentally increased nutrients have also been demonstrated to cause increases in lesion progression rates (Bruno et al. 2003).

Vibrio spp. infecting the endosymbiotic Symbiodinaceae cells have been implicated in the early stages of Caribbean yellow-band disease onset, with histological studies showing progressive degeneration of dinoflagellate cells, including swelling, vacuolisation, fragmentation, and loss of cellular integrity (Cervino et al. 2004). This results in breakdown of the symbiotic association between the coral host and Symbiodiniaceae, leading Cervino et al. (2004) to suggest the disease is principally a disease of the symbiont. Four *Vibrio* strains which were recovered from the mucus of diseased corals, reproduced disease signs when healthy corals were inoculated with all four *Vibrio* strains, but failed to cause disease when inoculated as individual isolates (Cervino et al. 2008). Later studies found these vibrios in both diseased and healthy tissues of affected corals (Cunning et al. 2009), so the specific causative agent (vibrios or other biotic and abiotic agents) remains to be elucidated.

4.2.6 Skeletal eroding band (SEB) and Caribbean ciliate infections (CCI)

SEB is associated with the folliculid ciliate *Halofolliculina corallasia* and was the first disease described from the Indo-Pacific (Antonius 1999). The disease presents as a dark band 1–10 cm wide resulting from high densities of the *H. corallasia* ciliate (Figure 4.1g). The band is located at the interface of recently denuded skeleton and apparently healthy coral tissues (Figure 4.2c). This condition is most commonly associated with acroporid and pocilloporid coral genera, though it has been recorded afflicting over 80 coral species found on reefs across the Indo-Pacific and the Red Sea (Antonius 1999; Page et al. 2016). Cróquer et al. (2006a) reported a similar ciliate infection on a number of coral species in the Caribbean, with the disease named Caribbean ciliate infection (CCI) due to potential different aetiologies of this condition and the Indo-Pacific skeletal eroding band (Cróquer et al. 2006; Rodríguez et al. 2009; Weil and Rogers 2011). However, recent morpho-molecular studies confirmed that both of these two conditions display similar macroscopic signs and fine scale skeletal

erosion patterns derived from a common ciliate phenotype unequivocally identified as *H. corallasia* (Montano et al. 2020). Though a species complex may exist between Indo-Pacific and Caribbean *H. corallasia* populations, both disease names should be synonymised to avoid confusion in reporting of these two diseases (Montano et al. 2020).

The ciliate *H. corallasia* has a free-living form (see Figure 4.2d) that migrates onto or immediately adjacent to living coral tissue before transitioning to a sessile form. In its sessile form, the ciliate penetrates living coral tissues and attaches to the coral skeleton, forming a lorica (sac-like housing) with a rounded posterior and cylindrical neck. The ciliate and its associated lorica has an average length of 220 μm and width of 95 μm, with the ciliate housed within the lorica and possessing two conspicuous retractable pericytostomial wings bearing feeding cilia. Often, only the neck of the lorica is visible above the coral skeletal surface. The presence of the ciliate compromises healthy coral tissues potentially through spinning and chemical secretions, but it also damages and erodes the coral skeleton (Cróquer et al. 2006; Page and Willis 2008; Page et al. 2016). The migration of the ciliates across the colony surface leaves empty lorica visible behind the advancing front of live ciliates. Injury of coral tissues facilitates colonisation by *H. corallasia*, though studies have shown that intact healthy coral tissues challenged with the ciliate fail to manifest disease signs or tissue loss (Page and Willis 2008). Hence, while *H. corallasia* is characteristic of the disease, additional factors that increase ciliate virulence or compromise host coral tissues may be required to promote halofolliculinid infestations that result in pathogenesis and coral tissue loss. SEB can sometimes be confused in field observations with BBD when the density of the invading ciliates is high (Page and Willis 2008). Lesions in SEB are often associated with other potential causative agents including cyanobacteria, and the contribution of these other agents is unclear in cases where ciliates occur at varying densities (Page et al. 2016).

4.2.7 Brown band disease (BrB)

BrB was first reported in 2003 from the Great Barrier Reef, though since has been recorded across

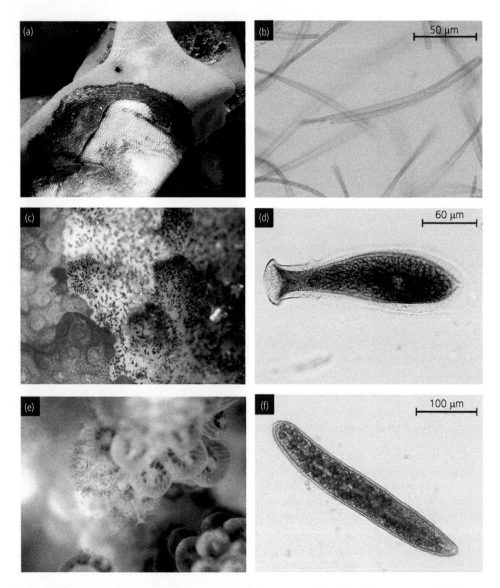

Figure 4.2 Images of common coral diseases and photomicrographs of microorganisms implicated in disease causation. (a) Macroscopic appearance of black band disease. (b) Photomicrograph of the dominant cyanobacterium (*Roseofilum reptotaenium*) associated with the black band disease microbial mat. (c) *Halofolliculina* bands associated with skeletal eroding band bordering healthy coral tissues. (d) Photomicrograph of a motile swarmer folliculind ciliate *Halofolliculina coralllasia* associated with skeletal eroding band. (e) Image of *Acropora tenuis* from the GBR with ciliates associated with brown band disease bordering healthy coral tissue. (f) Photomicrograph of the dominant protozoan scuticociliate (Class Oligohymenophorea; Genus *Porpostoma*) found in brown band disease lesions. Source: images supplied by Cathie Page and David Bourne.

the Indo-Pacific (Harvell et al. 2007; Nugues and Bak 2009; Page et al. 2009; Qiu et al. 2010; Onton et al. 2011; Weil et al. 2012) with similar disease signs also being reported from Caribbean reef systems (Randall et al. 2015). BrB is distinctively characterised by a brown band of variable width which separates apparently healthy tissues from exposed white coral skeleton (Figure 4.1h). There may also be a narrow zone of bleached tissue or denuded coral skeleton between the brown band and live apparently healthy tissue (Figure 4.2e). The brown band is comprised of a dense population of ciliates, which can consist of mixed populations, though generally is dominated by a protozoan scuticociliate of the class Oligohymenophorea (Bourne et al. 2008). Morphological work further identified the dominant ciliate as *Porpostoma guamense*, which was characterised from infected *Acropora* colonies from Guam (Lobban et al. 2011). Ciliates within the band can reach densities of ~ 120 cells ml^{-1} and actively feed on coral tissues, progressing at rates > 2 cm day^{-1} (Lobban et al. 2011). The ability to feed on living tissue means *P. guamense* is an ectoparasite of coral polyps and ingests Symbiodiniaceae cells along with the coral tissues (Figure 4.2f). These Symbiodinaceae cells remain actively photosynthesising, thereby benefiting the ciliate through a mixotrophic energy acquisition (Ulstrup et al. 2007). The corallivorous gastropod *Drupella* sp. was demonstrated to be an effective vector for BrB to spread between colonies (Nicolet et al. 2013). Additionally, predation by marine fishes and other invertebrates increased the occurrence of BrB through feeding scars, which create wounds sufficiently extensive to facilitate colonisation of coral tissues by the ciliates (Chong-Seng et al. 2011; Katz et al. 2014).

Ciliate species consistent with those found in BrB have been identified at the lesion interface of other diseases including WS (Indo-Pacific) and WBD (Caribbean) and are implicated in conferring the macroscopic visible signs of these diseases (Sweet and Bythell 2012; Sweet et al. 2014). However, *Philaster* sp. ciliates appear to be secondary colonisers within lesions resulting from alternate aetiologies. Other broad studies on WS have not consistently observed ciliates at the lesion interface (Pollock et al. 2017; Smith et al. 2020). Hence, while ciliates are linked to some WS cases, these are unlikely the primary causative agents, instead being histophagic, feeding on damaged tissues and endosymbionts. However, at high densities ciliates may play a role in pathogenesis through removal of coral tissues by their feeding activities resulting in denuded skeleton. Nicolet et al. (2013) reported visual differences in WS and BrB disease with the role of ectoparasite *Porpostoma guamense* and other distinct histophagic ciliates potentially contributing to these gross differences.

4.2.8 Stony coral tissue loss disease (SCTLD)

SCTLD was first reported in reefs off Florida in 2014, and subsequently spread through almost the entirety of the Florida Reef Tract (Precht et al. 2016; Walton et al. 2018) and is now also reported in many other regions of the wider Caribbean (Jamaica, Mexico, US Virgin Islands, Dominican Republic, Turks & Caicos Islands, Belize, Puerto Rico, and Grand Bahama (Alvarez-Filip et al. 2019; Weil et al. 2019)). The disease causes extensive coral mortality on already degraded reefs, compounded by observations that SCTLD disease signs have been reported for a large number (~ 24) of important reef-building coral genera (Muller et al. 2020). The environmental and ecological drivers of the disease outbreak are poorly understood, though the first reports coincided with a summer bleaching event across the Florida Reef Tract, combined with localised dredging operations which increased sedimentation deposition onto reefs in the vicinity of the Port of Miami, Florida (Miller et al. 2016; Walton et al. 2018). Disease transmission can occur by direct contact between coral colonies or through the water column (Aeby et al. 2019), with modelling studies reporting the epizootic followed a contagion model with spread up to 100 m day^{-1} and potentially facilitated by water currents (Muller et al. 2020).

Signs of SCTLD vary within and among affected coral species (Aeby et al. 2019), displaying focal or multifocal diffuse areas of tissue loss, and lesion boundaries adjacent to bleached tissues in some case studies (see Figure 4.1i). The rate of tissue loss can vary from acute (3–4 cm day^{-1}) to subacute, influenced by host species, individual host genotype, region, and time of year. A detailed case definition for SCTLD is available (NOAA 2018), with gross morphology of tissue loss distributed basally, peripherally, or both. Tissues bordering the lesions have indistinct bands (1–5 cm) of pallor, progressing to normal pigmentation away from denuded skeleton. Histological investigations report that tissue lesion pathology first affects the gastrodermal basal body wall, displaying signs of degradation, fragmentation, and swelling associated with disintegration of the mesoglea and subsequently progresses towards the oral surface, manifesting as tissue necrosis. Landsberg et al. (2020) histologically characterised SCTLD lesions from eight coral species, with all displaying lytic necrosis originating in the gastrodermis of the basal body wall, which progresses to the surface body wall as the lesions advance. The host cells display degenerative changes including disintegration of the mesoglea layers, degradation and fragmentation of the gastrodermal cells, plus mucocyte hypertrophy. In addition, endosymbiotic Symbiodiniaceae cells displayed a range of changes including necrosis, peripheral nuclear chromatin condensation, cytoplasmic vacuolation, deformation and degradation of chloroplasts, highlighting that SCTLD lesions result as a consequence of the disruption to the coral host and symbiont physiology leading to tissue sloughing from the underlying skeleton (Landsberg et al. 2020).

The aetiologic diagnosis of SCTLD is currently unknown, with studies showing the application of antibiotics or human wound treatment patches containing antiseptics and natural antioxidants can slow or halt lesion progression, suggesting some involvement of bacteria in the disease (Neely et al. 2020; Contradi et al. 2020). Higher relative abundance of sequences affiliated with the bacterial orders Rhodobacterales and Rhizobiales (Rosales et al. 2020) and an unclassified genus of the Flavobacteriales (Meyer et al. 2019) have been detected in disease lesions, though it is unknown if these taxa have a direct role in lesion progression or are secondary colonisers of compromised tissues. An obvious host-cell inflammatory response was not observed in histopathology studies, which may indicate that bacteria, fungi and parasites are not primary causes of the disease (Landsberg et al. 2020). Other aetiological causes under investigation include viral pathogens, toxicants, metabolic dysfunction and other environmental factors. Due to the ecological impact of the SCTLD disease outbreak across such a broad geographic and taxonomic range, and its occurrence on coral reefs already under intense pressures, significant research efforts are currently focussed on understanding this disease. However, a number of management failures in the early stage of the outbreak may have prevented effective management to stop spread and mitigate its impacts (Precht 2019).

4.3 Control and treatment of coral diseases

Control and treatment of coral diseases in open reef environments in which there are few physical barriers to disease spread is challenging, with prevention of outbreaks the best long-term option. This is best achieved through mitigation of anthropogenic stressors on coral reef ecosystems, which, if unchecked, may result in widespread disease outbreaks impacting coral community assemblages (Harvell et al. 1999, 2002, 2007). The biggest threat currently facing coral reefs is the global anthropogenic warming of seawater that is likely to contribute to the increased virulence of coral pathogens, and/or decreased coral host immunity, and ultimately deterioration of the functional coral holobiont, resulting in coral bleaching, disease outbreaks, and ultimately high coral mortality (Bruno et al. 2007; Muller et al. 2008; Heron et al. 2010; Brodnicke et al. 2019). Unfortunately, given current climate modelling projections, global seawater temperatures will continue to rise, driving further coral disease outbreaks. Difficulty in unambiguously identifying the causative agents of many coral diseases has hindered the development of potential control and treatment strategies. However, there have been a number of approaches trialled at different scales, and current global concern regarding declining reef ecosystem health is driving exploration of new approaches for mitigating coral diseases and building resilience in coral populations (Boström-Einarsson et al. 2018; National Academies of Sciences 2018, 2019).

The first approach to treat and manage a coral disease was conducted on BBD-infected colonies in the 1980s (Hudson 2000). The partially successful treatment involved using an aspirator device to remove the microbial mat infecting the colony and the pressing of modelling clay into the coral skeleton at the site post-aspiration to seal the wound and reduce reinfection (Hudson 2000). Epoxy resin has similarly been used to mechanically block progression of coral lesions (Miller et al. 2014), with chemical treatments such as chlorine also included in the resin (Aeby et al. 2015). Randall et al. (2018) explored three mitigation strategies for YBD, including shading, underwater aspiration of the disease tissue followed by sealing with modelling clay, and creating a firebreak through chiselling a trench between the lesion and adjacent apparently healthy tissues. The firebreak method was identified as the best for slowing the spread of the disease, though long-term benefits were unclear (Randall et al. 2018). Removal of the region or branch of a coral colony that displays disease signs or removal of whole colonies from the population can also be undertaken to reduce disease spread (National Academies of Sciences 2018). In aquarium settings, the screening of corals for disease agents prior to their introduction to aquaria, and the removal of any colony displaying disease signs is the best recommended approach to prevent transmission through the closed system (Sweet et al. 2011). Antibiotic treatments have also been trialled successfully in aquarium settings to both control disease and help identify if bacteria are implicated as biotic agents of the disease (Sheridan et al. 2013; Sweet et al. 2014). All these approaches, however, are small scale, labour intensive, and expensive, resulting in limited applicability at broader reef ecosystem scales.

Phage therapy has been trialled to control coral disease progression and transmission in both closed experimental systems and open reef environments. The concept of phage therapy is based upon the principle that every bacterium has one or many bacteriophages that have evolved (or coevolved) to infect and/or lyse that bacterium (Keen 2015). If a coral bacterial pathogen is known, then a phage can be isolated, grown in high abundance, and added to control the pathogen and prevent disease progression and/or transmission. The approach has been used widely in both human and agricultural systems (Doss et al. 2017). Bacteriophages targeting the coral pathogen *V. coralliilyticus* have been isolated and used effectively in small-scale experimental aquarium systems to stop progression of WP-like lesions on *Favia favus* in the Red Sea (Efrony et al. 2007, 2009; Cohen et al. 2013). A proof-of-concept field study was also conducted in the Red Sea and was shown to be effective in mitigating disease impacts (Atad et al. 2012). While the approach offers promise, there are many challenges to applying phage therapy to control coral diseases in open reef ecosystems, with an extensive risk/benefit analyses required for each application to ensure the benefits

of disease control are not offset by other deleterious impacts to coral reef ecosystems (National Academies of Sciences 2018).

Manipulation of the coral microbiome, essentially through addition of probiotics is emerging as another potential option to increase coral resilience to environmental stress and disease. The addition of beneficial microorganisms for corals may enhance nutrient cycling (including acquisition of heterotrophic nutrients), biological control of potential pathogens, supply essential trace nutrients, metals or vitamins, and mitigate the effects of reactive oxygen species (Peixoto et al. 2017). Rosado et al. (2018) demonstrated the potential for these microbes to minimise the impacts of environmental stressors. A range of bacterial species displaying potential beneficial traits were isolated and applied as a cocktail to corals in aquaria. These were subsequently subjected to bacterial challenge with the coral pathogen *V. coralliilyticus* and heat stress to simulate a bleaching event. Inoculated corals displayed improved health compared to controls, measured as lower bleaching metrics (Rosado et al. 2018). Heterotrophic feeds may provide an efficient route to deliver probiotics to corals, with recent studies demonstrating rotifers ingested beneficial microbes, which were subsequently ingested by *Pocillopora damicornis* corals (Assis et al. 2020). This approach could be applicable to coral aquaculture facilities, which are becoming more numerous and of increased size in response to increased reef restoration efforts (Barton et al. 2017).

Manipulation of the cell-to-cell chemical signalling (quorum sensing) of the coral microbiome may also hold promise for disease treatment. Quorum sensing through acyl homoserine lactones (AHLs) in bacterial communities can control the coordinated expression of virulence genes (Teplitski and Ritchie 2009), and hence disrupting (or 'quenching') AHL signals may allow the host to prevent pathogenesis. Interference of quorum sensing as disease biocontrol has been investigated in agriculture and other model systems (reviewed in Teplitski and Ritchie 2009). BBD microbial communities have demonstrated AHL-based quorum sensing activities, and hence may play a role in pathogenesis (Zimmer et al. 2014). A proof-of-concept study in corals conducted by Certner and Vollmer (2017) showed that microbial communities isolated from WBD-infected corals that were exposed to a quorum sensing inhibitor (i.e. antagonist of AHLs) were unable to establish disease signs on *A. cervicornis* (Certner and Vollmer 2018). The same microbial communities without the inhibitor treatment established WBD signs within 2 days. While the application of quorum quenching to disease biocontrol *ex situ* may be promising, the application *in situ* in open ocean systems remains a challenge.

4.4 Future directions

Diagnosis of coral diseases and the identification of their causative agents has proved challenging since the first reported cases emerged in reef ecosystems over 50 years ago (Montilla et al. 2019). The limited success in conclusively assigning causative agents to many coral diseases highlights the complex interactions that occur between the coral host, its diverse and complex microbiome,

and the surrounding environment (Bourne et al. 2016b). There remains a critical need for systematic morphological descriptions of coral diseases at the gross and cellular levels to enable comparative studies across large geographical regions (Work and Aeby 2006; Bourne et al. 2009). In addition, the links between host immunity, homeostasis of the microbial symbiont community, primary *vs.* opportunistic pathogens, the role of dysbiosis in disease onset, and how all these factors are influenced by environmental pressures, requires further elucidation (Vega Thurber et al. 2020). Genomic tools offer new ways to explore these questions as well as the interactions between holobiont members and their surrounding environment. However, genomics is only part of the toolkit to help elucidate disease causation through a deductive approach and should not be the focus at the expense of classical biomedical approaches to standardise disease case definitions (see Chapter 3 for more detailed discussion of this point).

4.5 Summary

- Novel coral diseases continue to be reported from coral reefs globally, however inconsistencies in disease nomenclature and the failure to apply systematic morphological descriptions of diseases at both the gross and cellular levels results in proliferation of diseases assumed to have narrow geographic distributions.
- The causation of most coral diseases remains poorly understood and remains an urgent research priority considering the increasingly emerging links between environmental stress and disease outbreaks.
- Dysbiosis is increasingly linked to many coral diseases through a shift in the symbiotic microbiome allowing proliferation of opportunistic bacteria/pathogens.
- Reefs globally are declining at an alarming rate and disease outbreaks have historically, and will continue to, contribute to this loss of coral cover and coral reef ecosystem functioning.
- Ongoing coral declines have led to research focussed on methods to mitigate further anthropogenic contributions to environmental impacts on coral reefs, build coral resilience, and increase disease resistance in the face of a rapidly changing environment.

Suggested further reading

Vega Thurber, R.L., Mydlarz, L.D., Brandt, M., Harvell, C.D., Weil, E., Raymundo, L., Willis, B., Langevin, S., Tracy, A.M., and Littman, R. 2020. Deciphering coral disease dynamics: Integrating host, microbiome, and the changing environment. *Frontiers in Ecology and Evolution* 8: 402. (Most recent review detailing the latest understanding of factors contributing to coral diseases.)

Harvell, D. 2019. *Ocean outbreak: Confronting the rising tide of marine disease*. California: University of California Press. (Details disease outbreaks in iconic marine animals including corals and how these diseases are driven by warming seas and other human impacts)

Mera, H. and Bourne, D.G. 2018. Disentangling causation: Complex roles of coral-associated microorganisms in disease. *Environmental Microbiology* 20: 431–449. (Short review detailing the complexity of elucidating microbial processes in coral diseases)

Bourne, D.G., Morrow, K.M., and Webster, N.S. 2016. Insights into the coral microbiome: Underpinning the health and resilience of reef ecosystems. *Annual Review of Microbiology* 70: 317–340. (Comprehensive review that details our current understanding of the coral microbiome and how it underpins coral health)

Woodley, C.M. Downs, C.A., Bruckner, A.W., Porter, J.W., and Galloway, B. 2016. *Disease of Corals*. Hoboken, New Jersey: Wiley-Blackwell. (Collection of chapters focused on coral disease globally)

Birkeland, C. 2015. *Coral Reefs in the Anthropocene*. Dordrecht, Netherlands: Springer-Netherlands. (Broad collection of chapters discussing importance of coral reefs and impacts that influence reefs today including disease)

Rohwer, F., Youle, M., and Vosten, D. 2010. *Coral Reefs in the Microbial Seas*. Granada Hills: Plaid Press. (General reading that highlights how microbes underpin reef health and the impact of human activity on these ecosystems)

Bourne, D.G., Garren, M., Work, T.M., Rosenberg, E., Smith, G., and Harvell, C.D. 2009. Microbial disease and the coral holobiont. *Trends in Microbiology* 17: 554–562 (Review detailing the role of microorganisms in coral diseases and the challenges to elucidate complex disease processes in corals)

Rosenberg, E. and Loya Y. 2004. *Coral Health and Disease*. Heidelberg, Germany: Springer-Verlag. (The first comprehensive book that provides in-depth discussion of the impact and causes of coral diseases globally)

Sweet, M., Burian, A., and Bulling, M. 2021. Corals as canaries in the coalmine: Towards the incorporation of marine ecosystems into the 'one health' concept. *Journal of Invertebrate Pathology* 107538.

Acknowledgements

The authors thank Greta Aeby (Department of Biological and Environmental Sciences, Qatar University) for supplying coral disease images.

References

Aeby, G.S. 2005. Outbreak of coral disease in the Northwestern Hawaiian Islands. *Coral Reefs* 24: 481–489

Aeby, G.S., Callahan, S., Cox, E.F., Runyon, C., Smith, A., Stanton, F.G., Ushijima, B., and Work, T. M. 2016. Emerging coral diseases in Kāneohe Bay, O Ahu, Hawai I (USA): Two major disease outbreaks of acute *Montipora* white syndrome. *Diseases of Aquatic Organisms* 119: 189–198

Aeby, G.S., Ross, M., Williams, G.J., Lewis, T.D., and Work, T.M. 2010. Disease dynamics of *Montipora* white syndrome within Kaneohe Bay, Oahu, Hawaii: Distribution, seasonality, virulence, and transmissibility. *Diseases of Aquatic Organisms* 91: 1–8

Aeby, G.S. and Santavy, D.L. 2006. Factors affecting susceptibility of the coral *Montastraea faveolata* to black-band disease. *Marine Ecology Progress Series* 318: 103–110

Aeby, G.S., Ushijima, B., Campbell, J.E. et al. 2019. Pathogenesis of a tissue loss disease affecting multiple species of corals along the Florida Reef Tract. *Frontiers in Marine Science* 6: 678

Aeby, G.S., Williams, G.J., Franklin, E.C., Kenyon, J., Cox, E.F., Coles, S., and Work, T.M. 2011. Patterns of coral disease across the Hawaiian archipelago: Relating disease to environment. *PloS One* 6: e20370

Aeby, G.S., Work, T.M., Runyon, C.M. et al. 2015. First record of black band disease in the Hawaiian archipelago: Response, outbreak status, virulence, and a method of treatment. *PLoS One* 10: e0120853

Ainsworth, T.D., Kramasky-Winter, E., Loya, Y., Hoegh-Guldberg, O., and Fine, M. 2007a. Coral disease diagnostics: What's between a plague and a band? *Applied and Environmental Microbiology* 73: 981–992

Ainsworth, T.D., Kvennefors, E.C., Blackall, L.L., Fine, M., and Hoegh-Guldberg, O. 2007b. Disease and cell death in white syndrome of Acroporid corals on the Great Barrier Reef. *Marine Biology* 151: 19–29

Alvarez-Filip, L., Estrada-Saldívar, N., Pérez-Cervantes, E., Molina-Hernández, A. , and González-Barrios, F. J. 2019. A rapid spread of the stony coral tissue loss disease outbreak in the Mexican Caribbean. *PeerJ* 7: e8069

Antonius, A. 1973. New observations on coral destruction in reefs. 10th Meeting of the Association of Island Marine Laboratories of the Caribbean Vol. 10: p. 3. Puerto Rico: University of Puerto Rico.

Antonius, A. 1977. Coral mortality in reefs: A problem for science and management. Proceedings of the International Coral Reef Symposium, Miami, Florida. Vol. 2; pp. 617–623

Antonius, A. 1985. Coral diseases in the Indo-Pacific: A first record. *Marine Ecology* 6: 197–218

Antonius, A. 1999. *Halofolliculina corallasia*, a new coral-killing ciliate on Indo-Pacific reefs. *Coral Reefs* 18: 300

Arboleda, M.D. and Reichardt, W.T. 2010. *Vibrio* sp. causing *Porites* ulcerative white spot disease. *Diseases of Aquatic Organisms* 90: 93–104

Assis, J., Abreu, F., Villela, H. et al. 2020. Delivering beneficial microorganisms for corals: Rotifers as carriers of probiotic bacteria. *Frontiers in Microbiology* 11: 608506

Atad, I., Zvuloni, A., Loya, Y. , and Rosenberg, E. 2012. Phage therapy of the white plague-like disease of *Favia favus* in the Red Sea. *Coral Reefs* 31: 665–670

Barash, Y., Sulam, R., Loya, Y. , and Rosenberg, E. 2005. Bacterial strain BA-3 and a filterable factor cause a white plague-like disease in corals from the Eilat coral reef. *Aquatic Microbial Ecology* 40: 183–189

Barton, J.A., Willis, B.L., and Hutson, K.S. 2017. Coral propagation: A review of techniques for ornamental trade and reef restoration. *Reviews in Aquaculture* 9: 238–256

Beeden, R., Maynard, J.A., Marshall, P.A., Heron, S.F., and Willis, B.L. 2012. A framework for responding to coral disease outbreaks that facilitates adaptive management. *Environmental Management* 49: 1–13

Ben-Haim, Y., Thompson, F.L., Thompson, C.C., Cnockaert, M.C., Hoste, B., Swings, J., and Rosenberg, E. 2003. *Vibrio coralliilyticus* sp. nov., a temperature-dependent pathogen of the coral *Pocillopora damicornis*. *International Journal of Systematic and Evolutionary Microbiology* 53: 309–315

Beurmann, S., Ushijima, B., Videau, P., Svoboda, C.M., Smith, A.M., Rivers, O.S., Aeby, G.S., and Callahan, S.M. 2017. *Pseudoalteromonas piratica* strain OCN003 is a coral pathogen that causes a switch from chronic to acute *Montipora* white syndrome in *Montipora capitata*. *PLoS One* 12: e0188319

Worms Editorial Board. 2019 World Register of Marine Species.

Borger, J.L. 2003. Three scleractinian coral diseases in Dominica, West Indies: Distribution, infection patterns and contribution to coral tissue mortality. *Revista de Biologia Tropical* 51: 25–38

Borger, J.L. 2005. Dark spot syndrome: A scleractinian coral disease or a general stress response? *Coral Reefs* 24: 139–144

Boström-Einarsson, L., Ceccarelli, D., Babcock, R.C. et al. 2018. Coral restoration in a changing world—A global synthesis of methods and techniques. A report for the Reef Restoration and Adaptation Program, Subproject 1a—Review of existing technologies, pilots and new initiatives. (p. 85)

Bourne, D.G., Ainsworth, T.D., Pollock, F.J., and Willis, B.L. 2015. Towards a better understanding of white syndromes and their causes on Indo-Pacific coral reefs. *Coral Reefs* 34: 233–242

Bourne, D.G., Ainsworth, T.D., and Willis, B.L. 2016a. White syndromes of Indo-Pacific corals. In *Diseases of Coral*. Woodley, C.M., Downs, C.A., Bruckner, A.W., Porter, J.W., and Galloway, S.B. (eds.) pp. 300–315. Hoboken, New Jersey: Wiley Blackwell

Bourne, D.G., Morrow, K.M., and Webster, N.S. 2016b. Insights into the coral microbiome: Underpinning the health and resilience of reef ecosystems. *Annual Review of Microbiology* 70: 317–340

Bourne, D.G., Boyett, H.V., Henderson, M.E., Muirhead, A., and Willis, B.L. 2008. Identification of a ciliate (Oligohymenophorea: Scuticociliatia) associated with brown band disease on corals of the Great Barrier Reef. *Applied and Environmental Microbiology* 74: 883–888

Bourne, D.G., Garren, M., Work, T.M., Rosenberg, E., Smith, G., and Harvell, C.D. 2009. Microbial disease and the coral holobiont. *Trends in Microbiology* 17: 554–562

Bourne, D.G., Muirhead, A., and Sato, Y. 2011. Changes in sulfate-reducing bacterial populations during the onset of black band disease. *The ISME Journal* 5: 559–564.

Bourne, D.G., Van Der Zee, M.J., Botté, E.S., and Sato, Y. 2013. Sulfur-oxidizing bacterial populations within cyanobacterial dominated coral disease lesions. *Environmental Microbiology Reports* 5: 518–524

Brodnicke, O., Bourne, D., Heron, S., Pears, R., Stella, J., Smith, H., and Willis, B. 2019. Unravelling the links between heat stress, bleaching and disease: Fate of tabular corals following a combined disease and bleaching event. *Coral Reefs* 38: 591–603

Bruckner, A. and Bruckner, R. 1997. Outbreak of coral disease in Puerto Rico. *Coral Reefs* 16: 260–260

Bruckner, A.W. 2009. Progress in understanding coral diseases in the Caribbean. *Coral Health and Disease in the Pacific: Vision for Action.* pp. 126–161

Bruckner, A.W. 2016a. White syndromes of western Atlantic reef-building corals. In *Diseases of Coral.* Woodley, C.M., Downs, C.A., Bruckner, A.W., Porter, J.W., and Galloway, S.B. (eds.) pp. 316–332. Hoboken, New Jersey: Wiley Blackwell

Bruckner, A.W. 2016b. History of coral disease research. In *Diseases of Coral.* Woodley, C.M., Downs, C.A., Bruckner, A.W., Porter, J.W., and Galloway, S.B. (eds.) pp. 52–84. Hoboken, New Jersey: Wiley Blackwell

Bruckner, A.W., Aronson, R.B., Bruckner, R.J., and Precht, W.F. 2002. Endangered acroporid corals of the Caribbean: Large-scale dynamics of coral reef systems. *Coral Reefs* 21: 41–42

Bruckner, A.W. and Bruckner, R.J. 2006. Consequences of yellow band disease (YBD) on *Montastraea annularis* (species complex) populations on remote reefs off Mona Island, Puerto Rico. *Diseases of Aquatic Organisms* 69: 67–73

Bruckner, A.W., Bruckner, R.J., and Williams Jr, E.H. 1997. Spread of a black-band disease epizootic through the coral reef system in St. Ann's Bay, Jamaica. *Bulletin of Marine Science* 61: 919–928

Bruckner, A.W. and Hill, R.L. 2009. Ten years of change to coral communities off Mona and Desecheo Islands, Puerto Rico, from disease and bleaching. *Diseases of Aquatic Organisms* 87: 19–31

Bruckner, A.W. and Riegl, B. 2016. Yellow-Band Diseases. In *Diseases of Coral.* Woodley, C.M., Downs, C.A., Bruckner, A.W., Porter, J.W., and Galloway, S.B. (eds.) Hoboken, New Jersey: Wiley Blackwell. 376–386

Bruno, J.F., Petes, L. E., Harvell, C.D., and Hettinger, A. 2003. Nutrient enrichment can increase the severity of coral diseases. *Ecology Letters* 6: 1056–1061

Bruno, J.F., Selig, E.R., Casey, K.S. et al. 2007. Thermal stress and coral cover as drivers of coral disease outbreaks. *PLoS Biology* 5: 1220–1227

Bythell, J.C., Pantos, O., and Richardson, L. 2004. White plague, white band, and other 'white' diseases. In *Coral Health and Disease.* Rosenberg, E. and Loya, Y. (eds.) pp. 351–364. Heildelberg: Springer-Verlag

Cárdenas, A., Rodriguez-R, L.M., Pizarro, V., Cadavid, L.F., and Arévalo-Ferro, C. 2012. Shifts in bacterial communities of two Caribbean reef-building coral species affected by white plague disease. *The ISME Journal* 6: 502–512

Casamatta, D., Stanić, D., Gantar, M., and Richardson, L.L. 2012. Characterization of *Roseofilum reptotaenium* (Oscillatoriales, Cyanobacteria) gen. et sp. nov. isolated from Caribbean black band disease. *Phycologia* 51: 489–499

Casas, V., Kline, D.I., Wegley, L., Yu, Y., Breitbart, M., and Rohwer, F. 2004. Widespread association of a Rickettsiales-like bacterium with reef-building corals. *Environmental Microbiology* 6: 1137–1148

Certner, R.H., Dwyer, A.M., Patterson, M.R., and Vollmer, S.V. 2017. Zooplankton as a potential vector for white band disease transmission in the endangered coral, *Acropora cervicornis. PeerJ* 5: e3502

Certner, R.H. and Vollmer, S.V. 2018. Inhibiting bacterial quorum sensing arrests coral disease development and disease-associated microbes. *Environmental Microbiology* 20: 645–657

Cervino, J., Goreau, T.J., Nagelkerken, I., Smith, G.W., and Hayes, R. 2001. Yellow band and dark spot syndromes in Caribbean corals: Distribution, rate of spread, cytology, and effects on abundance and division rate of zooxanthellae. *Hydrobiologia* 460: 53–63

Cervino, J.M., Hayes, R.L., Polson, S.W., Polson, S.C., Goreau, T.J., Martinez, R.J., and Smith, G.W. 2004. Relationship of *Vibrio* species infection and elevated temperatures to yellow blotch/band disease in Caribbean corals. *Applied and Environmental Microbiology* 70: 6855–6864

Cervino, J.M., Thompson, F.L., Gomez-Gil, B., Lorence, E.A., Goreau, T.J., Hayes, R.L., Winiarski-Cervino, K.B., Smith, G.W., Hughen, K., and Bartels, E. 2008. The *Vibrio* core group induces yellow band disease in Caribbean and Indo-Pacific reef-building corals. *Journal of Applied Microbiology* 105: 1658–1671

Chong-Seng, K., Cole, A., Pratchett, M., and Willis, B. 2011. Selective feeding by coral reef fishes on coral lesions associated with brown band and black band disease. *Coral Reefs* 30: 473–481

Cohen, Y., Joseph Pollock, F., Rosenberg, E., and Bourne, D.G. 2013. Phage therapy treatment of the coral pathogen *Vibrio coralliilyticus. Microbiology Open* 2: 64–74

Contardi, M., Montano, S., Liguori, G., Heredia-Guerrero, J.A., Galli, P., Athanassiou, A., and Bayer, I.S. 2020. Treatment of coral wounds by combining an antiseptic bilayer film and an injectable antioxidant biopolymer. *Scientific Reports* 10: 1–10

Cróquer, A., Bastidas, C., Lipscomp, D., Rodríguez-Martínez, R.E., Jordan-Dahlgren, E., and Guzman, H.M. 2006a. First report of folliculinid ciliates affecting Caribbean scleractinian corals. *Coral Reefs* 25: 187–191

Cróquer, A., Bastidas, C., and Lipscomb, D. 2006b. Folliculinid ciliates: A new threat to Caribbean corals? *Diseases of Aquatic Organisms* 69: 75–78

Croquer, A., Pauls, S.M., and Zubillaga, A.L. 2003. White plague disease outbreak in a coral reef at Los Roques

National Park, Venezuela. *Revista de Biologia Tropical* 51: 39–45

Cunning, J., Thurmond, J., Smith, G., Weil, E., and Ritchie, K. 2009. A survey of vibrios associated with healthy and yellow band diseased *Montastraea faveolata*. Proceedings of the 11th international coral reef symposium, pp. 206–210 Ft. Lauderdale, Florida, 7-11 July 2008 Session number 7. Ft. Lauderdale, Florida: International Coral Reef Society

Denner, E.B.M., Smith, G.W., Busse, H.J., Schumann, P., Narzt, T., Polson, S.W., Lubitz, W., and Richardson, L.L. 2003. *Aurantimonas coralicida* gen. nov., sp nov., the causative agent of white plague type II on Caribbean scleractinian corals. *International Journal of Systematic and Evolutionary Microbiology* 53: 1115–1122

Diaz-Soltera, H. 1999. Endangered and threatened species: A revision of candidate species list under the Endangered Species Act. Fed Register. *Federal Register* 64: 33466–33468

Domart-Coulon, I.J., Traylor-Knowles, N., Peters, E. et al. 2006. Comprehensive characterization of skeletal tissue growth anomalies of the finger coral *Porites compressa*. *Coral Reefs* 25: 531–543

Doss, J., Culbertson, K., Hahn, D., Camacho, J., and Barekzi, N. 2017. A review of phage therapy against bacterial pathogens of aquatic and terrestrial organisms. *Viruses* 9: 50

Dustan, P. 1977. Vitality of reef coral populations off Key Largo, Florida: Recruitment and mortality. *Environmental Geology* 2: 51–58

Edmunds, P.J. 1991. Extent and effect of black band disease on a Caribbean reef. *Coral Reefs* 10: 161–165

Efrony, R., Atad, I., and Rosenberg, E. 2009. Phage therapy of coral white plague disease: Properties of phage BA3. *Current Microbiology* 58: 139–145

Efrony, R., Loya, Y., Bacharach, E., and Rosenberg, E. 2007. Phage therapy of coral disease. *Coral Reefs* 26: 7–13

Egan, S., and Gardiner, M. 2016. Microbial dysbiosis: Rethinking disease in marine ecosystems. *Frontiers in Microbiology* 7: 991

Garzón-Ferreira, J., Gil-Agudelo, D., Barrios, L., and Zea, S. 2001. Stony coral diseases observed in southwestern Caribbean reefs. *Hydrobiologia* 460: 65–69

Gibbin, E., Gavish, A., Domart-Coulon, I. et al. 2018. Using NanoSIMS coupled with microfluidics to visualize the early stages of coral infection by *Vibrio coralliilyticus*. *BMC Microbiology* 18: 1–10

Gignoux-Wolfsohn, S., Marks, C.J., and Vollmer, S.V. 2012. White band disease transmission in the threatened coral, *Acropora cervicornis*. *Scientific Reports* 2: 804

Gignoux-Wolfsohn, S.A., Precht, W.F., Peters, E.C., Gintert, B.E., and Kaufman, L.S. 2020. Ecology, histopathology, and microbial ecology of a white-band disease outbreak in the threatened staghorn coral *Acropora cervicornis*. *Diseases of Aquatic Organisms* 137: 217–237

Gil-Agudelo, D., Smith, G., and Weil, E. 2006. The white band disease type II pathogen in Puerto Rico. *Revista de Biologia Tropical* 54: 59–67

Gil-Agudelo, D.L., Fonseca, D.P., Weil, E., Garzon-Ferreira, J., and Smith, G.W. 2007. Bacterial communities associated with the mucopolysaccharide layers of three coral species affected and unaffected with dark spots disease. *Canadian Journal of Microbiology* 53: 465–471

Gil-Agudelo, D.L., Smith, G.W., Garzón-Ferreira, J., Weil, E., and Petersen, D. 2004. Dark spots disease and yellow band disease, two poorly known coral diseases with high incidence in Caribbean reefs. In *Coral Health and Disease*. Rosenberg, E. and Loya, Y. (eds.) pp. 337–349. Heildelberg: Springer-Verlag

Gladfelter, W.B. 1982. White-band disease in *Acropora palmata*: Implications for the structure and growth of shallow reefs. *Bulletin of Marine Science* 32: 639–643

Gladfelter, W.B., Gladfelter, E.H., Monahan, R.K., Ogden, J.C., and Dill, R.F. 1977. Environmental studies of Buck Island Reef National Monument. (p. 144) US National Park Service Report

Green, E.P. and Bruckner, A.W. 2000. The significance of coral disease epizootiology for coral reef conservation. *Biological Conservation* 96: 347–361

Haapkyla, J., Seymour, A.S., Trebilco, J., and Smith, D. 2007. Coral disease prevalence and coral health in the Wakatobi Marine Park, south-east Sulawesi, Indonesia. *Journal of the Marine Biological Association of the United Kingdom* 87: 403–414

Hanahan, D. and Weinberg, R.A. 2000. The hallmarks of cancer. *Cell* 100: 57–70

Harvell, C.D., Kim, K., Burkholder, J.M. et al.1999. Emerging marine diseases—Climate links and anthropogenic factors. *Science* 285: 1505–1510

Harvell, C.D., Mitchell, C.E., Ward, J.R. et al. 2002. Climate warming and disease risks for terrestrial and marine biota. *Science* 296: 2158–2162

Harvell, D., Altizer, S., Cattadori, I.M., Harrington, L., and Weil, E. 2009. Climate change and wildlife diseases: When does the host matter the most? *Ecology* 90: 912–920

Harvell, D., Jordan-Dahlgren, E., Merkel, S., Rosenberg, E., Raymundo, L., Smith, G., Weil, E., and Willis, B. 2007. Coral disease, environmental drivers, and the balance between coral and microbial associates. *Oceanography* 20: 172–195

Heron, S.F., Willis, B.L., Skirving, W.J., Eakin, C.M., Page, C.A., and Miller, I.R. 2010. Summer hot snaps and winter conditions: Modelling white syndrome outbreaks on Great Barrier Reef corals. *PLoS One* 5: e12210

Hobbs, J.-P. and Frisch, A.J. 2010. Coral disease in the Indian Ocean: Taxonomic susceptibility, spatial distribution and the role of host density on the prevalence of white syndrome. *Diseases of Aquatic Organisms* 89: 1–8

Hoegh-Guldberg, O., Poloczanska, E.S., Skirving, W., and Dove, S. 2017. Coral reef ecosystems under climate change and ocean acidification. *Frontiers in Marine Science* 4: 158.

Hogarth, W. 2006. Endangered and threatened species: Final listing determinations for elkhorn coral and staghorn coral. *Federal Register* 71: 26852–26861

Hudson, H. 2000. First aid for massive corals infected with black band disease: An underwater aspirator and post-treatment sealant to curtail re-infection. In *Diving for Science in the 21st Century*. Hallock, P. and French, L. (eds.) pp. 10–11. Portland: The American Academy of Underwater Sciences

Hughes, T.P., Barnes, M.L., Bellwood, D.R. et al. 2017a. Coral reefs in the Anthropocene. *Nature* 546: 82–90

Hughes, T.P., Kerry, J.T., Álvarez-Noriega, M. et al. 2017b. Global warming and recurrent mass bleaching of corals. *Nature* 543: 373–377

Jacobson, D.M., Smith, L.W., Dailer, M.L. et al. 2006. Fine scale temporal and spatial dynamics of a Marshall Islands coral disease outbreak: Evidence for temperature forcing. *EOS, Transactions, American Geophysical Union* 87: (36)

Joyner, J.L., Sutherland, K.P., Kemp, D.W. et al. 2015. Systematic analysis of white pox disease in *Acropora palmata* of the Florida Keys and role of *Serratia marcescens*. *Applied and Environmental Microbiology* 81: 4451–4457

Kaczmarsky, L.T. 2006. Coral disease dynamics in the central Philippines. *Diseases of Aquatic Organisms* 69: 9–21

Kaczmarsky, L.T. and Richardson, L.L. 2007. Transmission of growth anomalies between Indo-Pacific Porites corals. *Journal of Invertebrate Pathology* 94: 218–221

Katz, S.M., Pollock, F.J., Bourne, D.G., and Willis, B.L. 2014. Crown-of-thorns starfish predation and physical injuries promote brown band disease on corals. *Coral Reefs* 33: 705–716

Keen, E.C. 2015. A century of phage research: Bacteriophages and the shaping of modern biology. *BioEssays* 37: 6–9

Kim, K. 2016. Diseases of octocorals. In *Diseases of Coral*. Woodley, C.M., Downs, C.A., Bruckner, A. W., Porter, J.W., and Galloway, S.B. (eds.) pp. 410–415. Hoboken: Wiley Blackwell

Kuta, K.G. and Richardson, L.L. 1996. Abundance and distribution of black band disease on coral reefs in the northern Florida Keys. *Coral Reefs* 15: 219–223

Kuta, K. and Richardson, L. 2002. Ecological aspects of black band disease of corals: relationships between disease incidence and environmental factors. *Coral Reefs* 21: 393–398

Lajeunesse, T.C., Parkinson, J.E., Gabrielson, P.W. et al. 2018. Systematic revision of Symbiodiniaceae highlights the antiquity and diversity of coral endosymbionts. *Current Biology* 28: 2570–2580

Landsberg, J.H., Kiryu, Y., Peters, E.C. et al. 2020. Stony coral tissue loss disease in Florida is associated with disruption of host–zooxanthellae physiology. *Frontiers in Marine Science* 7: 576013

Lawrence, J.R., Chenier, M.R., Roy, R., Beaumier, D., Fortin, N., Swerhone, G.D., Neu, T.R., and Greer, C.W. 2004. Microscale and molecular assessment of impacts of nickel, nutrients, and oxygen level on structure and function of river biofilm communities. *Applied and Environmental Microbiology* 70: 4326–4339

Lesser, M.P., Bythell, J.C., Gates, R.D., Johnstone, R.W., and Hoegh-Guldberg, O. 2007. Are infectious diseases really killing corals? Alternative interpretations of the experimental and ecological data. *Journal of Experimental Marine Biology and Ecology* 346: 36–44

Lesser, M.P. and Jarett, J.K. 2014. Culture-dependent and culture-independent analyses reveal no prokaryotic community shifts or recovery of *Serratia marcescens* in *Acropora palmata* with white pox disease. *FEMS Microbiology Ecology* 88: 457–467

Lobban, C.S., Raymundo, L.M., and Montagnes, D.J.S. 2011. *Porpostoma guamensis* n. sp., a Philasterine Scuticociliate associated with brown-band disease of corals. *Journal of Eukaryotic Microbiology* 58: 103–113

Loya, Y., Bull, G. and Pichon, M. 1984. Tumor formations in scleractinian corals. *Helgolander Meersuntersuchungen* 37: 138–151

Luna, G.M., Bongiorni, L., Gili, C., Biavasco, F., and Danovaro, R. 2010. *Vibrio harveyi* as a causative agent of the white syndrome in tropical stony corals. *Environmental Microbiology Reports* 2: 120–127

McClanahan, T.R. and Muthiga, N.A. 1998. An ecological shift in a remote coral atoll of Belize over 25 years. *Environmental Conservation*, 122–130

Meyer, J.L., Castellanos-Gell, J., Aeby, G.S., Häse, C.C., Ushijima, B., and Paul, V.J. 2019. Microbial community shifts associated with the ongoing stony coral tissue loss disease outbreak on the Florida Reef Tract. *Frontiers in Microbiology* 10: 2244

Meyer, J.L., Paul, V.J., Raymundo, L.J., and Teplitski, M. 2017. Comparative metagenomics of the polymicrobial black band disease of corals. *Frontiers in Microbiology* 8: 618

Miller, M.W., Karazsia, J., Groves, C.E., Griffin, S., Moore, T., Wilber, P., and Gregg, K. 2016. Detecting sedimentation impacts to coral reefs resulting from dredging the Port of Miami, Florida USA. *PeerJ* 4: e2711

Miller, M.W., Lohr, K.E., Cameron, C.M., Williams, D.E., and Peters, E.C. 2014. Disease dynamics and potential mitigation among restored and wild staghorn coral, *Acropora cervicornis*. *PeerJ* 2: e541

Montano, S., Maggioni, D., Liguori, G., Arrigoni, R., Berumen, M.L., Seveso, D., Galli, P., and Hoeksema, B.W. 2020. Morpho-molecular traits of Indo-Pacific and Caribbean *Halofolliculina* ciliate infections. *Coral Reefs* 39: 375–386

Montano, S., Strona, G., Seveso, D., Maggioni, D., and Galli, P. 2016. Widespread occurrence of coral diseases in the central Maldives. *Marine and Freshwater Research* 67: 1253–1262

Montilla, L. M., Ascanio, A., Verde, A., and Croquer, A. 2019. Systematic review and meta-analysis of 50 years of coral disease research visualized through the scope of network theory. *PeerJ* 7: e7041

Muller, E., Rogers, C.S., Spitzack, A.S., and Van Woesik, R. 2008. Bleaching increases likelihood of disease on *Acropora palmata* (Lamarck) in Hawksnest Bay, St John, US Virgin Islands. *Coral Reefs* 27: 191–195

Muller, E.M., Sartor, C., Alcaraz, N.I., and Van Woesik, R. 2020. Spatial epidemiology of the stony-coral-tissue-loss disease in Florida. *Frontiers in Marine Science* 7: 163

Myers, R.L. and Raymundo, L.J. 2009. Coral disease in Micronesian reefs: A link between disease prevalence and host abundance. *Diseases of Aquatic Organisms* 87: 97–104

National Academies of Sciences, Engineering and Medicine 2018. A research review of interventions to increase the persistence and resilience of coral reefs (p. 230). Washington, DC: The National Academies Press

National Academies of Sciences, Engineering and Medicine 2019 2019. A decision framework for interventions to increase the persistence and resilience of coral reefs, (p. 200). Washington, DC: The National Academies Press

Neely, K.L., Macaulay, K.A., Hower, E.K., and Dobler, M.A. 2020. Effectiveness of topical antibiotics in treating corals affected by stony coral tissue loss disease. *Peer Journal* 8: p.e9289

Nicolet, K., Chong-Seng, K., Pratchett, M., Willis, B., and Hoogenboom, M. 2018. Predation scars may influence host susceptibility to pathogens: Evaluating the role of corallivores as vectors of coral disease. *Scientific Reports* 8: 1–10

Nicolet, K., Hoogenboom, M.O., Gardiner, N., Pratchett, M., and Willis, B. 2013. The corallivorous invertebrate *Drupella* aids in transmission of brown band disease on the Great Barrier Reef. *Coral Reefs* 32: 585–595

Nugues, M.M. 2002. Impact of a coral disease outbreak on coral communities in St. Lucia: What and how much has been lost? *Marine Ecology Progress Series* 229: 61–71

Nugues, M.M. and Bak, R.P.M. 2009. Brown-band syndrome on feeding scars of the crown-of-thorn starfish *Acanthaster planci*. *Coral Reefs* 28: 507–510

Onton, K., Page, C.A., Wilson, S.K., Neale, S., and Armstrong, S. 2011. Distribution and drivers of coral disease at Ningaloo reef, Indian Ocean. *Marine Ecology Progress Series* 433: 75–84

Page, C.A., Baker, D.M., Harvell, C.D. et al. 2009. Influence of marine reserves on coral disease prevalence. *Diseases of Aquatic Organisms* 87: 135–150

Page, C.A., Cróquer, A., Bastidas, C. et al. 2016. *Halofolliculina* ciliate infections on corals (skeletal eroding disease). In *Diseases of Coral*. Woodley, C.M., Downs, C.A., Bruckner, A.W., Porter, J.W., and Galloway, S.B. (eds.) pp. 361–375. Hoboken, New Jersey: Wiley Blackwell.

Page, C.A. and Stoddart, J.A. 2010. New records of five coral diseases from the Pilbara Region of Western Australia. *Coral Reefs* 29: 987–987

Page, C.A. and Willis, B.L. 2006. Distribution, host range and large-scale spatial variability in black band disease prevalence on the Great Barrier Reef, Australia. *Diseases of Aquatic Organisms* 69: 41–51

Page, C.A. and Willis, B.L. 2008. Epidemiology of skeletal eroding band on the Great Barrier Reef and the role of injury in the initiation of this widespread coral disease. *Coral Reefs* 27: 257–272

Palmer, C.V. and Baird, A.H. 2018. Coral tumor-like growth anomalies induce an immune response and reduce fecundity. *Diseases of Aquatic Organisms* 130: 77–81

Pandolfi, J.M. and Jackson, J.B.C. 2006. Ecological persistence interrupted in Caribbean coral reefs. *Ecology Letters* 9: 818–826

Pantos, O., Cooney, R.P., Le Tissier, M.D., Barer, M.R., O'Donnell, A.G., and Bythell, J.C. 2003. The bacterial ecology of a plague-like disease affecting the Caribbean coral *Montastrea annularis* Environmental Microbiology 5: 370–382

Patten, N.L., Harrison, P.L., and Mitchell, J.G. 2008. Prevalence of virus-like particles within a staghorn scleractinian coral (*Acropora muricata*) from the Great Barrier Reef. *Coral Reefs* 27: 569–580

Patterson, K.L., Porter, J.W., Ritchie, K.B. et al. 2002. The etiology of white pox, a lethal disease of the Caribbean elkhorn coral, *Acropora palmata*. *Proceedings of the National Academy of Sciences of the United States of America* 99: 8725–8730

Peixoto, R.S., Rosado, P.M., Leite, D.C.D.A., Rosado, A.S., and Bourne, D.G. 2017. Beneficial microorganisms for corals (BMC): Proposed mechanisms for coral health and resilience. *Frontiers in Microbiology* 8: 341

Peters, E.C. 1984. A survey of cellular reactions to environmental stress and disease in Caribbean scleractinian corals. *Helgolander Meersuntersuchungen* 37: 113–137

Peters, E.C., Halas, J.C., and Mccarty, H.B. 1986. Calicoblastic neoplasms in *Acropora palmata*, with a review of reports on anomalies of growth and form in corals. *Journal of the National Cancer Institute* 76: 895–912

Pollock, F.J., Wada, N., Torda, G., Willis, B.L., and Bourne, D.G. 2017. White syndrome-affected corals have a distinct microbiome at disease lesion fronts. *Applied and Environmental Microbiology* 83: e02799–16

Pollock, F.J., Wood-Charlson, E.M., Van Oppen, M.J.H., Bourne, D.G., Willis, B.L., and Weynberg, K.D. 2014. Abundance and morphology of virus-like particles associated with the coral *Acropora hyacinthus* differ between healthy and white syndrome-infected states. *Marine Ecology Progress Series* 510: 39–43

Porter, J.W., Dustan, P., Jaap, W.C. et al. 2001. Patterns of spread of coral disease in the Florida Keys. *Hydrobiologia* 460: 1–24

Porter, J.W., Torres, C., Sutherland, K.P. et al. 2011. Prevalence, severity, lethality, and recovery of dark spots syndrome among three Floridian reef-building corals. *Journal of Experimental Marine Biology and Ecology* 408: 79–87

Pratte, Z.A. and Richardson, L.L. 2016. Possible links between white plague-like disease, scleractinian corals, and a cryptochirid gall crab. *Diseases of Aquatic Organisms* 122: 153–161

Precht, W.F. 2019. Failure to respond to a coral disease outbreak: Potential costs and consequences. *Peer Journal* Preprints: 1–33, https://doi.org/10.7287/peerj.preprints.27860v2

Precht, W.F., Gintert, B.E., Robbart, M.L., Fura, R., and Van Woesik, R. 2016. Unprecedented disease-related coral mortality in Southeastern Florida. *Scientific Reports* 6: 31374

Preston, S. and Richards, Z. 2020. Biological consequences of an outbreak of growth anomalies on Isopora palifera at the Cocos (Keeling) Islands. *Coral Reefs* 40: 1–13

Qiu, D., Huang, L., Huang, H., Yang, J., and Lin, S. 2010. Two functionally distinct ciliates dwelling in *Acropora* corals in the South China Sea near Sanya, Hainan Province, China. *Applied and Environmental Microbiology* 76: 5639–5643

Randall, C.J., Jordán-Garza, A.G., and Van Woesik, R. 2015. Ciliates associated with signs of disease on two Caribbean corals. *Coral Reefs*, 34: 243–247

Randall, C.J., Whitcher, E.M., Code, T., Pollock, C., Lundgren, I., Hillis-Starr, Z., and Muller, E.M. 2018. Testing

methods to mitigate Caribbean yellow-band disease on *Orbicella faveolata*. *PeerJ* 6: e4800

Raymundo, L.J., Harvell, C.D., and Reynolds, T.L. 2003. *Porites* ulcerative white spot disease: Description, prevalence, and host range of a new coral disease affecting Indo-Pacific reefs. *Diseases of Aquatic Organisms* 56: 95–104

Raymundo, L.J., Rosell, K.B., Reboton, C.T., and Kaczmarsky, L. 2005. Coral diseases on Philippine reefs: Genus *Porites* is a dominant host. *Diseases of Aquatic Organisms* 64: 181–191

Reeves, L. 1994. Newly discovered: Yellow band disease strikes keys reefs. *Underwater USA* 11: 16

Renegar, D.-E.A., Blackwelder, P., Miller, J., Gochfeld, D., and Moulding, A.L. 2008. Ultrastructural and histological analysis of dark spot syndrome in *Siderastrea siderea* and *Agaricia agaricites*. Proceedings of the 11th International Coral Reef Symposium, Ft Lauderdale, Florida 7

Richards, Z.T. and Newman, S.J. 2019. Outbreak of growth anomalies in Isopora palifera at Cocos (Keeling) Islands. *Marine Biodiversity*, 49: 1071–1072

Richardson, L.L. 1997. Occurrence of the black band disease cyanobacterium on healthy corals of the Florida Keys. *Bulletin of Marine Science* 61: 485–490

Richardson, L.L. 1998. Coral diseases: What is really known? *Trends in Ecology & Evolution* 13: 438–443

Richardson, L.L. 2004. Black band disease. In *Coral Health and Disease*. Rosenberg, E. and Loya, Y. (eds.) pp. 325–336. Berlin: Springer-Verlag

Richardson, L.L., Sekar, R., Myers, J.L., Gantar, M., Voss, J.D., Kaczmarsky, L., Remily, E.R., Boyer, G.L., and Zimba, P.V. 2007. The presence of the cyanobacterial toxin microcystin in black band disease of corals. *FEMS Microbiology Letters* 272: 182–187

Richardson, L.L., Smith, G.W., Ritchie, K.B., and Carlton, R.G. 2001. Integrating microbiological, microsensor, molecular, and physiologic techniques in the study of coral disease pathogenesis. *Hydrobiologia* 460: 71–78

Ritchie, K.B. 2006. Regulation of microbial populations by coral surface mucus and mucus-associated bacteria. *Marine Ecology Progress Series* 322: 1–14

Ritchie, K.B. and Smith, G.W. 1998. Type II white-band disease. *Revista De Biologia Tropical* 46: 199–203

Roder, C., Arif, C., Bayer, T., Aranda, M., Daniels, C., Shibl, A., Chavanich, S., and Voolstra, C.R. 2014. Bacterial profiling of white plague disease in a comparative coral species framework. *The ISME Journal* 8: 31–39

Rodríguez, S., Croquer, A., Guzmán, H.M., and Bastidas, C. 2009. A mechanism of transmission and factors affecting coral susceptibility to *Halofolliculina* sp. infection. *Coral Reefs* 28: 67–77

Rodriguez-Martinez, R., Banaszak, A., and Jordan-Dahlgren, E. 2001. Necrotic patches affect *Acropora palmata* (Scleractinia: Acroporidae) in the Mexican Caribbean. *Diseases of Aquatic Organisms* 47: 229–234

Rosado, P.M., Leite, D.C., Duarte, G.A. et al. 2018. Marine probiotics: Increasing coral resistance to bleaching through microbiome manipulation. *The ISME Journal* 13: 921–936

Rosales, S.M., Clark, A.S., Huebner, L.K., Ruzicka, R.R., and Muller, E.M. 2020. *Rhodobacterales* and *Rhizobiales* are associated with stony coral tissue loss disease and its suspected sources of transmission. *Frontiers in Microbiology* 11: 681

Rosenberg, E., Koren, O., Reshef, L., Efrony, R., and Zilber-Rosenberg, I. 2007. The role of microorganisms in coral health, disease and evolution. *Nature Reviews Microbiology* 5: 355–362

Rützler, K. and Santavy, D.L. 1983. The black band disease of Atlantic reef corals: I. Description of the cyanophyte pathogen. *Marine Ecology* 4: 301–319

Rutzler, K., Santavy, D.L., and Antonius, A. 1983. The black band disease of Atlantic reef corals. III. Distribution, ecology, and development. *Marine Ecology* 4: 329–358

Santavy, D.L. and Peters, E.C. 1997. Microbial pests: Coral disease in the Western Atlantic. Proceedings of the 8th International Coral Reef Symposium, pp. 607–612

Sato, Y., Bourne, D.G., and Willis, B.L. 2009. Dynamics of seasonal outbreaks of black band disease in an assemblage of *Montipora* species at Pelorus Island (Great Barrier Reef, Australia). *Proceedings of the Royal Society B: Biological Sciences* 276: 2795–2803

Sato, Y., Civiello, M., Bell, S.C., Willis, B.L., and Bourne, D.G. 2016. Integrated approach to understanding the onset and pathogenesis of black band disease in corals. *Environmental Microbiology* 18: 752–765

Sato, Y., Ling, E.Y., Turaev, D., Laffy, P., Weynberg, K.D., Rattei, T., Willis, B.L., and Bourne, D.G. 2017. Unraveling the microbial processes of black band disease in corals through integrated genomics. *Scientific Reports* 7: 1–14

Sato, Y., Willis, B.L., and Bourne, D.G. 2010. Successional changes in bacterial communities during the development of black band disease on the reef coral, *Montipora hispida*. *The ISME Journal* 4: 203–214

Sato, Y., Willis, B.L., and Bourne, D.G. 2013. Pyrosequencing-based profiling of archaeal and bacterial 16S rRNA genes identifies a novel archaeon associated with black band disease in corals. *Environmental Microbiology* 15: 2994–3007

Sekar, R., Kaczmarsky, L.T., and Richardson, L.L. 2008. Microbial community composition of black band disease on the coral host *Siderastrea siderea* from three regions of the wider Caribbean. *Marine Ecology Progress Series* 362: 85–98

Séré, M.G., Schleyer, M.H., Quod, J.P., and Chabanet, P.C.R. 2012. *Porites* white patch syndrome: An unreported coral disease on Western Indian Ocean reefs. *Coral Reefs* 31: 739

Shapiro, O.H., Kramarsky-Winter, E., Gavish, A.R., Stocker, R., and Vardi, A. 2016. A coral-on-a-chip microfluidic platform enabling live-imaging microscopy of reef-building corals. *Nature Communications* 7: 1–10

Sheridan, C., Kramarsky-Winter, E., Sweet, M., Kushmaro, A., and Leal, M.C. 2013. Diseases in coral aquaculture: Causes, implications and preventions. *Aquaculture* 396–399: 124–135

Shnit-Orland, M. and Kushmaro, A. 2009. Coral mucus-associated bacteria: A possible first line of defense. *FEMS Microbiology Ecology* 67: 371–380

Smith, H.A., Conlan, J.A., Pollock, F.J. et al. 2020. Energy depletion and opportunistic microbial colonisation in white syndrome lesions from corals across the Indo-Pacific. *Scientific Reports* 10: 1–14

Solano, O.D., Navas Suárez, G., and Moreno-Forero, S.K. 1993. Blanqueamiento coralino de 1990 en el Parque Nacional Natural Corales del Rosario (Caribe, colombiano). *Boletín de Investigaciones Marinas y Costeras-INVEMAR* 22: 97–111

Spalding, M., Burke, L., Wood, S.A., Ashpole, J., Hutchison, J., and Zu Ermgassen, P. 2017. Mapping the global value and distribution of coral reef tourism. *Marine Policy* 82: 104–113

Squires, D.F. 1965. Neoplasia in a coral? *Science* 148: 503–505

Sudek, M., Work, T.M., Aeby, G., and Davy, S. 2012. Histological observations in the Hawaiian reef coral, *Porites compressa*, affected by *Porites* bleaching with tissue loss. *Journal of Invertebrate Pathology* 111: 121–125

Sunagawa, S., Desantis, T.Z., Piceno, Y.M. et al. 2009. Bacterial diversity and white plague disease-associated community changes in the Caribbean coral *Montastraea faveolata The ISME Journal* 3: 512–521

Sussman, M., Mieog, J. C., Doyle, J., Victor, S., Willis, B.L., and Bourne, D.G. 2009. *Vibrio* zinc-metalloprotease causes photoinactivation of coral endosymbionts and coral tissue lesions. *PLoS One* 4: e4511

Sussman, M., Willis, B. L., Victor, S., and Bourne, D.G. 2008. Coral pathogens identified for white syndrome (WS) epizootics in the Indo-Pacific. *PLoS One* 3: e2393

Sutherland, K.P., Berry, B., Park, A. et al. 2016a. Shifting white pox aetiologies affecting *Acropora palmata* in the Florida Keys, 1994–2014. *Philosophical Transactions of the Royal Society B: Biological Sciences* 371: 20150205

Sutherland, K.P., Lipp, E.K., and Porter, J.W. 2016b. Acroporid serratiosis. In *Diseases of Coral*. Woodley, C.M., Downs, C.A., Bruckner, A.W., Porter, J.W., and Galloway, S.B. (eds.) pp. 221–230. New Jersey: Hoboken

Sutherland, K.P., Porter, J.W., and Torres, C. 2004. Disease and immunity in Caribbean and Indo-Pacific zooxanthellate corals. *Marine Ecology Progress Series* 266: 273–302

Sutherland, K.P., Porter, J.W., Turner, J.W. et al. 2010. Human sewage identified as likely source of white pox disease of the threatened Caribbean elkhorn coral, *Acropora palmata*. *Environmental Microbiology* 12: 1122–1131

Sutherland, K.P. and Ritchie, K.B. 2004. White pox disease of the Caribbean elkhorn coral, *Acropora palmata*. In *Coral Health and Disease*. Rosenberg, E. and Loya, Y. (eds.) pp. 289–300. Berlin: Springer-Verlag

Sweet, M., Burn, D., Croquer, A., and Leary, P. 2013. Characterisation of the bacterial and fungal communities associated with different lesion sizes of dark spot syndrome occurring in the coral *Stephanocoenia intersepta*. *PLoS One* 8: e62580

Sweet, M. and Bythell, J. 2012. Ciliate and bacterial communities associated with white syndrome and brown band disease in reef-building corals. *Environmental Microbiology* 14: 2184–2199

Sweet, M., Jones, R., and Bythell, J. 2011. Coral diseases in aquaria and in nature. *Journal of the Marine Biological Association of the United Kingdom* 92: 791–801

Sweet, M.J. and Bulling, M.T. 2017. On the importance of the microbiome and pathobiome in coral health and disease. *Frontiers in Marine Science* 4: 9

Sweet, M.J., Croquer, A., and Bythell, J.C. 2014. Experimental antibiotic treatment identifies potential pathogens of white band disease in the endangered Caribbean coral *Acropora cervicornis*. *Proceedings of the Royal Society B: Biological Sciences* 281: 20140094

Teplitski, M. and Ritchie, K. 2009. How feasible is the biological control of coral diseases? *Trends in Ecology & Evolution* 24: 378–385

Thompson, F.L., Barash, Y., Sawabe, T., Sharon, G., Swings, J., and Rosenberg, E. 2006. *Thalassomonas loyana* sp nov, a causative agent of the white plague-like disease of corals on the Eilat coral reef. *International Journal of Systematic and Evolutionary Microbiology* 56: 365–368

Ulstrup, K.E., Kuhl, M., and Bourne, D.G. 2007. Zooxanthellae harvested by ciliates associated with brown band syndrome of corals remain photosynthetically competent. *Applied and Environmental Microbiology* 73: 1968–1975

Ushijima, B., Smith, A., Aeby, G.S., and Callahan, S.M. 2012. *Vibrio owensii* induces the tissue loss disease *Montipora* white syndrome in the Hawaiian Reef Coral *Montipora capitata*. *PLoS One* 7: e46717

Ushijima, B., Videau, P., Burger, A.H., Shore-Maggio, A., Runyon, C.M., Sudek, M., Aeby, G.S., and Callahan, S.M. 2014. *Vibrio coralliilyticus* strain OCN008 is an etiological agent of acute *Montipora* white syndrome. *Applied and Environmental Microbiology* 80: 2102–2109

Vargas-Angel, B. 2009. Coral health and disease assessment in the US Pacific Remote Islands Areas. *Bulletin of Marine Science* 84: 211–227

Vega Thurber, R.L., Mydlarz, L.D., Brandt, M. et al. 2020. Deciphering coral disease dynamics: Integrating host, microbiome, and the changing environment. *Frontiers in Ecology and Evolution* 8: 402

Vollmer, S.V. and Kline, D.I. 2008. Natural disease resistance in threatened staghorn corals. *PLoS One* 3: e3718

Walton, C.J., Hayes, N.K., and Gilliam, D.S. 2018. Impacts of a regional, multi-year, multi-species coral disease outbreak in Southeast Florida. *Frontiers in Marine Science* 5: 323

Weil, E. 2004. Coral Diseases in the wider Caribbean. In *Coral Health and Disease*. Rosenberg, E. and Loya, Y. (eds.) pp. 35–68. Berlin: Springer-Verlag

Weil, E., Cróquer, A., and Urreiztieta, I. 2009. Yellow band disease compromises the reproductive output of the Caribbean reef-building coral *Montastraea faveolata* (Anthozoa, Scleractinia). *Diseases of Aquatic Organisms* 87: 45–55

Weil, E., Hernández-Delgado, E., Gonzalez, M., Williams, S., Suleimán-Ramos, S., Figuerola, M., and Metz-Estrella, T. 2019. Spread of the new coral disease 'SCTLD' into the Caribbean: implications for Puerto Rico. *Reef Encounter* 34: 38–43

Weil, E., Irikawa, A., Casareto, B., and Suzuki, Y. 2012. Extended geographic distribution of several Indo-Pacific coral reef diseases. *Diseases of Aquatic Organisms* 98: 163–170

Weil, E. and Rogers, C.S. 2011. Coral reef diseases in the Atlantic-Caribbean. In *Coral Reefs: An Ecosystem in Transition*. Dubinsky, Z., and Stambler, N. (eds.) pp. 465–491. Springer, Dordrecht, https://doi.org/10.1007/978-94-007-0114-4_27

Weil, E., Rogers, C.S. and Croquer, A. 2015. Octocoral diseases in a changing ocean. In *Marine Animal Forests: The Ecology of Benthic Biodiversity Hotspots*. Rossi, S., Bramanti, L., Gori, A., and Orejas Saco Del Valle, C. (eds.) pp. 1–55. Cham: Springer International Publishing

Williams, E.H., Bartels, P.J., and Bunkley-Williams, L. 1999. Predicted disappearance of coral-reef ramparts: a direct result of major ecological disturbances. *Global Change Biology* 5: 839–845

Williams, D.E. and Miller, M.W. 2005. Coral disease outbreak: pattern, prevalence and transmission in

Acropora cervicornis. *Marine Ecology Progress Series* 301: 119–128

Williams, G.J., Aeby, G.S., Cowie, R.O., and Davy, S.K. 2010. Predictive modeling of coral disease distribution within a reef system. *PLoS One* 5: e9264

Willis, B.L., Page, C.A., and Dinsdale, E.A. 2004. Corals disease on the Great Barrier Reef. In *Coral Health and Disease*. Rosenberg, E. and Loya, Y. (eds.) pp. 69–104. Heildelberg: Springer-Verlag

Work, T.M. and Aeby, G.S. 2006. Systematically describing gross lesions in corals. *Diseases of Aquatic Organisms* 70: 155–160

Work, T.M. and Aeby, G.S. 2014. Microbial aggregates within tissues infect a diversity of corals throughout the Indo-Pacific. *Marine Ecology Progress Series* 500: 1–9

Work, T.M., Aeby, G.S., and Coles, S.L. 2008a. Distribution and morphology of growth anomalies in *Acropora* from the Indo-Pacific. *Diseases of Aquatic Organisms* 78: 255–264

Work, T.M., Richardson, L.L., Reynolds, T.L., and Willis, B.L. 2008b. Biomedical and veterinary science can increase our understanding of coral disease. *Journal of Experimental Marine Biology and Ecology* 362: 63–70

Work, T.M., Kaczmarsky, L.T., and Peters, E.C. 2015. Skeletal growth anomalies in corals. In *Diseases of Coral*. Woodley, C.M., Down, C.A., Bruckner, A.W., Porter, J.W., and Galloway, S.B. (eds.) pp. 291–299. Hoboken, New Jersey: Wiley-Blackwell

Work, T.M. and Rameyer, R.A. 2005. Characterizing lesions in corals from American Samoa. *Coral Reefs* 24: 384–390

Work, T.M., Russell, R., and Aeby, G.S. 2012. Tissue loss (white syndrome) in the coral *Montipora capitata* is a dynamic disease with multiple host responses and potential causes. *Proceedings of the Royal Society B: Biological Sciences* 279: 4334–4341

Zimmer, B.L., May, A.L., Bhedi, C.D. et al. 2014. Quorum sensing signal production and microbial interactions in a polymicrobial disease of corals and the coral surface mucopolysaccharide layer. *PLoS One* 9: e108541

Zverlov, V., Klein, M., Lücker, S., Friedrich, M.W., Kellermann, J., Stahl, D., Loy, A., and Wagner, M. 2005. Lateral gene transfer of dissimilatory (bi)sulfite reductase revisited *Journal of Bacteriology* 187: 2203–2208

CHAPTER 5

Diseases of sponges

Heidi M. Luter and Nicole S. Webster

5.1 Introduction

5.1.1 Sponge biology, ecology, and microbiology

Sponges (phylum Porifera) represent the oldest living metazoans with fossil records dating back to the Precambrian (Li et al. 1998). In fact, recent evidence supports sponges as the sister group to all other animals (Feuda et al. 2017). Sponges are an incredibly diverse phyla with an estimated 20,000 species, of which 9,292 are formally described, occupying both marine and freshwater environments spanning the tropics to the poles (Van Soest et al. 2020). Sponges form an integral component of benthic communities and through their immense filtering capabilities contribute to a number of important ecosystem processes including benthic-pelagic coupling and nutrient cycling, as well as bioerosion and consolidation processes (Bell 2008; de Goeij et al. 2013; Maldonado et al. 2015).

The sponge body plan is simplistic, lacking true tissues and organs and instead relying on a diverse assortment of highly specialised cells, which facilitate important physiological processes such as water circulation, feeding and reproduction (Simpson 1984). Sponges consist of three main cell layers: (1) the outer layer, or pinacoderm, is made up of squamous pinacocyte cells, (2) the inner layer, or choanoderm, is made up of flagellated choanocyte cells that function to pump water through the sponge, and (3) the middle layer, or mesohyl (mesenchyme), which supports a skeletal matrix of collagen fibres and spicules for most groups of sponges. The mesohyl also contains the totipotent archaeocyte cells which can differentiate into other cell types (Müller 2006), as well as most of the symbiotic microbes (Wilkinson 1978a). Although generally described as simplistic, mounting evidence suggests that sponges are more complex than previously thought, with compartmentalisation of choanocyte cells (de Goeij et al. 2009) and distinct types of epithelial tissues being described (Leys et al. 2009). In addition, sponges possess a functional immune system (Müller 2003; Degnan 2015) and a repertoire of genes associated with the nervous systems of higher animals, despite lacking neurons (Srivastava et al. 2010).

Sponges represent one of the most ancient metazoan-symbioses which has likely attributed to their evolutionary success. They host a diverse range of microorganisms including bacteria, Archaea, unicellular algae, fungi, and viruses (Webster and Taylor 2012; Thomas et al. 2016; Laffy et al. 2018) which in some species can account for up to 40% of the total sponge biomass (Vacelet and Donadey 1977). Sponge-associated microbes play a vital role in the physiology and ecology of their host. In particular, microbes can produce secondary metabolites that the sponge host can utilise for defence (Schmidt et al. 2000; Piel et al. 2004), they can provide increased structural rigidity (Wilkinson 1978b), they play a role in synthesising vitamins that the host is incapable of producing (Fan et al. 2012; Hentschel et al. 2012) and photosynthetic microbes (i.e. cyanobacteria) can provide > 50% of the energy requirements for the sponge host (Wilkinson 1983). In a recent metabolic reconstruction of sponge symbionts by metagenomics,

Heidi M. Luter and Nicole S. Webster, *Diseases of sponges*. In: *Invertebrate Pathology*. Edited by Andrew F. Rowley, Christopher J. Coates and Miranda M.A. Whitten, Oxford University Press. © Oxford University Press (2022). DOI: 10.1093/oso/9780198853756.003.0005 the sponge-associated microbiome was shown to exhibit extensive taxonomic redundancy for pathways involved in carbon fixation, B-vitamin synthesis, taurine metabolism, sulphite oxidation, and most components of nitrogen metabolism (Engelberts et al. 2020).

5.1.2 Sponge disease background

Some of the most devastating sponge disease outbreaks have occurred in populations of commercial sponge species in the Caribbean and Mediterranean. One of the earliest epidemics decimated 70–95% of *Hippospongia* spp. in the Caribbean (Galstoff 1942). On a similar scale, the lucrative Mediterranean sponge fishery lost 90% of their output in the 1980s (Vacelet 1994) with 60% lost in the Ligurian Sea (Gaino et al. 1992). The causative agents for these epidemics were never confirmed; however, fungi were consistently identified in diseased sponges in the Caribbean (Galstoff 1942), whereas bacterial infections were suspected in the Mediterranean (Gaino et al. 1992; Vacelet et al. 1994). A number of sponge disease incidents have since been reported, spanning a broad geographical range and including a wide range of symptoms which will be the focus of the remainder of this chapter.

5.2 Principal diseases

The main diseases reported for sponges are summarised in Table 5.1. While there are some instances where a sponge disease is fully characterised, many studies instead focus on describing the symptoms, which frequently includes generalised necrosis and therefore will be discussed collectively as one disease. It is important to highlight that although names have been assigned to diseases in sponges, the causative agents of disease have only been defined in three instances (sponge boring necrosis, sponge necrosis syndrome, and fatal fungal infection).

5.2.1 Sponge boring necrosis

Sponge boring necrosis was first reported in isolated individuals of *Rhopaloeides odorabile* on the Great

Barrier Reef in 1998. Diseased sponges displayed extensive levels of necrosis which led to exposure of the sponge skeletal fibres. An alphaproteobacterial strain (NW4327) isolated from diseased tissue was used in infection trials where it was found to successfully induce disease symptoms in healthy sponges and was subsequently re-isolated to confirm Koch's postulates as the primary pathogen of sponge boring necrosis (Webster et al. 2002). Further examination using transmission electron microscopy showed this bacterium apparently burrowing through the spongin fibres (Figure 5.1a–f), with subsequent studies confirming strain NW4327 as *Pseudoalteromonas agarivorans* that produces a collagenolytic enzyme capable of degrading sponge skeletal fibres (Mukherjee et al. 2009), as well as virulence genes related to collagenase activity (Choudhury et al. 2015).

5.2.2 Sponge necrosis syndrome

Sponge necrosis syndrome was shown to affect up to 35% of the *Callyspongia (Euplacella)* aff *biru* population in the Maldives, with rapid disease progression rates (0.34 ± 0.08 cm d^{-1}) observed (Sweet et al. 2015). Histological examination revealed the presence of symbiotic bacteria and fungi associated with all tissues, although fungal hyphae dominated the necrosed tissues. Infection trials were performed using two bacterial isolates and three fungal isolates, with successful disease transmission to healthy individuals only observed when using a combination of one Rhodobacteraceae-affiliated bacterium and one *Rhabdocline* sp. fungus (Sweet et al. 2015). Sponge necrosis syndrome is the first formally described polymicrobial disease in sponges and the second study to determine the causative agent of disease.

5.2.3 Fatal fungal infection

A fatal fungal infection was identified in *Chondrosia reniformis* from the Marine Protected Area of Portofino in the Ligurian Sea, with between 5 and 20% of sponges affected (Greco et al. 2019). Symptoms of the infection included a softening and greying of the tissues and the formation of fluid filled pouch (oedema), with tissue rotting occurring a few days

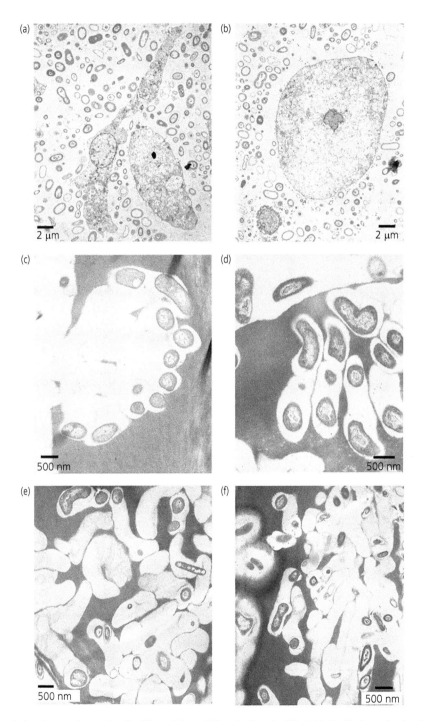

Figure 5.1 Transmission electron micrographs of healthy and diseased *Rhopaloeides odorabile*. (a, b). Healthy tissue showing the enormous diversity of symbiont morphotypes and characteristic sponge cells. (c, d) In the early stages of disease *Pseudoalteromonas agarivorans* enters the sponge collagen fibres and begins replicating. (e, f) In the advanced stages of disease extensive digestion of the sponge collagen fibres is evident.

after the first symptoms were observed. Histological examination revealed infected sponges had lost the organisation of choanocyte chambers characteristic of healthy sponges, and a proliferation of fungal hyphae morphologically consistent with the genus, *Aspergillus* was evident. Infection trials using the isolated fungus (identified as *Aspergillus tubingensis*) from diseased sponges successfully established the infection in healthy sponges, thereby confirming a fungal aetiology (Greco et al. 2019).

5.2.4 *Agelas* wasting syndrome

Between 2010 and 2015, *Agelas tubulata* populations in the Florida Keys showed an increased prevalence of this disease from 7 to 35%, with disease symptoms including localised lesions and shedding of the sponge skeleton (Deignan and Pawlik 2016). No specific microbial pathogen was identified, however, diseased samples displayed greater microbiome variability compared to healthy samples and also had a higher relative abundance of ammonia oxidising Archea belonging to the phylum Thaumarchaeota (Deignan et al. 2018).

5.2.5 Pustule disease

Pustule disease was described in populations of *Ircinia* spp. in the western Mediterranean and

African coasts, with severe outbreaks occurring in 2008–09 (Maldonado et al. 2010) and the summer of 2010 (Blanquer et al. 2016). During these outbreaks, sponges developed small pustules on their surface which eventually joined into larger necrotic lesions. Using transmission electron microscopy, Maldonado et al. (2010) identified the presence of a bacterium with a twisted rod morphology in all diseased samples. In addition, concentric barriers of collagen and phagocytic cells were visible, which the host appeared to be using to limit the spread of the bacterial infection. Later, Blanquer et al. (2016) characterised the microbiome changes associated with early disease in *I. fasciculata*, revealing increased bacterial diversity and shifts in the microbiome consistent with infection by opportunistic pathogens.

5.2.6 Brown lesion necrosis/brown spot syndrome

Outbreaks of brown lesion necrosis were reported in *Ianthella basta* populations in Papua New Guinea between 1996 and 2000 (Cervino et al. 2006). Similarly, *I. basta* populations from reefs on the Great Barrier Reef and Torres Strait suffered from brown spot syndrome (Figure 5.2), which affected 43 and 66% of the surveyed populations, respectively, between 2008 and 2010 (Luter et al. 2010a). Unlike populations in Papua New Guinea,

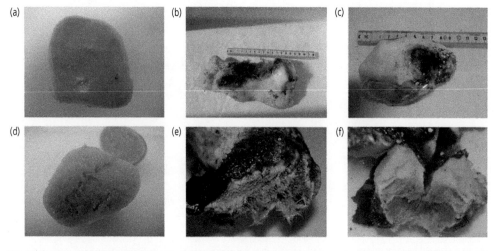

Figure 5.2 Photographs of healthy and diseased *Geodia barretti.* Healthy sponges were characterised by a smooth white outer cortex (a) and cream internal region with densely packed spicules and fibres comprising a rigid mesohyl (d). In contrast, diseased individuals (b, c) were affected by large brown-grey-black lesions, had high levels of fouling by algae and foraminiferans and the internal tissue revealed complete destruction of the mesohyl (e, f). Source: figure and legend modified from Luter et al. (2017) with permission.

where brown lesions were correlated with bacteria in the *Bacillus* and *Pseudomonas* genera (Cervino et al. 2006), comprehensive comparisons of healthy and diseased *I. basta* using various techniques did not identify any potential pathogens and brown spot syndrome could not be transmitted to heathy sponges (Luter et al. 2010b).

5.2.7 Sponge orange band (SOB) and bleaching

Sponge orange band (SOB) affects giant barrel sponge (*Xestospongia muta*) populations in the Caribbean and is synonymous with fatal bleaching (Cowart et al. 2006). As the name implies, a distinct orange band forms between the healthy and bleached/dead tissues, with examination revealing massive destruction of the pinacoderm and increased phagocytosis of cyanobacterial symbionts (Angermeier et al. 2011). The cause of SOB remains unknown, however, a consistent shift in the cyanobacterial community away from known sponge-specific symbionts (e.g. *Synechococcus/Prochlorococcus*) in healthy tissues towards lineages associated with coral disease (*Leptolyngbya* sp.) in diseased tissues has been reported (Angermeier et al. 2011).

There have also been a number of reports of sponge bleaching, however, whether these incidents lead to SOB and fatal bleaching remains unknown. In particular, sponge bleaching has been observed in the Florida Keys, Little Cayman Island, Cuba, Belize, Mexico, and Curaçao (Vicente 1990; Whalan 2018; Gammill and Fenner 2005; Nagelkerken et al. 2000; Paz 1997).

5.2.8 *Aplysina* red band syndrome (ARBS)

Aplysina red band syndrome (ARBS) affects sponges of the *Aplysina* genus in the Caribbean (Olson et al. 2006; Gochfeld et al. 2007; Olson et al. 2014). The syndrome was first documented in the Bahamas in 2004 where its prevalence in *A. cauliformis* populations increased from 3.2 to 12.2% over the course of the study period (July 2004–October 2005), with infected individuals displaying a rust-coloured leading edge typically transitioning into necrotic tissue (Olson et al. 2006). Similar to SOB, microbial profiling of diseased tissue revealed a dominance

of the cyanobacterium, *Leptolyngbya* sp. However, infection trials using this strain of the bacterium were unsuccessful in transmitting the syndrome, indicating it is likely not the sole causative agent of ARBS (Olson et al. 2014).

5.2.9 Black patch syndrome

Aplysina sponges have also displayed signs of disease in the Adriatic Sea, with over 40% of the *A. aerophoba* population affected by black patch syndrome (Webster et al. 2008b). Infected sponges developed black patches on the ectosome (outer, cortical layer), followed by severe necrosis that resulted in exposure of the skeletal fibres. Microbial analysis of healthy and diseased specimens revealed several sequences exclusively present in diseased tissues, including a bacterium with homology to a strain involved in black band disease in corals (Webster et al. 2008b).

5.2.10 Sponge white patch

Sponge white patch was first reported in Florida where up to 20% of the *Amphimedon compressa* population was impacted between November 2007 and July 2010 (Angermeier et al. 2012). Symptoms of this disease include white patches of variable sizes that appear sporadically along the sponge branches. Interestingly, transmission electron microscopy (TEM) revealed a spongin-boring bacterium similar to that found by Webster et al. (2002) in individuals affected by sponge boring necrosis, as well as a community shift in the associated microbiome (Angermeier et al. 2012). However, attempts to transmit sponge white patch from diseased to healthy individuals in the field were unsuccessful.

5.2.11 Uncharacterised pink necrosis and brown rot syndrome

Disease is not confined to marine sponges, with freshwater populations of *Lubomirskia baicalensis* in Lake Baikal, Russia also affected by a syndrome that turns the normally characteristic bright green sponge tissue pink (Denikina et al. 2016). Since the first report of *L. baicalensis* disease in

2011, a number of other symptoms have been reported from sponges of the Lubomirskiidae family including tissue necrosis, brown patches, and dirty-violet bacterial biofilms (Belikov et al. 2019), as well as bleaching (Kaluzhnaya and Itskovich 2015). Microbial profiling of healthy and diseased sponges was undertaken in 2011 and again in 2015 with varying results. For instance, in 2011 diseased sponges had a higher relative abundance of bacteria belonging to the Phylum Verrucomicrobia, which authors attributed to an increased production of methane in the lake (Denikina et al. 2016), whereas no such bacteria were identified in diseased samples from 2015, with samples instead hosting higher relative abundances of bacteria belonging to the phyla Bacteroidetes and Proteobacteria, than their healthy counterparts (Belikov et al. 2019). Interestingly, both studies noted the replacement of the symbiotic algae (*Choricystis* sp.) with a cyanobacterium, that in conjunction with opportunistic bacteria, may have attributed to the decline in host health.

The brown patches noted by Belikov and colleagues are generally consistent with brown rot syndrome, which reportedly affected a variable number of *L. baicalensis* across two sites in the lake (Kulakova et al. 2018). Kulakova et al. (2018) also undertook microbial profiling of healthy and diseased sponges, revealing increased relative abundances of potential pathogenic bacterial families, particularly the Oscillatoriaceae.

5.2.12 Generalised necrosis

More than half of the reported sponge diseases have not been named, and as such have been categorised under generalised necrosis, which includes all of the disease reports in commercial sponge species (Table 5.1). In addition, prevalence of disease was examined in a number of sponge species in Panama between 1984 and 1998 including *Iotrochota birotulata*, *Amphimedon compressa*, and *Aplysina fulva* which affected up to 4, 12, and 15% of the respective populations (Wulff 2007). Disease has also impacted the mangrove sponge, *Geodia papyracea*, with histological examination revealing infected sponges have an excess of cyanobacteria that are expelled via the formation of pseudogemmules (Rützler 1988). Interestingly, deep water

sponges of the same genus are also impacted by disease. Up to 20% of the *Geodia barretti* population surveyed in Norway displayed disease symptoms which included large brown-grey-black lesions, tissue degradation and increased fouling (Figure 5.3) (Luter et al. 2017). Microbial profiling revealed a reduction in the relative abundance of archaeal symbionts and increased relative abundances of *Bacteroidetes*, *Firmicutes* and *Deltaproteobacteria* in diseased tissues (Luter et al. 2017).

There have also been a number of anecdotal reports of sponge disease, including isolated incidents of *Carteriospongia foliascens* and *Coscinoderma matthewsi* on the Great Barrier Reef displaying disease-like symptoms (Figures 5.3b,c), as well as sponges from locations around Puerto Rico including *Svenzea zeai* and *Agelas conifera* at Mona Island and *Smenospongia conulosa* at Guanica Biosphere Reserve (Figures 5.3d–f; personal comments Jaaziel Garcia-Hernandez).

5.3 Control and treatment of diseases

Difficulty in determining the causative agents of sponge disease combined with the frequent inability to transmit infection from diseased to healthy individuals, suggests that microorganisms are not solely responsible for many of the sponge diseases reported to date. As such, no specific treatments have yet been developed. However, there is some evidence that environmental factors such as elevated seawater temperature may be linked to symptoms of sponge disease. For example, a number of previous sponge mortality events coincided with abnormally high seawater temperatures (Vicente 1990; Vacelet 1994; Vacelet et al. 1994; Cerrano et al. 2001). Additionally, one of the more recent disease epidemics, which affected 80–100% of the *Ircinia fasciculata* populations in the western Mediterranean, was linked to anomalous sea surface temperatures (Cebrian et al. 2011). Furthermore, 22% of *I. variabilis*, *Sarcotragus spinosulus*, and *Spongia officinalis* displayed disease symptoms directly correlated with calm seas and high seawater temperatures in the north Adriatic Sea (Di Camillo et al. 2012). Most recently, a mass mortality event was reported along the

Table 5.1 An overview of the principal diseases of sponges

Disease	Disease causing agent(s)	Host range	Geographical range	Pathology	Key references
Sponge boring necrosis	Alphaproteobacterial Strain NW4327: *Pseudoalteromonas agarivorans*	*Rhopaloeides odorabile*	Great Barrier Reef	Causative agent identified as *Pseudoalteromonas agarivorans* with collagenase activity allowing it to burrow through spongin fibres	Webster et al. 2002; Mukherjee et al. 2009; Choudhury et al. 2015
Sponge necrosis syndrome	Rhodobacteraceae affiliated bacterium and *Rhabdocline* affiliated fungus	*Callyspongia (Euplacella)* aff *biru*	Maldives	Polymicrobial disease—combination of a bacterium and fungus from the Rhodobacteraceae and *Rhabdocline* genera induce disease symptoms	Sweet e: al. 2015
Fatal fungal infection	*Aspergillus tubingensis*	*Chondrosia reniformis*	Ligurian Sea	Fungal infection caused by *Aspergillus tubingensis*. Once symptoms observed the sponge decays/dies within days	Greco et al. 2019
Agelas wasting syndrome	unknown	*Agelas tubulata*	Florida Keys	Dysbiosis, greater microbiome variability in diseased samples, including the abundance of Thaumarchaeota	Deignan and Pawlik 2016; Deignan et al. 2018
Pustule disease	unknown	*Ircinia fasciculata, Ircinia* spp.	Western Mediterranean and African coasts	Twisted rod bacteria identified by TEM*. Dysbiosis	Maldonado et al. 2010, Blanquer et al. 2016
Brown spot syndrome	unknown	*Ianthella basta*	Grea: Barrier Reef, Torres Strait	No dysbiosis or distinct pathogens detected	Luter et al. 2010a, 2010b
Brown lesion necrosis	unknown	*Ianthella basta*	Papua New Guinea	Culture based methods revealed diseased tissues correlated with bacterial strains within the *Bacillus* and *Pseudornonas* genera	Cervino et al. 2006
Sponge orange band	unknown	*Xestospongia muta*	Florida Keys, Bahamas	Dysbiosis, shift in cyanobacterial community associated with orange band	Cowart et al. 2006; Angermeier et al. 2011
Aplysina red band syndrome	unknown	*Aplysina cauliformis*	Bahamas	Dysbiosis, red colour attributed to cyanobacterial strain (*Leptolyngbya*) but no disease transmission	Olson et al. 2006; Gochfeld et al. 2C07; Olson et al. 2014

continued

Table 5.1 *Continued*

Disease	Disease causing agent(s)	Host range	Geographical range	Pathology	Key references
Black patch disease	unknown	*Aplysina aerophoba*	Slovenia, Adriatic Sea	Multiple sequences unique to diseased samples identified, including a Deltaproteobacteria implicated in coral black band disease	Webster et al. 2008
Sponge white patch	unknown	*Amphimedon compressa*	Florida Keys	Sponge-boring morphotype identified by TEM, dysbiosis	Angermeier et al. 2012
Uncharacterised pink necrosis	unknown	*Lubomirskia baicalensis, Lubomirskiidae* family	Lake Baikal, Russia	Dysbiosis, replacement of symbiotic algae with cyanobacterium and increased abundances of Verrumicrobia, Bacteroidetes and Proteobacteria	Belikov et al. 2019; Khanaev et al. 2018; Denikina et al. 2016
Brown rot syndrome	Unknown	*Lubomirskia baicalensis*	Lake Baikal, Russia	Dysbiosis, increase in abundances of potential pathogenic bacterial families, particularly the Oscillatoriaceae	Kulakova et al. 2018
Generalised necrosis	unknown	*Isodyctia palmata, Halichondria sitiens*	Kandalaksha Bay, White Sea, sub-Arctic	No microbial characterisation	Ereskovsky et al. 2019
Generalised necrosis	unknown	*Geodia barretti*	Norway	Dysbiosis, increased relative abundances of Bacteroidetes, Firmicutes and Deltaproteobacteria in diseased tissues	Luter et al. 2017
Generalised necrosis	unknown	*Crella cyathophora*	Red Sea	Diseased individuals enriched with a clade within the phylum Verrucomicrobia	Gao et al. 2015
Generalised necrosis	unknown	*Ircinia variabilis, Sarcotragus spinosulus, Spongia officinalis*	Adriatic Sea	No microbial characterisation, but SEM used to observe cyanobacterial mat	Di Camillo et al. 2013
Generalised necrosis	unknown	*Ircinia fasciculata, Sarcotragus spinosulum*	Mediterranean	TEM revealed degraded cyanobacteria associated with necrotic zones as well as an unidentified boring bacterium	Cebrian et al. 2011

Disease	Cause	Host sponge	Location	Notes	Reference
Generalised necrosis	unknown	*Xestospongia muta, Geodia* spp., *Ircinia* spp. *Verongula gigantea, Callyspongia plicifera*	Mexico	No microbial characterisation	Gammill and Fenner 2005
Generalised necrosis	unknown	*Xestospongia muta*	Curaçao	No microbial characterisation	Nagelkerken et al. 2000
Generalised necrosis	unknown	*Xestospongia muta*	Belize	No microbial characterisation	Paz 1997*
Generalised necrosis	unknown	*Ircinia spinosula and Ircinia* spp.	Mediterranean	No microbial characterisation	Corriero et al. 1996
Generalised necrosis	unknown	Commercial sponge spp.	Libya	No microbial characterisation	Gashout et al. 1989
Generalised necrosis	unknown	*Spongia and Hippospongia* spp., *Petrosia ficiformis, Ircinia variabilis, Anchinoe paupertas*	Sicily and Ligurian Coast	Ovoid bacteria filled the canaliculi of exposed skeletal fibres	Gaino et al. 1992; Rizzello et al. 1997
Generalised necrosis	unknown	*Hippospongia* spp., *Spongia* spp.	Mediterranean	Unidentified bacterium burrowing through tissues and white film postulated to be an *Oscillatoria* spp.	Economou and Konteatis 1988; Vacelet 1994; Vacelet et al. 1994
Generalised necrosis	unknown	*Geodia papyracea*	Belize	Tissue decay attributed to a cyanobacterium	Rützler 1988
Generalised necrosis	unknown	*Iotrochota* spp., *A. compressa, Aplysina fulva, Callyspongia vaginalis, Niphates* spp., *Xestospongia* spp., *Verongula rigida, Ircinia* spp.	Panama	No microbial characterisation	Wulff 2006a, 2006b; Wulff 2007
Generalised necrosis	unknown	*Hippospongia* spp.	Caribbean	Diseased tissues associated with fungi	Glastoff 1942; Storr 1964
Generalised necrosis	unknown	Commercial sponge spp.	British Honduras	Unbranched fungal filaments between live and dead tissue	Smith 1939, 1941
Generalised necrosis	unknown	Commercial sponge spp.	Cuba, Florida Keys, Bahamas	Diseased tissues associated with fungi	Galstoff et al. 1939; Smith 1941

This table represents a continuation from previous reviews on sponge diseases (see Luter and Webster 2017 and Webster 2007). * TEM = transmission electron microscopy

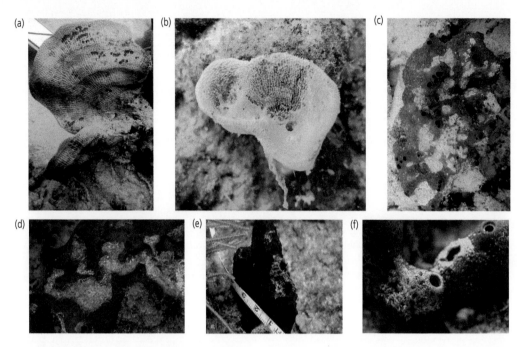

Figure 5.3 The elephant ear sponge, *Ianthella basta*, displaying brown spot syndrome (a). Anecdotal reports of disease in *Carteriospongia foliascens* (photo: Craig Humphrey) (b) *Coscinoderma matthewsi* (c) from the Great Barrier Reef and *Svenzea zeai* (d), *Smenospongia conulosa* (e) and *Agelas conifera* (f) from locations around Puerto Rico. Specifically, diseased *S. zeai* and *A. conifera* were observed at Mona Island and *S. conulosa* at Guanica Biosphere Reserve. Source: photographs courtesy of Jaaziel Garcia-Hernandez.

coasts of Kandalaksha Bay, White Sea, and sub-Arctic affecting *Isodyctia palmata* and *Halichondria sitiens* which coincided with a 6.5°C increase in average seawater temperature (Ereskovsky et al. 2019). It is unclear whether any of these events were caused by a specific pathogen or instead exceeded the physiological thresholds of the respective sponge species. However, numerous studies have shown that elevated temperatures can lead to microbial dysbiosis and adverse health outcomes for the sponge host (Lemoine et al. 2007; Webster et al. 2008a; Fan et al. 2013; López-Legentil et al. 2010; Ramsby et al. 2018). Also, elevated seawater temperatures impact the sponge viral community, with a 30-fold increase in retro-transcribing viruses from the Caulimoviridae and Retroviridae families in *R. odorabile* exposed to 32°C for 48h (Laffy et al. 2019), which the authors suggest contributes to the rapid decline in host health at higher temperature. Collectively, these studies indicate that limiting anthropogenic warming may reduce the likelihood of future mass sponge mortality events.

5.4 Future directions

The limited success in assigning specific microorganisms as causative agents of sponge disease highlights the complex interactions that occur within the sponge holobiont. Moving forward, it will be critical for researchers to start exploring links between host immunity, homeostasis of the symbiont community, primary *vs.* opportunistic pathogens, and the environmental pressures that increase disease susceptibility. Importantly, very little is known about the role of viruses in sponge disease and given that computational tools now make sponge viral community analysis tractable, sponge viromics should also become a priority area for the field of sponge disease.

5.5 Summary

- Only three studies have successfully identified the causative agent of disease in sponges, demonstrating the difficulty in assigning aetiology.
- Dysbiosis is commonly linked with disease in sponges e.g. shifts in the symbiotic microbiome give way to opportunistic bacteria/pathogens.
- Sponge mortalities are often correlated with elevated seawater temperatures.

Recommended further reading

Vanwonterghem, I. and Webster, N.S. 2020. Coral reef microorganisms in a changing climate. *iScience* 23: 100972 (Review on environmental driven dysbiosis in coral and sponge microbiomes)

Slaby, B.M., Franke, A., Rix, L., Pita, L., Bayer, K., Jahn, M.T., and Hentschel, U. 2019. Marine sponge holobionts in health and disease. In *Symbiotic Microbiomes of Coral Reefs Sponges and Corals*. Z. Li (ed.) p. 81: Netherlands: Springer (Review of the sponge holobiont including diseases)

Pita, L., Rix, L., Slaby, B.M., Franke, A., and Hentschel, U. 2018. The sponge holobiont in a changing ocean: From microbes to ecosystems *Microbiome* 6: 46 (Review on sponge-microbe symbiosis including links between environmental perturbations, dysbiosis and sponge diseases)

Luter, H.M. and Webster, N.S. 2017. Sponge disease and climate change. In *Climate Change, Ocean Acidification and Sponges Impacts Across Multiple Levels of Organization* J.L. Carballo and J. Bell (eds.), p. 525, Netherlands: Springer (Review of sponge diseases with a focus on the impacts of climate change)

Webster, N.S. 2007. Sponge disease: A global threat? *Environmental Microbiology* 9: 1363–1375 (Original review of sponge diseases)

References

Angermeier, H., Glockner, V., Pawlik, J.R., Lindquist, N.L., and Hentschel, U. 2012. Sponge white patch disease affecting the Caribbean sponge *Amphimedon compressa*. *Diseases of Aquatic Organisms* 99: 95–102

Angermeier, H., Kamke, J., Abdelmohsen, U.R., Krohne, G., Pawlik, J. R., Lindquist, N.L., and Hentschel, U. 2011. The pathology of sponge orange band disease affecting the Caribbean barrel sponge *Xestospongia muta*. *FEMS Microbiology Ecology* 75: 218–230

Belikov, S., Belkova, N., Butina, T., Chernogor, L., Martynova-Van Kley, A., Nalian, A., Rorex, C., Khanaev, I., Maikova, O., and Feranchuk, S. 2019. Diversity and shifts of the bacterial community associated with Baikal sponge mass mortalities. *PLoS One* 14: e0213926

Bell, J.J. 2008. The functional roles of marine sponges. *Estuarine, Coastal and Shelf Science* 79: 341–353

Blanquer, A., Uriz, M.J., Cebrian, E., and Galand, P.E. 2016. Snapshot of a bacterial microbiome shift during the early symptoms of a massive sponge die-off in the western Mediterranean. *Frontiers in Microbiology* 7: 752

Cebrian, E., Uriz, M.J., Garrabou, J., and Ballesteros, E. 2011. Sponge mass mortalities in a warming Mediterranean Sea: Are cyanobacteria-harboring species worse off? *PLoS One*, 6: e20211.

Cerrano, C., Magnino, G., Sarà, A., Bavestrello, G., and Gaino, E. 2001. Necrosis in a population of *Petrosia ficiformis* (Porifera, Demospongiae) in relation with environmental stress. *Italian Journal of Zoology* 68: 131–136

Cervino, J.M., Winiarski-Cervino, K., Polson, S.W., Goreau, T., and Smith, G.W. 2006. Identification of bacteria associated with a disease affecting the marine sponge *Ianthella basta* in New Britain, Papua New Guinea. *Marine Ecology Progress Series* 324: 139–150

Choudhury, J.D., Pramanik, A., Webster, N.S., Llewellyn, L.E., Gachhui, R., and Mukherjee, J. 2015. The pathogen of the Great Barrier Reef sponge *Rhopaloeides odorabile* is a new strain of *Pseudoalteromonas agarivorans* containing abundant and diverse virulence-related genes. *Marine Biotechnology* 17: 463–478

Corriero, G., Scalera-Liaci, L., and Rizzello, R. 1996. Osservazioni sulla mortalita di *Ircinia spinosula* (Schmidt) and *Ircinia* sp. (Porifera, Demospongiae) nell/ insenatura della Strea di Porto Cesareo. *Thalassia Salent* 22: 51–62

Cowart, J.D., Henkel, T.P., McMurray, S.E., and Pawlik, J.R. 2006. Sponge orange band (SOB): A pathogenic-like condition of the giant barrel sponge, *Xestospongia muta*. *Coral Reefs* 25: 513–513

De Goeij, J.M., De Kluijver, A., Van Duyl, F.C., Vacelet, J., Wijffels, R.H., De Goeij, A.F., Cleutjens, J.P., and Schutte, B. 2009. Cell kinetics of the marine sponge *Halisarca caerulea* reveal rapid cell turnover and shedding. *Journal of Experimental Biology* 212: 3892–3900

De Goeij, J.M., Van Oevelen, D., Vermeij, M.J., Osinga, R., Middelburg, J.J., De Goeij, A.F., and Admiraal, W. 2013. Surviving in a marine desert: The sponge loop retains resources within coral reefs. *Science N.Y.* 342: 108–110

Degnan, S.M. 2015. The surprisingly complex immune gene repertoire of a simple sponge, exemplified by the NLR genes: A capacity for specificity? *Developmental & Comparative Immunology* 48: 269–274

Deignan, L.K. and Pawlik, J.R. 2016. Demographics of the Caribbean brown tube sponge *Agelas tubulata* on Conch Reef, Florida Keys, and a description of *Agelas* wasting syndrome (AWS). Proceedings of the 13th International Coral Reef Symposium, 2016 Honolulu, HI, pp. 72–84

Deignan, L.K., Pawlik, J.R., and Erwin, P.M. 2018. *Agelas* wasting syndrome alters prokaryotic symbiont communities of the Caribbean brown tube sponge, *Agelas tubulata*. *Microbial Ecology* 76: 459–466

Denikina, N.N., Dzyuba, E.V., Bel'kova, N.L., Khanaev, I.V., Feranchuk, S.I., Makarov, M.M., Granin, N.G., and Belikov, S.I. 2016. The first case of disease of the sponge *Lubomirskia baicalensis*: Investigation of its microbiome. *Biology Bulletin* 43: 263–270

Di Camillo, C.G., Bartolucci, I., Cerrano, C., and Bavestrello, G. 2012. Sponge disease in the Adriatic Sea. *Marine Ecology* 34: 62–71

Economou, E. and Konteatis, D. 1988. Information on the sponge disease of 1986 in the waters of Cyprus. *Report of Department of Fisheries, Ministry of Agriculture and Natural Resources*. Cyprus: Ministry of Agriculture and Natural Resources, Republic of Cyprus

Engelberts, J.P., Robbins, S.J., De Goeij, J.M., Aranda, M., Bell, S.C., and Webster, N.S. 2020. Characterization of a sponge microbiome using an integrative genome-centric approach. *The ISME Journal* 14: 1100–1110

Ereskovsky, A., Ozerov, DA., Pantyulin, A.N., and Tzetlin, A.B. 2019. Mass mortality event of White Sea sponges as the result of high temperature in summer 2018. *Polar Biology* 42: 2313–2318

Fan, L., Liu, M., Simister, R., Webster, N.S., and Thomas, T. 2013. Marine microbial symbiosis heats up: The phylogenetic and functional response of a sponge holobiont to thermal stress. *ISME Journal* 7: 991–1002

Fan, L., Reynolds, D., Liu, M., Stark, M., Kjelleberg, S., Webster, N.S., and Thomas, T. 2012. Functional equivalence and evolutionary convergence in complex communities of microbial sponge symbionts. *Proceedings of the National Academy of Science, USA* 109: E1878–E1887

Feuda, R., Dohrmann, M., Pett, W., Philippe, H., Rota-Stabelli, O., Lartillot, N., Worheide, G., and Pisani, D. 2017. Improved modeling of compositional heterogeneity supports sponges as sister to all other animals. *Current Biology* 27: 3864–3870

Gaino, E., Pronzato, R., Corriero, G., and Buffa, P. 1992. Mortality of commercial sponges: Incidence in two Mediterranean areas. *Bolletino di Zoologia* 59: 79–85

Galstoff, P.S. 1942. Wasting disease causing mortality of sponges in the West Indies and Gulf of Mexico. *Proceedings of the VIII American Science Congress*. Department of State, Washington.

Galstoff, P.S., Brown, H.H., Smith, C.L., and Smith, F.G.W. 1939. Sponge mortality in the Bahamas. *Nature* 143: 807–808

Gammill, E.R. and Fenner, D. 2005. Disease threatens Caribbean sponges: Report and identification guide. *ReefBase Online Library*. http://www.reefbase.org/resource_center/publication/main.aspx?refid=24912

Gao, Z.M., Wang, Y., Tian, R.M., Lee, O.O., Wong, Y.H., Batang, Z.B., Al-Suwailem, A., Lafi, F.F., Bajic, V.B., and Qian, P.Y. 2015. Pyrosequencing revealed shifts of prokaryotic communities between healthy and disease-like tissues of the Red Sea sponge *Crella cyathophora*. *PeerJ* 3: e890

Gashout, S.F., Haddud, D.A., El-Zintani, A.A., and Elbare, R.M.A. 1989. Evidence for infection of Libyan sponge grounds. In International seminar on the combat of pollution and the conservation of marine wealth in the Mediterranean Sea. Gulf of Sirte, Marine Biological Resources Centre, pp. 100–113

Gochfeld, D.J., Schlöder, C., and Thacker, R.W. 2007. Sponge community structure and disease prevalence on coral reefs in Bocas del Toro, Panama. In *Porifera Research Biodiversity, Innovation and Sustainability*. Custódio, M.R., Lôbo-Hajdu, G., Hajdu, E., and Muricy, G. (eds.), pp. 335–343, Rio de Janeiro: Museu Nacional

Greco, G., Di Piazza, S., Gallus, L., Amaroli, A., Pozzolini, M., Ferrando, S., Bertolino, M., Scarfi, S., and Zotti, M. 2019. First identification of a fatal fungal infection of the marine sponge *Chondrosia reniformis* by *Aspergillus tubingensis*. *Diseases of Aquatic Organisms* 135: 227–239

Hentschel, U., Piel, J., Degnan, S.M., and Taylor, M.W. 2012. Genomic insights into the marine sponge microbiome. *Nature Reviews Microbiology* 10: 641–654

Kaluzhnaya, O.V. and Itskovich, V.B. 2015. Bleaching of Baikalian sponge affects the taxonomic composition of symbiotic microorganisms. *Russian Journal of Genetics* 51: 1153–1157

Kulakova, N.V., Sakirko, M.V., Adelshin, R.V., Khanaev, I.V., Nebesnykh, I.A., and Pérez, T. 2018. Brown rot syndrome and changes in the bacterial community of the Baikal sponge. *Lubomirskia baicalensis*. *Microbial Ecology* 75: 1024–1034

Laffy, P.W., Botte, E.S., Wood-Charlson, E.M., Weynberg, K.D., Rattei, T., and Webster, N.S. 2019. Thermal stress modifies the marine sponge virome. *Environmental Microbiology Reports* 11: 690–698

Laffy, P.W., Wood-Charlson, E.M., Turaev, D., Jutz, S., Pascelli, C., Botté, E.S., Bell, S.C., Peirce, T.E., Weynberg, K.D., Van Oppen, M.J.H., Rattei, T., and Webster, N.S. 2018. Reef invertebrate viromics: Diversity, host specificity and functional capacity. *Environmental Microbiology* 20: 2125–2141

Lemoine, N., Buell, N., Hill, A., and Hill, M. 2007. Assessing the utility of sponge microbial symbiont communities as models to study global climate change: A case study with *Halichondria bowerbanki*. In *Porifera Research: Biodiversity, Innovation and Sustainability*. Custódio, M.R., Lôbo-Hajdu, G., Hajdu, E., and Muricy, G. (eds.), pp. 419–425, Rio de Janeiro: Museu Nacional

Leys, S.P., Nichols, S.A., and Adams, E.D. 2009. Epithelia and integration in sponges. *Integrative and Comparative Biology* 49: 167–177

Li, C.-W., Chen, J.-Y., and Hua, T.-E. 1998. Precambrian sponges with cellular structures. *Science N.Y.* 279: 879–882

López-Legentil, S., Erwin, P.M., Pawlik, J.R., and Song, B. 2010. Effects of sponge bleaching on ammonia-oxidizing Archaea: Distribution and relative expression of ammonia monooxygenase genes associated with the barrel sponge *Xestospongia muta*. *Microbial Ecology* 60: 561–571

Luter, H.M. and Webster, N.S. 2017. Sponge disease and climate change. In *Climate Change, Ocean Acidification and Sponges Impacts Across Multiple Levels of Organization*. Carballo, J.L. and Bell, J. (eds.) The Netherlands: Springer

Luter, H.M., Bannister, R.J., Whalan, S., Kutti, T., Pineda, M.C., and Webster, N.S. 2017. Microbiome analysis of a disease affecting the deep-sea sponge *Geodia barretti*. *FEMS Microbiol Ecology* 93: fix074

Luter, H.M., Whalan, S., and Webster, N.S. 2010a. Prevalence of tissue necrosis and brown spot lesions in a common marine sponge. *Marine and Freshwater Research* 61: 484–489

Luter, H.M., Whalan, S., and Webster, N.S. 2010b. Exploring the role of microorganisms in the disease-like syndrome affecting the sponge *Ianthella basta*. *Applied and Environmental Microbiology* 76: 5736–5744

Maldonado, M., Aguilar, R., Bannister, R.J., Bell, J.J., Conway, K.W., Dayton, P.K., Díaz, C., Gutt, J., Kelly, M., Kenchington, E.L.R., Leys, S.P., Pomponi, S.A., Rapp, H.T., Rützler, K., Tendal, O.S., Vacelet, J., and Young, C.M. 2015. Sponge grounds as key marine habitats: A synthetic review of types, structure, functional roles, and conservation concerns. In *Marine Animal Forests: The Ecology of Benthic Biodiversity Hotspots*. Rossi, S., Bramanti, L., Gori, A., and Orejas Saco Del Valle, C. (eds.) pp. 1–13. Cham: Springer International Publishing

Maldonado, M., Sánchez-Tocino, L., and Navarro, C. 2010. Recurrent disease outbreaks in corneous demosponges of the genus *Ircinia*: Epidemic incidence and defense mechanisms. *Marine Biology* 157: 1577–1590

Mukherjee, J., Webster, N.S., and Llewellyn, L.E. 2009. Purification and characterization of a collagenolytic enzyme from a pathogen of the Great Barrier Reef sponge, *Rhopaloeides odorabile*. *PLoS One* 4: e7177

Müller, W.E.G. 2003. The origin of Metazoan complexity: Porifera as integrated animals. *Integrative and Comparative Biology* 43: 3–10

Müller, W.E.G. 2006. The stem cell concept in sponges (Porifera): Metazoan traits. *Seminars in Cell and Developmental Biology* 17: 481–491

Nagelkerken, I., Aerts, L., and Pors, L. 2000. Barrel sponge bows out. *Reef encounter*. Kansas: International Society for Reef Studies

Olson, J.B., Gochfeld, D.J., and Slattery, M. 2006. Aplysina red band syndrome: A new threat to Caribbean sponges. *Diseases of Aquatic Organisms* 71: 163–168

Olson, J.B., Thacker, R.W., and Gochfeld, D.J. 2014. Molecular community profiling reveals impacts of time, space, and disease status on the bacterial community associated with the Caribbean sponge *Aplysina cauliformis*. *FEMS Microbiology Ecology* 87: 268–279

Paz, M. 1997. New killer disease attacks giant barrel sponge. *San Pedro Sun*, 7 March. https://www.sanpedrosun.com/old/sponge.html

Piel, J., Hui, D., Wen, G., Butzke, D., Platzer, M., Fusetani, N., and Matsunaga, S. 2004. Antitumor polyketide biosynthesis by an uncultivated bacterial symbiont of the marine sponge *Theonella swinhoei*. *Proceedings of the National Academy of Sciences of the United States of America* 101: 16222–16227

Ramsby, B.D., Hoogenboom, M.O., Whalan, S., and Webster, N.S. 2018. Elevated seawater temperature disrupts the microbiome of an ecologically important bioeroding sponge. *Molecular Ecology* 27: 2124–2137

Rizzello, R., Corriero, G., and Scalera-Liaci, L. 1997. Extinction and recolonization of *Spongia officinalis* in the Marsala Lagoon. *Biolgia Marina Mediterranea* 4: 443–444

Rützler, K. 1988. Mangrove sponge disease induced by cyanobacterial symbionts: Failure of a primitive immune system? *Diseases of Aquatic Organisms* 5: 143–149

Schmidt, E.W., Obraztsova, A.Y., Davidson, SK., Faulkner, D.J., and Haygood, M.G. 2000. Identification of the antifungal peptide-containing symbiont of the marine sponge *Theonella swinhoei* as a novel δ-proteobacterium, 'Candidatus Entotheonella palauensis'. *Marine Biology* 136: 969–977

Simpson, T.L. 1984. *The Cell Biology of Sponges*. New York, NY, Springer-Verlag New York Inc.

Smith, F.G.W. 1939. Sponge mortality at British Honduras. *Nature* 144: 785–785

Smith, F.G.W. 1941. Sponge disease in British Honduras, and its transmission by water currents. *Ecology* 22: 415–421

Srivastava, M., Simakov, O., Chapman, J. et al. 2010. The *Amphimedon queenslandica* genome and the evolution of animal complexity. *Nature* 466: 720–726

Storr, J.F. 1964. Ecology of the Gulf of Mexico commercial sponges and its relation to the fishery. *Special Scientific Report*. Washington: US Fisheries and Wildlife Service

Sweet, M., Bulling, M., and Cerrano, C. 2015. A novel sponge disease caused by a consortium of microorganisms. *Coral Reefs* 34: 871–883

Thomas, T., Moitinho-Silva, L., Lurgi, M. et al. 2016. Diversity, structure and convergent evolution of the global sponge microbiome. *Nature Communications* 7: 1–12

Vacelet, J. 1994. Control of the severe sponge epidemic – Near East and Europe: Algeria, Cyprus, Egypt, Lebanon, Malta, Morocco, Syria, Tunisia, Turkey, Yugoslavia. *Technical Report: The struggle against the epidemic which is decimating Mediterranean sponges* FI: TCP/RAB/8853 FAO, Rome, pp. 39

Vacelet, J. and Donadey, C. 1977. Electron microscope study of the association between some sponges and bacteria. *Journal of Experimental Marine Biology and Ecology* 30: 301–314

Vacelet, J., Vacelet, E., Gaino, E., and Gallissian, M.F. 1994. Bacterial attack of spongin skeleton during the 1986–1990 Mediterranean sponge disease. In Van Soest,

R.W.M., Van Kempen, T.M.G. and Braeckan, J.C. (eds.) *Sponges in Time and Space*, pp. 355–362. Rotterdam, the Netherlands: A.A. Balkema

Van Soest, R.W.M., Boury-Esnault, N., Hooper, J.N.A. et al. 2020. World Porifera database. http://www.marinespecies.org/porifera/index.php

Vicente, V.P. 1990. Response of sponges with autotrophic endosymbionts during the coral-bleaching episode in Puerto Rico. *Coral Reefs* 8: 199–202

Webster, N.S., Negri, A.P., Webb, R.I., and Hill, R.T. 2002. A spongin-boring α-proteobacterium is the etiological agent of disease in the Great Barrier Reef sponge *Rhopaloeides odorabile*. *Marine Ecology Progress Series* 232: 305–309

Webster, N.S., Cobb, R.E., and Negri, A.P. 2008a. Temperature thresholds for bacterial symbiosis with a sponge. *The ISME Journal* 2: 830–842

Webster, N.S., Xavier, J.R., Freckelton, M., Motti, C.A., and Cobb, R. 2008b. Shifts in microbial and chemical patterns within the marine sponge *Aplysina aerophoba* during a disease outbreak. *Environmental Microbiology* 10: 3366–3376

Webster, N.S. and Taylor, M.W. 2012. Marine sponges and their microbial symbionts: Love and other relationships. *Environmental Microbiology* 14: 335–346

Whalan, S. 2018. Not just corals – sponges are bleaching too! *Frontiers in Ecology and the Environment* 16: 471–471

Wilkinson, C. 1983. Net primary productivity in coral reef sponges. *Science N.Y.* 219: 410–412.

Wilkinson, C.R. 1978a. Microbial associations in sponges. III. Ultrastructure of the *in situ* associations in coral reef sponges. *Marine Biology* 49: 177–185

Wilkinson, C.R. 1978b. Microbial associations in sponges. II. Numerical analysis of sponge and water bacterial populations. *Marine Biology* 49: 169–176

Wulff, J.L. 2006a. Rapid diversity and abundance decline in a Caribbean coral reef sponge community. *Biological Conservation* 127: 167–176

Wulff, J.L. 2006b. A simple model of growth form-dependent recovery from disease in coral reef sponges, and implications for monitoring. *Coral Reefs* 25: 419–426

Wulff, J.L. 2007. Disease prevalence and population density over time in three common Caribbean coral reef sponge species. *Journal of the Marine Biological Association of the UK* 87: 1715–1720

CHAPTER 6

Diseases of platyhelminths, acanthocephalans, and nematodes

Matt Longshaw and Andrew P. Shinn

6.1 Introduction

Parasites, commensals, and symbionts are, arguably, amongst the most successful forms of life on the planet. It should therefore come as no surprise when they themselves become parasitised by their compatriots and by other infectious agents. This was best summed up in Augustus De Morgan's 1872 poem *A Budget of Paradoxes* 'Great fleas have little fleas upon their backs to bite 'em, And little fleas have lesser fleas, and so *ad infinitum*'. In this chapter, the classes Monogenea, Trematoda, Cestoda, and Acanthocephala are wholly parasitic forms that occur in a range of vertebrate and invertebrate hosts whilst the phylum Nematoda and the class Turbellaria contain a number of parasitic and free-living species—each in turn are infected, to a greater or lesser extent by a range of agents. The bulk of these infections appear to be benign, with a smaller proportion being considered pathogenic. However, caution should be exercised with this view, since many of the publications do not explicitly set out to assess pathogenicity, rather the bulk of papers published appear to be of a faunistic nature, recording the presence of, and descriptions therein, of new and novel infections.

Within the chapter, we have tried, as far as possible, to verify and update where necessary the taxonomy of hosts and pathogens but recognise that in some cases this is in a state of flux. Accordingly, the reader should verify for themselves the current taxonomic classifications.

Finally, although we have tried to provide all known parasite and pathogen records for the taxa in the current chapter, it is likely, given the disparate nature of the literature, that we have missed some publications. However, we are, however content that we have been able to cover the bulk of the records and indeed the bulk of the different types of infections.

6.2 Phylum Platyhelminthes

6.2.1 Class Turbellaria

The class Turbellaria contains both free living and parasitic/commensal forms, with around 4500 species described worldwide from marine, freshwater and terrestrial environments. They typically reproduce sexually and are hermaphroditic with some species able to clone themselves or reproduce by budding. Although a wide range of infectious agents have been reported in turbellarians, including viruses, bacteria, apicomplexans, kinetoplastids, labyrinthulomycetes, haplosporidians, flagellates, ciliates, algae, orthonectids, dinoflagellates, digeneans, turbellarians, rotifers, nematodes, and copepods (see Table 6.1), over half of the infections reported are due to apicomplexans (mainly *Monocystella* spp.) and ciliates. Despite the wide diversity of infectious agents recorded, the number of pathogenic species is limited.

Matt Longshaw and Andrew P. Shinn, *Diseases of platyhelminths, acanthocephalans, and nematodes*. In: *Invertebrate Pathology*. Edited by Andrew F. Rowley, Christopher J. Coates and Miranda M.A. Whitten, Oxford University Press. © Oxford University Press (2022). DOI: 10.1093/oso/9780198853756.003.0006

Of the nine virus and virus-like infections reported from turbellarians, two are noted as pathogenic to their host. Reuter (1975) described the presence of a putative intranuclear Adenovirus in the brackish water *Gyratrix hermaphroditus* caught in Sweden. Virus particles measured 70 nm in diameter and were noted in a crystalline array in the nucleus and cytoplasm of the proboscis gland cells of the host. In heavily infected cells, nuclei are swollen, filled with viral particles that exclude other nuclear structures that show as intranuclear inclusion bodies. Additionally, there is a loss of endoplasmic reticulum and an accumulation of concentric membrane systems in infected cells. No data exists on whether the infection is terminal. A second pathogenic virus-like infection was noted in *Urastoma cyprinae*, a commensal of *Mytilus galloprovincialis* collected in Spain (Crespo-Gonzalez et al. 2008). Virus-like particles, arranged in paracrystalline arrays, were noted in the cytoplasm of the rhabdoid cells of the glandular system of the host. Infected cells contained several intracytoplasmic membranous vesicles which were considered as cytopathological by the authors. Virus particles were isometric and measured 24–30 nm in diameter.

Most bacterial infections in turbellarians have not been properly identified although most do not cause any apparent disease and could be considered incidental. Laboratory reared *Archaphanostoma agile* are susceptible to an unnamed, lethal, bacterial infection (Apelt 1969). Infected individuals left a trail of mucus behind them and changed to a purple-red colour before succumbing to the bacterial infection within 24 hours. Treatment of the culture water with p-amino-benzene-sulphonacetamide prevented disease outbreaks. A pathogenic bacterial infection of *Pterastericola fedotovi* commensal in *Pteraster militaris* from the White Sea coast of Russia was described by Karling (1970). The ovary of infected individuals was completely destroyed. The author considered that the infection was likely to have originated in the surrounding gut content of the starfish hyper-host.

Using a range of methods, including live observations, *in vitro* cultivation, transmission electron microscopy (TEM), and molecular biology, Schärer et al. (2007) described the thraustochytrid protist *Thraustochytrium caudivorum* as a pathogen of *Macrostomum lignano*. The protist actively invades host tissues and causes dissolution of the tail-plate and in extreme cases, dissolution of the entire worm. Eradication of the infection in the culture was achieved by bathing turbellarian eggs in the non-ionic surfactant Triton.

Infections with the ciliate *Tetrahymena* spp. can be lethal. The freshwater triclads *Crenobia alpina* and *Polycelis felina* infected with *Tetrahymena pyriformis* and *T. corlissi* become inactive and ultimately die as a result of infections (Wright 1981). Following death, the host tissues are consumed by the ciliates within 24 hours. Route of infection was considered to be via ingestion of free-swimming ciliates or through ingestion of infected food. Once in the gut, the ciliate appears to traverse the gut wall and invades the mesenchyme. *Dugesia gonocephala* infected by *Tetrahymena dugesiae* had open wounds which were considered by Rataj and Vd'ačný (2020) to be the likely route of entry by the ciliate, which was found within the mesenchyme. Infections by the peritrichous ciliate *Urceolaria mitra* are not normally noted as pathogenic to its turbellarian host. Bowen and Ryder (1994) reported an infection of *Polycelis tenuis* collected from a freshwater pond in Wales. Evidence of epithelial disturbance and erosion was noted, and where the parasite was attached to the ciliated surfaces of the planarian, it was suggested that host mobility may be impaired. Free-living *Stylochus* sp. from Australia exhibit black patches on the body surface when infected with the haplosporidian *Urosprodium cannoni* (Anderson et al. 1993).

An unusual infection of *Plagiostomum* sp. by the parasitic turbellarian *Oekiocolax plagiostomorum* collected in Greenland was noted by Reisinger (1929). Infected individuals were sluggish, with convex projections on the body surface, and with yellowish oval egg capsules of the parasite within the parenchyma. The parasite led to partial castration of its host by destroying the ovaries although testes and vitellaria appeared to be unaffected.

Complete suppression of gonadal development in female *Childia* (= *Paraphanostoma*) *macroposthium* and *C.* (= *P.*) *brachyposthium* by the orthonectid mesozoan *Rhopalura paraphanostomae* was noted by Westblad (1942). No effect by the orthonectid was noted in males of either host species. Similarly,

Table 6.1 List of turbellarian hosts and their infectious agents. Where A = Apicomplexa, Al = Algae, B = Bacteria, C = Ciliophora, Ces = Cestoda, Cop = Copepoda, D = Digenea, Dino = Dinophyceae, F = Flagellata, H = Haplosporidia, K = Kinetoplastida, L = Labyrinthulomycetes, N = Nematoda, O = Orthonectida, R = Rotifera, T = Turbellaria, V = Virus. Those recorded as pathogenic in the literature are highlighted in bold. For habitat FW = Freshwater, SW = seawater

Host	Parasite/pathogen	Habitat. Geographical location	Reference
Amaga (= *Geoplana*) *amagensis*	*Monocystella geoplanae* (A)	FW. USA	Desportes and Schrével (2013)
Amaga (= *Geoplana*) *becki*	*Monocystella geoplanae* (A)	FW. USA	Desportes and Schrével (2013)
Amaga bussoni (= *Geoplana olivacea*)	*Monocystella planariae* (A)	FW. Europe	Desportes and Schrével (2013)
Archaphanostoma agile	**Unknown (B)**	SW. Europe	Apelt (1969)
Bdellocephala punctata	Echinostomatid (D)	FW. Europe	Westblad (1942)
Caenoplana (= *Geoplana*) *steenstrupi*	*Monocystella planariae* (A)	FW. Europe	Desportes and Schrével (2013)
Childia (= *Paraphanostoma*) *brachyposthium*	***Rhopalura paraphanostomae* (O)**	SW. Europe	Westblad (1942)
Childia (= *Paraphanostoma*) *crissum*	Unknown (T)	SW. Europe	Westblad (1942)
Childia (= *Paraphanostoma*) *macroposthium*	***Rhopalura paraphanostomae* (O)**	SW. Europe	Westblad (1942)
Childia sp.	*Lepocreadium setiferoides* (D)	SW. USA	Curtis (2007)
Convoluta convoluta (= *paradoxa*)	*Chlorella* sp. (Al)		Keeble (1908)
Crenobia alpine	***Tetrahymena corlissi* (C), *Tetrahymena pyriformis* (C)**	FW. Europe	Wright (1981)
Cura foremanii	*Urceolaria mitra* (C)	FW. USA	Ball and Fernando (1968)
Cycloporus maculatus	*Monocystella* (= *Lankesteria*) *cyclopori* (A)	SW. Europe	Desportes and Schrével (2013)
Cylindrostoma fingalianum	Unknown (K)	SW. Sweden	Watson and Jondelius (1997)
Dendrocoelopsis piriformis	*Monocystella* sp. (A)	FW. USA	Carpenter (1982); Desportes and Schrével (2013)
Dendrocoelum (= *Dendrocoeles*) spp.	*Monocystella neodendrocoelum* (A)	FW. Europe	Desportes and Schrével (2013)
Dendrocoelum lacteum	*Cryptobia dendrocoeli* (K), *Monocystella planariae* (A), *Ophryoglena parasitica* (C), *Urceolaria mitra* (C)	FW. USA, Europe	Kahl (1931); Ball and Fernando (1968); Desportes and Schrével (2013); Stocchino et al. (2017)
Dendrocoelum obstinatum	Unknown (N)	FW. Europe	Stocchino et al. (2017)
Dendrocoelum sanctinaumae (= *Neodendrocoelum sanctinaumi*)	*Monocystella neodendrocoeli* (A)	FW. Europe	Desportes and Schrével (2013)
Dendrocoelum tismanae	*Monocystella spelaea* (A)	FW. Romania	Desportes and Schrével (2013)
Dugesia arabica	*Monocystella* sp. (A)	FW. Yemen	Harrath et al. (2013)

continued

Table 6.1 *Continued*

Host	Parasite/pathogen	Habitat. Geographical location	Reference
Dugesia gonocephala	*Haptophrya planariarum* (C), **Tetrahymena dugesiae (C)**, *Tetrahymena* sp. (C), *Urceolaria mitra* (C)	FW. Slovakia	Rataj and Vd'ačný (2019)
Dugesia japonica	Intranuclear crystalline structures (V)	FW. Japan	Kishida and Asai (1977)
Euplana gracilis	*Cercaria parvicaudata* (D), *Lepocreadium setiferoides* (D)	SW. USA	Curtis (2007); Stunkard (1950)
Fonticola macedonica	*Monocystella arndti* (A)	FW. Europe	Desportes and Schrével (2013)
Fonticola ochridana	*Monocystella arndti* (A)	FW. Europe	Desportes and Schrével (2013)
Fonticola sp.	*Urceolaria mitra* (C)	FW. USA	Ball and Fernando (1968)
Girardia tigrina	Extrachromosomal VLP (V), *Tetrahymena acanthophora* (C), *Tetrahymena nigricans* (C), **Tetrahymena sp. (C)**, *Urceolaria mitra* (C)	FW. Russia, Slovakia, USA	Ball and Fernando (1968); Rebrikov et al. (2002a b); Rataj and Vd'ačný (2020)
Graffiellus croceus (= *Macrorhynchus crocea*)	*Rhopalura variabeli* (O)	SW. Russia	Alexandrov and Sljusarev (1992)
Gyratrix hermaphroditus	**Adenovirus (V)**, Unknown (K)	SW. Sweden, Finland	Reuter (1975); Watson and Jondelius (1997)
Heterochaerus (= *Amphiscolops*) *langerhansi*	*Amphidinium klebsii* (Dino)	SW. USA	Taylor (1971)
Invenusta sp. or *Vannuccia* sp.	**Ciliocincta akkeshiensis (O)**	SW. Japan	Tajika (1979)
Kaburakia excelsa	*Pseudanthessius latus* (Cop)	SW. USA	Illg (1950)
Leptoplana sp.	*Monocystella* (= *Leidyana*) *leptoplanae* (A)	SW. UK	Bhatia and Setna (1924)
Leptoplana tremellaris	*Rhopalura leptoplanae* (O)	SW. Japan	Tajika (1979)
Macrostomum lignano	**Thraustochytrium caudivorum (L)**	SW. Europe	Schärer et al. (2007)
Meara sp.	Unknown (B)	SW. Norway, Sweden	Lundin (1998)
Meara stichopi	Unknown (B)	SW. Norway, Sweden	Lundin (1998)
Microstomum lineare	Echinostomatid (D)	SW. Europe	Westblad (1942)
Nemertoderma westbladi	Unknown (B)	SW. Norway, Sweden	Lundin (1998)
Neodendrocoelum suetinaumi	*Monocystella compacta* (A)	FW. Europe	Desportes and Schrével (2013)
Niobe zonata (= *Planaria limacina*)	*Lachmannella recurva* (C)	SW. Norway	Cepede (1910)
Notodactylus handschini	Cyanobacteria (B), Sessile rotifers (R), Stalked (C), Unknown (B)	FW. Australia	Jennings et al. (1992)
Otomesostoma auditivum	*Eucoccidium monoti* (A)	FW. Germany	Reisinger (1929)
Paracatenula sp.	Unknown (B)	SW. USA	Ott et al. (1982)
Paravortex cardii	*Urosporidium* sp. (H)	SW. Spain	Carballal et al. (2005)

continued

Table 6.1 *Continued*

Host	Parasite/pathogen	Habitat. Geographical location	Reference
Paravortex tapetis	Adenovirus (V), *Urosporidium* sp. (H)	SW. France	Noury-Shaïri et al. (1995); Carballal et al. (2005)
Phagocata (= *Planaria*) *albissima*	*Monocystella arndti* (A)	FW. Romania	Desportes and Schrével (2013)
Phagocata vitta	*Tetrahymena* sp. (C), Unknown (N)	FW. UK	Armitage and Young (1990); Rataj and Vd'ačný (2019)
Phrikoceros (= *Tytthosoceros*) *lizardensis*	*Pseudanthessius newmanae* (Cop)	SW. Australia	Humes (1997)
Plagiostomum lemani	*Monocystella neodendrocoeli* (A), *Monocystella plagiostomae* (A)	FW. Yugoslavia	Desportes and Schrével (2013)
Plagiostomum sp.	**Oekiocolax plagiostomorum (T)**	SW. Greenland	Reisinger (1929)
Planaria occulta	*Haptophrya* (= *Sieboldiellina*) *planariarum* (C)	FW. USA	Kenk (1969)
Planaria sp.	*Monocystella swarczewskyi* (A)	FW. Russia	Desportes and Schrével (2013)
Planaria torva	*Haptophrya planariarum* (C), *Monocystella planariae* (A), *Urceolaria mitra* (C)	FW. USA	Ball and Fernando (1968); Desportes and Schrével (2013)
Planocera sp.	*Peracreadium* sp. (D)	SW. Australia	Prudhoe (1945)
Polycelis felina	**Tetrahymena sp. (C),** *Urceolaria mitra* (C)	FW. USA, Europe	Ball and Fernando (1968); Wright (1981); Rataj and Vd'ačný (2019)
Polycelis nigra	Echinostomatid (D), *Urceolaria mitra* (C)	FW. USA, Europe	Ball and Fernando (1968); Kinne (1980)
Polycelis tenuis	**Urceolaria mitra (C)**	FW. UK	Bowen and Ryder (1994)
Procerodes (= *Gunda*) *segmentata*	*Steinella uncinata* (C)	SW. Poland	Sikora (1963)
Procerodes littoralis (= *ulvae*)	*Steinella uncinata* (C)	SW. Poland	Sikora (1963)
Procerodes lobata	*Steinella uncinata* (C)	SW. Poland	Sikora (1963)
Prorhynchus sp.	*Desmomonas prorhynchi* (F)	FW. Australia	Williams (1999)
Pseudobiceros sp. 1	*Pseudanthessius newmanae* (Cop)	SW. Australia	Humes (1997)
Pseudobiceros sp. 2	*Pseudanthessius newmanae* (Cop)	SW. Australia	Humes (1997)
Pseudoceros (= *Planaria*) *fusca*	*Monocystella planariae* (A),	FW. USA	Ball and Fernando (1968); Kinne (1980)
Pterastericola fedotovi	**Unknown (B)**	SW. Russia	Karling (1970)
Pterastericola vivipara	*Monocystella epibatis* (A)	SW. Australia	Cannon and Jennings (1988)
Rimacephalus (= *Dicotylus*) sp.	*Ophryoglena intestinalis* (C)	FW. Germany	Kahl (1931)
Schmidtea (= *Dugesia*) *lugubris*	*Monocystella planariae* (A), *Urceolaria mitra* (C)	FW. USA	Ball and Fernando (1968); Desportes and Schrével (2013)

continued

Table 6.1 *Continued*

Host	Parasite/pathogen	Habitat. Geographical location	Reference
Schmidtea (= *Dugesia*) *polychroa*	*Monocystella planariae* (A), *Urceolaria mitra* (C), Echinostomatid (D)	FW. USA, Europe	Ball and Fernando (1968); Kinne (1980); Desportes and Schrével (2013)
Sorocoelis sp.	*Monocystella planariae* (A), *Monocystella swarczewskyi* (A)	FW. Russia	Desportes and Schrével (2013)
Sphalloplana percoeca	*Monocystella* sp. (A)	FW. USA	Carpenter (1982)
Sphalloplana percoeca	Unknown (C)	FW. USA	Carpenter (1982)
Stylochus (*Stylochus*) *castaneus*	*Scolex polymorphus* (Ces)	SW. Italy	Kinne (1980)
Stylochus ellipticus	*Lepocreadium setiferoides* (D)	SW. USA	Curtis (2007)
Stylochus sp.	**Urosporidium cannoni (H)**	SW. Australia	Anderson et al. (1993)
Symsagittifera (= *Convoluta*) *roscoffensis*	Adenovirus (V)	SW. France	Oschman (1969)
Temnocephala brevicornis	*Ophiotaenia cohospes* (Ces)	FW. Uruguay	Cordero (1946)
Temnocephala chilensis	*Echinoparyphium megacirrus* (D)	FW. Argentina	Viozzi et al. (2005)
Temnocephala gargantua	Plerocercoid larvae (Ces), unknown (N)	FW. Uruguay	De León and Volonterio (2018)
Temnocephala iheringi	Adenovirus (V)	FW. Uruguay	Justine et al. (1991)
Temnocephala mexicana	Proteocephalid (Ces), *Rhaphidascaris* sp. (N)	FW. Mexico	De León and Volonterio (2018)
Temnocephala talicei	Unknown (N)	FW. Uruguay	De León and Volonterio (2018)
Temnohaswellia (= *Temnocephala*) *minor*	Unknown (B)	FW. Australia	Williams (1994)
Temnohaswellia (= *Temnocephala*) *novaezealandiae*	RLO (B)	FW. Australia	Williams (1991)
Temnosewellia rouxi	Plerocercoid larvae (Ces)	FW. Indonesia	De León and Volonterio (2018)
Temnosewellia semperi	Plerocercoid larvae (Ces), unknown (N)	FW. Sumatra	De León and Volonterio (2018)
Triclads	*Annelophrya sphaeronucleata* (C)	FW. Yugoslavia	Van der Velde (1978)
Troglocaridicola mrazeki	*Spirillum*-like *Rickettsia* sp. (B)	FW. Australia	Williams (1992)
Urastoma cyprinae	**Picornaviridae (V)**, Adenovirus (V), Chlamydia-like organisms (B), Mycoplasma-like organisms (B)	SW. Spain, France	Comps and Tige (1999); Crespo-Gonzalez et al. (2008)
Uteriporus vulgaris (=*Procerodes warreni*)	*Lepocreadium setiferoides* (D)	SW. USA	Curtis (2007)

infections of *Invenusta* sp. by the orthonectid *Ciliocincta akkeshiensis* also destroyed the reproductive organs of the host and transformed the pharynx into 'an accumulation of cells' (Tajika 1979).

6.2.2 Class Monogenea

Monogeneans are obligate parasites, mainly of aquatic vertebrates in marine and freshwater habitats. They reproduce mainly via oviparity with a lesser number of species being viviparous. In excess of 6,000 species and 750 genera have been described (Paladini et al. 2017). Despite the ubiquity of these parasites in aquatic habitats and the large number described, only a few infections have been reported in these hosts. This includes viruses, bacteria, fungi, ciliophorans, flagellates, kinetoplastids, microsporidians, and myxozoans (Table 6.2); of these, only four infections have been reported as pathogenic to their hosts.

An intracellular *Rickettsia*-like organism of *Euzetrema knoepffleri*, endoparasitic in the amphibian Corsican brook salamander, *Euproctus montanus*, led to cytoplasmic changes in the monogenean host. Fournier et al. (1975) suggested that the infection was present in all life stages of the parasite and was transmitted to the progeny through the gametes. *Pseudodiplodorchis americanus*, parasitic in another amphibian, *Scaphiopus couchii*, is hyperparasitised by bacteria and a microsporidian (Cable and Tinsley 1992). Pathologies associated with the microsporidian infection included a reduction in the number of cilia present in ciliated cells, loss of plasma membrane, loss of cell content of infected cells as well as complete loss of locomotory cells. Despite these infections, host invasion did not appear to be impaired (Cable and Tinsley 1992).

Myxozoans typically utilise two hosts in their life cycle, alternating between a vertebrate host such as a fish, and a free-living invertebrate host. A small number, all within the genus *Myxidium*, have been recorded as hyperparasites of monogeneans. Infected hosts showed marked pathological changes including loss of internal structures and parenchymal tissues being filled with myxospores; infections were considered lethal (Freeman and Shinn 2011).

6.2.3 Class Trematoda

6.2.3.1 Subclass Digenea

Digeneans are obligate parasites, utilising up to three hosts in their life cycles. Digenea occur in a wide range of terrestrial and aquatic hosts worldwide with an estimated > 10,000 described species.

A large number of infections have been reported in digeneans across a range of taxa including viruses, bacteria, algae, ciliophorans, haplosporidians, heterokonts, metamonads, flagellates, microsporidians, myxozoans, digeneans, nematodes, and copepods (Table 6.3). Of these, a number of pathogenic species have been recorded, with the bulk of these being microsporidians and haplosporidians.

Pathogenic *Urosporidium* spp. infections typically lead to a colour change of the infected digeneans (Howell 1967; Ormières et al. 1973; Cho et al. 2020). Furthermore, infections can be lethal, leading to reduced motility or castration (Zaika and Dolgikh 1963; Howell 1967; Kinne 1980). Opacity of the body, loss and replacement of host issues, and reduced fecundity, followed by death are the normal outcome of *Nosema* spp. and *Unikaryon* spp. infections of digeneans (Cort et al. 1960; Canning et al. 1974, 1983; Canning and Madhavi 1977; Canning and Olson Jr. 1980; Levron et al. 2005). These infections normally occur within the tissues of the digenean. In contrast, the microsporidian *Ovipleistophora dilpostomuri* of Lovy and Friend (2017) and the *Pleistophora* sp. of Paperna et al. (1978) infections appear restricted to the metacercarial cyst wall where they lead to hypertrophy of these tissues, followed by degeneration and ultimately death of the digenean host.

The myxozoan *Fabespora vermicola* is a pathogen of *Crassicutis archosargi*. Egg production in infected worms ceased as a result of infections, which obstructed reproductive ducts, destroyed the Mehlis' gland and led to necrosis of the gonads (Overstreet 1976). Plasmodia of *Myxobilatus paragasterostei* attached to the tegument of *Phyllodistomum folium* elicits a range of tissue changes including degeneration of parenchyma structures, a reduction in glycogen and a reduction in egg production along with mortality of the host (Dugarov et al. 2011). A *Hexamita* sp. infection of the (mainly female) reproductive organs of *Deropristis inflata* has been noted, with a particular focus in the eggs. Infected eggs have little to no protoplasm with maturation ceasing as a result (Hunninen and Wichterman 1938).

Table 6.2 List of monogenean hosts and their infectious agents along with their hyperhosts. Where B = Bacteria, C = Ciliophora, F = Flagellata, Fu = Fungi, M = Microsporidia, Myx = Myxozoa, K = Kinetoplastida, V = Virus. Those recorded as pathogenic in the literature are highlighted in bold. For habitat FW = Freshwater, SW = Seawater, Terr = Terrestrial

Host	Parasite/Pathogen	Hyperhost	Habitat. Geographical location	Reference
Ancyrocephalus mogurndae	Trichodina sp. (C)	Siniperca chuatsi	FW. China	Gao et al. (2001)
Cichlidogyrus halli (= halli typicus)	RLO? (B)	Sarotherodon niloticus, Coptodon (= Tilapia) zillii	FW. Egypt	El-Naggar and Kearn (1989)
Diclidophora merlangi	Mycoplasma-like (B)	Merlangius merlangus	SW. UK	Morris and Halton (1975)
Diplectanocotyla gracilis	**Myxidium incomptavermi (Myx)**	Megalops cyprinoides	SW. Malaysia	Freeman and Shinn (2011)
Diplectanum aequans	Reoviridae (V), Trichodina sp. (C)	Dicentrarchus labrax	SW. Israel, France	Mokhtar-Maamouri et al. (1976); Colorni and Diamant (2005)
Euzetrema knoepffleri	Unidentified (V) **RLO? (B)**	Euproctus montanus	Terr. France	Fournier et al. (1975); Justine and Bonami (1993)
Gyrodactylus avalonia	Unidentified (B)	Gasterosteus wheatlandi	FW. Canada	Cusack and Cone (1985)
Gyrodactylus colemanensis	Pseudomonas sp. (B), Vibrionaceae spp. (B)	Salvelinus fontinalis, Salmo salar, Oncorhynchus mykiss (= Salmo gairdneri)	FW. Canada	Cusack et al. (1988)
Gyrodactylus derjavinoides (= derjavini)	Unidentified (Fu)	Salmo trutta	FW. UK	Mennie et al. (2000)
Gyrodactylus salaris	Unidentified (B), Ichthyobodo necator (F)	Salmo salar	FW. Norway	Bakke et al. (2006)
Gyrodactylus salmonis	Unidentified (B)	Salvelinus fontinalis, Oncorhynchus mykiss, Salmo salar	FW. Canada	Cone and Odense (1984)
Gyrodactylus sp.	Unidentified (V)	Carassius auratus	FW. Australia	Jones and Whittington (1992)
Haliotrema sp.	Trichodina sp. (C)	Pterois miles	SW. Israel	Colorni and Diamant (2005)
Microcotyle sp.	Birnaviridae (V)	Abudefduf taurus (= analogus)	SW. Senegal	Justine and Bonami (1993)
Neobenedenia melleni	Amyloodinium ocellatum (Dino)	Sparus aurata	SW. Israel	Colorni (1994)
Platycephalotrema cf. ogawi (= Haliotrema sp.)	**Myxidium sp. (Myx)**	Platycephalus sp.	SW. Japan	Freeman and Shinn (2011)
Pseudodactylogyrus bini	Myxidium giardi? (Myx)	Anguilla anguilla	FW. Spain	Aguilar et al. (2004)
Pseudodiplorchis americanus	Unidentified (B), **Unidentified (M)**	Scaphiopus couchii	Terr. USA	Cable and Tinsley (1992)
Udonella murmanica	Cryptobia udonellae (K)	Caligus curtus	SW. Russia	Frolov and Kornakova (2001)
Zeuxapta seriolae	Unidentified (B)	Seriola hippos	SW. Australia	Rohde (1986)

Table 6.3 List of digenean hosts, their infectious agents along with their hyperhosts. Where Al = Algae, B = Bacteria, C = Ciliophora, Cop = Copepoda, D = Digenea, F = Flagellata, H = Haplosporidia, Het = Heterokonta, Met = Metamonada, M = Microsporidia, Myx = Myxozoa, N = Nematoda, V = Virus. Those recorded as pathogenic in the literature are highlighted in bold. For habitat FW = Freshwater, SW = Seawater, Terr = Terrestrial

Host	Parasite/Pathogen	Hyperhost	Habitat. Geographical location	Reference
Alcicornis (= *Bucephalus*) *longicornutus*	***Urosporidium constantae* (H)**	*Ostrea chilensis* (= *lutaria*)	SW. New Zealand	Howell (1967)
Allocreadium fasciatusi	***Nosema gigantica* (M), *Unikaryon allocreadii* (M)**	*Oryzias* (= *Aplocheilus*) *melastigma*	FW. India	Canning and Madhavi (1977)
Allopodocotyle pedicellata (= *chrysophrii*)	*Fabespora* sp. (Myx)	*Sparus aurata*	SW. France	Siau et al. (1981)
Amurotrema dombrowskajae	*Spironoura babei* (N)	*Spinibarbus* (= *Spinibarbichthys*) *denticulatus*	FW. Vietnam	Sey and Moravec (1986)
Apatemon sp.	*Nosema* sp. (M)	*Physa* sp., *Helisoma campanulatum*	FW. USA	Cort et al. (1960)
Artyfechinostomum (= *Echinostoma*) *malayanum*	***Nosema eurytremae* (M)**	*Indoplanorbis exustus*	FW. Malaysia	Canning et al. (1974); Colley et al. (1975)
Bacciger bacciger (= *Cercaria pectinata*)	***Urosporidium* (= *Anurosporidium*) *pelseneeri* (H)**	*Donax vittatus*, *Barnea candida*	SW. France	Caullery and Chappellier (1906); Kinne (1980)
Brachycoelium hospitale	*Gordius* sp. (N)	*Diemictylus viridescens*	FW. USA	Cort (1915)
Brachylaimus fuscatus	Rhabdoviridae (V)	*Ponsadenia duplocincta*	Terr. Kazakhstan	Zdarska et al. (1986)
Bunodera (= *Culaeatrema*) *inconstans*	Unidentified (B)	*Culaea inconstans*	FW. USA	Lasee and Sutherland (1993)
Cercaria dohema	*Nosema* sp. (M)	*Lymnaea stagnalis*	FW. USA	Cort et al. (1960)
Cercaria emarginatae	*Nosema* sp. (M)	*Ladislavella emarginata* (= *Stagnicola emarginata angulata*)	FW. USA	Cort et al. (1960)

continued

Table 6.3 *Continued*

Host	Parasite/Pathogen	Hyperhost	Habitat. Geographical location	Reference
Cercaria laruei	*Nosema* sp. (M)	*Ladislavella emarginata* (= *Stagnicola emarginata angulata*)	FW. USA	Cort et al. (1960)
Cercaria modicella	*Nosema* sp. (M)	*Galba* (= *Fossaria*) *obrussa*	FW. USA	Cort et al. (1960)
Cercaria sp.	*Microsporidium* sp. (M)	*Planorbis* (= *Tropidiscus*) sp.	FW. Germany	Sokolova and Overstreet (2020)
Cercaria yogena	*Nosema* sp. (M)	*Ladislavella emarginata* (= *Stagnicola emarginata angulata*)	FW. USA	Cort et al. (1960)
Clinostomum marginatum	*Achromobacter* sp. (B), *Edwardsiella tarda* (B), *Enterobacter agglomerans* (B)	*Ardea herodias*	Terr. USA	Aho et al. (1991)
Cotylurus flabelliformis	*Nosema* sp. (M)	*Lymnaea stagnalis*	FW. USA	Cort et al. (1960)
Crassicutis archosargi	**Fabespora vermicola (Myx)**	*Archosargus probatocephalus*	SW. USA	Overstreet (1976)
Derogenes varicus	Unknown (*Hatshekia* sp?) (Cop)	*Hippoglossus limandoides*	SW. UK	Lebour (1908)
Deropristis inflata	**Hexamita sp. (F)**	*Anguilla rostrata*	SW. USA	Hunninen and Wichterman (1938)
Diacetabulum sp.	*Unikaryon* sp. (M)	*Pachygrapsus transversus*	SW. USA	Sokolova and Overstreet (2020)
Digenea	*Microsporidium* sp. (M)	*Succinea* sp.	SW. USA	Sokolova and Overstreet (2020)
Diphterostomum brusinae	*Nosema diphterostomi* (M)	*Diplodus annularis*	SW. France	Levron et al. (2004)
Diplostomum flexicaudum	**Nosema strigeoideae (M)**	*Ladislavella* (= *Stagnicola*) *emarginata, Lymnaea stagnalis, Galba* (= *Fossaria*) *obrussa*	FW. USA	Cort et al. (1960); Hussey (1971)
Diplostomum helveticum (= *indistinctum*)	*Unikaryon diplostomi* (M)	Fish	FW. Russia	Shigina and Grobov (1972)

Diplostomum paraspathaceum	*Unikaryon diplostomi* (M)	Fish	FW. Russia	Shigina and Grobov (1972)
Diplostomum spathaceum	*Unikaryon diplostomi* (M)	Fish	FW. Russia	Shigina and Grobov (1972)
Echinoparyphium recurvatum	**Unikaryon slaptonleyi (M)**	*Peregriana* (= *Radix*) *peregra, Planorbis planorbis*	FW. UK	Canning et al. (1983)
Echinoparyphium spp.	*Pleistophora* sp. (M)	*Radix* (= *Lymnaea*) *rubiginosa*	FW. Malaysia	Sokolova and Overstreet (2020)
Echinostoma hystricosum	**Nosema eurytremae (M), Unikaryon piriformis (M)**	*Radix* (= *Lymnaea*) *rubiginosa*	FW. Malaysia	Canning et al. (1974)
Echinostoma revolutum (= *audyi*)	*Giardia muris* (Met), **Nosema eurytremae (M), Nosema sp. 1 (M), Nosema sp. 2 (M), Nosema vascicola (M), Unikaryon piriformis (M)**	*Mus musculus, Radix* (= *Lymnaea*) *rubiginosa, Indoplanorbis exustus*	FW. Malaysia Terr. USA	Dollfus (1945); Joe and Nasemary (1973); Canning et al. (1974); Palmieri et al. (1978)
Echinostome larvae	**Nosema echinostomi (M)**, *Tetracotyle* sp. (D)	*Ampullaceana balthica* (= *Lymnaea limosa*), *Radix auricularia*	FW. France, Russia	Dollfus (1946); Voronin (1974); Mekhraliev and Mikailov (1981)
Echinoparyphium dunni	**Nosema eurytremae (M), Nosema vascicola (M), Unikaryon piriformis (M)**	*Radix* (= *Lymnaea*) *rubiginosa, Indoplanorbis exustus*	FW. Malaysia	Canning et al. (1974)
Encyclometra colubrimurorum (= *bolongensis*)	*Pleistophora danilewskyi* (= *Glugea encyclometrae*) (M)	*Natrix natrix, Lacerta* sp., *Chalcides chalcides* (= *tridactylus*), *Emys orbicularis, Rana temporaria*	Terr. France	Guyénot and Naville (1924)
Eurytrema pancreaticum	*Nosema eurytremae* (= *Perezia helminthorum*) (M)	*Bradybaena similaris*	FW. Malaysia	Canning and Basch (1968); Colley et al. (1975)
Fasciola gignatica (= *gigantea*)	**Nosema sp. 1 (M), Nosema sp. 2 (M)**	*Radix* (= *Lymnaea*) *rubiginosa*	FW. USA	Palmieri et al. (1978)

continued

Table 6.3 *Continued*

Host	Parasite/Pathogen	Hyperhost	Habitat. Geographical location	Reference
Fasciola hepatica	*Anncaliia* (= *Nosema* = *Brachiola*) *algerae* (M), **Nosema eurytremae (M)**	*Biomphalaria glabrata, Galba* (= *Lymnaea*) *cubensis, Galba* (= *Lymnaea*) *truncatula*	FW. Malaysia	Canning et al. (1974); Costa and Bradley (1980); Franzen et al. (2006)
Genarchopsis goppo	*Tetracotyle* sp. (D)	*Semisulcospira niponica*	FW. Japan	Urabe (2001)
Gyliauchen nahaensis	Eubacteria (B), Nanobacteria (B), Spirochaeta (B)	*Siganus* sp.	SW. Australia	Hughes-Stamm et al. (1999)
Gymnophallus (= *Meiogymnophallus*) *minutus*	**Unikaryon legeri (M)**	*Cerstoderma edule*	SW. Europe	Canning and Nicholas (1974)
Gymnophallus nereicola	**Urosporidium jiroveci (H)**	*Abra segmentum* (= *ovata*)	SW. France	Ormières et al. (1973)
Gymnophallus strigatus	*Nosema legeri* (M)	Bivalves	SW. Europe	Dollfus (1946)
Hemiuridae	**Urosporidium tauricum (H)**	*Rissoa splendida*	SW. Ukraine	Zaika and Dolgikh (1963)
Heterophyes heterophyes	**Pleistophora sp. (M)**	*Chelon* (= *Liza*) *ramada*	SW. Egypt	Paperna et al. (1978)
Lecithocladium sp.	Haplosporidia sp. (H)	*Scomber japonicus*	SW. Japan	Dollfus (1946)
Lepocreadium manteri	**Nosema lepocreadii (M)**	*Leuresthes tenuis*	SW. USA	Canning and Olson Jr (1980)
Levinseniella (= *Monarrhenos*) *capitanea*	**Urosporidium crescens (H)**	*Callinectes sapidus*	SW. USA	Deturk (1940); Perkins (1971, 1979); Couch (1974); Pung et al. (2002)
Lintonium vibex	*Trichodina* (= *Cyclochaeta*) *domerguei* (C)	*Sphoeroides maculatus*	SW. USA	Linton (1940)
Megalodiscus (= *Diplodiscus*) *temperatus*	*Nyctotherus cordiformis* (C), *Opalina* cf. *obtrigonoidea* (Het), **Unidentified (M)**	*Ferressia novangliae, Lithobates* (= *Rana*) *clamitans, Lithobates* (= *Rana*) *pipiens*	FW. USA	Dollfus (1946); Smith (1959)

Megalophallus sp.	*Callinectes sapidus*	SW. USA	Deturk (1940); Perkins (1971, 1979); Couch (1974); Pung et al. (2002)
Mesostephanus appendicula-toides (= *appendiculatus*) (= *Cercaria rhionica*)	*Melaopsis praemorsa*	FW. Russia	Voronin (1974)
Microphallus (= *Spelotrema*) *nicolli*	*Callinectes sapidus*	SW. USA	Deturk (1940); Perkins (1971, 1979); Couch (1974); Pung et al. (2002)
Microphallus basodactylophal-lus	*Callinectes sapidus*	SW. USA	Deturk (1940); Perkins (1971, 1979); Couch (1974); Pung et al. (2002)
Microphallus (= *Spelotrema*) *carcini*	*Carcinus maenas*	SW. France, UK	Guyenot and Naville (1922); Dollfus (1946); Canning (1975)
Microphallus sp.	*Panopeus herbstii*	SW. USA	Sokolova and Overstreet (2020)
Microphallus turgidus	*Palaemonetes pugio*	SW. USA	Deturk (1940); Perkins (1971, 1979); Couch (1974); Pung et al. (2002)
Monorchidae	*Abra segmentum* (= *ovata*)	SW. France	Ormières et al. (1973)
Monorchis parvus (M)	*Diplodus annularis*	SW. France	Levron et al. (2005)
Urosporidium jiroveci? (H)			
Nosema monorchis (M)			
Notocotylus sp.	*Physella* (= *Physa*) *gyrina*	FW. USA	Hanelt (2009)
Nematophila grandis (= *Amphistoma grande*)	*Podocnemis expansa*	FW. Brazil	Travassos and Travassos (1934)
Neorenifer grandispinus	*Drymarchon corais*	Terr. USA	Manter (1943); Dollfus (1946)
Unidentified (V)	*Ondatra zibethicus*	Terr. Canada	Beverley-Burton and Sweeny (1972)
Unidentified (V)	*Felis catus*	Terr. USA	Byram et al. (1975)
Microsporidium dobrovolskyi (M)	*Natrix natrix*	Terr. Ukraine	Ginetsinskaya (1968)
Urosporidium astomatum nomen nudum (H)	*Donax variabilis*	SW. USA	Menke (1968)

Row labels at left (parasite taxa):

Urosporidium crescens (H)

Nosema rhionica (M)

Urosporidium crescens (H)

Urosporidium crescens (H)

Pleistophora spelotremae (M)

Unikaryon sp. (M)

Urosporidium crescens (H)

Urosporidium jiroveci? (H)

Nosema monorchis (M)

Paragordius varius (N)

Atractis trematophila (N)

Alaria marcianae (D)

Notocotylus urbanensis

Paragonimus kellicotti

Paralepoderma cloacicola

Parvatrema donacis

continued

Table 6.3 Continued

Host	Parasite/Pathogen	Hyperhost	Habitat. Geographical location	Reference
Parvatrema duboisi	**Urosporidium tapetis (H)**	Ruditapes philippinarum	SW. Korea	Le et al. (2015); Cho et al. (2020)
Petasiger segregatus (= Paryphostomum segregatum)	Cotylurus lutzi (D)	Biomphalaria glabrata	FW. Brazil	Basch (1969)
Phyllodistomum folium	**Myxobilatus paragasterostei (Myx)**, Myxobilatus sp. (Myx)	Leuciscus baicalensis, Leuciscus leuciscus	FW. UK	Longshaw et al. (2005); Dugarov et al. (2011)
Plagioporus sinitsini	Paragordius sp. (N)	Hypentelium nigricans	FW. USA	Fischthal (1942)
Plagiorchiidae	Nosema xiphidiocercariae (M)	Stagnicola (= Lymnaea) palustris	FW. Russia	Voronin (1974)
Podocotyloides magnatestis	Nosema podocotyloidis (M)	Parapristipoma octolineatum	SW. Senegal	Toguebaye et al. (2014)
Postharmostomum gallinum	Nosema eurytremae (= Perezia helminthorum) (M)	Bradybaena similaris	FW. Malaysia	Canning and Basch (1968)
Posthodiplostomum minimum	Nosema sp. (M), **Ovipleistophora diplostomuri (M)**	Physella (= Physa) parkeri, Physella (= Physa) magnalacustris, Physa sp., Lepomis macrochirus	FW. USA	Cort et al. (1960); Lovy and Friend (2017)
Preptetos (= Lepocreadium) trulla	Chilomastix sp. (F)	Ocyurus chrysurus	SW. USA	Dollfus (1946)
Prosorhynchoides haimeana (= Bucephalus cucullus)	Urosporidium sp. (H), Nosema dollfusi (M)	Crassostrea virginica	SW. USA	Sprague (1964); Andrews (1984)
Prosorhynchus squamatus	Helicosporidium sp. (Al)	Mytilus edulis	SW. Finland	Pekkarinen (1991, 1993)
Quinqueserialis quinqueserialis	Unidentified (V)	Ondatra zibethicus	Terr. Canada	Beverley-Burton and Sweeny (1972)
Rauschiella palmipedis (= Glypthelmins linguatula)	Microsporidium distomi (M)	Rhinella marina (= Bufo marinus)	Terr. Brazil	Dollfus (1946); Sokolova and Overstreet (2020)
Schistosoma japonicum	Salmonella typhimurium (B)	Mus musculus	Terr. USA	Tuazon et al. (1985)

Schistosoma mansoni	Campylobacter coli (B), Campylobacter jejuni (B), Anncaliia (Nosema = Brahiola) algerae (M)	Homo sapiens, Biomphalaria glabrata	Terr, FW.	Dike (1971); Lindbolm and Nilsson (1994); Franzen et al. (2006)
Stictodora lari	Urosporidium sp. (H)	Batillaria australis	SW. Australia	Reece et al. (2004)
Strigea sp.	Tetracotyle sp. (D)	Anisus spirorbis	FW. Russia	Mekhraliev and Mikailov (1981)
Telorchis ercolanii	Microsporidium ghigii (M) Pleistophora (= Glugea) danilewskyi (= encyclometrae) (M)	Natrix natrix, Lacerta sp., Chalcides chalcides (= tridactylus), Emys orticularis, Rana temporaria	Terr. Italy	Guyénot and Naville (1924); Sokolova and Overstreet (2020)
Tracheophilus sp.	**Nosema sp. 1, (M), Nosema sp. 2 (M)**	Radix (= Lymnaea) rubiginosa	FW. USA	Palmieri et al. (1978)
Unidentified digenean	Cotylurus flabelliformis (D)	Physid and planorbid snails	FW. USA	Campbell (1973)
Unidentified strigeoid	Nosema sp. (M)	Ladislavella emarginata (= Stagnicola emarginata angulata), Planorbella (= Helisoma) trivolvis, Helisoma campanulatum	FW. USA	Cort et al. (1960)

6.2.3.2 Subclass Aspidogastrea

Aspidogastreans, a sister group to the Digenea in the class Trematoda are obligate parasites of marine and freshwater hosts. Maturation can occur in molluscan and vertebrate hosts. To date, only one hyperparasitic infection of aspidogastreans has been confirmed, namely a picornavirus-like infection of *Cotylogaster occidentalis* from *Elliptio complanata*, visible as ovoid inclusions in the cytoplasm of live specimens. Particles measured 23 nm in diameter occurred in paracrystalline arrays with the tegument severely disrupted in areas containing these arrays. Large crystalline arrays containing hexagonal particles measuring approximately 20 nm in diameter have been noted in *Aspidogaster conchicola* by Halton (1972); whilst reminiscent of a viral infection, no virus particles have been definitively identified.

6.2.4 Class Cestoda

Cestodes are obligate parasites that typically utilise three hosts in their life cycle. Hyperparasitic infections of cestodes include viruses, microsporidians, haplosporidians, acanthocephalans, digeneans, cestodes, and nematodes. The majority of infections recorded are represented by microsporidians and other cestodes (Table 6.4). Few infections cause pathology in these hosts and appear to be limited to a few microsporidian infections and an acanthocephalan infection (Mackiewicz 1972; Aguilar et al. 2004; Poddubnaya et al. 2006).

6.3 Phylum Acanthocephala

Acanthocephalans are wholly parasitic, typically utilising at least two hosts in their life cycle. The numbers and diversity of infections in acanthocephalans reported to date are limited to sporadic reports of microsporidians, viruses, nematodes, and cestodes (Table 6.5). Of these, only microsporidians and an unusual swelling of the tegument have been noted as pathogenic to acanthocephalans. Lester and Wright (1978) reported the presence of a yellow mass in the tegument of *Metechinorhynchus salmonis* from the gut of salmonids in Lake Ontario, Canada. The authors concluded that the swellings in *M. salmonis* and those reported by

Amin (1975) were due to a build-up of glycogen due to a metabolic dysfunction. Two microsporidians have been noted in acanthocephalans with *Microsporidium acanthocephali* inducing pathological changes in *Acanthocephaloides propinquus*, including alteration of the tegument and the pseudocoelom (de Buron et al. 1990). In female acanthocephalans, the microsporidian damages or destroys the ovarian balls and fills the pseudocoelom with spores.

6.4 Phylum Nematoda

Pathogens of nematodes and their pathological effects are summarised in Table 6.6. For Fungi, the delineation between predaceous and endoparasitic fungal species is not clear. There is an enormous number of predaceous species that have been described but for the sake of brevity and for the purposes of this review, here we will only signpost the reader to species of fungi that have evolved specialised structures to capture nematodes (i.e. adhesive columns, knobs, and hyphae, or networks that entangle or impede the movement of nematodes facilitating the penetration of hyphae, or those that use non-constricting or constricting rings) and instead focus on records where the pathology on the nematode is recorded.

Viral infections of nematodes can often be readily diagnosed in live specimens by a lighter colouration due to loss of intestinal granules (Felix and Wang 2019). Amorphous cytoplasmic inclusions, pleiomorphic granular elements and degeneration of the hypodermal and oviduct cells were noted in *Trichosomoides crassicauda* infected by virus-like particles (Foor 1972). The authors were unable to unequivocally determine if the changes were due to the presence of the virus-like particles or formed as a result of aging or other external agents. A number of viral infections have been noted in *Caenorhabditis briggsae*, including Le Blanc and Melnik viruses which cause intestinal cell damage and Santeuil virus which reduces production of progeny (Felix et al. 2011; Frézal et al. 2019). Orsay virus infections in the intestinal cells of the related *Caenorhabditis elegans* were typified by the loss of storage granules with a fluid cytoplasm with low viscosity and

Table 6.4 List of cestode hosts and their infectious agents along with their hyperhosts. Where Aca = Acanthocephala, Ces = Cestoda, D = Digenea, H = Haplosporidia, M = Microsporidia, N = Nematoda, V = Virus. Those recorded as pathogenic in the literature are highlighted in bold. For habitat FW = Freshwater, SW = Seawater, Terr = Terrestrial

Host	Parasite/Pathogen	Hyperhost	Habitat. Geographical location	Reference
Bothriocephalus nigropunctatus	*Echinorhynchus cestodicola* (Aca)	*Sebastes norvegicus*	SW. USA	Dollfus (1946)
Catenotaenia dendritica	*Urosporidium charleyi* (H)	*Sciurus vulgaris*	Terr. France	Levron et al. (2004)
Catenotaenia pusilla	Unidentified (N)	*Mus musculus*	Terr. Poland	Dollfus (1946)
Choanoscolex abscissus	Plerocercoid larvae (Ces), Unidentified (N)	*Pseudoplatystoma corruscans, Pseudoplatystoma faciatus*	FW. Brazil	Rego and Gibson (1989)
Cotugnia digonopora	Unidentified (N)	*Gallus gallus domestica*	Terr. Poland	Dollfus (1946)
Cryptocotylepis (= Hymenolepis) anthocephalus	*Microsporidium (= Stempellia) moniezi* (M)	*Blarina brevicauda*	Terr. USA	Jones (1943)
Eubothrium salvelini	*Echinorhynchus salvelini* (Aca)	*Salvelinus (= Critivomer) namaycush*	FW. USA	Miller (1946)
Eutetrarhynchus ruficollis	*Microsporidium* sp (M)	*Mustelus* sp.	SW. France	Dollfus (1946)
Glaridacris confusus	**Nosema sp. (M)**	*Ictiobus bubalus*	FW. USA	Mackiewicz (1972)
Gyrocotyle sp.	*Gyrocotlye* spp. (Ces), *Gyrocotyloides nybelini* (Ces)	*Chimera monstrosa*	SW. Europe	Malmberg (1986)
Hepatoxylon trichuri (= Dibothriorhynchus attenuatus)	*Tentacularia coryphaenae* (Ces)	*Xiphias gladius*	SW. USA	Dollfus (1946)
Hunterella nodulosa	Unidentified (M)	Unknown (Catostomidae)	FW. USA	Mackiewicz (1972)
Hymenolepis anthocephalus	*Microsporidium moniezi* (M)	*Blarina brevicauda*	Terr. USA	Jones (1943)
Hymenolepis microstoma	Virus-like particles (V)	*Mus musculus*	Terr. Canada	Webb (1984)
Hymenolepis nana	*Nosema helminthorum* (M)	*Rattus rattus, Mus musculus*	Terr. No data	Bulla and Cheng (1977)
Hymenolepis sp.	*Typhlocoelum gambense* (D)	*Plectropterus gambensis*	Terr. South Africa	Dollfus (1946)
Inermicapsifer hyracis	Unidentified (Nem)	*Hyrax* sp.	Terr. East Africa	Baer (1924)
Isoglaridacris folius	Unidentified (M)	Unknown (poss. *Moxostoma erythrurum*)	FW. USA	Mackiewicz (1972)
Jauela glandicephalus	Plerocercoid larvae (Ces)	*Zungaro zungaro (= Paulicea luetkeni)*	FW. Brazil	Rego and Gibson (1989)
Khawia armeniaca	**Paratuzetia kupermani (M)**	*Potamothrix alatus paravanicus*	FW. Armenia	Poddubnaya et al. 2006
Khawia japonensis (= iowensis)	Unidentified (M)	Unknown (poss *Cyprinus carpio*)	FW. USA	Mackiewicz (1972)
Ligula colubri blumenbachii	Microsporidium sp. (M)	*Natrix (= Tropidonotus) natrix*	Terr. Italy	Poddubnaya et al. (2006)
Mariauxiella pimelodi	Cysticercoid (Ces)	*Pimelodus ornatus, Pimelodus* sp.	FW. Brazil, Paraguay	De Chambrier and Rego (1995)
Megathylacus jandia (= brooksi)	Plerocercoid larvae (Ces)	*Zungaro zungaro (= Paulicea luetkeni)*	FW. Brazil	Rego and Gibson (1989)
Moniezia benedeni	*Nosema helminthorum* (M)	*Ovis aries*	Terr. Europe	Bulla and Cheng (1977)
Moniezia expansa	*Nosema helminthorum* (M)	*Ovis aries*	Terr. Europe	Bulla and Cheng (1977)
Moniezia sp.	*Nosema helminthorum* (M)	*Bison bison*	Terr. USA	Bulla and Cheng (1977)

continued

Table 6.4 *Continued*

Host	Parasite/Pathogen	Hyperhost	Habitat. Geographical location	Reference
Monticellia macrocotyle sp. inq.	Unidentified (N)	*Pimelodus ornatus, Pseudo-platystoma corruscans (= Silurus macrocephalus)*	FW. Brazil	Rego and Gibson (1989)
Nomimoscolex arandasregoi sp. inq.	Plerocercoid larvae (Ces)	*Tachysurus* sp.	FW. Brazil	Rego and Gibson (1989)
Nomimoscolex sp.	*Unikaryon nomimoscolexi* (M)	*Clarotes laticeps*	FW. Senegal	Sene et al. (1997)
Oochoristica agamae	Unidentified (M)	ND	Terr. No data	Poddubnaya et al. (2006)
Ophiotaenia sp.	*Alaria marcianae* (D)	*Lithobates (= Rana) pipiens*	FW. USA	Schaefer and Etges (1969)
Peltidocotyle rugosa	Plerocercoid larvae (Ces)	*Zungaro zungaro (= Paulicea luetkeni)*	FW. Brazil	Rego and Gibson (1989)
Phyllobothrium sp.	*Tentacularia coryphaenae* (Ces)	Squids	SW. Atlantic	Gaevskaya (1978)
Phyllobothrium loliginis	*Trilocularia acanthiaevulgaris (= gracilis)* (Ces)	*Squalus acanthias*	SW. USA	Linton (1924)
Plerocercoides pancerii	**Microsporidium ghigii (M)**	*Natrix (= Tropidonotus) natrix*	Terr. Italy	Bulla and Cheng (1977)
Potamotrygonocestus sp.	Proteocephalidea (Ces)	*Paratrygon aiereba*	FW. Peru	Reyda and Olson (2003)
Proteocephalidea sp.	Plerocercoid larvae (Ces)	*Rhamdia quelen (= sapo)*	FW. Brazil	Rego and Gibson (1989)
Pseudodiorchis (= Dorchis) reynoldsi	*Microsporidium (= Stempellia) moniezi* (M)	*Blarina brevicauda*	Terr. USA	Jones (1943)
Rhinebothrium sp.	Proteocephalidea (Ces)	*Paratrygon aiereba*	FW. Peru	Reyda and Olson (2003)
Rhinebothroides sp.	Proteocephalidea (Ces)	*Potamotrygon motoro, Potamotrygon falkneri (= castexi)*	FW. Peru	Reyda and Olson (2003)
Rhodobothrium (= Anthobothrium) pulvinatum	*Trilocularia acanthiaevulgaris (= gracilis)* (Ces)	*Squalus acanthias*	SW. USA	Linton (1924)
Rudolphiella lobosa (= Corallobothrium lobosum)	Plerocercoid larvae (Ces)	*Luciopimelodus pati*	FW. Paraguay	Rego and Gibson (1989)
Scyphophyllidium (= Anindobothrium) guariticus	Proteocephalidea (Ces)	*Paratrygon aiereba*	FW. Peru	Reyda and Olson (2003)
Sparganum proliferum	Virus-like particles (V)	*Homo sapiens*	Terr. USA	Mueller and Strano (1974a)
Spirometra mansonoides	*Agamodistomum* sp. (D)	*Sus scrofa*	Terr. USA	Becklund (1962)
Spirometra spp.	Virus-like particles (V)	ND	Terr. USA	Mueller and Strano (1974b)
Staphylocystis (= Hymenolepis) bacillaris	*Nosema helminthorum* (M)	ND	Terr. Pakistan	Bulla and Cheng (1977)
Taenia crassiceps	Virus-like particles (V)	*Sus* sp.	Terr. Mexico	Laclette et al. (1990)
Taenia saginata	*Nosema helminthorum* (M)	*Homo sapiens, Ceratoppia bipilis, Xenillus tegeocranus*	Terr., SW. No data	Bulla and Cheng (1977)
Taenia solium	Virus-like particles (V)	*Sus* sp.	Terr. Mexico	Laclette et al. (1990)
Taenia taeniaeformis (= crassicolis)	*Uncinaria (= Dochmius) balsami* (N)	*Felis catus*	Terr. Italy	Dollfus (1946)
Travassiella avitellina	Plerocercoid larvae (Ces)	*Zungaro zungaro (= Paulicea luetkeni)*	FW. Brazil	Rego and Gibson (1989)
Triaenophorus nodulosus	**Acanthocephalus lucii (Aca)**	*Esox lucius*	FW. Europe	Aguilar et al. (2004)

Table 6.5 List of acanthocephalan hosts and their infectious agents along with their hyperhosts. Where Ces = Cestoda, M = Microsporidia, N = Nematoda, V = Virus. Those recorded as pathogenic in the literature are highlighted in bold. For habitat FW = Freshwater, SW = Seawater, Terr = Terrestrial

Host	Parasite/Pathology	Hyperhost	Habitat. Geographical location	Reference
Acanthocephaloides geneticus	*Microsporidium acanthocephali* (M)	*Arnoglossus laterna*	SW. France	Loubes et al. (1988); Lom (2002)
Acanthocephaloides propinquus	**Microsporidium acanthocephali (M), Microsporidium propinqui (M)**	*Gobius niger, Zosterisessor ophiocephalus*	SW. France	Loubes et al. (1988); de Buron et al. (1990)
Acanthocephalus parksidei	**Abnormal swelling**	*Semotilus atromaculatus, Catostomus commersonii, Lepomis cyanellus, Lepomis macrochirus, Oncorhynchus mykiss*	FW. USA	Amin (1975)
Metechinorhynchus salmonis	**Abnormal swelling**	*Coregonus hoyi, Oncorhynchus kisutch, O. tschawytscha*	FW. USA	Lester and Wright (1978)
Moniliformis moniliformis	*Hymenolepis diminuta* (Ces), Virus-like inclusions (V)	*Periplaneta americana*	Terr. Canada	Budziakowski et al. (1984); Taraschewski (2000)
Pomphorhynchus sp.	*Anguillicoloides crassus* (N)	*Neogobius melanostomus*	FW. Germany	Emde et al. (2014)
Yamagutisentis sp.	*Microsporidium acanthocephali* (M)	*Bugloglossidium luteum*	SW. France	Lom (2002)

the occasional presence of multi-membrane structures. The apical border of the intestine had extensive convolutions, intermediate filament disorganisation, as well as elongation of nuclei and nucleoli, and nuclear degeneration with concomitant fusion of the intestinal cells (Felix et al. 2011). Whilst infections did not change nematode longevity or brood size, progeny production was reduced.

Pathogenic bacterial infections invariably lead to death of the host (Stewart and Godwin 1963; Anderson et al. 1971; Darby 2005; Giblin-Davis et al. 2003; Zhang et al. 2016; Page et al. 2019). In addition, *Clostridium welchii* infections of *Ascaris lumbricoides* var. *suis* are associated with cuticular lesions (Manter 1929; Bird and Deutsch 1957), whilst *Salmonella typhimurium* infections of *Caenorhabditis elegans* lead to apoptosis of the gonad, *Microbacterium*

nematophilum reduces growth in the same host and *Bacillus thuringiensis* and *Cryptococcus neoformans* reduce host fecundity.

Fungal and fungal-like infections of nematodes have been exploited as biocontrol agents due to the negative impacts on the hosts. Fungi either reduce or impair mobility e.g. *Stylopage hadra* infections of *Acrobeles* spp. and *Acrobeloides* sp.; *Harposporium lilliputanum* infections of *Acrobeloides* sp. and *Eucephalobus* sp.; *Coprinus comatus* infections of *Meloidogyne arenaria* or *Pleurotus* spp. and *Resupinatus silvanus* infections of a number of unidentified nematodes (Drechsler 1935; Fowler 1970; Wood 1973; Thorn and Barron 1984; Luo et al. 2004), reduced egg output or survival e.g. *Pochonia chlamydosporia* infections in *Ascaridia galli, Ascaris lumbricoides, Enterobius vermicularis*, and

Table 6.6 List of nematode hosts and their infectious agents. Where A = Apicomplexa, Am = Amoeba, B = Bacteria, Dip = Diplomonada, F = Flagellata, Fu = Fungi, H = Haplosporidia, K = Kinetoplastida, Met= Metamonada, M = Microsporidia, V = Virus. Those recorded as pathogenic in the literature are highlighted in bold

Host	Parasite/Pathogen	Location	Reference
Acrobeles spp.	**Stylopage hadra (Fu)**	USA	Drechsler (1935)
Acrobeloides sp.	*Acrostalagmus obovatus* (Fu), **Harposporium lilliputanum (Fu),** **Stylopage hadra (Fu)**	New Zealand	Drechsler (1935); Fowler (1970); Wood (1973)
Actinolaimus sp.	*Adelea* spp. (A)	ND	Poinar and Hess (1988)
Ancylostoma spp.	**Duddingtonia flagrans (CG768) (Fu)**	Brazil	Kramer De Mello et al. (2014)
Aphelenchoides sp.	*Acrostalagmus obovatus* (Fu)	New Zealand	Fowler (1970)
Aphelenchoides besseyi	**Beauveria bassiana (Snef2598) (Fu),**	China	Zhao et al. (2013)
	Simplicillium chinense (Snef5) (Fu)		
Archromadora spp.	Amoebic infections—general (Am)	ND	Poinar and Hess (1988)
Ascaridia galli	**Pochonia chlamydosporia (= Verticillium chlamydosporium) (Fu)**	Denmark	Thapa et al. (2018)
Ascaris lumbricoides	*Dientamoeba fragilis* (Met), **Ponchonia chlamydosporium (Fu)**	Thailand Czech Republic, Brazil	Sukanahaketu (1977); Lysek and Štěrba (1991); Braga et al. (2007)
Ascaris lumbricoides var. *suis*	*Pseudomonas aeruginosa* (tentative ID) (B), **Clostridium welchii (B)**	UK?	Manter (1929); Bird and Deutsch (1957)
Ascaris mystax	*Nosema mystaci* (M)	Brazil	Lutz and Splendore (1908)
Ascaris suum	*Candida pseudotropicalis* (B), **Candida sp. (B)**, **Enterobacter (syn. Aerobacter) sp. (B)**, *Escherichia* sp. (B), **Escherichia coli (B)**, **Pseudomonas sp. (B)**, **Streptococcus sp. (B)**	USA	Stewart and Godwin (1963); Anderson et al. (1971)
Aspicularis tetraptera	*Trichomonas muris* (F)	USA	Theiler and Farber (1932)
Belonolaimus longicaudatus	**'Candidatus Pasteuria usgae' (B)**	USA	Giblin-Davis et al. (2003)
Brugia malayi	*Wolbachia* sp. (B)	No data	Fenn and Blaxter (2004)
Bursaphelenchus xylophilus	**Stropharia rugosoannulata (Fu)**	No data	Luo et al. (2006)
Caenorhabditis plicata	**Microbacterium nematophilum (B)**	Lab, USA	Darby (2005)

Species	Pathogens/parasites/symbionts	Location	References
Caenorhabditis sp.	**Beauveria bassiana (Snef2598) (Fu)**, **Simplicillium chinense (Snef5) (Fu)**	China	Zhao et al. (2013)
Caenorhabditis brenneri (wild)	*Pancytospora epiphaga* (M)	Colombia	Zhang et al. (2016)
Caenorhabditis briggsae	**Chryseobacterium nematophagum (B)**, **Le Blanc virus (V)**, **Melnik virus (V)**, **Santeuil virus (V)**, *Nematocida homosporus* (M), *Nematocida ausubeli* (M), *Nematocida major* (M), *Nematocida parisii* (M)	France	Felix et al. (2011); Zhang et al. (2016); Frezal et al. (2019); Page et al. (2019)
Caenorhabditis drosphilae	**Microbacterium nematophilum (B)**	Lab, USA	Darby (2005;
Caenorhabditis elegans	**Agrobacterium tumefaciens (B)**, **Burkholderia cenocepacia (B)**, **Burkholderia pseudomallei (B)**, **Burkholderia thailandensis (B)**, **Photorhabdus (syn. Xenorhabdus) luminescens (B)**, **Pseudomonas aeruginosa (B)**, **Salmonella typhimurium (B)**, **Serratia marcescens (B)**, **Shewanella frigidimarina (B)**, **Shewanella massilia (B)**, **Xenorhabdus nematophila (B)**, **Yersinia pestis (B)**, **Yersinia pseudotuberculosis (B)**, **Microbacterium nematophilum (B)**, **Bacillus thuringiensis (B)**, **Enterococcus faecalis (B)**, **Staphylococcus aureus (B)**, **Streptococcus agalactiae (Group B) (B)**, **Streptococcus dysgalactiae (B)**, **Streptococcus mitis (Viridans group) (B)**, **Streptococcus oralis (B)**, **Streptococcus pneumoniae (B)**, **Streptococcus pyogenes (Group A) (B)**, **Histoplasma capsulatum (B)**, **Candida albicans (B)**, *Cryptococcus neoformans (B)*, *Myzocytiopsis humicola* (Fu), *Drechmeria coniospora* (Fu), **Orsay virus (V)**, **Aeromonas hydrophila (B)**, **Erwinia carotovora (B)**, **Erwinia chrysanthemi (B)**, **Escherichia coli (B)**, *Nematocida displodere* (M), *Pancytospora epiphaga* (M), *Nematocida ausubeli* (M), *Nematocida homosporus* (M), *Nematocida majo* (M), **Nematocida parisii (M)**	France, Germany, Sweden, Colombia, UK, USA	Jansson (1994); Darby et al. (1999, 2002); Hodgkin et al. (2000); Marroquin et al. (2000); Aballay and Ausubel (2001); Gallagher and Manoil (2001); Garsin et al. (2001); O'Quinn et al. (2001); Couillault and Ewbank (2002); Jansen et al. (2002); Mylonakis et al. (2002); Kothe et al. (2003); Kurz et al. (2003); Bae et al. (2004); Bolm et al. (2004); Begun et al. (2005); Troemel et al. (2008); Johnson et al. (2009); Felix et al. (2011); Ardila-Garcia and Fast (2012); Zhang et al. (2016); Osman et al. (2018); Feistel et al. (2019);
Caenorhabditis remanei	*Nematocida ausubeli* (M)	India	Zhang et al. (2016)
Caenorhabditis tropicalis (wild)	*Nematocida major* (M)	Guadeloupe	Zhang et al. (2016)
Cephalobus sp.	*Adelea* spp. (A), **Stylopage hadra (Fu)**	No data	Drechsler (1935); Poinar and Hess (1988)
Chiloplacus sp.	*Acrostalagmus obovatus* (Fu)	New Zealand	Fowler (1970)

continued

Table 6.6 *Continued*

Host	Parasite/Pathogen	Location	Reference
Choerostrongylus pudendotectus	Swine influenza virus (V)	USA	Shope (1941)
Chromadora spp.	Amoebic infections—general (Am), Flagellates—general (F)	No data	Poinar and Hess (1988)
Coomansus (syn. *Mononchus*) *composticola*	*Legerella helminthorum* (A)	No data	Canning (1962)
Cooperia oncophora	*Giardia* sp. (Dip)	USA	Graham (1935)
Desmodora marci	**Nematocenator marisprofundi (M)**	Hydrate Ridge North, Hydrate Ridge South, and East Knoll	Sapir et al. (2014)
Diplogaster spp.	**Stylopage hadra** (Fu), Flagellates—general (F)	USA	Drechsler (1935); Poinar and Hess (1988)
Diploscapter spp.	**Stylopage hadra (Fu)**	USA	Drechsler (1935)
Dolichodorus heterocephalus	Virus-like particles (V)	USA	Zuckerman et al. (1973)
Dorylaimus sp.	*Adelea* spp. (A), Amoebic infections—general (Am)	No data	Poinar and Hess (1988)
Enterobius vermicularis	*Dientamoeba fragilis* (Met), **Pochonia chlamydosporia (Fu)**	Turkey Denmark Brazil	Girginkardeşler et al. (2008); Braga et al. (2009); Röser et al. (2013)
Eucephalobus sp.	*Acrostalagmus obovatus* (Fu), **Harposporium lilliputanum (Fu)**	New Zealand	Fowler (1970); Wood (1973)
Globodera sp.	*Pasteuria nishizawae* (B)	No data	Atibalentja et al. (2004); Atibalentja and Noel (2008)
Globodera pallida	**Catenaria auxiliaris (Fu),** Potato cyst nematode (PCN) picorna-like virus (PLV) (V)	General	Kerry (1980); Ruark et al. (2018)
Globodera rostochiensis	*Catenaria anguillulae* (Fu), 'Candidatus Paenicardinium endonii' (B), Potato cyst nematode (PCN) picorna-like virus (PLV) (V)	Not known	Kerry (1980); Atibalentja and Noel (2008); Ruark et al. (2018)
Hammerschmidtiella diesingi	*Streptomyces leidynematis* (Fu)	USA	Hoffman (1953)
Heterakis gallinarum	**Pochonia chlamydosporia** (Fu), *Histomonas meleagridis* (Met)	Denmark, UK	Lee (1969); Thapa et al. (2018)
Heterodera sp.	*Pasteuria nishizawae* (B)	No data	Atibalentja et al. (2004); Atibalentja and Noel (2008)

Host/nematode	Natural enemy	Location	Reference
Heterodera avenae	**Nematophthora gynophila (Fu), Catenaria auxiliaris (Fu)**	General	Kerry (1980)
Heterodera carotae	**Nematophthora gynophila (Fu)**	General	Kerry (1980)
Heterodera cruciferae	**Nematophthora gynophila (Fu)**	General	Kerry (1980)
Heterodera glycines	**Beauveria bassiana (Snef2598) (Fu)**, 'Candidatus Paenicardinium endonii' (B), SCN nyavirus (ScNV) (V), SCN phlebovirus (ScPV) (V), SCN rhabodovirus (ScRV) (V), SCN tenuivirus (ScTV) (V), SCN virus 5 (SbCNV-5) (V), SCN nyami-like virus (NLV) (V), Soybean cyst nematode (SCN) bunya-like virus (BLV) (V), **Simplicillium chinense (Snef5) (Fu), Hirsutella minnesotensis (Fu),**	China, USA	Chen et al. (2000); Atibalentja and Noel (2008); Bekal et al. (2011); Zhao et al. (2013); Ruark et al. (2018)
Heterodera goettingiana	'Candidatus Paenicardinium endonii' (B), Candidatus Pasteuria goettingianae (B), **Nematophthora gynophila (Fu)**	No data	Kerry (1980); Atibalentja et al. (2004); Atibalentja and Noel (2008)
Heterodera rostochiensis	**Grey sterile fungus (Fu), Rhizoctonia solani (Fu), Verticillium alboatrum (Fu)**	UK	Clarke et al. (1967); Ketudat (1969)
Heterodera schachtii	*Nematophthora gynophila* (Fu), *Pochonia rubescens* (syn. *Verticillium suchlasporium*) (Fu), **Catenaria auxiliaris (Fu)**, SCN nyami-like virus (NLV) (V)	General, USA	Kerry (1980); Lopez-Llorca and Robertson (1992); Ruark et al. (2018)
Heterodera trifolii	**Nematophthora gynophila (Fu)**, SCN nyami-like virus (NLV) (V), Soybean cyst nematode (SCN) bunya-like virus (BLV) (V)	General, USA	Kerry (1980); Ruark et al. (2018)
Heterorhabditis spp.	*Photorhabdus* spp. (B), *Xenorhabdus* spp. (B)	No data	Ciche et al. (2006)
immature anisakid similar to *Paranisakiopsis pectinis*	**Urosporidium spisuli (H)**	Virginia, N. Atlantic Ocean	Perkins et al. (1975)
Leidynema appendiculata	*Streptomyces leidynematis* (Fu)	USA	Hoffman (1953)
Longidorus elongatus	Raspberry ringspot virus (V), Tobacco rattle virus (V)	Scotland	Taylor and Robertson (1969, 1970)
Meloidogyne sp.	*Pasteuria penetrans* (B)	No data	Atibalentja et al. (2004); Atibalentja and Noel (2008)
Meloidogyne arenaria	**Coprinus comatus (Fu)**	No data	Luo et al. (2004)

continued

Table 6.6 *Continued*

Host	Parasite/Pathogen	Location	Reference
Meloidogyne incognita	**Beauveria bassiana (Snef2598) (Fu)**, **Simplicillium chinense (Snef5) (Fu)**, Candida albicans (various strains) (B), **Geotichum terrestre Y 2162 (Fu)**, **Issatchenkia scutulata Moh 76 (Fu)**, **Meyerozyma (syn. Pichia) gluilliermondii Moh10 (Fu)**, **Cryptococcus curvatus Y-0812 (Fu)**, **Pachytrichospora transvaalensis Y-1240 (Fu)**, **Acremonium strictum (Fu)**, **Acremonium strictum + Aspergillus terreus (Eurotiomycetes) (Fu)**, **Aspergillus terreus (Fu)**, **Viral disease (V)**	China, Egypt	Loewenberg et al. (1959); Hashem et al. (2008); Singh and Mathur (2010); Zhao et al. (2013)
Meloidogyne javanica	*Ponchonia chlamydosporium* (syn. *Verticillium chlamydosporium*) (Fu)	Spain	Lopez-Llorca et al. (2002)
Metastrongylus elongatus	Swine influenza virus (V)	USA	Shope (1941)
Metoncholaimus scissus	**Pleistophora sp. (M)**	USA	Hopper et al. (1970)
Monhystera spp.	Amoebic infections—general (Am)	No data	Poinar and Hess (1988)
Nematode not specified	*Nematoctonus concurrens* (Fu) *Nematoctonus leiosporus* (Fu), *Nematoctonus robustus* (Fu), *Arthrobotrys conoides* (Fu), *Arthrobotrys dactyloides* (Fu), *Arthrobotrys oligospora* (Fu), *Arthrobotrys robusta* (Fu), *Dactylaria gracilis* (Fu), *Genicularia cystosporia* (Fu), *Harposporium anguillulae* (Fu), *Monacrosporium cionopagum* (Fu), *Monacrosporidium eudermatum* (Fu), *Monacrosporium parvicollis* (Fu), *Acaulopage pectospora* (Fu), *Stylopage grandis* (Fu)	New Zealand	Fowler (1970); Wood (1973)
Nematodirus helvetianus	*Giardia* sp. (Dip)	USA	Bowman et al. (1990)
Neoaplectana agriotos	*Nosema mesnili* (M), *Pleistophora schubergi* (M)	No data	Veremtchuk and Issi (1970)
Neoaplectana glaseri	**Microsporidian gen. sp. unidentified (M)**, unidentified (Fu)	Brazil	Poinar (1988)
Odontophora rectangular	*Sporanauta perivermis* (M)	Canada	Ardila-Garcia and Fast (2012)
Onchocerca volvulus	*Wolbachia* sp. (B)	No data	Fenn and Blaxter (2004)
Oscheius sp. 3	*Enteropsectra breve* (M), *Enteropsectra longa* (M), *Nematocida ausubeli* (M), *Nematocida homosporus* (M)	France, Iceland, India	Zhang et al. (2016)
Oscheius tipulae	*Enteropsectra breve* (M), *Nematocida homosporus* (M), *Nematocida minor* (M), *Pancytospora philotis* (M)	France, Czech Republic	Zhang et al. (2016)

Host	Parasite/pathogen	Location	Reference
Oxyuris ornata	Unidentified gregarine (A)	Germany	Walter (1858)
Panagrellus redivivus	**Stropharia rugosoannulata (Fu)**, *Drechmeria coniospora* (Fu), *Coprinus comatus* (Fu)	No data	Jansson (1994); Luo et al. (2004, 2006)
Panagrellus spp.	**Duddingtonia flagrans (AC001, CG722) (Fu)**, **Monacrosporium thaumasium (NF34A) (Fu)**	Brazil	Silva et al. (2017)
Parafilaroides [decorum?]	*Brucella* sp. (B)	USA	Garner et al. (1997)
Paraphanolaimus spp.	Amoebic infections—general (Am)	No data	Poinar and Hess (1988)
Parascaris equorum	*Catenaria anguillulae* (Fu)	Ireland	Buckley and Clapham (1929)
Plectus sp.	*Acrostalagmus obovatus* (Fu)	New Zealand	Fowler (1970)
Pratylenchus sp.	*Pasteuria thornei* (B)	No data	Atibalentja et al. (2004); Atibalentja and Noel (2008)
Procephalobus sp.	*Nematocida ciargi* (M)	Spain	Zhang et al. (2016)
Protospirura muris	*Thelohania reniformis* (M)	USA	Kudo and Hetherington (1922)
Rhabditella typhae	*Nematocida homosporus* (M)	Portugal	Zhang et al. (2016)
Rhabditis myriophila	*Microsporidium rhabdophilum* (M)	USA	Poinar and Hess (1986)
Rhabditis sp.	**Harposporium lilliputanum (Fu)**, **Stylopage hadra (Fu)**	New Zealand USA	Drechsler (1935); Wood (1973)
Rhabditis terricola	**Nematoctonus campylosporus (Fu)**, **Nematoctonus concurrens (Fu)**, **Nematoctonus haptocladus (Fu)**, **Nematoctonus leiosporus (Fu)**, **Nematoctonus pachysporus (Fu)**, **Catenaria anguillulae (Fu)**, **Acrostalagmus goniodes (Fu)**, **Acrostalagmus obovatus (Fu)**, **Harposporium anguillulae (Fu)**, **Harposporium bysmatosporum (Fu)**, **Harposporium crissum (Fu)**, **Harposporium cycloides (Fu)**, **Harposporium diceraeum (Fu)**, **Harposporium helicoides (Fu)**, **Harposporium leptospira (Fu)**, **Harposporium lilliputanum (Fu)**, **Harposporium subuliforme (Fu)**, **Meria coniospora (Fu)**, **Haptoglossa heterospora (Fu)**, **Haptoglossa zoospora (Fu)**, **Lagenidium caudatum (Fu)**, **Myzocytium glutinosporum (Fu)**, **Myzocytium humicola (Fu)**, **Myzocytium intermedium (Fu)**, **Myzocytium lenticulare (Fu)**, **Myzocytium papillatum (Fu)**, **Myzocytium vermicolum (Fu)**, **Meristacrum asterospermum (Fu)**, **Rhopalomyces elegans (Fu)**	Canada	Barron (1977–1978)

continued

Table 6.6 *Continued*

Host	Parasite/Pathogen	Location	Reference
Romanomermis culicivorax	Virus-like particles (V)	USA	Poinar and Hess (1977)
Seinura sp.	*Acrostalagmus obovatus* (Fu)	New Zealand	Fowler (1970)
Steinerma spp.	*Photorhabdus* spp. (B), *Xenorhabdus* spp. (B)	No data	Ciche et al. (2006)
Strongyloides ratti	Swine influenza virus (V)	USA	Shotts et al. (1968)
Syphacia obvelata	*Trichomonas muris* (F)	USA	Theiler and Farber (1932); Becker (1933)
Thaumamermis cosgrovei	Isopod iridescent virus (V)	USA	Poinar et al. (1980); Hess and Poinar (1985)
Theristus (Daptonema) albigensis	Microsporidian gen. sp. unidentified (M)	USA	Hopper et al. (1970)
Torbrilus [?]	*Adelea* spp. (A)	No data	Poinar and Hess (1988)
Toxocara [syn. *limbata*] *leonina*	*Nosema helminthorum* (M)	Turkestan	Dollfus (1946)
Toxocara canis	**Paecilomyces lilacinus (Fu), Paecilomyces marquandii (Fu)**	Argentina	Basualdo et al. (2000)
Trichinella spiralis	Lymphocytic choriomeningitis virus (V)	Lab, USA	Syverton et al. (1947)
Trichodorus pachydermus	Raspberry ringspot virus (V), Tobacco rattle virus (V)	Scotland	Taylor and Robertson (1969, 1970)
Trichosomoides crassicauda	**Virus-like particles (V)**	USA	Foor (1972)
Trichostrongylus pergracilis	Unidentified amoeba (Am)	No data	Shipley (1909)
Trichuris trichiura	*Dientamoeba fragilis* (Met)	Worldwide	Dobell (1940)
Trilobus gracilis	*Leptomonas butschlii* (K)	No data	Wenyon (1926)

Species	Description	Location	Reference
Trilobus spp.	Amoebic infections—general (Am), Flagellates—general (F)	No data	Poinar and Hess (1988)
Tripyla spp.	*Adelea* spp. (A), Amoebic infections—general (Am)	No data	Poinar and Hess (1988)
Tylenchus sp.	*Acrostalagmus obovatus* (Fu), Amoebic infections—general (Am)	New Zealand	Fowler (1970); Poinar and Hess (1988)
Unidentified rhabditid nematode	**Podocrella (= Cordyceps) peltate (Fu)**	Costa Rica	Chaverri et al. (2005)
Unspecified	*Hohenbuehelia atrocaerulea* (Fu), *Hohenbuehelia grisea* (Fu), *Hohenbuehelia mastrucata* (Fu), *Hohenbuehelia petaloides* (Fu), *Hohenbuehelia portegna* **(Fu)**, **Pleurotus cornucopiae (Fu), Pleurotus cystidiosus (Fu), Pleurotus ostreatus (Fu), Pleurotus strigosus (Fu), Pleurotus subareolatus (Fu)**, *Resupinatus silvanus* (Fu)	Canada France	Thorn and Barron (1984)
Viannella sp.	*Giardia viscaciae* (Dip)	S. America	Thomson (1925)
Wilsonema sp.	*Acrostalagmus obovatus* (Fu)	New Zealand	Fowler (1971)
Wuchereria bancrofti	*Wolbachia* sp. (B)	No data	Fenn and Blaxter (2004)
Xiphinema americanum	*Candidatus Xiphinematobacter* spp. (B)	USA	Coomans et al. (2000); Atibalentja and Noel (2008)
Xiphinema brevicollum	*Candidatus Xiphinematobacter* spp. (B)	South Africa	Coomans et al. (2000); Atibalentja and Noel (2008)
Xiphinema rivesi	*Candidatus Xiphinematobacter* spp. (B)	USA	Coomans et al. (2000); Atibalentja and Noel (2008)

Heterakis gallinarum (see Sukanahaketu 1977; Lysek and Štěrba 1991; Braga et al. 2007, 2009; Girginkardeşler et al. 2008; Röser et al. 2013; Thapa et al. 2018), reduce fecundity e.g. *Paecilomyces lilacinus* and *Paecilomyces marquandii* infections of *Toxocara canis* (see Basualdo et al. 2000) or prevent cyst formation e.g. *Catenaria auxiliaris* infections of *Globodera pallida* (see Kerry 1980).

However, more often, the main outcome of fungal infections of nematodes is death e.g., *Duddingtonia flagrans* infections of *Ancylostoma* spp., and *Panagrellus* spp.; *Beauveria bassiana* infections of *Aphelenchoides besseyi*, *Caenarhabditis* sp., *Heterodera glycines*, and *Meloidogyne incognita*; *Simplicillium chinense* infections of *Aphelenchoides besseyi*, *Caenarhabditis* sp., *Heterodera glycines*, and *Meloidogyne incognita*; *Stropharia rugosoannulata* infections of *Bursaphelenchus xylophilus*, and *Panagrellus redivivus*; *Stylopage hadra* infections of *Cephalobus* sp., *Diplogaster* spp., *Diploscapter* spp., and *Rhabditis* sp.; *Hirsutella minnesotensis* infections of *Heterodera glycines*; *Monacrosporium thaumasium* infections of *Panagrellus* spp.; *Harposporium lilliputanum* infections of *Rhabditis* sp. and *Podocrella peltate* infections of unidentified rhabditid nematodes (Drechsler 1935; Wood 1973; Poinar and Hess 1988; Jansson 1994; Chen et al. 2000; Luo et al. 2004, 2006; Chaverri et al. 2005; Atibalentja and Noel 2008; Bekal et al. 2011; Zhao et al. 2013; Kramer De Mello et al. 2014; Silva et al. 2017; Ruark et al. 2018). In addition, a large number of fungal infections of *Meloidogyne incognita* and *Rhabditis terricola* lead to the death of their hosts (Loewenberg et al. 1959; Barron 1977–1978; Hashem et al. 2008; Singh and Mathur 2010; Zhao et al. 2013). Differential mortality of females has been noted for fungal infections of *Heterodera* spp. (Clarke et al. 1967; Ketudat 1969; Kerry 1980; Lopez-Llorca and Robertson 1992; Atibalentja et al. 2004, Atibalentja and Noel 2008; Ruark et al. 2018).

A number of microsporidian infections of nematodes have been reported as pathogenic to their hosts. *Nematocenator marisprofundi* from free-living deep sea *Desmodora marci* targets the body wall muscles causing cell lysis and subsequently degeneration of the muscle filaments in severe infections (Sapir et al. 2014). *Pleistophora* sp. infections of the free-living marine nematode *Metoncholaimus scissus* cause the nematode to change colour and become inactive (Hopper et al. 1970). Additionally, infected cells become hypertrophied and host eggs become infected, potentially acting as a route of infection to new hosts. An unidentified microsporidian infection of the parasitic nematode *Neoaplectana glaseri* from the beetle *Migdolus fryanus*, reduced fecundity with progeny being hyaline and smaller compared with uninfected juveniles (Poinar 1988). Curiously, infected third-stage juveniles possessed a normal, open intestine and developing gonads unlike uninfected individuals that normally have a collapsed intestine and rudimentary gonads. Death was a typical outcome in heavy infections in this and in infections of various nematodes with *Nematocida parisii* (see Ardila-Garcia and Fast 2012).

Infections by *Urosporidium spisuli* in immature anisakids from *Spisula solidissima* were responsible for clams being withheld from commercial sale due to the brown colouration of the nematodes as a result of the haplosporidian infection (Perkins et al. 1975). The authors did not consider that there were any further pathologies associated with the parasite nor were they of concern for human health.

6.5 Conclusions and discussion

This review has considered the range of parasites and pathogens reported from platyhelminths, acanthocephalans, and nematodes worldwide. Many of these reports are merely descriptive with limited information provided on potential pathogenicity. It is not clear, therefore, if the infections reported are pathogenic or benign due to the methods utilised. Despite the ubiquity of these hosts in most ecosystems and, in the case of parasitic forms, the wide diversity of hyperhosts infected, the number of hyperparasite records is comparatively sparse. More interestingly, the diversity of infections within each taxa varies markedly. For example, the dominant parasite group in turbellarians are the apicomplexans and the ciliates, in Monogenea the dominant group are bacteria, in Cestoda the dominant groups are microsporidians and other cestodes, in nematodes, the dominant groups are bacteria, fungi, and microsporidia, whilst in acanthocephalans and digeneans, the dominant groups are microsporidians. The hosts with the greatest diversity are the turbellarians, nematodes, and digeneans. This somewhat complex picture of parasite diversity may well reflect the efforts of scientists and availability of hosts rather than necessarily the true diversity of parasites and pathogens within these hosts. Furthermore, as a result of a need to find appropriate biocontrol methods, a large focus has been placed on classifying and describing the pathogens of nematodes of economic concern. It remains a possibility that if similar efforts were to be made with the other host taxa reviewed in this chapter, that similar diversities may be noted. In particular for the parasitic species, hyperparasitic infections may be overlooked due to a lack of specific observations by researchers who, unless they are describing new species or specific anatomical structures, are unlikely to focus much on internal features of these hosts.

More recent descriptive studies have utilised a range of techniques including molecular, light and electron microscopy to ensure taxonomic rigour and to describe, where appropriate, host-parasite interactions. Unfortunately, the older literature, whilst providing detailed morphological and pathological information of infections, suffers naturally from a lack of molecular data, and confused nomenclatural changes. There would therefore be merit in revisiting these infections through the collection of new material from the type hosts and more focused studies on the potential parasites and pathogens of the taxa reported in this chapter. It is imperative that future studies should utilise a range of techniques to characterise infections of these hosts and to re-examine previous infections to confirm their taxonomic classifications.

6.6 Summary

- Only a small proportion of the infections of Platyhelminthes, Acanthocephala, and Nematoda reported herein, appear to exert pathological effects with the bulk of infections reported appearing benign.
- The details for many infections, however, are lacking, many of which appear to result from serendipitous encounter, while the report of others seem to reflect sampling effort, researcher interest or are skewed by the medical, veterinary and/or agricultural importance of particular species, as major pathogens, or from their utility as model organisms (e.g. *C. elegans*) in medical research and development. A proportion of the reports result from the study of parasite species of economic and/or sanitary importance including those that pose a direct food safety risk or an issue of aesthetics with consumers.
- It follows that it is likely that the number of hyperparasite infections is grossly underestimated and that targeted studies that include observational methods as well as new methods such as eDNA techniques will increase the number of infections detected.
- Very few studies have purposefully set out to investigate the pathogenicity of infections in these taxa, as such the pathology of the parasite tissues themselves has been largely overlooked. Routinely, histopathology should be conducted on infected parasites alongside other methods to determine the impact of hyperparasitic infections on their hosts.
- Finally, there is a need to ensure taxonomic rigour in any studies, confirming the taxonomy of hyperhost, host and hyperparasite in future studies. Previously reported infections may benefit from reappraisal for impact as well as confirming their taxonomic position.

References

Aballay, A. and Ausubel, F.M. 2001. Programmed cell death mediated by ced-3 and ced-4 protects *Caenorhabditis elegans* from *Salmonella typhimurium*-mediated killing. *Proceedings of the National Academy of Sciences of the United States of America* 98: 2735–2739

Aguilar, A., Aragort, W., Alvarez, M.F., Leiro, J.M., and Sanmartin, M. 2004. Hyperparasitism by *Myxidium giardi* Cepede 1906 (Mxyozoa: Mxyosporea) in *Pseudodactylogyrus bini* (Kikuchi, 1929) Gussev, 1965 (Monogenea: Dactylogyridae), a parasite of the European eel *Anguilla anguilla* L. *Bulletin of the European Association of Fish Pathologists* 24: 287–292

Aho, J.M., Uglem, G.L., Moore, J.P., and Larson, O.R. 1991. Bacteria associated with the tegument of *Clinostomum marginatum* (Digenea). *The Journal of Parasitology* 77: 784–786

Alexandrov, K.E. and Sljusarev, G.S. 1992. A new species of orthonectids, *Rhopalura variabeli* sp. n. (Mesozoa) from the turbellarian *Macrorynchus crocea*. *Parazitologiia* 26: 347–351

Amin, O.M. 1975. Variability in *Acanthocephalus parksidei* Amin, 1974 (Acanthocephala: Echinorhynchidae). *The Journal of Parasitology* 61: 307–317

Anderson, T.J., Newman, L.J., and Lester, R.J.G. 1993. Light and electron microscope study of *Urosporidium cannoni* n. sp., a haplosporidian parasite of the polyclad turbellarian *Stylochus* sp. *Journal of Eukaryotic Microbiology* 40: 162–168

Anderson, W.R., Madden, P.A., and Tromba, F.G. 1971. Histopathologic and bacteriologic examination of cuticular lesions of *Ascaris suum*. *The Journal of Parasitology* 57(5): 1010–1014

Andrews, J.D. 1984. Epizootiology of diseases of oysters (*Crassostrea virginica*), and parasites of associated organisms in eastern North America. *Helgoländer Meeresuntersuchungen* 37: 149–166

Apelt, G. 1969. Fortpflanzungsbiologie, entwicklungszyklen und vergleichende frühentwicklung acoeler turbellarien. *Marine Biology* 4: 267–325

Ardila-Garcia, A.M. and Fast, N.M. 2012. Microsporidian infection in a free-living marine nematode. *Eukaryotic Cell* 11(12): 1544–1551

Armitage, M.J. and Young, J.O. 1990. A field and laboratory study of the parasites of the triclad *Phagocata vitta* (Duges). *Freshwater Biology* 24: 101–107

Atibalentja, N., Jakstys, B.P., and Noel, G.R. 2004. Life cycle, ultrastructure, and host specificity of the North American isolate of *Pasteuria* that parasitizes the soybean cyst nematode, *Heterodera glycines*. *Journal of Nematology* 36: 171–180

Atibalentja, N. and Noel, G.R. 2008. Bacterial endosymbionts of plant-parasitic nematodes. *Symbiosis* 46: 87–93

Bae, T., Banger, A.K., Wallace, A., Glass, E.M., Aslund, F., Schneewind, O., and Missiakas, D.M. 2004. *Staphylococcus aureus* virulence genes identified by *bursa aurealis* mutagenesis and nematode killing. *Proceedings of the National Academy of Sciences of the United States of America* 101: 12312–12317

Baer, J.G. 1924. On the occurrence of nematodes in the excretory duct of a cestode *Inermicapsifer hyracis*. *Journal of Helminthology* 2: 95–96

Bakke, T.A., Cable, J., and Østbø, M. 2006. The ultrastructure of hypersymbionts on the monogenean *Gyrodactylus salaris* infecting Atlantic salmon *Salmo salar*. *Journal of Helminthology* 80: 377–386

Ball, I.R. and Fernando, C.H. 1968. On *Urceolaria mitra* (Protozoa, Peritricha) epizoic on North American flatworms. *Canadian Journal of Zoology* 46: 981–985

Barron, G.L. 1977–1978. Nematophagous fungi: Endoparasites of *Rhabditis terricola*. *Microbial Ecology* 4(2): 157–163

Basch, P.F. 1969. *Cotylurus lutzi* sp. n. (Trematoda: Strigeidae) and its life cycle. *The Journal of Parasitology* 55: 527–539

Basualdo, J.A., Ciarmela, M.L., Sarmiento, P.L., and Minvielle, M.C. 2000. Biological activity of *Paecilomyces* genus against *Toxocara canis* eggs. *Parasitology Research* 86: 854–859

Becker, E.R. 1933. Two observations on helminths. *Transactions of the American Microscopical Society* 52(4): 361–362

Becklund, W.W. 1962. Occurrence of a larval trematode (Diplostomidae) in a larval cestode (Diphyllobothriidae) from *Sus scrofa* in Florida. *The Journal of Parasitology* 48: 286–286

Begun, J., Sifri, C.D., Goldman, S., Calderwood, S.B., and Ausubel, F.M. 2005. *Staphylococcus aureus* virulence factors identified by using a high-throughput *Caenorhabditis elegans*-killing model. *Infection and Immunity* 73: 872–877

Bekal, S., Domier, L.L., Niblack, T.L., and Lambert, K.N. 2011. Discovery and initial analysis of novel viral genomes in the soybean cyst nematode. *Journal of General Virology* 92: 1870–1879

Beverley-Burton, M. and Sweeny, P.R. 1972. Intranuclear, paracrystalline inclusions in various cells of *Quinqueserialis* and *Notocotylus urbanensis* (Trematoda: Notocotylidae). *Canadian Journal of Zoology* 50: 345–348

Bhatia, B.L. and Setna, S. 1924. On some new cephaline gregarines. *Parasitology* 16: 279–288

Bird, A.F. and Deutsch, K. 1957. The structure of the cuticle of *Ascaris lumbricoides* var. *suis*. *Parasitology* 47: 319–328

Bolm, M., Jansen, W.T., Schnabel, R., and Chhatwal, G.S. 2004. Hydrogen peroxide-mediated killing of

Caenorhabditis elegans: A common feature of different streptococcal species. *Infection and Immunity* 72: 1192–1194

Bowen, I. and Ryder, T.A. 1994. *Urceolaria mitra* (von Seib) epizoic on *Polycelis tenuis* (Ijima) an SEM study. *Cell Biology International* 18: 881–888

Bowman, D.D., Block, G., and Tanneberger, A. 1990. *Giardia* sp. in the intestine of *Nematodirus helvetianus* (Nematoda: Trichostrongyloidea) recovered at necropsy from a sheep (*Ovis aries*). *Transactions of the American Microscopical Society* 109(4): 422–424

Braga, F.R., Araújo, J.V., Campos, A.K., Carvalho, R.O., Silva, A.R., Tavela, A.O., and Maciel, A.S. 2007. Observação *in vitro* da ação dos isolados fúngicos *Duddingtonia flagrans, Monacrosporium thaumasium* e *Verticillium chlamydosporium* sobre ovos de *Ascaris lumbricoides* (Lineu, 1758). *Revista da Sociedade Brasileira de Medicina Tropical* 40(3): 356–358

Braga, F.R., Silva, A.R., Araujo, J.M., Ferreira, S.R., Araujo, J.V., and Frassy, L.N. 2009. Ação ovicida do fungo *Pochonia chlamydosporia* sobre ovos de *Enterobius vermicularis*. *Revista do Instituto Adolfo Lutz* 68(2): 318–321

Buckley, J.J.C. and Clapham, P.A. 1929. The invasion of helminth eggs by chytridiacean fungi. *Journal of Helminthology* 8(1): 1–14

Budziakowski, M.E., Mettrick, D.F., and Webb, R.A. 1984. Ultrastructural morphology of the nerve cells in the cerebral ganglion of the acanthocephalan *Moniliformis moniliformis*. *The Journal of Parasitology* 70: 719–734

Bulla, L. and Cheng, T.C. 1977. *Comparative Pathobiology. 2. Systematics of the Microsporidia*, 510pp. New York: Plenum Press

Byram, J.E., Ernst, S.C., Lumsden, R.D., and Sogandares-Bernal, F. 1975. Viruslike inclusions in the cecal epithelial cells of *Paragonimus kellicotti* (Digenea, Troglotrematidae). *The Journal of Parasitology* 61: 253–264

Cable, J. and Tinsley, R.C. 1992. Microsporidian hyperparasites and bacteria associated with *Pseudodiplorchis americanus* (Monogenea: Polystomatidae). *Canadian Journal of Zoology* 70: 523–529

Campbell, R.A. 1973. Studies on the biology of the life cycle of *Cotylurus flabelliformis* (Trematoda: Strigeidae). *Transactions of the American Microscopical Society* 92: 629–640

Canning, E.U. 1962. *Legerella helminthorum* n. sp. a coccidium parasitic in a nematode. *Archiv fur Protistenkunde* 105: 455–462

Canning, E.U. 1975. *The Microsporidian Parasites of Platyhelminthes: Their Morphology, Development, Transmission and Pathogenicity*. CIH Miscellaneous Publication 2: ii, 32pp.

Canning, E.U., Barker, R.J., Hammond, J.C., and Nicholas, J.P. 1983. *Unikaryon slaptonleyi* sp. nov. (Microspora: Unikaryonidae) isolated from echinostome and strigeid larvae from *Lymnaea peregra*: Observations on its morphology, transmission and pathogenicity. *Parasitology* 87: 175–184

Canning, E.U. and Basch, P.F. 1968. *Perezia helminthorum* sp.nov., a microsporidian hyperparasite of trematode larvae from Malaysian snails. *Parasitology* 58; 341–347

Canning, E.U., Foon, L.P., and Joe, L.K. 1974. Microsporidian parasites of trematode larvae from aquatic snails in West Malaysia. *Journal of Protozoology* 21: 19–25

Canning, E.U. and Madhavi, R. 1977. Studies on two new species of Microsporida hyperparasitic in adult *Allocreadium fasciatusi* (Trematoda, Allocreadiidae). *Parasitology* 75: 293–300

Canning, E.U. and Nicholas, J.P. 1974. Light and electron microscope observations on *Unikaryon legeri* (Microsporida, Nosematidae), a parasite of the metacercaria of *Meiogymnophallus minutus* in *Cardium edule*. *Journal of Invertebrate Pathology* 23: 92–100

Canning, E.U. and Olson, A. Jr. 1980. *Nosema lepocreadii* sp. n., a parasite of *Lepocreadium manteri* (Digenea: Lepocreadiidae) from the gut of the California grunion, *Leuresthes tenuis. The Journal of Parasitology* 66: 154–159

Cannon, L.R.G. and Jennings, J.B. 1988. *Monocystella epibatis* n.sp., a new aseptate gregarine hyperparasite of rhabdocoel turbellarians parasitic in the crown of thorns starfish, *Acanthaster planci* Linnaeus, from the Great Barrier Reef. *Archiv fur Protistenkunde* 136: 267–272

Carballal, M.J., Diaz, S., and Villalba, A. 2005. *Urosporidium* sp. hyperparasite of the turbellarian *Paravortex cardii* in the cockle *Cerastoderma edule. Journal of Invertebrate Pathology* 90: 104–107

Carpenter, J.H. 1982. Observations on the biology of cave planarians of the United States. *International Journal of Speleology*, 12: 9–26

Caullery, M. and Chappellier, A, 1906. *Anurosporidium pelseneeri* n. g., n. sp., haplosporidie infectant les sporocystes d'un trematode parasite de *Donax trunculus* L. *Compte Rendu de la Societe de Biologie Paris* 60: 325–328

Cépède, C. 1910. Recherches sur les infusoires astomes. *Archives de Zoologie Experimentale et Generale 5 Serie*, 3: 341–609

Chaverri, P., Samuels, G.J., and Hodge, K.T. 2005. The genus *Podocrella* and its nematode-killing anamorph *Harposporium. Mycologia* 97(2): 433–443

Chen, S., Liu, X.Z., and Chen, F.J. 2000. *Hirsutella minnesotensis* sp. nov., a new pathogen of the soybean cyst nematode. *Mycologia* 92(5): 819–824

Cho, Y.G., Kang, H.S., Le, C.T., Kwon, M.G., Jang, M.S., and Choi, K.S. 2020. Molecular characterization of

Urosporidium tapetis sp. nov., a haplosporidian hyperparasite infecting metacercariae of *Parvatrema duboisi* (Dollfus 1923), a trematode parasite of Manila clam *Ruditapes philippinarum* on the west coast of Korea. *Journal of Invertebrate Pathology* 175: 107454

Ciche, T.A., Darby, C., Ehlers, R.-U., Forst, S., and GoodrichBlair, H. 2006. Dangerous liaisons: The symbiosis of entomopathogenic nematodes and bacteria. *Biological Control* 38: 22–46

Clarke, A.J., Cox, P.M., and Shepherd, A.M. 1967. The chemical composition of the eggshells of the potato cyst-nematode, *Heterodera rostochiensis* Woll. *Biochemical Journal* 104: 1056–1060

Colley, F.C., Joe, L.K., Zaman, V., and Canning, E.U. 1975. Light and electron microscopical study of *Nosema eurytremae*. *Journal of Invertebrate Pathology* 26: 11–20

Colorni, A, 1994. Hyperparasitism of *Amyloodinium ocellatum* (Dinoflagellida: Oodinidae) on *Neobenedenia melleni* (Monogenea: Capsalidae). *Diseases of Aquatic Organisms* 19: 157–159

Colorni, A. and Diamant, A. 2005. Hyperparasitism of trichodinid ciliates on monogenean gill flukes of two marine fish. *Diseases of Aquatic Organisms* 65: 177–180

Comps, M. and Tige, G, 1999. Procaryotic infections in the mussel *Mytilus galloprovinciallis* and in its parasite the turbellarian *Urastoma cyprinae*. *Diseases of Aquatic Organisms* 38: 211–217

Cone, D.K. and Odense, P.H. 1984. Pathology of five species of *Gyrodactylus* Nordmann, 1832 (Monogenea). *Canadian Journal of Zoology* 62: 1084–1088

Coomans, A., Vandekerckhove, T.M., and Claeys, M. 2000. Transovarial transmission of symbionts in *Xiphinema brevicollum* (Nematoda: Longideridae). *Nematology*, 2: 443–449

Cordero, E.H. 1946. *Ophiotaenia cohospes* n. sp., de la tortuga fluvial *Hydromedusa tectifera* Cope, una larva plerocercoide en el parénquima de *Temnocephala brevicornis* Mont., y su probable metamorfosis. *Comunicaciones Zoologicas del Museo de Historia Natural de Montevideo* 2: 1–15

Cort, W.W. 1915. *Gordius* larvae parasitic in a trematode. *The Journal of Parasitology* 1: 198–199

Cort, W.W., Hussey, K.L., and Ameel, D.J. 1960. Studies on a microsporidian hyperparasite of strigeoid trematodes. I. Prevalence and effect on the parasitized larval trematodes. *The Journal of Parasitology* 46: 317–326

Costa, C.A.F. and Bradley, R.E. 1980. Hyperparasitism of intrasnail stages of *Fasciola hepatica* by a mosquito microsporidian parasite. *Journal of Invertebrate Pathology* 35: 175–181

Couch, J.A. 1974. Pathological effects of *Urosporidium* (Haplosporida) infection in microphallid metacercariae. *Journal of Invertebrate Pathology* 23: 389–396

Couillault, C. and Ewbank, J.J. 2002. Diverse bacteria are pathogens of *Caenorhabditis elegans*. *Infection and Immunity* 70: 4705–4707

Crespo-Gonzalez, C., Rodriguez-Dominguez, H., Soto-Bua, M., Segade, P., Iglesias, R., Arias-Fernandez, C., and Garcia-Estevez, J.M. 2008. Virus-like particles in *Urastoma cyprinae*, a turbellarian parasite of *Mytilus galloprovincialis*. *Diseases of Aquatic Organisms* 79: 83–86

Curtis, L.A. 2007. Larval trematode infections and spatial distributions of snails. *Invertebrate Biology* 126: 235–246

Cusack, R. and Cone, D.K. 1985. A report of bacterial microcolonies on the surface of *Gyrodactylus* (Monogenea). *Journal of Fish Diseases* 8: 125–127

Cusack, R., Rand, T., and Cone, D. 1988. A study of bacterial microcolonies associated with the body surface of *Gyrodactylus colemanensis* Mizelle and Kritsky, 1967 (Monogenea), parasitizing *Salmo gairdneri* Richardson. *Journal of Fish Diseases* 11: 271–274

Darby, C. 2005. Interactions with microbial pathogens. WormBook. The *C. elegans* Research Community, WormBook. http://www.wormbook.org/chapters/www_intermicrobpath/intermicrobpath.html [accessed Oct 2020]

Darby, C., Cosma, C.L., Thomas, J.H., and Manoil, C. 1999. Lethal paralysis of *Caenorhabditis elegans* by *Pseudomonas aeruginosa*. *Proceedings of the National Academy of Sciences of the United States of America*, 96: 15202–15207

Darby, C., Hsu, J.W., Ghori, N., and Falkow, S. 2002. *Caenorhabditis elegans*: Plague bacteria biofilm blocks food intake. *Nature* 417: 243–244

De Buron, I., Loubes, C., and Maurand, J. 1990. Infection and pathological alterations within the acanthocephalan *Acanthocephaloides propinquus* attributable to the microsporidian hyperparasite *Microsporidium acanthocephali*. *Transactions of the American Microscopical Society* 109: 91–97

De Chambrier, A. and Rego, A, 1995. *Mariauxiella pimelodi* n.g., n.sp. (Cestoda: Monticelliidae): A parasite of pimelodid siluroid fishes from South America. *Systematic Parasitology* 30: 57–65

De León, R.P. and Volonterio, O. 2018. A new species of *Temnocephala* (Platyhelminthes) with an unusual pharynx, including an amendment of the diagnosis of the genus. *Zootaxa* 4378: 323–336

Desportes, I, and Schrével, J. 2013. *Treatise on Zoology—Anatomy, Taxonomy, Biology. The Gregarines* (2 vols). Leiden, The Netherlands: Brill

Deturk, W.E. 1940. The occurrence and development of a hyper-parasite, *Urosporidium crescens* sp. nov. (Sporozoa, Haplosporidia), which infests the metacercariae of *Spelotrema nicolli*, parasitic in *Callinectes sapidus*. *Journal of the Elisha Mitchell Scientific Society* 56: 231–232

Dike, S.C. 1971. Ultrastructure of the esophageal region in *Schistosoma mansoni*. *American Society of Tropical Medicine and Hygiene* 20: 552–568

Dobell, C. 1940. Researches on the intestinal protozoa of monkeys and man. X. The life history of *Dientamoeba fragilis*: Observations, experiments, and speculations. *Parasitology* 32: 417–461

Dollfus, R.P. 1946. Parasites (animaux et végétaux) des helminthes: Hyperparasites, ennemis et prédateurs des helminthes parasites et des helminthes libres. Essai de compilation méthodique. *Encyclopédie Biologique*, 27, 462pp. Paris: Lechevalier

Drechsler, C. 1935. A new species of conidial phycomycete preying on nematodes. *Mycologia* 27(2): 206–215

Dugarov, Z.N., Batueva, M.D., and Pronina, S.V. 2011. Hyperparasitism by *Myxobilatus paragasterostei* Zaika, 1963 (Myxozoa: Myxosporea) in *Phyllodistomum folium* (Olfers, 1926) (Trematoda: Gorgoderidae), a parasite of the Siberian dace *Leuciscus baicalensis*. *Bulletin of the European Association of Fish Pathologists* 31: 31–35

El-Naggar, M.M. and Kearn, G.C. 1989. Haptor glands in the gill-parasitic, ancyrocephaline monogenean *Cichlidogyrus halli typicus* and the report of a possible prokaryotic symbiont. *International Journal for Parasitology* 19: 401–408

Emde, S., Rueckert, S., Kochmann, J., Knopf, K., Sures, B., and Klimpel, S. 2014. Nematode eel parasite found inside acanthocephalan cysts -A "Trojan horse" strategy? *Parasites & Vectors* 7: 504

Feistel, D.J., Elmostafa, R., Nguyen, N., Penley, M., Morran, L., and Hickman, M.A. 2019. A novel virulence phenotype rapidly assesses *Candida* fungal pathogenesis in healthy and immunocompromised *Caenorhabditis elegans* hosts. *mSphere* 4: e00697–18

Felix, M.A., Ashe, A., Piffaretti, J., Wu, G., Nuez, I., Belicard, T., Jiang, Y., Zhao, G., Franz, C.J., Goldstein, L.D., Sanroman, M., Miska, E.A., and Wang, D. 2011. Natural and experimental infection of *Caenorhabditis* nematodes by novel viruses related to nodaviruses. *PLoS Biology* 9(1): e1000586

Felix, M.A. and Wang, D. 2019. Natural viruses of *Caenorhabditis* nematodes. *Annual Review of Genetics* 53: 313–326

Fenn, K. and Blaxter, M. 2004. Are filarial nematode *Wolbachia* obligate mutualist symbionts? *Trends in Ecology and Evolution* 19: 163–166

Fischthal, J.H. 1942. A *Paragordius* larva (Gordiacea) in a trematode. *The Journal of Parasitology* 28: 167–167

Foor, W.E. 1972. Viruslike particles in a nematode. *The Journal of Parasitology* 58(6): 1065–1070

Fournier, A., Combes, C., and Vago, C. 1975. Mise en evidence de bacteries endocellulaires pathogenes chez le monogene *Euzetrema knoepffleri*. *Comptes Rendus Hebdomadaires des Séances de l'Académie des Sciences* 281: 1895–1896

Fowler, M. 1970. New Zealand predaceous fungi. *New Zealand Journal of Botany* 8: 283–302

Franzen, C., Nassonova, E.S., Schölmerich, J., and Issi, I.V. 2006. Transfer of the members of the genus *Brachiola* (Microsporidia) to the genus *Anncaliia* based on ultrastructural and molecular data. *Journal of Eukaryotic Microbiology* 53: 26–35

Freeman, M.A. and Shinn, A.P. 2011. Myxosporean hyperparasites of gill monogeneans are basal to the Multivalvulida. *Parasites & Vectors* 4: 220

Frézal, L., Jung, H., Tahan, S., Wang, D., and Félix, M.A. 2019. Noda-like RNA viruses infecting *Caenorhabditis* nematodes: Sympatry, diversity, and reassortment. *Journal of Virology* 93: e01170–19

Frolov, A.O. and Kornakova, E.E. 2001. *Cryptobia udonellae* sp. n. (Kinetoplastidea: Cryptobiida)—Parasites of the excretory system of *Udonella murmanica* (Udonellida). *Parazitologiya* 35: 458–459

Gaevskaya, A.V. 1978. Several cases of accidental hyperparasitism in cestodes. *Zoologicheskĭ Zhurnul* 57: 1262–1263

Gallagher, L.A. and Manoil, C. 2001. *Pseudomonas aeruginosa* PAO1 kills *Caenorhabditis elegans* by cyanide poisoning. *Journal of Bacteriology* 183: 6207–6214

Gao, Q., Nie, P., and Yao, W. 2001. Observation of *Ancyrocephalus mogurndae* (Monogenea: Ancyrocephalidae) from the gills of the mandarin fish, *Siniperca chuatsi* by scanning electron microscopy. *Acta Hydrobiologica Sinica* 255: 597–604

Garner, M.M., Lambourn, D.M., Jeffries, S.J., Hall, P.B., Rhyan, J.C., Ewalt, D.R., Polzin, L.M., and Cheville, N.F. 1997. Evidence of *Brucella* infection in *Parafilaroides* lungworms in a Pacific harbor seal (*Phoca vitulina richardsi*). *Journal of Veterinary Diagnostic Investigation* 9: 298–303

Garsin, D.A., Sifri, C.D., Mylonakis, E., Qin, X., Singh, K.V., Murray, B.E., Calderwood, S.B., and Ausubel, F.M. 2001. A simple model host for identifying Gram-positive virulence factors. *Proceedings of the National Academy of Sciences of the United States of America* 98: 10892–10897

Giblin-Davis, R.M., Williams, D.S., Bekal, S., Dickson, D.W., Brito, J.A., Becker, J.O., and Preston, J.F. 2003. '*Candidatus Pasteuria usgae*' sp. nov., an obligate endoparasite of the phytoparasitic nematode *Belonolaimus longicaudatus*. *International Journal of Systematic and Evolutionary Microbiology* 53: 197–200

Ginetsinskaya, T.A. 1968. *Trematodes; their Life-Cycles, Biology and Evolution*, 411pp., Nauka, Leningrad: Izdatelstvo

Girginkardeşler, N., Kurt, Ö., Kilimcioğlu, A.A., and Ü.Z. 2008. Transmission of *Dientamoeba fragilis*: Evaluation of the role of *Enterobius vermicularis*. *Parasitology International* 57: 72–75

Graham, G.L. 1935. *Giardia* infections in a nematode from cattle. *The Journal of Parasitology* 21: 127–128

Guyenot, E. and Naville, A. 1922. Un nouveau protiste, du genre *Dermocystidium*, parasite de la grenouille, *Dermocystidium ranae* nov. spec. *Revue Suisse de Zoologie*, 29: 133–145

Guyénot, E. and Naville, A. 1924. *Glugea encyclometrae* n. sp. et *G. ghigii* n. sp., parasites de *Platodes* et leur developpement dans l'hote vertebre (*Tropidonotus natrix* L.). *Revue Suisse de Zoologie* 31: 75–115

Halton, D.W. 1972. Ultrastructure of the alimentary tract of *Aspidogaster conchicola* (Trematoda: Aspidogastrea). *The Journal of Parasitology* 58: 455–467

Hanelt, B. 2009. Hyperparasitism by *Paragordius varius* (Nematomorpha: Gordiida) larva of monostome redia (Trematoda: Digenea). *The Journal of Parasitology* 95: 242–243

Harrath, A.H., Sluys, R., Aldahmash, W., Al-Razaki, A., and Alwasel, S. 2013. Reproductive strategies, karyology, parasites, and taxonomic status of *Dugesia* populations from Yemen (Platyhelminthes: Tricladida: Dugesiidae). *Zoological Science* 30: 502–508

Hashem, M., Omran, Y.A.M.M., and Sallam, N.M.A. 2008. Efficacy of yeasts in the management of root-knot nematode *Meloidogyne incognita*, in flame seedless grape vines and the consequent effect on the productivity of the vines. *Biocontrol Science and Technology* 18(4): 357–375

Hess, R.T. and Poinar, G.O. Jr. 1985. Iridoviruses infecting terrestrial isopods and nematodes. *Current Topics in Microbiology and Immunology* 116: 49–76

Hodgkin, J., Kuwabara, P.E., and Corneliussen, B. 2000. A novel bacterial pathogen, *Microbacterium nematophilum*, induces morphological change in the nematode *C. elegans*. *Current Biology* 10(24): 1615–1618

Hoffman, G.L. 1953. *Streptomyces leidnematis* n. sp., growing on two species of nematodes of the cockroach. *Transactions of the American Microscopical Society* 72(4): 376–378

Hopper, B.E., Meyers, S.P., and Cefalu, R. 1970. Microsporidian infection of a marine nematode *Metoncholaimus scissus*. *Journal of Invertebrate Pathology* 16: 371–377

Howell, M. 1967. The trematode, *Bucephalus longicornutus* (Manter, 1954) in the New Zealand mud-oyster, *Ostrea lutaria*. *Transactions and Proceedings of the Royal Society of New Zealand* 8: 221–237

Hughes-Stamm, S.R., Cribb, T.H., and Jones, M.K. 1999. Structure of the tegument and ectocommensal microorganisms of *Gyliauchen nahaensis* (Digenea: Gyliauchenidae), an inhabitant of herbivorous fish of the Great Barrier Reef, Australia. *The Journal of Parasitology* 85: 1047–1052

Humes, A.G. 1997. *Pseudanthessius newmanae*, new species (Copepoda: Poecilostomatoida: Pseudanthessiidae) from marine turbellarians in Australia. *Memoirs of the Queensland Museum* 42: 227–231

Hunninen, A.V. and Wichterman, R. 1938. Hyperparasitism: A species of *Hexamita* (Protozoa, Mastigophora) found in the reproductive systems of *Deropristis inflata* (Trematoda) from marine eels. *Journal of Parasitology* 24: 95–101.

Hussey, K.L. 1971. A microsporidan hyperparasite of strigeoid trematodes, *Nosema strigeoideae* sp. n. *The Journal of Protozoology* 18: 676–679

Illg, P.L. 1950. A new copepod, *Pseudanthessius latus* (Cyclopoida: Lichomolgidae), commensal with a marine flatworm. *Journal of the Washington Academy of Sciences* 40: 129–133

Jansen, W.T., Bolm, M., Balling, R., Chhatwal, G.S., and Schnabel, R. 2002. Hydrogen peroxide-mediated killing of *Caenorhabditis elegans* by *Streptococcus pyogenes*. *Infection and Immunity* 70: 5202–5207

Jansson, H.B. 1994. Adhesion of conidia of *Drechmeria coniospora* to *Caenorhabditis elegans* wild type and mutants. *Journal of Nematology*, 26: 430–435

Jennings, J.B., Cannon, L.R.G., and Hick, A.J. 1992. The nature and origin of the epidermal scales of *Notodactylus handschini* – an unusual temnocephalid turbellarian ectosymbiotic on crayfish from Northern Queensland. *The Biological Bulletin* 182: 117–128

Joe, L.K. and Nasemary, M. 1973. Transmission of *Nosema eurytremae* (Microsporida: Nosematidae) to various trematode larvae. *Zeitschrift für Parasitenkunde* 41: 109–117

Johnson, C.H., Ayyadevara, S., McEwan, J.E., and Shmookler Reis, R.J. 2009. *Histoplasma capsulatum* and *Caenorhabditis elegans*: A simple nematode model for an innate immune response to fungal infection. *Medical Mycology* 47: 808–813

Jones, A.W. 1943. A further description of *Stempellia moniezi* Jones, 1942, a microsporidian parasite (Nosematidae) of cestodes. *The Journal of Parasitology* 29: 373–378

Jones, M.K. and Whittington, I.D. 1992. Nuclear bodies in the egg cells of a *Gyrodactylus* species (Platyhelminthes, Monogenea). *Parasitology Research* 78: 534–536

Justine, J.L. and Bonami, J.R. 1993. Virus-like particles in a monogenean (Platyhelminthes) parasitic in a marine fish. *International Journal for Parasitology* 23: 69–75

Justine, J.L., De León, R.P., Mattei, X., and Bonami, J.R. 1991. Viral particles in *Temnocephala iheringi* (Platyhelminthes, Temnocephalidea), a parasite of the mollusc *Pomacea canaliculata*. *Journal of Invertebrate Pathology* 57: 287–289

Kahl, A. 1931. *Urtiere oder Protozoa I: Wimpertiere oder Ciliata (Infusoria 2. Holosticha. Tierwelt Deutschlands)*: Fischer Verlag, Jena.

Karling, T.G. 1970. On *Pterastericola fedotovi* (Turbellaria), commensal in sea stars. *Zeitschrift für Morphologie der Tiere* 67: 29–39

Keeble, F. 1908. The yellow-brown cells of *Convoluta paradoxa*. *Quarterly Journal of Microscopy Science* 52: 431–479

Kenk, R. 1969. Freshwater triclads (Turbellaria) of North America. I. The genus *Planaria*. *Proceedings of the Biological Society of Washington* 82: 539–558

Kerry, B. 1980. Biocontrol fungal parasites of female cyst nematodes. *Journal of Nematology* 12(4): 253–259

Ketudat, U. 1969. The effects of some soil-borne fungi on the sex ratio of *Heterodera rostochiensis* on tomato. *Nematologica* 15: 229–233

Kinne, O. 1980. *Diseases of Marine Animals*. Volume I. Chichester: John Wiley and Sons.

Kishida, Y. and Asai, E. 1977. Les cristaux intranucléaires des cellules pharyngiennes chez la planaire, *Dugesia japonica* Ichikawa et Kawakatsu. *Journal of Electron Microscopy* 26: 145–147

Kothe, M., Antl, M., Huber, B., Stoecker, K., Ebrecht, D., Steinmetz, I., and Eberl, L. 2003. Killing of *Caenorhabditis elegans* by *Burkholderia cepacia* is controlled by the cep quorum-sensing system. *Cellular Microbiology* 5: 343–351

Kramer De Mello, I.N., Braga, F.R., Avelar Monteiro, T.S., Freitas, L.G., Araujoa, J.M., Freitas Soares, F.E., and Araújo, J.V. 2014. Biological control of infective larvae of *Ancylostoma* spp. in beach sand. *Revista Iberoamericana de Micología* 31(2): 114–118

Kudo, R. and Hetherington, D.C. 1922. Notes on a microsporidian parasite of a nematode. *The Journal of Parasitology* 8(3): 129–132

Kurz, C.L., Chauvet, S., Andres, E., Aurouze, M., Vallet, I., Michel, G.P., Uh, M., Celli, J., Filloux, A., De Bentzmann, S., Steinmetz, I., Hoffmann, J.A., Finlay, B.B., Gorvel, J.P., Ferrandon, D., and Ewbank, J.J. 2003. Virulence factors of the human opportunistic pathogen *Serratia marcescens* identified by *in vivo* screening. *EMBO Journal* 22: 1451–1460

Laclette, J.P., Merchant, M.T., Damian, R.T., and Willms, K. 1990. Crystals of virus-like particles in the metacestodes of *Taenia solium* and *T. crassiceps*. *Journal of Invertebrate Pathology* 56: 215–221

Lasee, B.A. and Sutherland, D.R. 1993. Bacterial colonization of tegumental surfaces of *Culaeatrema inconstans*

Lasee et al. 1988 (Digenea) from the brook stickleback, *Culaea inconstans*. *Journal of Fish Diseases* 16: 83–85

Le, T.C., Kang, H.S., Hong, H.K., Park, K.J., and Choi, K.S. 2015. First report of *Urosporidium* sp., a haplosporidian hyperparasite infecting digenean trematode *Parvatrema duboisi* in Manila clam, *Ruditapes philippinarum* on the west coast of Korea. *Journal of Invertebrate Pathology* 130: 141–146

Lebour, M.V. 1908. *Fish Trematodes of the Northumberland Coast*. Northumberland Sea Fisheries Report for 1907, 1–62, Newcastle-upon-Tyne: Cail & Sons

Lee, D.L. 1969. The structure and development of *Histomonas meleagridis* (Mastigamoebidae: Protozoa) in the female reproductive tract of its intermediate host, *Heterakis gallinarum* (Nematoda). *Parasitology* 59: 877–884

Lester, R.J.G. and Wright, K.A. 1978. Abnormal swellings in the tegument of the acanthocephalan parasite *Metechinorhynchus salmonis*. *Journal of Invertebrate Pathology* 31: 271–274

Levron, C., Ternengo, S., Toguebaye, B.S., and Marchand, B. 2004. Ultrastructural description of the life cycle of *Nosema diphterostomi* sp. n., a microsporidia hyperparasite of *Diphterostomum brusinae* (Digenea: Zoogonidae), intestinal parasite of *Diplodus annularis* (Pisces: Teleostei). *Acta Protozoologica* 43: 329–336

Levron, C., Ternengo, S., Toguebaye, B.S., and Marchand, B. 2005. Ultrastructural description of the life cycle of *Nosema monorchis* n. sp. (Microspora, Nosematidae), hyperparasite of *Monorchis parvus* (Digenea, Monorchiidae), intestinal parasite of *Diplodus annularis* (Pisces, Teleostei). *European Journal of Protistology* 41: 251–256

Lindbolm, G.B. and Nilsson, L.A. 1994. Interaction between *Campylobacter jejuni/coli* and *Schistosoma mansoni*, a helminth parasite. *Bulletin of the Scandinavian Society for Parasitology* 4: 1–8

Linton, E. 1924. Notes on cestode parasites of sharks and skates. *Proceedings of the United States National Museum* 64: 1–114

Linton, E. 1940. Trematodes from fishes mainly from the Woods Hole region, Massachusetts. *Proceedings of the United States National Museum* 88: 1–172

Loewenberg, J.R., Sullivan, T., and Schuster, M.L. 1959. A virus disease of *Meloidogyne incognita*, the southern root knot nematode. *Nature* 184 (Suppl. 24): 1896

Lom, J. 2002. A catalogue of described genera and species of microsporidians parasitic in fish. *Systematic Parasitology* 53: 81–99

Longshaw, M., Frear, P.A., and Feist, S.W. 2005. Descriptions, development and pathogenicity of myxozoan (Myxozoa: Myxosporea) parasites of juvenile cyprinids (Pisces: Cyprinidae). *Journal of Fish Diseases* 28: 489–508

Lopez-Llorca, L.V., Olivares-Bernabeu, C., Salinas, J., Jansson, H.B., and Kolattukudy, P.E. 2002. Pre-penetration events in fungal parasitism of nematode eggs. *Mycological Research* 106: 499–506

Lopez-Llorca, L.V. and Robertson, W.M. 1992. Immunocytochemical localization of a 32-kDa protease from the nematophagous fungus *Verticillium suchlasporium* in infected nematode eggs. *Experimental Mycology* 16: 261–267

Loubes, C., Maurand, J., and De Buron, I. 1988. Premières observations sur deux microsporidies hyperparasites d'acanthocéphales de poissons marins et lagunaires. *Parasitology Research* 74: 344–351

Lovy, J. and Friend, S.E. 2017. Phylogeny and morphology of *Ovipleistophora diplostomuri* n. sp. (Microsporidia) with a unique dual-host tropism for bluegill sunfish and the digenean parasite *Posthodiplostomum minimum* (Strigeatida). *Parasitology* 144: 1898–1911

Lundin, K. 1998. Symbiotic bacteria on the epidermis of species of the Nemertodermatida (Platyhelminthes, Acoelomorpha). *Acta Zoologica* 79: 187–191

Luo, H., Li, X., Li, G.H., Pan, Y.B., and Zhang, K.Q. 2006. Acanthocytes of *Stropharia rugosoannulata* function as a nematode-attacking device. *Applied and Environmental Microbiology* 72: 2982–2987

Luo, H., Mo, M.H., Huang, X.W., Li, X., and Zhang, K.Q. 2004. *Coprinus comatus*: A basidiomycete fungus forms novel spiny structures and infects nematodes. *Mycologia* 96: 1218–1225

Lutz, A. and Splendore, A. 1908. Üeber Pebrine und vermandte Mikrosporidien. II Mitteilungen. Centr. Bakt., I Orig., *Zeitschrift fur Parasitenkunde* 46: 311–315

Lysek, H. and Št ěrba, J. 1991. Colonization of *Ascaris lumbricoides* eggs by the fungus *Verticillium chlamydosporium* Goddard. *Folia Parasitologica* 38: 255–259

Mackiewicz, J.S. 1972. Caryophyllidea (Cestoidea): A review. *Experimental Parasitology* 31: 417–512

Malmberg, G. 1986. The major parasitic platyhelminth classes—progressive or regressive evolution? *Hydrobiologia* 132: 23–29

Manter, H.W. 1929. A disease of *Ascaris lumbricoides*. The *Journal of Parasitology* 16: 101

Manter, H.W. 1943. One species of trematode, *Neorenifer grandispinus* (Caballero, 1938) attacked by another, *Mesocercaria marcianae* (La Rue, 1917). *The Journal of Parasitology* 29: 387–392

Marroquin, L.D., Elyassnia, D., Griffitts, J.S., Feitelson, J.S., and Aroian, R.V. 2000. *Bacillus thuringiensis* (Bt) toxin susceptibility and isolation of resistance mutants in the nematode *Caenorhabditis elegans*. *Genetics* 155: 1693–1699

Mekhraliev, A.A. and Mikailov, T.K. 1981. Cases of hyperparasitism of metacercarians *Tetracotyle* sp. in partenites of trematodes. *Parazitologiya* 15: 80–83

Menke, J.H. 1968. Urosporidium astomatum *n. sp., A Haplosporidian Infecting the Metacercariae of* Parvatrema donacis, *A Trematode Parasite of the* bivalve Donax variabilis. Texas A&M University, USA: PhD Thesis

Mennie, D.A.R., Collins, C., and Bruno, D.W. 2000. Colonisation of *Gyrodactylus derjavini* (Monogenea: Gyrodactylidae) by fungal-like hyphae. *Bulletin of the European Association of Fish Pathologists* 20: 215–216

Miller, R.B. 1946. Cestode 'parasitized' by acanthocephalan. *Science* 103: 762

Mokhtar-Maamouri, F., Lambert, A., Maillard, C., and Vago C 1976. Viral infection in a platyhelminth parasite. *Comptes Rendus Hebdomadaires des Seances de l'Academie des Sciences Serie D: Sciences Naturelles* 283: 1249–1251

Morris, G.P. and Halton, D.W. 1975. The occurrence of bacteria and mycoplasma-like organisms in a monogenean parasite, *Diclidophora merlangi*. *International Journal for Parasitology* 5: 495–498

Mueller, J.F. and Strano, A.J. 1974a. *Sparganum proliferum*, a *Sparganum* infected with a virus? *The Journal of Parasitology* 60: 15–19

Mueller, J.F. and Strano, A.J. 1974b. The ubiquity of type-C viruses in spargana of *Spirometra* spp. *The Journal of Parasitology* 60: 398

Mylonakis, E., Ausubel, F.M., Perfect, J.R., Heitman, J., and Calderwood, S.B. 2002. Killing of *Caenorhabditis elegans* by *Cryptococcus neoformans* as a model of yeast pathogenesis. *Proceedings of the National Academy of Sciences of the United States of America* 99: 15675–15680

Noury-Shaïri, N., Justine, J.L., and Bonami, J.R. 1995. Viral particles in a flatworm (*Paravortex tapetis*) parasitic in the commercial clam, *Ruditapes decussatus*. *Journal of Invertebrate Pathology* 65: 200–202

O'Quinn, A.L., Wiegand, E.M. and Jeddeloh, J.A. 2001. *Burkholderia pseudomallei* kills the nematode *Caenorhabditis elegans* using an endotoxin-mediated paralysis. *Cellular Microbiology* 3: 381–393

Ormières, R., Sprague, V., and Bartoli, P. 1973. Light and electron microscope study of a new species of *Urosporidium* (Haplosporida), hyperparasite of trematode sporocysts in the clam *Abra ovata*. *Journal of Invertebrate Pathology* 21: 71–86

Oschman, J.L. 1969. Endonuclear viruslike bodies in *Convoluta roscoffensis* (Turbellaria, Acoela). *Journal of Invertebrate Pathology* 13: 147–148

Osman, G.A., Fasseas, M.K., Koneru, S.L., Essmann, C.L., Kyrou, K., Srinivasan, M.A., Zhang, G., Sarkies, P., Felix, M.A., and Barkoulas, M. 2018. Natural infection of C.

elegans by an oomycete reveals a new pathogen-specific immune response. *Current Biology* 28: 640–648

Ott, J., Rieger, G., Rieger, R., and Enderes, F. 1982. New mouthless interstitial worms from the sulfide system: Symbiosis with prokaryotes. *Marine Ecology* 3: 313–333

Overstreet, R.M. 1976. Fabespora vermicola sp. n., the first myxosporidan from a platyhelminth. *The Journal of Parasitology* 62: 680–684

Page, A.P., Roberts, M., Felix, M.A., Pickard, D., Page, A., and Weir, W. 2019. The golden death bacillus *Chryseobacterium nematophagum* is a novel matrix digesting pathogen of nematodes. *BMC Biology* 17: 10

Paladini, G., Longshaw, M., Gustinelli, A., and Shinn, A.P. 2017. Parasitic diseases in aquaculture: Their biology, diagnosis and control. In *Diagnosis and Control of Diseases of Fish and Shellfish*, B. Austin and A. Newaj-Fyzul (eds.), pp. 37–107. Chichester, UK: John Wiley & Sons

Palmieri, J.R., Lai, P.C., Sullivan, J.T., and Cali, A. 1978. Effects of Microspora on snail and trematode tissue. *The Southeast Asian Journal of Tropical Medicine and Public Health* 9: 256–259

Paperna, I., Sabnai, I., and Castel, M. 1978. Microsporidian infection in the cyst wall of trematode metacercariae encysted in fish. *Annales de Parasitologie Humaine et Comparée* 53: 123–130

Pekkarinen, M. 1991. A coccidian hyperparasite in bucephalid trematode sporocysts in brackish water (Baltic Sea) *Mytilus edulis*. *Journal of Invertebrate Pathology* 57: 292–293

Pekkarinen, M. 1993. Bucephalid trematode sporocysts in brackish-water *Mytilus edulis*, new host of a *Helicosporidium* sp. (Protozoa: Helicosporida). *Journal of Invertebrate Pathology* 61: 214–216

Perkins, F.O. 1971. Sporulation in the trematode hyperparasite *Urosporidium crescens* De Turk, 1940 (Haplosporida: Haplosporidiidae): An electron microscope study. *The Journal of Parasitology* 57: 9–23

Perkins, F.O. 1979. Cell structure of shellfish pathogens and hyperparasites in the genera *Minchinia*, *Urosporidium*, *Haplosporidium*, and *Marteilia*—taxonomic implications. *Marine Fisheries Review* 41: 25–37

Perkins, F.O., Zwerner, D.E., and Dias, R.K. 1975. The hyperparasite, *Urosporidium spisuli* sp. n. (Haplosporea), and its effects on the surf clam industry. *The Journal of Parasitology* 61(5): 944–949

Poddubnaya, L.G., Tokarev, Y.S., and Issi, I.V. 2006. A new microsporidium *Paratuzetia kupermani* gen. et sp. n. (Microsporidia), a hyperparasite of the procercoid of the cestode *Khawia armeniaca* Chol. 1915 (Cestoda, Caryophyllidea). *Protistology* 4: 269–277

Poinar, G.O. Jr. 1988. A microsporidian parasite of *Neoaplectana glaseri* (Steinernemtidae: Rhabditida). *Revue de Nématologie* 11(3): 359–361

Poinar, G.O. Jr. and Hess, R. 1977. Virus like particles in the nematode *Romanomermis culicivorax* (Mermithidae) *Nature* 266(5599): 256–257

Poinar, G.O. Jr. and Hess, R. 1986. *Microsporidium rhabdophilum* sp. n. (Microsporida: Pansporoblastina), a parasite of the nematode, *Rhabditis myriophila* (Rhabditina: Rhabditidae). *Revue de Nématologie* 9: 369–375

Poinar, G.O. Jr. and Hess, R. 1988 Protozoan diseases. In *Diseases of Nematodes*. G.O. Poinar Jr. and H.H. Jansson (eds.), Volume 1, pp. 103–131, Boca Raton, FL: CRC Press

Poinar, G.O. Jr., Hess, R.T., and Cole, A. 1980. Replication of an iridovirus in a nematode (Mermithidae). *Intervirology* 14: 316–320

Prudhoe, S. 1945. XXXVI. Two notes on trematodes. *Annals and Magazine of Natural History* 12: 378–383

Pung, O.J., Khan, R.N., Vives, S.P., and Walker, C.B. 2002. Prevalence, geographic distribution, and fitness effects of *Microphallus turgidus* (Trematoda: Microphallidae) in grass shrimp (*Palaemonetes* spp.) from coastal Georgia. *The Journal of Parasitology* 88: 89–92

Rataj, M. and Vd'ačný, P. 2019. Living morphology and molecular phylogeny of oligohymenophorean ciliates associated with freshwater turbellarians. *Diseases of Aquatic Organisms* 134: 147–166

Rataj, M. and Vd'ačný, P. 2020. Multi-gene phylogeny of *Tetrahymena* refreshed with three new histophagous species invading freshwater planarians. *Parasitology Research* 119: 1523–1545

Rebrikov, D.V., Bogdanova, E.A., Bulina, M.E., and Lukyanov, S.A. 2002a. A new planarian extrachromosomal virus-like element revealed by subtractive hybridization. *Molecular Biology* 36: 813–820

Rebrikov, D.V., Bulina, M.E., Bogdanova, E.A., Vagner, L.L., and Lukyanov, S.A. 2002b. Complete genome sequence of a novel extrachromosomal virus-like element identified in planarian *Girardia tigrina*. *BMC Genomics* 3: 15

Reece, K.S., Siddall, M.E., Stokes, N.A., and Burreson, E.M. 2004. Molecular phylogeny of the Haplosporidia based on two independent gene sequences. *Journal of Parasitology* 90: 1111–1122

Rego, A. and Gibson, D. 1989. Hyperparasitism by helminths: New records of cestodes and nematodes in proteocephalid cestodes from South American siluriform fishes. *Memórias do Instituto Oswaldo Cruz* 84: 371–376

Reisinger, E. 1929. Zum ductus genito-intestinalis-problem. I. Über primäre geschlechtstrakt-darmverbindungen bei rhabdocoelen turbellarien.— Zugleich ein Beitrag zur europäischen und grönländischen Turbellarienfauna. *Zeitschrift für Morphologie und Ökologie der Tiere* 16: 49–73

Reuter, M. 1975. Viruslike particles in *Gyratrix hermaphroditus* (Turbellaria: Rhabdocoela). *Journal of Invertebrate Pathology* 25: 79–95

Reyda, F.B. and Olson, P.D. 2003. Cestodes of cestodes of Peruvian freshwater stingrays. *The Journal of Parasitology* 89: 1018–1024

Rohde, K. 1986. Ultrastructure of the pharynx and some parenchyma cells of *Zeuxapta seriolae* and *Paramicrocotyloides reticularis* (Monogenea: Polyopisthocotylea: Microcotylidae). *Australian Journal of Zoology* 34: 473–484

Röser, D., Nejsum, P., Carlsgart, A.J., Nielsen, H.V., and Stensvold, C.R. 2013. DNA of *Dientamoeba fragilis* detected within surface-sterilized eggs of *Enterobius vermicularis*. *Experimental Parasitology* 133: 57–61

Ruark, C.L., Gardner, M., Mitchum, M.G., Davis, E.L., and Sit, T.L. 2018. Novel RNA viruses within plant parasitic cyst nematodes. *PLoS ONE* 13(3): e0193881

Sapir, A., Dillman, A.R., Connon, S.A., Grupe, B.M., Ingels, J., Mundo-Ocampo, M., Levin, L.A., Baldwin, J.G., Orphan, V.J., and Sternberg, P.W. 2014. Microsporidia-nematode associations in methane seeps reveal basal fungal parasitism in the deep sea. *Frontiers in Microbiology* 5: 43

Schaefer, F.W. and Etges, F.J. 1969. Hyperparasitism of a larval cestode by a larval fluke. *The Journal of Parasitology* 55: 462

Schärer, L., Knoflach, D., Vizoso, D.B., Rieger, G., and Peintner, U. 2007. Thraustochytrids as novel parasitic protists of marine free-living flatworms: *Thraustochytrium caudivorum* sp. nov. parasitizes *Macrostomum lignano*. *Marine Biology* 152: 1095–1104

Sene, A., Ba, C.T., Marchand, B., and Toguebaye, B.S. 1997. Ultrastructure of *Unikaryon nomimoscolexi* n. sp. (Microsporida, Unikaryonidae), a parasite of *Nomimoscolex* sp. (Cestoda, Proteocephalidea) from the gut of *Clarotes laticeps* (Pisces, Teleostei, Bagridae). *Diseases of Aquatic Organisms* 29: 35–40

Sey, O. and Moravec, F. 1986. An interesting case of hyperparasitism of the nematode *Spironoura babei* Ha Ky, 1971 (Nematoda: Kathlaniidae). *Helminthologia* (Bratisl) 23: 173–177

Shigina, N. and Grobov, O. 1972. *Nosema diplostomi* sp. n. (Microsporidia: Nosematidae), a hypeparasite of trematodes of the genus *Diplostomum*. *Parazitologiya* 5: 469–475

Shipley, A.E. 1909. The thread-worms (Nematoda) of the red grouse (*Lagopus scoticus*). *Proceedings of the Zoological Society of London* 23: 335–350

Shope, R.E. 1941. Swine lungworm as reservoir and intermediate host for swine influenza virus; transmission of swine influenza virus by swine lungworm. *Journal of Experimental Medicine* 74: 41–68

Shotts, E.B., Foster, J.W., Brugh, M., Jordan, H.E., and McQueen, J.L. 1968. An intestinal threadworm as a reservoir and intermediate host for swine influenza virus. *Journal of Experimental Medicine* 127: 359–369

Siau, Y., Gasc, C., and Maillard, C. 1981. Premieres observations ultrastructurales d'une myxosporidie appartenant au genre *Fabespora*, parasite de trematode. *Protistologica* 17: 131–137

Sikora, J. 1963. Study on the parasitic ciliate *Steinella uncinata* (Schultze). *Acta Protozoologica* 1: 13–20

Silva, M.E., Silveira, W.F., Braga, F.R., and Araujo, J.V. 2017. Nematicide activity of microfungi (Orbiliales, Orbiliaceae) after transit through gastrointenstinal tract of '*Gallus domesticus*'. *Revista Brasileira de Saúde e Produção Animal* 18(1): 1–9

Singh, S. and Mathur, N. 2010. Biological control of root-knot nematode, *Meloidogyne incognita* infesting tomato. *Biocontrol Science and Technology* 20(8): 865–874

Smith, R.J. 1959. Ancylid snails: First intermediate host to certain trematodes with notes on ancylids as a new host for *Megalodiscus* and *Haematoloechus*. *Transactions of the American Microscopical Society* 78: 228–231

Sokolova, Y.Y. and Overstreet, R.M. 2020 Hyperparasitic spore-forming eukaryotes (Microsporidia, Haplosporidia, and Myxozoa) parasitizing trematodes (Platyhelminthes). *Invertebrate Zoology* 17: 93–117

Sprague, V. 1964. *Nosema dollfusi* n. sp. (Microsporidia, Nosematidae), a Hyperparasite of *Bucephalus cuculus* in *Crassostrea virginica*. *The Journal of Protozoology* 11: 381–385

Stewart, T.B. and Godwin, H.J. 1963. Cuticular lesions of *Ascaris suum* caused by *Pseudomonas* sp. *The Journal of Parasitology* 49(2): 231–234

Stocchino, G.A., Sluys, R., Kawakatsu, M., Sarbu, S.M., and Manconi, R. 2017. A new species of freshwater flatworm (Platyhelminthes, Tricladida, Dendrocoelidae) inhabiting a chemoautotrophic groundwater ecosystem in Romania. *European Journal of Taxonomy* 342: 1–21

Stunkard, H.W. 1950. Further observations on *Cercaria parvicaudata* Stunkard and Shaw, 1931. *The Biological Bulletin* 99: 136–142

Sukanahaketu, S. 1977. The presence of *Dientamoeba fragilis* in the *Ascaris lumbricoides* ova: The first report from Thailand. *Journal of the Medical Association of Thailand* 60: 265–268

Syverton, J.T., McCoy, O.R., and Koomen, J. 1947. The transmission of the virus of lymphocytic choriomeningitis by *Trichinella spiralis*. *Journal of Experimental Medicine* 85: 759–769

Tajika, K.I. 1979. A new species of the genus *Ciliocincta* Kozloff, 1965 (Mesozoa, Orthonectida) parasitic in a marine turbellarian from Hokkaido, Japan. *Journal of the*

Faculty of Science, Hokkaido University Series Zoology 21: 383–395

Taraschewski, H. 2000. Host-parasite interactions in Acanthocephala: A morphological approach. *Advances in Parasitology*, 46, 1–179

Taylor, C.E. and Robertson, W.M. 1969. The location of raspberry ringspot and tomato black ring viruses in the nematode vector, *Longidorus elongatus* (de Man) *Annals of Applied Biology* 64: 233–237

Taylor, C.E. and Robertson, W.M. 1970. Location of tobacco rattle virus in the nematode vector, *Trichodorus pachydermus* Steinhorst. *Journal of General Virology* 6: 179–182

Taylor, D.L. 1971. On the symbiosis between *Amphidinium klebsii* [Dinophyceae] and *Amphiscolops langerhansi* [Turbellaria: Acoela]. *Journal of the Marine Biological Association of the United Kingdom* 51: 301–313

Thapa, S., Thamsborg, S.M., Wang, R., Meyling, N.V., Dalgaard, T.S., Petersen, H.H., and Mejer, H. 2018. Effect of the nematophagous fungus *Pochonia chlamydosporia* on soil content of ascarid eggs and infection levels in exposed hens. *Parasites & Vectors* 11: 319

Theiler, H. and Farber, S.M. 1932. *Trichomonas muris*, parasitic in oxyurids of the white mouse. *The Journal of Parasitology* 19: 169

Thomson, J.G. 1925. A *Giardia* parasitic in a bursate nematode living in the Viscacha. *Protozoology* 1: 1–6

Thorn, R.G. and Barron, G.L. 1984. Carnivorous mushrooms. *Science* 224: 76–78

Toguebaye, B.S., Quilichini, Y., Diagne, P.M., and Marchand, B. 2014. Ultrastructure and development of *Nosema podocotyloidis* n. sp. (Microsporidia), a hyperparasite of *Podocotyloides magnatestis* (Trematoda), a parasite of *Parapristipoma octolineatum* (Teleostei). *Parasite* 21: 44

Travassos, L. and Travassos, L. 1934. *Atractis trematophila* n. sp., nematodeo parasito do ceco de um trematodeo Paramphistomoidea. *Memórias do Instituto Oswaldo Cruz* 28: 267–269

Troemel, E.R., Félix, M.A., Whiteman, N.K., Barrière, A., and Ausubel, F.M. 2008. Microsporidia are natural intracellular parasites of the nematode *Caenorhabditis elegans*. *PLoS Biology* 6(12): e309

Tuazon, C.U., Nash, T., Cheever, A., Neva, F., and Lininger, L. 1985. Influence of *Salmonella* bacteremia on the survival of mice infected with *Schistosoma mansoni*. *The Journal of Infectious Diseases* 151: 1166–1167

Urabe, M. 2001. Some rare larval trematodes of prosobranch snails, *Semisulcospira* spp., in the Lake Biwa drainage system, central Japan. *Parasitology International* 50: 191–199

Van Der Velde, G. 1978. *Lachmannella recurva* (Ciliata Astomida, Haptophryidae), a parasite of *Limapontia capitata* (Mollusca, Gastropoda). *Basteria* 42: 11–13

Veremtchuk, G.V. and Issi, I.V. 1970. Development of insect microsporidians in the entomopathogenic nematode, *Neoaplectana agriotos* (Nematodea: Steinernematidae). *Parasitologiya* 4: 3–7

Viozzi, G., Flores, V., and Rauque, C. 2005. An ectosymbiotic flatworm, *Temnocephala chilensis*, as second intermediate host for *Echinoparyphium megacirrus* (Digenea: Echinostomatidae) in Patagonia (Argentina *The Journal of Parasitology* 91: 229–231

Voronin, V.N. 1974. Microsporidian parasitism (Microsporidia, Nosematidae) in parthenogenetic generations and cercariae of trematodes from fresh-water molluscs. *Parazitologiia* 8: 359–364

Walter, G. 1858. Fernere beitrage zur anatomie und physiologie von *Oxyuris ornata*. *Zeitschrift für Wissenschaftliche Zoologie* 9(4): 485–495

Watson, N.A. and Jondelius, U. 1997. Spermiogenesis and sperm ultrastructure in *Cylindrostoma fingalianum* (Platyhelminthes, Prolecithophora) with notes on a protozoan (Kinetoplastid) symbiont closely associated with allosperm. *Invertebrate Reproduction and Development* 32: 273–282

Webb, R.A. 1984. Intranuclear bodies in the tissues of the scolex of the cestode *Hymenolepis microstoma*. *Canadian Journal of Zoology* 62: 107–111

Wenyon, C.M. 1926. *Protozoology*, pp. 1396. New York: William, Wood and Company

Westblad, E. 1942. Studien über skandinavische Turbellaria Acoela. II. *Arkiv för Zoologi*, 33A: 1–48

Williams, B.J. 1999. Description of a new flagellate protist *Desmomonas prorhynchi* gen. et sp. n. associated with problematical cell masses, parasitic in the turbellarian *Prorhynchus* sp. (Lecithoepitheliata). *Folia Parasitologica* 46: 248–256

Williams, J.B. 1991. Rickettsiae and giant lysosomes in the testes of *Temnocephala novaezealandiae* (Platyhelminthes: Temnocephaloidea). *Journal of Submicroscopic Cytology and Pathology* 23: 447–455

Williams, J.B. 1992. Ultrastructure of the intestinal epithelium and prey microorganisms of *Troglocaridicola mrazeki* (Platyhelminthes: Scutariellidae). *Journal of Submicroscopic Cytology and Pathology* 24: 473–481

Williams, J.B. 1994. Ultrastructural observations on *Temnocephala minor* (Platyhelminthes, Temnocephaloidea), including notes on endocytosis. *New Zealand Journal of Zoology* 21: 195–208

Wood, F.H. 1973. Nematode-trapping fungi from a tussock grassland soil in New Zealand. *New Zealand Journal of Botany* 2: 231–240

Wright, J.F. 1981. *Tetrahymena pyriformis* (Ehrenberg) and *T. corlissi* Thompson parasitic in stream-dwelling triclads (Platyhelminthes: Turbellaria). *The Journal of Parasitology* 67: 131–133

Zaika, V.E. and Dolgikh, A.V. 1963. A rare case of hyperparasitism of *Haplosporidium* of *Urosporidium tauricum* sp. n. in partenites of trematodes of the family Hemiuridae Luhe from the mollusc *Rissoa splendida* Eichw. *Zoologicheskiĭ Zhurnal* 42: 1727–1729

Zdarska, Z., Soboleva, T., and Weiser, J. 1986. Rhabdovirus-like particles in the sporocyst of the trematode *Brachylaimus fuscatus* from terrestrial mollusc *Ponsadenia duplocincta*. *Folia Parasitologica* 33: 277–279

Zhang, G., Sachse, M., Prevost, M.C., Luallen, R.J., Troemel, E.R., and Felix, M.A. 2016. A large collection of novel nematode-infecting Microsporidia and their diverse interactions with *Caenorhabditis elegans* and other related nematodes. *PLoS Pathogens* 12(12): e1006093

Zhao, D., Liu, B., Li, L.Y., Zhu, X.F., Wang, Y.Y., Wang, J.Q., Duan, Y.X., and Chen, L.J. 2013. *Simplicillium chinense*: A biological control agent against plant parasitic nematodes. *Biocontrol Science and Technology* 23(8): 980–986

Zuckerman, B.M., Himmelho, S., and Kisiel, M. 1973. Virus-like particles in *Dolichodorus heterocephalus*. *Nematologica* 19: 117–118

CHAPTER 7

Diseases of annelids

Jacqueline L. Stroud

7.1 Introduction

Annelids are soft bodied, segmented worms that inhabit terrestrial, freshwater, and marine habitats. Taxonomic methods have identified tens of thousands of species, whilst molecular methods have revealed significant levels of cryptic diversity (King et al. 2008). Annelids include ecosystem engineers that change soil and benthic communities through soil or sediment ingestion and defaecation processes. This can improve plant productivity in both terrestrial (Van Groenigen et al. 2014) and aquatic (Mermillod-Blondin and Lemoine 2010) ecosystems. The quantity and quality of annelids in ecosystems is important because they are an important food source. However, the consumption of annelids can lead to the trophic transfer of pollutants, and spread of parasitic diseases in wildlife, pets, fowl, swine, fish, and shrimp farming. Therefore, understanding annelid health is critical to both local and global ecosystem functioning.

Annelid immune systems have been characterised and are described in Chapter 1. Most laboratory research has focussed on the functioning of immunocompetent cells (coelomocytes). In both terrestrial and marine worms, pollution can impair the viability and activity of coelomocytes, reducing phagocytosis and encapsulation defensive mechanisms (Wieczorek-Olchawa et al. 2003; Browne et al. 2013). Laboratory studies show that introduced chemicals, particles, and pathogens cause histopathological alterations in the intestinal tract, which likely increases susceptibility to infections. Metal and pesticide pollution has been correlated

to monocystid gregarine infection levels in earthworms (Pizl 1985; Pižl and Sterzynska 1991). Terrestrial and aquatic annelids are reservoirs, sources and sinks for pathogenic organisms, but are harvested and translocated to new geographic regions around the world by the bait industry (Diggles 2011).

It is important to note that the lack of systematic data collection and paucity of records requires cautious interpretations of the scientific literature. For example, there is a curious mass parasitic infection of the *Lumbricus terrestris* earthworms in Germany. This was first noted 30 years ago, and is linked to this earthworm species being easy to collect and identify, easy to dissect due to its size, and the fact that Germany had the most parasitologists in the world (Sims and Gerard 1985). Annelids host many species, and these associations are poorly understood, with parasitism often speculated rather than proven through fitness cost measurements. Annelid health assessments are confounded by their regenerative abilities, common structural abnormalities and seasonal regression of sexual organs linked to a estivation (Stephenson 1930). In the absence of known diseases, this chapter brings together examples of infections and stressors in terrestrial, marine and freshwater environments, and Table 7.1 highlights some potential diseases in annelids. This review highlights that the current interpretation of soil biological monitoring (abundance of annelids indicates soil health) is incorrect because nothing is known about annelid health (e.g. fitness, pollutant accumulation, parasite-loads) which influences local ecosystem functions (trophic transfer of

Jacqueline L. Stroud, *Diseases of annelids*. In: *Invertebrate Pathology*. Edited by Andrew F. Rowley, Christopher J. Coates and Miranda M.A. Whitten, Oxford University Press. © Oxford University Press (2022). DOI: 10.1093/oso/9780198853756.003.0007

Table 7.1 An overview of potential diseases in annelids

Disease causing agent(s)	Host range	Geographical range	Pathology	Key references
Exophiala jeanselmei, a saprotrophic fungus (black yeast)	Earthworms *Octolasian tyrtaeum* and *Eisenia foetida*	Potentially global soils	Spherical black granules, consisting of brown to black hyphae and cell aggregates, observed in cocoon albumen & tissues of infected embryos of *O. tyrtaeum* and albumen of *E. foetida*. Some necrotic eggs.	Vakili (1993)
Monocystis sp., a gregarine protozoan	Earthworm *Lumbricus terrestris*	Potentially global soils	Parasitic castration, destroying testes and causing male ducts to atrophy	Stephenson (1930); Field and Michiels (2006)
Pathogens, metals, chemicals and particles	Earthworms	Global soils	Damage to gastrointestinal tract, body wall and reproductive organs	Kilic (2011); Vittori Antisari et al. (2016);
Gymnophallus choledochus, trematode	Polychaete *Diapatra neopolitana*	Estuary, Portugal	Ruptured muscle bundles, associated with setae and setal sack likely inhibiting movement	Rangel and Santos (2009)
Myxobolus cerebralis (Cnidaria: Myxozoa)	Oligochaete *Tubifex*	Cold waters	Temporal castration and reduced feeding activity linked to intestinal tract damage	Shirakashi and El-Matbouli (2009)
Cestodes	Tubificids	Potentially global	Potentially physically blocking gonad development	Courtney and Christensen (1991)
Australapatemon sp., trematode	Freshwater leech, *Erpobdella octoculata*	Lake, Finland	Sub-lethal effects on behaviour resulting in impaired hiding behaviour—which probably increases predation by next host (birds).	Karvonen et al. (2017)

pollutants, spread of fatal parasitic diseases to vertebrates). Similarly, there is a completely overlooked risk from the transport of diseased individuals and vectors around globe through the bait industry.

7.2 Principal diseases

There is no body of research on annelid diseases under laboratory or field conditions. Disease incidents are not easily detected in soft-bodied animals. For example, earthworms die and rapidly decay in their burrows. Similarly, the sampling methods for the small terrestrial enchytraeids are biased towards active and motile worms. Research

describes the immunocompetence of annelids, host-parasite interactions, and environmental pollutants that induce fitness costs. The inflicted pathology ranges from minor metabolic changes to major tissue damage resulting in mortality. Minor changes can compromise survival through reducing resilience to stress and/or increasing risks of predation (e.g. motility or behaviour changes).

7.2.1 Earthworms, infections, pollution, and parasites

Earthworms are hermaphrodites, and reproduction is through the production of egg capsules that are laid in the soil. These cocoons can remain viable for years and exhibit a delayed hatching strategy (Lowe and Butt 2014). However, cocoons are vulnerable to infection by the opportunistic soilborne pathogenic fungus *Exophiala jeanselmei* that can penetrate the cocoon layer, or might be carried by earthworms and directly transmitted to their eggs (Vakili 1993). *Exophiala jeanselmei* is a human pathogen as a cause of disease, and infection susceptibility is linked to those under physical or biological stress (Vakili 1993).

Microorganisms form an essential part of the earthworm diet, but earthworms also routinely consume pathogens. Two bacterial infections are reported in the early literature for anecic *L. terrestris* earthworms. Spherical bacteria invasions of the cytophores and spermatocytes which the author associated with parasitic castration, and *Spirochaeta lumbrici* associated with Kirloff's type bodies found in epithelial cells of the seminal vesicle walls (Stephenson 1930). However, many species have a diapause phase which is not well understood and leads to the loss of secondary sexual characteristics (Edwards and Bohlen 1996) which can confound castration interpretations (Stephenson 1930).

The gut has antimicrobial activities which may be linked to metabolites of symbiotic bacteria in the gut walls (Fiołka et al. 2010). The innate immune mechanism is linked to a protein called the Coelomic Cytolytic Factor. This is a pattern recognition molecule (Cooper and Roch 2003) and is analogous to the TNF-α (tumour-necrosis factor-α) in mammals. The model for the earthworm defence mechanism against infections is the sensing of microbes by receptors in the gut that leads to the release of coelomocytes from the lining of the coelom. These act as phagocytes and/or produce antimicrobial proteins and opsonins that reduce microbial abundance to control the infection (Dvořák et al. 2016).

Exposure to pollution impairs coelomocyte activity (Wieczorek-Olchawa et al. 2003). Further, pollution is directly linked to damage of the earthworm gastrointestinal tract. For example, incubating *E. andrei* earthworms in field-collected metal and radionuclide contaminated soil causes the intestinal epithelium layer to atrophy; becoming very thin, intervilli spaces disappear and villi fuse to form a continuous and thin layer of cells

(Lourenço et al. 2011). Particle pollutants, both microplastics (Rodriguez-Seijo et al. 2017) and nanoparticles (Adeel et al. 2019) can damage the gastrointestinal tract of earthworms. One study on silver and cobalt nanoparticles identified the presence of Ca/P calcifications, which are also found in human pathological tissues affected by cancer (Vittori Antisari et al. 2016).

It is important to note that earthworms are associated with a tolerance to pollution; their accumulation of pollutants is directly linked to the widespread poisoning of wildlife (Macdonald 1983). Metal pollution levels have been correlated with tissue necrosis in field collected *L. terrestris* earthworms, changes in chloragogenous cells (this is analogous to the liver in vertebrates) and the loss of structural integrity in the muscles (Kilic 2011). Disease incidence is unknown, but metal accumulation sites within the body are well established. For example, metal accumulation in the cerebral ganglion is linked to reducing burrowing efficacy, and metal accumulation in the sarcoplasm of the muscle cells reduces muscle activity due to the disruption of calcium transport (Bengtsson et al. 1983). Movement disorders likely influence earthworm susceptibility to predation (thus trophic transfer of pollutants (Macdonald 1983)).

Parasitic infections are curious: earthworm populations can be parasite-free or heavily infested (Stephenson 1930), and some infections have been linked to soil pollution. For example, a study on earthworms collected from German forests detected gregarines (protozoa belonging to the phylum Apicomplexa) in 20–30% of earthworms, which was linked to local air pollution. In comparison, infection rates were 0% in earthworms collected from unpolluted forests in Spain and Austria (Purrini 1983). Interestingly, enchytraeids were more sensitive than earthworms in this study, with infection rates of 60–80% by various bacteria, fungi, ciliates and gregarines in the polluted German forests compared with 5–20% infection rates in unpolluted forests in Spain and Austria (Purrini 1983).

Earthworms can be infected by a range of organisms including protozoa, bacteria, rotifers and nematodes (Stephenson 1930; Edwards and Bohlen 1996). The most well-studied earthworm parasite is the castrating gregarine protozoan *Monocystis* sp.

and parasite loads can impact host fitness in terms of growth (Field and Michiels 2005). The precise mode of infection has been subject to debate for nearly 100 years. It is thought to be via the alimentary canal, subsequently digestive juices cause the spores to open, and sporozoites move to the seminal vessels (Stephenson 1930). This is in agreement with more recent laboratory studies identifying oral transmission through the soil as the principal mode of infection (Field and Michiels 2006). The authors noted that parasite loads in wild populations are thirty times higher than laboratory infected earthworms, suggesting there is either a very slow infection development or very limited successful transfer of sporozoites through the gut wall and into the seminal vessels (Field and Michiels 2006). The infection may result in reproductive inefficiency due to hypertrophy of male funnel cells and parasitisation of sperm moralae (Stephenson 1930), that is, a paternity fitness cost (Field and Michiels 2006). Massive infections of *Monocystis* sp. lead to parasitic castration by destroying the testes and causing the male ducts to atrophy (Sims and Gerard 1985).

In terms of soil pollution, increased parasite infection rates are correlated to total zinc (but not lead, copper or cadmium) contamination (Pižl and Sterzynska 1991), and pesticides also directly increase the incidence of infections in soil dwelling earthworms (Pizl 1985). Pizl (1985) incubated uninfected earthworms with (and without) gregarine sporocysts in control and pesticide-treated soils for 26 weeks. Earthworms in untreated control soils displayed cyst infection rates of 64%, 36%, and 56% respectively for epigeic, endogeic, and anecic earthworms. In pesticide treated soils, however, the cyst infection rates were considerably elevated to 100% for epigeic, 75% for endogeic and 96% for anecic earthworms (Pizl 1985).

Nematodes are also commonly found infecting earthworms but there is little research into associated pathologies. Earthworms are often imported as live bait for freshwater fishing, which leads to nematodes being introduced into new geographic regions (Roepstorff et al. 2002). Nematode-infected earthworms can be ingested by birds and other animals, and in so doing transmit nematodes of commercial significance such as the gapeworm (*Syngamus trachealis*) which causes tracheo-bronchitis infections in fowl (Stephenson 1930) and lung worms (*Metastrongylus* spp.) that infect swine and cause parasitic bronchitis (López and Martinson 2017). Laboratory experiments in 1895 studying the regeneration of the central nervous system in earthworms found nematode abundance was linked to earthworm mortality rates (Stephenson 1930). However, more recent studies in *Dendrobaena* spp. tissues led to the possible detection of an RNA virus (Andrea et al. 2018). This is an interesting finding because there are currently no known earthworm viruses (Edwards and Bohlen 1996).

7.2.2 Polychaete parasites, infections, and abnormalities

Parasite biomass has been estimated to make up to 0.2–2.4% polychaete biomass in estuary habitats, indicating the significance of parasite ecology in aquatic ecosystem functioning (Kuris et al. 2008). A range of organisms have been identified in polychaetes, but it is unclear if this leads to disease. For example, an RNA virus OjRV1 has been isolated from the deep sea worm that colonises whale bones but it is remains to be established whether it is pathogenic (Urayama et al. 2018).

The most thoroughly researched polychaete parasites are the trematodes (Platyhelminthes) (Phelan et al. 2016; Rueckert et al. 2018). An analysis of polychaetes from a New Zealand harbour identified several trematode parasite species as well as other parasites. Whilst there was no relationship to host mortality, researchers speculated that the structural integrity of the host polychaete *Heteromastus filiformis* may be compromised by infection with the trematode Opecoelid E (family Opecoelidae) shown in Figure 7.1 (Peoples et al. 2012). In a Portugese estuary, a 100% infection prevalence of the trematode *Gymnophallus choledochus* was recorded in the polychaete *Diopatra neapolitana* (noting the specimen collection was by bait diggers, and the infection is suspected to impact mobility). It is thought that the branchiae (gills) are the entry point for *G. choledochus*, and this trematode has been found in association with the setae, setal sack (replaces setae) and muscle bundles causing host muscle rupture. This infection likely impacts host mobility and increases vulnerability to predation by aquatic birds

(a)　　　　　　　　　　　　　　(b)

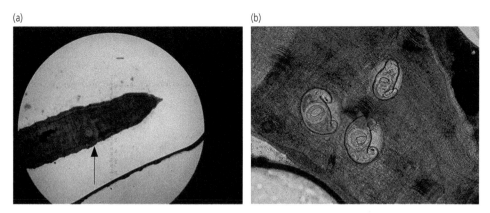

Figure 7.1 Parasite (metacercariae of an opecoelid trematode) associated with the (a) polychaete *Heteromastus filiformis*, an arrow indicates the general area of parasitized tissue in the worm (b) close-up of the parasite. Source: photo credit to Robert Peoples III.

(the definitive hosts of *G. choledochus*) (Rangel and Santos 2009). The infection by *G. choledochus* is likely to be sensitive to environmental change: in the North Sea it has seasonal host-switching pathways using polychaetes in the spring and summer, and molluscs (*Cerastoderma edule*) in the autumn and winter (Rangel and Santos 2009).

7.2.3 Oligochaete parasites, infections, pollution, and abnormalities

Helminth infections of the tubificids are common, but it is unclear if this leads to disease. Tubificid worms are also the definitive hosts for the actinosporean stage of several species of myxozoan parasites (currently classified as cnidarians). For example, 18% actinosporean infection rates have been detected in a population of the tubificid worm *Branchiura sowerbyi* living in the mud below a commercial carp farm (Xi et al. 2017). The myxosporean stage of myxozoan parasites infect fish (intermediate hosts) and are a common cause of fish diseases. The most studied myxozoan is *Myxobolus cerebralis* which can cause whirling disease in salmonid fish and hence jeopardises the cold-water aquaculture industry (Nehring et al. 2016). The oligochaete becomes infected by ingesting myxospores that develop for 2–3 months between its gut epithelial cells, and subsequently waterborne triactinomyxon spores are released with the worm faeces (Shirakashi and El-Matbouli 2009). Another

oligochaete, the sludge worm *Tubifex* has variable susceptibility to *M. cerebralis* infections (Nehring et al. 2016), however this is not expected to reduce the risk to salmonid hosts because exposure to myxospores increases the growth of both susceptible and resistant *T. tubifex* and their progeny biomass is unaffected (Elwell et al. 2006). Lineage III *T. tubifex* are highly susceptible to infections (Nehring et al. 2016), and whilst infection causes no impact on worm growth or survival, it does lead to reproductive inefficiency through temporal castration and reduced feeding activity linked to pathological damage of the intestinal tract (Shirakashi and El-Matbouli 2009). Metal pollution likely facilitates the development of the parasite in its annelid host, although the mechanism is unknown (Shirakashi and El-Matbouli 2010).

Other tubificid parasites include cestodes, which inhibit host sexual maturity or delay the breeding season. This is speculated to be caused by limited energy for reproductive growth or the parasite physically blocking gonad development (Courtney and Christensen 1991). Protozoan (ciliate) infections of tubificids are also common and large numbers of ciliates have been observed tightly packing the alimentary canal. Although this would be expected to interfere with feeding, no overt adverse effects have been detected and it has been suggested that they are commensal organisms rather than parasites (Stephenson 1930).

7.2.4 Leeches

Freshwater leeches and aquatic earthworms are the first non-vertebrate hosts of *Uranotaenia sapphirina* mosquitoes (Culicidae) to be described (Reeves et al. 2018).

Leeches are also vulnerable to trematode parasites. Trematode *Australapatemon* sp. infection rates in the leeches *Erpobdella octoculata* or *Helobdella stagnalis* range from 9–60% within the leech populations. These infections tend to fluctuate seasonally and are generally more prevalent in offshore sites compared with shore locations (Karvonen et al. 2017). Although no overt disease or increased mortality has been ascribed directly the parasite, it appears that it can disrupt the hiding behaviour of *E. octoculata* to increase its susceptibility to predation by water fowl, which are the next hosts in the parasite lifecycle (Karvonen et al. 2017).

7.3 Future directions

There is no research into annelid disease. There is a body of literature describing infections, parasites, and pollutants linked to annelid fitness suggesting that this is a productive area for future research. There are two urgent knowledge gaps to address.

There is a contemporary assumption that earthworm abundance is indicative of soil health. This bypasses the terrestrial annelid literature (1890–1990) describing the trophic transfer of pollutants and parasite transfer which cause fatal diseases in vertebrates. Advances will be made through research into the quantity (abundance) and quality (health) of annelids because they are an important food source for many animals. The second knowledge gap is linked to the 121,000 tonnes of annelids that are harvested per year for the bait industry (Cole et al. 2018). This industry is dependent on wild species collected from the environment so there is little knowledge of which annelid species are collected (let alone their disease or vector status) (Cole et al. 2018). Nematodes have been transmitted to new geographical regions through earthworms and detected in pigs (Roepstorff et al., 2002). The use of polychaetes and leeches as bait has resulted in the transmission of viruses to marine animals (Diggles, 2011). For example, polychaetes are a mechanical vector of white spot syndrome virus, a major shrimp pathogen (Vijayan et al., 2005; see Chapter 14). Therefore, understanding the incidence of annelid disease in species used for bait will help to protect ecosystem functioning.

Recommended further reading

Risks associated with the bait industry

Diggles, D.B.K. 2011. Risk Analysis. Aquatic animal diseases associated with domestic bait translocation. Final report prepared for the Australian Government Department of Agriculture, Fisheries and Forestry, Canberra.

Parasite-annelid interactions (now being reconfirmed in contemporary studies)

Stephenson, J. 1930. *The Oligochaeta*. Oxford: Clarendon Press.

References

Adeel, M., Ma, C., Ullah, S. et al. 2019. Exposure to nickel oxide nanoparticles insinuates physiological, ultrastructural and oxidative damage: A life cycle study on *Eisenia fetida*. *Environmental Pollution* 254: 113032

Andrea, A., Sara, F., Marina, P., Lorenzo, G., Steven, P., and Stefano, B. 2018. The earthworm *Dendrobaena veneta* (Annelida): A new experimental-organism for photobiomodulation and wound healing. *European Journal of Histochemistry* 62(1): 2867

Bengtsson, G., Nordström, S., and Rundgren, S. 1983. Population density and tissue metal concentration of lumbricids in forest soils near a brass mill. *Environmental Pollution Series A, Ecological and Biological* 30: 87–108

Browne, M.A., Niven, S.J., Galloway, T.S., Rowland, S.J., and Thompson, R.C. 2013. Microplastic moves pollutants and additives to worms, reducing functions linked to health and biodiversity. *Current Biology* 23: 2388–2392

Cole, V.J., Chick, R.C., and Hutchings, P.A. 2018. A review of global fisheries for polychaete worms as a resource for recreational fishers: Diversity, sustainability and research needs. *Reviews in Fish Biology and Fisheries* 28: 543–565

Cooper, E.L. and Roch, P. 2003. Earthworm immunity: A model of immune competence: The 7th international symposium on earthworm ecology Cardiff Wales 2002. *Pedobiologia* 47: 676–688

Courtney, C.C. and Christensen, B.M. 1991. The response of *Tubifex-tubifex* (Oligochaeta, Tubificidae) to a 2nd infection with *Gladiracris-catostomi* (Cestoidea, Caryophyllaeidae). *Journal of the Helminthological Society of Washington* 58: 118–121

Diggles, D.B.K. 2011. *Risk Analysis. Aquatic animal diseases associated with domestic bait translocation. Final report prepared for the Australian Government Department of Agriculture, Fisheries and Forestry, Canberra*

Dvořák, J., Roubalová, R., Procházková, P., Rossmann, P., Škanta, F., and Bilej, M. 2016. Sensing microorganisms in the gut triggers the immune response in *Eisenia andrei* earthworms. *Developmental and Comparative Immunology* 57: 67–74

Edwards, C.A. and Bohlen, P.J. 1996. *Biology and Ecology of Earthworms*. Netherlands: Springer

Elwell, L.C., Kerans, B.L., Rasmussen, C., and Winton, J.R. 2006. Interactions among two strains of *Tubifex* (Oligochaeta: Tubificidae) and *Myxobolus cerebralis* (Myxozoa). *Diseases of Aquatic Organisms* 68: 131–139

Field, S.G. and Michiels, N.K. 2005. Parasitism and growth in the earthworm *Lumbricus terrestris*: Fitness costs of the gregarine parasite *Monocystis* sp. *Parasitology* 130: 397–403

Field, S.G. and Michiels, N.K. 2006. Acephaline gregarine parasites *Monocystis* sp. are not transmitted sexually among their lumbricid earthworm hosts. *Journal of Parasitology* 6: 292–297

Fiołka, M.J., Zagaja, M.P., Piersiak, T.D., Wróbel, M., and Pawelec, J. 2010. Gut bacterium of *Dendrobaena veneta* (Annelida: Oligochaeta) possesses antimycobacterial activity. *Journal of Invertebrate Pathology* 105: 63–73

Karvonen, A., Faltynkova, A., Choo, J.M., and Valtonen, E.T. 2017. Infection, specificity and host manipulation of *Australapatemon* sp. (Trematoda, Strigeidae) in two sympatric species of leeches (Hirudinea). *Parasitology* 144: 1346–1355

Kilic, G.A. 2011. Histopathological and biochemical alterations of the earthworm (*Lumbricus terrestris*) as

biomarker of soil pollution along Porsuk River Basin (Turkey). *Chemosphere* 83: 1175–1180

King, R.A., Tibble, A.L., and Symondson, W.O.C. 2008. Opening a can of worms: Unprecedented sympatric cryptic diversity within British lumbricid earthworms. *Molecular Ecology* 17: 4684–4698

Kuris A.M., Hechinger, R.F., Shaw, J.C. et al. 2008. Ecosystem energetic implications of parasite and free-living biomass in three estuaries. *Nature* 454: 515–518

López, A. and Martinson, S.A. 2017. Respiratory system, mediastinum, and pleurae. In *Pathologic Basis of Veterinary Disease* (Sixth Edition), Zachary, J.F. (ed.), Chapter 9, pp. 471–560. Mosby.

Lourenço, J., Silva, A., Carvalho, F. et al. 2011. Histopathological changes in the earthworm *Eisenia andrei* associated with the exposure to metals and radionuclides. *Chemosphere* 85: 1630–1634

Lowe, C.N. and Butt, K.R. 2014. Cocoon viability and evidence for delayed hatching by the earthworm *Lumbricus terrestris* in a laboratory based study. *Zeszyty Naukowe* 17: 61–67

Macdonald, D.W. 1983. Predation on earthworms by terrestrial vertebrates. In *Earthworm Ecology: From Darwin to Vermiculture* Satchell, J.E. (ed.), pp. 393–414. Dordrecht, Netherlands: Springer

Mermillod-Blondin, F. and Lemoine, D.G. 2010. Ecosystem engineering by tubificid worms stimulates macrophyte growth in poorly oxygenated wetland sediments. *Functional Ecology* 24: 444–453

Nehring, R.B., Schisler, G.J., Chiaramonte, L., Horton, A., and Poole, B. 2016. Accelerated deactivation of *Myxobolus cerebralis* myxospores by susceptible and non-susceptible *Tubifex*. *Diseases of Aquatic Organisms* 121: 37–47

Peoples, R., Randhawa, H., and Poulin, R. 2012. Parasites of polychaetes and their impact on host survival in Otago Harbour, New Zealand. *Journal of the Marine Biological Association of the UK* 92: 449–455

Phelan, K., Blakeslee, A.M., Krause, M., and Williams, J.D. 2016. First documentation and molecular confirmation of three trematode species (Platyhelminthes: Trematoda) infecting the polychaete *Marenzelleria viridis* (Annelida: Spionidae). *Parasitology Research* 115: 183–194

Pizl, V. 1985. The effect of the herbicide zeatin-50 on the earthworm infection by monocystid gregarines *Pedobiologia* 28: 399–402

Pižl, V. and Sterzynska, M. 1991. The influence of urbanization on the earthworm infection by Monocystid gregarines. *Fragmenta Faunistica (Warsaw)* 35: 9–14

Purrini, K. 1983. Soil invertebrates infected by microorganisms. In *New Trends in Soil Biology. Proceedings of the VIII. Intl Colloquium of Soil Zoology*, Lebrun, P., Andre,

H.M., De Medths, A., Gregoire-Wino, C., and Wauthy, G. (eds.), pp. 167–178. Louvain-la-Neuve (Belgium): Dieu-Brichart

Rangel, L.F. and Santos, M.J. 2009. *Diopatra neapolitana* (Polychaeta: Onuphidae) as a second intermediate host of *Gymnophallus choledochus* (Digenea: Gymnophallidae) in the Aveiro Estuary (Portugal): Distribution within the host and histopathology. *Journal of Parasitology* 95: 1233–1236

Reeves, L.E., Holderman, C.J., Blosser, E.M. et al. 2018. Identification of *Uranotaenia sapphirina* as a specialist of annelids broadens known mosquito host use patterns. *Communications Biology* 1: 92–92

Rodriguez-Seijo, A., Lourenco, J., Rocha-Santos, T. et al. 2017. Histopathological and molecular effects of microplastics in *Eisenia andrei* Bouche. *Environmental Pollution* 220: 495–503

Roepstorff, A., Grønvold, J., Larsen, M. N., Kraglund, H.-O., and Fagerholm, H.-P. 2002. The earthworm *Lumbricus terrestris* as a possible paratenic or intermediate host of the pig parasite *Ascaris suum*. *Comparative Parasitology* 69: 206–210

Rueckert, S., Glasinovich, N., Diez, M.E., Cremonte, F., and Vázquez, N. 2018. Morphology and molecular systematic of marine gregarines (Apicomplexa) from Southwestern Atlantic spionid polychaetes. *Journal of Invertebrate Pathology* 159: 49–60

Shirakashi, S. and El-Matbouli, M. 2009. *Myxobolus cerebralis* (Myxozoa), the causative agent of whirling disease, reduces fecundity and feeding activity of *Tubifex tubifex* (Oligochaeta). *Parasitology* 136: 603–613

Shirakashi, S. and El-Matbouli, M. 2010. Effect of cadmium on the susceptibility of *Tubifex tubifex* to *Myxobolus cerebralis* (myxozoa), the causative agent of whirling disease. *Diseases of Aquatic Organisms* 89: 63–70

Sims, R.W. and Gerard, B.M. 1985. *Earthworms. Synopses of the British Fauna* (new series): No. 31, London: Brill and Backhuys

Stephenson, J. 1930. *The Oligochaeta*. Oxford: Clarendon Press

Urayama, S.I., Takaki, Y., Nunoura, T., and Miyamoto, N. 2018. Complete genome sequence of a novel RNA virus identified from a deep-sea animal, *Osedax japonicus*. *Microbes and Environments* 33: 446–449

Vakili, N.G. 1993. *Exophiala jeanselmei*, a pathogen of earthworm species. *Journal of Medical and Veterinary Mycology* 31: 343–346

Van Groenigen, J.W., Lubbers, I.M., Vos, H.M.J., Brown, G.G., De Deyn, G.B., and Van Groenigen, K.J. 2014. Earthworms increase plant production: A meta-analysis. *Scientific Reports* 4: 6365

Vijayan, K.K., Raj, V.S., Balasubramanian, C.P., Alavandi, S.V., Sekhar, V.T., and Santiago, T.C. 2005. Polychaete worms – a vector for white spot syndrome virus (WSSV). *Diseases of Aquatic Organisms* 63: 107–111

Vittori Antisari, L., Carbone, S., Gatti, A. et al. 2016. Effect of cobalt and silver nanoparticles and ions on *Lumbricus rubellus* health and on microbial community of earthworm faeces and soil. *Applied Soil Ecology* 108: 62–71

Wieczorek-Olchawa, E., Niklinska, M., Miedzobrodzki, J., and Plytycz, B. 2003. Effects of temperature and soil pollution on the presence of bacteria, coelomocytes and brown bodies in coelomic fluid of *Dendrobaena veneta*: The 7th international symposium on earthworm ecology · Cardiff · Wales · 2002. *Pedobiologia* 47: 702–709

Xi, B. W., Li, P., Liu, Q. C., Chen, K., Teng, T., and Xie, J. 2017. Description of a new Neoactinomyxum type actinosporean from the oligochaete *Branchiura sowerbyi* Beddard. *Systematic Parasitology* 94: 73–80

CHAPTER 8

Diseases of molluscs

Sharon A. Lynch, Andrew F. Rowley, Matt Longshaw, Shelagh K. Malham and Sarah C. Culloty

8.1 Introduction

The Phylum Mollusca is a large and diverse group of invertebrate protostomes of over 85,000 species including gastropods, cephalopods, and bivalves. Many molluscs reside in the marine environment but there are also examples in brackish and freshwater and a smaller number of terrestrial forms. Molluscs are generally free living with the exception of some gastropod parasites of echinoderms (Warén 1981). The anatomy and physiology of molluscs has been extensively studied and some, such as the mussel *Mytilus edulis* are widely used models for environmental biomonitoring (e.g. Świacka et al. 2019).

Molluscs are economically important as pests, food for humans, and some act as vectors of human diseases such as schistosomiasis. They are harvested for human consumption by traditional fishing but are increasingly cultured in many coastal communities worldwide. Bivalve molluscs including oysters, mussels, clams, scallops, and cockles are the principal products of this production. Approximately 89% of bivalve production worldwide comes from aquaculture (Wijsman et al. 2019) and this is set to rise further as capture fishing activities decline. In 2018, oyster production alone had an estimated value of US$7.2 billion with Pacific oyster (*Crassostrea gigas*) a key product valued at US$1.36 billion (FAO 2020). Similarly, clam and cockle production were worth US$9.72 billion, with the Manila clam, *Ruditapes (Venerupis) philippinarum* valued at US$6.9 billion in 2018 (FAO 2020). In the case of gastropods, whelks and abalone are the molluscs of main commercial importance. The common whelk, *Buccinum undatum* is found in Europe from Brittany and northwards to the Arctic and capture production reached over 41,000 tonnes in 2016 (FAO FishStat), however, its lengthy maturation period of 5–7 years makes them unsuitable candidates for farming. Up to 95% of abalones (*Haliotis* spp.) produced for human consumption are farmed in a number of countries including the USA, South Africa, and Japan. Their high market value of US$15–30 kg^{-1} (Mau and Jha 2018) makes them some of the most highly prized sea food.

This shift from capture fishing to aquaculture production has resulted in extensive research into the immune systems and diseases of bivalve and cephalopod molluscs. Similarly, as gastropods *Bulinus* and *Biomphalaria* are infected by schistosome larvae, they have also been subject to similar scrutiny (see Section 1.11.2 in Chapter 1). This chapter aims to provide a succinct overview of the principal diseases of molluscs with an emphasis on bivalves, gastropods, and cephalopods. As there is an extensive and growing literature in this field, we refer readers to key reviews for additional in-depth appraisal.

8.2 Principal diseases

Table 8.1 highlights the broad taxonomic diversity of main disease-causing agents of a number of commercially important molluscs ranging from a viral aetiology through to mortality associated with the presence of multicellular parasites. A common

Sharon A. Lynch et al., *Diseases of molluscs*. In: *Invertebrate Pathology*. Edited by Andrew F. Rowley, Christopher J. Coates and Miranda M.A. Whitten, Oxford University Press. © Oxford University Press (2022). DOI: 10.1093/oso/9780198853756.003.0008

Table 8.1 Examples of key diseases of some commercially important bivalve molluscs

Mollusc	Diseases	Causative agents	References
Sydney rock oyster, *Saccostrea glomerata*	1. Winter mortality syndrome 2. QX (Queensland Unknown) disease	1. *Bonamia roughleyi* 2. *Marteilia sydneyi*	1. Raftos et al. 2014 2. Perkins and Wolf 1976; Kleeman et al. 2002; Raftos et al. 2014

Pacific oyster, *Crassostrea (Magallana) gigas*	1. Pacific oyster mortality syndrome (POMS) in larvae and juveniles
	2. Vibriosis in adults
		1. Ostreid herpesvirus (OsHV-1μVar) and subsequent bacterial infections
		2. a. *Vibrio aestuarianus*
		   b. *V. splendidus*
		   c. *V. harveyi*
		3. *Tenacibaculum soleae* and other microbes including *V. aestuarianus*
			1. de Lorgeril et al. 2018; Delmotte et al. 2020
			2. a. Mandas et al. 2020
			   b. Lacoste et al. 2001; Le Roux et al. 2002
			   c. King et al. 2019
			3. Burioli et al. 2018

Eastern oyster, *Crassostrea virginica*	1. MSX (multinucleated sphere X) disease 2. 'Dermo' disease, Perkinsosis	1. *Haplosporidium nelsoni* 2. *Perkinsus marinus*	1. Burreson and Ford 2004 2. Villalba et al. 2004; Bower 2011
European flat oyster, *Ostrea edulis*	Bonamiasis	*Bonamia ostreae*	Culloty and Mulcahy 2007; Arzul and Carnegie 2015
Edible cockle, *Cerastoderma edule*	Mass mortality events associated with lack of burying ability	Disseminated neoplasia and trematode infection	Morgan et al. 2012
Grooved carpet shell clam, *Ruditapes decussatus*	Perkinsosis	*Perkinsus olseni* (= *P. atlanticus*)	Villalba et al. 2004
Manila clam, Japanese carpet shell clam, *Venerupis* [*Ruditapes*, *Tapes*] *philippinarum*	Perkinsosis Brown ring disease	*Perkinsus olseni* *Vibrio tapetis*	Choi et al. 2002; Villalba et al. 2004; Pretto et al. 2014 Paillard and Maes 1990, Borrego et al. 1996
Greenlip abalone, *Haliotis laevigata*	1. Perkinsosis 2. Abalone viral ganglioneuritis (AVG)	1. *Perkinsus olseni* 2. Abalone herpesvirus (AbHV) (Haliotid herpesvirus 1)	1. Goggin and Lester 1995; Bower 2000 2. Hooper et al. 2012; Chen et al. 2016

feature of many of these disease outbreaks is the physiological status of the host (e.g. gametogenesis, immune suppression) and adverse changes in the surrounding environment (e.g. temperature and salinity, presence of toxins, and anthropogenic chemicals) resulting in conditions likely to favour disease. Over the last 50 years there have been numerous reports of sudden mortality events in oysters (see Alfaro et al. 2019 for a review), cockles (Beukema and Dekker 2020), mussels (Charles et al. 2020a), and clams (Wei et al. 2019) during summer months often linked to higher seawater temperature (e.g. Garnier et al. 2007). For instance, a rise in the temperature of seawater from 20 to 25 °C, simulating heat waves causes greater mortality rates associated with changes in the microbiome of Pacific oysters with a higher abundance of potentially pathogenic vibrios (Green et al. 2019).

An important concept highlighted in Table 8.1. is the polymicrobial nature of some diseases. In the case of the Pacific oyster, *C. gigas* reared in Europe, Pacific Oyster Mortality Syndrome has been shown to be polymicrobial in nature, where oysters first become infected by the virus ostreid herpesvirus (OsHV-1 μVar) leaving them immunocompromised. Subsequent bacterial infections, often associated with vibrios, cause mortality in such populations (Petton et al. 2015; de Lorgeril et al. 2018). In

Table 8.2 Principal diseases of molluscs caused by viruses

Disease	Disease causing agent(s)	Host range	Geographic range	Pathology	Key references
Summer mortality disease/syndrome (SMS), Pacific oyster mortality syndrome (POMS) in larvae and juveniles	Ostreid herpesvirus-1 (OsHV-1) and variants including Ostreid herpesvirus microVar (OsHV-1 μVar)	Oysters (*Crassostrea gigas*, *C. angulata*, *C. sikamea*, *C. virginica*, *C. hongkongensis*, *Ostrea edulis*) Mussels (*Mytilus galloprovincialis*) Scallops (*Chlamys farreri*, *Patinopecten yessoensis*) Clams (*Venerupis phillipinarum*, *meretrix*, *Scapharca broughtonii*)	Europe (France, Ireland, Italy, The Netherlands, Spain, UK), North America, Australia, New Zealand, Korea, Japan	*Gross pathology*: cessation of feeding and swimming by larvae, which exhibit velar lesions, gaping in adults and pale digestive gland in spat and older oysters *Histopathology*: ulcerative and erosive lesions in the connective tissue of mantle, gills, labial palps and digestive tissue. Cytopathic effects including nuclear hypertrophy, nuclear chromatin margination and pyknosis. Inflammatory changes ranging from mild and localised, to severe and extensive	Renault et al. 1994; Davison et al. 2005; Friedman et al. 2005; Lynch et al. 2012; Bingham et al. 2013; Paul-Pont et al. 2014; Arzul et al. 2017
Abalone viral ganglioneuritis (AVG)	Abalone herpesvirus (AbHV) (Haliotid herpesvirus 1 [HaHV-1])	Ear abalone *Haliotis diversicolor supertexta*; Green-lip abalone, *H. laevigata*; Black-lip abalone *H. rubra*	Australia, Taiwan, possibly China	Spherical virus with an icosahedral core and an envelope of *ca*. 100 nm in diameter. *Histopathology*: Lesions include necrosis of the cerebral ganglia and nerve bundles in the muscle of the foot as well as in the muscular layers beneath the visceral organs. Haemocyte infiltration in the lamina propria of the digestive tract, as well as the loss of seminal tubules in the gonad	Corbeil 2020

affected animals, the virus multiplies within haemocytes and reduces the expression of antimicrobial peptides leaving the host susceptible to later bacterial infection. This results in bacterial dysbiosis (Lasa et al. 2019; Clerissi et al. 2020) with subsequent secondary infections by several species of vibrios including *V. aestuarianus*, *V. crassostreae*, *V. splendidus*, and *V. harveyi* (de Lorgeril et al. 2018; Alfaro et al. 2019).

8.3 Diseases caused by viruses

Table 8.2 summarises the main diseases of molluscs caused by viruses including a description of the pathology. A range of viral families, including Herpesviridae, Papovaviridae, Togaviridae, Retroviridae, Reoviridae, Birnaviridae, and Picornaviridae have been reported in marine molluscs (Renault and Novoa 2004; Meyers et al. 2009). Most of these viruses have not been characterised in molluscs and although birnaviruses have been detected in various marine molluscs, their infectivity is considered low or hypothetical (Arzul et al. 2017). Current knowledge of viruses in molluscs is mostly restricted to farmed host species such as *C. gigas*, the ear abalone *Haliotis diversicolor supertexta* and the Chinese scallop *Chlamys farreri* (Arzul et al. 2017; Table 8.2). Herpesviruses including ostreid herpes virus-1 (OsHV-1) and variants, in particular OsHV-1 microVar (μVar) have been associated with significant mortality events resulting in high losses in several marine mollusc species including *C. gigas* worldwide (de Lorgeril et al. 2018; Destoumieux-Garzón et al. 2020). The potential host range of these viruses has recently been extended to the cephalopod *Octopus vulgaris*, although its ability to replicate inside these animals is uncertain (Prado-Alvarez et al. 2021).

8.4 Diseases caused by bacteria

The main bacterial diseases of molluscs are summarised in Tables 8.3 and 8.4. Details of the gross- and histo-pathology of these can be found in these tables. The following text is purposely succinct as many of these diseases have been reviewed elsewhere (e.g. Gosling 2015; Travers et al. 2015; Dubert et al. 2017) and wherever possible we have sought to cover recent advances not described in these reviews.

8.4.1 Vibriosis

Vibrios are ubiquitous in the marine and estuarine environments, with near global distribution. They have been recognised as key pathogens of molluscs for over 60 years and they have traditionally been considered as opportunistic pathogens, often of early life stages of molluscs prior to the full development of the immune system. Adverse environmental conditions such as high (or low) temperature, lowered oxygen levels, presence of anthropogenic chemicals and salinity changes all favour heightened virulence of vibrios in this host–pathogen relationship.

There are over 15 different species of vibrios that may be pathogenic to molluscs and others that are normal members of the host's microbiota (Romalde et al. 2014). Due to the location of bivalves on/in the benthos and their filter feeding behaviour, these molluscs can be associated with a wide range and number of vibrios. Bivalve molluscs also harbour vibrios including *V. parahaemolyticus* and *V. vulnificus* in their tissues that can cause food poisoning outbreaks if eaten raw (Barker and Gangarosa 1974). Vibrios are difficult to identify using phenotypic and standard 16S rRNA analyses which do not permit accurate identification at the species level. Multilocus sequence analysis using an array of housekeeping genes, is used to give more precise identification (e.g. Sawabe et al. 2007; Gabriel et al. 2014; Rojas et al. 2019). Using these approaches, the list of 'classical' mollusc pathogens including *V. alginolyticus*, *V. (Listonella) anguillarum*, *V. harveyi*, *V. parahaemolyticus*, *V. splendidus*, and *V. vulnificus* has been extended over the last couple of decades and, in some cases, the original identification changed to reflect these new approaches. Table 8.4 summarises some of the more recent additions to this ever-increasing list of potential pathogens of molluscs. For a description of 'classical' vibrio mollusc pathogens see Travers et al. (2015).

Vibrio aestuarianus is an important opportunistic pathogen of oysters and increasingly involved in mortalities, especially in France (Garnier et al. 2007). It is capable of causing infections in oysters at all life history stages, either alone, in adults or in combination with earlier viral infections in juvenile stages. Recent outbreaks of vibriosis from 2012 onwards in adult oysters are thought to be as a result of the emergence of a genotype of *V. aestuarianus* with a *varS* gene that codes for the signal transduction histidine-protein kinase (Goudenège et al. 2015). This gene is necessary for controlling the pathway that produces a zinc metalloprotease virulence factor that previous studies have shown to adversely affect the defensive behaviour of haemocytes (Labreuche et al. 2010). This metalloprotease is cytotoxic towards haemocytes and its injection is toxic to susceptible bivalves.

Table 8.3 Principal diseases of molluscs caused by bacteria (based on Gosling 2015 and Travers et al. 2015)

Disease	Disease causing agent(s)	Host range	Geographic range	Pathology	Key references
Vibriosis	*Vibrio* (*Aliivibrio*, *Listonella*) spp. (see Table 8.4)	Wide range found in brackish and sea water	Global depending on species	Extracellular multiplication of bacteria, lysis of haemocytes, tissue destruction, septicaemia	See Table 8.4 for references
Brown ring disease	*Vibrio tapetis*	Clams (*Venerupis* [*Ruditapes, Tapes*] *philippinarum* and *V. decussatus*)	Europe (mainly France and Spain)	*Gross pathology*: Brown deposits (conchiolin) on inner shell *Histopathology*: Infects the extrapallial fluid and multiplies in haemocytes in acute infections resulting in septicaemia and tissue destruction	Paillard and Maes 1990; Borrego et al. 1996; Rahmani et al. 2019
Nocardiosis	*Nocardia crassostreae*	Pacific oyster, *C. gigas* (adults), mussel, *Mytilus gallprovincialis* and oyster, *Ostrea edulis*	USA, Japan Europe including the Mediterranean (Italy)	*Gross pathology*: Yellow-green lesions on mantle, gills, and adductor muscle in Pacific oysters *Histopathology*: Colonies of bacteria surrounded by haemocytic response to encapsulate these foci in tissues	Friedman et al. 1998; Friedman and Hedrick 1991; Englesma et al. 2008; Carella et al. 2013, 2018
Juvenile oyster disease (Roseovarius oyster disease)	*Roseovarius* (*Alliroseovarius*) *crassostreae*	Juvenile Eastern oyster, *Crassostrea virginica*	North-Eastern USA	*Gross pathology*: Emaciation, conchiolin deposits on inner shell surfaces, uneven valve margins, high mortality *Histopathology*: Lesions with haemocytic infiltration, epithelial cell degradation	Boettcher et al. 2000, 2015; Boardman et al. 2008
Rickettsiosis	Rickettsia-like bacteria in bivalves	Wide range including Pearl oysters, clams, and scallops	Global	*Histopathology*: Intracellular bacteria colonising cells in the mantle, digestive gland, gills, and connective tissue	Travers et al. 2015 (review)
Withering syndrome	*Xenohaliotis californiensis*	Abalone, *Haliotis cracherodii*	California and spread south into Mexico	*Histopathology*: Infection of epithelial cells in GI tract, degeneration of digestive gland, atrophy of pedal muscle	Friedman et al. 2000, 2002
	Francisella halioticida	Scallops, *Mizuhopecten* (*Patinopecten*) *yessoensis* and mussels (*Mytilus* spp.)	British Columbia, Canada, Brittany, France	*Gross pathology*: Pink pustules and/or pus in adductor muscle, pale digestive gland *Histopathology*: Haemocytic infiltration of necrotic affected muscle, intracellular bacteria within haemocytes (Figure 8.1)	Meyer et al. 2017; Kawahara et al. 2019; Charles et al. 2020b
	Endoziocomonas-like (additionally, some animals also have *Francisella* sp., *Anaplasma* sp. and *Mycoplasma* sp.)	Wide range of bivalves	Global	*Gross pathology*: None *Histopathology*: Microcolonies of bacteria in gills and digestive gland	Cano et al. 2020
Mycobacteriosis	*Mycobacterium* spp.	Wide range of gastropods (including snails and pen shells) and bivalves (including scallops and oysters)	Global? (terrestrial, freshwater and marine)	*Histopathology*: Inflammatory foci (nodular masses) with Ziehl-Neelsen acid-fast bacteria	Davidovich et al. 2020 (review)

Table 8.4 Some recently discovered diseases of molluscs caused by vibrios

Disease causing agent(s)	Host range	Geographic range	Pathology	Key references
Vibrio aestuarianus	Pacific oysters, *C. gigas* spat and adults	Europe mainly France, UK, Ireland, and Italy	*Histopathology*: General tissue damage and muscle necrosis, damage to tubules in digestive gland	Garnier et al. 2007
V. bivalvicida	1. Scallop, *Argopecten purpuratus*	1. Scallop hatchery in Chile	1. Disruption of velum, detachment of ciliary cells of velum, necrosis of digestive tissue	1. Rojas et al. 2019
	2. Carpet shell clam, *Ruditapes decussatus*	2. Hatchery in Spain		2. Dubert et al. 2016
V. coralliilyticus (*V. tubiashii*)	*C. gigas*, Kumamoto oysters, *C. sikamea*, clams, *Panope abrupta*, mussel, *Perna canaliculus*	Shellfish hatcheries in USA, Mexico. Potential spread to natural populations	*Gross pathology*: Deformed velum, loss of velar cells *Histopathology*: Abscesses below shell in extra-pallial cavity	Elston et al. 2008
V. crassostreae	*C. gigas*	Europe	Observed in oysters with POMS. Likely secondary invaders	Bruto et al. 2017
V. mediterranei	Noble pen shell (fan mussel), *Pinna nobilis*	Mortalities recorded in aquarium	*Histopathology*: Haemocyte infiltration in the digestive gland. Co-infection with mycobacteria?	Andree et al. 2020; Prado et al. 2020
V. pectenicida	Scallop, *Pecten maximus* (larvae)	Outbreaks of disease in hatchery (France)	Systemic infection of the soft tissues of the larvae resulting in tissue necrosis (due to production of exotoxins by the bacteria) and death	Lambert et al. 1998
V. ostreicida	Oyster, *Ostrea edulis*	Outbreaks of disease in hatchery (Spain)	Shown to be pathogenic in exposure experiments	Prado et al. 2015; Kumar et al. 2020

Infections caused by *V. aestuarianus* mainly occur in summer months, while winter temperatures favour the survival of the host (Parizadeh et al. 2018). However, oysters held at low temperatures although carrying undetectable levels of live *V. aestuarianius* following thermal stress exhibit vibrios in their haemolymph, perhaps implying that they persist within hosts at lower temperatures. Microcosm experiments showed 1 day after adding *V. aestuarianus* to water that these bacteria could be detected in the haemolymph of oysters, and by day 4 these were found in the gills, mantle, digestive gland, muscle, haemolymph, and palps showing the extensive colonisation of tissues (Parizadeh et al. 2018). In the same experiments, mutants of *A. aestuarianus* lacking the *varS* gene were unable

to colonise these tissues, reflecting the importance of this gene in virulence. Histology and *in situ* hybridisation (ISH) also revealed this progression of the disease from haemolymph to other tissues, accompanied by haemocyte recruitment and their subsequent lysis and agglutination (Parizadeh et al. 2018).

Vibrio crassostreae is a facultative pathogen of oysters. Virulent strains of this bacterium have a plasmid whose presence is essential for killing within the host (Bruto et al. 2017). These virulent strains are cytotoxic towards the host's haemocytes via a contact dependent mechanism that involves a novel protein coded by the gene, *r5.7* (Rubio et al. 2019). In contrast, a further oyster pathogen, *V. tasmaniensis* is an intracellular bacterium that inhabits the haemocytes of its host resulting in their destruction (Rubio et al. 2019).

Vibrio mediterranei is a pathogen of the noble pen shell, *Pinna nobilis* held under aquarium conditions (Prado et al. 2020). Populations of this large clam in the Mediterranean have been affected by disease caused by *Haplosporidium pinnae* (see Table 8.6) exacerbating its current status as 'critically endangered of extinction'. Small-scale experiments have shown that challenge of *P. nobilis* with this vibrio causes disease with characteristic pathology of this pathogen (Andree et al. 2020). This infection appears when aquarium temperatures exceed 24°C but its role as a pathogen in wild populations is unknown as are its virulence factors.

Finally, disease outbreaks caused by vibrios may well rely on the collaboration of several species of vibrio—rather than a single species, and avirulent strains of these bacteria may also assist in the disease process—suggesting that such infections can be polymicrobial in nature (Lemire et al. 2015).

8.4.2 Brown ring disease of clams

Brown ring disease is caused by strains of *Vibrio tapetis* (Borrego et al. 1996) and causes characteristic brown rings of dark organic deposits in the inner shell of clams (Paillard and Maes 1990). This disease can cause mortality in these hosts especially if the bacteria invade into the tissues resulting in generalised sepsis and host death. The causative agent can infect other animals by its release in faeces and pseudo-faeces (Mortensen et al. 2007).

8.4.3 Nocardiosis in oysters and mussels

Nocardia crassostreae is a Gram-positive bacterium belonging to the phylum Actinobacteria. It was first identified as a pathogen of the Pacific oyster, *C. gigas* (Friedman et al. 1988) but has since been found in the mussel, *Mytilus galloprovincialis* and the oyster, *Ostrea edulis* in the Gulf of Naples region of Italy (Carella et al. 2013, 2018). In *C. gigas*, the gross pathology characteristically shows yellow-green pustule-like lesions in the mantle, gills, and adductor muscle (Friedman and Hedrick 1991). Carella et al. (2018) suggest that this disease could be considered as an emerging condition

presumably based on its wider host range and geographic spread. Histological examination of infections shows microcolonies of *Nocardia* similar to those seen in other studies (Cano et al. 2020; see also Section 11.3.5).

8.4.4 Juvenile oyster disease (*Roseovarius* oyster disease)

The causative agent of this disease is *Roseovarius* (*Alliroseovarius*) *crassostreae* (Boettcher et al. 2000; Boardman et al. 2008). This is a disease of juvenile oysters only found in hatcheries. Like many bacterial diseases, outbreaks occur in summer months and mortality levels can be over 95% (see Travers et al. 2015 for details). Oysters with juvenile oyster disease show altered shell margins, poor growth rates, and deposits of pigment on the inner shell. High rates of mortality have been reported (Travers et al. 2015).

8.4.5 Rickettsia-like bacteria

Members of the order Rickettsiales (α-proteobacteria) are obligate intracellular parasites of both invertebrates and vertebrates. They are not able to be routinely cultured in artificial media and this results in difficulty in studying these, especially in invertebrate hosts where isolation, cultivation and experimental infections prove impossible to perform. The lack of molluscan cell lines for their propagation is a further problem exacerbating this situation. It is therefore not surprising that few potential rickettsial diseases of molluscs have been fully investigated especially with regard to fulfilling Koch's postulates.

There are several reports of rickettsia-like organisms (RLOs) in a wide range of bivalve molluscs (see Travers et al. 2015 for details and references therein) but these have not been identified and so their classification as rickettsia is at best tenuous (see Section 15.2.2 in Chapter 15 for further discussion of this problem). A recent comprehensive study has examined the nature of the bacteria found in microcolonies in the gills and other tissues of a wide range of molluscs using a combination of histology and 16S amplicon sequencing (Cano et al. 2020). They found that the most common operational taxonomic units (OTUs) belonged to *Endozoicomonas*-like bacteria belonging to the family Endozoicomonadaceae (γ-proteobacteria). These were found in a variety of bivalves, and a gastropod limpet from Europe, Australia, Asia, Africa, and South America. Other commonly observed OTUs were assigned to *Anaplasma*, *Mycoplasma* and *Francisella*-like sequences (Cano et al. 2020). The nature of the association between the bacteria in these microcolonies and their hosts (parasitic *vs.* symbiotic) is unclear.

The most investigated RLO infection in gastropods occurs in black abalones (*Haliotis cracherodii*) off the Californian coast in the USA. The causative agent of this condition, termed withering syndrome, is *Xenohaliotis californiensis* placed in the order Rickettsiales (i.e. true rickettsia) (Friedman et al. 2000, 2002). These rickettsia multiply within the epithelial cells of the GI tract and the pedal muscle undergoes atrophy reflected in the 'withering' name of the disease. This chronic condition is thought to be a major cause of the decline in the abalone fishery in coastal waters off California.

8.4.6 Other bacteria

Some other bacteria are also thought to be pathogens of various molluscs. These include mycobacteria, including *Mycobacterium marinum* that can also cause skin ulcers in humans from handling oysters (Davidovich et al. 2020), and *Francisella halioticida*, a newly reported pathogen of scallops and mussels (Meyer et al. 2017; Kawahara et al. 2019; Charles et al. 2020b; Figure 8.1).

8.5 Diseases caused by fungi including microsporidians

Reports of fungal infections in molluscs are few and their identification and role as primary pathogens is invariably unclear. For instance, most fungal infections of cephalopods have only been observed in aquarium-based animals raising the question whether such infections occur in the wild (Polgase 2019). Similarly, growth of fungi associated with eggs and larval stages of molluscs under culture conditions may reflect saprophytic activities on dead or dying animals. One fungal infection that has been explored is a form of shell disease in Pacific oysters, *C. gigas*, caused by *Ostracoblabe implexa* (Alderman and Jones 1971; Pirkova and Demenko 2008). In this condition, fungi can be observed growing in the shell margins, the hinge and the site of adductor muscle attachment resulting in a darkening of these areas. It is unclear, however, if this disease commonly affects live oysters as Pirkova and Demenko (2008) appeared to have mainly found the condition in dead animals.

(a)   (b)

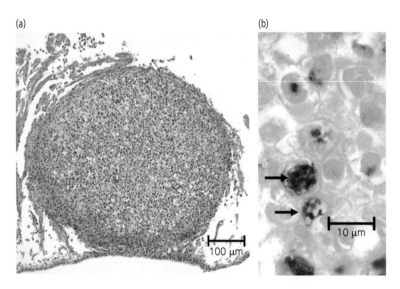

Figure 8.1 Infection of the scallop, *Mizuhopecten* (*Patinopecten*) *yessoensis* by the bacterium, *Francisella halioticida*. (a) Low power micrograph of a histological section showing a prominent focal lesion within the heart containing haemocytes and bacteria. (b) *In situ* hybridisation image counterstained with light green showing strong positive (blue-black) signal (unlabelled arrows) in *F. halioticida* bacteria apparently internalised in haemocytes within a lesion. Source: micrographs courtesy of Dr Gary Meyer (Fisheries and Oceans Canada, British Columbia, Canada).

8.5.1 Microsporidians

Microsporidia are single-celled, obligate intracellular, spore-forming eukaryotic parasites that were once considered to be protists but based on phylogenetics are now known to be most closely related to fungi (Keeling and Fast 2002; Hibbett et al. 2007) and for the purpose of this chapter are dealt with within the Kingdom Fungi. This phylum is present globally in terrestrial, marine, and freshwater ecosystems and consists of over 200 genera and *ca.* 1,300–1,500 species producing benign to lethal infections (Vávra and Lukeš 2013; Cali et al. 2017). A diverse range of both generalist and specialist Microsporidia infect vertebrate and invertebrate hosts (Stentiford et al. 2016), which can have significant effects on individual and population fitness. Aquatic organisms act as hosts for species in almost half of the known genera of Microsporidia and many of these parasites remain undescribed (Stentiford et al. 2013, 2016). Ten species are known to infect molluscs (Table 8.5).

The intracellular habitat is fundamental in their ability to evade recognition by the host immune system and to accumulate high parasite burdens within tissues of infected hosts (Stentiford et al. 2013; Szumowski and Troemel 2015). Microsporidians affect host survival but may also influence host phenotype, life history and behaviour (Stentiford et al. 2013). They can survive outside their hosts for up to several years facilitated by a spore life stage. Microsporidia spores are distinguished by a striking harpoon-like apparatus called the polar tube located at the anterior half of the spore (Larsson 2005; Winters and Faisal 2014). The long thread-like polar filament is surrounded by a polaroplast i.e. a lamella of membranes that extends from the polar tube as a coiled-up structure in the posterior half of the spore (Larsson 2005). When the spore is in the appropriate host and environment, it germinates and infection of the host occurs when the polar tube and filament evert to pierce the membrane of the host cell, which can happen in a matter of seconds (Xu and Weiss 2005). The polar tube then acts as a syringe to directly inject nuclei and sporoplasm (infective cytoplasm) into the host cell. The sporoplasm develops into a multinucleate plasmodium i.e. a cell-wall deficient form called a meront, which eventually differentiates to re-generate the spore form (1–40 µm

in diameter). The host cell eventually ruptures liberating the spores and allowing them to infect a new host (Lauckner 1983).

The microsporidian life cycle can vary with some species having a simple asexual life cycle (Ironside 2007), while others have a more complex one involving multiple hosts and both asexual and sexual reproduction. Both horizontal and vertical transmission life cycles are associated with microsporidians (Dunn and Smith 2001) with horizontal transmission characterised by a high parasite burden and associated pathogenicity, while vertical transmission is characterised by low virulence, which has potentially led to under-reporting of this important transmission route (Dunn and Smith 2001).

The genus *Steinhausia* is only found in molluscs (Sprague et al. 1972). *Steinhausia mytilovum* (*Steinhausia* sp.) is the causative agent of mussel egg disease. Its geographical range includes eastern United States (Sprague 1965; Figueras et al. 1991; Sunila et al. 2004), California, USA (Hillman 1991), Europe (Figueras et al. 1991; Villalba et al. 1997; Comtet et al. 2003; Rayyan and Chintiroglou 2003), Western Australia (Jones 1997; Jones and Creeper 2006), and Brazil (Matos et al. 2005). Host species include the blue mussel *Mytilus edulis*, the Mediterranean mussel *M. galloprovincialis*, and the Guiana swamp mussel *Mytella guyanensis*. Prevalence of infection varies from low (5.9% at one site in *M. galloprovincialis* in the Mediterranean; Comtet et al. 2004) to high (43% in *M. edulis* from Maryland, USA; Sprague 1965). Sporocysts of *Steinhausia* are spherical in shape (9–19 µm in diameter) and may contain 10 to 41 spherical spores (*ca.* 1–2 µm in diameter). *S. mytilovum* infects the cytoplasm and nucleus of mussel ova and can incite a moderate to severe diffuse-type haemocyte infiltration response inside infected gonadal follicles (De Vico and Carella 2012). Infected oocytes have been observed in apparently healthy follicles (Sunila et al. 2004; Jones and Creeper 2006). Mussel fecundity is believed to be inversely related to intensity of infection, as the nucleus of the ovum is distorted often resulting in the destruction of the egg (Jones and Creeper 2006). Horizontal transmission is thought to occur when loose spores are released along with intact eggs (Jones and Creeper 2006). A seasonal pattern in prevalence of infection has been

Table 8.5 Diseases of molluscs caused by microsporidians

Disease causing agent(s)	Host range	Geographic range	Pathology & location of infection	Key references
Bivalves				
Microsporidium rapuae	Mediterranean mussel, *Mytilus galloprovincialis*, Green lipped mussels, *Perna canaliculus*, Chilean oyster, *Ostrea lutaria* (=*Tiostrea chilensis, Ostrea lutaria*)	Europe, New Zealand	Gut connective tissue. Induces formation of oval cysts (20–70 μm) in connective tissue surrounding the gut. Cysts contain numerous spores surrounded by concentric layers of fibrous elements infiltrated by haemocytes	Jones 1981; Bower 2006a; Castinel et al. 2019
Steinhausia-like microsporidian in the ova of oysters	European flat oyster, *Ostrea edulis*, Sydney rock oyster, *Saccostrea glomerata*, Mangrove cupped oyster, *Crassostrea rhizophorae*, Hong Kong oyster, *C. hongkongensis*	France, Australia, Brazil, Southern China	All infections occur in the cytoplasm of mature ova. Occasionally, surrounding tissue exhibit an intense haemocyte infiltration response and some parasitised eggs become necrotic. The intensity of infection is usually low with generally fewer than 1% of the oocytes (= ovocytes) infected	Nascimento et al. 1986; Moss et al. 2007; Green et al. 2008; da Silva et al. 2012
Steinhausia mytilovum	*M. edulis, M. galloprovincialis*, Guiana swamp mussel, *Mytella guyanensis*, Common cockle, *Cerastoderma edule*	USA (Eastern United States, California), Europe (Italy, Spain, France and Greece), Western Australia, Brazil	All infections occur in the cytoplasm and nucleus of mature mussel ova. Moderate to severe diffuse-type haemocyte infiltration response. Infected oocytes (=ovocytes) have also been observed in apparently healthy follicles	Villalba et al. 1997; Sagristá et al. 1998; Carballal et al. 2001; Comtet et al. 2003, 2004; Sunila et al. 2004; Matos et al. 2005; Jones and Creeper 2006; De Vico and Carella 2012
Gastropods				
Microsporidium aplysiae	California sea hare, *Aplysia californica*	Northern California, USA to Gulf of California, Mexico	Nucleus and cytoplasm of neurons	Krauhs et al. 1979
M. novacastriensis	Grey garden slug, *Deroceras reticulatum*		Gut epithelium	Jones and Selman 1985
Microsporidium sp.	The ram's horn snail, *Biomphalaria straminea*	Not reported	Not reported	Richards 1973
Nosema sp.	Pond snail, *Lymnaea rubiginosa*	Malaysia	Hepatopancreas, heart, preputium, gizzard, mantle. Binucleate shizonts, sporonts, sporoblasts and spores (ovoid and ellipsoidal 4–6.3 μm in diameter) observed	Canning et al. 1974
Pleistophora husseyae	Freshwater snails, *Physa cubensis, P. heterostropha, Physa* sp., *Aplexa hypnorum*	Cuba, Eastern North America	All organs and tissues except radular cartilage	Michelson 1963
Sheriffia (= Steinhausia) brachynema	*Biomphalaria glabrata, B. helophilia, B. pfeifferi*	Central and South America	Gut epithelium, adjacent tissue, sometimes mantle	Richards and Sheffield 1971

observed with a positive correlation with cooler water temperatures (Matos et al. 2005). *Steinhausia* spp. have round spores with a short thick filament and no endospores that develop pressed against, and deforming, the host cell's nucleus. Although *Steinhausia* is pathogenic in gastropods, the other species are present in the oocytes/ova of bivalves but are not considered to be virulent. Few eggs are infected and usually with only a single spore (Cali et al. 2017).

Species known to affect the edible cockle *Cerastoderma edule* are *Steinhausia* sp. while the hyperparasite *Unikaryon legeri* infects the parasitic flatworms i.e. trematodes such as *Gymnophallus minutus* that infect *C. edule* (Longshaw and Malham 2013; Figure 8.6b). *U. legeri* infects the trematode metacercariae life stage and appear to be fatal; the microsporidian spores make the metacercariae enlarge to in some cases twice the normal size and alter their coloration to opaque/greyish white (Bowers and James 1967; Goater 1993). More than eighteen species are known to parasitise digeneans and eight of these species belong to the genus *Nosema* (Toguebaye et al. 2014). Hyperparasitism of *G. minutus* metacercariae by *U. legeri* is common in cockles on the south coast of Ireland (Fermer et al. 2010).

Within the gastropods, a detailed investigation of microsporidian presence in the freshwater snails *Bulinus* spp., *Biomphalaria* spp., *Lymnaea* spp., *Physa* sp., and *Planorbis* spp. using microsporidian specific primers found a wide range of prevalence ranging from 0% in *Physa* sp. collected in the UK to 83% in *Bulinus globosus* from Zimbabwe (McClymont et al. 2005). These data may imply a wide distribution and diversity of microsporidians in snails that has not to date been further elucidated.

8.6 Diseases caused by oomycetes and related forms

Oomycetes cause diseases in a number of aquatic animals including fish, crustaceans and molluscs. *Halioticida noduliformans* is the causative agent of abalone tubercle mycosis that has affected abalone production in several countries including Japan and South Africa (see Derevnina et al. 2016 for a review). These infections in abalone culture facilities in South

Africa have been reported to cause >90% mortality in spat and >30% death in older abalone (Macey et al. 2011).

8.7 Diseases caused by haplosporidans

The phylum Haplosporidia is a group of obligate protozoan parasites comprising 52 described species and several unnamed species, including ecologically and economically significant pathogens (Table 8.6). Four haplosporidian genera (*Haplosporidium, Minchinia, Urosporidium* and *Bonamia*) have been described based on morphology, development, ultrastructure (Azevedo and Hine 2017), and phylogenetic analysis based on small subunit ribosomal ribonucleic acid (SSU rRNA) (Reece et al. 2004; Arzul and Carnegie 2015). Haplosporidians are widely distributed in marine and freshwater invertebrates, although their status in Africa, much of Eurasia, and Central and South America is largely unknown (Azevedo and Hine 2017). Haplosporidian parasites of molluscs have been considered major pathogens of concern for aquatic animal health managers and shellfish industries globally, as significant mortalities have been associated with them in susceptible hosts (Culloty and Mulcahy 2007, Arzul and Carnegie 2015).

8.7.1 *Haplosporidium* spp.

Haplosporidium spp. are typically observed extracellularly in connective and epithelium tissues (Burreson and Ford 2004; Lynch et al. 2010; Arzul and Carnegie 2015) and are associated with abalone, false limpets, oysters, mussels and clams (Table 8.6). Multinucleated plasmodia (5 to > 50 µm in diameter depending on the number and size of nuclei present) are the most commonly observed life stage and sporulation is typical of this genus. During sporulation, plasmodia develop into sporocysts, with spore walls forming around each nucleus. Spores are subsequently released into the environment upon the death of the host (Arzul and Carnegie 2015; Azevedo and Hine 2017). Spores facilitate survival of *Haplosporidium* spp. during

Table 8.6 Diseases of molluscs caused by haplosporidans

Disease causing agent(s)	Host range	Geographic range	Pathology	Key references
Bivalves				
Bonamia ostreae	Oyster, *Ostrea edulis*	Europe, Atlantic and Pacific coasts, USA, Canada	Although many infected oysters appear normal, others may have yellow discolouration and/or extensive lesions (i.e. perforated ulcers) in the connective tissues of the gills, mantle and digestive gland. Pathology appears correlated to haemocyte destruction and diapedesis due to proliferation of *B. ostreae* (Figure 8.2)	Carnegie et al. 2000; Cochennec et al. 2000; Marty et al. 2006
B. exitiosa	Oysters, *Ostrea angasi*, *O. chilensis*, *O. edulis*	Australia, New Zealand, Europe	In *O. angasi* large focal necrotic lesions in gills. In *O. chilensis*, no gill lesions detected. Infected haemocytes are initially observed in connective tissue but as infection progresses, they can be found in all tissues either by tissue leakage, haemocyte diapedesis, or host decomposition	Hine et al. 2001; Abollo et al. 2008; Carnegie et al. 2014
B. perspora	Oyster, *O. stentina*	North Carolina, USA	*B. perspora* undergoing sporulation (spore formation) had sporogonic cell forms (sporonts, sporocysts and released spores) distributed throughout and sometimes completely filling the connective tissue of the oyster. Widespread disintegration of digestive diverticula with sporocysts and free spores occurring in the lumens of many digestive tubules. Haemocyte infiltration into infected tissue.	Carnegie et al. 2006
B. roughleyi	Oyster, *O. angasi*	Australia	A systemic intracellular infection in the haemocytes that is associated with focal abscess-type lesions in the gill, connective, and gonadal tissues and alimentary tract.	Carnegie et al. 2014
Bonamia sp. (unknown sp.)	Various species of oysters	Chile, USA, Europe	Pale digestive gland, severe ulcerations and/or abscess lesions especially in the gonad region, mantle (frequently near the adductor muscle) and gills and gaping.	Campalans et al. 2000; Hill et al. 2014
Haplosporidium armoricanum [= *Minchinia armoricana*]	Oyster, *O. edulis*	France (Atlantic and Mediterranean) Ireland, Spain (Galicia), the Netherlands,	Gill lesions are observed. Oysters in late stages of the infection are often thin and watery in appearance and have a characteristic brownish colour caused by masses of spores in the connective tissue of all organs.	Pichot et al. 1980; da Silva et al. 2005; Hine et al. 2007; Lynch et al. 2013
H. nelsoni	Oysters, *Crassostrea virginica*, *O. edulis*	Atlantic Coast, USA, Ireland	Infected juvenile oysters may have pale digestive glands, appear emaciated and show no new shell growth. In adult oysters, mantle recession has been reported from heavily infected *C. virginica*, accompanied by extensive fouling. Affected oysters are typically thin and watery with pale digestive diverticula.	Burreson 2005; Lynch et al. 2013
H. edule	Cockle, *Cerastoderma edule*	Galicia, Spain,	Infections are heaviest in the digestive gland and induce a haemocytic response especially when the plasmodial stage is dominant	Carballal et al. 2001; Azevedo et al. 2003
H. hinei	Pearl oyster, *Pinctada maxima*	W. Australia	Infection observed in the connective tissue of the digestive gland, the gut and digestive diverticulae. Presporulation and sporulation stages occur mainly in the connective tissue of the digestive gland (but not in the epithelia of the digestive tract) and mantle, and with fewer numbers in the heart, gills, foot and adductor muscle. Spores have been observed to be pleomorphic, but most are ovoid	Jones and Creeper 2006; Bearham et al. 2008a

continued

Table 8.6 *Continued*

Disease causing agent(s)	Host range	Geographic range	Pathology	Key references
H. pinnae	Fan mussel, *Pinna nobilis*	Mediterranean	Uninucleate cells observed throughout the connective tissue and haemolymph sinuses of the visceral mass. Binucleate cells and, rarely, multinucleate plasmodia detected in the connective tissue. Sporulation stages observed in the epithelium of the host digestive gland tubules	Catense et al. 2018
H. raabei	Mussel, *Dreissena polymorpha*	France, Germany, The Netherlands	Infection with multinucleate plasmodia and sporocysts were observed systemically in the connective tissue, the gills, gonad, and digestive gland. No evidence of haplosporidian cells in epithelia or muscle	Molloy et al. 2012
Minchinia occulta	Oyster, *Saccostrea cuccullata*	W. Australia	Focal infections in the connective tissue of the gills and as disseminated infections in the mantle, reproductive follicles, and around the digestive diverticulae. Sporulation confined to the connective tissue of the digestive gland	Bearham et al. 2008b
M. mercenariae	Clam, *M. mercenaria*	Virginia and New Jersey, USA	Multinucleated plasmodial and sporogonic stages (both 4.6–18.5 µm in diameter) replaced almost all of the connective tissue but were absent in the digestive epithelia. Spores observed	Ford et al. 2009
M. tapetis	Clams, *Ruditapes decussatus*, *Venerupis aureus*, *Ruditapes philippinarum*, Cockle, *Cerastoderma edule*	Europe (Ireland, France, Spain, Portugal)	First signs of infection are multinucleate plasmodia usually in the epithelia of the digestive tract. The pathogenicity of the plasmodial stage in clams is minimal but the sporulation stage in the connective tissue causes important lesions in the digestive gland and gills	Villalba and Navas 1988; Azevedo 2001; Albuixech-Martí et al. 2020
Minchinia sp. (unknown)	Clam, *Cyrenoida floridana*, mussels, *Mytlus edulis*	USA, Wales (UK) and Ireland	Multinucleated plasmodia observed in the connective tissues	Reece et al. 2004; Lynch et al. 2014, 2020
Gastropods				
Haplosporidium montforti	Abalone, *Haliotis tuberculata*	Galicia, Spain	Blotchy appearance of the epipodium. Dark foot pigmentation, loss of surface adherence and a limited ability to right themselves after they were set on their backs. These signs are not specific to infection with this parasite alone	Azevedo et al. 2006; Balseiro et al. 2006
H. patagon	Lesson's false limpet, *Siphonaria lessonii*	Patagonia, Argentina	Different stages of sporulation were observed as infections in the digestive gland. Oval or slightly subquadrate spores were observed	Ituarte et al. 2014
H. tuxtlensis	Striped false limpet, *Siphonaria pectinata*	Veracruz, Mexico	Detected in the visceral mass. Spores ellipsoidal (3.6 × 2.7 µm)	Vea and Siddall 2011

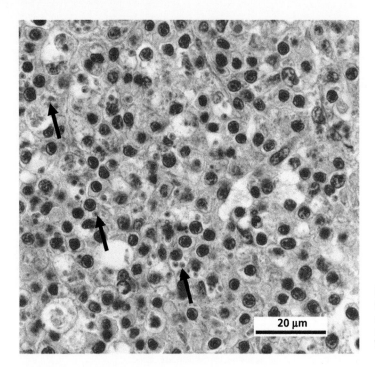

Figure 8.2 *Bonamia ostreae* (unlabelled arrows) seen in the haemocytes and connective tissue cells of the European oyster, *Ostrea edulis*. Source: micrograph courtesy of Matthew Green (Cefas, Weymouth, U.K)

periods of suboptimal conditions externally in the environment and possibly within the host. Experiments designed to demonstrate direct transmission between oysters with certain *Haplosporidium* spp., such as *Haplosporidium nelsoni* (the causative agent of multinucleated spore X (MSX) disease) in *Crassostrea virginica* have been unsuccessful. The involvement of an intermediate host, possibly plankton, to complete the life cycle of *H. nelsoni* is suspected (Powell et al. 1999; Burreson and Ford 2004; Arzul and Carnegie 2015). High abundances of *H. nelsoni* have been observed in the environment outside *C. virginica* even when oyster infection prevalence is low (Ford et al. 2009), which would indicate that this haplosporidian species is able to sustain itself outside the oyster host. A recent overview sought evidence for intermediate hosts of *H. nelsoni* and noted that although environmental DNA (eDNA) analyses showed positive signals in amphipods and polychaetes in particular, *in situ* hybridisation failed to show any evidence of parasites in these animals (Ford et al. 2018).

8.7.2 *Bonamia* spp.

Infection with *Bonamia* is typically observed in oysters globally and is intracellular, systemic, and associated with haemocyte infiltration into affected tissues. *Bonamia* spp. are the causative agent of the disease bonamiosis, also named microcell disease or haemocyte disease of flat oysters. Unicellular, binucleate (single cell life forms 2–6 μm in diameter) and plasmodial life stages are associated with *Bonamia* spp. (Figure 8.2). The plasmodial life stage is more commonly observed in *B. exitiosa* that infects several oyster species of the genus *Ostrea* (*O. angasi*, Chilean oyster *O. chilensis* and *O. edulis*) compared to *B. ostreae* (Arzul and Carnegie 2015). *B. perspora* is the only species for which spores have been observed thus far. *Bonamia* spp. are found intracellularly and extracellularly throughout connective tissues of infected oysters and are more abundant at the base of the epithelia of the gut and haemolymph sinuses (Carnegie et al. 2006). Direct transmission of *B. ostreae* between oysters has been observed in *O. edulis* and the Pacific oyster has been reported to act as a carrier of viable *B. ostreae* (Lynch et al. 2010).

Additionally, brittle stars have been found to successfully transmit *B. ostreae* to naive *O. edulis* (Lynch et al. 2007).

8.7.3 *Minchinia* spp.

Minchinia spp. have been detected in mussels, clams, cockles and oysters. Species belonging to the genus *Minchinia* have been detected in blue mussels *Mytilus edulis* (Lynch et al. 2014; Ward et al. 2018), Mediterranean mussels *Mytilus galloprovincialis* (Comps and Tige 1997), hard clams *Mercenaria mercenaria* (Ford et al. 2009), grooved carpet clam *Ruditapes decussatus* (Villalba et al. 1993; Azevedo et al. 2001), pullet carpet clams *Venerupis corrugata* (Villalba et al. 1993), banded carpet clam *Polititapes rhomboides* (Villalba et al. 1993), golden carpet clam *Polititapes aureus* (Navas et al. 1992), Florida marsh clam *Cyrenoida floridana* (Reece et al. 2004), and cockles *Cerastoderma edule* (Ramilo et al. 2018; Lynch et al. 2020; Albuixech-Martí et al. 2020) (Table 8.6). The main species are *M. mercenaria* (clams), *M. mercenaria*-like (cockles), *M. tapetis* (clams and cockles), and *Minchinia occulata* (oysters) (Table 8.6).

Life stages consist of uninucleate cells, binucleate cells, and multinucleate plasmodia life stages located in the connective tissue, digestive gland, gills and gonad. The dominant multinucleate plasmodial life stage encloses 3 to 14 nuclei (*M. mercenaria*) and 3 to 6 nuclei (*M. tapetis*). Spores of a *M. mercenaria*-like sp. have only been reported in cockles in Ireland (Lynch et al. 2020; Albuixech-Martí et al. 2020). Infection intensity with *M. mercenaria* is moderate or heavy in most records. Cockles with abundant plasmodia show a heavy inflammatory reaction, mainly in the digestive area (Lynch et al. 2020; Albuixech-Martí et al. 2020). *M. tapetis* appear in foci, mostly close to stomach branches or digestive primary ducts. Frequently, the parasites are surrounded by fibrous material. Cockles show heavy haemocytic infiltration around parasite foci. Sporogonic stages of *M. tapetis* have not been reported in cockles, although spores are known to occur in clam species (Azevedo 2001; Ford et al. 2009). Reported intensity of infection with *M. tapetis* is primarily light, less frequently moderate, and never heavy. The inflammatory reaction can cause some damage and host weakness. *Minchinia chitonis* has been detected in the chiton *Lepidochitona cinereus* (Polyplacophora) in Scotland and Northern Ireland (U.K.). Infection was reported to not be limited to certain tissues and was observed in the digestive gland, muscles and gills (Ball and Neville 1979; Baxter et al. 1989).

8.8 Diseases caused by mikrocytids

Mikrocytids are microcell (2–3 μm in diameter) parasites, similar to haplosporidians in some aspects of their morphology but placed in a separate sister family and order (Mikrocytidae, Mikrocytida respectively; Hartikainen et al. 2014). Mikrocytid infections of molluscs were first reported in diseased oysters from Denman Island in British Columbia, Canada in 1961 (Quayle 1961) but the causative agent was unknown. Farley et al. (1988) later identified and named the causative agent as *Mikrocytos mackini*. Infected animals display localised green lesions visible upon dissection on the body wall, mantle, labial palps, or adductor muscle (Abbott and Meyer 2014). Histopathology shows numerous single cell parasites in connective tissue with some haemocytic infiltration into affected tissues. Heavy infections may cause mortality in Pacific oysters. Since the original finding of Denman disease in oysters in Canada, it has been reported in other oyster species in several geographic regions (Table 8.7).

Improvement in molecular detection methods has resulted in better understanding of the evolution, species diversity and host range of mikrocytids. This has facilitated the identification of several other species of *Mikrocytos* including *M. boweri* in another oyster species, *Ostrea lurida* (Abbott et al. 2014), *M. mimicus* in *C. gigas* collected in the UK (Hartikainen et al. 2014), and *M. veneroïdes* and *M. donaxi* in clams, *Donax trunculus* in France (Figure 8.3a–c; Garcia et al. 2018).

Methods of transmission, the possibility as vectors/reservoirs, and host specificity of these parasites is currently unclear but the finding of mikrocytids in a range of other invertebrates including crustaceans suggest that they are more widely distributed in the marine environment than first envisaged (Hartikainen et al. 2014).

Table 8.7 Diseases of molluscs caused by mikrocytids

Disease	Disease causing agent(s)	Host range	Geographic range	Pathology	Key references
Denman disease	*Mikrocytos mackini*	Pacific oyster, *C. gigas* and Kumamoto oyster, *C. sikamea*	Canada and USA	*Gross pathology*: Focal greenish lesions *Histopathology*: Microcells (*ca.* 3 μm) in diameter in tissues, limited haemocytic infiltration	Quayle 1961; Farley et al. 1988; Abbott et al. 2011; Elston et al. 2012
	M. boweri	Olympia oyster, *Ostrea lurida*	Canada (and probably China and Europe in other oyster species)	Presence of parasite causes limited haemocytic infiltration into affected tissues. Microcells seen intracellularly	Abbott et al. 2014
	M. mimicus	Pacific oyster, *C. gigas*	U.K. (East coast)	*Gross pathology*: Mortality, gaping with limited intravalvular fluid. Focal green pustules. *Histopathology*: Intense focal haemocyte infiltration, necrosis. Microcells *ca.* 2–5 μm) in diameter within vesicular connective tissue cells and adductor muscle.	Hartikainen et al. 2014
	M. donaxi and *M. veneroïdes*	Wedge clam, *Donax trunculus*	Atlantic coast, France	*Gross pathology*: High levels of infection leading to mortality *Histopathology*: Microcells found both intra- and extra-cellularly in various tissues. No report of host response. (Figure 8.3a–c)	Garcia et al. 2018

8.9 Diseases caused by other protists

8.9.1 Marteiliosis

Marteiliosis is a paramyxid disease of bivalve molluscs caused by members of the genus *Marteilia* spp. This genus belongs to the phylum Cercozoa (Endomyxa), order Paramyxida. Recent studies have shown that *M. refringens* consists of two distinct species, *M. refringens* ('O' genotype) and *M. pararefringens* ('M' genotype) with distinct geographic distribution in Europe (Kerr et al. 2018). Since the late 1960s and early 1970s, *M. refringens* has been associated with mass mortalities of *O. edulis* from northern France and further south into Portugal, Greece, Morocco and Tunisia (Aber disease/Digestive gland disease; Berthe et al. 2004; Figure 8.4a). *M. pararefringens* has a more northern distribution including the UK and Scandinavia infecting mussels including *M. edulis* (Carrasco et al.

2017). *Marteilia* can range in size from 5–8 μm (early life stage) up to 40 μm (late stage, sporulation) in diameter. The morphology and cell cycle of *Marteilia* is unique in that it consists of a primary/stem cell and within this cell is a nucleus and multiple, between 3 and 15, secondary cells/sporoblasts created via internal cleavage (Figure 8.4b). Infections by all *Marteilia* spp. are initiated by a primary cell infecting the epithelial cells in the digestive tubules, and possibly the gills and labial palps (Kleeman et al. 2002; Bower 2006a, 2006b). Within each secondary cell there is a nucleus and up to six spores. Each spore contains a nucleus and another spore and so on. Sporulation occurs in the digestive tubules causing cellular damage in some cases (Kleeman et al. 2002; Carrasco et al. 2015). Mature spores are shed from the lumen of the digestive tubules and intestine and released in faecal matter to infect other susceptible bivalves. *M. refringens* may

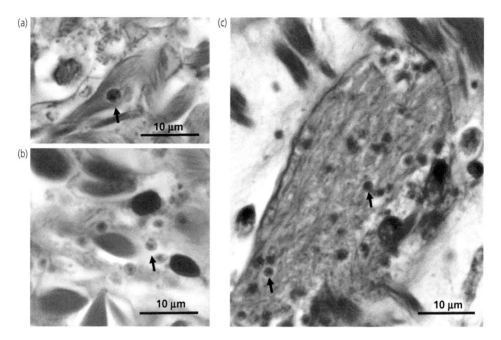

Figure 8.3 (a–c) Mikrocytid infections of wedge clams, *Donax trunculus*. (a) Intracellular *Mikrocytos donaxi* (arrow) in a myocyte. (b) Extra- and intra-cellular *Mikrocytos veneroïdes* (arrows) in muscle tissue. (c) Intracellular unicellular forms of *M. veneroïdes* (arrows) in neural ganglion. Source: micrographs courtesy of Céline Garcia (Ifremer, LGPMM Laboratory, Station Les Tremblade, France).

also be harboured by other invertebrates including copepods (*Paracartia latisetosa*) where putative life history stages have been found in the alimentary canal, digestive epithelium, and germinal site/gonadal tissue (Boyer et al. 2013; Arzul et al. 2014).

There are several other species of *Marteilia* that infect various bivalves. These include *M. sydneyi* that infects the Sydney rock oyster, *Saccostrea glomerata* (so-called QX disease) and possibly also the tropical black lip oyster *Saccostrea echinata* in Australia, and the rock oyster *Saccostrea forskali* in Thailand (Perkins and Wolf, 1976; Adlard and Ernst 1995, Taveekijakarn et al. 2002; Adlard and Nolan 2015), *M. cochillia* (genotype 'C') that infects the digestive gland of cockles, *Cerastoderma edule* in Spain (Carrasco et al. 2013; Darriba et al. 2020) and *M. tapetis* and *M. granula* (= *Eomarteilia*; Ward et al. 2016) that infect the Manila clam, *Ruditapes philippinarum* in Korea (Figure 8.4b; Kang et al. 2019) and Japan respectively (Itoh et al. 2014). *M. octospora* has been detected in the razor shell clam *Solen*

marginatus (Ruiz et al. 2016), however no associated mortalities were observed in that clam population in Galicia, Spain. Although *M. refringens* is a recognised pathogen of oysters, it has been reported in several clam species (Boyer et al. 2013). In the case of *M. sydneyi*, infections of rock oysters have been one of the key drivers of their decline as mortality levels of up to 95% have been found in some years (see review by Raftos et al. 2014). The earliest uninucleate stages of *M. sydneyi* are found in the epithelial cells in the gills and palps where initial replication occurs. The progeny of these cells spread via the haemolymph to the digestive gland resulting ultimately in sporulation. The haemocytes of the immune system of oysters can recognise and phagocytose *M. sydneyi* and low levels of phenoloxidase are associated with higher disease susceptibility (Raftos et al. 2014). Finally, like *M. refringens*, QX disease can be harboured by polychaete worms (*Nephtys australiensis*) where infective stages were observed in the alimentary canal epithelial lining (Adlard and Nolan 2015).

(a) (b)

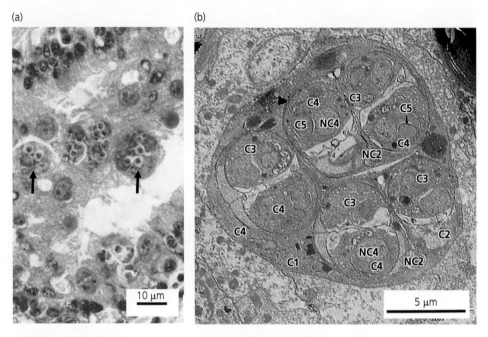

Figure 8.4 Marteiliosis in bivalve molluscs. (a) *Marteilia refringens* (unlabelled arrows) in the digestive gland epithelial cells of the European oyster, *Ostrea edulis*. These form characteristic nurse cells containing sporonts containing spores. Micrograph courtesy of Matthew Green (Cefas, Weymouth, UK) and Ana Grande and Francisco Ruano (Instituto Português do Mar e da Atmosfera, Portugal). (b) Transmission electron micrograph of *Marteilia tapetis* infecting the digestive tubules of the Manila clam *Ruditapes philippinarum*. C1: primary cell, C2: secondary cell, C3: tertiary cell, C4: fourth cell, C5: fifth cell, NC2, NC4: nucleus of C2 and C4. Source: micrograph courtesy of Hyun-Sil Kang (Southeast Sea Fisheries Research Institute, National institute of Fisheries Science, South Korea).

8.9.2 Perkinsosis and Dermo disease

Dermo disease is caused by the alveolate protist *Perkinsus marinus*, however, the initial observation of this parasite mis-identified it as a fungus belonging to the genus *Dermocystidium* (hence the name 'Dermo disease'). According to Andrews (1996) and Burreson and Ragone Calvo (1996) it first made its appearance in eastern oysters (*Crassostrea virginica*) in Chesapeake Bay in the early 1950s and since that time it has been one of the causes of the dramatic decline in this fishery.

Dermo disease is cyclical in its annual presence and severity of infection. At water temperatures of > 20°C it can kill oysters quickly and it over winters without causing immediate mortality only to reappear the next year when water temperatures increase. A combination of hot, dry summers with increased salinity and higher winter temperatures in Chesapeake Bay has strongly favoured the episodic nature of the disease and its severity. The geographic distribution of *P. marinus* infections has gradually increased northwards (Ford and Smolowitz 2007) presumably due to climate change. In its southern range it can be found in the Gulf of Mexico in *C. virginica* but with less impact on populations (Cáceres-Martínez et al. 2016).

There are three life history stages in *Perkinsus* spp., parasitising trophozoites, prezoosporangia, and zoospores (Ruano et al. 2015; Maeda et al. 2021), the latter two found outside the host. Dermo can be transmitted directly via faeces from dying or moribund oysters. Infectious stages gain entry to naïve oysters via the upper digestive tract as a result of the filter feeding activity of the oyster. Once inside its host, trophozoite stages are phagocytosed by haemocytes but these survive and multiply within these cells by inhibiting the respiratory burst

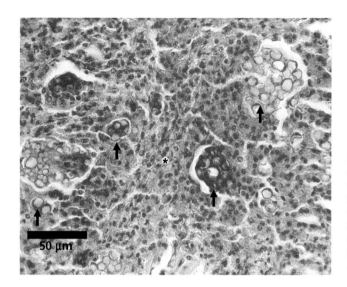

Figure 8.5 *Perkinsus olseni* (= *P. atlanticus*) infecting the carpet shell clam, *Ruditapes decussatus*. Note the characteristic signet ring shape (unlabelled arrows) of the trophozoites and infiltration of haemocytes into the tissues (*). Samples were taken from infected clams in the Algarve region of Portugal where this parasite can cause extensive mortality. Source: micrograph courtesy of Ana Grande and Francisco Ruano (Instituto Português do Mar e da Atmosfera, Portugal).

that normally kills ingested microbes. Recognition and binding of trophozoites to these phagocytes is achieved by galectins (Vasta et al. 2020; see Section 1.11.1 in Chapter 1). Insights into how *P. marinus* avoids the killing activities of host haemocytes comes from a transcriptomic analysis of this pathogen that revealed a number of antioxidant enzymes that may interfere with oxygen radical production and H_2O_2 (Joseph et al. 2010). Intrahaemocytic survival and multiplication results in the proliferation of the parasites leading to systemic infection and death (Ford et al. 2002).

Since the discovery and identification of *P. marinus* in the USA, other species of *Perkinsus* have been described from both bivalves and abalones causing perkinsosis often with high mortalities. *P. olseni* (= *P. atlanticus*; Figure 8.5) causes disease in abalones, *Haliotis rubra* in South Australia (Lester and Davis 1981), the Manila clam, *Venerupis* (*Ruditapes*) *philippinarum* in Japan (Waki et al. 2018) and Italy (Pretto et al. 2014), pearl oysters *Pinctada fucata* in India (Sanil et al. 2010), the carpet shell clam, *Ruditapes decussatus* in Galicia, Spain (Casas and Villalba 2012) and the short yellow clam, *Paphia malabarica* in India (Shamal et al. 2018). Similarly, *P. chesapeaki* has a wide host range including cockles, *Cerastoderma edule* on the Mediterranean coast (Carrasco et al. 2014) and Manila clams *V. philippinarum* in Spain (Ramilo et al. 2016).

8.9.3 Ciliates

Ciliate protists are often found living in association with molluscs, especially in the gills and alimentary canal. Many are probably harmless causing minor irritation at the most but some may have detrimental effects on their host and are hence classified as facultative parasites. There are reports of ciliates causing pathological conditions in both gastropods and bivalves. For example, *Tetrahymena limacis* and *T. rostrata* are thought to be parasitic to various species of terrestrial gastropods invading tissues including the digestive gland and causing mortality (Brooks 1968). Another species of *Tetrahymena*, *T. glochidiophila* when incubated with the glochidia of freshwater mussels of the genus *Lampsilis* spp. causes damage and loss of viability (Lynn et al. 2018; Prosser et al. 2018) but its importance in natural conditions is unknown. In the case of bivalves, infestations of both *C. virginica* and *Mytilus* spp. have been found. In the former case, ciliates invaded in the gills of oysters collected in Hampshire, USA causing the formation of hypertrophic lesions termed xenomas (McGurk et al. 2016). Host response to the presence of these ciliates within the xenomas was limited, perhaps suggesting immune evasion or a limited impact on the host. The temporal incidence of this infection occurred in late summer peaking in the autumn months. Significant numbers of the oysters infested with this unidentified ciliate also had co-infections with *P. marinus* and *H. nelsoni* (McGurk et al. 2016). The condition called 'mussel protozoan X' is also caused by ciliates recently identified as belonging to the order Rhynchodida, but not the same as any known genus and species (Fichi et al. 2018). These ciliates are found intracellularly within the epithelial cells that form the digestive tubules, in the surrounding intratubular spaces and in ovarian tissue. There is no apparent host response to their presence despite evidence of tissue damage. Although Fichi et al. (2018) found a high prevalence of infection albeit in a small sample size of *ca.* 30%, infection severity was low in the majority of mussels examined. The overall importance of this and other ciliate infestations is unclear.

8.10 Diseases caused by 'macroparasites'

In marine ecosystems, gastropod and bivalve species contain a species rich macroparasite community consisting of trematodes, turbellarians, nematodes, copepods and polychaetes (Tables 8.8–8.12). For example, in a study that catalogued the macroparasite community in ten mollusc species of a tidal basin in the Wadden Sea, 31 taxa of macroparasites were recorded (Thieltges et al. 2006). Some parasite species were observed in a single mollusc host species while others, trematode species in particular, were observed in several mollusc species. Highest parasite burdens were observed in gastropods; mud snails *Hydrobia ulvae* and common periwinkles, *Littorina littorea*, and in bivalves; common cockles *C. edule* and blue mussels *M. edulis* (Thieltges et al. 2006).

8.10.1 Trematodes

Trematodes (Phylum Platyhelminthes) are the most prevalent and abundant macroparasite group in coastal waters. They display a complex/heteroxenous life cycle with free-living and parasitic stages generally involving three host species consisting of a vertebrate primary host and invertebrate intermediate host species. The most deleterious stage is in the first intermediate host (a mollusc) where the parasite penetrates as miracidium larvae and asexually reproduces in

sporocysts/rediae to develop as cercariae larvae (de Montaudouin et al. 2000; Thielteges et al. 2006; Longshaw and Malham 2013). Metacercariae are the encysted juvenile trematodes and depending on the species, both sporocysts (that cause castration, weakness, and gaping of valves, eventual mortality) and metacercariae (that cause behavioural changes in the host and tissue destruction) are found in the tissues of several bivalve species. Various trematode species in the families Gymnophallidae, Echinostomatidae and Renicolidae have been observed in oyster, clam, and cockle species worldwide. Trematodes are flattened oval or worm-like animals, usually no more than a few centimetres in length, although species as small as 1 mm are known.

8.10.2 Turbellarians

Turbellaria (Phylum Platyhelminthes) exist as both endocommensals and parasites of molluscs. All belong to the Order Rhabdocoela and Family Graffillidae. The oval or pyriform shaped flatworms (< 2 mm long) are able to pass freely between the mantle cavity and alimentary canal (Figure 8.6e). They are generally found in low prevalences and intensities causing no apparent harm to the host. A common species is *Urastoma cyprinae* found on the gills of several bivalve species including mussels, oysters and clams where it can cause disorganisation of the gill filaments (including reduction of space between the lamellae and increase in the size of haemal sinuses within the lamellae), a heavy infiltration of haemocytes, and subsequent necrosis of adjacent gill tissue (Villalba et al. 1997; Nicole et al. 1999). Although typically found in the middle area of the four demibranchs, *U. cyprinae* have been observed over the entire gill surface and swimming freely in the pallial cavity (Villalba et al. 1997; Nicole et al. 1999). *U. cyprinae* has been observed in Eastern oysters *C. virginica*, and giant scallops *Placopecten magellanicus* in Atlantic Canada (Bower and Blackbourn 2003), in various clams, cockles (*C. edule*, *C. glaucum*), mussels (*Mytilus* spp.) and oysters (*O. edulis*, *C. gigas*) in Europe (Villalba et al. 1997; Trotti et al. 1998; Thieltges et al. 2006), in geoduck clams *Panopea generosa* in British Columbia, Canada, and Baja California

Table 8.8 Trematode parasites of molluscs

Genus/species	Host range	Geographic range	Pathology	References
Bucephalus minimus	Cockles (*Cerastoderma edule, C. glaucum*)	European coastline from Denmark, including Ireland and the UK, to Portugal	Sporocyst life stage invades the entire body with a negative effect on cockle fecundity, growth, condition, and survival. Severe castration has been reported in infected individuals	de Montaudouin et al. 2009; Magalhães et al. 2015
Gymnophallus choledochus	Cockles (*C. edule, C. glaucum*)	European coastline from Norway, including Ireland and the UK, to Portugal	The sporocyst life stage invades the entire body with a negative effect on cockle fecundity and survival. Initial stages of infection occur in the gonad and the digestive gland	Thieltges et al. 2006; de Montaudouin et al. 2009; Magalhães et al. 2020
Gymnophallus somateriae	Cockles (*C. edule, C. glaucum*)	Norway, Sweden and Denmark	Metacercariae infections occur between the shell and muscle resulting in calcareous concrements or muscle scars on the host's shell which may reduce burrowing behaviour	Lauckner 1992; Thieltges and Reise 2006; Thieltges et al. 2006
Himasthla elongata	Cockles (*C. edule, C. glaucum*)	European coastline from Norway, including Ireland and the UK, to Portugal	Infection of the foot with metacercariae is considered to impair the burrowing capability of cockles with effects on survival (Figure 8.6d)	Wegeberg and Jensen 1999; Thieltges and Reise 2006
Monorchis parvus	Cockles (*C. edule, C. glaucum*)	From Norway to Portugal	Sporocyst invades the entire body, causing damage to all tissue groups and castration	de Montaudouin et al. 2009; Magalhães et al. 2020
Parvatrema minutum	Cockles (*C. edule, C. glaucum*), Peppery furrow shell (*Scrobicularia plana*)	European coastline from Norway, including Ireland and the UK, to Portugal	Metacercariae are generally enclosed by epithelial tissue and appear as translucid pouches at the umbo. The outer mantle margin can also be infected, with metacercariae. Metacercariae abundance can reach < 2,000 per cockle	de Montaudouin et al. 2009; Gam et al. 2008; Fermer et al. 2010, 2011
Parvatrema affinis, Gymnophallus fossarum, Cercaria tapidis	Clams (*Venerupis* (= *Tapes*) *philippinarum, Tapes decussatus, Tapes aureus, Mya arenaria, Mercenaria mercenaria*)	Global, although each species probably has a confined distribution	Majority of species are innocuous. Certain species cause shell deformities, hyperplasia and metaplasia of the mantle epithelium tissue, castration, and deterioration of connective tissue adjacent to sporocysts	Lauckner 1983; Nago and Choi 2004
Unidentified trematode sp.	Abalone (Roe's abalone *Haliotis roei*, two donkey eared abalone *H. asinine*)	Australia in particular Western Australia	Infections with metacercariae usually systemically widespread within individual abalone. Much of the parenchymal tissues replaced with this life stage. The gonad appeared to be the initial site of infection. Infection with metacercariae appeared to be progressive and likely to lead to organ failure and death. Little inflammatory response associated with infection	Handlinger et al. 2006
Microphallus pygmaeus, Psilochasmus aglyptorchis, Renicola roscovita	Mud snail (*Hydrobia ulvae*), Common periwinkle (*Littorina littorea*)	Wadden Sea	Sporocysts/rediae/metacercariae life stages observed in the gonad and visceral mass	Thieltges et al. 2006
M. pygmaeus and *M. similis*	Periwinkles (*L. saxatlis* and *L. arcana*)	North Sea, UK	*M. similis*—Sporocysts in digestive gland cause damage to tissue. *M. pygmaeus* in digestive gland cause haemocyte infiltration	Bojko et al. 2016

Table 8.9 Turbellarians associated with molluscs

Genus/species	Host range	Geographic range	Pathology	References
Urastoma cyprinae	Oysters (Crassostrea virginica, C. gigas), Scallops (Placopecten magellanicus, Patinopecten caurinus), Clams (Panopea generosa, mercenaria, Tridacna maxima, Tridacna gigas), Cockles (Clinocardium nuttallii), Mussels (Mytilus galloprovincialis, M. edulis, M. californianus, Modiolus modiolus)	Australia, Brazil, European North Atlantic, the east coast of North America and north west coast of Mexico	Occur in the mantle cavity, gills and alimentary tract. Can cause disorganisation of the gill filaments (compressed and enlarged), a heavy infiltration of haemocytes and subsequent necrosis of adjacent gill tissue	Villalba et al. 1997; Cáceres Martínez et al. 1998; Nicole et al. 1999; Crespo-González et al. 2010
Paravortex gemellipara	Mussels (Modiolus plicatulus, M. edulis, Geukensia demissa, Ischadium recurvum, Mytilopsis leucophaeata, M. edulis platensis), oysters (C. virginica), Cockles (C. edule)	Prince Edward Island, Canada (Atlantic Ocean), North America (Atlantic Ocean, Gulf of Mexico), Chile, UK (English Channel)	Found primarily within the intestine and stomach, gills, kidney	Fleming et al. 1981; Brunel et al. 1998
Paravortex-like	Mussels (M. galloprovincialis)	Galicia, Spain, Portugal, Mediterranean and Black Sea	No host injury or haemocytic reaction observed	Villalba et al. 1997
Paravortex novel sp. P. mesodesma	Yellow clam Mesodesma mactroides	Uruguay	No direct damage to the host, however, these parasites are frequently observed inside the ducts of the digestive gland and may obstruct them.	Brusa et al. 2006
P. cardii	Cockles (C. edule, C. lamarki), Tellins (Macoma balthica)	English Channel (UK), North Sea (UK), Atlantic Ocean (UK, Spain), Baltic Sea	Occurs in the digestive gland, stomach, and intestine	Carballal et al. 2001; Zander and Reimer 2002
P. karlingi	Cockles (C. edule)	North Sea and Atlantic Ocean (UK)	Occurs in the intestine	Pike and Burt 1981
P. nicolli	Mussel (M. platensis)	Argentina	Occurs in the intestine	Szidat 1965
P. scrobiculariae	Abra ovata (= segmentum), Abra tenuis, Clams (R. decussatus, Scrobicularia plana)	Black Sea, English Channel and North Sea, Bristol Channel (UK), Mediterranean, Adriatic Sea	Occurs in the intestine and digestive gland	Gibbs 1982; Belofastova and Dimitrieva 1999
P. tapetis	Clam (R. decussatus)	Mediterranean	Occurs in the intestine and digestive gland	Noury-Sraïri et al.1995
Unidentified sp.	Cockle (C. edule), Clam (Mya arenaria)	Wadden Sea	No pathology description provided	Thieltges et al. 2006

in Mexico (Bower and Blackbourn 2003; Cáceres-Martínez et al. 2015), and in *C. gigas* in British Columbia, Canada (Bower and Blackbourn 2003). In Alaska, they have been observed on the gills, within the alimentary tract and kidney tissue of scallops, basket cockles, mussels and clams (Meyers and Burton 2009). Another turbellarian (*Paravortex* spp.) was reported as a symbiont in the intestinal lumen of < 3% of *M. galloprovincialis* from Galicia, Spain (Villalba et al. 1997). Robledo et al. (1994) suspected that turbellarians greatly reduce the feeding capacity in heavily infected mussels.

Table 8.10 Nematodes associated with molluscs

Genus/species	Host range	Geographic range	Pathology	References
Echinocephalus crassostreai	Oysters (*Crassostrea gigas*)	Hong Kong	Second and third larval stages occur primarily encysted in the gonad with minimal associated pathology	Cheng 1978
E. pseudouncinatus	Abalone (Pink *Haliotis corrugata*, Green *H. fulgens*)	Southern California, North America, and Gulf of California, Mexico	Blisters associated with encysted larval nematode results in a weakening of the foot muscle thus reducing its efficacy as a hold-fast organ	Milleman 1963; Bower 2000
E. sinensis	Oyster (*C. gigas*), Scallop (*Agropecten ventricosus*)	China	Green spots were observed in *C. gigas* infected with *E. sinensis* in the digestive diverticula, stomach, intestine, and mantle while yellow brown spots were observed in the adductor muscle of the Catarina scallop *A. ventricosus*	Ko et al. 1975; McLean 1983
Sulcascaris sulcata	Scallops (*Pecten* spp., *P. jacobeus*, *Agropecten gibbus*, *Aequipecten opercularis*) and clams	Australia, Western Atlantic, Northern Adriatic Sea, Mediterranean	Larvae (*ca.* 9–21 mm long) detected in the adductor muscle	Barber et al. 1987; Gosling 2015; Marcer et al. 2020
Nematode sp. (unidentified)	Oyster (*C. virginica*)	Louisiana, USA	Similar pathology observed as described for *E. crassostreai*	Gauthier et al. 1990
Nematode sp. (unidentified)	Cockle (*C. edule*)	Wadden Sea	No description of pathology	Thieltges et al. 2006

Table 8.11 Copepods and decapods associated with molluscs

Genus/species	Host range	Geographic range	Pathology	References
Copepoda				
Mytilicola intestinalis (Red worm disease)	Mussels (*Mytilus edulis*, *M. galloprovincialis*)	Adriatic Sea, Mediterranean Sea, North Sea including the UK and Ireland	Both copepod species are believed to be associated with poor growth and condition, extensive tissue damage and metaplasia of the gut wall and sporadic mortalities. In most infections there is no evidence of pathology caused by these parasites in oysters	Lauckner 1983; Thiesen 1987
M. orientalis	Oysters (*Crassostrea gigas*, *Ostrea edulis*)	Inland Sea of Japan, Pacific coast of North America, France, Netherlands, Ireland, Mediterranean		Lauckner 1983; Grizel 1985; Clantzig 1989; Holmes and Minchin 1995
Decapoda				
African pea crab, *Afropinnotheres monodi*	Cockle (*C. edule*), Mussels (*M. galloprovincialis*)	Portugal, North Africa (Atlantic) extending into the Mediterranean (Alboran Sea)	Located close to gills. Associated negative impact on cockle growth, condition and survival. Major effects occur when cockles are infected by soft females or more than one crab	Perez-Miguel et al. 2018, 2020; Cuesta et al. 2020
Pea crab, *Pinnotheres pisum*	Cockles (*C. edule*), mussels (*Mytilus* spp.)	Ireland, UK, Portugal, Canary Islands	Situated close to the gills, with low negative effect on cockle growth, condition and survival. Major negative effects occur when affected by soft females or more than one crab	Becker 2010; Triay-Portella et al. 2018

Table 8.12 Polychaetes associated with molluscs

Genus/species	Host range	Geographic range	Pathology	References
Mudworm, Blister worm, shell-boring polychaetes *Polydora* spp. (possibly *P. ciliata*, *P. limicola*, *P. ligni* and/or *P. websteri*), *P. hoplura*, *P. woodwicki*, *Dipolydora armata*, *Boccardia knoxi* and possibly other polychaete species in the family Sipionidae	Abalone *Haliotis kamtschatkana*, *H. diversicolor*, *H. midae*, *H. iris* and other species	Australia, New Zealand	Most infections are innocuous when at low intensity. Shell damage (internal deformities), weakening and a decrease in flesh weight observed with higher infestations and blistering of soft tissue	Ruck and Cook 1998; Diggles and Oliver 2005; Handlinger et al. 2006.
	Slipper limpet *Crepidula fornicata*	Wadden Sea	*P. ciliata* observed in the shell	Thieltges et al. 2006
	Oysters (*Crassostrea virginica*, *C. gigas*, *Ostrea edulis*, *Saccostrea glomerata*)	Global, although some species probably have limited distributions	Most infections are innocuous and are usually of low intensity with burrows (containing little or no mud) being confined to the shell. *P. websteri* and *P. ligni* in *C. virginica* may cause unsightly mud blisters in the shell and yellowish abscesses in the adductor muscle	Anderson 1990; Cácerez-Martinez 1998; Nell 2002
	Mussels (*Mytilus edulis*)	Global	A lower condition index and loss of market quality caused by *P. ciliata*. The burrows excavated by *P. ciliata* in blue mussel shells cause unsightly blisters containing compacted mud and debris, which also weaken the shell. Nacreous blisters may result in atrophy and detachment of the adductor muscle and possibly interfere with gamete production	Lauckner 1983; Pregenzer 1983
	Scallops (*Patinopecten yessoensis*, *Crassedoma giganteum*, *Placopecten magellanicus*, *P. websterii*)	Ubiquitous, although some species probably have limited distributions	In British Columbia, stunting, abnormally thickened shells and high mortalities caused by high intensities observed in *P. websterii*. The formation of mud blisters on the inner surface of the shell surface may impede adductor muscle attachment	Ball and McGladdery 2001

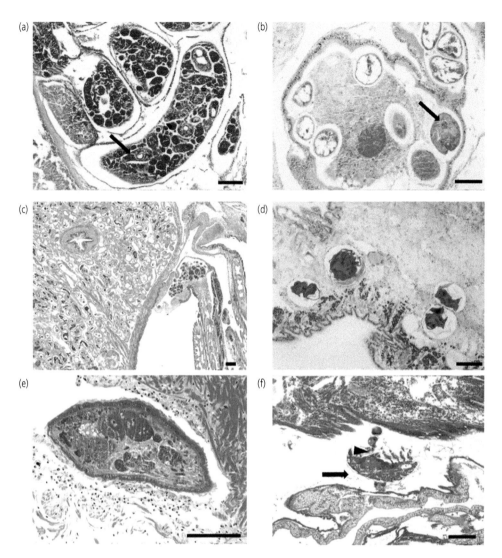

Figure 8.6 Macroparasite infections of bivalve molluscs. (a) *Bucephalus minimus* sporocysts (arrow denoting cercaria developing within) in edible cockle (*Cerastoderma edule*). (b) *Gymnophallus* (= *Meiogymnophallus*) *minutus* infection encapsulated by host tissues of edible cockle, hyperinfected with the microsporidian *Unikaryon legeri* (arrow). (c) Heavy digenean infection of Manila clam showing almost all tissues replaced by parasites. (d) *Himasthla* sp. infection in edible cockle. (e) Adult turbellarian (*Paravortex karlingi*) in tissues of edible cockle. (f) Copepod *Herrmannella rostrata* (arrow) attached loosely to gills of edible cockle host. Note presence of *Trichodina* sp. (arrowhead). Scale bars = 100 µm.

8.10.3 Nematodes

Nematode parasites are not frequently observed in marine molluscs and the biodiversity of this parasite group is poorly understood with most records from commercially important bivalve and gastropod species (McElwain et al. 2019). Their associations range from commensalism through to parasitism. *Echinocephalus* spp. have been observed in oysters and abalone, while *Sulcascaris sulcata* has been detected in scallops and clams (Barber et al. 1987; Gosling 2015; Marcer et al. 2020). The literature contains limited information about tissue damage or potential host responses associated with nematodes infecting marine bivalves. Pathological changes to infected tissues have mainly been reported as localised discolorations or cysts (McElwain et al. 2019). Green spots were observed in *C. gigas* infected with *E. sinensis* in the digestive diverticula, stomach, intestine and mantle (Ko et al. 1975) while yellow-brown spots were observed in the adductor muscle of the Catarina scallop *Agropecten ventricosus* (McLean 1983). A brown discolouration was observed in the adductor muscle of bivalve species infected with *S. sulcata* (Lester et al. 1980). Brownish spots indicated the presence of encapsulated nematodes in the stout razor clam *Tagelus plebeius* (McElwain et al. 2019). It is unclear if the tissue colourations are due to cellular damage, an immunological response or secretions from the parasite (McElwain et al. 2019).

8.10.4 Copepods

Parasitic copepods of bivalve molluscs are mostly within the order Cyclopoida. The more common species are *Mytilicola intestinalis* and *M. orientalis* that are obligate endoparasites inhabiting the alimentary tract (Lauckner 1983; Grizel 1985; Theisen 1987; Holmes and Minchin 1995). Individuals are 5 to 12 mm long, depending on the species, are relatively dedifferentiated, with reduced limbs and body segmentation (Meyers and Burton 2009). The geographic range of *M. intestinalis* is limited to Europe where it has been reported in mussels (*Mytilus* spp.), oysters (*C. gigas, O. edulis*), clams,

and cockles. Intestinal damage, poor growth, and mortalities in mussels have been associated with *Mytilicola* sp. infestations. Metaplasia of the mucosal epithelium in the alimentary tract is frequently observed with the detection of *Mytilicola* sp. In certain cases, the copepod's appendages result in tissue erosion, perforation, and encapsulation within the connective tissue. Blockage or obstruction of the ducts connecting the stomach to the digestive diverticulae has also been reported (Meyers and Burton 2009). Parasitic copepods of bivalves have a direct life cycle with separate sex parents producing fertilised eggs that are released directly into the water column with faeces where they hatch into free swimming larvae.

8.10.5 Polychaetes

Segmented worms belonging to class Polychaeta and family Spionidae are associated with gastropod and bivalve species, including abalone, slipper limpet, oysters, mussels, and scallops, worldwide (Cacerez-Martinez et al. 1998; Ball and McGladdery 2001; Handlinger et al. 2006; Thieltges et al. 2006). Polychaete worms construct burrows/tunnels (< 2 mm in diameter) on the shell surface or within the shell itself, which are lined with mud and debris resulting in the condition of 'mud blisters' in the soft tissues of the host. The internal shell surface is damaged resulting in the formation of yellow pustules or abscesses in the soft tissues that are in contact with these burrows. In severe cases, shell gaping in bivalves, overall poor body condition and mortalities occur (Meyers and Burton 2009). Mud worms have a simple direct life cycle with juvenile worms settling on the edge of the shell prior to building their burrow. Hermaphroditic adult worms produce egg capsules within the burrow which then hatch as larvae and enter the water column prior to locating a new host. Oysters are most tolerant of mud worms as they are able to secrete shell nacre more rapidly to wall off the burrows compared to other species such scallops, which may have their adductor muscle attachment weakened if burrows occur close by.

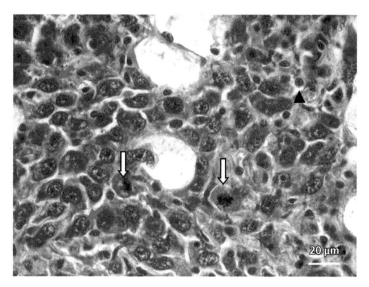

Figure 8.7 Histological appearance of disseminated neoplasia in *Mytilus galloprovincialis*. The neoplasic cells are larger than the normal haemocytes and some are in active division (large unlabelled arrows). Normal haemocytes (black arrowhead) and neoplasic cells (red unlabelled arrows). Scale bar = 20 μm. Source: micrograph courtesy of Dr Antonio Villalba Garcia (Centro de Investigacións Mariñas, Consellería do Mar, Xunta de Galicia, Spain).

8.11 Neoplasia

Disseminated neoplasia is a leukaemia-like condition of bivalve molluscs. Since the original observation of this condition in *Crassostrea virginica* and *C. gigas* in the 1960s (Farley, 1969), many species of bivalves including clams, oysters, mussels and cockles have been found with this condition (see review by Carballal et al. 2015 for details). The neoplastic cells characteristic of this disease are found in the haemolymph in large numbers. Unlike the host's haemocytes, these cells are generally larger with altered nuclear morphology, non-adhesive behaviour, and with no or little phagocytic activity (Figure 8.7; Carballal et al. 2015). Early studies showed that the disease can be horizontally transmitted between individuals of the same and other species. In some areas, the prevalence of this condition is high and mass mortalities and rapid declines of bivalve mollusc populations have been linked to its presence (e.g. Brousseau and Baglivo 1991; Morgan et al. 2012; Dairain et al. 2020).

Recent studies by Metzler and colleagues (e.g. Metzger et al. 2015, 2016, 2018; Yonemitsu et al. 2019) have elegantly revealed that disseminated neoplasia is an unusual form of cancer only found in a few other animals including the Tasmanian devil. The genotype of these neoplastic cells in bivalves does not match that of the host genotype (i.e. they are not derived from host cells as in most cancers). Transmission of this disease may occur over vast distances in the seas as evidenced by studies conducted by Yonemitsu et al. (2019). They found that disseminated neoplasia in *Mytilus edulis* in the northern latitudes and that in *M. chilensis* in South America, is genetically identical and that it arose from a third species of mussel, *M. trossulus*. It is therefore possible that this form of disseminated neoplasia could be traced back to a single bivalve mollusc sometime before the first observation of its presence in the late 1960s. Transmission of infected molluscs is probably aided by shipping activity or international movement of spat for aquaculture.

A second form of neoplasia in molluscs are gonadal neoplasms found in bivalve molluscs including clams and mussels (Carballal et al. 2015). This condition reduces the fecundity of affected individuals and probably also results in mortality. The aetiology of this condition is unclear.

8.12 Control and treatment

Control of infections in natural populations of molluscs is challenging and exacerbated by movement of non-native species, arrival of invasive species often carrying diseases, and the expansion and reach of human activities including aquaculture and industrial wastes. Improvements in rapid disease detection assays and satellite imagery may allow

Table 8.13 Approaches to the control and/or treatment of diseases of molluscs of commercial importance

Approach	Mechanism of action	Examples	References
Improved biosecurity	Isolation of host from pathogen(s)	Use of OIE lists of notifiable diseases and their potential pitfalls International control on movement of live and unprocessed animals	Carnegie et al. 2016 (review)
Improved water quality in hatcheries	Reduced microbial load and reduced contaminants	Recirculating aquaculture system (RAS) technology including ozonation of waste, UV treatment, control of pH and nitrogenous waste product removal	Xiao et al. 2019 (review of RAS technology)
Bacteriophages	Targeted lysis of known pathogenic bacteria	1. Depuration of oysters to remove *V. parahaemolyticus* 2. Control of *V. alginolyticus* in the larvae of oyster, *Saccostrea glomerata*	1. Rong et al. 2014 2. Le et al. 2020
Depuration	Removal of human pathogens from shellfish prior to consumption	Potential use of probiotics in depuration systems	Yeh et al. 2020 (review)
Selective breeding of resistant strains	Development of disease-resistant molluscan stocks through marker-assisted selection	Selective breeding in oysters has improved resistance against several diseases including (1) MSX, Dermo and ROD in *C. virginica*, (2) QX disease in *S. glomerata*, and (3) OsHV-1 infections in *C. gigas*, (4) Advances in genome editing	1. Guo et al. 2008; 2. Nell and Perkins 2006; 3. Dégremont 2011; 4. Houston et al. 2020 (review);
'Vaccination' and immune stimulation	Improve resistance to disease of hosts and their progeny	Trans-generational transfer of heightened disease resistance	Green et al. 2016
Probiotics	Inhibition of pathogen growth, reduction in virulence of pathogens, stimulation of immune system, change in microbial community in water and within host	Microbial probiotics added to water in hatcheries resulting in changes in microbial communities in water and in the mollusc (microbiome)	Stevick et al. 2019

modelling of disease progression at the population level and to improve prediction of disease outbreaks.

Bivalve hatcheries are important in the farming of shellfish to avoid depletion of natural populations and to maintain the quality of spat. The water quality of these is of importance to the growth and survival of these juvenile shellfish that are highly susceptible to infectious and non-infectious diseases. Water may be treated by ozonation, filtration and/or UV irradiation to reduce unwanted bacterial growth (Dubert et al. 2017). The quality and contamination of microalgal feed is also important for spat development (Elston et al. 2008). Antibiotics and other antimicrobial agents can improve growth and reduce the potential impact of bacteria such as vibrios, but their use is non-selective and may destroy other beneficial microbes. The continued use of many antimicrobial agents in some

aquaculture sectors, where antibiotic resistance has already been observed, may be subject to future legislation to limit their use and therefore alternate strategies are emerging. These include bacteriophage therapy, probiotics, and immune stimulants (Table 8.13). In the case of probiotics, phytoplankton including *Phaeobacter* spp. (so-called green water therapy) have been found to inhibit the growth of vibrios (Prado et al. 2010) while 'traditional' probiotic bacteria including aeromonads, *Pseudomonas*, *Bacillus* spp. and non-pathogenic vibrios may also have potential (e.g. Kesarcodi-Watson et al. 2012; Ma et al. 2019; Stevick et al. 2019; Sanchez-Ortiz et al. 2020). For instance, the action of the probiotic bacterium, *Bacillus pumilus* added to the water in an oyster hatchery resulted in an improvement in larval survival post-challenge with *V. coralliilyticus*. Detailed analyses of microbes in the tank biofilm, larvae, and water revealed changes in bacterial community structure including an increase in vibrio diversity potentially reducing the abundance of pathogenic vibrios (Stevick et al. 2019).

There is much scope for improvement of the disease resistance of bivalves using genetic approaches (Dégremont et al. 2015; Houston et al. 2020). Transcriptomes (all RNAs) have been sequenced in many mollusc species identifying genes related to immune response (Gómez-Chiarri et al. 2015). The transcriptomes of molluscs are sequenced after challenge trials with pathogens, and transcriptomic comparison between control and challenged molluscs can suggest the possible involvement of candidate genes in immune responses. The large numbers of candidate genes and their possible roles in molluscan immunity, including disease-resistant markers, genetic maps, and whole genome sequences, have contributed to the development of disease-resistant molluscan stocks through marker-assisted selection (Guo et al. 2008). Selective breeding in oysters has improved resistance against several diseases including multinucleated sphere X (MSX), *Dermocystidium marinum* (Dermo) and Roseovarius oyster disease (ROD) in *C. virginica* (Guo et al. 2008), Queensland unknown (QX) disease in Sydney rock oysters (Nell and Perkins 2006), and Ostreid herpes virus type 1 (OsHV-1) infections in *C. gigas* (Dégremont 2011).

The immune system of molluscs develops during organogenesis (Song et al. 2016) but early life history stages may be immune deficient. While immune stimulants and orally delivered 'vaccines' may have promise, these have both theoretical and practical constraints in terms of delivery and cost-benefit analysis. These are therefore unlikely to be important at least in the immediate future. A further potential approach comes from reports of transgenerational immune priming (a form of heightened resistance to disease passed from parent to offspring) that has been demonstrated in various invertebrates (see Section 1.7 in Chapter 1 for more details). A recent study found that larval progeny from Pacific oyster 'parents' treated with the immune stimulant, poly (I:C) had heightened survival when challenged with OsHV-1 (Green et al. 2016). This transfer of immunity from parent to offspring may have promise especially in the control of viral infections but the efficacy and duration of these events need to be further elucidated.

Finally, although depuration of shellfish and its feasibility as a process to reduce microbial contamination is outside the scope of this chapter, the reader is referred to a recent review (Yeh et al. 2020) for further details of advances in this procedure.

8.13 Future directions

8.13.1 Climate change and molluscan diseases

The marine environment is challenged by climate change globally and increasing temperature is acknowledged as being a key driver (Rhein et al. 2013). Bivalve filtration rates and the subsequent uptake of oxygen and food intake are strongly influenced by elevated seawater temperature, with a negative correlation observed (Eymann et al. 2020). It is recognised that marine species living near their thermal tolerance limits, for example *O. edulis* in Galicia, Spain, will be negatively impacted by warming (Eymann et al. 2020). At temperatures > 26°C, the gill tissue of *O. edulis* was observed to become partly anaerobic and cardiac dysfunction (arrhythmia) developed at 28°C followed by cardiac arrest at 30°C (Eymann et al. 2020). Additionally, a warmer marine environment may make oysters more susceptible to bacterial and viral infections (Zannella et al. 2017).

The effects of ocean acidification on shell production in juvenile and adult shelled molluscs is variable among species, however molluscan haemolymph may exhibit lower pH (acidosis) that has consequences for physiological processes such as respiration, excretion etc. and overall performance (Gazeau et al. 2013) including resistance to disease. Mackenzie et al. (2014) suggested that temperature, more than pH, may be the key driver affecting the immune response in *M. edulis*. In their study, both increases in temperature and/or lowered pH conditions led to changes in parasite abundance and diversity, pathological conditions, and bacterial incidence.

Most climate impact studies carried out to date have primarily measured the effects of a single stressor on a single mollusc species in one generation and at one stage of their life cycle. More long-term multigenerational studies are required to assess if species have the potential to adapt or acclimate under predicted climate change conditions at regional and local levels. Along with performance (% survival) at each stage of life, pathogen and parasite diversity, abundance and associations should also be assessed to identify potential future threats to mollusc populations.

8.13.2 Disease bottlenecks at hatcheries and therapies

Outbreaks of disease continue to be the main bottleneck for successful bivalve larval and spat production, most of them caused by species of the genus *Vibrio*. Both preventive and management measures that control pathogenic bacterial populations are recognised (Dubert et al. 2017). The over use and dependence on antibiotics has facilitated the development of resistant bacteria, which quickly transmit resistance genes via horizontal transfer

mechanisms, in the hatchery and marine environment (Cabello et al. 2013; Miranda et al. 2013). Phage therapy also represents a promising alternative strategy for prevention of disease outbreaks for *Vibrio* spp. associated with mortalities in aquaculture (Letchumanan et al. 2016) and in larval oyster and clam hatcheries. Future research is needed to better understand how environmental factors influence the efficiency of phage therapy in aquaculture systems, as salinity and organic matter have been shown to influence the efficacy of phage treatment (Silva et al. 2014).

RNA interference (RNAi) or clustered regularly interspaced short palindromic repeats (CRISPR) Cas9 protein-based therapies show great promise in combating various types of diseases caused by viral and parasitic agents. The CRISPR/Cas pathway can be used to degrade nonself RNA/DNA (Chin et al. 2017) and to reduce viral replication and titre (Gotesman et al. 2014). Such approaches may have merit in future studies.

8.14 Summary

- Molluscs are subject to a wide range of diseases caused by microbes through to macroparasites.
- Most is known about these diseases in bivalve molluscs of commercial importance including oysters, clams and mussels.
- Some disease states are polymicrobial in nature—e.g., Pacific oyster mortality syndrome caused by ostreid herpes virus-1 microVar (OsHV-1 µVar) and bacteria including *Vibrio aestuarianus*.
- Infections caused by vibrios are important in bivalve hatcheries where they can cause significant mortality.
- Molluscs, especially those in temperate regions, have been subject to epizootics in the last few decades probably associated with rapid changes in environmental conditions in summer months.
- Haplosporidians are a key group of pathogens causing significant mortalities and population declines—e.g. *Haplosporidium nelsoni* in the eastern oyster in the 1950s–1960s.
- Dermo disease caused by *Perkinsus marinus* caused destruction of oyster fisheries in the USA from 1950s onwards.
- The importance of nematodes, trematodes, and other 'macroparasites' in ecosystem health is not fully understood.
- Control and treatment of diseases is still in its infancy but there are new developments especially in rapid diagnostics and remote sensing that will assist in predictive modelling.

Suggested further reading

General

Cheng, T.C. 1967. Marine molluscs as hosts for symbioses with a review of known parasites of commercially important species. *Advances in Marine Biology* 5: 1–424

Gosling, E. 2015. Diseases and parasites. In *Marine Bivalve Molluscs*, 2nd edn, pp. 429–477. Chichester: Wiley

Viral diseases

Alfaro, A.C., Nguyen, T.V., and Merien, F. 2019. The complex interactions of *Ostreid herpesvirus* 1, *Vibrio* bacteria, environment and host factors in mass mortality outbreaks of *Crassostrea gigas*. *Reviews in Aquaculture* 11: 1148–1168

Bacterial diseases

Travers, M.-A., Boettcher Miller, K., Roque, A., and Friedman, C.S. 2015. Bacterial diseases in marine bivalves. *Journal of Invertebrate Pathology* 131: 11–31 (Excellent detailed review)

Cruz-Flores, R. and Cáceres-Martínez, J. 2020. Rickettsiales-like organisms in bivalves and marine gastropods: A review. *Reviews in Aquaculture* 12: 2010–2026

Microsporidia

Stentiford, G.D., Feist, S.W., Stone, D.M., Bateman, K.S., and Dunn, A.M. 2013. Microsporidia: Diverse, dynamic, and emergent pathogens in aquatic systems. *Trends in Parasitology* 29: 567–578

Stentiford, G.D., Becnel, J.J., Weiss, L.M. et al. 2016. Microsporidia—emergent pathogens in the global food chain. *Trends in Parasitology* 32: 336–348

Mikrocytids

Abbott, C.L. and Meyer, G.R. 2014. Review of *Mikrocytos* microcell parasites at the dawn of a new age of scientific discovery. *Diseases of Aquatic Organisms* 110: 25–32. (Excellent review of these parasites)

Haplosporidians

Arzul, I. and Carnegie, R.B. 2015. New perspective on the haplosporidian parasites of molluscs. *Journal of Invertebrate Pathology* 131: 32–42

Culloty, S.C. and Mulcahy, M.F. 2007. *Bonamia ostreae* in the native oyster *Ostrea edulis*: A Review. Marine Environment and Health Series No. 29 (Marine Institute Ireland ISSN NO:1649-0053) https://oar.marine.ie/handle/10793/269

Engelsma, M.Y., Culloty, S.C., Lynch, S.A., Arzul, I., and Carnegie, R.B. 2014. *Bonamia* parasites: A rapidly changing perspective on a genus of important mollusc pathogens. *Diseases of Aquatic Organisms* 110: 5–23

Marteiliosis

Carrasco, N., Green, T., and Itoh, N. 2015. *Marteilia* spp. parasites in bivalves: A revision of recent studies. *Journal of Invertebrate Pathology* 131: 53–57

Feist, S.W., Hine, P.M., Bateman, K.S., Stentiford, G.D., and Longshaw, M. 2009. *Paramarteilia canceri* sp. n. (Cercozoa) in the European edible crab (*Cancer pagurus*) with a proposal for the revision of the order Paramyxida Chatton, 1911. *Folia Parasitologica* 56: 73–85

Perkinsosis

Ruano, F., Batista, F.M., and Arcangeli, G. 2015. Perkinsosis in the clams *Ruditapes decussatus* and *R. philippinarum* in the Northeastern Atlantic and Mediterranean Sea: A review. *Journal of Invertebrate Pathology* 131: 58–67

Macroparasites

Longshaw, M. and Malham, S.K. 2013. A review of the infectious agents, parasites, pathogens and commensals of European cockles (*Cerastoderma edule* and *C. glaucum*). *Journal of the Marine Biological Association of the United Kingdom* 93: 227
Thieltges, D.W., Mouritsen, K.N., and Poulin, R. 2018. Ecology of parasites in mudflat ecosystems. In *Mudflat Ecology*, P.G. Beninger (ed.) pp. 213–242. Switzerland: Springer Nature (Wide ranging review of parasite ecology in mudflats)

Neoplasia

Carballal, M.J., Barber, B.J., Iglesias, D., and Villalba, A. 2015. Neoplastic diseases of marine bivalves. *Journal of Invertebrate Pathology* 131: 83–106

Control and treatment

Carnegie, R.B., Arzul, I., and Bushek, D. 2016. Managing marine mollusc diseases in the context of regional and international commerce: Policy issues and emerging diseases. *Philosophical Transactions of the Royal Society B* 371: 20,150,215
Dubert, J., Barja, J.L., and Romalde, J.L. 2017. New insights into pathogenic vibrios affecting bivalves in hatcheries: present and future prospects. *Frontiers in Microbiology* 8: 762
Dégremont, L., Garcia, C., and Allen, S.K. 2015. Genetic improvement for disease resistance in oysters: A review. *Journal of Invertebrate Pathology* 131: 226–241
Houston, R.D., Bean, T.P., Macqueen, D.J. et al. 2020. Harnessing genomics to fast-track genetic improvement in aquaculture. *Nature Reviews Genetics* 21: 389–409 (Wide ranging review of recent developments in genomics as applied to aquaculture species)

Acknowledgements

We thank the following people for their generous provision of micrographs: Frederico Batista, Kelly Bateman, and Matthew Green (Cefas, Weymouth, U.K.); Gary Meyer (Fisheries and Oceans Canada, British Columbia, Canada); Céline Garcia (Ifremer, LGPMM Laboratory, Station Les Tremblade, France); Ana Grande and Francisco Ruano (Instituto Português do Mar e da Atmosfera, Portugal); Hyun-Sil Kang (Southeast Sea Fisheries Research Institute, National institute of Fisheries Science, South Korea); and Antonio Villalba Garcia (Centro de Investigacións Mariñas, Consellería do Mar, Xunta de Galicia, Spain). We (SAL, AFR, SKM and SCC) acknowledge funding for the Bluefish project from EU ERDF Ireland-Wales Territorial Programme. This chapter is one of a series of reports from the UKRI-funded ARCH-UK project awarded to AFR.

References

Abbott, C.L., Gilmore, S.R., Lowe, G., Meyer, G., and Bower, S. 2011. Sequence homogeneity of internal transcribed spacer rDNA in *Mikrocytos mackini* and detection of *Mikrocytos* sp. in a new location. *Diseases of Aquatic Organisms* 93: 243–250

Abbott, C.L. and Meyer, G.R. 2014. Review of *Mikrocytos* microcell parasites at the dawn of a new age of scientific discovery. *Diseases of Aquatic Organisms* 110: 25–32

Abbott, C.L., Meyer, G.R., Lowe, G., Kim, E. and Johnson, S.C. 2014. Molecular taxonomy of *Mikrocytos boweri* sp. nov. from Olympia oysters *Ostrea lurida* in British Columbia, Canada. *Diseases of Aquatic Organisms* 110: 65–70

Abollo, E., Ramilo, A., Casas, S.M., Comesaña, P., Cao, A., Carballal, M.J. and Villalba, A. 2008. First detection of the protozoan parasite *Bonamia exitiosa* (Haplosporidia) infecting flat oyster *Ostrea edulis* grown in European waters. *Aquaculture* 274: 201–7

Adlard, R.D. and Ernst, I. 1995. Extended range of the oyster pathogen *Marteilia sydneyi*. *Bulletin of the European Association of Fish Pathology* 15: 119–21

Adlard, R.D. and Nolan, M.J. 2015. Elucidating the life cycle of *Marteilia sydneyi*, the aetiological agent of QX disease in the Sydney rock oyster (*Saccostrea glomerata*), *International Journal for Parasitology* 45: 419–26

Albuixech-Martí, S., Lynch, S.A., and Culloty, S.C. 2020. Biotic and abiotic factors influencing haplosporidian species distribution in the cockle *Cerastoderma edule* in Ireland. *Journal of Invertebrate Pathology* 174: 107425

Alderman, D. and Jones, G. 1971. Shell disease of oysters, *Fishery Investigations* 26: 1–19

Alfaro, A.C., Nguyen, T.V., and Merien, F. 2019. The complex interactions of *Osteid herpesvirus* 1, *Vibrio* bacteria, environment and host factors in mass mortality outbreaks of *Crassostrea gigas*. *Reviews in Aquaculture* 11: 1148–1168

Anderson, I.G. 1990. Diseases in Australian invertebrate aquaculture. In *Proceedings, Fifth International Colloquium on Invertebrate Pathology and Microbial Control*. Society for Invertebrate Pathology, 20–24 August 1990, pp. 38–48. Adelaide, Australia

Andree, K., Carrasco, N., Carella, F., Furones, D., and Prado, P. 2020. *Vibrio mediterranei*, a potential emerging pathogen of marine fauna: Investigation of pathogenicity using a bacterial challenge in *Pinna nobilis* and development of a species-specific PCR. *Journal of Applied Microbiology* 130: 617–631

Andrews, J.D. 1996. History of *Perkinsus marinus*, a pathogen of oysters in Chesapeake Bay 1950–1984. *Journal of Shellfish Research* 15: 13–16

Arzul, I. and Carnegie, R.B. 2015. New perspective on the haplosporidian parasites of molluscs. *Journal of Invertebrate Pathology* 131: 32–42

Arzul, I., Chollet, B., Boyer, S. et al. 2014. Contribution to the understanding of the cycle of the protozoan parasite *Marteilia refringens*. *Parasitology* 141: 227–240

Arzul, I., Corbeil, S., Morga, B. and Renault, T. 2017. Viruses infecting marine molluscs. *Journal of Invertebrate Pathology* 147: 118–135

Azevedo, C. 2001. Ultrastructural descriptions of the spore maturation stages of the clam parasite *Minchinia tapetis* (Vilela, 1951) (Haplosporida: Haplosporidiidae). *Systematic Parasitology* 49: 189–194

Azevedo, C., Balseiro, P., Casal, G. et al. 2006. Ultrastructural and molecular characterization of *Haplosporidium montforti* n. sp., parasite of the European abalone *Haliotis tuberculata*. *Journal of Invertebrate Pathology* 92: 23–32

Azevedo, C., Conchas, R.F., and Montes, J. 2003. Description of *Haplosporidium edule* n. sp. (Phylum Haplosporidia), a parasite of *Cerastoderma edule* (Mollusca, Bivalvia) with complex spore ornamentation. *European Journal of Protistology* 39: 161–167

Azevedo, C. and Hine, P.M. 2017. Haplosporidia. In *Handbook of the Protists*, Archibald J., Simpson A., and Slamovits C. (eds.) pp. 823–850. Cham, Switzerland: Springer International Publishing

Ball, M.C. and McGladdery, S.E. 2001. Scallop parasites, pests and diseases: Implications for aquaculture development in Canada. *Bulletin of the Aquaculture Association of Canada* 101–103: 13–18

Ball, S.J. and Neville, J.E. 1979. *Minchinia chitonis* (Lankester, 1885) Labbe, 1896, a haplosporidian parasite of the chiton *Lepidochitonoa cinereus*. *Journal of Molluscan Studies* 45: 340–344

Balseiro, P., Aranguren, R., Gestal, C., Novoa, B., and Figueras, A. 2006. *Candidatus* Xenohaliotis californiensis and *Haplosporidium montforti* associated with mortalities of abalone *Haliotis tuberculata* cultured in Europe. *Aquaculture* 258: 63–72

Barber, B.J., Blake, N.J., Moyer, M.A., and Rodrick, G.E. 1987. Larval *Sulcascaris sulcata* from calico scallops, *Argopecten gibbus*, along the Southeast coast of the United States. *Journal of Parasitology* 73: 476–480

Barker, W.H. and Gangarosa, E.J. 1974. Food poisoning due to *Vibrio parahaemolyticus*. *Annual Review of Medicine* 25: 75–81

Barnes, R.D. 1982. In: *Invertebrate Zoology*. pp. 230–235. Philadelphia, PA: Holt-Saunders International, Saunders College Publishing

Baxter, J.M., Hodgson, A.N., and Sturrock, M.G. 1989. Variations in infestation rates of *Lepidochitona cinereus* (Polyplacophora) by *Minchinia chitonis* (Sporozoa) in twelve populations in Scotland and Northern Ireland. *Marine Biology* 102: 107–117

Bearham, D., Spiers, Z., Raidal, S.R., Jones, J.B., Burreson, E.M., and Nicholls, P.K. 2008a. Spore ornamentation of *Haplosporidium hinei* n. sp. (Haplosporidia) in pearl oysters *Pinctada maxima* (Jameson, 1901). *Parasitology* 135: 521–527

Bearham, D., Spiers, Z., Raidal, S.R., Jones, J.B., and Nicholls, P.K. 2008b. Spore ornamentation of *Minchinia occulta* n. sp. (Haplosporidia) in rock oysters *Saccostrea cuccullata* (Born, 1778). *Parasitology* 135: 1271–1280

Becker, C. and Türkay, M. 2010. Taxonomy and morphology of European pea crabs (Crustacea: Brachyura: Pinnotheridae). *Journal of Natural History* 44: 1555–1575

Belofastova, I. and Dimitrieva E.V. 1999. The turbellarians of genus *Paravortex* (Rhabdocoela: Graffillidae)-parasites of the Black Sea bivalves. *Ecology of the Sea* 48: 76–78

Berthe, F.C.J., Le Roux, F., Adlard, R.D., and Figueras, A. 2004. Marteiliosis in molluscs: A review. *Aquatic Living Resources* 17: 433–448

Beukema, J.J. and Dekker, R. 2020. Winters not too cold, summers not too warm: Long-term effects of climate change on the dynamics of a dominant species in the Wadden Sea: The cockle *Cerastoderma edule* L. *Marine Biology* 167: 44

Bingham, P., Brangenberg, N., Williams, R., and Andel M.V. 2013. Marine and freshwater investigation into the first diagnosis of Ostreid herpesvirus type 1 in Pacific oysters. *Surveillance Wellington* 40: 20–24

Boardman, C.L., Maloy, A.P., and Boettcher, K.J. 2008. Localization of the bacterial agent of juvenile oyster disease (*Roseovarius crassostreae*) within affected eastern oysters (*Crassostrea virginica*). *Journal of Invertebrate Pathology* 97: 150–158

Boettcher, K.J., Barber, B.J., and Singer, J.T. 2000. Additional evidence that juvenile oyster disease is caused by a member of the *Roseobacter* group and colonization of nonaffected animals by *Stappia stellulata*-like strains. *Applied & Environmental Microbiology* 66: 3924–30

Boettcher, K.J., Geaghan, K.K., Maloy, A.P., and Barber, B.J. 2015. *Roseovarius crassostreae* sp. nov., a member of the Roseobacter clade and the apparent cause of juvenile oyster disease (JOD) in cultured Eastern oysters. *International Journal of Systematic and Evolutionary Microbiology* 55: 1531–1537

Bojko, J., Grahame, J.W., and Dunn, A.M. 2016. Periwinkles and parasites: The occurrence and phenotypic effects of parasites in *Littorina saxatilis* and *L. arcana* in northeastern England. *Journal of Molluscan Studies* 83: 69–78

Borrego, J.J., Castro, D., Luque, A., Paillard, C., Maes, P., Garcia, M.T., and Ventosa, A. 1996. *Vibrio tapetis* sp.

nov., the causative agent of Brown Ring disease affecting cultured clams. *International Journal of Systematic Bacteriology* 46: 480–484

Bower, S.M. 2000. Infectious diseases of abalone (*Haliotis* spp.) and risks associated with transplantation. In *Workshop on Rebuilding Abalone Stocks in British Columbia. Canadian Special Publication of Fisheries and Aquatic Sciences.* A. Campbell (ed.) Vol. 130: pp. 111–122. Ottowa, Canada: Fisheries and Oceans

Bower, S.M. 2006a. *Synopsis of Infectious Diseases and Parasites of Commercially Exploited Shellfish: Microsporidiosis of Dredge Oysters.* Ottowa Canada: Fisheries and Oceans

Bower, S.M. 2006b. Parasitic diseases of shellfish. In *Fish Diseases and Disorders, Volume 1: Protozoa and Metazoan Infections.* 2nd Edition, P.T.K. Woo (ed.) pp. 629–677. Wallingford: CABI

Bower, S.M. 2011. Synopsis of infectious diseases and parasites of commercially exploited shellfish: *Perkinsus marinus* ("dermo") disease of oysters. Fisheries and Oceans Canada

Bower, S.M. and Blackbourn, J. 2003. Geoduck clam (*Panopea generosa*): Anatomy, Histology, Development, Pathology, Parasites and Symbionts: Turbellaria of Geoduck Clams. https://www.mpo-dfo.gc.ca/science/aah-saa/species-especes/shellfish-coquillages/geopath/turbellaria-eng.html

Bowers, E.A. and James, B.L. 1967. Studies on the morphology, ecology and life-cycle of *Meiogymnophallus minutus* (Cobbold, 1859) comb. nov. (Trematoda: Gymnophallidae). *Parasitology* 57: 281–300

Boyer, S., Chollet, B., Bonnet, D., and Arzul, I. 2013. New evidence for the involvement of *Paracartia grani* (Copepoda, Calanoida) in the life cycle of *Marteilia refringens* (Paramyxea). *International Journal of Parasitology* 43: 1089–1099

Brooks, W.M. 1968. Tetrahymenid ciliates as parasites of the gray garden slug. *Hilgardia* 39: 205

Brousseau, D.J. and Baglivo, J.A. 1991. Field and laboratory comparisons of mortality in normal and neoplastic *Mya arenaria*. *Journal of Invertebrate Pathology* 57: 59–65

Brunel, P., Bossé, L., and Lamarche, G. 1998. Catalogue des invertébrés marins de l'estuaire et du golfe du Saint-Laurent. Conseil national de recherches du Canada. *Canadian Special Publication of Fisheries and Aquatic Sciences.*126: 405

Brusa, F., Ponce De León, R., and Damborenea, C. 2006. A new *Paravortex* (Platyhelminthes, Dalyellioida) endoparasite of *Mesodesma mactroides* (Bivalvia, Mesodesmatidae) from Uruguay. *Parasitology Research* 99: 566–571

Bruto, M., James, A., Petton, B. et al. 2017. *Vibrio crassostreae*, a benign oyster colonizer turned into a pathogen after plasmid acquisition. *ISME Journal* 11: 1043–1052

Burioli, E.A.V., Varello, K., Trancart, S. et al. 2018. First description of a mortality event in adult Pacific oysters in Italy associated with infection by a *Tenacibaculum soleae* strain. *Journal of Fish Diseases* 41: 215–221

Burreson, E. 2005. Shellfish: MSX disease still going strong. *Aquaculture Health International* 2: 13–14

Burreson, E.M. and Ford, S.E. 2004. A review of recent information on the Haplosporidia, with special reference to *Haplosporidium nelsoni* (MSX disease). *Aquatic Living Resources* 17: 499–517

Burreson, E.M. and Ragone Calvo, L.M. 1996. Epizootiology of *Perkinsus marinus* disease of oysters in Chesapeake Bay, with emphasis on data since 1985. *Journal of Shellfish Research* 15: 17–34

Cabello, F.C., Godfrey, H. P., Tomova, A. et al. 2013. Antimicrobial use in aquaculture re-examined: Its relevance to antimicrobial resistance and to animal and human health. *Environmental Microbiology* 15: 1917–1942

Cáceres-Martínez, J., Madero-López, L.H., Padilla-Lardizábal, G., and Vásquez-Yeomans, R. 2016. Epizootiology of *Perkinsus marinus*, parasite of the pleasure oyster *Crassostrea corteziensis*, in the Pacific coast of Mexico. *Journal of Invertebrate Pathology* 139: 12–18

Cáceres-Martínez, J., Vásquez-Yeomans, R., and Cruz-Flores, R. 2015. First description of symbionts, parasites, and diseases of the Pacific geoduck *Panopea generosa* from the Pacific coast of Baja California, Mexico. *Journal of Shellfish Research* 34: 751–756

Cáceres-Martínez, J., Vásquez-Yeomans, R., and Sluys, R. 1998. The turbellarian *Urastoma cyprinae* from edible mussels *Mytilus galloprovincialis* and *Mytilus californianus* in Baja California, NW Mexico. *Journal of Invertebrate Pathology* 72: 214–219

Cali, A., Becnel, J.J., and Takvorian, P.M. 2017. Microsporidia. In *Handbook of the Protists*. Archibald, J.M., Simpson, A.G.B., and Slamovits, C.H. (eds.) 2nd Edition, p. 1657. Halifax, Canada: Springer International Publishing

Campalans, M., Rojas, P., and Gonzalez, M. 2000. Haemocytic parasitosis in the farmed oyster *Tiostrea chilensis*. *Bulletin of the European Association of Fish Pathologists* 20: 31–33

Canning, E.U., Foon, L.P., and Joe, L.K. 1974. Microsporidian parasites of trematode larvae from aquatic snails in West Malaysia. *The Journal of Protozoology* 21: 19–25

Cano, I., Ryder, D., Webb, S.C. et al. 2020. Cosmopolitan distribution of *Endozoicomonas*-like organisms and other intracellular microcolonies of bacteria causing infection in marine mollusks. *Frontiers in Microbiology* 11: 577481

Carballal, M.J., Barber, B.J., Iglesias, D., and Villalba, A. 2015. Neoplastic diseases of marine bivalves. *Journal of Invertebrate Pathology* 131: 83–106

Carballal, M.J., Iglesias, D., Santamarina, J., Ferro-Soto, B., and Villalba, A. 2001. Parasites and pathologic conditions of the cockle *Cerastoderma edule* populations of the coast of Galicia (NW Spain). *Journal of Invertebrate Pathology* 78: 87–97

Carella, F., Aceto, S., Mangoni, O. et al. 2018. Assessment of the health status of mussels *Mytilus galloprovincialis* along the Campania coastal areas: A multidisciplinary approach. *Frontiers in Physiology* 9: 683

Carella, F., Carrasco, N., Andree, K.B., Lacuesta, B., Furones, D., and De Vico, G. 2013. Nocardiosis in Mediterranean bivalves: First detection of *Nocardia crassostreae* in a new host *Mytilus galloprovincialis* and in *Ostrea edulis* from the Gulf of Naples (Italy). *Journal of Invertebrate Pathology* 114: 324–328

Carnegie, R.B., Arzul, I., and Bushek, D. 2016. Managing marine mollusc diseases in the context of regional and international commerce: Policy issues and emerging concerns. *Philosophical Transactions of the Royal Society B – Biological Sciences* 371: 20150215

Carnegie, R.B., Barber, B.J., Culloty, S.C., Figueras, A.J., and Distel, D.L. 2000. Development of a PCR assay for detection of the oyster pathogen *Bonamia ostreae* and support for its inclusion in the Haplosporidia. *Diseases of Aquatic Organisms* 42: 199–206

Carnegie, R.B., Burreson, E.M., Hine, M. et al. 2006. *Bonamia perspora* n. sp. (Haplosporidia), a parasite of the oyster *Ostreola equestris*, is the first *Bonamia* species known to produce spores. *Journal of Eukaryotic Microbiology* 53: 232–245

Carnegie, R.B., Hill, K.M., Stokes, N.A., and Burreson, E.M. 2014. The haplosporidian *Bonamia exitiosa* is present in Australia, but the identity of the parasite described as *Bonamia* (formerly *Mikrocytos*) *roughleyi* is uncertain. *Journal of Invertebrate Pathology* 115: 33–40

Carrasco, N., Green, T., and Itoh, N. 2015. *Marteilia* spp. parasites in bivalves: A revision of recent studies. *Journal of Invertebrate Pathology* 131: 53–57

Carrasco, N., Hine, P.M., Durfort, M. et al. 2013. *Marteilia cochillia* sp nov., a new *Marteilia* species affecting the edible cockle *Cerastoderma edule* in European waters. *Aquaculture* 412: 223–30

Carrasco, N., Rojas, M., Aceituno, P., Andree, K.B., Lacuesta, B., and Furones, M.D. 2014. *Perkinsus chesapeaki* observed in a new host, the European common edible cockle *Cerastoderma edule*, in the Spanish Mediterranean coast. *Journal of Invertebrate Pathology* 117: 56–60

Carrasco, N., Voorbergen-Laarman, M., Lacuesta, B., Furones, D. and Engelsma, M.Y. 2017. Application of a competitive real time PCR for detection of *Marteilia refringens* genotype 'O' and 'M' in two geographical locations: The Ebro Delta, Spain and the Rhine-Meuse Delta, the Netherlands. *Journal of Invertebrate Pathology* 149: 51–55

Casas, S.M. and Villalba, A. 2012. Study of perkinsosis in the grooved carpet shell clam *Ruditapes decussatus* in Galicia (NW Spain). III. The effects of *Perkinsus olseni* infection on clam reproduction. *Aquaculture* 356–357: 40–47

Castinel, A., Webb, S., Jones, J.B., and Forrest, B.M. 2019. Disease threats to farmed green-lipped mussels *Perna canaliculus* in New Zealand: Review of challenges in risk assessment and pathway analysis. *Aquaculture Environment Interactions* 11: 291–304

Catanese, G., Grau, A., Valencia, J.M. et al. 2018. *Haplosporidium pinnae* sp. nov., a haplosporidan parasite associated with mass mortalities of the fan mussel, *Pinna nobilis*, in the western Mediterranean Sea. *Journal of Invertebrate Pathology* 157: 9–24

Charles, M., Bernard, I., Villalba, A. et al. 2020a. High mortality of mussels in northern Brittany – evaluation of the involvement of pathogens, pathological conditions and pollutants. *Journal of Invertebrate Pathology* 170: 107308

Charles, M., Villalba, A., Meyer, G. et al. 2020b. First detection of *Francisella halioticida* in mussels *Mytilus* spp. experiencing mortalities in France. *Diseases of Aquatic Organisms* 140: 203–208

Chen, I.W., Chang, P.H., Chen, M.S. et al. 2016. Exploring the chronic mortality affecting abalones in Taiwan: Differentiation of abalone herpesvirus-associated acute infection from chronic mortality by PCR and in situ hybridization and histopathology. *Taiwan Veterinary Journal* 42: 1–9

Cheng, T.C. 1978. Larval nematodes parasitic in shellfish. *Marine Fisheries Review* 40: 39–42.

Chin, W.X., Ang, S.K., and Chu, J.J.H. 2017. Recent advances in therapeutic recruitment of mammalian RNAi and bacterial CRISPR-Cas DNA interference pathways as emerging antiviral strategies. *Drug Discovery Today* 22: 17–30

Choi, K.S., Park, K.I., Lee, K.W., and Matsuoka, K. 2002. Infection intensity, prevalence, and histopathology of *Perkinsus* sp. in the Manila clam, *Ruditapes philippinarum* in Isahaya Bay, Japan. *Journal of Shellfish Research* 21: 119–25

Clantzig, S. 1989. Invertebrates d'introduction récente dans les lagunes méditerranéennes du Languedoc-Roussillon (France). *Bulletin de la Société Zoologique de France* 114: 151–152

Clerissi, C., De Lorgeril, J., Petton, B. et al. 2020. Microbiota composition and evenness predict survival rate of oysters confronted to Pacific oyster mortality syndrome. *Frontiers in Microbiology* 11: 311

Cochennec, N., Le Roux, F., Berthe, F., and Gerard, A. 2000. Detection of *Bonamia ostreae* based on small subunit ribosomal probe. *Journal of Invertebrate Pathology* 76: 26–32

Comps, M. and Tigé, G. 1997. Fine structure of *Minchinia* sp., a haplosporidian infecting the mussel *Mytilus galloprovincialis* L. *Systematic Parasitology* 38: 45–50

Comtet, T., Garcia, C., Le Coguic, Y., and Joly, J.P. 2003. Infection of the cockle *Cerastoderma edule* in the Baie des Veys (France) by the microsporidian parasite *Steinhausia* sp. *Diseases of Aquatic Organisms* 57: 135–139

Comtet, T., Garcia, C., Le Coguic, Y., and Joly, J.-P. 2004. First record of the microsporidian parasite *Steinhausia mytilovum* in *Mytilus* sp. (Bivalvia: Mytilidae) from France. *Diseases of Aquatic Organisms* 58: 261–264

Corbeil S. 2020. Abalone viral ganglioneuritis. *Pathogens* 9: 720

Crespo-González, C., Rodríguez-Domínguez, H., Segade, P., Iglesias, R., Arias, C., and García-Estévez, J.M. 2010. Seasonal dynamics and microhabitat distribution of *Urastoma cyprinae* in *Mytilus galloprovincialis*: Implications for its life cycle. *Journal of Shellfish Research* 29: 187–192

Cuesta, J.A., Perez-Miguel, M., González-Ortegón, E., Roque, D., and Drake, P. 2020. The prevalence of the pea crab *Afropinnotheres monodi* in mussels depending on the degree of habitat exposure: Implications for mussel culture. *Aquaculture* 520: 7347–7372

Culloty, S.C. and Mulcahy, M.F. 2007. *Bonamia ostreae* in the native oyster *Ostrea edulis*. Marine Environment and Health Series No. 29 (Marine Institute Ireland ISSN NO:1649-0053) https://oar.marine.ie/

Da Silva, P.M., Fuentes, J., and Villalba, A. 2005. Growth, mortality and disease susceptibility of oyster *Ostrea edulis* families obtained from brood stocks of different geographical origins, through on-growing in the Ria de Arousa (Galicia, NW Spain). *Marine Biology* 147: 965–977

Da Silva, P.M., Magalhães, A.R.M., and Barracco, M.A. 2012. Pathologies in commercial bivalve species from Santa Catarina State, southern Brazil. *Journal of the Marine Biological Association of the United Kingdom* 92: 571–579

Dairain, A., Engelsma, M.Y., Drent, J., Dekker, R., and Thieltges, D.W. 2020. High prevalences of disseminated neoplasia in the Baltic tellin *Limecola balthica* in the Wadden Sea. *Diseases of Aquatic Organisms* 138: 89–96

Darriba, S., Iglesias, D., and Carballal, M.J. 2020. *Marteilia cochillia* is released into seawater via cockle *Cerastoderma edule* faeces. *Journal of Invertebrate Pathology* 172: 107364

Davidovich, N., Morick, D., and Carella, F. 2020. Mycobacteriosis in aquatic invertebrates: A review of its emergence. *Microorganisms* 8: 1249

Davison, A.J., Trus, B.L., Cheng, N.Q. et al. 2005. A novel class of herpesvirus with bivalve hosts. *Journal of General Virology* 86: 41–53

De Lorgeril, J., Lucasson, A., Petton, B. et al. 2018. Immune-suppression by OsHV-1 viral infection causes fatal bacteraemia in Pacific oysters. *Nature Communications* 9: 4215

De Montaudouin, X., Kisielewski, I., Bachelet, G., and Desclaux, C. 2000. A census of macroparasites in an intertidal bivalve community, Arcachon Bay, France. *Oceanologica Acta* 23: 453–68

De Montaudouin, X., Thieltges, D.W., Gam, M. et al. 2009. Digenean trematode species in the cockle *Cerastoderma edule*: Identification key and distribution along the north-eastern Atlantic shoreline. *Journal of the Marine Biological Association of the United Kingdom* 89: 543–356

De Vico, G. and Carella, F. 2012. Morphological features of the inflammatory response in molluscs. *Research in Veterinary Science* 93: 1109–1115

Dégremont, L. 2011. Evidence of herpesvirus (OsHV-1) resistance in juvenile *Crassostrea gigas* selected for high resistance to the summer mortality phenomenon. *Aquaculture* 317: 94–98

Dégremont, L., Garcia, C., and Allen, S.K. 2015. Genetic improvement for disease resistance in oysters: A review. *Journal of Invertebrate Pathology* 131: 226–241

Delmotte, J., Chaparro, C., Galinier, R. et al. 2020. Contribution of viral genomic diversity to oyster susceptibility in the Pacific oyster mortality syndrome. *Frontiers in Microbiology* 11: 1579

Derevnina, L., Petre, B., Kellner, R. et al. 2016. Emerging oomycete threats to plants and animals. *Philosophical Transactions of the Royal Society B* 371: 20150459

Destoumieux-Garzón D., Canesi, L., Oyanedel, D., Travers, M.-A., Charrière, G.M., Pruzzo, C., and Vezzulli, L. 2020. Vibrio–bivalve interactions in health and disease. *Environmental Microbiology* 22: 4323–4341

Diggles, B.K. and Oliver, M. 2005. Diseases of cultured paua (*Haliotis iris*) in New Zealand. In *Diseases in Asian Aquaculture V*. Proceedings of the 5th Symposium on Diseases in Asian Aquaculture. P.J. Walker, R.G. Lester, and M.G. Bondad-Reantaso (eds.) pp. 275–287. Manila, Philippines: Fish Health Section, Asian Fisheries Society

Dubert, J., Romalde, J.L., Prado, S., and Barja, J.L. 2016. *Vibrio bivalvicida* sp nov., a novel larval pathogen for bivalve molluscs reared in a hatchery. *Systematic and Applied Microbiology* 39: 8–13

Dubert, J., Barja., J.L. and Romalde, J.L. 2017. New insights into pathogenic vibrios affecting bivalves in hatcheries: Present and future prospects. *Frontiers in Microbiology* 8: 762

Dunn, A.M. and Smith, J.E. 2001. Microsporidian life cycles and diversity: The relationship between virulence and transmission. *Microbes and Infection* 3: 381–388

Elston, R.A., Hasegawa, H., Humphrey, K.L., Polyak, I.K., and Häse, C.C. 2008. Re-emergence of *Vibrio tubiashii* in bivalve shellfish aquaculture: Severity, environmental drivers, geographic extent and management. *Diseases of Aquatic Organisms* 82: 119–134

Elston, R.A., Moore, J., and Abbott, C.L. 2012. Denman Island disease (causative agent *Mikrocytos mackini*) in a new host, Kumamoto oysters *Crassostrea sikamea*. *Diseases of Aquatic Organisms* 102: 65–71

Englesma, M.Y., Roozenburg, I., and Joly, J.P. 2008. First isolation of *Nocardia crassostreae* from Pacific oyster *Crassostrea gigas* in Europe. *Diseases of Aquatic Organisms* 80: 229–234

Eymann, C., Götze, S., Bock, C. et al. 2020. Thermal performance of the European flat oyster, *Ostrea edulis* (Linnaeus, 1758)-explaining ecological findings under climate change. *Marine Biology* 167: 17

FAO 2020. *The State of World Fisheries and Aquaculture 2020*. Rome: FAO

Farley, A.C., Wolf, P.W., and Elston, R.A.1988. A long-term study of 'microcell' disease in oysters with a description of a new genus, *Mikrocytos* (g.n.), and two new species, *Mikrocytos mackini* (sp. n.) and *Mikrocytos roughleyi* (sp. n.). *Fisheries Bulletin* 86: 581–593

Farley, C.A. 1969. Probable neoplastic disease of the hematopoietic system in oysters *Crassostrea virginica* and *Crassostrea gigas*. *National Cancer Institute Monographs* 31: 541–555

Fermer, J., Culloty, S.C., Kelly, T.C., and O'Riordan, R.M. 2010. Temporal variation of *Meiogymnophallus minutus* infections in the first and second intermediate host. *Journal of Helminthology* 84: 362

Fermer, J., Culloty, S.C., Kelly, T.C., and O'Riordan, R.M. 2011. Parasitological survey of the edible cockle *Cerastoderma edule* (Bivalvia) on the south coast of Ireland. *The Journal of the Marine Biological Association of the United Kingdom* 91: 923–928

Fichi, G., Carboni, S., Bron, J.E., Ireland, J., Leaver, M.J., and Paladini, G. 2018. Characterisation of the intracellular protozoan MPX in Scottish mussels, *Mytilus edulis* Linnaeus, 1758. *Journal of Invertebrate Pathology* 153: 99–108

Figueras, A.J., Jardón, C.F., and Caldas, J.R. 1991. Diseases and parasites of rafted mussels (*Mytilus galloprovincialis* Lmk): Preliminary results. *Aquaculture* 99: 17–33

Fleming, L.C., Burt, M.D.B., and Bacon, G.G. 1981. On some commensal Turbellaria of the Canadian east coast. *Hydrobiologia* 84: 131–137

Ford, S.E., Chintala, M.M., and Bushek, D. 2002. Comparison of *in vitro*- cultured and wild-type *Perkinsus marinus*. I. Pathogen virulence. *Diseases of Aquatic Organisms* 51: 187–201

Ford, S.E. and Smolowitz, R. 2007. Infection dynamics of an oyster parasite in its newly expanded range. *Marine Biology* 151: 119–133

Ford, S.E., Stokes, N.A., Burreson, E.M. et al. 2009. *Minchinia mercenariae* n. sp (Haplosporidia) in the hard clam *Mercenaria mercenaria*: Implications of a rare parasite in a commercially important host. *Journal of Eukaryotic Microbiology* 56: 542–551

Ford, S.E., Stokes, N.A., Burreson, E.M. et al. 2018. Investigating the life cycle of *Haplosporidium nelsoni* (MSX): A review. *Journal of Shellfish Research* 37: 679–693

Friedman C.S., Beaman B.L., Hedrick R.P., Beattie J.H. and Elston R.A. 1988. Nocardiosis of adult Pacific oysters, *Crassostrea gigas*. *Journal of Shellfish Research* 7: 216

Friedman, C.S., Andree, K.B., Beauchamp, K.A. et al. 2000. 'Candidatus *Xenohaliotis californiensis*' a newly described pathogen of abalone, *Haliotis* spp., along the west coast of North America. *International Journal of Systematic and Evolutionary Microbiology* 50: 847–855

Friedman, C.S., Beaman, B.L., Chun, J., Goodfellow, M., Gee, A, and Hedrick, R.P. 1998. *Nocardia crassostreae* sp. nov., the causal agent of nocardiosis in Pacific oysters. *International Journal of Systematic Bacteriology* 48: 237–246

Friedman, C.S., Biggs, W., Shields, J.D., and Hedrick, R.P. 2002. Transmission of withering syndrome in black abalone, *Haliotis cracherodii* Leach. *Journal of Shellfish Research* 21: 817–824

Friedman, C.S., Estes, R.M., Stokes, N.A. et al. 2005. Herpes virus in juvenile Pacific oysters *Crassostrea gigas* from Tomales Bay, California, coincides with summer mortality episodes. *Diseases of Aquatic Organisms* 63: 33–41

Friedman, C.S. and Hedrick, R.P. 1991. Pacific oyster nocardiosis: Isolation of the bacterium and induction of laboratory infections. *Journal of Invertebrate Pathology* 57: 109–120

Gabriel, M.W., Matsui, G.Y., Friedman, R., and Lovell, C.R. 2014. Optimization of multilocus sequence analysis for identification of species in the genus *Vibrio*. *Applied and Environmental Microbiology* 80: 5359–5365

Gam, M., Bazaïri, H., Jensen, K.T., and De Montaudouin, X. 2008. Metazoan parasites in an intermediate host population near its southern border: The common cockle (*Cerastoderma edule*) and its trematodes in a Moroccan coastal lagoon (Merja Zerga). *Journal of the Marine Biological Association of the United Kingdom* 88: 357–364

Garcia, C., Haond, C., Chollet, B. et al. 2018. Descriptions of *Mikrocytos veneroides* n. sp and *Mikrocytos donaxi* n. sp (Ascetosporea: Mikrocytida: Mikrocytiidae), detected during important mortality events of the wedge clam *Donax trunculus* Linnaeus (Veneroida: Donacidae), in France between 2008 and 2011. *Parasites and Vectors* 11: 119

Garnier, M., Labreuche, Y., Garcia, C., Robert, M., and Nicolas, J.L. 2007. Evidence for the involvement of pathogenic bacteria in summer mortalities of the Pacific oyster *Crassostrea gigas*. *Microbial Ecology* 53: 187–196

Gauthier, J.D., Soniat, T.M., and Rogers, J. S. 1990. A parasitological survey of oysters along salinity gradients in coastal Louisiana. *Journal of World Aquaculture Society* 21: 105–115

Gazeau, F., Parker, L.M., Comeau, S. et al. 2013. Impacts of ocean acidification on marine shelled molluscs. *Marine Biology* 160: 2207–2245

Gibbs, P.E. 1982. The turbellarian *Paravortex* parasitic in scrobiculariid bivalves: New records from the southwest England. *Journal of the Marine Biological Association of the United Kingdom* 62: 739–740

Goater, C.P. 1993. Population biology of *Meiogymnophallus minutus* (Trematoda: Gymnophallidae) in cockles from the Exe Estuary. *Journal of the Marine Biological Association of the UK* 73: 163–177

Goggin, C.L. and Lester, R.J.G. 1995. *Perkinsus*, a protistan parasite of abalone in Australia: A review. *Marine Fisheries Research* 46: 639–646

Gómez-Chiarri, M., Guo, X., Tanguy, A., He, Y., and Proestou, D. 2015. The use of -omic tools in the study of disease processes in marine bivalve mollusks. *Journal of Invertebrate Pathology* 131: 137–154

Gosling, E. 2015. Diseases and parasites. In *Marine Bivalve Molluscs*, 2nd edn, pp. 429–477. Hoboken, New Jersey, USA: Wiley

Gotesman, M., Soliman, H., Besch, R., and El-Matbouli, M. 2014. *In vitro* inhibition of cyprinid herpesvirus-3 replication by RNAi. *Journal of Virology Methods* 206: 63–6

Goudenège, D., Travers, M.A., Lemire, A. et al. 2015. Comparative genomics of *Vibrio aestuarianus*. *Environmental Microbiology* 17: 4189–4199

Green, T.J., Helbig, K., Speck, P., and Raftos, D.A. 2016. Primed for success: Oyster parents treated with poly(I:C) produce offspring with enhanced protection against Ostreid herpesvirus type I infection. *Molecular Immunology* 78: 113–120

Green, T.J., Jones, B.J., Adlard, R.D., and Barnes, A.C. 2008. Parasites, pathological conditions and mortality in QX-resistant and wild-caught Sydney rock oysters, *Saccostrea glomerata*. *Aquaculture* 280: 35–38

Green, T.J., Siboni, N., King, W.L., Labbat, M., Seymour, J.R., and Raftos, D.A. 2019. Simulated marine heat wave alters abundance and structure of *Vibrio* populations associated with the Pacific oyster resulting in a mass mortality event. *Microbial Ecology* 77: 736–747

Grizel, H. 1985. *Mytilicola orientalis* Mori, parasitism. (Parasitose à *Mytilicola orientalis* Mori.) Identification Leaflets for *Diseases and Parasites of Fish and Shellfish* 20: 4

Guo, X., Wang, Y., Wang, L., and Lee, J.H. 2008. Oysters. In *Genome Mapping and Genomics in Fishes and Aquatic Animals*. T.D. Kocher and C. Kole (eds.) pp. 163–175. Berlin, Germany: Springer

Handlinger, J., Bastianello, S., Callinan, R. et al. 2006. Abalone aquaculture subprogram: A national survey of diseases of commercially exploited abalone species to support trade and translocation issues and the development of health surveillance programs. Report number: FRDC Project No. 2002/201

Hartikainen, H., Stentiford, G.D., Bateman, K.S. et al. 2014. Mikrocytids are a broadly distributed and divergent radiation of parasites in aquatic invertebrates. *Current Biology* 24: 807–812

Hibbett, D.S., Binder, M., Bischoff, J.F. et al. 2007. A higher-level phylogenetic classification of the Fungi. *Mycological Research* 111: 509–47

Hill, K.M., Stokes, N.A., Webb, S.C. et al. 2014. Phylogenetics of *Bonamia* parasites based on small subunit and internal transcribed spacer region ribosomal DNA sequence data. *Diseases of Aquatic Organisms* 110: 33–54

Hillman, R.E. 1991. *Steinhausia mytilovum* (Minisporida, Chitridiopsidae) in *Mytilus* sp. in California – A new geographic record. *Journal of Invertebrate Pathology* 57: 144–145

Hine, P.M., Cochennec-Laureau, N., and Berthe, A.F. 2001. *Bonamia exitiosus* n. sp. (Haplosporidia) infecting flat oysters *Ostrea chilensis* in New Zealand. *Diseases of Aquatic Organisms* 47: 63–72

Holmes, J.M.C. and Minchin, D. 1995. Two exotic copepods imported into Ireland with the Pacific oyster *Crassostrea gigas* (Thunberg). *Irish Naturalist's Journal* 25: 17–20

Hooper, C., Slocombe, R., Day, R., and Crawford, S. 2012. Leucopenia associated with abalone viral ganglioneuritis. *Australian Veterinary Journal* 90: 24–28

Houston, R.D., Bean, T.P., Macqueen, D.J. et al. 2020. Harnessing genomics to fast-track genetic improvement in aquaculture. *Nature Reviews Genetics* 21: 389–409

Ironside, J.E. 2007. Multiple losses of sex within a single genus of Microsporidia. *BMC Evolutionary Biology* 7: 1–16

Itoh, N., Yamamoto, T., Kang, H.-S. et al. 2014. A novel paramyxean parasite, *Marteilia granula* sp. nov. (Cercozoa), from the digestive gland of Manila clam *Ruditapes philippinarum* in Japan. *Fish Pathology* 49: 181–193

Ituarte, C., Bagnato, E., Siddall, M., and Cremonte, F. 2014. A new species of Haplosporidium Caullery & Mesnil, 1899 in the marine false limpet *Siphonaria lessonii* (Gastropoda: Siphonariidae) from Patagonia. *Systematic Parasitology* 88: 63–73

Jones, A.A. and Selman, B.J. 1985. *Microsporidium novacastriensis* n. sp., a microsporidian parasite of the grey field slug, *Deroceras reticulatum*, 2. *The Journal of Protozoology* 32: 581–586

Jones, J.B. 1981. A new *Microsporidium* from the oyster *Ostrea lutaria* in New Zealand. *Journal of Invertebrate Pathology* 38: 67–70

Jones, J.B. 1997. *Steinhausia* sp. (Microspora: Chytridiopsidae) infecting ova of *Mytilus galloprovincialis* in western Australia. In *10th International Congress of Protozoology*. Monday 21 July–Friday 25 July 1997, Programme & Abstracts. Business Meetings & Incentives. M. Pascoe (ed). P. 112. Australia: The University of Sydney

Jones, J.B. and Creeper, J. 2006. Diseases of pearl oysters and other molluscs: A Western Australian perspective. *Journal of Shellfish Research* 25: 233–238

Joseph, S, Fernández-Robledo, J.A., Garner, M.J. et al. 2010. The alveolate *Perkinsus marinus*: Biological insights from EST gene discovery. *BMC Genomics* 11: 228

Kang, H.-S., Itoh, N., Limpanont, Y., Lee, H.-M., Whang, I., and Choi, K.-S. 2019. A novel paramyxean parasite, *Marteilia tapetis* sp. nov. (Cercozoa) infecting the digestive gland of Manila clam *Ruditapes philippinarum* from the southeast coast of Korea. *Journal of Invertebrate Pathology* 163: 86–93

Kawahara, M., Meyer, G.R., Lowe, G.J. et al. 2019. Parallel studies confirm *Francisella halioticida* causes mortality in Yesso scallops *Patinopecten yessoensis*. *Diseases of Aquatic Organisms* 135: 127–34

Keeling, P.J. and Fast, N.M. 2002. Microsporidia: Biology and evolution of highly reduced intracellular parasites. *Annual Reviews in Microbiology* 56: 93–116

Kerr, R., Ward, G.M., Stentiford, G.D. et al. 2018. *Marteilia refringens* and *Marteilia pararefringens* sp nov are distinct parasites of bivalves and have different European distributions. *Parasitology* 145: 1483–1492

Kesarcodi-Watson, A., Miner, P., Nicolas, J.-L., and Robert, R. 2012. Protective effect of four potential probiotics against pathogen-challenge of the larvae of three bivalves: Pacific oyster (*Crassostrea gigas*), flat oyster (*Ostrea edulis*) and scallop (*Pecten maximus*). *Aquaculture* 344–349: 29–34

King, W.L., Siboni, N., Kahlke, T., Green, T.J., Labbate, M., and Seymour, J.R. 2019. A new high throughput sequencing assay for characterizing the diversity of natural vibrio communities and its application to a Pacific oyster mortality event. *Frontiers in Microbiology* 10: 2907

Kleeman, S.N., Adlard, R.D., and Lester, R.J.G. 2002. Detection of the initial infective stages of the protozoan parasite *Marteilia sydneyi* in *Saccostrea glomerata* and their development through to sporogenesis. *International Journal of Parasitology* 32: 767–784

Ko R.C., Morton B., and Wong P.S. 1975. Prevalence and histopathology of *Echinocephalus sinensis* (Nematoda: Gnathostomatidae) in natural and experimental hosts. *Canadian Journal of Zoology* 53: 550–559

Krauhs, J.M., Long, J.L., and Baur Jr, P.S. 1979. Spores of a new microsporidan species parasitizing molluscan neurons. *The Journal of Protozoology* 26: 43–46

Kumar, R.S., Galvis, F., Wasson, B.J. et al. 2020. Draft genome sequence of *Vibrio ostreicida* Strain PP-203, the type strain of a pathogen that infects bivalve larvae. *Microbiology Resource Announcements* 9: e00913–20

Labreuche, Y., Le Roux, F., Henry, J. et al. 2010. *Vibrio aestuarianus* zinc metalloprotease causes lethality in the Pacific oyster *Crassostrea gigas* and impairs the host cellular immune defenses. *Fish & Shellfish Immunology* 29: 753–758

Lacoste, A., Kalabert, F., Malham, S., et al. 2001. A *Vibrio splendidus* strain is associated with summer mortality of juvenile oysters *Crassostrea gigas* in the Bay of Morlaix (North Brittany, France). *Diseases of Aquatic Organisms* 46: 139–145

Lambert, C., Nicolas, J.L., Cilia, V., and Corre, S. 1998. *Vibrio pectenicida* sp. nov. a pathogen of scallop (*Pectin maximus*) larvae. *International Journal of Systematic Bacteriology* 48: 481–487

Larsson, J.R. 2005. Fixation of microsporidian spores for electron microscopy. *Journal of Invertebrate Pathology* 90: 47–50

Lasa, A., Di Cesare, A., Tassistro, G. et al. 2019. Dynamics of the Pacific oyster pathobiota during mortality episodes in Europe assessed by 16S rRNA gene profiling and a new target enrichment next-generation sequencing strategy. *Environmental Microbiology* 21: 4548–62

Lauckner, G. 1983. Diseases of Mollusca: Bivalvia. In *Diseases of Marine Animals, 2*. Kinne, O. (ed.) Hamburg: 1038: Biologische Anstalt Helgoland

Le Roux, F., Gay, M., Lambert, C. et al. 2002. Comparative analysis of *Vibrio splendidus*-related strains isolated during *Crassostrea gigas* mortality events. *Aquatic Living Resources* 15: 251–258

Le, T.S., Southgate, P.C., O'Connor, W., Vu, S.V., and Kurtböke, D.I. 2020. Application of bacteriophages to control *Vibrio alginolyticus* contamination in oyster (*Saccostrea glomerata*) larvae. *Antbiotics (Basel)* 9: 415

Lemire, A., Goudenège, D., Versigny, T. et al. 2015. Populations, not clones, are the unit of vibrio pathogenesis in naturally infected oysters. *ISME Journal* 9: 1523–1531

Lester R.J.G., Blair, D., and Heald, D. 1980. Nematodes from scallops and turtles from Shark Bay, western Australia. *Australian Journal of Marine and Freshwater Research* 31: 713–717

Lester, R.J.G. and Davis, G.H.G. 1981. A new *Perkinsus* species (Apicomplexa, Perkinsea) from the abalone *Haliotis ruber*. *Journal of Invertebrate Pathology* 37: 181–187

Letchumanan, V., Chan, K.G., Pusparajah, P. et al. 2016. Insights into bacteriophage application in controlling *Vibrio* species. *Frontiers in Microbiology* 19: 1114

Longshaw, M. and Malham, S.K. 2013. A review of the infectious agents, parasites, pathogens and commensals of European cockles (*Cerastoderma edule* and *C. glaucum*). *Journal of the Marine Biological Association of the United Kingdom* 93: 227

Lynch, S.A., Abollo, E., Ramilo, A., Cao, A., Culloty, S.C., and Villalba, A. 2010. Observations raise the question if the Pacific oyster, *Crassostrea gigas*, can act as either a carrier or a reservoir for *Bonamia ostreae* or *Bonamia exitiosa*. *Parasitology* 137: 1515–1526

Lynch, S.A., Armitage, D.V., Coughlan, J., Mulcahy, M.F., and Culloty, S.C. 2007. Investigating the possible role of benthic macroinvertebrates and zooplankton in the life cycle of the haplosporidian *Bonamia ostreae*. *Experimental Parasitology* 115: 359–368

Lynch, S.A., Carlsson, J., O' Reilly, A., Cotter, E., and Culloty, S.C. 2012. A previously undescribed ostreid herpes virus 1 (OsHV-1) genotype detected in the Pacific oyster, *Crassostrea gigas*, in Ireland. *Parasitology* 139: 1526–1532

Lynch, S.A., Lepée-Rivero, S., Kelly, R. et al. 2020. Detection of haplosporidian protistan parasites supports an increase to their known diversity, geographic range and bivalve host specificity. *Parasitology* 147: 584–592

Lynch, S.A., Morgan, E., Carlsson, J. et al. 2014. The health status of mussels, *Mytilus* spp., in Ireland and Wales with the molecular identification of a previously undescribed haplosporidian. *Journal of Invertebrate Pathology* 118: 59–65

Lynch, S.A., Villalba, A., Abollo, E., Engelsma, M., Stokes, N.A., and Culloty, S.C. 2013. The occurrence of haplosporidian parasites, *Haplosporidium nelsoni* and *Haplosporidium* sp., in oysters in Ireland. *Journal of Invertebrate Pathology* 112: 208–212

Lynn, D.H., Doerder, F.P., Gillis, P.L., and Prosser, R.S. 2018. *Tetrahymena glochidiophila* n. sp., a new species of *Tetrahymena* (Ciliophora) that causes mortality to glochidia larvae of freshwater mussels (Bivalvia). *Diseases of Aquatic Organisms* 127: 125–36

Ma, Y.X., Liu, J.C., Li, M., Tao, W., Yu, Z.C., and Liu, Y.B. 2019. The use of *Pseudoalteromonas* sp. F15 in larviculture of the Yesso scallop, *Patinopecten yessoensis*. *Aquaculture Research* 50: 1844–1850

Macey, B.M., Christison, K.W., and Mouton, A. 2011. *Halioticida noduliformans* isolated from cultured abalone (*Haliotis midae*) from South Africa. *Aquaculture* 315: 187–195

Mackenzie, C.L., Lynch, S.A., Culloty, S.C. and Malham, S.K. 2014. Future oceanic warming and acidification alter immune response and disease status in a commercial shellfish species, *Mytilus edulis* L. *PLoS ONE* 9: e99712

Maeda, K., Yoshinaga, T., and Itoh, N. 2021. Development of a simple host-free medium for efficient prezoosporulation of *Perkinsus olseni* trophozoites cultured *in vitro*. *Parasitology International* 80: 102186

Magalhães, L., Daffe, G., Freitas, R., and De Montaudouin, X. 2020. *Monorchis parvus* and *Gymnophallus choledochus*: Two trematode species infecting cockles as first and second intermediate host. *Parasitology* 147: 643–658

Magalhães, L., Freitas, R., and De Montaudouin, X. 2015. Review: *Bucephalus minimus*, a deleterious trematode parasite of cockles *Cerastoderma* spp. *Parasitology Research* 114: 1263–1278

Mandas, D., Salati, F., Polinas, M. et al. 2020. Histopathological and molecular study of Pacific oyster tissues provides insights into *V. aestuarianus* infection related to oyster mortality. *Pathogens* 9: 492

Marcer, F., Tosi, F., Franzo, G. et al. 2020. Updates on ecology and life cycle of *Sulcascaris sulcata* (Nematoda: Anisakidae) in Mediterranean grounds: Molecular identification of larvae infecting edible scallops. *Frontiers in Veterinary Science* 7: 64

Marty, G.D., Bower, S.M., Clarke, K.R. et al. 2006. Histopathology and a real-time PCR assay for detection of *Bonamia ostreae* in *Ostrea edulis* cultured in western Canada. *Aquaculture* 261: 33–42

Matos, E., Matos, P., and Azevedo, C. 2005. Observations on the intracytoplasmic microsporidian *Steinhausia mytilovum*, a parasite of mussel (*Mytella guyanensis*) oocytes from the Amazon River estuary. *Brazilian Journal of Morphological Sciences* 22: 183–186

Mau, A. and Jha, R. 2018. Aquaculture of two commercially important molluscs (abalone and limpet): Existing knowledge and future prospects. *Reviews in Aquaculture* 10: 611–625

McClymont, H.E., Dunn, A.M., Terry, R.S., Rollinson, D., Littlewood, T.J., and Smith, J.E. 2005. Molecular data suggest that microsporidian parasites in freshwater snails are diverse. *International Journal for Parasitology* 35: 1071–1078

McElwain, A., Warren, M.B., Pereira, F.B., Ksepka, S. P., and Bullard, S.A. 2019. Pathobiology and first report of larval nematodes (*Ascaridomorpha* sp.) infecting freshwater mussels (*Villosa nebulosa*, Unionidae), including an inventory of nematode infections in freshwater and marine bivalves. *International Journal for Parasitology Parasites and Wildlife* 10: 41–58

McGurk, E.S., Ford, S., and Bushek, D. 2016. Unusually abundant and large ciliate xenomas in oysters, *Crassostrea virginica*, from Great Bay, New Hampshire, USA. *Journal of Invertebrate Pathology* 137: 23–32

McLean, N. 1983. An echinocephalid nematode in the scallop *Argopecten aequisulcatus* (Mollusca: Bivalvia). *Journal of Invertebrate Pathology* 42: 273–276

Metzger, M., Villalba, A., Carballal, M., Sherry, J., Reinisch, C., Muttray, A.F., Baldwin, S.A., and Goff, S.P. 2016. Widespread transmission of independent cancer lineages within multiple bivalve species. *Nature* 534: 705–709

Metzger M.J., Paynter, A.N., Siddall, M.E., and Goff, S.P. 2018. Horizontal transfer of retrotransposons between bivalves and other aquatic species of multiple phyla. *Proceedings of the National Academy of Sciences U.S.A.* 115: E4227–E4235

Metzger, M.J., Villalba, A., Carballal, M.J., Iglesias, D., Sherry, J., Reinisch, C., Muttray, A.F., Baldwin, S.A., and Goff, S.P. 2016. Widespread transmission of independent cancer lineages within multiple bivalve species. *Nature* 534: 705–709

Metzger, M.J., Reinisch, C., Sherry, J., and Goff, S.P. 2015. Horizontal transmission of clonal cancer cells causes leukemia in soft-shell clams. *Cell* 161: 255–263

Meyer, G.R., Lowe, G.J., Gilmore, S.R., and Bower, S.M. 2017. Disease and mortality among Yesso scallops *Patinopecten yessoensis* putatively caused by infection with *Francisella halioticida*. *Diseases of Aquatic Organisms* 125: 79–84

Meyers, T. and Burton, T. 2009. Diseases of wild and cultured shellfish in Alaska. Alaska Department of Fish and Game, https://www.adfg.alaska.gov/static/species/disease/pdfs/shellfish_disease_book.pdf

Meyers, T.R., Burton, T., Evans, W., and Starkey, N. 2009. Detection of viruses and virus-like particles in four species of wild and farmed bivalve molluscs in Alaska, USA, from 1987 to 2009. *Diseases of Aquatic Organisms* 88: 1–12

Michelson, E.H. 1963. *Plistophora husseyi* sp. n., a microsporidian parasite of aquatic pulmonate snails. *Journal of Insect Pathology* 5: 28–38

Miranda C., Rojas R., Garrido M., Geisse J., and González G. 2013. Role of shellfish hatchery as a reservoir of antimicrobial resistant bacteria. *Marine Pollution Bulletin* 74: 334–43

Molloy, D.P., Giambérini, L., Stokes, N.A., Burreson, E.M., and Ovcharenko, M.A. 2012. *Haplosporidium raabei* n. sp. (Haplosporidia): A parasite of zebra mussels, *Dreissena polymorpha* (Pallas, 1771). *Parasitology* 139: 463–477

Morgan, E., O'Riordan, R.M., Kelly, T.C., and Culloty, S.C. 2012. Influence of disseminated neoplasia, trematode infections and gametogenesis on surfacing and mortality in the cockle *Cerastoderma edule*. *Diseases of Aquatic Organisms* 98: 73–84

Mortensen, S., Arzul, I., Miossec, L., et al. 2007. Molluscs and crustaceans 5.3.7 Brown ring disease (infections with *Vibrio tapetis*). In *Review of disease interactions and pathogen exchange between farmed and wild finfish and shellfish in Europe*. R. Raynard, T. Wahli, I. Vatsos, and S. Mortensen (eds.) pp. 341–347. Oslo: VESO on behalf of DIPNET

Moss, J.A., Burreson, E.M., Cordes, J.F. et al. 2007. Pathogens in *Crassostrea ariakensis* and other Asian oyster species: Implications for non-native oyster introduction to Chesapeake Bay. *Diseases of Aquatic Organisms* 77: 207–223

Nago, T.T.T. and Choi, K.-S. 2004. Seasonal changes of *Perkinsus* and Cercaria infections in the Manila clam *Ruditapes philippinarum* from Jeju, Korea. *Aquaculture* 239: 57–68

Nascimento, I.A., Smith, D.H., Kern, F.I. and Pereira, S.A. 1986. Pathological findings in *Crassostrea rhizophorae* from Todos os Santos Bay, Bahia, Brazil. *Journal of Invertebrate Pathology* 47: 340–349

Navas, J., Castillo, M., Vera, P., and Ruiz-Rico, M. 1992. Principal parasites observed in clams, *Ruditapes decussatus* (L), *Ruditapes-philippinarum* (Adamset-Reeve), *Venerupis pullastra* (Montagu) and *Venerupis aureus* (Gmelin), from the Huelva coast (SW Spain). *Aquaculture* 107: 193–9

Nell, J.A. 2002. Farming triploid oysters. *Aquaculture* 210: 1–4

Nell, J.A. and Perkins, B. 2006. Evaluation of the progeny of third-generation Sydney rock oyster *Saccostrea glomerata* (Gould, 1850) breeding lines for resistance to QX disease *Marteilia sydneyi* and winter mortality *Bonamia roughleyi*. *Aquaculture Research* 37: 693–700

Nicole, T.B., Boghen, A.D., and Allard, J. 1999. Attraction of *Urastoma cyprinae* (Turbellaria: Urastomidae) to the eastern oyster *Crassostrea virginica*. *Diseases of Aquatic Organisms* 37: 139–144

Noury-Sraïri, N., Justine, J.-L. and Bonami, J.-R. 1995. Viral particles in a flatworm (*Paravortex tapetis*) parasitic in the commercial clam, *Ruditapes decussatus*. *Journal of Invertebrate Pathology* 65: 200–202

Paillard, C. and Maes, P. 1990. Etiologie de la maladie de l'anneau brun chez *Tapes philippinarum*: Pathogénicité d'un *Vibrio* sp. *Comptes Rendus de l'Académie des Sciences. Série 3, Sciences de la vie* 310: 15–20

Parizadeh, L., Tourbiez, D., Garcia, C. et al. 2018. Ecologically realistic model of infection for exploring the host damage caused by *Vibrio aestuarianus*. *Environmental Microbiology* 20: 4343–4355

Paul-Pont, I., Evans, O., Dhand, N.K., Rubio, A., Coad, P., and Whittington, R.J. 2014. Descriptive epidemiology of mass mortality due to Ostreid herpesvirus-1 (OsHV-1) in commercially farmed Pacific oysters (*Crassostrea gigas*) in the Hawkesbury River estuary, Australia. *Aquaculture* 422–423: 146–159

Perez-Miguel, M., Cuesta, J.A., Navas, J.I., García Raso, J.E., and Drake, P. 2018. The prevalence and effects of the African pea crab *Afropinnotheres monodi* on the condition of the mussel *Mytilus galloprovincialis* and the cockle *Cerastoderma edule*. *Aquaculture* 491: 1–9

Perez-Miguel, M., Drake, P., and Cuesta, J.A. 2020. Temperature effect on the African pea crab *Afropinnotheres monodi*: Embryonic and larval developments, fecundity and adult survival. *Journal of Experimental Marine Biology and Ecology* 527: 151380

Perkins, F.O. and Wolf, P.H. 1976. Fine structure of *Marteilia sydneyi* sp. n.- Haplosporidian pathogen of Australian oysters. *Journal of Parasitology* 62: 528–538

Petton, B., Bruto, M., James, A., Labreuche, Y., Alunno-Bruscia, M., and Le Roux, F. 2015. *Crassostrea gigas* mortality in France: The usual suspect, a herpes virus, may not be the killer in this polymicrobial opportunistic disease. *Frontiers in Microbiology* 6: 686

Pichot, Y., Comps, M., Tige, G., Grizel, H., and Rabouin, M.A. 1980. Recherches sur *Bonamia ostreae* gen. n., sp. n., parasite nouveau de l'huître plate *Ostrea edulis*. *Revue des travaux de l'Institut des pêches maritimes* 43: 131–140

Pike, A.W. and Burt, M.D.B. 1981. *Paravortex karlingi* sp. nov. from *Cerastoderma edule* L., in Britain. *Hydrobiologia* 84: 23–30

Pirkova, A.V. and Demenko, D.P. 2008. Cases of shell disease in the giant oyster *Crassostrea gigas* (Bivalvia) cultivated in the Black Sea. *Russian Journal of Marine Biology* 34: 309–315

Polgase, J.L. 2019. Cephalopod diseases caused by fungi and Labyrinthulomycetes. In *Handbook of Pathogens and*

Diseases in Cephalopods. C. Gestal et al. (eds.) pp. 113–122. Cham, Switzerland: Springer

Powell, E.N., Klinck, J.M., Ford, S.E., Hofmann, E.E., and Jordon, S.J. 1999. Modeling the MSX parasite in eastern oyster (*Crassostrea virginica*) populations. III. Regional application and the problem of transmission. *Journal of Shellfish Research* 18: 517–537

Prado, P., Carrasco, N., Catanese, G. et al. 2020. Presence of *Vibrio mediterranei* associated to major mortality in stabled individuals of *Pinna nobilis* L. *Aquaculture* 519: 734899

Prado, P., Dubert, J., Romalde, J.L., Toranzo, A.E., and Barja, J.L. 2015. *Vibrio ostreicida* sp. nov., a new pathogen of bivalve larvae. *International Journal of Systematic and Evolutionary Microbiology* 64: 1641–1646

Prado, S., J.L. Romalde, J.L., and Barja, J.L. 2010. Review of probiotics for use in bivalve hatcheries. *Veterinary Microbiology* 145: 187–197

Prado-Alvarez, M., García-Fernández, P., Faury, N., Azevedo, C., Morga, B. and Gestal, C. 2021. First detection of OsHV-1 in the cephalopod *Octopus vulgaris*. Is the octopus a dead-end for OsHV-1? *Journal of Invertebrate Pathology* 107553

Pregenzer, C. 1983. Survey of metazoan symbionts of *Mytilus edulis* (Mollusca: Pelecypoda) in Southern Australia. *Australian Journal of Marine and Freshwater Research* 34: 387396

Pretto, T., Zambon, M., Civettini, M. et al. 2014. Mass mortality in Manila clams (*Ruditapes philippinarum*) farmed in the Lagoon of Venice, caused by *Perkinus olseni*. *Bulletin of the European Association of Fish Pathologists* 34: 43–53

Prosser, R.S., Lynn, D.H., Salerno, J., Bennett, J., and Gillis, P.L. 2018. The facultatively parasitic ciliated protozoan, *Tetrahymena glochidiophila* (Lynn, 2018), causes a reduction in viability of freshwater mussel glochidia. *Journal of Invertebrate Pathology* 157: 25–31

Quayle, D.B. 1961. Denman Island oyster disease and mortality, 1960. *Fisheries Research Board Canada* 713: 1–9

Raftos, D.A., Kuchel, R., Aladaileh, S., and Butt, D. 2014. Infectious microbial diseases and host defense responses in Sydney rock oysters. *Frontiers in Microbiology* 5: 135

Rahmani, A., Corre, E., Richard, G. et al. 2019. Transcriptomic analysis of clam extrapallial fluids reveals immunity and cytoskeleton alterations in the first week of Brown Ring Disease development. *Fish & Shellfish Immunology* 93: 940–948

Ramilo, A., Abollo, E., Villalba, A., and Carballal, M.J. 2018. A *Minchinia mercenariae*-like parasite infects cockles *Cerastoderma edule* in Galicia (NW Spain). *Journal of Fish Diseases* 41: 41–8

Ramilo, A., Pintado, J., Villalba, A., and Abollo, E. 2016. *Perkinsus olseni* and *P. chesapeaki* detected in a survey of perkinsosis of various clam species in Galicia (NW Spain) using PCR–DGGE as a screening tool. *Journal of Invertebrate Pathology* 133: 50–58

Rayyan, A. and Chintiroglou, C.C. 2003. *Steinhausia mytilovum* in cultured mussel *Mytilus galloprovincialis* in the Thermaikos Gulf (north Aegean Sea, Greece). *Diseases of Aquatic Organisms* 57: 271–273

Reece, K.S., Siddal, M.E., Stokes, N.A., and Burreson, E.M. 2004. Molecular phylogeny of the haplosporidia based on two independent gene sequences. *International Journal of Parasitology* 90: 1111–1122

Renault, T., Le Deuff, R.M., Cochennec, N., and Maffart, P. 1994. Herpesvirus associated with mortalities among Pacific oyster, *Crassostrea gigas*, in France—Comparative study. *Revue Médecine Vétérinaire* 145: 735–742

Renault, T. and Novoa, B. 2004. Viruses infecting bivalve molluscs. *Aquatic Living Resources* 17: 397–409.

Rhein, M., Rintoul, S.R., Aoki, S. et al. 2013. Observations: Ocean. In Climate change 2013: *The Physical Science Basis. Contribution of working Group I to the fifth assessment report of the intergovernmental panel on climate change*. T.F. Stocker et al. (eds.) pp. 255–316. Cambridge: Cambridge University Press

Richards, C.S. 1973. A potential intermediate host of *Schistosoma mansoni* in Grenada. *Journal of Parasitology* 59: 111

Richards, C.S. and Sheffield H.G. 1971. Unique host relations and ultrastructure of a new microsporidian of the genus *Coccospora* infecting *Biomphalaria glabrata*. *Proceedings of the IV International Colloquium on Insect Pathology* 4: 439–452

Robledo, J.A.F., Ráceres-Martínez, J.C., Sluys, R., and Figueras, A. 1994. The parasitic turbellarian *Urastoma cyprinae* (Platyhelminthes: Urastomidae) from blue mussel *Mytilus galloprovincialis* in Spain: Occurrence and pathology. *Diseases of Aquatic Organisms* 18: 203–210

Rojas, R., Miranda, C.D., Romero, J., Barja, J.L., and Dubert, J. 2019. Isolation and pathogenic characterization of *Vibrio bivalvicida* associated with a massive larval mortality event in a commercial hatchery of scallop *Argopecten purpuratus* in Chile. *Frontiers in Microbiology* 10: 855

Romalde, J.L., Diéguez, A.L., and Balboa, S. 2014. New *Vibrio* species associated with molluscan microbiota: A review. *Frontiers in Microbiology* 4: 413

Rong, R., Lin, H., Wang, J., Khan, M.N., and Li, M. 2014. Reductions of *Vibrio parahaemolyticus* in oysters after bacteriophage application during depuration. *Aquaculture* 418: 171–176

Ruano, F., Batista, F.M., and Arcangeli, G. 2015. Perkinsosis in the clams *Ruditapes decussatus* and *R. philippinarum* in the Northeastern Atlantic and Mediterranean Sea: A review. *Journal of Invertebrate Pathology* 131: 58–67

Rubio, T., Oyanedel, D., Labreuche, Y. et al. 2019. Species-specific mechanisms of cytotoxicity toward immune cells determine the successful outcome of *Vibrio* infections. *Proceedings of the National Academy of Science U.S.A.* 116: 14238–14247

Ruck, K.R. and Cook, P.A. 1998. Sabellid infestations in the shells of South African molluscs: Implications for abalone mariculture. Journal of Shellfish Research Conference: 3rd International Symposium of Abalone Biology, Fisheries and Culture, Volume 17: 693–699

Ruiz, M., López, C., Lee, R.-S., Rodríguez, R., and Darriba, S. 2016. A novel paramyxean parasite, *Marteilia octospora* n. sp. (Cercozoa) infecting the Grooved Razor Shell clam *Solen marginatus* from Galicia (NW Spain). *Journal of Invertebrate Pathology* 135: 34–42

Sagristà, E., Bozzo, M.G., Bigas, M., Poquet, M., and Durfort, M. 1998. Developmental cycle and ultrastructure of *Steinhausia mytilovum*, a microsporidian parasite of oocytes of the mussel, *Mytilus galloprovincialis* (Mollusca, Bivalvia). *European Journal of Protistology* 34: 58–68

Sanchez-Ortiz, A.C., Mazon-Suastegui, J.M. Flores-Miranda, M.D. et al. 2020. Probiotic bacterium and microalga interaction on rearing Kumamoto oyster *Crassostrea sikamea* spat. *Current Microbiology* 77: 2758–2765

Sanil, N.K., Vijayan, K.K., Kripa, V., and Mohamed, K.S. 2010. Occurrence of the protozoan parasite, *Perkinsus olseni* in the wild and farmed Pearl Oyster, *Pinctada fucata* (Gould) from the Southeast coast of India. *Aquaculture* 299: 8–14

Sawabe, T., Kita-Tsukamoto, K., and Thompson, F.L. 2007. Inferring the evolutionary history of vibrios by means of multilocus sequence analysis. *Journal of Bacteriology* 189: 7932–7936

Shamal, P., Zacharia, P.U., Binesh, C.P., Pranav, P. et al. 2018. *Perkinsus olseni* in the short neck yellow clam, *Paphia malabarica* (Chemnitz, 1782) from the southwest coast of India. *Journal of Invertebrate Pathology* 159: 113–120

Silva Y. J., Costa L., Pereira C. et al. 2014. Phage therapy as an approach to prevent *Vibrio anguillarum* infections in fish larvae production. *PLoS ONE* 9: e114197

Song, X., Wang, H., Xin, L. et al. 2016. The immunological capacity in the larvae of Pacific oyster *Crassostrea gigas*. *Fish & Shellfish Immunology* 49: 461–469

Sprague, V. 1965. Observations on *Chytridiopsis mytilovum* (Field), formerly *Haplosporidium mytilovum* Field, (Microsporida?). *Journal of Protozoology* 12: 385–9

Sprague, V., Ormieres, R., and Manier, J.F. 1972. Creation of a new genus and a new family in the Microsporida. *Journal of Invertebrate Pathology* 20: 228–231

Stentiford, G.D., Becnel, J.J., Weiss, L.M. et al. 2016. Microsporidia–emergent pathogens in the global food chain. *Trends in Parasitology* 32: 336–48

Stentiford, G.D., Feist, S.W., Stone, D.M., Bateman, K.S., and Dunn, A.M. 2013. Microsporidia: Diverse, dynamic, and emergent pathogens in aquatic systems. *Trends in Parasitology* 29: 567–578

Stevick, R.J., Sohn, S., Modak, T.H. et al. 2019. Bacterial community dynamics in an oyster hatchery in response to probiotic treatment. *Frontiers in Microbiology* 10: 1060

Sunila, I., Williams, L., Russo, S., and Getchis, T. 2004. Reproduction and pathology of blue mussels, *Mytilus edulis* (L.) in an experimental longline in Long Island Sound, Connecticut. *Journal of Shellfish Research* 23: 731–740

Świacka, K., Maculewicz, J., Smolarz, K., Szaniawska, A., and Caban, M. 2019. Mytilidae as model organisms in the marine ecotoxicology of pharmaceuticals—A review. *Environmental Pollution* 254B: 113082

Szidat, L. 1965. Los parásitos de los mitílidos y los daños por ellos causados II. Los parásitos de *Mytilus edulis* platensis (Orb.) (mejillón del plata). *Comunicaciones del Museo Argentino de Ciencias Naturales 'Bernardino Rivadavia'* 1: 1–16

Szumowski, S.C. and Troemel, E.R. 2015. Microsporidia–host interactions. *Current Opinion in Microbiology* 26: 10–16

Taveekijakarn, P., Nash, G., Somsiri, T., and Putinaowarat, S. 2002. *Marteilia*-like species: First report in Thailand. *The Aquatic Animal Health Research Institute Newsletter* 11: 1–2

Theisen, B.F. 1987. *Mytilicola intestinalis* Steuer and the condition of its host *Mytilus edulis* L. *Ophelia* 27: 77–86

Thieltges, D.W., Krakau, M., Andresen, H., Fottner, S., and Reise, K. 2006. Macroparasite community in molluscs of a tidal basin in the Wadden Sea. *Helgoland Marine Research* 60: 307

Thieltges, D.W. and Reise, K. 2006. Metazoan parasites in intertidal cockles *Cerastoderma edule* from the northern Wadden Sea. *Journal of Sea Research* 56: 284–293

Toguebaye, B.S., Quilichini, Y., Diagne, P.M., and Marchand, B. 2014. Ultrastructure and development of *Nosema podocotyloidis* n. sp. (Microsporidia), a hyperparasite of *Podocotyloides magnatestis* (Trematoda), a parasite of *Parapristipoma octolineatum* (Teleostei). *Parasite* 21: 44

Travers, M.-A., Boettcher Miller, K., Roque, A., and Friedman, C.S. 2015. Bacterial diseases in marine bivalves. *Journal of Invertebrate Pathology* 131: 11–31

Triay-Portella, R., Perez-Miguel, M., González, J.A., and Cuesta, J.A. 2018. On the presence of *Pinnotheres pisum* (Brachyura, Pinnotheridae) in the Canary Islands (NE Atlantic), its southernmost distribution limit. *Crustaceana* 91: 1397–1402

Trotti, G.C, Baccarani, E.M., Giannetto, S., Giuffrida, A., and Paesanti, F. 1998. Prevalence of *Mytilicola intestinalis* (Copepoda: Mytilicolidae) and *Urastoma cyprinae* (Turbellaria: Hypotrichinidae) in marketable mussels *Mytilus galloprovincialis* in Italy. *Diseases of Aquatic Organisms* 32: 145–149

Vasta, G.R., Feng, C., Tasumi, S. et al. 2020. Biochemical characterization of oyster and clam galectins: Selective recognition of carbohydrate ligands on host hemocytes and *Perkinsus* parasites. *Frontiers in Chemistry* 8: 98

Vávra, J. and Lukeš, J. 2013. Microsporidia and 'the art of living together'. *Advances in Parasitology* 82: 253–319

Vea, I.M. and Siddall, M.E. 2011. Scanning electron microscopy and molecular characterization of a new *Haplosporidium* species (Haplosporidia), a parasite of the marine gastropod *Siphonaria pectinata* (Mollusca: Gastropoda: Siphonariidae) in the Gulf of Mexico. *Journal of Parasitology* 97: 1062–1066

Villalba, A., Lopez, M., and Carballal, M. 1993. Parasites and pathological conditions of 3 clam species, *Ruditapes decussatus, Venerupis pullastra*, and *Venerupis rhomboides*, in the Galician Rias. *Proceedings of 4th National Congress on Aquaculture*, pp. 551–556. Centro Investigaciones Marinas, Spain

Villalba, A., Mourelle, S.G., Carballal, M.J., and Lopez, C. 1997. Symbionts and diseases of farmed mussels *Mytilus galloprovincialis* throughout the culture process in the Rias of Galicia (NW Spain). *Diseases of Aquatic Organisms* 31: 127–139

Villalba, A. and Navas, J.I. 1988. Occurrence of *Minchinia tapetis* and a *Perkinsus*-like parasite in cultured clams, *Ruditapes decussatus* and *R. philippinarum*, from south Atlantic coast of Spain. Preliminary results. In *Third International Colloquium on Pathology in Marine Aquaculture*. F.O. Perkins and T.C. Cheng (eds.) pp. 57–58, 2–6 Oct. 1988, Gloucester Point, VA: Virginia Institute of Marine Science, Gloucester Point

Villalba, A., Reece, K.S., Ordáz, M.C., Casas, S.M., and Figueras, A. 2004. Perkinsosis in molluscs: A review. *Aquatic Living Resources* 17: 411–432

Waki, T., Takahashi, M., Eki, T. et al. 2018. Impact of *Perkinsus olseni* infection on a wild population of Manila clam *Ruditapes philippinarum* in Ariake Bay, Japan. *Journal of Invertebrate Pathology* 153: 134–144

Ward, G., Feist, S, Noguera, P. et al. 2018. Detection and characterisation of haplosporidian parasites of the blue mussel *Mytilus edulis* including description of the novel parasite *Minchinia mytili* n. sp. *Diseases of Aquatic Organisms* 133: 57–68

Ward, G.M., Bennett, M., Bateman, K. et al. 2016. A new phylogeny and environmental DNA insight into paramyxids: An increasingly important but enigmatic clade of protistan parasites of marine invertebrates. *International Journal for Parasitology* 46: 605–619

Warén, A. 1981. Eulimid gastropods parasitic on echinoderms in the New Zealand region. *New Zealand Journal of Zoology* 8: 313–324

Wegeberg, A.M. and Jensen, K.T. 1999. Reduced survivorship of *Himasthla* (Trematoda, Digenea) — infected cockles (*Cerastoderma edule*) exposed to oxygen depletion. *Journal of Sea Research* 42: 325–331

Wei, Z.X., Xin, L.S., Zhang, W.W. Bai, C.M., Wang, C.M., and Li, C.H. 2019. Isolation and characterization of *Vibrio harveyi* as a major pathogen associated with mass mortalities of ark clam, *Scapharca broughtonii*, in summer. *Aquaculture* 511: 734248

Wijsman, J.W.M., Troost, K., Fang, J., and Roncarti, A. 2019. Global production of marine bivalves. Trends and challenges. In *Goods and Services of Marine Bivalves*. A.C. Smaal Et Al. (eds.) pp. 7–26, Cham, Switzerland: Springer Open

Winters, A.D. and Faisal, M. 2014. Molecular and ultrastructural characterization of *Dictyocoela diporeiae* n. sp. (Microsporidia), a parasite of *Diporeia* spp. (Amphipoda, Gammaridea). *Parasite* 21: 9

Xiao, R., Wei, Y., An, D. et al. 2019. A review on the research status and development trend of equipment in water treatment processes of recirculating aquaculture systems. *Reviews in Aquaculture* 11: 863–895

Xu, Y. and Weiss, L.M. 2005. The microsporidian polar tube: A highly specialised invasion organelle. *International Journal of Parasitology* 35: 941–953

Yeh, H., Skubel, S.A., Patel, H., Cai She, D., Bushek, D., and Chikindas, M.L. 2020. From farm to fingers: An exploration of probiotics for oysters, from production to human consumption. *Probiotics and Antimicrobial Proteins* 12: 351–364

Yonemitsu, M.A., Giersch, R.M., Polo-Prieto, M. et al. 2019. A single clonal lineage of transmissible cancer identified in two marine mussel species in South America and Europe. *Elife* 8: e47788

Zander, C.D. and Reimer, L.W. 2002. Parasitism at the ecosystem level in the Baltic Sea. *Parasitology* 124: 119–135

Zannella, C., Mosca, F., Mariani, F. et al. 2017. Microbial diseases of bivalve mollusks: Infections, immunology and antimicrobial defense. *Marine Drugs* 15: E182

Diseases of Arthropods

CHAPTER 9

Diseases of chelicerates

Christopher J. Coates

9.1 Introduction

The Subphylum Chelicerata is one of the oldest and second most speciose groups among the Phylum Arthropoda (Lozano-Fernandez et al. 2019)—with members often referred to as 'living fossils' (intact remains are ~ 450 to 510 million years old). Extant chelicerates inhabit both terrestrial and aquatic ecosystems, with their evolutionary relatedness proving difficult to resolve due to rapid radiation and extinction events among/within groups and two independent putative water-to-land transitions. Common chelicerates like spiders, scorpions, ticks, and mites tend to be mistaken for insects, just as horseshoe crabs are misidentified as crustaceans. The Pantopoda (sea spiders) and Xiphosura (horseshoe crabs) are the aquatic chelicerate orders, which consist of ~ 1,300 and 4 species, respectively. The remaining members—in excess of 110,000 known species—are terrestrial air-breathers grouped broadly as the arachnids (Sharma 2018). The chelicerate body plan is usually bi-segmented (tagmata) into the anterior prosoma and the posterior opisthosoma with jointed limbs, and as is the case with all arthropods, covered entirely in a chitinous cuticle. The defining morphological feature of this subphylum is the chelicerae ('jaws' and pedipalps), which are appendages that appear immediately before the mouth in the forms of articulated fangs or feeding pincers (Dunlop and Lamsdell 2017).

The biological and commercial importance of this enigmatic group cannot be overstated; chelicerates represent vectors of devastating human diseases: (ticks; Paules et al. 2018), agricultural pests (mites),

wielders of poisons and venoms (spiders, scorpions), and critical for the detection of endotoxin contaminants in pharmaceuticals and on medical devices (horseshoe crabs; Krisfalusi-Gannon et al. 2018). Because of their many uses, the vast majority of research has been associated with human health impacts, leading to a knowledge deficit with respect to disease of chelicerates *in situ*. Against this background, recent evidence positions spiders as rich sources of endosymbionts and pathobionts (e.g. Goodacre et al. 2006; Rosenwald et al. 2020), and more broadly, efforts are focussed on developing integrated pest management strategies using entomopathogenic, acaropathogenic, and araneogenous fungi for controlling terrestrial chelicerates such as mites and ticks (e.g. Fernandes and Bittencourt 2008; Carr and Roe 2016). Outside the scope of this text is the vectorial role of acarids in spreading animal and plant diseases, the pathobiology of arachnid bites and stings (puncture wounds), and bacteria-induced food poisoning associated with arachnophagy (human consumption of spiders and scorpions; e.g. Grabowski and Klein 2017).

In this chapter, I review the diseases and pathobiology of chelicerates, wherein much information has been accrued from captive settings, such as scorpions from zoological collections and horseshoe crabs in aquaria.

9.2 Bacterial diseases of terrestrial chelicerates

It should come as no surprise that microbial intrusion of the solid or liquid tissues of arachnids is

Christopher J. Coates, *Diseases of chelicerates*. In: *Invertebrate Pathology*. Edited by Andrew F. Rowley, Christopher J. Coates and Miranda M.A. Whitten, Oxford University Press. © Oxford University Press (2022). DOI: 10.1093/oso/9780198853756.003.0009 detected as non-self by immune cells (haemocytes), resulting in inflammation and the production of antimicrobial factors (Fukuzawa et al. 2008; Tonk et al. 2015; also see Chapter 1). Bacterial septicaemia and tissue damage (e.g. necrosis) triggers haemocyte infiltration and swelling of the affected area (Figure 9.1a–d), which can lead to further deterioration (Figure 9.1). Haemocytes will target the bacteria and form cellular aggregates called capsules/nodules to prevent further spread (Figure 9.1). Newton and Smolowitz (2018) provided histopathological evidence of such responses in the goliath bird-eating spider and green bottle blue tarantula; focal damage in the digestive gland, clot formation, haemocyte aggregation, and melanisation in a damaged limb (scab-like

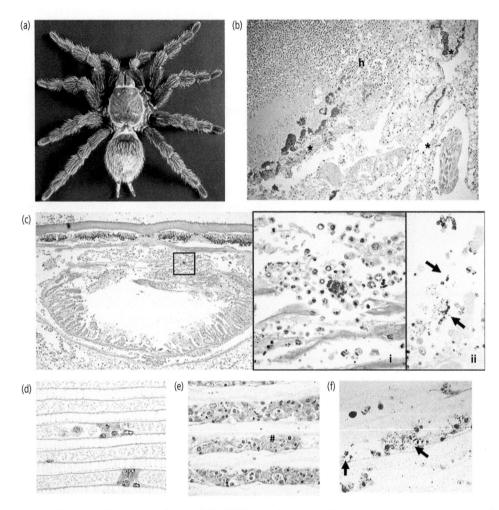

Figure 9.1 Bacterial septicaemia-induced mortality of an arachnid in the absence of external gross lesions. A green bottle blue tarantula died acutely of bacterial sepsis (a). Histopathology identified extensive tissue necrosis (*) and haemocyte (h) infiltration of the digestive gland (b), with large colonies of intralesional bacteria. Haemocyte reactivity was also apparent in the pericardial sinus (c) and heart (i) with internalised bacteria evident on Gram's staining consistent with bacterial sepsis (ii). Normal book lung lamellae (d) of a Goliath bird-eating spider, consisting of thin chitinous layers separated by haemolymph channels containing haemocytes and separated by chitinous pillars. In photomicrograph (e) haemocyte infiltration and swelling of book lung lamellae. The haemolymph spaces are filled with and thrombosed by haemocyte aggregates with intracellular bacteria visualised with Gram's staining (f), black arrows. The book lungs may be a focal site of bacterial dissemination in sepsis. Source: Dr Alisa (Harley) Newton (Disney's Animals, Science and Environment, Florida USA) kindly provided the original images, which were taken during her time at the San Diego Zoo.

appearance), and expansion of a septic pericardium and ventricular lumen by circulating phagocytic haemocytes (small, intracellular bacilli were visible; Figure 9.1c, 9.1f). It is likely that the book lungs act as the main site for bacterial dissemination in sepsis (Figure 9.1d and 9.1e). Williams (1987) provided a long list of bacterial genera associated with scorpions, including *Bacillus*, *Lactobacillus*, *Pseudomonas*, *Staphylococcus*, *Acinetobacter*, *Moraxella*, *Escherichia*, *Proteus*, and *Rickettsia*. Like so many invertebrates, arachnids can be asymptomatic carriers and reservoirs of disease (notably bacteria and viruses; Figure 9.1), yet it is now apparent that terrestrial chelicerates are rich sources of intra- and inter-cellular endosymbionts. Although not covered in this chapter, the reader should be aware that there is much literature on the association of microbes with spider bites (dermonecrotic lesions), and mechanical transfer of bacteria through envenomation (e.g. *Clostridium perfringens*, Monteiro et al. 2002; Gaver-Wainwright et al. 2011; Dunbar et al. 2020).

9.2.1 Endosymbionts

Advances in high-throughput sequencing have facilitated in-depth assessments of complex microbial communities associated with invertebrates, although it remains unclear to what extent they are beneficial (e.g. nutrient-providing mutualists), benign, or causing disease/trauma among/within hosts (*Rickettsiella* being an exception—see the next section). In 2012, Zug and Hammerstein calculated that ~ 40% of known terrestrial arthropods are infected with *Wolbachia*—one of the most common reproductive parasites of metazoans. While prevalence can be low among individual populations or species, incidence estimates rise to > 70% when the other most common endosymbiotic bacterial genera, *Cardinium*, *Rickettsia* and *Spiroplasma* are taken into account (Duron et al. 2008a; Zug and Hammerstein 2012; Weinert et al. 2015). In most cases, these intracellular endosymbionts are transferred vertically from mother to offspring, and many can modulate the reproductive behaviour and outputs of their hosts in severe ways:

• forced feminisation, asexuality (parthenogenesis), and killing males results in sex ratio distortion in favour of females,

• cytoplasmic incompatibility—embryonic mortality can occur when a *Wolbachia*-positive male mates with an uninfected female (Werren et al. 2008),

• reproductive isolation by limiting dispersal, and consequently gene flow (e.g. *Rickettsia*, Goodacre et al. 2009).

Duron et al. (2008b) were the first to provide evidence of maternal transfer/inheritance of *Cardinium hertigii* and *Wolbachia* among spiders. Horizontal transfer across species does occur also. *Cardinium* bacteria are repeatedly recorded in spiders, and it is estimated that at least 20% of all species are infected—alongside harvestmen, scorpions and acarids (Duron et al. 2008a and 2008b; Martin and Goodacre 2009; Stefanini and Duron 2012; Bryson Jr 2014). Like *Wolbachia*, *Cardinium* also causes cytoplasmic incompatibility in hosts, but the mechanism appears to have evolved separately (Penz et al. 2012). *Cardinium* transmission occurs via the egg cytoplasm of infected matrilines. Arachnid hosts can harbour multiple endosymbionts concurrently (Curry et al. 2015), with some evidence suggesting that spider mites (*Tetranychus piercei*) co-infected with *Wolbachia* and *Cardinium* are more likely to produce non-viable offspring (Zhu et al. 2012).

A recent study by White et al. (2020) exemplifies the richness and diversity of endosymbiotic bacteria in agricultural spiders, wherein the authors examined 267 infected individuals covering 14 species to reveal 27 unique bacterial operational taxonomic units across seven genera (including those mentioned previously and *Rickettsiella* for example). In fact, the dwarf spider *Idionella rugosa* had up to five different symbionts per individual. Vanthournout and Hendrickx (2015) interrogated the microbiome of another Dwarf spider, *Oedothorax gibbosus*, to reveal communities replete with endosymbionts (e.g. novel *Rhabdochlamydia* and *Acinetobacter*).

Martin and Goodacre (2009) used PCR to confirm the presence of *Cardinium* in a parthenogenetic scorpion, *Liocheles australasiae*. Bryson Jr. (2014) screened 41 species of vaejovid scorpions from North America for the presence of *Cardinium*, *Rickettsia*, *Spiroplasma*, and *Wolbachia* using a similar PCR approach but found no measurable evidence of infection in any of the 61 specimens using leg, gonad, or whole instar tissue extracts. Bryson

Jr. (2014) cautioned that such targeted approaches taken in previous works are prone to error (false positives), as he retrieved amplicons of the expected size for *Rickettsia* (n = 8) but sequencing revealed a diversity of non-specific target amplification. This point extends beyond endosymbiont detection in arachnids—PCR should form part of a multi-method screen for disease in invertebrates (see Chapter 3). Nevertheless, the evidence for endosymbiont presence and diversity is incontrovertible, which could have previously unknown repercussions for the ecology and evolution of these beasties (Stefanini and Duron 2012). To date, three bacterial genera, *Cardinium*, *Wolbachia*, and *Rickettsiella*, have been validated experimentally as reproductive parasites of arachnids (Table 9.1).

In 1982, Hess and Hoy used electron microscopy to record the presence of two pervasive intracellular bacteria, types A and B, in the ovaries/eggs, midgut, and Malpighian tubule of the phytoseiid mite *Metaseiulus occidentalis*. Later, Weeks and Breeuwer (2003) confirmed one of these to be *Cardinium*. Just like spiders and scorpions, many ticks and mites are common hosts to endosymbionts (e.g. *Schineria* and *Spiroplasma*), counting several genera that interfere with reproductive performance through cytoplasmic incompatibility and sex-ratio distortion (Reeves et al. 2006; Zhang et al. 2016; Table 9.1). Acarids are prolific arbovectors transferring many disease-causing agents to humans, livestock, and agricultural stocks, yet there is little evidence of mite or tick-specific bacterial pathogens. The obligate intracellular bacterium *Rhabdochlamydia helvetica* is a purported pathogen of ticks (e.g. *Ixodes ricinus*) due to the abundance of the bacterium in the invertebrate, the small genome (a sign of a specialised parasite), and limited prevalence (although clear evidence of host damage and reactivity are lacking; Pillonel et al. 2019).

Members of the genus *Rickettsiella* (Family Coxiellaceae; γ-proteobacteria) are intracellular pathobionts and symbionts observed in chelicerates, such as, spiders (Morel 1977; Rosenwald et al. 2020), scorpions (Morel 1976; Han et al. 2020), ticks (Anstead and Chilton 2014) and mites (De Luna et al. 2009). In fact, *Rickettsiella* spp. have been detected across a broad geographic distribution in many arthropod hosts including crustaceans (Romero et al. 2000; Cordaux et al. 2007) and insects (Adams et al. 1997; Leclerque and Kleespies 2008; Kleespies et al. 2011)—but should not be confused with the genus *Rickettsia* (α-proteobacteria). Symbiotic *Rickettsiella* spp. are restricted to specific tissue types (e.g. ovaries), and do not cause pathology or cellular aberrations in hosts. Pathogenic strains target all vital organs and are immunogenic, most likely due to the abundance of lipopolysaccharides (LPS) and peptidoglycans in the cell membranes. *Rickettsiella* accumulate as micro-colonies within host intracellular vacuoles, particularly haemocytes—using them as vehicles to disseminate throughout the haemocoel.

In an early study, Morel (1976) observed rickettsiella-like bacteria in wild-caught common yellow scorpions *Buthus occitanus* (5% of animals collected were diseased). Examination of tissues provided evidence of replication within the hepatopancreatic cells with bacteria eventually spilling into the gut and haemocoel where they were ingested and destroyed by haemocytes. Experimental infection of seemingly healthy scorpions with tissue homogenates killed 90 to 100% between two and eight months post-inoculation. Within one month, micro-colonies up to 7.5 μm in size were observed, progressing up to 20 μm after two further weeks and in some cases the bacteria burst out of the hepatopancreas into the caecal lumen. Several weeks before dying, the infected scorpions stopped feeding and some developed secondary infections of the gut. The causative agent was named *Porochlamydia buthi* (Morel 1976), but was later reclassified to *Rikettsiella chironami* based on a genetic assessment (Williams 1987). Another rickettsiella-like infection was determined in spiders from the suborder Mesothelae (Haupt, 2002). Depending on the spider genera, bacterial prevalence ranged from ~ 1.5–4.5% (n = 549). Bacterial cells and colonies—similar to those described previously—were found exclusively inside hepatopancreatic cells. As the infection progressed, all intermediate cells filled with these bacteria until the organelles (including the nucleus) and subcellular structures were no longer visible. Even at such an advanced stage, the infection did not appear to spread to the midgut. The use of nutrients by the proliferating bacteria prevents the spider

Table 9.1 Bacterial (endo)symbionts and pathobionts of terrestrial chelicerates (not exhaustive)

Bacterial agent	Host	Location/Source	Pathology	References
Actinetobacter *Cardinium* *Rhabdochlamydia* *Rickettsia* *Wolbachia*	Dwarf spider, *Oedothorax gibbosus*	Belgium	Detected and profiled using illumina reads from 16S rRNA (V3–V4 regions)	Varthournout and Hendrickx (2015)
Cardinium hertigii *Wolbachia* sp.	Spiders: representatives of the Araneidae, Linyphiidae, Lycosidae, Pholcidae, Salticidae and Tetragnathidae	Across Northern Europe	**Cytoplasmic incompatibility.** Transfer *via* maternal inheritance confirmed. Approximately 22% and 37% of specimens were infected with *Cardinium* and *Wolbachia*, respectively	Duron et al. (2008b); Stefanini and Duron (2012)
Cardinium sp. *Wolbachia* sp.	Mites: -*Bryobia sarothamni* - *Eotetranychus suginamensis* - *Tetranychus* species	-France, -Japan -China, Japan	**Cytoplasmic incompatibility**	Gotoh et al. (2007); Ros and Breeuwer (2009); Natamura et al. (2009); Zhang et al. (2016)
Rickettsiella chironami	Scorpion, *Buthus occitanus*	Montpellier, France	Bacteria reside and replicate within the hepatopancreatic tissues. Later stages of infection include the spread of bacteria to muscle, gut and haemolymphatic tissues. Septicaemia tends to be fatal.	Morel (1976)
Rickettsiella scorpionisepticum	Scorpion, *Pandinus imperator*	Zoo collection, Denver Colorado (USA); private collection in Saskatoon, Saskatchewan (Canada)	Haemocyte infiltration of many tissue types, nodulation, and phagocytic cells replete with bacteria. Some tissue necrosis. See Figure 9.2.	Han et al. (2020)
Rickettsiella sp. *Wolbachia* and *Rickettsia* spp.	Spider, *Mermessus fradeorum*	Original collection took place at the University of Kentucky's Spindletop Research Farm (USA)	Identified using NGS, V4 region of bacterial 16S. **Cytoplasmic incompatibility** (CI) observed in infected hosts. This is the first record for a γ-proteobacterium inducing CI	Rosenwald et al. (2020); Curry et al. (2015)

continued

Table 9.1 *Continued*

Bacterial agent	Host	Location/Source	Pathology	References
Rickettsiella spp. (F) *Cardinium* spp. (F/UK) *Schineria* spp. (F) *Spiroplasma* spp. (F/UK)	Mite, *Dermanyssus gallinae*	Poultry farms across France (16 surveyed) and UK (1 surveyed)	Presence confirmed *via* PCR	De Luna et al. (2009)
Rickettsiella species	Spider, *Pisaura mirabilis*	-	Presence confirmed in digestive and interstitial cells	Morel (1977)
Rickettsiella species	Trapdoor spiders, *Heptathela*, *Ryuthela* and *Liphistius* species	Collected in Kyushu, Okinawa and Malaysia	Prevalence varied from 1.5–4.5%. Bacterial morphotypes—mostly rods—visible in the intermediate cells of the hepatopancreas. Late stage infection involves the dissolution of the intermediate cells as the spider dies, and the spilling of bacteria into the surrounding soil	Haupt (2002)
Rickettsiella species [*R. grylli* from crickets] [*R. ixodidis*] [*R. popilliae*]	Ticks: *Ixodes angustus* [88%] *Ixodes sculptus* [43%] *Ixodes kingi* [4%]	Western Canada	Presence confirmed *via* PCR	Anstead and Chilton (2014)

from moulting. As the tissue putrefies, the spiders become pale and the opisthosoma inflates—gentle prodding of the cuticle leads to rupture and seepage of a milky-while liquid (Haupt 2002).

Two days prior to *in exitus*, seemingly infected emperor scorpions (*Pandinus imperator*) did not respond to stimuli (including tail lifting), were lethargic and weak, but had no gross external signs

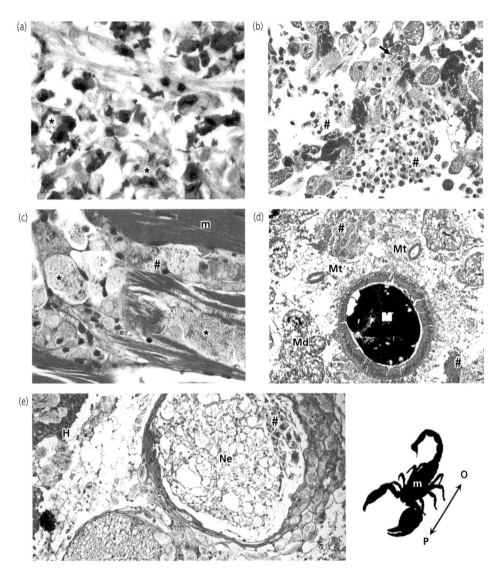

Figure 9.2 Histopathology of *Rickettsiella scorpionisepticum* infection of the emperor scorpion *Pandinus imperator*. Fite's acid fast staining of haemocytes from the mesosoma of a scorpion (a). Staining confirmed the presence of small intracellular bacteria (denoted by *). Aggregation and swelling of haemocytes with intracytoplasmic bacteria (denoted by #) in the intrahaemocoelic spaces of an infected scorpion (b). The digestive gland is infiltrated by haemocytes, and some large, granular eosinophilic cells are present (arrow). Bundles of striated muscle (m) from the mesosoma, depicting a moyocyte infiltrated by haemocytes containing bacteria (#) and extrahaemocytic bacteria (*) (c). A section through the prosoma reveals the midgut (m), midgut diverticula (Md), Malpighian tubules (Mt) and several instances of haemocyte aggregates (#) or coalescing nodules (d). Haemocyte infiltration of a degenerating nerve (Ne) within the prosoma of a compromised scorpion (e). Apparent haematopoietic tissue (H) is also visible. Inset—scorpion silhouette outlining the mesosomal tergites (m), prosoma (p) and opisthosoma (o). Source: Dr Sushan Han (Colorado State University, USA) kindly provided the original images for a–e.

of disease. Necropsy of several specimens revealed haemocyte infiltration of many tissues including the midgut (and diverticula), heart, haemolymph vasculatures, nerves, and muscle (Han et al. 2020; Figure 9.2a–9.2e). Phagocytic haemocytes were replete with bacterial morphotypes (~ 0.6–1.3 µm), elementary and intermediate bodies, degenerated microbial fragments, and cytoplasmic vacuoles. Using electron micrographs, PCR-mediated amplification of the 16S rRNA region, and a multi-gene (*gidA, rspA, sucB*) phylogeny, the authors diagnosed *Rickettsiella scorpionisepticum* n. sp. as the causative agent (Han et al. 2020).

9.3 Fungal diseases of terrestrial chelicerates

Some entomopathogenic and plant pathogenic fungi also target arachnids (e.g. *Beauveria, Ophiocordyceps* species; see Chapter 12), with those specific to spiders referred to as araneopathogenic (Evans and Samson 1987; Evans 2013). *Aspergillus* and *Fusarium* isolates have been collected from the scorpion *Androctonus australis* (Nentwig 1985). Shrestha et al. (2019) performed an exhaustive review of arachnid-targeting fungi and categorised ~ 90 hypocrealean species (phylum Ascomycota) that infect at least 22 spider and harvestmen families. Despite their broad geographic distribution across temperate/tropical regions (e.g. Canada, Strongman 1991; Panama, Nentwig 1985; Evans 2013; Hughes et al. 2016) and perceived importance to spider ecology, spider-fungal antibiosis is a surprisingly overlooked topic. Of the many fungal genera identified, *Akanthomyces, Hevansia* (Figure 9.3f), *Cordyceps, Gibellula* (Figure 9.3a), and *Torrubiella* are most closely associated with mycosis-induced mortality of spiders—with the latter three also targeting scorpions (Gysin and LeCoroiler, 1969). For example, a 2014 study from Brazil identified at least nine *Gibellula* species including *G. dimorpha* found in 80 parasitised spiders collected, with the Pholcidae (~ 27%, cellar spiders) and Anyphaenidae (~ 25%, ghost spiders) families most frequently affected (Costa, 2014).

In a manner similar to insects, fungal spores adhere to and penetrate the intact or damaged exoskeleton in order to reach the haemocoel and

the nutrients stored therein (Figures. 9.3 and 9.4). Fungi colonise the outer surfaces of the spider in a mycelium with visible fruiting bodies of varying sizes bursting out of the mycosed cadavers (Figure 9.3); for example, *Gibellula* sp. infection of ghost spiders (e.g. *Macrophyces pacoti*) can lead to white-yellow body discolouration (Brescovit et al. 2019) and the formation of lilac-coloured synnemata (Hughes et al. 2016). The exoskeleton of spiders and insects are not the same—spiders have an unhardened opisthosoma with a mesocuticle layer in the absence of an exocuticle (Foelix 2011; Figure 9.5)—suggesting those microbes circumventing the different cuticle barriers have evolved specifically to do so. Replication of infectious fungi within the haemolymph takes the form of budding-yeast or creeping hyphae—releasing toxic chemicals and proteins to oppose the cellular and humoral defences of the spider (a strategy similar to entomopathogenic Ascomycota; see Chapter 12 and Butt et al. 2016). Systemic mycosis eventually leads to the replacement of internal tissues with hyphal masses, e.g. the effacement of the book lungs, and large clusters of melanised cellular debris and lipofuscin-like materials (Figure 9.4).

Several historical accounts of fungal disease within spider populations reveal high prevalence in some areas: 80 spiders infected with *Gibellula pulchra* from ten trees across forest sites in Ghana, 120 spiders infected with *G. pulchra* from a year-long survey in the same region (Samson and Evans 1973; Evans 1974); ~ 70 spiders infected with *Cordyceps thaxteri* attached to foliage in the mountains of North Carolina (samples from July—August 1887; Mains 1939). The hypocrealean fungus *Purpreocillium (Nomuraea) atypicola* is considered an anamorph of *Cordyceps* and a major regulator of trapdoor spider populations (Coyle et al. 1990). During fieldwork in Argentina, 50 *Actinopus* sp. burrows were studied, 17 of which contained the fungus *P. atypicola* usually in the form of a white mycelial mass covering the host (examples of other species can be viewed in Figure 9.3). Fungal projections (fruiting synnemata) were growing out from each of the dead spiders and lifting the trapdoor for aerial conidia dispersal (Coyle et al. 1990). Approximately 18% of trapdoor spiders (*Latouchia* sp.) collected in Malaysia were infected with the

Figure 9.3 Parasitic fungi and mycosed cadavers of spiders. In images (a–e), the fungi on display belong to the genus *Gibellula*, and in image (f) a *Hevansia* species is depicted. (b) This specimen is a representative of ~ 100 spiders collected from Bala Lake (Llyn Tegid) in Wales on a single winter's day. (a, c–f) These specimens were gathered from a small, university forest reserve in Brazil. The tree-like and flask-like structures are asexual synnemata and sexual perithecia, respectively. Source: Harry C. Evans (Centre for Agriculture and Bioscience International, Surrey UK) kindly provided the original images.

same fungus (Haupt 2002), and it has been recorded on mycosed spiders in Brazil, Japan, Panama, Sri Lanka, and Ghana (Nentwig 1985; Greenstone et al. 1987; Coyle et al. (1990) and references therein). As these spiders live within the moist soil, death by mycosis is very common as are opportunistic and secondary infections (saprotrophic). Interestingly, the production of synnemata by *Purpreocillium* sp. is restricted to infections of burrow-dwelling spiders, whereas hunting or web-building spiders are enveloped in mononematous conidiophores (Coyle et al. 1990). Hughes et al. (2016) suggest these

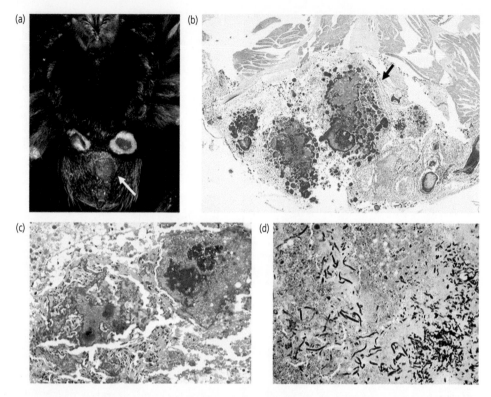

Figure 9.4 Opportunistic fungal infections associated with ulcerative lesions of the arachnid carapace. Large ulcers (white arrow) affect the book lung ostia and ventral opisthosoma of a Mexican red-legged tarantula (a). Tissue histopathology reveals severe haemocyte infiltration effacing (black arrow) the book lungs (b), several grossly melanised nodules/capsules, and perhaps lipofuscin (c), higher magnification of b). Staining of tissue sections with Grocott's methenamine silver reveals abundant intralesional fungal hyphae (d). Source: Dr Alisa (Harley) Newton (Disney's Animals, Science and Environment, Florida USA) kindly provided the original images, which were taken during her time at the San Diego Zoo.

outbreaks are akin to 'graveyard events' described for fungi-infected ants (*Ophiocordyceps unilateralis*) and appear to manipulate ants to leave their nest and die under leaves, and in some cases, the graveyards contain several thousand cadavers (Pontoppidan et al. 2009). Mycosed spiders are found predominantly on the underside of elevated foliage—again, similar to many insects—and may represent a key stage of disease progression where the fungus manipulates the behaviour of the spider to increase the likelihood of conidial transfer to new hosts. Trapdoor spiders are an exception to this, where parasitised hosts die *in situ* and use phototrophic stromata to push the sporulating tissues (perithecia) above ground.

Although rare, there have been reports of fungi targeting harvestmen, notably the suborders Laniatores and Eupnoi (Shrestha et al. 2019). In 1901, Möller described an infection of Neotropical harvestmen (*Gonyleptes* sp.) with *Torrubiella gonylepticida*. In Britain, Leatherdale (1970) identified a member of the harvestmen family Phalangiidae with an *Ophiocordyceps verrucosa* infection. Much later, Barbosa et al. (2016) found an *Acanthogonyleptes* sp. parasitised with *T. aranicida* in Brazil.

The acaropathogenic fungus *Hirsutella thompsonii* (Order Hypocreales)—also a pathogen of spiders—is the primary cause of epizootics among eriophyoid mites. Records of mass mortality events exist as early as 1924 when numbers of the citrus rust mite *Phyllocoptruta oleivora* collapsed across Florida, USA (Speare and Yothers 1924; Fisher 1950; Agrawal et al. 2015). In addition to *P. oleivora*, several isolates of *H. thompsonii* have been recovered from blueberry bud mites *Acalitus vaccinia*, and coconut fruit mites *Aceria guerreronis*, and *Eriophyes* spp. (Samson

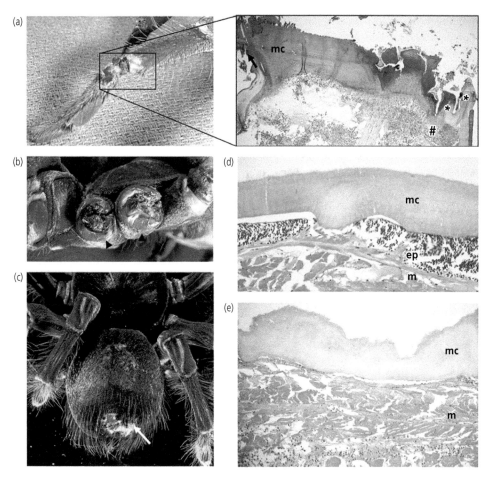

Figure 9.5 Trauma to the carapace of a goliath bird-eating spider. Routine observations of arachnids reveal limb fractures (a), leg amputations (b), and erosions/ulcers on the opisthosoma (c). Inset (a), a histological section (black box) has been prepared through the highlighted lesion. In response to injury, haemocytes appear to aggregate (#) to assist haemolymph clotting, in addition to enhanced melanin deposition (*). The mesocuticle (mc) layer is deformed and brittle. Photomicrographs (d) and (e) represent normal and compromised areas of the opisthosoma, respectively. In contrast to the healthy tissue barrier depicted in (d) with a continuous laminated mesocuticle (mc), and distinct epidermal (ep) and subjacent muscle (m) layers, the compromised opisthosoma in (e) is irregular, the epidermis is no longer visible, and setae are lost. Source: Dr Alisa (Harley) Newton (Disney's Animals, Science and Environment, Florida USA) kindly provided the original images, which were taken during her time at the San Diego Zoo.

et al. 1980; Weibelzahl and Liburd 2009). Members of the *Hirsutella* genus are highly infectious and widespread (Fernandes and Bittencourt 2008), and therefore exploited as biopesticides in attempts to control agricultural pests—*H. tompsonii* produces several toxic factors including the protein hirsutellin A that is a potent acaricide and insecticide (Herrero-Galán et al. 2013; Wang et al. 2018). Members of the fungal order Entomophthorales (notably *Conidiobolus*, *Entomophora*, and *Neozygites* species)

are the most frequently encountered pathogens of mites—conidia adhere to the cuticle, develop infection pegs, cross the tissue barriers, and ramify in the haemocoel until all internal tissues are replaced (Poinar Jr and Poinar 1998; Balazy et al. 2009). Fungal infection of mites leads to lethargy/quiescence and usually discolouration (darker for hypocrealean and lighter for entomophthoralean fungi). Once considered unicellular protozoans, the microsporidia are now classified as fungi with several genera pathogenic toward mites, e.g. *Nosema*, (Bjørnson et al. 1996). Microsporidian spores are ingested, and from the midgut the parasite emits a polar filament that facilitates the intracellularisation of protoplasm (e.g. fat body, haemocytes). Spores develop inside host cells and accumulate in the haemocoel prior to their release upon death of the acarid.

9.4 Parasites and pests of spiders and scorpions

The animal/pet exotic trade has assisted in spreading parasitic mites and pests beyond traditional spatial ranges (Pizzi 2009; Masan et al (2012). Across the literature a plethora of macro-parasites (ecto and endo) have been reported for arachnids (Table 9.2). Parasitic worms, especially nematodes are present in captive and wild-caught spiders and scorpions (Table 9.2). Panagrolaimidae nematodes accumulate in the oral structures of spiders causing damage and leading to anorexia, whereas mermithid nematodes despoil the haemolymph of nutrients until they burst out of the moribund arachnid. Of primary concern are nematode infections and mite infestations of tarantulas, which are rife in Africa, America, and Asia (Pizzi et al. 2003). Mite deutonymphs (Astigmata, Heterostigmata) are prevalent among captive spiders and usually phoretic, whereas Prostigmata mite larvae are principally parasitic (Pizzi 2009). As mites accumulate on the spider, especially the soft intersegmental membranes (pleurites), the book lungs become blocked and dysfunctional. Some mites are saprophytic, feeding on the decaying/necrotic flesh of nematode-infected tarantulas (Baker, 1992; Pizzi, 2009 and 2012). In fact, mites and humpbacked flies, *Megaselia scalaris* (Family Phoridae) are supposed vectors of nematodes to spiders (Pizzi et al. 2003; Masan et al. 2012).

9.5 Disease of aquatic chelicerates

9.5.1 Pathobiology of horseshoe crabs

9.5.1.1 In the wild

The Atlantic (or American) horseshoe crab *Limulus polyphemus* is a contender for the most famous marine invertebrate—largely because of the value of its copper-based haemolymph (~ \$15,000 per litre). Frederick Bang and Jack Levin's discovery of the clotting apparatus in *L. polyphemus* from the mid-twentieth century (Bang 1956; Levin and Bang 1964) has led to the eradication of rabbits for pyrogen (endotoxin) screening. The single immune cell type—the amebocytes—are isolated and processed for their clotting components to produce the pictogram-level sensitive test for Gram-negative bacterial endotoxins known as the *Limulus* (or *Tachypleus*) amebocyte lysate (LAL/TAL). The importance of *Limulus* extends beyond biomedicine; horseshoe crab adults and their eggs play critical roles in maintaining migratory bird populations along the east coast of North America, bioturbation of sediment, bait for eels and whelks, fertilizer, and previously as livestock feed and a model for the study of vision and embryogenesis (reviewed by Walls et al. 2002). In addition to *L. polyphemus*, there are three other extant species of horseshoe crab: *Tachypleus gigas* (Indo-specific), *T. tridentatus* (tri-spine), *Carcinoscorpius rotundicauda* (mangrove), which all inhabit coastal waters of south-eastern Asia. There are few historical case reports of disease in horseshoe crabs across the literature. Beyond their industrial value, concern for the health of these animals has risen due to declining populations in the wild, and because of their importance in display and touch-tank aquaria (education).

As with most solid surfaces immersed in the marine environment, the carapace of the long-lived horseshoe crabs are subject to colonisation by a plethora of macro- and micro-organisms: diatoms, coelenterates, cyanobacteria, flatworms, annelids, bryozoans, isopods, amphipods, gastropods, green algae, mussels, oysters, barnacles, tunicates, pelecypods, and polychaetes (Leibovitz 1986; Key Jr et al. 1996 and 2000; Patil and Anil 2000; Grant 2001; Botton 2009; Shuster 2009; Tan et al. 2011; Table 9.3). While the majority of these residents have little to no effect on the day-to-day life of the horseshoe crab, certain opportunistic fungi, bacteria and algae secrete bioactives that can degrade the shell (Nolan and Smith 2009; Braverman et al. 2012; Shinn et al. 2015). The accumulation

Table 9.2 Macro-parasites and pests of arachnids

Agent	Host	Pathologic features	References (not exhaustive)
Insects: -Flies (dipterans) Acroceridae, Phoridae -Wasps (hymenopterans) Pompilidae, Ichneumonidae, and Sphecidae	Diverse range of spiders, notably tarantulas, in the wild	Deposited Acroceridae larvae make their way to the book lungs and penetrate the lamellae. The 4th instar feeds intensively and cause fatality through opisthosomal bursting. Pompilid wasps sting a spider to paralyse it, and then deposit an egg. The developing larva eats the spider from the inside out. Ichneumonid wasps do not paralyse but place eggs on legs or abdomen, where the larva will cross into the haemocoel and feed on the haemolymph.	Eason et al. (1967); Pizzi (2009, 2012); Foelix (2011)
Parasitic mites (examples): -*Clinotrombium metae* -*Eutrombuium lockleii* -*Ljunghia bristowi* -*Ljunghia pulleinei*	Spiders (examples): -*Ceraticelus emertoni* -*Oxyopes salticus* -*Liphistius malayanus* -*Selenocosmia stirlingi*	Ectoparasites (mostly Prostigmata) that attach to the soft region (pleuron) of the prosoma—lateral moult structures (proximal leg joints). 89% of hosts were parasitised by a single mite larva, with one host having nine larvae. Infestation can occlude book lungs.	Finnegan (1933); Welbourn and Young (1988); Pizzi (2009); Masan et al (2012)
Parasitic nematodes: -Rhabditida -Thelastomatidae	Scorpions, e.g. *Centruroides exilicauda*; *Euscorpius italicus*	Nematodes and embryondated ova seen in few scorpions. Laboratory infection of harmful *C. exilicauda* with rhabditid nematodes to limited success. Some evidence of a *Strongyloides*-like worm in the faeces of scorpions.	Poinar Jr and Stockwell (1988); Barus and Koubkova (2002); Gouge and Snyder (2005); Frye (2012)
Parasitic nematodes: -Allantonematidae	Mites: *Cosmolaelaps cuneifer* *Euryparasitus terribilis*	Reports are rare. Nematodes penetrate the host cuticle, and reside in the body cavity where the female produces eggs and juveniles.	Warren (1941); Poinar Jr and Poinar (1998)
Parasitic worms: -Horsehair (Gordian) worms -Mermithid and Panagrolaimid nematodes	At least 51 species of spider and harvestmen	Swollen abdomen (e.g. greatly enlarged, distorted opisthosoma); altered epigynum (female genital structure), malformed palpi and legs, and poorly developed male secondary sexual characteristics. Infection leads to the reduction/absence of the digestive gland, muscle atrophy, and reproductive system decay. Host mortality when parasites emerge Panagrolaimid nematodes accumulate in the mouths of captive spider and may lead to secondary infections (*Bacillus* and *Proteus* spp.). Anorexia and posture changes are sings of infection, with a thick, white oral discharge often a sign of fatality.	Poinar Jr (1985, 1987); Cokendolpher (1993); Poinar Jr et al. (2000); Pizzi et al. (2003); Pizzi (2009)

of damage over time allows noxious and chitinoclastic agents to gain a foothold and further weaken these abrasions, gradually digesting through the shell and exposing the underlying tissues to the microbiologically harsh environment—such circumstances can prove fatal (Harrington et al. 2008; Braverman et al. 2012; LaDouceur et al. 2019).

When horseshoe crabs reach maturity *ca.* 9–11 years old, they no longer moult (i.e. terminal anecdysis; Shuster and Sekiguchi 2003), therefore these animals are more susceptible to colonisation or infestation with epibionts such as barnacles, encrusting bryozoans, and limpets. Interestingly, Heres et al. (2020) observed a scleractinian coral

(*Cladocora arbuscula*) growing on the opisthosoma of *L. polyphemus*. Heavy and persistent infestations can reduce mobility and cause respiratory difficulties (Table 9.3). Despite the hardiness of the exoskeleton, documented mortalities of adult horseshoe crabs are regularly associated with cuticular damage, fractures, pitting, thinning, dysecdysis and erosion (Figure 9.6)—some are caused by conspecifics. A series of hypodermal glands line the outer cuticle layer of the carapace, which secrete an anti-fouling agent. This surfactant-like substance helps to keep the carapace clean of commensals and disease-causing agents (Harrington and Armstrong 1999; Harrington et al. 2008). Although the complete composition of this dermal exudate remains undetermined, it might exert its effect by two means: (1) the continuous production and secretion of a mucus-like material entraps and lifts off

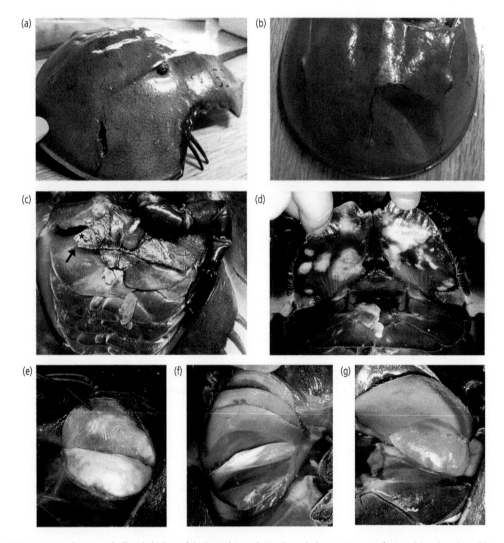

Figure 9.6 Carapace damage and gill pathobiology of the horseshoe crab *Limulus polyphemus.* Carapace fracture (a) and crushing (b) are common features of injured horseshoe crabs—usually associated with moulting. Damage (tearing; black arrow) to the gill cover (operculum; c). The black asterisk denotes the presence of green algae. Unknown fungal infection on the underside of the gill cover (d). Views of the gills depicting extensive blood (haemolymph) clotting (e), emphysema (f) and oedema (g; swollen, or ballooning). Source: images d–g are courtesy of Professor Andrew Shinn (Benchmark,Thailand).

microscopic settlers preventing their contact with the solid surface and the formation of a structural bond, or (2) the cytolytic activities kill colonisers directly.

Green algae (chlorophycophytal) and fungi (e.g. *Fusarium* spp.) are the most common aetiological agents associated with 'shell disease' among wild and captive horseshoe crabs, respectively (Leibovitz and Lewbart 1987; Smith and Berkson 2005; Braverman et al. 2012; Tuxbury et al. 2014; Table 9.3). Both culprits form mats across the dorsum of the host and the leaflets of the ventral book gills. Mild cases are denoted by focal distortions and shell discolouration in the form of white/pale (fungal) and dark green (algal) regions. Once established in post-moult crabs, algae and fungi use rhizoidal and hyphal structures to breach the cuticle and form pits that develop into ulcerative lesions. Bacterial consortia are often associated with these carapace lesions but are considered superficial in most cases (LaDouceur et al. 2019). Extensive coverage of the carapace deforms the lateral eyes, which can affect negatively an animal's ability to sense/find a mate (Duffy et al. 2006). Eventually, the disease progresses through the full thickness of the cuticle/dermal layers to the sub-adjacent viscera (Braverman et al. 2012). Cyanobacteria such as *Oscillatoria* spp. can cover the surfaces of the gills and manifest as dark discolouration, swelling and putrefaction (Leibovitz 1986; Table 9.3).

Host immune cells, namely the amebocytes, are competent phagocytes but their key role is to release cytoplasmic-derived granules containing biocidal and biostatic factors that thwart intruders and seal wounds (Mürer et al. 1975; Bursey 1977; Armstrong and Rickles 1982; Coates et al. 2012a). The amebocytes are triggered by the presence of fungal β-glucans and algal lipopolysaccharide-like ligands in the haemocoel (Söderhäll et al. 1985; Conrad et al. 2006). Amebocytes migrate towards compromised tissues causing swelling, dilation of lamellar tissues, vacuolisation, and an increase in epithelial subcuticular spaces. Amebocytes tend to degranulate and aggregate to form 'coagula' (similar to haemocyte nodules described in Chapter 1) in infected gill lamellae, which can sometimes occupy the entire lumen (Newton and Smolowitz 2018; LaDouceur et al. 2019). Haemolymph clotting is triggered primarily by lipopolysaccharides (endotoxins) from Gram-negative bacteria, a phenomenon that was originally misidentified as a disease of the horseshoe crab rather than a mechanism of immunity (Bang 1956). Despite such a robust cellular response, crabs are not completely resistant to bacteria—some can be found in the haemolymph (Brandin and Pistole 1985), and septicaemia does occur in debilitated animals.

Several species of turbellarid worms are found among the horseshoe crab gills, and on the appendages including the base of the telson (see Table 9.3). Between the gill lamellae, the marine triclad flatworm *Bdelloura candida* deposits encapsulated eggs (Huggins and Waite 1993). Such eggs are not displaced easily due to an adhesive plaque and stem (Figure 9.7). These ectoparasites are likely to be haematophagous—accessing haemolymph nutrients from small lesions made in the thin gill tissue. Long after the '*Limulus* leech' has emerged (Figure 9.7c), the eggshell will remain *in situ* and can contribute to the development of focal necrosis, damage and inflammation of the gills (Groff and Liebovitz 1982). Such compromised tissues leave the animal vulnerable to opportunistic infections. Inspections of effete horseshoe crabs have also encountered digenetic trematodes, free-living nematodes, and some protozoans (Table 9.3).

9.5.1.2 In captivity

The main cause of mortality in captive adult horseshoe crabs is the syndrome panhypoproteinemia (Smith and Berkson 2005). Within one month of capture and beyond, animals display clear signs of reduced soluble protein levels. Smith et al. (2002) established a reference interval for total protein of seemingly healthy individuals, 34 to 112 mg mL^{-1}. Oxygenated horseshoe crab haemolymph (blood) is blue due to the binding of dioxygen between two copper atoms present in the active site of the respiratory pigment haemocyanin, which constitutes > 90% of total protein (Coates et al. 2012b). It is unsurprising that a reduction of protein concentration leads to pigment loss, and in turn, the haemolymph becomes pale and clear (Smith and Berkson 2005; Coates and Nairn 2014). Nolan and Smith (2009) proposed several reasons why this occurs, including nutritional deficiency, protein-losing enteropathy, and nephropathy.

Table 9.3 Disease and pathobiology of wild and captive horseshoe crabs

Disease	Causative agent(s)	Host(s) & location(s)	Pathology/Syndrome	References
Wild-caught				
Carapace lesions (shell disease)	Cyanobacteria (*Oscillatoria* spp.) *Beggiatoa* spp.	*L. polyphemus*	Bacteria are usually found on/among the book gills, but most of them do not agitate the animal. Rupturing of the thin chitinous leaflets can be caused by filamentous cyanobacteria like *Oscillatoria* spp. Late stage infections involve deep tissue penetration of the circulatory sinuses, oedema (Figure 9.6), necrosis, and death. Other bacteria include *Flavobacterium* sp. *Leucothrix* sp. (Figure 9.8), *Pseudomonas* sp. and *Vibrio* sp.	Leibovitz (1986); Leibovitz and Lewbart (2003)
	Green algae (*Chlorophytal*)	*L. polyphemus*	Dorsal carapace is covered by green algae, affecting the eyes (ocelli) and the arthrodial membrane. Algae insert themselves within the chitinous lamina and eventually penetrate through to the soft epithelial and sub-adjacent tissues. Shell and moulting deformities, ocular degeneration, respiratory failure, and cardiac haemorrhage are all linked to late stages of this disease. (See Figure 9.6)	Leibovitz and Lewbart (1987); Braverman et al. (2012)
Gill disease	Branchiomycosis	*L. polyphemus*	Fungal infection of the book gills and operculum (Figure 9.6), that can manifest as white/discoloured regions and lead to mechanical blockage of lamellae and multifocal necrosis. (Also observed in captive animals)	Leibovitz and Lewbart (2003)
	Turbellarid worms -*Bdelloura candida* -*B. parasitica* -*B. propinqua* -*B. wheeleri* -*Syncoelidium pellucidum*	*B. candida* has been recovered from the book gills of *L. polyphemus* along the east coast of the USA, from Maine to Florida	Gravid *B. candida* deposits encapsulated eggs at the base of the gill leaflets. These eggs are not displaced easily due to an adhesive plaque and stem (Figure 9.7). Long after the triclad has emerged (Figure 9.7), the eggshell will remain *in situ* and can contribute to the development of necrotic lesions and damage. There is also evidence of haemocyte aggregation and melanisation.	Groff and Liebovitz (1982); Landy and Leibovitz (1983); Huggins and Waite (1993); Riesgo et al. (2017); Brianik et al. (2021)
Infestations (commensals, epibionts)	Consortia	-*C. rotundicauda* -*L. polyphemus* -*T. gigas*	Bryozoans, barnacles, tube-building polychaetes, molluscs (slipper limpets, blue mussels, oysters) most commonly use horseshoe crabs as substrates	Patil and Anil, (2000); Botton, 2009; Shuster, 2009)
	Encrusting bryozoans (and others)	-*T. gigas* (West Bengal, India)	Diverse organisms were seen to infest *T. gigas* collected from inshore regions of West Bengal (1972): sea anemones (*Metridium*), polychaete (*Gattyana*), barnacles (*Balanus*, *Chthamalus*), amphipod (*Cheiriphotis*) isopod (*Cleantis*), bivalves (*Ostrea*), brachyuran (*Thalamita*), bryozoan (*Membranipora*)	Rao and Rao (1972); Key et al. (1996, 2000); Tan et al. (2011)
		-*T. gigas* (Singapore)	Approximately, 77% of *T. gigas* collected from the seas adjacent to Singapore (1996) were fouled by bryozoans (*Electra angulata*, *Biflustra savartii*)	

	T. gigas (Pantai Balok, Malaysia)	Out of 161 *T. gigas* collected (2011), ~17% had acorn barnacles (*Balanus*), ~10% had pedunculated barnacles (*Octolasmis*), and ~1% had conical or flat slipper limpets (*Calyptraea* and *Crepidula*, respectively). *In extremis*, epibionts carpeted the entire dorsal carapace, and some of the ventral surfaces leading to mobility issues	Betton (1981); Turner et al. (1988); Deaton et al (1989);
	Mussels	*L. polyphemus*	Fast growing blue mussel (*Mytilus edulis*) on dorsal carapace. Ribbed mussel (*Geukensia demissa*) on ventral book gills	
Parasites	Ciliated (*Pananophrys* spp.) and flagellated (*Hexamita* spp.) protozoans Digenetic trematodes (*Microphallus* sp.)	*L. polyphemus*	The trematode *Microphallus limuli* uses the horseshoe crab as a second intermediate host. Encysted metacercaria can be found in brain, connective tissues, muscles, and eyes of juvenile and inter-moult crabs. The final host, Herring gulls (*Larus argentatus*), ingest the cysts from the tissues of up-turned crabs	Stunkard (1950 and 1951); Johnson (1977); Leibovitz and Lewbart (2003); Smith and Berkson (2005); LaDouceur et al. (2C19).
	Nematodes (*Grathponema* and *Monhysteria* spp.)		Varieties of organisms observed on horseshoe crabs are known to cause disease in other animals, e.g. Paramoebiasis in blue crabs (*Callinectes sapidus*), Hexamitiasis in fresh/salt-water fish. Free-living nematodes have been reported to invade the carapace of juvenile and compromised animals	

Captive animals

Fusariosis	*Fusarium solani* complex	*L. polyphemus* (National Aquarium, Baltimore, Maryland USA)	Over a period of 3 years. 39 horseshoe crabs displaying lesions on the dorsal/ventral carapace and gills were assessed. The extent of damage varied from multifocal (superficial) discolouration to ulcers and pitting—some penetrating the full thickness of the carapace. Tissue histology revealed several fungal morphotypes in 97% of crabs, including hyphae, associated with lesions. Amebocyte infiltration was visible in compromised haemocoel tissues. Gonads, connective tissue, muscle and the hepatopancreas were infected in some cases. Additionally, black discolouration and necrosis were observed in mycotic gills.	Tuxbury et al. (2014)
Lesions: mycotic dermatitis and branchitits, bacterial tubulitis and enteritis	Identity was not confirmed in all cases: -*Fusarium* sp. -Gram-negative rods (n = 10) -Gram-negative cocci (n = 4) -Gram-positive rods (n = 1) -Acid-fast bacilli (n = 1)	*L. polyphemus*	Archived tissue samples of *L. polypehmus* retrieved from nine states (Maryland, California, Washington, Nevada, Oklahoma, Ohio, Nebraska, Texas, Louisiana) determined bacterial and fungal infections in 88% and 59% of animals, respectively. Tissues displaying lesions (in descending order) were the compound eyes (85%), body wall (83%), gills (65%), hepatopancreas (61%), chitinous gut (50%), non-chitinous gut (36%), heart (35%) and brain (13%). Lesions were absent from the coxal gland and gonads. Fungi were primarily responsible for ulcerative (body wall lesions), whereas bacteria targeted internal soft tissues. The presence of bacteria and/or fungi was accompanied by amebocyte-mediated inflammation in the majority of cases.	LaDouceur et al. (2019)

continued

Table 9.3 Continued

Disease	Causative agent(s)	Host(s) & location(s)	Pathology/Syndrome	References
Mycosis	Identity was not determined	*L. polyphemus* (Ripley's Aquarium of the Smokies, Gatlinburg, Tennessee, USA)	Irregularly expansile lesions, soft tissues of the gills, joint spaces of the legs and proximal telson (n = 10). Also visible, 1–2 mm circular, discoloured red-brown ulcerated lesions of the ventral carapace. Treatment of animals was attempted with intravascular administration of ftraconazole.	Allender et al. (2008)
Mycosis and bacteriosis	*Aspergillus niger* (fungus) and *Shewanella putrefaciens* (bacterium)	*C. rotundicauda* and *T. gigas*	Horseshoe crabs were collected from Banting (Selangor, Malaysia) and maintained in a laboratory. In captivity, *T. gigas* eggs and trilobites (larvae) were blighted by several microbial infections that manifested as discolouration (red, grey, black) and spore formation on the surface. Those severely affected ceased developing and died. A range of fungi (*Aspergillus*, *Gliocladium*, and *Penicillium* species) and bacteria (*Bacillus cereus*, *Corynebacterium* sp. and *Enterococcus faecalis*) were identified on *T. gigas* and adult *C. rotundicauda*, but the most damaging were *A. niger* and *S. putrefaciens*.	Faizul et al. (2015)
Non-infectious	Panhypoproteinemia	*L. polyphemus* and *T. tridentatus*	Gradual decrease in soluble protein levels, and immune cell numbers (amebocytes). Haemolymph becomes pale with reduced clotting potential. Respiratory difficulties, anorexia, lethargy and hypovolemia are likely co-morbidities	Smith and Berkson (2005); Coates et al. (2012b); Kwan et al. (2014)
Parasites	Various -Paramoebidae -Trematodes -Nematode larvae -Arthropods	*L. polyphemus*	Pathological features range from non-invasive cyanobacterial presence on the gills to encysted (sometimes degraded) metacercariae with little to no inflammation (including the brain), to myocellular necrosis and amebocyte infiltration linked to the presence of amoeba (and bacteria).	LaDouceur et al. (2019); Tuxbury et al. (2014)
Protozoan infestation	*Zoothamnium duplicatum*	Outbreak recorded in juvenile *L. polyphemus* under commercial-scale, hatchery conditions	Gross coverage of the exoskeleton, including mechanical blockage of the cuticular hypodermal glands (Figure 9.8). Complete colonisation of the gill lamellae, occluding the lumen and progressing to the proximal gill chamber (Figure 9.8). Amebocyte infiltration of the tissue in response to the protozoan. Over 95% of 2nd/3rd instars died within two months. The addition of sand to aquaria and a freshwater bath for 1 hour with 10 ppm chlorine reduced epibiont coverage and growth (without measureable side-effects).	Shinn et al. (2015) Chen et al. (1989)
		Asian horseshoe crab	An unidentified protozoan infection of the haemolymph was documented in an Asian horseshoe crab	

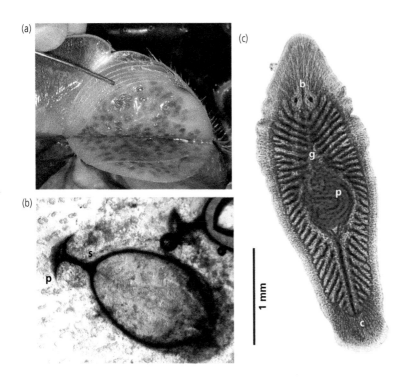

Figure 9.7 Association of the turbellarid *Bdelloura candida* with the horseshoe crab *Limulus polyphemus*. Gross appearance of egg capsules attached to the surface of gill leaflets of an adult horseshoe crab (a). An intact egg casing with the adhesive plaque (p) and stem (s) visible (b). Preparation of a mature worm—harvested from an egg—stained with Mayer's paracarmine (c). Inset: b, brain (cerebral ganglion, and two eyespots); c, caudal adhesive disc; g, gut; p, pharyngeal cavity and mouth. Source: Professor Andrew Shinn (Benchmark, Thailand) kindly provided the original images.

Moribund animals are anorexic and lethargic, and when haemolymph is extracted, the time needed to clot increases substantially. Coates et al. (2012b) monitored several haemolymph indicators in captive horseshoe crabs held at different temperatures, $8\,^\circ$C to $23\,^\circ$C. After 56 days, haemocyanin levels had decreased dramatically, with those held at/above room temperature ($\geq 18\,^\circ$C) were most severely affected, ~ 69% reduction. Similarly, amebocyte numbers also halved over the duration of the experiment, declining from ~ 3.5×10^7 mL^{-1} to < 1.5×10^7 mL^{-1}. As haemocyanin-derived phenoloxidase activity and the circulating amebocytes are integral components of horseshoe crab biological defences (Coates et al. 2011, 2012a, and 2013), their loss will leave the animal immunocompromised and more susceptible to opportunists. The knock-on effects of lowered acute phase protein levels are decreased oxygen tension (haemocyanin deficiency) and the loss of oncotic pressure (oedema is often observed in the gills; Figure 9.6).

Captivity-associated stress and morbidity in aquatic animals, including horseshoe crabs (Coates et al., 2012b) are linked to opportunistic infections and environmental factors such as salinity, temperature, pH, poor water quality, and ammonia toxicity (Nolan and Smith, 2009). Recently, Friel et al. (2020) interrogated the microbiomes of three wild-caught and three captive horseshoe crabs (> 2 years) across four body parts (cloaca, carapace, gills, oral cavity), by sequencing 16S rRNA gene amplicons. Distinct bacterial profiles were found between the wild-caught and captive samples, with the former displaying more diverse microbial populations dominated by α- and γ-*proteobacteria*, and *Bacteroidia*. For those crabs, microbial richness declined after 1 month in captivity. Aeromonads and pseudomonads represented > 80% of bacteria identified from the bodies of long-term captive animals. Friel et al. (2020) posited that these bacteria, which are known opportunists capable of producing chitin-degrading proteins, might enable or exacerbate shell disease of horseshoe crabs. In an independent study, necropsy examinations of diseased horseshoe crabs identified *Pseudomonas putrefaciens* and *Aeromonas hydrophila* from the haemolymph (LaDouceur et al. 2019). Many bacterial genera including *Vibrio* and *Pseudoalteromonas* have been recovered from the gills and surfaces of wild horseshoe crabs and those reared *ex situ* (C.

rotundicauda, L. polyphemus, T. gigas), yet it remains unclear whether some are symbionts contributing to health, opportunistic commensals, or pathobionts causing trauma (Smith et al. 2011; Thompson et al. 2011; Ismail et al. 2015; Friel et al. 2020)

Fungal infections (dermatitis, branchitis) are the scourge of horseshoe crabs held in aquaria and touch tanks (Leibovitz and Lewbart 2003; Densmore et al. 2005; Allender et al. 2008; Faizul et al. 2015;

LaDouceur et al. 2019). Over a three-year period at the National Aquarium of Baltimore, 80 horseshoe crabs died (Tuxbury et al. 2014). Circular, tan lesions on the dorsal prosoma/opisthosoma including the eyes and the ventral exoskeleton including the book gills were visible on all animals. Affected book gills were disfigured with friable, missing and blackened/necrotic leaflets. The carapace had localised to extensive erosion and pitting, and in

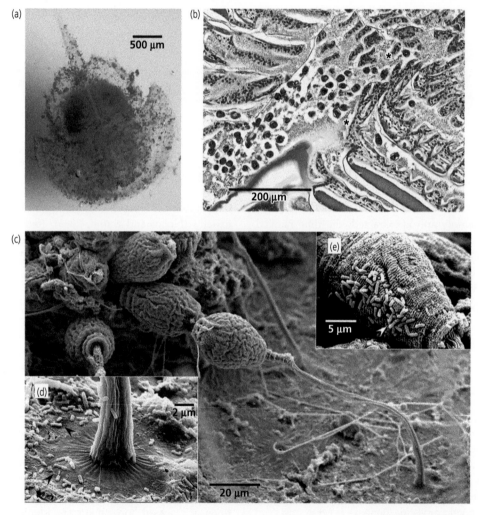

Figure 9.8 Infestation of juvenile horseshoe crabs *Limulus polyphemus* with the ciliate protozoan *Zoothamnium duplicatum*. Light micrograph of a second instar *L. polyphemus* stained with neutral red (a). Growth of *Z. duplicatum* has carpeted the entire carapace. Histopathology of the (proximal) gill chamber removed from infested *L. polyphemus* (b). Gill lamellae are congested with *Z. duplicatum*, and immune cells (amebocytes) have infiltrated the tissue (denoted by an asterisk, *), presumably in response to the parasite. Scanning electron micrograph of branching *Z. duplicatum* zooids on the carapace of horseshoe crabs (c). Rod shaped bacteria (perhaps *Leucothrix* spp.; arrowheads) associated with the anchoring disc (d) and anterior stalk (e) of *Z. duplicatum*.

some cases, muscle myositis and atrophy of the appendages. Moreover, lesions contained biofilms consisting of Gram-positive/negative bacterial rods/cocci. Histopathology screening of 39 individuals revealed fungi to be the causative agent of this cutaneous infection in 97% of cases. Culturing and molecular screening (ITS region) of fungi were dominated by members of the *Fusarium solani* complex, with some evidence of *Hortaea werneckii* and *Stemphylium* sp. (Tuxbury et al. 2014).

An outbreak of the vorticellid *Zoothamnium duplicatum* decimated ~ 40,000 (~ 96%) 2nd/3rd instars in a commercial-scale hatchery for *L. polyphemus* (Shinn et al. 2015). Colony growth was extensive and blanketed the entire carapace, obstructing the hypodermal glands, penetrating the lamellae, and accumulating in the proximal gill chamber (Figure 9.8). Mucilaginous and filamentous rod-shaped bacteria akin to *Leucothrix* sp. were visible on the zooids and anchoring discs of *Z. duplicatum* (Figure 9.8). The addition of sterile sand and immersion in chlorine baths (10 ppm for 1 hour) proved successful in treating and discouraging re-growth. Lining aquaria with fine sands permits crabs to burrow and dislodge epiphytes (self-cleaning) and can help juveniles to shed their moults (Smith and Berkson 2005).

Most recently, LaDouceur et al. (2019) performed an in-depth, broad histological survey of archived *L. polyphemus* tissues held in zoos or aquaria from nine American states. Systemic infections contained large amorphous coagulates of amebocyte debris among degenerated/necrotic tissue. Fungi were most often found progressing through the carapace or gills from the external surfaces and bacteria were associated with inflammatory foci of many internal organs (e.g. the interstitium of the hepatopancreas, enteritis; Table 9.3). Approximately 41% of horseshoe crabs were diagnosed with parasites distributed among the viscera such as the brain and heart, with encysted trematodes visible in 39% of hepatopancreatic tissues.

9.5.2 Pathobiology of sea spiders

Of the two extant marine chelicerate groups—the pycnogonids (sea spiders) and xiphosurans (horseshoe crabs)—sea spiders are more diverse and geographically widespread. Unusually, sea spiders lack a body cavity (coelom), which is understood to have been lost over time as they represent a basal arthropod lineage. Sea spiders themselves are parasitic and raptorial (carnivorous), feeding directly on polychaete worms, bryozoans, jellyfish, sponges, sea anemones, echinoderms, and gastropods using their proboscis to puncture tissues and gorge on the body fluids. Larval stages of several species, e.g. *Phoxichilidium femoratum* and *Pycnogonum littorale*, use unsuspecting marine invertebrates as hosts, e.g. hydrozoans, during developmental cycles where they can form 'cysts' and remain ecto-parasitic for several weeks (King and Crapp 1971; Arnaud and Bamber 1988). In some instances, pycnogonids have evolved to be endoparasites as is the case for *Ascorhynchus endoparasiticus* living inside the pallial cavity of the gastropod *Scaphander punctostriatus* (Arnaud 1978; Arnaud and Bamber 1988 and references cited therein).

Unfortunately, insufficient records exist with respect to their pathobiology. As is the case for horseshoe crabs (see Section 9.5.1), the pycnogonids are liable to colonisation by aquatic epibionts (Key et al., 2013; Lane et al., 2016 and 2018), where the hard exoskeleton acts as a refugium for bryozoans, sponges, bivalves, hydroids, and brachiopods to name a few (Wyer and King, 1973). Key et al. (2013) scrutinised encrusting bryozoans on predatory pycnogonids and determined that 35% of bryozoan colonies spread across two body parts, which may decrease flexibility (locomotion) and enhance susceptibility to predation. Lane et al. (2016) concluded that gross colonisation (~ 75%) of sea spider exoskeletons (*Achelia chelata* and *Achelia gracilipes*) with algae and diatoms are not costly. In a second study on several Antarctic species, Lane et al. (2018) calculated a decrease in oxygen diffusion coefficient across the cuticle when encrusting bryozoan coverage was extensive—but such cases are rare.

Pipe (1982) surveyed epibionts on *P. femoratum* from the North Sea and noted the presence of peritrich and suctorian ciliates on the dorsal surfaces of leg segments in addition to filamentous *Leucothrix*-like bacteria (but very little fouling on the ventrum). It is not surprising to find bacteria on the surface of a sea spider, especially considering their microbiologically rich surroundings, but offers little insight into putative microbe-pycnogonid antibiosis. Most recently, Conway (2015) identified a flavivirus (ssRNA) signal in a cDNA library prepared from *Endeis spinosa*—again, this does not affirm an infection and may represent an old genomic integration event of viral nucleic acids.

9.6 On the lack of viruses

Across all the chelicerates, there are few characterisations of viruses. In 1955, Muma studied some effete rust mites (*Panonychus citri*) in Florida that had a black resin-like material seeping from the anus, which was later attributed to a baculovirus infection of the midgut epithelial cells where bacilliform virions were seen replicating within the nuclei. A key diagnostic factor of this disease is the presence of birefringent crystals in the midgut caeca (Smith and Cressman 1962; Reed and Desjardins 1978). An additional Baculoviridae was discovered in European red mites (*P. ulmi*; Steinhaus 1960; Bird 1967), where the virus replicates inside fat body cells and manifests as dark red colouration. This disease was a common cause of epizootics among mite populations in California and Arizona (Reed 1981). Transmission between conspecifics is via a faecal-oral route, and sometimes shedding of virions occurs from reproductive openings onto leaves. Liu (1991) described virus-like agents in the cytoplasm of tracheal mites (*Acarapis woodi*) infesting Scottish honey bees, which appeared to damage some of the host cells. Another midgut-based virus—this time rod-shaped—was determined to be the cause of 'red death' in predatory mites (*Metaseiulus occidentalis*), in which paralysis, bloated and shiny cuticles accompanied the distinct red discolouration (Poinar Jr and Poinar 1998). Nuclei of the midgut cells were hypertrophied and/or displaced in viremic mites. In the same greenhouse at the University of California (Berkeley), other icosahedral and hexagonal shaped viruses were found in epidermal and midgut tissues of *T. urticae* and *M. occidentalis* using electron microscopy—linked to lowered egg production and death (Poinar Jr and Poinar 1998).

Morel (1978) diagnosed a baculovirus infection of the hepatopancreas as the cause of death of a nursey web spider (*Pisaura mirabilis*). Pizzi (2012) reckoned iridoviruses of spiders were asymptomatic and wrote about anecdotal evidence of imperial scorpions becoming infected with an iridovirus from eating affected crickets, yet further highlighted the paucity of viral pathobiology. In earlier works, Lewis (1979) detailed the presence of a virus associated with a rickettsia-like infection of tick (*Ixodes ricinus*) ovaries, and Williams (1987) noted viruses as a cause of infectious disease in scorpions alongside fungal and rickettsial agents.

The most infamous virus-chelicerate interaction is that of the parasitic mite, *Varroa destructor*, and the RNA-based deformed wing virus (DWV). The mites feed on honey bees, and in the process, transfer DWV as well as other viruses and microsporidia, which can have devastating consequences for the insect colonies (Bowen-Walker et al., 1999; Ramsey et al., 2019). However, there is an absence of evidence to suggest DMV is detrimental to the carrier mites, and their relationship is treated as symbiotic. High-throughput sequencing has enabled much progress in the study of arachnid microbiome(s), so perhaps this approach can be applied to resolve their virome (e.g., Debat 2017; Li et al. 2015; Guo et al. 2020).

9.7 Summary

- There is a paucity of contemporary data for infections of scorpions in the wild.
- The reader should err on the side of caution when considering those aetiologies found only in capture settings—notably horseshoe crabs—as many remain undetected in the wild and are unlikely to influence the long-term ecology of the animal.
- There is a knowledge deficit with regards pathologic conditions of sea spiders.
- Ostensibly, the book lungs and book gills of chelicerates are the primary source of pathogens taking over the haemocoel.
- For horseshoe crabs and spiders, evidence suggests 'skirmishes on the cuticle' are likely to determine disease outcomes.

- Interestingly, arachnids are targeted by unique genera of hypocrealean fungi (in contrast to those that target insects).
- Probing the bacterial microbiome of arachnids and horseshoe crabs will undoubtedly reveal further details on the widespread, intimate relationships with endosymbionts and the consequences of captivity-induced dysbiosis, respectively.
- To date, three bacterial genera, *Cardinium*, *Wolbachia*, and *Rickettsiella*, have been validated experimentally as reproductive parasites of arachnids.

Suggested further reading

Bruin, J. and van der Geest, L.P. 2009. *Diseases of mites and ticks*. Springer

Evans, H.C. 2013. Fungal pathogens of spiders. In *Spider Ecophysiology*, pp. 107–121. Berlin, Heidelberg: Springer

Hughes, D.P., Araújo, J.P.M., Loreto, R.G., Quevillon, L., De Bekker, C., and Evans, H.C. 2016. From so simple a beginning: The evolution of behavioral manipulation by fungi. *Advances in Genetics* 94: 437–469

LaDouceur, E.E., Mangus, L., Garner, M.M., and Cartoceti, A.N. 2019. Histologic findings in captive American horseshoe crabs (*Limulus polyphemus*). *Veterinary Pathology* 56: 932–939

Newton, A.L. and Smolowitz, R. 2018. Invertebrates. In *Pathology of Wildlife and Zoo Animals*, pp. 1019–1052. Academic Press, https://www.sciencedirect.com/book/9780128053065/pathology-of-wildlife-and-zoo-animals#book-des-cription

Acknowledgements

I would like to thank my colleague Dr Christopher B Cunningham (Swansea University) for commenting on a draft version of this chapter. I should also like to extend my gratitude to Andy Shinn, Harley Newton, Harry Evans and Sushan Han for providing such beautiful images to complement the text.

References

Adams, J.R., Clark, T.B., Tompkins, G.J., Neel, W.W., Schroder, R.F., and Schaefer, P.W. 1997. Histopathological Investigations on *Rickettsiella*-like sp. and Nonoccluded Viruses Infecting the Pecan Weevil *Curculio caryae*, the Squash Beetle *Epilachna borealis*, and the Mexican Bean Beetle *Epilachna varivestis*. *Journal of Invertebrate Pathology* 69(2): 119–124

Agrawal, Y., Khatri, I., Subramanian, S., and Shenoy, B.D. 2015. Genome sequence, comparative analysis, and evolutionary insights into chitinases of entomopathogenic fungus *Hirsutella thompsonii*. *Genome Biology and Evolution* 7(3): 916–930

Allender, M.C., Schumacher, J., Milam, J., George, R., Cox, S., and Martin-Jimenez, T. 2008. Pharmacokinetics of intravascular itraconazole in the American horseshoe crab (*Limulus polyphemus*). *Journal of Veterinary Pharmacology and Therapeutics* 31: 83–86

Anstead, C.A. and Chilton, N.B. 2014. Discovery of novel *Rickettsiella* spp. in ixodid ticks from Western Canada. *Applied and Environmental Microbiology* 80: 1403–1410

Armstrong, P.B. and Rickles, F.R. (1982). Endotoxin-induced degranulation of the Limulus amebocyte. *Experimental Cell Research*, 140(1): 15–24

Arnaud, F. (1978). A new species of Ascorhynchus (Pycnogonida) found parasitic on an opisthobranchiate mollusc. *Zoological Journal of the Linnean Society*, 63(1–2): 99–104

Arnaud, F. and Bamber, R.N. (1988). The biology of Pycnogonida. In *Advances in Marine Biology*, Vol. 24, pp. 1–96. Academic Press

Baker, A.S. 1992. Acari (mites and ticks) associated with other arachnids. J.E. Cooper, P. Pearce-Kelly, D.L. Williams (eds.), Arachnida: Proceedings of a Symposium on Spiders and Their Allies. pp. 126–131. London: Chrion Press

Bałazy, S., Miętkiewski, R., Tkaczuk, C., Wegensteiner, R., and Wrzosek, M. 2009. Diversity of acaropathogenic fungi in Poland and other European countries. In *Diseases of Mites and Ticks*, pp. 53–70. Dordrecht: Springer

Bang, F.B. 1956. A bacterial disease of *Limulus polyphemus*. *Bulletin of the Johns Hopkins Hospital* 164: 542–549

Barbosa, B.C., Maciel, T.T., Abegg, A.D., Borges, L.M., Rosa, C.M., and Vargas-Peixoto, D. 2016. Arachnids Infected by arthropod-pathogenic fungi in an urban fragment of Atlantic Forest in southern Brazil. *Nature Online* 14: 11–114

Baruš, V. and Koubková, B. 2002. The first species of *Thelastoma* Leidy, 1849 (Nematoda: Thelastomatidae) parasitising the scorpion *Euscorpius italicus* (Chactidae: Scorpionidea). *Systematic Parasitology* 53(2): 141–146

Bird, F.T. 1967. A virus disease of the European red mite *Panonychus ulmi* (Koch). *Canadian journal of microbiology* 13(8): 1131–1131

Bjørnson, S., Steiner, M.Y., and Keddie, B.A. 1996. Ultrastructure and pathology of *Microsporidium phytoseiulin*. sp. infecting the predatory mite, Phytoseiulus persimilis Athias-Henriot (Acari: Phytoseiidae). *Journal of Invertebrate Pathology* 68(3): 223–30

Botton, M.L. 1981. The gill books of the horseshoe crab (*Limulus polyphemus*) as a substrate for the blue mussel (*Mytilus edulis*). *Bulletin of the New Jersey Academy of Science* 26: 26–28

Botton, M.L. 2009. The ecological importance of horseshoe crabs in estuarine and coastal communities: A review and speculative summary. In *Biology and Conservation of Horseshoe Crabs*, John T. Tanacredi, Mark L. Botton, David Smith (eds.) pp. 45–63. Boston, MA: Springer

Bowen-Walker, P.L., Martin, S.J., and Gunn. A. 1999. The Transmission of Deformed Wing Virus between Honeybees (*Apis mellifera* L.) by the Ectoparasitic Mite *Varroa jacobsoni* Oud. *Journal of Invertebrate Pathology* 73: 101–106

Brandin, E.R. and Pistole, T.G. 1985. Presence of microorganisms in hemolymph of the horseshoe crab *Limulus polyphemus*. *Applied and Environmental Microbiology* 49(3): 718–720

Braverman, H., Leibovitz, L., and Lewbart, G.A. 2012. Green algal infection of American horseshoe crab (*Limulus polyphemus*) exoskeletal structures. *Journal of Invertebrate Pathology* 111: 90–93

Brescovit, A.D., Villanueva-Bonilla, G.A., Sobczak, J.C.M., Nóbrega, F.A.D.S., Oliveira, L.F.M., Arruda, I.D.P., and Sobczak, J.F. 2019. Macrophyes pacoti n. sp. (Araneae: Anyphaenidae) from Brazilian Atlantic Forest, with notes on an araneopathogenic fungus. *Zootaxa* 4629(2): 294–300

Brianik, C.J., Bopp, J., Piechocki, C., Liang, N., O'Reilly, S., Ceratto, R.M., Allam, B. 2021. Infection prevalence, intensity and tissue damage caused by the parasitic flatworm, *Bdelloura candida*, in the American horseshoe crab (*Limulus polyphemus*). Research Square 10.21203/rs.3.rs-842746/v1.

Bryson R.W. Jr. 2014. Bacterial endosymbiont infections in 'living fossils': A case study of North American vaejovid scorpions. *Molecular Ecology Resources* 14(4): 789–793

Bursey, C.R. (1977). Histological response to injury in the horseshoe crab, *Limulus polyphemus*. *Canadian Journal of Zoology* 55(7): 1158–1165

Butt, T.M., Coates, C.J., Dubovskiy, I.M., Ratcliffe, N.A. 2016. Entomopathogenic fungi: new insights into host–pathogen interactions. *Advances in Genetics* 94: 307–364.

Carr, A.L. and Roe, M. 2016. Acarine attractants: Chemoreception, bioassay, chemistry and control. *Pesticide Biochemistry and Physiology* 131: 60–79

Chen, I.J., Hong, S.J., Chen, Y.M., and Yang, Y.C. 1989. Cultivation of horseshoe crab amebocytes. *The Kaohsiung Journal of Medical Sciences* 5(9): 516–521

Coates, C.J., Bradford, E.L., Krome, C.A., and Nairn, J. 2012b. Effect of temperature on biochemical and cellular properties of captive *Limulus polyphemus*. *Aquaculture* 334: 30–38

Coates, C.J., Kelly, S.M., and Nairn, J. 2011. Possible role of phosphatidylserine–hemocyanin interaction in the innate immune response of *Limulus polyphemus*. *Developmental & Comparative Immunology* 35: 155–163

Coates, C.J. and Nairn, J. 2014. Diverse immune functions of hemocyanins. *Developmental & Comparative Immunology* 45(1): 43–55

Coates, C.J., Whalley, T., and Nairn, J. 2012a. Phagocytic activity of *Limulus polyphemus* amebocytes in vitro. *Journal of Invertebrate Pathology*, 111: 205–210

Coates, C.J., Whalley, T., Wyman, M., and Nairn, J. 2013. A putative link between phagocytosis-induced apoptosis and hemocyanin-derived phenoloxidase activation. *Apoptosis* 18: 1319–1331

Cokendolpher, J.C. 1993. Pathogens and parasites of Opiliones (Arthropoda: Arachnida). *Journal of Arachnology* 120–146

Conrad, M.L., Pardy, R.L., Wainwright, N., Child, A., and Armstrong, P.B. 2006. Response of the blood clotting system of the American horseshoe crab, *Limulus polyphemus*, to a novel form of lipopolysaccharide from a green alga. *Comparative Biochemistry and Physiology Part A: Molecular & Integrative Physiology* 144(4): 423–428

Conway, M.J. 2015. Identification of a flavivirus sequence in a marine arthropod. *PloS one* 10(12): e0146037

Cordaux, R., Paces-Fessy, M., Raimond, M., Michel-Salzat, A., Zimmer, M., and Bouchon, D. 2007. Molecular characterization and evolution of arthropod-pathogenic Rickettsiella bacteria. *Applied and Environmental Microbiology* 73(15): 5045–5047

Costa, P.P. 2014. *Gibellula spp. associadas a aranhas da Mata do Paraíso, Viçosa-MG* (Doctoral dissertation, Universidade Federal de Viçosa).

Coyle, F.A., Goloboff, P.A., and Samson, R.A. 1990. *Actinopus* trapdoor spiders (Araneae, Actinopodidae) killed by the fungus, *Nomuraea atypicola* (Deuteromycotina). *Acta Zoologica Fennica* 190: 89–93

Curry, M.M., Paliulis, L.V., Welch, K.D., Harwood, J.D., and White, J.A. 2015. Multiple endosymbiont infections and reproductive manipulations in a linyphiid spider population. *Heredity* 115(2): 146–152

De Luna, C.J., Moro, C.V., Guy, J.H., Zenner, L., and Sparagano, O.A. 2009. Endosymbiotic bacteria living inside the poultry red mite (*Dermanyssus gallinae*). In *Control of Poultry Mites (Dermanyssus)*, Olivier A.E. Sparagano (ed.) pp. 105–113. Dordrecht: Springer

Deaton, L.E., and Kempler, K.D. 1989. Occurrence of the Ribbed Mussel, *Geukensia demissa*, on the book gills of a horseshoe crab, *Limulus polyphemus*. *The Nautilus (Philadelphia, PA)*: 103(1)

Debat, H.J. 2017. An RNA virome associated to the golden orb-weaver spider *Nephila clavipes*. *Frontiers in Microbiology* 8: e2097.

Densmore, C., Crawford, E., Smith, D., and Dykstra, M. 2005. Probable branchiomycosis among a captive population of horseshoe crabs, *Limulus polyphemus*, from Delaware Bay. In *Proceedings of the 30th Eastern Fish Health Workshop*, Shepherdstown, WV

Duffy, E.E., Penn, D.J., Botton, M. L., Brockmann, H.J., and Loveland, R.E. 2006. Eye and clasper damage influence male mating tactics in the horseshoe crab, Limulus polyphemus. *Journal of Ethology* 24(1): 67–74

Dunbar, J.P., Khan, N.A., Abberton, C.L., Brosnan, P., Murphy, J., Afoullouss, S. et al. 2020. Synanthropic spiders, including the global invasive noble false widow Steatoda nobilis, are reservoirs for medically important and antibiotic resistant bacteria. *Scientific Reports* 10(1): 1–11

Dunlop, J.A. and Lamsdell, J.C. 2017. Segmentation and tagmosis in Chelicerata. *Arthropod Structure & Development* 46(3): 395–418

Duron, O., Bouchon, D., Boutin, S., Bellamy, L., Zhou, L., Engelstädter, J., and Hurst, G.D. 2008a. The diversity of reproductive parasites among arthropods: Wolbachia do not walk alone. *BMC biology* 6(1): 27

Duron, O., Hurst, G.D., Hornett, E.A., Josling, J.A., and Engelstädter, J.A.N. 2008b. High incidence of the maternally inherited bacterium *Cardinium* in spiders. *Molecular Ecology* 17(6): 1427–1437

Eason, R.R., Peck, W.B., and Whitcomb, W.H. 1967. Notes on spider parasites, including a reference list. *Journal of the Kansas Entomological Society* 40: 422–434

Evans, H.C. 1974. Natural control of arthropods, with special reference to ants (Formicidae), by fungi in the tropical high forest of Ghana. *Journal of Applied Ecology* 37–49

Evans, H.C. 2013. Fungal pathogens of spiders. In *Spider Ecophysiology*, Wolfgang Nentwig (eds.) pp. 107–121. Berlin, Heidelberg: Springer

Evans, H.C. and Samson, R.A. 1987. Fungal pathogens of spiders. *Mycologist* 1(4): 152–159

Faizul, M.I.M., Eng, I.I.T., Christianus, A., and Abdel-Hadi, Y.M. 2015. Bacteria and fungi identified on horseshoe crabs, *Tachypleus gigas* and *Carcinoscorpius rotundicauda* in the laboratory. In *Changing Global Perspectives on Horseshoe Crab Biology, Conservation and Management*, Ruth H. Carmichael, Mark L. Botton, Paul K.S. Shin, Siu Gin Cheung (eds.) pp. 303–311. Cham: Springer

Fernandes, E.K.K. and Bittencourt, V.R.E.P. 2008. Entomopathogenic fungi against South American tick species. In *Diseases of Mites and Ticks*, pp. 71–93. Dordrecht: Springer

Finnegan, S. 1933. A new Species of Mite parasitic on the Spider *Liphistius malayanus* Abraham, from Malaya. *Proceedings of the Zoological Society of London* 103(2): 413–417

Fisher, F.E. 1950. Two new species *of* Hirsutella Patouillard. *Mycologia* 42(2): 290–297

Foelix, R. 2011. *Biology of Spiders*. USA: Oxford University Press

Friel, A.D., Neiswenter, S.A., Seymour, C.O., Bali, L.R., McNamara, G., Leija, F., Jewell, J., and Hedlund, B.P. 2020. Microbiome shifts associated with the introduction of wild Atlantic horseshoe crabs (*Limulus polyphemus*) into a touch-tank exhibit. *Frontiers in Microbiology* 11: 1398

Frye F.L. (2012). Chapter 12 Scorpions. In *Invertebrate Medicine*, 2nd Edition, Lewbart, G. A. (ed.). John Wiley & Sons

Fukuzawa, A.H., Vellutini, B.C., Lorenzini, D.M., Silva Jr, P.I., Mortara, R.A., Da Silva, J. M., and Daffre, S. 2008. The role of hemocytes in the immunity of the spider *Acanthoscurria gomesiana*. *Developmental & Comparative Immunology* 32(6): 716–725

Gaver-Wainwright, M.M., Zack, R.S., Foradori, M.J., and Lavine, L.C. 2011. Misdiagnosis of spider bites: Bacterial associates, mechanical pathogen transfer, and hemolytic potential of venom from the hobo spider, *Tegenaria agrestis* (Araneae: Agelenidae). *Journal of Medical Entomology* 48(2): 382–388

Goodacre, S. L., Martin, O. Y., Bonte, D., Hutchings, L., Woolley, C., Ibrahim, K. et al. 2009. Microbial modification of host long-distance dispersal capacity. *BMC Biology* 7(1): 1–8

Goodacre, S.L., Martin, O.Y., Thomas, C.G., and Hewitt, G.M. 2006. Wolbachia and other endosymbiont infections in spiders. *Molecular Ecology* 15(2): 517–527

Gotoh, T., Noda, H., and Ito, S. 2007. Cardinium symbionts cause cytoplasmic incompatibility in spider mites. *Heredity* 98(1): 13–20

Gouge, D.H. and Snyder, J.L. 2005. Parasitism of bark scorpion *Centruroides exilicauda* (Scorpiones: Buthidae) by entomopathogenic nematodes (Rhabditida: Steinernematidae; Heterorhabditidae). *Journal of Economic Entomology* 98(5): 1486–1493

Grabowski, N.T. and Klein, G. (2017). Bacteria encountered in raw insect, spider, scorpion, and centipede taxa including edible species, and their significance from the food hygiene point of view. *Trends in Food Science & Technology* 63: 80–90

Grant, D. 2001 Living on Limulus. In *Limulus in the Limelight: A Species 350-Million Yeas in the Making and in Peril?* Tanacredi J.T. (ed.) pp. 135–145. New York: Kluwer Academic/Plenum Publishers

Grbic, M., Khila, A., Lee, K.Z., Bjelica, A., Grbic, V., Whistlecraft, J., Verdon, L., Navajas, M., and Nagy, L. 2007. Mity model: *Tetranychus urticae*, a candidate for chelicerate model organism. *Bioessays* 29: 489–496

Greenstone, M.H., Ignoffo, C.M., and Samson, R.A. 1987. Susceptibility of spider species to the fungus *Nomuraea atypicola*. *The Journal of Arachnology* 15(2): 266–268

Groff, J.M. and Liebovitz, L. 1982. A gill disease of *Limulus polyphemus* associated with triclad turbellarian infection. *The Biological Bulletin* 163: 392

Guo L., Lu, X., Liu, X., Li, P., Wu, J., Xing, F. et al. 2020. Metatranscriptomic analysis reveals the virome and viral genomic evolution of medically important mites. *Journal of Virology* 95(7): e01686-20.

Gysin, J. and Le Corroiler, Y. 1969. Mycoses chez le scorpion 'Androctonus australis Hector'. *Archives Institut Pasteur d'Aalgérié* 47: 83–92

Han, S., Armién, A.G., Hill, J.E., Fernando, C., Bradway, D.S., Stringer, E. et al. 2020. Infection with a Novel Rickettsiella species in Emperor Scorpions (*Pandinus imperator*). *Veterinary Pathology* 57(6): 858–870

Harrington, J.M. and Armstrong, P.B. 1999. A cuticular secretion of the horseshoe crab, *Limulus polyphemus*: A potential anti-fouling agent. *The Biological Bulletin* 197: 274–5

Harrington, J.M., Leippe, M., and Armstrong, P.B. 2008. Epithelial immunity in a marine invertebrate: A cytolytic activity from a cuticular secretion of the American horseshoe crab, *Limulus polyphemus*. *Marine Biology* 153: 1165–1171

Haupt, J. 2002. Fungal and rickettsial infections of some East Asian trapdoor spiders. European Arachnology. Toft S. and Scharff N. (eds.) pp. 45–49. Arhaus University Press, Arhaus ISBN 8,779,340,016

Heres, B.M., Kilcollins, R.F., and Crowley, C.E. 2020. Novel epibiont coral found on *Limulus polyphemus* (Atlantic horseshoe crab) in Northwestern Florida. *Southeastern Naturalist* 19(3): N45

Herrero-Galán, E., García-Ortega, L., Olombrada, M., Lacadena, J., Del Pozo, Á. M., Gavilanes, J.G., and Oñaderra, M. 2013. Hirsutellin A: A paradigmatic example of the insecticidal function of fungal ribotoxins. *Insects* 4(3): 339–356

Hess, R.T. and Hoy, M.A. 1982. Microorganisms associated with the spider mite predator *Metaseiulus* (= *Typhlodromus*) *occidentalis*: Electron microscope observations. *Journal of Invertebrate Pathology* 40(1): 98–106

Huggins, L.G. and Waite, J.H. 1993. Eggshell formation in *Bdelloura candida*, an ectoparasitic turbellarian of the horseshoe crab *Limulus polyphemus*. *Journal of Experimental Zoology* 265: 549–557

Hughes, D.P., Araújo, J.P.M., Loreto, R.G., Quevillon, L., De Bekker, C., and Evans, H.C. 2016. From so simple a beginning: The evolution of behavioral manipulation by fungi. *Advances in Genetics* 94: 437–469

Ismail, N., Faridah, M., Ahmad, A., Alia'm, A.A., Khai, O.S., Sofa, M.F.A.M., and Manca, A. 2015. Marine bacteria associated with horseshoe crabs, *Tachypleus gigas* and *Carcinoscorpius rotundicauda*. In *Changing Global Perspectives on Horseshoe Crab Biology, Conservation and Management*, Ruth H. Carmichael, Mark L. Botton, Paul K.S. Shin, Siu Gin Cheung (eds.) pp. 313–320. Cham: Springer

Johnson, P.T. 1977. Paramoebiasis in the blue crab, Callinectes sapidus. *Journal of Invertebrate Pathology* 29(3): 308–320

Key Jr, M.M., Jeffries, W.B., Voris, H.K., and Yang, C.M. 1996. Epizoic bryozoans, horseshoe crabs, and other mobile benthic substrates. *Bulletin of Marine Science* 58(2): 368–384

Key Jr, M.M., Jeffries, W.B., Voris, H.K., Yang, C.M., and Jackson, J.B.C. 2000. Bryozoan fouling pattern on the horseshoe crab *Tachypleus gigas* (Müller) from Singapore. In *Proceedings of the 11th International Bryozoology Association Conference*, pp. 265–271. Balboa: Smithsonian Tropical Research Institute

Key, M.M., Knauff, J.B., and Barnes, D.K. 2013. Epizoic bryozoans on predatory pycnogonids from the south Orkney Islands, Antarctica: 'If you can't beat them, join them'. In *Bryozoan Studies, 2010*, pp. 137–153. Berlin, Heidelberg: Springer

King, P.E. and Crapp, G.B. 1971. Littoral pycnogonids of the British Isles. *Field Studies* 3(3): 455–480

Kleespies, R.G., Marshall, S.D., Schuster, C., Townsend, R.J., Jackson, T.A., and Leclerque, A. 2011. Genetic and electron-microscopic characterization of Rickettsiella bacteria from the manuka beetle, *Pyronota setosa*

(Coleoptera: Scarabaeidae). *Journal of Invertebrate Pathology* 107(3): 206–211

Krisfalusi-Gannon, J., Ali, W., Dellinger, K., Robertson, L., Brady, T.E., Goddard, M.K. et al. 2018. The role of horseshoe crabs in the biomedical industry and recent trends impacting species sustainability. *Frontiers in Marine Science* 5: 185

Kwan, B.K., Chan, A.K., Cheung, S.G., and Shin, P.K. 2014. Hemolymph quality as indicator of health status in juvenile Chinese horseshoe crab Tachypleus tridentatus (Xiphosura) under laboratory culture. *Journal of Experimental Marine Biology and Ecology* 457: 135–142

LaDouceur, E.E., Mangus, L., Garner, M.M. and Cartoceti, A.N. (2019). Histologic findings in captive American horseshoe crabs (*Limulus polyphemus*). *Veterinary Pathology*, 56: 932–939

Landy, R.B. and Leibovitz, L.A. 1983. A preliminary study of the toxicity and therapeutic efficacy of formalin in the treatment of triclad turbellarid worm infestations. Limulus polyphemus. In *Proceedings of the Annual Meeting of the Society of Invertebrate Pathology*. NY: Ithaca

Lane, S.J., Shishido, C.M., Moran, A.L., Tobalske, B.W., and Woods, H.A. 2016. No effects and no control of epibionts in two species of temperate pycnogonids. *The Biological Bulletin* 230(2): 165–173

Lane, S.J., Tobalske, B.W., Moran, A.L., Shishido, C.M., and Woods, H.A. 2018. Costs of epibionts on Antarctic sea spiders. *Marine Biology* 165(8): 137

Leatherdale, D. 1970. The arthropod hosts of entomogenous fungi in Britain. *Entomophaga* 15(4): 419–435

Leclerque, A. and Kleespies, R.G. 2008. Genetic and electron-microscopic characterization of *Rickettsiella tipulae*, an intracellular bacterial pathogen of the crane fly, *Tipula paludosa*. *Journal of Invertebrate Pathology* 98(3): 329–334

Leibovitz, L. (1986). Cyanobacterial diseases of the horseshoe-crab (*Limulus polyphemus*). *The Biological Bulletin* 171(2): 482–483

Leibovitz, L. and Lewbart, G.A. 1987. A green algal (Chlorophycophytal) infection of the dorsal surface of the exoskeleton, and associated organ structures, in the horseshoe-crab, Limulus polyphemus. *The Biological Bulletin* 73(2): 430–430

Leibovitz, L. and Lewbart, G.A. (2003). Diseases and symbionts: Vulnerability despite tough shells. In *The American Horseshoe Crab*, Shuster C.N., Barlow R.B., and Brockmann H.J. (eds.) pp. 245–275. Cambridge, MA: Harvard University Press

Levin, J. and Bang, F.B. 1964. A description of cellular coagulation in the Limulus . *Bulletin of the Johns Hopkins Hospital* 115: 337–345

Lewis, D. 1979. The detection of rickettsia-like microorganisms within the ovaries of female *Ixodes ricinus* ticks. *Zeitschrift fur Parasitenkunde* 59(3): 295–298

Li, C.X., Shi, M., Tian, J.H., Lin, X.D., Kang, Y.J. et al. 2015. Unprecedented genomic diversity of RNA viruses in arthropods reveals the ancestry of negative-sense RNA viruses. *eLife* 4: e05378.

Liu, T.P. 1991. Virus-like particles in the tracheal mite *Acarapis woodi* (Rennie). *Apidologie* 22(3): 213–219

Lozano-Fernandez, J., Tanner, A.R., Giacomelli, M., Carton, R., Vinther, J., Edgecombe, G.D., and Pisani, D. 2019. Increasing species sampling in chelicerate genomic-scale datasets provides support for monophyly of Acari and Arachnida. *Nature Communications* 10: 1–8

Mains, E.B. 1939. Entomogenous species of *Akanthomyces*, *Hymensotilbe* and *Insecticola* in North America. *Mycologia* 42: 566–589

Mains, E.B. 1954. Species of *Cordyceps* on spiders. *Bulletin of the Torrey Botanical Club* 81(6): 492–500

Martin, O.Y. and Goodacre, S.L. 2009. Widespread infections by the bacterial endosymbiont Cardinium in arachnids. *The Journal of Arachnology* 37(1): 106–108

Masan, P., Simpson, C., Perotti, M.A., and Braig, H.R. 2012. Mites parasitic on Australasian and African spiders found in the pet trade; a redescription of *Ljunghia pulleinei* Womersley. *PLoS One* 7(6): e39019

Möller, A. 1901. Botanische Mittheilungen aus den Tropen von Schimper AFW, Heft 9. *Phycomyceten und Ascomyceten*, untersuchungen aus Brasilien

Fischer, Jena, Monteiro, C.L.B., Rubel, R., Cogo, L.L., Mangili, O.C., Gremski, W., and Veiga, S.S. 2002. Isolation and identification of *Clostridium perfringens* in the venom and fangs of *Loxosceles intermedia* (brown spider): Enhancement of the dermonecrotic lesion in loxoscelism. *Toxicon* 40(4): 409–418

Morel, G. 1976. Studies on *Porochlamydia buthi*, sp. n., an intracellular pathogen of the scorpion *Buthus occitanus*. *Journal of Invertebrate Pathology* 28(2): 167–175

Morel, G. 1977. Study of a *Rickettsiella* (Rickettsia) pathogen of the spider *Pisaura mirabilis*. *Annales de Microbiologie* 128: 49–59

Morel, G. 1978. Les maladies microbiennes des Arachnides (Acariens exceptés). *Symposium of the Zoological Society of London* 42: 477–481

Muma, M.H. 1955. Factors contributing to the natural control of citrus insects and mites in Florida. *Journal of Economic Entomology* 48(4): 432–438

Mürer, E.H., Levin, J., and Holme, R. 1975. Isolation and studies of the granules of the amebocytes of *Limulus polyphemus*, the horseshoe crab. *Journal of Cellular Physiology* 86(3): 533–542

Nakamura, Y., Kawai, S., Yukuhiro, F., Ito, S., Gotoh, T., Kisimoto, R. et al. 2009. Prevalence of Cardinium bacteria in planthoppers and spider mites and taxonomic revision of 'Candidatus Cardinium hertigii' based on detection of a new Cardinium group from biting midges. *Applied and Environmental Microbiology* 75(21): 6757–6763

Nentwig, W. 1985. Parasitic fungi as a mortality factor of spiders. *The Journal of Arachnology* 13(2): 272–274

Newton, A.L. and Smolowitz, R. 2018. Invertebrates. In *Pathology of Wildlife and Zoo Animals*, pp. 1019–1052. Academic Press

Nolan, M.W. and Smith, S.A. 2009. Clinical evaluation, common diseases, and veterinary care of the horseshoe crab, *Limulus polyphemus*. In *Biology and Conservation of Horseshoe Crabs*, John T. Tanacredi, Mark L. Botton, David Smith (eds,). pp. 479–499. Boston, MA: Springer

Patil, J.S. and Anil, A.C. 2000. Epibiotic community of the horseshoe crab *Tachypleus gigas*. *Marine Biology* 136: 699–713

Paules, C.I., Marston, H.D., Bloom, M.E., and Fauci, A.S. 2018. Tickborne diseases—confronting a growing threat. *New England Journal of Medicine* 379(8): 701–703

Penz, T., Schmitz-Esser, S., Kelly, S.E., Cass, B.N., Müller, A., Woyke, T. et al. (2012). Comparative genomics suggests an independent origin of cytoplasmic incompatibility in *Cardinium hertigii*. *PLoS Genetics* 8(10): e1003012

Pillonel, T., Bertelli, C., Aeby, S., De Barsy, M., Jacquier, N., Kebbi-Beghdadi, C. et al. 2019. Sequencing the obligate intracellular *Rhabdochlamydia helvetica* within its tick host *Ixodes ricinus* to investigate their symbiotic relationship. *Genome Biology and Evolution* 11(4): 1334–1344

Pipe, A.R. 1982. Epizoites on marine invertebrates: With particular reference to those associated with the pycnogonid *Phoxichilidium tubulariae* Lebour, the amphipod *Caprella linearis* (L.) and the decapod *Corystes cassivelaunus* (Pennant). *Chemistry in Ecology* 1(1): 61–74

Pizzi, R. 2009. Parasites of tarantulas (Theraphosidae). *Journal of Exotic Pet Medicine* 18(4): 283–288

Pizzi, R. 2012. Spiders. In *Invertebrate Medicine*, 2nd Edition. Lewbart, G.A. (ed.) Chapter 11, pp. 187–221. New York: John Wiley & Sons

Pizzi, R., Carta, L., and George, S. 2003. Oral nematode infection of tarantulas. *Veterinary Record* 152(22): 695–695

Poinar G.O. Jr. 1985. Mermithid (Nematoda) parasites of spiders and harvestmen. *Journal of Arachnology* 13: 121–128

Poinar G.O. Jr. 1987. Nematode parasites of spiders. In *Ecophysiology of Spiders*, Wolfgang Nentwig (ed.) pp. 299–308. Berlin, Heidelberg: Springer

Poinar G.O. Jr., Čurčic, B.P., Karaman, I.M., Cokendolpher, J.C., and Mitov, P.G. 2000. Nematode parasitism of harvestmen (Opiliones: Arachnida). *Nematology* 2(6): 587–590

Poinar G.O. Jr. and Stockwell, S.A. 1988. A new record of a nematode parasite (Mermithidae) of a scorpion. *Revue de Nématologie* 11(3): 361–364

Poinar G.O. Jr. and Poinar, R. 1998. Parasites and pathogens of mites. *Annual Review of Entomology* 43(1): 449–469

Pontoppidan, M.B., Himaman, W., Hywel-Jones, N.L., Boomsma, J.J., and Hughes, D.P. 2009. Graveyards on the move: The spatio-temporal distribution of dead *Ophiocordyceps*-infected ants. *PLoS one* 4(3): e4835

Rama Rao, K.V. and Surya Rao, K.V. 1972. Studies on Indian king crabs (Arachnida, Xiphosura). *Proceedings of the Indian National Science Academy B* 38: 206–211

Ramsey, S. D., Ochoa, R., Bauchan, G., Gulbronson, C., Mowery, J.D., Cohen, A. et al. (2019). Varroa destructor feeds primarily on honeybee fat body tissue and not hemolymph. *Proceedings of the National Academy of Sciences*, 116(5): 1792–1801

Reed, D.K. 1981. Control of mites by non-occluded viruses. *Microbial control of pests and plant diseases, 1970–1980*

Reed, D.K. and Desjardins, P.R. 1978. Isometric virus-like particles from citrus red mites, Panonychus citri. *Journal of Invertebrate Pathology* 31(2): 188–193

Reeves, W.K., Dowling, A.P., and Dasch, G.A. (2006). Rickettsial agents from parasitic dermanyssoidea (Acari: Mesostigmata). *Experimental & Applied Acarology* 38(2–3): 181–188

Riesgo, A., Burke, E.A., Laumer, C., and Giribet, G. 2017. Genetic variation and geographic differentiation in the marine triclad *Bdelloura candida* (Platyhelminthes, Tricladida, Maricola), ectocommensal on the American horseshoe crab *Limulus polyphemus*. *Marine Biology* 164(5): 111

Romero, X., Turnbull, J.F., and Jiménez, R. (2000). Ultrastructure and cytopathology of a rickettsia-like organism causing systemic infection in the redclaw crayfish, *Cherax quadricarinatus* (Crustacea: Decapoda), in Ecuador. *Journal of Invertebrate Pathology* 76(2): 95–104

Ros, V.I.D. and Breeuwer, J.A.J. 2009. The effects of, and interactions between, *Cardinium* and *Wolbachia* in the doubly infected spider mite 'i' Bryobia sarothamni. *Heredity* 102(4): 413–422

Rosenwald, L.C., Sitvarin, M.I., and White, J.A. 2020. Endosymbiotic Rickettsiella causes cytoplasmic incompatibility in a spider host. *Proceedings of the Royal Society B* 287: 20201107

Samson, R.A. and Evans, H.C. 1973. Notes on Entomogenous Fungi from Ghana: I. The Genera *Gibellula* and Pseudogibellula. *Acta Botanica Neerlandica* 22(5): 522–528

Samson, R.A., McCoy, C.W., and O'Donnell, K.L. 1980. Taxonomy of the acarine parasite Hirsutella thompsonii. *Mycologia* 72(2): 359–377

Sharma, P.P. (2018). Chelicerates. *Current Biology* 28: R774–R778

Shinn, A.P., Mühhölzl, A.P., Coates, C.J., Metochis, C., and Freeman, M.A. 2015. *Zoothamnium duplicatum* infestation of cultured horseshoe crabs (*Limulus polyphemus*). *Journal of Invertebrate Pathology* 125: 81–86

Shrestha, B., Kubátová, A., Tanaka, E., Oh, J., Yoon, D.H., Sung, J.M., and Sung, G. H. 2019. Spider-pathogenic fungi within Hypocreales (Ascomycota): Their current nomenclature, diversity, and distribution. *Mycological Progress* 18(8): 983–1003

Shuster, C.N. 2009. Public participation in studies on horseshoe crab populations. In *Biology and Conservation of Horseshoe Crabs*, John T. Tanacredi, Mark L. Botton, David Smith (eds.) pp. 585–594. Boston, MA: Springer

Shuster, C.N. and Sekiguchi, K. (2003). Growing up takes about ten years and eighteen stages. In *The American Horseshoe Crab*. Shuster C.N., Barlow R.B., and Brockman H.J. (eds.) pp. 103–132. Cambridge, MA: Harvard University Press .

Smith, K.M. and Cressman, A.W. 1962. Birefringent crystals in virus-diseased citrus red mites. *Journal of Insect Pathology* 4(2): 229

Smith, S.A. and Berkson, J. 2005. Laboratory culture and maintenance of the horseshoe crab (*Limulus polyphemus*). *Lab Animal* 34: 27–34

Smith, S.A., Berkson, J.M., and Barratt, R.A. 2002. Horseshoe crab (*Limulus polyphemus*) hemolymph, biochemical and immunological parameters. *Proceedings of the International Association of Aquatic Animal Medicine* 33: 101–102

Smith, S.A., Scimeca, J.M., and Mainous, M.E. 2011. Culture and maintenance of selected invertebrates in the laboratory and classroom. *ILAR Journal* 52(2): 153–164

Söderhäll, K., Levin, J., and Armstrong, P.B. 1985. The effects of β1, 3-glucans on blood coagulation and amebocyte release in the horseshoe crab, Limulus polyphemus. *The Biological Bulletin* 169(3): 661–674

Speare, A.T. and Yothers, W.W. 1924. Is there an entomogenous fungus attacking the citrus rust mite in Florida? *Science*, 60(1541), 41–42

Stefanini, A. and Duron, O. 2012. Exploring the effect of the *Cardinium* endosymbiont on spiders. *Journal of Evolutionary Biology* 25(8): 1521–1530

Steinhaus, E. 1960. Possible virus disease in European red mite. *California Agriculture* 14(2): 3

Strongman, D.B. 1991. *Gibellula pulchra* from a spider (Salticidae) in Nova Scotia, Canada. *Mycologia* 83(6): 816–817

Stunkard, H.W. 1950. Microphallid metacercariae encysted in *Limulus polyphemus. The Biological Bulletin* 99(2)

Stunkard, H.W. 1951. Observations on the morphology and life-history of *Microphallus limuli* n. sp. (Trematoda: Microphallidae). *The Biological Bulletin* 101(3): 307–318

Tan, A.N., Christianus, A., and Satar, M.A. 2011. Epibiont infestation on horseshoe crab *Tachypleus gigas* (Müller) at Pantai Balok in Peninsular Malaysia. *Our Nature* 9: 9–15

Thompson, J., Ben Ahmeid, A.A., Pickett, C., and Neil, D. 2011. Identification of cost-effective measures to improve holding conditions and husbandry practices for the horseshoe crab *Limulus polyphemus*. Project Report. Glasgow, UK: University of Glasgow

Tonk, M., Cabezas-Cruz, A., Valdés, J.J., Rego, R. O., Grubhoffer, L., Estrada-Pena, A. et al. 2015. *Ixodes ricinus* defensins attack distantly-related pathogens. *Developmental & Comparative Immunology* 53(2): 358–365

Turner, L.L., Kammire, C., and Sydlik, M A. (1988). Preliminary report, composition of communities resident on limulus carapaces. *The Biological Bulletin*, 175(1), 312–312.

Tuxbury, K.A., Shaw, G.C., Montali, R.J., Clayton, L.A., Kwiatkowski, N.P., Dykstra, M.J. and Mankowski, J.L. 2014. *Fusarium solani* species complex associated with carapace lesions and branchitis in captive American horseshoe crabs *Limulus polyphemus*. *Diseases of Aquatic Organisms* 109: 223–230

Vanthournout, B. and Hendrickx, F. 2015. Endosymbiont dominated bacterial communities in a dwarf spider. *PLoS One* 10(2): e0117297

Walls, E.A., Berkson, J., and Smith, S.A. (2002). The horseshoe crab, *Limulus polyphemus*: 200 million years of existence, 100 years of study. *Reviews in Fisheries Science* 10(1): 39–73

Wang, L., Zhang, S., Li, J. H., and Zhang, Y.J. 2018. Mitochondrial genome, comparative analysis and evolutionary insights into the entomopathogenic fungus *Hirsutella thompsonii*. *Environmental Microbiology* 20(9): 3393–3405

Weeks, A.R. and Breeuwer, J.A. 2003. A new bacterium from the Cytophaga-Flavobacterium-Bacteroides phylum that causes sex ratio distortion. *Insect Symbiosi*, 165–176

Weibelzahl, E. and Liburd, O.E. 2009. Epizootic of *Acalitus vaccinii* (Acari: Eriophyidea) caused by *Hirsutella thompsonii* on southern highbush blueberry in north-central Florida. *Florida Entomologist* 601–617

Weinert, L.A., Araujo-Jnr, E.V., Ahmed, M.Z., and Welch, J.J. 2015. The incidence of bacterial endosymbionts in terrestrial arthropods. *Proceedings of the Royal Society B: Biological Sciences* 282(1807): 20150249

Welbourn, W.C. and Young, O.P. 1988. Mites parasitic on spiders, with a description of a new species of *Eutrombidium* (Acari, Eutrombidiidae). *Journal of Arachnology* 373–385

Werren, J.H., Baldo, L. and Clark, M.E. 2008. *Wolbachia*: Master manipulators of invertebrate biology. *Nature Reviews Microbiology* 6(10): 741–751

White, J.A., Styer, A., Rosenwald, L.C., Curry, M.M., Welch, K. D., Athey, K.J., and Chapman, E.G. 2020. Endosymbiotic bacteria are prevalent and diverse in agricultural spiders. *Microbial Ecology* 79(2): 472–481

Williams, D.L. 1987. Studies in arachnid disease. In *Arachnida: Proceedings of a one day symposium on Spiders and Their Allies held at the Zoological Society of London, 21 November 1987.* Cooper J.E., Pearce-Kelly P., and Williams D.L. (eds.) pp. 116–125. Keighley, England: Chiron

Wyer, D.W. and King, P.E. 1973. Relationships between some British littoral and sublittoral bryozoans and pycnogonids. *Living and fossil Bryozoa* 642

Zhang, Y.K., Chen, Y.T., Yang, K., and Hong, X.Y. 2016. A review of prevalence and phylogeny of the bacterial symbiont *Cardinium* in mites (subclass: Acari). *Systematic and Applied Acarology* 21(7): 978–990

Zhu, L.Y., Zhang, K.J., Zhang, Y.K., Ge, C., Gotoh, T., and Hong, X. Y. 2012. *Wolbachia* strengthens *Cardinium*-induced cytoplasmic incompatibility in the spider mite *Tetranychus piercei* McGregor. *Current Microbiology* 65(5): 516–523

Zug, R. and Hammerstein, P. 2012. Still a host of hosts for *Wolbachia*: Analysis of recent data suggests that 40% of terrestrial arthropod species are infected. *PloS one* 7(6): e38544

CHAPTER 10

Viral diseases of insects

Vera I.D. Ros, Delphine Panziera, Remziye Nalcacioglu, Jirka Manuel Petersen, Eugene Ryabov, and Monique M. van Oers

10.1 Introduction

Insects harbour a wide range of viruses and the recent revolution of large-scale sequencing technologies has only further increased the number of virus species recognised. These viruses include plant and animal viruses vectored by insects but also a wide range of viruses that are specific to insects. In this chapter, we describe the main virus families known to cause disease in wild and cultured insect species. We focus on insects in terrestrial environments, although terrestrial insects may have an aquatic phase in their life cycle. Viral infections in insects may show very high mortality rates, but viruses may also adopt a persistent or latent phase (see Box 10.1) and only cause a disease outbreak under certain circumstances, such as population overcrowding or other unfavourable conditions for insect health. Viruses may preferably transmit horizontally, so between individual insects within a population but there are also clear examples of viruses being transmitted vertically (so from parent to offspring). Also, the infection routes are species dependent, although for most insect viruses, horizontal transmission occurs via oral uptake of contaminated food. Sexual transmission, infection via trachea, wounds, cannibalism of infected conspecifics, and vectoring by other arthropod species may also happen (e.g. Varroa mites feeding on bee larvae, or parasitoids injecting viruses into the parasitised hosts).

We discuss the properties of the key virus families by dividing these into groups based on the nature of their genomes and replication strategy, since that, in many cases, also reflects their evolutionary relationships. We illustrate the impact that viruses may have on particular groups of insects by providing examples of viral infections in insects in their natural environments and in insects reared for pollination, for food and feed, as well as for biological control purposes. The global production of insects is estimated at 100,000 tons for 2023 and a further 10-fold increase is foreseen for 2030, with matching annual investment. Edible insects form an important protein source in human diets and the global edible insect market was validated at 112 million US$ in 2019, with an estimated gain of over 40% by 2026 (Ahuja and Mamtani 2020). Pest insects such as caterpillars are grown to culture entomopathogens and parasitoids to be used in integrated pest management. Likewise, fruit flies and insects that serve as vectors for human or veterinary diseases are reared for sterile insect technology (SIT) applications. In all these cases, successful insect farming relies heavily on culturing large colonies of a single insect species. Such high-density monoculture conditions in artificial environments easily trigger viral disease outbreaks, accompanied with extensive economic losses for the producers.

Per virus family, we reflect on viral features and the associated pathology, as well as on ecological aspects if sufficient data are available. Due to the broadness of the topic, we refer to overview papers when available. Clear behavioural changes caused by viruses are also indicated, but for an in-depth discussion we refer to Han et al. (2015). Figure 10.1 shows the molecular features of the various taxonomical groups of viruses causing disease in insects. The set-up of Figure 10.2 allows the reader to identify the kind of viruses found thus far in

Vera I.D. Ros et al., *Viral diseases of insects*. In: *Invertebrate Pathology*. Edited by Andrew F. Rowley, Christopher J. Coates and Miranda M.A. Whitten, Oxford University Press. © Oxford University Press (2022). DOI: 10.1093/oso/9780198853756.003.0010

Box 10.1 Some terminology on insect viruses

In insect pathology **covert** virus infections (also known as occult, sublethal, inapparent, silent, or asymptomatic infections), are described as being either persistent or latent. **Persistent infections** involve low levels of viral genome replication as well as expression of a range of viral genes, and virus particles (virions) may be produced. **Latent infections** do not involve viral genome replication or virion production and exhibit minimal gene expression to maintain latency. Covert infections are characterised by the absence of visible signs of disease, although persistent infections may negatively affect host fitness, referred to as sublethal disease.

Horizontal transmission is from one individual to another within or between generations through environmental contamination. **Vertical transmission** goes from parent to offspring, which may occur via contamination of the egg surface or the virus may be within the eggs (Cory and Myers 2003; Williams et al. 2017b).

---

various orders of insects. Figures 10.3 to 10.6 present examples of disease symptoms, while Tables 10.1 and 10.2 summarise the characteristics regarding host, pathology, and transmission for a number of ribonucleic acid (RNA) and deoxyribonucleic acid (DNA) viruses used as examples. The application of pathogenic insect viruses in biocontrol programmes is a separate section in this chapter. A brief viral disease management section is included for apiculture and insect mass rearing, despite the fact that knowledge gaps and lack of available diagnostic tools and service centres often form serious obstacles in this respect. We conclude with a future outlook, where we especially reflect on the large number of insect viruses being discovered these days and the difficulties encountered to understand which role all these viruses play in ecosystems.

10.2 Major groups of RNA viruses causing disease in insects

10.2.1 Insect viruses with positive sense RNA genomes

10.2.1.1 Introduction

RNA viruses in insects vary widely in their structural properties and genome characteristics and as a consequence belong to a large variety of taxonomic groups (Figure 10.1; Table 10.1). As new RNA viruses are being discovered at high speed, the RNA virus taxonomy is also highly dynamic. For this book chapter, we selected some prominent examples and refer to Ryabov (2017) for an excellent overview of RNA viruses in invertebrate species. Many insect-infecting viruses have single-stranded (ss) positive (+) sense RNA genomes (meaning the RNA can be directly translated into protein). The majority of these viruses are classified in families within the orders *Picornavirales* and *Nidovirales* or belong to one of the tetravirus families (for these viruses see the separate sections to follow). Insect specific (+) strand RNA viruses are also represented in the genus *Flavivirus* (family *Flaviviridae*). These insect-specific flaviviruses are in general not associated with disease symptoms, although they may affect the replication of vertebrate flaviviruses of medical importance vectored by mosquitoes (see for a review Blitvich and Firth 2015). Eilat virus, first found in the Negev desert, is an example of a mosquito (*Anopheles coustani*) restricted member of the family *Togaviridae* (genus *Alphavirus*). A close relative is Taï Forest virus in *Culex decens* mosquitoes detected in Ivory Coast (Hermanns et al. 2017).

10.2.1.2 Insect viruses in the order *Picornavirales*

Viruses in the large order *Picornavirales* have a (+) sense RNA genome with a small (3–4 kDa) viral protein that is genome-linked (VPg) to the 5′ end, and a polyA-tail at the 3′ end. The genomic RNA is translated via a cap-independent translation mechanism into one or two polyproteins that give rise to functional proteins through auto-proteolytical activities. Three main viral protein (VP) subunits of approximately 25 kDa each assemble into small,

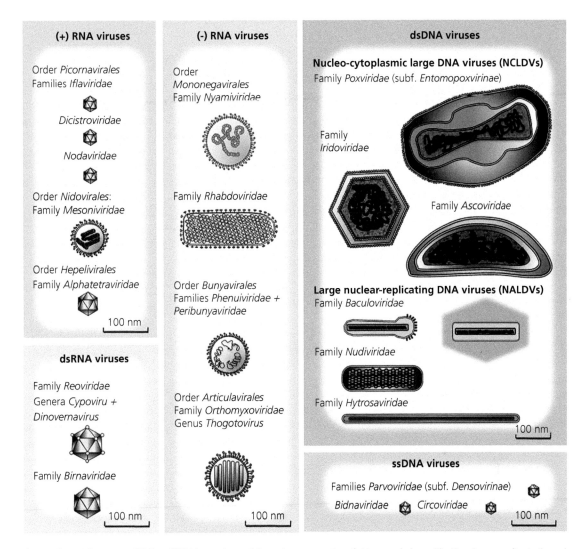

Figure 10.1 Major groups of RNA and DNA insect viruses. Schematic representation of virion morphology with virion sizes according to the scale shown in each panel.

icosahedral, non-enveloped particles of approximately 30 nm (Figure 10.1); (Le Gall et al. 2008). The viruses in the families *Dicistroviridae* and *Iflaviridae* have been exclusively found in invertebrate species and frequently occur in insects of economic importance, including silkworms and pollinator species (Figure 10.2). Picornavirus infections in the western honeybee (*Apis mellifera*) are illustrated by the following examples chosen for their dominant worldwide prevalence in honeybee colonies (Beaurepaire et al. 2020; Chen and Siede 2007).

The acute bee paralysis virus (ABPV) complex (family *Dicistroviridae*) formed by ABPV, Kashmir bee virus (KBV), and Israeli acute bee paralysis virus (IAPV) trigger a series of both physical and behavioural symptoms in infected honeybees, including gradual darkening and loss of thorax and abdomen hairs, trembling, inability to fly, and rapid death (Chen and Siede 2007).

Sacbrood virus (SBV) (family *Iflaviridae*) prevents *A. mellifera* larvae from pupating due to the accumulation of virus-rich ecdysial fluid beneath the skin

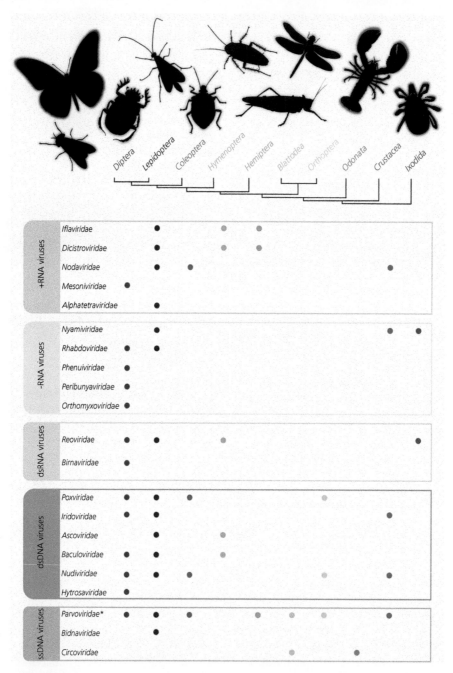

Figure 10.2 Overview of the main virus families causing disease in insects and their occurrence in orders of insects, crustaceans and ticks (order Ixodida). Only arthropod orders known to be infected by members of the listed virus families are shown. For an in-depth overview of viral diseases of crustaceans, we refer to Chapter 14 of this book. The phylogenetic tree reflects the relationship between the various orders of arthropods displayed (based on Misof et al. 2014). *The insect-infecting viruses classified in the family *Parvoviridae* are better known as densoviruses (subfamily *Densovirinae*).

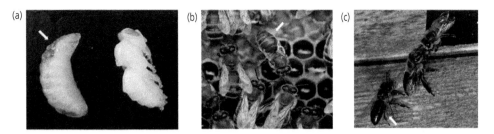

Figure 10.3 Typical symptoms of viral infections in honeybees. Honey bee pupae showing symptoms of Sacbrood virus (SBV) infection (a), and adults with underdeveloped wings as a consequence of Deformed wing virus (DWV) infection (b), or with a black, hairless abdomen as a result of Acute bee paralysis virus (ABPV) infection (c). In all images the insects that display visible signs of infection (arrows) are compared to normal individuals of the same age. Source: photos were taken by Eugene Ryabov (a) and Bram Cornelissen (b and c).

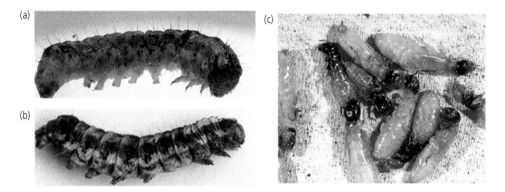

Figure 10.4 Images of insects infected with iridoviruses and entomopoxviruses. Non-infected *Helicoverpa armigera* caterpillars (a) are compared to conspecifics with a patent insect iridescent virus 6 (IIV6) infection (b), showing the typical iridescent appearance. *Neotermes jouteli* termite cadavers showing a dark-head syndrome (c), most likely due to an entomopoxvirus infection. Source: figures were used with permission from Choevenc et al., 2013 and Gencer et al. 2020.

(Figure 10.3A). SBV is also present in the hypopharyngeal glands of honeybee workers and causes precocious foraging activity, as well as a shortened lifespan.

Black queen cell virus (BQCV) (family *Dicistroviridae*) affects adult honeybees as well as queen larvae, displaying symptoms resembling SBV due to the accumulation of ecdysial fluid. The infected pupae turn brown and die leaving a dark colour on the queen cells walls. BQCV has also the ability to replicate in bumblebees *(Bombus sp.)*.

Other iflaviruses pathogenic to honeybees are deformed wing virus A (DWV-A) (Lanzi et al. 2006) and Varroa destructor virus 1 (nowadays called DWV-B) (Ongus et al. 2004), and recombinants between these two strains (Dalmon et al. 2017; Moore et al. 2011). While DWV is often present in colonies as a covert infection, its successful adaptation to vector transmission through the invasive mite species *Varroa destructor* has led to an increase in prevalence and virulence in honeybee colonies worldwide. DWV causes wing and abdomen deformities when transmitted to developing honeybee pupae by these *Varroa* mites (Figure 10.3B) and impaired cognitive functions in honeybee workers. DWV is also known to transmit horizontally to other species of insects that share their habitat with *A. mellifera* (Yañez et al. 2020), and thus, poses a potential threat to wild insect communities. Bees can also be infected with slow bee paralysis virus (SBPV) (family *Iflaviridae*) affecting the nervous system and causing paralysis of the anterior legs, followed by death within 24 to 48 hours after the first symptoms (Beaurepaire et al. 2020).

Iflaviruses have also been reported in lepidopteran species and are transmitted horizontally as well as vertically. Examples are Spodoptera exigua iflavirus 1 (SeIV-1) and 2 (SeIV2) that naturally infect *S. exigua* populations as non-lethal covert infections, and that occur individually as well as mixed (Jakubowska et al. 2014; Virto et al. 2014). These iflaviruses were found to increase the susceptibility of *S. exigua* caterpillars to baculovirus infections (see Section 10.3); (Carballo et al. 2020; Jakubowska et al. 2016). Iflavirus infections in *Helicoverpa armigera* negatively affect host fitness (Yuan et al. 2020), as evidenced by prolonged development times and higher mortality levels of larvae and pupae.

Nora viruses have been found in *Drosophila melanogaster*, wasps, honeybees and lepidopteran species, and these together potentially form a new family within the order *Picornavirales* (Jakubowska et al. 2014; Marzoli et al. 2020; Yang et al. 2019). While mostly non-symptomatic, Nora virus infection was found to impair geotaxis in *D. melanogaster* (Rogers et al. 2020).

10.2.1.3 Nodaviruses

Nodaviruses are small (appr. 30 nm), non-enveloped bipartite (+) sense RNA viruses with icosahedral symmetry. Viruses in the genus *Alphanodavirus* infect insects and well-known members include Nodamura virus, Flock House Virus (FHV), Black beetle virus, and Pariacoto virus (Yong et al. 2017). FHV originated from the grass grub *Costelytra zealandica* but is also frequently present as persistent infection in *Drosophila melanogaster* cell lines (Dasgupta et al. 1994). Wuhan nodavirus was isolated from *Pieris rapea* larvae (Liu et al. 2006). The *Trichoplusia ni* derived H5 cell line, frequently used in baculovirus expression technology, is also known to be latently infected with an alphanodavirus, whose replication is boosted by a baculovirus infection (Li et al. 2007). Mosinovirus (MoNV) infecting mosquitoes has a capsid protein of unknown origin that is not related to that of other nodaviruses (Schuster et al. 2014). These viruses are often persistently infecting insects, likely transmit vertically, and clear disease symptoms have not been reported to our knowledge.

10.2.1.4 Chronic bee paralysis virus

Chronic bee paralysis virus (CBPV) is an unclassified RNA virus carrying two (+) sense RNA fragments (Olivier et al. 2008). Genome analysis reveals significant distance from any defined family, although the RNA-dependent RNA polymerase (RdRP) of CBPV shares conserved domains with RdRPs of nodaviruses (see Section 10.2.1.3) and plant-infecting viruses in the family *Tombusviridae*. CBPV causes severe disease in honeybee colonies as it replicates in the nervous system and/or the mandibular and hypopharyngeal glands causing two types of symptoms. The type 1 syndrome is defined by worker bees displaying abnormal trembling of the wings or the whole body making them completely unable to fly. These bees either cluster together on top of the hive, or are seen crawling on the ground in front of the hive in masses of sometimes thousands of individuals (Beaurepaire et al. 2020; Chen and Siede 2007). Infected workers often also have bloated abdomens caused by a distension of the honey sac and die within a week. The type 2 syndrome is characterised by the presence of hairless bees ('black robbers') in the colony. These are at first able to fly but are attacked by healthy workers (like robber bees would be). They start trembling after a few days and die rapidly. The two types of symptoms can co-occur within a single colony. Generally, CBPV persists in colonies as a covert infection, but may rapidly convert into overt infections, causing significant colony losses. It is thought that environmental factors such as nutritional or thermal stressors might be triggers of CBPV outbreaks. CBPV has a worldwide distribution, with a recently reported increase in prevalence in Asia, Europe, and North America. In England and Wales, government honeybee health inspection services recorded an exponential increase of cases between 2007 and 2017, suggesting CBPV-associated disease might be considered as emerging (Budge et al. 2020).

10.2.1.5 Viruses in the order *Nidovirales* that infect insects

The viruses in the order *Nidovirales* are enveloped and have relatively large, monopartite, (+) sense

(a)

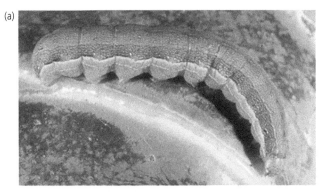

(b)

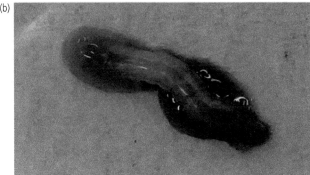

Figure 10.5 Liquefaction due to baculovirus infection in lepidopteran insects. A non-infected beet army worm feeding on leaf material (a), compared to a liquefied conspecific due to a Spodoptera exigua MNPV infection (b), 4–5 days post infection.
Source: figure (a) was used with permission from https://entomology.k-state.edu/extension/insect-information/crop-pests, photo (b) was taken by Vera Ros.

RNA genomes that are capped and polyadenylated. Sub-genomic mRNAs and ribosomal frameshifting assist to express all open reading frames (ORFs). The resulting primary translation products often need proteolytic cleavage to obtain functional proteins. The insect-infecting viruses in this taxon fall under the family *Mesoniviridae*. These viruses form spherical particles, 60–80 nm in diameter, with trimeric glycoprotein spikes on the surface (Figure 10.1). The envelope is structurally supported from within by the matrix protein. Mesonivirus genomes are 20 kb in size and associated with the nucleocapsid protein. Mesoniviruses have a wide geographical distribution and occur in various genera of mosquitoes, including those vectoring human diseases (Lauber et al. 2012; Vasilakis et al. 2014). Recent data indicate that these viruses can be transmitted both vertically and horizontally (Newton et al., 2020; Ye et al., 2020). No clear pathological effects have been described for these viruses, and to what level these viruses affect the fitness and vectoring capacity of infected mosquitoes needs further study.

10.2.1.6 Three families of tetraviruses

Tetraviruses are non-enveloped viruses with a typical T = 4 icosahedral symmetry (hence the name) and a diameter of 40 nm. Their genomes are 6–8 kb in length. Tetraviruses typically infect the midgut of caterpillars via the oral route, but isolates vary considerably in pathogenicity, causing symptoms ranging from unapparent to lethal infections (Dorrington et al. 2020). Tetraviruses belong to either of three families (*Alphatetraviridae*, *Carmotetraviridae*, and *Permutotetraviridae*) based on distinct properties of their RdRP (ICTV 2020; Walter et al. 2010). The RdRP of alphatetraviruses has typical alpha-like virus properties (van der Heijden and Bol 2002) and their genome has 5′ cap and 3′ tRNA-like structures, and is either monopartite (in the genus *Betatetravirus*) or bipartite (genus *Omegatetravirus*). The genomic RNA is translated into a polyprotein, which is cleaved to give individual proteins. Viruses in the family *Alphatetraviridae* are for instance Nudaurelia capensis β virus and Helicoverpa armigera stunt virus (HaSV), which is

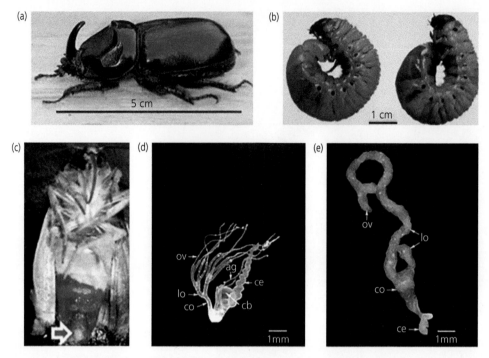

Figure 10.6 Symptoms of nudivirus infections. *Oryctes rhinoceros* beetle (a) with third instar larva with heavy signs of Oryctes rhinoceros nudivirus (OrNV) infection, including prolapse of the rectum (b, left), compared to a healthy specimen (b, right). *Helicoverpa zea* female infected with Helicoverpa zea nudivirus (HzNV-2) showing a virus-filled 'waxy plug' (c). Ovary tissue was isolated from normal (d) and infected, female pupae showing malformations (e). ov = ovarioles, la = lateral oviduct, co = common oviducts, ag = accessory glands, ce = cervix bursa, cb = corpus bursa. Source: images (a and b) were re-used from Huger 2005 with permission from Alois H. Huger; Images (c-e) were re-used with permission from Rallis and Burand 2002, and from Burand et al. 2012.

used for biocontrol applications (Brooks et al. 2002). Viruses in the family *Carmotetraviridae* have RdRps with similar properties as plant tombusviruses and also have non-segmented RNA genomes. Providence virus, persistently infecting a *Helicoverpa armigera* midgut cell line belongs to this family (Pringle et al. 2003). Permutotetraviruses on the other hand, have RdRP that are distantly related to those of birnaviruses (Zeddam et al. 2010) (see also Section 10.2.3.2), and which contain permuted domains. Their non-segmented genomes have a VPg protein at the 5′ end and a pseudoknot at the 3′ end. Examples of permutoetraviruses are Thosea asigna virus and Euprosterna elaeasa virus, both infecting slug moths of the family Limacodidae (Sugiharti et al. 2010).

10.2.1.7 Negeviruses

Negeviruses form a diverse group of non-segmented, (+) sense RNA viruses that await formal classification (Vasilakis et al. 2013). Negevirus particles appear spherical and 45–55 nm in diameter. They have been isolated from mosquitoes and sand flies (family Phlebotominae) (Nunes et al. 2017). Unpublished data also indicate the presence of a negevirus in tsetse flies (family Glossinidae) (Abd-Alla, personal communication). So far, no data on mode of transmission or pathology have been reported for these viruses. Nor is it known whether vector competence of mosquitoes and tsetse flies for human pathogens is affected.

10.2.2 Insect viruses with negative sense RNA genomes

10.2.2.1 Viruses in the order *Monanegavirales*

The order *Mononegavirales* harbours two families with insect-infecting viruses with non-segmented, (-) sense RNA genomes, the *Rhabdoviridae* and *Nyamiviridae*. Rhabdovirus particles are rod-shaped and enveloped (Figure 10.1; Table 10.1). Dipteran

Table 10.1 Taxonomically classified insect viruses with RNA genomes with selected examples and references.[1,2]

Virus order	(sub) Family	Prominent examples	Original host(s) [Order: Family][3]	Transmission routes	Described pathology	References
Picornavirales	*Dicistroviridae*	Acute bee paralysis virus (ABPV) complex	*Apis mellifera* (Hymenoptera: Apidae)	Oral-faecal, *Varroa* mites, vertically	Trembling and affected behaviour, eventually paralysis and death	Beaurepaire et al. 2020
		Black queen cell virus (BQCV)	*Apis mellifera*	Oral-faecal, via shared resources, vertically	Queen development failure at the pupal stage, shortened lifespan in adult workers	
		Cricket paralysis virus (CrPV)	Crickets (Orthoptera: Gryllidae)	Via egg surface, faeces, cannibalism	Nymphs with paralyzed hind-legs, disorientation, mortality. May also remain covert	Reinganum et al. 1970; Wigley and Scotti 1983
		Rhipalosiphum padi virus (RhPV)	*Rhipalosiphum padi* (Hemiptera: Aphidae)	Transovarially and horizontally via plants	Reduced lifespan	Gildow and D'Arcy 1988
	Iflaviridae	Sacbrood virus (SBV)	*Apis mellifera*	Oral-faecal, environment	Pupation failure, impaired foraging and reduced lifespan in adults	Beaurepaire et al. 2020
		Deformed wing virus (DWV)	*Apis mellifera*	Varroa mites, oral-faecal, vertically, environment	Development problems with occasional wing deformities, shortened adult lifespan, impaired learning	
		Helicoverpa armigera iflavirus (HaIV)	*Helicoverpa armigera* (Lepidoptera: Noctuidae)	Horizontally and vertically	Mainly in fatbody, prolonged development with higher mortality	Yuan et al. 2020
		Spodoptera exigua iflavirus (SeIV)	*Spodoptera exigua* (Lepidoptera: Noctuidae)	Vertically	Covert infections, increased susceptibility to baculoviruses	Carballo et al. 2020; Jakubowska et al. 2016
		Glossina pallidipes iflavirus (GpIV)	*Glossina pallidipes* (Diptera: Glossinidae)	Probably vertically	Covert infections	Abd Alla, Personal communication

RNA (+) sense

continued

Table 10.1 *Continued*

<table>
<thead>
<tr><th></th><th>Virus order</th><th>(sub) Family</th><th>Prominent examples</th><th>Original host(s) [Order: Family][3]</th><th>Transmission routes</th><th>Described pathology</th><th>References</th></tr>
</thead>
<tbody>
<tr><td rowspan="12">RNA (+) sense</td><td rowspan="5">*Nodamuvirales*</td><td rowspan="5">*Nodaviridae*</td><td>Flock House virus (FHV)</td><td>*Costelytra giveni* (Coleoptera: Scarabaeidae)</td><td>Unknown, most data from cell culture studies</td><td>Unknown, most data from cell culture studies</td><td>Dasgupta et al. 1994</td></tr>
<tr><td>Nodamura virus (NOV)</td><td>*Culex tritae-niorhynchus* (Diptera: Culicidae)</td><td>Unclear, may be via vertebrates</td><td>Inapparent infections</td><td>Yong et al. 2017; Schneeman et al. 1998</td></tr>
<tr><td>Wuhan nodavirus (WhNV)</td><td>*Pieris rapae* (Lepidoptera: Pieridae)</td><td>Unclear</td><td>No clear pathology</td><td>Liu et al. 2006</td></tr>
<tr><td>Black beetle virus (BBV)</td><td>*Heteronychus arator* (Coleoptera: Scarabaeidae)</td><td>Unclear</td><td>Mortality in black beetles, paralysis in wax moth larvae</td><td>Schneeman et al. 1998</td></tr>
<tr><td>*Mesoniviridae*</td><td>Nam Dinh virus (NDiV)</td><td>*Culex* mosquitoes (Diptera: Culicidae)</td><td>Vertically and horizontally</td><td>Inapparent infections</td><td>Newton et al. 2020; Ye et al. 2020.</td></tr>
<tr><td>*Nidovirales*</td><td>*Alphatetraviridae*</td><td>Helicoverpa armigera stunt virus (HaSV)</td><td>*Helicoverpa armigera* (Lepidoptera: Noctuidae)</td><td>Orally</td><td>Specific for midgut cells, pathology limited to first three larval stages</td><td>Brooks et al. 2002</td></tr>
<tr><td>*Hepelivirales*</td><td>*Carmotetraviridae*</td><td>Providence virus</td><td>Cell line derived from *Helicoverpa armigera*</td><td>Unknown</td><td>Persistent infections in cell culture</td><td>Pringle et al. 2003</td></tr>
<tr><td>*Tolivirales*</td><td>*Permutotetraviridae*</td><td>Thosea asigna virus (TaV)</td><td>*Thosea asigna* (Lepidoptera: Limacodidae)</td><td>Orally</td><td>Feeding termination and liquefying, infectivity higher in the early larval stages</td><td>Zeddam et al. 2010; Sugiharti et al. 2010</td></tr>
<tr><td>Floating family</td><td>*Flaviviridae*</td><td>Culex flavivirus (CxFV)</td><td>*Culex* spp. (Diptera: Culicidae)</td><td>Vertical transmission via transovarial route</td><td>No clear symptoms, competitive interaction with dual-host flaviviruses</td><td>Blitvich and Firth 2015</td></tr>
<tr><td>*Amarillovirales*</td><td>*Togaviridae*</td><td>Eilat virus (EILV)</td><td>*Anopheles coustani* (Diptera: Culicidae)</td><td>Orally</td><td>Little pathology in cell cultures, possible interaction with dual-host alphaviruses</td><td>Nasar et al. 2012</td></tr>
<tr><td>*Martelivirales*</td><td></td><td></td><td></td><td></td><td></td><td></td></tr>
</tbody>
</table>

Genome	Order	Family	Virus	Host	Transmission	Effect	Reference
RNA (-) sense viruses	Mononegavirales	Nyamiviridae	Orinoco virus (ONCV)	Pasiphila rectangulata (Lepidoptera: Geometridae)	Unknown	NGS data only	Dietzgen et al. 2017
		Rhabdoviridae	Ceratitis capitata sigma virus (CCapSV)	Ceratitis capitata (Diptera: Tephritidae)	Biparental vertical transmission	Not apparent	Longdon et al. 2017
			Drosophila melanogaster sigmavirus (DMelSV)	Drosophila melanogaster (Diptera: Drosophilidae)	Vertically	Decreased viability and fitness, longer development time. Increase in male reproductive success	Rittschof et al. 2013
			Spodoptera frugiperda rhabdovirus (SfRV)	Spodoptera frugiperda (Lepidoptera: Noctuidae)	Unknown	Not apparent	Schroeder et al. 2019.
	Bunyavirales	Phenuiviridae	Badu virus (BADUV)	Culex spp. (Diptera: Culicidae)	Unknown	Fitness costs and effect on mammalian bunyaviruses unknown	Hobson-Peters et al. 2016
		Peribunyaviridae	Herbert virus (HEBV)	Mainly Culex spp. (Diptera: Culicidae)	Unknown	Fitness costs unclear, and effect on mammalian bunyaviruses unknown	Marklewitz et al. 2013
	Articulavirales	Orthomyxoviridae	Sinu virus (SINUV)	Stinging mosquitoes but species unknown (Diptera: Culicidae)	Unknown	Not apparent	Contreras-Gutierrez et al. 2017
RNA (ds)	Reovirales	Reoviridae	Cypovirus 1 (CPV-1)	Bombyx mori (Lepidoptera: Bombicidae)	Orally	Limited to midgut epithelial cells	Belloncik et al. 1989
	Floating family	Birnaviridae	Drosophila X virus (DXV)	Drosophila melanogaster (Diptera: Drosophilidae)	Contact transmission	Reduced survival time and sensitivity to anoxia	Teninges et al. 1979

[1] For a more extensive overview table of bee viruses see Beaurepaire et al. (2020).

[2] For an overview of the insect orders known to be infected by viruses in these families see Figure 10.2.

[3] An individual virus might (experimentally) infect a broader host range than indicated here. The order and family to which the host insects belong are indicated unless mentioned higher up in the table.

species can naturally be infected by sigmaviruses (family *Rhabdoviridae*, genus *Sigmavirus*) (Langdon et al. 2017). Sigmaviruses have been described in various *Drosophila* species, the Mediterranean fruit fly *Ceratitis capitata*, and in the speckled wood butterfly *Pararge aegeria* (Longdon et al. 2017). These viruses are vertically transmitted by females and, to a lesser extent, also by males. Sigmaviruses decrease egg to adult viability, increase development time, reduce fitness in crowded conditions, but may increase male reproductive success (Rittschof et al. 2013). An insect specific rhabdovirus (RV) chronically infects cultured *Spodoptera frugiperda* (Sf9) cells (Ma et al. 2014) without causing cytopathic effects. Sf-RV is, however, a complication for the biotechnological application of these cells, e.g. for vaccine production. SF-RV was also detected in wild *S. frugiperda* populations and also has the ability to infect other lepidopterans (Schroeder et al. 2019). Viruses belonging to the family *Nyamiviridae* (genus *Orinovirus*) have been detected in moths by NGS studies, only. These viruses may be enveloped and spherical as found for related nyamiviruses (Dietzgen et al. 2017).

10.2.2.2 Bunyaviruses and thogotoviruses

Insect viruses with (-) sense RNA genomes and enveloped spherical particles may belong to the order *Bunyavirales* (families *Phenuiviridae* or *Peribunyaviridae*), or to the family *Orthomyxoviridae* (genus *Thogotovirus*). The genomes of these viruses are segmented and present in the virions as nucleoprotein complexes. Co-packaging of the RDRP (L protein) in the virions allows the first rounds of genome replication. Bunyaviruses and thogotoviruses initiate viral mRNA synthesis via host mRNA cap-snatching (see review by Olschewski et al. 2020). Examples of insect-specific viruses in the family *Phenuiviridae* are Badu virus (genus *Phasivirus*), found with high prevalence in *Culex* spp. mosquitoes on Badu Island (Hobson-Peters et al. 2016), and viruses in the genus *Goukovirus*, named after the Gouléako virus in mosquitoes in Western Africa (Marklewitz et al. 2011). A second clade of mosquito-restricted bunyaviruses is classified in the family *Peribunyaviridae* (genus *Herbevirus*), named after Herbert virus, which was also isolated in Western Africa (Marklewitz et al. 2013). Sinu virus,

isolated in Colombia, is an example of a mosquito-specific thogotovirus (Contreras-Gutierrez et al. 2017). What effect these insect-specific bunyaviruses and thogotoviruses have on the fitness of their insect host and on their ability to vector related viruses, remains to be analysed in most cases.

10.2.3 Insect viruses with double stranded RNA

10.2.3.1 Cypoviruses

Within the family *Reoviridae* the cytoplasmic polyhedrosis viruses (CPVs) form a specific clade (genus *Cypovirus*). Cypoviruses have multiple (10–12) linear, double stranded (ds) RNA genome segments. The virions have an icosahedral capsid (55–69 nm), with 20 nm long hollow spikes and are highly stable (Belloncik et al. 1989). The dsRNA genome segments can be transcribed into mRNA, within the virion. As their name implies, these viruses occlude thousands of progeny virions in cytoplasmic polyhedral-shaped occlusion bodies that protect the viruses after their release into the environment. Cypoviruses are transmitted orally, and infections are limited to midgut epithelial cells. Symptoms are often chronic, mimicking starvation, and include a reduction in feeding and in larval growth. CPVs have been found in insects belonging to various orders, including Lepidoptera, Diptera, Hymenoptera, Coleoptera, and Neuroptera (Zhao et al. 2018). The Aedes pseudoscutellaris reovirus, which is characterised by nine genome fragments and is non-occluded (Attoui et al. 2005) is so far the sole member of the genus *Dinovernavirus*.

10.2.3.2 Birnaviruses

The members of the family *Birnaviridae* have non-enveloped icosahedral particles and a bi-segmented, linear dsRNA genome (5.9–6.9 kbp in length). The largest segment translates into a polyprotein that is cleaved to obtain four to five functional proteins. Three entomobirnavirus isolates have been described: Drosophila X virus, Espirito Santo Virus (ESV), and Culex Y virus (CYV). DXV was isolated from *Drosophila* cell cultures and causes a reduction in survival time and a sensitivity to anoxia (Teninges et al. 1979). ESV and CYV appear to be closely related, with the main difference that ESV, isolated from a human patient, replicated only in the presence of Dengue-2 virus in *Culex* cells (Vancini et al. 2012), while CYV isolated from naturally occurring *Culex pipiens* mosquitoes in Germany did not need a helper virus to replicate (Marklewitz et al. 2012). ESV and CYV now both fall under the species *Mosquito X virus*.

10.3 DNA viruses of insects

10 3.1 Insect-infecting small DNA viruses with single-stranded genomes

10.3.1.1 Common features

The small DNA viruses of insects are characterised by small (18–27 nm), icosahedral, non-enveloped particles. These viruses are extremely stable in their environment. Their genomes typically carry *rep* (replication initiator/non-structural (NS) proteins) and *cap* (capsid proteins) genes. Their linear or circular genome is packaged as ssDNA. These single-stranded genomes first need to be converted into dsDNA before transcription can occur by cellular RNA polymerase needed for this complementary strand synthesis and further replication is provided by dividing cells. Consequently, tissues with dividing cells are affected. The ssDNA viruses with linear genomes have developed special features to make sure that their genome is not shortened with every replication cycle, which involves terminal hairpin structures that serve as primers, or, alternatively, a protein-primed mechanism (Tijssen and Bergoin 1995). Insect ssDNA viruses belong to the families *Parvoviridae* (subfamily *Densovirinae*), *Circoviridae* (usually from the genus *Cyclovirus*), or *Bidnaviridae* (Figures 10.1–10.2; Table 10.2). An excellent overview of the characteristics of the various invertebrate-infecting viruses classified in these families has been presented before, in Tijssen et al. 2017.

10.3.1.2 Densoviruses

The viruses in the subfamily *Densovirinae* (family *Parvoviridae*) have linear ssDNA genomes with complex terminal palindromes, for some genera in the form of inverted terminal repeats that fold into terminal structures to allow priming of the complementary DNA strand (Bando et al. 1990; Pénzes et al.

2020). Many densoviruses have their *rep* and *cap* genes in ambisense orientation of each other, hence the term ambidensovirus in the name of most genera in this subfamily. Among the *rep*-gene encoded proteins is the NS1 protein, an endonuclease essential for replication. Many densoviruses have been found in lepidopteran species, but they also occur in insects in the orders Orthoptera, Hemiptera, Hymenoptera, Diptera and Blattodea, as well as in aquatic arthropods (Tijssen and Bergoin, 1995; ICTV, 2020). Reared crickets (order Orthoptera) are frequently infected with Acheta domesticus densovirus (AdDV), which regularly causes devastating disease outbreaks that hamper cricket mass production considerably (Szelei et al. 2011). Examples in lepidopteran insects include an ambidensovirus infecting the wax moth *Galleria mellonella* (family Pyralidae) (Tijssen et al. 2003), another mass-reared insect for food and feed, as well as the Papilio polyxenus iteradensovirus isolated from the black swallowtail *P. polyxenus* (family Papilionidae), which naturally occurs in North America (Yu et al. 2012).

Another example is the Junonia coenia densovirus (JcDV), isolated from the common buckeye *J. coenia* (family Nymphalidae). This virus infects orally but bypasses the midgut without replication and then infects haemocytes and trachea, resulting in hypoxia (Mutuel et al. 2010). JcDV-derived vectors are being used for transgene delivery in cultured lepidopteran cells (Bossin et al. 2003). This strategy makes use of the tendency of this virus to integrate its DNA into the host genome (comparable to Adeno-associated virus (AAV) vectors (subfamily *Parvovirinae*) used for gene therapy). This integration also occurs naturally and numerous endogenous viral elements (EVEs) with homology to parvoviruses/densoviruses have been reported in genomes of both vertebrates and invertebrates. These EVEs have recently been classified in the subfamily *Hamaparvovirinae*. Whether EVEs play a role in the insects' life cycle, for instance in providing antiviral immunity, is only beginning to be unravelled. Free (non-integrated) densoviruses may, however, also have mutual relationships with their insect hosts: asexual clones of the rosy apple aphid only develop wings when infected with Dysaphis plantaginea densovirus (DplDV). These wings allow the aphids to seek for new plants to colonise and feed on, although at a fitness cost for the infected individuals (Ryabov et al. 2009). For a more complete overview of densovirus biology we refer to Pénzes et al. (2020).

10.3.1.3 Bidnaviruses

These linear ssDNA viruses have a bi-segmented genome, but only one genome segment of either positive or negative polarity ends up in their capsids, leading to four possible capsids (Tijssen et al. 2017). The replication mechanism of bidnaviruses does not involve a cleavage step by a viral endonuclease (as seen in the densoviruses). Instead, these viruses encode a type B protein-primed DNA polymerase. The corresponding *polyB* gene may find its origin in a large DNA transposon, in which an ancestral densovirus genome has been integrated (Krupovic and Koonin 2014). Such an event may also explain why the terminal inverted repeats of bidnaviruses do not contain the typical hairpins found in densovirus genomes. Currently, *Bombyx mori bidensovirus* (BmBDV) is the only approved species in the family *Bidnaviridae* (ICTV, 2020). BmBDV infects mainly midgut columnar cells, causing these cells to collapse. This results in loss of appetite and physical decline of the host (Kumar et al. 2019).

10.3.1.4 Circular Rep-encoding ssDNA (CRESS-DNA) viruses

The circular Rep-encoding ssDNA (CRESS-DNA) viruses form a remarkably diverse group of arthropod-infecting viruses (Rosario et al. 2018). Many CRESS-DNA viruses have been found in typical insect predators. Dragonflies, for instance, form a rich source of potential insect infecting CRESS-DNA viruses, as these insects tend to accumulate viruses from their consumed prey (Rosario et al. 2012). Viruses of the genus *Cyclovirus* (family *Circoviridae*) are among the frequently detected CRESS-DNA viruses in dragonflies. These viruses have small circular ssDNA genomes in a range of 1.7 to 1.9 kb. The genome has an ambisense nature with the two main ORFs orientated inversely. These ORFs encode the putative capsid and the rolling circle replication initiator (Rep) proteins. The Rep-proteins of CRESS-DNA viruses belong to the HUH endonuclease superfamily (Chandler

et al. 2013), members of which also appear in bacteria and bacteriophages. Cycloviruses are further characterised by a stem loop structure between the 5′ ends of the two main ORFs with a conserved nonanucleotide loop sequence that marks the origin of DNA replication (Delwart and Li 2012). CRESS-DNA viruses may have evolved from a combination of a bacterial plasmid that encoded Rep for a rolling circle-replication mechanism and cDNA copies of capsid genes of eukaryotic positive-sense RNA viruses (Kazlauskas et al. 2019; Krupovic and Koonin 2014). There seem to have been at least three of such evolutionary events, which may explain the high diversity among these viruses. The cycloviral genomes obtained from dragonflies do not separate into a distinct invertebrate infecting clade, but group together with cycloviruses found in human and bat faeces, suggesting that cycloviruses in mammalian faeces may originate from (accidentally) consumed invertebrates, as discussed by Tijssen et al. (2017). A CRESS-DNA virus found in reared house cricket populations in the USA and Japan, and also more recently in Jamaican field crickets (*Gryllis assimilis*) in the USA (Pham et al. 2013a), has arbitrarily been named Acheta domesticus volvovirus (AdVVV). The virus was discovered in samples of died-off crickets that were negative for AdDV (Pham et al. 2013b) and may therefore induce mortality. The small AdVVV genome measures close to 2.5 kb and has ORFs in both the sense and antisense orientation. Like cycloviruses, AdVVV contains a nine-base loop sequence, that likely serves as replication origin. However, AdVVV appears to be the sole member of a novel virus family.

10.3.2 Nucleo-cytoplasmic large DNA viruses of insects

10.3.2.1 Common features and taxonomical positions

Nucleo-cytoplasmic large DNA viruses (NCLDVs) constitute an apparently monophyletic group of viruses that infect animals and diverse unicellular eukaryotes (Colson et al. 2013). Therefore, these viruses have recently been placed in the phylum *Nucleocytoviricota* (ICTV 2020). They typically have

Table 10.2 Taxonomically classified insect viruses with DNA genomes with selected examples and references[1]

	Higher order taxon	(sub) Family	Prominent examples	Original host(s) (Order: Family)[2]	Transmission routes	Described pathology	References
DNA (ss)	Family *Parvoviridae*	*Denso-viirinae*	Acheta domestica densovirus (AdDV)	*Acheta domestica* (Orthoptera: Gryllidae)	Orally, cannibalism, egg surface	Covert infections as well as environmentally triggered lethal infections	Szelei et al. 2011, Maciel- Vergara unpublished data
			Junonia coenia densovirus (JcDNV)	*Junonia coenia* (Lepidoptera: Nymphalidae)	Orally	Replication in haemocytes and visceral tracheae which may lead to hypoxia	Mutuel et al. 2010
	Poliivirales	*Bidnaviridae*	*Bombyx mori* bidensovirus (BmBDV)	*Bombyx mori* (Lepidoptera: Bombicidae)	Orally	Targets mainly midgut columnar cells. Loss of appetite and physical decline	Kumar er al. 2019
	Cirlivirales	*Circoviridae*	Dragonfly associated cyclovirus 1	Unknown	Unknown	Not infecting dragonflies, consumed with diet containing infected insects	Rosario et al. 2012
DNA (ds)	Phylum *Nucleocyto-viiricota*	*Ascoviridae*	Spodoptera frugiperda ascovirus 1a (SfAV-1a)	*Spodoptera frugiperda* (Lepidoptera: Noctuidae)	By parasitoid wasps	Milky white haemolymph containing refractile virion vesicles	Asgari et al. 2017
		Iridoviridae	Chilo iridescent virus (CIV/IIV6)	*Chilo suppressalis* (Lepidoptera: Crambidae)	Horizontally via cannibalism or predation on IIV-infected tissues or vector mediated	Overt infections giving an iridescent colour as well as covert infections	Muttis et al. 2020; Williams 2008
		Entomopox-virinae	Melolontha melolontha entomopoxvirus (MmEPV)	*Melolontha melolontha* (Coleoptera: Scarabaeidea)	Orally	Apart from general morbidity, symptoms of infected larvae differ from host to host	Arif and Kurstak 1991
			Melanoplus sanguinipes entomopoxvirus (MsEPV)	*Melanoplus sanguinipes* (Orthoptera: Acrididae)	Orally	Inhibition of pigment production, developmental delay, and reduced food consumption	E'landson 2008

continued

Table 10.2 *Continued*

	Higher order taxon	(sub) Family	Prominent examples	Original host(s) (Order: Family) [2]	Transmission routes	Described pathology	References
DNA (ds)	*Naldaviricetes*	*Baculoviridae*	Culex nigripalpus (CuniNPV)	*Culex nigripalpus* (Diptera: Culicidae)	Horizontally via contaminated water	Lethargy of larvae, surface floating	Andreadis et al. 2003
			Neodiprion sertifer nucleopolyhedrovirus (Nese NPV)	*Neodiprion sertifer* [Hymeoptera, Diprionidae]	Faecal-oral route	Midgut infection, infectious diarrhea, mortality	Hajek and van Frankenhuyzen 2017
			Autographa californica nucleopolyhedrovirus (AcMNPV), Bombyx mori (Bm) NPV), Cydia pomonella granulovirus (CpGV)	Various moth spp. (Lepidoptera: Noctuidae, Bombycidae and Tortricidae)	Orally, as well as vertically	Broad tissue tropism, often lethal infections, behavioural changes, covert infections also reported	Williams et al. 2017a
		Nudiviridae	Helicoverpa zea nudivirus 2 (HzNV-2)	*Helicoverpa zea* (Lepidoptera: Noctuidae)	Sexual transmission as well as oral infection. Vertically via ovaries.	Reproductive tissue. Malformed gonads of both sexes. Infectious waxy plug at the tip of the female abdomen. Overt and covert infections	Burand et al. 2004; Rallis and Burand 2002
			Gryllus bimaculatus nudivirus (GbNV)	*Gryllus bimaculatus* (Orthoptera: Gryllidae)	Horizontally via oral route, cannibalism	Hypertrophied fat body cells. Reduced growth, lethargy, swelling and viscid, opalescent haemolymph. Death often in last instar	Huger 1985
			Oryctes rhinoceros nudivirus (OrNV)	*Oryctes rhinoceros* (Coleoptera: Scarabaeidea)	Horizontally during mating or via faeces	Hypertrophied midgut and fat body cells. Discolorations, oviduct malformations	Huger 2005; Zelazny 1976
		Hytrosaviridae	Glossina pallidipes salivary gland hypertrophy virus (GpSGHV)	*Glossina pallidipes* (Diptera: Glossinidae)	Vertically via milk glands and horizontally via feeding resources	Hypertrophied salivary glands, sterility in F1 leading to colony collapse, may also be covert	Abd Alla et al. 2009
			Musca domestica gland hypertrophy virus (MdSGHV)	*Musca domestica* (Diptera: Muscidae]	Horizontally via oral-faeces route	Female sterility, effect on vitellogenesis	Lietze et al. 2012; Kariithi et al. 2017b

[1] For an overview of the insect orders known to be infected by viruses in these families see Figure 10.2.

[2] An individual virus might (experimentally) infect a broader host range than indicated here. The order and family to which the host insects belong are indicated unless mentioned higher up in the table.

large virions and linear or circular dsDNA genomes of considerable size. Replication of NCLDVs either occurs exclusively within the cytoplasm or involves both the nucleus and the cytoplasm of the host cell. The insect-infecting NCLDVs belong to the families *Ascoviridae, Iridoviridae*, and the more distantly related *Poxviridae* (Figures 10.1 and 10.4; Table 10.2).

10.3.2.2 Ascoviruses

Ascoviruses, belonging to the family *Ascoviridae*, are large, enveloped, double-stranded DNA viruses with circular genomes of 100–200 kbp. Virions contain an internal lipid membrane surrounding the DNA/protein core. They can be bacilliform, ovoidal, or allantoid in shape, and measure about 130 nm in diameter and 200–400 nm in length. Members of this relatively small family are classified into two genera: *Ascovirus* and *Toursvirus*. While the *Ascovirus* genus includes three species, the *Toursvirus* genus so far includes only one species (ICTV 2020). Viral replication starts in the host cell nucleus. Subsequently, host cell DNA is degraded, the nucleus ruptures, and the cell breaks up into a cluster of virion-containing vesicles that turn the host haemolymph milky white (Asgari et al. 2017). Some members of the *Ascovirus* genus are embedded in occlusion body structures (Cheng et al. 2000; Federici et al. 1990), which may be functionally similar to that of baculoviruses (see Section 3.3). Ascoviruses infect mainly lepidopteran larvae and are rarely transmitted via oral ingestion (Govindarajan and Federici 1990). Most ascoviruses are transmitted by parasitoid wasps and cause a chronic, fatal disease with a cytopathology resembling that of apoptosis (Bideshi et al. 2006) (see Chapter 13 for a more general description of parasitoids). Phylogenetic studies showed that ascoviruses emerged relatively recently from an ancestral invertebrate iridovirus lineage (Piegu et al. 2015).

10.3.2.3 Invertebrate iridescent viruses

The invertebrate iridescent viruses (IIVs) belong to the subfamily *Betairidovirinae* within the family *Iridoviridae*. The members of the *Iridoviridae* classified in the genera *Iridovirus* and *Chloriridovirus* infect agriculturally and medically important pest insects, while those in the genus *Decapodiridovirus* infect isopods, penaeid shrimp, and crayfish (Chinchar et al. 2017b). Invertebrate iridescent virus 6 (IIV6) is the best studied IIV of the genus *Iridovirus*. The common name is Chilo iridescent virus, referring to *Chilo suppressalis* (Lepidoptera: Crambidae). IIV3 and IIV9 infect mosquitoes naturally and fall in the genus *Chloriridovirus*. Most IIVs can be propagated *in vivo* in the larvae of *Galleria mellonella*. Besides, *in vitro* propagation of IIVs can be performed in different cell lines from dipteran, lepidopteran, coleopteran, and homopteran species (Williams 2008). IIVs display low infectivity following oral administration. However, cannibalism or predation of IIV-infected tissues is considered a likely mechanism of transmission for several species (Carter 1973). Vector-mediated transmission of IIVs by parasitic mermithid nematodes was also demonstrated (Muttis et al. 2020). IIVs can produce either covert or patent infections. Covert IIV infections are not lethal and not obvious to the naked eye, but may cause extended development time, reduced fecundity and longevity (Marina et al. 2003). Patent disease causes insects to develop an obvious iridescent colour that typically ranges from violet, blue, green, or orange, and is usually fatal in the larval or pupal stages (Figure 10.4A).

The genome of IIVs comprises a single molecule of linear dsDNA (163–220.2 kbp) that is terminally redundant and circularly permutated (Goorha and Murti 1982). The terminal repetition adds another 10–30% to the size of the genome. The putative ORFs, ranging in number from 126 to 215, are closely packed, lack introns, and are generally non-overlapping (Chinchar et al. 2017a). Twenty-six ORFs appeared to be conserved (core genes) across all iridovirid species (Eaton et al. 2010). Viral genome replication starts in the nucleus and is followed by genome concatamerisation and subsequent cleavage, particle assembly and maturation in the cytoplasm. Virions accumulate in the cytoplasm within paracrystalline arrays (hence the iridescence) or acquire an envelope by budding from the plasma membrane (Liu et al. 2016). IIV virions have an icosahedral shape and vary between 120–180 nm in diameter. The DNA/protein core is surrounded by an internal lipid membrane, together enclosed by a protein capsid and in the case of those particles that bud out of cells, an outer viral envelope. The virion structure is formed of 12 penta- and 20 tri-symmetrons that both predominantly comprise hexavalent capsomers composed of the major and minor capsid proteins (Yan et al. 2009). The outer surface of the capsid is covered by flexible fibrils emanating from the surface of the virion. Proteomic analysis of IIV particles indicates the presence of approximately 54–64 virion-associated proteins (Ince et al. 2010; Wong et al. 2011). At the start of the infection, the viral DNA/protein core is released into the cytoplasm following receptor-mediated endocytosis in the case of enveloped virions or after entry at the plasma membrane for the naked viruses, and is transported into the nucleus (Braunwald et al. 1985; Chitnis et al. 2008). One or more virion-associated proteins are needed for the initiation of IIV gene transcription (Cerutti et al. 1989), which occurs in a regulated temporal cascade. IIV6 transcripts possess generally short 5' untranslated regions and lack poly A tails (Dizman et al. 2012; Nalçacioğlu et al. 2003). Biochemical and *in silico* evidence suggest the existence of viral microRNAs (miRNA) that modulate viral gene expression (Wong et al. 2011).

10.3.2.4 Entomopoxviruses

Entomopoxviruses (EPVs) have been reported from about 40 different insect species, mainly belonging to the orders Lepidoptera, Coleoptera, Orthoptera, and Diptera (Arif and Kurstak 1991; Granados 1981), but also Blattodea (Radek and Fabel 2000), and Hymenoptera (Lawrence 2002; Viljakainen et al. 2018). Most EPVs primarily infect the larvae of insects and have a restricted host range. EPV transmission occurs mainly through the oral route with the ingestion of spheroids. Transovarial transmission of an EPV in *Diachasmimorpha longicaudata* parasitoid wasps representing a mutualistic interaction was also reported (Coffman et al. 2020). EPVs infect primarily the fat body (Roberts and Granados 1968), but also haemocytes, and other tissues of the insect might be affected (Perera et al. 2010). Symptoms of EPV-infected larvae include a sluggish and whitish body, colour change of cuticle and haemolymph, white-spotted or mottled exterior, or a decrease in feeding, but differ from host to host (see Figure 10.4B for an example). EPV

infections generally develop slowly and can take several weeks to kill the insect (Ishii et al. 2002).

Insect-infecting poxviruses belong to the subfamily *Entomopoxvirinae*, which comprises three genera reflecting the host range and virion morphology of their members. Entomopoxviruses (EPVs) infect coleopteran (genus *Alphaentomopoxvirus*), lepidopteran and orthopteran (*Betaentomopoxvirus*) or dipteran (*Gammaentomopoxvirus*) insects (Skinner et al. 2011). *Amsacta moorei entomopoxvirus* (AMEV) is the best-studied betaentomopoxvirus, and one of the EPVs with a fully sequenced viral genome (Bawden et al. 2000). The virions have a brick-shaped or ovoid structure and consist of three major components: a centrally positioned electron-dense core containing the genome, one or two lateral bodies, and an outer lipoprotein envelope. The virion sizes between 150–250 nm in diameter and 250–400 nm in length. EPVs contain a linear dsDNA genome around 225–380 kb in length with a high AT content. The genome has inverted terminal repeats (ITRs) that range from 5.6 kb to 24 kb. ORF numbers in the sequenced EPVs vary between 267 and 334 of which 49 ORFs are conserved across all species of poxviruses. Besides these 49 core genes, EPVs also include a number of genes that exhibit similarity to host genes (Bratke and McLysaght 2008; Dall et al. 2001; Hughes and Friedman 2003) and secondary structure alignment comparing proteins of EPVs and baculoviruses revealed 33 clusters of homologous genes shared by these two virus families (Theze et al. 2015).

Two different infectious virus morphologies exist: the intracellular mature virus (IMV) and the extracellular enveloped virus (EEV). The majority of EPVs produce proteinaceous matrices called spheroids, comprised of spheroidin protein, at the late stage of infection. Mature virions are occluded in these spheroids for protection against harsh environmental conditions. Late in the infection, spindle structures consisting of fusolin protein are also produced by EPVs. The spindles can either assemble in the endoplasmic reticulum of infected cells or occur embedded within the crystalline lattice of the spheroids (Bergoin et al. 1976). Purified spindles are not infectious but enhance virus infectivity by disrupting the peritrophic membrane (Mitsuhashi et al. 2007). Especially, the N-terminus of the fusolin protein plays an essential role in this process (Takemoto et al. 2008). Spheroids dissolve in the alkaline environment of the insect midgut (Bilimoria and Arif 1979). Released virions attach to the midgut epithelium and appear to fuse with the cell membrane allowing entry of the viral core and lateral bodies into the cytoplasm (Granados 1973). Heparin-binding glycosyltransferase, one of the entry-fusion complex proteins described extensively for vertebrate poxviruses (Moss 2016) was also identified in AMEV (Inan et al. 2018). EPV replication occurs entirely in the cytoplasm and virus particles therefore contain a DNA-dependent RNA polymerase. Transcription in EPVs is believed to be regulated in a cascaded fashion and in a similar way as for other members of the *Poxviridae* (Yang et al. 2010).

10.3.3 Large nuclear DNA viruses of arthropods

10.3.3.1 Common features

This is a clade of viruses with rod-shaped enveloped nucleocapsids and large circular dsDNA genomes that replicate and assemble in the nucleus of infected host cells and are exclusively pathogenic to arthropods (Williams et al. 2017a). This clade includes members of the families *Baculoviridae*, *Nudiviridae*, and *Hytrosaviridae* (Figures 10.1, 10.2, 10.5, and Table 10.2). Nudi- and baculoviruses first evolved around 310 million years ago (Mya), in the Paleozoic Era during the Carboniferous Period, with the first insects (Thezé et al. 2011). Bracoviruses, from the paraphyletic family *Polydnaviridae* are endogenised viruses derived from nudiviruses. More distantly related to these viruses are the hytrosaviruses. An evolutionary connection also appears to exist with the whispoviruses (family *Nimaviridae*), infecting Crustaceans (Wang and Jehle 2009). Due to their evolutionary relationships the class *Naldaviricetes* has been installed for these four families. The viruses within these families share a set of four genes coding for *per os* infectivity factors (PIFs) (Boogaard et al. 2018; Wang et al. 2017), as well as *odv-e66*, which has a role in bridging the peritrophic membrane (Hou et al. 2019). These findings suggest that these viruses apply an evolutionarily conserved mechanism for midgut infection.

Noteworthy is the vast variability in length that rod-shaped dsDNA viruses can establish. The variety in length of rod-shaped virions can be explained with a general mechanism that adapts the packaging and particle formation to match diverse viral genomes sizes (Velamoor et al. 2020).

10.3.3.2 Baculoviruses

The name 'baculovirus' refers to the rod-shaped (baculum, Latin for stick) nucleocapsids of these viruses. The family *Baculoviridae* comprises insect viruses with 80–180 kbp circular dsDNA genomes (Ros 2020; Williams et al. 2017a). Four genera are distinguished (Jehle et al. 2006): *Alphabaculovirus* (lepidopteran-specific nucleopolyhedroviruses [NPVs]) with 56 recognised species, *Betabaculovirus* (lepidopteran-specific granuloviruses [GVs]) with 26 species, *Gammabaculovirus* (hymenopteran-specific NPVs) with two official species and *Deltabaculovirus* (dipteran-specific NPVs) with a single species (ICTV 2020). Lepidopteran-specific NPVs are further divided into Group I and Group II alphabaculoviruses, which differ considerably in gene content. Group I alphabaculoviruses form a monophyletic clade within Group II alphabaculoviruses (Zanotto et al. 1993) and have the budded virus Fusion protein (F) replaced with GP64 (Blissard and Theilmann 2018; Wang et al. 2016). Baculoviruses are unique among viruses in having a biphasic replication cycle with two morphologically and biochemically distinct virion types, budded viruses (BVs) and occlusion derived viruses (ODVs) (e.g. see reviews by Harrison and Hoover 2012; Rohrmann 2019). Both virion types contain cylindrical, rod-shaped nucleocapsids, which are 250 to 300 nm in length and have a diameter of 30 to 60 nm. BVs always contain a single nucleocapsid and are produced during early stages of infection. BVs are responsible for cell-to-cell transmission within the host. ODVs are formed later in the infection and are embedded in a proteinaceous crystalline matrix forming viral occlusion bodies (OBs) that are either polyhedral (NPV) or granular shaped (GVs). OBs are released from the host and are responsible for horizontal spread of the virus from host to host. ODVs may have one or several nucleocapsids within a single envelope, depending on the species of baculovirus, forming single (S) or multiple (M) NPVs, respectively.

Baculoviruses infect the larval stages of holometabolous insects from the orders Lepidoptera, Hymenoptera (family Diprionidae), and Diptera (only in the mosquito *Culex nigripalpus*) (e.g. Ros 2020; Williams et al. 2017a). Baculoviruses infections are often lethal, resulting in total liquefaction of the host (Figure 10.5). For some GVs, gammabaculoviruses and deltabaculoviruses infections are limited to midgut epithelial cells of the host. Most baculovirus species infect one or a few related host species. However, some baculovirus species have a very broad host range, with the well-characterised Autographa californica MNPV (AcMNPV, group I *Alphabaculovirus*) infecting more than 90 lepidopteran species. Insect hosts get infected when consuming OB contaminated food. The OBs dissolve in the alkaline midgut environment, releasing the ODVs. The ODV envelope binds and fuses with the membrane of the midgut cells releasing the nucleocapsids into the cytoplasm (Slack and Arif 2006). This binding and fusion process is enabled through a set of conserved viral proteins called *per os* infectivity factors (PIFs). Currently, ten baculoviral *pif* genes are known (Boogaard et al. 2019; Javed et al. 2017). In the nucleus, the nucleocapsids are dissembled and the viral DNA is replicated. New nucleocapsids are produced. These are transported to the cell membrane, where they bud out of the cell forming BVs. BVs spread from cell to cell, infecting different tissues. At the very late stage of infection, ODVs and OBs are produced. Cells are then lysed, and the larval exoskeleton is degraded, resulting in release of the OBs into the environment (Figure 10.5B).

Depending on the species, baculoviruses encode between 89 and 181 ORFs (Ros 2020). These ORFs are distributed over both strands of DNA. A set of 38 core genes is shared among all baculovirus species, encoding genes required for basic processes including DNA replication, gene transcription, virion formation, and oral infectivity. Additional sets of genes are shared within the genera, or among closely related baculoviruses within these genera (van Oers 2020). Baculovirus genomes are also characterised by the presence of several homologous repeat regions (*hr's*), which function as origins of replication and enhancers of transcription. Transcription of baculovirus genes is regulated in a temporal manner, including immediate early, delayed early, late, and very late genes.

The detrimental effect of baculoviruses on insect hosts has been known for centuries, as the collapse of reared caterpillar colonies was a recurrent problem in the silk industry in China and Europe, caused by Bombyx mori NPV (BmNPV, a group I alphabaculovirus). Furthermore, baculoviruses are well-known for altering caterpillar behaviour, inducing hyperactivity (enhanced locomotion) and tree-top disease (climbing to elevated positions prior to death) (Gasque et al. 2019). These virus-induced behavioural changes aid the spread of the virus by increasing the area over which the virus is distributed, and by enhancing the visibility of the dead cadavers to predators. Several viral genes are likely to be involved in inducing these behavioural manipulations, although the exact mechanisms depend on the virus-host combination studied (Ros et al. 2015). While horizontal transmission, via the ingestion of OBs, is the major route of baculovirus transmission, baculoviruses can also be transmitted vertically, from parents to offspring. It is becoming clear that baculoviruses also cause covert infections, which are seemingly symptomless (Williams et al. 2017b). Such covert infections may allow virus persistence over time, for example to pertain in periods of low host density with limited chances of horizontal transmission.

10.3.3.3 Nudiviruses

The name 'nudivirus' originates from Latin (nudi = naked, uncovered) and is based on the early assumption that nudiviruses would represent non-occluded baculoviruses. Sequence data and phylogenetic analyses, however, revealed that these viruses form a distinct virus family, referred to as *Nudiviridae*. Members of the *Nudiviridae* are officially grouped into four genera (Harrison et al. 2020; ICTV 2021). Nudiviruses infect a broad range of hosts from different insect orders, including Orthoptera, Coleoptera, Lepidoptera, and Diptera. Identified hosts include (among others) crickets (Gryllus bimaculatus nudivirus, GbNV), palm rhinoceros beetles (Oryctes rhinoceros nudivirus, OrNV), corn earworm moths (Heliothis zea nudivirus, HzNV-1, and HzNV-2), and fruit flies (Drosophila innubila nudivirus, DiNV; Kallithea virus, KV). OrNV (genus *Alphanudivirus*) is being applied successfully as a viral biopesticide against coconut palm tree-infesting *O. rhinoceros*) (Huger 2005). A genus *Gammanudivirus* has been installed for nudiviruses infecting the giant tiger prawn *Penaeus monodon* (PmNV) (Yang et al. 2014b) and the European lobster *Homarus gammarus* (Holt et al. 2019). The genus *Deltanudivirus*, on the other hand, harbours Tipula oleracea nudivirus (ToNV) in crane fly (Bézier et al. 2015). The increasing availability of full genome sequences of nudiviruses from terrestrial arthropods, as well as marine and fresh water crustacean species, such as the recently published *Dikerogammarus haemobaphes* nudivirus infecting the demon shrimp (Allain et al. 2020) and the brown shrimp-infecting Crangon crangon nudivirus (Van Eynde et al. 2018, Bateman et al., 2021), may urge a reorganisation of the nudivirus family in the near future.

The infection of nudiviruses is not exclusively bound to certain developmental stages or host tissues. Asymptomatic infections have been observed in larvae and adults. While lethal infections usually occur in larvae, adults often become chronically infected. A typical sign of OrNV infection in larvae is the prolapse of the rectum (Figure 10.6A). Sterility and malformations due to nudivirus infections have also been described (Bézier et al. 2015), as exemplified by a female *Helicoverpa zea* moth infected with HzNV-2 (Figure 10.6C-E) with malformed ovary tissue and a virus-filled 'waxy plug' around the reproductive openings that promotes sexual transmission of the virus (Burand et al. 2012). GbNV is pathogenic to various species of field crickets, including those reared for food and feed (Huger 1985). These examples emphasise both the potential that nudiviruses harbour as biocontrol agents and the impact nudivirus infections may have on commercially reared insects. Furthermore, nudiviruses can durably integrate into their host's genome, which gave rise to endogenous nudivirus species, such as Nilaparvata lugens endogenous nudivirus (Cheng et al. 2014) or Fopius arisanus endogenous nudivirus (Burke 2019). This ability of genome integration, combined with comparative phylogenetic analyses of endogenous viral elements found in braconid wasps created the theory that bracoviruses (family *Polydnaviridae*), originated from an ancestral nudivirus that integrated into the genome of a braconid ancestor wasp ~ 100 Mya (Bézier 2009; Zhang et al. 2020).

The enveloped virions of nudiviruses contain cylindrical nucleocapsids of ellipsoidal or rodshaped form with variable length and width. Contrary to the assumption that nudiviruses are usually non-occluded, some nudiviruses do assemble their virions into occlusion bodies, made up of proteins with similarity to the polyhedrin in baculoviruses (ToNV) or with distinct properties from the baculoviral OBs (PmNV) (Bézier et al. 2015; Yang et al. 2014b). The genome of nudiviruses is a circular molecule of dsDNA of 96–232 kbp encoding approximately 89–155 proteins. The ORF content and the order can differ significantly (Harrison et al. 2020). The 32 ORFs present in all nudivirus genomes make up the set of core genes (Bézier et al. 2015), many of which are homologues to genes found in baculoviruses and hytrosaviruses. The core genes involved in transcription (*lef-4*, *lef-5*, *lef-8*, *lef-9*, and *p47*) are highly conserved between nudiviruses and baculoviruses. Eight nudivirus core genes are homologous to baculovirus *per os* infectivity factors. Furthermore, nudiviruses and baculoviruses share a few core genes involved in viral packaging, assembly, and morphogenesis (*p33*, *vp39* *vp91*, *vlf-1*, *38k*, and *p6.9*), DNA replication, repair, and recombination (*dnapol*, *helicase*), and one (*ac81*) of unknown function (Bézier et al. 2015; Holt et al. 2019; Yang et al. 2014b). However, due to the ongoing discovery and full sequencing of nudivirus genomes, the exact number of nudivirus core genes might be redefined in the near future.

10.3.3.4 Hytrosaviruses

Hytrosaviruses replicate in the salivary glands of adult flies leading to strongly enlarged salivary glands (see Abd-Alla et al. 2007 for an image). The best-studied examples of hytrosaviruses are Glossina pallidipes salivary gland hypertrophy virus (GpSGHV) in tsetse and Musca domestica SGHV (MdSGHV) in houseflies. The narcissus bulb fly *Meridon sequestris* can also become infected by an SGHV (Abd-Alla et al. 2009). The enveloped, rodshaped virions are relatively long (500–1000 nm) and with a diameter of 50–100 nm and carry the circular dsDNA genome with a size of 120–190 kbp (Abd-Alla et al. 2009). Salivary gland hyperplasia in *Glossina pallidipes* and the associated virus were first recognised in the late 1970s (Jaenson, 1978). The virus was studied in more detail when outbreaks leading to colony collapse occurred in tsetse mass rearing facilities (for the release of sterile insects). Disease symptoms induced by GpSGHV also include gonadal anomalies, which may result in sterility, and altered sperm production. The exact pathologies depend on the strain of GpSGHV (Kariithi et al. 2017a). Infection of milk glands in females promotes the vertical transmission of the virus to the next generation. In tsetse production facilities flies may carry GpSGHV in a covert, asymptomatic state. Also, in sub-Saharan Africa, SGHV infections are common in *Glossina* species, but mostly also remain asymptomatic (Meki et al. 2018b). The underlying mechanism that triggers the switch from covert to overt infections is still under research but may involve differential levels of viral and host encoded miRNAs in symptomatic versus asymptomatic flies (Meki et al. 2018a). Also, the microbiome composition and secondary infections with other viruses may affect the outcome of a GPSGHV infection. Infections of houseflies with MdSGHV leads to apparent disease symptoms and 100% female sterility by preventing egg production (Kariithi et al. 2017b; Prompiboon et al. 2010). Such infections occur with a prevalence of 0.5% and 10% in housefly populations (Lietze et al. 2012). For both viruses altered mating behaviour has been reported. MdSGHV is considered as a potential biocontrol agent, while the vertically transmitted GpSGHV is unsuitable for tsetse vector control. We further refer to an in-depth comparison available in literature (Kariithi et al. 2017a).

10.3.4 Filamentous viruses of Hymenoptera

Two apparently non-related filamentous viruses (FV) with large dsDNA genomes have been found in hymenopteran insects. The FV infecting the parasitoid *Leptopilina boulardi* (LbFV) replicates in the nucleus of cells lining the oviduct of the female parasitoid wasps and assembles into nucleocapsids of

over 1 mm in length (Varaldi et al. 2006). The nucleocapsids acquire an envelope when they are transported to the cytoplasm. They are injected with the eggs into *Drosophila* larvae. Infection with LbFV promotes superparasitism, whereby the female wasps are stimulated to lay eggs in already parasitised fly larvae (Varaldi et al. 2006). This behaviour leads to horizontal transmission of LbFV and allows the coexistence of multiple parasitoid species (Patot et al. 2012). A transcriptome study (Varaldi and Lepetit 2018) revealed interesting leads on how this virus manipulates parasitoid behaviour and how superparasitism is regulated genetically. LbHV shares several features with the nuclear-associated large DNA viruses previously described, but there is insufficient data available to understand whether they share an evolutionary history.

The second large enveloped FV infects *Apis mellifera* (AmFV). The nearly 500 kbp dsDNA genome is encapsidated by two major nucleoproteins, and the resulting nucleocapsid is folded and wrapped in a tri-laminate envelope (Gauthier et al. 2015). AmFV prevalence appears to be high throughout the year with a peak in spring (Hartmann et al. 2015). Severely affected honeybees are characterised by milky-white haemolymph. AmFV sequences have also been found in *Varroa* mites, but it is unclear if these mites may serve as vector for AmFV (Gauthier et al. 2015). Classification of this virus is difficult: the AmFV genome encodes 28 proteins with significant similarity with NCLDV proteins, while 13 proteins (including PIF 0–3 and BRO proteins) find their closest known orthologues in baculoviruses.

10.4 Use of insect viruses in biocontrol of insect pests

Most of our current understanding of insect viruses derives from research on their potential as biological control agents. The use of entomopathogenic viruses in crop protection is steadily increasing (Lacey et al. 2015). Insect viruses are naturally occurring pathogens that are often host specific, making them highly suitable as environmentally friendly alternatives to chemical insecticides in the management of pest insects. Candidate viruses must be able to easily infect insects, be virulent enough to kill the insects (or at least to stop them from eating), and be sufficiently fast acting to limit feeding damage (Erlandson 2008). Of the known insect viruses, baculoviruses are most widely used as viral biological control agents and these are discussed in more detail in the next section. Another successful example is the ongoing use of Oryctes rhinoceros nudivirus (OrNV) to control rhinoceros beetle larvae on oil palms in Asia (Ramle et al. 2005). It is highly pathogenic to the larvae, which hide in burrows in oil and coconut palm trunks. The control programme also involves using a pheromone to attract adults that can subsequently be infected and used to disseminate and horizontally transmit the virus via their faeces to natural habitats including breeding sites and feeding burrows (Huger 2005).

Other RNA and DNA viruses have also been explored for their potential as biological control agents, but their application is often constrained due to limited virulence leading to chronic infections (cypoviruses, entomopoxviruses, iridoviruses), low pathogenicity in larger instars (Helicoverpa armigera stunt virus (HaSV)), low oral infectivity (iridoviruses, ascoviruses), lack of a large-scale production system (HaSV, cypoviruses, densoviruses), or close relatedness to vertebrate-infecting viruses (densoviruses, entomopoxviruses, iridoviruses). Among the RNA viruses, Helicoverpa armigera stunt virus (HaSV; *Alphatetraviridae*) can be used against *Helicoverpa* species (Christian et al. 2005; Erlandson, 2008). HaSV is highly pathogenic to early instar larvae, causing rapid cessation of feeding, slowing of larval growth, and shrinkage of the larval body (stunting) and eventually death. However, it has hardly any effect on late instar larvae. Although HaSV is robust and can be easily stored, large-scale production is very challenging, which currently limits further development of this virus for biological control (Christian et al. 2005). The potential of cypoviruses (*Reoviridae*), which mainly infect lepidopteran hosts, has also been explored. The CPV-based product Matsukemin was used in Japan against the pine caterpillar *Dendrolimus spectabilis* between 1974 and 1995 (Erlandson, 2008; Eberle et al. 2012). The downside of CPVs is that they often cause chronic infections. Infected larvae may still pupate and develop into adults, and larvae may fully recover from viral infection by sloughing off

infected midgut cells. Therefore, CPVs may be used in systems where some crop damage can be tolerated. An exception is the Norape argyrrhorea CPV, which is highly virulent towards *N. argyrrhorea*, an important oil-palm pest in South America, and causes rapid larval feeding cessation and mortality (Zeddam et al. 2003). However, large-scale production is challenging since the virus is restricted to the midgut and this may limit commercialisation.

Among the DNA viruses, densoviruses (family *Parvoviridae*) infect a range of invertebrates. The virulence of densoviruses varies and depends on the virus species. Some strains infecting economically important pest insect species have been explored for their potential as biological control agents, including Galleria mellonella densovirus against waxmoth infestations of bee hives and Periplaneta fuliginosa densovirus against the smoky-brown cockroach *P. fuliginosa* (Erlandson, 2008; Jiang et al. 2008; Prasad and Srivastava, 2016). Large-scale production is limiting commercialisation of densovirus products, although progress has been made for mosquito densoviruses by optimising *in vivo* production in mosquito larvae (Sun et al. 2019). In many countries, commercialisation is further limited due to the homology between densoviruses and vertebrate parvoviruses. Although densoviruses are restricted to invertebrates, extensive safety tests would be required for registration of densovirus products as biological control agents (Erlandson 2008). The same applies to entomopoxviruses (*Poxviridae*), although there is substantial evidence that entomopoxviruses are solely restricted to insects and are well separated from vertebrate poxviruses. Entomopoxviruses have been investigated for their potential use to control orthopteran (grasshopper and locust) pests. Melanoplus sanguinipes entomopoxvirus has a broad host species range and is infectious upon ingestion, however disease develops slowly, and the virus is deemed insufficiently virulent (Erlandson 2008). Despite the creation of a scorpion toxin-encoding recombinant IIV with increased insecticidal activity, IIVs remain of limited interest as biocontrol agents due to their broad host range, limited mortality, low oral infectivity, and the existence of vertebrate relatives in fish and amphibians (Hernandez et al. 2000; Kleespies et al. 1999; Nalcacioglu et al. 2016). Ascoviruses (family *Ascoviridae*), on the other hand cause high mortality among economically important insect pests, and therefore have potential to be used as biological control agents (Chen et al. 2020; Hamm et al. 1986). Although ascoviruses exhibit poor oral infectivity (being transmitted by parasitoid wasps) their potential as biocontrol agent may be enlarged by combining the virus with *Bacillus thuringiensis* (Yu et al. 2020).

Baculoviruses comprise the most well-characterised insect viruses, and they are used extensively for biocontrol of pest insects. Most baculovirus species are highly host specific, infecting one or a few related host species. Baculoviruses have no negative effects on non-target organisms (including natural enemies) and the embedding of the virions into OBs gives makes them very persistent outside the host, allowing easy handling, application, and long-term storage. More than 60 baculovirus-based pesticides have been utilised worldwide to control diverse insect pests, mostly from the genera *Alphabaculovirus* (lepidopteran-specific NPVs) and *Betabaculovirus* (lepidopteran-specific GVs), (Beas-Catena et al. 2014), but a gammabaculovirus-based product to control the sawfly *Neodiprion sertifer* is also available (Hajek and Van Frankenhuyzen 2017). Baculoviruses are used to control certain globally important lepidopteran pest species, which are often pests to a wide range of crops (Lacey et al. 2015). Effective baculovirus products include Anticarsia gemmatalis MNPV (AgMNPV) to control the velvet been caterpillar *A. gemmatalis*; Cydia pomonella GV (CpGV) to control the codling moth *C. pomonella* in apples, pears and walnuts; Helicoverpa armigera NPV (HearNPV) to control the cotton bollworm *H. armigera*, a highly polyphagous pest worldwide, and Spodoptera exigua MNPV (SeMNPV) to control the beet armyworm *S. exigua*, a highly polyphagous and widely distributed pest both in the field and in greenhouses (Beas-Catena et al. 2014; Lacey et al. 2015). In addition, baculoviruses are successfully used to control forest insect species in North America and Europe, including gypsy moths and sawflies.

Drawbacks for the more extensive use of baculoviruses as biological control agents include their slower speed of kill relative to other (chemical) insecticides, their narrow host specificity and relatively high production costs (large-scale production still relies on *in vivo* systems) which may hinder commercialisation (Beas-Catena et al. 2014; Lacey et al. 2015). The susceptibility of baculoviruses to ultraviolet (UV) light leads to solar degradation, which necessitates frequent re-application of the virus (Lacey et al. 2015). Resistance against baculoviruses has been observed in certain European *C. pomonella* populations. To avoid the development of resistance, alternative baculovirus isolates are used, and baculovirus application is alternated with other biological strategies. To improve the speed of kill, baculoviruses have been genetically modified to delete specific genes or to add genes encoding insect hormones, enzymes, or insect-specific toxins (Beas-Catena et al. 2014; Erlandson 2008; Lacey et al. 2015), although there are substantial regulatory barriers that hinder worldwide application of genetically modified products.

A major issue hindering commercialisation of potential insect viruses as biological control agents remains the high cost related to mass production of those viruses. Production of commercial insect viruses is dependent on often expensive *in vivo* systems using living insects, which are also difficult to rear in a high-quality disease-free state. To further enhance the potential of other non-baculoviral insect viruses as biological control agents, a huge advance in fundamental knowledge on viral taxonomy, pathology, ecology, and mass-production systems is also required (Lacey et al. 2015). New viral products need to be explored and technological advances are needed to allow mass production of viral control products. The search for strains with an increased speed of action is an ongoing process. In addition, a better understanding of the interactions between viral products and other biological control agents is needed. Nonetheless, insect viruses (especially baculoviruses) have a promising future as biopesticides and can be successfully implemented in integrated pest management strategies.

10.5 Viral disease management in reared insects

Viruses have been problematic in insect rearing worldwide long before viruses themselves where discovered. Early Neolithic farmers in North Africa already kept honeybees (Roffer-Salque et al. 2016) and *Bombyx mori* (the silkworm or silk moth) was domesticated for silk production around 7,500 years ago (Eilenberg and Bruun Jensen 2018; Yang et al. 2014a). Several diseases have been found in domesticated honeybees and *B. mori* larvae caused by a range of pathogens including viruses. BmNPV, for instance, has frequently led to complete collapse of silkworm colonies. Many outbreaks in insect production facilities, however, are caused by previously unknown, emerging pathogens, such as the densovirus targeting house cricket colonies and the hytrosavirus responsible for the collapse of tsetse fly rearings (Abd-Alla et al. 2009). At the moment, there is no effective control measurement for AdDV (or any other densovirus), other than thoroughly cleaning the facilities and starting over. A complication is that healthy looking crickets can also harbour AdDV (as detected by quantitative PCR) and appear to carry this covert infection over to their offspring (Maciel-Vergara et al. unpublished data). Efforts are ongoing to find out exactly how the virus is transmitted to the offspring, and if egg sanitation might provide a workable solution. Whether the introduction of an artificial EVE in the reared cricket genome may be an option to reduce AdDV titres has to our knowledge not been analysed so far and may in practice not be the final solution as this would need acceptance of the entrance of a genetically modified organism (GMO) into the food chain.

A good example of collected knowledge leading to an effective adaptation in an insect rearing strategy is the changing of the blood feeding regime implemented by FAO/IAEA in their tsetse fly production lines (Abd-Alla et al. 2013). Previously, trays with tsetse fly batches were placed in a random order on blood containers covered with artificial membranes. The shared blood meal served as an efficient medium for horizontal transmission of the virus, where a few infected flies could transmit to almost the whole colony, leading to severe effects on the reproduction capacity. By changing the feeding policy to a strict order of trays (clean feeding system) it was possible to drastically reduce the amount of virus over time, and eventually a virus-free tsetse colony was obtained. Moreover, combining the clean feeding system with the addition of the antiviral drug valacyclovir in the blood meal to suppress viral infection provided a faster and effective control of GpSGHV (Abd-Alla et al.

2014). Crucial here was also the fact that living flies could be screened by PCR for the presence of the virus without sacrificing them, by excising only one middle leg, which allowed the start of a colony with (almost) clean flies (Abd-Alla et al. 2007). These GpSGHV management strategies led to the development of a step-by-step standard operational procedure to detect and manage GpSGHV in tsetse fly factories that is now followed by several rearing facilities.

Management of viral diseases is also crucial in apiculture. Social bees belonging to the genera *Apis* and *Bombus* represent the vast majority of managed pollinator species used in agriculture worldwide (Gallai et al. 2009). They live in densely populated colonies in which closely related individuals (usually siblings) have frequent contact within a confined nest environment, thus elevating infection risks. Additionally, individuals forage for nectar and pollen in a wide area (on average in a radius of 6 km around the colony, but up to more than 10 km, in which they get exposed to a multitude of external pathogens through intra and inter species interactions) (Yañez et al. 2020). Modern intensive beekeeping practices, involving large-scale movements of colonies and trade of queens, are responsible for the fast spread of *Varroa* mites and their associated viruses (SBPV, ABPV, and DWV) (Beaurepaire et al. 2020). Allowed to reach higher virulence through vector transmission, virus prevalence has increased wherever *Varroa destructor* invaded, and a spill over of viruses has been reported within pollinator communities. Chemical control of *Varroa* mites may reduce the prevalence of several viruses and it is the vector rather than the viruses that is currently managed in beekeeping practices: *Varroa* control is mandatory in all countries where *Varroa* is present, and strict regulations apply to the importation of live bees. Within the European Union, introduction of cages of queens is only authorised when certified free of exotic parasites by veterinarian authorities (EU council directive 92/65/EEC of 13 July 1992). It is important to note that no viral species belongs to the list of exotic pests restricting the trade of honeybee queens and germplasm, although it is known that honeybee queens can carry several viruses at the same time (Chen and Siede 2007) and transmit these to their offspring.

We so far have only fragmentary ideas of which viruses and stress factors are responsible for disease outbreaks in mass-reared insects. We also do not know how these viruses often manage to remain in the population in an invisible status over many generations before they suddenly cause disease. For every insect species and group of viruses the measures to be taken can be different, and detailed knowledge on the viruses and their transmission strategy as well as discriminative diagnostic tools are crucial to prevent virus outbreaks in industrial insect rearing facilities. A limiting factor in virus disease control is the fact that there are clear restrictions in the measures that can be taken, as many of the reared insects will end up in the food chain or are even directly meant for human consumption, meaning that the choice of chemical disinfectants that can be used safely, is limited. In general, effective management of viral diseases in insect production systems includes an early diagnosis (preferably based on a screening for multiple pathogens at the same time), the use of clean production facilities with optimal rearing conditions, the supply of clean food sources and substrates, the separation of production stocks, and the maintenance of genetic variation in the insect population (Eilenberg et al. 2015; Eilenberg and Bruun Jensen 2018). A clear problem here is that standardised diagnostic services for insect rearing companies (like those existing for vertebrate animal husbandry) are almost non-existent at the moment.

10.6 Future Outlook

Despite the large variety in insect infecting viruses as outlined in Section 2, it also has become clear that we have only discovered a very small fraction of the insect-infecting viruses. One of the main reasons for this is that we have only examined viruses that cause major disease symptoms and only in insects that are well-known to us as agricultural pests, as vectors of medical importance, or as being beneficial like pollinators, and silkworms. For many taxonomic groups of insects in a variety of geographical ranges, we have hardly paid any attention to the virome, let alone know about transmission and pathology of potential viruses. It is therefore not surprising that outbreaks of previously unknown viruses are frequently seen when novel insect species are reared on a large scale, for example to serve as food and feed or for waste management purposes. Metagenomics using Next Generation Sequencing (NGS) have proven to be powerful tools, not only to screen an organism's virome, but also to analyse large collections of samples for viral DNA/RNA sequences. The metagenomic/transcriptomic discoveries revealed an overwhelming number of unknown viral sequences in insects (and other invertebrates), several even with novel genome properties that set the corresponding viruses aside from known viral families (Paez-Espino et al. 2016; Porter et al. 2019; Shi et al. 2016; Wu et al. 2020). A useful tool to detect replicating viruses in insects is deep-sequencing of small RNA species that are generated as a result of antiviral RNA interference; see for example Göertz et al. (2019). Despite its merits, the clear limitation of NGS is that it only reveals genetic data that hint towards the existence of so far unknown viruses (and mostly only if they show some detectable homology to already-known viruses (Obbard 2018)).

The major challenge is to decipher the biological properties of these new viruses, which in some cases, also may turn out to be endogenous viral elements instead (see also Obbard 2018). As a consequence, there is a quickly growing gap between the fast-expanding information in viral genome sequence databases and the relatively slowly progressing build-up of knowledge on virus biology and virus-host interactions. This includes, for example, incomplete information on virus host ranges, pathogenesis (if any), and routes of transmission. This kind of biological data is essential for understanding of the role that insect viruses play in ecosystems: whether these viruses are a potential risk due to their pathogenetic features, whether they co-exist in a neutral relationship with the insect hosts, or even may provide the host with additional traits that increase the fitness of the insect under certain circumstances (Roossinck and Bazan 2017). This kind of information is of broad interest for the protection of natural ecosystems and for the establishing of resilient agricultural settings. In addition, it is of growing importance to obtain such knowledge for viruses of cultured insects in order to mitigate, or even prevent, viral disease outbreaks in the insect mass-rearing industry.

10.7 Conclusions

In recent years, it became absolutely clear that we have only seen a very tiny 'tip of the iceberg' regarding the insect-infecting viruses that likely exist globally. Hence, it is obvious that insect viruses with new traits will be discovered and that novel viral families will be recognised in the future as follow-up studies on metagenomic analyses or when novel species of insects gain economic importance, for example in the food and feed industry as bio-control agents, or in waste management. One of the clear challenges is the development of diagnostic tools and services for mass reared insects. On the other hand, it also becomes obvious that viruses may play various roles in ecosystems, and can either have pathogenic, mutualistic, or neutral relationships with their hosts, and in addition may influence the impact of other pathogens and the vector competence of mosquitoes for vertebrate viruses. The exact nature of these interactions with the host and with other components of the microbiome needs to be identified for many (newly) discovered insect viruses.

Acknowledgements

The authors are grateful to Peter Tijssen and Irene Meki for critical reading of sections before submission. Gabriela Maciel Vergara and Adly Abd-Alla are acknowledged for sharing unpublished data. Jody Hobson-Peters and Hanna-Isadora Huditz are acknowledged for sharing knowledge on mesoniviruses and negeviruses, respectively. Jirka Manuel Petersen is financially supported by the EU Horizon 2020 Research and Innovation Framework Programme INSECT DOCTORS (MSCA-ITN-2019, Project 859850). Vera Ros is supported by a VIDI-grant of the Dutch Research Council (NWO; VI.Vidi.192.041).

Recommended further reading

Extended review on bee viruses:

Beaurepaire, A., Piot, N., Doublet, V., Antunez, K., Campbell, E., Chantawannakul, P.,Chejanovsky, N., Gajda, A., Heerman, M., Panziera, D., Smagghe, G., Yanez, O., de Miranda, J. R., and Dalmon, A. 2020. Diversity and global distribution of viruses of the western honey bee, *Apis mellifera. Insects* 11(4)

Overview of biological control using insect viruses:

Lacey, L.A., Grzywacz, D., Shapiro-Ilan, D.I., Frutos, R., Brownbridge, M., and Goettel, M. S. 2015. Insect pathogens as biological control agents: Back to the future. *Journal of Invertebrate Pathology* 132:1–41

Overviews of the main groups of insect viruses (with RNA, small and large DNA genomes):

Ryabov, E.V. 2017. Invertebrate RNA virus diversity from a taxonomic point of view. *Journal of Invertebrate Pathology* 147: 37–50

Tijssen, P., Pénzes, J.J., Yu, Q., Pham, H., and Bergoin, M. 2017. Diversity of small, single-stranded DNA viruses of invertebrates and their chaotic evolutionary past. *Journal of Invertebrate Pathology* 147: 23–36

Williams, T., Bergoin, M., and van Oers, M.M. 2017a. Diversity of large DNA viruses of invertebrates. *Journal of Invertebrate Pathology* 147: 4–22

References

Abd-Alla, A., Bossin, H., Cousserans, F., Parker, A., Bergoin, M., and Robinson, A. 2007. Development of a non-destructive PCR method for detection of the salivary gland hypertrophy virus (SGHV) in tsetse flies. *Journal of Virological Methods* 139: 143–149

Abd-Alla, A.M., Kariithi, H.M., Mohamed, A.H., Lapiz, E., Parker, A.G., and Vreysen, M. J. 2013. Managing hytrosavirus infections in *Glossina pallidipes* colonies: Feeding regime affects the prevalence of salivary gland hypertrophy syndrome. *PLoS ONE* 8: e61875

Abd-Alla, A.M., Marin, C., Parker, A.G., and Vreysen, M.J. 2014. Antiviral drug valacyclovir treatment combined with a clean feeding system enhances the suppression of salivary gland hypertrophy in laboratory colonies of *Glossina pallidipe*s. *Parasite Vectors* 7: 214

Abd-Alla, A.M., Vlak, J.M., Bergoin, M., Maruniak, J.E., Parker, A., Burand, J.P., Jehle, J. A., Boucias, D.G., and Hytrosavirus Study Group of the ICTV. 2009. *Hytrosaviridae*: A proposal for classification and nomenclature of a new insect virus family. *Archives of Virology* 154: 909–918

Ahuja, K. and Mamtani, K. 2020. Edible insects market size by product (beetles, caterpillars, grasshoppers, bees, wasps, ants, scale insects & tree bugs), by application (flour, protein bars, snacks), industry analysis report, regional outlook, application potential, price trends, competitive market share & forecast, 2020 – 2026. https://www.gminsights.com/industry-analysis/edible-insects-market, October 2020

Allain, T.W., Stentiford, G.D., Bass, D., Behringer, D.C., and Bojko, J. 2020. A novel nudivirus infecting the invasive demon shrimp *Dikerogammarus haemobaphes* (Amphipoda). *Science Reports* 10: 14816

Andreadis, T.G., Becnel, J.J., and White, S.E. 2003. Infectivity and pathogenicity of a novel baculovirus, CuniNPV from *Culex nigripalpus* (Diptera: Culicidae) for thirteen species and four genera of mosquitoes. *Journal of Medical Entomology* 40: 512–517.

Arif, B. and Kurstak, E. 1991. The entomopoxviruses. In *Virus of Invertebrates*, E. Kurstak (eds.) pp. 179–195. Marcel Dekker Inc., New York.

Asgari, S., Bideshi, D.K., Bigot, Y., Federici, B.A., Cheng, X.W., and ICTV Report Consortium. 2017. ICTV Virus Taxonomy Profile: *Ascoviridae*. *Journal of General Virology* 98: 4–5

Attoui, H., Mohd Jaafar, F., Belhouchet, M., Biagini, P., Cantaloube, J. F., De Micco, P., and De Lamballerie, X. 2005. Expansion of family *Reoviridae* to include nine-segmented dsRNA viruses: Isolation and characterization of a new virus designated Aedes pseudoscutellaris reovirus assigned to a proposed genus (*Dinovernavirus*). *Virology* 343: 212–223

Bando, H., Choi, H., Ito, Y., and Kawase, S. 1990. Terminal structure of a densovirus implies a hairpin transfer replication which is similar to the model for AAV. *Virology* 179: 57–63

Bateman, K.S., Kerr, R., Stentiford, G.D., Bean, T.P., Hooper, C., Van Eynde, B., Delbare, D., Bojko, J., Christiaens, O., Taning, C.N.T., Smagghe, G., van Oers, M.M., van Aerle, R. 2021. Identification and full characterisation of two novel crustacean infecting members of the family *Nudiviridae* provides support for two subfamilies. *Viruses* 13: 1694

Bawden, A.L., Glassberg, K.J., Diggans, J., Shaw, R., Farmerie, W., and Moyer, R.W. 2000. Complete genomic sequence of the Amsacta moorei entomopoxvirus: Analysis and comparison with other poxviruses. *Virology* 274: 120

Beas-Catena, A., Sánchez-Mirón, A., García-Camacho, F., Contreras-Gómez, A., and Molina-Grima, E. 2014. Baculovirus biopesticides: An overview. *Journal of Animal & Plant Sciences* 24: 362–373

Beaurepaire, A., Piot, N., Doublet, V., Antunez, K., Campbell, E., Chantawannakul, P., Chejanovsky, N., Gajda, A., Heerman, M., Panziera, D., Smagghe, G., Yanez, O., De Miranda, J.R., and Dalmon, A. 2020. Diversity and global distribution of viruses of the Western honeybee, *Apis mellifera*. *Insects* 11: 239

Belloncik, S., Maramorosch, K., Murphy, F.A., and Shatkin, A.J. 1989. Cytoplasmic polyhedrosis viruses—*Reoviridae*. *Advances in Virus Research* 37: 173–209

Bergoin, M., Devauchelle, G., and Vago, C. 1976. Fusiform inclusions associated with the entomopoxvirus of the beetle *Melolontha melolontha*. *Journal of Ultrastructural Research* 55: 17–30

Bézier, A. 2009. Polydnavirus hidden face: The genes producing virus particles of parasitic wasps. *Journal of Invertebrate Pathology* 101: 194–203

Bézier, A., Thézé, J., Gavory, F., Gaillard, J., Poulain, J., Drezen, J.-M., and Herniou, E. A. 2015. The genome of the nucleopolyhedrosis-causing virus from *Tipula oleracea* sheds new light on the *Nudiviridae* family. *Journal of Virology* 89: 3008–3025

Bideshi, D.K., Demattei, M.-V., Rouleux-Bonnin, F., Stasiak, K., Tan, Y., Bigot, S., Bigot, Y., and Federici, B.A. 2006. Genomic sequence of spodoptera frugiperda ascovirus 1a, an enveloped, double-stranded DNA insect virus that manipulates apoptosis for viral reproduction. *Journal of Virology* 80: 11791–11805

Bilimoria, S.L. and Arif, B.M. 1979. Subunit protein and alkaline protease of entomopoxvirus spheroids. *Virology* 96: 596–603

Blissard, G.W. and Theilmann, D.A. 2018. Baculovirus entry and egress from insect cells. *Annual Review of Virology* 5: 113–139

Blitvich, B.J. and Firth, A.E. 2015. Insect-specific flaviviruses: A systematic review of their discovery, host range, mode of transmission, superinfection exclusion potential and genomic organization. *Viruses* 7: 1927–1959

Boogaard, B., Evers, F., Van Lent, J.W.M., and van Oers, M.M. 2019. The baculovirus Ac108 protein is a *per os* infectivity factor and a component of the ODV entry complex. *Journal of General Virology* 100: 669–678

Boogaard, B., van Oers, M.M., and Van Lent, J.W.M. 2018. An advanced view on baculovirus *per os* infectivity factors. *Insects* 9: 84

Bossin, H., Fournier, P., Royer, C., Barry, P., Cérutti, P., Gimenez, S., Couble, P., and Bergoin, M. 2003. Junonia coenia densovirus-based vectors for stable transgene expression in Sf9 cells: Influence of the densovirus sequences on genomic integration. *Journal of Virology* 77: 11060

Bratke, K.A. and McLysaght, A. 2008. Identification of multiple independent horizontal gene transfers into poxviruses using a comparative genomics approach. *BMC Evolutionary Biology* 8: 67

Braunwald, J., Nonnenmacher, H., and Tripier-Darcy, F. 1985. Ultrastructural and biochemical study of frog virus 3 uptake by BHK-21 cells. *Journal of General Virology* 66; 283–293

Brooks, E.M., Gordon, K.H., Dorrian, S.J., Hines, E.R., and Hanzlik, T.N. 2002. Infection of its lepidopteran host by the Helicoverpa armigera stunt virus (*Tetraviridae*). *Journal of Invertebrate Pathology* 80: 97–111

Budge, G.E., Simcock, N.K., Holder, P.J., Shirley, M.D.F., Brown, M A., Van Weymers, P. S.M., Evans, D.J., and Rushton, S P. 2020. Chronic bee paralysis as a serious emerging threat to honeybees. *Nature Communications* 11; 2164

Burand, J.P., Kim, W., Afonso, C.L., Tulman, E.R., Kutish, G.F., Lu, Z., and Rock, D.L. 2012. Analysis of the genome of the sexually transmitted insect virus Helicoverpa zea nudivirus 2. *Viruses* 4: 28–61.

Burand, J.P., Rallis, C.P., and Tan, W 2004. Horizontal transmission of Hz-2V by virus infected *Helicoverpa zea* moths. *Journal of invertebrate Pathology* 85: 128–131

Burke, G.R. 2019. Common themes in three independently derived endogenous nudivirus elements in parasitoid wasps. *Current Opinion in Insect Science* 32; 28–35

Carballo, A., Williams, T., Murillo, R., and Caballero, P. 2020. Iflavirus covert infection increases susceptibility to nucleopolyhedrovirus disease in *Spodoptera exigua*. *Viruses* 12; 509

Carter, J.B. 1973. The mode of transmission of Tipula iridescent virus: II. Route of infection. *Journal of Invertebrate Pathology* 21: 136–143

Cerutti, M., Cerutti, P., and Devauchelle, G. 1989. Infectivity of vesicles prepared from Chilo iridescent virus inner membrane: Evidence for recombination between associated DNA fragments. *Virus Research* 12: 299–313

Chandler, M., De La Cruz, F., Dyda, F., Hickman, A.B., Moncalian, G., and Ton-Hoang, B. 2013. Breaking and joining single-stranded DNA: The HUH endonuclease superfamily. *Nature Review Microbiology* 11: 525–538

Chen, G., Liu, H., Mo, B. C., Hu, J., Liu, S. Q., Bustos-Segura, C., Xue, J., and Wang, X. 2020. Growth and development of *Helicoverpa armigera* (Lepidoptera: Noctuidae) larvae infected by Heliothis virescens ascovirus 3i (HvAV-3i). *Frontiers in Physiology* 11: 93

Chen, Y.P. and Siede, R. 2007. Honeybee viruses. *Advances in Virus Research* 70; 33–80

Cheng, R.L., Xi, Y., Lou, Y.H., Wang, Z., Xu, J.Y., Xu, H.J., and Zhang, C.X. 2014. Brown planthopper nudivirus DNA integrated in its host genome. *Journal of Virology* 88: 5310–5318

Cheng, X.-W., Carner, G.R., and Arif, B.M. 2000. A new ascovirus from *Spodoptera exigua* and its relatedness to the isolate from *Spodoptera frugiperda*. *Journal of General Virology* 81: 3083–3092

Chinchar, V.G., Hick, P., Ince, I.A., Jancovich, J.K., Marschang, R., Qin, Q., Subramaniam, K., Waltzek, T.B., Whittington, R., Williams, T., Zhang, Q.Y., and ICTV Report Consortium. 2017a. ICTV Virus Taxonomy Profile: Iridoviridae. *Journal of General Virology* 98: 890–891

Chinchar, V.G., Waltzek, T.B., and Subramaniam, K. 2017b. Ranaviruses and other members of the family *Iridoviridae*: Their place in the virosphere. *Virology* 511: 259 271

Chitnis, N.S., D'Costa, S.M., Paul, E.R., and Bilimoria, S.L. 2008. Modulation of iridovirus-induced apoptosis by endocytosis, early expression, JNK, and apical caspase. *Virology* 370: 333–342

Chouvenc, T., Mullins, A.J., Efstathion, C.A, and Su N.Y. 2013. Virus-like symptoms in a termite (Isoptera: Kalotermitidae) field colony. *Florida Entomologist* 96: 1612–1614

Christian, P.D., Murray, D., Powell, R., Hopkinson, J., Gibb, N.N., and Hanzlik, T.N. 2005. Effective control of a field population of *Helicoverpa armigera* by using the small RNA virus Helicoverpa armigera stunt virus (*Tetraviridae: Omegatetravirus*). *Journal of Economical Entomology* 98: 1839–1847

Coffman, K.A., Harrell, T.C., and Burke, G.R. 2020. A mutualistic poxvirus exhibits convergent evolution with other heritable viruses in parasitoid wasps. *Journal of Virology* 94: e02059–e02119

Colson, P., De Lamballerie, X., Yutin, N., Asgari, S., Bigot, Y., Bideshi, D.K., Cheng, X.W., Federici, B.A., Van Etten, J.L., Koonin, E.V., La Scola, B., and Raoult, D. 2013. Megavirales, a proposed new order for eukaryotic nucleocytoplasmic large DNA viruses. *Archives of Virology* 158: 2517–2521

Contreras-Gutierrez, M.A., Nunes, M.R.T., Guzman, H., Uribe, S., Suaza Vasco, J.D., Cardoso, J.F., Popov, V.L., Widen, S.G., Wood, T.G., Vasilakis, N., and Tesh, R.B. 2017. Sinu virus, a novel and divergent orthomyxovirus related to members of the genus *Thogotovirus* isolated from mosquitoes in Colombia. *Virology* 501: 166–175

Cory, J.S. and Myers, J.H. 2003. The ecology and evolution of insect baculoviruses. *Annual Review of Ecology, Evolution and Systemics* 34: 239–272

Dall, D., Luque, T., and O'Reilly, D. 2001. Insect-virus relationships: Sifting by informatics. *Bioessays* 23: 184–193

Dalmon, A., Desbiez, C., Coulon, M., Thomasson, M., Le Conte, Y., Alaux, C., Vallon, J., and Moury, B. 2017. Evidence for positive selection and recombination hotspots in Deformed wing virus (DWV). *Science Reports* 7: 41045

Dasgupta, R., Selling, B., and Rueckert, R. 1994. Flock house virus: A simple model for studying persistent infection in cultured *Drosophila* cells. *Archives of Virology Supplement* 9: 121–132

Delwart, E. and Li, L. 2012. Rapidly expanding genetic diversity and host range of the *Circoviridae* viral family and other Rep encoding small circular ssDNA genomes. *Virus Research* 164: 114–121

Dietzgen, R.G., Ghedin, E., Jiang, D., Kuhn, J.H., Song, T., Vasilakis, N., Wang, D., and ICTV Report Consortium 2017. ICTV virus taxonomy profile: *Nyamiviridae*. *Journal of General Virology* 98: 2914–2915

Dizman, Y.A., Demirbag, Z., Ince, I.A., and Nalcacioglu, R. 2012. Transcriptomic analysis of Chilo iridescent virus immediate early promoter. *Virus Research* 167: 353–357

Dorrington, R.A., Jiwaji, M., Awando, J.A., and Bruyn, M.M. 2020. Advances in tetravirus research: New insight into the infectious virus lifecycle and an expanding host range. *Current Issues in Molecular Biology* 34: 145–162

Eaton, H.E., Ring, B.A., and Brunetti, C.R. 2010. The genomic diversity and phylogenetic relationship in the family *Iridoviridae*. *Viruses* 2: 1458–1475

Eberle, K.E., Jehle, J.A., and Hüber, J. 2012. Microbial control of crop pests using insect viruses. In *Integrated Pest Management: Principles and Practice*, D.P.A.S. Abrol (ed.) pp. 281–98. Wallingford, UK:CABI Publishing.

Eilenberg, J. and Bruun Jensen, A. 2018. Prevention and management of diseases in terrestrial invertebrates. In *Ecology of Invertebrate Diseases*, A.E. Hacek and D.I. Shapiro-Ilan (eds.) pp. 495–526. Hoboken, NJ: John Wiley & Sons.

Eilenberg, J., Vlak, J.M., Nielsen-leroux, C., Cappellozza, S., and Jensen, A.B. 2015. Diseases in insects produced for food and feed. *Journal of Insects as Food and Feed*, 1: 87–102

Erlandson, M. 2008. Insect pest control by viruses. In *Encyclopedia of Virology*, 3rd edition edn, B.W.J. Mahy, and M.H.V. Van Regenmortel (eds.) pp. 125–33. Academic Press, Amsterdam

Federici, B.A., Vlak, J.M., and Hamm, J.J. 1990. Comparative study of virion structure, protein composition and genomic DNA of three ascovirus isolates. *Journal of General Virology* 71: 1661–1668

Gallai, N., Salles, J.-M., Settele, J., and Vaissière, B. E. 2009. Economic valuation of the vulnerability of world agriculture confronted with pollinator decline. *Ecological Economics* 68; 810–821

Gasque, S.N., van Oers, M.M., and Ros, V.I.D. 2019. Where the baculoviruses lead, the caterpillars follow: Baculovirus-induced alterations in caterpillar behaviour. *Current Opinion in Insect Science* 33: 30–36

Gauthier, L., Cornman, S., Hartmann, U., Cousserans, F., Evans, J.D., De Miranda, J.R., and Neumann, P. 2015. The Apis mellifera filamentous virus genome. *Viruses* 7: 3798–3815

Gencer, D., Yesilyurt, A., Güllü, M., Demir, I., and Nalcacioglu, R. 2020. Insecticidal activities of wild type and recombinant invertebrate iridescent viruses on five common pests. *Turkish Journal of Entomology* 44: 365–373

Gildow, F.E. and D'Arcy, C.J. 1988. Barley and oats as reservoirs for an aphid virus and the influence on barley yellow dwarf virus transmission. *Phytopathology* 78: 811–816

Göertz, G.P., Miesen, P., Overheul, G.J., Van Rij, R.P., van Oers, M.M., and Pijlman, G.P. 2019. Mosquito small RNA responses to West Nile and insect-specific virus infections in *Aedes* and *Culex* mosquito cells. *Viruses* 1, 271, doi: 10.3390/v11030271

Goorha, R. and Murti, K.G. 1982. The genome of frog virus 3, an animal DNA virus, is circularly permuted and terminally redundant. *Proceedings of the National Academy of Science* 79: 248–252

Govindarajan, R. and Federici, B.A. 1990. Ascovirus infectivity and effects of infection on the growth and development of noctuid larvae. *Journal of Invertebrate Pathology* 56: 291–299

Granados, R.R. 1973. Entry of an insect poxvirus by fusion of the virus envelope with the host cell membrane. *Virology* 52: 305–309

Granados, R.R. 1981. The entomopoxviruses. In *Pathogenesis of Invertebrate Microbial Diseases*, E.W. Davidson (ed.) p. 101. Montclair: Allanheld Osmun

Hajek, A.E. and Van Frankenhuyzen, K. 2017. Use of entomopathogens against forest pests. In *Microbial Control of Insect and Mite Pests*, L.A. Lacey (ed.) pp. 313–30. Amsterdam, NL: Academic Press.

Jamm, J.J., Pair, S.D., and Marti, O.G. 1986. Incidence and host range of a new ascovirus isolated from fall armyworm, *Spodoptera frugiperda* (Lepidoptera: Noctuidae). *Florida Entomologist* 69: 524–541

Han, Y., van Oers, M.M., Van Houte, S., and Ros, V.I.D. 2015. Host manipulations by parasites and viruses In *Host Manipulations by Parasites and Viruses*, H. Mehlhorn (ed.) pp. 149–74. Springer, Cham, Switzerland.

Harrison, R. and Hoover, K. 2012. Baculoviruses and other occluded insect viruses. In *Insect Pathology*, F.K. Vega (ed.) pp. 73–131. Amsterdam: Elsevier

Harrison, R.L., Herniou, E.A., Bézier, A., Jehle, J.A., Burand, J.P., Theilmann, D.A., Krell, P.J., van Oers, M.M., Nakai, M., and ICTV Report Consortium. 2020. ICTV Virus Taxonomy Profile: *Nudiviridae*. *Journal of General Virology* 101: 3–4

Hartmann, U., Forsgren, E., Charriere, J.D., Neumann, P., and Gauthier, L. 2015. Dynamics of Apis mellifera filamentous virus (AmFV) infections in honeybees and relationships with other parasites. *Viruses* 7: 2654–2667

Hermanns, K., Zirkel, F., Kopp, A., Marklewitz, M., Rwego, I.B., Estrada, A., Gillespie, T.R., Drosten, C., and Junglen, S. 2017. Discovery of a novel alphavirus related to Eilat virus. *Journal of General Virology* 98: 43–49

Hernandez, O., Maldonado, G., and Williams, T. 2000. An epizootic of patent iridescent virus disease in multiple species of blackflies in Chiapas, Mexico. *Medical and Veterinary Entomology* 14: 458–462

Hobson-Peters, J., Warrilow, D., McLean, B.J., Watterson, D., Colmant, A.M., Van Den Hurk, A.F., Hall-Mendelin, S., Hastie, M.L., Gorman, J.J., Harrison, J.J., Prow, N.A., Barnard, R.T., Allcock, R., Johansen, C.A., and Hall, R.A. 2016. Discovery and characterisation of a new insect-specific bunyavirus from *Culex* mosquitoes captured in northern Australia. *Virology* 489: 269–281

Holt, C.C., Stone, M., Bass, D., Bateman, K.S., Van Aerle, R., Daniels, C.L., Van Der Giezen, M., Ross, S.H., Hooper, C., and Stentiford, G.D. 2019. The first clawed lobster virus Homarus gammarus nudivirus (HgNV n. sp.) expands the diversity of the *Nudiviridae*. *Science Reports* 9: 10086

Hou, D., Kuang, W., Luo, S., Zhang, F., Zhou, F., Chen, T., Zhang, Y., Wang, H., Hu, Z., Deng, F., and Wang, M. 2019. Baculovirus ODV-E66 degrades larval peritrophic membrane to facilitate baculovirus oral infection. *Virology* 537: 157–164

Huger, A.M. 1985. A new virus disease of crickets (Orthoptera: Gryllidae) causing macronucleosis of fatbody. *Journal of Invertebrate Pathology* 45: 108–111

Huger, A.M. 2005. The Oryctes virus: Its detection, identification, and implementation in biological control of the coconut palm rhinoceros beetle, *Oryctes rhinoceros* (Coleoptera: Scarabaeidae). *Journal of Invertebrate Pathology* 89: 78–84

Hughes, A.L. and Friedman, R. 2003. Genome-wide survey for genes horizontally transferred from cellular organisms to baculoviruses. *Molecular Biology and Evolution* 20: 979–987

ICTV 2020. Virus Taxonomy by the International Committee on the Taxonomy of Viruses, Release 2020 https://talk.ictvonline.org/taxonomy/

Inan, C., Muratoglu, H., Arif, B.M., and Demirbag, Z. 2018. Amsacta moorei entomopoxvirus encodes a functional heparin-binding glycosyltransferase (AMV248). *Virus Genes* 54: 438–445

Ince, I.A., Boeren, S.A., van Oers, M.M., Vervoort, J.J., and Vlak, J.M. 2010. Proteomic analysis of Chilo iridescent virus. *Virology* 405: 253–258

Ishii, T., Takatsuka, J., Nakai, M., and Kunimi, Y. 2002. Growth characteristics and competitive abilities of a nucleopolyhedrovirus and an entomopoxvirus in larvae of the smaller tea tortrix, *Adoxophyes honmai* (Lepidoptera: Tortricidae). *Biological Control* 23: 96–105

Jaenson, T.G. 1978. Virus-like rods associated with salivary gland hyperplasia in tsetse, *Glossina pallidipes*. *Transactions of the Royal Society of Tropical Medicine and Hygiene* 72: 234–238

Jakubowska, A.K., D'Angiolo, M., Gonzalez-Martinez, R.M., Millan-Leiva, A., Carballo, A., Murillo, R., Caballero, P., and Herrero, S. 2014. Simultaneous occurrence of covert infections with small RNA viruses in the lepidopteran *Spodoptera exigua*. *Journal of Invertebrate Pathology* 121: 56–63

Jakubowska, A.K., Murillo, R., Carballo, A., Williams, T., Van Lent, J.W.M., Caballero, P., and Herrero, S. 2016. Iflavirus increases its infectivity and physical stability in association with baculovirus. *Peer Journal* 4: e1687

Javed, M.A., Biswas, S., Willis, L.G., Harris, S., Pritchard, C., van Oers, M.M., Donly, B.C., Erlandson, M.A., Hegedus, D.D., and Theilmann, D.A. 2017. Autographa californica multiple nucleopolyhedrovirus AC83 is a *per os* infectivity factor (PIF) protein required for occlusion-derived virus (ODV) and budded virus nucleocapsid assembly as well as assembly of the PIF complex in ODV envelopes. *Journal of Virology* 91: e02115–16

Jehle, J.A., Blissard, G.W., Bonning, B.C., Cory, J.S., Herniou, E.A., Rohrmann, G.F., Theilmann, D.A., Thiem, S.M., and Vlak, J.M. 2006. On the classification and nomenclature of baculoviruses: A proposal for revision. *Archives of Virology* 151: 1257–1266

Jiang, H., Zhou, L., Zhang, J.M., Dong, H.F., Hu, Y.Y., and Jiang, M.S. 2008. Potential of Periplaneta fuliginosa densovirus as a biocontrol agent for smoky-brown cockroach, *P. fuliginosa*. *Biological Control* 46: 94–100

Kariithi, H.M., Meki, I.K., Boucias, D.G., and Abd-Alla, A.M. 2017a. Hytrosaviruses: Current status and perspective. *Current Opinion in Insect Science* 22: 71–78

Kariithi, H.M., Yao, X., Yu, F., Teal, P.E., Verhoeven, C.P., and Boucias, D.G. 2017b. Responses of the housefly,

Musca domestica, to the hytrosavirus replication: Impacts on host's vitellogenesis and immunity. *Frontiers in Microbiology* 8: 583

Kazlauskas, D., Varsani, A., Koonin, E.V., and Krupovic, M. 2019. Multiple origins of prokaryotic and eukaryotic single-stranded DNA viruses from bacterial and archaeal plasmids. *Nature Communication* 10: 3425

Kleespies, R.G., Tidona, C.A., and Darai, G. 1999. Characterization of a new iridovirus isolated from crickets and investigations on the host range. *Journal of Invertebrate Pathology* 73: 84–90

Krupovic, M. and Koonin, E.V. 2014. Evolution of eukaryotic single-stranded DNA viruses of the *Bidnaviridae* family from genes of four other groups of widely different viruses. *Science Rep* 4: 5347

Kumar, H., Sun, Z., Cao, G., Xue, R., Hu, X., and Gong, C. 2019. Bombyx mori bidensovirus infection alters the intestinal microflora of fifth instar silkworm (*Bombyx mori*) larvae. *Journal of Invertebrate Pathology* 163: 48–63.

Lacey, L.A., Grzywacz, D., Shapiro-Ilan, D.I., Frutos, R., Brownbridge, M., and Goettel, M. S. 2015. Insect pathogens as biological control agents: Back to the future. *Journal of Invertebrate Pathology* 132: 1–41

Lanzi, G., De Miranda, J.R., Boniotti, M.B., Cameron, C.E., Lavazza, A., Capucci, L., Camazine, S.M., and Rossi, C. 2006. Molecular and biological characterization of deformed wing virus of honeybees (*Apis mellifera* L.). *Journal of Virology* 80: 4998–5009

Lauber, C., Ziebuhr, J., Junglen, S., Drosten, C., Zirkel, F., Nga, P.T., Morita, K., Snijder, E. J., and Gorbalenya, A.E. 2012. *Mesoniviridae*: A proposed new family in the order *Nidovirales* formed by a single species of mosquito-borne viruses. *Archives of Virology* 157: 1623–1628

Lawrence, P.O. 2002. Purification and partial characterization of an entomopoxvirus (DLEPV) from a parasitic wasp of tephritid fruit flies. *Journal of Insect Science* 2: 10

Le Gall, O., Christian, P., Fauquet, C.M., King, A M., Knowles, N.J., Nakashima, N., Stanway, G., and Gorbalenya, A.E. 2008. *Picornavirales*, a proposed order of positive-sense single-stranded RNA viruses with a pseudo-T = 3 virion architecture. *Archives of Virology* 153: 715–727

Li, T.C., Scotti, P.D., Miyamura, T., and Takeda, N. 2007. Latent infection of a new alphanodavirus in an insect cell line. *Journal of Virology* 81: 10890–10896

Lietze, V.U., Geden, C.J., Doyle, M.A., and Boucias, D.G. 2012. Disease dynamics and persistence of Musca domestica salivary gland hypertrophy virus infections in laboratory house fly (*Musca domestica*) populations. *Applied and Environmental Microbiology* 78, 311–317

Liu, C., Zhang, J., Yi. F., Wang, J., Wang, X., Jiang, H., Xu, J., and Hu, Y. 2006. Isolation and RNA1 nucleotide sequence determination of a new insect nodavirus from *Pieris rapae* larvae in Wuhan city, *China*. *Virus Research* 120: 28–35

Liu, Y., Tran, B.N., Wang, F., Ounjai, P., Wu, J., and Hew, C. L. 2016. Visualization of assembly intermediates and budding vacuoles of Singapore grouper iridovirus in grouper embryonic cells. *Science Reports* 6: 18696

Longdon, B., Day, J.P., Schulz, N., Leftwich, P.T., De Jong, M.A., Breuker, C.J., Gibbs, M., Obbard, D.J., Wilfert, L., Smith, S.C.L., McGonigle, J.E., Houslay, T.M., Wright, LI., Livraghi, L., Evans, L.C., Friend, L.A., Chapman, T., Vontas, J., Kambouraki, N., and Jiggins, F.M. 2017. Vertically transmitted rhabdoviruses are found across three insect families and have dynamic interactions with their hosts. *Proceedings of the Royal Society B*, 284, 20162381

Ma, H., Galvin, T.A., Glasner, D.R., Shaheduzzaman, S., and Khan, A.S. 2014. Identification of a novel rhabdovirus in *Spodoptera frugiperda* cell lines. *Journal of Virology* 88: 6576–6585

Maciel-Vergara, G. and Ros, V.I.D. 2017. Viruses of insects reared for food and feed. *Journal of Invertebrate Pathology* 147: 60–75

Marina, C.F., Ibarra, J.E., Arredondo-Jimenez, J.I., Fernandez-Salas, I., Valle, J., and Williams, T. 2003. Sublethal iridovirus disease of the mosquito *Aedes aegypti* is due to viral replication not cytotoxicity. *Medicine Veterinary Entomology* 17: 187–194

Marklewitz, M., Gloza-Rausch, F., Kurth, A., Kummerer, B.M., Drosten, C., and Junglen, S. 2012. First isolation of an entomobirnavirus from free-living insects. *Journal of General Virology* 93; 2431–2435

Marklewitz, M., Handrick, S., Grasse, W., Kurth, A., Lukashev, A., Drosten, C., Ellerbrok, H., Leendertz, F.H., Pauli, G., and Junglen, S. 2011. Gouleako virus isolated from West African mosquitoes constitutes a proposed novel genus in the family *Bunyaviridae*. *Journal of Virology* 85: 9227–9234

Marklewitz, M., Zirkel, F., Rwego, I.B., Heidemann, H., Trippner, P., Kurth, A., Kallies, R., Briese, T., Lipkin, W.I., Drosten, C., Gillespie, T.R., and Junglen, S. 2013. Discovery of a unique novel clade of mosquito-associated bunyaviruses. *Journal of Virology* 87: 12850–12865

Marzoli, F., Forzan, M., Bortolotti, L., Pacini, M.I., Rodriguez-Flores, M.S., Felicioli, A., and Mazzei, M. 2020. Next generation sequencing study on RNA viruses of *Vespa velutina* and *Apis mellifera* sharing the same foraging area. *Transboundary and Emerging Diseases* 00(00): 1–13. https://doi.org/10.1111/tbed.13878

Meki, I.K., Ince, I.A., Kariithi, H.M., Boucias, D.G., Ozcan, O., Parker, A.G., Vlak, J.M., van Oers, M.M., and Abd-Alla, A.M.M. 2018a. Expression profile of *Glossina pallidipes* microRNAs during symptomatic and asymptomatic infection with Glossina pallidipes salivary gland hypertrophy virus (*Hytrosavirus*). *Frontiers in Microbiology* 9: 2037

Meki, I.K., Kariithi, H.M., Ahmadi, M., Parker, A.G., Vreysen, M.J.B., Vlak, J.M., van Oers, M.M., and Abd-Alla, A.M.M. 2018b. Hytrosavirus genetic diversity and ecoregional spread in *Glossina species*. *BMC Microbiology* 18: 143

Misof, B., Liu, S., Meusemann, K., Peters, R.S., Donath, A., Mayer, C. et al. 2014. Phylogenomics reveals the timing and pattern of insect evolution. *Science* 346: 763–767

Mitsuhashi, W., Kawakita, H., Murakami, R., Takemoto, Y., Saiki, T., Miyamoto, K., and Wada, S. 2007. Spindles of an entomopoxvirus facilitate its infection of the host insect by disrupting the peritrophic membrane. *Journal of Virology* 81: 4235–4243

Moore, J., Jironkin, A., Chandler, D., Burroughs, N., Evans, D.J., and Ryabov, E.V. 2011. Recombinants between Deformed wing virus and Varroa destructor virus-1 may prevail in *Varroa destructor*-infested honeybee colonies. *Journal of General Virology* 92: 156–161

Moss, B. 2016. Membrane fusion during poxvirus entry. *Seminars Cell Development Biology* 60: 89–96

Muttis, E., Micieli, M.V., Bonica, M.B., Ghiringhelli, P.D., and Garcia, J.J. 2020. Mosquito iridescent virus: New records from nature and infections using *Strelkovimermis spiculatus* (Mermithidae) as a vector under laboratory vonditions. *Neotropical Entomology* 49: 268–274

Mutuel, D., Ravallec, M., Chabi, B., Multeau, C., Salmon, J.M., Fournier, P., Ogliastro, M., 2010. Pathogenesis of Junonia coenia densovirus in *Spodoptera frugiperda*: A route of infection that leads to hypoxia. *Virology* 403: 137–144

Nalçacioğlu, R., Marks, H., Vlak, J M., Demirbag, Z., and van Oers, M.M. 2003. Promoter analysis of the Chilo iridescent virus DNA polymerase and major capsid protein genes. *Virology* 317: 321–329

Nalçacioglu, R., Muratoglu, H., Yesilyurt, A., van Oers, M.M., Vlak, J.M., and Demirbag, Z. 2016. Enhanced insecticidal activity of Chilo iridescent virus expressing an insect specific neurotoxin. *Journal of Invertebrate Pathology* 138: 104–111

Nasar, F., Palacios, G., Gorchakov, R.V., Guzman, H., Da Rosa, A.P., Savji, N., Popov, V.L., Sherman, M.B., Lipkin, W.I., Tesh, R.B., and Weaver, S.C. 2012. Eilat virus, a unique alphavirus with host range restricted to insects by RNA replication. *Proceedings of the National Academy of Sciences USA* 109: 14622–14627.

Newton, N.D., Colmant, A.M.G, O'Brien, C.A., Ledger, E., Paramitha, D., Bielefeldt-Ohmann, H., Watterson, D., McLean, B.J., Hall-Mendelin, S., Warrilow, D., Van Den Hurk, A.F. Liu, W., Hoare, C., Kizu, J.R., Gauci, P.J., Haniotis, J., Doggett, S.L., Shaban, B., Johansen, C.A., Hall, R.A., and Hobson-Peters, J. 2020. Genetic, morphological and antigenic relationships between

mesonivirus isolates from Australian mosquitoes and evidence for their horizontal transmission. *Viruses* 12: 1159

Nunes, M.R.T., Contreras-Gutierrez, M.A., Guzman, H., Martins, L.C., Barbirato, M.F., Savit, C. et al. 2017. Genetic characterization, molecular epidemiology, and phylogenetic relationships of insect-specific viruses in the taxon *Negevirus Virology* 504: 152–167

Obbard, D.J. 2018. Expansion of the metazoan virosphere: Progress, pitfalls, and prospects. *Current Opinion in Virology* 31: 17–23

Olivier, V., Blanchard, P., Chaouch, S., Lallemand, P., Schurr, F., Celle, O., Dubois, E., Tordo, N., Thiery, R., Houlgatte, R., and Ribiere, M. 2008. Molecular characterisation and phylogenetic analysis of Chronic bee paralysis virus, a honeybee virus. *Virus Research* 132: 59–68

Olschewski, S., Cusack, S., and Rosenthal, M. 2020. The cap-snatching mechanism of bunyaviruses. *Trends in Microbiology* 28: 293–303

Ongus, J.R., Peters, D., Bonmatin, J.M., Bengsch, E., Vlak, J.M., and Van Oers, M.M. 2004. Complete sequence of a picorna-like virus of the genus *Iflavirus* replicating in the mite *Varroa destructor*. *Journal of General Virology* 85: 3747–3755

Paez-Espino, D., Eloe-Fadrosh, E.A., Pavlopoulos, G.A., Thomas, A.D., Huntemann, M., Mikhailova, N., Rubin, E., Ivanova, N.N., and Kyrpides, N.C. 2016. Uncovering Earth's virome. *Nature* 536: 425–430

Patot, S., Allemand, R., Fleury, F., and Varaldi, J. 2012. An inherited virus influences the coexistence of parasitoid species through behaviour manipulation. *Ecology Letters* 15: 603–610

Pénzes, J., Pham, H., Yu, Q., Bergoin, M., and Tijssen, P. 2020. Parvoviruses of Invertebrates (*Parvoviridae*). In *Encyclopedia of Virology*, 4th edn. D. Bamford and M. Zuckerman (eds.). Vol. 4, pp. 835–848. Amsterdam, NL: Elsevier Ltd.

Perera, S., Li, Z., Pavlik, L., and Arif, B. 2010. Entomopoxviruses. In *Insect Virology*, A.S. and K.N. Johnson (eds.) pp. 83–115. Norfolk, United Kingdom: Caister Academic Press

Pham, H.T., Bergoin, M., and Tijssen, P. 2013b. *Acheta domesticus* volvovirus, a novel single-stranded circular dna virus of the house cricket. *Genome Announcements* 1: e00079–13

Pham, H.T., Iwao, H., Bergoin, M., and Tijssen, P. 2013a. New volvovirus isolates from *Acheta domesticus* (Japan) and *Gryllus assimilis* (United States). *Genome Announcements* 1: e00328–13.

Piegu, B., Asgari, S., Bideshi, D., Federici, B.A., and Bigot, Y. 2015. Evolutionary relationships of iridoviruses and divergence of ascoviruses from invertebrate iridoviruses in the superfamily Megavirales. *Molecular Phylogenetics and Evolion* 84: 44–52

Porter, A.F., Shi, M., Eden, J.S., Zhang, Y.Z., and Holmes, E.C. 2019. Diversity and evolution of novel invertebrate DNA viruses revealed by meta-transcriptomics. *Viruses* 11: 1092

Prasad, V. and Srivastava, S. 2016. Insect Viruses. In *Ecofriendly Pest Management for Food Security*, Omkar (ed.) pp. 411–442. San Diego: Academic Press

Pringle, F.M., Johnson, K.N., Goodman, C.L., McIntosh, A.H., and Ball, L.A. 2003. Providence virus: A new member of the *Tetraviridae* that infects cultured insect cells. *Virology* 306: 359–370

Prompiboon, P., Lietze, V.U., Denton, J.S., Geden, C.J., Steenberg, T., and Boucias, D.G. 2010. Musca domestica salivary gland hypertrophy virus, a globally distributed insect virus that infects and sterilizes female houseflies. *Applied and Environmental Microbiology* 76: 994–998

Radek, R. and Fabel, P. 2000. A new entomopoxvirus from a cockroach: Light and electron microscopy. *Journal of Invertebrate Pathology* 75: 19–27

Rallis, C.P. and Burand, J.P. 2002. Pathology and ultrastructure of Hz-2V infection in the agonadal female corn earworm, *Helicoverpa zea*. *Journal of Invertebrate Pathology* 81: 33–44

Ramle, M., Wahid, M.B., Norman, K., Glare, T.R., and Jackson, T.A. 2005. The incidence and use of Oryctes virus for control of rhinoceros beetle in oil palm plantations in Malaysia. *Journal of Invertebrate Pathology* 89: 85–90

Reinganum, C, O'Loughlin, G.T., and Hogan, T.W. 1970. A nonoccluded virus of the field crickets *Teleogryllus oceanicus* and *T. commodus* (Orthoptera: Gryllidae); *Journal of Invertebrate Pathology* 16: 214–220

Rittschof, C.C., Pattanaik, S., Johnson, L., Matos, L.F., Brusini, J., and Wayne, M.L. 2013. Sigma virus and male reproductive success in *Drosophila melanogaster*. *Behavioral Ecology and Sociobiology* 67: 529–540

Roberts, D.W. and Granados, R.R. 1968. A poxlike virus from *Amsacta moorei* (Lepidoptera: Arcxtiidae). *Journal of Invertebrate Pathology* 12: 141–143

Roffet-Salque, M., Regert, M., Evershed, R., Outram, A.K., Cramp, L.J.E., Orestes Decavallas, O., Dunne, J., Gerbault, P., Mileto, S., Mirabaud, S., Pääkkönen, M., Smyth, J., Šoberl, L., and Whelton, H L. 2016. Widespread exploitation of the honeybee by early Neolithic farmers. *Nature* 534: S17–S18

Rogers, A., Towery, L., McCown, A., and Carlson, K.A. 2020. Impaired geotaxis as a novel phenotype of Nora virus infection of *Drosophila melanogaster*. *Scientifica (Cairo)* 2020: 1804510

Rohrmann, G.F. 2019. Baculovirus Molecular Biology. 4th edition. Bethesda (MD): National Center for Biotechnology Information. https://www.ncbi.nlm.nih.gov/books/NBK543458/

Roossinck, M.J. and Bazan, E.R. 2017. Symbiosis: Viruses as intimate partners. *Annual Review of Virology* 4: 123–139

Ros, V. I. D. 2020. Baculoviruses: General features. In *Encyclopedia of Virology*, 4th edn. D. Bamford and M. Zuckerman (eds.) 4: pp. 739–746. Amsterdam, NL, Elsevier Ltd.

Ros, V.I.D., Van Houte, S., Hemerik, L., and van Oers, M.M. 2015. Baculovirus-induced tree-top disease: How extended is the role of egt as a gene for the extended phenotype? *Molecular Ecology* 24: 249–258

Rosario, K., Dayaram, A., Marinov, M., Ware, J., Kraberger, S., Stainton, D., Breitbart, M., and Varsani, A. 2012. Diverse circular ssDNA viruses discovered in dragonflies (Odonata: Epiprocta). *Journal of General Virology* 93: 2668–2681

Rosario, K., Mettel, K.A., Benner, B.E., Johnson, R., Scott, C., Yusseff-Vanegas, S.Z., Baker, C.C.M., Cassill, D.L., Storer, C., Varsani, A., and Breitbart, M. 2018. Virus discovery in all three major lineages of terrestrial arthropods highlights the diversity of single-stranded DNA viruses associated with invertebrates. *Peer Journal* 6: e5761

Ryabov, E.V. 2017. Invertebrate RNA virus diversity from a taxonomic point of view. *Journal of Invertebrate Pathology* 147: 37–50

Ryabov, E.V., Keane, G., Naish, N., Evered, C., and Winstanley, D. 2009. Densovirus induces winged morphs in asexual clones of the rosy apple aphid *Dysaphis plantaginea*. *Proceedings of the National Academy of Sciences, USA* 106: 8465

Schneemann, A., Reddy, V., and Johnson, J.E. 1998. The structure and function of nodavirus particles: A paradigm for understanding chemical biology. *Advances in Virus Research* 50: 381–446

Schroeder, L., Mar, T.B., Haynes, J.R., Wang, R., Wempe, L., and Goodin, M.M. 2019. Host range and population survey of Spodoptera frugiperda rhabdovirus. *Journal of Virology* 93: e02028-18Schuster S., Zirkel, F., Kurth, A., Van Cleef, K.W., Drosten, C., Van Rij, R.P., and Junglen, S. 2014. A unique nodavirus with novel features: Mosinovirus expresses two subgenomic RNAs, a capsid gene of unknown origin, and a suppressor of the antiviral RNA interference pathway. *Journal of Virology* 88: 13447–13459

Shi, M., Lin, X.D., Tian, J.H., Chen, L.J., Chen, X., Li, C.X., Qin, X.C., Li, J., Cao, J.P., Eden, J.S., Buchmann, J., Wang, W., Xu, J., Holmes, E.C., and Zhang, Y.Z. 2016. Redefining the invertebrate RNA virosphere. *Nature* 540: 539–543

Skinner, M.A., Buller, R.M., Damon, I.K., Lefkowitz, E.J., McFadden, G., McInnes, C.J., Mercer, A.A., Moyer, R.W., and Upton, C. 2011. *Poxviridae*. In *Virus Taxonomy: Ninth Report of the International Committee on Taxonomy of Viruses*, A.M.Q. King, M.J. Adams, E.B. Carstens, and E.J. Lefkowitz (eds.) pp. 291–309. New York: Elsevier Academic Press.

Slack, J. and Arif, B.M. 2006. The baculoviruses occlusion derived virus: Virion structure and function. In *Advances in Virus Research*, K. Maramorosch and A.J. Shatkin (eds) 69: pp. 99–165. Amsterdam, NL: Academic Press

Sugiharti, M., Ono, C., Ito, Y., Asano, S., Sahara, K., Pujiastuti, Y., and Bando, H., 2010. Isolation of the Thosea asigna virus (TaV) from the epizootic *Setothosea asigna* larvae collected in South Sumatra and a study on its pathogenicity to Limacodidae larvae in Japan. *Journal of Insect Biotechnology and Sericology* 79: 117–124

Sun, Y., Dong, Y., Li, J., Lai, Z., Hao, Y., Liu, P., Chen, X., and Gu, J. 2019. Development of large-scale mosquito densovirus production by *in vivo* methods. *Parasites & Vectors* 12: 255

Szelei, J., Woodring, J., Goettel, M.S., Duke, G., Jousset, F.X., Liu, K.Y., Zadori, Z., Li, Y., Styer, E., Boucias, D.G., Kleespies, R.G., Bergoin, M., and Tijssen, P. 2011. Susceptibility of North American and European crickets to Acheta domesticus densovirus (AdDNV) and associated epizootics. *Journal of Invertebrate Pathology* 106: 394–399

Takemoto, Y., Mitsuhashi, W., Murakami, R., Konishi, H., and Miyamoto, K. 2008. The N-terminal region of an entomopoxvirus fusolin is essential for the enhancement of peroral infection, whereas the C-terminal region is eliminated in digestive juice. *Journal of Virology* 82: 12406–12415

Teninges, D., Ohanessian, A., Richard-Molard, C., and Contamine, D. 1979. Isolation and biological properties of Drosophila X virus. *Journal of General Virology* 42: 241–254

Thezé, J., Bezier, A., Periquet, G., Drezen, J.M., and Herniou, E.A. 2011. Paleozoic origin of insect large dsDNA viruses. *Proceedings of the National Academy of Sciences* 108: 15931–15935

Thezé, J., Takatsuka, J., Nakai, M., Arif, B., and Herniou, E.A. 2015. Gene acquisition convergence between entomopoxviruses and baculoviruses. *Viruses* 7: 1960–1974

Tijssen, P. and Bergoin, M. 1995. Densonucleosis viruses constitute an increasingly diversified subfamily among the parvoviruses. *Seminars in Virology* 6: 347–355

Tijssen, P., Li, Y., El-Far, M., Szelei, J., Letarte, M., and Zadori, Z. 2003. Organization and expression strategy of the ambisense genome of densonucleosis virus of *Galleria mellonella*. *Journal of Virology* 77: 10357–10365

Tijssen, P., Pénzes, J.J., Yu, Q., Pham, H., and Bergoin, M. 2017. Diversity of small, single-stranded DNA viruses of invertebrates and their chaotic evolutionary past. *Journal of Invertebrate Pathology* 147: 23–36

Van Der Heijden, M.W. and Bol, J.F. 2002. Composition of alphavirus-like replication complexes: Involvement of virus and host encoded proteins. *Archives of Virology* 147: 875–898

Van Eynde, B., Christiaens, O., Delbare, D., Cooreman, K., Bateman, K.S., Stentiford, G. D., Dullemans, A.M., van Oers, M.M., and Smagghe, G. 2018. Development and application of a duplex PCR assay for detection of Crangon bacilliform virus in populations of European brown shrimp (*Crangon crangon*). *Journal of Invertebrate Pathology* 153: 195–202

van Oers, M. 2020. Baculovirus: Molecular biology and replication. In *Encyclopedia of Virology*, 4th edn, D. Bamford and M. Zuckerman (eds.), Vol. 4, pp. 747–758. Amsterdam, NL: Elsevier Ltd.

Vancini, R., Paredes, A., Ribeiro, M., Blackburn, K., Ferreira, D., Kononchik, J.P., Jr., Hernandez, R., and Brown, D. 2012. Espirito Santo virus: A new birnavirus that replicates in insect cells. *Journal of Virology* 86: 2390–2399

Varaldi, J. and Lepetit, D. 2018. Deciphering the behaviour manipulation imposed by a virus on its parasitoid host: Insights from a dual transcriptomic approach. *Parasitology* 145: 1979–1989

Varaldi, J., Petit, S., Bouletreau, M., and Fleury, F. 2006. The virus infecting the parasitoid *Leptopilina boulardi* exerts a specific action on superparasitism behaviour. *Parasitology* 132: 747–756

Vasilakis, N., Forrester, N.L., Palacios, G., Nasar, F., Savji, N., Rossi, S.L., Guzman, H., Wood, T.G., Popov, V., Gorchakov, R., Gonzalez, A.V., Haddow, A.D., Watts, D.M., Da Rosa, A.P., Weaver, S.C., Lipkin, W.I., and Tesh, R.B. 2013. *Negevirus*: A proposed new taxon of insect-specific viruses with wide geographic distribution. *Journal of Virology* 87: 2475–2488

Vasilakis, N., Guzman, H., Firth, C., Forrester, N.L., Widen, S.G., Wood, T.G., Rossi, S. L., Ghedin, E., Popov, V., Blasdell, K.R., Walker, P.J., and Tesh, R.B. 2014. Mesoniviruses are mosquito-specific viruses with extensive geographic distribution and host range. *Virological Journal* 11: 97

Velamoor, S., Mitchell, A., Humbel, B.M., Kim, W., Pushparajan, C., Visnovsky, G., Burga, L.N., and Bostina, M. 2020. Visualizing nudivirus assembly and egress. *Microbiology Journal* 11: e01333–20

Viljakainen, L., Holmberg, I., Abril, S., and Jurvansuu, J. 2018. Viruses of invasive Argentine ants from the European main supercolony: Characterization, interactions and evolution. *Journal of General Virology* 99: 1129–1140

Virto, C., Navarro, D., Tellez, M.M., Herrero, S., Williams, T., Murillo, R., and Caballero, P. 2014. Natural populations of *Spodoptera exigua* are infected by multiple viruses that are transmitted to their offspring. *Journal of Invertebrate Pathology* 122: 22–27

Walter, C.T., Pringle, F.M., Nakayinga, R., De Felipe, P., Ryan, M.D., Ball, L.A., and Dorrington, R.A. 2010. Genome organization and translation products of Providence virus: insight into a unique tetravirus. *Journal of General Virology* 91: 2826–2835

Wang, Q., Bosch, B.J., Vlak, J.M., van Oers, M.M., Rottier, P.J., and Van Lent, J.W.M. 2016. Budded baculovirus particle structure revisited. *Journal of Invertebrate Pathology* 134: 15–22

Wang, X., Liu, X., Makalliwa, G.A., Li, J., Wang, H., Hu, Z., and Wang, M. 2017. Per os infectivity factors: A complicated and evolutionarily conserved entry machinery of baculovirus. *Science China Life Sciences* 60: 806–815

Wang, Y. and Jehle, J.A. 2009. Nudiviruses and other large, double-stranded circular DNA viruses of invertebrates: New insights on an old topic. *Journal of Invertebrate Pathology* 101: 187–193

Wigley, P.O.J. and Scotti, P.D. 1983 The seasonal incidence of cricket paralysis virus in a population of the New Zealand small field cricket, *Pteronemobius nigrovus* (Orthoptera: Gryllidae), *Journal of Invertebrate Pathology* 41: 378–380

Williams, T. 2008. Natural invertebrate hosts of iridoviruses (*Iridoviridae*). *Neotropical Entomology* 37: 615–632

Williams, T., Bergoin, M., and Van Oers, M.M. 2017a. Diversity of large DNA viruses of invertebrates. *Journal of Invertebrate Pathology* 147: 4–22

Williams, T., Virto, C., Murillo, R., and Caballero, P. 2017b. Covert infection of insects by baculoviruses. *Frontiers in Microbiology*, 8: 1337

Wong, C.K., Young, V.L., Kleffmann, T., and Ward, V.K. 2011. Genomic and proteomic analysis of invertebrate iridovirus type 9. *Journal of Virology* 85: 7900–7911

Wu, H., Pang, R., Cheng, T., Xue, L., Zeng, H., Lei, T., Chen, M., Wu, S., Ding, Y., Zhang, J., Shi, M., and Wu, Q. 2020. Abundant and diverse RNA viruses in insects revealed by RNA-seq analysis: Ecological and evolutionary implications. *Microbiology Systems* 5: e00039-20

Yan, X., Yu, Z., Zhang, P., Battisti, A.J., Holdaway, H.A., Chipman, P.R., Bajaj, C., Bergoin, M., Rossmann, M.G., and Baker, T.S. 2009. The capsid proteins of a large, icosahedral dsDNA virus. *Journal of Molecular Biology* 385: 1287–1299

Yañez, O., Piot, N., Dalmon, A., De Miranda, J.R., Chantawannakul, P., Panziera, D., Amiri, E., Smagghe, G., Schroeder, D., and Chejanovsky, N. 2020. Bee viruses: Routes of infection in Hymenoptera. *Frontiers in Microbiology* 11: 943

Yang, S.Y., Han, M.J., Kangm L.F., Li, Z.W., Shen, Y.H., and Zhang, Z. 2014a. Demographic history and gene flow during silkworm domestication. *BMC Evolutionary Biology* 14: 185.

Yang, X., Xu, P., Yuan, H., Graham, R.I., Wilson, K., and Wu, K. 2019. Discovery and characterization of a novel picorna-like RNA virus in the cotton bollworm *Helicoverpa armigera*. *Journal of Invertebrate Pathology* 160: 1–7

Yang, Y.T., Lee, D.Y., Wang, Y., Hu, J.M., Li, W.H., Leu, J.H., Chang, G.D., Ke, H.M., Kang, S.T., Lin, S.S., Kou, G H., and Lo, C.F. 2014b. The genome and occlusion bodies of marine *Penaeus monodon* nudivirus (PmNV, also known as MBV and PemoNPV) suggest that it should be assigned to a new nudivirus genus that is distinct from the terrestrial nudiviruses. *BMC Genomics* 15: 628

Yang, Z., Bruno, D. P., Martens, C. A., Porcella, S. F., and Moss, B. 2010. Simultaneous high-resolution analysis of vaccinia virus and host cell transcriptomes by deep RNA sequencing. *Proceedings of the National Academy of Sciences* 107: 11513–11518

Ye, G., Wang, Y., Liu, X., Dong, Q., Cai, Q., Yuan, Z., and Han, X. 2020. Transmission competence of a new mesonivirus, Yichang virus, in mosquitoes and its interference with representative flaviviruses. *PLoS Neglected Tropical Diseases* 14(11): e0008920.

Yong, C.Y., Yeap, S.K., Omar, A.R., and Tan, W.S. 2017. Advances in the study of nodavirus. *Peer Journal* 5: e3841

Yu, H., Yang, C. J., Li, N., Zhao, Y., Chen, Z.M., Yi, S. J., Li, Z.Q., Adang, M.J. and Huang, G.H. 2020. Novel strategies for the biocontrol of noctuid pests (Lepidoptera) based on improving ascovirus infectivity using *Bacillus thuringiensis*. *Insect Science*, https://doi.org/10.1111/1744-7917.12875

Yu, Q., Hajek, A.E., Bergoin, M., and Tijssen, P. 2012. *Papilio polyxenes* densovirus has an iteravirus-like genome organization. *Journal of Virology* 86: 9534–9535

Yuan, H., Xu, P., Xiao, Y., Yang, L., Yang, X., and Wu, K. 2020. Infection of cotton bollworm by Helicoverpa armigera iflavirus decreases larval fitness. *Journal of Invertebrate Pathology* 173: 107384

Zanotto, P.M., Kessing, B.D., and Maruniak, J.E. 1993. Phylogenetic interrelationships among baculoviruses: Evolutionary rates and host associations. *Journal of Invertebrate Pathology* 62: 147–164

Zeddam, J.-L., Arroyo Cruzado, J., Luna Rodriguez, J., Ravallec, M., and Candiotti Subilete, E. 2003. A cypovirus from the South American oil-palm pest *Norape argyrrhorea* and its potential as a microbial control agent. *Biological Control* 48: 101–112

Zeddam, J.L., Gordon, K.H., Lauber, C., Alves, C.A., Luke, B.T., Hanzlik, T.N., Ward, V. K., and Gorbalenya, A.E. 2010. Euprosterna elaeasa virus genome sequence and evolution of the *Tetraviridae* family: Emergence of bipartite genomes and conservation of the VPg signal with the dsRNA *Birnaviridae* family. *Virology* 397: 145–154

Zelazny, B. 1976. Transmission of a baculovirus in populations of *Oryctes rhinoceros*. *Journal of Invertebrate Pathology* 27: 221–227

Zhang, Y., Wang, J., and Han, G.-Z. 2020. Chalcid wasp paleoviruses bridge the evolutionary gap between bracoviruses and nudiviruses. *Virology* 542: 34–39

Zhao, Y., Sun, J., Labropoulou, V., Swevers, L., and Smagghe, G. 2018. Beyond baculoviruses: Additional biotechnological platforms based on insect RNA viruses. In *Advances in Insect Physiology*, Volume 55, G. Schmagge (Ed): Pp. 123–62. Academic Press, London.

CHAPTER 11

Bacterial diseases of insects

Heba Abdelgaffar, Trevor Jackson, and Juan Luis Jurat-Fuentes

11.1 Introduction: Classification and pathology commonalities

Bacteria are highly diverse and ubiquitous prokaryotes defined as unicellular microorganisms lacking a membrane separating the genetic material from the cytoplasm and other cell organelles. Through ~ 479 million years of potential coevolution (Misof et al. 2014), bacteria have evolved a broad range of tight relationships with their insect hosts. Many of these interactions are pathogenic, fulfilling Koch's postulates of isolation from a diseased host and subsequent infection with the isolated bacterium resulting in the same disease in a healthy host, and these are the focus of this chapter. The emphasis of this chapter will be on bacterial disease caused by biocontrol agents (Ruiu 2015) or those affecting insect rearing (Eilenberg et al. 2015). Bacterial entomopathogens symbiotic with nematodes (*Photorhabdus* spp. and *Xenorhabdus* spp.) are discussed in Chapter 13.

The ability to induce disease in insects is mostly dictated by the presence in the bacterial genome of genes encoding virulence factors. Thus, acquisition of foreign virulence genes from the environment (transformation), bacteriophages (transduction), or other bacteria (conjugation) can increase virulence or create new pathogens. Pathogenicity islands are a good example of mobile genetic elements containing large clusters of genes involved in different aspects of bacterial virulence (Schmidt and Hensel 2004). Examples of insect virulence genes in pathogenicity islands include the toxin complex (Tc) in *Yersinia* spp. (Fuchs et al. 2008;

Hurst et al. 2011) and multiple insecticidal toxins in nematode-associated *Photorhabdus* spp. (Waterfield et al. 2004) and *Xenorhabdus* spp. (Brown et al. 2004). Genetic information relevant to entomopathogenesis is also stored in self-replicating plasmids, which can be shared through conjugation. The most notable example is the pBtoxis plasmid in *Bacillus thuringiensis* (*Bt*) susbp. *israelensis*, which harbours six insecticidal protein genes (Berry et al. 2002).

Entomopathogenic bacteria (Table 11.1) are found in all three major groups of Eubacteria, including bacteria with Gram-positive (Firmicutes) and Gram-negative (Gracilicutes) type cell wall, as well as bacteria lacking a cell wall (Tenericutes). Traditionally, bacterial taxonomy has used phenotypic characterisation and DNA-DNA hybridisation values in differentiating species (Wayne et al. 1987), yet advances in genetic sequencing technologies have allowed more detailed comparisons, including the development of species-specific DNA barcodes (Paul et al. 2020). For this chapter, we have selected the Genome Taxonomy Database classification (Parks et al. 2020), based on nucleotide identity to representative species genomes, as reference for taxonomy and nomenclature of bacterial entomopathogens. Throughout the chapter, bacterial strains within a species are defined as the descendants of a single isolate in pure culture.

Most bacterial entomopathogens have obligate or facultative relationships with their host. Obligate bacterial entomopathogens, such as *Paenibacillus popilliae* infecting scarab larvae, complete their life cycle in the host. In contrast, facultative

Heba Abdelgaffar, Trevor Jackson, and Juan Luis Jurat-Fuentes, *Bacterial diseases of insects*. In: *Invertebrate Pathology*. Edited by Andrew F. Rowley, Christopher J. Coates and Miranda M.A. Whitten, Oxford University Press. © Oxford University Press (2022). DOI: 10.1093/oso/9780198853756.003.0011

Table 11.1 An overview of bacterial diseases of insects.

Disease	Disease causing agent(s)	Host range	Geographical range	Pathology	Key references
Sudden death or 'sotto' disease	*Bacillus thuringiensis*	Lepidoptera, Coleoptera, Diptera, Orthoptera	Worldwide	Paralysis, disruption of the midgut epithelium and septicaemia	(Yu et al. 1997; Knowles and Ellar 1987; Fast and Angus 1965)
	Lysinibacillus sphaericus	Dipteran larvae (*Culex, Anopheles, Simulium* spp.)	Worldwide	Paralysis, disruption of the midgut epithelium and septicaemia	(Sharma et al. 2020; Silva-Filha et al. 1999; Nicolas et al. 1993)
Milky disease	*Paenibacillus popilliae*	Coleopteran larvae	Worldwide	Penetration of the midgut, colonisation of the haemocoel	(Zhang et al. 1997; Splittstoesser et al. 1973)
American foulbrood	*Paenibacillus larvae*	Honeybee	Worldwide	Bacterial proliferation in the gut followed by host death by bacteraemia and degradation of tissues	(Garcia-Gonzalez et al. 2014)
European foulbrood	*Melissococcus plutonius*	Honeybee	Worldwide	Asymptomatic gut colonisation followed by tissue damage and potential infection by secondary invaders	(Bailey 1983)
	Brevibacillus laterosporus	Lepidoptera, Diptera, Coleoptera	Worldwide	Altered microvilli followed by cytoplasmic vacuolisation and leakage of cellular contents into the gut lumen and disruption of the epithelial barrier	(Marche et al. 2017)
	Clostridium bifermentans	Diptera	Worldwide	Protein complex produced during sporulation targets gut cells	(Qureshi et al. 2014)
Amber disease	*Serratia entomophila*	*Costelytra giveni*	New Zealand	Toxin production, antifeeding, gut enzyme disruption, chronic starvation until death.	(Jackson et al. 2001; Hurst et al. 2007)
	Yersinia entomophaga	Coleoptera Lepidoptera, Orthoptera	New Zealand	Toxin production, gut reaction, pore formation, invasion of haemocoel	(Hurst et al. 2011; Piper et al. 2019)
	Pseudomonas entomophila	Diptera, Coleoptera Lepidoptera	Unknown	Colonisation of midgut, production of toxins, degradation of the midgut epithelium	(Vodovar et al. 2005; Dieppois et al. 2015)
	Burkholderia rinojensis	Lepidoptera, Hemiptera Acarina	Unknown	Heat stable toxin	(Cordova-Kreylos et al. 2013)
	Rickettsiella pyronotae	Coleoptera	New Zealand	Passes through gut barrier, colonises fat body, haemocytes and other tissues. Chronic disease.	(Kleespies et al. 2011)

entomopathogenic bacteria in the genus *Bacillus* can also grow outside the host. Opportunistic or 'potential' pathogens depend on factors weakening the insect and disrupting the epithelial gut barrier to reach and proliferate in the haemolymph to cause disease.

The most common point of entry for entomopathogenic bacteria is the oral cavity (*per os*), although infection may also occur through compromised integument (Maciel-Vergara et al. 2018) or trachea, and the egg. Upon entry, bacteria must overcome the immune response and colonise the host to cause disease. There is evidence for transgenerational immune priming of eggs (Dhinaut et al. 2018). In addition, successful colonisation of the gut lumen may require overcoming filtering structures (Lanan et al. 2016; Kuraishi et al. 2011) and the highly stringent physicochemical conditions (e.g. pH, ionic strength, redox potential) in the insect gut (Appel 2017). Ingestion of the pathogen typically causes symptoms including cessation of feeding and paralysis, which increases interactions with the host gut cells but also can produce defensive diarrhoea and vomiting to limit exposure to the pathogen.

In order to invade the main body cavity (haemocoel) and access the nutrient-rich haemolymph, bacterial pathogens need to disrupt or cross the gut epithelial barrier. This property has probably driven evolutionary selection in bacteria to develop insecticidal toxins and other virulence factors targeting gut cells. It is interesting to note that although pore formation in the epithelial membrane is a common step for bacterial pathogenesis in insects (Figure 11.1), the pore-forming toxins produced to disrupt the gut epithelium are structurally and functionally diverse (Crickmore et al. 2020b). While there are examples of bacterial toxins killing the insect host without systemic infection (Nishitsutsuji-Uwo and Endo 1980), invasion of the haemocoel leading to bacterial proliferation and lethal septicaemia are commonly observed. Changes in colouration, consistency, and melanisation are often observed during late stages of disease. Some bacterial entomopathogens, including members of the *Bacillus*, *Lysinibacillus*, and *Paenibacillus* genera, undergo sporulation during the late stages of pathogenesis. The spore is a highly resistant life stage and can persist in the environment, which is highly advantageous as suitable host availability is unpredictable.

Insecticidal sprays based on bacterial entomopathogens and transgenic crops producing toxin genes from these bacteria have been successfully commercialised for pest control. These bacterial insecticides are highly specific and induce rapid cessation of feeding and death, maintaining insect populations below economic thresholds.

Most bacteria undergo horizontal transmission from an infected to a new host. Inevitably, host health plays a major role in susceptibility to bacterial disease. Crowding, poor nutrition, adverse environmental conditions, and cannibalism can favour horizontal transmission and infection, leading to epizootics. In fact, most bacterial entomopathogens were originally identified from diseased or dead insects closely associated with environments with those conditions, such as mass rearing or pest outbreaks with a high density of hosts. In environments where hosts are not readily available, bacterial entomopathogens can survive in highly resistant life forms (spores) or be transported by other organisms (such as symbiotic nematodes) until a suitable host is available. Disease onset and transmission is also dependent on the microbiome of the host. Gut microflora can inhibit colonisation by pathogenic bacteria (Engel and Moran 2013) and may compete for resources and hinder growth after host death (Raymond et al. 2007).

The identification of bacterial pathogens needs microscopic observation (400–1000 × magnification and phase contrast objectives) of the haemolymph or tissue smears to detect refringent spores, vegetative bacteria (rods/cocci), and in some cases parasporal bodies containing insecticidal proteins. Confirmation may be obtained by amplification and sequencing of 16S ribosomal RNA genes with specific primers.

11.2 Gram-positive entomopathogens

11.2.1 *Bacillus thuringiensis (Bt)*

Originally identified in 1901 as the causative agent for the 'sotto' (in Japanese) or 'sudden death' disease in silkworm (*Bombyx mori*), the *Bt* bacterium was later formally described as causing 'flaccid

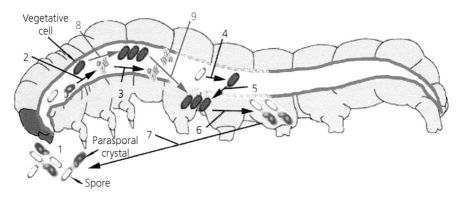

Figure 11.1 Schematic view of a generalised pathogenic process for spore-forming entomopathogenic bacteria. The host ingests spores by themselves or inside sporangia with parasporal crystals composed of insecticidal proteins (1). These parasporal crystals are solubilised (2) and the insecticidal proteins it contains are processed to activated toxins, which bind to receptors on the brush border membrane of the gut to disrupt the epithelium (3). Changing physicochemical conditions in the gut activate germination of spores into vegetative cells (4), which can cross the disrupted gut barrier (5) to invade the haemocoel and proliferate to cause septicaemia. Successful proliferation in the haemocoel usually involves the production of additional toxins and pathogenic factors. Upon consumption of resources in the insect cadaver, vegetative cells sporulate and produce parasporal bodies (6), which are released to the environment (7) to restart the infective cycle. In some cases (*Paenibacillus larvae*), spores germinate upon entering the gut (8) and vegetative cells colonise and proliferate in the gut before disrupting the gut epithelial barrier and invading the haemocoel (9).

syndrome' in larvae of the Mediterranean flour moth (*Ephestia küehniella*) in the German state of Thuringia (Berliner 1915). This description included mention of a parasporal inclusion, which was later demonstrated to solubilise in alkaline solutions similar to the digestive fluids in lepidopteran larval gut, and thus release the proteins responsible for toxicity (Angus 1954). Production of this proteinaceous parasporal crystal is typically observed in *Bt* (Figure 11.2A) but is not a signature for species identification from other members of the *Bacillus cereus sensu lato* group (Ehling-Schulz et al. 2019). These inclusions are composed of a number of different Cry (from crystal) and/or Cyt (from cytolytic) proteins, which constitute the main Bt virulence factors. These pesticidal proteins are named and classified according to amino acid sequence identity and structural similarity into 391 holotype sequences, with Cry proteins folding in a three domain structure (three-domain Cry toxins) being the largest group (271 proteins) (Crickmore et al. 2020b).

Bt isolates were historically characterised by serotyping of the H flagellar antigen in vegetative cells (de Barjac and Bonnefoi 1968), which allowed classification into 85 serotypes or subspecies (Jurat-Fuentes and Jackson 2012; Lecadet et al.

1999). However, this method is now discontinued due to sparse availability of antisera and issues with agglutination and cross-reactivity with *B. cereus*. Alternative DNA-based methods have been proposed for classifying *Bt* isolates, including polymorphisms in flagellin genes (Yu et al. 2002), fingerprinting of extragenic palindromic sequences (Reyes-Ramírez and Ibarra 2005) and multi-locus sequence typing (MLST) analysis (Wang et al. 2018).

The generalised infective cycle for *Bt* (Figure 11.1) commences with the ingestion of spores, the most common life stage outside of a host, together with crystalline inclusions. Proteases in the alkaline and reducing conditions of the gut fluids solubilise the Cry and/or Cyt protein crystals, processing them to activated toxins. Emerging preliminary evidence suggests the possibility that Cry protoxins may not be processed as quickly as predicted from *in vitro* assays and that they may be toxic through a distinct mode of action compared to processed (activated) Cry toxins (Qi et al. 2020). The Cry protoxin occurs in two main forms, based on size (Figure 11.2B). For the larger protoxins (130–140 kDa), processing removes about 500 and 29–43 amino acids from the C- and N-terminus, respectively, while in smaller Cry protoxins (67–75 kDa) activation only removes

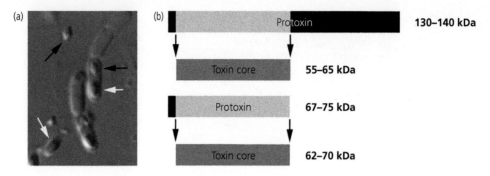

Figure 11.2 Microscopic morphology of *Bacillus thuringiensis* cells and schematic view of the two main types of Cry three-domain insecticidal proteins based on protoxin size. (a) Spores (white arrows) and parasporal crystals (black arrows) observed in a differential interference contrast (DIC) micrograph detailing the late sporulation phase of a *B. thuringiensis* subsp. *kurstaki* strain HD73 culture. Note that spores and crystals appear inside a sporangium and in free form. (b) Schematic alignment of larger (130–140 kDa) and shorter (67–75 kDa) Cry protoxins and their activation to a toxin core. Protoxins in the larger group include Cry1A and Cry1F proteins, while Cry2A, Cry3A, and Cry11A belong to the shorter group. The relative length of protoxins, toxin cores, and cleaved regions are approximate and not drawn to scale.

the N-terminal peptide (Zalunin et al. 2015). Fragments of DNA encoding Cry genes remain associated with activated Cry toxins and may influence toxin stability and mode of action (Guo et al. 2011). On the other hand, Cyt protoxins are processed to 23 kDa activated toxins (Tetreau et al. 2020).

The majority of Cry proteins (89%) have a three-domain structural pattern (Adang et al. 2014), suggesting commonalities in their mode of action. In comparison, the structures of resolved Cyt toxins present a single domain topology (Soberón et al. 2013). Depending on the insect host, activated Cry and Cyt toxins must cross the peritrophic matrix, a chitin-rich semi-permeable protective structure surrounding the food bolus, to interact with the midgut epithelium. Retention of Cry toxins by the peritrophic matrix may affect toxicity through toxins binding to its surface (Rees et al. 2009), and disruption of the matrix with chitinases enhances Cry toxicity (Sampson and Gooday 1998). While Cry toxins bind to protein receptors on the brush border membrane of midgut cells (Adang et al. 2014), Cyt toxins present affinity for membrane lipids (Soberón et al. 2013). Although the specific midgut proteins serving as receptors depend on the Cry toxin and host, the most accepted model supports the hypothesis that binding to these receptors is conducive to the formation of a cation-selective Cry toxin pore that disrupts homeostasis and leads to cell death by colloid osmotic lysis (Knowles and Ellar 1987).

Information from Cry-resistant insect strains and functional tests supports a critical role for cadherin-like and ATP binding cassette (ABC) transporters in Cry intoxication (Jurat-Fuentes et al. 2021; Heckel 2020). Interaction of Cry toxin monomers with cadherin receptors seems important for further processing resulting in oligomerisation (Gómez et al. 2002). Engineered Cry toxins capable of cadherin-independent oligomerisation overcome resistance to Cry toxins in insects with altered cadherin expression (Soberón et al. 2007). While the specific role of ABC transporters in Cry intoxication is still under study, the current hypothesis suggests that these transmembrane pumps may facilitate prepore oligomer insertion in the membrane (Heckel 2020).

About 10% of the described Cry proteins are distinct from the three-domain folding model and structurally resemble binary toxins, epsilon toxin-mosquitocidal toxin (ETX-MTX) proteins, and other known toxins (Adang et al. 2014). Herein we refer to these proteins using their old Cry toxin naming and, if changed, we also present the most current nomenclature (Crickmore et al. 2020a). The Cry34Ab1/Cry35Ab1 binary proteins expressed in transgenic maize to control rootworms are renamed as aegerolysin-related pesticidal protein (Gpp34Ab1) and toxin-10 pesticidal protein (Tpp35Ab1), respectively. The hypothesised mode of action of this binary crystal toxin includes membrane permeation leading to disruption of cell homeostasis. The role for each of the partner proteins in binding to receptors and cytotoxicity remains unclear (Narva et al. 2017). The crystal proteins structurally resembling ETX-MTX toxins are beta pore forming toxins and are renamed as ETX/MTX2-related pesticidal proteins (Mpp), such as Cry51Aa2 with activity against hemipterans (Baum et al. 2012). There is evidence that some of these crystal proteins containing ETX-MTX domains, such as Cry64Ba and Cry64Ca (renamed as Mpp64Ba and Mpp64Ca, respectively), are co-expressed and possibly act as a binary toxin (Liu et al. 2018).

The Cyt toxins are commonly described as insecticidal and haemolytic (i.e. lysing red blood cells), yet activity against insects presents unique features. For instance, Cyt1Aa binding to non-saturated membrane lipids, subsequent oligomerisation, membrane insertion and pore formation appear exclusive to intoxication of insect cells (Onofre et al. 2020). Formation of non-cytolytic aggregates enabling anchoring of Cry toxins (Pérez et al. 2007) or cytolytic porous oligomers seems to depend on the local concentration of Cyt1Aa toxin in the enterocyte membrane (Tetreau et al. 2020). Combining Cry and Cyt toxins can enhance effects. For instance, binding of Cry to Cyt toxins results in synergism of Cry toxicity in *B. thuringiensis* susbsp. *israelensis* (*Bti*) against mosquito larvae (Crickmore et al. 1995) and delays the evolution of resistance to *Bti* and its Cry toxins (Wirth et al. 2005).

Different types of hosts have been defined based on symptomatology of *Bt* infection, including increased haemolymph pH and potassium concentration (Type I), observable gut paralysis (Types I and II but not Type III) and requiring the presence of spores for mortality (Type III) (Jurat-Fuentes and Jackson 2012). Gut damage results in extensive melanisation (Figure 11.3A). Widespread enterocyte (gut epithelial cell) death after Cry and/or Cyt intoxication results in disruption of the midgut epithelial barrier and mixing of haemolymph and gut contents, lowering the midgut pH and activating *Bt* spore germination. Gut disruption results in the first set of disease symptoms approximately 15–30 min post-ingestion, including cessation of feeding and paralysis (Fast and Angus 1965). Gut

paralysis prevents passage of the germinating *Bt* cells, which invade the haemocoel.

Vegetative cells invading the haemocoel produce a number of pathogenic factors, including Vip and secreted insecticidal protein (Sip, now renamed as Mpp) toxins, chitinases, proteases, phospholipases, and iron sequestering systems to overcome the host and degrade its tissues. Antimicrobials such as bacteriocins are also secreted to inhibit growth of bacterial competitors (Salazar-Marroquín et al. 2016). Among the Vip toxins, the Vip1 and Vip2 proteins are active against Coleoptera and Hemiptera and are predicted to function as a binary toxin, with Vip1 being responsible for binding to receptors and Vip2 entering and killing the cell through its ADP-ribosyltransferase activity against actin (Chakroun et al. 2016). This is reflected in the new pesticidal protein nomenclature (Crickmore et al. 2020b), where Vip1 and Vip2 are renamed as the Vip protein binding (Vpb1) and Vip protein active (Vpa2) component, respectively. On the other hand, the Vip3 family contains 14 holotype proteins active against Lepidoptera and present no similarity to Vip1/2 proteins (Crickmore et al. 2020b). Increased used of transgenic plants producing Vip3Aa for control of lepidopteran larvae has fuelled interest in elucidating the mode of action of Vip3 proteins in midgut cells, although it is possible the toxin targets alternative cell types in the haemocoel. The Vip3Aa protein (approx. 88 kDa) is activated by proteolysis to yield fragments from the N—(19-kDa) and C-terminus (approx. 65-kDa) that remain tightly associated. Binding of the activated Vip3Aa toxin to receptors on the membrane of midgut cells is conducive to pore formation (Lee et al. 2003). Candidate Vip3Aa receptors have been identified from cultured insect cells, including intracellular (ribosomal protein S2) and membrane (scavenger receptor class C and fibroblast growth factor receptor) proteins (Singh et al. 2010; Jiang et al. 2018a, 2018b). There is evidence supporting internalisation of Vip3Aa and activation of apoptosis in *Spodoptera exigua* larvae (Hernández-Martínez et al. 2017). As with Cry toxins, histopathology of Vip3Aa intoxication shows lysed and swollen cells leaking cellular material (Yu et al. 1997). While there is no information available on the mode of action of Sip toxins, they are active against selected coleopteran larvae

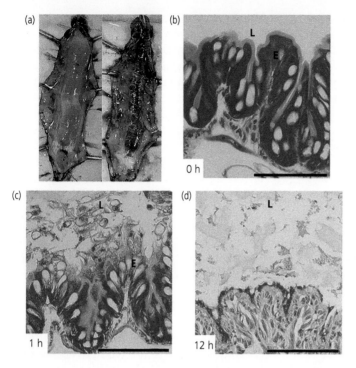

Figure 11.3 Infection of a lepidopteran host by *Bacillus thuringiensis*. (a) Melanisation of the gut in larvae of *Heliothis virescens* after 30 minutes since ingestion of Cry1Ac protoxin (right), compared with an untreated larva (left). (b-d) Histological sections stained with haematoxylin and eosin (H&E) detailing the midgut epithelium of *H. virescens* larvae before (b) and after treatment with Cry1Ac protoxin for 1 (c) and 12 (d) hours. The gut lumen (L) is at the top of each panel and the scale bar = 100 μm. Note the healthy midgut epithelium (E) in the control panel (b) compared to extensive damage and shedding of cell contents and dying cells into the lumen after 1 hour (c). After 12 hours (d), only the connective tissue and muscle layer remain of the midgut epithelium, with cell debris observed in the gut lumen.

and probably contribute to overcoming the host during the exponential growth phase. During this time of tissue damage and bacterial proliferation, approximately 2 h post-ingestion, the host appears very sluggish or paralyzed.

While gut microbiota may influence the outcome, under normal conditions proliferation of vegetative *Bt* cells is responsible for septicaemia and ultimately host death (Raymond et al. 2010), usually within 3–4 hours since ingestion. Host death, exhaustion of nutrients, and cell density in the cadaver signals initiation of *Bt* sporulation controlled by regulatory transcription factors. Synthesis of parasporal bodies is concomitant with formation and maturation of the spore in the mother cell compartment (8–12 h), until the sporangium liberates the mature spore and parasporal body into the environment.

Once released from the host cadaver, *Bt* spores are very persistent in the environment. However, effective horizontal transmission of *Bt* has been difficult to demonstrate experimentally and field epizootics are rare. This inconsistency may be explained by environmental factors affecting viability of crystal toxins or by interactions with social 'cheats', *Bt* cells

that do not produce toxins yet exploit benefits of toxin-producing cells, thus hindering the onset of natural outbreaks (Raymond et al. 2012). In contrast, *Bt* epizootics have been reported in mass rearing of silkworms and other lepidopteran larvae (Lietze et al. 2010). In these cases and considering the resilience of *Bt* spores, complete cleaning of the rearing facility and removal of dead insects would be necessary to curb the spread of the disease.

11.2.2 Lysinibacillus sphaericus

Also known as *Bacillus sphaericus* but later reclassified as *Lysinibacillus sphaericus*, this bacterium is characterised by the localisation of its spherical spore at one end of the sporangium. The bacterium causes disease in mosquitoes (especially *Culex* and *Anopheles* larvae), black flies and non-biting midges, with suggested potential as a brood pathogen in stingless bees (Fünfhaus et al. 2018). It is persistent even in polluted waters, which has promoted its common use as a biopesticide in mosquito control programmes.

The main virulence factors produced by *L. sphaericus* are binary (Bin) proteins that are co-transcribed from the same operon and stored in parasporal bodies. The two components of the binary (Bin) toxins, BinA and BinB (renamed as Tpp1 and Tpp2, respectively) fulfil complementary roles in the intoxication process. Only Bin A presents some mosquitocidal activity when tested individually (Nicolas et al. 1993), and thus BinB may be viewed as a synergist of BinA. Both Bin components are proteolytically activated in the digestive fluids of susceptible larvae to 41.9 kDa BinA and 43–46 kDa Bin B proteins, before associating in heterodimers (Surya et al. 2016). The Bin B protein binds to glycosylphosphatidylinositol (GPI) anchored α-glucosidases (maltases) as receptors on the membrane of gastric caecae and posterior midgut cells (Silva-Filha et al. 1999). Binding of Bin B to these receptors is hypothesised to internalise the Bin toxin through lipid raft-dependent endocytosis into target cells (Sharma et al. 2020). Relevance of Bin B binding for toxicity is highlighted by the lack of Bin B binding to non-susceptible *A. aegypti* (Ferreira et al. 2014) and cases of resistance to Bin toxin involving unavailability of Bin B receptors (Silva-Filha 2017). Once internalised, the Bin toxin kills the target cell by yet unknown mechanisms. The Cyt1Aa toxin from *Bti* synergises Bin and Bin A toxicity against Bin-resistant or refractory mosquito larvae through a mechanism that requires functional activity of Cyt1Aa but it probably does not involve binding between Bin and Cyt1Aa (Nascimento et al. 2020). A chimeric fusion of Cyt1Aa with Bin A displays high mosquitocidal activity, supporting that Bin A does not require interaction with Bin B for internalisation and cytotoxicity (Bideshi et al. 2017).

Some of the additional mosquitocidal virulence factors that may be produced by *L. sphaericus* strains include a binary toxin composed of a three-domain Cry toxin (Cry48Aa), a binary toxin (Cry49Aa, now renamed as Tpp49Aa), and mosquitocidal toxins (Mtx). The Cry48Aa/Cry49Aa binary toxin is produced as a parasporal crystal and overcomes resistance to Bin toxin (de Melo et al. 2009), supporting a distinct mode of action. In contrast, Mtx proteins are produced and remain associated with *L. sphaericus* vegetative cells. The mode of action of Mtx proteins varies depending on the protein (Berry 2012), with proteins in the Mtx1 family displaying

ADP-ribosyltransferase activity and proteins in the ETX/MTX2 structural family resembling the pore-forming *Clostridium* epsilon toxin family. While production of Mtx proteins is not directly associated with high mosquitocidal activity, they can synergise Bin and Cry11Aa (from *Bti*) activity (Wirth et al. 2007).

Striking similarities can be identified between *L. sphaericus* and *Bt* pathogenesis. Disruption of the gut epithelial barrier by Bin or Cry48Aa/Cry49Aa proteins allows spores and vegetative cells to invade the haemocoel, as done by Cry and Cyt proteins in *Bt* pathogenesis. The host tissues are then used as nutrient resource to support vegetative growth and production of Mtx proteins, akin to production of Vip and Sip toxins in *Bt*, to target unknown cell types. Exhaustion of host resources stimulates sporulation and subsequent release of infective bacteria from the host cadaver to recycle in the environment (Becker et al. 1995).

11.2.3 Milky disease bacteria

The *Paenibacillus* genus contains entomopathogens causing milky disease in larvae of scarab beetles (Figure 11.4A). The general inability of the milky disease bacteria to grow and sporulate on standard media has challenged taxonomic definition and initial classification was based on morphotype. Molecular genetics methods have been used to examine differences between isolates and support the contention of localised evolution and specificity based on morphological and pathological data (Dingman 2009). Draft sequences of *P. popilliae* and *P. lentamorbus* have been published which will further add to our understanding of this group (Grady et al. 2016). The best-studied bacterium causing this disease is *P. popilliae*, originally described and formerly known as *Bacillus popilliae*. This bacterium was originally morphologically characterised by a distinctive resting stage with a spore and parasporal body contained within a thick sporangium, giving the cells a footprint-like appearance under light microscopy (Type A), although some isolates lack the parasporal body (Type B) (Figure 11.4B). As an obligate pathogen, the *P. popilliae* bacterium is highly fastidious and only found as a spore when outside its natural hosts, larvae of the scarabaeid beetles. The

(a)

(b)

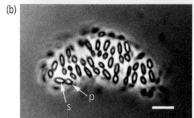

(c)

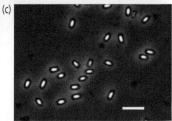

Figure 11.4 Symptoms and causative agents for milky disease in larvae of *Costelytra giveni*. (a) Live larvae of *C. giveni* with milky disease (right) and healthy (left). Note milky appearance of haemolymph exuding from cut leg of the diseased compared to clear haemolymph in the healthy larva. (b) Haemolymph sample of milky spore Type A with spore (s) and parasporal body (p). (c) Milky spore Type B lacking a parasporal body. Scale bar = 5 μm.

observation that host specificity may be overcome by injection of vegetative cells into the haemocoel (Milner 1981) suggests that the ability to traverse the gut epithelial barrier is critical in determining host range.

Pathogenesis is initiated with ingestion of *P. popilliae* spores and subsequent germination in the physicochemical conditions of the host midgut. Vegetative cells enter the midgut epithelium by phagocytosis, where they may proliferate before invading the haemocoel (Splittstoesser et al. 1973). Once reaching the haemocoel, the vegetative cells proliferate without causing toxaemia, allowing the host to remain active and continue feeding through a long period of infection. Several waves of vegetative propagation and subsequent sporulation in the cells occur in the haemolymph. In the late disease stages the high density of spores results in the typical whitish haemolymph colouration that gives name to the milky disease (Figure 11.4A). The host remains active until a few days before death, when they become sluggish and become brownish through melanisation, except for regions where the white haemolymph is visible. The cause

of host death in milky disease seems to involve host starvation caused by the successive waves of *P. popilliae* proliferation in the haemolymph. Once the host dies, spores liberate into the soil where they remain until ingested by a new host.

Although their role in pathogenicity is unclear, a number of Cry proteins in the Cry18 and Cry43 families have been identified from strains of *P. popilliae* and other milky disease bacteria (Yokoyama et al. 2004). These toxins are not required for the milky disease progression but together with the action of secreted chitinases may assist in crossing the gut epithelial barrier (Zhang et al. 1997).

11.2.4 Foulbrood diseases

Honeybee larvae affected by foulbrood disease die before completing the pupation stage, greatly affecting viability of colonies. Typically, infected brood changes colour from white to light tan/black and emits a foul odour giving name to the disease. The causative agent of American foulbrood (AFB), *Paenibacillus larvae*, is the most destructive and best studied of the brood disease bacterial entomopathogens (Fünfhaus et al. 2018). This bacterium displays peritrichous (distributed all over the whole cell surface) flagella and is considered an obligate pathogen, as spores are the only infectious form and young honeybee larvae the only known hosts. The AFB disease can be caused by any of four *P. l. larvae* strains differing in their enterobacterial repetitive intergenic consensus (ERIC) sequences (ERIC I to IV, respectively). This classification also agrees with phenotypic and virulence differences, with higher virulence and faster speed of kill by ERIC II-IV (6–7 days) compared to ERIC I (12–14 days).

The infective process of AFB starts with ingestion of spore-contaminated food by larvae. Spores rapidly germinate and vegetative cells proliferate to high numbers in the larval midgut, without disrupting the epithelium (Yue et al. 2008). During this phase, the bacterium produces a number of peptide and non-ribosomal antibiotics to eliminate the honeybee gut microbiome (Müller et al. 2015) and a lytic polysaccharide monooxygenase (*Pl*CBP49) that degrades the chitin in the peritrophic matrix to reach the midgut epithelium (Garcia-Gonzalez et al. 2014). Early instars are most susceptible to the bacteria as the peritrophic matrix thickens after 36 h post-hatch and is harder to degrade. Once *P. l. larvae* can directly interact with the midgut cells, it switches to an invasive destructive stage through the action of secreted virulence factors (Erban et al. 2019). Characterised toxins include binary proteins such as the Plx1 toxin displaying mono-ADP-ribosyltransferase activity targeting RNA/DNA, or Plx2 and C3larvin toxins disrupting actin through Rho-ADP ribosylation (Fünfhaus et al. 2013; Krska et al. 2015). The specific role of these and other toxins in pathogenesis varies depending on the ERIC genotype (Erban et al. 2019). Surface layer (S-layer) proteins probably protect the bacterium against biotic and abiotic stress, mediate adhesion to host cells and peritrophic matrix, and are known to participate in virulence (Poppinga et al. 2012). The process resulting in haemocoel invasion remains elusive but breaching of the gut epithelial barrier coincides with host death from bacteraemia (Djukic et al. 2014). The bacterium continues to proliferate in the haemocoel using nutrients from the degradation of the host cadaver until nutrient depletion and sporulation. Diseased brood appears a light tan or caramel colour and produces the characteristic 'dead fish' foul odour. Worker bees are not susceptible to infection but they spread it by acquiring and dispersing spores when cleaning contaminated cells. Adult hygienic behaviour can decrease levels of spore contamination and explain lower susceptibility to AFB in Asian (*Apis cerana*) compared to European (*Apis mellifera*) honeybee (Chen et al. 2000).

Melissococcus plutonius is the causative agent of the European foulbrood (EFB) disease in both Asian and European honeybees. Cells of *M. plutonius* are non-motile, shaped as lanceolate cocci and do not form a spore. Not much is known about *M. plutonius* pathogenesis, other than infection is from ingestion of contaminated food and that saprophytes acting as 'secondary invaders', may participate in decomposition of the host cadaver. Observations from *in vitro* infections support the hypothesis that virulence depends on *M. plutonius* genotype and is influenced by host background but not by infection with secondary invaders (Lewkowski and Erler 2019). Affected larvae appear curled upwards in their cells until the decomposed cadavers dry down to brownish flakes.

11.2.5 *Brevibacillus laterosporus*

Characterised by a canoe-shaped parasporal body attached to one side of the spore (Hannay 1957), *B. laterosporus* presents a wide range of infectivity towards coleopteran, dipteran and lepidopteran insects (Ruiu 2013). The parasporal crystal is composed of insecticidal proteins (Zubasheva et al. 2010) but production of the crystal is not necessary for pathogenicity (Ruiu et al. 2007). Strains not producing these crystals synthesise spore surface proteins (Marche et al. 2017), chitinases (Prasanna et al. 2013), collagenase-like proteases, antimicrobials, and a number of insecticidal proteins homologous to Cry75 (renamed as Mpp75), Vip1–4, and ETX/MTX2 toxin genes (Marche et al. 2018; Glare et al. 2020). These virulence factors induce alteration of microvilli, cytoplasmic vacuolisation and changes to the mitochondria in gut cells, progressing to leakage of cellular contents into the gut lumen and disruption of the epithelial barrier

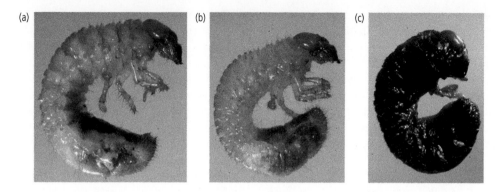

Figure 11.5 Characteristics of infection of a scarab larva (*Costelytra giveni*) with the Gram-negative entomopathogen *Serratia entomophila*. (a) Healthy larva. (b) Larva infected by *S. entomophila* causing amber disease, showing apparent transparency due to gut clearance. (c) Dead larva showing septicaemia and melanisation after colonisation of the haemocoel by bacteria.

(Ruiu et al. 2012). Infection in *Musca domestica* larvae results in reduced feeding and movement, with death following within 72 h.

11.2.6 *Clostridium bifermentans*

Isolates of the anaerobic bacterium *C. bifermentans* show mosquitocidal activity. The best-studied isolate, *C. bifermentans* subsp. *malaysia*, is highly toxic to larvae of *Anopheles* but has lower toxicity to *Aedes* and *Culex* mosquitoes (Thiery et al. 1992). Highest levels of toxicity occur during sporulation and greatly decrease with sporangium lysis, probably due to released proteases lysing virulence factors (Nicolas et al. 1990). The main virulence factor for activity against *Aedes* in *C. bifermentans* subsp. *malaysia* is a complex of four proteins, including two Cry-like proteins (Cry16Aa, Cry17Aa) and two proteins with similarity to the aegerolysin family of proteins (Cbm17.1 and Cbm17.2), encoded in a 'Cry toxin' operon (Qureshi et al. 2014). The virulence factors targeting *Anopheles* mosquitoes are unknown but are not located in the Cry operon.

11.3 Gram-negative entomopathogens

11.3.1 *Serratia* spp.

The genus *Serratia* (Enterobacteriaceae) comprises ubiquitous species, with several strains and biotypes frequently isolated from insects (Grimont et al. 1979). Among these, the red-pigmented *Serratia marcescens* is identified as a 'potential' pathogen (Jeong et al. 2010; Pineda-Castellanos et al. 2015) with a lethal dose of just a few cells per insect once in the haemocoel but limited invasive capacity. Intrahaemocoelic infection with *S. marcescens* activates immune responses in crickets and surprisingly increases egg laying as a response to infection (Adamo 1999). Infections with red bacteria (probably *S. marcescens*) have been reported in mass rearing of the house cricket (*Acheta domesticus*) for food and feed (Eilenberg et al. 2015). Since the bacterium does not sporulate, disease spread is stalled by thorough cleaning and elimination of diseased individuals.

Probably the best characterised entomopathogenic *Serratia* spp. infection is that of amber disease caused by isolates of *Serratia entomophila* in the New Zealand grass grub, *Costelytra giveni* (formerly *C. zealandica*) (Grimont et al. 1988). Larval amber disease symptoms include cessation of feeding, gut clearance, amber colouration, and eventual death (Figure 11.5). The virulence determinants of the disease reside on a 155-kb plasmid designated as pADAP (amber disease-associated plasmid), which include a *S. entomophila* pathogenicity (*sep*) and an antifeeding (*afp*) gene cluster needed for amber disease. After ingestion of the pathogenic bacteria from the soil, the (*afp*) gene cluster expresses a phage-tail like bacteriocin responsible for cessation of feeding by the infected larva (Hurst et al. 2007). Active *sep* genes are arranged in a virulence-associated Tc (Toxin complex or ABC) cluster (Hurst et al. 2000) and secretion of Sep toxins leads to clearance of organic matter and digestive fluids from the midgut through the hindgut and anus, leaving larvae with the translucent amber colouration characteristic of amber disease (Jackson et al. 1993). As the bacteria colonise the midgut, *Serratia* toxins degrade the cytoskeletal network and prevent secretion of digestive proteinases (Gatehouse et al. 2008). Denied nutrition, larval fat bodies are auto consumed and tissues weaken until bacteria finally invade the haemocoel after days or weeks of inactivity of the larva, causing death through septicaemia (Jackson et al. 2001).

Amber disease is unusual as it appears to be unique to *C. giveni* in New Zealand, where natural epizootics are common (Jackson et al. 2018). The pADAP plasmid has only been found in New Zealand bacterial isolates and substantial screening efforts have not found other insects susceptible to the pADAP bearing strains (Jackson 2003). In newly cultivated soils and pastures free from grass grub, most *Serratia* isolates are non-pathogenic. As the grass grub populations build up, *Serratia* spp. causing amber disease enter the population, spread and cause an epizootic leading to population collapse. This impact of natural epizootics of amber disease on grass grub populations stimulated the development of a selected *S. entomophila* strain as a commercial biocontrol (Jackson 2007). Interestingly, a strain of *S. proteamaculans* has recently been discovered showing a broader host range with activity against a range of scarab beetle larvae (Hurst et al. 2018).

11.3.2 *Chromobacterium* spp.

Pigmented *Chromobacterium* spp. (Betaproteobacteria) are generally soil and water associated organisms but prospecting efforts have identified entomopathogenic isolates of *C. subtsugae* (Martin et al. 2007a; Martin et al. 2007b), *C. vaccinii* (Farrar et al. 2018b), and *C. sphagni* (Farrar et al. 2018a). Pathogenicity appears associated with insecticidal toxins expressed in culture supernatants and filtrates and an extract of *C. subtsugae* fermentation has been commercialised as the

insecticide (Grandevo®). In another example, an isolate of *C. piscinae* produces a Gram-negative insecticidal protein (GNIP1Aa) with highly specific activity against the Western corn rootworm (*Diabrotica virgifera virgifera*) when purified or expressed in transgenic maize (Sampson et al. 2017). This toxin is the only bacterial member of the membrane attack complex/perforin (MACPF) superfamily for which insecticidal activity has been identified, as MACPF proteins are characterised as pore-forming toxins targeting pathogens and infected cells (Zaitseva et al. 2019).

Genomic comparisons have predicted a range of secondary metabolites, chitinases and other genes that could contribute to insecticidal activity of *Chromobacterium spp.* (Blackburn et al. 2016; Vöing et al. 2015). Notably, genomes of most *Chromobacterium* spp. compared harbour a pathogenicity island including genes encoding type III secretion systems (T3SSs) required for the secretion of effector proteins (Batista and da Silva Neto 2017).

The *Chromobacterium* sp. Panama (Csp_P) strain isolated from the midgut of the mosquito *Aedes aegypti* reduces the lifespan of adult and larval mosquitoes (Ramirez et al. 2014). After ingestion, the bacteria colonise the mosquito midgut activating the immune response and inhibiting the normal microflora, significantly reducing larval and adult survival. Moreover, the bacterium produces stable bioactive factors with anti-*Plasmodium* and anti-dengue activities, reducing vector competence. A dried, nonlive extract of *Chromobacterium* (Csp_P) has been developed as an effective mosquito biopesticide (Caragata et al. 2020).

11.3.3 *Yersinia* spp.

As with other Gram-negative bacteria, *Yersinia* isolates are often only weakly pathogenic to insects and variable in effect (Figure 11.6). Oral pathogenicity relates to the presence of toxin complex (Tc or A_5BC) gene homologues, which are common among insecticidal *Yersinia* isolates (Fuchs et al. 2008). Thus, the highly pathogenic isolate *Yersinia entomophaga* MH96 identified from New Zealand grass grub (*C. giveni*) larvae, contains a 32 kb pathogenicity island encoding an insecticidal Tc complex, Yen-Tc,

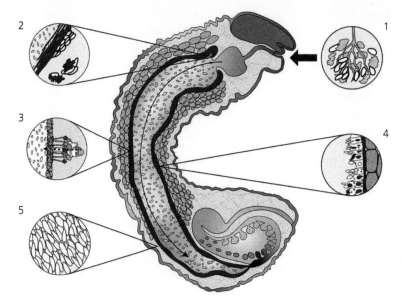

Figure 11.6 Generalised characterisation of disease process for Gram-negative entomopathogenic bacteria. 1; Ingestion from soil or contaminated material. 2; Colonisation of membranes and particulate matter. Release of toxins and response. 3; Pore formation and delivery of toxins to the cells. 4; Midgut cellular disruption. 5; Invasion and growth of bacteria in the haemocoel causing septicaemia. Factors vary for different species of Gram-negative bacteria (see text).

and two chitinase proteins forming a composite Tc molecule (Hurst et al. 2011). The Yen-Tc complex consists of a pore-forming TcA component that allows cytotoxic TcB and TcC elements to enter the cell (Piper et al. 2019).

After ingestion of the bacteria, the Yen-Tc toxin complex including active chitinases is secreted into the larval gut and binds to midgut cells (Busby et al. 2013). The larva slows feeding, followed by a violent regurgitation and clearance of the gut contents in what appears to be an attempt to purge the bacteria (Hurst et al. 2014). After colonisation of the insect gut, *Y. entomophaga* MH96 rapidly degrades the gut epithelial membranes before invading the haemocoel, causing septicaemia and death within a few days.

While the Tc genes contain the main determinants of disease, sequence analysis has shown that the *Y. entomophaga* MH96 genome contains a diverse array of toxins and secretion systems, which may contribute to the virulence of the bacterium (Hurst et al. 2016). The Tc (ABC) toxin comprises a conserved structural scaffold capable of accommodating diverse structural motifs (Piper et al. 2019), possibly explaining the wide range of insect pests (coleopteran, lepidopteran, and orthopteran species) susceptible

to *Yersinia* spp. and the potential for discovery of novel insect pathologies within this genus.

11.3.4 *Pseudomonas* spp.

Bacteria from the family Pseudomonadaceae are frequently isolated from dead and diseased insects and some strains have been found to display the characteristics of potential pathogens, causing disease in laboratory conditions (Bucher 1963). Thus, *Pseudomonas entomophila* originally isolated from fruit fly (*Drosophila melanogaster*) (Vodovar et al. 2005) is highly pathogenic to the original host and able to kill insects from at least three different Orders (Dieppois et al. 2015). Upon ingestion, the bacterium accumulates in the insect's crop blocking further food uptake. Genomic analysis shows that *P. entomophila* encodes a number of virulence factors including proteases, secondary metabolites and toxins (Dieppois et al. 2015). The secreted pore-forming toxin monalysin targets gut cells (Opota et al. 2011) and its damage to gut cells blocks the epithelium defence response (Chakrabarti et al. 2012). Degradation of the midgut epithelium allows bacteria to enter the haemocoel leading to septicaemia and death. Large bacterial doses can overcome the immune system of the host resulting in faster death.

11.3.5 *Burkholderia rinojensis*

Bacteria of the genus *Burkholderia* (Betaproteobacteria) are known as insect symbionts, antagonists and even as biocontrol agents against plant pathogens. However, *Burkholderia rinojensis* A396 was isolated from soil and showed activity against caterpillars and mites (Cordova-Kreylos et al. 2013). The bacterium appears to produce heat-stable toxins and heat-treated whole cell cultures have been developed as a biopesticide that is active by exposure and ingestion, inducing exoskeleton degradation and moulting interference in lepidopteran larvae and hemipteran pests.

11.3.6 *Rickettsiella* spp.

The genus *Rickettsiella* are obligate intracellular bacteria that can be found as pathogens and symbionts of insects and ticks. Insect pathogens in the genus were originally defined by pathotype, *Rickettsiella popilliae*, *R. gryllyi*, and *R. chironomi*, in a classification that has been largely supported by whole genome analysis (Leclerque 2008). The disease process was described for infection of scarab beetle larvae (Kleespies et al. 2011). The infective cells ingested during feeding, pass through the midgut epithelium to enter the haemocoel, where they gain entry through endocytosis into cells in the fat body (the insect liver) and the haemolymph. Pleiomorphic forms develop within infected cells producing protein crystals and masses of small *Rickettsiella* as the disease develops. Eventually, infected cells lyse releasing vast numbers of *Rickettsiella* and crystals into the haemolymph, producing a white to blue colouration of the infected larvae. *Rickettsiella* diseases of insects are chronic, and the infected host will lose vigour over time, becoming sluggish and ceasing feeding before death. Infection can lead to epizootics of disease in an insect population as described for an outbreak of *Rickettsiella pyronotae* in the Manuka beetle (*Pyronota* spp.) (Marshall et al. 2017).

Care should be taken when working with entomopathogenic *Rickettsiella* due to concerns over the potential for inflammation and infection induced in vertebrates (Delmas and Timon-David 1985).

11.4 Summary, research needs, and future directions

Bacteria are ubiquitous and many have evolved to become opportunist or obligate pathogens of a wide range of insects. Stress and crowding conditions favour bacterial disease onset and transmission.

Bacteria from many genera have been developed as commercial agents for control of insect pests and mites. Bacterial genes have been incorporated into plants to provide resistance to insect pests.

Spore-forming bacteria in the Bacillaceae family are well characterised as potent pathogens in larvae of Lepidoptera (moths and caterpillars), Coleoptera (beetles and grubs), and Diptera (flies and mosquitoes).

Insecticidal proteins produced as parasporal bodies or secreted by vegetative cells are the main virulence factors allowing bacterial entomopathogens to disrupt the midgut epithelial barrier and reach the main body cavity to cause septicaemia.

In some cases, the bacterial entomopathogen follows a saprophytic lifestyle in the gut of the host until reaching a critical population mass that triggers a shift to an invasive destructive stage resulting in septicaemia. Expanded understanding of diseases induced by bacterial entomopathogens increases our ability to identify control methods to protect beneficial insects from disease and guide the development of entomopathogens as biological control agents for crop protection and insect pest control. Interest in the discovery of bacterial entomopathogens and characterisation of their pathogenesis and virulence factors has been mostly driven by the latter aspect. This explains why the best-studied entomopathogens are bacteria with features that are beneficial for field application, such as production of a highly stable life stage (spore) and potent virulence factors (Cry, Bin, Vip, Tc, and other toxins) for use against pest species. Other research efforts have focussed on impact on a limited range of beneficial insects (e.g. honeybee). Consequently, we probably only have a limited view into the world of insect diseases caused by bacteria.

Biodiscovery and screening programmes have resulted in the identification of new bacterial isolates causing disease in insects, some of them with potential for commercialisation in crop protection (Marrone 2019). However, successful commercialisation of entomopathogenic bacteria does not only depend on their pesticidal potential but also on an understanding of the manner in which they can be used. Challenges to their adoption include misunderstandings of their unique mode of action and how it can benefit integrated programmes rather than focusing on immediate effects on insect populations when considering cost and efficacy (Marrone 2019).

The advent of high throughput sequencing technologies with continuously reducing costs have synergised the discovery of new pesticidal genes in bacteria with broader and/or higher activity. Genome sequencing projects have thus provided relevant insights into the lifestyle and ensemble of virulence factor genes harboured by bacterial entomopathogens, frequently outpacing the characterisation of their mode of action. More than a thousand pesticidal proteins have been identified thus far from bacterial entomopathogens, classified in sixteen structural classes (Crickmore et al. 2020a). Aside from the well characterised three-domain Cry, binary Bin, and Cyt proteins, the mode of action of most of these pesticidal proteins remains obscure and needs investigation. Structural similarities can provide relevant clues towards characterisation of their mode of action (Berry and Crickmore 2017) but experimental evidence is ultimately needed to fully understand toxicity.

The role of bacterial diseases in beneficial insects such as honeybees and other insects providing services to ecosystems is not well understood. In some cases a causal bacteria-disease relationship remains to be demonstrated (Fünfhaus et al. 2018). A deeper understanding of these interactions is needed to manage diseases and ensure continued delivery of the services. Similar information is needed for bacterial entomopathogens that can affect mass production of insects for human food and animal feed or for release or rearing programmes (Francuski and Beukeboom 2020).

Recommended further reading

Berry, C. 2012. The bacterium, *Lysinibacillus sphaericus*, as an insect pathogen. *Journal of Invertebrate Pathology* 109: 1–10 (A complete review on *L. sphaericus*, including virulence factors, genomics, and field use.)

Bravo, A., Gill, S.S., and Soberón, M. 2019. *Bacillus thuringiensis*: Mechanisms and use, in *Encyclopedia of Microbiology*, 4th edn, Schmidt, T.M. (ed.) pp. 307–332. Cambridge, MA, USA: Academic Press (Complete review on *B. thuringiensis*, including virulence factors, genomics and applications.)

Crickmore, N., Berry, C., Panneerselvam, S., Mishra, R., Connor, T.R., and Bonning, B.C. 2020. A structure-based nomenclature for *Bacillus thuringiensis* and other bacteria-derived pesticidal proteins. *Journal of Invertebrate Pathology* 107438, https://doi.org/10.1016/j.jip.2020.107438 (Description of the recently proposed nomenclature for pesticidal proteins from bacteria.)

Fünfhaus, A., Ebeling, J., and Genersch, E. 2018. Bacterial pathogens of bees. *Current Opinion in Insect Science* 26, 89–96 (Detailed review of entomopathogenic bacteria affecting bees.)

Glare, T.R., Jurat-Fuentes, J.L., and O'Callaghan, M. 2017. Basic and applied research: Entomopathogenic bacteria. In *Microbial Control of Insect and Mite Pests*, 1st edn, Lacey, L.A. (ed.) Chapter 4, pp. 47–67. Academic Press, Cambridge, MA, USA (Detailed review on the use of entomopathogenic bacteria for pest control.)

Jackson, T.A., Berry, C., and O'Callaghan, M. 2018. Bacteria. In *Ecology of Invertebrate Diseases*, Hajek A.E. and Shapiro-Ilan D. (eds.) pp. 287–326. Oxford, UK: John Wiley & Sons Ltd. (Detailed review of the ecology of bacterial entomopathogens.)

Jurat-Fuentes, J.L., Heckel, D., and Ferré, J. 2021. Mechanisms of resistance to insecticidal proteins from *Bacillus thuringiensis*. In *Annual Review of Entomology*, 66, 121–140 (The most recent in-depth review on mechanisms of resistance to insecticidal proteins from *B. thuringiensis*.)

Jurat-Fuentes, J.L. and Jackson, T.A. 2012. Bacterial entomopathogens. In *Insect Pathology*, 2nd edn, Vega F.E. and Kaya H.K. (eds.) pp. 265–349. Cambridge, MA, USA: Academic Press (Systematic in-depth treatise on bacterial entomopathogens.)

Ruiu, L. 2013. *Brevibacillus laterosporus*, a pathogen of invertebrates and a broad-spectrum antimicrobial species. *Insects* 4, 476–492 (The most up to date and complete review of *B. laterosporus* as entomopathogen and producer of antimicrobials.)

Acknowledgements

J.L. Jurat-Fuentes acknowledges support from Agriculture and Food Research Initiative Foundational Program competitive grant 2018-67013-27820 and Hatch Multistate NC-246 from the US Department of Agriculture National Institute of Food and Agriculture. T.A Jackson is grateful for support from AgResearch, New Zealand.

References

Adamo, S.A. 1999. Evidence for adaptive changes in egg laying in crickets exposed to bacteria and parasites. *Animal Behaviour* 57: 117–124

Adang, M., Crickmore, N., and Jurat-Fuentes, J.L. 2014. Diversity of *Bacillus thuringiensis* crystal toxins and mechanism of action. In *Advances in Insect Physiology Vol. 47: Insect Midgut and Insecticidal Proteins*, Dhadialla T.S., and Gill S. (eds.) pp. 39–87. San Diego, CA: Academic Press

Angus, T.A. 1954. A bacterial toxin paralyzing silkworm larvae. *Nature* 173: 545–546

Appel, H.M. 2017. The chewing herbivore gut lumen: Physicochemical conditions and their impact on plant nutrients, allelochemicals, and insect pathogens. In *Insect-Plant Interactions*, Bernays, E.A. (ed) pp. 225–240. Boca Raton, FL: CRC Press

Bailey, L. 1983. *Melissococcus pluton*, the cause of European foulbrood of honey bees (*Apis* spp.). *Journal of Applied Bacteriology* 55(1): 65–69

Batista, J.H. and Da Silva Neto, J.F. 2017. *Chromobacterium violaceum* pathogenicity: Updates and insights from genome sequencing of novel *Chromobacterium* species. *Frontiers in Microbiology* 8: 2213

Baum, J.A., Sukuru, U.R., Penn, S.R. et al. 2012. Cotton plants expressing a hemipteran-active *Bacillus thuringiensis* crystal protein impact the development and survival of *Lygus hesperus* (Hemiptera: Miridae) nymphs. *Journal of Economic Entomology* 105: 616–624

Becker, N., Zgomba, M., Petric, D., Beck, M., and Ludwig, M. 1995. Role of larval cadavers in recycling processes of *Bacillus sphaericus*. *Journal of the American Mosquito Control Association* 11: 329–334

Berliner, E. 1915. Uber die schlaffsucht der mehlmottenraupe (*Ephestia kuhniella*, Zell.) und ihren erreger *B. thuringiensis* n. sp. *Zeitschrift Für Angewandte Entomologie* 2: 29–56

Berry, C. 2012. The bacterium, *Lysinibacillus sphaericus*, as an insect pathogen. *Journal of Invertebrate Pathology* 109: 1–10

Berry, C. and Crickmore, N. 2017. Structural classification of insecticidal proteins—Towards an *in silico* characterisation of novel toxins. *Journal of Invertebrate Pathology* 142: 16–22

Berry, C., O'Neil, S., Ben-Dov, E., Jones, A.F., Murphy, L., Quail, M.A., Holden, M.T., Harris, D., Zaritsky, A., and Parkhill, J. 2002. Complete sequence and organization of pBtoxis, the toxin-coding plasmid of *Bacillus thuringiensis* subsp. *israelensis*. *Applied and Environmental Microbiology* 68: 5082–5095

Bideshi, D.K., Park, H.-W., Hice, R.H., Wirth, M.C., and Federici, B.A. 2017. Highly effective broad spectrum chimeric larvicide that targets vector mosquitoes using a lipophilic protein. *Scientific Reports* 7: 11282

Blackburn, M.B., Sparks, M.E., and Gundersen-Rindal, D.E. 2016. The genome of the insecticidal *Chromobacterium subtsugae* PRAA4-1 and its comparison with that of *Chromobacterium violaceum* ATCC 12472. *Genomics Data* 10: 1–3

Brown, S.E., Cao, A.T., Hines, E.R., Akhurst, R.J., and East, P.D. 2004. A novel secreted protein toxin from the insect pathogenic bacterium *Xenorhabdus nematophila*. *Journal of Biological Chemistry* 279: 14595–14601

Bucher, G.E. 1963. Nonsporulating bacterial pathogens. In *Insect Pathology, An Advanced Treatise*. Steinhaus, E.A. (ed.) pp. 117–147. New York: Academic Press

Busby, J.N., Panjikar, S., Landsberg, M.J., Hurst, M.R., and Lott, J.S. 2013. The BC component of ABC toxins is an RHS-repeat-containing protein encapsulation device. *Nature* 501: 547–550

Caragata, E.P., Otero, L.M., Carlson, J.S., Borhani Dizaji, N., and Dimopoulos, G. 2020. A nonlive preparation of *Chromobacterium* sp. Panama (Csp_P) is a highly effective larval mosquito biopesticide. *Applied and Environmental Microbiology* 86(11): e00240-20, doi:10.1128/AEM.00240-20

Chakrabarti, S., Liehl, P., Buchon, N., and Lemaitre, B. 2012. Infection-induced host translational blockage inhibits immune responses and epithelial renewal in the *Drosophila* gut. *Cell Host & Microbe* 12: 60–70

Chakroun, M., Banyuls, N., Bel, Y., Escriche, B., and Ferré, J. 2016. Bacterial vegetative insecticidal proteins (VIP) from entomopathogenic bacteria. *Microbiology and Molecular Biology Reviews* 80: 329–350.

Chen, Y.-W., Wang, C.-H., An, J., and Kai-Kuang, H. 2000. Susceptibility of the Asian honeybee, *Apis cerana*, to American foulbrood, *Paenibacillus larvae*. *Journal of Apicultural Research* 39: 169–175

Cordova-Kreylos, A.L., Fernandez, L.E., Koivunen, M., Yang, A., Flor-Weiler, L., and Marrone, P.G. 2013. Isolation and characterization of *Burkholderia rinojensis*., a non-*Burkholderia cepacia* complex soil bacterium with insecticidal and miticidal activities. *Applied and Environmental Microbiology* 79: 7669–7678

Crickmore, N., Berry, C., Panneerselvam, S., Mishra, R., Connor, T.R., and Bonning, B.C. 2020a. Bacterial Pesticidal Protein Resource Center [Online]. Available: https://www.bpprc.org

Crickmore, N., Berry, C., Panneerselvam, S., Mishra, R., Connor, T.R., and Bonning, B.C. 2020b. A structure-based nomenclature for *Bacillus thuringiensis* and other bacteria-derived pesticidal proteins. *Journal of Invertebrate Pathology* Jul: 107438. doi: 10.1016/j.jip.2020.107438

Crickmore, N., Bone, E.J., Williams, J.A., and Ellar, D.J. 1995. Contribution of the individual components of the δ-endotoxin crystal to the mosquitocidal activity of *Bacillus thuringiensis* subsp *israelensis*. *FEMS Microbiology Letters* 131: 249–254

De Barjac, H. and Bonnefoi, A. 1968. A classification of strains of *Bacillus thuringiensis* Berliner with a key to their differentiation. *Journal of Invertebrate Pathology* 11: 335–347

De Melo, J.V., Jones, G.W., Berry, C. et al. 2009. Cytopathological effects of *Bacillus sphaericus* Cry48Aa/Cry49Aa toxin on binary toxin-susceptible and -resistant *Culex quinquefasciatus* larvae. *Applied and Environmental Microbiology* 75: 4782–4789

Delmas, F. and Timon-David, P. 1985. Effect of invertebrate rickettsiae on vertebrates: Experimental infection of mice by *Rickettsiella grylli*. *Comptes Rendus de l Academie des Sciences. Serie III, Sciences de la Vie* 300: 115–117

Dhinaut, J., Chogne, M., and Moret, Y. 2018. Trans-generational immune priming in the mealworm beetle protects eggs through pathogen-dependent mechanisms imposing no immediate fitness cost for the offspring. *Developmental and Comparative Immunology* 79: 105–112

Dieppois, G., Opota, O., Lalucat, J., and Lemaitre, B. 2015. *Pseudomonas entomophila*: A versatile bacterium with entomopathogenic properties. In *Pseudomonas: Volume 7: New Aspects of Pseudomonas Biology*, Ramos, J.-L., Goldberg, J.B., and Filloux A. (eds.) pp. 25–49. Dordrecht, Netherlands: Springer

Dingman, D.W. 2009. DNA fingerprinting of *Paenibacillus popilliae* and *Paenibacillus lentimorbus* using PCR-amplified 16S-23S rDNA intergenic transcribed spacer (ITS) regions. *Journal of Invertebrate Pathology* 100: 16–21

Djukic, M., Brzuszkiewicz, E., Fünfhaus, A. et al. 2014. How to kill the honey bee larva: Genomic potential and virulence mechanisms of *Paenibacillus larvae*. *PLoS One* 9: e90914

Ehling-Schulz, M., Lereclus, D., and Koehler, T.M. 2019. The *Bacillus cereus* group: *Bacillus* species with pathogenic potential. *Microbiology Spectrum* 7(3): 10.1128/microbiolspec.GPP3-0032–2018

Eilenberg, J., Vlak, J.M., Nielsen-leroux, C., Cappellozza, S., and Jensen, A.B. 2015. Diseases in insects produced for food and feed. *Journal of Insects as Food and Feed* 1: 87–102

Engel, P. and Moran, N.A. 2013. The gut microbiota of insects—diversity in structure and function. *FEMS Microbiology Reviews* 37: 699–735

Erban, T., Zitek, J., Bodrinova, M., Talacko, P., Bartos, M., and Hrabak, J. 2019. Comprehensive proteomic analysis of exoproteins expressed by ERIC I, II, III and IV *Paenibacillus larvae* genotypes reveals a wide range of virulence factors. *Virulence* 10: 363–375

Farrar, R.R., Jr., Gundersen-Rindal, D., Kuhar, D., and Blackburn, M.B. 2018a. Insecticidal activity of a recently described bacterium, *Chromobacterium sphagni. Journal of Entomological Science* 53: 333–338

Farrar, R.R., Jr., Gundersen-Rindal, D.E., Kuhar, D., and Blackburn, M.B. 2018b. Insecticidal activity of *Chromobacterium vaccinii*. *Journal of Entomological Science* 53: 339–346

Fast, P.G. and Angus, T.A. 1965. Effects of parasporal inclusions of *Bacillus thuirngiensis* var. *sotto* Ishiwata on the permeability of the gut wall of *Bombyx mori* (Linnaeus) larvae. *Journal of Invertebrate Pathology* 7: 29–32

Ferreira, L.M., Romao, T.P., Nascimento, N.A., Costa, Mda, C., Rezende, A.M., De-melo-neto, O.P., and Silva-Filha, M.H. 2014. Non conserved residues between Cqm1 and Aam1 mosquito alpha-glucosidases are critical for the capacity of Cqm1 to bind the Binary toxin from *Lysinibacillus sphaericus*. *Insect Biochemistry and Molecular Biology* 50: 34–42

Francuski, L. and Beukeboom, L.W. 2020. Insects in production—an introduction. *Entomologia Experimentalis et Applicata* 168; 422–431

Fuchs, T.M., Bresolin, G., Marcinowski, L., Schachtner, J., and Scherer, S. 2008. Insecticidal genes of *Yersinia* spp.: Taxonomical distribution, contribution to toxicity towards *Manduca sexta* and *Galleria mellonella*, and evolution. *British Medical Council Journal of Microbiology* 8: 214

Fünfhaus, A., Ebeling, J., and Genersch, E. 2018. Bacterial pathogens of bees. *Current Opinion in Insect Science* 26: 89–96

Fünfhaus, A., Poppinga, L., and Genersch, E. 2013. Identification and characterization of two novel toxins expressed by the lethal honeybee pathogen *Paenibacillus larvae*, the causative agent of American foulbrood. *Environmental Microbiology* 15: 2951–2965

Garcia-Gonzalez, E., Poppinga, L., Fünfhaus, A., Hertlein, G., Hedtke, K., Jakubowska, A., and Genersch, E. 2014. *Paenibacillus larvae* chitin-degrading protein PlCBP49 is a key virulence factor in American foulbrood of honeybees. *PLoS Pathogens* 10: e1004284

Gatehouse, H.S., Marshall, S.D., Simpson, R.M., Gatehouse, L.N., Jackson, T.A., and Christeller, J.T. 2008. *Serratia entomophila* inoculation causes a defect in exocytosis in *Costelytra zealandica* larvae. *Insect Molecular Biology* 17: 375–385

Glare, T.R., Durrant, A., Berry, C., Palma, L., Ormskirk, M.M., and Cox, M.P. 2020. Phylogenetic determinants of toxin gene distribution in genomes of *Brevibacillus laterosporus*. *Genomics* 112: 1042–1053

Gómez, I., Sánchez, J., Miranda, R., Bravo, A., and Soberón, M. 2002. Cadherin-like receptor binding facilitates proteolytic cleavage of helix alpha-1 in domain I and oligomer pre-pore formation of *Bacillus thuringiensis* Cry1Ab toxin. *FEBS Letters* 513: 242–246

Grady, E.N., MacDonald, J., Liu, L., Richman, A., and Yuan, Z.C. 2016. Current knowledge and perspectives of *Paenibacillus*: A review. *Microbial Cell Factories* 15: 203

Grimont, P.A.D., Grimont, F., and Lysenko, O. 1979. Species and biotype identification of *Serratia* strains associated with insects. *Current Microbiology* 2: 139–142

Grimont, P.A.D., Jackson, T.A., Ageron, E., and Noonan, M.J. 1988. *Serratia entomophila* sp. nov. associated with amber disease in the New Zealand grass grub *Costelytra zealandica*. *International Journal of Systematic Bacteriology* 38: 1–6

Guo, S., Li, J., Liu, Y., Song, F., and Zhang, J. 2011. The role of DNA binding with the Cry8Ea1 toxin of *Bacillus thuringiensis*. *FEMS Microbiology Letters* 317: 203–210

Hannay, C.L. 1957. The parasporal body of *Bacillus laterosporus* Laubach. *The Journal of Biophysical and Biochemical Cytology* 3: 1001–1010

Heckel, D.G. 2020. How do toxins from *Bacillus thuringiensis* kill insects? An evolutionary perspective. *Archives of Insect Biochemistry and Physiology* 104(2): e21673

Hernández-Martínez, P., Gomis-Cebolla, J., Ferré, J., and Escriche, B. 2017. Changes in gene expression and apoptotic response in *Spodoptera exigua* larvae exposed to sub-lethal concentrations of Vip3 insecticidal proteins. *Scientific Reports* 7: 16245, https://doi.org/10.1038/s41598-017-16406-1

Hurst, M.R., Beard, S.S., Jackson, T.A., and Jones, S.M. 2007. Isolation and characterization of the *Serratia entomophila* antifeeding prophage. *FEMS Microbiology Letters* 270: 42–48

Hurst, M.R., Beattie, A., Altermann, E., Moraga, R.M., Harper, L.A., Calder, J., and Laugraud, A. 2016. The draft genome sequence of the *Yersinia entomophaga* entomopathogenic type strain MH96T. *Toxins (Basel)* 8: 143

Hurst, M.R., Glare, T.R., Jackson, T.A., and Ronson, C.W. 2000. Plasmid-located pathogenicity determinants of *Serratia entomophila*, the causal agent of amber disease of grass grub, show similarity to the insecticidal toxins of *Photorhabdus luminescens*. *Journal of Bacteriology* 182: 5127–5138

Hurst, M.R., Jones, S.A., Binglin, T., Harper, L.A., Jackson, T.A., and Glare, T.R. 2011. The main virulence determinant of *Yersinia entomophaga* MH96 is a broad host-range toxin complex active against insects. *Journal of Bacteriology* 193: 1966–1980

Hurst, M.R., Van Koten, C., and Jackson, T.A. 2014. Pathology of *Yersinia entomophaga* MH96 towards *Costelytra zealandica* (Coleoptera; Scarabaeidae) larvae. *Journal of Invertebrate Pathology* 115: 102–107

Hurst, M.R.H., Beattie, A., Jones, S.A., Laugraud, A., Van Koten, C., and Harper, L. 2018. *Serratia proteamaculans* strain AGR96X encodes an antifeeding prophage (Tailocin) with activity against grass grub (*Costelytra giveni*) and manuka beetle (*Pyronota* spp.) larvae. *Applied and Environmental Microbiology* 84: e02739–17

Jackson, T.A. 2003. Environmental safety of inundative application of a naturally occurring biocontrol agent,

Serratia entomophila. In *Environmental Impacts of Microbial Insecticides: Need and Methods for Risk Assessment*, Hokkanken H. and Hajek A. (eds.) pp. 169–176. Dordrecht: Kluwer Academic Publishers

Jackson, T.A. 2007. A novel bacterium for control of grass grub. In *Biological Control: A Global Perspective*, Vincent C., Goettel M.S., and Lazarovits G. (eds.) pp. 160–184. Cambridge, MA: CAB International

Jackson, T.A., Berry, C., and O'Callaghan, M. 2018. Bacteria. In *Ecology of Invertebrate Diseases*, Hajek A.E. and Shapiro-Ilan D. (eds.) pp. 287–326. Oxford, UK: John Wiley & Sons Ltd.

Jackson, T.A., Boucias, D.G., and Thaler, J.O. 2001. Pathobiology of amber disease, caused by *Serratia* spp., in the New Zealand grass grub, *Costelytra zealandica*. *Journal of Invertebrate Pathology* 78: 232–243

Jackson, T.A., Huger, A.M., and Glare, T.R. 1993. Pathology of amber disease in the New Zealand grass grub *Costelytra zealandica* (Coleoptera: Scarabaeidae). *Journal of Invertebrate Pathology* 61: 123–130

Jeong, H., Mun, H., Oh, H., Kim, S., Yang, K., Kim, I., and Lee, H. 2010. Evaluation of insecticidal activity of a bacterial strain, *Serratia* sp. EML-SE1 against diamondback moth. *Journal of Microbiology* 48: 541–545

Jiang, K., Hou, X.-y., Tan, T.-t. et al. 2018a. Scavenger receptor-C acts as a receptor for *Bacillus thuringiensis* vegetative insecticidal protein Vip3Aa and mediates the internalization of Vip3Aa via endocytosis. *PLoS Pathogens* 14: e1007347

Jiang, K., Hou, X., Han, L., Tan, T., Cao, Z., and Cai, J. 2018b. Fibroblast growth factor receptor, a novel receptor for vegetative insecticidal protein Vip3Aa. *Toxins* 10: 546.

Jurat-Fuentes, J.L., Heckel, D., and Ferré, J. 2021. Mechanisms of resistance to insecticidal proteins from *Bacillus thuringiensis*. *Annual Review of Entomology* 66: 121–140, https://doi.org/10.1146/annurev-ento-052620-073348

Jurat-Fuentes, J.L. and Jackson, T.A. 2012. Bacterial entomopathogens. In *Insect Pathology*, 2nd edn, Vega, F.E. and Kaya H.K. (eds.) pp. 265–349. Cambridge, MA, USA: Academic Press

Kleespies, R.G., Marshall, S.D., Schuster, C., Townsend, R.J., Jackson, T.A., and Leclerque, A. 2011. Genetic and electron-microscopic characterization of *Rickettsiella* bacteria from the manuka beetle, *Pyronota setosa* (Coleoptera: Scarabaeidae). *Journal of Invertebrate Pathology* 10: 206–211

Knowles, B.H. and Ellar, D.J. 1987. Colloid-osmotic lysis is a general feature of the mechanism of action of *Bacillus thuringiensis* delta-endotoxins with different insect specificities. *Biochimica et Biophysica Acta—General Subjects* 924: 509–518

Krska, D., Ravulapalli, R., Fieldhouse, R.J., Lugo, M.R., and Merrill, A.R. 2015. C3larvin toxin, an ADP-ribosyltransferase from *Paenibacillus larvae*. *Journal of Biological Chemistry* 290: 1639–1653

Kuraishi, T., Binggeli, O., Opota, O., Buchon, N., and Lemaitre, B. 2011. Genetic evidence for a protective role of the peritrophic matrix against intestinal bacterial infection in *Drosophila melanogaster*. *Proceedings of the National Academy of Science USA* 108: 15966–15971

Lanan, M.C., Rodrigues, P.A.P., Agellon, A., Jansma, P., and Wheeler, D.E. 2016. A bacterial filter protects and structures the gut microbiome of an insect. *The ISME Journal* 10: 1866–1876

Lecadet, M.M., Frachon, E., Dumanoir, V.C., Ripouteau, H., Hamon, S., Laurent, P., and Thiery, I. 1999. Updating the H-antigen classification of *Bacillus thuringiensis*. *Journal of Applied Microbiology* 86: 660–672

Leclerque, A. 2008. Reorganization and monophyly of the genus *Rickettsiella*: All in good time. *Applied and Environmental Microbiology* 74: 5263–5264

Lee, S.H., Dimock, K., Gray, D.A., Beauchemin, N., Holmes, K.V., Belouchi, M., Realson, J., and Vidal, S.M. 2003. Maneuvering for advantage: The genetics of mouse susceptibility to virus infection. *Trends in Genetics* 19: 447–457

Lewkowski, O. and Erler, S. 2019. Virulence of *Melissococcus plutonius* and secondary invaders associated with European foulbrood disease of the honeybee. *Microbiologyopen* 8: e00649

Lietze, V.-U., Schneider, G., Prompiboon, P., and Boucias, D.G. 2010. The detection of *Bacillus thuringiensis* in mass rearing of *Cactoblastis cactorum* (Lepidoptera: Pyralidae). *The Florida Entomologist* 93: 385–390

Liu, Y., Wang, Y., Shu, C., Lin, K., Song, F., Bravo, A., Soberón, M., and Zhang, J. 2018. Cry64Ba and Cry64Ca, two ETX/MTX2-type *Bacillus thuringiensis* insecticidal proteins active against hemipteran pests. *Applied and Environmental Microbiology* 84: e01996–e01917

Maciel-Vergara, G., Jensen, A.B., and Eilenberg, J. 2018. Cannibalism as a possible entry route for opportunistic pathogenic bacteria to insect hosts, exemplified by *Pseudomonas aeruginosa*, a pathogen of the giant mealworm *Zophobas morio*. *Insects* 9: 88

Marche, M.G., Camiolo, S., Porceddu, A., and Ruiu, L. 2018. Survey of *Brevibacillus laterosporus* insecticidal protein genes and virulence factors. *Journal of Invertebrate Pathology* 155: 38–43

Marche, M.G., Mura, M.E., Falchi, G., and Ruiu, L. 2017. Spore surface proteins of *Brevibacillus laterosporus* are involved in insect pathogenesis. *Scientific Reports* 7: 43805

Marrone, P.G. 2019. Pesticidal natural products—status and future potential. *Pest Management Science* 75: 2325–2340

Marshall, S.D.G., Townsend, R.J., Kleespies, R.G., Van Koten, C., and Jackson, T.A. 2017. An epizootic of *Rickettsiella* infection emerges from an invasive scarab pest outbreak following land use change in New Zealand. *Annals of Clinical Cytology and Pathology* 3: 1058–1060

Martin, P.A., Gundersen-Rindal, D., Blackburn, M., and Buyer, J. 2007a. *Chromobacterium subtsugae* sp. nov., a betaproteobacterium toxic to Colorado potato beetle and other insect pests. *International Journal of Systematic and Evolutionary Microbiology* 57: 993–999

Martin, P.A., Hirose, E., and Aldrich, J.R. 2007b. Toxicity of *Chromobacterium subtsugae* to southern green stink bug (Heteroptera: Pentatomidae) and corn rootworm (Coleoptera: Chrysomelidae). *Journal of Economic Entomology* 100: 680–684

Milner, R.J. 1981. A novel milky disease organism from Australian scarabaeids: Field occurrence, isolation, and infectivity. *Journal of Invertebrate Pathology* 37: 304–309

Misof, B., Liu, S., Meusemann, K., Peters, R.S., Donath, A. et al. 2014. Phylogenomics resolves the timing and pattern of insect evolution. *Science* 346: 763–767

Müller, S., Garcia-Gonzalez, E., Genersch, E., and Süssmuth, R.D. 2015. Involvement of secondary metabolites in the pathogenesis of the American foulbrood of honey bees caused by *Paenibacillus larvae*. *Natural Product Reports* 32: 765–778

Narva, K.E., Wang, N.X., and Herman, R. 2017. Safety considerations derived from Cry34Ab1/Cry35Ab1 structure and function. *Journal of Invertebrate Pathology* 142: 27–33

Nascimento, N.A., Torres-Quintero, M.C., Molina, S.L. et al. 2020. Functional *Bacillus thuringiensis* Cyt1Aa is necessary to synergize *Lysinibacillus sphaericus* binary toxin (Bin) against Bin-Resistant and -refractory mosquito species. *Applied and Environmental Microbiology* 86: e02770–02719

Nicolas, L., Hamon, S., Frachon, E., Sebald, M., and De Barjac, H. 1990. Partial inactivation of the mosquitocidal activity of *Clostridium bifermentans* serovar *malaysia* by extracellular proteinases. *Applied Microbiology and Biotechnology* 34: 36–41

Nicolas, L., Nielsen-Leroux, C., Charles, J.F., and Delécluse, A. 1993. Respective role of the 42- and 51-kDa components of the *Bacillus sphaericus* toxin overexpressed in *Bacillus thuringiensis*. *FEMS Microbiology Letters* 106: 275–279

Nishitsutsuji-Uwo, J. and Endo, Y. 1980. Mode of action of *Bacillus thuringiensis* endotoxn: Relative roles of spores and crystals in toxicity to *Pieris*, *Lymantria* and *Ephestia* larvae. *Applied Entomology and Zoology* 15: 416–424

Onofre, J., Pacheco, S., Torres-Quintero, M.C., Gill, S.S., Soberón, M., and Bravo, A. 2020. The Cyt1Aa toxin from *Bacillus thuringiensis* inserts into target membranes via different mechanisms in insects, red blood cells, and lipid liposomes. *Journal of Biological Chemistry* 295: 9606–9617

Opota, O., Vallet-Gély, I., Vincentelli, R. et al. 2011. Monalysin, a novel ß-pore-forming toxin from the *Drosophila* pathogen *Pseudomonas entomophila*, contributes to host intestinal damage and lethality. *PLoS Pathogens* 7: e1002259.

Parks, D.H., Chuvochina, M., Chaumeil, P.-A., Rinke, C., Mussig, A.J., and Hugenholtz, P. 2020. A complete domain-to-species taxonomy for Bacteria and Archaea. *Nature Biotechnology* 38(9): 1079–1086, https://doi.org/10.1038/s41587-020-0501-8

Paul, B., Kavia Raj, K., Murali, T.S., and Satyamoorthy, K. 2020. Species-specific genomic sequences for classification of bacteria. *Computers in Biology and Medicine* 123: 103874

Pérez, C., Muñoz-Garay, C., Portugal, L.C., Sánchez, J., Gill, S.S., Soberón, M., and Bravo, A. 2007. *Bacillus thuringiensis* ssp. *israelensis* Cyt1Aa enhances activity of Cry11Aa toxin by facilitating the formation of a pre-pore oligomeric structure. *Cellular Microbiology* 9: 2931–2937

Pineda-Castellanos, M.L., Rodriguez-Segura, Z., Villalobos, F.J., Hernandez, L., Lina, L., and Nunez-Valdez, M.E. 2015. Pathogenicity of isolates of *Serratia marcescens* towards larvae of the scarab *Phyllophaga blanchardi* (Coleoptera). *Pathogens* 4: 210–228

Piper, S.J., Brillault, L., Rothnagel, R., Croll, T.I., Box, J.K., Chassagnon, I., Scherer, S., Goldie, K.N., Jones, S.A., Schepers, F., Hartley-Tassell, L., Ve, T., Busby, J.N., Dalziel, J.E., Lott, J.S., Hankamer, B., Stahlberg, H., Hurst, M.R.H., and Landsberg, M.J. 2019. Cryo-EM structures of the pore-forming A subunit from the *Yersinia entomophaga* ABC toxin. *Nature Communications* 10: 1952

Poppinga, L., Janesch, B., Funfhaus, A. et al. 2012. Identification and functional analysis of the S-layer protein SplA of *Paenibacillus larvae*, the causative agent of American Foulbrood of honeybees. *PLoS Pathogens* 8: e1002716

Prasanna, L., Eijsink, V.G.H., Meadow, R., and Gåseidnes, S. 2013. A novel strain of *Brevibacillus laterosporus* produces chitinases that contribute to its biocontrol potential. *Applied Microbiology and Biotechnology* 97: 1601–1611

Qi, L., Qiu, X., Yang, S., Li, R., Wu, B., Cao, X., He, T., Ding, X., Xia, L., and Sun, Y. 2020. Cry1Ac protoxin and its activated toxin from *Bacillus thuringiensis* act differently during the pathogenic process. *Journal of Agricultural and Food Chemistry* 68: 5816–5824

Qureshi, N., Chawla, S., Likitvivatanavong, S., Lee, H.L., and Gill, S.S. 2014. The cry toxin operon of *Clostridium bifermentans* subsp. *malaysia* is highly toxic to *Aedes* larval mosquitoes. *Applied and Environmental Microbiology* 80: 5689–5697

Ramirez, J.L., Short, S.M., Bahia, A.C., Saraiva, R.G., Dong, Y., Kang, S., Tripathi, A., Mlambo, G., and Dimopoulos, G. 2014. *Chromobacterium* Csp_P reduces malaria and dengue infection in vector mosquitoes and has entomopathogenic and *in vitro* anti-pathogen activities. *PLoS Pathogens* 10(10): e1004398, https://doi.org/10.1371/journal.ppat.1004398

Raymond, B., Davis, D., and Bonsall, M.B. 2007. Competition and reproduction in mixed infections of pathogenic and non-pathogenic *Bacillus* spp. *Journal of Invertebrate Pathology* 96: 151–155

Raymond, B., Johnston, P.R., Nielsen-Leroux, C., Lereclus, D., and Crickmore, N. 2010. *Bacillus thuringiensis*: An impotent pathogen? *Trends in Microbiology* 18: 189–194

Raymond, B., West, S.A., Griffin, A.S., and Bonsall, M.B. 2012. The dynamics of co-operative bacterial virulence in the field. *Science* 337: 85–88

Rees, J.S., Jarrett, P., and Ellar, D.J. 2009. Peritrophic membrane contribution to Bt Cry delta-endotoxin susceptibility in Lepidoptera and the effect of calcofluor. *Journal of Invertebrate Pathology* 100: 139–146

Reyes-Ramírez, A. and Ibarra, J.E. 2005. Fingerprinting of *Bacillus thuringiensis* type strains and isolates by using *Bacillus cereus* group-specific repetitive extragenic palindromic sequence-based PCR analysis. *Applied and Environmental Microbiology* 71: 1346–1355

Ruiu, L. 2013. *Brevibacillus laterosporus*, a pathogen of invertebrates and a broad-spectrum antimicrobial species. *Insects* 4: 476–492

Ruiu, L. 2015. Insect pathogenic bacteria in integrated pest management. *Insects* 6: 352–367

Ruiu, L., Floris, I., Satta, A., and Ellar, D.J. 2007. Toxicity of a *Brevibacillus laterosporus* strain lacking parasporal crystals against *Musca domestica* and *Aedes aegypti*. *Biological Control* 43: 136–143

Ruiu, L., Satta, A., and Floris, I. 2012. Observations on house fly larvae midgut ultrastructure after *Brevibacillus laterosporus* ingestion. *Journal of Invertebrate Pathology* 111: 211–216

Salazar-Marroquín, E.L., Galán-Wong, L.J., Moreno-Medina, V.R., Reyes-López, M.Á., and Pereyra-Alférez, B. 2016. Bacteriocins synthesized by *Bacillus thuringiensis*: Generalities and potential applications. *Reviews in Medical Microbiology* 27: 95–101

Sampson, K., Zaitseva, J., Stauffer, M., Vande Berg, B., Guo, R., Tomso, D., McNulty, B., Desai, N., and Balasubramanian, D. 2017. Discovery of a novel insecticidal protein from *Chromobacterium piscinae*, with activity against Western corn rootworm, *Diabrotica virgifera*. *Journal of Invertebrate Pathology* 142: 34–43

Sampson, M.N. and Gooday, G.W. 1998. Involvement of chitinases of *Bacillus thuringiensis* during pathogenesis in insects. *Microbiology* 144: 2189–2194

Schmidt, H. and Hensel, M. 2004. Pathogenicity islands in bacterial pathogenesis. *Clinical Microbiology Reviews* 17: 14–56

Sharma, M., Aswal, V.K., Kumar, V., and Chidambaram, R. 2020. Small-angle neutron scattering studies suggest the mechanism of BinAB protein internalization. *International Union of Crystallography Journal* 7: 166–172

Silva-Filha, M.H., Nielsen-Leroux, C., and Charles, J.F. 1999. Identification of the receptor for *Bacillus sphaericus* crystal toxin in the brush border membrane of the mosquito *Culex pipiens* (Diptera: Culicidae). *Insect Biochemistry and Molecular Biology* 29: 711–721

Silva-Filha, M.H.N.L. 2017. Resistance of mosquitoes to entomopathogenic bacterial-based larvicides: Current status and strategies for management. In *Bacillus thuringiensis and Lysinibacillus sphaericus: Characterization and use in the Field of Biocontrol*. Fiuza, L.M. Polanczyk, R.A., and Crickmore, N. (eds.) pp. 239–257. Cham: Springer International Publishing

Singh, G., Sachdev, B., Sharma, N., Seth, R., and Bhatnagar, R.K. 2010. Interaction of *Bacillus thuringiensis* vegetative insecticidal protein with ribosomal S2 protein triggers larvicidal activity in *Spodoptera frugiperda*. *Applied and Environmental Microbiology* 76: 7202–7209

Soberón, M., López-Díaz, J.A., and Bravo, A. 2013. Cyt toxins produced by *Bacillus thuringiensis*: A protein fold conserved in several pathogenic microorganisms. *Peptides* 41: 87–93

Soberón, M., Pardo-López, L., López, I., Gómez, I., Tabashnik, B.E., and Bravo, A. 2007. Engineering modified Bt toxins to counter insect resistance. *Science* 318: 1640–1642

Splittstoesser, C.M., Tashiro, H., Lin, S.L., Steinkraus, K.H., and Fiori, B.J. 1973. Histopathology of the European chafer, *Amphimallon majalis*, infected with *Bacillus popilliae*. *Journal of Invertebrate Pathology* 22: 161–167

Surya, W., Chooduang, S., Choong, Y.K., Torres, J., and Boonserm, P. 2016. Binary toxin subunits of *Lysinibacillus sphaericus* are monomeric and form heterodimers after *in vitro* activation. *PLoS One* 11: e0158356

Tetreau, G., Banneville, A.S., Andreeva, E.A. et al. 2020. Serial femtosecond crystallography on *in vivo*-grown crystals drives elucidation of mosquitocidal Cyt1Aa bioactivation cascade. *Nature Communications* 11(1): 1153, https://doi.org/10.1038/s41467-020-14894-w

Thiery, I., Hamon, S., Gaven, B., and De Barjac, H. 1992. Host range of *Clostridium bifermentans* serovar.

malaysia, a mosquitocidal anaerobic bacterium. *Journal of the American Mosquito Control Association* 8: 272–277

Vodovar, N., Vinals, M., Liehl, P., Basset, A., Degrouard, J., Spellman, P., Boccard, F., and Lemaitre, B. 2005. *Drosophila* host defense after oral infection by an entomopathogenic *Pseudomonas* species. *Proceedings of the National Academy of Sciences of the United States of America* 102: 11414–11419

Vöing, K., Harrison, A., and Soby, S.D. 2015. Draft genome sequence of *Chromobacterium vaccinii*, a potential biocontrol agent against mosquito (*Aedes aegypti*) larvae. *Genome Announcements* 3(3): e00477–15, https://doi.org/10.1128/genomeA.00477-15

Wang, K., Shu, C., Soberón, M., Bravo, A., and Zhang, J. 2018. Systematic characterization of *Bacillus* Genetic Stock Center *Bacillus thuringiensis* strains using Multi-Locus Sequence Typing. *Journal of Invertebrate Pathology* 155: 5–13

Waterfield, N.R., Daborn, P.J., and Ffrench-Constant, R.H. 2004. Insect pathogenicity islands in the insect pathogenic bacterium *Photorhabdus*. *Physiological Entomology* 29: 240–250

Wayne, L.G., Brenner, D.J., Colwell, R.R., Grimont, P.A.D., Kandler, O., Krichevsky, M.I., Moore, L.H., Moore, W.E.C., Murray, R.G.E., Stackebrandt, E., Starr, M.P., and Truper, H.G. 1987. Report of the ad-hoc-committee on reconciliation of approaches to bacterial systematics. *International Journal of Systematic Bacteriology* 37: 463–464

Wirth, M.C., Park, H.W., Walton, W.E., and Federici, B.A. 2005. Cyt1A of *Bacillus thuringiensis* delays evolution of resistance to Cry11A in the mosquito *Culex quinquefasciatus*. *Applied and Environmental Microbiology* 71: 185–189

Wirth, M.C., Yang, Y., Walton, W.E., Federici, B.A., and Berry, C. 2007. Mtx toxins synergize *Bacillus sphaericus* and Cry11Aa against susceptible and insecticide-resistant *Culex quinquefasciatus* larvae. *Applied and Environmental Microbiology* 73: 6066–6071

Yokoyama, T., Tanaka, M., and Hasegawa, M. 2004. Novel cry gene from *Paenibacillus lentimorbus* strain Semadara inhibits ingestion and promotes insecticidal activity in *Anomala cuprea* larvae. *Journal of Invertebrate Pathology* 85: 25–32

Yu, C.G., Mullins, M.A., Warren, G.W., Koziel, M.G., and Estruch, J.J. 1997. The *Bacillus thuringiensis* vegetative insecticidal protein Vip3A lyses midgut epithelium cells of susceptible insects. *Applied and Environmental Microbiology* 63: 532–536

Yu, J., Tan, L., Liu, Y., and Pang, Y. 2002. Phylogenetic analysis of *Bacillus thuringiensis* based on PCR amplified fragment polymorphisms of flagellin genes. *Current Microbiology* 45: 139–143

Yue, D., Nordhoff, M., Wieler, L.H., and Genersch, E. 2008. Fluorescence in situ hybridization (FISH) analysis of the interactions between honeybee larvae and *Paenibacillus larvae*, the causative agent of American foulbrood of honeybees (*Apis mellifera*). *Environmental Microbiology* 10: 1612–1620

Zaitseva, J., Vaknin, D., Krebs, C. et al. 2019. Structure–function characterization of an insecticidal protein GNIP1Aa, a member of an MACPF and β-tripod families. *Proceedings of the National Academy of Sciences* 116: 2897–2906

Zalunin, I.A., Elpidina, E.N., and Oppert, B. 2015. The role of proteolysis in the biological activity of Bt insecticidal crystal proteins. In *Bt Resistance: Characterization and Strategies for GM Crops Producing Bacillus thuringiensis Toxins*. Soberón, M., Gao, A., and Bravo A. (eds.) pp. 107–118. Wallingford, Oxfordshire, England: CABI International

Zhang, J., Hodgman, T.C., Krieger, L., Schnetter, W., and Schairer, H.U. 1997. Cloning and analysis of the first cry gene from *Bacillus popilliae*. *Journal of Bacteriology* 179: 4336–4341

Zubasheva, M.V., Ganushkina, L.A., Smirnova, T.A., and Azizbekyan, R.R. 2010. Larvicidal activity of crystal-forming strains of *Brevibacillus laterosporus*. *Applied Biochemistry and Microbiology* 46: 755–762

CHAPTER 12

Fungal and oomycete diseases of insects

Almudena Ortiz-Urquiza

12.1 Introduction

The white muscardine disease or 'mal del segno' in silkworms (*Bombyx mori*), caused by the fungus *Beauveria bassiana* (Balsamo) Vuillemin, was one the first fungal diseases to be characterised in insects. This fungal pathogen, along with other insect pathogens (e.g. microsporidians and viruses), constituted the breeding ground of the socioeconomic changes that led to the decline of the European sericulture industry in the nineteenth century. The identification of *B. bassiana* as the causative agent of the white muscardine disease in silkworms and the experimental work on the transmission of this infectious agent from infected insect to healthy insect helped scientists like Agostino Bassi (1773–1856) and Louis Pasteur (1822–1895) to overthrow the spontaneous generation theory and champion the germ theory, which later extrapolated to human diseases. In fact, Bassi was the first to prove that an infectious agent can be transmitted from animal to animal through the inoculation of contagious structures.

Agostino Bassi is considered the father of insect pathology due to his studies on the muscardine disease in the silkworm published in 1835 under the title *Del Mal del Segno, Calcinaccio o Moscardino*. In his manuscript, Bassi reported that the disease affecting the silkworm was caused by a vegetable parasite, which later was named *Botrytis bassiana* in honour of Bassi (currently *Beauveria bassiana*) by the naturalist Giuseppe Balsamo-Crivello (1800–1874) (Mazzarello et al. 2013). Nevertheless, fungal diseases

affecting insects were observed prior to Bassi's work. The entomologists William Kirby (1750–1850) and William Spense (1783–1860) published in 1826 that insects were natural hosts of 'parasitic assailants'—now classified within the Kingdom Fungi (Kirby and Spense 1826; Steinhaus 1956). These initial reports, together with early observations regarding the occurrence of natural insect epizootics, assisted to devise the idea of microbial pest control and the belief that causative agents of insect diseases could be used for pest control. The concept of using microbial pest control was proposed in several instances by Agostino Bassi, Louise Pasteur, and the American entomologist John L. LeConte (1825–1883) and pioneered in the field by Hermann A. Hagen (1817–1893) and Elie Metchnikoff (1845–1914), among others (Steinhaus 1956). Today, fungus-(and oomycete-) based pesticides account for *ca.* 15% of the biopesticide market, with their compound annual growth rate continuing to rise (Mishra et al. 2014).

The aim of this chapter is to provide an up-to-date taxonomic context for insect pathogenic fungi and oomycetes. It also discusses the ecological diversity, the evolutionary history and the varied infection strategies and host-ranges of these insect pathogens.

12.2 Insect-fungal/oomycete interactions

With different ecological adaptations as obligate or facultative parasites, insect pathogenic fungi and oomycetes regulate insect populations naturally. In

Almudena Ortiz-Urquiza, *Fungal and oomycete diseases of insects*. In: *Invertebrate Pathology*. Edited by Andrew F. Rowley, Christopher J. Coates and Miranda M.A. Whitten, Oxford University Press. © Oxford University Press (2022). DOI: 10.1093/oso/9780198853756.003.0012 this scenario, the host and parasite densities are inter-dependent so, in general, the incidence of the parasitic disease diminishes as hosts become sparse (Fisher et al. 2012). However, facultative pathogens (e.g. facultative fungi and oomycetes) can survive outside the host as saprophytes and this lifestyle makes them less dependent on host density (Naranjo-Ortiz and Gabaldón 2019a). Thus, facultative insect pathogens can cause greater mortalities in the insect populations than obligate pathogens.

Fungal and oomycete diseases can also affect insect behaviour and fitness. Changes in host behaviour usually facilitate the spread of the disease (Roy et al. 2006: Shang et al. 2015). For instance, species of the complex *Ophiocordyceps unilateralis* s.l. cause ant workers to leave the nest and start wandering in search of an optimum microclimate site, where the fungus can grow and sporulate. Before dying, the ants climb onto the vegetation about 25 cm above the ground, and grip on a leaf, twig, or branch by biting it firmly (Evans et al. 2011; de Bekker et al. 2014; Figure 12.1). The ant will stay attached to the vegetation after its death due to a major atrophy of the mandibles, most likely caused by extensive fungal colonisation of these tissues (Hughes et al. 2011). After the ant's death, the fungus will grow a stroma from the back of the ant's head, and perithecia containing ascospores will develop on one side of the stroma (Figure 12.1). Sometimes, however, infections originate behavioural changes in the host aimed to suppress the pathogen. Examples of these are the behavioural fever (i.e. increase of body temperature) in the desert locust, *Schistocerca gregaria*, in response to *Metarhizium acridum* infection, sanitary care in social insects (i.e. allogrooming or avoidance of the nest if infected), and premature moulting and burrowing in larvae (Oi and Pereira 1993; Ortiz-Urquiza and Keyhani 2013; Fan et al. 2012; Shang et al. 2015). Generally, fungal and oomycete infections decrease fitness in insects, with reduced lifespan, fecundity, and fertility observed in various insect species upon infection with sublethal concentrations (Quesada-Moraga et al. 2006; Scholte et al. 2006; Ondiaka et al. 2015).

Insect pathogenic fungi and oomycetes infect their host via penetration of the cuticle, although oral infection can be an alternate route for some species (Araújo and Hughes 2016). These pathogens use a combination of mechanical forces and hydrolytic and assimilatory enzymes to breach the insect integument and exploit the insect resources within the haemocoel including carbohydrates like trehalose. The ability to secrete chitinases and proteases to digest the procuticle, consisting of microfibres of chitin embedded in a matrix of proteins, is a common feature of pathogenic fungi and oomycetes and is central for successful infection (St Leger et al. 1987; Screen et al. 2001; Humber 2008). In addition, other factor contributing to entomopathogenicity in fungi and oomycetes is the production of trehalase, enzymes that catalyse the hydrolysis of trehalose into glucose (Zhao et al. 2006; Olivera et al. 2016; Litwin et al. 2020). Trehalose is the most abundant carbohydrate in the insect haemolymph, where this disaccharide fuels processes requiring high metabolic rates (e.g. insect flight) and buffers glucose fluctuation by acting as a glucose sink (Matsushita and Nishimura 2020). In the entomopathogenic fungus, *M. acridum*, the acidic trehalase ATM1 has been reported to contribute to fungal proliferation in the insect haemocoel and virulence (Jin et al. 2015).

Once inside the haemocoel, some fungi like *Metarhizium* spp. and *Beauveria* spp. (Hypocreales, Ascomycota) undergo a dimorphic transition and grow as yeast-like cells namely hyphal bodies, other fungi (e.g. species within Entomophthoromycota) produce protoplasts (viz. cells without cell walls), and some species of oomycetes and chytrids show hyphal growth when proliferating within the haemocoel (Araújo and Hughes 2016). As the haemocoel and internal organs are invaded, fungi and oomycetes work their way out of the insect and sporulate on its surface. Some entomopathogenic fungal species, particularly those within the phylum Entomophthoromycotina, sporulate while the host is still alive. However, most species of entomopathogenic fungi and all species of insect pathogenic oomycetes kill their host before sporulating on them (Goettel et al. 2005). Sporulating insects or insect corpses facilitate the dispersal of the spores and secondary infections (Figure 12.2).

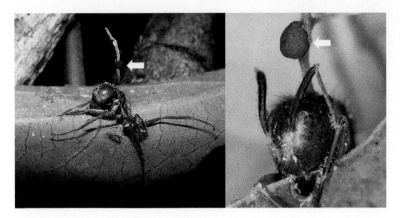

Figure 12.1 Ants infected with *Ophiocordyceps unilateralis* (complex). The so-called 'zombie ant' is biting (death grip) into vegetation and the fungal stroma arises from the dorsum. White arrows point to the ascoma. Specimens depicted were collected in Mata do Paraiso—a small forest reserve belonging to the University of Viçosa in Minas Gerais. Source: Harry C. Evans (Centre for Agriculture and Bioscience International, Surrey UK) provided the original images.

12.3 Insect pathogenic fungi - diversity and ecology

Fungi participate in important ecological processes such as decomposition and nutrient cycling and can be found living in long-term associations with plants, algae, animals, and other organisms. Most fungal species produce hyphae; cylindrical multinucleated cells with a common cytoplasm encased in a chitinous cell wall. Fungal hyphae, which operate like a single harmonised unit, are often compartmentalised by perforated septa (Richards et al. 2017). Having a hyphal thallus that exhibits indefinite apical growth allows fungi to search for nutrients. In addition, fungi possess a refined secretome enabling them to obtain nutrients from highly polymerised and hydrophobic compounds that are not easily assimilable by other microbes (e.g. cellulose, lignin, long chain hydrocarbons) (Pedrini et al. 2013; Boddy and Hiscox 2017; Hiscox, O'Leary and Boddy 2018; Naranjo-Ortiz and Gabaldón 2019a). The large number of genes encoding for hydrolytic and assimilatory enzymes and membrane transporters in fungi is reminiscent of the ancestral parasitic lifestyle of fungi, involving breaking into other organisms to plunder nutrients (Naranjo-Ortiz and Gabaldón 2019a).

Apart from some parasitic species, most of the extant fungi are terrestrial saprophytes which feed on organic substrate by osmotrophy (i.e. uptake of dissolved organic compound by osmosis) (Torruella et al. 2018). However, ecologically speaking, the lifestyle of ancestral fungi, including the last common fungal ancestral (LCFA) was very different.

Omics analyses of marine and terrestrial organisms and pre-date fossil evidence of terrestrial biota suggest that LCFA was likely a marine parasite of microalgae, with phagotrophic capabilities (i.e. feeding by engulfing) and both amoeboid and flagellar mobility (Del Campo et al. 2015; Richards et al. 2017; Spatafora et al. 2017; Torruella et al. 2018).

Pathogenicity towards insects has arisen several times during fungal evolution yielding either specialised or broad host range insect pathogens (Ortiz-Urquiza and Keyhani 2013; Zheng et al. 2013). Broad host range in insect pathogenic fungal species is associated to the loss of sexuality and genome features (e.g. genome restructuring, evidence of multiple independent events of lateral gene transfer and protein family expansions) that enable compatible interactions with a large number of hosts (i.e. a generalist lifestyle). Specialised fungal pathogens, on the other hand, are characterised by fast-evolving proteins and retention of sexual reproduction (Hu et al. 2014; Wang et al. 2016; Zhang et al. 2019). With different strategies to colonise the insect body, insect pathogenic fungi infect their hosts (and also disseminate in the environment) via sexual and asexual spores. Sexual spores can be motile (e.g. zoospores—present in zoosporic fungi) and non-motile. Asexual spores or mitotic spores are called conidia regardless of taxon and are normally released passively. Infection starts by adhesion of a spore (sexual or asexual) to the outermost layer of the insect cuticle or epicuticle, which is rich in lipids (Nelson et al. 1995; de Renobales et al. 1991; see Section 1.10 and Chapter 1, Figure 1.2 for further details of this process). Adhesion to the

Figure 12.2 Entomopathogenic fungi and insect cadavers. (a) *Beauveria* cf. *bassiana* (Ascomycota, Cordycipitaceae) infection of a weevil. (b) *Metarhizium* sp. (Ascomycota, Clavicipitaceae) infection of a coleopteran. (c) *Erynia* sp. (Entomophthoramycota, Entomophthoraceae) infection of a dipteran. (d) *Ophiocordyceps* cf. *gracilis* infection of a ghost moth larva (Hepialidae). The white arrow points to an emerging perithecial neck containing ascospores. (e) *Hirsutella* sp. (now classified within the family Ophicordycipitaceae) infection of a dipteran. Specimens depicted in panels (a–e) were collected in Mata do Paraiso—a small forest reserve belonging to the University of Viçosa in Minas Gerais. (f) *Entomophthora* cf. *muscae* (Entomophthoramycota, Entomophthoraceae) of a dipteran (Calliphoridae)—observed in an allotment in the UK. Source: Harry C. Evans (Centre for Agriculture and Bioscience International, Surrey UK) kindly provided the original images.

epicuticle appears to be controlled by a two-step process, encompassing non-specific passive adsorption of fungal spores followed by the consolidation of the attachment (Boucias et al. 1988; Boucias and Pendland 1991; Ortiz-Urquiza and Keyhani 2013).

After spore attachment, interaction with the insect epicuticle triggers the expression of assimilatory and hydrolytic enzymes and other factors that promotes germination of the spores and penetration of the insect integument (Ortiz-Urquiza and Keyhani

2013; Ortiz-Urquiza and Keyhani 2016). Once inside the haemocoel, insect pathogenic fungi encounter the host's defences. If they overcome the cellular and humoral insect responses, depending on the species, fungi will proliferate within the haemocoel as hyphal filaments, yeast-like cells, or protoplast (Araújo and Hughes 2016; see Section 1.10 in Chapter 1 for a description of immune avoidance/suppression in fungi).

12.4 Taxonomy of fungal pathogens of insects

Fungi are the one of the largest and most diverse group of eukaryotes (Choi and Kim 2017), with about 2.2 to 3.8 million species, of which only *ca.* 120,000 species have been named to date (Hawksworth and Lücking 2017). Most recent classifications of fungi describe nine phyla, with four of them belonging to the group of zoosporic fungi (i.e. Cryptomycota, Microsporidia, Chytridyomycota, and Blastoclamycota), and the other five being part of the non-flagellated fungi (viz., Zoopagomycota, Mucoromycota, Glomeromycota, Ascomycota, and Basidiomycota) (Spatafora et al. 2017; Ahrendt et al. 2018).

12.4.1 Phylum Cryptomycota (or Rozellomycota)

To date, no insect pathogenic species have been described within the Phylum Cryptomycota. This phylum includes only a handful of described taxa (e.g. *Rozella* spp.) and taxa derived from environmental sapling (Spatafora et al. 2017). *Rozella* is the best-known genus of Cryptomycota and is commonly found in environmental samples.

12.4.2 Phylum Microsporidia

With approximately 1,300 species described and grouped into 160 genera, 93 microsporidian genera are insect pathogens (Becnel and Andreadis 2014). Microsporidia are obligate intracellular parasites of protists and animals, including vertebrates and invertebrates. The phylogenetic placement of Microsporidia within fungi has been a controversial topic since the discovery of microsporidians in 1857 by Carl Wilhelm von Nägeli (1817–1891). In the past,

they have been placed in many polyphyletic groups. They have been classified as Schizomycete fungi (i.e. an artificial group that included yeast and bacteria), Sporozoa, and Cnidosporidia (Keeling 2009). Phylogenetic analyses place Microsporidia as a sister group of the phylum Cryptomycota within the Kingdom Fungi (Keeling and Doolittle 1996; Hirt et al. 1999; Bouzat et al. 2000; James et al. 2006a; Hibbett et al. 2007). The scientific community has agreed to keep microsporidians under the classification of the Code of Zoological Nomenclature, despite being fungi whose nomenclature follows the Botanical Code (Weiss 2005; Redhead et al. 2009). Classification of Microsporidia within the Botanical Code would entail to re-assign formal botanical names to approximate 1,000 species of microsporidians that were named based on the assumption that they were protists. Microsporidia have small genomes with fast-rate nucleotide substitutions and lack Golgi apparatus and mitochondria (Keeling et al. 2005; Corradi and Keeling 2009; Solter et al. 2012). The latter makes Microsporidia depend on the host to obtain ATP, which is imported from the host into the microsporidial cell or the vestigial mitochondria, the mitosomes (Williams et al. 2008; Tsaousis et al. 2008).

Insects are the type host of ~ 90–93 microsporidian genera, including *Nosema* spp. Unlike the great majority of entomopathogenic fungi, which infect via penetration of the cuticle, microsporidians infect insects primary by cannibalism of infected animals or ingestion of spores disseminated in the environment from faeces. Additional infecting routes include transovarial transmission (i.e. Microsporidia infect the eggs as ovaries and associated structures in the female are infected) and parasitoid oviposition activities (Solter et al. 2012). Venereal transmission in insect adults through Microsporidia-infected males does not play a major role in the horizontal transmission of the pathogen (Solter et al. 1991; Goertz et al. 2007). Upon ingestion, spores germinate extending a polar tube that punctures the midgut epithelial cells and injects the spore content into the cytoplasm of the host cell, where the infection is circumspect. Microsporidia can cause systemic infection or display tissue tropism, with some species only infecting the fat body or the gut. Symptoms associated with microsporidian infection encompasses tissue lesions (i.e. dark spots on the larval integument also called pepper (pe'brine) disease (Pasteur 1870), abnormal development (e.g. stunted growth, deformed wings), physiological effects (e.g. short-ened lifespan, decrease fertility), and altered behaviour (Becnel and Andreadis 2014; Mishra et al. 2018). Severe infections result in targeted cells filled with spores that lead to host death (Solter et al. 2012).

Two major clades in Microsporidia group most of the insect pathogenic microsporidians, name-ly Amblyospora/Parathelohania and Nosema/Vairimorpha clades. The Amblyospora/Parathe-lohania clade encompasses the greatest number of mosquito pathogens, 122 of the approximately 150 microsporidian species described as mosquito pathogens. Microsporidia within this clade/group use an intermediate host, a copepod, to complete their life cycles, although this does not apply to all the species within the clade (e.g. *Edhazardia aedis* and *Culicospora magna*) (Andreadis 2007). Species of *Amblyospora* and *Parathelohania* are host-specific, with *Amblyospora* spp. chiefly associated to *Aedes* and *Culex* mosquitoes and *Parathelohania* spp. with *Anopheles* spp. (Solter et al. 2012). The Nosema/Vairimorpha clade are pathogens of ter-restrial insects, particularly Lepidoptera, although notably examples in Hymenoptera include *Nosema vespula* (in wasps), *N. ceranae* and *N. apis* (in hon-eybees; Box 12.1), *N. bombi* (in bumble bees), and *Vairimorpha invictae* (in fire ants).

12.4.3 Phylum Chytridiomycota *sensu lato*

Chrytidiomycota *sensu lato* includes the rede-fined phyla Chitridiomycota *sensu stricto*, Blatocladiomycota, Monoblepharidomycota, and Neocallimastogomycota (Radek et al. 2017). No insect pathogenic species have been described amongst the Monoblepharidomycota and Neocal-limastogomycota. Within Chytriodyomycota *sensu lato*, one genus, *Nephridiophaga* has been described as an insect parasite. *Nephridiophaga* species are unicellular chytrids that infect the Malpighian tubules of insects. In particular, the species *N. maderae* and *N. blatellae* are parasites of cockroaches (Radek et al. 2017). One genus of the phylum Chytridiomycota *sensu stricto*, *Myiophagus*, has been reported as an insect pathogen. *Myiophagus* species have been reported infecting scale insects and pupae of Diptera (Petch 1948; Karling 1948; Muma and Clancy 1961).

Three genera, *Catenaria*, *Coelomycidium*, and *Coelo-momyces*, in the phylum Blastocladiomycota include insect pathogenic species (Vega et al. 2012). *Catenar-ia* species are primary known for infecting nema-todes but some species also can infect small flies and coleopteran larvae (Vega et al. 2012; Araújo and Hughes 2016). Species of *Coelomycidium* can attack scale insects, larvae of Coleoptera, and larvae of Diptera (i.e. black flies) (Tanada and Kaya 2012; Jitklang et al. 2012). *Coelomomyces* species are obli-gate parasites that often require two hosts (e.g. a mosquito larva and a copepod) to complete their life cycle (Scholte et al. 2004). *Coelomomyces* spp. infect mostly aquatic Diptera, including Culicidae, Phy-chodidae, Chironomidae, Simuliidae, and Taban-idae (Leão and Pedroso 1965; Roberts 1970; Chap-man 1974; Scholte et al. 2004). *Coelomomyces* is high-ly pathogenic of mosquitoes and is found infect-ing egg, larvae, and adult mosquitoes, although larval infection is the most common type (Araújo

---

> ## Box 12.1 *Nosema*-Insect Antibiosis
>
> Nosemosis in beehives and colonies is an important condi-tion. Bees encounter spores through the ingestion of feeds, and epithelial cells of the GI tract become infected lead-ing to dysentery. Infected bees may show lethargy and foraging flights become shortened. Colonies contaminat-ed with *Nosema* spp. are substantially smaller (reduced brood rearing) and produce less honey. The faecal drop-pings from infected bees contain spores that remain viable for in excess of one year, providing an environmental source of further infection. *N. ceranae* causes immunosuppression in *Apis mellifera* (Li et al. 2018) and *Melipona colimana* (Macias-Macias et al. 2020).

and Hughes 2016). Usually, larvae infected with *Coelomomyces* (e.g. *C. psorophorae and C. stegomyiae* var. *stegomyiae*) die and form resting spores (i.e. sporangia), although occasionally larvae survive and after pupation, infected adults emerge. During ovary maturation in females, hyphae migrate from the haemocoel into the ovaries and after the blood meal, these hyphae form resting sporangia. Infected female adult mosquitoes mate but no eggs are laid. Instead, resting sporangia are disseminated in every attempt of oviposition (Lucarotti 1992; Lucarotti and Shoulkamy 2000). Such naïve action ensures the introduction and the constant presence of the pathogen in breeding sites (e.g. tree holes, water containers).

12.4.4 Phylum Zoopagomycotina

Originally classified as Zygomycota, based on sexual reproduction that occurs through zygospores, this phylum represents the first-divergent species of non-flagellated fungi (Hibbett et al. 2007; Spatafora et al. 2016). Zoopagomycotina includes three subphyla: Zoopagomycotina, Entomophthoromycotina, and Kickxellomycotina (Hibbett et al. 2007; Spatafora et al. 2016; Naranjo-Ortiz and Gabaldón 2019a). Species within these three subphyla are generally associated with animals, protists, and other fungi (primary within the phylum Mucoromycotina) as parasites or commensals and rarely interact with plants (Benny et al. 2014; Spatafora et al. 2016).

Four fungal orders comprise the subphylum Kickxellomycotina: Dimargaritales, Kickxellales, Asellariales, and Harpellales (Hibbett et al. 2007). Dimargaritales and Kickxellales species are saprobes or mycoparasites, while Asellariales and Harpellales populate the gut microbiome of insects, isopods, and Collembola, probably functioning as commensals (Guardia Valle and Cafaro 2008; Spatafora et al. 2017). Some species of Harpellales, however, are parasite of insects. In these fungi, parasitism seems to have evolved secondarily from commensalism as a strategy to ensure fungal environmental transmission in aquatic insect larvae (e.g. blackflies, midges, mosquitoes, mayflies, stoneflies, caddisflies, and beetles) and greater rates of colonisation (White et al. 2006; Naranjo-Ortiz and Gabaldón 2019b). Examples of parasitic

Harpellales include *Smittium morbosum*, which infects through the gut and kills mosquitos by preventing the larvae from moulting (Wang et al. 2012) and *Genistellospora homothalica*, *Pennella simulii*, and *Harpella melusinae*, which produce chlamydospores (thick-walled large resting spores) that invade the ovaries of female insects (e.g. blackflies) and are dispersed during oviposition (White et al. 2006). In addition, within Kicksellales, two *incertae sedis* genera, *Zygnemomyces* and *Ballocephala* (formerly Entomophtoromycotina, order Entophthorales, family Meristacreceae) (Humber 2012; Gryganskyi et al. 2013) can infect insects but are better known for affecting tardigrades (Humber 2016).

The subphylum Entomophthoromycotina consists of a monophyletic group of species within the early divergent terrestrial fungi and it can be considered as a separate fungal phylum, the Entomopthtoromycota (Humber 2012). This phylum/sub phylum encompasses about 250 species, which are mostly arthropod pathogens or soil-dwelling saprobes (Gryganskyi et al. 2013). Entomophthoromycotina (or Entomophthoromycota) comprises three orders: (i) Basidiobolales—consisting of primary saprophytic genera, (ii) Neozygitales, and (iii) Entomopthtorales—both including saprophytic and invertebrate pathogenic genera (Humber 2012). Entomopathogenic species in Entomophthoromycotina are distributed across 16 genera: *Neozygtes*, *Thaxlerosporium* (order Neozygitales, family Neozygitaceae), *Ancyclistes*, *Conidiobolus*, *Macrobiophthora* (order Entophthorales, family Ancylistaceae), *Completoria* (order Entophthorales, family Completoriaceae), *Erynia*, *Eryniopsis*, *Pandora*, *Strongwellsea*, *Batkoa*, *Entomophaga*, *Entomophthora*, *Massospora* (order Entophthorales, family *Entomophthoraceae*), *Meristacrum* (order Entophthorales, family *Meristacreceae*), and *Tarichium* (order and family uncertain as it is assigned to both Entomophtorales and Neozygitales) (Gryganskyi et al. 2013; Araújo and Hughes 2016).

Species of Entomophthoromycotina (or Entomophthoromycota) produces homothallic zygospores that act as survival structures, and forcibly discharged infective conidia on infected (alive) insects or cadavers, that, in some instance, can form secondary conidia (Gryganskyi et al. 2013; Eilenberg, Lovett and Humber 2020). Some species of

Entomophthoramycotina (e.g. *Strongwellsea* spp. *Massospora* spp, *Entomophthora* spp., *Erynia* spp. (Figure 12.2c,f), and *Entomophaga* spp.) produce infective conidia and sometimes zygospores, on the still-living host. This is part of a 'twisted plan of dispersal' where the fungus hijacks the host and use it to actively spread the fungal propagules as the insect flies or wanders around. Notable examples of such ultimate form of dispersion have been reviewed in Roy et al. (2006); Araújo and Hughes (2016) and include *Strongwellsea canstrans* whose infection is mainly circumscribed to the abdomen of the adult of the cabbage and seedcorn maggots (*Hylemia brassicae* and *H. platura*, Diptera: Antomyiidae). *Strongwellsea canstrans*-infected female and male flies present large holes in the abdomen containing abundant fungal masses and conidiophores producing conidia. Despite these large lesions, the flies act normally until the fungal infection is so widespread that it causes castration and subsequent death. *Massospora cicadina* represents another similar case. Infection of cicadas by this fungus results in extensive fungal growth in the cicada's abdomen causing it to fall apart. The head and the thorax remain intact and the ability to flight is retained, turning the insect into a living disseminator of conidia or resting spores. *Erynia kansana*-infected flies also have been observed missing abdominal parts as in the case of *M. cicadina* (MacLeod and Müller-Kögler 1973). Calyptrate flies parasitised by *Entomophaga* spp. release body exudates from the disintegrating abdomens. These body fluids adhere flies to riverbeds and attract other flies, which become infected as conidia are released from the insect body (Hutchison 1962).

12.4.5 Phylum Mucoromycota

The phylum Mucoromycota are zygosporic fungi consisting of two subphyla, Mortierellomycotina and Mucoromycotina. Fungi in this phylum, unlike species in Zoopagomycota, are associated with plants either as rhizosphere-competent or decomposers (Spatafora et al. 2017). Mortierellomycotina includes one order, Mortierellales and 13 genera with more than 100 named species (Wagner et al. 2013), does not contain any insect pathogenic species. Mucoromycotina comprises three orders: Mucorales, Umbelosidales, and Edogonales

(Spatafora et al. 2016). Only one species, *Sporodiniella umbellate* (order Mucorales) is entomopathogenic and has been found infecting hemipterans and lepidopterans (Evans and Samson 1977; Samson et al. 1988; Chien and Hwang 1997). Nevertheless, environmental sampling has revealed that these two subphyla are ubiquitous in most environments, suggesting that most of their biodiversity remains un-investigated (Tedersoo et al. 2014; Ziaee et al. 2016; Tedersoo et al. 2017).

12.4.6 Phylum Basidiomycota

Thirty-two per cent of the known fungal species belong to the Basidiomycota (Dai et al. 2015). Most of these are major decomposers of organic matter but this phylum also includes plant and animal pathogens, saprophytic fungi that grow as yeasts, and mycoparasites (Naranjo-Ortiz and Gabaldón 2019a). There are more than 30,000 species distributed into 16 classes and 52 orders. Four subphyla form the phylum Basidiomycota: Agaromycotina (e.g. mushroom, jelly fungi, bracket fungi), Pucciniomycotina (e.g. rusts), Ustilagomycotina (e.g. smuts), and Wallemiomycotina (Zhao et al. 2017).

Only two subphyla, the Agaromycotina and Pucciniomycotina, are known to include insect pathogens. *Fibularhizoctonia* (order Atheliales) and *Trechispora* (Incertae cedis), of the subphylum Agaromycotina, are associated with eggs of the subterranean termites from the genera *Reticulitermes*, *Coptotermes*, and *Nasutitermes* (Isoptera: Rhinotermitidae). These fungi, which parasitise the eggs, form sclerotia (i.e. compact masses of hardened fungal mycelia) that mimic the termite eggs (Matsuura 2006, Matsuura and Yashiro 2010). Termite workers mistake this sclerotia for eggs and carry them to the incubation chamber with the rest of the eggs. However, as workers carry and care for the eggs, they cover these sclerotia with fungistatic salivary secretions that prevent their germination. Inactivated sclerotia may protect termite eggs from other fungal infections and/or other pathogens (Matsuura et al. 2000). Antifungal compounds localised in the fungal cell wall are common in fungi, including entomopathogens, where such compounds are released in the surrounding microenvironment to antagonise competitors (Tong et al. 2020; Marx 2004). The apparent symbiotic relationship between these sclerotia-forming fungi and termites, as reported by Matsuura (2010), seems to be the result of a co-evolutionary interaction where the host has outrun the pathogen (Pedrini et al. 2015) and the pathogen's efforts to increase dispersal and infection through parallel evolution have backfired.

Ninety per cent of the species in the Puccinomycotina are plant pathogens belonging to the group rust fungi, the order Uredinales. The remaining fungi include plant pathogens, mycoparasites, saprotrophs, and entomopathogens (Aime et al. 2006). The order Septobasidiales contains two families, Pachnocybaceae and Septobasidiaceae. While the only genera in *Pachnocybaceae* is found in wood, all five genera of the Septobasidiaceae family are parasites of scale insects (Hemiptera: Coccoidea), with most of the species placed in the genera *Septobasidium*, *Auriculoscypha*, and *Uredinella* (Couch 1938; Henk and Vilgalys 2007). *Septobasidium* spp. grow as mycelial mats that cover the scale insects. Some species *Septobasidium* can produce large structures, which are observed to protrude from the branch or the leaf where the insect feed and others form mycelial arrangements, namely 'insect houses', where insects dwell (Henk 2005; Henk and Vilgalys 2007). The relationship between *Septobasidium* spp. and insects is a peculiar one, where the fungus grows over a group of scale insects but only infects some of them. Thus, Coach (1931) hypothesised that the relationship between *Septobasidium* spp. and scale insects is parasitic for some individuals but mutualistic for most of them, as these mycelial mats and structures could protect the scale insects from parasitic wasps. Unfortunately, there are no published records of such a hypothesis because *Septobasidium* spp. have been largely overlooked (Gómez and Henk 2016). The Septobasidiaceae genera *Auriculoscypha* and *Uredinella* parasite scale insects similarly to *Setobasidium* spp. However, *Uredinella* spp. form thin mycelial mats on top of infected insects (Henk and Vilgalys 2007).

12.4.7 Phylum Ascomycota

With two thirds of the described fungal species, the phylum Ascomycota includes the largest number of species within the Kingdom Fungi (Lutzoni et al. 2004; Schoch et al. 2009; McLaughlin and Spatafora 2014; Box 12.2). Ascomycete fungi group into three subphyla, namely Taphrinomycotina, Saccharomycotina, and Pezizomycotina. The first two of these do not contain any insect pathogenic species. Although some species within the Pezizomycotina remain unclassified, most of those described are currently organised in 67 orders grouped into 13 classes: Arthoniomycetes, Coniocybomycetes, Lichinomycetes, Lecanoromycetes, Dothideomycetes, Eurotiomycetes, Geoglossomycetes, Laboulbeniomycetes, Leothiomycetes, Orbiliomycetes, Pezizomycetes, Sordariomycetes, and Xylonomycetes (Hibbett et al. 2007; Adl et al. 2012; Spatafora et al. 2017). The classes Arthoniomycetes, Coniocybomycetes, Lichinomycetes, and Lecanoromycetes consist of only lichenised fungi, the rest of the Pezizomycota classes show varied lifestyles and include mycorrhizal, plant pathogens, endophytes, animal parasites and symbionts, mycoparasites, amoebophagous, endolichenic, or endolythic (Stajich et al. 2009; Spatafora et al. 2017; Naranjo-Ortiz and Gabaldón 2019a). Five orders within the Pezizomycotina include insect pathogens, namely, Laboulneniales (class Laboulbeomyctes), Pleosporales, Myringiales (class Dothideomycetes), Ascosporales (class Eurotiomycetes), and Hypocreales (class Sordariomycetes) (Vega et al. 2012; Araújo and Hughes 2016).

Species of the order Laboulbeniales do not show an asexual state and do not produce mycelium. These fungi are obligate haustorial parasites of insects, and as such, they are not normally considered as fungal pathogens (Weir and Blackwell 2005). Within the Pleosporales, species of the genus *Podonectria* infect scale insects (Kobayasi 1977). The associated anamorphs to this genus are *Tetracrium* and *Tetranachrium* (Kobayasi 1977; Roberts and Humber 1981). The order Myringiales include several species which are also pathogens of scale insects (Alexopoulos et al. 1996).

The genus *Ascosphaera* in the family Ascosphaeraceae, includes 30 species of parasites of bees that can exploit the bees' detritus and resources (e.g. wax, honey, cocoons, larval faeces) or parasitise the brood (Klinger et al. 2013). *Ascosphaera* spp. exhibit a remarkable similarity with the grains of pollen (McManus and Youssef 1984; Araújo and Hughes

> ### Box 12.2 Important ascomycete entomopathogens
>
> *Metarhizium* spp. have been widely studied in terms of their mechanisms of pathogenicity (see Shin et al. 2020 for a review) and their interaction with host defences (e.g. Grizanova et al. 2019). They and their products have been developed as biological control agents to deal with a wide variety of insect pests. *M. brunneum* (*anisopliae*) causes green muscardine disease in > 200 insect species and was
>
> originally described as a pathogen of silkworms. A second important ascomycete pathogen is the soil dweller *Beauveria bassiana*, the cause of white muscardine disease (Ortiz-Urquiza and Keyhani 2016). This fungus has also been developed as a biological control agent of agricultural pests (e.g. thrips, whiteflies; see Arthurs and Dara 2019 for a review)

2016). Bees mix the pollen with nectar to form bee-bread, which is fed to the brood. We could argue that parallel evolution in *Ascosphaera* spp. resulting in mimicry of the pollen grains accounts for part of the fungus dispersal and pathogenic success, as infection of *Ascosphaera* spp. in larvae occurs via ingestion. Upon infection, *Ascosphaera* spp. develop in the body of bee larva, producing abundant mycelia in the insect cavity and sporulating on the larva (Vojvodic et al. 2012).

The three families of the order Hypocreales, Clavicipitaceae, Cordypitaceae, and Ophicordycipitaceae encompass a number of insect pathogenic genera. The family Clavicipitaceae includes the entomopathogenic genera *Aschersonia*, *Hypocrella*, *Regiocrella*, *Metacordyceps*, and *Metarhizium* (Sung et al. 2007). In Hyprocreales insect pathogenic fungi, host preference is not family or genus-dependent, although some species are restricted to a single host or a handful of closely related insect species. The family Cordycipitaceae, originally included in Clavicipitaceae, includes most *Cordyceps* species and it is characterised by the production of brightly coloured and fleshy stromata. This family has been restructured into 11 generic names: *Akanthomyces*, *Ascopolyporus*, *Beauveria*, *Cordyceps*, *Engyodontium*, *Gibellula*, *Hyperdermium*, *Parengyodontium*, *Simplicillium*, *Hevansia*, and *Blackwellomyces* (Kepler et al. 2017) in compliance with the 'one fungus one name' policy that does not longer permit separate names for pleomorphic fungi (Taylor 2011). The Ophiocodycipitacea family was proposed by Sung et al. (2007) to accommodate species from the former Clavicipitiacea family showing low phylogenetic support within the Clavicipitiaceae or Cordypitaceae. The family Ophicordycipitaceae is recognised by producing darkly pigmented tough stromata displaying aperithecial apices. Reorganisation of this family by Quandt et al. (2014), in an attempt to eliminate the dual nomenclature system, resulted into 6 genera, namely *Polycephalomyces*, *Purpureocillium*, *Harposporium*, *Drenchmeria*, *Tolypocladium*, and *Ophycordyceps*.

12.5 Insect pathogenic oomycetes

Best known as plant pathogens, oomycetes are heterotrophic eukaryotes, which share many ecological roles and functional traits with fungi, including nutrient absorption and hyphal growth, see Table 9.1. Differences in their cell wall composition, made mostly of glucans and cellulose instead of chitin, diploid vegetative state as opposed to haploid or dikaryotic and mitochondria with tubular cristae in lieu of flattened have revealed that oomycetes are more closely related to the majority of marine algae and other heterotrophic protists than fungi (Beakes et al. 2012). Oomycetes have developed saprophytic and parasitic lifestyles, and as parasites, they can infect plants, algae, protists, fungi, and animals, including arthropods, nematodes, and vertebrates (Beakes et al. 2014). Genetic similarities with the parasitic alveolates of the phylum Apicomplexa suggest that these fungus-like organisms evolved from marine parasites and they made it onto the land hitchhiking onto their animal hosts or via host switching from seaweeds to coastal plants (Beakes et al. 2012).

Classified as Stramenopiles or Heterokonts, insect pathogenic oomycetes produce zoospores, which are motile asexual single nucleated wall less spores with two flagella. Although infection through the digestive track is possible, infection starts with a primary or secondary zoospore that encysts, germinates, and penetrates the cuticle (Shen et al. 2019). Once in the haemocoel, oomycete species produce mycelia that colonise the insect causing its death. After the death of the insect, oomycetes make their way out of the insect body and produce zoosporangia that releases more zoospores that can cause secondary infections. Simultaneously, hyphae on the insect cuticle can undergo meiosis and differentiate into haploid gametes (male and female). These gametes remain attached to the parental hyphae, namely antheridium an oogonium. Subsequently, an oospore is produced by the fusion of the gametes through a fertilisation tube. Oospores are double-walled resting structures and thus, unlike zoospores, are more resistant to harsh environmental conditions (Ascunce et al. 2016, 2017).

With fewer species than fungi, oomycetes are divided into two major fungal-like classes and several early divergent classes without mycelial structure (viz. holocarpic). The two major fungal-like classes encompass three and four orders respectively, namely Saprolegniomycetes (orders: Atkinsiellales s. lat., Leptomitales, and Saprolegniales) and Peronosporomycetes (orders: Rhipidiales, Albuginales, and Peronosporales s. lat.). While the majority of plant pathogens belong to the orders Peronosporales s. lat. and Albuginales (class Peronosporomycetes), most of the animal pathogens group within the orders Saprolegniales and Leptomitales (class Saprolegniomycetes) (Beakes and Thines 2017).

12.5.1 Order Atkinsiellales (Class Saprolegniomycetes)

This order consists of three families, Atkinsiellaceae, Cryptocolaceae, and Lagenismatacae, and contains mostly saprophytic oomycetes and a handful of parasites of marine crustaceans, algae and terrestrial invertebrates (Beakes and Thines 2017; See Chapter 16 for details of oomycetes that infect crustaceans). Within the order Atkinsiellales only two species, *Atkinsiella entomophaga* and *Crypticola clavulifera* have been reported to parasite insects (Martin 1977; Beakes and Thines 2017; Mendoza, Vilela and Humber 2018; Frances et al.

1989; Table 12.1). *A. entomophaga*, currently *Crypticola entomomophaga* (Dick 1998) is a holocarpic freshwater species reported to infect egg masses of midges (Diptera) and caddisflies (Trichoptera) (Martin 1977). *C. clavulifera* is also an holocarpic species found to naturally infect larvae of the mosquito *Forcipomyia marksae* (Diptera) (Frances et al. 1989), although laboratory insect bioassays have shown that *C. clavulifera* can also infect other insect larvae, including the mosquitoes *Aedes aegypti*, *Ae. notoscriptus*, *Anopheles farauti*, *Culex annulirostris*, *Cx. Quirquefasciatus*, and the midge *Chriornmus tepperi* (Frances 1991).

12.5.2 Order Leptomitales (Class Saprolegniomycetes)

The most recent classification (Dick 2001) includes four families in the order Leptomitales: the Apodachyllaceae, Ducellariaceae, Leptollegniellaceae, and Leptomitaceae. However, molecular data suggest that the family Apodachyllaceae should be included within the Leptomitaceae (Beakes and Thines 2017). To date, no molecular data are available for members of the families Ducellariaceae and Leptollegniellaceae, so these two families remain *incertae sedis* (i.e. unplaced) (Beakes and Thines 2017; Table 12.1). Leptomitales encompass holocarpic genera such as *Apodachlya*, *Leptomitus*, *Aphanodictyon*, *Brevilegniella*, *Ducellaria*, *Leptolegniella*, *Eurychosmopsis*, *Nemathophthora*, *Pythiella*, *Aphanomycopsis*, and *Leptolegnia* (Table 12.1). Within the last two genera, *Aphanomycopsis sexualis*, *Leptolegnia chapmanii* and *L. caudata* are insect pathogens. While *A. sexualis* has been found infecting eggs of the non-biting midge *Glyptotendipes lobiferus*, *L. caudata and L. chapmanii* are natural enemies of *Anopheles* and *Aedes* mosquito larvae (Martin 1975; Mc Innis and Zattau 1982; Seymour 1984; Bisht et al. 1996; Gutierrez et al. 2017). In particular, *L. chapmanii* shows high specificity and virulence for mosquito larvae, including *Aedes*, *Anopheles*, *Culex*, and *Ochlerotatus*, with infected larvae found in aquatic habitats such as tree holes, artificial containers, freshwater/brackish floodplains and ponds (Lastra et al. 1999; López Lastra et al. 2004; Gutierrez et al. 2017).

Table 12.1 Insect pathogenic fungal-like oomycetes within the taxonomic framework proposed by Dick (2001) and Beakes and Thines (2017).

Class	Order	Family	Genus	Insect pathogenic species	Host range	Key references
Saprolegniomycetes	Atkinsiellales s. lat.	Atkinsiellaceae	~	~	~	Martin (1977)
		Crypticolaceae	*Crypticola*	*Crypticola (Atkinsiella) entomophaga*	Eggs of midges and caddisflies	Frances, Sweeney and Humber (1989); Frances (1991)
				Crypticola clavulifera	Mosquito larvae (*Forcipomyia marksae, Aedes aegypti, Ae. notoscriptus, Anopheles farauti, Cx. annulirostris, Cx. quinquefasciatus*) Midge larvae (*Chironomus tepperi*)	
		Lagenismataceae	~	~	~	~
	Leptomitales	Leptomitaceae	~	~	~	~
		(*Incertae sedis* families) Ducellariaceae Leptolegniallaceae	*Leptolegnia*	*Leptolegnia chapmanii*	Mosquito larvae *Aedes* spp., *Anopheles* spp., *Culex* spp. and *Ochlerotatus* spp.	Seymour (1984);López Lastra et al. (2004); Gutierrez et al. (2017)
				Leptolegnia caudata	Mosquito larvae *Aedes* spp. *Anopheles* spp.	Bisht, Joshi and Khulbe (1996)
			Aphanomycopsis	*Aphanomycopsis sexualis*	Eggs of the non-biting midges (*Glyptotendipes lobiferus*)	Martin (1975)
	Saprolegniales	Verrucalvaceae	*Aphanomyces*	*Aphanomyces laevis*	Mosquito larvae *Ae. culcifacies and Cx. quinquefasciatus*	Bisht et al. 1996; Patwardhan et al. (2006)
		Achlyaceae	~	~	~	~

continued

Table 12.1 Continued

Class	Order	Family	Genus	Insect pathogenic species	Host range	Key references
		Saprolegniaceae	Saprolegnia	Saprolegnia delica	Larvae of caddisflies Rhyacophila siltala and Rhyacophila dorsalis Nymphs of some stoneflies and mayflies (e.g. Ephemera danica)	Sarowar et al. (2013)
				Saprolegnia hypogyna	Larvae of caddisflies Rhyacophila siltala and Rhyacophila dorsalis	Sarowar et al. (2013)
				Saprolegnia diclina	Larvae of caddisflies Rhyacophila siltala and Rhyacophila dorsalis Nymphs of stoneflies	Sarowar et al. (2013)
			Couchia	Couchia circumplexa	Eggs of the midge Polypedilum simulans	Martin (1981)
				Couchia amphora	Eggs of the midge Polypedilum simulans	Martin (2000)
				Couchia limnophila	Eggs of the midge Glyptotendipes lobiferus	Martin (2000)
Peronosporomycetes	Rhipidiales	Rhicipidiacea	?	?	?	?
	Albuginales	Albuginaceae	?	?	?	?
	Peronosporales s. lat.	Salispiliaceae	?	?	?	?
		Peronosporacea	?	?	?	?
		Pythiaceae s. lat.	Lagenidium	Lagenidium giganteum	Mosquito larvae of Ae. aegypti, An. gambiae, Cx. quinquefasciatus, Cx. trasalis, and Ochlerotatus spp. Larvae of the biting midge Culicoides molestus	Rueda, Patel and Axtell (1991); Golkar et al. (1993); Woodring, Kaya and Kerwin (1995); Wright and Easton (1996); Scholte et al. (2004)

Pythium	Pythium carolinianum	Mosquito larvae Cx. quinquefaciatus and Ae. albopictus	Su et al. (2001)
	Pythium sierrensis	Mosquito larvae of An. freeborni, C. incidens, Cx. tarsalis, Oc. californica, Oc sierrensis, Oc. triseriatus and Uranotaenia anhydor	Clark et al. (1966); Scholte et al. (2004)
	Pythium flevoense	Mosquito larvae of Oc. sierrensis	Washburn et al. (1988)
	Pythium guiyangense	Mosquito larvae of Cx. pipiens quinquefasciatus, Cx. tritaeniorhynchus, Cx. pipiens pallens, Cx. mimulus, Cx. tianpinggensis, Cx. theileri, Cx. pseudovishnui, Cx. minor, Ae. elsiae, Ae. novoniveus, Ae. formosensis, Ae. albopictus, Ae. aegypti, and An. sinensis.	Su (2008)
	Pythium insidiosum	Mosquito larvae of Ae. aegypti	Vilela et al. (2018)

12.5.3 Order Saprolegniales (Class Saprolegniomycetes)

This order includes eucarpic (i.e. thallus differentiated into vegetative and reproductive structures) species classified into three families, namely Verrucalvaceae, Achlyaceae, and Saprolegniaceae (Beakes and Thines 2017; Rocha et al. 2018; Masigol et al. 2020). The insect pathogenic genus *Aphanomyces* belongs to the Verrucalvaceae family. *Aphanomyces* species are commonly found in mosquito insectaries causing temporal epizootics (Seymour and Briggs 1985; Scholte et al. 2004). *A. laevis* has been reported to infect larvae of the mosquito *A. culcifaciens* and *C. quinquefasciatus* (Bisht et al. 1996; Patwardhan et al. 2006). The family Achlyaceae incorporates the genus *Achlya* s. str., *Brevilegnia*, *Dyctyuchus*, and *Thraustotheca* (Beakes et al. 2014; Choi et al. 2019), which comprise primary saprotrophic and pathogens of fish and crustaceans. No pathogenic species in Achlyaceae have been discovered yet. Within the family Saprolegniaceae, several species of the genus *Saprolegnia* (e.g. *S. delica*, *S. hypogyna* and *S. diclina*) have been isolated from river insects including immature and adult stages of caddisflies (Trichoptera), stoneflies (Plecoptera), and mayflies (Ephemeroptera) (Sarowar et al. 2013). *Saprolegnia* spp. cause devastating diseases in amphibians, fish, crustaceans, and other aquatic animals and it is been proposed that aquatic insects can act as a disease reservoir of *Saprolegnia* for aquatic vertebrates (Sarowar et al. 2013). Also, amongst the Saprolegniaceae, three species of the genus *Couchia*, namely *C. circumplexa*, *C. amphora*, and *C. limnophila*, have been observed parasitising eggs of midges (Diptera: Chironomidae), including the species *Polypedilum simulans* and *Gypotendipes lobiferus* (Martin 1981, 2000).

12.5.4 Order Rhipidiales (Class Peronosporomycetes)

Circumscribed to only one family, the Rhicipidiacea, the order Rhipidiales includes the saprophytic genera *Araiospora*, *Aqualinderella*, *Mindeniella*, *Nellymyces*, *Rhipidium*, and *Sapromyces* (Bennett et al. 2018). To date, no insect pathogenic species of Rhipidiales have been reported.

12.5.5 Order Albuginales (Class Peronosporomycetes)

Albuginales are obligate biotrophs plant pathogens of flowering plants (Thines 2014) and so far, no insect or animal pathogenic species have been described within this order.

12.5.6 Order Peronosporales s. lat. (Class Peronosporomycetes)

Peronosporales are currently classified into three families: Salispiliaceae, Peronosporacea, and Pythiaceae s. lat. (Beakes et al. 2014). Important genera amongst the Pythiaceae family include saprophytic and plant and animal pathogenic genera such as *Myzocytiopsis*, *Saligenidium* (or marine *Lagenidium*), *Lagenidium*, *Lagena*, *Pythiogeton*, and *Pythium* (Blackwell 2011; Hatai 2012; Rocha et al. 2014; Ho 2018).

Although, the genus *Lagenidium* in the Pythiaceae family is a primary pathogen of invertebrates (e.g. crustaceans, nematodes and insects) (Glenn and Hc 1978; Kerwin and Washino 1986; Kerwin and Washino 1987; Nakamura et al. 1995), some species of have been reported to also infect mammals (Grooters 2003; Reinprayoon et al. 2013). To date, only one species of *Lagenidium*, *L. giganteum*, has been identified as a pathogen of aquatic insect larvae such as *Aedes*, *Ochlerotatus*, and *Culex* mosquitoes, the biting midge *Culicoides molestus* and to a lesser extent, *Anopheles* larvae (Golkar et al. 1993; Wright and Easton 1996; Scholte et al. 2004; Jaronski and Mascarin 2016). *Lagenidium* is a facultative insect parasite and can be found in freshwater habitats, infecting mosquitoes or growing saprophytically on vegetation debris (Scholte et al. 2004; Kerwin 2007). In nature, infection rates of *L. giganteum* depend on water temperature and salinity levels as these two factors affect zoosporogenesis (Merriam and Axtell 1982; Jaronski and Axtell 1983). Given its persistence in the environment, which can last up to a year and its compatibility with other biological control agents (e.g. *Bacillus thurigiensis*), *L. giganteum* is available commercially in the USA as a mosquito control under names like Laginex ® (Jaronski and Axtell 1983; Orduz and Axtell 1991).

Pathogenic oomycetes produce a plethora of protein effectors to manipulate the host immune system to achieve successful infection. Transcriptomic and phylogenetic analyses have shown that the insect pathogen, *L. giganteum*, expresses effectors typical of oomycete phytopathogens, which suggests that *L. giganteum* has evolved from a plant-pathogenic common ancestor (Quiroz Velasquez et al. 2014).

Pythium (family Pythiaceae) species are primarily soil-borne, although they can also be found in freshwater and marine habitats growing saprophytically on detritus or parasitising plant, algae, fungi, and animals (van der Plaats-Niterink 1981). Four species of *Pythium* have been identified as pathogens of mosquito larvae, namely *P. carolinianum*, *P. sierrensis*, *P. flevoense*, and *P. guiyangense* (Saunders et al. 1988; Su et al. 2001; Su 2008; Shen et al. 2019). In addition, pathogen of mammals,

P. insidiosum has also been found infecting larvae of *Aedes egyptii*, which indicates that this species may use hosts other than mammals to complete its life cycle in lakes and other freshwater bodies (Vilela et al. 2018). Recent phylogenetic analyses have revealed that the insect pathogenic species of *Pythium*, such as *P. guiyangense*, are more related to plant pathogens than mammalian pathogens. For a plant pathogen, adaptation to a mosquito pathogenic lifestyle includes, amongst other genetic features, loss of genetic enzymes able to degrade the plant cell wall and Nep1-like protein effectors inducing plant cell death and expansion of kinases (Shen et al. 2019). Despite the pathogenicity of these *Pythium* species against mosquitoes, very little attention has been paid to its mode of action and suitability as a biocontrol agent for mosquito larvae.

12.6 Summary

Fungal insect disease outcomes follow a common series of events:

Contact (spore adhesion) → Consolidation (differentiation of infection structures) → Cuticle degradation/penetration → Haemocoel colonisation (hyphae and spores) → Cadaver formation and sporulation:

- Fungi and oomycetes represent diverse and important pathogens in insects, and unlike other agents of disease, can circumvent the cuticle-based defences.
- Intact fungi, propagules, and their products are potent insecticides, which are designed to target arbovectors of zoonotic disease and agricultural pests.
- Microsporidians, notably *Nosema* are economically important pathogens of hymenopterans, representing key ecological regulators of bee populations in commercial settings and *in situ*.
- Haemocoelic histopathology is dominated by the rapid reproduction and spread of fungal morphotypes, the congestion of tissues (e.g. gastrointestinal tract, Malpighian tubules), and their eventual replacement. Haemocyte infiltration of compromised tissues and the formation of melanin-rich cellular aggregates are common features of the doomed insect.

Recommended further reading

General

Cameron, S.A. and Sadd, B.M. 2020. Global trends in bumble bee health. *Annual Review of Entomology* 65: 209–232 (Review of bumble bee diseases including *Nosema*)

Samson, R.A., Evans, H.C., and Latgé, J.P. 2013. *Atlas of Entomopathogenic Fungi*. Springer Science & Business Media

Vega, F.E. and Kaya, H.K. 2012. *Insect Pathology*, 2nd edn, New York: Academic Press (several chapters on fungal diseases in insects)

Mechanisms of pathogenicity

Shin, T.Y, Lee, M.R., Park, S.E., Lee, S.J., Kim, W.J., and Kim, J.S. 2020. Pathogenesis-related genes of entomopathogenic fungi. *Archives of Insect Biochemistry and Physiology* 105: e21747

Ortiz-Urquiza, A. and Keyhani, N.O. 2016. Molecular genetics of *Beauveria bassiana* infection of insects. In *Genetics and Molecular Biology of Entomopathogenic Fungi*, Lovett B. and St Leger R.J. (eds.) pp. 165–249. New York: Academic Press

Valero-Jimenez, C.A., Wiegers, H., Zwaan, B.J., Koenraadt, C.J.M., and van Kan, J.A.L 2016. Genes involved in virulence of the entomopathogenic fungus *Beauveria bassiana*. *Journal of Invertebrate Pathology* 133: 41–49

Wang, V.C.S. and Wang, S.B. 2017. Insect pathogenic fungi: genomics, molecular interactions, and genetic improvements. *Annual Review of Entomology* 62: 73–90

Microsporidians

Grupe, A.C. II and Quandt, C.A. 2020. A growing pandemic: A review of *Nosema* parasites in globally distributed domesticated and native bees. *PLoS Pathogens* 16: e1008580 (Brief overview of *Nosema* spp. infected bees globally)

Oomycetes

Kerwin, J.L. 2007. Oomycetes: *Lagenidium giganteum*. *Journal of the American Mosquito Control Association* 23: 50–57

Scholte, E.J., Knols, B.G.J., Samson, R.A. and Takken, W. 2004. Entompathogenic fungi for mosquito control: A review. *Journal of Insect Science* 4: 19 (Extensive review on *Lagenidium*)

References

Adl, S.M., Simpson, A.G., Lane, C.E. et al. 2012. The revised classification of eukaryotes. *Journal of Eukaryotic Microbiology* 59: 429–493

Ahrendt, S.R., Quandt, C., Ciobanu, D. et al. 2018. Leveraging single-cell genomics to expand the fungal tree of life. *Nature Microbiology* 3: 1417–1428

Aime, M. C., Matheny, P.B., Henk, D.A. et al. 2006. An overview of the higher level classification of Pucciniomycotina based on combined analyses of nuclear large and small subunit rDNA sequences. *Mycologia* 98: 896–905

Alexopoulos, C.J., Mims, C.W., and Blackwell, M. 1996. *Introductory Mycology*. John Wiley and Sons.

Andreadis, T.G. 2007. Microsporidian parasites of mosquitoes. *Journal of the American Mosquito Control Association* 23: 3–29

Araújo, J.P. and Hughes, D.P. 2016. Diversity of entomopathogenic fungi: Which groups conquered the insect body? In *Genetics and Molecular Biology of Entomopathogenic Fungi*. Brian Lovett and Raymond St. Leger (eds.) *Advances in Genetics*, vol. 94, pp. 1–39. Cambridge, Massachusetts: Academic Press Elservier,

Arthurs, S. and Dara, S.K. 2019. Microbial biopesticides for invertebrate pests and their markets in the United States. *Jounal of Invertebrate Pathology* 165: 13–21

Ascunce, M.S., Huguet-Tapia, J.C., Braun, E.L., Ortiz-Urquiza, A, Keyhani, N.O., and Goss, E.M. 2016 Whole genome sequence of the emerging oomycete pathogen *Pythium insidiosum* strain CDC-B5653 isolated from an infected human in the USA. *Genomics Data* 7: 60

Ascunce, M.S., Huguet-Tapia, J.C., Ortiz-Urquiza, A., Keyhani, N.O., Braun, E.L., and Goss, E.M. 2017. Phylogenomic analysis supports multiple instances of polyphyly in the oomycete peronosporalean lineage. *Molecular Phylogenetics and Evolution* 114: 199–211

Beakes, G.W., Glockling, S.L., and Sekimoto, S. 2012. The evolutionary phylogeny of the oomycete 'fungi'. *Protoplasma* 249: 3–19

Beakes, G.W., Honda, D., and Thines, M. 2014. 3 Systematics of the Straminipila: Labyrinthulomycota, Hyphochytriomycota, and Oomycota. In *Systematics and Evolution. The Mycota (A Comprehensive Treatise on Fungi as Experimental Systems for Basic and Applied Research)*, Vol. 7A, McLaughlin D. and Spatafora J. (eds.) Berlin, Heidelberg: Springer

Beakes, G.W. and Thines, M. 2017. Hyphochytriomycota and Oomycota. In *Handbook of the Protists*, J.M. Archibald, A.G.B. Simpson, and C.H. Slamovits (eds.) pp. 435–505. Cham: Springer International Publishing

Becnel, J. and Andreadis, T. 2014. Microsporidia in insects. In *Microsporidia: Pathogens of Opportunity*, 1st edn, Weiss L.M. and Becnelpp J.J. (eds.) pp. 521–570. Chichester: J. Wiley and Sons

Bennett, R.M., Devanadera, M.K., Dedeles, G.R., and Thines, M. 2018. A revision of Salispina, its placement in a new family, Salispinaceae (Rhipidiales), and description of a fourth species, S. hoi sp. nov. *IMA Fungus*, 9: 259–269

Benny, G.L., Humber, R.A. and Voigt, K. 2014. 8 Zygomycetous fungi: Phylum Entomophthoromycota and Subphyla Kickxellomycotina, Mortierellomycotina, Mucoromycotina, and Zoopagomycotina. In *Systematics and Evolution. The Mycota (A Comprehensive Treatise on Fungi as Experimental Systems for Basic and Applied Research)*, vol. 7A, D. McLaughlin and J. Spatafora (eds.) pp. 209–250. Berlin & Heidelberg: Springer

Bisht, G.S., Joshi, C., and Khulbe, R.D. 1996. Watermolds: Potential biological control agents of malaria vector *Anopheles culicifacies*. *Current Science* 70: 393–395

Blackwell, M. 2011. The Fungi: 1, 2, 3 … 5.1 million species? *American Journal of Botany* 98: 426–438

Boddy, L. and Hiscox, J. 2018. Fungal ecology: Principles and mechanisms of colonization and competition by saprotrophic fungi. In *The Fungal Kingdom*, J. Heitman et al. (eds.) pp. 293–308. Washington D.C.: American Society for Microbiology

Boucias, D. and Pendland, J. 1991. Attachment of mycopathogens to cuticle. In *The Fungal Spore and Disease Initiation in Plants and Animals*. G.T. Cole and H.C. Hoch (eds.) pp. 101–127. New York: Plenum Press

Boucias, D.G., Pendland, J.C., and Latge, J.T. 1988. Nonspecific factors involved in attachment of entomopathogenic deuteromycetes to host insect cuticle. *Applied and Environmental Microbiology*, 54: 1795–1805

Bouzat, J.L., McNeil, L.K., Robertson, H.M. et al. 2000. Phylogenomic analysis of the alpha proteasome gene family from early-diverging eukaryotes. *Journal of Molecular Evolution* 51: 532 543

Chapman, H.C. 1974. Biological control of mosquito larvae. *Annual Review of Entomology* 19: 33–59

Chien, C.-Y. and Hwang, B.-C. 1997. First record of the occurrence of *Sporodiniella umbellata* (Mucorales) in Taiwan. *Mycoscience* 38: 343–346

Choi, J. and Kim, S.H. 2017 A genome Tree of Life for the Fungi kingdom. *Proceedings of the National Academy of Sciences USA* 114: 9391–9396

Clark, T.B., Kellen, W.R., Lindegren J.E., and Sanders, R.D. 1966 Pythium sp. (Phycomycetes: Pythiales) pathogenic to mosquito larvae. *Journal of Invertebrate Pathology* 8: 351–354

Corradi, N. and Keeling, P. 2009. Microsporidia: A journey through radical taxonomical revisions. *Fungal Biology Reviews* 23: 1–8

Couch, J.N. 1931. Memoirs: the biological relationship between *Septobasidium retiforme* (B. & C.) Pat. and *Aspidiotus osborni* New. and Ckll. *Journal of Cell Science*, 2: 383–438

Couch, J.N. 1938. *The Genus Septobasidium*. North Carolina: University of North Carolina Press

Dai, Y.-C., Cui, B.K., Si, J. et al. 2015. Dynamics of the worldwide number of fungi with emphasis on fungal diversity in China. *Mycological Progress* 14: 62

De Bekker, C., Quevillon, L.E., Smith, P.B. et al. 2014. Species-specific ant brain manipulation by a specialized fungal parasite. *BMC Ecology and Evolution* 14: 166

De Renobales, M., Nelson, D.R., and Blomquist, G.J. 1991. Cuticular lipids. In *Physiology of the Insect Epidermis*. K. Binnington and A. Retnakaran (eds.) pp. 240-251. East Melbourne, Victoria, Australia: CSIRO Publications

Del Campo, J., Mallo, D., Massana, R., et al. 2015. Diversity and distribution of unicellular opisthokonts along the European coast analysed using high-throughput sequencing. *Environmental Microbiology* 17: 3195–3207

Dick, M.W. 2001. *Straminipilous FungiSystematics of the Peronosporomycetes Including Accounts of the Marine Straminipilous Protists, the Plasmodiophorids and Similar Organisms*. Dordrecht: Kluwer Academic Publishers

Eilenberg, J., Lovett, B., and Humber, R.A. 2020. Secondary conidia types in the insect pathogenic fungal genus *Strongwellsea* (Entomophthoromycotina: Entomophthorales) infecting adult Diptera. *Journal of Invertebrate Pathology* 174: 107399

Evans, H.C., Elliot, S.L., and Hughes, D.P. 2011. *Ophiocordyceps unilateralis*: A keystone species for unraveling ecosystem functioning and biodiversity of fungi in tropical forests? *Communicative & Integrative Biology* 4: 598–602

Evans, H.C. and Samson, R.A. 1977. *Sporodiniella umbellata*, an entomogenous fungus of the Mucorales from cocoa farms in Ecuador. *Canadian Journal of Botany* 55: 2981–2984

Fan, Y., Pereira, R.M., Kilic, E., Casella, G., and Keyhani, N.O. 2012. Pyrokinin β-neuropeptide affects necrophoretic behavior in fire ants (*S. invicta*), and expression of β-NP in a mycoinsecticide increases its virulence. *PLOS One* 7: e26924

Fisher, M. C., Henk, D.A., Briggs, C.J. et al. 2012. Emerging fungal threats to animal, plant and ecosystem health. *Nature* 484: 186–194

Frances, S.P. 1991. Pathogenicity, host range and temperature tolerance of *Crypticola clavulifera* (Oomycetes: Lagenidiales) in the laboratory. *Journal of the American Mosquito Control Association* 7: 504–506

Frances, S. P., Sweeney, A.W., and Humber, R.A. 1989. *Crypticola clavulifera* gen. et sp. nov. and *Lagenidium giganteum*: Oomycetes pathogenic for dipterans infesting leaf axils in an Australian rain forest. *Journal of Invertebrate Pathology* 54: 103–111

Glenn, F.E. and Hc, C. 1978. A natural epizootic of the aquatic fungus *Lagenidium giganteum* in the mosquito *Culex territans*. *Mosquito News* 38: 522–524

Goertz, D., Solter, A., Linde, L.F., and Linde, A. 2007. Horizontal and vertical transmission of a *Nosema* sp. (Microsporidia) from *Lymantria dispar* (L.) (Lepidoptera: Lymantriidae). *Journal of Invertebrate Pathology* 95: 9–16

Goettel, M., Eilenberg, J., and Glare, T. 2005. Entomopathogenic fungi and their role in regulation of insect populations. *Comprehensive Molecular Insect Science* 6: 361–406

Golkar, L., Lebrun, R.A., Ohayon, H., Gounon, P., Papierok, B., and Brey, P.T. 1993. Variation of larval susceptibility to *Lagenidium giganteum* in three mosquito species. *Journal of Invertebrate Pathology* 62: 1–8

Gómez, L. and Henk, D. 2016. Validation of the species of *Septobasidium* (Basidiomycetes) described by John N. Couch. *Lankesteriana* 4: 75–96

Grizanova, E.V., Coates, C.J., Dubovskiy, I.M., and Butt, T.M. 2019. *Metarhizium brunneum* infection dynamics differ at the cuticle interface of susceptible and tolerant morphs of *Galleria mellonella*. *Virulence* 10: 999–1012

Grooters, A.M. 2003. Pythiosis, lagenidiosis, and zygomycosis in small animals. *Veterinary Clinics: Small Animal Practice* 33: 695–720

Gryganskyi, A.P., Humber, R.A., Smith, M.E., Hodge, K., Huang, B., Voigt, K., and Vilgalys, R. 2013. Phylogenetic lineages in Entomophthoromycota. *Persoonia* 30: 94–105

Guardia Valle, L. and Cafaro, M. 2008. First report of zygospores in Asellariales and new species from the Caribbean. *Mycologia* 100: 122–131

Gutierrez, A.C., Rueda Páramo, M.E., Falvo, M.L., López Lastra, C.C., and García, J.J. 2017. *Leptolegnia chapmanii* (Straminipila: Peronosporomycetes) as a future biorational tool for the control of *Aedes aegypti* (L.). *Acta Tropica* 169: 112–118

Hatai, K. 2012. Diseases of fish and shellfish caused by marine fungi. In *Biology of Marine Fungi. Progress in Molecular and Subcellular Biology*, vol 53. Raghukumar, C. (ed.) pp. 15–52. Heidelberg and Berlin: Springer

Hawksworth, D.L. and Lücking, R. 2017. Fungal diversity revisited: 2.2 to 3.8 million species. *Microbiology Spectrum* 5 (4)

Henk, D.A. 2005. New Species of *Septobasidium* from Southern Costa Rica and the Southeastern United States. *Mycologia* 97: 908–913

Henk, D.A. and Vilgalys, R. 2007. Molecular phylogeny suggests a single origin of insect symbiosis in the Pucciniomycetes with support for some relationships within the genus *Septobasidium*. *American Journal of Botany* 94: 1515–1526

Hibbett, D.S., Binder, M., Bischoff, J.F. et al. 2007. A higher-level phylogenetic classification of the Fungi. *Mycological Research* 111: 509–547

Hirt, R.P., Logsdon, Jr. J.M., Healy, B., Dorey, M.W., Doolittle, W.F., and Embley, T.M. 1999. Microsporidia are related to Fungi: Evidence from the largest subunit of RNA polymerase II and other proteins. *Proceedings of the National Academy of Sciences of the United States of America USA* 96: 580–585

Hiscox, J., O'Leary, J., and Boddy, L. 2018. Fungus wars: Basidiomycete battles in wood decay. *Studies in Mycology* 89: 117–124

Ho, H.H. 2018. The taxonomy and biology of *Phytophthora* and *Pythium*. *Journal of Bacteriology and Mycology Open Access* 6: 40–45

Hu, X., Xiao, G., Zheng, P. et al. 2014. Trajectory and genomic determinants of fungal-pathogen speciation and host adaptation. *Proceedings of the National Academy of Sciences USA* 111: 16796

Hughes, D.P., Andersen, S.B., Hywel-Jones, N.L., Himaman, W., Billen, J., and Boomsma, J.J. 2011. Behavioral mechanisms and morphological symptoms of zombie ants dying from fungal infection. *BMC Ecology* 11: 13

Humber, R.A. 2008. Evolution of entomopathogenicity in fungi. *Journal of Invertebrate Pathology* 98: 262–266

Humber, R.A. 2012. Entomophthoromycota: A new phylum and reclassification for entomophthoroid fungi. *Mycotaxon* 120: 477–492

Humber, R.A. 2016 Entomophthoromycota: A new overview of some of the oldest terrestrial fungi. In *Biology of Microfungi*, Li, D.W. (ed.) pp. 127–145. Cham: Springer

Hutchison, J.A. 1962. Studies on a new Entomophthora attacking calyptrate flies. *Mycologia* 54: 258–271

James, T., Letcher, P., Longcore, J. et al. 2006b. A molecular phylogeny of the flagellated fungi (Chytridiomycota) and description of a new phylum (Blastocladiomycota). *Mycologia* 98: 860–871

James, T.Y., Kauff, F., Schoch, C.L. et al. 2006a. Reconstructing the early evolution of Fungi using a six-gene phylogeny. *Nature* 443: 818–822

Jaronski, S. and Axtell, R. 1983. Persistence of the mosquito fungal pathogen *Lagenidium giganteum* (Oomycetes: Lagenidiales) after introduction into natural habitats. *Mosquito News* 43(3): 332–337

Jaronski, S. and Mascarin, G. 2016. Mass production of fungal entomopathogens. In *Microbial Control of Insect and Mite Pests From Theory to Practice*, Lacey, L.A. (ed.) pp. 141–155, London: Academic Press

Jin, K., Peng, G., Liu, Y., and Xia, Y. 2015. The acid trehalase, ATM1, contributes to the *in vivo* growth and virulence of the entomopathogenic fungus, *Metarhizium acridum*. *Fungal Genetics and Biology* 77: 61–67

Jitklang, S., Ahantarig, A., Kuvangkadilok, C., Baimai, V., and Adler, P. 2012. Parasites of larval black flies (Diptera: Simuliidae) in Thailand. *Songklanakarin Journal of Science and Technology* 34: 597–599

Karling, J.S. 1948. Chytridiosis of scale insects. *American Journal of Botany* 35: 246–254

Keeling, J. P., Fast, M.N., Law, S.L., Williams, A.P.B., and Slamovits, H.C. 2005. Comparative genomics of Microsporidia. *Folia Parasitologica* 52: 8–14

Keeling, P. 2009. Five questions about Microsporidia. *PLoS Pathogens* 5: e1000489

Keeling, P. and Doolittle, W. 1996. Alpha-tubulin from early-diverging eukaryotic lineages and the evolution of the tubulin family. *Molecular Biology and Evolution*, 13: 1297–1305

Kepler, R.M., Luangsa-Ard, J.J., Hywel-Jones, N.L. et al. 2017. A phylogenetically-based nomenclature for Cordycipitaceae (Hypocreales). *IMA Fungus* 8: 335–353

Kerwin, J.L. 2007. Oomycetes: *Lagenidium giganteum*. *Journal of the American Mosquito Control Association* 2: 50–57

Kerwin, J.L. and Washino, R.K. 1986. Ground and aerial application of the sexual and asexual stages of *Lagenidium giganteum* (Oomycetes: Lagenidiales) for mosquito control. *Journal of the American Mosquito Control Association* 2: 182–189

Kerwin, J.L. and Washino, R.K. 1987. Ground and aerial application of the asexual stage of *Lagenidium giganteum* for control of mosquitoes associated with rice culture in the Central Valley of California. *Journal of the American Mosquito Control Association* 3: 59–64

Kirby, W. and Spense, W. 1826. Diseases of insects. In *An introduction to Entomology: or Elements of the Natural History of Insects*, pp. 197–232. William Kirby and William Spence (eds.) London: Longman, Hurst, Rees, Orne and Brown

Klinger, E. G., James, R.R., Youssef, N.N., and Welker, D.L. 2013. A multi-gene phylogeny provides additional insight into the relationships between several *Ascosphaera* species. *Journal of Invertebrate Pathology* 112: 41–48

Kobayasi, Y. 1977. Two new species of Podonectria (Clavicipitaceae). *Zootaxa* 3821: 146–150

Lastra, C.C.L., Steciow, M.M., and García, J.J. 1999. Registro más austral del hongo *Leptolegnia chapmanii* (Oomycetes: Saprolegniales) como patógeno de larvas de mosquitos (Diptera: Culicidae). *Revista Iberoamericana de Micología* 16: 143–145

Leão, A.E.A. and Pedroso, M.C. 1965. Nova espécie do gênero *Coelomomyces* parasito de ovos de Phlebotomus. *Mycopathologia et Mycologia Applicata*, 26: 305–307

Li, W., Chen, Y., and Cook, S.C. 2018. Chronic *Nosema ceranae* infection inflicts comprehensive and persistent immunosuppression and accelerated lipid loss in host *Apis mellifera* honey bees. *International Journal of Parasitology* 48: 433–444

Litwin, A., Nowak, M., and Różalska, S. 2020. Entomopathogenic fungi: Unconventional applications. *Reviews in Environmental Science and Bio/Technology*, 19: 23–42

López-Lastra, C.C., Scorsetti, A.C., Marti, G.A., and García, J.J. 2004. Host range and specificity of an Argentinean isolate of the aquatic fungus *Leptolegnia chapmanii* (Oomycetes: Saprolegniales), a pathogen of mosquito larvae (Diptera: Culicidae). *Mycopathologia*, 158: 311–315

López-Lastra, C.C., Toledo, A.V., Manfrino, R.G., and Gutierrez, A.C. 2019. Southernmost records of Entomophthoromycotina. Updated review of Entomophthoralean fungal insect pathogens of Argentina. *Caldasia* 41: 349–357

Lucarotti, C.J. 1992. Invasion of *Aedes aegypti* ovaries by *Coelomomyces stegomyiae*. *Journal of Invertebrate Pathology* 60: 176–184

Lucarotti, C.J. and Shoulkamy, M.A. 2000. *Coelomomyces stegomyiae* infection in adult female *Aedes aegypti* following the first, second, and third host blood meals. *Journal of Invertebrate Pathology* 75: 292–295

Lutzoni, F., Kauff, F., Cox, C.J., et al. 2004. Assembling the fungal tree of life: Progress, classification, and evolution of subcellular traits. *American Journal of Botany* 91: 1446–1480

Macías-Macías, J.O., Tapia-Rivera, J.C., De La Mora, A. et al. 2020. *Nosema ceranae* causes cellular immunosuppression and interacts with thiamethoxam to increase mortality in the stingless bee *Melipona colimana*. *Scientific Reports* 10: 17021

MacLeod, D.M. and Müller-Kögler, E. 1973. Entomogenous fungi: *Entomophthora* species with pear-shaped to almost spherical conidia (Entomophthorales: Entomophthoraceae). *Mycologia* 65: 823–893

Martin, W.W. 1975. *Aphanomycopsis sexualis*, a new parasite of midge eggs. *Mycologia* 67: 923–933

Martin, W.W 1977. The development and possible relationships of a new *Atkinsiella* parasitic in insect eggs. *American Journal of Botany* 64: 760–769

Martin, W.W. 1981. *Couchia circumplexa*, a water mold parasitic in midge eggs. *Mycologia* 73: 1143–1157

Martin, W.W. 2000. Two new species of *Couchia* parasitic in midge eggs. *Mycologia* 92: 1149–1154

Marx, F. 2004. Small, basic antifungal proteins secreted from filamentous ascomycetes: A comparative study regarding expression, structure, function and potential application. *Applied Microbiology and Biotechnology* 65: 133–142

Masigol, H., Khodaparast, S.A., Mostowfizadeh-Ghalamfarsa, R. et al. 2020. Taxonomical and functional diversity of Saprolegniales in Anzali Lagoon, Iran. *Aquatic Ecology* 54: 323–336

Matsushita, R. and Nishimura, T. 2020. Trehalose metabolism confers developmental robustness and stability in *Drosophila* by regulating glucose homeostasis. *Communications Biology* 3: 170

Matsuura, K. 2006. Termite-egg mimicry by a sclerotium-forming fungus. *Proceedings of the Royal Society B. Biological Sciences* 273: 1203–1209

Matsuura, K., Tanaka, C., and Nishida, T. 2000. Symbiosis of a termite and a sclerotium-forming fungus: Sclerotia mimic termite eggs. *Ecological Research* 15: 405–414

Matsuura, K. and Yashiro, T. 2010. Parallel evolution of termite-egg mimicry by sclerotium-forming fungi in distant termite groups. *Biological Journal of the Linnean Society* 100: 531–537

Mazzarello, P., Garbarino, C., and Cani, V. 2013. *Bassi, Agostino. eLS.* Chichester: John Wiley & Sons, Ltd

Mc Innis, T. and Zattau, W.C. 1982. Experimental infection of mosquito larvae by a species of the aquatic fungus *Leptolegnia*. *Journal of Invertebrate Pathology* 39: 98–104

McLaughlin, D.J. and Spatafora, J. 2014. *Systematics and Evolution: Part A.* Heidelberg and Berlin: Springer

McManus, W.R. and Youssef, N.N. 1984. Life cycle of the chalk brood fungus, *Ascosphaera aggregata*, in the alfalfa leafcutting bee, *Megachile rotundata*, and its associated symptomatology. *Mycologia* 76: 830–842

Mendoza, L., Vilela, R., and Humber, R. 2018. Taxonomic and phylogenetic analysis of the Oomycota mosquito larvae pathogen *Crypticola clavulifera*. *Fungal Biology* 122: 847–855

Merriam, T.L. and Axtell, R.T. 1982. Salinity tolerance of two isolates of *Lagenidium giganteum* (Oomycetes: Lagenidiales), a fungal pathogen of mosquito larvae. *Journal of Medical Entomology* 19: 388–393

Mishra, J., Tewari, S., Singh, S., and Arora, N. 2014 Biopesticides: Where we stand? In *Plant Microbes Symbiosis: Applied Facets*, Arora N.K. (ed) pp. 37–75. India: Springer

Mishra, S., Jaiswal, K., and Kashyap, D. 2018. Microsporidia as an entomopathogen: A Review. *International Journal of Advanced Scientific Research and Management* 3: 68–75

Muma, M.H. and Clancy, D.W. 1961. Parasitism of purple scale in Florida citrus groves. *The Florida Entomologist* 44: 159–165

Nakamura, K., Nakamura, M., Hatai, K., and Zafran, K. (1995) *Lagenidium* infection in eggs and larvae of mangrove crab (*Scylla serrata*) produced in Indonesia. *Mycoscience* 36: 399–404

Naranjo-Ortiz, M.A. and Gabaldón, T. 2019a. Fungal evolution: Major ecological adaptations and evolutionary transitions. *Biological Reviews of the Cambridge Philosophical Society* 94: 1443–1476

Naranjo-Ortiz, M.A. and Gabaldón, T. 2019b. Fungal evolution: Diversity, taxonomy and phylogeny of the Fungi. *Biological Reviews* 94: 2101–2137

Nelson, D.R., Blomquist, G.M., and Hamilton, R.J. 1995. Waxes: Chemistry, molecular biology and functions. In *Insect Waxes*. Richard J. Hamilton (ed.), pp. 1–90: Dundee, Scotland: Oily Press

Oi, D. and Pereira, R. 1993. Ant behavior and microbial pathogens (Hymenoptera: Formicidae). *Florida Entomologist* 76: 63–74

Olivera, I.E., Fins, K.C., Rodriguez, S.A., Abiff, S.K., Tartar, J.L., and Tartar, A. 2016. Glycoside hydrolases family 20 (GH20) represent putative virulence factors that are shared by animal pathogenic oomycetes but are absent in phytopathogens. *BMC Microbiology* 16: 232

Ondiaka, S.N., Masinde, E.W., Koenraadt, C.J.M., Takken T., and Mukabana, W.R. 2015. Effects of fungal infection on feeding and survival of *Anopheles gambiae* (Diptera: Culicidae) on plant sugars. *Parasites & Vectors* 8: 35

Orduz, S. and Axtell, R.C. 1991. Compatibility of *Bacillus thuringiensis* var. *israelensis* and *Bacillus sphaericus* with the fungal pathogen *Lagenidium giganteum* (Oomycetes: Lagenidiales). *Journal of the American Mosquito Control Association* 7: 188–193

Ortiz-Urquiza, A., and Keyhani, N.O. 2016. Molecular genetics of *Beauveria bassiana* infection of insects. In *Genetics and Molecular Biology of Entomopathogenic Fungi*, Lovett B. and St Leger R.J. (eds.) pp. 165–249. New York: Academic Press

Ortiz-Urquiza, A. and Keyhani, O.N. 2013. Action on the surface: Entomopathogenic fungi versus the insect cuticle. *Insects* 4(3): 357–374

Pasteur, L. 1870. *Études sur la maladie des vers à soie: moyen pratique assuré de la combattre et d'en prévenir le retour.* Paris: Gauthier-Villars, successeur de Mallet-Bachelier

Patwardhan, A., Gandhe, R., Ghole, V., and Mourya, D. 2006. Larvicidal activity of the fungus *Aphanomyces* (Oomycetes: Saprolegniales) against *Culex quinquefasciatus*. *The Journal of Communicable Diseases* 37: 269–274

Pedrini, N., Ortiz-Urquiza, A., Huarte-Bonnet, C., Fan, Y., Juárez, M.P., and Keyhani, N.O. 2015. Tenebrionid secretions and a fungal benzoquinone oxidoreductase form competing components of an arms race between a host and pathogen. *Proceedings of the National Academy of Sciences USA* 112: P. E3651

Pedrini, N., Ortiz-Urquiza, A., Zhang, S., and Keyhani, N. 2013. Targeting of insect epicuticular lipids by the entomopathogenic fungus *Beauveria bassiana*: Hydrocarbon oxidation within the context of a host-pathogen interaction. *Frontiers in Microbiology* 4: 24

Petch, T. 1948. A revised list of British entomogenous fungi. *Transactions of the British Mycological Society* 31: 286–304

Quandt, C.A., Kepler, R.M., Gams, W. et al. 2014. Phylogenetic-based nomenclatural proposals for Ophiocordycipitaceae (Hypocreales) with new combinations in Tolypocladium. *IMA Fungus* 5: 121–134

Quesada-Moraga, E., Ruiz-García, A., and Santiago-Alvarez, C. 2006. Laboratory evaluation of entomopathogenic fungi *Beauveria bassiana* and *Metarhizium anisopliae* against puparia and adults of *Ceratitis capitata* (Diptera: Tephritidae). *Journal of Economic Entomology* 99: 1955–1966

Quiroz Velasquez, P.F., Abiff, S.K., K. Fins, K.C. et al. 2014. Transcriptome analysis of the entomopathogenic Oomycete *Lagenidium giganteum* reveals putative virulence factors. *Applied and Environmental Microbiology* 80: 6427

Radek, R., Wurzbacher, C., Gisder, S. et al. 2017. Morphologic and molecular data help adopting the insect-pathogenic nephridiophagids (Nephridiophagidae) among the early diverging fungal lineages, close to the Chytridiomycota. *MycoKeys* 25: 31–50

Redhead, S., Kirk, P., Keeling, P., and Weiss, L. 2009. Proposals to exclude the phylum Microsporidia from the Code. *Taxon* 58: 10–11

Reinprayoon, U., Permpalung, N., Kasetsuwan, N., Plongla, R., Mendoza, L., and Chindamporn, A. 2013. *Lagenidium* sp. ocular infection mimicking ocular pythiosis. *Journal of Clinical Microbiology* 51: 2778–2780

Richards, T.A., Leonard, G., and Wideman, J.G. 2017. What defines the 'Kingdom' fungi? *Microbiology Spectrum* 5(3)

Roberts, D.W. 1970. *Coelomomyces, Entomophthora, Beauveria*, and *Metarrhizium* as parasites of mosquitoes. *Miscellaneous Publications of the Entomological Society of America* 7: 140–155

Roberts, D.W. and Humber, R.A. 1981. Entomogenous fungi. In *Biology of Conidial Fungi*, Cole G.T. and Kendrick B. (eds.) pp. 201–236. New York: Academic Press

Rocha, J.R.S., Sousa, N.D.C., Santos, L.A., Pereira, A.A., Negreiros, N.C., Sales, P.C.L., and Trindade Júnior, O.C. 2014. The genus *Pythiogeton* (Pythiogetonaceae) in Brazil. *Mycosphere* 5: 623–634

Rocha, S.C.O., Lopez-Lastra, C.C., Marano, De Souza, J.I., Rueda-Páramo, M.E., and Pires-Zottarelli, C.L.A. 2018. New phylogenetic insights into Saprolegniales (Oomycota, Straminipila) based upon studies of specimens isolated from Brazil and Argentina. *Mycological Progress* 17: 691–700

Roy, H. E., Steinkraus, D.C., Eilenberg, J., Hajek, A.E., and Pell, J.K. 2006. Bizarre interactions and endgames: Entomopathogenic fungi and their arthropod hosts. *Annual Review of Entomology* 51: 331–357

Rueda, L., Patel, K., and Axtell, R. 1991. Efficacy of encapsulated *Lagenidium giganteum* (Oomycetes: Lagenidiales) against *Culex quinquefasciatus* and *Aedes aegypti* larvae in artificial containers. *Journal of the American Mosquito Control Association*, 6: 694–699

Samson, R.A., Evans, H.C., and Latgé, J.P. 1988. *Atlas of Entomopathogenic Fungi.* New York: Springer Verlag

Sarowar, M., Berg, A., McLaggan, D., Young, M., and Van West, P. 2013. *Saprolegnia* strains isolated from river insects and amphipods are broad spectrum pathogens. *Fungal Biology* 117: 752–763

Saunders, G.A., Washburn, J.O., Egerter, D.E., and Anderson, J.R. 1988. Pathogenicity of fungi isolated from field-collected larvae of the western treehole mosquito, *Aedes sierrensis* (Diptera: Culicidae). *Journal of Invertebrate Pathology* 52: 360–363

Schoch, C.L., Sung, G.H., López-Giráldez, F. et al. 2009. The Ascomycota tree of life: A phylum-wide phylogeny clarifies the origin and evolution of fundamental reproductive and ecological traits. *Systematic Biology* 58: 224–239

Scholte, E.J., Knols, B.G., and Takken, W. 2006. Infection of the malaria mosquito *Anopheles gambiae* with the entomopathogenic fungus *Metarhizium anisopliae* reduces blood feeding and fecundity. *Journal of Invertebrate Pathology* 91: 43–49

Scholte, E.-J., Knols, B.G.J., Samson, R.A., and Takken, W. 2004. Entomopathogenic fungi for mosquito control: A review. *Journal of Insect Science* 4: e19

Screen, S.E., Hu, G., and St Leger, R.J. 2001. Transformants of *Metarhizium anisopliae* sf. *anisopliae* overexpressing chitinase from *Metarhizium anisopliae* sf. *acridum* show early induction of native chitinase but are not altered in pathogenicity to *Manduca sexta*. *Journal of Invertebrate Pathology* 78: 260–266

Seymour, R. and Briggs, J.D. 1985. Occurrence and control of *Aphanomyces* (Saproleginales: Fungi) infections in laboratory colonies of larval *Anopheles*. *Mosquito News* 1: 100–102

Seymour, R.L. 1984. *Leptolegnia chapmanii*, an oomycete pathogen of mosquito larvae. *Mycologia* 76: 670–674

Shang, Y., Feng, P., and Wang, C. 2015. Fungi that infect insects: Altering host behavior and beyond. *PLoS Pathogens* 11: e1005037

Shen, D., Tang, Z., Wang, C. et al. 2019 Infection mechanisms and putative effector repertoire of the mosquito pathogenic oomycete *Pythium guiyangense* uncovered by genomic analysis. *PLoS Genetics*, 15: e1008116

Shin, T.Y, Lee, M.R., Park, S.E., Lee, S.J., Kim, W.J., and Kim, J.S. 2020. Pathogenesis-related genes of entomopathogenic fungi. *Archives of Insect Biochemistry and Physiology* 105: e21747

Solter, L., Becnel, J., and Oi, D. 2012. Microsporidian entomopathogens. In *Insect Pathology*, 2nd edn, Vega F.E. and Kaya H.K. (eds.) pp. 221–263. New York: Academic Press

Solter, L.F., Maddox, J.V., and Onstad, D.W. 1991. Transmission of *Nosema pyrausta* in adult European corn borers. *Journal of Invertebrate Pathology* 57: 220–226

Spatafora, J.W., Aime, M.C., Grigoriev, I.V., Martin, F., Stajich, J.E., and Blackwell, M. 2017. The Fungal Tree of Life: from molecular systematics to genome-scale phylogenies. *Microbiology Spectrum* 5(5)

Spatafora, J.W., Chang, Y., Benny, G.L. et al. 2016. A phylum-level phylogenetic classification of zygomycete fungi based on genome-scale data. *Mycologia* 108: 1028–1046

St Leger, R.J., Cooper, R.M., and Charnley, A.K. 1987. Production of cuticle-degrading enzymes by the entomopathogen *Metarhizium anisopliae* during infection of cuticles from *Calliphora vomitoria* and *Manduca sexta*. *Microbiology* 133: 1371–1382

Stajich, J. E., Berbee, M.L., Blackwell, M. et al. 2009. The fungi. *Current Biology* 19: R840–R845

Steinhaus, E. 1956. Microbial control—the emergence of an idea. A brief history of insect pathology through the nineteenth century. *Hilgardia* 26: 107–160

Su, X. 2008. A report of the mosquito host range of *Pythium guiyangense* Su. *Nature Precedings* :1

Su, X. Q., Zou, F.H., Guo, Q., Huang, J., and Chen, T.X. 2001. A report on a mosquito-killing fungus, *Pythium carolinianum*. *Fungal Diversity* 7: 129–133

Sung, G.-H., Hywel-Jones, N.L., Sung, J.-M., Luangsa-Ard, J.J., Shrestha, B., and Spatafora, J.W. 2007. Phylogenetic classification of *Cordyceps* and the clavicipitaceous fungi. *Studies in Mycology* 57: 5–59

Tanada, Y. and Kaya, H.K. 2012. *Insect Pathology*, 1st edn, London: Academic Press

Taylor, J.W. 2011. One fungus = one name: DNA and fungal nomenclature twenty years after PCR. *IMA Fungus* 2: 113–120

Tedersoo, L., Bahram, M., Põlme, S. et al. 2014. Global diversity and geography of soil fungi. *Science*, 346: 1256688

Tedersoo, L., Bahram, M., Puusepp, R., Nilsson, R.H., and James, T.Y. 2017. Novel soil-inhabiting clades fill gaps in the fungal tree of life. *Microbiome* 5: 42

Thines, M. 2014. Phylogeny and evolution of plant pathogenic oomycetes—a global overview. *European Journal of Plant Pathology* 138: 431–447

Thines, M. and Spring, O. 2005. A revision of Albugo (Chromista, Peronosporomycetes). *Mycotaxon* 92: 443–458

Tong, S., Li, M., Keyhani, N.O. et al. 2020. Characterization of a fungal competition factor: Production of a conidial cell-wall associated antifungal peptide. *PLoS Pathogens* 16: e1008518

Torruella, G., Grau-Bové, X., Moreira, D. et al. 2018. Global transcriptome analysis of the aphelid *Paraphelidium tribonemae* supports the phagotrophic origin of fungi. *Communications Biology* 1: 231

Tsaousis, A. D., Kunji, E.R., Goldberg, A.V., Lucocq, J.M., Hirt, R.P., and Embley, T.M. 2008. A novel route for ATP acquisition by the remnant mitochondria of Encephalitozoon cuniculi . *Nature* 453: 553–556

Van Der Plaats-niterink, A.J. 1981. *Monograph of the Genus* Pythium. The Netherlands: Centraalbureau voor Schimmelculutures in Baarn

Vega, F.E., Meyling, N.V., Luangsa-Ard, J.J. and Blackwell, M. 2012. Fungal Entomopathogens. In *Insect Pathology*, 2nd Edn, Vega, F., and Kaya, H.K., (eds.) pp. 171–220. San Diego, CA: Academic Press

Vilela, R., Montalva, C., Luz, C., Humber, R.A., and Mendoza, L. 2018. *Pythium insidiosum* isolated from infected mosquito larvae in central Brazil. *Acta Tropica* 185: 344–348

Vojvodic, S., Boomsma, J.J., Eilenberg, J., and Jensen, A.B. 2012. Virulence of mixed fungal infections in honey bee brood. *Frontiers in Zoology* 9: 1–6

Wagner, L., Stielow, B., Hoffmann, K. et al. 2013. A comprehensive molecular phylogeny of the Mortierellales (Mortierellomycotina) based on nuclear ribosomal DNA. *Persoonia* 30: 77–93

Wang, Yan, Tretter, Eric and Lichtwardt, Robert and White, Merlin. 2012. Overview of 75 years of Smittium research, establishing a new genus for Smittium culisetae, and prospects for future revisions of the 'Smittium' Clade. *Mycologia.* 105: 10.3852/11-311

Wang, J.B., St. Leger, R.J. and Wang, C. 2016. Advances in genomics of entomopathogenic fungi. In *Advances in Genetics*, Lovett B. and St. Leger R.J. (eds.) pp. 67–105. London: Academic Press

Washburn, J.O., Egerter, D.E., Anderson, J.R., and Saunders, G.A. 1988. Density Reduction in Larval Mosquito (Diptera: Culicidae) Populations by Interactions between a Parasitic Ciliate (Ciliophora: Tetrahymenidae) and an Opportunistic Fungal (Oomycetes: Pythiaceae) Parasite. *Journal of Medical Entomology* 25: 307–314

Weir, A. and Blackwell, M. 2004. Fungal biotrophic parasites of insects and other arthropods. In *Insect–Fungal Associations: Ecology and Evolution*, Vega F.E. and Blackwell M. (eds.) pp. 119– 145. Oxford: Oxford University Press

Weiss, L.M. (2005) The first united workshop on Microsporidia from invertebrate and vertebrate hosts. *Folia Parasitol (Praha)* 52: 1–7

White, M.M., Lichtwardt, R.W., and Colbo, M.H. 2006. Confirmation and identification of parasitic stages of obligate endobionts (Harpellales) in blackflies (Simuliidae) by means of rRNA sequence data. *Mycological Research* 110: 1070–1079

Williams, B.A., Haferkamp, I., and Keeling, P.J. 2008. An ADP/ATP-specific mitochondrial carrier protein in the microsporidian *Antonospora locustae*. *Journal of Molecular Biology* 375: 1249–1257

Woodring, J.L., Kaya, H.K., and Kerwin, J.L. 1995. *Lagenidium giganteum* in Culex tarsalis larvae: Production of infective propagules. *Journal of Invertebrate Pathology* 66: 25–32

Wright, P.J. and Easton, C.S 1996. Natural incidence of *Lagenidium giganteum* Couch (Oomycetes: Lagenidiales) infecting the biting midge *Culicoides molestus* (Skuse) (Diptera: Ceratopogonidae). *Australian Journal of Entomology* 35: 131–134

Zhang, Q., Chen, X., Xu, C. et al. 2019. Horizontal gene transfer allowed the emergence of broad host range entomopathogens. *Proceedings of the National Academy of Sciences USA* 116: 7982

Zhao, H., Charnley, A.K., Wang, Z. et al. 2006. Identification of an extracellular acid trehalase and its gene involved in fungal pathogenesis of *Metarizium anisopliae*. *Journal of Biochemistry* 140: 319–327

Zhao, R.-L., Li, G.-J., Sanchez-Ramirez, S. et al. 2017. A six-gene phylogenetic overview of Basidiomycota and allied phyla with estimated divergence times of higher taxa and a phyloproteomics perspective. *Fungal Diversity* 84: 75–99

Zheng, P., Xia, Y., Zhang, S., and Wang, C. 2013. Genetics of *Cordyceps* and related fungi. *Applied Microbiology and Biotechnology* 97: 2797–2804

Ziaee, A., Zia, M., Bayat, M., and Hashemi, J. 2016. Identification of Mucorales isolates from soil using morphological and molecular methods. *Current Medical Mycology* 2: 13

CHAPTER 13

Parasitic diseases of insects

Miranda M.A. Whitten

13.1 Introduction

Parasites of insects represent an exceptionally diverse group of organisms with equally varied life-cycles and adaptations. The following sections are organised by parasite taxonomic group and examine parasites' impact on insect health, their transmission, mechanisms of pathogenesis and their interactions with (and evasion of) host immune defences. A surprisingly common, and sometimes spectacular, adaptation among these parasites is their capacity to alter the behaviour of their hosts to enhance their transmissibility. Parasites are often chronically present in the gut and other tissues of insects, resulting in a general loss of fitness at the individual or population level. However, other parasite species are powerfully virulent and can cause significant epizootics among social and mass-reared commercially important species, either by direct pathogenesis or through the involvement of parasite-derived bacteria and viruses. As discussed in the following sections, this presents obvious challenges for the management of outbreaks but also opportunities to exploit such organisms as biocontrol agents.

The infection and infestation of insects by parasites is a subject well represented in the published literature but it is understandably dominated by studies on commercially important insects such as honeybees and pests such as mosquitoes. The aim of this chapter is therefore not only to update our current understanding of insect parasites but also to highlight lesser-known parasites and neglected areas of research, such as the fascinating and often highly virulent entomopathogenic ciliates and amoebae. Our modern appreciation of the vital role of insects in ecosystems and the expanding commercial exploitation of diverse insect taxa requiring mass rearing (e.g. as biocontrol agents or for entomophagy), is certain to create new challenges and stimulate further research into previously neglected or unexplored areas of insect parasitology.

This chapter focuses primarily but not exclusively on the protozoan, helminth, and arthropod parasites of insects. However, parasites of medical or veterinary significance that are vectored by insects are outside the scope of this chapter.

13.2 Principal parasites and diseases

13.2.1 Protozoan infections

Protozoa (protists) are unicellular eukaryotic animals roughly 30 to 75 μm in length. Despite the availability of molecular techniques, classification still relies to a large extent on the mode of movement: protozoa exhibit amoeboid movement, gliding motility, or employ flagellae, or rows of beating cilia. Monoxenous protozoa are defined as those that utilise a single insect host, whereas insects represent just one of several hosts for a heteroxenous parasite. Lifecycles are often complex and will be discussed in detail only where it is directly relevant to insect pathology. The principal diseases of insects caused by protozoa are summarised in Tables 13.1–4.

Miranda M.A. Whitten, *Parasitic diseases of insects*. In: *Invertebrate Pathology*. Edited by Andrew F. Rowley, Christopher J. Coates and Miranda M.A. Whitten, Oxford University Press. © Oxford University Press (2022). DOI: 10.1093/oso/9780198853756.003.0013

13.2.1.1 Gregarines (Phylum Apicomplexa)

Apicomplexan parasites form oocysts containing infective sporozoites; these replicate intracellularly and invade host cells with the assistance of secretions from the apical complex—a cluster of secretory organelles from which the phylum derives its name. Apicomplexans include important human pathogens such as species of *Cryptosporidium* and *Plasmodium* but the gregarine subclass (Gregarinasina) contains parasites exclusively of invertebrates. Gregarines are extremely common parasites among all the invertebrate taxa and they have adapted to a wide range of habitats and insect hosts (summarised in Table 13.1). Gregarines divide into two orders: the eugregarines (which usually have characteristic septate bodies; Figure 13.1) and the neogregarines (which are usually aseptate). Eugregarines often chronically infect the gut and gut-associated tissues, with heavy parasite burdens

inflicting sublethal or subpathogenic effects on fitness when combined with additional stressors such as food deprivation (e.g. Tsubaki and Hooper 2004). In contrast, neogregarines usually cause systemic infections, with overt pathogenicity linked to destruction of the fat body.

The vast majority of gregarines are monoxenous and transmitted by the faecal-oral route. After ingestion infective sporozoites attach to the midgut and/or hindgut epithelial cells with a specialised apical structure called an epimerite or mucron. Some species penetrate the cell completely. They next develop into trophozoite feeding forms that rapidly increase in size, eventually detaching and pairing up as gamonts in side-to-side or head-to-tail arrangements (syzygy). These paired forms are unmistakable in dissected gut preparations when viewed under low-power magnification (Figure 13.1). Subsequently, gametocysts form

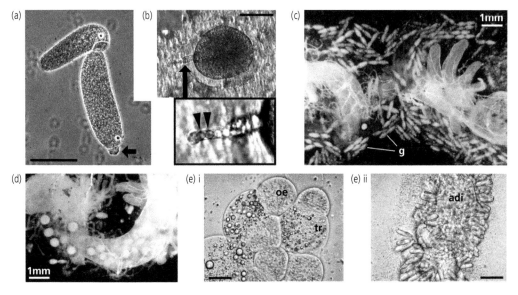

Figure 13.1 Gregarine infections in insects. (a) Eugregarine gamonts of unknown species in syzygy in the gut lumen of a mealworm larva *Tenebrio molitor*. Arrow indicates the epimerite. Scale bar: 50 μm. Source: Reuben James, Swansea University. (b) Dehiscing eugregarine gametocyst with apparent spore duct (arrow) from the faecal pellet of a buffalo worm larva (*Alphitobius diaperinus*). Oocysts can just be seen inside the spore duct [arrowheads in inset box]. Scale bar: 50 μm. Source: Chiara Marchisio, Swansea University. (c) Multitudes of eugregarine *Gregarina* sp. gamonts (g) in the midgut and gastric caeca of the cockroach *Blattella germanica*, with many appearing in syzygy. Source: from (Lopes and Alves 2005), with permission from Elsevier. (d) Spherical gametocysts of *Gregarina* sp. in the last portion of the midgut of the cockroach *Blattella germanica*. Source: from (Lopes and Alves 2005), with permission from Elsevier. (e) Bumblebee *Bombus pascuorum* female infected with the neogregarine *Apicystis cryptica* sp. n. (i) Healthy fat body globule with large oenocytes (oe) and trophocytes (tr). (ii) Hypertrophied fat body cells. Parasite oocysts are arrayed side-by-side near the cell periphery. Infected adipocytes (adi) are enlarged but other fat body cells are degraded. Scale bars: 25 μm. Source: from (Schoonvaere et al. 2020), with permission from Elsevier.

in the hindgut, which are also very characteristic (Figure 13.1).

Often, large numbers of eugregarine trophozoites simply cause an obstruction in the gut that limits nutrient absorption, eventually reducing overall fitness (Åbro 1971; Åbro 1996; Lucarotti 2000) or immune competence (Siva-Jothy and Plaistow 1999; Siva-Jothy et al. 2001). The gut epithelium can also be damaged and perforated, as in damselflies infected with *Hoplorhynchus* sp. (Åbro 1971; Åbro 1996) and in mosquito larvae infected with *Ascogregarina* spp. (Lantova and Volf 2014). In the latter example, the Malpighian tubules of pupae and adults also fill with gametocysts, causing distension and local tissue damage. The negative impact of *Ascogregarina* on certain *Aedes* mosquitoes may be sufficient at the population level to warrant its development as a biopesticide agent (Lantova and Volf 2014). So too might *Gregarina blattarum*, which can act synergistically with the fungus *Metarhizium* to kill cockroaches (Lopes and Alves 2005; see also Chapter 12, this volume).

Serious neogregarine infections are generally (but not exclusively) characterised by invasion of the haemocoel and consumption of the fat body (fat body hypertrophism) resulting in a range of pathologies including developmental problems, sluggish behaviour, and premature death (Table 13.1). Neogregarines *Apicystis bombi* and *Apicystis cryptica* sp. n. are currently spreading among bumblebee and honeybee populations as emerging diseases (Maharramov et al. 2013; Schoonvaere et al. 2020). Solitary bees (*Osmia* spp.) may also act as *Apicystis* reservoirs (Ravoet et al. 2014). *Apicystis bombi* was first discovered in Italy in 1988 and has since been recorded in nearly twenty *Bombus* species as well as solitary bees and the honeybee *Apis mellifera*. It is important to note that *A. cryptica* sp. n. was recently discovered in a metagenomic survey of *Bombus pascuorum* and may historically have been misidentified as *A. bombi* (Maharramov et al. 2013). Severe infections are associated with fat body destruction in infected workers and increased sucrose sensitivity, increased worker mortality, and reduced colony success. It is likely that the spread of the infection is assisted by commercial pollination services (Schoonvaere et al. 2020).

Neogregarine outbreaks can also reduce the efficiency of insect mass breeding programmes. Quick, reliable, and cheap diagnostics have been successfully adopted to regularly screen the health of some mass-reared species. For example, sweet potato weevil larvae (which are pests reared in Sterile Insect Technique programmes) exhibit characteristic signs of *Farinocystis* spp. infection (Table 13.1). After careful verification with microscopy and PCR, it has been established that a quick visual inspection provides a sufficiently reliable neogregarine diagnostic (Tsurui et al. 2015). Furthermore, the spores of some neogregarines, such as *Mattesia* spp. can often be seen through the intact sternites of live insects (Alfazairy et al. 2020).

13.2.1.2 Coccid disease—Coccidia (Phylum Apicomplexa)

Coccidia are apicomplexans that share many features with gregarines. Most coccidia are pathogens of vertebrates but a few are entomopathogens, albeit very under-researched. Of the seven genera described as entomopathogens (*Adelina, Barrouxia, Chagasella, Ganapatiella, Ithania, Legerella, Rasajeyna*), only species in the *Adelina* genus are commonly recorded (Table 13.1). Host insects are mainly larvae of lepidopteran or coleopteran species, including several that are regarded as stored product pests but also the economically important silkworm *Bombyx mori*. Transmission is primarily faecal-oral (or via cannibalism), beginning with the ingestion of oocysts that release sporocysts from which sporozoites develop (Yarwood 1937). In lepidopterans, the infection appears to be localised to the midgut and results in a form of diarrhoeal disease during which the larvae fail to develop. In beetles, sporozoites penetrate the gut and infect the fat body, in which oocysts, sporocysts, and sporozoites can be observed. Female beetles lose fat reserves and may die before oviposition, resulting in a loss of fecundity of up to 90% (King et al. 1981; Malone and Dhana 1988). *Adelina tenebrionis*, which causes coccid disease in tenebrioid beetles, is a potential bio-control agent (Alfazairy et al. 2020).

13.2.1.3 Amoebiasis—Amoebozoa supergroup

The Amoebozoa supergroup includes several genera responsible for amoebic disease (amoebiasis)

Table 13.1 Diseases of insects caused by parasitic protozoa from the phylum Apicomplexa (not exhaustive).

Apicomplexan parasite	Host order(s)	Host examples	Pathology & notes	Key References
Subclass Gregarinasina, order Eugregarinorida				
Gregarina spp. including *niphandrodes*, *typographi*, *cuneata*, *blattarum* von Siebold	Coleoptera, Blattodea	e.g. bark beetles (Scolitinae). *Tenebrio molitor*, cockroaches *Blatta* spp. & *Periplaneta* spp.	Infects midgut lumen. Sub-lethal or sub-pathogenic depending on parasite load. Synergistic action with fungus *Metharhizium anisopliae* or insecticide treatments.	Lopes and Alves (2005); Holuša et al. (2013); Yahaya et al. (2017)
Leidyana canadensis	Lepidoptera	e.g. eastern hemlock looper (*Lambdina fiscellaria*)	Gut infection of larvae. Occluded, swollen gut visible in live caterpillars. Reduced fitness.	Lucarotti (2000)
Leidyana apis	Hymenoptera	Honeybee (*Apis mellifera*)	Infects adult midgut epithelium. Heavy losses in apiculture.	Théodorides (1956)
Ascogregarina barretti & other spp.	Diptera, Siphonaptera	Mosquito e.g. *Aedes* spp., bat flies (Nycteribiidae), hump-backed flies (Phoridae), fleas	In mosquito, infects gut & Malpighian tubules. Compromised female development & fecundity when food is limited.	Walker et al. (1987); Lantova and Volf (2014)
Ascogregarina culicis & taiwanensis	primarily Diptera	Mosquitoes: Culicidae, *Aedes aegypti*, *Ae. albopictus*	Infects Malpighian tubules & midgut. Sublethal infection, reduced adult longevity. Associated with the ability of *Ae. aegypti* mosquitoes to transmit arboviruses vertically. Proposed as a biocontrol agent.	Blackmore et al. (1995); Lantova and Volf (2014)
Psychodiella spp.	Diptera	Sandflies (Psychodidae)	When the fly matures, the gregarines migrate to the female accessory glands; oocysts are released with the eggs during oviposition.	Lantova and Volf (2014)
Hoplorhynchus spp.	Odonata	Damselflies e.g. *Pyrrhosoma nymphula* & *Calopteryx virgo*	Infects midgut & hindgut. Gut blockage & perforation. Adult longevity compromised when environmental conditions are suboptimal.	Åbro (1971); Hecker et al. (2002); Tsubaki and Hooper (2004)
Diplocystis tipulae	Diptera	Cranefly (*Tipula paludosa*)	Infects haemocoel. Reduces fat reserves, retards larval growth.	Er and Gökçe (2005)
Subclass Gregarinasina, order Neogregarinorida				
Apicystis cryptica sp. n.	Hymenoptera	Bumblebee (*Bombus pascuorum*)	Emerging disease, widespread. Infects hypopharangeal gland and fat body. Hypertrophism of fat body. Can be fatal. Historically may have been confused with *Apicystis bombi*.	Schoonvaere et al. (2020)

continued

Table 13.1 *Continued*

Apicomplexan parasite	Host order(s)	Host examples	Pathology & notes	Key References
Apicystis (formerly *Mattesia*) *bombi*	Hymenoptera	Bumblebee (*Bombus terrestris* & others spp.), honeybee (*Apis mellifera*), solitary bees (*Osmia* spp.)	Emerging disease, widespread. Fat body hypertrophism in infected workers. Increased sucrose sensitivity, worker mortality, reduced colony success.	Maharramov et al. (2013); Ravoet et al. (2014); Plischuk et al. (2017)
Ophryocystis e.g. *elektroscirrha*	Coleoptera & Lepidoptera	Monarch & queen butterflies (*Danaus* spp.), *Tenebrio* spp.	Infects larval gut. In lab studies, high parasite density compromises larval survival, adult size & adult lifespan.	Altizer and Oberhauser (1999); Altizer et al. (2000)
Farinocystis spp.	Coleoptera	West Indian sweet potato weevil (*Euscepes postfasciatus*)	Infects gut & haemocoel. Late instar larvae appear pale/opaque. Swollen body & head, sluggish, swollen fat body, death. 'Sticky, flabby' cuticle. Problematic because hosts are mass-reared for SIT.	Tsurui et al. (2015)
Mattesia spp.	Coleoptera, Hymenoptera & Lepidoptera	e.g. spruce bark beetle (Scoltinae), fire ants (*Solenopsis* spp.)	Infects larval fat body & haemolymph. Moulting problems, deformities, premature death. In sublethal infections, fecundity & longevity are compromised.	Jouvenaz and Anthony (1979); Yaman and Radek (2015)
Mattesia geminate	Hymenoptera	Tropical fire ant (*Solenopsis geminata*)	Develops in larval hypodermis. Disrupts eye development. Severe cuticular melanisation. Pupa turns black & dies. Host species specific, i.e. cannot infect *Solenopsis invicta*.	Jouvenaz and Anthony (1979); Buschinger and Kleespies (1999)
Class Conoidasida, order Coccidia				
Adelina spp. including *sericesthis* & *tenebrionis*	Coleoptera & Lepidoptera, Orthoptera, Diptera, Blattodea	Scarab beetle, black beetle (*Heteronychus arator*), greenheaded leaf roller (*Planotortrix excessana*), tortrix moths, silkworm (*Bombyx mori*)	Coccid disease. In silkworm, larval midgut infected. Larvae develop poorly & are shrunken: diarrhoea, brown mucus around the anus. Cysts detected in faeces. In beetles, fat body of adults & larvae infected. Females lose fat reserves & can die before oviposition.	Malone and Dhana (1988); Alfazairy et al. (2020)

Note: the location/source of the parasite is assumed to be widespread unless noted otherwise in the table above.

in insects, many of which are collectively termed the 'malamoebae'. Most, but not all, genera of entomopathogenic amoebae develop in two forms: infective cysts that tolerate the external environment and an obligate endoparasitic trophozoite feeding stage that exhibits the classic features of an amoeba, i.e. the ability to form pseudopodia and to feed by phagocytosis. Research on entomopathogenic amoebae has fallen out of favour since the 1980s, despite their significant virulence in mass-reared insects and their potential as biopesticides. Many older descriptions of entompathogenic amoebae have unfortunately not been supplemented for more than a century (e.g. *Malpighiella refringens* in the rat flea; Minchin 1910). Their true diversity and incidence is therefore likely to be greatly under-reported and the taxonomic classification of many genera is *incertae sedis* (Adl et al. 2019) since, due to the paucity of molecular analyses, it is still based on microscopic observations of cyst morphology (e.g. Abdel Rahman et al. 2015).

Host species come from a wide range of insect orders including the Coleoptera, Hymenoptera (including honeybees), Siphonaptera, Diptera, Zygentoma, Archaeognatha, and most particularly the Orthoptera (Table 13.2). Lange and Lord (2012) provide an excellent review covering the majority of these species. While *Malpighamoeba mellificae* is thought to only infect adult *Apis mellifera* honeybees (King and Taylor 1936), *Malamoeba locustae* notably has a very broad host range including 50 orthopteran species and its transmission between other insect orders is probable (e.g. Larsson 1976).

Entomopathogenic amoebae infect the midgut and, in most cases, the Malpighian tubules (Figure 13.2). Transmission is via the faecal-oral route, usually by ingesting food contaminated with infective cysts. *Malamoeba locustae* is the best-described species and a major pathogen of locusts and grasshoppers, in which the lifecycle was carefully elucidated by King and Taylor (1936). Primary trophozoites emerging from ingested cysts infect the lumen and epithelium of the gastric caeca and midgut (Taylor and King 1937) and then develop amoeboid secondary trophozoites that migrate to Malpighian tubules about 1 week post infection (Braun et al. 1988). Here, cysts develop en

masse and after another week they begin to exit via the hindgut in the faeces.

Lightly infected insects appear asymptomatic. In heavy infections, however, cysts pack the entire lumen of the Malpighian tubules which take on a pale, opaque or 'glassy' appearance with significant hypertrophy (longer, larger surface area with a thinner epithelium; Rossi et al. 2019); (Figure 13.2). The brush border is damaged and sometimes the Malpighian tubule perforates with consequent infection of the body cavity (Taylor and King 1937). Melanin deposition and encapsulation reactions are often evident in the fat body and muscles (Larsson 1976). Typically, parasitised insects become sluggish and anorexic, then experience tetanic twitching, 'coma', and finally death. Older insects (nymphs and adults) are most severely affected. In Orthoptera, the final moult can be compromised, leading to adult deformities, and egg formation can be impaired (Taylor and King 1937). Malpighian tubules are the functional equivalent of kidneys but also have roles in metabolism, immunity, and detoxifying self-derived toxins, plant secondary metabolites, and xenobiotics. Heavy parasitisation will therefore disrupt all of these functions and this is a particular problem for phytophagous insects in arid habitats, which experience greater energy costs associated with reabsorbing vital fluids and ions (Rossi et al. 2019).

Taylor and King (1937) calculated that 2–3 million *M. locustae* cysts are expelled in a 24 h period from a single lab-infected *Melanoplus* grasshopper, so it is easy to see why amoebae pose a major outbreak risk to captive breeding systems. Amoebic disease afflicts cultured silkworms, although the precise identity of the amoeba remains unknown (James and Li 2012). Somewhat controversially, *Malpighamoeba mellificae* is implicated in amoebiasis in honeybee (*Apis mellifera*) workers (Austrian Agency for Health and Food Security et al. 2014), however the UK Animal and Plant Health Agency point out that notable symptoms may only arise when a bee is co-infected with *Nosema apis* (Bailey 1968, Animal & Plant Health Agency, UK, accessed 2020). (See Chapter 12 for details of nosemosis.)

Although some gross pathological changes to the Malpighian tubules may be seen macroscopically, traditionally a diagnosis must be made by

Table 13.2 Examples of amoebic diseases of insects.

Pathogen	Host order(s)	Host examples	Pathology & notes	Key References
Supergroup Amoebozoa				
Malamoeba locustae	Orthoptera, Zygentoma (formerly Thysanura)	Broad host range including locusts, grasshoppers, silverfish	Infects Malpighian tubules, midgut, gastric caeca. Causes moulting problems, delayed development, anorexia, sluggishness, twitching, reproductive failure. Co-infections with gregarines are common.	King and Taylor (1936); Larsson (1976); Ernst and Baker (1982); Hinks and Ewen (1986); Abdel Rahman et al. (2015)
Malamoeba scolyti	Coleoptera	Bark beetles (e.g. *Ips typographus* & *Dryocoetes autographus*)	Gut infection & destruction of Malphigian tubule epithelium (tubules appear thin and milk-coloured). Significant mortality, especially in pupae.	Purrini and Žižka (1983); Michalková et al. (2012)
Malpighiella refringens incertae sedis	Siphonaptera	Rat flea (*Ceratophyllus fasciatus*)	Infects Malpighian tubules. Sublethal effects.	Minchin (1910)
Dobellina mesnili nov. gen. incertae sedis (formerly *Entamoeba mesnili*)	Diptera	Winter crane fly (*Trichocera sp.*)	Infects gut lumen of overwintering larvae. Dense parasite mats form. Estimated up to 92% of larvae are infected in the UK.	Bishop and Tate (1939)
Entamoeba spp. (precise identity unknown)	Lepidoptera	Silkworm (*Bombyx* spp.)	Silkworm amoebic disease. Infects the gut of late instar larvae. Cysts develop in gut epithelial cells, which swell & collapse. Larval growth slows, the posterior end shrinks, faeces turn green. Upon death, cadaver turns black & becomes mummified.	James and Li (2012)
Vahlkampfia sp.	Archaeognatha	Bristletail (*Promesomachilis hispanica*)	Midgut infection with destruction of epithelium brush border.	Larsson et al. (1992)
Malpighamoeba locustae	Orthoptera	Desert locust (*Schistocerca gregaria*)	Infects gut and Malpighian tubules. Impaired osmoregulation; tubules appear swollen (packed with cysts) and cloudy. Reduced feeding. Can be fatal.	Rossi et al. (2019)
Malpighamoeba mellificae	Hymenoptera	Honeybee (*Apis mellifera*) (?)	Pathology in adult bees, mainly workers. Bees become sluggish, shake wings without flying. Faeces become yellow and more fluid (like amoebic dysentery) with a strong unpleasant odour. (Note: often co-infects with Nosema, which may cause some of the previously given symptoms).	Liu (1985)

Note: the location/source of the parasite is assumed to be widespread unless noted otherwise in the previous table.

microscopy (trophozoites in the midgut and gastric caeca; trophozoites and cysts in the Malpighian tubules; cysts in the hindgut and faeces). Trophozoites are roughly 5–10 µm in diameter and amoeboid, and cysts are of a similar size but more conspicuous due to their oval shape and the refractile nature of their thick walls (Figure 13.2c–d).

Fixation and staining can obviously affect shape and size and it is possible to confuse amoeba cysts with other parasites.

Field trials with amoebae as biocontrol agents against orthopterans have been limited. When Taylor and King (1937) scattered cyst-laden faeces from lab insects along roadsides in the USA, the local

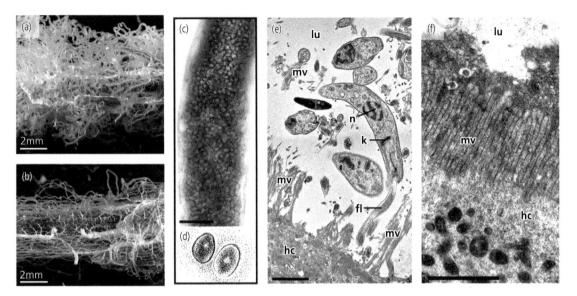

Figure 13.2 Amoebic and trypanosome infections of insects. (a) Swollen, cloudy Malpighian tubules from a *Malpighamoeba*-infected locust. Source: from (Rossi et al. 2019). (b) The Malpighian tubules from an uninfected locust are thinner and more transparent. Source: from (Rossi et al. 2019). (c) Detail of the lumen of an infected locust Malpighian tubule filled with *Malpighamoeba* cysts. Source: from (Rossi et al. 2019). Scale bar: 50 μm. (d) Diagram of cysts at high magnification. Source: from Miranda M.A. Whitten. (e) TEM showing degradation of the brush border of a firebug (*Pyrrhocoris apterus*) infected with the trypanosomatid *Blastocrithidia papi*. Microvilli (mv) of the host epithelial cell (hc) are largely absent or in fragmented groups. Note the nucleus (n), kinetoplast (k) and flagellum (f) of a trypanosomatid in the gut lumen (lu). Source: from (Frolov et al. 2018). Scale bar: 2 μm. (f) TEM of the brush border structure of an intact Malpighian tubule from a healthy firebug for comparison. Source: from (Frolov et al. 2018). Scale bar: 2 μm.

grasshopper infection rate rose modestly from 0.3% to 4.7%. Bearing in mind that species specificity was not taken into account, this was an encouraging first attempt. Lange (2002) indicated that seasonal inoculative releases of *Malamoeba locustae* cysts in Argentina could aid transmission and allow long-term persistence. Clearly, however, there is not yet any feasible hope of using amoebae to control locust plagues.

13.2.1.4 Trypanosomiasis (class Kinetoplastida)

Trypanosomatids (class Kinetoplastida, family Trypanosomatidae) are flagellated protozoa that possess a kinetoplast—a unique organelle containing mitochondrial DNA situated close to the origin of the single flagellum. Several trypanosomatids cause disease in humans, livestock, and crops after being vectored by insects (e.g. sleeping sickness, n'gana, phytomonas). Many more related trypanosomatid species are directly pathogenic in insects.

The entompathogenic trypanosomatids belong to the genera *Lotmaria*, *Crithidia*, *Blastocrithidia*, *Leptomonas*, *Rhynchoidomonas*, *Herpetomonas*, and *Trypanosoma* and they primarily target hemipteran and hymenopteran hosts (Table 13.3) and their distribution is effectively worldwide. Classically, trypanosomatids have been categorised according to six major morphotypes based on the flagellated stage of development but this chapter does not dwell on classification as it complicated and molecular identification methods are triggering a major taxonomic revision. There are several reviews on the subject (e.g. Maslov et al. 2013; Wheeler et al. 2013).

Disease in insects is usually inflicted by the monoxenous (single-host) trypanosomatids, with variable species-specific impacts on their health. These protozoa infect insects via oral-faecal transmission (and cannibalism). After ingestion, the trypanosomatid either remains unattached in the gut lumen, attaches to the gut epithelium in a thick carpet (often in the hindgut, with invasion of the

Table 13.3 Trypanosomatid diseases of insects caused by protozoa from the class Kinetoplastea (not exhaustive).

Pathogen	Host order(s)	Host examples	Pathology & notes	Key References
Lotmaria passim	Hymenoptera	Honeybee (*Apis mellifera*)	The predominant trypanosomatid in *Apis mellifera* (global). Infects adult hindgut. Reduced worker longevity. Effects on bee & colony health are poorly understood but possibly a cause of winter mortality.	Schwarz et al. (2015); Gómez-Moracho et al. (2020); Quintana et al. 2021
Crithidia mellificae	Hymenoptera	Honeybee (*Apis mellifera*), Africanized honeybee	Infects adult hindgut lumen & epithelium. Reduced worker longevity. Strongly implicated in colony winter loss. Pathology undergoing re-evaluation because prior to 2014 it was not realized that *Lotmaria passim* co-infects with *Crithidia mellificae*.	Ravoet et al. (2013); Gómez-Moracho et al. (2020)
cand. *Crithidia mexicana* sp. nov.	Hymenoptera	Bumblebee (*Bombus* spp.)	Recently identified in bumblebees from Mexico. Pathological effects not yet studied; may represent an emerging parasite.	Gallot-Lavallée et al. (2016)
Crithidia expoeki	Hymenoptera	Bumblebee (*Bombus* spp.)	Partially castrates founding queens, greatly reduces fitness.	Schmid-Hempel et al. (2018)
Crithidia bombi	Hymenoptera	Bumblebee (*Bombus* spp.)	Infects gut. Hibernating queens lose mass & are partially castrated. Significantly reduces colony-founding success, colony size, male production & overall fitness, by up to 40%. Impairs learning ability. Mortality of food-stressed workers increased by up to 50%.	Brown et al. (2000); Gegear et al. (2006); Schmid-Hempel et al. (2018)
Crithidia fasciculata	Diptera	Anopheline mosquito (*Anopheles* spp.)	Forms dense mats in hindgut. Likely sub-pathogenic linked to energy cost of supporting infection.	Brooker (1971)
Blastocrithidia papi	Hemiptera	Firebug (*Pyrrhocoris apterus*)	Infects midgut, hindgut; degrades Malpighian tubule brush border. Disrupted peristalsis, constipation, then defaecation at oviposition, which coats eggs with infective cysts.	Frolov et al. (2018)
Blastocrithidia triatomae	Hemiptera	Several Reduviid bug genera e.g. *Triatoma* (but not *Rhodnius* spp.)	Infects gut. Aberrations in food ingestion, digestion, excretion, moulting, immune response. Oxygen deprivation to hindgut & Malpighian tubules. Developmental deformities, reduced longevity.	Schaub and Schnitker (1988); Schaub (1994); Schaub (2009)
Trypanosoma rangeli	Hemiptera	Reduviid bug (*Rhodnius* spp.)	Invades gut, haemocoel, haemocytes, fat body & salivary glands. Immunosuppression, sickness syndrome affecting multiple organs in later instars. Discoloured opaque haemolymph. Difficulties feeding & moulting, deformities, compromised development, increased mortality.	Grewal (1957); Tobie (1970); Añez and East (1984); Schaub (2006)
Leptomonas wallacei	Hemiptera	Lygaeid bug (*Oncopeltus fasciatus*)	Infects midgut. Delayed nymphal development. Reduced lifespan when exacerbated by stress, especially in females. Fewer eggs laid; fewer eggs eclose. Egg reabsorption. Morphological abnormalities.	Vasconcellos et al. (2019)

Note: the location/source of the parasite is assumed to be widespread unless noted otherwise in the previous table.

Malpighian tubules), or traverses the epithelium to invade the haemocoel. The microvilli are commonly damaged (Figure 13.2e), and occasionally the epithelium is so badly degraded that the basal lamina is exposed. There are only a few examples of heteroxenous entomopathogenic species, the most notable being *Trypanosoma rangeli* (discussed in the following section).

Many of the most overt trypanosome-related pathologies have been described from observations of infections in the blood-sucking reduviid bugs (Hemiptera) and these host-parasite relationships were intensively studied by Schaub and colleagues (e.g. Schaub and Schnitker 1988; Schaub and Loesch 1989; Schaub 1994; Eichler and Schaub 1998). *Trypanosoma rangeli* is pathogenic exclusively to insects from the genus *Rhodnius* in which it invades the gut, haemocoel, and salivary glands. On the other hand, the monoxenous trypanosomatid *Blastocrithidia triatominae* is pathogenic to insects within several other reduviid genera but interestingly the *Rhodnius* spp. are refractory to *B. triatominae*. Both *B. triatominae* and *T. rangeli* infections manifest as sickness syndromes affecting multiple organs that worsen in later instars. Typical symptoms include damage to the gut epithelium or Malpighian tubule brush border, constipation or altered peristalsis, developmental delay, or reduced longevity and fecundity. Infected insects are sluggish and exhibit aberrant behaviours including problems with feeding. That *Rhodnius* species alone—out of many other reduviid bugs—should be refractory to *B. triatominae* is something of a puzzle. A convincing hypothesis is that the *R. prolixus* symbiotic gut bacterium (*Rhodococcus rhodnii*) confers refractoriness, since this symbiont is lacking in the other reduviids that are permissive to *B. triatominae*. Eichler and Schaub (1998) observed that *B. triatominae* limits tracheole development in permissive hosts and this symptom can be reversed by B-vitamin supplementation. Underdevelopment of the tracheoles is also exhibited by *R. prolixus* when reared aposymbiotically (i.e. without its symbiotic bacteria). Eichler and Schaub (1998) then further established that *R. rhodnii* supplies B-vitamins to *R. prolixus*; this is in effect an example of a symbiont synthesising probiotics for its host.

Trypanosomatids also compromise fecundity. In the hemipteran firebug *Oncopeltus fasciatus*, *Leptomonas* infection of the midgut leads to reduced longevity (especially in stressed females), morphological abnormalities, and reduced fecundity with egg reabsorption (Vasconcellos et al. 2019). A thick carpet of parasites forms on the gut wall that interferes with its function and reduces the available microvillar surface area. *Blastocrithidia papi* disrupts firebug peristalsis and triggers defaecation behaviour concomitant with oviposition. This coats the eggs with infective cysts and in turn allows parasite transmission to the newly hatched offspring (Vasconcellos et al. 2019).

The host interactions in trypanosomatids are highly variable. Mixed infections are common and this has led to confusion in ascribing a particular disease to a particular trypanosomatid species. For example, two trypanosomatid species are now known to infect the hindgut of the honeybee *Apis mellifera*: *Lotmaria passim* which was discovered in 2015 (Schwarz et al. 2015) when molecular tools became available to differentiate it from *Crithidia mellificae* with which it often co-infects. It is now becoming apparent that of the two species, *L. passim* is more prevalent (e.g. Quintana et al. 2021) and probably the more virulent (Gómez-Moracho et al. 2020), and it may be the major cause of colony winter mortality.

In bumblebees, *Crithidia bombi* gut infections—which are very common in the wild—increase the mortality of food-stressed workers up to 1.5-fold and reduce the reproductive output of queens. The interaction between *Crithidia bombi* and the bumblebee has been studied intensively and it is an important model system of host-parasite evolutionary ecology (for a review see Schmid-Hempel et al. 2019). At the population level, *C. bombi* profoundly impacts colony size and colony-founding success, especially when combined with other stressors (Brown et al. 2000; Brown et al. 2003). The parasite mainly colonises the hindgut and infective cells will be voided in faeces within 3–4 days of infection. There is neither vertical transmission nor a cyst stage, so the parasite must overwinter within the gut of a mated daughter (new queen) as she hibernates. Laboratory-based experiments indicate that founding queens become partially castrated by the spring, with resource allocation diverted to the fat body at the expense of the reproductive system (Brown et al. 2000).

Bumblebees transmit *Crithidia bombi* via faecally contaminated surfaces (on flowers or in the nest), or via infected larvae and nectar (reviewed by Schmid-Hempel et al. 2019). Heavily parasitised bumblebees have impaired memory and difficulty in learning how to manipulate new types of flowers, possibly due to excess nitric oxide generated as part of the immune response (Gegear et al. 2006). Nevertheless, bumblebees adopt prophylactic avoidance behaviours towards flowers that are already contaminated with *C. bombi* (Fouks and Lattorff 2011) and infected bees even appear to self-medicate since they spend considerably longer foraging flowers with high nectar iridoid glycoside secondary metabolites (e.g. catalpol), which suppress *Crithidia* infections (Richardson et al. 2015).

Understandably, there is mounting anxiety about the role of commercial bumblebees' transportation for pollination in the spread of these diseases. Considering their propensity to cause co-infections, molecular methods such as ITS1 RFLP (Ravoet et al. 2015) are now replacing the microscopic examination of faeces and dissected gut to discriminate trypanosomatid species.

13.2.1.5 Ciliatosis—Ciliates (Phylum Ciliophora)

Ciliatosis was first described in 1921 by Lamborn in Kuala Lumpur, who noted an abundance of ciliated protozoa—*Lambornella stegomyiae*—in unusually pale, moribund larvae of *Aedes albopictus* mosquitoes (Lamborn, 1921). Two families and four genera of ciliates contain confirmed entomopathogens (Table 13.4): *Lambornella*, *Tetrahymena*, *Chilodonella*, and *Ophryoglena* and some are clearly able to synchronise with the lifecycle of their hosts. They are commonly found in aquatic breeding sites, especially in paddy fields, tree holes and ponds (Washburn and Anderson 1986; Das 2003). It is notable that dipteran larvae are common hosts, including many species of mosquitoes, midges and blackfly that vector other diseases. Ciliates are therefore regarded as potential biocontrol agents. Of the four species known to infect mosquitoes (*Lambornella stegomyiae*, *L. clarki*, *Tetrahymena pyriformis*, and

Chilodonella uncinata), the latter two are candidates for commercial development as bio-larvicides.

Ciliates undergo a shift from a free-living bactivorous organism (trophont) to a fast-moving host-seeking parasitic one (theront) (Batson and Beale 1983). *Lambornella* spp. and *Ch. uncinata* switch to theronts when mosquito larvae enter their aquatic environment, as might happen when a female mosquito lays her eggs. The switch may be a defensive strategy to avoid predation by the mosquito larvae (Mercer and Anderson 1994; Arshad and Sulaiman 1995; Zaritsky et al. 1992). If the infected host survives to adulthood (as in some *Lambornella*, *Chilodonella*, and *Ophryoglena* infections) ciliates can also use insects as a way to disseminate to new habitats (Das 2013). Ciliatosis commences when theronts form invasion cysts that either actively penetrate the cuticle (e.g. *Ch. uncinata* and *L. clarki*) or enter passively through natural openings (e.g. *Tetrahymena* spp.). Opportunistic cuticular invasion during moulting and via pre-existing damage (wounds) are also likely (Corliss 1960). Chironomid midge larvae are particularly susceptible in this regard because cuticle abrasions and wounds often arise from attacks by co-infecting mermithid nematodes (see Section 13.2.2.1); (Golini and Corliss 1981). Very few cysts are needed to set up successful systemic infections (reviewed by Washburn et al. 1991). An endoparasitic replicative form then squeezes out through a small hole made at the base of the cyst and penetrates into the haemocoel where it replicates systemically (Figure 13.3). Interestingly, there are no reports of ingested ciliates being able to invade through the gut (Corliss 1960; Zaritsky et al. 1992).

The endoparasitic ciliate indiscriminately consumes host tissues including haemocytes, muscle, and fat body (Batson 1985; Gaino and Rebora 2000; Das 2003). The host cuticle tends to become more translucent and ciliates can be visualised clearly through the cuticle of moribund alder fly and mosquito larvae moving rapidly in the body cavity (Figure 13.3), head, and even inside the antennae. Ciliates are identifiable in haemolymph samples as pear-shaped ('pyriform') cells, 50–100 μm long with many tiny fast-beating cilia, and inclusions in the cytoplasm make the cells appear opaque under transmitted light (Batson 1985; Figure 13.3). Authors describing these infections tend to use verbs like

Table 13.4 Parasitic diseases of insects caused by ciliated protozoa, and an algal species.

Pathogen	Host order(s)	Host examples	Pathology & notes	Key References
Phylum Ciliophora: Order Hymenostomatida				
Tetrahymena pyriformis complex	Diptera	Many species including mosquito *Aedes* spp.	Infects aquatic larvae: systemic infection. Large ciliates can be seen through cuticle. Has potential in biocontrol for delivering Bti spores.	Zaritsky et al. (1992)
Tetrahymena (1) *dimorpha* (2) *sialidos*	(1) Diptera, (2) Megaloptera	(1) Black fly (*Simulium equinum* & *S. ornatum*), (2) alder fly larvae (*Sialis lutaria*)	Infects haemocoel & most organs (not gut) of all developmental stages. Dramatic increase in parasites in adults; death 2–3 days after eclosion.	(1) Batson and Beale (1983); (2) Batson (1985)
Tetrahymena chironomi	Diptera	Chironomid midges (e.g. *Chironomus plumosus*)	Facultative pathogen infecting haemocoel, head, Malpighian tubules & anal papillae. Slow-moving larvae die before pupation.	Corliss (1960); Golini and Corliss (1981)
Lambornella spp. (e.g. *clarki, stegomyiae*)	Diptera	Mosquitoes (e.g. *Aedes sierrensis, Anopheles* spp).	Penetrates larval cuticle. Infects ovaries & causes parasitic castration. Female oviposits parasite instead of eggs.	Egerter et al. (1986); Mercer et al. (2010)
Ophryoglena sp.	Ephemeroptera	Mayfly (*Baetis, Caenis* & *Centroptilum* spp.)	Systemic infection of nymph haemocoel. Destroys several tissues & parasitically castrates females. Often lethal.	Gaino and Rebora (2000); Ball and Turner (2003)
Phylum Ciliophora: Order Chlamydodontida				
Chilodonella uncinata	Diptera	Mosquitoes (*Anopheles, Aedes, Culex* spp.)	Systemic infection of larvae (head capsule, antennae, body cavity, saddle, anal gills & siphon). Extensive tissue damage. Death before pupation. Potential biocontrol agent.	Das (2003); Das (2019)
Phylum Chlorophyta: Order Chlorellales				
Helicosporidium parasiticum	Broad host range including Coleoptera, Diptera, Lepidoptera	e.g. Black fly (*Simulium jonesi*), corn earworm (*Helicoverpa zea*), culicine mosquitoes, carpet beetles	Infects midgut, haemocoel, fat body, muscle of larval host. Developmental deformities, reduction in size, mortality, reduced longevity & fecundity. Insects often fail to reach adulthood.	Kellen and Lindegren (1973); Hembree (1979); Boucias et al. (2001); Yaman and Radek (2007)

Note: the location/source of the parasite is assumed to be widespread unless noted otherwise in the previous table.

'astonishing' to describe the sheer number of the parasites. The time course over which the host tissues are consumed varies considerably from a few weeks to very many months, depending on the species but eventually the insect becomes moribund, and the ciliates convert back into trophonts and escape from the disintegrating host tissues.

Some ciliate pathogens (e.g. *Ch. uncinata* and *T. chironomi*) kill their hosts with great rapidity prior to pupation (Das 2003, Das 2019; Das and Tuli 2019) and these offer the most potential as biocontrol agents, as well as posing a considerable risk to captive mosquito rearing facilities (Dr Bina Pani Das, personal communication). In laboratory infections, the mortality rate can be extremely high (for example 99.5% of *Aedes albopictus* infected with *Lambornella stegomyiae*; Arshad and Sulaiman 1995). Nevertheless, a few adult mosquitoes may emerge with sub-lethal infections, albeit often with limited fertility, to aid dispersal of the ciliate.

Some hosts like the alderfly (*Sialis lutaria*) have much longer lifecycles that involve overwintering.

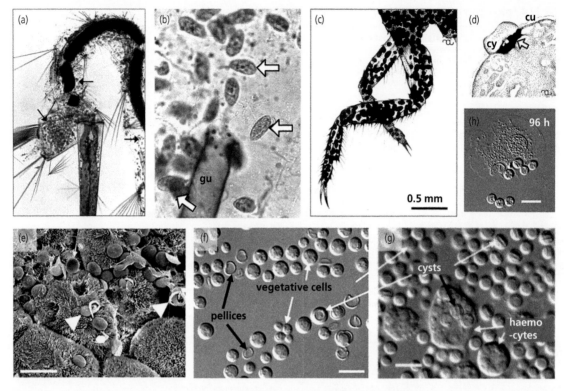

Figure 13.3 Ciliate and helicosporidial (algal) infections of insects. (a) Photomicrograph of a dead 4th stage mosquito larva (*Culex tritaeniorhynchus*) infected with numerous *Chilodonella uncinata* ciliates (arrows) visible through the excessively transparent cuticle (Das 2003). (b) Detail of the same mosquito larva at higher magnification showing ciliates (arrows) and the degenerating host gut (gu) (Das 2003). (c) Legs of an infected alderfly showing a heavy infection with *Tetrahymena sialidos* ciliates visible though the cuticle. Source: redrawn by Miranda M.A. Whitten from (Batson 1985), with permission from The Royal Society. (d) Failed *Lambornella clarki* cyst (cy) on the outside of a mosquito cuticle (cu). Note the melanised body underneath (arrow.) Source: redrawn by Miranda M.A. Whitten from (Corliss and Coats 1976), with permission from John Wiley & Sons Inc. (e-h) *Spodoptera exigua* (beet armyworm) larvae after oral challenge with *Helicosporidium* cysts. Source: from (Bläske-Lietze and Boucias 2005), with permission from Elsevier. (e) SEM of the midgut lumen 4 hours post ingestion, showing intact spheroid cysts and dehiscing cysts (arrowheads) releasing invasive filamentous cells. (f, g) haemolymph sampled at 168 hours (f) and 240 hours (g), showing freely circulating vegetative cells, pellicles, and cysts; (h) cyst development inside a haemocyte. Scale bars: 2 μm.

The lifecycle of the ciliate *Tetrahymena sialidos* has evolved to synchronise with its host by limiting replication of the endoparasitic stage over winter, then explosively peaking the following summer. The death of the (now adult) host coincides with the emergence of the next generation of insect larvae, so that ciliates can simply emerge from one corpse and quickly infect a new victim (Batson 1985). A similar situation occurs in *Tetrahymena dimorpha* infections of blackflies (Batson and Beale 1983) and in *Ch. uncinata* infections of adult female *Culex tritaeniorhynchus* in Northern India in which transovarial transmission occurs in the overwintering hosts. Parasitic castration of adult females is another feature in some ciliatoses. Systemic *Ophryoglena* sp. infection of several mayfly species compromises egg development, specifically by destroying follicle cells, and this results in the failure of the synthesis of egg membranes. Adult females that survive infection for long enough nevertheless experience ovarian distension due to the number of parasites in the tissue and this triggers an oviposition behaviour in which ciliate masses are 'laid' instead of eggs (Ball and Turner 2003). Parasitic castration is an effective manipulative dispersal mechanism that is utilised by several types of entomopathogen and it also occurs in mosquitoes infected by *L. clarki* and *Ch. uncinata* (Egerter et al. 1986; Das 2013).

Field studies by Das (2013) indicate that the high incidence of *Ch. uncinata* in Northern India provides a 'natural check' on the abundance of mosquito vectors of Japanese encephalitis (JE), despite the presence of other risk factors for JE outbreaks. Effective wide-scale control of vector mosquito species employing bio-larvicides requires appropriate formulation and easy application. *Chilodonella uncinata* has many useful properties such as high virulence, resistance to desiccation, simple laboratory culture and transport and dispersal via 'infusion bags' designed for use in paddy fields and other aquatic environments (Das 2019; Das and Tuli 2019). *Tetrahymena pyriformis* has also been scrutinised as a vehicle to deliver *Bacillus thuringiensis* subsp. *Israelensi* (Bti) toxin via its food vacuole; a kind of 'living microencapsulation' (Manasherob et al. 1998; see also Chapter 11, this volume).

An immune response against theront cysts is sometimes seen at the cuticle, which may appear 'peppered' with melanised wounds indicating encapsulation and healing triggered by failed invasions (Batson 1985; Corliss and Coats 1976); (Figure 13.3d). However, there is no obvious systemic immune response, suggesting that ciliates either suppress or quickly overwhelm an insect's immune capacity. Batson (1985) speculated that the constantly motile nature and size of the ciliates would prevent effective cellular immunity but this publication predates much of what we now know about humoral defences in insects (Chapter 1) and the subject is in need of further attention. Indeed, with the exception of the aforementioned studies aimed at biocontrol and the development of protocols for *in vitro* culture (Strüder-Kypke et al. 2001), insect ciliatosis has received surprisingly little research in recent decades.

13.2.1.6 Algae

Helicosporidium parasiticum (Phylum Chlorophyta: Order Chlorellales) is the only known species of entomopathogenic alga, though there are probably several infective strains. Its appearance does not resemble an alga as it lacks chlorophyll and other pigments, and this fact hampered its proper classification until the availability of molecular analyses. It is an obligate endoparasite with a wide host range that includes members of the Coleoptera, Diptera, and Lepidoptera with a worldwide distribution (Kellen and Lindegren 1973); (Table 13.4).

Transmission of *H. parasiticum* is horizontal, via the faecal-oral route (Boucias et al. 2001; Bläske-Lietze et al. 2006). Infective barrel-shaped 3.5 µm diameter cysts are ingested (usually by a larva) and upon entering the midgut the cysts attach to the peritrophic matrix and dehisce to release a corkscrew-shaped filamentous cell that uncoils from three non-motile ovoid cells (Boucias et al. 2001; reviewed by Tartar 2013); (Figure 13.3). While the ovoid cells quickly degenerate, the filamentous cell breaches the peritrophic matrix and midgut epithelium and invades the haemocoel within 1–2 days (Boucias et al. 2001). Light microscopy of haemolymph samples or body squashes show a range of characteristic vegetative cell clusters and pellicles, and (at later stages of infection) empty pellicles and barrel-shaped cysts (Figure 13.3). The vegetative cell stages appear to evade host cellular immune defences and eventually the haemolymph fills with overwhelming numbers of parasites. At death, the cuticle disintegrates and releases cysts into water or aquatic microenvironments such as plant sap (Bläske-Lietze et al. 2006). In Lepidoptera, infected larvae sometimes manage to pupate but the adults exhibit wing deformities, lower fecundity, and reduced longevity (Bläske and Boucias 2004).

Helicosporodium spp. were once considered as a potential biocontrol agent because it infects several species of arbovirus-transmitting mosquito species (Hembree 1979). Unfortunately, the parasite is not sufficiently species-specific and the tables have now turned: it is an emergent problem in the mass-rearing of beneficial insects such as *Rhizophagous grandis* (a coleopteran predator of the pest bark beetle *Dendroctonus micans*) (Yaman and Radek 2007).

13.2.2 Helminth (worm) infections

Helminths are a diverse group of unsegmented metazoan worms with elongated, flat or round bodies. They all develop through egg, larval (juvenile) and adult stages but the lifecycles vary dramatically by species. Many helminths are parasites of medical significance, some of which are vectored by insects. However these parasites usually have only subtle impacts on the insect host and will not be discussed in this section. The principal helminth diseases of insects are summarised in Table 13.5 and are caused by the nematodes (roundworms), nematomorphs (hairworms or gordiid worms), and trematodes (flukes).

13.2.2.1 Nematodes and their bacteria

Nematodes are non-segmented roundworms, with a complete gut and a nervous system, and they are usually dioecous. Apart from the fact that all nematodes require water for reproduction, it is very hard to make generalisations about nematodes that parasitise insects. Some have broad host ranges, while others are species-specific; some kill very rapidly with the aid of symbiotic bacteria while others do not rely on bacteria and create sub-lethal chronic infections synchronised to the host's lifecycle; some manipulate host behaviour and there are multiple strategies for host seeking and host invasion. Nematodes can have phoretic (hitch-hiking), facultative, or obligate relationships with insects.

This section aims to give a broad overview of nematodes including some lesser-known species and those that have a bearing on the survival of mass-reared insects, with examples summarised in Table 13.5. Detailed lifecycles will not be discussed in this section. Readers wishing for more depth, particularly on the entomopathogenic Heterorhabditidae and Steinernematidae, are directed to Lewis and Clarke (2012) and Bhat et al. (2020) for up-to-date lists of entomopathogenic nematodes and their geographical distributions.

While nematodes from many genera can be considered entomoparasitic, the entomopathogenic nematodes (EPNs) are a special guild of nematodes whose members have evolved with symbiotic gut bacteria, all of which are lethal pathogens to many insects (Figure 13.5g,–h). Disease is caused primarily by the bacteria (through a combination of toxic bacterial metabolites, immune suppressants, and overwhelming multiplication) but also to varying extents by the nematode itself (excretory-secretory products that destroy tissues and subvert host immunity). These virulence factors will be explored more fully below. In nematology, 'entomopathogenic' therefore has a very specific definition: an EPN rapidly kills its host within 5 days of infection with the aid of a bacterial partner—this is not necessarily an obligate relationship but the worm must pass on the bacteria to future generations—and there is evolutionary selection for increasing virulence (Dillman et al. 2012). Most known EPNs belong to the Heterorhabditidae and Steinernematidae families and they form an 'insecticidal complex' with Gram-negative bioluminescent *Photorhabdus* spp. and *Xenorhabdus* spp. symbiotic bacteria respectively (Gammaproteobacteria: family Enterobacteriaceae).

There has been interest in a possible new group of EPNs from the family Rhabditidae, whose intestinal mutualistic bacteria are *Serratia* species. Certainly, several *Serratia* species are entomopathogenic (e.g. Patil et al. 2012); (see Chapter 11). EPN status is tentatively ascribed to *Heterorhabditidoides* (*Oscheius*) *safricana*, which rapidly kills *Galleria mellonella* larvae in the laboratory (Serepa-Dlamini and Gray 2018.) These nematodes may be mid-way in their evolution between scavengers and true EPNs and will doubtless receive much further study. It should be noted that experiments exploring 'EPN potential' are usually performed with model 'pan-susceptible' host insects (waxmoth or mealworm larvae, i.e. *Galleria mellonella* or *Tenebrio molitor*). It is rare to find EPNs in their natural hosts in the wild because their bodies disintegrate so quickly, so it is normal practice to use model insects as live bait to search for candidate EPNs in soil (Kaya and Stock 1997). This means that the identity of the true host, or host range, is frequently unknown.

The rest of the (non-EPN) entomoparasitic nematodes are mainly represented by the mermithids, ascaradid, and other rhabditid species. Mermithidae is a large family comprising at least ten genera, which parasitise a wide range of insects and frequently cause epizootics (Nickle 1972). *Romanomermis* species such as *culicivorax* are possibly the most studied since these have potential as biocontrol agents against culicine mosquitoes. Entomoparasitic nematodes are under evolutionary selection for decreasing virulence and do not utilise mutualistic bacteria but together with nematomorph worms (Section 13.2.3.2) they share the ability to

Table 13.5 Examples of nematode, trematode and nematomorph worm infections of insects (not exhaustive).

Worm	Host order(s)	Host examples	Pathology	Key References
Phylum Nematoda: Class Chromadorea (Phasmids), order Rhabditida, family Steinernematidae (EPNs)				
Steinernema (~ 100 spp.) and endosymbiont *Xenorhabdus spp.* (~ 26 spp.)	Broad host range including Thysanoptera, Coleoptera, Diptera, Orthoptera, Anoplura, Hymenoptera, Blattodea, Hemiptera	Thrips, sciarid flies, weevils, fungus gnats, termites, crickets, scarab beetles, leaf miner flies, many other soil insects	Highly virulent. Worm releases bacterial symbiont inside host, which rapidly multiply & kill insect within 1–2 days. Core venom proteins are both tissue-damaging & immune-modulating. Characteristic brown cadaver.	Peters (1996); Chang et al. (2019)
Phylum Nematoda: Class Chromadorea (Phasmids), order Rhabditida, family Heterohabditidae (EPNs)				
Heterorhabditis (~ 21 spp.) and endosymbiont *Photorhabdus spp.* (~ 19 spp.)	Broad host range including Coleoptera, Lepidoptera, Diptera	Phlebotomine sandfly, scarab beetles, weevils, noctuid moths	Highly virulent. Bacterial symbiont kills larvae & suppresses the growth of microbial competitors. Characteristic red cadaver that disintegrates into sticky, gummy masses.	Peters (1996); Clarke (2008); Dillman et al. (2012)
Phylum Nematoda: Class Chromadorea (Phasmids), order Rhabditida, family Rhabditidae (EPNs?)				
Oscheius safricana and mutualist bacterium *Serratia marcescens*	(Lepidoptera) host range unknown	e.g. wax moth (*Galleria mellonella*)	Kills wax moth larvae; currently unclear whether *Oscheius* spp. are entomopathogenic or necromenic (scavengers).	Serepa-Dlamini and Gray (2018)
Phylum Nematoda: Class Chromadorea (Phasmids), order Rhabditida, family Sphaerulariidae				
Deladenus (Beddingia) siricidicola	Hymenoptera	Pine woodwasp (*Sirex noctilio*)	Infects larvae, invades haemocoel, ovaries & eggs. Parasitic castration. Forces nemapositing behaviour.	Bedding (2009)
Sphaerularia bombi	Hymenoptera	Bumblebees (*Bombus* spp.)	Infects queen during overwintering. Develops in haemocoel but penetrates gut to emerge from anus. Queen fails to found colonies; parasitic castration. Prevalence up to 50% in some locations.	Colgan et al. (2020)
Phylum Nematoda: Class Enoplea (Aphasmids), order Mermithida				
Mermis nigrescens	Wide host range: Hymenoptera, Orthoptera, Coleoptera, Dermaptera, Lepidoptera	Honeybee (*Apis* spp.) bumblebee (*Bombus* spp.), grasshoppers, earwigs, beetles, moth larvae	Infects gut & haemocoel. Hatched larvae penetrate gut. Very large nematode. Induces lethal positive hydrotaxis. Occurs widely in N. & S. America, Europe, Asia.	Baylis (1947); Tripodi and Strange (2018); Herbison et al. (2019)
Allomermis solenopsi	Hymneoptera	Fire ant (*Solenopsis invicta*)	Infects haemocoel. Distended abdomen. Death upon emergence of large mature worm. Used in USA for biocontrol of invasive fire ants.	Poinar et al. (2007)
Romanomermis iyengari & culicivorax	Diptera	Mosquito (*Anopheles gambiae* & over 90 other spp.)	Enters via larval cuticle; develops in haemocoel. Kills within 10 days. Used in biocontrol of mosquito larvae.	Galloway and Brust (1979); Abagli et al. (2019)

continued

Table 13.5 *Continued*

Worm	Host order(s)	Host examples	Pathology	Key References
Phylum Nematoda: Class Chromadorea (Phasmids), order Ascaridida, family Thelastomatidae				
Hammerschmidtiella diesigni, *Leidynema* spp., *Thelastoma* spp.	Blattodea	Cockroaches (e.g. *Leucophaea maderae*)	Infects hindgut. Reduces fat body content, lesions in hindgut tissues, increased mortality in mass rearing systems	Taylor (1968); Gałęcki and Sokół (2019)
Phylum Nematomorpha (hairworms): Class Gordioidea				
Gordionus spp.	Coloeoptera	Ground beetles (~ 79 spp.)	Worms develop in haemocoel of adult host. Induce positive hydrotaxis. Death often occurs due to destructive emergence of mature worm and/or drowning.	Ernst et al. (2016)
Spinochordodes spp.	Orthoptera	Crickets, grasshoppers, katydids		Thomas et al. (2002)
Paragordius tricuspidatus	Orthoptera	Crickets		Thomas et al. (2002); Thomas et al. (2003)
Phylum Platyhelminthes: Class Trematoda, order Plagiorchiida				
Lancet liver fluke (*Dicrocoelium dendriticum*)	Hymenoptera	Various ant spp. e.g. *Formica fusca*	Cysts fill the haemocoel. Induces 'suicidal' behavioural manipulation.	Libersat et al. (2018); Martín-Vega (2018)

Note: the location/source of the parasite is assumed to be widespread unless noted otherwise in the previous table.

manipulate host behaviour with often lethal consequences for the insect.

All nematode lifecycles progress from an egg to juvenile stages (usually four larval instars) to adults. Any stage can be infective to insects, including the eggs, as determined by the nematode species. If the infective stage is a larva, it is called an 'infective juvenile' (IJ) or 'pre-parasite' and usually forages for insects in soil or on damp surfaces. For all EPNs, the IJ is a third stage larva. The EPN IJs use two strategies to locate hosts: cruise or ambush. Cruisers like *Steinernema glaseri* and *Heterorhabditis bacteriophora* forage over large distances for slow-moving insect larvae and rely heavily on chemoreception for seeking and evaluating suitable hosts (e.g. Grewal et al. 1994). Ambushers such as *Steinernema carpocapsae* rely much less on chemoreceptive cues and tend to launch surprise attacks on dense populations of fast-moving insects. Infective worms penetrate the host via the mouth, anus, spiracles, or through gaps in the cuticle (e.g. membranes between segments). Proteolytic enzymes or a tooth-like stylet are employed to breach barriers such as the gut

epithelium or the cuticle. Eggs are, for example, the infective stage of the mermithid *Mermis nigrescens*. The eggs are anchored with byssus threads to plant surfaces where they can be ingested by host insects (Figure 13.4e); (Nickle, 1972). After hatching in the gut, *M. nigrescens* larvae penetrate the gut wall and invade the haemocoel. Some other mermithid species are infective as larvae.

The events that take place *after* host penetration divide very clearly along the lines of 'entomopathogenic' versus 'entomoparasitic' infections and will be discussed separately. In the case of EPNs, exposure to insect haemolymph triggers release of their symbiotic bacteria by defaecation (*Xenorhabdus* spp.) or regurgitation (*Photorhabdus*) (Martens and Goodrich-Blair 2005; Ciche et al. 2008). The bacteria rapidly multiply and kill the host and the nematode consumes the 'soup' of bacteria and deteriorating insect tissues. What virulence factors are responsible for such rapid killing? The IJs produce 'excretory-secretory proteins' (ESPs) also referred to as venom proteins; a set of 52 core venom proteins have been identified in *S. feltiae* and

S. carpocapsae that are induced when IJs encounter a suitable host (Chang et al. 2019). The major roles of ESPs are in suppressing the host immune response and in damaging host tissues. *Steinernema* spp. worms produce a range of proteases, which have triple roles since they can aid in initial penetration of the host's barriers, in suppressing various arms of the immune response such as melanisation and encapsulation, and finally in helping to break down the tissues of the cadaver (reviewed by Cooper and Eleftherianos 2016; for a review of insect immune responses to EPNs see Castillo et al. 2011).

The genomes of key *Photorhabdus* and *Xenorhabdus* species exhibit many pathogenicity islands containing multiple toxin genes (reviewed by Sajnaga and Kazimierczak 2020). Toxins include PirA/PirB and 'Makes Caterpillars Floppy' (Mcf) from *Photorhabdus* species, XaxA/XaxB from *Xenorhabdus* species, and Tc (expressed by many entomopathogenic bacteria including various *Serratia* species; see Chapter 11 for details). Insect larvae quickly stop feeding and die when injected with these toxins (reviewed by Lewis and Clarke 2012). *Xenorhabdus* bacteria also protect the *Heterorhabditis* worms by suppressing host insect encapsulation and AMP synthesis activities and *Photorhabdus* species have been shown to suppress melanisation by synthesis of an 'ST' molecule.

A nutrient-rich insect cadaver would ordinarily be quickly spoiled by a whole community of decomposer microbes, or eaten by scavengers, and either outcome would of course be disastrous for the developing nematodes. To prevent this from happening, a cocktail of EPN-derived excretory-secretory products and symbiont-derived secondary metabolites is synthesised to inhibit soil microbes and/or prevent predation by arthropods. Secondary metabolite gene clusters account for 6% of the genome of *Photorhabdus* (listed by Stock et al. 2017) and include antifungal factors and broad-spectrum antibacterials. *Photorhabdus luminescens* infected cadavers release volatiles to deter scavengers (Baur et al. 1998) while attracting insect larvae that are suitable as future hosts for the maturing nematodes (Baiocchi et al. 2017; Zhang et al. 2019). Thus, EPNs and their symbionts, in addition to being biological control agents, also represent

treasure troves of other potentially exploitable molecules.

Often several generations of EPNs will be able to develop in the cadaver before resources are depleted and IJs emerge to seek a fresh host. Steinernematid nematode infections result in characteristically brown insect cadavers, whereas those killed by heterorhabditid nematodes turn red and finally disintegrate into sticky, gummy masses; the red colour very likely acts as a further deterrent against visual predators such as birds. As mentioned previously, extensive melanisation is not a feature in EPN infections and so black corpses are usually associated with other causes of death and any nematodes found in such bodies may be an incidental finding.

In the case of entomoparasitic (non-EPN) nematodes, the infection is a more protracted affair. Many of these nematodes compromise the reproductive capacity of their hosts through parasitic castration and typically (though not always) mortally damage the host upon emergence due to their large size (especially the mermithids) (e.g. Poinar 1981; Nickle 1972; Figure 13.4). Entomoparasitic nematodes can manipulate the host to perform self-destructive behaviours that improve the chance of successful parasite transmission—attributes that are also desirable in biological control agents. Some mermithid species infect aquatic insects, in which case the IJs simply penetrate the cuticle and develop inside the haemocoel before killing the host upon emergence. A movie prepared by Rutgers University Entomology Department of showing the mermithid *Romanomermis iyengari* emerging from a mosquito larva can be viewed at https://www.youtube.com/watch?v=m8HaZPV5wIs (last viewed 1 October 2021).

Mermithids trigger a water-seeking behaviour (positive hydrotaxis) in terrestrial insect hosts. The final-stage larva forces its way out once the insect has located a suitably damp environment in which the parasite can mate and lay eggs. *Mermis nigrescens*, for example, forces earwigs to enter water, a behaviour in which they would not normally engage (Herbison et al. 2019). This worm species is also exceptionally large (the adult females are 10 cm long), so worm emergence usually kills the host. The mermithid *Allomermis solenopsi* is used in

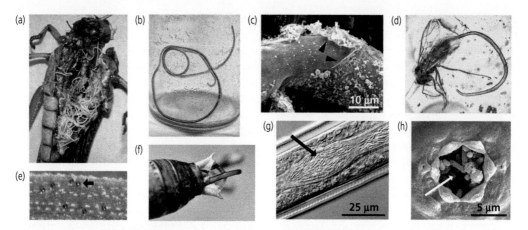

Figure 13.4 Nematode and nematomorph infections of insects. (a) Mermithid nematodes filling the body cavity of an adult blue-winged grasshopper from a zoo invertebrate breeding centre. (Attard et al. 2008), with permission from the American Association of Zoo Veterinarians. (b) An adult female *M. nigrescens* Dujardin in a glass vial. Note the numerous minute eggs adhering to the sides of the vial and the dark line of eggs within the adult's body. Source: from John Capinera, University of Florida. (c) SEM of a live mermithid evading the encapsulation response in a permissive insect host. A large fragment of the surface coat has been shed (coat edges indicated by arrowheads), clearing immune products away to show the underlying nematode epicuticle. Source: from (Shamseldean et al. 2007), with permission from Brill. (d) The fossil mermithid *Heydenius formicinus* emerging from a *Prenolepis henschei* ant in Baltic amber approximately 40 million years old. Source: from (Poinar 2012). (e) Infective eggs (arrow) of the grasshopper mermithid *Mermis nigrescens* adhering to grass foliage; these will be ingested by host insects. Source: from John Capinera, University of Florida. (f) A gordian (nematomorph) worm emerging from the posterior abdominal segment (probably the anus) of the bush cricket *Antaxius pedestris*. Source: from Gilles San Martin, Walloon Agricultural Research Centre. (g) Anterior end of a *Heterorhabditis megidis* infective juvenile nematode, with *Photorhabdus temperata* rods visible (arrow) inside the lumen of the intestine. Source: from (Sajnaga and Kazimierczak 2020). (h) SEM *Photorhabdus temperata* rods (arrow) inside the mouth of the entomopathogenic nematode *Heterorhabditis zealandica*. Source: from (Sajnaga and Kazimierczak 2020).

the biological control of imported fire ant workers (*Solenopsis invicta*) in the USA (Poinar et al. 2007) and represents just one of many *Allomermis* species to infect *Solenopsis* spp. Parasitism of ants by mermithids is an ancient relationship that has endured for at least 40 million years, as evidenced by beautiful examples in fossil amber (e.g. Poinar 2012; Figure 13.4). Infected ants exhibit characteristically distended abdomens.

As mentioned previously, nematodes also affect insect fecundity and reproductive behaviours. Females of the pest woodwasp *Sirex* are forced to 'nemaposit' juvenile worms instead of eggs following infection with the rhabditid nematode *Deladenus* (*Beddingia*) *siricidicola* (Bedding 2009). Parasitic castration is also achieved by the rhabditid nematode *Sphaerularia bombi* in bumblebee queens by suppressing development of the corpora allata, a key endocrine organ that assists ovary development through the synthesis of juvenile hormone.

Infected queens do not develop eggs and so do not found colonies (Colgan et al. 2020).

Like EPNs, entomoparasitic nematodes develop in the insect haemocoel and they are thus exposed to the full arsenal of the immune system. Although nematodes will be efficiently melanotically encapsulated in non-permissive hosts (Poinar et al. 1979; See Chapter 1), studies with *Romanomermis* mermithids have revealed an avoidance strategy in permissive hosts. These worms are in fact partially encapsulated, but they secrete and repeatedly shed 'disposable' outer coats, shrugging off the capsule of accumulating haemocytes before it becomes too confining or dangerous. The shed layers also act like decoys for the immune system (Shamseldean et al. 2007; Figure 13.4c). During this time the worm grows slowly, presumably as an energy trade-off, but eventually the host immune response becomes depleted and the worm then continues to develop unimpeded (Shamseldean et al. 2007). Why they are unable to do this in non-permissive hosts is presently still unclear.

Many nematodes (and particularly the EPNs) are of course excellent biological control agents and EPNs have been used for this purpose since at least the 1930s. However, these same attributes make nematode outbreaks very difficult to control in insect mass-rearing programmes, especially in damp conditions. In a recent survey, for example, *Steinernema* spp. were detected in 29% of commercial cricket farms in the EU along with many other parasites (Gałęcki and Sokół 2019). It is important to avoid feeding insects with plants with attached mermithid eggs or plants from fields treated with EPNs and to avoid importing insects of an unknown health status. Nematode species are usually confirmed by molecular sequencing of the 18S ribosomal DNA or mitochondrial cytochrome c oxidase subunit I (COI) sequences.

13.2.2.2 Nematomorphs

Nematomorphs (hairworms or gordiid worms) are obligate parasites of arthropods with a free-living aquatic adult stage and they occur worldwide. They belong to the phylum Nematomorpha, whose members are poorly classified. Species from three genera are known to parasitise insects: *Spinochordodes* and *Paragordius* species, which infect orthopteran hosts and *Gordionus* species, which have a broad host range among the Coleoptera (Table 13.5). In many respects hairworms closely resemble the mermithids (Section 13.2.3.1), not least because both types of worm have convergently evolved to induce water-seeking behaviours in their hosts, which is necessary due to the reliance of the adult worms on a wet environment.

As obligate parasites, all hairworms require at least one, and usually two, hosts (Hanelt et al. 2005). Larval hairworms first enter an aquatic ('paratenic' or 'transport') host such as a larval fly, usually via ingestion, in which they form benign cysts that endure through metamorphosis (Hanelt and Janovy 2004). The paratenic host is then eaten by an omnivorous or carnivorous terrestrial insect (the definitive host) and the hairworm cysts re-activate in the gut. The hairworm develops until it fills most of the available space (Figure 13.4) whereupon it compels the insect to seek and enter water, especially at night, so that the mature worm can escape from the host's haemocoel (Thomas et al. 2002; Thomas et al. 2003). An excellent video of *Paragordius varius* worms exiting a cricket (*Acheta domesticus*) produced by the Hanelt laboratory at the University of New Mexico can be viewed at https://sites.google.com/site/haneltlab/movies (last viewed 1 October 2021). Provided the insect avoids drowning and predation by fish, death is not an inevitable outcome and there are records of insects recovering sufficiently to successfully oviposit after egress of mature gordiids.

Behavioural manipulation in nematodes and hairworms are poorly understood processes. Neurotransmitters such as taurine are found in higher concentrations in infected crickets and as taurine is involved in regulating brain osmoprotection it is hypothesised to induce an insatiable thirst in the host (Thomas et al. 2003 and references therein).

13.2.2.3 Trematodes

Trematodes (Phylum Platyhelminthes) rarely cause notable infections in insects. A single example will be described in this section: the generalist parasite, *Dicrocoelium dendriticum* (lancet liver fluke) (Table 13.5). As part of a very complicated lifecycle, *D. dendriticum* can infect a wide range of ant species as second intermediate hosts. Infection begins when ants ingest snail 'slime balls' containing fluke larvae; these penetrate the gut, migrate through the haemocoel and then encyst as metacercariae in the abdomen. A single larva, however, takes an exceptional route and comes to lie adjacent to the suboesophageal ganglion, where direct physical contact is established between the parasite and the brain tissue (Martín-Vega et al. 2018). Infected ants are compelled to climb to the tip of a blade of grass each evening, clamping on with their mandibles to await being eaten by grazing animals (the definitive hosts). During attachment, ants do not feed or defend themselves and if the ant survives the night it returns to the nest but it makes further attempts the next night on the same plant (Libersat et al. 2018; Martín-Vega et al. 2018). The molecular basis of this fascinating manipulation is yet to be established.

13.2.3 Arthropod parasitoids and infestations

Arthropod parasites of insects are dominated by the mites, wasps, and flies, and examples are summarised in Table 13.6.

13.2.3.1 Mites

Parasites are highly represented among the large red velvet mites (order Trombidiidae) and families within the order Mesostigmata. Dozens of mite species parasitise bees, some relatively harmlessly but a handful are extremely serious and can contribute to the collapse of a whole colony of social bees (Sammataro et al. 2000), especially if bees are subjected to several stressors simultaneously that combine to lower immunity (Goulson et al. 2015). Some mites are also vectors of bacterial and viral diseases (Wilfert et al. 2016; see also Chapters 10 and 11).

The best-known bee mite is *Varroa destructor*, an obligate ectoparasitic mite of immature and adult *Apis* spp. bees that vectors several viral pathogens including deformed wing virus (Wilfert et al. 2016). *Varroa destructor* underwent an evolutionary host-shift from the Asian honeybee (*Apis cerana*) to the European honeybee (*Apis mellifera*), probably facilitated by the commercial exchange of queens about 60 years ago (Wilfert et al. 2016) and it is now an almost cosmopolitan pest. The life cycle has two phases: one in which they feed on adult bees, called the phoretic phase and a reproductive phase that takes place within a sealed honeycomb cell, in which the mites lay eggs on developing bee larvae (capped brood). Female mites travel on worker bees into the hive whereupon they lay eggs into the brood cells just before they are sealed.

Until very recently it was assumed that *V. destructor* feeds on haemolymph but it is now understood that it primarily ingests fat body tissue (Ramsey et al. 2019). Since the fat body plays a critical role in the immune response, hormone regulation, synthesis of wax precursors, pesticide detoxification, and other metabolic roles, this finding helps explain why *Varroa* has such a profound negative impact on the health of its bee hosts. At the time of writing, it is not yet known if other important bee mites such as the tracheal mite *Acarapis woodi* (Figure 13.5), *Varroa jacobsoni*, *Tropilaelaps*, or *Locustacarus* spp. also consume fat body. Varoosis can lead to impaired development, decreased protein and lipid content, desiccation, impaired metabolism, precocious foraging, greater winter mortality, immunodeficiency, decreased longevity, and reduced pesticide tolerance (reviewed by Ramsey et al. 2019). The *Varroa* cryptic species complex contains at least four species, the other major pest being *jacobsoni* which shares the host *Apis cerana* with *A. destructor*. Other mites important to the health of social and solitary bees are listed in Table 13.6 and a dedicated website for the identification of bee mites can be found at BeeMiteID (http://idtools.org/id/mites/beemites). For a discussion of *Varroa* mite control measures, see Section 10.5.

A surprising array of water mites (family Arrenuridae) parasitise mosquitoes and the aquatic stages of other insects. The majority of water mites thus far studied belong to two genera: *Parathyas* (Hydryphantidae) and *Arrenurus* (Arrenuridae) (Table 13.6). Water mite larvae are small exoparasites of the Odonata, Hemiptera and Diptera (especially adult midges and mosquitoes; see Figure 13.5), which significantly reduce host longevity and fecundity (Smith and McIver 1984). Water mites pierce the host cuticle with their blade-like chelicerae and consume haemolymph. As they tend to infect a high percentage of the local mosquito population, water mites are considered to have potential as biological control agents.

13.2.3.2 Wasps and flies

Parasitism has evolved in a great many representatives of the Hymenoptera and Diptera, and also in the Strepsiptera (which are all obligate endoparasites), Coleoptera, Lepidoptera and a few other orders. Their insect hosts tend to belong to these same orders, as well as the Hemiptera (Table 13.6). Respectively, idiobiont (ecto-) and koinobiont (endo-) parasitoids lay eggs either on, or in, their insect hosts. The successful development of their offspring relies on altering the physiology and/or behaviour of the parasitised host. The host typically dies after successful emergence and pupation of the parasite. Cases of superparasitism (i.e. the co-existence of several parasitoid species in the

Table 13.6 Examples of arthropod parasites and parasitoids of insects (not exhaustive).

Insect family	Parasite/parasitoid example	Host order(s)	Host examples	Pathology & notes	Key References
order Hymenoptera					
family Braconidae (Parasitoid wasps)	~ 46,000 species of braconid parasitoid wasp (e.g. *Aphidus*, *Cotesia*, *Glyptapanteles*, *Dinocampus* spp.)	Lepidoptera, Hemiptera, Coleoptera	Very wide lepidopteran host range; coccinellid beetles, aphids, etc.	Can be endoparasitic or exoparasitic; all developmental stages of host can be attacked. Usually cause a combination of host immunsuppression, arrested development, behavioural manipulation & eventually death. Partially achieved through mutualism with polydnavirus and/or virus-like particle and/or venom. Effects too diverse to generalize (see references & main text for details). Several species are used as biological control agents.	Lee et al (2009); Rodriguez et al. (2C13); Moreau and Asgari (2015); Hughes and Libersat (2018)
Ichneumonidae (Parasitoid wasps)	~ 40,000 species of ichneumonid parasitoid wasp (e.g. *Nasonia*, *Megarhyssa*, *Ichneumon* spp.).	Diptera, Coleoptera, Lepidoptera	Very wide host range including blowflies, flesh flies, beetles, butterflies, moths, weevils etc.		
Figitidae (Parasitoid wasps)	~ 1500 species of figitid parasitoid wasp (e.g. *Leptopilina* spp.).	Diptera, Neuroptera	Fruit flies *Drosophila* spp., muscid flies, lacewings. etc.		
order Diptera					
family Phoridae (Parasitoid flies)	*Pseudacreon* spp. including *tricuspis*	Hymenoptera	Several ant genera e.g. *Lasius*, *Crematogaster* & *Solenopsis* (fire ants)	Develops in haemocoel & head, usually fatal. Head decapitated shorty before adult fly emerges.	Porter (1998); Chen and Fadamiro (2018)
family Tachinidae (Parasitoid flies)	*Exorista bombycis*, *Crossocosmia sericaria*, *Ctenophorocera pavida*, *Blepharipa zebina* (ujiflies/uzi flies)	Lepidoptera	Silkworm *Bombus* spp.	Invades cuticle & haemocoel. Melanotic lesions, discoloured haemolymph, failure to pupate, death. Emergence of final stage maggot kills host. Causes mass losses of cocoons in India, China & elsewhere.	Narayanaswamy and Govindan (2000)

continued

Table 13.6 *Continued*

Insect family	Parasite/parasitoid example	Host order(s)	Host examples	Pathology & notes	Key References
order Strepsiptera					
family Myrmecolacidae	e.g. *Stichotrema dallatorre-anum*	Zygentoma, Diptera, Mantodea, Orthoptera, Blattodea, Heteroptera, Hymenoptera	~34 insect families	Adult female occupies a significant volume of the host abdomen. Host dies upon emergence of adult male or after all 1st instar larvae emerge. Possibly also parasitic castration.	Kathirithamby et al. (2003)

Arachnids	Parasite example	Host order(s)	Host examples	Pathology	Key References
order Mesostigmata					
family Laelapidae	*Varroa destructor* (& also vectored viruses)	Hymenoptera	Honeybee (adults, pupae, larvae) *Apis cerana, Apis mellifera*	Varoosis. Ectoparasite of adults & brood; feeds on fat body. Serious pest of *A. mellifera* worldwide; contributes to colony collapse in winter (synergistic effect with other stressors). Impaired development, decreased protein & lipid content, desiccation, impaired metabolism, winter mortality, immunodeficiency, decreased longevity & reduced pesticide tolerance. Mites also vector entomopathogenic viruses.	Ritter (1981); Ramsey et al. (2019)
	Varroa jacobsoni	Hymenoptera	Honeybee *Apis cerana*	Varoosis in adult bees & brood; severe in older larvae & pupae. Similar effects to *V. destructor*. Pupae may not develop into adults, or emerge underweight with deformities.	De Jong et al. (1982)
	Tropilaelaps clareae, Euvarroa sp.	Hymenoptera	Honeybees e.g. *Apis mellifera, Apis dorsata*	External parasitic mites in Middle East & SE Asia. Feeds on haemolymph of developing bees inside capped brood cells. Irregular brood pattern, dead or malformed wingless bees at the hive entrance.	Sammataro et al. (2000)

Order Trombidiformes (red velvet mites)

family Pyemotidae	Pyemotes ventricosus & other spp. in the ventricosus group	Silkworm (Bombyx sp.), bees including Apis spp.	Lepidoptera, Hymenoptera	Generalist ectoparasite. Mite pierces the cuticle, injects neurotoxin, ingests haemolymph. Host paralyzed, dies.	Tomalski et al. (1988)
family Podapolipidae	Locustacarus spp. (tracheal mite)	Bumblebee Bombus, beetles	Hymenoptera, Coleoptera	Tracheal mite: invades metastomal air sacs, pierces tracheal walls, ingests hemolymph. Diarrhoea, lethargy & cessation of foraging. Widespread.	Otterstatter and Whidden (2004)
family Tarsonemidae	Acarapis woodi	Honeybees & bumblebees (Apis spp. & Bombus spp.)	Hymenoptera	Tracheal mite: pierces tracheal wall to ingest haemolymph; compromises air exchange. Bees have disjointed wings, difficulty flying, distended abdomens. Worse in cold climates (but can be asymptomatic).	Cepero et al. (2015)
family Erythraeidae	Leptus spp.	Honey bee Apis mellifera (also emergent in Africanised honeybees)	Hymenoptera	Ectoparasite of adult workers. Leptus spp. can also transmit Spiroplasma (responsible for May disease).	Martin and Correia-Oliveira (2016)
family Arrenuridae	Water mites e.g. Arrenurus, Thyas Parathyas & Euthyas spp.	> 100 mosquito species (Aedes, Anopheles & Culex spp.), dragonflies, midges, water bugs	Odonata, Hemiptera, Diptera	Obligate external parasite of adult host. Heavy infestations associated with reduced host fecundity & longevity. Considered as biocontrol agents.	Smith and McIver (1984); Simmons and Hutchinson (2016); Atwa et al. (2017)

Note: the location/source of the parasite is assumed to be widespread unless noted otherwise in the previous table.

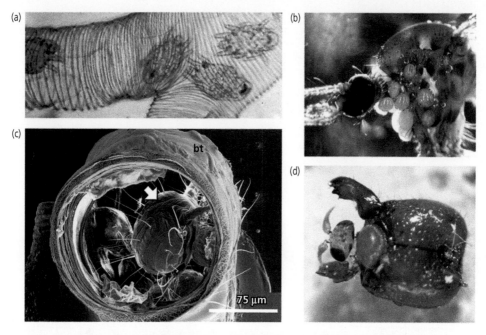

Figure 13.5 Arthropod parasites of insects. (a) Honeybee tracheal mites (*Acarapis woodi*) visible in a bee trachea (approx 150 μm in length). Source: image in the Public Domain; Released by USDA. (b) An *Aedes* sp. adult mosquito heavily infected with *Arrenurus danbyensis* water mites. Source: from (Atwa et al. 2017), with permission from Elsevier. (c) An SEM of an adult female *Acarapis woodi* mite (arrow) inside a honeybee trachea (bt). Source: by Pavel Klimov, Bee Mite ID idtools.org/id/mites/beemites Public Domain image, https://commons.wikimedia.org/w/index.php?curid=58419329). (d) An adult phorid fly *Pseudacteon litoralis* emerging from a decapitated ant head. Source: from Sanford D. Porter, USDA-ARS.

same host) can also occur, apparently facilitated by filamentous viruses (see Chapter 10, Section 3.4).

Endoparasitoid wasps inject one or more eggs into their host, often along with a quantity of venom and polydnaviruses (PDVs). The PDVs from braconid and ichneumonid endoparasitoid wasps are bracoviruses and ichnoviruses respectively and although the morphologies are dissimilar and they evolved independently, they share many similar features in terms of function and transmission. The PDVs are integrated into and dispersed in multiple loci in the wasp genome; they replicate only in the pupa and female ovarian calyx cells. Thus when wasp females oviposit, the proviral genome is present in the cells of the embryo, while the secreted fluid surrounding the eggs contains a suspension of encapsidated PDV virions that subsequently infect host cells. These virions contain multiple circular dsDNAs containing a large proportion of noncoding DNA and also symbiotically-incorporated virulence genes acquired over a long evolutionary period from multiple eukaryote sources (by horizontal gene transfer), including genes more recently acquired from the parasitoid wasp (reviewed in Herniou et al. 2013; Strand and Burke 2015). Interestingly, the replication machinery is permanently integrated into the wasp genome and is absent from the packaged virion, which crucially means that they can infect but not replicate, in the parasitised host. The virion DNA rapidly integrates into the parasitised host cell genome. Furthermore, virulence genes are only transcribed inside infected host cells and not in the wasp, and the gene products essentially prevent the host immune system from killing the wasp eggs and larvae and trigger other physiological alterations such as retarded growth that support development of the wasp at the expense of the host (Herniou et al. 2013; Strand and Burke 2015).

The proteomes of bracoviruses can be accessed via UniProt and indicate several virulence genes including multiple I-kappa-B-like proteins that suppress the host immune response through NF-kappa-B inactivation, mucin-like proteins that inhibit haemocyte adhesion and phagocytosis, and protease inhibitors that deregulate the melanisation cascade (e.g. Beck and Strand 2007) but also some proteins for which no function has yet been ascribed.

Few ichnovirus genes are known, and their exact role, if any, in immunomodulation is not yet known. One ichneumonid wasp, *Venturia canescens*, has lost its association with PDVs and instead delivers virus-like particles (VLPs) to its host, the caterpillar of the Mediterranean flour moth *Ephestia kuehniella*. These consist of wasp-derived virulence proteins—not DNA—packaged into a nudiviral lipid envelope and they are released from calyx cells into the oviduct lumen where they become attached to the wasp egg. Once in the parasitised caterpillar host, VLPs act as liposomes to deliver immunosuppressive proteins directly to host cells (Pichon et al. 2015). It is not presently known whether all parasitoid wasps employ VLPs, but a very similar situation to *V. canescens* occurs in the braconid *Fopius arisanus*, a commercially-used parasitoid of tephritid fruit flies (Burke et al. 2018).

Some species of endoparasitoid wasp lay eggs containing specialised 'teratocyte' cells, which are formed during embryogenesis and are released when the eggs hatch inside the host (reviewed in Strand 2014). These eggs tend to be tiny, so the embryo emerges into the host in a very underdeveloped state and starts feeding on host haemolymph. Transcriptomic and proteomic evidence suggests that teratocyte secretory proteins (TSPs) include several immunosuppressive factors that prevent recognition of the wasp offspring and/or destruction by the host immune response. It is noteworthy that antimicrobial TSPs are also synthesised as 'protective supplements' for the host's disarmed immune system, presumably because a microbial infection would be equally disastrous to both host and parasitoid. Other TSPs have roles in retarding host growth (Burke and Strand 2014).

A further strategy employed by both endo- and ecto-parasitoid wasps is a venom, often injected in more than one location in the host's body to ensure tissue-specific action. Ectoparasitoid venoms usually incorporate paralytic neurotoxins to prevent the host from attempting to remove the parasitoid, and sometimes include factors that retard host development. Endoparasitoid venoms interfere with haemocyte activity and melanotic pathways, appear to act synergistically with PDVs, and may also act as antimicrobial factors and chemical castrators. An excellent review of parasitoid venoms can be found in Moreau and Asgari (2015).

Host behavioural manipulation is also a hallmark of parasitoid attacks mediated via venoms containing neurotoxins. An example is the *Thyrinteina leucocerae* caterpillar parasitised by *Glyptapanteles* spp. braconid wasps. Even when reduced to a moribund state following emergence of the wasp larvae from its body, a parasitised caterpillar will forego feeding and use the last of its energy to defend the emergent braconids while they pupate (Grosman et al. 2008). Although detailed discussions of behavioural manipulation are beyond the scope of this chapter (see Hughes and Libersat (2018) and Libersat et al. (2018) for recent reviews), it is also important to note that host behaviour can also be manipulated by non-PDV viruses transmitted by parasitoid wasps and these can replicate in the host brain (Dheilly et al. 2015).

Several species of phorid flies and all of the tachinid flies are parasitoids. Most tachinids are endoparasites, gaining entry to the host either by direct egg injection, or sometimes by being ingested as eggs. *Pseudacteon* flies (phorid, humpback, or 'decapitating' flies) are koinobiont, solitary parasitoids of worker ants. Eggs are injected into the ant haemocoel (Porter 1998) and once hatched the larva migrates to the head where it consumes mostly haemolymph. Just prior to pupation it releases a chemical (probably an enzyme) that degrades the membranes holding the ant's exoskeleton together. The larva then consumes the contents of the ant's head, which falls free of the body (Figure 13.5d) and finally the fly pupates inside the remains of the head. These flies are currently under evaluation for the biological control of imported fire ants in the USA (e.g. Chen and Fadamiro 2018).

Some species of tachinids and strepsiptera that develop in their host's haemocoel appear to evade immune destruction by becoming ensheathed in a camouflaging 'cloak' of host-derived cells, while leaving the host's immune system functional. The sheath surrounding a tachinid is composed of several layers of host haemocytes and fat body cells, which seem to deposit in an organised sequence (Yamashita et al. 2019). The molecular factors underpinning these interactions are yet to be elucidated. Although at first glance this appears similar to encapsulation, in fact melanin deposition is absent and the larva emerges unscathed. Other tachinids escape the haemocoel into other tissues. As they penetrate the host cuticle, strepsipteran larvae wrap themselves into an invagination of the basal layer of the host epidermis, which is in direct contact with the haemolymph (a 'host-derived epidermal bag')—a crude yet highly effective camouflage tactic that has allowed them to parasitise an extremely wide range of hosts (Kathirithamby et al. 2003). Parasitoid tachinids are a major problem in sericulture (e.g. Ujifly or uzifly species are responsible for 10–30% of cocoon crop losses in India and China Narayanaswamy and Govindan 2000), while others are used in biological control (e.g. against the Asian corn borer moth).

13.3 Future directions and conclusions

Some likely areas of major research expansion are in:

- Our understanding of host immune suppression mechanisms by parasitoid wasps, especially in species other than the well-studied ichneumonid/braconid species. For example, the VLPs, renamed Mixed Strategy Extracellular Vesicles (MSEVs), are immune-suppressive particles made in venom glands of figitid wasps *Leptopilina* spp. Comparative proteomic studies of the MSEV protein mixtures and analyses of the wasp genome are starting to uncover the identity of MSEV proteins, many of which are unannotated/novel (Wey et al. 2020). It is expected that CRISPR-cas9 mediated allele disruption will greatly assist functional genomics experiments to identify the roles of these genes.
- The role of teratocytes and non-coding miRNAs in parasitoid wasps. Recent work with *Cotesia vestalis*, a braconid wasp that endoparasitises *Plutella xylostella* caterpillars, shows that both teratocytes and PDVs produce non-coding miRNAs. Teratocyte-derived miRNA (Cve-miR-281-3p) and PDV-derived miRNA (Cve-miR-novel22-5p-1) specifically arrest host growth by modulating expression of its ecdysone receptor with the net effect that the insect's development is retarded (Wang et al. 2018). Many other parasitoid/PDV miRNAs await identification and functional investigation.
- An acceleration in the discovery and licencing of parasite and parasitoid candidates for the biological control of pest insects and a new generation of recombinant biocontrol agents with enhanced virulence (e.g. Karabörklü et al. 2017). This will probably be driven by multiple factors such as new legislation restricting the use of various chemical insecticides, the evolution of insecticide resistance, and increased pressure to enhance agricultural productivity.
- A reappraisal of the key players in parasite co-infections. The probable historical misidentification of *Apicystis cryptica* sp. n. as *Apicystis bombi* came to light in a metagenomic survey of bumblebees (Maharramov et al. 2013). Parasite co-infections are extremely common in insects (probably much more common than is generally recognised) and the causative organisms may be indistinguishable from one another with traditional identification techniques. It is only with molecular techniques that some parasites can be properly differentially diagnosed and these complex relationships in disease states can finally be teased apart.

This chapter has also highlighted some areas of insect parasitism in which research has stalled, or indeed has never been brought into the modern scientific era. We still know almost nothing about how ciliates invade the insect cuticle or how they interact with the host immune system, which is all the more surprising considering their virulence towards mosquitoes. There is also surprisingly little in the published scientific literature on parasites in sericulture (at least that is available in English). There is a major imbalance between research devoted to parasites that are of commercial use in biological control, or those which affect bees, compared with the vast majority of parasites of insects. While this is understandable, it leaves us poorly equipped to deal with emerging diseases in vulnerable ecosystems, especially those under pressure from climate change or adverse anthropogenic influences. We are well aware that terrestrial and freshwater ecosystems will cease to function without insects but the loss of parasite biodiversity engenders very little outcry. This is a short-sighted view considering the vital role that parasites also play in insect ecology. Recent conservative estimates suggest that 5–10% of parasite species are committed to extinction by 2070, either directly due to climate change or indirectly through host co-extinction (Carlson et al. 2017).

Recommended further reading

Protozoa

Das, B.P. 2019. *Chilodonella uncinata*—as potential protozoan biopesticide for mosquito vectors of human diseases. *Acta Scientific Microbiology* 2(12): 90–96 (Overview of ciliates infecting mosquitoes and potential use as biopesticides.)

Egerter, D.E., Anderson, J.R., and Washburn, J.O. 1986. Dispersal of the parasitic ciliate *Lambornella clarki*: Implications for ciliates in the biological control of mosquitoes. *Proceedings of the National Academy of Sciences of the United States of America* 83(19): 7335–7339 (Transmission strategy, effects on host reproduction, and pathological features.)

King, R.L. and Taylor, A.B. 1936. *Malpighamœba locustae*, n. sp. (Amoebidae), a protozoan parasitic in the Malpighian tubes of grasshoppers. *Transactions of the American Microscopical Society* 55(1): 6–10 (Detailed description of amoebic infection and resulting pathology, still relevant despite the publication year.)

Lamborn, W.A. 1921. A protozoon pathogenic to mosquito larvae. *Parasitology* 13(3): 213–215 (Seminal article describing mosquito ciliatosis.)

Lantova, L. and Volf, P. 2014. Mosquito and sand fly gregarines of the genus *Ascogregarina* and *Psychodiella* (Apicomplexa: Eugregarinorida, Aseptatorina)—Overview of their taxonomy, life cycle, host specificity and pathogenicity. *Infection, Genetics and Evolution* 28: 616–627 (Pathological effects of eugregarines on mosquito species.)

Schaub, G.A. 1994. Pathogenicity of trypanosomatids on insects. *Parasitology Today* 10(12): 463–468 (Authoritative overview of host-parasite interactions concentrating on trypanosomatid infections in hemipterans.)

Tartar, A. 2013. The non-photosynthetic algae *Helicosporidium* spp.: Emergence of a novel group of insect pathogens. *Insects* 4(3): 375–91(Possibly the only review paper available for this parasite.)

Nematodes and other worms

Bhat, A.H., Chaubey, A.K., and Askary, T.H. 2020. Global distribution of entomopathogenic nematodes, *Steinernema* and *Heterorhabditis*. *Egyptian Journal of Biological Pest Control* 30(1): 31 (Detailed lists of EPNs and their geographical locations.)

Dillman, A.R., Chaston, J.M., Adams, B.J. et al. 2012. An entomopathogenic nematode by any other name. *PLOS Pathogens* 8(3): e1002527 (A detailed discussion of the defining characteristics of an EPN.)

Hughes, D.P. and Libersat, F. 2018. Neuroparasitology of parasite–insect associations. *Annual Review of Entomology* 63(1): 471–487 (Up-to-date review of insect behavioural manipulation by parasites and pathogens, using a spatial approach.)

Lewis, E.E. and Clarke, D.J. 2012. Nematode Parasites and Entomopathogens. In *Insect Pathology*, 2nd edn, Vega F.E. and Kaya, H.K (eds.) pp. 395–424. San Diego: Academic Press (Detailed chapter on nematodes that infect insects, with an emphasis on EPNs.)

Poinar, G. 2012. Nematode parasites and associates of ants: Past and present. *Psyche: A Journal of Entomology* Article ID 192,017 (Excellent images and good sections on mermithids and general overview of the nematodes.)

Thomas, F., Ulitsky, P., Augier, R., et al. 2003. Biochemical and histological changes in the brain of the cricket *Nemobius sylvestris* infected by the manipulative parasite *Paragordius tricuspidatus* (Nematomorpha). *International Journal for Parasitology* 33(4), 435–443 (A rare study showing changes in the brain of insects infected by nematomorphs.)

Parasitic arthropods and parasitoids

Kathirithamby, J., Ross, L.D., and Johnston, J.S. 2003. Masquerading as self? Endoparasitic Strepsiptera (Insecta) enclose themselves in host-derived epidermal bag. *Proceedings of the National Academy of Sciences of the United States of America* 100(13): 7655–7659 (A window into the bizarre life cycle of a barely known order of parasitic insects.)

Ramsey, S.D., Ochoa, R., Bauchan, G. et al. 2019. *Varroa destructor* feeds primarily on honey bee fat body tissue and not hemolymph. *Proceedings of the National Academy of Sciences* 116(5): 1792–1801 (The dietary preferences, and pathological impact, of *Varroa* examined through superb microscopy.)

Strand, M.R. 2014. Teratocytes and their functions in parasitoids. *Current Opinion in Insect Science* 6: 68–73 (Overview of teratocytes and future research avenues.)

Strand, M.R. and Burke, G.R. 2015. Polydnaviruses: From discovery to current insights, *Virology* 02/07, 479–480, 393–402 (A timeline-based review of key discoveries in PDV research.)

References

Abagli, A.Z., Alavo, T.B.C., Perez-Pacheco, R., and Platzer, E.G. 2019. Efficacy of the mermithid nematode, *Romanomermis iyengari*, for the biocontrol of *Anopheles gambiae*, the major malaria vector in sub-Saharan Africa. *Parasites & Vectors* 12(1): 253

Abdel Rahman, K.M., El-Shazly, M.M., and Ghazawy, N.A. 2015. The incidence of the protozoa *Malamoeba locustae* and *Gregarina granhami* in three acridian communities during summer month. *Donnish Journal of Entomology and Nematology* 1(1): 001–008

Åbro, A. 1971. Gregarines: Their effects on damselflies (Odonata: Zygoptera). *Entomologica Scandinavica* 2: 294–300

Åbro, A. 1996. Gregarine infection of adult *Calopteryx virgo* L. (Odonata: Zygoptera). *Journal of Natural History* 30(6): 855–859

Adl, S.M., Bass, D., Lane, C.E. et al. 2019. Revisions to the classification, nomenclature, and diversity of eukaryotes. *The Journal of Eukaryotic Microbiology* 66(1): 4–119

Alfazairy, A.A., El-Abed, Y.M.G., Karam, H.H., and Ramadan, H.M. 2020. Morphological characteristics of local entomopathogenic protozoan strains isolated from insect cadavers of certain stored-grain pests in Egypt. *Egyptian Journal of Biological Pest Control* 30(1): 7

Altizer, S.M. and Oberhauser, K.S. 1999. Effects of the protozoan parasite *Ophryocystis elektroscirrha* on the fitness of monarch butterflies (*Danaus plexippus*). *Journal of Invertebrate Pathology* 74(1): 76–88

Altizer, S.M., Oberhauser, K.S., and Brower, L.P. 2000. Associations between host migration and the prevalence of a protozoan parasite in natural populations of adult monarch butterflies. *Ecological Entomology* 25(2): 125–139

Añez, N. and East, J.S. 1984. Studies on *Trypanosoma rangeli* Tejera, 1920 II. Its effect on feeding behaviour of triatomine bugs. *Acta Tropica* 41(1): 93–95

Animal & Plant Health Agency (UK) (no date) Amoeba Disease BeeBase http://www.nationalbeeunit.com/index.cfm?pageid=193. Accessed 25-04-2020

Arshad, H.H. and Sulaiman, I. 1995. Infection of *Aedes albopictus* (Diptera: Culicidae) and *Ae. aegypti* with *Lambornella stegomyiae* (Ciliophora: Tetrahymenidae). *Journal of Invertebrate Pathology* 66(3): 303–306

Attard, L.M., Carreno, R.A., Paré, J.A., Peregrine, A.S., Dutton, C.J., and Mason, T.R. 2008. Mermithid nematode infection in a colony of blue-winged grasshoppers (*Tropidacris collaris*). *Journal of Zoo and Wildlife Medicine* 39: 488–492

Atwa, A.A., Bilgrami, A.L., and Al-Saggaf, A.I.M. 2017. Host–parasite interaction and impact of mite infection on mosquito population. *Revista Brasileira de Entomologia* 61(2): 101–106

Bailey, L. 1968. The measurement and interrelationships of infections with *Nosema apis* and *Malpighamoeba mellificae* of honey-bee populations. *Journal of Invertebrate Pathology* 12(2): 175–179

Baiocchi, T., Lee, G., Choe, D.-H., and Dillman, A.R. 2017. Host seeking parasitic nematodes use specific odors to assess host resources. *Scientific Reports* 7(1) 6270

Ball, S.L. and Turner, S.A. 2003. Ciliate presence and the incidence of castration in the obligately parthenogenetic mayfly, *Centroptilum triangulifer* (Ephemeroptera: Baetidae). *Northeastern Naturalist* 10(4): 421–424

Batson, B.S. 1985. A paradigm for the study of insect-ciliate relationships: *Tetrahymena sialidos* sp. nov. (Hymenostomatida: Tetrahymenidae), parasite of larval *Sialis lutaria* (Linn.) (Megaloptera: Sialidae). *Philosophical Transactions of the Royal Society of London. B, Biological Sciences* 310; 123–144

Batson, B.S. and Beale, G.H. 1983. *Tetrahymena dimorpha* sp. nov. (Hymenostomatida: Tetrahymenidae), a new ciliate parasite of Simuliidae (Diptera) with potential as a model for the study of ciliate morphogenesis. *Philosophical Transactions of the Royal Society of London. B, Biological Sciences* 301(1106): 345–363

Baur, M.E., Kaya, H.K., and Strong, D.R. 1998. Foraging ants as scavengers on entomopathogenic nematode-killed insects. *Biological Control* 12(3): 231–236

Baylis, H.A. 1947. The larval stages of the nematode *Mermis nigrescens*. *Parasitology* 38(1/2) 10–16

Beck, M.H. and Strand, M.R. 2007. A novel polydnavirus protein inhibits the insect prophenoloxidase activation pathway. *Proceedings of the National Academy of Sciences of the United States of America* 104(49): 19267–19272

Bedding, R.A. 2009. Controlling the pine-killing woodwasp, *Sirex noctilio*, with nematodes. In *Use of Microbes for Control and Eradication of Invasive Arthropods. Progress in Biological Control*, vol. 6. Hajek, A.E. Glare, T.R., and O'Callaghan M. (eds.) pp. 213–235. Dordrecht: Springer

Bhat, A.H., Chaubey, A.K., and Askary, T.H. 2020. Global distribution of entomopathogenic nematodes, Steinernema and Heterorhabditis, *Egyptian Journal of Biological Pest Control*, 30(1): 31

Bishop, A. and Tate, P. 1939. The morphology and systematic position of *Dobellina mesnili* nov.gen. (Entamoeba mesnili Keilin, 1917). *Parasitology* 31(4): 501–511

Blackmore, M.S., Scoles, G.A., and Craig, G.B. 1995. Parasitism of *Aedes aegypti* and *Ae. albopictus* (Diptera: Culicidae) by *Ascogregarina* spp. (Apicomplexa: Lecudinidae) in Florida. *Journal of Medical Entomology* 32: 847–852

Bläske, V.-U. and Boucias, D.G. 2004. Influence of *Helicosporidium* spp. (Chlorophyta: Trebouxiophyceae) infection on the development and survival of three noctuid species. *Environmental Entomology* 33(1): 54–61

Bläske-Lietze, V.-U. and Boucias, D.G. 2005. Pathogenesis of *Helicosporidium* sp. (Chlorophyta: Trebouxiophyceae) in susceptible noctuid larvae. *Journal of Invertebrate Pathology* 90(3): 161–168

Bläske-Lietze, V.-U., Shapiro, A.M., Denton, J.S., Botts, M., Becnel, J.J., and Boucias, D.G. 2006. Development of the insect pathogenic alga *Helicosporidium*. *The Journal of Eukaryotic Microbiology* 53(3): 165–176

Boucias, D.G. Becnel, J.J., White, S.E., and Bott, M., 2001. *In vivo* and *in vitro* development of the protist *Helicosporidium* sp. *Journal of Eukaryotic Microbiology* 48(4): 460–470

Braun, L., Ewen, A.B., and Gillott, C. 1988. The life cycle and ultrastructure of *Malameba locustae* (King and Taylor) (Amoebidae) in the migratory grasshopper *Melanoplus sanguinipes* (f.) (Acrididae). *The Canadian Entomologist* 120(8–9): 759–772

Brooker, B.E. 1971. Flagellar attachment and detachment of *Crithidia fasciculata* to the gut wall of *Anopheles gambiae*. *Protoplasma* 73(2): 191–202

Brown, M.J.F., Loosli, R., and Schmid-Hempel, P. 2000. Condition-dependent expression of virulence in a trypanosome infecting bumblebees. *Oikos* 91(3): 421–427

Brown, M.J.F., Schmid-Hempel, R., and Schmid-Hempel, P. 2003. Strong context-dependent virulence in a host–parasite system: Reconciling genetic evidence with theory. *Journal of Animal Ecology* 72(6): 994–1002

Burke, G.R., Simmonds, T.J., Sharanowski, B.J., and Geib, S.M. 2018. Rapid viral symbiogenesis via changes in parasitoid wasp genome architecture. *Molecular Biology and Evolution* 35(10): 2463–2474

Burke, G.R. and Strand, M.R. 2014. Systematic analysis of a wasp parasitism arsenal. *Molecular Ecology* 23(4): 890–901

Buschinger, A. and Kleespies, R. 1999. Host range and host specificity of an ant-pathogenic gregarine parasite, *Mattesia geminata* (Neogregarinida: Lipotrophidae). *Entomologia Generalis* 24(2): 93–104

Carlson, C.J., Burgio, K.R., Dougherty, E.R. et al. 2017. Parasite biodiversity faces extinction and redistribution in a changing climate. *Science Advances* 3(9): e1602422

Castillo, J.C., Reynolds, S.E., and Eleftherianos, I. 2011. Insect immune responses to nematode parasites. *Trends in Parasitology* 27(12): 537–547

Cepero, A., Martín-Hernández, R., Prieto, L. et al. 2015. Is *Acarapis woodi* a single species? A new PCR protocol to evaluate its prevalence. *Parasitology Research* 114(2): 651–658

Chang, D.Z., Serra, L., Lu, D., Mortazavi, A., and Dillman, A.R. 2019. A core set of venom proteins is released by entomopathogenic nematodes in the genus *Steinernema*. *PLoS Pathogens* 15(5): e1007626

Chen, L. and Fadamiro, H.Y. 2018. Pseudacteon phorid flies: Host specificity and impacts on solenopsis fire ants. *Annual Review of Entomology* 63(1): 47–67

Ciche, T.A., Kim, K., Kaufmann-Daszczuk, B., Nguyen, K.C.Q., and Hall, D.H. 2008. Cell invasion and matricide during *Photorhabdus luminescens* transmission by *Heterorhabditis bacteriophora* nematodes. *Applied and Environmental Microbiology* 74(8): 2275–2287

Clarke, D.J. 2008. Photorhabdus: A model for the analysis of pathogenicity and mutualism. *Cellular Microbiology* 10(11): 2159–2167

Colgan, T.J., Carolan, J.C., Sumner, S., Blaxter, M.L., and Brown, M.J.F. 2020. Infection by the castrating parasitic nematode *Sphaerularia bombi* changes gene expression in *Bombus terrestris* bumblebee queens. *Insect Molecular Biology* 29(2): 170–182

Cooper, D. and Eleftherianos, I. 2016. Parasitic nematode immunomodulatory strategies: Recent advances and perspectives. *Pathogens* 5(3): 58

Corliss, J.O. 1960. *Tetrahymena chironomi* sp.nov., a ciliate from midge larvae, and the current status of facultative parasitism in the genus *Tetrahymena*. *Parasitology* 50 (1–2): 111–153.

Corliss, J.O. and Coats, D.W. 1976. A new cuticular cyst-producing tetrahymenid ciliate, *Lambornella clarki* n. sp., and the current status of ciliatosis in culicine mosquitoes. *Transactions of the American Microscopical Society* 95(4): 725–739.

Das, B.P. 2003. *Chilodonella uncinata*—a protozoa pathogenic to mosquito larvae, *Current Science. Current Science Association*, 85(4): 483–489

Das, B.P. 2013. Mosquito surveillance tools used and methodology followed in ecological study on JE vectors in Northern India. In *Mosquito Vectors of Japanese Encephalitis Virus from Northern India*, Das, B.P. (ed.) pp. 17–23. India: Springer Briefs in Animal Sciences, Springer

Das, B.P. 2019. *Chilodonella uncinata*—As potential protozoan biopesticide for mosquito vectors of human diseases. *Acta Scientific Microbiology* 2(12): 90–96

Das, B.P. and Tuli, N.R. 2019. Field evaluation of *Chilodonella uncinata* formulation against *Aedes aegypti* in desert coolers and cemented tanks in Delhi. *International Journal of Mosquito Research*, 6(2): 39–45

De Jong, D., De Jong, P.H., and Gonçalves, L.S. 1982. Weight loss and other damage to developing worker honeybees from infestation with *Varroa jacobsoni*. *Journal of Apicultural Research* 21(3): 165–167

Dheilly, N.M., Maure, F., Ravallec, M. et al. 2015. Who is the puppet master? Replication of a parasitic wasp-associated virus correlates with host behaviour manipulation. *Proceedings of the Royal Society B: Biological Sciences* 282(1803): 20142773. doi: 10.1098/rspb.2014.2773

Dillman, A.R., Chaston, J.M., Adams, B.J. et al. 2012. An entomopathogenic nematode by any other name. *PLOS Pathogens* 8(3): p.e1002527

Egerter, D.E., Anderson, J.R., and Washburn, J.O. 1986. Dispersal of the parasitic ciliate *Lambornella clarki*: Implications for ciliates in the biological control of mosquitoes. *Proceedings of the National Academy of Sciences of the United States of America* 83(19): 7335–7339

Eichler, S. and Schaub, G.A. 1998. The effects of aposymbiosis and of an infection with *Blastocrithidia triatomae* (Trypanosomatidae) on the tracheal system of the reduviid bugs *Rhodnius prolixus* and *Triatoma infestans*. *Journal of Insect Physiology* 44(2): 131–140

Er, M.K. and Gökçe, A. 2005. Effect of *Diplocystis lipulae* Sherlock (Eugregarinida: Apicomplexa), a coelomic gregarine pathogen of tipulids, on the larval size of *Tipula paludosa* Meigen (Tipulidae: Diptera). *Journal of Invertebrate Pathology* 89(2): 112–115

Ernst, C.M., Hanelt, B., and Buddle, C.M. 2016. Parasitism of ground beetles (Coleoptera: Carabidae) by a new species of hairworm (Nematomorpha: Gordiida) in arctic Canada. *Journal of Parasitology* 102(3): 327–335

Ernst, H.P. and Baker, G.L. 1982. *Malameba locustae* (King and Taylor) (Protozoa: Amoebidae) in field populations of Orthoptera in Australia. *Journal of the Australian Entomological Society* 21(0): 295–296

Fouks, B. and Lattorff, H.M.G. 2011. Recognition and avoidance of contaminated flowers by foraging bumblebees (*Bombus terrestris*). *PLOS One* 6(10): e26328

Frolov, A.O., Malysheva, M.N., Ganyukova, A.I., Yurchenko, V., and Kostygov, A.Y. 2018. Obligate development of *Blastocrithidia papi* (Trypanosomatidae) in the Malpighian tubules of *Pyrrhocoris apterus* (Hemiptera) and coordination of host-parasite life cycles. *PLoS One* 13(9): e0204467

Gaino, E. and Rebora, M. 2000. *Ophryoglena* sp. (Ciliata: Oligohymenophora) in *Caenis luctuosa* (Ephemeroptera: Caenidae). *Acta Protozoologica* 39: 225–231

Gałęcki, R. and Sokół, R. 2019. A parasitological evaluation of edible insects and their role in the transmission of parasitic diseases to humans and animals. *PLoS one* 14(7): e0219303

Gallot-Lavallée, M., Schmid-Hempel, R., Vandame, R., Vergara, C.H., and Schmid-Hempel, P. 2016. Large scale patterns of abundance and distribution of parasites in Mexican bumblebees. *Journal of Invertebrate Pathology* 133: 73–82

Galloway, T.D. and Brust, R.A. (1979) Review of the genus *Romanomermis* (Nematoda: Mermithidae) with a description of *R. communensis* sp.n. from Canada. *Canadian Journal of Zoology* 57(2): 281–289

Gegear, R.J., Otterstatter, M.C., and Thomson, J.D. 2006. Bumble-bee foragers infected by a gut parasite have an impaired ability to utilize floral information. *Proceedings Biological Sciences* 273(1590): 1073–1078

Golini, V.I. and Corliss, J.O. 1981. A note on the occurrence of the Hymenostome ciliate tetrahymena in chironomid larvae (Diptera: Chironomidae) from the Laurentian Great Lakes. *Transactions of the American Microscopical Society* 100(1): 89–93

Gómez-Moracho, T., Buendía-Abad, M., Benito, M. et al. 2020. Experimental evidence of harmful effects of *Crithidia mellificae* and *Lotmaria passim* on honey bees. *International Journal for Parasitology* 50: 1117–1124

Goulson, D., Nicholls, E., Botías, C., and Rotheray, E.L. 2015. Bee declines driven by combined stress from parasites, pesticides, and lack of flowers. *Science* 347(6229): 1255957

Grewal, M.S. (1957) Pathogenicity of *Trypanosoma rangeli* Tejera, 1920 in the invertebrate host. *Experimental Parasitology* 6(2): 123–130

Grewal, P.S., Lewis, E.E., Gaugler, R., and Campbell, J.F. 1994. Host finding behaviour as a predictor of foraging strategy in entomopathogenic nematodes. *Parasitology* 108: 207–215

Grosman, A.H., Janssen, A., De Brito, E.F. et al. 2008. Parasitoid increases survival of its pupae by inducing hosts to fight predators. *PLoS One* 3(6) p. e2276.

Hanelt, B. and Janovy Jr, J. 2004 Life cycle and paratenesis of American gordiids (Nematomorpha: Gordiida). *Journal of Parasitology* 90(2): 240–244

Hanelt, B., Thomas, F., and Schmidt-Rhaesa, A. 2005. Biology of the Phylum Nematomorpha. *Advances in Parasitology* 59: 243–305

Hecker, K.R., Forbes, M.R., and Léonard, N.J. 2002. Parasitism of damselflies (*Enallagma boreale*) by gregarines: Sex biases and relations to adult survivorship. *Canadian Journal of Zoology* 80(1): 162–168

Hembree, S.C. (1979) Preliminary report of some mosquito pathogens from Thailand. *Mosquito News* 39(3): 575–582

Herbison, R.E.H., Evans, S., Doherty, J.-F., and Poulin, R. 2019. Let's go swimming: Mermithid-Infected earwigs exhibit positive hydrotaxis. *Parasitology* 146(13): 1631–1635

Herniou, E.A., Huguet, E., Thézé, J., Bézier, A., Periquet, G., and Drezen, J.-M. et al. 2013. When parasitic wasps hijacked viruses: Genomic and functional evolution of polydnaviruses. *Philosophical Transactions of the Royal Society B: Biological Sciences* 368(1626): 20130051

Hinks, C.F. and Ewen, A.B. 1986. Pathological effects of the parasite *Malameba locustae* in males of the migratory grasshopper *Melanoplus sanguinipes* and its interaction with the insecticide, cypermethrin. *Entomologia Experimentalis et Applicata* 42(1): 39–44

Holuša, J., Lukášová, K., Wegensteiner, R., Grodzki, W., Pernek, M., and Weiser, J. 2013. Pathogens of the bark beetle *Ips cembrae*: Microsporidia and gregarines also known from other *Ips* species. *Journal of Applied Entomology* 137(3): 181–187

Hughes, D.P. and Libersat, F. 2018. Neuroparasitology of parasite–insect associations. *Annual Review of Entomology* 63(1): 471–487

James, R.R. and Li, Z. 2012. From silkworms to bees: Diseases of beneficial insects. In *Insect Pathology*, 2nd edn, Vega, F.E. and. Kaya, H.K (eds.) pp. 425–459. San Diego: Academic Press

Jouvenaz, D.P. and Anthony, D.W. 1979 *Mattesia geminata* sp. n. (Neogregarinida: Ophrocystidae) a parasite of the tropical fire ant, *Solenopsis geminata* (Fabricius). *The Journal of Protozoology* 26(3): 354–356

Karabörklü, S., Azizoglu, U., and Azizoglu, Z.B. 2017. Recombinant entomopathogenic agents: A review of biotechnological approaches to pest insect control. *World Journal of Microbiology & Biotechnology* 34(1): 14

Kathirithamby, J., Ross, L.D., and Johnston, J.S. 2003. Masquerading as self? Endoparasitic Strepsiptera (Insecta) enclose themselves in host-derived epidermal bag. *Proceedings of the National Academy of Sciences of the United States of America* 100(13): 7655–7659

Kaya, H.K. and Stock, S.P. 1997. Techniques in insect nematology. In *Manual of Techniques in Insect Pathology*, Lacey, L.A. (ed.) pp. 281–324. London: Academic Press

Kellen, W.R. and Lindegren, J.E. 1973 New host records for *Helicosporidium parasiticum*. *Journal of Invertebrate Pathology* 22(2): 296–297

King, P.D., Mercer, C.F., and Meekings, J.S. 1981. Ecology of black beetle, *Heteronychus arator* (Coleoptera: Scarabaeidae)—population studies. *New Zealand Journal of Agricultural Research* 24(1): 87–97

King, R.L. and Taylor, A.B. 1936 *Malpighamœba locustae*, n. sp. (Amoebidae), a protozoan parasitic in the Malpighian tubes of grasshoppers. *Transactions of the American Microscopical Society* 55(1): 6–10

Lamborn, W.A. 1921. A protozoon pathogenic to mosquito larvae. *Parasitology* 13(3): 213–215

Lange, C.E. 2002. La amebiasis debilitativa de los ortópteros y su potencial para el control biológico de acridios (Orthoptera: Acridoidea) en la Argentina. *Revista de Investigaciones Agropecuarias* 31(3): 25–38 (in Spanish)

Lange, C.E. and Lord, J.C. 2012. Protistan Entomopathogens. In *Insect Pathology*, 2nd edn, Vega F.E. and Kaya H.K. (eds.) pp. 367–394. San Diego: Academic Press

Lantova, L. and Volf, P. 2014. Mosquito and sand fly gregarines of the genus *Ascogregarina* and *Psychodiella* (Apicomplexa: Eugregarinorida, Aseptatorina)—Overview of their taxonomy, life cycle, host specificity and pathogenicity. *Infection, Genetics and Evolution* 28: 616–627

Larsson, J.I.R., De Roca, C.B., and Gaju-Ricart, M. 1992. Fine structure of an amoeba of the genus *Vahlkampfia* (Rhizopoda, Vahlkampfiidae), a parasite of the gut epithelium of the bristletail *Promesomachilis hispanica* (Microcoryphia, Machilidae). *Journal of Invertebrate Pathology* 59(1): 81–89

Larsson, R. 1976. Insect pathological investigations on Swedish thysanura: Observations on *Malamoeba locustae* (Protozoa, Amoebidae) from *Lepisma saccharina* (Thysanura, Lepismatidae). *Journal of Invertebrate Pathology* 28(1): 43–46

Lee, M.J., Kalamarz, M.E., Paddibhatla, I., Small, C., Rajwani, R., and Govind, S. 2009. Virulence factors and strategies of *Leptopilina* spp.: Selective responses in *Drosophila* hosts. *Advances in Parasitology* 70: 123–145

Lewis, E.E. and Clarke, D.J. 2012. Nematode parasites and entomopathogens. In *Insect Pathology*, 2nd edn, Vega, F.E. and Kaya, H.K. (eds.), Chapter 11, pp. 395–424. San Diego: Academic Press

Libersat, F., Kaiser, M., and Emanuel, S. 2018. Mind control: How parasites manipulate cognitive functions in their insect hosts. *Frontiers in Psychology* 9: 572

Liu, T.P. 1985. Scanning electron microscopy of developmental stages of *Malpighamoeba mellificae* Prell in the honey bee. *The Journal of Protozoology* 32(1): 139–144

Lopes, R.B. and Alves, S.B. 2005. Effect of *Gregarina* sp. parasitism on the susceptibility of *Blattella germanica* to some control agents. *Journal of Invertebrate Pathology* 88(3): 261–264

Lucarotti, C.J. 2000. Cytology of *Leidyana canadensis* (Apicomplexa: Eugregarinida) in *Lambdina fiscellaria* larvae (Lepidoptera: Geometridae). *Journal of Invertebrate Pathology* 75(2): 117–125

Maharramov, J., Meeus, I., Maebe, K. et al. 2013. Genetic variability of the Neogregarine *Apicystis bombi*, an etiological agent of an emergent bumblebee disease. *PLoS one* 8(12): e81475

Malone, L.A. and Dhana, S. 1988. Life cycle and ultrastructure of *Adelina tenebrionis* (Sporozoea: Adeleidea) from *Heteronychus arator* (Coleoptera: Scarabaeidae). *Parasitology Research* 74(3): 201–207

Manasherob, R., Ben-Dov, E., Zaritsky, A., and Barak, Z. 1998. Germination, growth, and sporulation of *Bacillus thuringiensis* subsp. *israelensis* in excreted food vacuoles of the Protozoan *Tetrahymena pyriformis*. *Applied and Environmental Microbiology* 64(5): 1750–1758

Martens, E.C. and Goodrich-Blair, H. 2005. The *Steinernema carpocapsae* intestinal vesicle contains a subcellular structure with which *Xenorhabdus nematophila* associates during colonization initiation. *Cellular Microbiology* 7(12): 1723–1735

Martin, S.J. and Correia-Oliveira, M.E. 2016. The occurrence of ecto-parasitic *Leptus* sp. mites on Africanized honey bees. *Journal of Apicultural Research* 55(3): 243–246

Martín-Vega, D., Garbout, A., Ahmed, F. et al. 2018. 3D virtual histology at the host/parasite interface: Visualisation of the master manipulator, *Dicrocoelium dendriticum*, in the brain of its ant host. *Scientific Reports* 8(1): 8587

Maslov, D.A., Votýpka, J., Yurchenko, V., and Lukeš, J. 2013. Diversity and phylogeny of insect trypanosomatids: All that is hidden shall be revealed. *Trends in Parasitology* 29(1): 43–52

Mercer, D.R. and Anderson, J.R. 1994. Tannins in treehole habitats and their effects on *Aedes sierrensis* (Diptera: Culicidae) production and parasitism by *Lambornella clarki* (Ciliophora: Tetrahymenidae). *Journal of Medical Entomology* 31(1): 159–167.

Mercer, D.R., Washburn, J.O., and Anderson, J.R. 2010. Vertical oviposition and *Lambornella clarki* (Ciliophora: Tetrahymenidae) dispersal by *Aedes sierrensis* (Diptera: Culicidae) in California. *Journal of Vector Ecology* 35(1): 20–27

Michalková, V., Krascsenitsová, E., and Kozánek, M. 2012. On the pathogens of the spruce bark beetle *Ips typographus* (Coleoptera: Scolytinae) in the Western Carpathians. *Biologia* 67(1): 217–221

Minchin, E.A. (1910). On some parasites observed in the rat flea (*Ceratophyllus fasciatus*). Festschr. 60 Geburstag R. Hertwigs, 1, 289e302

Moreau, J.S. and Asgari, S. 2015. Venom proteins from parasitoid wasps and their biological functions. *Toxins* 7(7): 2385–2412.

Narayanaswamy, T.K. and Govindan, R. 2000. Mulberry silkworm ujifly, *Exorista bombycis* (Louis) (Diptera: Tachinidae). *Integrated Pest Management Reviews* 5(4): 231–240

Nickle, W.R. 1972. A Contribution to our knowledge of the Mermithidae (Nematoda). *Journal of Nematology* 4(2): 113–146

Österreichische Agentur für Gesundheit und Ernährungssicherheit GmbH, Institut für Saat und Pflanzgut, Pflanzenschutzdienst und Bienen Abteilung Bienenkunde und Bienenschutz. 2014. Amöbenruhr, Malpighamoebiose, Merkblatt 5. (In German). https://www.ages.at/download/0/0/fcd311a4e94b8e 47666a8bdc2a6f813686e08ae0/fileadmin/AGES2015/ Themen/Krankheitserreger_Dateien/Bienen/Infor mationsblatt_5_Am%C3%B6benruhr_Nov_2014.pdf (Retrieved18/06/2020.)

Otterstatter, C.M. and Whidden, L.T. 2004. Patterns of parasitism by tracheal mites (*Locustacarus buchneri*) in natural bumble bee populations. *Apidologie* 35(4): 351–357

Patil, C.D., Patil, S. V, Salunke, B.K., and Salunkhe, R.B. 2012. Insecticidal potency of bacterial species *Bacillus thuringiensis* SV2 and *Serratia nematodiphila* SV6 against larvae of mosquito species *Aedes aegypti*, *Anopheles stephensi*, and *Culex quinquefasciatus*. *Parasitology Research* 110(5): 1841–1847

Peters, A. 1996. The natural host range of *Steinernema* and *Heterorhabditis* spp. and their impact on insect populations. *Biocontrol Science and Technology* 6(3): 389–402

Pichon, A., Bézier, A., Urbach, S. et al. 2015. Recurrent DNA virus domestication leading to different parasite virulence strategies. *Science Advances* 1(10): e1501150

Plischuk, S., Antúnez, K., Haramboure, M., Minardi, G.M., and Lange, C.E. 2017. Long-term prevalence of the protists *Crithidia bombi* and *Apicystis bombi* and detection of the microsporidium *Nosema bombi* in invasive bumble bees. *Environmental Microbiology Reports* 9(2): 169–173

Poinar, G. 2012. Nematode parasites and associates of ants: Past and present. *Psyche* 192017, doi:10.1155/2012/192017

Poinar, G.O. 1981. Distribution of *Pheromermis pachysoma* (Mermithidae) determined by paratenic invertebrate hosts. *Journal of Nematology* 13(3): 421–424

Poinar, G.O., Hess, R.T., and Petersen, J.J. 1979. Immune Responses of Mosquitoes against *Romanomermis culicivorax* (Mermithida: Nematoda). *Journal of Nematology* 11(1): 110–116

Poinar, G.O.J., Porter, S.D., Tang, S., and Hyman, B.C. 2007. *Allomermis solenopsi* n. sp. (Nematoda: Mermithidae) parasitising the fire ant *Solenopsis invicta* Buren (Hymenoptera: Formicidae) in Argentina. *Systematic Parasitology* 68(2): 115–128

Porter, S.D. 1998. Biology and behavior of pseudacteon decapitating flies (diptera: phoridae) that parasitize

Solenopsis fire ants (Hymenoptera: Formicidae). *Florida Entomologist* 81(3): 292–309

Purrini, K. and Žižka, Z. 1983. More on the life cycle of *Malamoeba scolyti* (Amoebidae: Sarcomastigophora) parasitizing: The bark beetle *Dryocoetes autographus* (Scolytidae, Coleoptera). *Journal of Invertebrate Pathology* 42(1): 96–105

Quintana, S., Plischuk, S., Brasesco, C. et al. 2021. *Lotmaria passim* (Kinetoplastea: Trypanosomatidae) in honey bees from Argentina. *Parasitology International* 81: 102244

Ramsey, S.D., Ochoa, R., Bauchan, G. et al. 2019. *Varroa destructor* feeds primarily on honey bee fat body tissue and not hemolymph. *Proceedings of the National Academy of Sciences* 116(5): 1792–1801

Ravoet, J., De Smet, L., Meeus, I., Smagghe, G., Wenseleers, T., and De Graaf, D.C. 2014. Widespread occurrence of honey bee pathogens in solitary bees. *Journal of Invertebrate Pathology* 122: 55–58

Ravoet, J., Maharramov, J., Meeus, I. et al. 2013. Comprehensive bee pathogen screening in Belgium reveals *Crithidia mellificae* as a new contributory factor to winter mortality. *PLOS one* 8(8): e72443

Ravoet, J., Schwarz, R.S., Descamps, T. et al. 2015. Differential diagnosis of the honey bee trypanosomatids *Crithidia mellificae* and *Lotmaria passim*. *Journal of Invertebrate Pathology* 130: 21–27

Richardson, L.L., Adler, L.S., Leonard, A.S. et al. 2015. Secondary metabolites in floral nectar reduce parasite infections in bumblebees. *Proceedings of the Royal Society B: Biological Sciences* 282(1803): 20142471.

Ritter, W. 1981. Varroa disease of the honeybee *Apis mellifera*. *Bee World* 62(4): 141–153

Rodriguez, J.J., Fernández-Triana, J.L., Smith, M.A. et al. 2013. Extrapolations from field studies and known faunas converge on dramatically increased estimates of global microgastrine parasitoid wasp species richness (Hymenoptera: Braconidae). *Insect Conservation and Diversity* 6(4): 530–536

Rossi, M., Ott, S.R., and Niven, J.E. 2019. *Malpighamoeba* infection compromises fluid secretion and P-glycoprotein detoxification in Malpighian tubules. *Scientific Reports* 10: 15953

Sajnaga, E. and Kazimierczak, W. 2020. Evolution and taxonomy of nematode-associated entomopathogenic bacteria of the genera *Xenorhabdus* and *Photorhabdus*: An overview. *Symbiosis* 80(1): 1–13

Sammataro, D., Gerson, U., and Needham, G. 2000. Parasitic mites of honey bees: Life history, implications, and impact. *Annual Review of Entomology* 45: 519–548

Schaub, G.A. 1994. Pathogenicity of Trypanosomatids on insects. *Parasitology Today* 10(12): 463–468

Schaub, G.A. 2006. Parasitogenic alterations of vector behaviour. *International Journal of Medical Microbiology* 296(Suppl 1): 37–40

Schaub, G.A. 2009. Interactions of Trypanosomatids and Triatomines. *Advances in Insect Physiology* 37: 177–242

Schaub, G.A. and Loesch, P. 1989. Parasite/host-interrelationships of the trypanosomatids *Trypanosoma cruzi* and *Blastocrithidia triatomae* and the reduviid bug *Triatoma infestans*: Influence of starvation of the bug. *Annals of Tropical and Medical Parasitology* 83(3): 215–223

Schaub, G.A. and Schnitker, A. 1988. Influence of *Blastocrithidia triatomae* (Trypanosomatidae) on the reduviid bug *Triatoma infestans*: Alterations in the Malpighian tubules. *Parasitology Research* 75 (2): 88–97

Schmid-Hempel, P., Aebi, M., Barribeau, S. et al. 2018. The genomes of *Crithidia bombi* and *C. expoeki*, common parasites of bumblebees. *PLOS One* 13(1): e0189738

Schmid-Hempel, P., Wilfert, L., and Schmid-Hempel, R. 2019. Pollinator diseases: The *Bombus–Crithidia* system. In *Wildlife Disease Ecology: Linking Theory to Data and Application (Ecological Reviews)*, Wilson, K. Fenton, A. and Tompkins, D. (eds.) pp. 3–31. Cambridge: Cambridge University Press

Schoonvaere, K., Brunain, M., Baeke, F., De Bruyne, M., De Rycke, R., and De Graaf, D.C. 2020. Comparison between *Apicystis cryptica* sp. n. and *Apicystis bombi* (Arthrogregarida, Apicomplexa): Gregarine parasites that cause fat body hypertrophism in bees. *European Journal of Protistology* 73: 125688

Schwarz, R.S., Bauchan, G.R., Murphy, C.A., Ravoet, J., De Graaf, D.C., and Evans, J.D. 2015. Characterization of two species of trypanosomatidae from the honey bee *Apis mellifera*: *Crithidia mellificae* Langridge and McGhee, and *Lotmaria passim* n. gen., n. sp. *The Journal of Eukaryotic Microbiology* 62(5): 567–583

Serepa-Dlamini, M.H. and Gray, V.M. 2018. A new species of entomopathogenic nematode *Oscheius safricana* n. sp. (Nematoda: Rhabditidae) from South Africa. *Archives of Phytopathology and Plant Protection* 51(5–6): 309–321

Shamseldean, M., Platzer, E., and Gaugler, R. 2007. Role of the surface coat of *Romanomermis culicivorax* in immune evasion. *Nematology* 9(1): 17–24

Simmons, T.W. and Hutchinson, M.L. 2016. A Critical Review of All Known Published Records for Water Mite (Acari: Hydrachnidiae) and Mosquito (Diptera: Culicidae) Parasitic Associations From 1975 to Present. *Journal of Medical Entomology* 53(4): 737–752

Siva-Jothy, M.T. and Plaistow, S.J. 1999. A fitness cost of eugregarine parasitism in a damselfly. *Ecological Entomology* 24(4): 465–470

Siva-Jothy, M.T., Tsubaki, Y., Hooper, R.E., and Plaistow, S.J. 2001. Investment in immune function under chronic and acute immune challenge in an insect. *Physiological Entomology* 26(1): 1–5

Smith, B.P. and McIver, S.B. 1984. The impact of *Arrenurus danbyensis* Mullen (Acari: Prostigmata; Arrenuridae) on a population of *Coquillettidia perturbans* (Walker) (Diptera: Culicidae). *Canadian Journal of Zoology* 62: 1121–1134

Stock, S.P., Kusakabe, A., and Orozco, R.A. 2017. Secondary metabolites produced by *Heterorhabditis* symbionts and their application in agriculture: What we know and what to do next. *Journal of Nematology* 49(4): 373–383

Strand, M.R. 2014. Teratocytes and their functions in parasitoids. *Current Opinion in Insect Science* 6: 68–73

Strand, M.R. and Burke, G.R. 2015. Polydnaviruses: From discovery to current insights. *Virology* 479–480: 393–402

Strüder-Kypke, M.C., Wright, A.-D.G., Jerome, C.A., and Lynn, D.H. 2001. Parallel evolution of histophagy in ciliates of the genus *Tetrahymena*. *BMC Evolutionary Biology* 1(1): 5

Tartar, A. 2013. The Non-Photosynthetic Algae *Helicosporidium* spp.: Emergence of a novel group of insect pathogens. *Insects* 4 (3): 375–391

Taylor, A.B. and King, R.L. (1937) Further studies on the parasitic amebae found in grasshoppers. *Transactions of the American Microscopical Society* 56(2): 172–176

Taylor, R.L. (1968) Tissue damage induced by an oxyuroid nematode, *Leidynema* sp., in the hindgut of the Madeira cockroach, *Leucophaea maderae*. *Journal of Invertebrate Pathology* 11(2): 214–218

Théodorides, J. (1956) A propos des Grégarines d'Hyménoptères Apoidea. *Annales de Parasitologie Humaine et Comparee* 31: 315–316

Thomas, F., Schmidt-Rhaesa, A., Martin, G., Manu, C., Durand, P., and Renaud, F. (2002) Do hairworms (Nematomorpha) manipulate the water seeking behaviour of their terrestrial hosts? *Journal of Evolutionary Biology* 15(3): 356–361

Thomas, F., Ulitsky, P., Augier, R. et al. 2003. Biochemical and histological changes in the brain of the cricket *Nemobius sylvestris* infected by the manipulative parasite *Paragordius tricuspidatus* (Nematomorpha). *International Journal for Parasitology* 33(4): 435–443

Tobie, E.J. (1970) Observations on the development of *Trypanosoma rangeli* in the hemocoel of *Rhodnius prolixus*. *Journal of Invertebrate Pathology* 15(1): 118–125

Tomalski, M.D., Bruce, W.A., Travis, J., and Blum, M.S. 1988. Preliminary characterization of toxins from the straw itch mite, *Pyemotes tritici*, which induce paralysis in the larvae of a moth. *Toxicon* 26(2): 127–132

Tripodi, A.D. and Strange, J.P. 2018. Rarely reported, widely distributed, and unexpectedly diverse: Molecular characterization of mermithid nematodes (Nematoda:

Mermithidae) infecting bumble bees (Hymenoptera: Apidae: *Bombus*) in the USA. *Parasitology* 145(12): 1558–1563

Tsubaki, Y. and Hooper, R.E. 2004. Effects of eugregarine parasites on adult longevity in the polymorphic damselfly *Mnais costalis* Selys. *Ecological Entomology* 29(3): 361–366

Tsurui, K., Ohishi, T., Kumano, N., Teruya, K., Toyosato, T., and Shiromoto, K. 2015. Symptoms shown by late-stage larvae of the West Indian sweet potato weevil *Euscepes postfasciatus* (Coleoptera: Curculionidae) infected with *Farinocystis* sp. (Neogregarinida: Lipotrophidae). *Applied Entomology and Zoology* 50: 245–254

Vasconcellos, L.R.C., Carvalho, L.M.F., Silveira, F.A.M. et al. 2019. Natural infection by the protozoan *Leptomonas wallacei* impacts the morphology, physiology, reproduction, and lifespan of the insect *Oncopeltus fasciatus*. *Scientific Reports* 9(1): 17468

Walker, E.D., Poirier, S.J., and Veldman, W.T. 1987. Effects of *Ascogregarina barretti* (Eugregarinida: Lecudinidae) infection on emergence success, development time, and size of *Aedes triseriatus* (Diptera: Culicidae) in microcosms and tires. *Journal of Medical Entomology* 24(3) 303–309

Wang, Zhi-zhi, Ye, X., Shi, M., et al. 2018. Parasitic insect-derived miRNAs modulate host development. *Nature Communications* 9(1): 2205

Washburn, J.O. and Anderson, J.R. 1986. Distribution of *Lambornella clarki* (Ciliophora: Tetrahymenidae) and other mosquito parasites in California treeholes. *Journal of Invertebrate Pathology* 48(3): 296–309

Washburn, J.O., Anderson, J.R., and Mercer, D.R. 1991. Parasitism of newly hatched *Aedes sierrensis* (Diptera: Culicidae) larvae by *Lambornella clarki* (Ciliophora: Tetrahymenidae) following habitat flooding. *Journal of Invertebrate Pathology* 58(1): 67–74

Wey, B., Heavner, M.E., Wittmeyer, K.T., Briese, T., Hopper, K.R., and Govind, S. 2020. Immune suppressive extracellular vesicle proteins of *Leptopilina heterotoma* are encoded in the wasp genome, *G3 (Bethesda, Md.). Genetics Society of America* 10(1): 1–12.

Wheeler, R.J., Gluenz, E., and Gull, K. 2013. The limits on trypanosomatid morphological diversity. *PLoS ONE* 8(11): e79581

Wilfert, L., Long, G., Leggett, H.C. et al. 2016. Deformed wing virus is a recent global epidemic in honeybees driven by *Varroa* mites. *Science* 351(6273): 594–597

Yahaya, Z.S., Izzaudin, N.A.I., and Razak, A.F.A. 2017. Parasitic *Gregarine blattarum* found infecting American cockroaches, *Periplaneta americana*, in a population in Pulau Pinang, Malaysia. *Tropical Life Sciences Research* 28(1): 145–149

Yaman, M. and Radek, R. 2007. Infection of the predator beetle *Rhizophagus grandis* Gyll. (Coleoptera, Rhizophagidae) with the insect pathogenic alga *Helicosporidium* sp. (Chlorophyta: Trebouxiophyceae). *Biological Control* 41(3): 384–388

Yaman, M. and Radek, R. 2015. *Mattesia weiseri* sp. nov., a new neogregarine (Apicomplexa: Lipotrophidae) pathogen of the great spruce bark beetle, *Dendroctonus micans* (Coleoptera: Curculionidae, Scolytinae). *Parasitology Research* 114(8): 2951–2958

Yamashita, K., Zhang, K., Ichiki, R.T., Nakamura, S., and Furukawa, S. 2019. Novel host immune evasion strategy of the endoparasitoid *Drino inconspicuoides*. *Bulletin of Entomological Research* 109(5): 643–648

Yarwood, E.A. 1937. The life cycle of *Adelina cryptocerci* sp.nov., a coccidian parasite of the roach *Cryptocercus punctulatus*. *Parasitology* 29: 370–390

Zaritsky, A., Ben-Dov, E., Zalkinder, V., and Barak, Z. 1992. Digestibility by and pathogenicity of the protozoa *Tetrahymena pyriformis* to larvae of *Aedes aegypti*. *Journal of Invertebrate Pathology* 59(3): 332–334

Zhang, X., Machado, R.A., Doan, C. Van, Arce, C.C., Hu, L., and Robert, C.A. 2019. Entomopathogenic nematodes increase predation success by inducing cadaver volatiles that attract healthy herbivores. *eLife* 8. doi: 10.7554/eLife.46668

CHAPTER 14

Viral diseases of crustaceans

Arun K. Dhar, Roberto Cruz-Flores, and Kelly S. Bateman

14.1 Introduction

In 1974, Couch described the first viral infection in pink shrimp, *Penaeus duorarum*, by an occluded rod-shaped virus and this pathogen is now commonly known as Baculovirus penaei (BP) or tetrahedral baculovirus (Couch 1974). Since the initial report of this disease, almost 30 viruses have been discovered in penaeid shrimp and other farmed crustaceans. The emergence of viral diseases in farmed crustaceans coincided with the growth of penaeid shrimp farming from a nascent industry in early 70s to a major global industry.

In 2018, global production of finfish, crustaceans and molluscs was estimated to 179 million metric tonnes and 82 million tonnes (*ca.* 46% with an estimated value of US$ 250 billion) of which originated from aquaculture. Marine shrimp production dominates the crustacean sector (9.4 million tonnes in 2018) and penaeid shrimp accounts for over 60% of the total production of crustacean species. Two penaeid shrimp, *Penaeus* (also named *Litopenaeus* by some) *vannamei*, and *P. monodon*, and a freshwater species, *Macrobrachium rosenbergii* are the predominant species that are farmed worldwide; and *P. vannamei* represents about 75% of farmed shrimp worldwide. It is now widely recognised that diseases are major limiting factors in the growth and expansion of shrimp and other crustacean farming (Flegel 2012; Lightner et al. 2012). It is estimated that disease cause *ca.* 40% of production loss annually mainly due to viral diseases (Stentiford et al. 2012). As the shrimp industry becomes more industrialised and intensive farming

are increasingly practiced with both native and non-native species, it is inevitable that new diseases will continue to emerge. This highlights the need for an on-going disease surveillance, developing strategies to prevent the spread of diseases and economic losses caused by disease outbreaks worldwide (Shinn et al. 2018). Most of our current knowledge of viral diseases of crustaceans comes from studies on commercially important species. Therefore, this chapter aims to provide a comprehensive summary of viral diseases impacting these crustaceans including those produced by aquaculture-based methods (e.g. shrimp) and those caught from the wild (e.g. lobsters and crabs).

14.2 Diseases caused by DNA viruses

14.2.1 Infectious hypodermal and hematopoietic necrosis virus (IHHNV)

Infectious hypodermal and hematopoietic necrosis virus (IHHNV), also known as *Penaeus stylirostris* densovirus (*Pst*DNV), was first reported in juveniles and sub-adults of the blue shrimp (*P. stylirostris*) in Hawaii in the early 1980s causing mass mortalities (Lightner et al. 1983). Subsequently, the disease was reported in the Pacific white shrimp (*P. vannamei*) where it did not cause mortalities but resulted in growth retardation and deformities in the appendages referred to as runt deformity syndrome (RDS) (Kalagayan et al. 1991). The virus infects all life stages (eggs, larvae, post-larvae, juveniles, and adults) of *P. vannamei* and is transmitted vertically. Horizontal transmission occurs when

Arun K. Dhar, Roberto Cruz-Flores, and Kelly S. Bateman, *Viral diseases of crustaceans*. In: *Invertebrate Pathology*. Edited by Andrew F. Rowley, Christopher J. Coates and Miranda M.A. Whitten, Oxford University Press. © Oxford University Press (2022). DOI: 10.1093/oso/9780198853756.003.0014 healthy shrimp cannibalize on dead or moribund animals or through contaminated water (Dhar et al. 2014, 2019). In the black tiger shrimp, *P. monodon*, infections by IHHNV are mostly asymptomatic but causes growth reduction (Flegel 2006; Dhar et al. 2014). Many commercially available genetic lines of *P. vannamei* shrimp appeared to be tolerant or resistant to IHHNV. Shrimp farmers from Latin America where *P. vannamei* is widely farmed claim IHHNV infection does not cause economic losses. However, published reports from Australia and India indicate that the disease continues to cause economic losses in *P. monodon* and *P. vannamei* culture (Jagadeesan et al. 2019; Sellars et al. 2019). Even in experimental infections, it was shown that IHHNV can cause high-level mortality in crayfish (*Procambarus clarkii*) (Chen et al. 2018). This indicates that IHHNV still remains as an economically important viral pathogen in crustacean aquaculture.

IHHNV virions are icosahedral, non-enveloped measuring 22–23 nm in size and contain a single-stranded DNA genome of ~ 4.1 kb. The virus has been classified under the family *Parvoviridae*, subfamily *Densovirinae*. The IHHNV genome contains three major open reading frames, left, middle, and right; and encodes nonstructural proteins NS1, NS2, and the viral capsid protein. Among the three IHHNV genes, the capsid protein gene is the most variable and hence it is used for delineating genetic relationship among IHHNV isolates. Phylogenetic analysis using the IHHNV capsid protein sequence from geographical isolates around the world showed that IHHNV genotypes can be categorised into three lineages. The first clade, Type 1 genotype is formed by Southeast Asian isolates including isolates from Thailand, Philippines, and Vietnam. The second clade, Type 2 genotype is composed of isolates from the Americas and East Asia, and the third clade, Type 3 genotype contains isolates from the Indo-Pacific region. While Types I and II represent infectious forms of the virus, Type III genotype represents a genome integrated non-infectious form of the virus (Dhar et al. 2019).

IHHNV replicates in tissues of ectodermal (i.e. epidermis, hypodermal epithelium of fore- and hindgut, nerve cord, and nerve ganglia) and mesodermal (i.e. hematopoietic organs, antennal gland, gonads, lymphoid organ, and connective tissue) origins and produces intranuclear, haloed,

eosinophilic Cowdry type A intranuclear inclusion bodies (large inclusions of nucleic acid and protein which push chromatin to margins of the nucleus, contain virions and stain eosinophilic to basophilic) (Figure 14.1 a, b). In IHHNV tolerant *P. vannamei* shrimp, inclusion bodies may not be seen but the viral nucleic acid can be detected by *in situ* hybridisation using IHHNV-specific probe and by polymerase chain reaction-based molecular diagnostics (Dhar et al. unpublished).

14.2.2 Hepatopancreatic parvovirus (HPV)

Hepatopancreatic parvovirus (HPV) was first reported in a commercial farm in Singapore in 1982 in *Penaeus merguiensis* (Lightner and Redman 1985). Subsequently, the virus spread throughout Southeast Asia, Australia, Africa, Middle East, and Latin America (Safeena and Rai 2012). The susceptible species of shrimp included *P. orientalis*, *P. semisulcatus*, *P. esculentus*, *P. penicillatus*, *P. monodon*, *P. vannamei*, *P. stylirostris*, and freshwater prawn, *M. rosenbergii*, in both captive and wild populations (Safeena and Rai 2012).

HPV infection causes non-specific clinical signs, including necrosis and atrophy of the hepatopancreas, poor growth rates, anorexia, and reduced preening with a concurrent increase in surface, and gill fouling by epicommensal organisms (Dhar et al. 2014). Occasionally, there is a visible opacity of the tail muscles (Lightner and Redman 1985). More virulent strains of HPV are associated with stunted growth and increased mortality (Dhar et al. 2014). The disease is insidious, as the virus allows for a seemingly normal growth through the larval and post-larval stages, with mortalities occurring mostly at juvenile stages (Lightner et al. 1985). The viral load is suspected to be linked with the severity of HPV infection, and selective breeding for a low HPV titre trait may help to develop genetically resistant lines of shrimp (Phuthaworn et al. 2016).

The virus transmission occurs horizontally via contaminated water and cannibalism and is facilitated by high-density stocking (OIE 2007a; Tang et al. 2008). Horizontal transmission may also occur through carrier hosts. For example, HPV was detected in the Australian mud crab (*Scylla serrata*) (Owens et al. 2010). It was also demonstrated experimentally that Australian red-claw crayfish (*Cherax*

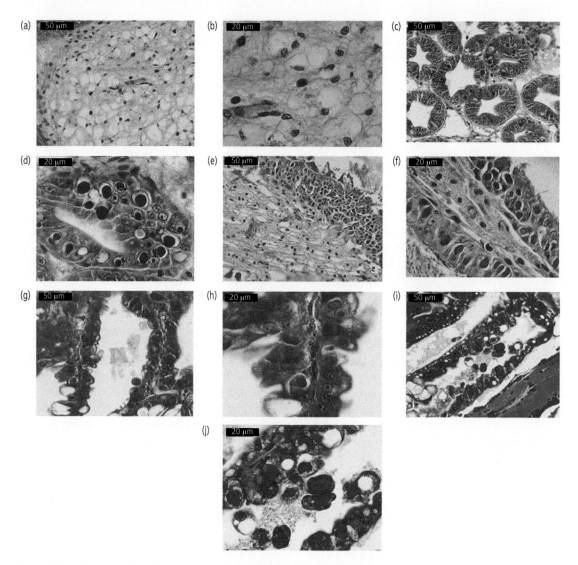

Figure 14.1 (a–j) Histopathology of DNA viruses that infect shrimp species of commercial importance. (a) Several eosinophilic intranuclear Cowdry Type A inclusion bodies in then connective tissues of *P. stylirostris* formed by IHHNV. (b) High magnification of three Cowdry Type A inclusion bodies of IHHNV. (c) Several basophilic inclusion bodies formed by HPV in the tubules of the hepatopancreas of *P. monodon*. (d) High magnification of a hepatopancreas tubule with a large number of basophilic inclusion bodies formed by HPV in *P. monodon*. (e) A large number of basophilic intranuclear inclusion bodies in the stomach epithelia of *P. vannamei* caused by WSSV. (f) High magnification of the stomach epithelia of shrimp infected with WSSV with a large number of inclusion bodies. (g) Hepatopancreas tubule of *P. vannamei* heavily infected with multiple spherical eosinophilic intranuclear inclusion bodies caused by PmNV. (h) High magnification where multiple spherical eosinophilic inclusion bodies are observed. (i) Hepatopancreas of *P. vannamei* larvae infected with PvSNPV, several eosinophilic tetrahedral inclusions are observed including some in the lumen. (j) High magnification of the eosinophilic tetrahedral inclusion bodies.

quadricarinatus) might be a short-term (up to 30 days) carrier of HPV (La Fauce et al. 2007). Brine shrimp (*Artemia franciscana*) at different life stages (from nauplius to adult), were also found to be able to transmit HPV to *P. monodon* in experimental challenges. It is unknown if *Artemia* is a reservoir, host or only a mechanical vector, although high viral loads detected in *Artemia* may suggest that the HPV might replicate in it.

The virions of HPV are icosahedral measuring 22 nm and contain a single-stranded DNA genome. HPV genome is ~ 6.3 kb and contains three open reading frames (ORFs): two partially overlapping ORFs, ORF1, and ORF2 representing two non-structural genes and a separate ORF, ORF3 representing a structural gene (Dhar et al. 2014, 2019). ORF1 encodes a non-structural protein of unknown function. ORF2 encodes a non-structural protein NS1. ORF3 encodes a viral capsid protein (VP). The VP gene has the highest genetic variation among the HPV isolates ranging from 18 to 31% (Tang et al. 2008). HPV infection is restricted to hepatopancreas. Upon infection, HPV produces basophilic intranuclear inclusion bodies within the E- and F-cells along the distal portion of the hepatopancreatic tubule that are considered pathognomonic of HPV (Figure 14.1 c,d). It is interesting to note that all three polypeptides of HPV have functional nuclear localisation signals (NLS), whereas, in IHHNV, only the NS1 protein contains NLS (Owens 2013; Shike et al. 2000).

HPV seems to be more genetically diverse compared to IHHNV (La Fauce et al. 2007; Tang et al. 2008). A Bayesian phylogenetic analysis based on the nucleotide sequence of VP revealed four genotypes of HPV, with isolates from Tanzania and Madagascar forming one subclade, Thailand, Indonesia, and India forming the second subclade, Australia and New Caledonia forming the third, and Korea and China forming the forth subclade. Overall, a common characteristic of the trees is the clustering of the isolates based on their geographic distribution (Dhar et al. 2014). According to the International Committee on Taxonomy of Viruses (ICTV), Hepatopancreatic parvovirus is now named *Decapod hepandensovirus 1* which is a single species within the *Hepadensovirus* genus, belonging to the subfamily *Densovirinae* and family *Parvoviridae* (formerly genus *Brevidensovirus*, subfamily *Densovirinae*, family *Parvoviridae*) (Cotmore et al. 2019).

14.2.3 Spawner isolated mortality virus (SMV)

Spawner isolated mortality virus (SMV) was first reported from a research facility in Townsville, Queensland, Australia, in 1993. The virus was reported to cause mortalities in broodstock of *P. monodon* and is similar or possibly identical to mid-crop mortality syndrome of *P. monodon* in grow out ponds (Owens et al. 1998). In addition to *P.*

monodon, the virus infects *P. esculentus*, *P. japonicus*, *P. merguiensis*, and *Metapenaeus ensis* in laboratory challenges, and also infects red-claw crayfish (*Cherax quadricarinatus*) (Owens et al. 1998, 2003). The virus infects cells of endodermal origin in multiple organs including the distal ends of hepatopancreatic tubules, midgut and hindgut caecae and folds, the terminal ampoule and the medial vas deferens in male reproductive organ, as well as in the ovary and the lymphoid organ. The virions are non-enveloped, icosahedral particles measuring 20 nm in diameter. Phylogenetic analyses revealed that SMV does not cluster with IHHNV and HPV suggesting that shrimp parvoviruses appear to have a diverse origin.

14.2.4 Lymphoidal parvo-like virus (LPV)

Parvovirus infections in the lymphoid organ of cultured *P. monodon*, *P. merguiensis*, *P. esculentus*, and the hybrid of *P. monodon* and *P. esculentus* were first reported from Australia (Owens et al. 1991). The virus infects the lymphoid organ, nervous tissue, antennal gland, and haematopoietic tissues. In virus infected shrimp, the lymphoid organ is hypertrophied and contains multinucleated 'giant cells' (Owens et al. 1991). Infected cells display basophilic, spherical-shaped intranuclear inclusion bodies, hypertrophied nucleus and marginated chromatin. In the lymphoid organ of heavily infected animals, paracrystalline arrays of LPV particles, measuring 18–20 nm in diameter, can be seen on the edge of the large electron-dense intranuclear inclusions (Owens et al. 1991).

14.2.5 *Penaeus monodon* metallodensovirus

A new potential pathogen of *P. monodon* from Vietnam has been recently described and termed *Penaeus monodon* metallodensovirus (PmMDV). This virus has been found to be the first member of a third divergent linage within the family *Parvoviridae*, subfamily *Hamaparvovirinae* (Pénzes et al. 2020). The genome of PmMDV is 4,374 nt flanked by ITRs of 416 nt which fold into a regular T-shaped hairpins. Four open reading frames have been annotated, with no significant sequence similarity with other Parvoviruses, except for SF3 helicase domain in it nonstructural protein (Pénzes et al. 2020). PmMDV is described as an example of convergent evolution among invertebrate parvoviruses concerning the host-derived capsid structure. Currently, there is limited information on the pathogenicity of this virus, the susceptible hosts, and its geographic distribution. Considering the availability of the nucleotide sequence of PmMDV, efforts should be put forward to develop diagnostic tools to determine to what extent this potential pathogen is found in farmed populations of shrimp worldwide and if it is associated with mortalities.

14.2.6 Circoviruses

(a) Shrimp hepatopancreas associated circular DNA virus (shrimp CDV): A circovirus was reported from apparently healthy northern pink shrimp *P. duorarum* captured from the Gulf of Mexico (Ng et al. 2013). The genome sequence of shrimp CDV was determined using metagenomic analysis of semi-purified viral DNA extracted from the hepatopancreas tissue. The Shrimp CDV genome is circular and consists of 1,956 nt, containing a putative replication initiator protein (Rep) and a capsid protein. Shrimp CDV Rep protein has 21 to 34% identity to the cognate protein of other circoviruses and other unclassified ssDNA viruses. A putative DNA hairpin structure containing a conserved nona-nucleotide motif (AGGTATTAC) similar to other circoviruses was found in the short intergenic region of shrimp CDV. However, the pathogenicity of Shrimp CDV has not been determined and it is not known if this virus represents an ingested environmental contaminant (Ng et al. 2013).

(b) Penaeus monodon circovirus (PmCV-1): The genome of a circular Rep-encoding single-stranded DNA was sequenced from a purified virus isolated from diseased *Penaeus monodon* shrimps in Vietnam (Pham et al. 2014). The viral genome was found to be 1,777 nt long containing three ORFs encoding 266, 255, and 146 amino acids (aa). The polypeptide containing 266 aa showed about 30% sequence identity with the putative Rep protein of a circovirus and

contains rolling circle replication and superfamily 3 helicase motifs, the 255 aa containing protein shared 25% similarity to a circovirus capsid protein, and the smallest of the three proteins did not show any similarity to any sequence in the NCBI database. A 13-nt inverted repeat forming a stem loop structure containing a canonical nonanucleotide, TAATATTAC, which is a signature sequence in the non-coding region between Rep and cap protein in circoviruses and cycloviruses was found in *PmCV-1*. (Pham et al. 2014). While the genome structure of *PmCV-1* was published, the biophysical properties, morphology and infectivity of the purified virus were not reported.

14.2.7 White spot syndrome virus (WSSV)

White spot disease (WSD) caused by white spot syndrome virus (WSSV) was arguably the most serious disease to affect shrimp aquaculture until the relatively recent emergence of acute hepatopancreatic necrosis disease, a bacterial disease of penaeid shrimp (see Section 15.2.1 in Chapter 15). WSSV initially surfaced in Taiwan and China between 1991 and 1992 and subsequently caused widespread catastrophic losses in shrimp aquaculture at a global scale (Lightner 1996; Lo et al. 1996; Lightner et al. 2012). The total economic losses attributed to WSD have been estimated by Lightner et al. (2012) to be around US$ 8–15 billion since its emergence. Furthermore, Flegel et al. (2008) and Stentiford et al. (2012) speculate that the losses caused by WSSV have been increasing yearly by US$ 1 billion. However, these estimations are nearly a decade old and with the ever-rising production of shrimp aquaculture they might not accurately reflect the actual economic impact of WSSV; and therefore a more recent calculation of the current impact of WSSV is long overdue.

White spot syndrome virus belongs to the family, *Nimaviridae* and is the sole member of the genus *Whispovirus* (Lightner et al. 2012; Wang et al. 2019; Dey et al. 2019). The WSSV virions are large (80 – 120 × 250 – 380 nm) rod-shaped to elliptical, show a trilaminar envelope and possess a tail-like appendage (Durand et al. 1997; Tsai et al. 2006; Wang et al. 2019). The genome of WSSV is contained in the rod-shaped nucleocapsid that is covered by the envelope (Durand et al. 1997). WSSV is composed of highly coiled circular dsDNA that may vary in length between 281–312 kbp depending on the geographical isolate (Van Hulten et al. 2001; Oakey and Smith 2018; Oakey et al. 2019). Approximately 180 open reading frames have been annotated on the WSSV genome, however, most lack identifiable homologues in online databases (Van Hulten et al. 2001; Dey et al. 2019). Currently, to our knowledge, based on the morphological and genomic properties of WSSV, it is the only Girus (giant virus) known to infect crustaceans (Van Etten et al. 2010).

The host range of WSSV is extremely wide including decapods and non-decapods (crabs, prawns, lobsters, crayfish, copepods, and arthropods) (Lo et al. 1996; Oidtmann and Stentiford 2011). Presently, *ca.* 98 species of crustaceans have been identified as hosts suitable for WSSV (Stentiford et al. 2009). The large repertoire of hosts for WSSV plays a fundamental role in the transmission and occurrence of WSD outbreaks in shrimp producing facilities. The transmission of WSSV may occur via horizontal and vertical routes (Lo et al. 1996). Most commonly, transmission occurs by the oral route when organisms feed on infected or contaminated food via the water column, when the virus is filtered through the gill. The virus is also transmitted vertically from an infected broodstock to offspring via oocytes (Sánchez-Paz 2010; Lightner et al. 2012; Dey et al. 2019).

It has been generally reported that WSSV presents a delay before the onset of clinical signs and mortality after shrimp have been exposed to the pathogen. Once clinical signs develop, usually there is 90–100% mortality within 3–8 days (Corbel et al. 2001). Clinical signs include lethargy, reduced food intake, reddish discolouration of the body and appendages, loose cuticle, pale hepatopancreas, low response to stimulus, and thinning and delayed clotting of haemolymph (Wang et al. 2000; Dey et al. 2019). The presence of the classical circular white spots (from which the name of the disease was derived) can be detected in the cuticle of the cephalothorax and tail of the affected shrimp (Lo et al. 1996). While the calcified circular white spots are hallmark of WSD and are routinely used as presumptive diagnostic sign for the disease at a pond

side level, this sign does not appear in all infected shrimp. At a histological level, an early stage infection is characterised by eosinophilic inclusion bodies in the infected cells and during advance stages of infection the hypertrophied nuclei of the infected cells contain inclusion bodies that are more basophilic (Figure 14.1 e,f); (Lightner 1996; Lo et al. 1996). WSSV infects all tissues of ectodermal and mesodermal origin (Lightner 1996). Target tissues include: haemolymph, gills, stomach, body cuticular epithelium, haematopoietic tissues, lymphoid organ, antennal glands, foregut, testes, and ovaries (Rajan et al. 2000; Escobedo-Bonilla et al. 2008).

14.2.8 Shrimp haemocyte iridescent virus (Decapod Iridescent Virus 1 (DIV1)

Since its formal description by Qiu et al. (2017), Shrimp Haemocyte Iridescent Virus (SHIV), later termed Decapod iridescent virus 1 (DIV1), has kept the shrimp aquaculture industry uneasy in recent years due to its reported pathogenicity and its ability to infect several species of crustaceans. SHIV was initially isolated and characterised from diseased *P. vannamei* presenting massive die-offs in a farm in Zhejian Province in China in 2014 (Qiu et al. 2017). However, it is important to mention that between July and November of 2014 *Cherax quadricarinatus* presenting a lethal infection in farms in Fujian, China were used to isolate and perform a preliminary characterisation *Cherax quadricarinatus* iridovirus (CQIV). CQIV was later recognised as an additional strain of DIV1 (Xu et al. 2016; ICTV 2020a). Therefore, the report by Xu et al. (2016) can be considered the first report of DIV1 in farmed crustaceans.

DIV1 belongs to the family *Iridoviridae*, subfamily *Betairidovirinae*, genus *Decapodiridovirus*. Two isolates are currently recognised, SHIV and CQIV (Xu et al. 2016; Qiu et al. 2017). The DIV1 virions are large, present icosahedral morphology with a diameter of 150–158 nm (Xu et al. 2016; Chen et al. 2019; Qiu et al. 2019, 2017; ICTV 2020a). Both genomes from SHIV and CQIV have been sequenced and consist of linear dsDNA of ~ 165 kbp with 170 ORF (Qiu et al. 2018a, 2018b;ICTV 2020a). Phylogenetic analysis of SHIV and CQIV indicate that these isolates cluster apart from existing genera. However, the results are inconclusive with one analysis placing SHIV with members of the *Betairidovirinae* and the other analysis placing CQIV with members of *Alphairidovirinae*. Due to the lack of an identifiable DNA methyltransferase gene and the principal host species being invertebrates, SHIV and CQIV are classified as *Betairidovirinae* (ICTV 2020a).

Several species of crustaceans are susceptible to DIV1. These include *P. vannamei*, *C. quadricarinatus*, *M. rosenbergii*, *M. nipponense*, *M. superbum*, *Exopalaemon carnicauda*, and *Procambarus clarkii* (Xu et al. 2016; Chen et al. 2019; Qiu et al. 2017, 2019). In *P. vannamei*, clinical signs include empty stomach and gut, slight loss of colour on the surface of the hepatopancreas, soft shell, and reduced swimming ability (Qiu et al. 2017). While in *M. rosenbergii* affected animals exhibit a characteristic white triangle area under the carapace at the base of the rostrum, hepatopancreatic atrophy with colour fading and yellowing in this section, empty stomach and guts (Qiu et al. 2019). Histologically, *P. vannamei* show basophilic inclusion bodies and karyopyknosis in haematopoietic tissue and haemocytic infiltration in the gills, hepatopancreas, and periopods (Qiu et al. 2017). Interestingly, inclusion bodies in *M. rosenbergii* and *M. nipponense* are described as 'dark eosinophilic inclusions mixed with basophilic tiny staining' and are observed in haematopoietic tissue, haemocytes in the hepatopancreas sinus, and in the gills (Qiu et al. 2019). It appears that the susceptibility of each cultured crustacean species varies, with *P. vannamei* being the most susceptible; and *M. rosenbergii* and *E. carnicauda* being less susceptible (Chen et al. 2019; Qiu et al. 2019, 2017).

Considering the potential of DIV1 to infect a wide host repertoire and the high mortality it causes in *P. vannamei*, it appears that the virus infection may results in major economic losses to the shrimp industry like WSSV and AHPND caused in the past. Currently, DIV1 has only been reported in cultured crustaceans from China and great efforts should be put forward to limit the spread of this virus. Diagnostic methods such as nested-polymerase chain reaction (PCR), qPCR and *in situ* hybridisation are available for DIV and rigorous testing of shrimp and shrimp products should be performed on any

sample to be exported to a DIV1-free location to prevent further spread of the virus (Qiu et al. 2018a, 2020).

14.2.9 *Penaeus monodon* nudivirus

The *Penaeus monodon* nudivirus (PmNV), widely known as Monodon Baculovirus of shrimp (MBV), is the second shrimp virus (BP being the first) to be identified causing diseases problems in cultured shrimp and to be morphologically characterised. In 1977, laboratory-reared *P. monodon* showing disease signs were studied by histology and electron microscopy examination (Lightner and Redman 1981). These analyses led to the identification of occluded viral particles in the nuclei of the hepatopancreatic epithelial cells of *P. monodon*. These finding represented the second reported virus disease in penaeid shrimp and the first reported from *P. monodon*.

Penaeus monodon nudivirus is the causative agent of spherical baculovirosis in shrimp (Rajendran et al. 2012). Based on ultrastructural studies PmNV was originally classified as a Baculovirus (Lightner and Redman 1981; Brock et al. 1983; Lightner et al. 1983). Later, Yang et al. (2014) sequenced the complete genome of PmNV and studied the chemical properties of the polyhedrin protein. Based on these studies, they clarified the taxonomic position of PmNV assigning it to the family *Nudiviridae* and proposed a new genus for PmNV, the *Gammanudivirus*. PmNV are rod-shaped, enveloped, 265 to 282 nm in length and 68 to 77 nm in diameter with unique appendages at each extremity (Mari et al. 1993). The genome of PmNV is composed of circular dsDNA of 119,638 bp in length and 115 ORF were annotated (Yang et al. 2014).

PmNV has been observed in all life stages of *P. monodon* and is widely considered the most common virus of this species. Although wild *P. monodon* seems to be the primary host for PmNV, this pathogen has also been recorded in *P. penicillatus*, *P. indicus*, *P. semisulcatus*, *P. mergiensis*, *P. kerathurus*, *P. esculentus*, *P. vannamei*, *Metapenaeus ensis*, *M. lysianassa*, and *M. rosenbergii* (Brock et al. 1983; Vijayan et al. 1995; Gangnonngiw et al. 2010; Rajendran et al. 2012; Afsharnasab et al. 2014). Gross signs of infection include lethargy, reduced feeding and preening activities, and retarded growth (Rajendran et al. 2012). Larval stages are reported to be more susceptible to PmNV. Histologically, PmNV infected shrimp present prominent multiple, spherical intranuclear occlusion bodies (OB) in the cells of hepatopancreas and anterior midgut, these inclusions stain eosinophilic with hematoxylin and eosin (H&E) staining (Figure 14.1g–h). The major mode of transmission of the virus is horizontal via occlusion bodies (OB) released through faecal matter.

While PmNV is no longer considered a serious pathogen by many and it has been delisted as a reportable agent by the World Organisation for Animal Health (OIE), it is still capable of causing losses in hatchery operations and in larger animals that are stressed. Shrimp affected by PmNV can become more susceptible to secondary bacterial, fungal or viral infections that might lead to serious diseases and mortality. Therefore, it is still fundamental to produce PmNV free lines of broodstock and use adequate disinfection hatchery protocols when using PmNV positive stock for production.

14.2.10 *Baculovirus penaei* (also known as *Penaeus vannamei* single nucleopolyhedrovirus, PvSNPV)

Nuclear polyhedrosis viruses (NPVs) in penaeid shrimp are categorised based on the shape of the occlusion bodies (OBs) that include tetrahedral intranuclear OB of the *Baculovirus penaei* (BP)-type and rounded intranuclear OBs of the Monodon baculovirus (MBV)-type. *Baculovirus penaei* is the first virus discovered in penaeid shrimp, *P. duorarum*, wild pink shrimp (Couch 1974). Later the virus was described in cultured shrimp, *P. vannamei* causing high larval mortality in the hatcheries. *Baculovirus penaei* is widely distributed in the Americas including Hawaii, the Atlantic and Caribbean coasts of South and Central America, and the Gulf of Mexico (Brock and Lightner 1990). *Baculovirus penaei*-type viruses have been reported from 15 penaeid species and in the related penaeids, *Trachypenaeus similis* and *Protschypene precipua*. Since all BP-type viruses have a single enveloped nucleocapsid (S subtype), the proposed generic name of these viruses is SNPV. Based on morphological data obtained by transmission electron microscopy on many BP-isolates, Lightner and colleagues proposed the existence of

four different BP-type strains. These include Gulf of Mexico strain originally described in wild pink shrimp, *P. duorarum* (PdSNPV) by Couch in 1974, Hawaii strain known to infect native *P. marginatus* (PmSNPV), Pacific strain isolated from a population of *P. vannamei* imported to Hawaii from Ecuador (PmSNPV), and a fourth strain of BP from the western Atlantic off the coast of Brazil (Lightner et al. 1985; Durand et al. 1998).

Baculovirus penaei is an enteric virus and affects mucosal epithelial cells of the hepatopancreas tubules and the anterior midgut. In infected cells, the nucleus is hypertrophied, chromatin marginalised with or without pathognomonic tetrahedral occlusion bodies can be seen. Tetrahedral occlusion bodies can also be seen in wet mount squashes of the hepatopancreas or faeces and in H&E-stained histological sections eosinophilic occlusion bodies are distinctly evident (Figure 14.1i–j). The virions are rod-shaped, enveloped particles measuring 312–320 nm in length and 75–87 nm in diameter. The nucleocapsids are 306–312 nm in length and 62–68 nm in diameter with a cross-hatched surface arranged in a helical pattern and with a trilaminar structure capping both extremities of the virion (Bonami et al. 1995). The viral genome contains a large circular dsDNA with a molecular mass of 75×10^6 Da (Summers 1977). *Baculovirus penaei* was an OIE-listed disease until 2012.

14.2.11 *Panulirus argus* Virus 1

Panulirus argus virus 1 (*Pa*V1) is a DNA virus reported to infect spiny lobsters (*Panulirus argus*) from the Florida Keys (Shields and Behringer 2004). The virus has since been found to be widespread across the Caribbean region (Behringer et al. 2011). All life stages have been shown to be susceptible to infection, including the planktonic larval stages (Moss et al. 2011), however the disease appears most prevalent and causes high levels of mortalities in the smallest juvenile lobsters (Behringer et al. 2011). Infected juveniles appear lethargic and possess a milk coloured haemolymph which does not clot (Shields and Behringer 2004). The virus has been shown to infect the fixed phagocytes and connective tissue cells of the hepatopancreas and haemocytes. In heavily infected animals the hepatopancreas is reported to be atrophied and infected cells have been described in connective tissues cells surrounding most organs (Shields and Behringer 2004; Li et al. 2008). Affected cells display enlarged nuclei with marginalised chromatin and contain virions with an electron dense icosahedral nucleocapsid surrounded by an envelope, virions measuring ~182 nm in diameter. Interestingly, healthy lobsters have been shown to avoid diseased individuals (Behringer et al. 2006; Behringer and Butler 2010), this process is thought to reduce transmission of infection (Behringer et al. 2011). PaV1 is dsDNA virus with genome of 70,866 bp and has recently been characterised within a new family of DNA viruses *Mininucleoviridae* (Subramaniam et al. 2020). A related virus has recently been found in shore crabs, *Carcinus maenas*, termed CMv1 (Figure 14.3c, d; Subramaniam et al. 2020).

14.2.12 Bi-Facies Virus (BFV)

Herpes-like virus (HLV) was described infecting the blue crab (*Callinectes sapidus*) but was later renamed bi-facies virus (BFV) due to the unusual nuclear development which was observed (Johnson 1988a, 1976a). Initially described in the haemocytes of a juvenile female blue crab, the infection has been shown to cause disease and mortality in both wild and captured crabs (Johnson 1984), current prevalence of infection and mortality is unknown. Moribund blue crabs were shown to have milk coloured haemolymph which did not clot, mortality is thought to be due to the destruction of circulating haemocytes (Johnson 1978). Nuclei within haemocytes and haematopoietic tissues were reported to be enlarged with granular Feulgen positive contents, some nuclei containing Cowdry Type A inclusions. Gill epithelium and connective tissue cells were also affected in some crabs. Two types of icosahedral virions have been described, development stages of both types occurring together. Type A particle show both inner and outer envelopes and measure 197 × 233 nm. Type B particles lack the outer envelope and measure 174 × 191 nm. Short rod-shaped electron dense cores surrounded by an electron dense sphere were reported in both virion types. BFV shares ultrastructural features similar to those of *C. maenas* virus 1 (CmV1; Figure 14.2c,d) which was initially named herpes like virus (Bojko

et al. 2018). It is possible that bi-facies virus will also belong to the newly described viral family, *Mininucleoviridae* (Subramaniam et al. 2020), however viral sequence data would be needed to confirm this potential classification. Sparks and Morado (1986) described a viral infection in the blue King crab *Paralithodes platypus* and highlighted the ultrastructural similarities to BFV. The infection was isolated from the antennal gland epithelial cells, severe damage of this organ was reported with affected cells displaying enlarged nuclei with marginalised chromatin and containing an amorphous granular substance. Hexagonal nucleocapsids containing rod shaped electron dense core were reported to be surrounded by a toroidal structure and an electron dense capsid, measuring 140 × 165 nm in diameter. Infection was thought to be fatal due to the level of damage present within the antennal gland and bladder. Current prevalence of this virus is unknown.

14.2.13 *Homarus gammarus* nudivirus (HgNV)

Homarus gammarus nudivirus (HgNV) is the first viral infection to be described within clawed lobsters (Holt et al. 2019). The infection was described in juvenile European lobsters (*Homarus gammarus*) during a study to evaluate the growth of hatchery reared lobsters in sea-based container culture systems The lobsters showed no external signs of infection, and it is not known whether this infection results in mortalities, to date the infection has not been described in larger lobsters. Histological assessment of tissues revealed enlarged nuclei with marginalised chromatin within the hepatopancreatic epithelial cells (Figure 14.2a). Enveloped virions were shown to accumulate at the nuclear membrane, contained a rod-shaped electron dense nucleocapsid and measured 180 × 136 nm. In some cases, the nucleocapsid appearing 'u' or 'v' shaped within the envelope (Figure 14.2b). HgNV has a circular dsDNA genome of 107,063 bp and has been characterised within the family *Nudiviridae* (Holt et al. 2019).

14.2.14 Baculoviral midgut gland necrosis virus (BMN)

Baculoviral midgut gland necrosis virus (BMN) was identified in *P. japonicus* following reports of

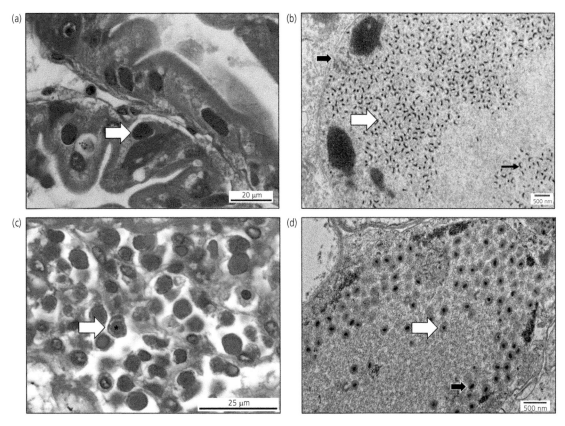

Figure 14.2 (a–d) Histology and electron microscopy of DNA viruses that infect lobster and crab species of commercial and ecological importance. (a) *Homarus gammarus* Nudivirus (HgNV) infection within the hepatopancreas of juvenile European lobsters *H. gammarus*. Infections can be seen within multiple epithelial cells of the hepatopancreas tubules, infected nuclei are enlarged, with marginated chromatin and possess an eosinophilic inclusion body (arrow). H&E Stain. Scale bar = 20 μm. (b) Nucleus from a HgNV infected cell containing rod-shaped virions. Virions accumulate at the periphery of the nuclear membrane (black arrow), rod shaped HgNV virions (white arrow) possess an electron dense nucleocapsid surrounded by a trilaminar membrane (envelope). In some cases, the rod shaped nucleocapsid appears to bend within the envelope forming a 'u' or 'v' shape (line arrow). TEM. Scale bar = 500 nm. (c) *Carcinus maenas* virus 1 (CmV1) infection within the haemocytes of the shore crab *C. maenas*. Nuclei are enlarged with marginalised chromatin (arrow) and contain an eosinophilic inclusion body (*). H&E Stain. Scale bar = 25 μm. (d) Enlarged nuclei within haemocyte containing CmV1 virions. Cross-sectioned viral particles clearly display an electron dense sphere within the capsid (black arrow) surrounded by an envelope (white arrow). TEM. Scale bar = 500 nm.

high-level, sudden onset mortalities in larvae and post-larvae (Sano et al. 1981, 1984). The condition has also been known as mid-gut gland cloudy disease, with affected shrimp displaying cloudy pale hepatopancreas tissues. Extensive cellular necrosis was reported within the midgut, cells displaying enlarged nuclei with marginalised chromatin. Rod shaped nucleocapsids surrounded by a closely fitting envelope with lateral extension at one end, measuring 310 × 72 nm were observed within these nuclei. Although initially thought to be a Baculovirus infection it is likely to belong to the

Nudiviridae family as highlighted by (Bateman and Stentiford 2017). Genomic information is lacking for this virus.

14.2.15 Baculo A and Baculo B

Baculo A was reported from the hepatopancreatic epithelial cells of blue crabs *C. sapidus* (Johnson 1976b). The virus is not thought to be associated with disease; however it was noted that the infection may prevent cells from performing normal functions (Johnson and Lightner 1988). Infected nuclei within the hepatopancreas were shown to be up to 2.3 times the normal size with marginalised chromatin. Virions consisted of rod-shaped nucleo-capsids surrounded by an envelope with a lateral protrusion at one end and measured 260–300 nm in length and 60 nm in diameter. No genetic infor-mation is available for this infection but given the size and shape of the virions and location of infec-tion it is likely this virus will belong to the family *Nudiviridae*. Baculo B was also reported in blue crabs but within the haemocytes and haematopoietic tis-sues (Johnson 1983, 1984). It is not known whether this virus infection is associated with disease as the virus was reported in both sick and normal crabs. Infected nuclei appeared hypertrophied (up to 1.6 times normal size) with marginalised chromatin. Electron dense rod-shaped nucleocapsids with one squared end and one rounded end were reported to be surrounded by an envelope, virions measuring 85–100 nm x 370–390 nm. Johnson (1988b) highlight-ed similarities between this virus and that described in *C. maenas*, rod-shaped virus of *Carcinus maenas* (RVCM). Genomic information is lacking for this virus.

14.2.16 *Cancer pagurus* bacilliform virus (CpBV)

A non-occluded bacilliform virus was described in the hepatopancreas of juvenile edible crabs (*Cancer pagurus*) (Bateman and Stentiford 2008). Enlarged nuclei with marginalised chromatin and containing a fine granular eosinophilic matrix were report-ed in the F (fibrillar) and R (reserve) cells of the hepatopancreas. Virions measuring 210 × 60 nm contain an electron dense rod shaped nucleocap-sid surrounded by a close-fitting envelope with a distinctive lateral expansion at one end (see Figure 3.1 in Chapter 3). A tail-like appendage coming from the end of nucleocapsid was observed with-in this expansion of the envelope. As highlight-ed by Bateman and Stentiford (2017) there have been multiple rod-shaped viral particles described in the hepatopancreas of various species of crus-tacea over recent years, the majority of which have been described as intranuclear bacilliform virus-es. As these viruses lack an occlusion body it has been suggested that these viruses may belong to the family *Nudiviridae* (Bateman and Stentiford 2017)

however genomic information is currently lacking to enable full characterisation. Several pathogens of commercial and ecological significance have been described from this species and it has been shown that the prevalence of these pathogens can vary between the different life stages sampled (Bateman et al. 2011). To date the virus has only been observed in juvenile edible crabs, it has not been shown in larger crabs and it is not known if this infection results in mortalities.

14.2.17 *Scylla* Baculovirus (SBV)

Scylla Baculovirus was described in the hepatopan-creatic epithelial cells of the mud crab (*Scylla scyl-la*) (Anderson and Prior 1992). Although initially described as a Baculovirus owing to size and shape of virions and location of infection, it is likely to be an intranuclear bacilliform virus or nudivirus infec-tion. Nuclei in the R cells of the hepatopancreas were reported to be enlarged with marginalised chromatin and contain basophilic inclusion. Rod shaped nucleocapsids surrounded by a close-fitting envelope with an apical expansion at one end and measure 253 × 44 nm. Owens et al. (2010) report-ed the presence of intranuclear inclusions within the hepatopancreas in both wild and farmed mud crabs when investigating disease as a potential bottleneck in the culture of this species. Prevalence of infec-tion was reported to be 9% in wild broodstock and 13% in progeny larvae and it was suggested that viruses may affect hatchery production. Genomic information is lacking for this virus.

14.2.18 *Cherax quadricarinatus* Bacilliform Virus (CqBV) and *Cherax destructor* Bacilliform Virus (CdBV)

Intranuclear bacilliform viruses have been described from multiple species of freshwater crayfish (Edgerton 1996; Longshaw 2011), here we highlight infections reported in two species of commercially exploited crayfish, the redclaw crayfish (*C. quadricarinatus*) and the common yabby (*C. destructor*). Both virus infections are reported in the hepatopancreatic epithelial cells, infected cells possessing enlarged nuclei with marginalised chromatin and contain eosinophilic inclusions.

Although shown to be highly prevalent within crayfish populations, the intensity of infection was reported to be low (Anderson and Prior 1992; Edgerton 1996). *Cherax quadricarinatus* Bacilliform Virus (CqBV) was first described in farmed populations of red claw crayfish in Australia (Anderson and Prior, 1992) and *Cherax destructor* Bacilliform Virus (CdBV) was described from farmed populations of the common yabby in Australia (Edgerton 1996). Both infections possess rod-shaped electron dense nucleocapsids surrounded by a closely fitting envelope with lateral expansion at one end. CqBV virions measure 262×103 nm and CdBV virions measure 304 × 68 nm (Edgerton 1996). There is no genomic information for these viral infections.

14.2.19 *Chionoecetes opilio* Bacilliform Virus (CoBV)

Kon et al. (2011) reported an intranuclear bacilliform virus in snow crabs (*Chionoecetes opilio*) from the Sea of Japan. Infected crabs were reported to have a milk-coloured haemolymph. Diseased animals were observed in wild and reared populations when the water temperature was between 2–3°C. Enlarged nuclei with marginalised chromatin and basophilic inclusions were observed in the interstitial connective tissues of multiple organs. Enveloped rod-shaped virions measuring 338 × 144 nm were reported to be morphologically similar to WSSV, crabs were negative for WSSV via PCR. Kawato et al. (2018) reported WSSV sequences within the genomes of various crustacean species, they used this sequence data along with sequences from *Chionoecetes opilio* Bacilliform Virus (CoBV) and showed that these viruses had conserved genes with those reported for WSSV and likely belong to the family *Nimaviridae*.

14.2.20 *Cherax quadricarinatus* parvo-like virus (PV) and *Cherax destructor* systemic parvo-like virus (CdSPV)

Edgerton et al. (2000) reported a parvovirus infection in the gills from moribund *C. quadricarinatus* which had been relocated from a holding tank at a farm. Hypertrophied nuclei with marginalised chromatin were observed within the gill epithelium. Virus-like particles 20 nm in diameter were observed within the affected nuclei. *Cherax quadricarinatus* parvovirus (CqPV) was identified following a mass mortality event in farmed *C. quadricarinatus* (Bowater et al. 2002). Moribund crayfish were weak, anorexic, and lethargic, and parvovirus infections were identified in the gills, cuticular epithelium and epithelial cells of the foregut, midgut and hindgut. Virions measuring 19.5 nm were reported within the affected hypertrophied nuclei in these tissues. Bowater et al. (2002) highlighted the similarities between CqPV and that described by Edgerton et al. (2000), it is possible that these descriptions are two separate isolations of the same virus. Bochow et al. (2015) provided sequence data of the virus described by Bowater et al. (2002) and proposed the virus be named *Cherax quadricarinatus* Densovirus and be placed within the genus *Ambidensovirus* of the subfamily *Densovirinae*.

Cherax destructor systemic parvo-like virus (CdSPV) was discovered in a single moribund *C. destructor* during a disease screen of a farmed population (Edgerton et al. 1997). Intranuclear Cowdry Type A inclusions were reported in hypertrophied nuclei distributed systemically throughout the crayfish tissues, with the highest concentrations noted in the gills. Gill tissues showed extensive necrotic foci characterised by pyknotic nuclei, haemocytic infiltration and melanisation. Hepatopancreas and midgut tissues were also reported to be necrotic with haemocytic infiltration and melanotic foci common in the interstitial tissues. Electron dense hexagonal particles measuring 18–25 nm were visualised in aggregates between the viroplasm and inner nuclear membranes of affected cells. There is limited information available on the impact of these viruses in the aquaculture industry.

14.2.21 Penaeid haemocytic rod-shaped virus (PHRV)

Penaeid haemocytic rod-shaped virus (PHRV) was identified in the haemocytes of a hybrid shrimp species (*P. esculentus* crossed with *P. monodon*)

(Owens 1993). Enveloped virions measuring 588 × 119 nm contained rod-shaped electron dense nucleocapsids, some nucleocapsids being reported to be 'u' or 'v' shaped similar to that reported for rod shaped virus of *C. maenas* (RVCM) (Johnson 1988b). Moribund shrimp were also shown to be co-infected with PstDNV, Owens (1993) suggested that PHRV may be a latent infection only observed due to the concurrent infection. No genomic information is available for this infection.

14.3 Diseases caused by RNA viruses

14.3.1 Taura syndrome virus (TSV)

Taura syndrome (TS), caused by Taura syndrome virus (TSV) first emerged in Ecuador from an unknown source in 1991, and by 1993 the disease was recognised as a major disease affecting penaeid shrimp (Hasson et al. 1995). The disease is now widely distributed in the shrimp-farming regions of the Americas, South-East Asia, and the Middle East (Lightner et al. 2012). TSV susceptible hosts where viral infectivity has been shown unequivocally include Pacific white shrimp (*P. vannamei*), blue shrimp (*P. stylirostris*), giant tiger prawn (*P. monodon*), northern white shrimp (*P. setiferus*), greasyback shrimp (*Metapenaeus ensis*), and northern brown shrimp (*P. aztecus*). In addition, there are a number of other species of crustaceans where TSV has been detected but active viral infection has not been demonstrated yet. These include fleshy prawn (*P. chinensis*), giant river prawn (*M. rosenbergii*), northern pink shrimp (*P. duorarum*), kuruma prawn (*P. japonicus*), southern white shrimp (*P. schmitti*), gulf killifish, blue crab, swamp crab, copepods, and the barnacles (*Chelonibia patula* and *Octolasmis muelleri*). There are three main stages of infection, acute, transition, and chronic which are grossly distinguishable. During the acute phase, infected animals display pale reddish body colouration and pleopods and the tail fan turn red, and hence the farmers called the disease 'red tail' when the disease first appeared in Ecuador. Shrimp that survive acute phase infection undergo a transition phase and show multifocal, irregularly shaped melanised cuticular lesions. The animals undergoing a transition phase that may or may not show soft cuticles and red-chromatophore expansion. These animals may display normal feeding behaviour. Upon moulting, these shrimp enter into a chronic phase without showing any obvious clinical signs and may carry the infection life-long. Histologically, the acute phase infection is characterised by multifocal areas of necrosis in the cuticular epithelium of the general body surface, appendages, gills, hindgut, foregut, and occasionally cells of the subcuticular connective tissues, and adjacent striated muscle fibres underneath the affected cuticular epithelium. Affected cells display an increased eosinophilia of the cytoplasm and pyknotic or karyorrhectic nuclei (Figure 14.3 a,b). Cytoplasmic remnants of necrotic cells along with pyknotic and karyorrhectic nuclei that give a characteristic 'peppered' or 'buckshot-riddled' appearance are considered as pathognomonic for this disease during the acute phase. During the chronic phase, histologically the only sign of infection is the presence of numerous lymphoid organ spheroids (Hasson et al. 1995).

After Taura syndrome emerged in *P. vannamei* from Ecuador during 1992 through 1994, some *P. stylirostris* lines were found to possesses resistance against TSV. Around this time, like in Ecuador, shrimp farmers in Mexico were also experiencing crop failure in *P. vannamei*. Following the discovery that some genetic lines of *P. stylirostris* are TS resistant, farmers in western Mexico switched from TSV-susceptible *P. vannamei* to TSV-resistant *P. stylirostris*. However, in 1998–1999, a new 'strain' of TSV (Type B) emerged and caused massive epizootics in *P. stylirostris* (Erickson et al. 2002; Robles-Sikisaka et al. 2002; Aldana Aranda et al. 2011). The Ecuadorian strain of TSV was termed as Type-A.

Taura syndrome virions are non-enveloped, 31–32 nm in diameter containing a single-stranded RNA of ~ 10.2 kb (Robles-Sikisaka et al. 2001; Mari et al. 2002). The TSV genome contains two ORFs which are separated by a 207-nucleotide intergenic region. The ORF1 encodes a 234 kDa nonstructural polyprotein containing conserved motifs of a helicase, protease, and RNA-dependent RNA polymerase (Mari et al. 2002). The ORF2 encodes three structural proteins, VP2, VP1, and VP3, with a predicted molecular mass of 36.4, 54.6, and 21.1 kDa, respectively (Mari et al. 2002). TSV is currently classified as a member of the *Dicistroviridae* family, genus *Aparavirus* in the order *Picornavirales* (Valles et al. 2017). A phylogenetic analysis using the TSV capsid protein VP1 showed that there are four distinct lineages of TSV: Mexico, Southeast Asia, Belize/Nicaragua, and Venezuela/Aruba (Wertheim et al. 2009). Since then new genotypes of TSV has been reported from Columbia (Aranguren et al. 2013) and Saudi Arabia (Tang et al. 2012).

14.3.2 Yellow head virus (YHV) and Gill-associated virus (GAV)

Yellow head virus (YHV) and Gill-associated virus (GAV) form the yellow head complex and are classified by the ICTV as a single species (Cowley et al. 2012; OIE 2009a). Yellow head disease was initially reported in *P. monodon* from central Thailand in 1990 from which the pathogen rapidly spread along the eastern and western coasts of the Gulf of Thailand to southern farming regions (Chantanachookin et al. 1993; Walker and Sittidilokratna 2008). Both YHV and GAV are reported to be pathogenic to shrimp, however, disease by GAV is less severe than disease caused by YHV (Walker and Sittidilokratna 2008).

Both YHV and GAV are positive-sense single-stranded RNA (ssRNA) viruses and are members of the order *Nidovirales*, family *Roniviridae*, genus *Okavirus* (Cowley et al. 2012; Munro and Owens 2007; Walker and Sittidilokratna 2008). There are currently ten recognised genotypes within the yellow head complex (Cowley et al. 2015). The YHV and GAV genomes are 26,662 nt and 26,235 nt in length, present a similar structure with the main differences occurring in UTRs (Cowley and Walker 2002, Walker and Sittidilokratna 2008). These viral genomes are comprised of four long ORFs designated ORF1a, ORF1b, ORF2, and ORF3. Phylogenetic analysis has delineated ten distinct genotypes in the yellow head complex based on a partial sequence of the ORF1b (Walker and Sittidilokratna 2008; OIE 2009a; Cowley et al. 2015). The morphology of YHV and GAV is indistinguishable, virions are enveloped, bacilliform in shape with rounded ends, and 40–60 nm x 150–200 nm in dimensions (Cowley et al. 2012; Munro and Owens 2007).

The host range of YHV and GAV is extremely wide and are reviewed by Munro and Owens (2007). However, YHV and GAV are considered endemic in *P. monodon* populations across its natural geographic range (Walker and Sittidilokratna 2008). The disease is usually characterised by a pale to yellowish colouration of the cephalothorax (from where the disease gets its name) and gills due to the underlying yellow hepatopancreas that is observed through the translucent carapace of the prawn and also a generally pale or bleached appearance of the organism (Chantanachookin et al. 1993; Cowley et al. 1999). Shrimp are susceptible to YHV infection from late post-larval stages but mortality in ponds usually occurs in early-to-late stages (Walker and Sittidilokratna 2008). YHV and GAV infect tissue of ectodermal and mesodermal origin. Histologically, in severe infections, generalised cell degeneration with prominent nuclear condensation pyknosis and karyorrhexis, and basophilic perinuclear cytoplasmic inclusions in affected tissues are observed (Figure 14.3c,d); (Chantanachookin et al. 1993; Walker and Sittidilokratna 2008). The infection by YHV and GAV can be experimentally propagated horizontally by injection, ingestion, immersion, and cohabitation. Vertical transmission has also been reported (Chantanachookin et al. 1993; Cowley et al. 2002).

14.3.3 Mourilyan virus (MoV)

Mourilyan virus (MoV) was initially identified in diseased *P. monodon* from Australia in 1996 (Cowley et al. 2005a. In a study targeted toward GAV, three MoV clones were identified in a cDNA library randomly amplified from dsRNA purified from lymphoid organ RNA of *P. monodon* that had been experimentally inoculated with infectious material of GAV diseased prawns (Cowley et al. 2005a). The authors of this study correctly assumed that the MoV cDNAs were derived from a viral RNA species that was co-purified with the larger GAV dsRNA.

The MoV genome is composed of four segments of negative sense single-stranded RNA (Cowley, McCulloch, Rajendran et al. 2005; Cowley, McCulloch, Spann et al. 2005). The segments have been termed L (6319 nt), M (2987 nt), S1 (1079–1557 nt), and S2 (1364 nt). Basic local alignment search tool (BLAST) searches using the ORFs encoded in the MoV genome reveal similarity with Uukuniemi virus and viruses of the genus *Phlebovirus* and *Bunyaviridae*. TEM analyses have revealed spherical to ovoid particles of ~ 100 x 85 nm, that may vary in electron density and appear to be enveloped (Cowley et al. 2005a, 2005b).

To date, four penaeid species have been reported to harbour MoV mainly *P. monodon*, *P. japonicus*, *P. esculentus* and *P. merguiensis* (Cowley et al. 2005a, 2005b; OIE 2007b). No obvious gross signs have been associated with MoV infected shrimp. In heavy infections ISH signal for MoV can be observed in all tissue of mesodermal and ectodermal origin including: haematopoietic tissues, gills, epicardium, heart, cuticular epithelia, foregut, midgut and antennal gland, and ventral/peripheral nerves (Cowley et al. 2005b; OIE 2007b). Histologically, lymphoid organ spheroids are the most obvious pathology caused by MoV (OIE 2007b).

MoV is one of the less well characterised viruses of cultured shrimp. In general, studies have shown that the main host species, *P. monodon* and *P. japonicus*, can tolerate high levels of infection (Oanh et al. 2011; Sellars et al. 2006). Due to its lower impact on shrimp culture, in comparison to other viruses, it has not been the focus of many studies. However, efforts should still be made to limit the spread of this pathogen and develop Specific Pathogen Free/ Specific Pathogen Resistant (SPF/SPR)lines of shrimp.

14.3.4 Infectious myonecrosis virus (IMNV)

Infectious myonecrosis disease caused by Infectious Myonecrosis Virus (IMNV) was first described in north-east Brazil where it caused significant disease and mortalities (Lightner et al. 2012). IMNV first drew the attention of farmers during 2002 when cultured *P. vannamei* showed extensive necrotic areas in the skeletal muscle tissues and shrimp ponds reach cumulative mortalities of 70% at harvest time (Poulos et al. 2006).

Phylogenetic analysis based on the RdRp of IMNV suggest it is a member of the family *Totiviridae* (Poulos et al. 2006). Currently, the International Committee of Taxonomy of Viruses (ICTV) has listed IMNV as related viruses which may be members of the family *Totiviridae* but have not been

approved as species (ICTV 2020b). The genome of IMNV consists of dsRNA of 7,560 bp with two identified non-overlapping ORFs (Poulos et al. 2006). The ORF1 encodes a putative RNA binding protein and a capsids protein, while ORF2 encodes a putative RdRp. The IMNV particle is icosahedral in shape and 40 nm in diameter. Recent studies have found deletion mutants of IMNV in Indonesia that might be associated with reduced disease (Mai et al. 2019).

The main host of IMNV is *P. vannamei*, however, experimental infections have shown that *P. stylirostris* and *P. monodon* are also susceptible to the virus (OIE 2009b). The main signs of disease include extensive necrotic areas in the skeletal muscle tissues, primarily in the distal abdominal segments and the tail fan. The colour of the muscle lesions is generally white and opaque in appearance (Poulos et al. 2006). Histologically, lesions are characterised by coagulative muscle necrosis, often accompanied by fluid accumulation in between muscle fibers, haemocytic infiltration, and fibrosis (Figure 14.3e,f). Lymphoid organ spheroids are commonly observed as well as dark cytoplasmic basophilic inclusion bodies in muscle cells, haemocytes, and connective tissue cells.

Losses to IMNV have been significant in Brazil, where they have been estimated to exceed $100 million between 2002 and 2006 alone (Lightner et al. 2012). IMNV has now spread to South-East Asia. In Indonesia losses due to IMNV are estimated at US$1 billion. While Brazil and Indonesia have been the countries most affected by IMNV, the appearance of new genotypes more adaptable to new environments could further the spread of this virus to shrimp-farming nations that have proven resilient up to now.

14.3.5 Nodaviruses: *Macrobrachium rosenbergii* nodavirus, *Penaeus vannamei* nodavirus (PvNV) and *Farfantepenaeus duorarum* nodavirus (FdNV)

The white tail disease (WTD), caused by *Macrobrachium rosenbergii* nodavirus, was first reported in the Guadeloupe Island in the French West

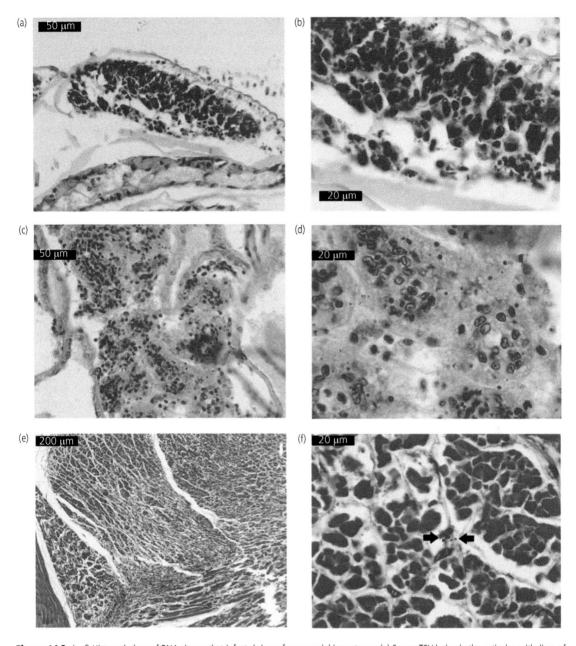

Figure 14.3 (a–f) Histopathology of RNA viruses that infect shrimp of commercial importance. (a) Severe TSV lesion in the cuticular epithelium of *P. vannamei* where the classical 'Buckshot riddle' appearance is observed. (b) High magnification of the TSV caused lesion where nuclear pyknosis and karyorrhexis are evident as well as small cytoplasmic spherical basophilic and eosinophilic inclusion bodies. (c) Lymphoid organ of a YHV infected *P. monodon* showing general disorganisation and loss of definition in the tubules. (d) High magnification of the lymphoid organ of a YHV infected shrimp some pyknotic nuclei are observed accompanied by small cytoplasmic basophilic inclusion bodies. (e) Coagulative necrosis of the skeletal muscle (*P. vannamei*) accompanied by haemocyte infiltration and fibrosis caused by IMNV. (f) High magnification of perinuclear basophilic inclusion bodies (black arrows) adjacent to coagulative necrosis lesions.

Indies in 1997 causing mass mortality of post-larvae specimens in *M. rosenbergii* hatchery (Arcier et al. 1999). Subsequently, high mortality due to WTD was reported in post-larvae of penaeid shrimp including Indian shrimp (*P. indicus*), Kuruma prawn (*P. japonicus*), and black tiger shrimp (*P. monodon*) (Sudhakaran et al. 2006; Ravi et al. 2009). Later, the disease was reported in juvenile *P. vannamei* but was not associated with high mortality (Senapin et al. 2012).

The aetiological agent of WTD is a non-enveloped, icosahedral virus, *Macrobrachium rosenbergii* nodavirus (MrNV), measuring 30 ± 3 nm in diameter belonging to the family *Nodaviridae* (Bonami et al. 2005). The viral genome consists of two positive-sense, single-stranded RNA fragments. RNA-1 (3202 bases) encodes an RNA-dependent RNA polymerase (RdRp) and a B2 protein, whereas RNA-2 (1175 bases) encodes the viral capsid protein. In WTD-affected shrimp, MrNV is often accompanied by a smaller virus, extra small virus (XSV). XSV is also a non-enveloped, icosahedral measuring 15 nm in diameter, and contains a single-stranded, positive sense RNA (796 bases) that encodes a single capsid protein (Bonami et al. 2005; Bonami and Sri Widada 2011). It was hypothesised that XSV is a satellite virus that depends on the RdRp of MrNV for its replication (Qian et al. 2003).

Unlike most shrimp viruses that do not infect fish or insect cells, MrNV was found to replicate in fish SS1, mosquito C6/36, and insect Sf9 cells. However, replication of the virus in these cells appears to be limited, although cytopathic effects were observed in the infected cells. Jariyapong et al. (2018) transfected Sf9 cells with *in vitro* transcribed MrNV RNA1 and RNA2, and produced mature virion that were infectious to shrimp, although the virions were bigger in size, 40–50 nm in diameter compared to the wild-type virus that are 30 ± 3 nm in diameter (Jariyapong et al. 2018). Recently, recombinant MrNV and XSV were expressed in Sf9 cells using a baculovirus expression system that produced virions similar in size to wild-type virions. Both MrNV and XSV were purified from recombinant baculovirus infected Sf9 cells, and it was demonstrated that MrNV alone can cause clinical signs of WTD (Gangnonngiw et al.

2020). In addition, it was shown when healthy shrimp were challenged with purified XSV, the viral RNA could replicate without the presence of MrNV suggesting a RdRp of host origin can support replication of XSV in virus-challenged shrimp. This study has opened up avenues for the application of reverse genetic approaches in studying viral pathogenesis in shrimp especially considering the lack of an immortal cell line of shrimp origin or any other crustaceans.

During 2004 to 2006, *P. vannamei* shrimp samples collected in Belize displayed clinical signs, white, opaque lesions in the tails and histopathology similar to IMNV. But these samples were tested negative by reverse transcriptase-polymerase chain reaction (RT-PCR) for IMNV (Tang et al. 2007). In affected farms, the disease caused in a 50% reduction in production. The purified virus from affected shrimp measured 19–27 nm in diameter and the viral genome contained two single-stranded positive-sense RNAs: RNA1 (3111 bp) and RNA2 (1183 bp). The genome organisation of the virus was found to be similar to MrNV, and phylogenetic analyses using the predicted amino acid sequence of the RNA-dependent RNA polymerase revealed that *Pv*NV is a member of the genus *Alphanodavirus* and closely related to MrNV (Tang et al. 2011). Interestingly, in laboratory bioassay using SPF *P. vannamei*, the virus did not cause mortality. It was reported that when the stocking density was high (> 50 m^{-2}), or when the temperature was > 32 $^\circ$C, survival in *Pv*NV-infected shrimp decreased to 40% (Tang et al. 2007). Faeces of seabirds consuming PvNV infected shrimp, barnacles, and zooplankton were found to serve as a passive carrier of the virus. Histopathology of affected shrimp revealed multifocal necrosis and haemocytic fibrosis in the skeletal muscle. In addition, basophilic, cytoplasmic inclusions were found in striated muscle, lymphoid organ, and connective tissues. Transmission electron microscopy showed cytoplasmic inclusions containing para-crystalline arrays of virions (Tang et al. 2011).

Another Nodavirus has been described in apparently healthy *F. duorarum* from the Gulf of Mexico (Ng et al. 2013). While analyzing large-scale RNA sequence data derived from hepatopancreas tissue, these authors identified two partial sequences encoding RdRp and the capsid proteins that showed similarity to MrNV and PvNV. While both MrNV and PvNV are known to affect the musculature of the shrimp and cause muscle necrosis, it is not known whether FdNV would also affect the muscle of *F. duorarum*.

14.3.6 Covert mortality syndrome virus

Covert mortality syndrome caused by Covert mortality nodavirus (CMNV) produced major disease outbreaks starting in 2009 in China. The disease was named covert mortality disease since affected shrimped preferred to hide at the bottom of the ponds instead of swimming to the surface. Shrimp farmers observed that mortality was more evident from 60–80 days post-stocking, with cumulative mortalities reaching as high as 80% (Zhang et al. 2014).

Genetic information of CMNV is limited to a clone of 1185 bp of the viral genome. Phylogenetic analysis using the amino acid sequence of the RdRp of CMNV strongly suggest that CMNV is new species within the genus *Alphanodavirus*. The size of virus may vary, two distinct virus-like particles have been observed by TEM, a larger virus-like particle of 32 nm in diameter and a smaller virus-like particle of 19 nm in diameter (Zhang et al. 2014).

The principal hosts for CMNV are *P. vannamei*, *P. chinensis*, *P. japonicus*, *P. monodon*, and *M. rosenbergii* (Zhang et al. 2017). Potential vectors and reservoirs include: brine shrimp, barnacles, rotifers, amphipods, *Crassostrea gigas*, hermit crab *Diogenes edwardsii*, *Meretrix lusoria*, *Ocypode cordimundus*, *Parathemisto guadichaudi*, and fiddler crab (Liu et al. 2018). Clinical signs of the affected shrimp include hepatopancreatic atrophy with colour fading, empty stomach and guts, soft shell, slow growth and in many cases uneven whitish muscle lesions in the abdominal segments. Histologically, the muscle fibres display muscle fragmentation tending towards coagulative, muscular lysis, myonecrosis, and haemocytic in filtration (Zhang et al. 2014). Vacuolation in the cytoplasm of epithelial cells in the tubules in the hepatopancreas and eosinophilic inclusions were observed within the tubular epithelia (Zhang et al. 2014). Recently, CMNV was found

to jump species barriers and cause disease in the economically important Japanese flounder (Wang et al. 2019).

14.3.7 Laem Singh virus (LSNV)

The virus was originally described from Laem Singh district in Chanthaburi Province in Thailand in 2006 and found to be associated with retarded growth in *P. monodon*, called monodon slow growth syndrome (MSGS) (Sritunyalucksana et al. 2006). The virus has also been detected from apparently healthy *P. monodon* and *P. vannamei* (which appears to be resistant). MSGS was the major reason for the switch from indigenous *P. monodon* to non-native *P. vannamei*, unaffected by LSNV infection in Thailand from around 2002 onwards. The virus has been reported from Thailand, Vietnam, Malaysia, Indonesia, India, Sri Lanka, and East Africa (Flegel 2012).

In *P. monodon*, the virus has been attributed to retinopathy which is characterised by abnormally enlarged haemolymphatic vessels, haemocytic infiltration, and rupture of the membrane that separates the fasticulated zone from the overlying row of reticular cells (Pratoomthai et al. 2008). The virus was also found within the lymphoid organ, heart, and connective tissues of the hepatopancreas. During the late stage of infection, lymphoid organ spheroids could be seen in the infected animals (Sritunyalucksana et al. 2006).

Recently, LSNV virus has been characterised. The virions are icosahedral, 23–33 nm in diameter and the genome contains two, single-stranded RNAs. The RNA1 is 2206 nt and contains two ORFs. The ORF1 encodes a protease and ORF2 encodes a RdRp. The RNA2 is 1,846 nucleotides long and contains a single ORF that encodes a viral capsid (Taengchaiyaphum et al. 2020). The phylogenetic analyses using RdRp and viral capsid protein sequence revealed that LSNV is most closely related to insect and other arthropod-borne viruses infecting plants in the family *Luteoviridae in* the *Sobemovirus* group (Taengchaiyaphum et al. 2020). Interestingly, both RNA1 and RNA2 shows 99% similarity to corresponding RNAs of Wenzhou virus 9 (WZSV 9) reported from China, and it has been proposed that these two viruses are different isolates of the same virus (Taengchaiyaphum et al. 2020).

14.3.8 *Macrobrachium rosenbergii* Golda Virus

Since 2011, larval stages of *M. rosenbergii* have been experiencing mass mortalities (reaching 100%) in Bangladesh. Recent studies by Hooper et al. (2020) have determined the aetiological agent as a potential new member of the order *Nidovirales*. The 29 kb single-stranded positive sense genome of *Macrobrachium rosenbergii* Golda Virus (MrGV) presents similarities in particular protein motif sequences to YHV (Hooper et al. 2020). Currently, MrGV is widespread in southern Bangladesh and has not been reported in other parts of the world. Due to its devastating impact in *M. rosenbergii* culture in Bangladesh, biosecurity measures in particular PCR testing of post-larva coming from this region should be carried out to mitigate the global spread of this novel virus.

14.3.9 *Penaeus japonicus* reovirus, *Penaeus monodon* reovirus, *Penaeus vannamei* reovirus

Three reovirus infections have been described within penaeid shrimps. *Penaeus japonicus* reovirus was reported within experimental tanks following mass mortalities after 15 days culture (Tsing and Bonami 1987). Affected shrimp were noted to possess altered behaviour, lack of burrowing in the sand, and were shown to have a reddish colouration of the telson, uropods, and hepatopancreas. Paraspherical non-enveloped virions measuring 61 nm in diameter were observed within the R (reserve) and F (fibrillar) cells of the hepatopancreas. A reo-like virus was identified from the hepatopancreas of *P. monodon* following mortalities in grow out ponds in Malaysia (Nash et al. 1988). Infected shrimp were shown to accumulate at the edge of the pond. Hepatopancreas tissues were vacuolated and displayed necrosis of the epithelial cells, non-enveloped paraspherical virions measuring 50–70 nm were identified within the cytoplasm of affected cells. It is not known whether this viral infection was the causative agent of the mortality event as shrimp were shown to be co-infected with *Penaeus monodon* nudivirus (PmNV), rickettsia, and a Gramnegative bacterium. Krol et al. (1990) reported a reo-like virus in larval stages of *P. vannamei* and was shown to occur concurrently with an experimentally induced PvSNPV infection. Reo-like virus was shown to develop within the epithelial cells of the anterior midgut and within R and F cells within the hepatopancreas. Shrimp with reo-like infection were also shown to have PvSNPV infections but not all PvSNPV infected shrimp showed signs of reo-like viral infection, occasionally both viral infections were shown to occur within the same cells. Non-enveloped paraspherical virions, measuring 40–60 nm in diameter, were identified within a filamentous matrix which resembled a virogenic stroma in the cytoplasm of infected cells (Krol et al. 1990). It is not known whether this infection was introduced during the experimental exposure to PvSNPV or manifested due to stress of infection with the more pathogenic PvSNPV infection. Genomic data are not available for this virus and it is not known whether this is the same, similar, or different to the other reo-like viruses reported in other penaeid shrimp.

14.3.10 *Callinectes sapidus* reovirus 1 (CsRV1)

Callinectes sapidus reovirus 1 was initially described in juvenile blue crabs, *C. sapidus* and was previously known as Reo-like virus (RLV) (Johnson 1977; Johnson and Bodammer 1975). The viral infection has been associated with mortalities of blue crabs in soft shell production facilities (Flowers et al. 2018; Spitznagel et al. 2019), infected crabs appearing lethargic, lack of moulting, and occasionally displaying tremors and paralysis (Johnson 1977). The virus affects the haemocytes, haematopoietic tissues, connective tissues, and the nervous system with eosinophilic to basophilic inclusions reported within the cytoplasm of affected cells (Johnson 1977; Tang et al. 2011). Non-enveloped icosahedral particles measuring 55 nm in diameter were identified within these inclusions, sometimes forming paracrystalline arrays (Bowers et al. 2010). Flowers et al. (2016) sequenced 12 segments of the CsRV1 genome and highlighted that this virus is similar to those described in the mud crab (MCRV) and Chinese mitten crab (WX-2012) (Deng et al. 2012;

Shen et al. 2015). The virus has been shown to be present in blue crab populations along the Atlantic coast of the USA (Flowers et al. 2016; Zhao et al. 2020). Management of culture conditions such as salinity and temperature has been shown to reduce mortalities within culture facilities, mortality being shown to be two-fold higher in flow through systems as opposed to recirculating aquaculture systems (Spitznagel et al. 2019).

14.3.11 Mud crab reovirus (MCRV)

Mud crab reovirus (MCRV) has been described in the mud crab *Scylla serrata* (Weng et al. 2007) and shown to be the causative agent of sleeping disease. Affected crabs were lethargic, showed loss of appetite and a grey colouration, moribund crabs displayed atrophied hepatopancreas and loose gills. Icosahedral viral particles measuring 70 nm in diameter were identified in eosinophilic inclusion bodies in the cytoplasm of connective tissue cells in the hepatopancreas, gills, and gut. Deng et al. (2012) sequenced 12 genome segments, identified the genome has a total length of 24.464 kbp and tentatively placed this virus within the Reoviridae family, suggesting the erection of a new genus *Crabreovirus*. This virus is thought to be highly pathogenic and is associated with large economic losses of cultured crabs in China.

14.3.12 *Eriocheir sinensis* reovirus (EsRV905, EsRV816, and EsRV WX-2012)

Zhang et al. (2004) reported a reovirus infection (EsRV905) from farmed Chinese mitten crabs (*Eriocheir sinensis*). The virus infects the connective tissues cells of the gill, gut and hepatopancreas with cytoplasmic inclusions shown to contain virions measuring 55 nm in diameter with electron dense cores and 23 kbp genome. Following infection trials, the virus was shown to cause 30% mortality but was not associated with Tremor Disease. A second reovirus (EsRV816) was isolated from Chinese mitten crabs by Zhang and Bonami (2012). The virus was shown to infect the same tissues as EsRV905, but viral particles were reported to be slightly larger, 60 nm in diameter, genome of 23 kbp. Shen et al. (2015) presented the near full genome of a

third reovirus from Chinese mitten crabs and identified 12 gene sequences which were similar to those reported for SsRV, this dsRNA virus of 23.914 kb was named EsRV WX-2012. It is not clear whether EsRV905, EsRV816, and EsRV WX-2012 are the same virus isolated at different times or three different reovirus infections, further work is needed to clarify this.

14.3.13 *Cherax quadricarinatus* reovirus

Edgerton et al. (2000) described a reovirus infection in a single moribund crayfish (*C. quadricarinatus*) sampled from a holding tank. Eosinophilic cytoplasmic inclusions were observed in the hepatopancreas, and these were shown to contain non-enveloped icosahedral particles measuring 35–40 nm in diameter. Genomic data are not available for this infection and the impact upon the aquaculture industry is unknown.

14.3.14 *Cherax quadricarinatus* giardiavirus-like virus

Cherax quadricarinatus giardia-like virus was described in the hepatopancreas of *C. quadricarinatus* (Edgerton et al. 1994). The infection was reported to cause low-level mortalities in juvenile crayfish (Edgerton et al. 1994; Edgerton and Owens 1997), affected cells displaying hypertrophied nuclei containing less developed eosinophilic or well-developed basophilic inclusion bodies. Hexagonal virions measuring 25 nm in diameter in highly ordered paracrystalline arrays were identified within these inclusion bodies. The virus was described to be morphologically similar to *Giardia lamblia* virus and was placed within the *Totiviridae* family. Genomic data are not available for this infection and the impact upon the aquaculture industry is unknown.

14.3.15 *Eriocheir sinensis* ronivirus (EsRNV)

Eriocheir sinensis ronivirus (EsRNV) is also known as sighs disease due the sound caused by the slow extrusion of bubbles at night (Zhang and Bonami 2007; Bonami and Zhang 2011). The viral infection is associated with black gill syndrome and has been associated with mortalities in farmed crabs. Infection corresponds to a multifocal to generalised necrosis of connective tissues of the gill, heart, hepatopancreas and gut, with nuclei showing pyknosis and karyorrhexis. Enveloped rod-shaped virions measuring 60–110 x 24–42 nm were identified within the cytoplasm of affected cells, non-mature particles were occasionally associated with the endoplasmic reticulum, similar to that described by Spann et al. (1997) in GAV infections. Tissue tropism, histopathology and ultrastructural characteristics suggest this virus is similar to viruses within the family *Roniviridae* and genus *Okavirus*, such as YHV, GAV, however genome sequencing is needed to clarify this relationship.

14.3.16 Mud crab dicistrovirus 1

Mud crab dicistrovirus 1 was identified in the mud crab *S. serrata* and is reported to be pathogenic to crabs under culture conditions (Guo et al. 2013). Infected crabs displayed lack of appetite, no response to stimuli and were shown to alter behaviours, remaining on the sand in daylight. The virus was shown to infect multiple tissues, virus particles approximately 30 nm in diameter were isolated from purified tissues. Nucleotide sequence analysis indicated a linear positive-sense ssRNA genome of 10,415 nucleotides. Phylogenetic analysis highlighted a close genetic relationship between this virus and TSV. The virus has been placed within the *Aparavirus* genus in the *Dicistroviridae* family (Guo et al. 2013).

14.3.17 Chequa iflavirus and Athtabvirus

Chequa iflavirus was identified in farmed *C. quadricarinatus*, increased mortalities were reported in stressed crayfish following transportation and translocation of the animals (Sakuna et al. 2017). Icosahedral virions measuring 14 nm were reported in muscle tissues that appeared fractured with haemocytic infiltration. The complete genome of a positive sense ssRNA virus with a 9,933 bp genome was identified with NextGen sequencing. Genome sequences highlighted similarities between this virus and *Iflaviruses* reported in insects and the virus was tentatively placed within the

order *Picornavirales*. A second viral infection was also identified within this same sequence data, a bunya-like virus named Athtabvirus (Sakuna et al. 2018). Athtabvirus is thought to be related to Mourilyan and Whenzhou Shrimp virus 1 from penaeid shrimp (Cowley et al. 2005b; Li et al. 2015). No histological or ultrastructural descriptions of Athtab virus infection are available but the virus is thought to infect the muscle, nerve, heart, gill, hepatopancreas, and antennal gland tissues. Both viruses were implicated in the mortality event of the crayfish.

14.3.18 *Cancer pagurus* systemic bunya-like virus (CpSBV)

A systemic bunya-like virus infection was reported from the edible crab, *C. pagurus*, during experimental studies (Corbel et al. 2003). Large amounts of virions were isolated from haemolymph samples of affected crabs. Enveloped round virions measuring 60–70 nm in diameter were described to have a tail like structure 70–110 nm in length and 25–35 nm in diameter. Virions were identified in the gills, hepatopancreas, and heart tissues, particles present free in the plasma between the cells or accumulated in cytoplasmic vesicles. Genome was suggested to comprise of three segments of ssRNA. The impact of this virus on wild populations is unknown.

14.3.19 *Eriocheir sinensis* bunya-like virus (EsBV)

A Bunya-like virus infection has been associated with tremor disease which is known to cause mortalities and impact production in farmed Chinese mitten crabs, *E. sinensis* (Huang et al. 2019). No histological or ultrastructural descriptions of this virus infection are available. The genome was identified via next generation sequencing, negative-sense ssRNA genome composed of 6.7 kbp and showed the greatest similarity to the Wenling crustacean virus 9 (Shi et al. 2016). These viruses were suggested to belong to the family *Peribunyaviridae*.

14.3.20 Lymphoid organ vacuolisation virus (LOVV)

Lymphoid organ vacuolisation virus (LOVV) affects the lymphoid organ of *P. vannamei* (Bonami et al. 1992). Affected shrimp showed no obvious signs of disease but histological analysis of the lymphoid organ revealed multifocal degenerative lesions of the parenchymal cells. Virions were identified within cytoplasmic inclusion bodies, spherical enveloped virions containing hexagonal nucleocapsids measuring 52–54 nm. Bonami et al. (1992) suggested that this virus belongs to the Family *Togaviridae* based on size, shape, and structure. Pathogenicity of this virus is unknown and there is no genomic information available.

14.3.21 Wenzhou shrimp virus 1 and 2

Two penaeid shrimp viruses were isolated from the Wenzhou region in China following a study to sequence all negative stranded RNA viruses from a selection of arthropods (Li et al. 2015). Wenzhou shrimp virus 1 was identified in *P. monodon* and Wenzhou shrimp virus 2 in *P. monodon* and *E. carinicauda*. It is not known whether these viruses are associated with any mortalities or pathology.

14.4 Diagnosis of viral diseases in farmed crustaceans

The 'gold standard' methods for the detection of all viral pathogen at most shrimp life stages are PCR-based methods (PCR, nested-PCR, RT-PCR, nested RT-PCR, qPCR, and RT-qPCR) (OIE 2007a, 2009a, 2017, 2018, 2019). An exception can be made for the detection of WSSV in larval and post-larval stages where the PCR-based methods are not recommended (OIE 2019). Histopathology, antibody-based assays, sequencing, and *in situ* hybridisation provide good detection capability in most cases but are limited by the time required for diagnosis. However, these test in combination with a PCR-based method provide a confirmatory diagnosis. In addition, a new generation of diagnostic techniques has been currently proposed harnessing the CRISPR/Cas machinery to detect viral nucleic acids (Chaijarasphong et al. 2019; Sullivan

et al. 2019). These techniques, when combined with a lateral flow colourimetric assay provide a fully field-deployable, next-generation diagnostic (Sullivan et al. 2019). Additionally, pond-side detection devices like the Gene Drive are already under evaluation in Thailand. These devices could provide farmers the opportunity to make real-time management decisions without the need to relay on delayed results from a distant diagnostic laboratory (Flegel 2019; Minardi et al. 2019). Finally, pond site NGS analysis might be possible in the near future with the use of technologies such as the MinIon from Oxford NanoPore that could potentially allow farmers to simultaneously detect multiple pathogens, and could even identify new and currently unknown pathogens.

In general, electron microscopy is still used as the first technique for virus detection (Dey et al. 2019). However, recently, with the advent of next generation sequencing (NGS) technologies, a genomic approach is being increasingly employed to genetically characterise a virus before conducting the ultrastructural characterisation. This approach was recently utilised by Dong et al. (2020) to characterise a novel Hepe-like virus from farmed *M. rosenbergii*. In addition, Cruz-Flores et al. (2020) have proposed to use NGS using nucleic acids directly derived Davidson´s-fixed paraffin embedded tissue sections that present pathological lesions to detect and characterise viral pathogens. These genomic approaches will expedite the genetic characterisation of the pathogen thus decreasing the time needed to develop molecular diagnostic tool that can be ultimately employed to screen crustaceans and limit the spread of emerging pathogens.

14.5 Disease control and management

Viral disease management in crustacean aquaculture is complex in nature due to the large number of variables modulating a disease outcome. While best management practices might reduce the possibility of a viral outbreak the risk will always be there. Biosecurity strategies such as stocking ponds with SPF/SPR lines of animals, quarantine of exotic bloodstock before introducing into the breeding programme, the exclusion of vectors and carrier organisms, the use of greenhouses, and disinfection at the start of a production cycle have all proven effective at reducing the possibility of a viral outbreak. New approaches such as shrimp polyculture and biofloc technology have been reported to not only provide a food source but also act as an immune booster that provides some level of protection when a viral outbreak occurs (Bunting 2005; Dey et al. 2019). However, there is a gap between laboratory research and development of products that provide consistent performance and commercially feasible.

Development of therapeutics against shrimp viral diseases is an area where significant progress has been made, yet nothing is available for commercial applications (see also Chapter 1, this volume). So far, a variety of approaches have been proven to work in laboratory settings, these include: vaccination (reviewed in Flegel 2019; Chapter 1), the use of host-derived antimicrobial peptides, the use of immunostimulants as feed additives (reviewed by Dey et al. 2019), RNAi (reviewed by Dey et al. 2019; Flegel 2019), probiotics (reviewed by Wang et al. 2019), and the selection or induction of endogenous viral elements (EVE) to provide heritable tolerance in shrimp (reviewed by Flegel 2019). Future research and development in these areas with a targeted goal on developing therapeutics for large-scale applications will completely redefine how diseases are managed in shrimp aquaculture in the future.

The future success of shrimp farming will depend on farming shrimp that are genetically superior for growth and or disease resistance in a biosecure production facility. While major progress has been made in developing SPF stocks and biosecurity protocols to ensure freedom from diseases in farms, success in developing SPR lines against most viral diseases has been limited (Wyban 2019). So far, breeding for disease resistance has been most successful against Taura syndrome disease. Due to the availability of Taura syndrome resistant lines of shrimp for commercial breeding purposes, the threat of Taura syndrome has been largely eliminated in shrimp farming regions worldwide (Moss et al. 2012). In contrast to developing disease resistant lines against Taura syndrome disease, there has been little progress in developing resistant lines against white spot disease, the most important microbial disease in crustaceans. Selected lines of *P. vannamei* from Panama were identified that showed significant resistance to WSSV infections in laboratory challenge studies (Cuéllar-Anjel et al. 2012) and microsatellite markers associated with WSSV susceptibility/resistance were reported (Chakroborty et al. 2015). Low heritability (h^2) and negative genetic correlations between body weights and WSSV resistance in *P. vannamei* (Trang et al. 2019) are challenges in developing WSSV-resistant lines. Perhaps precision breeding, also known as genome editing, may enable to expedite the development of WSSV resistant lines in shrimp in the near future.

14.6 Future directions

- It is inevitable that diseases will continue to pose a serious threat to growth and profitability of shrimp and other crustacean aquaculture worldwide. Therefore, early detection of diseases will help to contain the spread and prevent economic losses. While the laboratory-based diagnostic has served shrimp and other crustacean industry very well over the past four decades, there is a growing need to commercialise point-of-care diagnostics especially for farmers in many developing nations where the infrastructure is lacking. Development of point-of-care diagnostics will enable a farmer to make a decision to harvest the crop before a large-scale mortality occurs and the farmer suffer a complete loss of harvest. This is an area of research which will certainly gain attention in the coming years. There are already some platform technologies in the market but none of these has not been adopted widely primarily because of cost and utility of the platform.

- Increasingly new diseases are identified for which more than one aetiological agents are associated with clinical outcomes and disease outbreaks. In these cases, having a knowledge of pathobiome is going to be more critical than detecting the presence or absence of one or two pathogen. This is another area which will certainly gain attention especially in studying complex diseases with unknown aetiology.

- For long-term growth and sustainability of shrimp farming, disease prevention alone is not enough. While prevention via biosecurity remains as a cornerstone in avoiding disease introduction in a production system, these measures are not enough to prevent large-scale mortalities and disease outbreaks that are reoccurring periodically. Development and commercial availability of disease resistant lines of shrimp will be a key factor in mitigating economic losses caused by viral diseases in the near future. Combining with the development of diseases resistant lines, there is an urgent need to develop therapeutics against viral diseases in crustaceans. While there are many publications describing the utility of RNA interference-based therapeutics, so far there is no commercially available therapeutics. One of the bottlenecks in developing therapeutics that can be applied at a pond level is the lack of a robust and commercially feasible oral delivery system of the therapeutic molecules. This is another area which is certainly going to receive major attention in the years to come.

- Finally, like the historical developments in terrestrial farming such as poultry, the long-term sustainability of commercial farming of crustacean is likely to be an enclosed production system as opposed to an open system as practiced now. Farming crustacean in indoor system/highly controlled outdoor environment will enable the farmers to have a complete control starting with stocking ponds/raceways with disease-free post-larvae, preventing the introduction of vectors carrying diseases, monitoring diseases on-site, and taking steps to ensure the implementation of all other production management criteria to have a successful harvest in a sustainable manner.

14.7 Summary

- Viral diseases of cultured crustaceans have by far had the largest economic impact of all diseases (bacterial, fungal, and protistan) of farmed crustaceans.
- Viruses like IHHNV, WSSV, TSV, and YHV have caused the collapse of the shrimp industry on several occasions and have changed how crustaceans are farmed around the world.
- New emerging-OIE listed agents like DIV1 should be given special attention to limit the spread of the pathogen and avoid a new panzootic that could once again severely affect crustacean farming.

- Currently, there are no effective treatments or therapeutics for viral diseases of crustaceans, therefore early diagnosis and intensive monitoring remain the cornerstones of managing and controlling disease outbreaks of viral origin.

Acknowledgements

Funding for this research was provided by the College of Agriculture & Life Sciences in The University of Arizona to Arun K. Dhar. A special thanks to Dr Hung Nam Mai for providing critical comments on several sections of the book chapter.

References

Afsharnasab, M., A, M.M., Azaritakami, G., and Sharifrohani, M. 2014. Ultrastructural and pathogenesis of Monodon baculovirus in SPF shrimp, *Litopenaeus vannamei* imported to Iran. *Iranian Journal of Fisheries Sciences* 13: 640–652

Aldana Aranda, D., Frenkiel, L., Brulé, T., Montero, J., and Baqueiro Cárdenas, E. 2011. Occurrence of Apicomplexa-like structures in the digestive gland of *Strombus gigas* throughout the Caribbean. *Journal of Invertebrate Pathology* 106: 174–178

Anderson, I.G. and Prior, H.C. 1992. Baculovirus freshwater infections in the mud crab, *Scylla serrata*, and a crayfish, *Cherax quadricarinatus*, from Australia. *Journal of Invertebrate Pathology* 60: 265–273

Aranguren, L.F., Salazar, M., Tang, K., Caraballo, X., and Lightner, D. 2013. Characterization of a new strain of Taura syndrome virus (TSV) from Colombian shrimp farms and the implication in the selection of TSV resistant lines. *Journal of Invertebrate Pathology* 112: 68–73

Arcier, J.M., Herman, F., Lightner, D.V., Redman, R.M., Mari, J., and Bonami, J.R. 1999. A viral disease associated with mortalities in hatchery-reared postlarvae of the giant freshwater prawn *Macrobrachium rosenbergii*. *Diseases of Aquatic Organisms* 38: 177–181

Bateman, K. and Stentiford, G. 2008. Cancer pagurus bacilliform virus (CpBV) infecting juvenile European edible crabs *C. pagurus* from UK waters. *Diseases of Aquatic Organisms* 79: 147–151

Bateman, K.S., Hicks, R.J., and Stentiford, G.D. 2011. Disease profiles differ between non-fished and fished populations of edible crab (*Cancer pagurus*) from a major commercial fishery. *ICES Journal of Marine Science* 68: 2044–2052

Bateman, K.S. and Stentiford, G.D. 2017. A taxonomic review of viruses infecting crustaceans with an emphasis on wild hosts. *Journal of Invertebrate Pathology* 147: 86–110

Behringer, D.C. and Butler, M.J. 2010. Disease avoidance influences shelter use and predation in Caribbean spiny lobster. *Behavioral Ecology and Sociobiology* 64: 747–755

Behringer, D.C., Butler, M.J., and Shields, J.D. 2006. Avoidance of disease by social lobsters. *Nature* 441: 421–421

Behringer, D.C., Butler IV, M.J., Shields, J.D., and Moss, J. 2011. Review of Panulirus argus virus 1—a decade after its discovery. *Diseases of Aquatic Organisms* 94: 153–160

Bochow, S., Condon, K., Elliman, J., and Owens, L. 2015. Marine genomics first complete genome of an Ambidensovirus; Cherax quadricarinatus densovirus, from freshwater crayfish *Cherax quadricarinatus*. *Marine Genomics* 24: 305–312

Bojko, J., Stebbing, P.D., Dunn, A.M. et al. 2018. Green crab *Carcinus maenas* symbiont profiles along a North Atlantic invasion route. *Diseases of Aquatic Organisms* 128: 147–168

Bonami, J., Bruce, L., Poulos, B., Mari, J., and Lightner, D. 1995. Partial characterization and cloning of the genome of PvSNPV (= BP-type virus) pathogenic for *Penaeus vannamei*. *Diseases of Aquatic Organisms* 23: 59–66

Bonami, J.R., Lightner, D.V., Redman, R.M., and Poulos, B.T. 1992. Partial characterization of a togavirus (LOVV) associated with histopathological changes of the lymphoid organ of penaeid shrimps. *Diseases of Aquatic Organisms* 14: 145–152

Bonami, J.R., Shi, Z., Qian, D., and Widada, J.S. 2005. White tail disease of the giant freshwater prawn, *Macrobrachium rosenbergii*: Separation of the associated virions and characterization of MrNV as a new type of nodavirus. *Journal of Fish Diseases* 28: 23–31

Bonami, J.R. and Sri Widada, J. 2011. Viral diseases of the giant fresh water prawn *Macrobrachium rosenbergii*: A review. *Journal of Invertebrate Pathology* 106: 131–142

Bonami, J.R. and Zhang, S. 2011. Viral diseases in commercially exploited crabs: A review. *Journal of Invertebrate Pathology* 106: 6–17

Bowater, R.O., Wingfield, M., Fisk, A., Condon, K.M.L., Reid, A., Prior, H. and Kulpa, E.C. 2002. A parvo-like virus in cultured redclaw crayfish *Cherax quadricarinatus* from Queensland, Australia. *Diseases of Aquatic Organisms* 50: 79–86

Bowers, H.A., Messick, G.A., Hanif, A. et al. 2010. Physicochemical properties of double-stranded RNA used to discover a reo-like virus from blue crab *Callinectes sapidus*. *Diseases of Aquatic Organisms* 93: 17–29

Brock, J., Lightner, D., and Bell, T.A. 1983. A review of four viruses (BP, MBV, BMN, and IHHNV) diseases of penaeid shrimp with particular reference to clinical significance, diagnosis and control in shrimp aquaculture. In *71st international council for the exploration of the sea*. 1–8

Brock, J. and Lightner, D.V. 1990. Diseases of Crustacea. In *Diseases of Marine Animals*, Vol III, Kinne O. (ed.) pp. 245–349. Hamburg: Inter-Research

Bunting, S. 2005. Low impact aquaculture. Colchester: www.essex.ac.uk/ces/esu/occ-papers.shtm, 1–22

Chaijarasphong, T., Thammachai, T., Itsathitphaisarn, O., Sritunyalucksana, K., and Suebsing, R. 2019. Potential application of CRISPR-Cas12a fluorescence assay coupled with rapid nucleic acid amplification for detection of white spot syndrome virus in shrimp. *Aquaculture* 512: 734340

Chakrabarty, U., Dutta, S., Mallik, A., Mondal, D., and Mandal, N. 2015. Identification and characterisation of microsatellite DNA markers in order to recognise the WSSV susceptible populations of maringiant black tiger shrimp, *Penaeus monodon*. *Veterinary Resources* 46: 1–10. https://doi.org/10.1186/s13567-015-0248-2

Chantanachookin, C., Boonyaratpalin, S., Kasornchandra, J. et al. 1993. Histology and ultrastructure reveal a new granulosis-like virus in *Penaeus monodon* affected by yellow-head disease. *Diseases of Aquatic Organisms* 17: 145–157

Chen, B-K., Dong, Z., Pang, N., Nian, Y., and Yan, D. 2018. A novel real-time PCR approach for detection of infectious hypodermal and haematopoietic necrosis virus (IHHNV) in the freshwater crayfish *Procambarus clarkii*. *Journal of Invertebrate Pathology* 157: 100–103

Chen, X., Qiu, L., Wang, H., Zou, P., Dong, X., Li, F., and Huang, J. 2019. Susceptibility of *exopalaemon carinicauda* to the infection with shrimp hemocyte iridescent virus (SHIV 20141215), a strain of decapod iridescent virus 1 (DIV1). *Viruses* 11: 1–15

Corbel, V., Coste, F., and Bonami, J.R. 2003. CpSBV, a systemic virus of the edible crab, *Cancer pagurus* (L.). *Journal of Fish Diseases* 26: 121–126

Corbel, V., Zuprizal, Z., Shi, C., Huang, Sumartono, Arcier, J.M., and Bonami, J.R. 2001. Experimental infection of European crustaceans with white spot syndrome virus (WSSV). *Journal of Fish Diseases* 24: 377–382

Cotmore, S.F., Agbandje-mckenna, M., Canuti, M. et al. 2019. ICTV virus taxonomy profile: Parvoviridae. *Journal of General Virology* 100: 367–368

Couch, J.A. 1974. An enzootic nuclear polyhedrosis virus of pink shrimp: Ultrastructure, prevalence, and enhancement. *Journal of Invertebrate Pathology* 24: 311–331

Cowley, J., Dimmock, C., Wongteerasupaya, C., Boonsaeng, V., Panyim, S., and Walker, P. 1999. Yellow head virus from Thailand and gill-associated virus from Australia are closely related but distinct prawn viruses. *Diseases of Aquatic Organisms* 36: 153–157

Cowley, J., Moody, N.J.G., Mohr, P.G., Rao, M., and Cowley, J. 2015. *Aquatic Animal Health Subprogram: Viral presence, prevalence and disease management in wild populations of the Australian Black Tiger prawn (Penaeus monodon).* Queensland Australia: Fisheries Research and Development Corporation

Cowley, J.A., Hall, M.R., Cadogan, L.C., Spann, K.M., and Walker, P.J. 2002. Vertical transmission of gill-associated virus (GAV) in the black tiger prawn *Penaeus monodon. Diseases of Aquatic Organisms* 50: 95–104

Cowley, J.A., McCulloch, R.J., Rajendran, K.V., Cadogan, L.C., Spann, K.M., and Walker, P.J. 2005. RT-nested PCR detection of Mourilyan virus in Australian *Penaeus monodon* and its tissue distribution in healthy and moribund prawns. *Diseases of Aquatic Organisms* 66: 91–104

Cowley, J.A., McCulloch, R.J., Spann, K.M., Cadogan, L.C., and Walker, P.J. 2005. Preliminary molecular and biological characterization of Mourilyan virus (MoV): a new bunya-related virus of peaneid prawns. *Diseases in Asian Aquaculture* V, 113–124

Cowley, J.A. and Walker, P.J. 2002. The complete genome sequence of gill-associated virus of *Penaeus monodon* prawns indicates a gene organisation unique among nidoviruses. *Archives of Virology* 147: 1977–1987

Cowley, J.A., Walker, P.J., Flegel, T.W. et al. 2012. Roniviridae. In *Virus Taxonomy. Ninth Report of the International Committee on Taxonomy of Viruses*, pp. 829–834. UK: Elsevier.

Cruz-Flores, R., Mai, H.N., Kanrar, S., Aranguren Caro, L.F., and Dhar, A.K. 2020. Genome reconstruction of white spot syndrome virus (WSSV) from archival Davidson's-fixed paraffin embedded shrimp (*Penaeus vannamei*) tissue. *Scientific Reports* 10: 13425

Cuéllar-Anjel, J., White-Noble, B., Schofield, P., Chamorro, R., and Lightner, D. V. 2012. Report of significant WSSV-resistance in the Pacific white shrimp, Litopenaeus vannamei, from a Panamanian breeding program. *Aquaculture* 368–369: 36–39.

Deng, X., Lü, L., Ou, Y. et al. 2012. Sequence analysis of 12 genome segments of mud crab reovirus (MCRV). *Virology* 422: 185–194

Dey, B.K., Dugassa, G.H., Hinzano, S.M., and Bossier, P. 2019. Causative agent, diagnosis and management of white spot disease in shrimp: A review. *Reviews in Aquaculture* 12: 822–865

Dhar, A.K., Cruz-Flores, R., Caro, L.F.A., Siewiora, H.M., and Jory, D. 2019. Diversity of single-stranded DNA containing viruses in shrimp. *VirusDisease* 30: 43–57

Dhar, A.K., Robles-Sikisaka, R., Saksmerprome, V., and Lakshman, D.K. 2014. Biology, genome organization, and evolution of Parvoviruses in marine shrimp. *Advances in Virus Research* 89: 85–139

Dong, X., Hu, T., Liu, Q., Li, C., Sun, Y., Wang, Y., Shi, W., Zhao, Q., and Huang, J. 2020. A novel Hepe-like virus from farmed giant freshwater prawn *Macrobrachium rosenbergii. Viruses* 12: 3

Durand, S., Lightner, D.V., and Bonami, J.R. 1998. Differentiation of BP-type baculovirus strains using *in situ* hybridization. *Diseases of Aquatic Organisms* 32: 237–239

Durand, S., Lightner, D. V., Redman, R.M., and Bonami, J.R. 1997. Ultrastructure and morphogenesis of White Spot Syndrome Baculovirus (WSSV). *Diseases of Aquatic Organisms* 29: 205–211

Edgerton, B. 1996. A new bacilliform virus in Australian *Cherax destructor* (Decapoda : Parastacidae) with notes on *Cherax quadricarinatus* bacilliform virus (= Cherax baculovirus). *Diseases of Aquatic Organisms* 27: 43–52

Edgerton, B. and Owens, L. 1997. Age at first infection of *Cherax quadricarinatus* by Cherax quadricarinatus bacilliform virus and Cherax Giardiavirus-like virus, and production of putative virus-free crayfish. *Aquaculture* 152: 1–12

Edgerton, B., Owens, L., Glasson, B., and Beer, S. 1994. Description of a small dsRNA virus from freshwater crayfish *Cherax quadricarinatus. Diseases of Aquatic Organisms* 18: 63–69

Edgerton, B., Webb, R., and Wingfield, M. 1997. A systemic parvo-like virus in the freshwater crayfish *Cherax destructor. Diseases of Aquatic Organisms* 29: 73–78

Edgerton, B.F., Ian, G. and Elizabeth, C. 2000. Description of a presumptive hepatopancreatic reovirus, and a putative gill parvovirus, in the freshwater crayfish *Cherax quadricarinatus. Diseases of Aquatic Organisms* 41: 83–90

Erickson, H.S., Zarain-Herzberg, M., and Lightner, D. V. 2002. Detection of Taura syndrome virus (TSV) strain differences using selected diagnostic methods: Diagnostic implications in penaeid shrimp. *Diseases of Aquatic Organisms* 52: 1–10.

Escobedo-Bonilla, C.M., Alday-Sanz, V., Wille, M., Sorgeloos, P., Pensaert, M.B., and Nauwynck, H.J. 2008. A review on the morphology, molecular characterization, morphogenesis and pathogenesis of white spot syndrome virus. *Journal of Fish Diseases* 31: 1–18

Flegel, T.W. 2006. Detection of major penaeid shrimp viruses in Asia, a historical perspective with emphasis on Thailand. *Aquaculture* 258: 1–33

Flegel, T.W. 2012. Historic emergence, impact and current status of shrimp pathogens in Asia. *Journal of Invertebrate Pathology* 110: 166–173

Flegel, T.W. 2019. A future vision for disease control in shrimp aquaculture. *Journal of the World Aquaculture Society* 50: 249–266

Flegel, T.W., Lightner, D.V, Lo, C.H.U.F., and Owens, L. 2008. Shrimp disease control: Past, present and future. *Diseases in Asian Aquaculture* VI: 355–378

Flowers, E.M., Johnson, A.F., Aguilar, R., and Schott, E.J. 2018. Prevalence of the pathogenic crustacean virus Callinectes sapidus reovirus 1 near flow-through blue crab aquaculture in Chesapeake Bay, USA. *Diseases of Aquatic Organisms* 129: 135–144

Flowers, E.M., Simmonds, K., Messick, G.A., Sullivan, L., and Schott, E.J. 2016. PCR-based prevalence of a fatal reovirus of the blue crab, *Callinectes sapidus* (Rathbun) along the northern Atlantic coast of the USA. *Journal of Fish Diseases* 39: 705–714

Gangnonngiw, W., Bunnontae, M., Phiwsaiya, K., Senapin, S., and Dhar, A.K. 2020. In experimental challenge with infectious clones of *Macrobrachium rosenbergii* nodavirus (MrNV) and extra small virus (XSV), MrNV alone can cause mortality in freshwater prawn (*Macrobrachium rosenbergii*). *Virology* 540: 30–37

Gangnonngiw, W., Laisutisan, K., Sriurairatana, S. et al. 2010. Monodon baculovirus (MBV) infects the freshwater prawn *Macrobrachium rosenbergii* cultivated in Thailand. *Virus Research* 148: 24–30

Guo, Z.X., He, J.G., Xu, H.D., and Weng, S.P. 2013. Pathogenicity and complete genome sequence analysis of the mud crab dicistrovirus-1. *Virus Research* 171: 8–14

Hasson, K.W., Lightner, D.V., Poulos, B.T. et al. 1995. Taura syndrome in *Penaeus vannamei*: Demonstration of a viral etiology. *Diseases of Aquatic Organisms* 23: 115–126

Holt, C.C., Stone, M., Bass, D. et al. 2019. The first clawed lobster virus Homarus gammarus nudivirus (HgNV n. sp.) expands the diversity of the Nudiviridae. *Scientific Reports* 9: 1–15

Hooper, C., Debnath, P., Biswas, S. et al. 2020. A novel RNA virus, Macrobrachium rosenbergii Golda Virus (MrGV), linked to mass mortalities of the larval giant freshwater prawn in Bangladesh. *Viruses* 12: 11–20

Huang, P., Zhang, X., Ame, K.H., Shui, Y., Xu, Z., Serwadda, A., and Shen, H. 2019. Genomic and phylogenetic characterization of a bunya-like virus from the freshwater Chinese mitten crab *Eriocheir sinensis*. *Acta Virologica* 63: 433–438

ICTV 2020a. Genus: Decapodiridovirus [online]. *International Committee on Taxonomy of Viruses*. Available from: https://talk.ictvonline.org/ictv-reports/ictv _online_report/dsdna-viruses/w/iridoviridae/1301/ genus-decapodiridovirus#:~:text=Biology, appeared major targets for infection. [Accessed 10 Sep 2020]

ICTV 2020b. Totiviridae Family: Totiviridae Virion properties Morphology [online]. *Family: Totiviridae*. Available from: https://talk.ictvonline.org/ictv-reports/ictv_ 9th_report/dsrna-viruses-2011/w/dsrna_viruses/ 191/totiviridae.

Jagadeesan, V., Praveena, P.E., Otta, S.K., and Jithendran, K.P. 2019. Classical runt deformity syndrome cases in farmed *Penaeus vannamei* along the east coast of India. *Journal of Coastal Research* 86: 107

Jariyapong, P., Pudgerd, A., Weerachatyanukul, W., and Hirono, I. 2018. Construction of an infectious Macrobrachium rosenbergii nodavirus from cDNA clones in Sf9 cells and improved recovery of viral RNA with AZT treatment. *Aquaculture* 483: 111–119

Johnson, P. 1976a. A herpeslike virus from the blue crab, *Callinectes sapidus*. *Journal of Insect Science* 27: 419–420

Johnson, P. 1976b. A baculovirus from the blue crab, *Callinectes sapidus*. *Proclamation of International Colloquium on Invertebrate Pathology* 27: 419–420

Johnson, P. 1988a. Development and morphology of an unusual nuclear virus of the blue crab *Callinectes sapidus*. *Diseases of Aquatic Organisms* 4: 67–75

Johnson, P. 1988b. Rod-shaped nuclear viruses of crustaceans: Hemocyte-Infecting species. *Diseases of Aquatic Organisms* 5: 111–122

Johnson, P. and Lightner, D. 1988. Rod-shaped nuclear viruses of crustaceans: Gut-infecting species. *Diseases of Aquatic Organisms* 5: 123–141

Johnson, P.T. 1977. A viral disease of the blue crab, *Callinectes sapidus*: Histopathology and differential diagnosis. *Journal of Invertebrate Pathology* 29: 201–209

Johnson, P.T. 1978. Viral diseases of the blue crab, *Callinectes sapidus*. *Marine Fisheries Review* 40: 13–15

Johnson, P.T. 1983. Diseases caused by viruses, rickettsias, bacteria and fungi. In *The Biology of Cristaceans*, Vol. 6, Provenzano, A.J.Z. (ed.) pp. 1–78 New York: Academic Press

Johnson, P.T. 1984. Viral diseases of marine invertebrates. *Helgoländer Meeresuntersuchungen* 37: 65–98

Johnson, P.T. and Bodammer, J.E. 1975. A disease of the blue crab, *Callinectes sapidus*, of possible viral etiology. *Journal of Invertebrate Pathology* 26: 141–143

Kalagayan, H., Godin, D., Kanna, R., Hagino, G., Sweeney, J., Wyban, J., and Brock, J. 1991. IHHN virus as an etiological factor in Runt-Deformity Syndrome (RDS) of juvenile *Penaeus vannamei* cultured in Hawaii. *Journal of the World Aquaculture Society* 22: 235–243

Kawato, S., Shitara, A., Wang, Y., Nozaki, R., Kondo, H., and Hirono, I. 2018. Crustacean genome exploration reveals the evolutionary origin of White Spot Syndrome Virus. *Journal of Virology* 93: 1–23

Kon, T., Isshiki, T., and Miyadai, T. 2011. Milky hemolymph syndrome associated with an intranuclear bacilliform virus in snow crab *Chionoecetes opilio* from the Sea of Japan. *Fisheries Science* 77: 999–1007

Krol, R., Hawkins, W., and Overstreet, R. 1990. Reo-like virus in white shrimp *Penaeus vannamei* (Crustacea: Decapoda): Co-occurrence with Baculovirus penaei in experimental infections. *Diseases of Aquatic Organisms* 8: 45–49

La Fauce, K.A., Elliman, J., and Owens, L. 2007. Molecular characterisation of hepatopancreatic parvovirus (PmergDNV) from Australian *Penaeus merguiensis*. *Virology* 362: 397–403

Li, C., Shields, J.D., Ratzlaff, R.E., and Butler, M.J. 2008. Pathology and hematology of the Caribbean spiny lobster experimentally infected with Panulirus argus virus 1 (PaV1). *Virus Research* 132: 104–113

Li, C.X., Shi, M., Tian, J.H. et al. 2015. Unprecedented genomic diversity of RNA viruses in arthropods reveals the ancestry of negative-sense RNA viruses. *eLife* 2015: 1–26

Lightner, D.V. 1996. *A Handbook of Shrimp Pathology and Diagnostic Procedures for Diseases of Cultured Penaeid Shrimp*. Lousiana, USA: World Aquaculture Society

Lightner, D.V. and Redman, R.M. 1981. A baculovirus-caused disease of the penaeid shrimp, *Penaeus monodon*. *Journal of Invertebrate Pathology* 38: 299–302

Lightner, D.V. and Redman, R.M. 1985. A parvo-like virus disease of penaeid shrimp. *Journal of Invertebrate Pathology* 45: 47–53

Lightner, D.V, Redman, R.M., Bell, T., and Brock, J. 1983. Detection of IHHN virus in *Penaeus stylirostris* and *P. vannamei* imported into Hawaii. *Journal of World Mariculture Society* 225: 212–225

Lightner, D.V., Redman, R.M., and Bell, T.A. 1983. Observations on the geographic distribution, pathogenesis and morphology of the baculovirus from *Penaeus monodon* Fabricius. *Aquaculture* 32: 209–233

Lightner, D.V., Redman, R.M., Pantoja, C.R. et al. 2012. Historic emergence, impact and current status of shrimp pathogens in the Americas. *Journal of Invertebrate Pathology* 110: 174–183

Lightner, D.V., Redman, R.M., Williams, R.R. et al. 1985. Recent advances in penaeid virus disease investigations. *Journal of the World Mariculture Society* 16: 267–274

Liu, S., Wang, X., Xu, T.T., Li, X., Du, L., and Zhang, Q. 2018. Vectors and reservoir hosts of covert mortality nodavirus (CMNV) in shrimp ponds. *Journal of Invertebrate Pathology* 154: 29–36

Lo, C.F., Ho, C.H., Peng, S.E. et al. 1996. White spot syndrome baculovirus (WSBV) detected in cultured and captured shrimp, crabs and other arthropods. *Diseases of Aquatic Organisms* 27: 215–225

Longshaw, M. 2011. Diseases of crayfish: A review. *Journal of Invertebrate Pathology* 106: 54–70

Mai, H.N., Hanggono, B., Caro, L.F.A., Komaruddin, U., Nur'aini, Y.L., and Dhar, A.K. 2019. Novel infectious myonecrosis virus (IMNV) genotypes associated with disease outbreaks on *Penaeus vannamei* shrimp farms in Indonesia. *Archives of Virology* 164: 3051–3057

Mari, J., Bonami, J., Poulos, B., and Lightner, D.V. 1993. Preliminary characterization and partial cloning of the genome of a baculovirus from *Penaeus monodon* (PmSNPV = MBV). *Diseases of Aquatic Organisms* 16: 207–215

Minardi, D., Bateman, K.S., Kuzdzal, A. et al. 2019. Testing of a pond-side molecular diagnostic tool for the detection of white spot syndrome virus in shrimp aquaculture. *Journal of the World Aquaculture Society* 50: 18–33

Moss, J., Butler, M.J., Behringer, D.C., and Shields, J.D. 2011. Genetic diversity of the Caribbean spiny lobster virus, Panulirus argus virus 1 (PaV1), and the discovery of PaV1 in lobster postlarvae. *Aquatic Biology* 14: 223–232

Moss, S.M., Moss, D.R., Arce, S.M., Lightner, D.V., and Lotz, J.M. 2012. The role of selective breeding and biosecurity in the prevention of disease in penaeid shrimp aquaculture. *Journal of Invertebrate Pathology* 110: 247–250

Munro, J. and Owens, L. 2007. Yellow head-like viruses affecting the penaeid aquaculture industry. *Aquaculture Research* 38: 893–908

Nash, M., Nash, G., Anderson, I.G., and Shariff, M. 1988. A reo-like virus observed in the tiger prawn, *Penaeus monodon* Fabricius, from Malaysia. *Journal of Fish Diseases* 11: 531–535

Ng, T.F.F., Alavandi, S., Varsani, A., Burghart, S., and Breitbart, M. 2013. Metagenomic identification of a nodavirus and a circular ssDNA virus in semi-purified viral nucleic acids from the hepatopancreas of healthy

Farfantepenaeus duorarum shrimp. *Diseases of Aquatic Organisms* 105: 237–242

Oakey, H.J. and Smith, C.S. 2018. Complete genome sequence of a white spot syndrome virus associated with a disease incursion in Australia. *Aquaculture* 484: 152–159

Oakey, J., Smith, C., Underwood, D. et al. 2019. Global distribution of white spot syndrome virus genotypes determined using a novel genotyping assay. *Archives of Virology* 164: 2061–2082

Oanh, D.T.H., Van Hulten, M.C.W., Cowley, J.A., and Walker, P.J. 2011. Pathogenicity of gill-associated virus and Mourilyan virus during mixed infections of black tiger shrimp (*Penaeus monodon*). *Journal of General Virology* 92: 893–901

Oidtmann, B. and Stentiford, G.D. 2011. White Spot Syndrome Virus (WSSV) concentrations in crustacean tissues—a review of data relevant to assess the risk associated with commodity trade. *Transboundary and Emerging Diseases* 58: 469–482

OIE 2007a. Hepatopancreatic parvovirus disease. In *OIE Aquatic Animal Disease Cards*, (2007), pp. 1–4. Paris, France: Office International des Epizooties

OIE 2007b. Mourilyan virus. *OIE Aquatic Animal Disease Cards*, 4–5: OIE

OIE 2009a. Yellow head disease. In *Manual of Diagnostic Tests for Aquatic Animals*, pp. 20–23. Paris, France: Office International des Epizooties

OIE 2009b. Infectious myonecrosis. In *Manual of Diagnostic Tests for Aquatic Animals*, pp. 138–147. Paris, France: Office International des Epizooties

OIE 2017. Infection with infectious hypodermal and haematopoietic necrosis virus. In *Manual of Diagnostic Tests for Aquatic Animals*, pp. 1–18. Paris, France: OIE

OIE 2018. Infection with Taura virus. In *Manual of Diagnostic Tests for Aquatic Animals*, pp. 1–18. Paris, France: Office International des Epizooties

OIE 2019. Infection with white spot syndrome virus. In *Manual of Diagnostic Tests for Aquatic Animals*, pp. 1–16. Paris, France: Office International des Epizooties

Owens, L. 1993. Description of the first haemocytic rod-shaped virus from a penaeid prawn. *Diseases of Aquatic Organisms* 16: 217–221

Owens, L. 2013. Bioinformatical analysis of nuclear localisation sequences in penaeid densoviruses. *Marine Genomics* 12: 9–15

Owens, L., De Beer, S., and Smith, J. 1991. Lymphoidal parvovirus-like particles in Australian penaeid prawns. *Diseases of Aquatic Organisms* 11: 129–134

Owens, L., Haqshenas, G., McElnea, C., and Coelen, R. 1998. Putative spawner-isolated mortality virus associated with mid-crop mortality syndrome in farmed *Penaeus monodon* from northern Australia. *Diseases of Aquatic Organisms* 34: 177–185

Owens, L., Liessmann, L., La Fauce, K., Nyguyen, T., and Zeng, C. 2010. Intranuclear bacilliform virus and hepatopancreatic parvovirus (PmergDNV) in the mud crab *Scylla serrata* (Forskal) of Australia. *Aquaculture* 310: 47–51

Owens, L., McElnea, C., Snape, N., Harris, L., and Smith, M. 2003. Prevalence and effect of spawner-isolated mortality virus on the hatchery phases of *Penaeus monodon* and *P. merguiensis* in Australia. *Diseases of Aquatic Organisms* 53: 101–106

Pénzes, J.J., Pham, H.T., Chipman, P. et al. 2020. Molecular biology and structure of a novel penaeid shrimp densovirus elucidate convergent parvoviral host capsid evolution. *Proceedings of the National Academy of Sciences of the United States of America*, 117: 20211–20222

Pham, H.T., Yu, Q., Boisvert, M., Van, H.T., Bergoin, M., and Tijssen, P. 2014. A circo-like virus isolated from *Penaeus monodon* shrimps. *Genome Announcements* 2: 4–5

Phuthaworn, C., Nguyen, N.H., Quinn, J., and Knibb, W. 2016. Moderate heritability of hepatopancreatic parvovirus titre suggests a new option for selection against viral diseases in banana shrimp (*Fenneropenaeus merguiensis*) and other aquaculture species. *Genetics Selection Evolution* 48: 64

Poulos, B.T., Tang, K.F.J., Pantoja, C.R., Bonami, J.R., and Lightner, D.V. 2006. Purification and characterization of infectious myonecrosis virus of penaeid shrimp. *Journal of General Virology* 87: 987–996

Pratoomthai, B., Sakaew, W., Sriurairatana, S., Wongprasert, K., and Withyachumnarnkul, B. 2008. Retinopathy in stunted black tiger shrimp *Penaeus monodon* and possible association with Laem-Singh virus (LSNV). *Aquaculture* 284: 53–58

Qian, D., Shi, Z., Zhang, S., Cao, Z., Liu, W., Li, L., Xie, Y., Cambournac, I., and Bonami, J.R. 2003. Extra small virus-like particles (XSV) and nodavirus associated with whitish muscle disease in the giant freshwater prawn, *Macrobrachium rosenbergii*. *Journal of Fish Diseases* 26: 521–527

Qiu, L., Chen, M., Wan, X., Li, C. and Zhang, Q. 2017. Characterization of a new member of Iridoviridae, Shrimp hemocyte iridescent virus (SHIV), found in white leg shrimp (*Litopenaeus vannamei*). *Scientific Reports* 7: 11834

Qiu, L., Chen, M.M., Wan, X.Y., Zhang, Q.L., Li, C., Dong, X., Yang, B., and Huang, J. 2018a. Detection and quantification of shrimp hemocyte iridescent virus by TaqMan probe based real-time PCR. *Journal of Invertebrate Pathology* 154: 95–101

Qiu, L., Chen, M.M., Wang, R.Y., Yuan, X., Chen, W., Qing, L., Zhang, L., and Dong, X. 2018b. Complete genome sequence of shrimp hemocyte iridescent virus (SHIV)

isolated from white leg shrimp, *Litopenaeus vannamei*. *Archives of Virology* 163: 781–785

Qiu, L., Chen, X., Guo, X.-M. et al. 2020. A TaqMan probe based real-time PCR for the detection of Decapod iridescent virus 1. *Journal of Invertebrate Pathology* 173: 107367

Qiu, L., Chen, X., Zhao, R., Li, C., Gao, W., Zhang, Q., and Huang, J. 2019. Description of a natural infection with Decapod Iridescent Virus 1 in farmed giant freshwater prawn, *Macrobrachium rosenbergii*. *Viruses* 11: 354

Rajan, P.R., Ramasamy, P., Purushothaman, V., and Brennan, G.P. 2000. White spot baculovirus syndrome in the Indian shrimp *Penaeus monodon* and *P. indicus*. *Aquaculture* 184: 31–44

Rajendran, K.V., Makesh, M., and Karunasagar, I. 2012. Monodon baculovirus of shrimp. *Indian Journal of Virology* 23: 149–160

Ravi, M., Nazeer Basha, A., Sarathi, M. et al. 2009. Studies on the occurrence of white tail disease (WTD) caused by MrNV and XSV in hatchery-reared post-larvae of *Penaeus indicus* and *P. monodon*. *Aquaculture* 292: 117–120

Robles-Sikisaka, R., Garcia, D.K., Klimpel, K.R., and Dhar, A.K. 2001. Nucleotide sequence of 3′-end of the genome of Taura syndrome virus of shrimp suggests that it is related to insect picornaviruses. *Archives of Virology* 146: 941–952.

Safeena, M.P. and Rai, P. 2012. Molecular biology and epidemiology of hepatopancreatic parvovirus of penaeid shrimp. *Indian Journal of Virology* 23: 191–202

Sakuna, K., Elliman, J., and Owens, L. 2017. Discovery of a novel Picornavirales, Chequa iflavirus, from stressed redclaw crayfish (*Cherax quadricarinatus*) from farms in northern Queensland, Australia. *Virus Research* 238: 148–155

Sakuna, K., Elliman, J., Tzamouzaki, A., and Owens, L. 2018. A novel virus (order Bunyavirales) from stressed redclaw crayfish (*Cherax quadricarinatus*) from farms in northern Australia. *Virus Research* 250: 7–12

Sánchez-Paz, A. 2010. White spot syndrome virus: An overview on an emergent concern. *Veterinary Research* 41: 6

Sano, T., Nishimura, T., Fukuda, H., Hayashida, T.I., and Momoyama, K. 1984. Baculoviral mid-gut gland necrosis (BMN) of kuruma shrimp (*Penaeus japonicus*) larvae in Japanese intensive culture systems. *Helgoländer Meeresuntersuchungen* 264: 255–264

Sano, T., Nishimura, T., Oguma, K., Momoyama, K., and Takeno, N. 1981. Baculovirus infection Penaeus of cultured Japonicus kuruma shrimp in Japan. *Fish Pathology* 15: 185–191

Sellars, M.J., Cowley, J.A., Musson, D. et al. 2019. Reduced growth performance of Black Tiger shrimp (*Penaeus monodon*) infected with infectious hypodermal and hematopoietic necrosis virus. *Aquaculture* 499: 160–166

Sellars, M.J., Keys, S.J., Cowley, J.A., McCulloch, R.J., and Preston, N.P. 2006. Association of Mourilyan virus with mortalities in farm pond-reared *Penaeus* (*Marsupenaeus*) *japonicus* transferred to maturation tank systems. *Aquaculture* 252: 242–247

Senapin, S., Jaengsanong, C., Phiwsaiya, K. et al. 2012. Infections of MrNV (Macrobrachium rosenbergii nodavirus) in cultivated whiteleg shrimp *Penaeus vannamei* in Asia. *Aquaculture* 338–341: 41–46.

Shen, H., Ma, Y., and Hu, Y. 2015. Near-full-length genome sequence of a novel reovirus from the Chinese mitten crab, *Eriocheir sinensis*. *Genome Announcements* 3: 8–9

Shi, M., Lin, X.D., Tian, J.H. et al. 2016. Redefining the invertebrate RNA virosphere. *Nature* 540: 539–543

Shields, J.D. and Behringer, D.C. 2004. A new pathogenic virus in the Caribbean spiny lobster *Panulirus argus* from the Florida Keys. *Diseases of Aquatic Organisms* 59: 109–118

Shike, H., Dhar, A.K., Burns, J.C., Shimizu, C., Jousset, F.X., Klimpel, K.R., and Bergoin, M. 2000. Infectious hypodermal and hematopoietic necrosis virus of shrimp is related to mosquito Brevidensoviruses. *Virology* 277: 167–177

Shinn, A.P., Pratoomyot, J., Griffiths, D., Trong, T.Q., Vu, N.T., Jiravanichpaisal, P., and Briggs, M. 2018. Asian shrimp production and the economic costs of disease. *Asian Fisheries Science* 31: 29–58

Spann, K.M., Cowley, J.A., Walker, P.J., and Lester, R.J.G. 1997. A yellow-head-like virus from *Penaeus monodon* cultured in Australia. *Diseases of Aquatic Organisms* 31: 169–179

Sparks, A.K. and Morado, J.F. 1986. A herpes-like virus disease in the blue king crab, *Paralithodes platypus*. *Diseases of Aquatic Organisms* 1: 115–122

Spitznagel, M.I., Small, H.J., Lively, J.A., Shields, J.D., and Schott, E.J. 2019. Investigating risk factors for mortality and reovirus infection in aquaculture production of soft-shell blue crabs (*Callinectes sapidus*). *Aquaculture* 502: 289–295

Sritunyalucksana, K., Apisawetakan, S., Boon-nat, A., Withyachumnarnkul, B., and Flegel, T.W. 2006. A new RNA virus found in black tiger shrimp *Penaeus monodon* from Thailand. *Virus Research* 118: 31–38

Stentiford, G.D., Bonami, J.R., and Alday-Sanz, V. 2009. A critical review of susceptibility of crustaceans to Taura syndrome, Yellowhead disease and White Spot Disease and implications of inclusion of these diseases in European legislation. *Aquaculture* 291: 1–17

Stentiford, G.D., Neil, D.M., Peeler, E.J. et al. 2012. Disease will limit future food supply from the global crustacean fishery and aquaculture sectors. *Journal of Invertebrate Pathology* 110: 141–157

Subramaniam, K., Behringer, D.C., Bojko, J. et al. 2020. A new family of DNA viruses causing disease in crustaceans from diverse aquatic biomes. *mBio* 11: 1–14

Sudhakaran, R., Syed Musthaq, S., Haribabu, P., Mukherjee, S.C., Gopal, C., and Sahul Hameed, A.S. 2006. Experimental transmission of *Macrobrachium rosenbergii* nodavirus (MrNV) and extra small virus (XSV) in three species of marine shrimp (*Penaeus indicus, Penaeus japonicus* and *Penaeus monodon*). *Aquaculture* 257: 136–141

Sullivan, T.J., Dhar, A.K., Cruz-Flores, R., and Bodnar, A.G. 2019. Rapid, CRISPR-Based, field-deployable detection of white spot syndrome virus in shrimp. *Scientific Reports* 9: 1–7

Summers, M., 1977. *Characterization of shrimp baculovirus*, 1st edn. Couch, J (ed), Ecological Research Series. U.S environmental protection agency, Gulf Breeze, Florida

Taengchaiyaphum, S., Srisala, J., and Sanguanrut, P. 2020. Full genome characterization of Laem Singh virus (LSNV) in shrimp *Penaeus monodon*. *bioRxiv* 9: 1–16

Tang, K.F.J., Navarro, S.A., Pantoja, C.R., Aranguren, F.L., and Lightner, D.V. 2012. New genotypes of white spot syndrome virus (WSSV) and Taura syndrome virus (TSV) from the Kingdom of Saudi Arabia. *Diseases of Aquatic Organisms* 99: 179–185

Tang, K.F.J., Pantoja, C.R., and Lightner, D.V. 2008. Nucleotide sequence of a Madagascar hepatopancreatic parvovirus (HPV) and comparison of genetic variation among geographic isolates. *Diseases of Aquatic Organisms* 80: 105–112

Tang, K.F.J., Pantoja, C.R., Redman, R.M., and Lightner, D.V. 2007. Development of *in situ* hybridization and RT-PCR assay for the detection of a nodavirus (PvNV) that causes muscle necrosis in *Penaeus vannamei*. *Diseases of Aquatic Organisms* 75: 183–190

Tang, K.F.J., Pantoja, C.R., Redman, R.M., Navarro, S.A., and Lightner, D.V. 2011. Ultrastructural and sequence characterization of *Penaeus vannamei* nodavirus (PvNV) from Belize. *Diseases of Aquatic Organisms* 94: 179–187

Trang, T.T., Hung, N.H., Ninh, N.H., Knibb, W., and Nguyen, N.H. 2019. Genetic variation in disease resistance against white spot syndrome virus (WSSV) in *liptopenaeus vannamei*. *Frontiers in Genetics* 10: 1–10

Tsai, J.-M., Wang, H.-C., Leu, J.-H. et al. 2006. Identification of the nucleocapsid, tegument, and envelope proteins of the shrimp white spot syndrome virus virion. *Journal of Virology* 80: 3021–3029

Tsing, A. and Bonami, J.R. 1987. A new viral disease of the tiger shrimp, *Penaeus japonicus* Bate. *Journal of Fish Diseases* 10: 139–141

Valles, S.M., Chen, Y., Firth, A.E. et al. 2017. ICTV virus taxonomy profile: Dicistroviridae. *Journal of General Virology* 98: 355–356

Van Etten, J.L., Lane, L.C. and Dunigan, D.D. 2010. DNA viruses: The really big ones (Giruses). *Annual Review of Microbiology* 64: 83–99

Van Hulten, M.C.W., Witteveldt, J., Peters, S. et al. 2001. The white spot syndrome virus DNA genome sequence. *Virology* 286: 7–22

Vijayan, K., Alavandi, S.V., Rajendran, K.V., and Alagarswami, K. 1995. Prevalence and histopathology of monodon baculovirus (MBV) infection in *Penaeus monodon* and *P. Indicus* in shrimp farms in the South-East coast of India. *Asian Fisheries Science* 8: 267–272

Walker, P.J. and Sittidilokratna, N. 2008. Yellow head virus. In *Encyclopedia of Virology*, 3rd edn, Mahy W.J. and Van Regenmortel M.H.V. (eds.) pp. 476–483. UK: Elsevier

Wang, A., Ran, C., Wang, Y. et al. 2019. Use of probiotics in aquaculture of China—a review of the past decade. *Fish & Shellfish Immunology* 86: 734–755

Wang, C., Liu, S., Li, X., Hao, J., Tang, K.F.J., and Zhang, Q. 2019. Infection of covert mortality nodavirus in japanese flounder reveals host jump of the emerging alphanodavirus. *Journal of General Virology* 100: 166–175

Wang, H.C., Hirono, I., Maningas, M.B.B., Somboonwiwat, K., and Stentiford, G. 2019. ICTV virus taxonomy profile: Nimaviridae. *Journal of General Virology* 100: 1053–1054

Wang, Q., Nunan, L.M., and Lightner, D.V. 2000. Identification of genomic variations among geographic isolates of white spot syndrome virus using restriction analysis and Southern blot hybridization. *Diseases of Aquatic Organisms* 43: 175–181

Weng, S.P., Guo, Z.X., Sun, J.J., Chan, S.M., and He, J.G. 2007. A reovirus disease in cultured mud crab, *Scylla serrata*, in southern China. *Journal of Fish Diseases* 30: 133–139

Wertheim, J.O., Tang, K.F.J., Navarro, S.A., and Lightner, D.V. 2009. A quick fuse and the emergence of Taura syndrome virus. *Virology* 390: 324–329

Wyban, J., 2019. Selective breeding of *Penaeus vannamei*: Impact on world aquaculture and lessons for future. *Journal of Coastal Residences* 86: 1–5.

Xu, L., Wang, T., Li, F., and Yang, F. 2016. Isolation and preliminary characterization of a new pathogenic iridovirus from redclaw crayfish *Cherax quadricarinatus*. *Diseases of Aquatic Organisms* 120: 17–26

Yang, Y., Lee, D., Wang, Y. et al. 2014. The genome and occlusion bodies of marine Penaeus monodon nudivirus (PmNV, also known as MBV and PemoNPV) suggest that it should be assigned to a new nudivirus genus that is distinct from the terrestrial nudiviruses. *BMC Genomics* 15: 628

Zhang, Q., Liu, Q., Liu, S. et al. 2014. A new nodavirus is associated with covert mortality disease of shrimp. *Journal of General Virology* 95: 2700–2709

Zhang, Q., Xu, T., Wan, X. et al. 2017. Prevalence and distribution of covert mortality nodavirus (CMNV) in cultured crustacean. *Virus Research* 233: 113–119

Zhang, S. and Bonami, J.R. 2007. A roni-like virus associated with mortalities of the freshwater crab, *Eriocheir sinensis* Milne Edwards, cultured in China, exhibiting 'sighs disease' and black gill syndrome. *Journal of Fish Diseases* 30: 181–186

Zhang, S. and Bonami, J.R. 2012. Isolation and partial characterization of a new reovirus in the chinese mitten crab, *Eriocheir sinensis* H. Milne Edwards. *Journal of Fish Diseases* 35: 733–739

Zhang, S., Shi, Z., Zhang, J., and Bonami, J.R. 2004. Purification and characterization of a new reovirus from the Chinese mitten crab, *Eriocheir sinensis. Journal of Fish Diseases* 27: 687–692

Zhao, M., Behringer, D.C., Bojko, J. et al. 2020. Climate and season are associated with prevalence and distribution of trans-hemispheric blue crab reovirus (Callinectes sapidus reovirus 1). *Marine Ecology Progress Series* 647: 123–133

CHAPTER 15

Bacterial diseases of crustaceans

Andrew F. Rowley

15.1 Introduction

Crustaceans are a large group of arthropods with over 67,000 known species that inhabit a wide range of habitats including freshwater, estuarine, and sea water (Brusca and Brusca 2003). Additionally, there are also a small number of species of isopods that inhabit the terrestrial environment. The large body size range and variable lifestyle of crustaceans, ranging from free-living to parasitic forms, including the ectoparasitic copepods and the endoparasitic sacculinids, exemplify their successful radiation. Therefore, it is not surprising that with this diversity of habitat, form and mode of existence, they are subject to a wide range of bacterial diseases.

It is becoming increasingly clear that the tissues of crustaceans and other invertebrates are not always free from bacteria. For example, it is not uncommon to find small numbers (> 10^5 cfu.ml) of culturable bacteria in the haemolymph of various crustaceans (Tubiash et al. 1975; Smith et al. 2014) yet their presence does not necessarily imply that these are potential disease-causing agents and that the host is diseased. Indeed, such bacteria in the haemolymph of invertebrates are probably part of a normal microbiome in which they may function to interfere with the growth of pathogens (Wang and Wang 2015; Sumithra et al. 2019)

Members of the Crustacea have economic importance to humankind and play key roles in our food security ranging from prey to marine fishes of commercial importance (e.g. krill and zooplankton) to food for human consumption (e.g. shrimp, crabs and lobsters). They may also be crucial components of the diets in some communities across the globe as well as a major source of income. There is increasing dependence of production of some crustaceans via aquaculture and this has resulted in the appearance of new microbial diseases accompanied by a dramatic increase in our knowledge of their nature and approaches to control these.

This chapter reviews the main bacterial diseases of crustaceans. While emphasis is placed on crustaceans of economic importance, it endeavours to provide a balanced overview of all crustaceans regardless of the perceived commercial status.

15.2 Principal diseases

The principal diseases of crustaceans caused by bacteria are summarised in Table 15.1.

15.2.1 Vibriosis

Vibrios are widespread in the marine environment and although many species are halophiles, some species are also present in estuarine and freshwater environments. They are members of the family Vibrionaceae (γ-proteobacteria). Several species of vibrios are pathogenic to both humans and animals including strains of *V. parahaemolyticus* that cause human gastroenteritis as a result of eating raw or contaminated shellfish and fish (Letchumanan et al. 2014).

Vibrios are frequently considered to be opportunistic pathogens of shrimp mainly because disease outbreaks often occur in degraded environmental conditions associated with build-up of waste

Andrew F. Rowley, *Bacterial diseases of crustaceans*. In: *Invertebrate Pathology*. Edited by Andrew F. Rowley, Christopher J. Coates and Miranda M.A. Whitten, Oxford University Press. © Oxford University Press (2022). DOI: 10.1093/oso/9780198853756.003.0015

Table 15.1 An overview of the principal bacterial diseases of crustaceans

Disease	Disease causing agent(s)	Host range	Geographical range	Pathology	Key references
Vibriosis	Vibrios including *V. alginolyticus*, *V. campbellii*, *V. harveyi*, and *V. parahaemolyticus* (See Table 15.2)	Widespread across many aquatic crustaceans	Global in aquatic environments depending on species	Bacterial multiplication in tissues, high levels of mortality especially in juvenile hosts. High levels of mortality especially in juvenile hosts. May be exacerbated by poor environmental conditions including high stocking density, nitrogenous waste products, low levels of oxygenation and elevated temperatures during shrimp and crab culture	Austin and Zhang 2006; Prachumwat et al. 2019; Tran et al. 2013
Rickettsiosis (including milky diseases)	*Rickettsia*, *Wolbachia* and rickettsia-like organisms (RLOs)	RLOs: Wide across aquatic crustaceans *Wolbachia*: Isopods and amphipods	RLOs: Global in aquatic environments *Wolbachia*: mainly terrestrial	RLOs: Bacteria multiply intracellularly in a range of cell types including haemocytes and fixed phagocytes (Figures 15.2d–f). Haemolymph milky in colour due to presence of large numbers of bacteria (Figure 15.2a) *Wolbachia*: Intracellular endosymbionts, showing vertical transmission. Influences host reproduction (e.g. feminisation of males)	Edgerton and Prior 1999; Eddy et al. 2007; Federici et al. 1974; Nunan et al. 2010; Cordaux et al. 2004, 2011, 2012
	Rickettsiella isopodorum, *R. armadillidii*, *R. grylli* (?) and *Aquirickettsiella gammari*	Isopods	Europe, USA	Bacteria develop intracellularly, haemolymph becomes milky in colour, mortality	Bojko et al. 2018; Kleespies et al. 2014; Wang and Chandler 2016
Tremor disease	*Spiroplasma* spp. e.g. *S. eriocheiris*	Chinese mitten crab (*Eriocheir sinensis*); crayfish (*Procambarus clarkii*); giant river prawn (*Macrobrachium rosenbergii*); shrimp (*Litopenaeus = Penaeus vannamei*)	China	Rickettsia-like disease. External: Weakness, tremor of pereopods resulting in mortality. Epizootic in nature mainly in summer months. Histopathology: Intracellular inclusions (multiplying bacteria?) in intratubular cells of the hepatopancreas. Bacteria found in muscle, thoracic ganglion and connective tissue	Ning et al. 2019; Srisala et al. 2018; Wang and Gu 2002; Wang et al. 2004; 2005, 2011
Shell disease syndromes (See Table 15.4)	Disease caused by changes in microbial populations (dysbiosis). Probably no single pathogen involved.	Mainly crabs and lobsters and a few other marine crustaceans	Global	External: Lesions on cuticle, appendages and gills (Figures 15.3 and 15.4). In severe cases, may cause secondary infections in internal tissues. Variable levels mortality. Most forms of the disease are of low prevalence and low severity. Several different forms of the disease in American lobsters including an emerging condition, epizootic shell disease in American lobsters (Figures 15.5a–c)	Castro et al. 2012; Gomez-Chiarri and Cobb 2012; Shields 2013; Vogan et al. 2008

continued

Table 15.1 Continued

Disease	Disease causing agent(s)	Host range	Geographical range	Pathology	Key references
Various including red body-type disease caused by *Proteus penneri*	Various Gram-negative bacteria including *Aeromonas* spp., *Proteus penneri*	Various species of shrimp		Variable levels of mortality, some may be localised infections often associated with degraded environmental conditions. Not all species are true primary pathogens with independent invasive ability	e.g. Cao et al. 2014
Gaffkaemia (red tail disease)	*Aerococcus viridans* var. *homari*	Lobsters (*Homarus americanus, H. gammarus*)	N. America and Europe	Histopathology: Bacteria multiply in hepatopancreas, muscle and haemocytes (Figures 15.6a,b) leading to sepsis. Host dies due to lack of haemostasis	Cawthorn 2011; Stewart et al. 1966
	Streptococcus penaeicida	Shrimp (*L. vannamei*)	Latin America, Guatemala	External: Affected shrimp show lethargic behaviour before coming moribund. Histopathology: Histology shows extensive bacteraemia and many free streptococci in lymphoid gland and haemolymph spaces in tissues. Bacteria elicit a host response with melanised and vacuolated haemocytes in gills and lymphoid gland.	Hasson et al. 2009; Morales-Covarrubias et al. 2018
	Bacillus licheniformis CG-B52	Pacific white shrimp (*L. vannamei*)	Colombian Caribbean coast	High levels of mortality, disease thought to be induced by handling stress	Gálvez et al. 2016
White muscle disease	*Lactobacillus* spp. (including *L. garvieae*)	Giant freshwater prawn, *M. rosenbergii*	Taiwan	Histopathology: Necrosis in muscle accompanied by haemocyte infiltration leading to nodule (granuloma) formation.	Cheng and Chen 1998; Chen et al. 2001; Wang et al. 2008
Mycobacteriosis	*Mycobacterium gordonae*	Redclaw crayfish, *Cherax quadricarinatus*	Commercial hatchery in Israel	Pathology: Low level of mortality. Animals are lethargic and lay on one side. Histopathology: Nodules containing bacteria in gills and hepatopancreas. Co-infection with viral disease	Davidovich et al. 2019
	Aquimarina hainanensis	Crab and shrimp larvae including *Scylla serrata*	Hatchery in Japan	Necrotic and melanised lesions in appendages, highly mortality	Midorikawa et al. 2020

products, reduced oxygen tension or other pollutants. Shrimp under these conditions are undoubtedly immunocompromised (e.g. Le Moullac and Haffner 2000; Wang and Chen 2006) and the poor environmental conditions cause increases in the number of vibrios in shrimp farms and hatcheries. It is also clear, however, that vibriosis is not always linked to environmental degradation implying that vibrios can act as primary pathogens with the ability to invade and colonise otherwise healthy hosts.

Of all diseases caused by vibrios, most is known about infection of larval and juvenile shrimp caused by *V. harveyi* (luminous and non-luminous vibriosis) and *V. parahaemolyticus* (acute hepatopancreatic necrosis disease (AHPND) and early mortality syndrome (EMS)) in aquaculture grow-out ponds (Table 15.2). Other species of vibrios shown in Table 15.2 are probably of less global importance in disease outbreaks.

AHPND was first reported in China in 2009 as an emergent condition. This disease is now widespread in several countries including Vietnam (2010), Thailand (2012), Malaysia (2011), Mexico (2013), the Philippines (2014) and the USA in 2017 (Dhar et al. 2019; Lee et al. 2015; Nunan et al. 2014) where it has caused dramatic loss of shrimp production (Flegel 2019). Tran et al. (2013) were the first authors to show that AHPND is caused by virulent strains of *V. parahaemolyticus* (VP$_{AHPND}$). These strains have an extrachromosomal plasmid (pVA1) that codes for a binary cytotoxin, PirA^{vp} and PirB^{vp} (Lee et al. 2015). The pathology of AHPND is highly characteristic with extensive, toxin-driven damage to the cells of the hepatopancreatic tubules with sloughing and necrosis (Figure 15.1a–c; Lee et al. 2015; Prachumwat et al. 2019). Pathology appears to be dependent on the presence of both toxins as one toxin alone has limited effect of the integrity of cells in the hepatopancreatic tubules (Prachumwat et al. 2019). The action of these toxins is similar to those produced by certain bacterial pathogens of insects (Tang and Lightner 2014). Recent reports have demonstrated that other species of vibrios, notably *V. campbellii*, *V. owensii* and *V. punensis* harbouring the pVA1 plasmid, can also cause outbreaks of AHPND (Dong et al. 2017a,b; Restrepo et al. 2018). It has been suggested that this has been acquired by plasmid transfer in shrimp farms

affected by *V. parahaemolyticus* (Dong et al. 2017c). However, a recent study found that archived isolates of V. *campbellii* collected from before the first outbreak of AHPND in China exhibited sequences 99% identical to that of *pir^{vp}A* and *pir^{vp}B* genes (Wangman et al. 2018)—perhaps ruling out this assertion. Strains of *V. parahaemolyticus* carrying a mutant pVA plasmid that are unable to code for Pir^{vp} toxins are moderately pathogenic and these may contribute to disease outbreaks attributed to early mortality syndrome (EMS) where there is no evidence of the presence of AHPND (Phiwsaiya et al. 2017).

Luminous vibriosis, so-called because affected shrimp glow in the dark, is mainly caused by *V. harveyi* (Austin and Zhang 2006). Outbreaks of disease occur in hatcheries and grow-out ponds in several species of shrimp. The condition can result in poor growth, high mortality and affected animals show systemic bacteraemia in late stages of infection. Infections caused by *V. harveyi* have also been recorded in crabs mainly during hatchery production (e.g. Zhang et al. 2014).

While there is an increasing knowledge of the mechanisms of pathogenicity in clinical isolates of vibrios (e.g. Klein et al. 2018; Zhang and Orth 2013) understanding of the equivalent mechanisms in environmental isolates, such as those causing vibriosis in crustaceans, is less well developed. For vibrios to be successful pathogens, they need to attach to their host and its tissues using adhesins, agglutinins, and chitin-binding proteins, and then release factors that facilitate the invasion and colonisation of tissues. They also need strategies to overcome the cellular and humoral host defences. Table 15.3 shows a variety of virulence factors known to be utilised by vibrios in general. There is extensive literature on potential virulence factors produced by a number of different species of vibrios. Many of these studies have taken a traditional approach where bacteria are grown in liquid culture and the cell free supernatant removed to collect these extracellular products (ECPs). Crude preparations of ECPs contain a wide range of active molecules including enzymes such as proteases, lipases, and chitinases, together with haemolysins and non-enzymatic cytotoxins (Table 15.3). A weakness in this approach is that it is difficult to assess which of these factors are

Table 15.2 Examples of diseases of crustaceans caused by vibrios

Species	Disease	Host range	Geographic range	Pathology	Key references
V. alginolyticus	Emulsification disease	Swimming crabs (*Portunus trituberculatus*)	China	Hepatopancreas becomes 'emulsified' and haemolymph milky in consistency due the presence of bacteria. Bacteria cause dysbiosis in gut microbiota leading to further pathology	Shi et al. 2019; Wang et al. 2006
V. alginolyticus	Zoea-2 syndrome	Shrimp *Litopenaeus* (*Penaeus*) *vannamei*	India	Impairment of moulting, high mortality, damage to hepatopancreas reported in shrimp hatcheries. Causative agent not fully proven	Kumar et al. 2017
V. campbellii	Luminous vibriosis and acute hepatopancreatic necrosis disease (AHPND)	Shrimp including *L. vannamei*	China	Strains with *PirA/B* genes show characteristic histopathology of AHPND	Dong et al. 2017a, 2017b; Gomez-Gil et al. 2004;
V. cholerae (non-O1 strains)		(1) Blue crabs (*Callinectes sapidus*) (2) Oriental river prawn, *Macrobrachium nipponense*	(1) Louisiana, USA (2) China	(1) Bacteria found in haemolymph of blue crabs by RT-PCR, no report of any pathology, true infection? (2) Commercial hatchery, high mortality	(1) Sullivan and Neigel 2018; (2) Li et al. 2019a
V. damsela (= *Photobacterium damselae* subsp. *damselae*)		Shrimp (*Penaeus monodon*)	Taiwan, India	Opportunistic pathogen causing poor growth, anorexia and necrotic musculature	Rivas et al. 2013; Song et al. 1993
V. fluvialis or *Photobacterium indicum*?	Limp lobster disease	American lobster (*Homarus americanus*)	(1) Gulf of Maine, USA (2) Temporary storage in pounds	(1) Lethargy, weakness, lack of sensory response, sepsis. Field collected. (2) Condition with similar pathology observed in holding facility but causative agent reported as *P. indicum*. Thought to be an opportunistic infection	(1) Tall et al. 2003; (2) Basti et al. 2010
V. harveyi	Luminous vibriosis	Wide range including crabs (*P. trituberculatus*), rock lobsters (*Jasus verreauxi*) and various species of shrimp		Bacterial multiplication in hepatopancreas leading to septicaemia	Diggles et al. 2000; Liu et al. 1996; Zhang et al. 2014
V. harveyi	Non-luminous vibriosis (e.g. bacterial white tail disease)	Shrimp (*L. vannamei*)	China	Muscle necrosis in tail region	Zhou et al. 2012

Species	Disease/Syndrome	Host	Location	Pathology	References
V. natriegens		Swimming crabs (*P. triberculatus*)	China	High mortality in juvenile crabs	Bi et al. 2016
V. nigripulchritudo	Summer syndrome	Shrimp including *Marsupenaeus japonicus* and *Litopenaeus stylirostris*	New Caledonia, Madagascar and Japan	High mortality in juvenile shrimp. Virulent strains produce nigritoxin	Goarant et al. 2006; Goudenège et al. 2013; Le Roux et al. 2010; Sakai et al. 2007; Walling et al. 2010
V. owensii	AHPND (in some strains)	Spiny lobster (*Panulirus ornatus*) and shrimp including *P. monodon* and *L. vannamei*	Australia, China	Originally observed in spiny lobsters and shrimp in Australia. Recently found that strains of *V. owensii* that have *PirAB* genes can cause AHPND with its characteristic histopathology	Cano-Gómez et al. 2009; Liu et al. 2018
V. parahaemolyticus	AHPND (in some strains) Early mortality syndrome (EMS; in some strains)	Shrimp including *L. vannamei*	SE Asia and Mexico	Strains associated with AHPND produce toxins (PirA and PirB) that damage hepatopancreatic tubule cells (Figure 15.1c) resulting in their necrosis and encapsulation. High levels of mortality. Strains that cannot produce the toxins associated with AHPND outbreaks are moderately pathogenic.	Aguirre-Guzmán et al. 2010; Kumar et al. 2021; Lee et al. 2015; Tran et al. 2013
V. penaecida	Syndrome 93	Shrimp (*Penaeus stylirostris*)	New Caledonia	Epizootic condition, darkened cuticle, muscle necrosis, reduced clotting ability of haemolymph	Costa et al. 1998
V. punensis	AHPND-like	Shrimp including. *L. vannamei*	South America	Pathology like that of other outbreaks of AHPND	Restrepo et al. 2018
V. vulnificus		(1) Blue crabs (*Callinectes sapidus*) (2) *Macrobrachium rosenbergii*	(1) Louisiana, USA (2) China	(1) Bacteria found in haemolymph by RT-PCR, no report of any pathology, true infection? (2) Infection of zoea in hatcheries. Zoea show poor growth leading to mortality	(1) Sullivan and Neigel 2018; (2) Li et al. 2019b

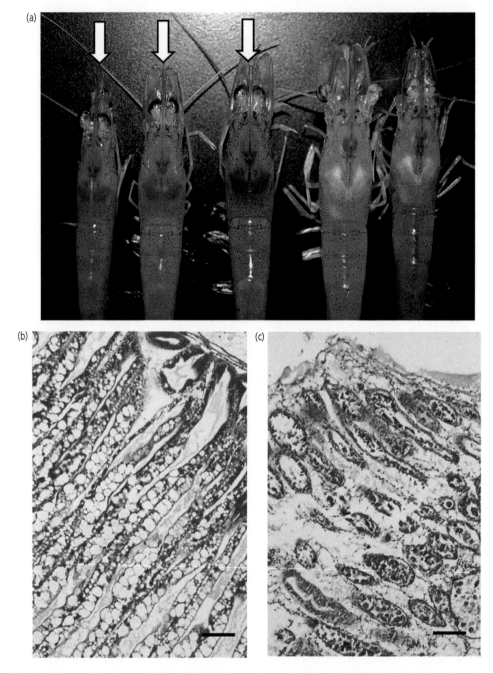

Figure 15.1 (a–c) Characteristic symptoms of shrimp infected with acute hepatopancreatic necrosis disease caused by *V. parahaemolyticus*. (a) Gross changes in infected shrimp (arrows) compared with the two uninfected animals on the right-hand side of the figure. (b) Section through hepatopancreas from an uninfected shrimp. Note normal tubule structure. (c) Section through hepatopancreas of shrimp affected by acute hepatopancreatic necrosis disease. Note extensive disruption of normal tubule structure with sloughing of epithelial cells that form the tubules. Scale bars = 100 μm. Source: micrographs courtesy of Professor Kallaya Sritunyalucksana, Research Unit National Center for Genetic Engineering and Biotechnology (BIOTEC) National Science and Technology Development Agency (NSTDA), Bangkok, Thailand.

key determinants of pathogenicity. It is clear that not all of the virulence factors listed in Table 15.3 are of central importance to different strains of vibrios that are crustacean pathogens, and their profile will differ between species/strains of vibrios and dissimilar hosts. For instance, haemolysins are important virulence factors in *V. parahaemolyticus* infections in humans (Wang et al. 2015), yet these may be of limited importance in the mechanisms of pathogenicity in those strains of *V. parahaemolyticus* responsible for outbreaks of AHPND (Li et al. 2017). Similarly, siderophores that are involved in the acquisition of iron required by microbes may be essential in strains of *V. harveyi* pathogenic to fish but not in invertebrates (Owens et al. 1996), presumably because of the lack of haemoglobin in many invertebrates. Finally, key factors found to be important in the pathogenesis of vibrios in humans have not been examined in crustacean diseases. One such factor is the multivalent adhesion molecule 7 (MAM7) that binds *V. parahaemolyticus* to host cells. This molecule has not been reported to be important in strains of *V. parahaemolyticus* pathogenic to crustaceans but Krachler et al. (2011) demonstrated that it was a key virulence factor in a nematode (*Caenorhabditis elegans*) disease model suggesting that its role stretches beyond just mammalian cells alone.

There are several reports of the presence of cytotoxins in ECPs released by strains of vibrios that are pathogenic towards crustaceans (Table 15.3; Harris and Owens 1999; Lee et al. 2015). These are distinct to other toxins such as haemolysins, yet with the notable example of the PirA/PirB binary toxin of *V. parahemolyticus*, *V. campbellii*, and *V. owensii*, our understanding of their structure and molecular activities is limited. Studies have found that both toxins are required to bring about the symptoms of AHPND (Kumar et al. 2019; Prachumwat et al. 2019) but there are probably other factors produced by these causative agents (Kumar et al. 2019) that are part of pathogenesis. A 32 kDa 'bacteriocin-like substance' produced by *V. harveyi* that is toxic to closely related vibrios, has been studied by Prasad et al. (2005). This may give the producers an advantage both within the host and in the surrounding environment but as this factor has not been characterised, its importance in infections is unknown.

Similarly, the two toxins, T1 and T2 of *V. harveyi* ECPs, appear to be distinct to the PirA/B binary toxin in terms of their mass (Harris and Owens 1999) but again their mode of action is unknown.

Secretion systems are important in the delivery of microbial virulence factors to host cells (see Defoirdt 2018; de Souza Santos et al. 2015; Wang et al. 2015 for reviews). Type III secretion systems carry out this process in several strains of vibrios known to be pathogenic to crustaceans (Ruwandeepika et al. 2012). *V. parahaemolyticus* has two separate Type III systems, T3SS1 and T3SS2 (de Souza Santos et al. 2015). T3SS1 occurs in two subtypes, T3SS1a and T3SS1b (Li et al. 2017). Strains of *V. parahaemolyticus* responsible for AHPND have either of these two subtypes. In more recent years a second secretion system has been elucidated, the Type VI, that is found in both clinical and environmental isolates (Wang et al. 2015). Li et al. (2017) found that AHPND-causing strains of *V. parahaemolyticus* display an antibacterial Type VI secretion system that was not found in non-AHPND strains. Similarly, Fu et al. (2018) characterised one strain of *V. parahaemolyticus* isolated from AHPND outbreaks in China and found both T6SS1 and T6SS2 virulence determinants. They also found some evidence that T6SS1 is involved in antibacterial activity. Overall, both Li et al. (2017) and Fu et al. (2018) considered that the presence of this system gives these isolates a competitive advantage over other pathogens that facilitates their ability to cause infections in shrimp.

Bacteriophages are commonly found associated with pathogenic strains of vibrios. They may act as vehicles for the transfer of genetic material and aid in the movement of some virulence factors. The association between bacteriophages and vibrios associated with crustacean diseases has been investigated especially in the 'Harveyi clade' including *V. harveyi*, *V. campbellii* and *V. owensii* (Busico-Salcedo and Owens 2013; Munro et al. 2003). Infecting avirulent and bacteriophage-free strains of *V. harveyi* with a bacteriophage from a virulent strain of the same species, induces virulence with the associated up-regulation of expression resulting in production of several proteins including a haemolysin (Munro et al. 2003). Similarly, virulence of members of the Harveyi clade towards nauplii of *Penaeus*

Table 15.3 Examples of known and potential virulence factors in vibrios pathogenic to crustaceans (Adapted from Austin and Zhang, 2006; Ruwandeepika et al. 2012 and Defoirdt 2014)

Virulence factor	Function	Importance in pathogenicity in crustaceans	References
Adhesins including chitin binding proteins	Attachment to cells with and without a chitinous boundary	Little information available but pathogens may have homologues of the chitin- binding protein of *V. cholerae* and *V. parahaemolyticus* that binds both chitin on zooplankton and epithelial cells from human GI tract	Gode-Potratz et al. 2011; Kim et al. 2005
Agglutinins	Adhesion of host immune cells and pathogens?	Extracellular products of *V. harveyi* 'lysogenised' with a bacteriophage contain an agglutinin that crosslinks the haemocytes of some species of shrimp	Intraprasong et al. 2009; Khemayan et al. 2014
Bacteriophages	Transformation of avirulent to virulent strains by transfer of ability to form virulence factors (lysogenic conversion)	Bacteriophages associated with various members of the Harveyi clade (e.g. *V. harveyi*, *V. campbellii* and *V. owensii*). Results in elevated levels of key virulence factors including agglutinins, haemolysins and chitinases	Ruangpan et al. 1999; Oakley et al. 2002; Munro et al. 2003; Intraprasong et al. 2009; Khemayan et al. 2014; Busico-Salcedo and Owens 2013
Capsules & slime	Adhesion of pathogens to surfaces (including biofilm generation). Protection against host defences (e.g. antiphagocytic).	Little information available but biofilms are important in aquaculture facilities in pathogen resistance to desiccation and resistance to antimicrobial chemicals	Ruwandeepika et al. 2012
Chitinases	Breakdown of insoluble chitin to soluble products. Important in chitin digestion in cuticle and in cuticle-lined gut allowing access of bacteria into tissues	Importance in shell disease in the digestion of the chitin-rich endocuticle. May play a role in bacterial invasion over cuticle-lined areas of gut	Vogan et al. 2008; Zha et al. 2018a; 2018b
Flagella (motility), pili (adhesion)	Colonisation of host	Potential role in crustacean diseases unknown	Zhang and Orth 2013
Haemolysins	Lysis of cells including vertebrate erythrocytes by damage to cell membrane	Virulence genes (*vhh*) found in isolates causing disease in various crustaceans. Strains unable to produce haemolysin are avirulent in *Artemia* model	Li et al. 2019a, 2019b; Ruwandeepika et al. 2011
Lipases including phospholipases	Hydrolysis of lipid and phospholipid	Little information available but lipases are produced by many species of vibrio	Ruwandeepika et al. 2012
Lipopolysaccharides	Toxicity, bacterial protection	Lethal toxin of *V. harveyi* strain pathogenic to shrimp	Montero and Austin 1999

Virulence factor	Function	Notes	References
Proteases (cysteine proteases, metalloproteases, serine proteases)	Digestion of proteins in tissues aiding invasive behaviour of pathogens	Several proteases have been characterised in disease causing strains of *V. harveyi*	Liu and Lee 1999
Quorum sensing systems	Bacteria-bacteria communication	Regulation of expression of virulence factors including Type III secretion system, toxins and chitinases. Loss of luminescence in *V. harveyi* after inhibition of quorum sensing	Defoirdt et al. 2008, 2010; Liu et al 2018; Manefield et al. 2010; Ruwandeepika et al. 2015
Secretion systems (including Types III and VI)	Delivery of virulence factors into host cells (Type III). Interference with bacterial growth allowing competitive advantage (Type VI)	Both Type III and VI systems thought to be of importance in strains of *V. parahaemolyticus* associated with AHPND/EMS	Li et al. 2017; Fu et al. 2018; Li et al. 2019a
Siderophores (e.g. vibriobactin, vulnibactin)	Iron binding and transportation into bacteria	(1) Important in pathogenic strains of *V. harveyi* in fish and other vertebrates, but not in invertebrates? (2) Potential role in pathogenesis of shell disease in spiny lobster?	(1) Owens et al. 1996; (2) Zha et al. 2018a
Toxins	Cytotoxic activity distinct to that of other factors (e.g. haemolysins, proteolytic enzymes)	(1) PirA/PirB binary toxin is a major virulence factor in strains of vibrios associated with AHPND. (2) Toxins 'T1' and 'T2' produced by some *V. harveyi* strains. Both cause mortality upon injection into shrimp. Mechanism of action unknown but appear to unlike PirA/B in mass. (3) Nigritoxin, an 83 kDa protein, produced by virulent strains of *V. nigripulchritudo*	(1) Lee et al. 2015; (2) Harris and Owens 1999; (3) Goudenège et al. 2013

monodon is linked to the presence of a bacterio-phage and virulence factors including chitinases and haemolysins (Busico-Salcedo and Owens 2013).

Quorum sensing systems have been shown to play important roles in several processes including biofilm formation, colonisation and virulence (Miller and Bassler 2001; Rutherford and Bassler 2012). They are mediated by a variety of small signal molecules. The quorum sensing system of *V. harveyi* has been extensively studied (e.g. Mukherjere and Bassler 2019; Papenfort and Bassler 2016) where it consists of three separate autoinducers (Harveyi autoinducer-1, autoinducer-2 and Cholerae autoinducer-1) that are products of synthases. Quorum sensing systems regulate the expression of key virulence genes as reviewed by Ruwandeepika et al. (2012). The expression of virulence genes in some bacteria is controlled by quorum sensing which involves N-acylhomoserine lactone (AHLs) signalling molecules produced by these bacteria. Torres et al. (2018) explored the role of AHLs as communication molecules in several species of vibrio including the pathogen, *V. owensii* using brine shrimp (*Artemia salina*) as a model assay system. They demonstrated that virulence of *V. owensii* towards brine shrimp was linked to the production of N-acylhomoserine-like molecules. Similarly, Ruwandeepika et al. (2011) were able to show that the quorum sensing master regulator of Gram-negative bacteria, luxR, plays an important role in virulence in vibrios in the Harveyi clade (e.g. *V. harveyi* and *V. campbellii*) towards brine shrimp. Strains of vibrios with varying virulence from low to high, positively correlated with *luxR* expression. In terms of virulence factor production, avirulent strains showed low expression of the haemolysin gene, *vhh*, while highly virulent strains exhibited 50-fold higher expression of *vhh*. These observations imply that haemolysins are important determinants of virulence in members of the Harveyi clade at least in their pathogenicity towards *Artemia*.

15.2.2 Rickettsia, *Wolbachia* and rickettsia-like organisms (RLOs)

Rickettsia are obligate intracellular parasites that are members of the O. Rickettsiales (α-proteobacteria). There are several reports of diseases of crustaceans caused by Rickettsia-like organisms (RLOs). These include diseases in the amphipod *Crangonyx floridanus* (Federici et al. 1974), the redclaw crayfish, *Cherax quadricarinatus* (Edgerton and Prior 1999; Romero et al. 2000), and 'milky diseases' in the European green crab, *Carcinus maenas* (Figures 15.2a–f; Eddy et al. 2007; Nunan et al. 2010) the edible crab, *Cancer pagurus* (Thrupp et al. 2016), the black tiger shrimp, *Penaeus monodon* (Nunan et al. 2010), and the spiny lobster, *Panulirus* sp. (Nunan et al. 2010). The identification of the causative agent(s) of such conditions as rickettsia-like in older literature is often tentative and based on their intracellular location and a failure to culture them in a range of defined culture media. In terms of pathology, some of the diseases are strikingly similar with intracellular bacteria seen within fixed phagocytes in the interstitial spaces of the hepatopancreas (Figures 15.2c–f), and in phagocytic haemocytes in circulation and in solid tissues (e.g. Eddy et al. 2007, Nunan et al. 2010). In other RLO infections, intracellular bacteria have been observed in cells that form the tubules of the hepatopancreas (e.g. Edgerton and Prior 1999). In studies where authors have attempted to identify the causative agent, some have been found to be unrelated to members of the Rickettsiales (e.g. Eddy et al. 2007; Nunan et al. 2010; Thrupp et al. 2016) but in one case the putative disease-causing agent has been identified as a novel bacterium, *Candidatus* Hepatobacter penaei placed within the Rickettsiales (Nunan et al. 2013). This latter bacterium is the causative agent of a key disease of shrimp culture called necrotising hepatopancreatitis that affects shrimp culture in the Americas. Further genome sequencing using a multilocus approach of this agent has confirmed the association of this agent within the Rickettsia but places it in the order Holosporales (Leyva et al. 2018).

Members of the genus *Wolbachia* are intracellular endosymbionts of arthropods including isopods and amphipods (Bouchon et al. 1998; Cordaux et al. 2012). These bacteria belong to the order Rickettsiales (α-proteobacteria) and are predominantly vertically transmitted via infected eggs. They have been referred to as 'reproductive parasites' as they alter the reproduction of their arthropod hosts by a variety of mechanisms (Cordaux et al. 2004, 2011, 2012). Levels of infection in populations

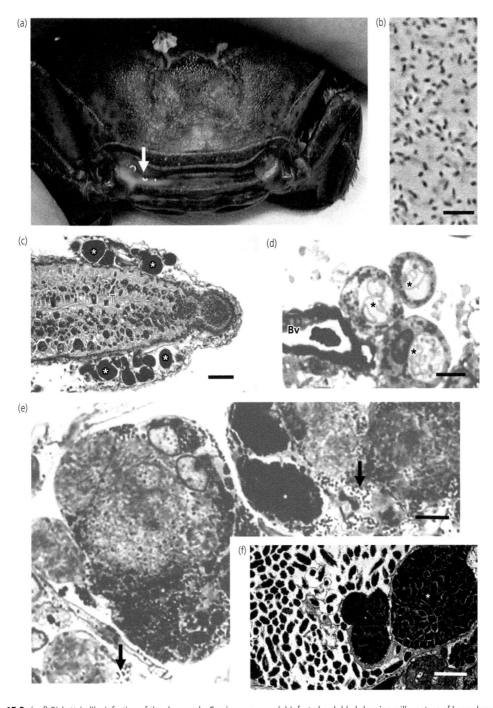

Figure 15.2 (a–f) Rickettsia-like infection of the shore crab, *Carcinus maenas*. (a) Infected crab bled showing milky nature of haemolymph (unlabelled arrow). (b) Phase contrast micrograph of infected haemolymph showing extensive bacterial septicaemia. Scale bar = 5 μm. (c) Histological section of hepatopancreas from infected crab showing swollen nature of fixed phagocytes (*) in the intertubular space. Tubule (T). Scale bar = 50 μm. (d) Plastic section of fixed phagocytes (*) in the hepatopancreas from an uninfected animal. Blood vessel (Bv). Scale bar = 10 μm. (e) Thick plastic section of swollen fixed phagocytes (*) from infected crab containing large numbers of bacteria. Note bacteria free in haemolymph (unlabelled arrow). Scale bar = 10 μm. (f) Transmission electron micrograph showing bacteria in (*) and around fixed phagocytes. Scale bar = 2 μm.

can be extremely high; for example, as many as 46% of woodlice have been observed to be infected with *Wolbachia* (Werren and O'Neil 1997). A recent report that *Wolbachia* in its isopod host, *Armadillidium vulgare,* can influence host microbiota (Dittmer and Bouchon 2018) suggests that it may have additional impact other than just reproductive fitness.

15.2.3 *Rickettsiella* and *Aquirickettsiella*

The genus *Rickettsiella* (Family Coxiellaceae, γ-proteobacteria) includes intracellular pathogens of arthropods mainly found in the terrestrial environment. These bacteria are not to be confused with the members of the genus *Rickettsia* that belong to the order Rickettsiales (α-proteobacteria) despite some similarities in their pathology. The taxonomy of some *Rickettsiella* living in association with crustaceans is currently unclear but there are probably one or two species that infect crustaceans, namely *R. isopodorum* (= *R. armadillidii*) and *R. grylli* than infect isopods including woodlice *Porcellio scaber* and *A. vulgare* (Cordaux et al. 2007; Kleespies et al. 2014; Vago et al. 1970; Wang and Chandler 2016). *R. armadillidii* causes the haemolymph of susceptible crustaceans to become milky in colour and the host finally dies (Vago et al. 1970). Some of the RLOs that remain unidentified, such as that described by Federici et al. (1974), could be members of the genus *Rickettsiella* or within the wider group, rather than true Rickettsia.

A recent addition to this group is a novel bacterium that the authors have designated as '*Candidatus* Aquirickettsiella gammari' found infecting freshwater amphipods, *Gammarus fossarum*, collected in Poland (Bojko et al. 2018). These bacteria develop in the hepatopancreatic tubules as their main site of development with secondary foci in the musculature, nervous system, gonad and haemocytes. The annotated genome sequence of A. gammari has given some insight into possible mechanisms of pathogenicity employed by this organism (Bojko et al. 2018).

15.2.4 Shell disease syndromes

Shell disease is a cuticular disease of crustaceans including crabs, lobsters, shrimp, and crayfish with a global distribution (Table 15.4). One of the original descriptions of this condition was by Hess (1937) in the American lobster, *Homarus americanus* collected in Nova Scotia. Shell disease is also referred to as black spot, rust spot shell disease, bacterial shell disease, box burnt disease, or brown spot disease (Andersen et al. 2000; Gomez-Chiarri and Cobb 2012; Vogan et al. 2008). Affected animals have cuticular lesions either on the carapace, appendages or occasionally on the cuticle-lined gills (Figures 15.3a–d). These lesions form progressively and become melanised as a result of activation of the cuticle-based phenoloxidase system in response to the damage (Figures 15.3a,b,d). The causative agents of the disease are unclear but although initial reports suggested it may be a fungal and bacterial condition (Sindermann 1989), it is probably bacterial in nature in most cases with the exception of burn spot disease in freshwater crayfish (*Astacus astacus*) caused by the fungus, *Fusarium avenaceum* (Makkonen et al. 2013; see Chapter 16).

Shell disease lesions contain a mixture of chitinolytic and non-chitinolytic bacteria and 'traditional' phenotypic taxonomic identification using a combination of morphology and reactivity in biochemical tests, has tentatively identified these bacteria as belonging to *Vibrio, Aeromonas, Pseudomonas, Pseudoalteromonas, Alteromonas, Flavobacterium, Spirillum, Moraxella, Pasteurella,* and *Photobacterium* genera (Getchell 1989; Vogan et al. 2002; 2008; Zha et al. 2018b). Approaches using culture independent methods, likely to give a more accurate overview of the bacterial community in the lesions, are few and mainly related to a form of shell disease termed epizootic shell disease seen in the American lobster, *H. americanus* (e.g. Chistoserdov et al. 2012; Feinmark et al. 2017; Quinn et al. 2017).

The prevalence and severity of shell disease is highly variable depending on site and species examined (Table 15.4). For example, in shore crabs, *Carcinus maenas* and European lobsters *Homarus gammarus* the prevalence of infection is < 5% (Davies et al. 2015; Wootton et al. 2012) and severity is also low with the lesions superficial and pinpoint in nature (Figure 15.3d). At the other end of the disease spectrum are edible crabs, *Cancer pagurus*, where the prevalence of the condition in juvenile onshore crabs in Gower, UK, ranges from 40 to 63%

Table 15.4 Examples of enzootic/endemic shell disease syndrome—pathology, prevalence, severity and geographical distribution

Genus & species (common name)	Location	Features	References
Argulus (ectoparasite of fish)	Taken from infected fish held under aquarium conditions	Small lesions found to be associated with bacteria	Rushton-Mellor and Whitfield 1993
Callinectes sapidus (blue crab)	Various sites in USA including Albemarle-Pamlico estuary, USA	Variable severity but crabs reported with severe lesions. Termed Pamlico River shell disease	Rosen 1967; Noga et al. 2000
Cancer borealus (Jonah crab)	Rhode Island Sound, USA	Variable disease prevalence (> 80%) in males linked to moult cycle	Truesdale et al. 2019
Cancer irroratus (Atlantic rock crab)	Eastern seaboard, USA including New Jersey, Maine, Massachusetts, Maryland, Delaware and Virginia	Prevalence ranging from 26 to 64%. Higher levels of disease in areas of environmental degradation?	Sawyer 1991
Cancer pagurus (edible or brown crab)	Gower Peninsula and Isle of Man, UK	Prevalence ranged from 15–25% (Isle of Man) and > 63% (Gower). Extensive lesions penetrating through (cuticle (Figure 15.3a))	Vogan et al. 1999; King et al. 2014
Carcinus maenas (European green crab)	South Wales, UK	Pin-point lesions mainly on walking legs and ventral surface. Low prevalence (< 5%) and low severity	Unpublished observations of author
Crangon crangon (brown shrimp)	Poole Harbour, UK	Variable size lesions. > 80% shrimp affected	Dyrynda 1998
Jasus edwardsii (spiny lobster)	New Zealand and Australia	Specialist form of shell disease called tail fan necrosis where the lesions are localised in this region. Found in lobsters in the wild and at higher levels after holding animals at sea	Zha et al. 2018b, 2018c
Homarus americanus (American lobster)	New England, USA	Winter impoundment shell disease. No prevalence data given. Pathology showed a range of severity including host response in underlying tissues	Smolowitz et al. 1992
H. americanus	New York Bight, USA	Lesions found in lobsters collected from vicinity of sewage sludge at higher levels than found in clean sites (no data given of prevalence). Lobsters held for > 6 weeks in aquaria with sewage sludge developed lesions but no evidence of any controls in experiment.	Young and Pearce 1975
Homarus gammarus (European lobster)	Lundy, UK	Small, superficial lesions mainly on claws (Figure 15.3d). Low prevalence and low severity. Disease likely to result from crowding stress and fighting behaviour	Wootton et al. 2012; Davies et al. 2015
Panulirus argus (spiny lobster)	Florida Keys and the Dry Tortugas, USA	Wild caught lobsters with lesions on uropods and telsons	Porter et al. 2001
Palinurus elephas (European spiny lobster)	Sicily, Italy	Uropod ulceration, damage on ventral surface. Disease caused by abrasion in aquarium facility	Mancuso et al. 2010
Penaeus indicus (Indian white shrimp)	India	Brown-spot disease	Sahul Hameed 1994
Scylla serrata (mud crab)	Queensland, Australia	Prevalence reported to be 22%. Many of these had 'rust spots' unlike the classical eroded forms of shell disease. Cause of this condition unknown but unlikely to be bacterial in nature	Andersen et al. 2000

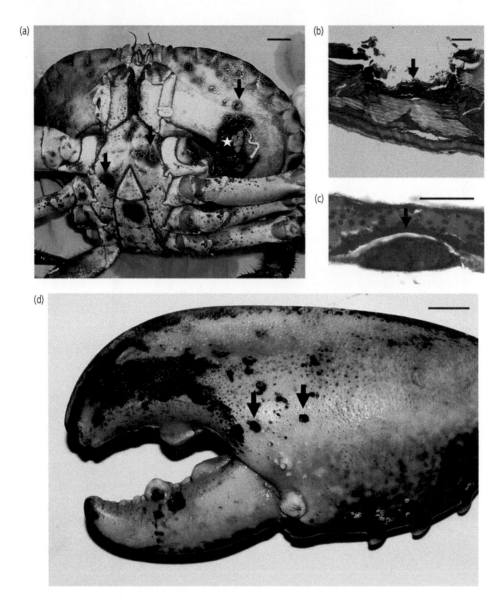

Figure 15.3 (a–d) (a) Shell disease lesions (unlabelled arrows) on the ventral surface of an edible crab, *Cancer pagurus*. The area marked with an asterisk is a hole in the cuticle probably resulting from fighting injury. The area around it exhibits shell disease. Scale bar = 1 cm. (b) Histological section through the cuticle of *C. pagurus* showing erosion of the cuticle in an area of shell disease. Note melanisation of cuticle (unlabelled arrow). Scale bar = 100 μm. (c) Lesion in the gill of *C. pagurus* with melanised scab and reformed, swollen epidermis (unlabelled arrow). Scale bar = 50 μm. (d) Claw of a European lobster, *H. gammarus*, with pinpoint lesions (arrows) probably resulting from conflict or feeding abrasion. Scale bar = 1 cm.

(Vogan et al. 1999; Powell and Rowley, 2005) with large deep lesions particularly on the ventral carapace (Figure 15.3a). It is very unlikely that crustaceans with low severity infections are at major risk of disease. However, in those animals with deep lesions, where bacteria may gain entry to underlying tissues, there is the possibility of death resulting from a variety of secondary infections (Vogan et al. 2001). The most likely outcome of such conditions in many crustaceans is a failure to successfully survive moulting as a result of fusion between the original, damaged cuticle and the newly formed underlying fresh cuticle.

It is a widely held view that shell disease may be linked to degraded environmental conditions either in the wild or in holding tanks and impounds (Figure 15.4; Vogan et al. 2008). For instance, Young and Pearce (1975) sampled lobsters (*H. americanus*) and crabs (*Cancer irroratus*) from an area affected by sewage sludge and contaminated dredging spoils in the New York Bight and they reported increased disease prevalence from affected areas but gave no quantitative data or method or analysis to support their viewpoint. Prolonged storage of crustaceans in either aquaria or as found during impoundment, may also result in heightened levels of disease but this is most likely to be a consequence of both increased fighting behaviour resulting in damaged limbs together with environmental degradation. Finally, Powell and Rowley (2005) examined the prevalence and severity of shell disease in pre-recruit *C. pagurus* in one site in Gower Peninsula, UK before and after the closure of a nearby sewage outfall, and found no relation between pre- and post-closure, suggesting that pollution from raw sewage was not an environmental driver of this

condition. Instead, observations of crabs on-site in this location revealed a back-burrowing behaviour into rocky crevices and sandy substrates that probably results in abrasion of the cuticle hence initiating shell disease (Vogan and Rowley 2002). Overall, although environmental factors are probably of importance in the development of classical shell disease, behaviours such as fighting and abrasion also play important roles in the pathogenesis of this condition (Vogan and Rowley 2002).

Although few of the surveys of shell disease reviewed in Table 15.4 have involved any seasonal studies observing changes in prevalence and severity, it is probable that the condition shows few changes in these criteria that may link this to climate change. However, studies have shown a new form of shell disease, termed epizootic shell disease (ESD), that affects lobsters (*H. americanus*) of the eastern coast off the USA is an emerging disease (Castro et al. 2006). This condition was first observed in 1996 near Rhode Island, USA but it has spread from this location along the coast since that time (Castro and Angell 2000; Castro et al. 2012). The pathology of ESD is distinct to other forms of shell disease found in *H. americanus* implying a different causative nature (Smolowitz et al. 2005). The disease

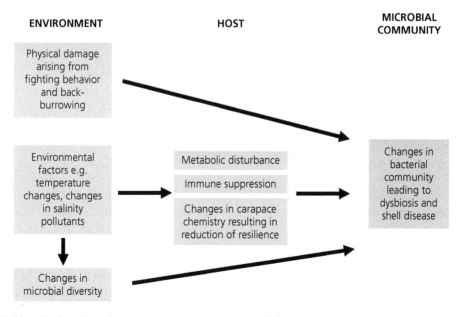

Figure 15.4 Schematic of the relationship between environment factors and shell disease.

(a)

(b)

(c)

Figure 15.5 (a–c) Epizootic shell disease in the American lobster, *Homarus americanus*. (a) Uninfected lobster. (b) Dorsal surface of severely infected lobster with epizootic shell disease showing extensive pitting of the carapace. (c) Claw of infected lobster in Figure 15.5b. Scale bars = 2 cm. Source: micrographs courtesy of Professor J.D. Shields (Virginia Institute of Marine Sciences, USA).

can cause rapid and extensive erosion of the dorsal area of lobster carapaces (Figures 15.5a–c) and population assessment shows that it results in mortality (Hoenig et al. 2017) hence affecting the viability of lobster fisheries in this area.

The bacteria associated with ESD are not fully characterised. Chistoserdov et al. (2005) utilised a culture dependent evaluation of bacterial communities in diseased and non-diseased areas of the carapace of *H. americanus* and found two groups of bacteria commonly isolated in lesions; one within the Flavobacteriaceae and another similar

to *Pseudoalteromonas gracilis*. Later studies taking a non-culture dependent approach, demonstrated the presence of *Aquimarina homaria* (now renamed *A. macrocephali* subsp. *homaria*; Quinn et al. 2017) a novel species in the Flavobacteriaceae and *Thalassobius* sp. in lesions from lobsters with ESD (Chistoserdov et al. 2012). Attempts to infect healthy lobsters with these bacteria suggest that *A. macrocephali* subsp. *homaria* may act as a primary pathogen but only if lobsters are first abraded (Quinn et al. 2012). The current view is that ESD, probably like other forms of shell disease, is an example of a dysbiotic condition of the microbial community resulting from changes in the environmental surroundings of lobsters in this region (Figure 15.4). Several environmental factors have been proposed as being the trigger for ESD, namely alkyphenols (Laufer et al. 2012, 2013) and elevated temperature (e.g. Glenn and Pugh 2006). The currently held viewpoint is that temperature is a principal driver of ESD (Groner et al. 2018; Shields 2019; Tanaka et al. 2017). Overall, ESD is a serious condition affecting the sustainability of lobster fishery in this part of the USA (Dellinger 2012; Hoenig et al. 2017). Accidental release of imported American lobsters with ESD into European waters could pose a risk to native lobsters (*H. gammarus*) but there is currently no evidence of such a possibility arising from aquarium-based studies (Davies et al. 2014; Whitten et al. 2014).

15.2.5 Gaffkaemia (red tail disease)

Gaffkaemia is an important disease of lobsters, especially those held under impoundment conditions. The disease was first reported by Sniezko and Taylor (1947). Since its first discovery in the USA, there have been periodic reports of its presence in native European lobsters (*H. gammarus*) in Europe mainly in holding facilities (Stebbing et al. 2012). It is widely believed that the source of this disease agent in Europe can be traced back to its original presence in the USA (Stebbing et al. 2012). This most likely occurred by the import of live (diseased) American lobsters for commercial purposes and their accidental release. Very few outbreaks of this disease have been found in the wild but such lobsters may harbour the bacteria and show clinical symptoms of disease when held under poor environmental conditions such as those in some holding facilities including pounds.

The causative agent of gaffkaemia is a Gram-positive bacterium, *Aerococcus viridans* var. *homari* that has a characteristic tetrad appearance in liquid culture and in the host's tissues (Figures 15.6a,b). The disease has few characteristic symptoms, but late-stage infections cause a lack of haemolymph coagulation resulting from a dramatic decline in haemocyte numbers (Stewart et al. 1966). Thus, infected lobsters when damaged will bleed to death and release the bacteria into surrounding waters.

Both virulent and avirulent strains of *A. viridans* var. *homari* have been identified and long-term culture of virulent strains results in loss of virulence that can be regained by culture in cell-free lobster haemolymph (Stewart et al. 2004). Clark and Greenwood (2011) reported that virulent strains of *A. viridans* var. *homari* produce several proteins that appear to be linked to virulence. One of these is a homologue of chaperonin 60 that in other bacteria is a virulence factor (Greenwood et al. 2005; Clark and Greenwood 2011).

15.2.6 Other diseases caused by Gram-positive bacteria

Although most bacterial diseases of crustaceans are caused by Gram-negative bacteria, there are some Gram-positive pathogens belonging to the Family Bacillaceae (*Bacillus licheniformis*), the Family Streptococcaceae (*Lactobacillus* spp. and *Streptococcus penaei*), and the Family Spiroplasmataceae (*Spiroplasma* spp. including *S. eriocheiris*) (Table 15.1). With the exception of spiroplasmas that cause tremor disease in a range of crustaceans, the importance of these bacteria in disease outbreaks, particularly during shrimp production, is probably localised. Of some interest, however, is the findings of Hasson et al. (2009) that *S. penaei* remains viable after several series of freezing and thawing of infected shrimp. This is an important observation pertinent to disease transmission caused by the global trade of fresh and frozen product.

Bacteria belonging to the genus *Mycobacterium* have been reported to be opportunistic pathogens of crustaceans including redclaw crayfish, *Cherax quadricarinatus*, under hatchery conditions (Figure 15.6c; Davidovich et al. 2019). This condition was found to only occur as a co-infection with the viral pathogen, *C. quadricarinatus* bacilliform virus and result in limited mortality so its importance as a crustacean pathogen is probably limited.

15.3 Control and treatment of bacterial diseases

This section considers various approaches to controlling the spread of disease and how to deter

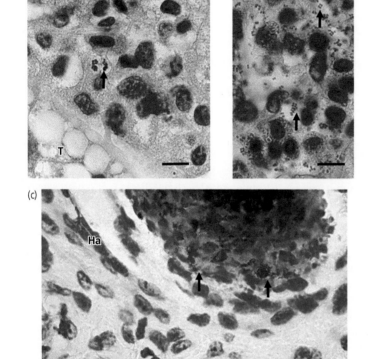

Figure 15.6 (a–c) Histopathological appearance of *Aerococcus viridans* var. *homari* in the intertubular space of the hepatopancreas of the American lobster, *Homarus americanus*.
(a) Apparent intracellular bacteria in phagocytes (unlabelled arrow) in the intratubular space.
(b) Large numbers of extracellular bacteria in the haemocoel spaces of the hepatopancreas. Note characteristic tetrad forms (unlabelled arrows) and eosinophilic granular haemocytes in the haemocoelic space. Tubule cell (T). Scale bars = 10 μm. Source: micrographs courtesy of Dr K. Fraser Clark. (c) *Mycobacterium gordonae* (unlabelled arrows) infecting the gill of redclaw crayfish, *Cherax quadricarinatus*.
Haemocytes (Ha). Scale bar = 10 μm. Source: micrograph courtesy of Drs Tobia Pretto and Nadav Davidovich.

these should they appear. It focuses on bacterial diseases of shrimp simply because this is the principal crustacean aquaculture product.

15.3.1 Antibacterial chemicals

Animal production is highly reliant on antibiotic use but there is increasing evidence that the use of such products can have profound environmental effects and result in potential for the spread of antibiotic resistance, ultimately putting at risk disease control in human and veterinary medicine (Heuer et al. 2009; Lulijwa et al. 2020; Santos and Ramos 2018). In a recent review of antibiotic use in agriculture and aquaculture by Done et al. (2015), it was noted that of the 51 antibiotics commonly used in aquaculture and agriculture, 76% of these are also important in treating human diseases. There is also increasing evidence that vibrios causing disease in crustaceans are increasingly resistant to antibiotics (e.g. Dong et al. 2017c; Fu et al. 2018; Han et al. 2015). For example, some of the strains of vibrios causing AHPND have multiple antibiotic resistance genes (*tet 35* and *tet 34*) and antibiotic sensitivity testing reveals their resistance to a wide range of antibiotics including tetracycline, streptomycin and β lactam antibiotics including ampicillin (Dong et al. 2017c; Fu et al. 2018). Similarly, other strains of *V. parahaemolyticus* involved in AHPND are resistance to tetracyclines and ampicillin related to the plasmid mediated *tetB* gene (Han et al. 2015). Treatment of shrimp with antibiotics can alter the microbiota of the gut (Zeng et al. 2019) potentially resulting in other health problems. Overall, overreliance on treating bacterial diseases with antibiotics is not a sustainable practice regardless of what region of the world is considered (see review by Cabello et al. 2013 for further details).

15.3.2 Bacteriophage therapy

Bacteriophage (phage) therapy harnesses the lytic ability of bacteriophages for various bacterial pathogens of humans and animals. Although the first report of phage therapy dates back to just over a century ago (see Doss et al. 2017 for a review of phage therapy in general) it has not been widely used in human and veterinary medicine but the increasing threat of antibiotic resistance in bacterial pathogens may change this.

In terms of treating bacterial diseases of commercially important crustaceans, such as shrimp and crabs, there is potential for these to replace antibiotics if they can be proven to be both effective and safe. In theory, resistance to phages in target bacteria should be far less of a problem than that seen towards antibiotics and unwanted environmental effects should be minimal. Because phages have a narrower range of target species than many antibiotics, this could be a useful attribute in terms of unwanted effects on non-target bacteria. There are several reports of trials in shrimp hatcheries using various phages to control vibrio growth (see Rao and Lalitha 2015 and Kalatzis et al. 2018 for reviews). Most of these show improvements in shrimp health following phage application. For example, Raghu Patil et al. (2014) isolated several lytic bacteriophages from clams and oysters and one of these was pathogenic against a wide range of strains of *V. harveyi*. They attempted to determine if application of this phage could protect post-larval shrimp in a realistic hatchery-like environment, however, survival was only increased from 40 to 60%.

A more promising approach may be to use 'cocktails' of phages that are more efficient at lysing pathogens. Mateus et al. (2014) found synergism between three types of phages in killing *V. parahaemolyticus* but this approach was in a culture system that may not replicate the same conditions found in hatcheries. Similar results have been found using multiple phages in improving hatching success and survival rates in brine shrimp (a food source commonly used in aquaculture). Hatching rates improved from *ca.* 50 to 100% with survival improving from 40–50% to 85–89% in the presence of phages (Quiroz-Guzmán et al. 2018). Finally, there are several reports of potential phage therapy against vibrios associated with AHP-ND/EMS that show potential promise (e.g. Angulo et al. 2019; Jun et al. 2016, Zermeño-Cervantes et al. 2018). Although there is much potential from a range of phage therapy trials, important hurdles still exist before their general introduction an aquaculture environment including determining optimal modes of delivery, ensuring purity of phage preparations and gaining regulatory approval for their use (Kalatzis et al. 2018).

15.3.3 Interference of virulence factor expression using quorum sensing inhibition

Quorum sensing interference may represent an alternate approach to our reliance of antibiotics (Defoirdt 2014; Manefield et al. 2010). For instance, Manefield et al. (2010) studied an algal-derived halogenated furanone and its ability to affect the growth and virulence of *V. harveyi*. They found that bacterial luminescence and toxin T1 levels were reduced in the presence of the inhibitor. Development of safe and effective agents, however, is still in its infancy especially in relation to bacterial diseases of aquaculture-produced crustaceans. A recent review by Liu et al. (2018) critically assessed whether such approaches would be practical in an aquaculture setting and concluded that further research was needed using appropriate experimental design. Therefore, effective therapy in the field may be some way from practical fruition.

15.3.4 Natural products

Large numbers of products extracted from plants, micro- and macro-algae have been shown to have growth enhancing properties for aquaculture species, immune stimulating properties and direct or indirect antimicrobial activities (Citarasu 2010;

Reverter et al. 2017). In some cases, the active factors are unknown as the experimental approach uses crude extracts from whole plants or algae while in other cases known compounds are tested. For example, a methanol extract of a mixture of herbs, including neem, Aloe vera, guava, basil, and coriander, increased survival following challenge of black tiger shrimp (*Penaeus monodon*) with *V. harveyi* (AftabUddin et al. 2017) yet the active ingredients in this 'cocktail' of crude extracts was unknown. In terms of known compounds, the prebiotics including non-digestible oligosaccharides, chitin, and inulin are key examples of this latter group (see reviews by Akhter et al. 2015 and Ringø et al. 2010 and for further details). Addition of powdered chitin to diets of European shore crabs, *Carcinus maenas*, resulted in higher survival rates but the outcome of challenge of these animals with *V. alginolyticus* was unaffected by additional of this substance to the diets (Powell and Rowley 2007). The authors suggested that chitin in the diets could reduce the presence of chitin-binding pathogens in the gut by acting as a purgative. Furthermore, addition of chitin to diets of the freshwater prawn, *Macrobrachium rosenbergii* has apparent immune-stimulatory activity (Kumar et al. 2015).

Glucans, such as β1,3 and β 1,6 glucans are potent immunostimulatory molecules in both vertebrates and invertebrates (Meena et al. 2013). These products are extracted from a number of sources including yeast and macroalgae and are available commercially as feed supplements (e.g. Macrogard®, ImmunoMax-FS, EcoActiva™). In crustaceans, glucans can stimulate immune factors including phenoloxidase and heightened haemocyte respiratory burst activity (e.g. Chang et al. 2011) and shrimp receiving this dietary supplement show enhanced resistance to *vibrio* challenge (e.g. Chang et al. 2011; Shivanda Murthy et al. 2009). Long-term (i.e. constant) administration of immunostimulatory molecules, such as glucans, could result in 'immune fatigue' (Smith et al. 2003) and so their use needs careful monitoring for effectivity.

15.3.5 Probiotics

There has been extensive research on the application of probiotics to control bacterial diseases of crustaceans, subject to production via aquaculture and a number of reviews describing progress in the field exist (e.g. Hai 2015; Hoseinfar et al. 2018; Kumar et al. 2016; Ninawe and Selvin 2009; Ringø et al. 2021) to which the reader is pointed for further information. There are several interpretations of the term probiotic but a recent revision of the original definition based on that by the WHO and the FAO is 'live microorganisms that, when administered in adequate amounts, confer a health benefit on the host' (Hill et al. 2014). A more aquaculture-facing definition has been proposed by Merrifield et al. (2010) as 'any microbial cell provided via the diet or rearing water that benefits the host fish, fish farmer or fish consumer, which is achieved, in part at least, by improving the microbial balance of the fish'. However, to conform with the globally accepted concept, probiotics need to be viable at the point of delivery and this may not always be achieved in a farm-based scenario.

Organisms that are candidates as probiotics for use in crustacean aquaculture include bacteria, fungi, and microalgae (Hai 2015). These either target creating changes in the rearing water by improving water quality, or by changing the gut microbiome to improve the animal's overall health (Dittmann et al. 2017). In the case of bacteria, research has either focussed on testing existing probiotics designed for human and agricultural animals or bacteria isolated from the host's gut or its environment. Most studies have examined Gram positive bacteria such as lactic acid bacilli including *Lactobacillus* spp. (e.g. Chiu et al. 2007) and *Pediococcus acidilactici* (e.g. Castex et al. 2008) and members of the Bacteriaceae for example *Bacillus subtilis* (Tseng et al. 2009; Zokaeifar et al. 2012, 2014). Gram negative bacteria including aeromonads and vibrios have also attracted some attention because of their ability to suppress the growth of pathogenic vibrios *in vitro* (e.g. Newaj-Fyzul et al. 2014; Thompson et al. 2010). A smaller number of studies have examined the potential of bacteria isolated from the alimentary canals of crustaceans looking for those that inhibit the growth and colonisation of the alimentary canal by potential pathogens (see Li, E et al. 2018 and Ninawe and Selvin 2009 for reviews). The approach generally taken is to test the inhibitory activity of isolated bacteria against known shrimp pathogens using standard *in vitro* assays. Although several potential probiotics strains have been identified including *Bacillus aquimaris* SH6 (Ngo et al. 2016), none of these appear to have made the transition to widespread commercial use.

There is extensive literature showing that addition of various putative probiotic bacteria to diets of shrimp improves growth rate (Castex et al. 2008; Zokaeifar et al. 2012), immune reactivity (Chiu et al. 2007; Tseng et al. 2009) and disease resistance (Balcazar and Rojas-Luna 2007; Castex et al. 2008; Tseng et al. 2009). The observed increase in growth rate is probably brought about by a combination of production of digestive enzymes by the probiotics and changes in the overall gut microbiome leading to enhanced digestion and removal of harmful bacteria. For example, application of *P. acidilactici* (Bactocell™) in feed achieved higher survival rates in shrimp affected by vibriosis caused by *V. nigripulchritudo* (Castex et al. 2008). This probiotic was found to cause a decrease in the number of viable vibrios recovered from the haemolymph and the hepatopancreas but, like many other probiotics, this effect needed constant application of *P. acidilactici* in the diets as it failed to colonise the alimentary canal.

A second approach is to reduce the presence of vibrios in hatchery water using microalgae—the so-called green water culture system (see review by Dash et al. 2017). These microalgae include *Nannochlorosis* sp., *Chaetoceros*, *Isochrysis*, and *Spirulina*, some of which have antibacterial activity that may be associated with the production of long chain fatty acids acting as a general disinfectant.

15.3.6 Immunisation

The immune system of crustaceans, like that of other invertebrates, is devoid of lymphocytes and specific antibodies and so it is considered to be non-specific in nature (see Chapter 1, and Rowley and Pope 2012). There is increasing evidence that the invertebrate immune system does have a degree of specificity and memory that are both key requirements for the development of effective vaccines although it achieves this by various mechanisms unique to these animals (Rowley and Powell 2007). Initial studies showed that shrimp (*P. monodon*) vaccinated with formalin-inactivated *Vibrio* spp. showed modest improvement in survival post-challenge with live, virulent *V. alginolyticus* (Teunissen et al. 1998)

but these observations were without a mechanistic explanation and were not designed to prove specificity of the protective response (a hallmark of classical vaccination). A potential mechanistic explanation of these authors' findings resides in enhancement of the cellular defence process of phagocytosis (Pope et al. 2011; Rowley and Powell 2007) and this may involve pathogen-specific isoforms of the Down syndrome cell adhesion molecule (Armitage et al. 2015; Hung et al. 2013; Li X-J et al. 2018; see Chapter 1 Text Box 1.2). Whether this approach to disease control would be effective and practical in an aquaculture-based environment is unclear.

15.3.7 Improvement in environmental conditions

Outbreaks of vibriosis in hatcheries and in postlarval shrimp are common in water containing high levels of waste products especially at high temperature with lower oxygen tension. It is well known that shrimp, under these conditions are immunocompromised. Environmental conditions can be improved using recirculating aquaculture systems (RAS) but these are expensive in capital and recurrent costs. Biofloc technology is a more cost-effective approach. It is characterised by the generation of floccules consisting of small porous particles of solid materials including feed, faeces and detritus together with attached organisms such as heterotrophic bacteria, phytoplankton, protozoans and fungi (Avnimelech 2012; Crab et al. 2012). These systems have minimal water change and need robust oxygenation. It is thought that Biofloc systems reduce disease outbreaks by lowering the numbers of bacterial pathogens in the water and by enhancement of the immune system of shrimp (Aguilera et al. 2019; Kim et al. 2013). Furthermore, the Biofloc system can reduce the virulence of AHPND causing strains of *V. parahaemolyticus* (Kumar et al. 2020)

15.4 Future directions

Since the publication of the more recent reviews of bacterial diseases of crustaceans (e.g. Wang 2011) great advances have been made in what is a fast-moving research field. Looking ahead, here are some likely developments:

15.4.1 The crustacean microbiome

Many bacterial pathogens of crustaceans invade via the alimentary canal (see Chapter 1) and so a deeper understanding of the microbial community in the gut is important. The recent advances in methodology such as next generation sequencing, are bringing about significant advances in our understanding of the interaction between bacterial pathogens and gut microbiome, largely using shrimp as model animals. There is increasing evidence from 16S rRNA sequencing that some diseases affecting the alimentary canal of crustaceans have a complex interaction with the normal microbiota in this region and this is providing greater insight into disease pathogenesis. It is now well known that the gut microbiota helps to prevent pathogen colonisation and so changes to this brought about by life history stage, diet and/or physical and chemical environmental factors, can have profound implications to the health and disease status of animals (e.g. Holt et al. 2020). 16S rRNA amplification and Illumina sequencing has made important advances in our understanding of the interaction between bacterial pathogens and the microbiota in the gut utilising shrimp as a model organism (Li et al. 2018; Xiong 2018; Xiong et al. 2018a,b). Healthy shrimp show marked changes in bacterial operational taxonomic units (OTU)s from larvae through to adults (Xiong et al. 2017, 2018). For example, larval OTUs are dominated by members of the Rhodobacteriaeceae while adults show different cohorts of OTUs including Actinomycetales, Marinicellaceae and Flavobacteriaceae. In the case of disease, such shrimp show less microbial diversity in the alimentary canal and a reduction in commensal bacteria (Xiong et al. 2018b; Zhu et al. 2016). There is also a marked increase in taxa including Rhodobacteriaceae, *Vibrio* spp. and Flavobacteriaceae in diseased shrimp (Xiong et al. 2017), some of which may switch from avirulent to virulent states.

Arising from these studies is the concept of polymicrobial pathogenesis of crustacean disease where the interaction between invading and resident bacteria results in disease that may not have a single causative agent (Dai et al. 2018; Yu et al. 2018). An example of such a condition may be white faeces syndrome of shrimp where the faecal strands have an unusual whitish appearance

and the diseased midgut becomes white-yellow in colour (Sriurairatana et al. 2014). This is a serious condition that has dramatically affected shrimp production. An initial suggestion was that this syndrome is caused by the microsporidian parasite, *Enterocytozoon hepatopenaei* but we now know that this is not the only agent involved in the condition (Tangprasittipap et al. 2013). Sequencing of 16S rRNA gene amplicons has revealed several changes in bacterial microbiota in affected shrimp with increases in *Candidatus* Bacilloplasma and *Phascolarctobacterium* and decreases in *Paracoccus* and *Lactococcus* (Hou et al. 2018), without revealing a single cause. Overall, this syndrome is probably another example of a dysbiotic condition, like epizootic shell disease syndrome in American lobsters and such conditions are likely to be of increasing importance, particularly in a changing environment. Most recently, Kumar et al. (2022) concluded that this condition is caused by the microsporidian and another unidentified agent. A better understanding of gut microbiome communities in aquaculture species such as shrimp, crabs and lobsters, will also be of increasing importance in future development of more effective probiotics (Li et al. 2018).

The bacterial communities found in the haemolymph of apparently healthy crustaceans is not well understood. For example, could these be present in a defensive remit to antagonise with the activities of potential pathogens or are they simply a transient population of chance invaders? The long-term dynamics of these bacterial populations need to be established to facilitate a clearer appreciation of the significance of such microbes to the health of the host.

15.4.2 Climate change and emerging bacterial diseases of crustaceans

There are few examples of bacterial diseases of crustaceans that are linked to climate change. Epizootic shell disease (ESD) of lobsters is one condition where there is sound evidence from laboratory and field observations of how increased water temperature is a key driver of this condition (Barris et al. 2018; Groner et al. 2018; Shields 2019). Lobsters harbouring ESD held under aquarium conditions for 5–6 months had a faster progression of lesions at higher temperatures (Barris et al. 2018). Groner et al. (2018) employed a 34-year mark-recapture dataset to investigate the relationship between temperature, changes in moulting behaviour (termed moulting phenology) and ESD in the Long Island Sound. They found a strong link between these factors and reported that the prevalence of ESD was higher in hotter than cooler summers. It is not known whether elevated temperatures merely leave lobsters susceptible to ESD due to changes in the timing of moulting alone or as a result of changes in the microbial composition in surrounding waters.

15.5 Summary

- Bacteria are important pathogens of many crustaceans and under aquaculture conditions these can result in infections causing significant economic losses.
- Infections such acute hepatopancreatic necrosis disease, caused by vibrios can plague shrimp production.
- Over reliance on the use of antibacterial chemicals to control these diseases results in serious environmental and human health problems.
- The influence of the host's microbiome on the prevalence and severity of bacterial diseases in crustaceans is poorly understood.

Recommended further reading

Vibriosis

de Souza Santos, M., Salomon, D., Li P., Krachler, A-M., and Orth, K. 2015. *Vibrio parahaemolyticus* virulence determinants. In *The Comprehensive Sourcebook of Bacterial Protein Toxins*. Alouf, J., Ladant, D., and Popoff, M.R. (eds.) pp. 230–260. Waltham, MA: Elsevier

de Souza Valente, C., and Wan, A.H.L. 2021. Vibrio and major commercially important vibriosis diseases in decapod crustaceans. *Journal of Invertebrate Pathology* 181: 107527

Defoirdt, T., Boon, N., Sogeloos, P., Verstraete, W., and Bossier, P. 2008. Quorum sensing and quorum quenching in *Vibrio harveyi*: Lessons learned from *in vivo* work. *The ISME Journal* 2: 19–26

Kumar, R., Ng, T.H., and Wang, H.C. 2020. Acute hepatopancreatic necrosis disease in penaeid shrimp. *Reviews in Aquaculture* 12: 1867-1880

Prachumwat, A., Taengchaiyaphum, S., Mungkongwongsiri, N., Aldama-Cano, D.J., Flegel, T.W., and Sritunyalucksana, K. 2019. Update on early mortality syndrome/acute hepatopancreatic necrosis disease by April 2018. *Journal of the World Aquaculture Society* 50: 5–17 (Excellent review on AHPND in shrimp)

Ruwandeepika, H.A., Sanjeewa, P., Jayaweera, T. et al. 2012. Pathogenesis, virulence factors and virulence regulation of vibrios belonging to the *Harveyi* clade. *Reviews in Aquaculture* 4: 59–74 (Comprehensive review of virulence mechanisms of vibrios pathogenic to fish and shellfish)

Rickettsia and RLOs

Bouchon, D., Zimmer, M., and Dittmer, J. 2016. The terrestrial isopod microbiome: An all-in-one toolbox for animal-microbe interactions of ecological relevance. *Frontiers in Microbiology* 23: 1472 (Timely account focussing on the interaction between isopod microbiomes)

Shell disease syndrome

Castro, K.M., Cobb, J.S., Gomez-Chiarri, M., and Tlusty, M. 2012. Epizootic shell disease in American lobsters *Homarus americanus* in southern New England: Past, present and future. *Diseases of Aquatic Organisms* 100: 149–158

Gomez-Chiarri, M. and Cobb, J.S. 2012. Shell disease in the American lobsters, *Homarus americanus*: A synthesis of research from the New England Lobster Research Initiative: Lobster shell disease. *Journal of Shellfish Research* 31: 583–590 (An overview of the research carried out on epizootic shell disease and their main findings)

Shields, J.D. 2013. Complex etiologies of emerging diseases in lobsters (*Homarus americanus*) from Long Island Sound. *Canadian Journal of Fisheries and Aquatic Science* 70: 1576–1587 (Review of emerging diseases including epizootic shell disease in American lobsters)

Shields, J.D. 2019. Climate change enhances disease processes in crustaceans: Case studies in lobsters, crabs, and shrimp. *Journal of Crustacean Biology* 39: 673–683 (Key evaluation of evidence of how climate change can influence disease processes in crustaceans)

Vogan, C.L., Powell, A., and Rowley, A.F. 2008. Shell disease in crustaceans—just chitin recycling gone wrong? *Environmental Microbiology* 10: 826–835 (Review of enzootic shell disease with reference to chitinolytic activities)

Diagnosis, control and treatment of bacterial diseases

Angulo, C., Loera-Muro, A., Trujillo, and Luna-Gonzalez, A. 2019. Control of AHPND by phages: A promising biotechnological approach. *Reviews in Aquaculture* 11: 989–1004

Cabello, F.C., Godfrey, H.P., Tomova, A. et al. 2013. Antimicrobial use in aquaculture re-examined: Its relevance to antimicrobial resistance and to animal and human health. *Environmental Microbiology* 15: 1917–1942 (Excellent review on antibiotic use in aquaculture in general and its environmental and animal health consequences)

Defoirdt, T. 2014. Virulence mechanisms of bacterial aquaculture pathogens and antivirulence therapy for aquaculture. *Reviews in Aquaculture* 6: 100–114

Kalatzis, P.G., Castill, D., Katharios, P., and Middelboe, M. 2018. Bacteriophage interactions with marine pathogenic vibrios: Implications for phage therapy. *Antibiotics* 7: 15 (Review of principles and trials of bacteriophage therapy)

Kumar, V., Roy, S., Meena, D. K., and Sarkar, U.K. 2016. Application of probiotics in shrimp aquaculture: Importance, mechanisms of action, and methods of administration. *Reviews in Fisheries Science & Aquaculture* 24: 342–368

Li, E., Xu, C., Wang, X. et al. 2018. Gut microbiota and its modulation for healthy farming of Pacific white shrimp *Litopenaeus vannamei*. *Reviews in Fisheries Science and Aquaculture* 26: 381–399 (Review of gut microbiota and their role in gut health in relation to bacterial probiotics)

Flegel, T.W. 2019. A future vision for disease control in shrimp aquaculture. *Journal of the World Aquaculture Society* 50: 249–266 (Review of approaches to the treatment and control of bacterial and viral diseases of shrimp)

Ringø, E., Van Doan, H., Lee, S., and Song, S.K. 2021. Lactic acid bacteria in shellfish: Possibilities and challenges. *Reviews in Fisheries Science and Aquaculture* 28: 139-169

Rowley, A.F. and Pope, E.C. 2012. Vaccines and crustacean aquaculture—a mechanistic explanation. *Aquaculture* 334–337: 1–11 (Review of potential mechanisms and actions of vaccination)

Vaiyapuri, M., Pailla, S., Badireddy, M.R. et al. 2021. Antimicrobial resistance in vibrios in shrimp aquaculture: Incidence, identification schemes, drivers and mitigation measures. *Aquaculture Research* 52: 2923–2941

Xiong, J. 2018. Progress in the gut microbiota in exploring shrimp disease pathogenesis and incidence. *Applied Microbiology and Biotechnology* 102: 7343–7350 (Review of potential novel diagnostic approaches to bacterial diseases of shrimp)

Acknowledgements

I am grateful to the following people who provided images to illustrate this chapter: Professor Jeffery Shields (Virginia Institute of Marine Science, USA), Dr K. Fraser Clark (Dalhousie University, Canada) Dr Tobia Pretto (Instituto Zooprofilattico Sperimentale de Venezie, Padova, Italy), Dr Nadav Davidovich, and Professor Kallaya Sritunyalucksana, (Research Unit National Center for Genetic Engineering and Biotechnology (BIOTEC) National Science and Technology Development Agency (NSTDA), Bangkok, Thailand).

Thanks are also given to past and current collaborators (Drs. Fiona Eddy, Adam Powell, Claire Vogan, Amanda Smith, Emma Wootton and Ed Pope) for the provision of unpublished micrographs. Original work referred to in this chapter was partially funded by the Interreg Ireland-Wales, Bluefish and Susfish projects, and the UKRI (BB/P017215/1) ARCH-UK. This chapter is part of a series of position papers produced by the ARCH-UK project.

References

AftabUddin. S., Siddique, M.A.M., Romkey, S.S., and Shelton, W.J. 2017. Antibacterial function of herbal extracts on growth, survival and immunoprotection in the black tiger shrimp *Penaeus monodon*. *Fish & Shellfish Immunology* 65: 52–58

Aguilera, D., Escalante-Herrera, K., Gaxiola, G. et al. 2019. Immune response of the Pacific white shrimp, *Litopenaeus vannamei*, previously reared in biofloc and after an infection assay with *Vibrio harveyi*. *Journal of the World Aquaculture Society* 50: 119–136

Aguirre-Guzmán, G., Sánchez-Martínez, J.G., Pérez-Castañeda, R., Palacios-Monzón, A., Trujillo-Rodríguez, T. and De La Cruz-hernández, N.I. 2010. Pathogenicity and infection route of *Vibrio parahaemolyticus* in American white shrimp, *Litopenaeus vannamei*. *Journal of the World Aquaculture Society* 41: 464–470

Akhter, N., Wu, B., Memon, A.M., and Mohsin, M. 2015. Probiotics and prebiotics associated with aquaculture: A review. *Fish and Shellfish Immunology* 45: 733–741

Andersen, L.E., Norton, J.H., and Levy, N.H. 2000. A new shell disease in the mud crab *Scylla serrata* from Port Curtis, Queensland (Australia). *Diseases of Aquatic Organisms* 43: 233–239

Angulo, C., Loera-Muro, A., Trujillo, E. and Luna-González, A. 2019. Control of AHPND by phages: A promising biotechnological approach. *Reviews in Aquaculture* 11: 989–1004

Armitage, S.A.O., Peuß, R., and Kurtz, J. 2015. Dscam and pancrustacean immune memory—a review of the evidence. *Developmental and Comparative Immunology* 48: 315–323

Austin, B. and Zhang, X.-H. 2006. *Vibrio harveyi*: A significant pathogen of marine vertebrates and invertebrates. *Letters in Applied Microbiology* 43: 119–124

Avnimelech, Y. 2012. *Biofloc Technology—A Practical Guide Book 2*: The World Aquaculture Society, Baton Rouge, USA

Balcazar, J.L. and Rojas-Luna, T. 2007. Inhibitory activity of probiotic *Bacillus subtilis* UTM 126 against vibrio species confers protection against vibriosis in juvenile shrimp (*Litopenaeus vannamei*). *Current Microbiology* 55: 409–412

Barris, B.N., Shields, J.D., Small, H., Huchin-Mian, J.P., O'Leary, P., Shawver, J.V., Glenn, R.P., and Pugh, T.L. 2018. Laboratory studies on the effect of temperature on epizootic shell disease in the American lobster, *Homarus americanus*. *Bulletin of Marine Science* 94: 887–902

Basti, D., Bricknell, I., Hoyt, K., Chang, E.S., Halteman. W., and Bouchard, D. 2010. Factors affecting post-capture survivability of lobster *Homarus amercanus*. *Diseases of Aquatic Organisms* 90: 153–166

Bi, K., Zhang, X., Yan, B., Gao, H., Gao, X., and Sun, J. 2016. Isolation and molecular identification of *Vibrio natriegens* from diseased *Portunus triberculatus* in China. *Journal of the World Aquaculture Society* 47: 854–861

Bojko, J., Dunn, A.M., Stebbing, P.D. et al. 2018. 'Candidatus Aquirickettsiella gammari' (Gammaproteobacteria: Legionellales: Coxiellaceae): A bacterial pathogen of the freshwater crustacean *Gammarus fossarum* (Malacostraca: Amphipoda). *Journal of Invertebrate Pathology* 156: 41–153

Bouchon, D., Rigaud, T., and Juchault, P. 1998. Evidence for widespread *Wolbachia* infection in isopod crustaceans: Molecular identification and host feminization. *Proceedings of the Royal Society of London B* 265: 1081–1090

Brusca, R.C. and Brusca, G.J. 2003. *Invertebrates*, 2nd edn. Sunderland, MA: Sinauer

Busico-Salcedo, N. and Owens, L. 2013. Virulence changes to Harveyi clade bacteria infected with bacteriophage from *Vibrio owensii*. *Indian Journal of Virology* 24: 180–187

Cabello, F.C., Godfrey, H.P., Tomova, A. et al. 2013. Antimicrobial use in aquaculture re-examined: Its relevance to antimicrobial resistance and to animal and human health *Environmental Microbiology* 15: 1917–1942

Cano-Gómez, A., Goulden, E., Owens, L., and Høj, L. 2009. *Vibrio owensii* sp. nov., isolated from cultured crustaceans in Australia. *FEMS Microbiology Letters* 302: 175–181

Cao, H., He, S., Lu, L., Yang, X., and Chen, B. 2014. Identification of *Proteus penneri* isolate as the causal agent of red body disease of the cultured white shrimp *Penaeus vannamei* and its control with *Bdellovibrio bacteriovorus*. *Antonie van Leeuwenhoek* 105: 423–430

Castex, M., Chim, L., Pham, D. et al. 2008. Probiotic *P. acidilactici* application in shrimp *Litopenaeus stylirostris* culture subject to vibriosis in New Caledonia. *Aquaculture* 275: 182–193

Castro, K.M. and Angell, T.E. 2000. Prevalence and progression of shell disease in American lobster, *Homarus americanus*, from Rhode Island waters and the offshore canyons. *Journal of Shellfish Research* 19: 691–700

Castro, K.M., Cobb, J.S., Gomez-Chiarri, M., and Tlusty, M. 2012. Epizootic shell disease in American lobsters *Homarus americanus* in southern New England: Past, present and future. *Diseases of Aquatic Organisms* 100: 149–158

Castro, K.M., Factor, J.R., Angell, T., and Landers, D.F., Jr. 2006. The conceptual approach to lobster shell disease revisited. *Journal of Crustacean Biology* 26: 646–660

Cawthorn, R.J. 2011. Diseases of American lobsters (*Homarus americanus*): A review. *Journal of Invertebrate Pathology* 106: 71–78

Chang, J., Zhang, W., Mai, K. et al. 2011. Effects of dietary β-glucan and glycyrrhizin on non-specific immunity and disease resistance of white shrimp, *Litopenaeus vannamei* (Boone) challenged with *Vibrio alginolyticus*. *Aquaculture Research* 42: 1101–1109

Chen, S.C., Lin, Y.D., Liaw, L.L., and Wang, P.C. 2001. *Lactococcus garvieae* infection in the giant freshwater prawn, *Macrobrachium rosenbergii* confirmed by polymerase chain reaction and 16S rDNA sequencing. *Diseases of Aquatic Organisms* 45: 45–52

Cheng, W and Chen, J.C. 1998. Isolation and characterization of an Enterococcus-like bacterium causing muscle necrosis and mortality in *Macrobrachium rosenbergii* in Taiwan. *Diseases of Aquatic Organisms* 34: 93–101

Chistoserdov, A.Y., Mirasol, F., Smolowitz, R., and Hsu, A. 2005. Culture-dependent characterization of the microbial community associated with epizootic shell disease lesions in American lobster, *Homarus americanus*. *Journal of Shellfish Research* 24: 741–747

Chistoserdov, A.Y., Quinn, R.A., Gubbala, S.L., and Smolowitz, R. 2012. Bacterial communities associated with lesions of shell disease in the American lobster, *Homarus americanus* Milne-Edwards. *Journal of Shellfish Research* 31: 449–462

Chiu, C.-H., Guu, Y.-K., Liu, C.-H., Pan T.-M., and Cheng, W. 2007. Immune responses and gene expression in white shrimp, *Litopenaeus vannamei*, induced by *Lactobacillus plantarum*. *Fish and Shellfish Immunology* 23: 364–377

Citarasu, T. 2010. Herbal biomedicines: A new opportunity for aquaculture industry. *Aquaculture International* 18: 403–414

Clark, K.F. and Greenwood, S.J. 2011. *Aerococcus viridans* expression of Cpn60 is associated with virulence during infection of the American lobster, *Homarus americanus* Milne Edwards. *Journal of Fish Diseases* 34: 831–843

Cordaux, R., Bouchon, D., and Greve, P. 2011. The impact of endosymbionts on the evolution of host sex-determination mechanisms. *Trends in Genetics* 27: 332–341

Cordaux, R., Michel-Salzat A., Frelon-Raimond M., Rigaud T., and Bouchon, D. 2004. Evidence for a new feminizing *Wolbachia* strain in the isopod *Armadillidium vulgare*: Evolutionary implications. *Heredity* 93: 78–84

Cordaux, R., Paces-Fessy, M., Raimond, M., Michel-Salzat, A., Zimmer, M., and Bouchon, D. 2007. Molecular characterization and evolution of arthropod-pathogenic *Rickettsiella* bacteria. *Applied and Environmental Microbiology* 73: 5045–5047

Cordaux, R., Pichon, S., Hatira, H.B. et al. 2012. Widespread *Wolbachia* infection in terrestrial isopods and other crustaceans. *ZooKeys* 176: 123–131

Costa, R., Mermoud, I., Koblavi, S. et al. 1998. Isolation and characterization of bacteria associated with a *Penaeus stylirostris* disease (Syndrome 93) in New Caledonia. *Aquaculture* 164: 297–309

Crab, R., Defoirdt, T., Bossier, P., and Verstraete, W. 2012. Biofloc technology in aquaculture: Beneficial effects and future challenges. *Aquaculture* 356–357: 351–356

Dai, W., Yu, W., Xuan, L., Tao, Z., and Xiong, J. 2018. Integrating molecular and ecological approaches to identify potential polymicrobial pathogens over a shrimp disease progression. *Applied Microbiology and Biotechnology* 102: 3755–3764

Dash, P., Avunje, S., Tandel, R.S., Sandeep, R.S., and Panigrahi, A. 2017. Biocontrol of luminous vibrios in shrimp aquaculture: A review of current approaches and future perspectives. *Reviews in Fisheries Science and Aquaculture* 25: 245–255

Davidovich, N., Pretto, T., Blum, S.E., Baider, Z., Grossman, R., Kaidar-Shwartz, H., Dveyrin, Z., and Rorman, E. 2019. *Mycobacterium gordonae* infecting redclaw crayfish *Cherax quadricarinatus*. *Diseases of Aquatic Organisms* 135: 169–174

Davies, C.E., Johnson, A.F., Wootton, E.C. et al. 2015. Effects of population density and body size on disease ecology of the European lobster in a temperate marine conservation zone. *ICES Journal of Marine Science* 72: i128–i138

Davies, C.E., Whitten, M.M.A., Kim, A. et al. 2014. A comparison of the structure of American (*Homarus americanus*) and European (*Homarus gammarus*) lobster cuticle with particular reference to shell disease susceptibility. *Journal of Invertebrate Pathology* 117: 33–41

De Souza Santos, M., Salomon, D., Li, P., Krachler, A.-M., and Orth, K. 2015. *Vibrio parahaemolyticus* virulence determinants. In *The Comprehensive Sourcebook of Bacterial Protein Toxins*, Alouf, J., Ladant, D., and Popoff M.R. (eds.) pp. 230–260. Waltham, MA: Elsevier

Defoirdt, T. 2014. Virulence mechanisms of bacterial aquaculture pathogens and antivirulence therapy for aquaculture. *Reviews in Aquaculture* 6: 100–114

Defoirdt, T. 2018. Quorum-sensing systems as targets for antivirulence therapy. *Trends in Microbiology* 26: 313–328

Defoirdt, T., Boon, N., Sorgeloos, P., Verstraete, W., and Bossier, P. 2008. Quorum sensing and quorum quenching in *Vibrio harveyi*: Lessons learned from *in vivo* work. *The ISME Journal* 2: 19–26

Defoirdt, T., Ruwandeepika, H.A.D., Karunasagar, I., Boon, N., and Bossier, P. 2010. Quorum sensing negatively regulates chitinase in *Vibrio harveyi*. *Environmental Microbiology Reports* 2: 44–49

Dellinger, L. 2012. A fisherman's perspective. *Journal of Shellfish Research* 31: 581–582

Dhar, A.K., Piamsomboon, P., Caro, L.F.A., Kanrar, S. Adami, R. Jr., and Juan, Y.-S. 2019. First report of acute hepatopancreatic necrosis disease (AHPND) occurring in the USA. *Diseases of Aquatic Organisms* 132: 241–247

Diggles, B.K., Moss, G.A., Carson, J., and Anderson, C.D. 2000. Luminous vibriosis in rock lobsters *Jasus verreauxi* (Decapoda: Palinuridae) phyllosoma larvae associated with infection by *Vibrio harveyi*. *Diseases of Aquatic Organisms* 43: 127–137

Dittmann, K.K., Rasmussen, B.B., Castex, M., Gram, L., and Benton-Tilia, M. 2017. The aquaculture microbiome at the centre of business creation. *Microbial Biotechnology* 10: 1279–1282

Dittmer, J. and Bouchon, D. 2018. Feminizing *Wolbachia* influence microbiota composition in the terrestrial isopod *Armadillidium vulgare*. *Scientific Reports* 8: 6998

Done, H.Y., Venkatesan, A.K., and Halden, R.U. 2015. Does the growth of aquaculture create antibiotic resistance threats different to those associated with land animal production in agriculture? *AAPS Journal* 17: 513–524

Dong, X., Wang, H., Xie, G. et al. 2017a. An isolate of *Vibrio campbellii* carrying the pir^{VP} gene causes acute hepatopancreatic necrosis disease, *Emerging Microbes & Infections* 6: 1–3

Dong, X., Wang, H., and Xie, G. 2017b. Complete genome sequence of *Vibrio campbellii* strain 20130629003S01 isolated from shrimp with acute hepatopancreatic necrosis disease. *Gut Pathogens* 9: 31

Dong, X., Bi, D., Wang, H. et al. 2017c. $pirAB^{vp}$-bearing *Vibrio parahaemolyticus* and *Vibrio campbellii* pathogens isolated from the same AHPND-affected pond possess highly similar pathogenic plasmids. *Frontiers in Microbiology* 8: 1859

Doss, J., Culbertson, K., Hahn, D., Camacho, J., and Barekzi, N. 2017. A review of phage therapy against bacterial pathogens of aquatic and terrestrial organisms. *Viruses* 9: 50

Dyrynda, E.A. 1998. Shell disease in the common shrimp *Crangon*: Variations within an enclosed estuarine system. *Marine Biology* 132: 445–252

Eddy, F., Powell, A., Gregory, S. et al. 2007. A novel bacterial disease of the European shore crab, *Carcinus maenas* molecular pathology and epidemiology. *Microbiology* 153: 2839–2849

Edgerton, B.F. and Prior, H.C. 1999. Description of a hepatopancreatic rickettsia-like organisms in the redclaw crayfish *Cherax quadricarinatus*. *Diseases of Aquatic Organisms* 36: 77–80

Federici, B.A., Hazard, E.I., and Anthony, D.W. 1974. Rickettsia-like organism causing disease in a crangonid amphipod from Florida. *Applied and Environmental Microbiology* 28: 885–886

Feinmark, S.G., Martínez, A.U., Bowen, J.L., and Tlusty, M.F. 2017. Fine-scale transition to lower bacterial diversity and altered community composition precedes shell disease in laboratory-reared juvenile American lobster. *Diseases of Aquatic Organisms* 124: 41–54

Flegel, T.W. 2019. A future vision for disease control in shrimp aquaculture. *Journal of the World Aquaculture Society* 50: 249–266

Fu, S., Wang, L., Tian, H., Wei, D., and Liu, Y. 2018. Pathogenicity and genomic characterization of *Vibrio parahaemolyticus* strain PB1937 causing shrimp acute hepatopancreatic necrosis disease in China. *Annals of Microbiology* 68: 175–184

Gálvez, E.J.C., Carrillo-Castro, K., Zarate, L., Guiza, L., Pieper, D.H., Garcia-Bonilla, E., Salazar, L., and Junca, H. (2016). Draft genome sequence of *Bacillus licheniformis* CG-B52, a highly virulent bacterium of Pacific white shrimp (*Litopenaeus vannamei*), isolated from a Colombian Caribbean aquaculture outbreak. *Microbiology Resource Announcements* 4: e00321-16

Getchell, R.G. 1989. Bacterial shell disease in crustaceans: A review. *Journal of Shellfish Research* 8: 1–6

Glenn, R.P. and Pugh, T.L. 2006. Epizootic shell disease in American lobster (*Homarus americanus*) in Massachusetts coastal waters: Interactions of temperature, maturity, and intermolt duration. *Journal of Crustacean Biology* 26: 639–645

Goarant, C., Ansquer, D., Herlin, J., Domalain, D., and Decker, S.D. 2006. 'Summer Syndrome' in *Litopenaeus stylirostris* in New Caledonia: Pathology and epidemiology of the etiological agent. *Aquaculture* 253: 105–113

Gode-Potratz, C.J., Kustusch, R.J., Breheny, P.J., Weiss, D.S., and McCarter, L.L. 2011. Surface sensing in *Vibrio parahaemolyticus* triggers a programme of gene expression that promotes colonization and virulence. *Molecular Microbiology* 79: 240–263

Gomez-Chiarri, M. and Cobb, J.S. 2012. Shell disease in the American lobsters, *Homarus americanus*: A synthesis of research from the New England Lobster Research Initiative: Lobster shell disease. *Journal of Shellfish Research* 31: 583–590

Gomez-Gil, B., Soto-Rodríguez, S., Garcia-Gasca, A. et al. 2004. Molecular identification of *Vibrio harveyi*-related isolates associated with diseased aquatic organisms. *Microbiology* 150: 1769–1777

Goudenège, D., Labreuche, Y., Krin, E. et al. 2013. Comparative genomics of pathogenic lineages of *Vibrio nigripulchritudo* identifies virulence-associated traits. *The ISME Journal* 7: 1985–1996

Greenwood, S.D.J., Keith, I.R., Despres, B.M., and Cawthorn, R.J. 2005. Genetic characterization of the lobster pathogen *Aerococcus viridans* var. *homari* determined by 16S rRNA gene sequence and randomly amplified polymorphic DNA analyses. *Diseases of Aquatic Organisms* 63: 237–246

Groner, M. L., Shields, J.D., Landers, D.F. Jr., Swenarton, J., and Hoenig, J.M. 2018. Rising temperatures, molting phenology, and epizootic shell disease in the American lobster. *The American Naturalist* 192: E163–E177

Hai, N.V. 2015. The use of probiotics in aquaculture. *Journal of Applied Microbiology* 119: 917–935

Han, J.E., Mohney, L.L., Tang, K.F.J., Pantoja, C.R., and Lightner, D.V. 2015. Plasmid mediated tetracycline resistance of *Vibrio parahaemolyticus* associated with acute hepatopancreatic necrosis disease (AHPND) in shrimps. *Aquaculture Reports* 2: 17–21

Harris, L.J. and Owens, L. 1999. Production of exotoxins by two luminous *Vibrio harveyi* strains known to be primary pathogens of *Penaeus monodon* larvae. *Diseases of Aquatic Organisms* 38: 11–22

Hasson, K.W., Wyld, E.M., Fan, Y. et al. 2009. Streptococcosis in farmed *Litopenaeus vannamei*: A new emerging bacterial disease of penaeid shrimp. *Diseases of Aquatic Organisms* 86: 93–106

Hess, E.A. 1937. A shell disease in lobsters (*Homarus americanus*) caused by chitinovorous bacteria. *Journal of the Biological Board of Canada* 3: 358–362

Heuer, O.E., Kruse, H., Grave, K., Collignon, P., Karunasagar, I., and Angulo, F.J. 2009. Human health consequences of use of antimicrobial agents in aquaculture. *Clinical Infectious Diseases* 49: 1248–1253

Hill, C., Guarner, F., Reid, G. et al. 2014. Expert consensus document. The International Scientific Association for Probiotics and Prebiotics consensus statement on the scope and appropriate use of the term probiotic. *Nature Reviews Gastroenterology & Hepatology* 11: 506–514

Hoenig, J. M., Groner, M.L., Smith, M.W. et al. 2017. Impact of disease on the survival of three commercially fished species. *Ecological Applications* 27: 2116–2127

Holt, C.C., Van Der Giezen, M., Daniels, C., Stentiford, G.D., and Bass, D. 2020. Spatial and temporal axes impact ecology of the gut microbiome in juvenile European lobster (*Homarus gammarus*). *The ISME Journal* 14: 531–543

Hosenifar, S.H., Sun, Y.-Z., Wang, A., and Zhou, Z. 2018. Probiotics as means of disease control in aquaculture, a review of current knowledge and future perspectives. *Frontiers in Microbiology* 9: 2429

Hou, D., Huang, Z., Zeng, S. et al. 2018. Intestinal bacterial signatures of white feces syndrome in shrimp. *Applied Microbiology and Biotechnology* 102: 3701–3709

Hung, H.-Y., Ng, T.H., Lin, J.-H., Chiang, Y.-A., Chuang, Y.-C., and Wang, H.-C. 2013. Properties of *Litopenaeus vannamei* Dscam (LvDscam) isoforms related to specific pathogen recognition. *Fish and Shellfish Immunology* 35: 1272–1281

Intaraprasong, A., Khemayan, K., Pasharawipas, T., and Flegel, T.W. 2009. Species-specific virulence of *Vibrio harveyi* for black tiger shrimp is associated with bacteriophage-mediated hemocyte agglutination. *Aquaculture* 29: 185–192

Jun, J.W., Han, J.E., Tang, K.F.J. et al. 2016. Potential application of bacteriophage pVp-1: Agent combating *Vibrio parahaemolyticus* strains associated with acute hepatopancreatic necrosis disease (AHPND) in shrimp. *Aquaculture* 457: 100–103

Kalatzis, P.G., Castill, D., Katharios, P., and Middelboe, M. 2018. Bacteriophage interactions with marine pathogenic vibrios: Implications for phage therapy. *Antibiotics* 7: 15

Khemayan, K., Prachumwat, A., Sonthayanon, B., Intaraprasong, A., Sriurairatana, S., and Flegel, T.W. 2014. Complete genome sequence of virulence-enhancing siphophage VHS1 from *Vibrio harveyi*. *Applied and Environmental Microbiology* 78: 2790–2796

Kim, S.-K., Pang, Z., Seo, A.-C., Cho, Y.-R., Samocha, T., and Jang, I.-K. 2013. Effect of bioflocs on growth and immune activity of Pacific white shrimp, *Litopenaeus vannamei* postlarvae. *Aquaculture Research* 45: 362–371

King, N.G., Duncan, P.F., Kennington, K., Wootton, E.C., and Jenkins, S.R. 2014. Characterisation of shell disease syndrome in brown crab, *Cancer pagurus*, in a discrete Irish Sea fishery. *Journal of Crustacean Biology* 34: 40–46

Kirn, T.J., Jude, B.A., and Taylor, R.K. 2005. A colonization factor links *Vibrio cholerae* environmental survival and human infection. *Nature* 438: 863–866

Kleespies, R.G., Federici, B.A., and Leclerque, A. 2014. Ultrastructural characterization and multilocus sequence analysis (MLSA) of 'Candidatus Rickettsiella isopodorum', a new lineage of intracellular bacteria infecting woodlice (Crustacea: Isopoda). *Systematic and Applied Microbiology* 37: 351–359

Klein, S., Pipes, S., and Lovell, CR. 2018. Occurrence and significance of pathogenicity and fitness islands in environmental vibrios. *AMB Express* 8: 177

Kratchler, A.M., Ham, H., and Orth, K. 2011. Outer membrane adhesion factor multivalent adhesion molecule 7 initiates host cell binding during infection by Gram-negative pathogens. *Proceedings of the National Academy of Sciences, USA* 108: 11614–11619

Kumar, B.T.N., Murthy, H.S., Patil, P., Doddamani, P.L., and Patil, R. 2015. Enhanced immune response and resistance to white tail disease in chitin-diet fed freshwater prawn, *Macrobrachium rosenbergii*. *Aquaculture Reports* 2: 34–38

Kumar, T.S., Vidya, R., Kumar, S., Alavandi, S.V., and Vijayan, K.K. 2017. Zoea-2 syndrome of *Penaeus vannamei* in shrimp hatcheries. *Aquaculture* 479: 759–767

Kumar, T.S., Makesh, M., Alavandi, S.V., and Vijayan, K.K. 2022. Clinical manifestations of white feces syndrome (WFS), and its association with *Enterocytozoon hepatopenaei* in *Penaeus vannamei* grow-out farms: A pathological investigation. *Aquaculture* 547: 737463

Kumar, V., Nguyen, D.V., Baruah, K., and Bossier, P. 2019. Probing the mechanism of VP$_{AHPND}$ extracellular proteins toxicity purified from *Vibrio parahaemolyticus* AHPND strain in germ-free *Artemia* test system. *Aquaculture* 504: 414–419

Kumar, V., Roy, S., Meena, D.K., and Sarkar, U.K. 2016. Application of probiotics in shrimp aquaculture: Importance, mechanisms of action, and methods of administration. *Reviews in Fisheries Science & Aquaculture* 24: 342–368

Kumar, V., Wille, M., Lourenco, T.M., and Bossier, P. 2020. Biofloc-based enhanced survival of *Litopenaeus vannamei* upon AHPND-causing *Vibrio parahaemolyticus* challenge is partially mediated by reduced expression of its virulence genes. *Frontiers in Microbiology* 11: 1270

Laufer, H., Chen M, Baclaski, B. et al. 2013. Multiple factors in marine environments affecting lobster survival, development, and growth, with emphasis on alkylphenols: A perspective. *Canadian Journal of Fisheries and Aquatic Sciences* 70: 588–600

Laufer, H., Chen, M., Johnson, M., Demir, N., and Bobbitt, J.M. 2012. The effect of alkylphenols on lobster shell hardening. *Journal of Shellfish Research* 31: 555–562

Le Moullac, G. and Haffner, P. 2000. Environmental factors affecting immune responses in Crustacea. *Aquaculture* 191: 121–131

Le Roux, F., Labreuche, Y., Davis, B.M. et al. 2010. Virulence of an emerging pathogenic lineage of *Vibrio nigripulchritudo* is dependent on two plasmids. *Environmental Microbiology* 13: 296–306

Lee, C.-T., Chen, I.-T., Yang, Y.-T. et al. 2015. Key virulence factors in AHPND *V. parahaemolyticus* becomes virulent by acquiring a plasmid that expresses the deadly toxin.

Proceedings of the National Academy of Sciences USA 112: 10798–10803

Letchumanan, V., Chan, K.G., and Lee, L.H. 2014. *Vibrio parahaemolyticus*: A review on the pathogenesis, prevalence, and advance molecular identification techniques. *Frontiers in Microbiology* 5: 705

Leyva, J.M., Martinez-Porchas, M., Hernández-López, J., Vargas-Albores, F., and Gollas-Galván, T. 2018. Identifying the causal agent of necrotizing hepatopancreatitis in shrimp: Multilocus sequence analysis approach. *Aquaculture Research* 49: 1795–1802

Li, E., Xu, C., Wang, X. et al. 2018. Gut microbiota and its modulation for healthy farming of Pacific white shrimp *Litopenaeus vannamei*. *Reviews in Fisheries Science and Aquaculture* 26: 381–399

Li, P., Kinch, L.N., Ray, A. et al. 2017. Acute hepatopancreatic necrosis disease-causing *Vibrio parahaemolyticus* strains maintain an antibacterial type VI secretion system with versatile effector repertoires. *Applied Environmental Microbiology* 83: e00737–17

Li, X., Yang, H., Gao, X., Zhang, H., Chen, N., Miao, Z., Liu, X., and Zhang, X. 2019a. The pathogenicity characterization of non-O1 *Vibrio cholerae* and its activation on immune system in freshwater shrimp *Macrobrachium nipponense*. *Fish and Shellfish Immunology* 87: 507–514

Li, X., Zou, Y., Jiang, Q. et al. 2019b. Virulence properties of *Vibrio vulnificus* isolated from diseased zoea of freshness shrimp *Macrobrachium rosenbergii*. *Microbial Pathogenesis* 127: 166–171

Li, X.-J., Yang, L., Li, D., Zhu, Y.-T., Wang, Q., and Li, W.-W. 2018. Pathogen-specific binding soluble Down syndrome cell adhesion molecule (Dscam) regulates phagocytosis via membrane-bound Dscam in crab. *Frontiers in Immunology* 9: 801

Liu, L., Xaio, J., Zhang, M. et al. 2018. A *Vibrio owensii* strain as a causative agent of AHPND in cultured shrimp, *Litopenaeus vannamei*. *Journal of Invertebrate Pathology* 153: 156–164

Liu, P.C. and Lee, K.K. 1999. Cysteine protease is a major exotoxin of pathogenic luminous *Vibrio harveyi* in the tiger prawn, *Penaeus monodon*. *Letters in Applied Microbiology* 28: 428–430

Liu, P.C., Lee, K.K., Yii, K.C., Kou, G.H., and Chen, S.N. 1996. Isolation of *Vibrio harveyi* from diseased kuruma prawns *Penaeus japonicus*. *Current Microbiology* 33: 129–132

Liu, Y., Qin, Q.W. and Defoirdt, T. 2018. Does quorum sensing interference affect the fitness of bacterial pathogens in the real world? *Environmental Microbiology* 20: 3918–3926

Lulijwa, R., Rupia, E.J., and Alfaro, A.C. 2020. Antibiotic use in aquaculture, policies and regulation, health and environmental risks: A review of the top 15 major producers. *Reviews in Aquaculture* 12: 664–677

Makkonen, J., Jussila, J., Koistinen, L., Paaver, T., Hurt, M., and Kokko, H. 2013. *Fusarium avenaceum* causes burn spot disease syndrome in noble crayfish (*Astacus astacus*). *Journal of Invertebrate Pathology* 113: 184–190

Mancuso, M., Costanzo, M.T., Maricchiolo, G., Gristina, M., Zaccone, R., Cuccu, D., and Genovese, L. 2010. Characterization of chitinolytic bacteria and histological aspects of Shell Disease Syndrome in European spiny lobsters (*Palinurus elephas*) (Fabricius 1787). *Journal of Invertebrate Pathology* 104: 242–244

Manefield, M., Harris, L., Rice, S.A., De Nys, R., and Kjellebreg, S. 2010. Inhibition of luminescence and virulence in the black tiger prawn (*Penaeus monodon*) pathogen *Vibrio harveyi* by intercellular signal antagonists. *Applied and Environmental Microbiology* 66: 2079–2084

Mateus, L., Costa, L., Silva, Y.J., Pereira, C., Cunha, A., and Almeida, A. 2014. Efficiency of phage cocktails in the inactivation of *Vibrio* in aquaculture. *Aquaculture* 424–425: 167–173

Meena, D.K., Das, P., Kumar, S., et al. 2013. Beta-glucan: An ideal immunostimulant in aquaculture (a review): *Fish Physiology and Biochemistry* 39: 431–457

Merrifield, D.L., Dimitroglou, A., Foey, A. et al. 2010. The current status and future focus of probiotic and prebiotic applications for salmonids. *Aquaculture* 302: 1–18

Midorikawa, Y., Shimizu, T., Sanda, T., Hamasaki, K., Dan, S., Lal, M.T.B.M., Kato, G., and Sano, M. 2020. Characterization of *Aquimarina hainanensis* isolated from diseased mud crab *Scylla serrata* in a hatchery. *Journal of Fish Diseases* 43: 541–549

Miller, M.B. and Bassler, B.L. 2001. Quorum sensing in bacteria. *Annual Reviews in Microbiology* 55: 165–199

Montero, A.B. and Austin, B. 1999. Characterization of extracellular products from an isolate of *Vibrio harveyi* recovered from diseased post-larval *Penaeus vannamei*. *Journal of Fish Diseases* 22: 377–386

Morales-Covarrubias, M.S., Del Carmen Bolan-Mejía, M., Vela Alonso AI, Fernandez-Garayzabal, J.F., and Gomez-Gil, B. 2018. *Streptococcus penaeicida* sp. nov., isolated from a diseased farmed Pacific white shrimp (*Penaeus vannamei*). *International Journal of Systematic and Evolutionary Microbiology* 68: 1490–1495

Mukherjee, S. and Bassler, B.L. 2019. Bacterial quorum sensing in complex and dynamically changing environments. *Nature Reviews Microbiology* 17: 371–382

Munro, J., Oakey, H.J., Bromage, E., and Owens, L. 2003. Experimental bacteriophage-mediated virulence in strains of *Vibrio harveyi*. *Diseases of Aquatic Organisms* 54: 187–194

Newaj-Fyzul, A., Al-Harbi, A.H., and Austin, B. 2014. Developments in the use of probiotics for disease control in aquaculture, *Aquaculture* 431: 1–11

Ngo, H.T., Nguyen, T.T.N., Nguyen Q.M. et al. 2016. Screening of pigmented *Bacillus aquimaris* SH6 from the intestinal tracts of shrimp to develop a novel feed supplement for shrimp. *Journal of Applied Microbiology* 121: 1357–1372

Ninawe, A.S. and Selvin, J. 2009. Probiotics in shrimp aquaculture: Avenues and challenges. *Critical Reviews in Microbiology* 35: 43–66

Ning, M., Xiu Y., Yuan M., Bi, J., Hou, L., Gu, W., Wang W., and Meng Q. 2019. *Spiroplasma eriocheiris* invasion into *Macrobrachium rosenbergii* hemocytes Is mediated by pathogen enolase and host lipopolysaccharide and β-1, 3-glucan binding protein. *Frontiers in Immunology* 10: 1852

Noga, E.J., Smolowitz, R., and Khoo, L.H. 2000. Pathology of shell disease in the blue crab, *Callinectes sapidus* Rathbun, (Decapoda: Portunidae). *Journal of Fish Diseases* 23: 389–399

Nunan, L.M., Lightner, D., Pantoja, C., and Gomez-Jimenez, S. 2014. Detection of acute hepatopancreatic disease (AHPND) in Mexico. *Diseases of Aquatic Organisms* 111: 81–86

Nunan, L.M., Pantoja, C.R., Gomez-Jimenez, S., and Lightner, D.V. 2013. '*Candidatus* Hepatobacter penaei,' an intracellular pathogenic enteric bacterium in the hepatopancreas of the marine shrimp *Penaeus vannamei* (Crustacea: Decapoda). *Applied and Environmental Microbiology* 79: 1407–1409

Nunan, L.M., Poulos, B.T., Navarro, S., Redman, R.M., and Lightner, D.V. 2010. Milky hemolymph syndrome (MHS) in spiny lobsters, penaeid shrimp and crabs. *Diseases of Aquatic Organisms* 91: 105–112

Oakey, H.J., Cullen, W.R., and Owens, L. 2002. The complete nucleotide sequence of the *Vibrio harveyi* bacteriophage VHML. *Journal of Applied Microbiology* 9: 1089–1098

Owens, L., Austin, D.A., and Austin, B. 1996. Effect of strain origin on siderophore production in *Vibrio harveyi* isolates. *Diseases of Aquatic Organisms* 27: 157–160

Papenfort, K. and Bassler, B.L. 2016. Quorum sensing signal-response systems in Gram-negative bacteria. *Nature Reviews in Microbiology* 14: 576–588

Phiwsaiya, K., Charoensapsri, W., Taengphu, S. et al. 2017. A natural *Vibrio parahaemolyticus* ΔpirA^{Vp} pirB^{Vp}+ mutant kills shrimp but produces neither PirVp toxins nor Acute Hepatopancreatic Necrosis disease lesions. *Applied and Environmental Microbiology* 83: e00680

Pope, E.C., Powell, A., Roberts, E.C., Shields, R.J., Wardle, R., and Rowley, A.F. 2011. Enhanced cellular immunity in shrimp (*Litopenaeus vannamei*) after 'vaccination'. *PLoS ONE* 6: e20960

Porter, L., Butler, M.W., and Reeves, R.H. 2001. Normal bacterial flora of the Caribbean spiny lobster, *Panulirus argus* and its possible role in shell disease. *Journal of Marine and Freshwater Research* 52: 1401–1405

Powell, A. and Rowley, A.F. 2005. Unchanged prevalence of shell disease in the edible crab *Cancer pagurus* four years after decommissioning of a sewage outfall at Langland Bay, UK. *Diseases of Aquatic Organisms* 68: 83–87

Powell, A. and Rowley, A.F. 2007. The effect of dietary chitin supplementation on the survival and immune reactivity of the shore crab, *Carcinus maenas*. *Comparative Biochemistry and Physiology A Molecular and Integrative Physiology* 147: 122–128

Prachumwat, A., Taengchaiyaphum, S., Mungkongwongsiri, N., Aldama-Cano, D.J., Flegel, T.W., and Sritunyalucksana, K. 2019. Update on early mortality syndrome/acute hepatopancreatic necrosis disease by April 2018. *Journal of the World Aquaculture Society* 50: 5–17

Prasad, S., Morris, P.C., Hansen, R., Meaden, P.G. and Austin, B. (2005). A novel bacteriocin-like substance (BLIS) from a pathogenic strain of *Vibrio harveyi*. *Microbiology* 151: 3051–3058

Quinn, R.A., Hazra, S., Smolowitz, R., and Chistoserdov, A.Y. 2017. Real-time PCR assay for *Aquimarina macrocephali* subsp. *homaria* and its distribution in shell disease lesions of *Homarus americanus*, Milne-Edwards, 1837, and environmental samples. *Journal of Microbiological Methods* 139: 61–67

Quinn, R.A., Metzler, A., Smolowitz, R.M., Tlusty, M., and Chistoserdov, A.Y. 2012. Exposures of *Homarus americanus* shell to three bacteria isolated from naturally occurring epizootic shell disease lesions. *Journal of Shellfish Research* 31: 485–493

Quiroz-Guzmán, E., Peña-Rodriguez, E., Vázquez-Juárez, R., Barajas-Sandoval, D.R., Balcázar, J.L. and Martínez-Díaz, S.F. 2018. Bacteriophage cocktails as an environmentally-friendly approach to prevent *Vibrio parahaemolyticus* and *Vibrio harveyi* infections in brine shrimp (*Artemia franciscana*) production. *Aquaculture* 492: 273–279

Raghu Patil, J., Desai, S.N., Roy, P., Durgaiah, M., Saravanan, R.S., and Vipra, A. 2014. Simulated hatchery system to assess bacteriophage efficacy against *Vibrio harveyi*. *Diseases of Aquatic Organisms* 12: 113–119

Rao, B.M. and Lalitha, K.V. 2015. Bacteriophages for aquaculture: Are they beneficial or inimical? *Aquaculture* 437: 146–154

Restrepo, L., Bayot, B., Arciniegas, S., Bajaña, L., Betancourt, B., Panchana, F., and Muñoz, A.R. 2018. PirVP genes causing AHPND identified in a new *Vibrio* species (*Vibrio punensis*) within the commensal *Orientalis* clade. *Scientific Reports* 8: 13080

Reverter, M., Tapissier-Bontemp, N., Sasal, P., and Saulnier, D. 2017. Use of medicinal plants in aquaculture. In *Diagnosis and Control of Diseases of Fish and Shellfish*, Austin B. and Newaj-Fyzul A. (eds.) pp. 223–261. NY: Wiley

Ringø, E., Olsen, R.E., Gifstad, T.Ø. et al. 2010. Prebiotics in aquaculture: A review. *Aquaculture Nutrition* 6: 117–136

Ringø, E., Van Doan, H., Lee, S. and Song, S.K. 2021. Lactic acid bacteria in shellfish: Possibilities and challenges. *Reviews in Fisheries Science & Aquaculture* 28: 139–169

Rivas, A.J., Lemos, M.L., and Osorio, C.R. 2013. Photobacterium damselae subsp. *damselae*, a bacterium pathogenic for marine animals and humans. *Frontiers in Microbiology* 4: 283

Romero, X., Turnbull, J.F., and Jimenez, R. 2000. Ultrastructure and cytopathology of a rickettsia-like organism causing systemic infection in the redclaw crayfish, *Cherax quadricarinatus* (Crustacea: Decapoda), in Ecuador. *Journal of Invertebrate Pathology* 76: 95–104

Rosen, B. 1967. Shell disease of the blue crab, *Callinectes sapidus*. *Journal of Invertebrate Pathology* 9: 348–353

Rowley, A.F. and Pope, E.C. 2012. Vaccines and crustacean aquaculture—a mechanistic explanation. *Aquaculture* 334–337: 1–11

Rowley, A.F. and Powell, A. 2007. Invertebrate immune systems—specific, quasi-specific, or nonspecific? *Journal of Immunology* 179: 7209–7214

Ruangpan, L., Danayadol, Y., Direkbusarakom, S., Siurairatana, S., and Flegel, T.W. 1999. Lethal toxicity of *Vibrio harveyi* to cultivated *Penaeus monodon* induced by a bacteriophage. *Diseases of Aquatic Organisms* 35: 195–201

Rushton-Mellor, S. and Whitfield, P.J. 1993. Transmission and scanning electron microscopic studies of crustacean shell disease in fish lice of the genus *Argulus* (Crustacea: Branchiura). *Journal of Zoology (London)* 229: 397–404

Rutherford, S.T. and Bassler, B.L. 2012. Bacterial quorum sensing: Its role in virulence and possibilities for its control. *Cold Spring Harbor Perspectives in Medicine* 2: a012427

Ruwandeepika, H.A.D., Bhowmick, P.P., Karunasagar, I., Bossier, P., and Defoirdt, T. 2011. Quorum sensing regulation of virulence gene expression in *Vibrio harveyi in vitro* and *in vivo* during infection of gnotobiotic brine shrimp larvae. *Environmental Microbiology Reports* 3: 597–602

Ruwandeepika, H.A.D., Jayaweera, T.S.P., Bhowmick, P.B., Karunasagar, I., Bossier. P., and Defoirdt, T. 2012. Pathogenesis, virulence factors and virulence regulation of vibrios belonging to the *Harveyi* clade. *Reviews in Aquaculture* 4: 59–74

Ruwandeepika, H.A.D., Karunasagar, I., Bossier, P., and Defoirdt, T. 2015. Expression and quorum sensing regulation of Type III secretion system genes of *Vibrio harveyi* during Infection of gnotobiotic brine shrimp. *PLoS ONE* 10: e0143935

Sahul Hameed, A.S. 1994. Experimental transmission and histopathology of brownspot disease in shrimp (*Penaeus indicus*) and lobster (*Panulirus homarus*). *Journal of Aquaculture in the Tropics* 9: 311–322

Sakai, T., Hirae, T., Yuasa, K. et al. 2007. Mass mortality of cultured kuruma prawn *Penaeus japonicus* caused by *Vibrio nigripulchritudo*. *Fish Pathology* 42: 141–147

Santos, L. and Ramos, F. 2018. Antimicrobial resistance in aquaculture: Current knowledge and alternatives to tackle the problem. *International Journal of Antimicrobial Agents* 52: 135–143

Sawyer, T.K. 1991. Shell disease in the Atlantic rock crab, *Cancer irroratus* Say, 1817, from the northeastern United States. *Journal of Shellfish Research* 10: 495–497

Shi, C., Xia, M., Li, R. et al. 2019. *Vibrio alginolyticus* infection induces coupled changes of bacterial community and metabolic phenotype in the gut of swimming crab. *Aquaculture* 499: 251–259

Shields, J.D. 2013. Complex etiologies of emerging diseases in lobsters (*Homarus americanus*) from Long Island Sound. *Canadian Journal of Fisheries and Aquatic Science* 70: 1576–1587

Shields, J.D. 2019. Climate change enhances disease processes in crustaceans: Case studies in lobsters, crabs, and shrimp. *Journal of Crustacean Biology* 39: 673–683

Shivanda Murthy, H., Li, P., Lawrence, A.L., and Gatlin, D.M. 2009. Dietary β-glucan and nucleotide effects on growth, survival and immune responses of Pacific white shrimp, *Litopenaeus vannamei*. *Journal of Applied Aquaculture* 21: 160–168

Sindermann, C.J. 1989. The shell disease syndrome in marine crustaceans. NOAA Technical Memorandum NMFS-F/NEC-64

Smith, A.L., Whitten, M.M.A., Hirschle, L. et al. 2014. Bacterial septicaemia in prerecruit edible crabs, *Cancer pagurus* L. *Journal of Fish Diseases* 37: 729–737

Smith, V.J., Brown, J.H., and Hauton, C. 2003. Immunostimulation in crustaceans: Does it really protect against infection? *Fish and Shellfish Immunology* 15: 71–90

Smolowitz, R., Chistoserdov, A.Y., and Hsu, A. 2005. A description of the pathology of epizootic shell disease in the American lobster, *Homarus americanus* H. Milne Edwards 1837. *Journal of Shellfish Research* 24: 749–756

Smolowitz, R.M., Bullis, R.A., and Abt, A.D. 1992. Pathological cuticular changes of winter impoundment shell disease preceding and during intermolt in the American lobster, *Homarus americanus*. *The Biological Bulletin (Woods Hole)* 183: 99–112

Snieszko, S.F. and Taylor, C.C. 1947. A bacterial disease of the lobster (*Homarus americanus*). *Science* 105: 500

Song, Y.-L., Cheng, W., and Wang, C.-H. 1993. Isolation and characterization of *Vibrio damsela* infectious for cultured shrimp in Taiwan. *Journal of Invertebrate Pathology* 61: 24–31

Srisala, J., Pukmee, R., McIntosh, R. et al. 2018. Distinctive histopathology of *Spiroplasma eriocheiris* in the giant river prawn *Macrobrachium rosenbergii*. *Aquaculture* 493: 93–99

Sriurairantana, S., Boonyawiwat, V., Gannonngiw, W., Laosutthipong, C., Hiranchan, J., and Flegel, T.W. 2014. White feces syndrome in shrimp arises from transformation, sloughing and aggregation of hepatopancreatic microvilli into vermiform bodies superficially resembling gregarines. *PLoS One* 9: e99170

Stebbing, P.D., Pond, M.J., Peeler, E, Small, H.J., Greenwood, S.J., and Verner-Jeffreys, D. 2012. Limited prevalence of gaffkaemia (*Aerococcus viridans* var. *homari*) isolated from wild-caught European lobsters *Homarus gammarus* in England and Wales. *Diseases of Aquatic Organisms* 100: 159–167

Stewart, J.E., Cornick, J.W., Spears, D.I., and McLeese, D.W. 1966. Incidence of *Gaffkya homari* in natural lobster (*Homarus americanus*) populations of the Atlantic region of Canada. *Journal of the Fisheries Research Board of Canada* 23: 1325–1330

Stewart, J.E., Cornick, J.W., Zwicker, B.M., and Arie, B. 2004. Studies on the virulence of *Aerococcus viridans* var. *homari*, the causative agent of gaffkemia, a fatal disease of homarid lobsters. *Diseases of Aquatic Organisms* 60: 149–155

Sullivan, T.J. and Neigel, J.E. 2018. Effects of temperature and salinity on prevalence and intensity of infection of blue crabs, *Callinectes sapidus*, by *Vibrio cholerae*, *V. parahaemolyticus*, and *V. vulnificus* in Louisiana. *Journal of Invertebrate Pathology* 151: 82–90

Sumithra, G., Reshma, K.J., Christo, J.P., Anusree, V.N., Drisya, D., Kishor, T.G., Revathi, D.N., and Sanil, N.K. 2019. A glimpse towards cultivable hemolymph microbiota of marine crabs: Untapped resource for aquatic probiotics/antibacterial agents. *Aquaculture* 501: 119–127

Tall, B.D., Fall, S., Pereira, M.R. et al. 2003. Characterization of *Vibrio fluvialis*-like strains implicated in Limp Lobster Disease. *Applied and Environmental Microbiology* 69: 7435–7446

Tanaka, K.R., Belknap, S.L., Homola, J.J., and Chen, Y. 2017. A statistical model for monitoring shell disease in inshore lobster fisheries: A case study in Long Island Sound. *PLoS ONE* 12: e0172123

Tang, K.F. and Lightner, D.V. 2014. Homologues of insecticidal toxin complex genes within a genomic island in the marine bacterium *Vibrio parahaemolyticus*. *FEMS Microbiology Letters* 361: 34–42

Tangprasittipap, A., Srisala, J., Chouwdee, S. et al. 2013. The microsporidian *Enterocytozoon hepatopenaei* is not the cause of white feces syndrome in whiteleg shrimp *Penaeus* (*Litopenaeus*) *vannamei*. *BMC Veterinary Research* 9: 139

Teunissen, O.S.P., Faber, R., Booms, G.H.R., Latscha, T., and Boon, J.H. 1998. Influence of vaccination on vibriosis resistance of the giant black tiger shrimp *Penaeus monodon* (Fabricius). *Aquaculture* 164: 359–366

Thompson, J., Gregory, S., Plummer, S., Shields, R., and Rowley, A.F. 2010. An *in vitro* and *in vivo* assessment of the potential of *Vibrio* spp. as probiotics for the Pacific White shrimp, *Litopenaeus vannamei*. *Journal of Applied Microbiology* 109: 1177–1187

Thrupp, T.J., Whitten, M.M.A., and Rowley, A.F. 2016. A novel bacterial infection of the edible crab, *Cancer pagurus*. *Journal of Invertebrate Pathology* 133: 83–86.

Torres, M., Reina, J.C., Fuentes-Monteverde, J.C. et al. 2018. AHL-lactonase expression in three marine emerging pathogenic *Vibrio* spp. reduces virulence and mortality in brine shrimp (*Artemia salina*) and Manila clam (*Venerupis philippinarum*). *PLoS ONE* 13: e0195176

Tran, L.H., Nunan, L., Redman, R.M. et al. 2013. Determination of the infectious nature of the agent of acute hepatopancreatic necrosis syndrome affecting penaeid shrimp. *Diseases of Aquatic Organisms* 105: 45–55

Truesdale, C.L., McManus, M.C., and Collie, J.S. 2019. Growth and molting characteristics of Jonah crab (*Cancer borealis*) in Rhode Island Sound. *Fisheries Research* 211: 13–20

Tseng, D.-Y., Ho, P.-L., Huang, S.-Y. et al. 2009. Enhancement of immunity and disease resistance in the white shrimp, *Litopenaeus vannamei*, by the probiotic, *Bacillus subtilis* E20. *Fish and Shellfish Immunology* 26: 339–344

Tubiash, H.S., Sizemore, R.K., and Colwell, R.R. 1975. Bacterial flora of the hemolymph of the blue crab, *Callinectes sapidus*: Most probable numbers. *Applied and Environmental Microbiology* 29: 388–392

Vago, C., Meynadier, G., Juchault, P., Legrand, J.J., Amargier, A., and Duthoit, J.L. 1970. A rickettsial disease of isopod crustaceans. *Comptes Rendus de l'Academie des Science D*, 271: 2061–2063

Vogan, C.L., Costa-Ramos, C., and Rowley, A.F. 2001. A histological study of shell disease syndrome in the edible crab *Cancer pagurus*. *Diseases of Aquatic Organisms* 47: 209–217

Vogan, C.L., Costa-Ramos, C., and Rowley, A.F. 2002. Shell disease syndrome in the edible crab, *Cancer pagurus*—isolation, characterization and pathogenicity of chitinolytic bacteria. *Microbiology* 148: 743–754

Vogan, C.L., Llewellyn, P.J., and Rowley, A.F. 1999. Epidemiology and dynamics of shell disease in the edible crab *Cancer pagurus*: A preliminary study of Langland Bay, Swansea, UK. *Diseases of Aquatic Organisms* 35: 81–87

Vogan, C.L., Powell, A., and Rowley, A.F. 2008. Shell disease in crustaceans—just chitin recycling gone wrong? *Environmental Microbiology* 10: 826–835

Vogan, C.L. and Rowley, A.F. 2002. Dynamics of shell disease in the edible crab *Cancer pagurus*: A comparative study between two sites on the Gower Peninsula, South Wales, UK. *Diseases of Aquatic Organisms* 52: 151–157

Walling, E., Vourney, E., Ansquer, D., Beliaeff, B., and Goarant, C. 2010. *Vibrio nigripulchritudo* monitoring and strain dynamics in shrimp pond sediments. *Journal of Applied Microbiology* 108: 2003–2011

Wang, F.-I. and Chen, J.-C. 2006. The immune response of tiger shrimp *Penaeus monodon* and its susceptibility to *Photobacterium damselae* subsp. *damselae* under temperature stress. *Aquaculture* 258: 34–41

Wang, G.L., Shan, J., Chen, Y., and Li, Z. 2006. Study on pathogens and pathogenesis of emulsification disease of *Portunus trituberculatus*. *Advances in Marine Science* 24: 526–531

Wang, P.C., Lin, Y.D., Liaw, L.L., Chern, R.S., and Chen, S.C. 2008. *Lactococcus lactis* subspecies *lactis* also causes white muscle disease in farmed giant freshwater prawns *Macrobrachium rosenbergii*. *Diseases of Aquatic Organisms* 79: 9–17

Wang, R., Zhong, Y., Gu, X., Yuan, J., Saeed, A.F., and Wang, S. 2015. The pathogenesis, detection, and prevention of *Vibrio parahaemolyticus*. *Frontiers in Microbiology* 6: 144

Wang, W. 2011. Bacterial diseases of crabs: A review. *Journal of Invertebrate Pathology* 106: 18–26

Wang, W., Gu, W., Ding, Z., Ren, Y., Chen, J., and Hou, Y. 2005. A novel *Spiroplasma* pathogen causing systemic infection in the crayfish *Procambarus clarkii* (Crustacea; Decapoda), in China. *FEMS Microbiology Letters* 249: 131–137

Wang, W., Gu, W., Gasparich, Ge., Bi, K., Ou, J., Meng, Q., Liang, T., Feng, Q., Zhang, J., and Zhang, Y. 2011. *Spiroplasma eriocheiris* sp. nov., associated with mortality in the Chinese mitten crab, *Eriocheir sinensis*. *International Journal of Systematic and Evolutionary Microbiology* 61: 703–708

Wang, W. and Gu, Z. 2002. Rickettsia-like organism associated with tremor disease and mortality of the Chinese mitten crab *Eriocheir sinensis*. *Diseases of Aquatic Organisms* 48: 149–153

Wang, W., Wen, B., Gasparich, G.E., Zhu, N., Rong, L., Chen, J., and Xu, Z. 2004. A spiroplasma associated with tremor disease in the Chinese mitten crab (*Eriocheir sinensis*). *Microbiology* 150: 3035–3040

Wang, X.-W. and Wang, J.-X. 2015. Crustacean hemolymph microbiota: Endemic, tightly controlled, and utilization expectable. *Molecular Immunology* 68: 404–411

Wang, Y. and Chandler, C. 2016. Candidate pathogenicity islands in the genome of 'Candidatus Rickettsiella isopodorum', an intracellular bacterium infecting terrestrial isopod crustaceans. *Peer Journal* 4: e2806

Wangman, P.J, Longyant, S., Taengchaiyaphum, S., Senapin, S., Sithigorngul, P., and Chaivisuthangku-ra, P. 2018. PirA & B toxins discovered in archived shrimp pathogenic *Vibrio campbellii* isolated long before EMS/AHPND outbreaks. *Aquaculture* 497: 494–502

Werren, J.H. and O'Neill, S.L. 1997. The evolution of heritable symbionts. In *Influential Passengers: Inherited Microorganisms and Invertebrate Reproduction*, O'Neill, S.L., Hoffmann, A.A., and Werren, J.H. (eds.) pp. 1–41. Oxford: Oxford University Press

Whitten, M.M.A., Davies, C.E., Kim, A. et al. 2014. Cuticles of European and American lobsters harbor diverse bacterial species and differ in disease susceptibility. *MicrobiologyOpen* 3: 395–409

Wootton, E.C., Woolmer, A.P., Vogan, C.L. et al. 2012. Increased disease calls for a cost-benefits review of marine reserves. *PLoS One* 7: e51615

Xhang, X.-J., Bai, X.-S., Yan, B.-L., Bi, K.-R., and Qin, L. 2014. *Vibrio harveyi* as a causative agent of mass mortalities of megalopa in the seed production of swimming crab *Portunus triberculatus*. *Aquaculture International* 22: 661–672

Xiong, J. 2018. Progress in the gut microbiota in exploring shrimp disease pathogenesis and incidence. *Applied Microbiology and Biotechnology* 102: 7343–7350

Xiong, J., Yu, W., Dai, W., Zhang, J., Qiu, Q., and Ou, C. 2018a. Quantitative prediction of shrimp disease incidence via the profiles of gut eukaryotic microbiota. *Applied Microbiology and Biotechnology* 102: 3315–3326

Xiong, J., Dai, W., Qiu, Q., Zhu, J., Yang, W., and Li, C. 2018b. Response of host-bacterial colonization in shrimp to developmental stage, environment and disease. *Molecular Ecology* 27: 3686–3699

Xiong, J., Zhu, J., Dai, W., Dong, C., Qiu, Q., and Li, C. 2017. Integrating gut microbiota immaturity and disease-discriminatory taxa to diagnose the initiation and severity of shrimp disease. *Environmental Microbiology* 19: 1490–1501

Young, J.S. and Pearce, J.B. 1975. Shell disease in crabs and lobsters from New York Bight. *Marine Pollution Bulletin* 6: 101–105

Yu, W., Wu, J.-H., Zhang, J., Yang, W., Chen, J., and Xiong, J. 2018. A meta-analysis reveals universal gut bacterial signatures for diagnosing the incidence of shrimp disease. *FEMS Microbiology Ecology* 94: 5

Zeng, S., Hou, D., Liu, J., Ji, P., Wang, S., He, J., and Huang, Z. 2019. Antibiotic supplement in feed can perturb the intestinal microbial composition and function in Pacific white shrimp. *Applied Microbiology and Biotechnology* 103: 3111–3122

Zermeño-Cervantes, L.A., Makarov, R., Lomelí-Ortega, C.O., Martínez-Díaz, S.F. and Cardona-Félix, C.S. 2018. Recombinant LysVPMS1 as an endolysin with broad lytic activity against *Vibrio parahaemolyticus* strains associated to acute hepatopancreatic necrosis disease. *Aquaculture Research* 49: 1723–1726

Zha, H., Jeffs, A., Dong, Y., and Lewis, G. 2018a. Potential virulence factors of bacteria associated with tail fan necrosis in the spiny lobster, *Jasus edwardsii*. *Journal of Fish Diseases* 41: 817–828

Zha, H., Jones, B., Lewis, G., Dong, Y., and Jeffs, A. 2018b. Pathology of tail fan necrosis in the spiny lobster, *Jasus edwardsii*. *Journal of Invertebrate Pathology* 154: 5–11

Zha, H., Jeffs, A., Dong, Y., and Lewis, G. 2018c. Characteristics of culturable bacteria associated with tail fan necrosis in the spiny lobster, *Jasus edwardsii*. *Bulletin of Marine Science* 94: 979–994

Zhang, L. and Orth, K. 2013. Virulence determinants for *Vibrio parahaemolyticus* infection. *Current Opinion in Microbiology* 16: 70–77

Zhang, X.-J., Bai, X.-S., Yan, B.-L., Bi, K.-R. and Qin, L. 2014. *Vibrio harveyi* as a causative agent of mass mortalities of megalopa in the seed production of swimming crab *Portunus triberculatus*. *Aquaculture International* 22: 661–672

Zhou, J., Fang, W., Yang, X., Zhou, S., Hu, L., Li, X., Qi, X., Su, H., and Xie, L. 2012. A nonluminescent and highly virulent *Vibrio harveyi* strain is associated with "bacterial white tail disease" of *Litopenaeus vannamei* shrimp. *PLoS ONE* 7: e29961

Zhu, J., Dai, W., Qiu, Q., Dong, C., Zhang, J., and Xiong, J. 2016. Contrasting ecological processes and functional compositions between intestinal bacterial community in healthy and diseased shrimp. *Microbial Ecology* 72: 975–985

Zokaeifar, H., Babaei, N., Saad, C.R., Kamarudin, M.S., Sijam, K., and Balcazar, J.L. 2014. Administration of *Bacillus subtilis* strains in the rearing water enhances the water quality, growth performance, immune response, and resistance against *Vibrio harveyi* infection in juvenile white shrimp, *Litopenaeus vannamei*. *Fish and Shellfish Immunology* 36: 68–74

Zokaeifar, H., Balcázar, J.L., Saad, C.R. et al. 2012. Effects of *Bacillus subtilis* on the growth performance, digestive enzymes, immune gene expression and disease resistance of white shrimp, *Litopenaeus vannamei*. *Fish and Shellfish Immunology* 33: 683–689

CHAPTER 16

Fungal and oomycete diseases of crustaceans

Andrew F. Rowley, Jenny Makkonen, and Jeffrey D. Shields

16.1 Introduction

True fungi, including microsporidians, belong to the Kingdom Fungi (Lee et al. 2008). Less is known about the diversity and role of fungi in the aquatic than the terrestrial environments. For example, according to Amend et al. (2019) *ca.* 1100 species of marine-specific fungi have been identified, yet this may represent only *ca.* 10% of the total species present (Jones 2011). Fungi in aquatic environments are found in a number of associations with animals, plants, and algae ranging from commensalism to parasitism. Recent studies have highlighted their association with diseases of coral (see Chapter 4) and amphibians where infections caused by the chytrid *Batrachochytrium dendrobatidis* have resulted in global declines of these hosts (e.g. O'Hanlon et al. 2018; Cohen et al. 2019). Crustaceans are also subject to diseases caused by true fungi but here they are frequently considered as secondary invaders or opportunistic pathogens.

Oomycetes, sometimes called 'water moulds' were originally thought to be basal fungi due to their shared morphology. However, molecular taxonomy has shown that they are a distinct lineage currently placed within the monophyletic Stamenopiles-Alveolata-Rhizaria supergroup (McCarthy and Fitzpatrick 2017). They are important pathogens of plants and animals including fish and arthropods. Examples include the fish pathogen, *Saprolegnia parasitica*, and the closely related agent of crayfish plague, *Aphanomyces astaci* (Phillips et al. 2007).

This aim of this chapter is to provide a succinct overview of infections of crustaceans caused by both fungi, including microsporidians, and oomycetes with particular emphasis on the pathology of these interactions.

16.2 Principal diseases

The principal diseases of crustaceans caused by fungi and oomycetes are summarised in Tables 16.1–16.3. Reports of their identification, especially of true fungi, based on morphological and growth characteristics alone, may prove to be unreliable. Older literature reporting oomycetes as members of the Kingdom Fungi and/or being 'fungal-like' can result in confusion of their current taxonomic status.

16.3 Diseases caused by true fungi

16.3.1 Black gill disease

This condition has been reported in a number of crustaceans including the Kuruma shrimp (*Penaeus japonicus*; Rhoobunjongde et al. 1991), the black tiger shrimp, *Penaeus monodon* (Khoa et al., 2004), the Pacific white shrimp, *Litopenaeus* (*Penaeus*) *vannamei* (Karthikeyan et al. 2015), the American lobster, *Homarus americanus* (Lightner and Fontaine 1975), and the rock lobster, *Panulirus ornatus* (Nha et al. 2009). Such infections are brought about by several species of *Fusarium* and possibly *Aspergillus awamori* (Karthikeyan et al. 2015). The gills appear black and damaged due to extensive haemocytic infiltration

Andrew F. Rowley, Jenny Makkonen and Jeffrey D. Shields, *Fungal and oomycete diseases of crustaceans*. In: *Invertebrate Pathology*. Edited by Andrew F. Rowley, Christopher J. Coates and Miranda M.A. Whitten, Oxford University Press. © Oxford University Press (2022). DOI: 10.1093/oso/9780198853756.003.0016 to encapsulate developing fungal hyphae in the gill lamellae. Such nodules are heavily melanised (see Chapter 1 for details of encapsulation responses in invertebrates). In heavy infections, death of the host can occur due to blockage of the gills resulting from this extensive haemocytic response to the presence of the pathogen (Solangi and Lightner 1976).

16.3.2 Burn spot disease (a form of shell disease)

There are several reports of burn spot disease in crustaceans caused by fungi (Table 16.1). In most cases, the causative agent is reported as belonging to the genus *Fusarium* sp. (see Sindermann 1989 for review). Areas of cuticular damage with associated melanisation represent sites of active penetration by fungal hyphae. Makkonen et al. (2013) isolated *Fusarium avenaceum* from cuticular lesions of the freshwater crayfish, *Astacus astacus*, from Estonia. They characterised the fungus using ITS and EF1alpha-gene sequences and were able to prove Koch's postulates by showing that infection by the fungus caused the symptoms of burn spot disease (Figure 16.1). It is unclear whether this disease is localised in the cuticle or penetrates underlying tissues within the haemocoel.

16.3.3 *Metschnikowia bicuspidata* infections

Infections caused by this yeast-like fungus have been reported in the giant freshwater prawn (Chen et al. 2007), the Chinese mitten crab (Bao et al., 2021) and *Daphnia* spp. (Codreanu and Codreanu-Balcescu 1981; Shocket et al. 2018). Similar infections have also been reported from amphipods (Messick et al. 2004). This fungus infects *Daphnia* by ingestion of spores during feeding. It passes through the gut barrier and infects the host's tissues. Spores formed by infected hosts are released into the water upon their death to infect new hosts (Dallas et al. 2018). Infected hosts have reduced fecundity, and mortality occurs *ca.* 14 days post-association. The association between *M. bicuspidata* and its hosts has proven to be a useful model to explore host—parasite interactions within an ecological context (Hall et al. 2009, 2010; Overholt et al. 2012; Dallas et al. 2018; Shocket et al. 2018). The relationship between environmental temperature and parasitism has demonstrated

that this driver affects the rate of ingestion of spores (i.e. feeding rates) by the host rather than altering the virulence of the parasite (Shocket et al. 2018). *M. bicuspidata* spores are susceptible to UV inactivation and hence infection rates are reduced in the summer months during periods of high UV radiation. Lakes with high UV transparency are therefore less likely to harbour outbreaks of this disease (Overholt et al. 2012). Other chytrids have also been shown to affect natural populations of Cladocera (e.g., Johnson et al. 2006, 2009).

16.3.4 Lethargic crab disease

This is a disease of mangrove land crabs, *Ucides cordatus*, found in areas of the Brazilian coast (Table 16.1). This condition has spread both north and south from its original centre (Boeger et al. 2005, 2007) and has caused a reduction in landings of this economically important host species. Systemic infections develop quickly, with hyphae, conidia and yeast cells present in the tissues. Affected animals display symptoms of lethargy and reduced walking activity perhaps linked to neural damage. The main cause of this condition is a black yeast, *Exophiala cancerae* (Orélis-Ribeiro et al. 2011) although Vicente et al. (2012) also found another fungus, *Fonsecaea brasiliensis*, associated with the condition. This second pathogenic fungus appears to be less virulent and may be a secondary invader to *E. cancerae*. As in many fungal infections, there is evidence of extensive host response to the presence of the pathogen that is overwhelmed in most cases resulting in mortality. Boeger et al. (2007) suggested that some crabs may be resistant to infection depending on their level of immune competence.

16.3.5 *Ophiocordyceps*-like infections of edible crabs (*Cancer pagurus*)

A yeast-like pathogen has been discovered in edible crabs collected off the south coast of the UK. The original description of this infection noted that all infected specimens had co-infections with the dinoflagellate parasite, *Hematodinium*, that is common in both species (Stentiford et al. 2003; see also Chapter 17). This resulted in the assumption that the yeast-like agent is a secondary invader

Table 16.1 An overview of the principal fungal diseases of crustaceans

Disease	Disease causing agent(s)	Host range	Geographical range	Pathology	Key references
Black gill disease	Various species of *Fusarium*, including *F. solani*, and *Aspergillus awamori*	Wide including shrimp and lobsters	Widespread	Characteristic haemocytic response in the gills resulting in melanised cellular capsules formed by haemocytes surrounding the fungi	Lightner and Fontaine 1975; Solangi and Lightner 1976; Khoa et al. 2004; Karthikeyan et al. 2015
Burn spot disease	*Fusarium* spp. including *F. avenaceum*	Crayfish	Finland	Fungus penetrates through cuticle leading to formation of melanised lesions (Figure 16.1)	Sindermann 1989; Makkonen et al. 2013
Eroded swimmeret syndrome	*Fusarium* spp. and *Aphanomyces astaci* (oomycete)	Crayfish (*Pacifastacus leniusculus*)	Sweden, Finland	Progression erosion of swimmerets with melanisation. Only found in females. Causes reduction in fecundity	Edsman et al. 2015
	Gilbertella persicaria	Black tiger shrimp, *Penaeus monodon*	India	Degeneration of muscle and hepatopancreas, gill damage leading to potential respiratory dysfunction. Poor growth	Karthikeyan and Gopalakrishnan 2015
	Metschnikowia bicuspidata	*Daphnia* spp., Chinese mitten crabs, *Eriocheir sinensis* and freshwater prawns, *Macrobrachium rosenbergii*	Europe, China, USA, Taiwan	In *Daphnia*, spores taken up upon feeding, cross the gut barrier and cause haemocoelic infection. In *M. rosenbergii*, causes necrosis of muscle, milky haemolymph and swollen cephalothorax	Bao et al 2021; Codreanu and Codreanu- Balcescu 1981; Chen et al. 2007; Shocket et al. 2018

Disease	Fungus	Host	Location	Description	References
Lethargic crab disease	*Exophiala cancerae* and *Fonsecaea brasiliensis*	Mangrove-land crab, *Ucides cordatus*	Brazil	General necrosis of tissues including heart, hepatopancreas and gills. Strong haemocytic response to presence of yeast-like fungi in all tissues	Boeger et al 2005, 2007; Vicente et al. 2012
	Ophiocordyceps-like	Crabs, *Cancer pagurus* and *Necora puber*	UK (South coast of England and South Wales)	Widespread infection of all tissues with large numbers of yeast and spore-like forms in the haemolymph (Figure 16.2a). Strong haemocytic response to presence of fungus (Figure 16.2d). Some affected crabs have co-infections with the dinoflagellate parasite, *Hematodinium*	Stentiford et al. 2003; Smith et al. 2013
	Hypocrealean fungus	European shore crabs, *Carcinus maenas*	U.K. (South Wales)	Widespread infection of all tissues with large numbers of yeast and spore-like forms in the haemolymph. Most affected crabs have co-infections with other pathogens and parasites	Davies et al. 2020a
Black mat syndrome	*Trichomaris invadens*	Tanner, (*Chionoecetes bairdi*) and snow (*C. opilio*) crabs	North America	Widespread infection of tissues, fungus may inhibit moulting	Hibbits et al. 1981

(a)

(b)

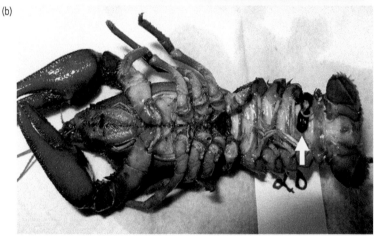

Figure 16.1 Characteristic melanised lesions (unlabelled arrows) on the carapace of the noble crayfish, *Astacus astacus* following infection with *Fusarium avenaceum*. (a) Dorsal surface view. (b) Ventral surface view.

resulting from an already weakened host. Smith et al. (2013) also found this yeast-like infection in juvenile *C. pagurus* collected from South Wales, UK but only at low prevalence (< 10%). They noted that although some of the crabs also had co-infections with *Hematodinium*, there were some individuals that only had the yeast-like agent, showing that it is a primary pathogen (Smith et al. 2013; Smith and Rowley 2015). To explore this putative relationship further, they took groups of crabs with and without natural *Hematodinium* infections and challenged both with equal numbers of the yeast-like pathogen. The presence of *Hematodinium* slowed down the multiplication of the fungus rather than enhanced it and such crabs died faster than the

non-*Hematodinium*-containing crabs suggesting that the fungus may allow this dinoflagellate to more rapidly overwhelm its host. This 'arrangement' allows both *Hematodinium* and the fungus to 'gain' from this association. Initial studies suggest that this yeast-like fungus is a member of the genus *Ophiocordyceps* closely related to entomopathogenic *O. gracilis* and *O. entomorrhiza* (Smith et al. 2013). Infected crabs show extensive numbers of fungal elements in the haemolymph (Figure 16.2b–d). Other tissues including the gills and hepatopancreas are replete with fungi both free and within heavily melanised nodules (Figure 16.2d). These fungi can apparently escape the nodules to cause further septicaemia (Smith et al. 2013).

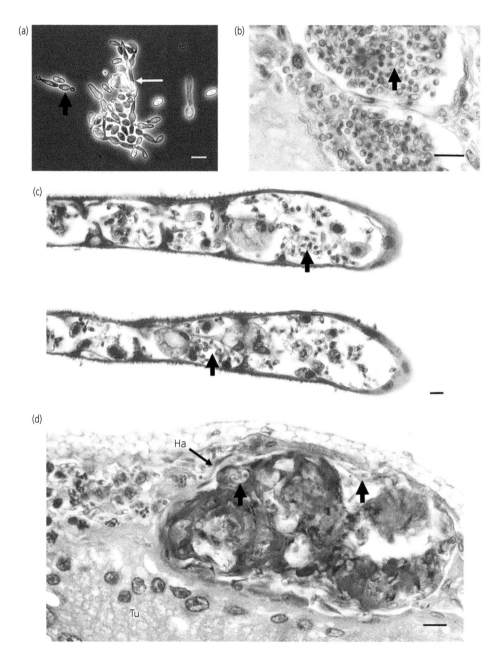

Figure 16.2 Infection of edible crabs, *Cancer pagurus* by an *Ophiocordyceps*-like fungal pathogen. (a) Micrograph showing various forms of the infecting fungus in culture with both yeast-like (large unlabelled arrow) and hyphal (small unlabelled arrow) forms. (b) Large numbers of fungi (unlabelled arrow) in an infected crab in haemolymph 'vessels' running through the hepatopancreas. (c) Fungi (unlabelled arrows) in the terminal gill filaments. (d) Nodule formation is response to the presence of fungi in circulation. Note melanisation (brown colour) in the centre of the nodule and fungi within this structure (unlabelled arrows). The thin encapsulating layer of haemocytes (Ha) around the core of the nodule is apparently incomplete such that fungi can escape this structure. Hepatopancreatic tubule (Tu). Scale bars = 10 µm. Source: micrographs courtesy of Dr Amanda Smith.

16.3.6 Mycosis of the European shore crab, *Carcinus maenas*

Shore crabs surveyed in South Wales, UK have a low prevalence (< 0.5%) of a fungal disease caused by an unidentified, and potentially new, hypocrealean fungus (Davies et al. 2020a). While the pathology and morphology of this disease-causing agent is similar to that found in diseased edible crabs surveyed in the same location (Smith et al. 2013) it is taxonomically distinct. Shore crabs infected by the fungus often had co-infections with either *Hematodinium* sp. or a newly discovered haplosporidium, *Haplosporidium carcini* (Davies et al. 2020b).

16.3.7 *Trichomaris invadens* infections of snow and tanner crabs

The ascomycete *Trichomaris invadens* infects the shell of crabs, *Chionoecetes bairdi and C. opilio*, causing a condition known as black mat syndrome. The fungus is naturally coloured black and its encrusting hyphae give the carapace of infected crabs a blackened appearance (Sparks and Hibbits 1979). Hyphae can penetrate through the cuticle into the epidermis and organs, possibly leading to death (Sparks 1982). Prevalence levels can be quite high in some host populations (Hicks 1982; Dick et al. 1998). Infected crabs may be able to moult out of the disease but the fungus may inhibit moulting.

16.4 Diseases caused by microsporidian fungi

Microsporidians are highly derived parasites, often with exquisite host specificity (Smith 2009). All taxa are obligate, intracellular, or even intranuclear parasites that produce small (< 6 μm) unicellular spores with a highly resistant, chitin-rich spore wall. In some taxa, the life cycle includes merogony and sporogony but not gamogony, while in others, there is sexual reproduction via gamogony. The mature spore is the infectious stage and possesses a specialised polar tube that can be quickly fired into a naïve host cell (Franzen 2004). The sporoplasm then invades the host cell or nucleus and begins vegetative merogony. Several genera in crustaceans such as *Abelspora* and *Endoreticulatus*, live in a host-derived parasitophorous vacuole and others such as *Ameson* live directly in the cytoplasm or nucleoplasm. In addition, several genera, e.g., *Thelohania* and *Pleistophora*, develop spores inside a characteristic sporophorous vesicle (SPV), with eight or 32 spores arranged in a packet formed by the SPV; but other genera, such as *Ameson, Abelspora, Nadelspora*, and *Ormieresia*, develop spores freely in the cytoplasm of the host cell. Microsporidians are typically monoxenous, having one host but a few taxa are heteroxenous, having more than one host, such as several species of *Amblyospora* that alternate between mosquitoes and copepods (Andreadis 1985), and *Desmozoon lepeophtherii*, which uses salmon and *Lepeoptherius salmonis*, the 'salmon louse', as well as other fish-copepod systems (Table 16.2; Stentiford et al. 2018). Recent advances in molecular techniques have facilitated descriptions of new taxa in crabs, lobsters, and other crustaceans (Stentiford et al. 2010, 2014), as well as highlighted new phylogenetic associations (Bacela-Spychalska et al. 2018; Vávra et al. 2017).

Of *ca*. 200 genera of microsporidians, more than 60 infect crustaceans (Table 16.2). They are frequently observed in crustaceans but they are difficult to diagnose, not only because of their size but because they require ultrastructural and molecular characterisation for proper identification (Figure 16.3a–d). Nonetheless, at least 20 genera have been described from decapods, nine genera from amphipods, two genera and species from isopods, 20 genera from copepods, 14 genera from cladocerans, five genera from anostracans, and one genus from ostracods. Surprisingly, none have been reported from barnacles, mysids, and euphausiids. The fact that only two species, *Mrazekia argoisi* and *Octosporea porcellioi*, infect isopods is puzzling, because these well-studied hosts are common in terrestrial and aquatic habitats, often feeding on detritus, and so one would expect them to have a higher diversity of these parasites. In contrast, the amphipods, which are often found in similar habitats with isopods, have a rich diversity of microsporidians (Table 16.2).

Notably, microsporidians in crustaceans are often discovered by accident. For example, an undescribed species of *Ameson*, was found during a study of the reproductive anatomy of its crab host

Table 16.2 Microsporidian genera that infect crustaceans, including number of known species, host taxa, and site of infection. Taxa in parentheses are alternate or definitive hosts.

Parasite	Number of species	Host taxa	Site of infection
Abelspora	1	Decapoda	Hepatopancreas
Agglomerata	3	Cladocera, Copepoda	Adipose cells, hypodermis
Agmasoma	1	Decapoda	Muscle
Alfvenia	3	Copepoda	Adipose tissue
Amblyospora	6	Copepoda, Amphipoda, (+mosquitoes)	Adipose tissue
Ameson	4	Decapoda	Muscle
Anncaliia[1]	1	Amphipoda	Fat body, myocytes
Anostracospora	1	Anostraca	Gut epithelium
Areospora	1	Decapoda	Muscle
Berwaldia	2	Cladocera	Adipose and connective tissue
Binucleata	1	Cladocera	Epithelial cells
Cucumispora	3	Amphipoda	Muscle
Cougourdella	1	Copepoda	?
Desmozoon	1	Copepoda	Hypodermis
Dictyocoela	8	Amphipoda	Muscle, gonadal cells
Duboscquia	1	Copepoda, (+mosquitoes)	?
Endoreticulatus	1	Anostraca	Intestinal epithelium
Enterocytospora	1	Anostraca	Intestinal epithelium
Enterocytozoon	1	Decapoda	Hepatopancreas
Enterospora	2	Decapoda	Hepatopancreas
Facilispora	1	Copepoda	Connective tissue
Fibrillanosema	1	Amphipoda	Gonadal tissue
Flabelliforma	2	Ostracoda, Copepoda	Adipose tissue, hypodermis, ovaries
Globulispora[2]	1	Cladocera	Intestinal epithelium
Glugoides	1	Cladocerans	Gut epithelium
Gurleya	1	Cladocerans	Hypodermal tissue
Gurleyides	1	Cladocerans	Adipose tissue
Hamiltosporidium	2	Cladocera	Fat bodies, hypodermis, ovaries
Hepatospora	1	Decapoda	Hepatopancreas
Hyalinocysta	2	Copepoda, (+Diptera)	?
Holobispora	1	Copepoda	?
Inodosporus	2	Decapoda, (+fish)	Muscle
Lanatospora	3	Copepoda	Adipose tissue
Larssonia	1	Cladocera	Adipose tissue
Marssoniella	1	Copepoda	Ovarian tissue
Mrazekia	2	Isopoda, Copepoda	Adipose tissue
Myospora	1	Decapoda	Muscle
Nadelspora	1	Decapoda	Muscle
Nelliemelba	1	Copepoda	Muscle
Norlevinea	1	Cladocera	Ovarian
Nosema	5	Anostraca, Amphipoda	Muscle
Nosemoides	2	Copepoda	Gut epithelium
Octosporea	2	Cladocera, Isopoda, Amphipoda	Muscle
Ordospora	1	Cladocera	Gut epithelium
Ormieresia	1	Decapoda	Muscle
Ovipleistophora	1	Decapoda, (+fish)	Muscle
Parahepatospora[3]	1	Decapoda	Hepatopancreatic epithelia
Parathelohania	1	Copepoda, (+mosquitoes)	Systemic
Paranucleospora		Copepoda	Epithelial cells
Perezia	4 spp., 1 sp. in crustaceans	Decapoda	Muscle

continued

Table 16.2 *Continued*

Parasite	Number of species	Host taxa	Site of infection
Pleistophora	31 spp., 5 spp. in crustaceans	Cladocera, Amphipoda, Decapoda	Muscle
Pseudoberwaldia[4]	1	Cladocera	Adipose tissue
Pyrotheca	1 sp. in crustaceans	Copepoda	?
Stempellia	1 sp. in crustaceans	Copepoda	?
Thelohania	14	Decapoda	Muscle
Trichotuzetia	1	Copepoda	Systemic
Triwangia	1	Decapoda	Connective tissue
Tuzetia	4	Copepoda, Amphipoda, Decapoda	Adipose tissue
Vairimorpha	2	Decapoda	Muscle
Vavraia	3	Anostraca, Decapoda	Muscle

[1] Tokarev et al. (2018); [2] Vávra et al. (2016); [3] Bojko et al. (2017); [4] Vávra et al. (2019). Source: compiled from Bronnvall and Larsson 1995; Stentiford et al. 2012; Becnel et al. 2014; Bojko and Ovarchenko 2019 and WoRMS Editorial Board 2020, with several recent additions indicated with footnotes to their authorities

(Walker and Hinsch 1972). In contrast, those crustaceans used as common laboratory models, have a well-described fauna, including 14 genera from *Daphnia* spp. and five genera from *Artemia* spp.

Given their specialised lifestyle, microsporidians have a predilection for a specific tissue type (Figure 16.3a–d). In crustaceans, most species infect the musculature of their hosts, including skeletal and cardiac muscles (Figure 16.3b-d). Meronts of the muscle-dwelling parasites grow along individual muscle fibres, eventually lysing them, with spores released into the haemolymph where they infect new muscle fibres (Weidner 1970). Heavily infected crustaceans are lethargic, often with an altered colour. Those living in muscles cause it to become friable, with whitish streaks or milky white coloration of the muscle fibres. Those that infect other tissues may not develop systemic infections and so can be more difficult to find or diagnose properly. Those genera that infect epithelial cells, e.g. *Enterospora*, *Hepatospora*, and possibly *Abelspora* tend to cause focal infections that may be self-limiting and do not lead to outward signs of disease; however, they can cause focal destruction of infected tissues (e.g. Stentiford et al. 2007).

Microsporidians use horizontal and vertical transmission to infect new hosts. Horizontal transmission occurs through the ingestion of spores, typically in the carcass of an infected host, by cannibalism or predation on an infected host, or by filter feeding or detrital feeding of spores released

through the faeces of an infected host. Vertical transmission is accomplished through infection of the ovary with transovarial spread into progeny (Terry et al. 2004). A few taxa are capable of both types of transmission (Haag et al. 2020). Horizontal transmission is usually associated with highly pathogenic infections whereas vertical transmission is associated with more benign infections (Mangin et al. 1995). Transmission appears to be either horizontal through faecal waste, as with *Glugoides intestinalis* and *Ordospora* sp., or vertically through transovarial transmission as with *Tuzetia* spp. (Mangin et al. 1995), or both, as with *Hamiltosporidium* spp. (Haag et al. 2020). These are ideal parasites to study the evolution and pathogenicity of different transmission pathways because their hosts can be maintained in laboratory studies.

Microsporidian infections can have direct and indirect effects on host populations. In *Artemia* spp., they are known to have a high prevalence in solar salterns (Martinez et al. 1992) and they can reduce female fecundity, in effect acting as partial castrators (Rode et al. 2013a). An introduced species of *Artemia* had a higher prevalence of *Enterospora artemiae*, whereas a native species had a higher prevalence of *Anostracaspora rigaudi* (Rode et al. 2013a) and this has implications to disease ecology as both parasites are capable of spill over into new hosts (Rode et al. 2013b; Lievens et al. 2019). Direct negative effects on host populations have been shown in *Daphnia magna* and other cladocerans.

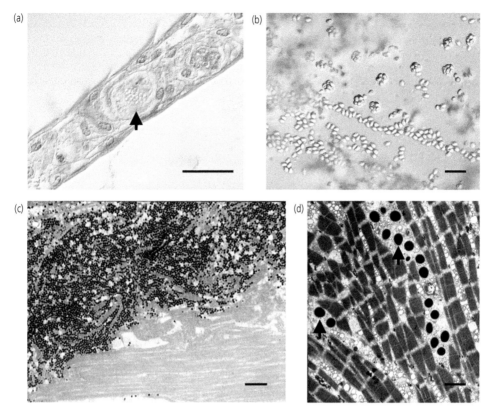

Figure 16.3 (a) Disseminated infection of *Pleistophora*-like microsporidian (arrowhead) in the gill of the blue crab, *Callinectes sapidus*. H&E, Nomarski. Scale bar = 20 µm. (b) Infection of *Thelohania* sp. in a muscle smear from *C. sapidus*. Scale bar = 20 µm. (c) *Ameson* sp. in the musculature of the crab, *Portunus armatus*. Giemsa. Scale bar = 10 µm. (d) Electron micrograph showing spores (arrowheads) of *Ameson* sp. along muscle fibres of *P. armatus*. Scale bar = 1 µm.

Infections can affect host fitness and exert strong control over laboratory populations (e.g. Stirnadel and Ebert 1997; Ben Ami et al. 2013; Haag et al. 2020).

Microsporidian infections in amphipods have also been well studied (for review, see Bojko and Ovarchenko 2019). Several genera use vertical transmission (Dunn et al. 1993; Galbreath et al. 2010; Terry et al. 2004) and are known to affect host sexual systems through distortion of the sex ratio through feminisation of males (Dunn and Rigaud 1998; Haine et al. 2007; Mautner et al. 2007). Those genera that use horizontal transmission are generally pathogenic and, in some cases, may exert significant ecological effects on their amphipod hosts (Fielding et al. 2005; MacNeil et al. 2011; Bacela-Spychalska et al. 2014). Microsporidians have also been used to investigate host preferences in a species complex

(Quiles et al. 2019), host specificity and invasion threat (Bacela-Spychalska et al. 2012, 2018; Wilkinson et al. 2011), parasite speciation in cryptic host species (Quiles et al. 2019; Park et al. 2020), competition between hosts (Bunke et al. 2019), and several other ecological features.

Microsporidians in decapods are known mostly as pathogens affecting commercially important species or as parasites in well-studied ecologically important species. Indeed, infected crabs and shrimp are often referred to colloquially as 'cotton' crab or shrimp because of the chalky white coloration and mealy texture of the flesh. In many respects they are not as well-known as those using amphipods and this may be because most taxa have lower prevalence levels in their host populations. For example, only two microsporidians, *Ameson hernnkindi* and *Myospora metanephrops*, have been

> ## Box 16.1 Global declines in crayfish are caused by multiple drivers including disease
>
> There are ca. 590 species of freshwater crayfish mainly found in the Northern Hemisphere. It has been estimated that ca. 32% of these are at risk of extinction (Richman et al. 2015) due to a variety of reasons including habitat loss, pollution, over-harvesting, climate change, and the introduction of non-indigenous crayfish species. In Europe, the introduction of the North American signal crayfish, *Pacifastacus* *leniusculus* is the main cause of declines of native species including the noble crayfish, *Astacus astacus* (Holdich et al. 2014). Not only has this brought crayfish plague to these countries but the invaders have altered the ecosystem due to their physical activities. In the UK alone, it is estimated that this costs £2 million p.a. in attempts to control and manage these invasive crayfish (Holdich et al. 2014).

described from lobster hosts and these had very low prevalence levels in these commercially exploited species (Stentiford et al. 2010; Small et al. 2019). This is because infections were identified from rarely captured, lethargic, obviously diseased hosts, which typically do not enter pots or traps. Another reason these parasites may not be well known is that their hosts are rarely examined for parasites. For example, a survey of *Eriocheir sinensis* reported prevalence levels of *Hepatospora eriocheir* as high as 70%, with overt signs of infection, yet this parasite was only recently described (Stentiford et al. 2011; Wang and Chen 2007). Because of their seemingly low prevalence, most studies in decapods have been at the level of taxonomic and systematic characterisations; however, the primary pathophysiological and biochemical studies on infections in crustaceans have been done in crabs (Vernice and Sprague 1970; Findley et al. 1981; Vivarès and Cuq 1981; Ding et al. 2018).

16.5 Diseases caused by oomycetes

16.5.1 Crayfish plague (krebspest)

The causative agent of crayfish plague is the oomycete *Aphanomyces astaci* that has contributed to substantial declines in crayfish native to Europe (Holdich et al. 2009; Box 16.1). North American crayfish act as reservoirs of the disease and their transportation into Europe over the last two centuries has resulted in the displacement and infection of native crayfish populations. Particularly badly affected are most species of European crayfish including the noble crayfish, *Astacus astacus*, leaving them classified as endangered (Holdich et al. 2009).

The basis of susceptibility *vs.* resistance to *A. astaci* is partially understood and is linked to the interaction between the oomycete hyphae and the host's cuticular defences. North American crayfish can carry *A. astaci* on the body surface but the hyphae of this parasite are normally disarmed within the cuticle because they activate the host's prophenoloxidase (proPO) cascade resulting in the formation of a variety of antimicrobial factors (Söderhäll and Ajaxon 1982; Cerenius et al. 2003; see Chapter 1 for details of this system). The constant overexpression of the proPO system is the main mechanism underlying the increased resistance of the North American crayfish species against the crayfish plague. In these crayfish species, proPO is continuously overexpressed, inhibiting the pathogen growth and, hence, the infection development. In native European crayfish species, the activation of the proPO system is often too inefficient and slow to successfully combat the disease (Cerenius et al. 2003).

Immunocompromised North American crayfish can, however, become diseased by *A. astaci* reflecting the importance of the cuticular immune system in defending such animals. When most European crayfish are exposed to the zoospores of *A. astaci*, they penetrate the cuticle with varying levels of activation of the proPO system depending on the host and the genotype of parasite (Figure 16.4.a, b). Overall, susceptible hosts infected by highly virulent strains of *A. astaci* often show little melanisation (a product of the proPO system; see Chapter 1 for details of this system and its products) in the cuticle around penetrating hyphae while resistant crayfish actively

(a)

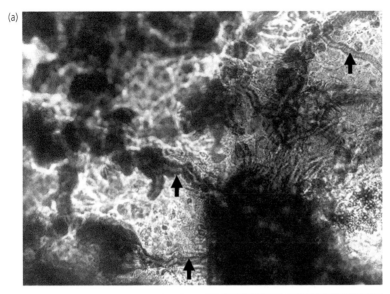

(b)

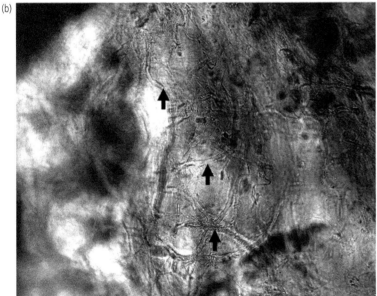

Figure 16.4 Infection of noble crayfish, *Astacus astacus*, by different strains of *Aphanomyces astaci*. (a) Melanised hyphae (unlabelled arrows) in a cleaned abdomen cuticle preparation of a low virulent strain of *A. astaci*. (b) Non-melanised hyphae (unlabelled arrows) in an acute infection caused by a virulent strain of *A. astaci*. Source: micrographs courtesy of Dr Satu Viljamaa-Dirks, OIE Reference Laboratory for Crayfish Plague, Finnish Food Authority, Kuopio, Finland.

suppress hyphal penetration with the deposition of this pigment and associated bioactive intermediates.

Susceptible crustaceans to *A. astaci* are not just found in Europe alone. For example, Australian species of crayfish are susceptible to crayfish plague (Unestam, 1975) and freshwater crabs including *Potamon potamios* from Turkey can become infected by this parasite (Svoboda et al. 2014). Amphipods and isopods sharing the same environment do not appear to be susceptible to crayfish plague

(Svoboda et al. 2014, 2017). Signal crayfish, *P. leniusculus*, can host species of *Aphanomyces* that are closely related to *A. astaci* but show no virulence towards this host. However, their effects on other crayfish species in the aquatic environment remain unknown (Makkonen et al. 2019).

In recent years, the world-wide distribution of *A. astaci* has been highlighted in several studies. It has been detected in invasive North American crayfish in Japan (Mrugała et al. 2017), Indonesia (Putra et al. 2018) and South America (Peiró et al. 2016). No reports exist yet of *A. astaci* population collapses among its native hosts in North America (Makkonen et al. 2019) because these hosts are resistant to its transmissive stage. However, North American crayfish species do act as carriers of the disease agent outside their original distribution range in the wild (e.g. Svoboda et al. 2017) and in captivity (Panteleit et al. 2017).

The transmission of crayfish plague is horizontal via biflagellate zoospores that are chemotactically attracted to the host (Cerenius and Söderhäll 1984a). The zoospores are capable of encystment and later re-emergence that extends their survival in water (Cerenius and Söderhäll 1984b). *A. astaci* undergoes repeated zoospore emergence for at least three generations (Cerenius and Söderhäll 1984b), which has been proposed to represent an adaptation to the parasitic lifestyle. When the zoospores of *A. astaci* find a host, they encyst and germinate. Germination is often successful in wounds, body openings and in the soft cuticle of the crayfish (Unestam and Weiss 1970). When the acute stage of the disease is reached, the hyphae grow out of the cuticle and form sporangia. Primary spores are formed in sporangia and then released from the hyphal tip. Secondary swimming zoospores are then released from the primary cysts into the ambient water (Cerenius and Söderhäll 1984b).

Currently, *A. astaci* strains have been mainly isolated from infected European crayfish or North American crayfish stocked into European freshwaters. There are six genotypes of *A. astaci* recognised (Huang et al. 1994; Svoboda et al. 2017; Oidtmann 2018; Panteleit et al. 2019; Table 16.4). Groups A-E originate from a variety of North American crayfish but all are found in Europe albeit with variable virulence. As *A. astaci* Group A was probably the first established genotype in Europe back in the nineteenth century, this form shows lower virulence (e.g. Makkonen et al. 2012, 2014; Becking et al. 2015) probably due to the lengthier co-habitation period with native European crayfish. Recently Panteleit et al. (2019) demonstrated that some North American crayfish species carry a novel genotype of *A. astaci*. This latest genotype, named as F, was isolated from North American *Faxonius rusticus* individuals, which were introduced more recently into Europe.

The development of molecular biology methods for *A. astaci* research (e.g. Oidtmann et al. 2006; Vrålstad et al. 2009) and diagnostics have highly accelerated our understanding of crayfish plague and provided improved methods for rapid diagnostics of affected crayfish populations.

16.5.2 *Lagenidium* spp.

Lagenidium spp. are oomycete pathogens and saprophytes of a variety of aquatic animals including crustaceans (Table 16.3). For example, larvae of the Dungeness crab, *Cancer magister*, can become infected with this oomycete under aquarium conditions resulting in significant levels of mortality (Armstrong et al. 1976). Similarly, there are many reports of damage to brooded eggs (particularly those at the periphery of the egg mass) and larvae in marine crustaceans under culture conditions (e.g. Lee et al. 2016). Kuris (1990) has suggested that this association is likely to be uncommon under wild conditions and this may reflect unsuitable environmental conditions in hatcheries.

16.5.3 *Haliphthoros* spp., *Halocrusticida* spp. and *Halioticida* sp.

These oomycetes are largely responsible for infection of eggs and larvae of a wide range of crustaceans (Table 16.3). A recent study of a novel oomycete, *Halioticida noduliformans*, found its presence in eggs from the European lobster, *Homarus gammarus* but only in specimens held in captivity (Holt et al. 2018). Whether such infections are of importance in lobsters sampled directly from the wild is unclear.

16.6 Generalised pathology of fungal and oomycete infections in crustaceans

Tables 16.1 and 16.3 summarise the main results of histopathological investigation of these conditions. Despite the diversity of hosts and pathogens, most infections reveal clear similarities. Various forms of fungal elements (e.g. hyphae, conidia and yeast-like forms) are disseminated throughout the tissues via the haemolymph that is replete with these pathogens at a late stage of infection. In many cases there is clear evidence of a marked immune response to their presence with phagocytosis and

Table 16.3 An overview of the principal diseases of crustaceans caused by oomycetes

Disease	Disease causing agent(s)	Host range	Geographical range	Pathology	Key references
Crayfish plague	*Aphanomyces astaci*	Mainly freshwater and semi-terrestrial forms including crayfish	Principally Europe and USA but also Australia	Affected animals show behavioural changes including increased daylight appearance and increasing paralysis. Zoospores form hyphae that penetrate through the cuticle resulting in variable levels of host response (melanisation; Figure 16.3). Rapid mortality follows in susceptible species such as *astacus* as a result of multiplication in tissues	Söderhäll and Ajaxon 1982; Alderman 1996; Cerenius et al. 2003; Svoboda et al. 2014; Oidtmann 2018
	Lagenidium spp.	Wide range of marine crustaceans	Mainly in crabs and lobsters in captivity and/or under aquaculture conditions	Infection of brooded eggs and free larvae leading to variable levels of mortality	Armstrong et al. 1976; Kuris 1990; Ramasamy et al. 1996; Lee et al. 2016
	Halioticida noduliformans	European lobster, *Homarus gammarus*		Infection of brooded eggs leading to discoloration. Eggs show reduced yolk and hyphae penetrate from the damaged egg. Gills show invasion by hyphae resulting in host response. Not found in lobsters sourced from the wild.	Holt et al. 2018
Haliphthoros 'fungus' disease of lobsters	*Haliphthoros* spp. including *H. milfordensis*	Lobsters, *Homarus americanus*, *H. gammarus*, rock lobster, *Jasus edwardsii*, mud crabs, *Scylla tranquebarica* and larval shrimp		Damage to cuticle, blackened gills in some species. Infection/contamination of eggs and larvae of some crustaceans resulting in significant mortality	Fisher 1988; Diggles 2001; Chukanhom et al. 2003; Lee et al. 2017

nodule/capsule formation (e.g. Boeger et al. 2007; Smith et al. 2013). These defence reactions are accompanied by melanisation as a result of the activation of the proPO system (Figure 16.2d; see Chapter 1 for a description of this host defence mechanism). Despite these attempts to wall off the pathogens, they break free to parasitise the host. Hence diseases, such as black gill disease, simply reflect the location of this cellular defence response in the gill lamellae and not necessarily a common

Table 16.4 Summary of genotype groups of *Aphanomyces astaci*.

Genotype nomenclature	Crayfish source	Comments
A (*As*)	*Astacus astacus* and *A. leptodactylus*	Present in Europe since 1860s (Huang et al. 1994). Less virulent than other genotypes (Makkonen et al. 2014)
B (*Ps I*)	*Pacifastacus leniusculus* (USA) and *A. astacus* (Sweden) (strain I)	Highly virulent towards *A. astacus* (Makkonen et al. 2014; Viljamaa-Dirks et al. 2016)
C (*Ps II*)	*P. leniusculus* (Canada)	
D (*Pc*)	*Procambarus clarkii* (Spain)	
E (*Or*)	*Orconectes limosus*	Not recognised in Oidtmann 2018
F	*Faxonius rusticus*	Panteleit et al. 2019

Based on, and updated from Svoboda et al. (2017) and Oidtmann (2018)

disease-causing agent, i.e. it is not pathognomonic for the agent.

Another noted common finding is the presence of other co-infections including other species of fungi (e.g. Vicente et al. 2012) or eukaryotic pathogens such as *Hematodinium* (Stentiford et al. 2003; Smith et al. 2013). These relationships may suggest complex interactions between fungi/oomycetes and other disease-causing agents possibly related to an immunocompromised host.

16.7 Control and treatment

There are few treatments available to deal with fungal and oomycete disease outbreaks in rearing facilities other than observing good hygiene. In the case of crayfish plague, Oidtmann (2018) concluded that the main approach is to maintain biosecurity to avoid introduction of infected animals or contaminated fishing equipment (Alderman 1996) into areas where native crayfish are present. The reason for this is that once introduced, spread of the disease is inevitable with obvious consequences to native species. Continued importation of non-native crayfish for the aquarium trade and aquaculture places this approach at risk because people often release their 'pets' into natural areas, thereby releasing the pathogen (Holdich et al. 2009).

Chemical treatment of fungal and oomycete infections of crustaceans held in captivity has often relied on toxic chemicals such as formalin and malachite green although Rantamäki et al. (1992) found that magnesium chloride can control crayfish plague transmission. A search for more environmentally safe alternatives is underway including extracts from seaweeds (e.g. Saito and Lal 2019).

16.8 Future directions

With the exception of crayfish, the interaction between the crustacean immune system and pathogenic fungi and oomycetes is poorly understood. The role of mycotoxins and other secondary metabolites in interfering with both phagocytosis and encapsulation in insects is well established (see Chapters 1 and 12) and such interactions are moreover likely to occur in crustacean hosts. Many pathogenic fungi and oomycetes elicit a strong inflammatory response in their crustacean hosts resulting in phagocytosis and encapsulation but they are resistant to killing and emerge from phagocytes and multicellular capsules to cause septicaemia. Many studies (e.g. Cerenius et al. 2003) have investigated the recognition pathway and defensive response that queues proPO in crayfish, largely because of the ease of working with these species but also because of the importance of *A. astaci* as a pathogen in these hosts. Nonetheless, features such as transmission, susceptibility, resistance and virulence remain largely unknown. Studies on the early infection and resistance dynamics would greatly facilitate our understanding of the nature of fungal and oomycete infections in crustacean hosts.

Some taxa, particularly the oomycetes, are important egg parasites of crustaceans, yet little work has been done to show their role in nature, where they may exert strong regulation over host populations (e.g., Redfield and Vincent 1979; Burns 1985). In freshwater systems, these appear to play a major role in host populations but their role in estuarine and marine systems is unknown. Egg parasites may not seem important in r-selected hosts but they can impose mortality rates far higher than any other predator in some ecosystems.

Fungi and oomycete parasites are not just facultative saprophytes, some have extraordinary host specificity (e.g. microsporidians, oomycetes) and the nature of this specificity has largely been ignored. Several Microsporidia have direct life cycles, hence an exploration of their host preferences and infection dynamics would be a major contribution. One avenue to their discovery and perhaps to their host specificity is further elaboration of the mycobiome, the fungal community present in an organism. Knowledge of 'mycobiome' community structure within both healthy and diseased crustaceans may shed more light on these relationships but they lag far behind that of bacterial microbiomes (see Chapter 15). In addition, the potential interaction between fungi and other disease agents, both prokaryotic and eukaryotic, is poorly appreciated although the role of co-infection and infection as a stressor may be of increasing importance under climate change (e.g. Edsman et al. 2015).

16.9 Summary

- Fungi and oomycetes are important pathogens of crustaceans particularly early life history stages held in poor environmental conditions.
- Microsporidian infections are widespread in crustaceans and can have significant impacts on populations via vertical transmission and alteration of host mating systems.
- Microsporidians are relatively common in crustaceans, yet they are often difficult to detect and identify. They can impose significant impacts on some host populations, yet they are extraordinarily rare in others.
- Crayfish plague continues to cause major changes in populations of native crayfish in Europe and other parts of the world. It is a major threat to freshwater systems.
- Although there is often a strong cellular immune response exhibited to the presence of pathogenic fungi and oomycetes in their hosts, these parasites appear resistant to the killing mechanisms resulting in sepsis and death.

Recommended further reading

Holdich, D.M, Reynolds, J.D., Souty-Grosset, C., and Sibley, P.J. 2009. A review of the ever-increasing threat to European crayfish from non-indigenous crayfish species. *Knowledge and Management of Aquatic Systems* 11: 394–395. (Excellent detailed overview of crayfish declines in Europe and role of all factors in this decline)

Oidtmann, B. 2018. Infection with *Aphanomyces astaci* (crayfish plague). In *OIE—Manual of Diagnostic Tests for Aquatic Animals*, Chapter 2.2.2. (12/10/2018) (http://www.oie.int/index.php?id =2439&L=0&htmfile=chapitre_aphanomyces_astaci .htm) (Excellent overview of all aspects of this disease including practical diagnostic methods.)

Stentiford, G.D., Feist, S.W., Stone, D.M., Bateman, K.S., and Dunn, A.M. 2012. Microsporidia: diverse, dynamic, and emergent pathogens in aquatic systems *Trends in Parasitology* 29: 567–578. (Wide-ranging review of microsporidian parasites in aquatic hosts)

Svoboda, J., Mrugala, A., Kozubíková-Balcarová, E., and Petrusek, A. 2017. Hosts and transmission of the crayfish plague pathogen *Aphanomyces astaci*: a review. *Journal of Fish Diseases* 40: 127–140. (Recent review on crayfish plague)

Acknowledgements

We are grateful to Drs Amanda Smith and Satu Viljamaa-Dirks (OIE Reference Laboratory for Crayfish Plague, Finnish Food Authority, Kuopio, Finland) for generously providing previously unpublished images for this chapter. A.F. Rowley also wishes to acknowledge funding from UKRI (ARCH-UK Network Grant) and the Ireland—Wales 2014–2020 European Territorial Co-operation programme project, Bluefish, that supported some of the original studies of this author reviewed in this chapter.

References

Alderman, D.J. 1996. Geographical spread of bacterial and fungal diseases of crustaceans. *Revue Scientifique et Technique de l' Office International des Epizootics* 15: 603–632

Amend, A., Burgaud, G., Cunliffe, M., Edgcomb, V.P., Ettinger, VC., Gutiérrez, M.H., Heitman, J., Hom, E.F.Y., Ianiri, G., Jones, A.C., Kagami, M., Picard, K.T., Quandt, C.A., Raghukumar, S., Riquelme, M., Stajich, J., Vargas-Muñiz, J., Walker, A.K. et al. 2019. Fungi in the marine environment: Open questions and unsolved problems. *MBio* 10: e11089–e11118

Andreadis, T.G. 1985. Experimental transmission of a microsporidian pathogen from mosquitoes to an alternate copepod host. *Proceedings of the National Academy of Sciences USA* 82: 5574–5577

Armstrong, D.A., Buchanan, D.V., and Caldwell, R.S. 1976. A mycosis caused by *Lagenidium* sp. in laboratory-reared larvae of the Dungeness crab, *Cancer magister*, and possible chemical treatments. *Journal of Invertebrate Pathology* 28: 329–336

Bacela-Spychalska, K., Rigaud, T., and Wattier, R.A. 2014. A co-invasive microsporidian parasite that reduces the predatory behaviour of its host *Dikerogammarus villosus* (Crustacea, Amphipoda). *Parasitology* 141: 254–258

Bacela-Spychalska, K., Wattier, R.A., Genton, C., and Rigaud, T. 2012. Microsporidian disease of the invasive amphipod *Dikerogammarus villosus* and the potential for its transfer to local invertebrate fauna. *Biological Invasions* 14: 1831–1842

Bacela-Spychalska, K., Wróblewski, P., Mamos, T., Grabowski, M., Rigaud, T., Wattier, R., Rewicz, T., Konopacka, A., and Ovcharenko, M. 2018. Europe-wide reassessment of Dictyocoela (Microsporidia) infecting native and invasive amphipods (Crustacea): Molecular versus ultrastructural traits. *Scientific Reports* 8: 8945

Bao, J., Jiang, H., Shen, H., Xing, Y., Feng, C., Li, X.,and Braquart-Varnier, C. 2021. First description of milky disease in the Chinese mitten crab *Eriocheir sinensis* caused by the yeast *Metschnikowia bicuspidata*. *Aquaculture* 534: 735984

Becking, T., Mrugała, A., Delaunay, C., Svoboda, J., Raimond, M., Viljamaa-Dirks, S., Petrusek, A., Grandjean, F., and Braquart-Varnier, C. 2015. Effect of experimental exposure to differently virulent *Aphanomyces astaci* strains on the immune response of the noble crayfish *Astacus*. *Journal of Invertebrate Pathology* 132: 115–124

Becnel, J.J., Takvorian, P.M., and Cali, A. 2014. Checklist of available generic names for microsporidia with type species and type hosts. In *Microsporidia: Pathogens of Opportunity*, Weiss, L.M. and Becnel J.J. (eds.) pp. 671–686. New York: John Wiley and Sons

Ben-Ami, F. and Routtu, J. 2013. The expression and evolution of virulence in multiple infections: The role of specificity, relative virulence and relative dose. *BMC Evolutionary Biology* 13: 97

Boeger, W.A., Pie, M.R., Ostrensky, A., and Patella, L. 2005. Lethargic crab disease: Multidisciplinary evidence supports a mycotic etiology. *Memoires of the Institute Oswaldo Cruz* 100: 161–167

Boeger, W.A., Pie, M.R., Vicente, V., Ostrensky, A., Hungria, D., and Castilho, G.G. 2007. Histopathology of the mangrove land crab *Ucides cordatus* (Ocypodidae) affected by lethargic crab disease. *Diseases of Aquatic Organisms* 78: 73–81

Bojko, J., Clark, F., Bass, D., Dunn, A.M., Stewart-Clark, S., Stebbing, P.D., and Stentiford, G.D. 2017. *Parahepatospora carcini* n. gen., n. sp., a parasite of invasive *Carcinus maenas* with intermediate features of sporogony between the Enterocytozoon clade and other Microsporidia. *Journal of Invertebrate Pathology* 143: 124–134

Bojko, J. and Ovcharenko, M. 2019. Pathogens and other symbionts of the Amphipoda: Taxonomic diversity and pathological significance. *Diseases of Aquatic Organisms* 136: 3–36

Bronnvall, A. and Larsson, J.R. 1995. Description of *Lanatospora tubulifera* sp. n. (Microspora, Tuzetiidae) with emended diagnosis and new systematic position for the genus *Lanatospora*. *Archiv für Protistenkunde* 146: 69–78

Bunke, M., Dick, J.T., Hatcher, M.J., and Dunn, A.M. 2019. Parasites influence cannibalistic and predatory interactions within and between native and invasive amphipods. *Diseases of Aquatic Organisms* 136: 79–86

Burns, C.W. 1985. Fungal parasitism in a copepod population: The effects of *Aphanomyces* on the population dynamics of *Boeckella dilatata* Sars. *Journal of Plankton Research* 72: 201–205

Cerenius, L., Bangyeekhun, E., Keyser, P., Söderhäll, I., and Söderhäll, K. 2003. Host prophenoloxidase expression in freshwater crayfish is linked to increased resistance to the crayfish plague fungus, *Aphanomyces astaci*. *Cellular Microbiology* 5: 353–357

Cerenius, L. and Söderhäll, K. 1984a. Chemotaxis in *Aphanomyces astaci*, an arthropod-parasitic fungus. *Journal of Invertebrate Pathology* 43: 278–281

Cerenius, L. and Söderhäll, K. 1984b. Repeated zoospore emergence from isolated spore cysts of *Aphanomyces astaci*. *Experimental Mycology* 8: 370–377

Chen, S.-C., Chen, Y.-C., Kwang, J., Manopo, I., Wang, P.C., Chaung, H.C., Liaw, L.L., and Chiu, S.H. 2007. *Metschnikowia bicuspidata* dominates in Taiwanese cold-weather yeast infections of *Macrobrachium rosenbergii*. *Diseases of Aquatic Organisms* 75: 191–199

Chukanhom, K., Borisutpeth, P., Van Khoa, L., and Hatai, K. 2003. *Haliphthoros milfordensis* isolated from black tiger prawn larvae (*Penaeus monodon*) in Vietnam. *Mycoscience* 44: 123–127

Codreanu, R. and Codreanu-Balcescu, D. 1981. On two *Metschnikowia* yeast species producing hemocoelic infections in *Daphnia magna* and *Artemia salina* (Crustacea, Phyllopoda) from Romania. *Journal of Invertebrate Pathology* 37: 22–27

Cohen, J.M., Civitello, D.J., Venesky, M.D., McMahon, T.A., and Rohr, J.R. 2019. An interaction between climate change and infectious disease drove widespread amphibian declines. *Global Change Biology* 25: 927–937

Dallas, T.A., Krkoš;ek, M., and Drake, J.M. 2018. Experimental evidence of a pathogen invasion threshold. *Royal Society Open Science* 5: 171975

Davies, C.E., Malkin, S.H., Thomas, J.E., Batista, F.M., Rowley, A.F., and Coates, C.J. 2020a. Mycosis is a disease state encountered rarely in shore crabs, *Carcinus maenas*. *Pathogens* 9: 462

Davies, C.E., Bass, D., Ward, G.M., Batista, F.M., Malkin, S.H., Thomas, J.E., Bateman, K., Feist, S.W., Coates, C.J., and Rowley, A.F. 2020b. Diagnosis and prevalence of two new species of haplosporidians infecting shore crabs *Carcinus maenas*: *Haplosporidium carcini* n. sp., and *H. cranc* n. sp. *Parasitology* 147: 1229–1237

Dick, M.H., Donaldson, W.E., and Vining, I.W. 1998. Epibionts of the tanner crab *Chionoecetes bairdi* in the region of Kodiak Island, Alaska. *Journal of Crustacean Biology* 18: 519–527

Diggles, B.K. 2001. A mycosis of juvenile spiny rock lobster, *Jasus edwardsii* (Hutton,1875) caused by *Haliphthoros* sp., and possible methods of chemical control. *Journal of Fish Diseases* 24: 99–110

Ding, Z., Pan, J., Huang, H., Jiang, G., Chen, J., Zhu, X., Wang, R., and Xu, G. 2018. An integrated metabolic consequence of *Hepatospora eriocheir* infection in the Chinese mitten crab *Eriocheir sinensis*. *Fish & Shellfish Immunology* 72: 443–451

Dunn, A.M., Adams, J., and Smith, J.E. 1993. Transovarial transmission and sex ratio distortion by a microsporidian parasite in a shrimp. *Journal of Invertebrate Pathology* 61: 248–252

Dunn, A.M. and Rigaud, T. 1998. Horizontal transfer of parasitic sex ratio distorters between crustacean hosts. *Parasitology* 117: 15–19

Edsman, L., Nyström, P., Sandström, A., Stenberg, M., Kokko, H., Tiitinen, V., Makkonen, J., and Jussila J. 2015. Eroded swimmeret syndrome in female crayfish *Pacifastacus leniusculus* associated with *Aphanomyces astaci* and *Fusarium* spp. infections. *Diseases of Aquatic Organisms* 112: 219–228

Fielding, N.J., MacNeil, C., Robinson, N., Dick, J.T.A., Elwood, R.W., Terry, R.S., Ruiz, Z., and Dunn, A.M. 2005. Ecological impacts of the microsporidian parasite *Pleistophora mulleri* on its freshwater amphipod host *Gammarus duebeni celticus*. *Parasitology* 131: 331–336

Findley, A.M., Blakeney, E.W. Jr., and Weidner, E.H. 1981. *Ameson michaelis* (Microsporida) in the blue crab, *Callinectes sapidus*: Parasite-induced alterations in the biochemical composition of host tissues. *Biological Bulletin, Woods Hole* 161: 115–125

Fisher, W.S. 1988. Fungus (*Haliphthoros*) disease of lobsters. In *Disease Diagnosis and Control in North American Aquaculture. Developments in* Aquaculture *and Fisheries Science*, Sindermann, C.J. and Lightner, D.V. (eds.) Vol. 17, pp. 251–254. Amsterdam: Elsevier

Franzen, C. 2004. Microsporidia: How can they invade other cells? *Trends in Parasitology* 20: 275–279

Galbreath, J.G.S., Smith, J.E., Becnel, J.J., Butlin, R.K., and Dunn, A.M. 2010. Reduction in post-invasion genetic diversity in *Crangonyx pseudogracilis* (Amphipoda: Crustacea): A genetic bottleneck or the work of hitchhiking vertically transmitted microparasites? *Biological Invasions* 12: 191–209

Haag, K.L., Pombert, J.F., Sun, Y., De Albuquerque, N.R.M., Batliner, B., Fields, P., Lopes, T.F., and Ebert, D. 2020. Microsporidia with vertical transmission were likely shaped by nonadaptive processes. *Genome Biology and Evolution* 12: 3599–3614

Haine, E.R., Motreuil, S., and Rigaud, T. 2007. Infection by a vertically-transmitted microsporidian parasite is associated with a female-biased sex ratio and survival advantage in the amphipod *Gammarus roeseli*. *Parasitology* 134: 1363–1367

Hall, S.R., Simonis, J.L., Nisbet, R.M., Tessier, A.J., and Cáceres, C.E. 2009. Resource ecology of virulence in a planktonic host-parasite system: An explanation using dynamic energy budgets. *The American Naturalist* 174: 149–162

Hall, S.R., Smyth, R., Becker, C.R., Duffy, M.A., Knight, C.J., MacIntyre, S., Tessier, A.J., and Cáceres C.E. 2010. Why are *Daphnia* in some lakes sicker? Disease ecology, habitat structure, and the plankton. *BioScience* 60: 363–375

Hibbitts, J., Hughes, G.C., and Sparks, A.K. 1981. *Trichomaris invadens* gen. et sp. nov., an ascomycete parasite of the tanner crab (*Chionoecetes bairdi* Rathbun Crustacea; Brachyura). *Canadian Journal of Botany* 59: 2121–2128

Hicks, D.M. 1982. Abundance and distribution of black mat syndrome on stocks of Tanner crab, *Chionoecetes bairdi*, in the northwestern Gulf of Alaska. In *Proceedings of the International Symposium on the genus* Chionoecetes, pp. 563–579. Alaska: University of Alaska

Holdich, D.M., James, J., Jackson, C., and Peay, S. 2014. The North American signal crayfish, with particular reference to its success as an invasive species in Great Britain. *Ethology, Ecology and Evolution* 26: 232–262

Holdich, D.M, Reynolds, J.D., Souty-Grosset, C., and Sibley, P.J. 2009. A review of the ever-increasing threat to European crayfish from non-indigenous crayfish species. *Knowledge and Management of Aquatic Systems* 11: 394–395

Holt, C., Foster, R., Daniels, C.L. Van Der Giezen, M., Feist, S.W., Stentiford, G.D., and Bass, D. 2018. *Halioticida noduliformans* infection in eggs of lobster (*Homarus gammarus*) reveals its generalist parasitic strategy in marine invertebrates. *Journal of Invertebrate Pathology* 154: 109–116

Huang, T., Cerenius, L., and Söderhall, K. 1994. Analysis of genetic diversity in the crayfish plague fungus, *Aphanomyces astaci*, by random amplification of polymorphic DNA. *Aquaculture* 126: 1–9

Johnson, P.T., Ives, A.R., Lathrop, R.C., and Carpenter, S.R. 2009. Long-term disease dynamics in lakes: Causes and consequences of chytrid infections in *Daphnia* populations. *Ecology* 90: 132–144

Johnson, P.T., Longcore, J.E., Stanton, D.E., Carnegie, R.B., Shields, J.D., and Preu, E.R. 2006. Chytrid infections of *Daphnia pulicaria*: Development, ecology, pathology and phylogeny of *Polycaryum laeve*. *Freshwater Biology* 51: 634–648

Jones, E.B.G. 2011. Fifty years of marine mycology. *Fungal Diversity* 50: 73–112

Karthikeyan, V. and Gopalakrishnan, A. 2014. A novel report of phytopathogenic fungi *Gilbertella persicaria* infection on *Peneus monodon*. *Aquaculture* 430: 224–229

Karthikeyan, V., Selvakumar, P., and Gopalakrishnan, A. 2015. A novel report of fungal pathogen *Aspergillus awamori* causing black gill infection on *Litopenaeus vannamei* (Pacific white shrimp). *Aquaculture* 444: 36–40

Khoa, L.V., Hatai, K., and Aoki, T. 2004. *Fusarium incarnatum* isolated from black tiger shrimp, *Penaeus monodon* Fabricius, with black gill disease cultured in Vietnam. *Journal of Fish Diseases* 27: 507–515

Kuris A. 1990. A review of patterns and causes of crustacean brood mortality. In *Crustacean Egg Production*, Wenner A. and Kuris, A. (eds.) pp. 117–139. Rotterdam: AA Balkema

Lee, S.C., Corradi, N., Byrnes III, E.J., Torres-Martinez, S., Dietrich, F.S., Keeling, P.J., and Heitman, J. 2008. Microsporidia evolved from ancestral sexual fungi. *Current Biology* 18: 1675–1679

Lee, Y.N., Hatai, K., and Kurata, O. 2015. *Haliphoros sabahensis* sp. nov. isolated from mud crab *Scylla tranquebarica* eggs and larvae in Malaysia. *Fish Pathology* 52: 31–37

Lee, Y.N., Hatai, K., and Kurata, O. 2016. First report of *Lagenidium thermophilum* isolated from eggs and larvae of mud crab (*Scylla tranquebarica*) in Sabah, Malaysia. *Bulletin of the European Association of Fish Pathology* 36: 111–117

Lievens, E.J., Rode, N.O., Landes, J., Segard, A., Jabbour-Zahab, R., Michalakis, Y., and Lenormand, T. 2019. Long-term prevalence data reveals spillover dynamics in a multi-host (*Artemia*), multi-parasite (Microsporidia) community. *International Journal for Parasitology* 49: 471–480

Lightner, D.V. and Fontaine, C.T. 1975. A mycosis of the American lobster, *Homarus americanus* caused by *Fusarium* sp. *Journal of Invertebrate Pathology* 25: 239–245

MacNeil, C. and Dick, J.T. 2011. Parasite-mediated intraguild predation as one of the drivers of co-existence and exclusion among invasive and native amphipods (Crustacea). *Hydrobiologia* 665: 247–256

Makkonen, J., Jussila, J., Koistinen, L., Paaver, T., Hurt, M., and Kokko, H. 2013. *Fusarium avenaceum* causes burn spot diseases syndrome in noble crayfish (*Astacus astacus*). *Journal of Invertebrate Pathology* 113: 184–190

Makkonen, J., Jussila, J., Kortet, R., Vainikka, A., and Kokko, H. 2012. Differing virulence of *Aphanomyces astaci* isolates and elevated resistance of noble crayfish *Astacus astacus* against crayfish plague. *Diseases of Aquatic Organisms* 102: 129–136

Makkonen, J., Kokko, H., Gökmen, G., Ward, J., Umek, J., Kortet, R., Petrusek, A., and Jussila, J. 2019. The signal crayfish (*Pacifastacus leniusculus*) in Lake Tahoe (USA) hosts multiple *Aphanomyces* species. *Journal of Invertebrate Pathology* 166: 107218

Makkonen J., Kokko, H., Vainikka, A, Kortet, R., and Jussila, J. 2014. Dose-dependent mortality of the noble crayfish (*Astacus astacus*) to different strains of the crayfish plague (*Aphanomyces astaci*). *Journal of Invertebrate Pathology* 115: 86–91

Mangin, K.L., Lipsitch, M., and Ebert, D. 1995. Virulence and transmission modes of two microsporidia in *Daphnia magna*. *Parasitology* 111: 133–142

Martinez, M.A., Vivarès, C.P., De Medeiros Rocha, R., Fonseca, A.C., Andral, B., and Bouix, G. 1992. Microsporidiosis on *Artemia* (Crustacea, Anostraca): Light and electron microscopy of *Vavraia anostraca* sp. nov. (Microsporidia, Pleistophoridae) in the Brazilian solar salterns. *Aquaculture* 107: 229–237

Mautner, S.I., Cook, K.A., Forbes, M.R., McCurdy, D.G. and Dunn, A.M. 2007. Evidence for sex ratio distortion by a new microsporidian parasite of a corophiid amphipod. *Parasitology* 134: 1567–1573

McCarthy, C.G.P. and Fitzpatrick, D.A. 2017. Phylogenomic reconstruction of the oomycete phylogeny derived from 37 genomes. *mSphere* 2: e00095–17

Messick, G.A., Overstreet, R.M., Nalepa, T.F., and Tyler, S. 2004. Prevalence of parasites in amphipods *Diporeia* spp. from Lakes Michigan and Huron, USA. *Diseases of Aquatic Organisms* 59: 159–170

Mrugała, A., Kawai, T., Kozubíková-Balcarová, E., and Petrusek, A. 2017. *Aphanomyces astaci* presence in Japan: A threat to the endemic and endangered crayfish species *Cambaroides japonicus*? *Aquatic Conservation* 27: 103–114

Nha V.V., Hoa, D.T., and Khoa, L.V. 2009. Black gill disease of cage-cultured ornate rock lobster *Panulirus ornatus*. *Aquaculture Asia* 14: 35–37

O'Hanlon, S.J., Rieux, A., Farrer, R.A., Rosa, R.A., Waldman, B. et al. 2018. Recent Asian origin of chytrid fungi causing global amphibian declines. *Science* 360: 621–627

Oidtmann, B. 2018. Infection with *Aphanomyces astaci* (crayfish plague). In *OIE—Manual of Diagnostic Tests for Aquatic Animals* Chapter 2.2.2. (12/10/2018) (http://www.oie.int/index.php?id=2439&L=0&htmfile=chapitre_aphanomyces_astaci.htm)

Oidtmann, B., Geiger, S., Steinbauer, P., Culas, A., and Hoffmann, R.W. 2006. Detection of *Aphanomyces astaci* in North American crayfish by polymerase chain reaction. *Diseases of Aquatic Organisms* 72: 53–64

Orélis-Ribeiro, R., Boeger, W.A., Vicente, V.A., Chammas, M., and Ostrensky, A. (2011). Fulfilling Koch's postulates confirms the mycotic origin of Lethargic Crab Disease. *Antonie Van Leeuwenhoek* 99: 601-608.

Overholt, E.P., Hall, S.R., Williamson, C.E., Meikle, C.K., Duffy, M.A., and Cáceres, C.E. 2012. Solar radiation decreases parasitism in *Daphnia*. *Ecology Letters* 15: 47–54

Panteleit, J., Horvarth, T., Jussila, J., Makkonen, J., Perry, W., Schultz, R., Theissinger, K., and Schrimpf, A., 2019. Invasive rusty crayfish *(Faxonius rusticus)* populations in North America are infected with the crayfish plague disease agent (*Aphanomyces astaci*). *Freshwater Science* 38: 425–433

Panteleit, J., Keller, N.S., Kokko, H., Jussila, J., Makkonen, J., Theissinger, K., and Schrimpf, A. 2017. Investigation of ornamental crayfish reveals new carrier species of the crayfish plague pathogen (*Aphanomyces astaci*). *Aquatic Invasions* 12: 77–83

Park, E., Jorge, F. and Poulin, R. 2020. Shared geographic histories and dispersal contribute to congruent phylogenies between amphipods and their microsporidian parasites at regional and global scales. *Molecular Ecology* 29: 3330–3345

Peiró, D.F., Almerão, M.P., Delaunay, C., Jussila, J., Makkonen, J., Bouchon, D., Araujo, P.B., and Souty-Grosset, C. 2016. First detection of the crayfish plague pathogen *Aphanomyces astaci* in South America: A high potential risk to native crayfish. *Hydrobiologia* 781: 181–190

Phillips, A.J., Anderson, V.L., Robertson, E.J., Secombes, C.J., and Van West, P. 2007. New insights into animal pathogenic oomycetes. *Trends in Microbiology* 16: 13–19

Putra, M.D., Bláha, M., Wardiatno, Y., Krisanti, M., Yonvitner, Jerikho, R., Kamal, M.M., Mojžišová, M., Bystřický, P.K., Kouba, A., Kalous, L., Petrusek, A., and Patoka, J. 2018. *Procambarus clarkii* (Girard, 1852) and crayfish plague as new threats for biodiversity in Indonesia. *Aquatic Conservation* 2018: 1–7

Quiles, A., Bacela-Spychalska, K., Teixeira, M., Lambin, N., Grabowski, M., Rigaud, T., and Wattier, R.A. 2019. Microsporidian infections in the species complex *Gammarus roeselii* (Amphipoda) over its geographical range: Evidence for both host–parasite co-diversification and recent host shifts. *Parasites and Vectors* 12: 327

Ramasamy, P., Rajan, P. R., Jayakumar, R., Rani, S., and Brennan, G.P. 1996. *Lagenidium callinectes* (Couch, 1942) infection and its control in cultured larval Indian tiger prawn, *Penaeus monodon* Fabricius. *Journal of Fish Diseases* 19: 75–82

Rantamäki, J., Cerenius, L., and Söderhäll, K. 1992. Prevention of transmission of the crayfish plague fungus (*Aphanomyces astaci*) to the freshwater crayfish *Astacus astacus* by treatment with MgCl$_2$. *Aquaculture* 104: 11–18

Redfield, G.W. and Vincent, W.F. 1979. Stages of infection and ecological effects of a fungal epidemic on the eggs of a limnetic copepod. *Freshwater Biology* 9: 503–510

Rhoobunjongde, W., Hatai, K., Wada, S., and Kubota, S.S. 1991. *Fusarium moniliforme* (Sheldon) isolated from gills of Kuruma prawn *Penaeus japonicus* (Bate) with black gill disease. *Nippon Suisan Gakkaishi* 57: 629–635

Richman, N.I., Böhm, M., Adams, S. et al. 2015. Multiple drivers of decline in the global status of freshwater crayfish (Decapoda: Astacidea). *Philosophical Transactions of the Royal Society B* 370: 20140060

Rode, N.O., Lievens, E.J., Segard, A., Flaven, E., Jabbour-Zahab, R., and Lenormand, T. 2013a. Cryptic microsporidian parasites differentially affect invasive and native *Artemia* spp. *International Journal for Parasitology* 43: 795–803

Rode, N.O., Lievens, E.J., Flaven, E., Segard, A., Jabbour-Zahab, R., Sanchez, M.I., and Lenormand, T. 2013b. Why join groups? Lessons from parasite-manipulated *Artemia*. *Ecology Letters* 16: 493–501

Saito, H. and Lal, T.M. 2019. Antimycotic activity of seaweed extracts (*Caulerpa lentillifera* and *Eucheuma cottonii*) against two genera of marine oomycetes, *Lagenidium* spp. and *Haliphthoros* spp. *Biocontrol Science* 24: 73–80

Shocket, M.S., Strauss, A.T., Hite, J.L. Šljivar, M., Civitello, D.J., Duffy, M.A., Cáceres, C.E., and Hall, S.R. 2018. Temperature drives epidemics in a zooplankton-fungus disease system: A trait-driven approach points to transmission via host foraging. *The American Naturalist* 191: 435–451

Sindermann, C.J. 1989. The shell disease syndrome in marine crustaceans. *NOAA Technical Memorandum NMFS-F/NEC-64*

Small, H.J., Stentiford, G.D., Behringer, D.C., Freeman, M.A., Atherley, N.A.M., Reece, K.S., Bateman, K.S., and Shields, J.D. 2019. Characterization of microsporidian *Ameson herrnkindi* sp. nov. infecting Caribbean spiny lobsters *Panulirus argus*. *Diseases of Aquatic Organisms* 136: 209–218

Smith, A.L., Hamilton, K.M., Hirschle, L., Wootton, E.C., Vogan, C.L., Pope, E.C., Eastwood, D.C., and Rowley, A.F. 2013. Characterization and molecular epidemiology of a fungal infection of edible crabs (*Cancer pagurus*) and interaction of the fungus with the dinoflagellate parasite *Hematodinium*. *Applied and Environmental Microbiology* 79: 783–793

Smith, A.L. and Rowley, A.F. 2015. Effects of experimental infection of juvenile edible crabs *Cancer pagurus* with the parasitic dinoflagellate *Hematodinium* sp. *Journal of Shellfish Research* 34: 511–519

Smith, J.E. 2009. The ecology and evolution of microsporidian parasites. *Parasitology* 136: 1901–1914

Söderhäll, K. and Ajaxon, A. 1982. Effect of quinones and melanin on mycelial growth of *Aphanomyces* spp. and extracellular protease of *Aphanomyces astaci*, a parasite on crayfish. *Journal of Invertebrate Pathology* 39: 105–109

Solangi, M.A. and Lightner, D.V. 1976. Cellular inflammatory response of *Penaeus aztecus* and *P. setiferus* to the pathogenic fungus, *Fusarium* sp., isolated from the California brown shrimp, *P. californiensis*. *Journal of Invertebrate Pathology* 27: 77–86

Sparks, A.K. 1982. Observations on the histopathology and probable progression of the disease caused by *Trichomaris invadens*, an invasive ascomycete, in the Tanner crab, *Chionoecetes bairdi*. *Journal of Invertebrate Pathology* 40: 242–254

Sparks, A.K. and Hibbits, J. 1979. Black mat syndrome, an invasive mycotic disease of the Tanner crab, *Chionoecetes bairdi*. *Journal of Invertebrate Pathology* 34: 184–191

Stentiford, G.D. and Bateman, K.S. 2007. *Enterospora* sp., an intranuclear microsporidian infection of hermit crab *Eupagurus bernhardus*. *Diseases of Aquatic Organisms* 75: 73–77.

Stentiford, G.D., Bateman, K.S., Dubuffet, A., Chambers, E., and Stone, D.M. 2011. Hepatospora eriocheir (Wang and Chen, 2007) gen. et comb. nov. infecting invasive Chinese mitten crabs (*Eriocheir sinensis*) in Europe. *Journal of Invertebrate Pathology* 108: 156–166

Stentiford, G.D., Bateman, K.S., Feist, S.W., Oyarzún, S., Uribe, J.C., Palacios, M., and Stone, D.M. 2014. *Areospora rohanae* n. gen. n. sp. (Microsporidia; Areosporiidae n. fam.) elicits multi-nucleate giant-cell formation in southern king crab (*Lithodes santolla*). *Journal of Invertebrate Pathology* 118: 1–11

Stentiford, G.D., Bateman, K.S., Small, H.J., Moss, J., Shields, J.D., Reece, K.S., and Tuck, I. 2010. *Myospora metanephrops* (ng, n. sp.) from marine lobsters and a proposal for erection of a new order and family (Crustaceacida; Myosporidae) in the Class Marinosporidia (Phylum Microsporidia). *International Journal for Parasitology* 40: 1433–1446

Stentiford, G.D., Evans, M., Bateman, K., and Feist, S.W. 2003. Co-infection by a yeast-like organism in *Hematodinium*-infected European edible crabs *Cancer pagurus* and velvet swimming crabs *Necora puber* from the English Channel. *Diseases of Aquatic Organisms* 54: 195–202

Stentiford, G.D., Feist, S.W., Stone, D.M., Bateman, K.S., and Dunn, A.M. 2012. Microsporidia: Diverse, dynamic, and emergent pathogens in aquatic systems. *Trends in Parasitology* 29: 567–578

Stentiford, G.D., Ross, S., Minardi, D., Feist, S.W., Bateman, K.S., Gainey, P.A., Troman, C., and Bass, D. 2018. Evidence for trophic transfer of *Inodosporus octospora* and *Ovipleistophora arlo* n. sp. (Microsporidia) between crustacean and fish hosts. *Parasitology* 145: 1105–1117

Stirnadel, H.A. and Ebert, D. 1997. Prevalence, host specificity and impact on host fecundity of microparasites and epibionts in three sympatric *Daphnia* species. *Journal of Animal Ecology* 66: 212–222

Svoboda, J., Mrugala, A., Kozubíková-Balcarová, E., and Petrusek, A. 2017. Hosts and transmission of the crayfish plague pathogen *Aphanomyces astaci*: a review. *Journal of Fish Diseases* 40: 127–140

Svoboda, J., Strand, D.A., Vrålstad, T. Grandjean, F., Edsman, L., Kozák, P., Kouba, A., Fristad, R.F., Bahadir Koca, S., and Petrusek, A. 2014. The crayfish plague pathogen can infect freshwater-inhabiting crabs. *Freshwater Biology* 59: 918–929

Terry, R.S., Smith, J.E., Sharpe, R.G., Rigaud, T., Littlewood, D.T.J., Ironside, J.E., Rollinson, D., Bouchon, D., MacNeil, C., Dick, J.T., and Dunn, A.M. 2004. Widespread vertical transmission and associated host sex–ratio distortion within the eukaryotic phylum Microspora. *Proceedings of the Royal Society of London, Series B: Biological Sciences* 271: 1783–1789

Tokarev, Y.S., Sokolova, Y.Y., Vasilieva, A.A., and Issi, I.V. 2018. Molecular and morphological characterization of *Anncaliia azovica* sp. n. (Microsporidia) infecting *Niphargogammarus intermedius* (Crustacea, Amphipoda) from the Azov Sea. *Journal of Eukaryotic Microbiology* 65: 296–307

Unestam, T. 1975. Defence reactions and susceptibility of Australian and New Guinean freshwater crayfish to European-crayfish-plague fungus. *Australian Journal of Experimental Biology and Medical Science* 53: 349–359

Unestam, T. and Weiss, D.W. 1970. The host-parasite relationship between freshwater crayfish and the crayfish disease fungus *Aphanomyces astaci*: Responses to infection by a susceptible and a resistant species. *Journal of General Microbiology* 60: 77–90

Vávra, J., Fiala, I., Krylová, P., Petrusek, A., and Hyliš, M. 2019. Establishment of a new microsporidian genus and species, *Pseudoberwaldia daphniae* (Microsporidia, Opisthosporidia), a common parasite of the *Daphnia longispina* complex in Europe. *Journal of Invertebrate Pathology* 162: 43–54

Vávra, J., Hyliš, M., Fiala, I., and Nebesářová, J. 2016. *Globulispora mitoportans* n.g., n. sp., (Opisthosporidia: Microsporidia) a microsporidian parasite of daphnids with unusual spore organization and prominent mitosome-like vesicles. *Journal of Invertebrate Pathology* 135: 43–52

Vávra, J., Hyliš, M., Fiala, I., Sacherová, V., and Vossbrinck, C.R. 2017. Microsporidian genus *Berwaldia* (Opisthosporidia, Microsporidia), infecting daphnids (Crustacea, Branchiopoda): Biology, structure, molecular phylogeny and description of two new species. *European Journal of Protistology* 61: 1–12

Vernick, S.H. and V. Sprague. 1970. *In vitro* muscle lysis accompanying treatment with extract of crab muscle infected with *Nosema* sp. *Journal of Parasitology* 56: 352–353

Vicente, V.A., Orelis-Ribeiro, Najafzadeh, M.J., Sun, J., Guerra, R.S., Miesch, S., Ostensky, A., Meis, J.F., Klaassen, C.H., De Hoog, G.S., and Boeger, W.A. 2012. Black yeast-like fungi associated with Lethargic Crab Disease (LCD) in the mangrove-land crab, *Ucides cordatus* (Ocypodidae). *Veterinary Microbiology* 158: 109–112

Viljamaa-Dirks, S., Heinikainen, S., Virtala, A.-M.K., Torssonen, H., and Pelkonen, S. 2016. Variation in the hyphal growth rate and the virulence of two genotypes of the crayfish plague organisms *Aphanomyces astaci*. *Journal of Fish Diseases* 39: 753–764

Vivarès, C.P. and Cuq, J.L. 1981. Physiological and metabolic variations in *Carcinus mediterraneus* (Crustacea: Decapoda) parasitized by *Thelohania maenadis* (Microspora: Microsporida): An ecophysiopathological approach. *Journal of Invertebrate Pathology* 37: 38–46

Vrålstad, T., Knutsen, A.K., Tengs, T., and Holst-Jensen, A. 2009. A quantitative TaqMan® MGB real-time polymerase chain reaction-based assay for detection of the causative agent of crayfish plague *Aphanomyces astaci*. *Veterinary Microbiology* 137: 146–155

Walker, M.H. and Hinsch, G.W. 1972. Ultrastructural observations of a microsporidian protozoan parasite in *Libinia dubia* (Decapoda). *Zeitschrift für Parasitenkunde* 39: 17–26

Wang, W. and Chen, J. 2007. Ultrastructural study on a novel microsporidian, *Endoreticulatus eriocheir* sp. nov. (Microsporidia, Encephalitozoonidae), parasite of Chinese mitten crab, *Eriocheir sinensis* (Crustacea, Decapoda). *Journal of Invertebrate Pathology* 94: 77–83

Weidner, E. 1970. Ultrastructural study of microsporidian development. *Zeitschrift für Zellforschung und Mikroskopische Anatomie* 105: 33–54

Wilkinson, T.J., Rock, J., Whiteley, N.M., Ovcharenko, M.O., and Ironside, J.E. 2011. Genetic diversity of the feminising microsporidian parasite *Dictyocoela*: New insights into host-specificity, sex and phylogeography. *International Journal for Parasitology* 41: 959–966

WoRMS Editorial Board 2020. World Register of Marine Species. Available from http://www.marine species.orgatVLIZ. Accessed several times in 2020.

CHAPTER 17

Parasites of crustaceans

Jeffrey D. Shields

17.1 Introduction

Crustaceans serve as hosts to disparate taxa of parasites ranging from unusual protistans, to the more classically studied helminths, and to the bizarrely adapted parasitic crustaceans. The parasites of decapods, particularly shrimp, crabs, and lobsters are the best known because of the commercial importance of these hosts, however, other parasites, particularly those using copepods, cladocerans and amphipods, have also received attention because of the ecological importance of these hosts and their use in laboratory models. Notable studies have contributed to our understanding of vertical transmission, the influence of cryptic host species complexes, as well as the emergence of new pathogens. Nonetheless, due to the obscure nature of many parasite taxa, few have received more than cursory taxonomic descriptions and systematic placement. There is no doubt that many more will be found.

This chapter reviews the major parasitic taxa that use Crustacea as hosts. Although there have been several outstanding reviews of specific taxa and comprehensive reviews of the parasites within specific host taxa, until now there have been no systematic reviews attempting to cover the diverse range of parasites in the entire host subphylum. Notable reviews include systematic treatment of the parasites of cladocerans (Green 1974), copepods (Ho and Perkins 1985), euphausiids (Gómez-Gutiérrez and Morales-Ávila 2016), amphipods (Bojko and Ovcharenko 2019), and specific decapods, such as freshwater crayfish (Edgerton et al. 2002; Longshaw

2016), brachyuran crabs (Shields and Overstreet 2007; Shields et al. 2015), anomuran hermit crabs (Williams and McDermott 2004; McDermott et al. 2010) and lobsters (Shields et al. 2006; Shields 2012). Several earlier reviews coalesced the literature and offered seminal insights into many of the parasites and pathogens of crustaceans (Couch 1983; Johnson 1983, Overstreet 1983). The focus of this review will be on eukaryotic parasites, organisms that have an intimate but negative relationship with their host. For in-depth treatment of the fouling organisms and commensals that live on crustaceans, see Overstreet (1983) and Fernandez-Leborans (2010).

17.2 Principal parasites

17.2.1 Protozoa (Protistans)

The higher-level systematics of the protozoa have been revamped over the last two decades, stabilising around the scheme used by the Tree of Life and incorporated into the World Register of Marine Species (Ruggiero et al. 2015; WoRMS Editorial Board 2020). Although I present the systematics of the parasitic taxa found in crustaceans, the emphasis is on specific parasites and not on their taxonomic positions or systematic relationships. Crustaceans have a high diversity of parasitic protists that range broadly over many types of host-parasite associations. I have, therefore, not covered fouling organisms, particularly those on decapods; however, I have included parasitic representatives, such as apostome ciliates, several peritrich ciliates,

Jeffrey D. Shields, *Parasites of crustaceans*. In: *Invertebrate Pathology*. Edited by Andrew F. Rowley, Christopher J. Coates and Miranda M.A. Whitten, Oxford University Press. © Oxford University Press (2022). DOI: 10.1093/oso/9780198853756.003.0017 and others that have more specialised relationships. I have also not covered the diverse assemblage of protists associated with the egg clutches of their hosts. For a review of egg mortality or other negative effects from this specialised fouling community, see Kuris (1991). The parasitic protozoans of crustaceans have been reviewed in the past by Sprague and Couch (1971); Couch (1983); Meyers (1990), and Bradbury (1994).

17.2.1.1 Kingdom Chromista: Infrakingdom Alveolata: Phylum Myzozoa: Subphylum Dinozoa: Infraphylum Dinoflagellata

Dinoflagellates are usually thought of as harmful algal bloom-forming species or as primary producers in aquatic and marine systems Indeed, over 2,000 species have been described and only *ca.* 7% (140 species in 35 genera) are parasitic. The parasitic species occur in both aquatic and marine habitats and show distinctive host preferences, infecting protists, crustaceans, other invertebrates, and fishes (Coats 1999). The classical work by Chatton (1920) is still highly relevant in terms of the biology of these organisms Three orders within the Dinoflagellata contain representatives that parasitise crustaceans: Coccidiniales (formerly Chytriodinales), Blastodiniales, and Syndiniales. The latter two are entirely parasitic, with highly derived morphologies, and life history stages. For a detailed review of these groups, see Shields (1994) and Skovgaard et al. (2012).

Members of Chytriodiniales are egg parasites of copepods, euphausiids, and possibly shrimp (Figure 17.1a–d). Four genera are known from crustaceans. They have not been well studied but recent molecular work has addressed their life cycles and phylogeny (Gómez et al. 2009a). Prevalence on host eggs can reach 50% and the effect on their host populations is thought to be damaging (Cachon and Cachon 1987). In terms of distribution, *Dissodinium pseudolunula* has been recorded over a broad area of the northeast Atlantic Ocean and North Sea and when at high densities, it is thought to negatively affect host populations (John and Reid 1983). *Chytriodinium* parasites have been shown to have a widespread distribution in world oceans; thus, they may represent a significant but understudied

component of the microplankton, exerting significant control over many pelagic copepods (Strassert et al. 2018).

Blastodiniales are parasites of copepods. They live in the foregut of their hosts and make the infected copepod appear opaque or black (Figures 17.1e–h). The life cycle was elucidated by Chatton (1920). Most species lack chloroplasts or have vestigial chloroplasts but those with chloroplasts are capable of modest photosynthesis (Pasternak et al. 1984; Skovgaard et al. 2012). These parasites are known to reduce host fecundity (Chatton 1920; Jepps 1937; Sewell 1951). They may also feminise male copepods (Chatton 1920; Skovgaard et al. 2012).

Blastodinium spp. have important ecological consequences in affected copepod populations. Prevalence levels were 58% in nearshore populations of *Calanus finmarchicus* off Norway but 5% in offshore populations (Fields et al. 2015). Infected females showed signs of starvation and the parasite was thought to affect recruitment depending on spatial extent of infections. In the Mediterranean Sea, 16% of copepods were infected, with infections found mainly in corycaeid and calanoid copepods (Alves-De Souza et al. 2011). Infections were also found mostly in copepods at the deep chlorophyll maximum. Mortality rates were estimated at around 15% per day (Skovgaard and Saiz 2006). Epizootics have been reported in the North Sea, Mediterranean Sea, and the southeast Pacific Ocean (for review, see Shields 1994). Tumour-like anomalies on copepods are likely wounds from the parasitic mass as it can protrude through the cuticle of the infected host (Skovgaard 2004).

Syndiniales are internal parasites of many marine organisms, including radiolarians, algae, crustaceans, and fish eggs. Members of the order have unusual life cycles and are known to produce two types of dinospores, a micro- and macro-dinospore, which may or may not be gametes (Chatton 1920; Coats 1999). All members of the order lack chloroplasts (Chatton 1920; Jepps 1937; Gornik et al. 2015). Several species use copepods, amphipods, and decapod crustaceans as hosts and they range from egg parasites of copepods and shrimp, such as *Actinodinium* and *Trypanodinium*, to internal parasites of copepods, amphipods, and

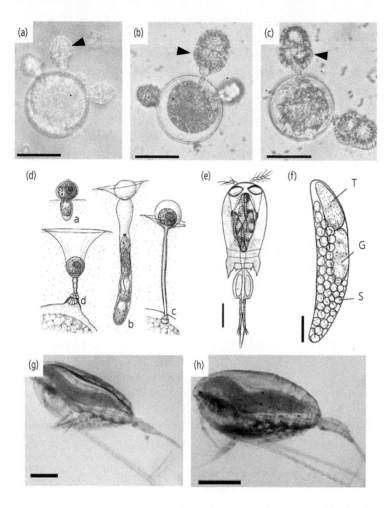

Figure 17.1 (a–c) *Syltodinium listii* (arrowheads) infecting copepod eggs. Note the change in size of the parasites as they consume the eggs. Bars = 50 μm. (d) Stages of infection of *Chytriodinium parasiticum* as it penetrates into a copepod egg. Lower case letters refer to the development of the peduncle through the embryonic coat. (e) *Blastodinium navicula* in the gut of *Corycaeus geisbrechti*. Bar = 100 μm. (f) *Blastodinium pruvoti* from a copepod showing trophocyte (T), gonocyte (G), and sporocytes (S). Bar = 50 μm. (g) *Blastodinium hyalinum* in gut of *Calanus* sp. (h) *Blastodinium* cf. *contortum* in gut of *Temora stylifera*. Bar = 500 μm. Sources: (a–c) from Gómez et al. 2019, with permission; (d) from Dogiel 1906; (e–h) from Skovgaard et al. 2012, with permission.

decapods, such as *Syndinium* and *Hematodinium*. Those infecting embryos kill their host when they release their dinospores. Those developing as internal parasites live in the haemocoel of their host and eventually kill their host because they release dinospores through perforations in the gills, cuticle, or gut wall.

Species of *Syndinium* can occur at high prevalence levels in their host populations. Prevalence levels up to 30% have been reported (Ianora et al. 1987) and they can impose significant mortality risk to their infected hosts, with peaks up to 15–42% per day (Kimmerer and McKinnon 1990). Given the importance of copepods as primary consumers in food webs, it is surprising that so few studies have been published on this group of organisms They likely impose significant mortality

and nutrient recycling in most marine planktonic systems.

Species of *Hematodinium* are significant pathogens of decapods, particularly commercially important crabs and lobsters. They have received the most attention of the parasitic protists in crustaceans. Important reviews include those by Meyers (1990); Shields (1994); Stentiford and Shields (2005); Morado (2011); and Small (2012). Due to a lack of distinctive morphological characters and a limited number of comparative DNA sequences, there are only two described species and one well-known undescribed species. *Hematodinium perezi* is the best known and occurs over a broad geographic range, occurring from Europe, eastern North America, and China (Small et al. 2012; Li et al. 2013). *Hematodinium* spp. are broad host generalists. At least 38 decapods and several amphipods are known to harbour infections (Stentiford and Shields 2005; Pagenkopp et al. 2012; Small 2012). *Hematodinium*-like infections are known from several cold-water decapods and this parasite has a circumboreal distribution (Hamilton et al. 2010; Ryazanova et al. 2010; Small et al. 2012; Small 2012). Infected tanner and snow crabs develop a condition known as Bitter Crab Disease, or Bitter Crab Syndrome, that affects their flavour, rendering them unmarketable (Meyers et al. 1987; Taylor and Khan 1995).

The life cycles of *H. perezi* and *Hematodinium* sp. from *N. norvegicus* have been examined in detail using *in vitro* culture (Appleton and Vickerman 1998; Li et al. 2011a). These parasites have asexual division (budding and merogony) in several life history stages and thus are capable of very rapid proliferation in their hosts (Figures 17.2a–e). Dinospores are the likely infectious stage (Shields 1994; Frischer et al. 2006; Shields et al. 2017). Dinospores from cultures and those released from sporulating crabs survive up to 7 days at 21–23°C but do not form cysts (Li et al. 2011a; Coffey et al. 2012; Huchin Mian et al. 2018). Sexual reproduction has not been observed but the parasites are haplo-diploid with strong evidence for sexuality in their life cycle (Pagenkopp et al. 2014).

Using sentinel studies with caged crabs, transmission of *H. perezi* and infection of juvenile blue crabs occurs rapidly and can develop quickly into heavy infections (7–14 d) (Shields et al. 2017). Transmission is strongly influenced by environmental conditions and the incidence, or rate of new infections, can be extraordinarily high (Huchin-Mian et al. 2017). Moulting is not a host factor in transmission (Huchin-Mian et al. 2017). Cannibalism and scavenging (as trophic transfer) have been ruled out as modes of transmission (Li et al. 2011b) and sexual transmission is not supported by the evidence (Stentiford and Shields 2005).

Several host systems show distinct seasonal or annual cycles of infection with *Hematodinium* spp. (Meyers et al. 1987; Messick and Shields 2000; Stentiford et al. 2001; Shields et al. 2005, 2007; Davies et al. 2019). For example, *H. perezi* is a temperate species with sharp autumnal peaks in prevalence that overlap with settlement of juvenile crabs (Messick and Shields 2000; Lycett and Pitula 2017;

Small et al. 2019). Infections in juvenile edible crabs, *Cancer pagurus*, by *Hematodinium* sp. also show peaks in prevalence in relation to seasonality (Smith et al. 2015). Low and high temperatures contribute to the mortality of infected crabs (Huchin-Mian et al. 2018). Spore densities have been quantified in environmental waters but show little relationship with prevalence of infection in crabs (Pitula et al. 2012; Hanif et al. 2013). Hydrographic features that constrain water masses appear to retain spores and promote epidemics (Shields 1994; Stentiford and Shields 2005).

Hematodinium spp. are significant pathogens in fisheries and cultures of commercial species and thus warrant attention in infected populations. Recent studies have investigated host responses to infection (e.g. Rowley et al. 2015; Li et al. 2016, 2019), risk analysis to other species (Davies and Rowley 2015), behavioural effects (Butler et al. 2014; Stentiford et al. 2015) and potential model hosts for infection (O'Leary and Shields 2017). Additional studies are needed on their life cycles, transmission pathways, and disease ecology. For example, modelling studies indicate that these parasites can hinder efforts to rebuild important snow crab stocks (Siddeek et al. 2010) but more work is needed in this area.

17.2.1.2 Infraphylum Protalveolata: Class Ellobiopsea: Order Ellobiopsida (*Ellobiopsis, Ellobiocystis, Thalassomyces*)

The ellobiopsids are a small clade of protists now classified in the Infraphylum Protalveolata, as a sister taxon to the dinoflagellates (Silberman et al. 2004; Gómez et al. 2009b). They are comprised of at least three genera, *Ellobiopsis, Ellobiocystis*, and *Thalassomyces*, with two other genera unresolved in their taxonomic affiliations. With the exception of *Rhizellobiopsis*, they all infect crustaceans. Much of their taxonomy was undertaken by Chatton (1920), Boschma (1949, 1959), and Wing (1975).

Species of *Ellobiocystis* are harmless commensals, attaching externally on or near the mouthparts and legs of their crustacean hosts and presumably absorbing food from the host's feeding activity. *Ellobiopsis* causes localised damage at the site of attachment but penetration into the

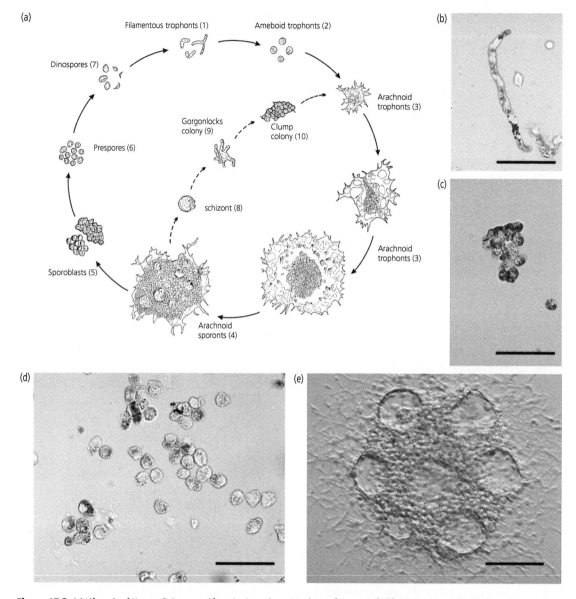

Figure 17.2 (a) Life cycle of *Hematodinium perezi* from *in vitro* culture. Numbers refer to specific life history stages observed in culture. (b) Filamentous trophont in dissection smear stained with 0.3% neutral red. (c) Clump colony from inoculum preparation stained with neutral red. (d) Amoeboid trophonts from haemolymph smear showing variable neutral red uptake. Bars = 50 μm. (e) Arachnoid sporont from culture showing schizonts developing within sporont. Bar = 100 μm. Source: (a) From Li et al. 2011a and author's collection, with permission.

host can cause parasitic castration (Chatton 1920). *Thalassomyces, Rhizellobiopsis,* and *Parallobiopsis* are mesoparasites, in that they have an external body attached to rootlets that penetrate into host tissues, that presumably function in nutrient absorption. They castrate their hosts and may feminise males. *Thalassomyces* have been reported projecting

from the eyestalks of their shrimp hosts (Wing 1975, Shields, personal observation), where they presumably interfere with neuronal control of host behaviour (Figure 17.3a). These intriguing parasites can range up to 30% prevalence and likely exert control over their host populations when at high prevalence levels.

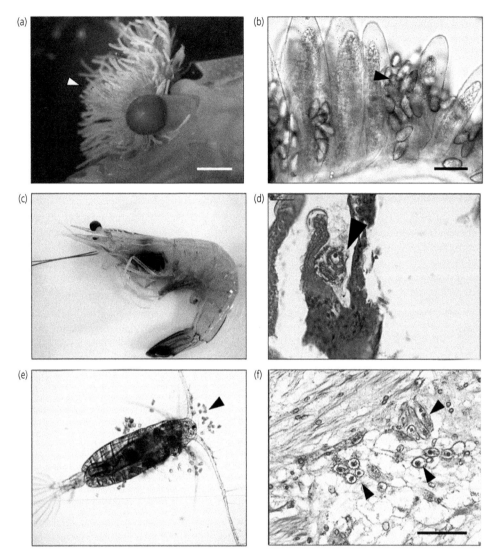

Figure 17.3 (a) *Thalassomyces californiensis* (arrowhead) growing from the eyestalks of *Pasiphaea emarginata*. Bar = 2 mm. (b) Tomites of *Synophrya histolytica* (arrowhead) in the gills of a hermit crab. Wet mount. Bar = 25 μm. (c) *Penaeus setifer* exhibiting shrimp black gill disease due to putative infection by *Hyalophysa lynni*. (d) Gill lamella from shrimp showing an encysted trophont of *H. lynni* (arrowhead) with incipient melanisation (m). H&E. Bar = 50 μm. (e) Heavy infestation of *Zoothamnion intermedium* (arrowhead) on *E. affinis*. (f) *Orchitophrya stellarum* in the glial cells around the optic nerve of *Callinectes sapidus*. Ciliates can be seen in cross and longitudinal section (arrowheads). H&E. Bar = 50 μm. Sources: (a) author's collection; (c) and (d) from Tuckey et al. 2021, and author's collection, with permission; (e) courtesy L. Safi; (f) author's collection.

17.2.1.3 Phylum Ciliophora: Class Phyllopharyngea: Subclass Suctoria, Subclass Chonotrichia, Class Heterotrichea and Class Oligohymenophorea: Subclass Apostomatia, subclass Scuticociliatia, and subclass Peritrichia

Ciliophora, or ciliates, are common epibionts on the external surfaces and gills of virtually all crustaceans. Indeed, they have a high diversity on these hosts with several subclasses using crustaceans almost exclusively as basiobionts (hosts on which the epibionts live). The hard cuticle of crustaceans provides a convenient, mobile surface for sites of attachment and improved sites for feeding activities (e.g. attached to the mouthparts of their hosts). Several ciliates use crustaceans as facultative surfaces or as obligate hosts in commensal relationships (see reviews of Bradbury 1994; Morado and Small

1995; Fernández-Leborans 2010). One subclass, Chonotrichia, is comprised of three orders with at least 43 genera comprised of well over 115 species (Jankowski 1973; Fernández-Leborans 2010). The order, Chilodochonida, uses decapods, the Exogemmida uses isopods, amphipods, and copepods, and the Cryptogemmida uses amphipods, copepods, and leptostracans. Although many of the epibionts in the subclass Chonotrichia are obligate commensals, they are not considered pathogens. They are sometimes found as epibionts on parasitic copepods and as such may be considered hyperparasites, or parasites of parasites. In addition, several chonotrichs show remarkable site specificity on their amphipod hosts, a feature of niche specialisation that has not received much ecological attention other than the work by Jankowski (1973). Similarly, the subclass Suctoria is comprised of three orders that infest crustaceans, with at least 34 genera comprised of at least 125 species that are commensal on the gills, mouthparts, and exoskeleton of their crustacean hosts (Morado and Small 1995; Fernández-Leborans and Tato-Porto 2000a, b). Several of these may show site fidelity on their hosts as well. Members of Heterotrichida are mostly free living and include well-known taxa such as *Stentor* and *Stylonicha*. A few taxa in this subclass are obligate symbionts such as the folliculinids on hermit crabs (Basile et al. 2004) and those that are parasitic on terrestrial isopods (Delgery et al. 2006).

The class Oligohymenophorea has three subclasses whose taxa have developed more specialised relationships with their crustacean hosts. Apostomatia is a highly specialised subclass of ciliates that uses crustacean hosts. The taxon is comprised of three orders, with 6 families, 15 genera, and at least 37 species (Bradbury 1994; Landers 2004; Landers et al. 2006; Chantangsi et al. 2013; Ohtsuka et al. 2015). The 'apostomes' show several adaptations for a symbiotic lifestyle, including lacking a 'mouth' or buccal cavity, having an unusual arrangement of their kineties and ciliature, and having tightly synchronised life cycles with the moult cycle of their hosts. They have four different life history strategies: (1) those that feed on host exuvial fluids; (2) those that invade the haemocoelom of their hosts and feed on haemolymph; (3) those that invade the soft tissues of their host; and (4) those that live in the renal organs, intestines, and liver of cephalopods

(Lynn and Strüder-Kypke 2019). Those that feed on host exuvia tend to be broad host generalists on crustaceans (Bradbury 1994). The life cycle involves a phoront that encysts on the new host cuticle and undergoes division as a tomont, which then excysts as a trophont to feed osmotrophically on exuvial fluids in the host's moulted exuvium. The trophont then swims to the new host instar where it encysts as a phoront until the next molt (Bradbury and Trager 1967). In most cases, the apostomes are commensals, encysting externally on host surfaces; however, those that invade host tissues are truly parasitic and can cause significant pathology therein (see below).

The pathogenic apostomes are capable of penetrating the thin gill cuticle of their host and invading the soft tissues to feed osmotrophically on haemolymph or organ and tissue constituents (Figure 17.3b). Hosts are susceptible after moulting, before the cuticle completes tanning. *Synophrya hypertrophica* is a pathogenic apostome with a broad host range. The trophont is histophagous, tunnels through the thin gill cuticle to burrow into the soft connective tissues of the gill lamellae (Johnson and Bradbury 1976). Invaded tissues become necrotic and melanised by an intensive host response to the injury, and the melanised areas appear as black spots in the gills and tissues of the infected host. Prevalence levels can be high in affected host populations. For example, *Ovalipes stephensoni* had a prevalence of 94.5% but the affected gill area was not extensive enough to impair respiration; hence it was not considered an important pathogen (Haefner and Spacher 1985). This parasite is a broad host generalist infecting several brachyurans, anomurans, and shrimp hosts but it is limited to high salinity waters (Johnson and Bradbury 1976; Haefner and Spacher 1985; Bradbury 1994).

A similar parasite, *Hyalophysa lynni*, causes 'shrimp black gill disease' in penaeid shrimp, particularly *Penaeus setiferus* (Frischer et al. 2017; Landers et al. 2020). The parasite encysts externally on the gills of their shrimp hosts and elicit a strong melanisation response, albeit physical penetration into the host tissues may not occur (Figure 17.3 c,d); (Frischer et al. 2018; Tuckey et al. 2021). This parasite may affect the swimming physiology of their hosts making them more susceptible to predation (Gooding et al. 2020). Another apostome, *Vampyrophrya pelagica* lives and feeds on calanoid copepod hosts (Chatton and Lwoff 1935; Grimes and Bradbury 1992). It feeds on injured copepods or those eaten by predators such as chaetognaths and other crustaceans and can survive digestion by these predators to infect new copepod hosts. Indeed, excystation of the trophonts is stimulated by predation or by injuries sustained by the copepod host. Prevalence levels, host preferences, and seasonal dynamics in the Sea of Japan show distinctive patterns in host usage, with shifts in host species associated with seasonality (Ohtsuka et al. 2004).

Other apostomes are more virulent, and act as parasitoids, ultimately killing their hosts. Species of *Collinia* and *Metacollinia* infect amphipods and species of *Pseudocollinia* infect euphausiids (Gómez-Gutiérrez et al. 2012; Lynn et al. 2014). These parasites live in the haemocoels of their hosts where they rapidly amplify and burst through the host cuticle to release swimming tomites, the presumptive infectious stage (Capriulo and Small 1986; Gómez-Gutiérrez et al. 2006; Lynn and Strüder-Kypke 2019). *Pseudocollinia* infects primarily female hosts (Gómez-Gutiérrez et al. 2006; Gómez-Gutiérrez et al. 2012). The free-swimming tomites have a novel association with an undescribed bacterium. The tomites and bacteria associate in long filaments that cause the tomites to encyst as phoronts. The filaments are likely eaten by euphausiids which are then infected (Gómez-Gutiérrez et al. 2012). Host pathophysiology indicates that the ciliates feed on fatty acids (Gómez-Gutiérrez et al. 2015). These parasites have been found causing epidemics in populations of midwater euphausiids resulting in mass mortalities (Gómez-Gutiérrez et al. 2003).

Much like the chonotrichs, Peritrichia is a clade of epibionts that occur on many crustaceans, including freshwater crayfish, copepods, amphipods, and crabs. It is comprised of two orders, containing 25 genera. At least 270 species have been reported from crustaceans (Morado and Small 1995; Fernández-Leborans and Tato-Porto 2000a). They live on the exoskeleton, gills, mouthparts, and limbs of their hosts (Figure 17.3e). Common genera include *Vorticella*, *Epistylis*, *Operculariella*, and *Zoothamnion*. Their biology is tightly linked to the habitat of their hosts as well as their host moult cycles; because of this, they have been used as indicator species for water

quality and prevailing environmental conditions (Henebry and Ridgeway 1979; Khan and Thulin 1991). Peritrichs are usually harmless commensals but high intensity infections can foul the gills and body and interfere with respiration, excretion, feeding activity, or motility (Visse 2007; Jones et al. 2016). High intensity infestations can affect host survival and increase the stress level of an infested hosts.

Scuticociliatia includes free-living and parasitic taxa, with some taxa known for inquilinism (internal commensalism) in echinoderms. In crustaceans, they are facultative parasites of amphipods, crabs, and lobsters. At least eight taxa have been reported in crustaceans in the literature but they are notoriously difficult to identify due to relatively plastic morphological characters that can vary over time in culture. Although infections can be spectacular, with high densities, host castration, limb loss, and death, there have only been few studies on their taxonomy and biology (for reviews, see Morado and Small 1995; Wiackowski et al. 1999). In decapods, they are typically associated with injured hosts, and outbreaks are associated with crowded laboratory or commercial settings (Grolière and Leglise 1977; Armstrong et al. 1981; Cawthorn et al. 1996; Small et al. 2013). They are facultative parasites and require a portal of entry for transmission as they cannot survive digestive fluids (Loughlin et al. 1998; Miller et al. 2013). In crabs and lobsters, they are associated with cold winter temperatures (Cawthorn et al. 1996; Small et al. 2013; Miller et al. 2013).

Although scuticociliates are facultative parasites, their ease of culture and manipulation, and infectivity via needle injection make them suitable for experimental studies. Infections in crabs and lobsters progress quickly, often killing their hosts within 7–10 days. Infected hosts exhibit lethargy and reduced clotting ability within 2–3 days of infection. They have an extraordinary growth rate and can phagocytise host haemocytes, which contributes to a decline in host defences (Cain and Morado 2001; Miller et al. 2013). Ciliates can invade connective tissues, heart, muscle, thoracic ganglion, and haemopoietic tissues (Messick and Small 1996). Infiltration of the nerves results in infection-induced autotomy (Figure 17. 3f) (Miller et al. 2013).

Anophyroides haemophila causes 'bumper car' disease in the American lobster, *Homarus americanus*

(Cawthorn et al. 1996). The haemolymph of infected lobsters often turns milky white and exhibits coagulopathies (Sherburne and Bean 1991; Cawthorn 1997). Lobsters inoculated with the ciliate develop granulomatous lesions in the affected tissues (Athanassopoulou et al. 2004). Ciliates occur systemically in the haemolymph and prefer oxygen-rich tissues such as gills but in heavy infections they can be found in most tissues. A *Mesanophrys*-like scuticociliate in *Nephrops norvegicus* secretes proteases that probably contribute to its invasiveness and ability to adapt to the parasitic life style (Small et al. 2005).

Anophryoides haemophila was a relatively unknown pathogen until the early 1970s when Aiken et al. (1973) noted a prevalence of 20% in lobster impoundments. The parasite was reported in winter samples in later years (Aiken and Waddy 1983). One outbreak reached 100% prevalence in impoundments from Maine (Sherburne and Bean 1991). This is primarily a disease of crowded conditions found in lobster impoundments as wild lobsters usually have a low prevalence (Lavallée et al. 2001).

17.2.1.4 Phylum Myzozoa: Subphylum Apicomplexa: Infraphylum Sporozoa: Class Conoidasida (gregarines), Order Eucoccidiorida (coccidians: *Aggregata*)

Apicomplexa is an entirely parasitic subphylum of intracellular protists. All members of the clade have an apical complex at some point in their life cycle, and this organelle is used for host cell invasion. The taxon is highly derived, with a high diversity in vertebrate hosts, including well-known genera that infect humans (e.g. *Plasmodium*), pets (e.g. *Toxoplasma*), and livestock (e.g. *Eimeria*, *Babesia*). Gregarines and coccidians can have direct (monoxenous) or indirect (heteroxenous) life cycles. Heteroxenous gregarines in crustaceans use molluscs as intermediate hosts and crustaceans as definitive hosts. In contrast, heteroxenous coccidians use crustaceans as intermediate hosts and cephalopods as definitive hosts. Although striking, the evolution of this pattern has not been studied.

In their definitive hosts, gregarines live in the digestive tract or epithelia of organs with a lumen. They infect the connective tissues of their intermediate hosts. They are common in annelids,

molluscs, and arthropods but have been reported in many invertebrates. Although gregarines are known mainly as insect and annelid parasites, at least 17 genera with 134 species infect crustaceans, including barnacles, amphipods, and many decapods (Bradbury 1994; Clopton 2002). Only one heteroxenous life cycle has been fully characterised, that for *Nematopsis ostrearum*, which uses oysters and xanthid mud crabs (Prytherch 1940 but see Clopton 2002). The giant lobster gregarine, *Porospora gigantea* uses molluscs and lobsters as hosts (Sprague and Couch 1971). It can be quite common in lobsters (Figure 17.4a). These parasites show a remarkable host specificity, typically at the species or genus level (Bradbury 1994) but there have been few infection studies to demonstrate their life cycles in crustaceans (Clopton 2002).

Coccidians are intracellular parasites of invertebrates and vertebrates. They show a high degree of tissue specificity in their hosts, living in blood cells, gut epithelia, fat bodies, or salivary glands. One genus, *Aggregata*, with at least 20 species, is known from crustaceans. The majority of the species use brachyuran crabs (Bradbury 1994) but a few have been reported in penaeid and oplophorid shrimp (Sardella and Martorelli 1997). Those using crustaceans are heteroxenous. The life cycle of *Aggregata eberthi* includes merogony in the crustacean host, with gamogony and sporogony in the cephalopod definitive host (Porchet-Hennere and Richard 1971; Gestal et al. 2002). The description of *A. bathytherma* from an octopus from a deep-sea hydrothermal vent hints at the potential crustaceans that may serve as intermediate hosts in that ecosystem (Gestal et al. 2010). Other than molecular work on species from cephalopd definitive hosts (e.g. Castellanos-Martínez et al. 2013), little has been done on these parasites in their crustacean hosts.

17.2.1.5 Kingdom Chromista: Infrakingdom Heterokonta: Phylum Oomycota: Class Peronosporea: Order Pythiales (Lagenidium, Pythium), Order Saprolegniales (Saprolegnium, Halipthoraceae, Olpidium), and Order Hyphochytriales

The oomycetes are fungus-like water moulds that were formerly included in the kingdom Fungi and regarded as 'lower fungi'. They are now grouped

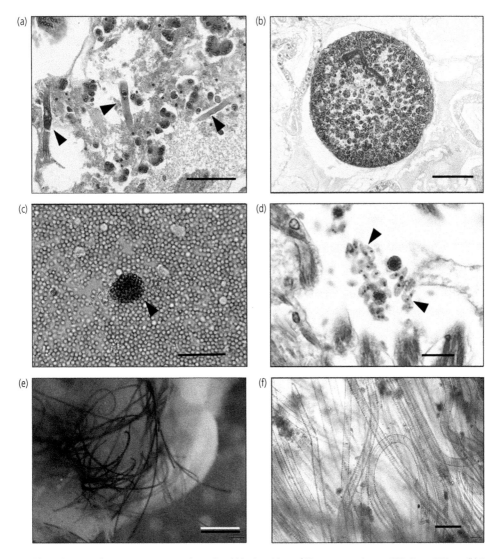

Figure 17.4 (a) Trophozoites of *Psorosporea gigantea* (arrowheads) in the midgut of *Homarus americanus*. H&E. Bar = 100 μm. (b) A hypertrophied microphallid metacercaria infected by the hyperparasite *Urosporidium crescens*. H&E. Bar = 150 μm. (c) Tissue smear showing freed black spores and a packet of spores in a sporophorous vacuole (arrowhead) of *U. crescens* from an infected metacercaria in *Uca pugnax*. Bar = 20 μm. (d) Trophozoites of *Paramoeba perniciosa* (arrowheads) in a haemal sinus of *Callinectes sapidus*. H&E. Bar = 20 μm. (e) *Enterobryus* sp. (Eccrinales) infection in the foregut of a fiddler crab, *Uca pugnax*. Bars = 100 μm. (f). Higher magnification of a *Enterobryus* sp. (Eccrinales) from the foregut of *U. pugnax* stained with neutral red. Bar = 20 μm. Source: author's collection.

within the kingdom Chromista as protists (e.g. Keeling et al. 2005). These are dealt with in Chapter 16.

17.2.1.5.1 *Phylum Bigyra: Subphylum Sagenista: Class Labyrinthulea: Order Thraustochytrida*

The thraustochytrids are related to slime moulds. They have been reported in the egg masses, or clutches, of *Cancer* spp. on the West Coast of the USA (Shields 1990) as well as *Callinectes sapidus* from the East Coast of the USA (Rogers-Talbert 1948). They are usually saprophytic feeding on dead eggs at the periphery of the clutch. An unidentified thraustochytrid in *Scylla serrata* caused significant egg mortality in an experimental Australian hatchery (Kvingedal et al., 2006). Several protists can be found in the egg clutches of crustaceans but few studies have examined the nature of these organisms and their relationship with the host (e.g. Shields 1990).

17.2.1.6 Kingdom Chromista: Infrakingdom Rhizaria: Phylum Retaria (= Cercozoa in part): Class Ascetosporea: Order Paradinida, Order Paramyxida, and Order Haplosporidia

The Phylum Cercozoa has undergone revisions with several classes moved to the phylum Retaria (Cavalier-Smith et al. 2018). Retaria now contains several classes of free-living foraminifera, as well as Ascetosporea, whose members are entirely parasitic. Ascetosporea contains three orders with taxa that parasitise crustaceans: Paradinida, Paramyxida, and Haplosporidia. The first order is poorly known and the latter two include well-known parasites of bivalves (see Chapter 8), with a few taxa infecting crustaceans.

Paradinida were once classified as dinoflagellate affiliates but are now in the Ascetosporea (Cavalier-Smith et al. 2018). Species of *Paradinium* are internal parasites of copepods. Their life cycle was described by Chatton (1920). Notably, orange gonospheres attach to the caudal rami of their host making them very distinctive. Copepod hosts may be castrated, and males may be feminised (see Shields 1994). These parasites are widespread and their environmental DNA has been identified from many regions (Cleary and Durbin 2016; Ward et al. 2018).

Paramyxida is a taxon of obligate parasites with a unique cell-within-a-cell development. Endogeny occurs in primary and secondary cells to form tertiary cells, which are the infectious stage. Their cells possess haplosporosomes, an organelle that is unique to Paramyxida and Haplosporidia. The order is comprised of five genera, two of which use crustaceans as hosts. *Paramarteilia orchestiae* is a parasite of amphipods (Ginsburger-Vogel and Desportes 1979; Ward et al. 2016). This parasite causes feminisation and intersexuality with female-skewed sex ratios in amphipod hosts (Ginsburger-Vogel 1991). Parasite-induced sexual changes can affect host populations and have been shown to be caused by *P. orchestiae* and not by microsporidian infections (Short et al. 2012; Pickup and Ironside 2018). *Paramarteilia canceri* is a parasite of *Cancer pagurus* and occurs in systemic infections in the haemocoel but also occurs intracellularly in mesodermal tissues (Feist et al. 2009). It may also infect *Maja squinado* (Ward et al. 2016). Notably, *Marteilia refringens*, an important pathogen of bivalves in France, has been found using copepods in their life cycle (Audemard et al. 2004; Carrasco et al. 2007). The parasite has some host specificity in copepods and appears to use transovarial and trans-stadial transmission but the role of copepods in the life cycle needs further study (Boyer et al. 2013; Arzul et al. 2014).

Haplosporidia are obligate parasites that undergo sporogony to develop ornate, operculated spores from a multinucleate plasmodium; however, not all taxa form spores (Carnegie et al. 2006). Unlike Microsporidia, Haplosporidia lack a polar filament. The order consists of 45 known species that are primarily parasites of molluscs but several taxa infect crustaceans (Burreson and Ford 2004). Although most taxa in Haplosporidia produce spores, the transmissive stage remains unknown and no life cycles have been established for any members of the order. Several species of *Haplosporidium* infect amphipods (e.g. Larsson 1987; Winters and Faisal 2014; Urrutia et al. 2019). They develop in connective tissue, fat cells, muscle, and digestive tissue, and then develop into systemic infections in the haemocoel. Individual amphipods develop lethal infections and there is circumstantial evidence that haplosporidian infections have led to the decline in amphipod populations in the Great Lakes, USA (Winters and Faisal 2014; Cave and Strychar 2014).

Several species of *Haplosporidium* also infect brachyuran crabs. *Haplosporidium louisiana* infects xanthid mud crabs from the east and Gulf coasts of the USA (Sprague 1963; Perkins 1975). Sporogony produces large black spores which can give the muscles a black appearance in heavy infections. As with those in amphipods, they infect mesodermally-derived tissues and later become systemic infections. Recently, at least four species in the order have been described from *Carcinus maenas* (Stentiford et al. 2013; Hartikainan et al. 2014; Davies et al. 2020). This is surprising considering that this crab has been the focus of many parasitological investigations since the early 1900s and highlights the role of molecular techniques and systematic surveys in finding new pathogens. To wit, *Paramikrocytos canceri*, an intracellular parasite infecting antennal gland epithelia of *Cancer pagurus* (Hartikainen et al., 2014), has a high prevalence ranging from 15–70% (Bateman et al. 2011; Thrupp et al. 2013). In a survey of locally abundant crab species, the parasite was found only in *C. pagurus* (Edwards et al. 2019). Thus, there is clearly much to be learned by continued surveys and re-examination of such well-studied hosts.

Although not strictly a parasite of crustaceans, *Urosporidium crescens* is found in brachyuran crabs on the east and Gulf coasts of the USA (Figure 17.4b,c). It is a hyperparasite that infects metacercariae of microphallid trematodes, particularly *Microphallus* spp. and *Levinseniella* spp., that use *Callinectes sapidus*, *Uca* spp., and grapsids as a second intermediate hosts. The hyperparasite causes infected metacercariae to swell in size and become visibly obvious. It does not infect crab tissues per se but its spores are often observed in crab haemolymph. The infected metacercariae are known colloquially as 'pepper crab', 'pepper-spot', or 'buck-shot' and consumers often seek advice about whether the crabs are safe to eat (Shields and Overstreet 2007). Although pepper crabs are safe to eat, as they show dead and dying worms, the uninfected metacercariae may be a health risk to humans who eat raw or poorly cooked crab hosts.

Haplosporidia have also been reported from penaeid and caridean shrimp but none have been adequately described (Dyková et al. 1988; Bower and Meyer 2002; Nunan et al. 2007; Utari et al. 2012). Localised outbreaks with prevalence as high as 30% and low host survival have occurred sporadically in penaeid cultures (Utari et al. 2012). An outbreak in Indonesia was thought to arise from infected postlarval shrimp. Prevalence of an undescribed species in a natural population of *Pandalus platyceros* off British Columbia, Canada, was 20% (Bower and Meyer 2002). Thus, as with *Carcinus*

maenas, comprehensive parasitology surveys will likely uncover more of these parasites in shrimp and other crustacean hosts.

17.2.1.7 Kingdom Protozoa: Phylum Euglenozoa: Class Kinetoplastea

Members of the Kinetoplastea are flagellate parasites, with important pathogens in vertebrates, including humans, fishes, and invertebrates. Two kinetoplasteans have been reported from crustaceans. An undescribed kinetoplastid flagellate was identified from the haemolymph of 34% of the blue crabs, *C. sapidus*, tested during the development of a novel diagnostic technique (Troedsson et al. 2008). DNA sequences showed that the flagellate was related to the free-living kinetoplastid *Procryptobia sorokini*. The other known kinetoplastean is *Perkinsiella amoebae*, which is the parasome, an organelle, of *Neoparamoeba* spp. (Dyková et al. 2003). Given the expansion of molecular techniques to many studies, this taxon will no doubt see some additions in crustaceans.

17.2.1.8 Kingdom Protozoa: Phylum Amoebozoa: Subphylum Lobosa: Class Discosea: Order Dactylopodida (*Paramoeba, Neoparamoeba*)

Only two parasitic amoebae are known from crustaceans and they both infect decapods. *Paramoeba perniciosa* infects *Callinectes sapidus*, and other crabs, in the coastal lagoons of the eastern and Gulf coasts of the USA (Sprague and Beckett 1966; Sprague et al. 1969). It is likely *Neoparamoeba pemaquidensis sensu* Dyková et al. (2007) but it has not been examined with molecular tools. It has two distinct morphologies: a large trophozoite (to 25 µm) lives in connective tissues and nerves and a smaller trophozoite (to 7 µm) develops in systemic infections (Figure 17.4d); (Johnson 1977). The amoeba can cause mortalities and during one outbreak losses were estimated at 30% of the population (Sprague and Beckett 1966; Sprague et al. 1969; Newman and Ward 1973). The route of transmission remains unknown but cyst stages have not been observed and cannibalism did not result in infections (Newman and Ward 1973; Couch 1983). Heavy infections impose significant pathophysiological changes on host crabs (Pauley et al. 1975).

Neoparamoeba pemaquidensis is a facultative parasite of lobsters, *Homarus americanus*. The amoeba is a common soil constituent but it contributed to a lobster mortality in Long Island Sound, USA (Mullen et al. 2004). The mortality coincided with several environmental stressors and possibly contaminants (see Shields 2013 for review). The weakened lobsters were attacked by the amoeba and 90–95% of the dead and dying lobsters were infected with it (Mullen et al. 2004). Although prevalence levels were high during the stressful conditions, they were very low in subsequent years (Mullen et al. 2005). Surveys of *Carcinus maenas* from the North Sea and North Atlantic found evidence of *Neoparamoeba* infections in populations from the Faroe Islands and Nova Scotia but not from other regions (Bojko et al. 2018). Three different amoebae were identified, and their prevalence was thought to be related to salmon aquaculture as salmon are also hosts for *N. pemaquidensis*. Both *P. perniciosa* and species of *Neoparamoeba* possess a symbiotic nucleosome, termed the Nebenkörper or parasome. In *Neoparamoeba*, it has been identified as a symbiotic kinetoplastid, *Perkinsiella amoebae* (Dyková et al. 2003; Young et al. 2014).

17.2.1.9 Kingdom Protozoa: Phylum Choanozoa: Class Ichthyosporea (formerly Mesomycetozoea): Order Eccrinales, Order Amoebidiales, and Order Ichthyophonida

Ichthyosporea is a clade of protists near the divergence of animals and fungi (Cafaro 2005; Lohr et al. 2010; Glockling et al. 2013). Three orders contain symbionts in crustaceans: Eccrinales, Amoebidiales, and Ichthyphonida. Eccrinales are benign commensals that live internally in the hindgut of arthropod hosts (Figure 17. 4e,f). There are least 47 genera known and at least 15 genera use crustaceans (Misra 1998). They specialise on isopods, amphipods, and a few decapods, mainly those in freshwater or intertidal areas. Members of Amoebidiales are ectosymbionts on the exoskeletons of aquatic arthropods including cladocerans, amphipods, copepods, and isopods as well as aquatic insect larvae. One genus,

Amoebidium, with five species infects crustaceans. The symbionts attach externally to their hosts and undergo development in synchrony with their host's moult cycle (Lichtwardt 1986). Ichthyophonidans include *Psorospermium haecklii*, a crayfish symbiont with uncertain pathology (see Edgerton et al. 2002 for review) and *Caullerya mesnili*, an important pathogen of cladocerans. The latter can be found at high prevalence levels and has been shown to affect populations of *Daphnia* in laboratory experiments (Bittner et al. 2002; Lohr et al. 2010).

17.2.2 Helminths

17.2.2.1 Kingdom Animalia: Phylum Platyhelminthes: Order Rhabdocoela (Temnocephala), and Order Fecampiida (Fecampia)

Turbellarian worms include free-living, symbiotic, and parasitic taxa. The best-known examples from crustaceans are symbionts of hermit crabs (Williams and McDermott 2004; McDermott et al. 2010) and freshwater crayfish (Edgerton et al. 2002). For example, Temnocephalidae is a family of highly modified symbionts with a specialised relationship with crayfish. The worms possess oral tentacles and a posterior 'sucker' and live externally on the exoskeleton and egg masses of their hosts. The clade has a distribution throughout the southern continents, except for Africa, and that was used to support the original concept of Gondwanaland and 'continental drift' (for review, see Gelder 1999). Although these symbionts are primarily found on crayfish, they also use freshwater shrimps and crabs (Cannon and Joffe 2001; Peralta et al. 2005).

Fecampiida are protelean parasitoids of barnacles, isopods, amphipods, and decapods (Southward 1950; Christensen and Hurley 1977; Kuris et al. 2002). They can castrate infected hermit crabs and shrimp (Bellon-Humbert 1983). A single worm fills the haemocoel of its infected host and eventually kills it when it emerges as a mature worm (Southward 1950). Fecampiids are cryptic species living inside their crustacean hosts. Their cocoons are more often observed than their parasitic or adult stages (Sluys and van Ginkel 1989). They have been reported from abyssal to intertidal habitats but they are a very cryptic clade, in part, because of their parasitism.

17.2.2.2 Kingdom Animalia: Phylum Platyhelminthes: Superclass: Neodermata: Class Trematoda

Flukes, or trematodes, are members of a large class of parasitic worms. They can have complex, multi-host life cycles and their first intermediate host is, with few exceptions, a mollusc. They use crustaceans as a second intermediate host and virtually every major systematic group has been reported with cysts (metacercarial stages) of these worms. Trematodes are often trophically transmitted parasites, with the second intermediate host eaten by the definitive host to complete the life cycle. Larval trematodes known as cercariae encyst as metacercariae in the tissues of their crustacean host. Although trematodes exhibit high host specificity for their molluscan first intermediate hosts (Cribb et al. 2001), they have less specificity in their choice of second intermediate hosts, often using other invertebrates, amphibians, or fishes. Vertebrates are the definitive hosts, or the host where maturation and sexual reproduction occur. Trematodes in crustaceans were reviewed by Overstreet (1983), those in copepods by Ho and Perkins (1985). Several trematodes use crustaceans as second intermediate hosts. For example, *Derogenes varicus*, a broadly distributed parasite of marine fishes, uses pelagic copepods as intermediate hosts (Køie 1979) and the well-known families Paragonimidae and Microphallidae use freshwater and estuarine shrimps and crabs as intermediate hosts.

Species of *Paragonimus westermani* form a species complex of closely related lung flukes that are pathogenic in humans and other crab-eating mammals (e.g. Blair et al. 1999). These parasites are endemic in Southeast Asia and parts of Africa, and closely related species occur in North and South America (Vieira et al. 1992; Procop 2009). They use felines, mustelids, and other carnivorous mammals as definitive hosts. They can also use a broad range of decapods, including crayfish, freshwater shrimps, and crabs, as intermediate hosts (Blair et al. 1999). Crustaceans get infected via penetration of the cercariae through the gills or by eating infected

snails (Shibahara 1991). Humans obtain infections by eating poorly cooked crab or shrimp. One such delicacy is known as drunken shrimp, as the hosts are marinated in wine and eaten raw. The Chinese and Japanese mitten crabs, *Eriocheir sinensis* and *E. japonica*, are important vectors in Asia, and prevalence levels can be quite high in these hosts (Cho et al. 1991; Odermatt et al. 2007; Kim et al. 2009; Doanh et al. 2012). This is but one reason why crustaceans should be cooked before being eaten.

The metacercariae of Microphallidae are often found in amphipods and decapods (Figures 17.5a,b). The primary hosts are snails in intertidal and nearshore estuarine and marine habitats. Because of the accessibility of the intermediate hosts and the neotenic development of adult characters, the life cycles of many microphallids are known. For example, *Microphallus primas*, a common digenetic trematode in the hepatopancreas of *C. maenas*, enters the crab host as a cercarial stage which uses a penetration cyst and stylet to penetrate the thin cuticle of the crab's gill (Saville and Irwin 2005). Once inside the crab, the cercariae migrate into connective tissues and encyst as metacercariae (Figure 17.5b). The latter excyst in the stomach of a shore bird and develop rapidly into adults (Stunkard 1957; Threlfall 1968). Other microphallids are host generalists, using mustelids, marsh rats, birds, and fishes as definitive hosts.

Microphallids are often found in common shore crabs, such as *Hemigrapsus* spp., *Uca* spp., and *Carcinus* spp. Their presence has been used to make inferences about the introductions of these crabs into non-native regions, the likelihood of parasite introductions or acquisition of new hosts, and their acquisition by climant migrants (Blakeslee et al. 2009; McDermott 2011; Zetlmeisl et al. 2011; Johnson et al. 2020).

A few trematodes have been reported from lobsters. *Thulakiotrema genitale*, a microphallid, encysts as a metacercaria in gonads of *Panulirus cygnus* (Deblock et al. 1991). The definitive host is unknown, but is likely a turtle or fish. *Cymatocarpus solearis*, a brachycoelid, is often found in the abdominal muscles of *P. argus* from Mexico (Gómez del Prado-Rosas et al., 2003; Briones-Fourzan et al. 2016). Cysts are visible through the lightly sclerotised ventral abdomen. The green sea turtle, *Caretta*

caretta, is the definitive host (Caballero 1959). *Stichocotyle nephropis*, an aspidogastrid trematode, infects the wall of the digestive tract of *Nephrops norvegicus* (MacKenzie 1963) and *H. americanus* (Montreuil 1954; Brattey and Campbell 1986). Lobsters probably get it from eating infected molluscan hosts. Rays are definitive hosts (Odhner, 1898; Linton, 1940; MacKenzie 1963; Symmonds 1972).

Trematode metacercariae can occur in high intensity infections that affect host behaviour, morbidity and mortality (Robaldo et al. 1999; Gonzalez 2016; Violante-González et al. 2016). Tissue necrosis is often noted in high intensity infections in many metacercarial infections (Stentiford and Feist 2005). Infections in nervous tissue may result in nerve damage and additional morbidities (Sparks and Hibbits 1981). Trematode infections in the nerves and in the haemocoel can affect the behaviour of their amphipod hosts making them more susceptible to parasite-induced trophic transfer (Gates et al. 2017; Johnson and Heard 2017). Infections in nervous tissue cause an increase in enzymes associated with inflammation that may explain altered behaviours (Helluy and Thomas 2010). These parasites can also lower fecundity and increase the respiration rate of infected hosts (Gates et al. 2017; Arundell et al. 2019).

Parasite ecology can be investigated with trematodes in crustaceans because of the abundance of infected snails and suitable hosts, particularly amphipods. Infection studies have been used to examine aspects of the host-parasite association in terms of host suitability, infection dynamics, development time, and other aspects of disease ecology. For example, more rapid development in snails and increased transmission to amphipods has been associated with temperature in the laboratory (Koprivnikar et al. 2014; Mouritsen et al. 2018).

17.2.2.3 Kingdom Animalia: Phylum Platyhelminthes: Superclass: Neodermata: Class Cestoda

Tapeworms, or cestodes, are an entirely parasitic clade that live in the intestines or coeloms of their vertebrate definitive hosts and absorb nutrients through their highly absorptive tegument. They have both indirect or direct life cycles. Arthropods

serve as the first intermediate hosts for several clades of cestodes. Crustaceans eat cestode eggs released from the faeces of infected fishes. The eggs hatch as coracidium or onchosphere larvae depending on the order. They penetrate through the gut wall and develop as a larva in the haemocoel or tissues of their crustacean host (Figure 17.5c). Fishes eat the infected crustaceans and the parasite burrows through the gut wall to develop as a metacestode in the body cavity of the fish second intermediate host. When ingested by a suitable definitive host, the metacestodes evaginate their scolex upon digestion and use it to attach to the gut wall of their definitive host, where they grow and mature.

The orders Tetraphyllidea, Trypanorhyncha, and Lecanicephalidea use crustaceans as intermediate hosts and elasmobranchs as definitive hosts. Pseudophyllidea use copepods as intermediate hosts and fishes and mammals as definitive hosts. Cyclophyllidea use many arthropods as intermediate hosts, including branchiopods (brine shrimp, e.g. Georgiev et al. 2007), and birds and mammals as definitive hosts. Very few life cycles have been worked out for marine cestodes because of the difficulties in working with large elasmobranch hosts in experimental settings (Butler 1987; Jensen and Bullard 2010). Moreover, many of these taxa have a very high host specificity in their first intermediate hosts adding to the logistical difficulty of working on their life cycles (see review by Ho and Perkins 1985); nonetheless, molecular genomics should lead to a better understanding of cestode host specificity and their use of different intermediate and definitive hosts (e.g. Morales-Avila et al. 2019).

Freshwater copepods serve as the first intermediate host for *Diphyllobothrium latum*, the broad fish tapeworm. This tapeworm uses fish-eating mammals, including humans, as definitive hosts. Humans get infected by eating raw or poorly cooked freshwater fishes, particularly salmonids and coregonids. The host-parasite association has been examined for another pseudophyllid, *Schistocephalus* spp., in their copepod host. Studies have included methods for tracking infections using fluorescent microscopy (Kurtz et al. 2002), detailing responses of infected copepods to stimuli (Jakobsen et al. 1998), as well as showing that altered copepod behaviours facilitate parasite-induced trophic transfer (Benesh 2019).

Cestode infections are relatively common in branchiopods, copepods, and shrimp, presumably because these hosts are suspension feeders, which enhances the probability of their ingesting parasite eggs. Decapods tend to have fewer cestode infections (see Shields et al. 2006; Shields et al. 2015 for reviews), but some species can develop high intensity infections (Shields 1992; Gurney et al. 2004; Zetlmeisl et al. 2011). Although little work has established whether infections are fatal, heavily infected crabs show alterations in their physiology (Gurney et al., 2006) and infected brine shrimp show changes in their coloration, parasitic castration, and changes in their lipid levels (Amat et al. 1991), as well as behavioural changes that may facilitate parasite-induced trophic transfer (Sánchez et al. 2007, 2009). Indeed, brine shrimp, *Artemia* spp., make excellent models for studying the effects of host-parasite associations in the first intermediate host because they have a well-described cyclophyllidean fauna that can be identified in the cysticercoid stage (e.g. Vasileva et al. 2009).

17.2.2.4 Kingdom Animalia: Phylum Nemertea: Class Hoplonemertea

Ribbon worms, or nemerteans, possess a protrusible proboscis used in prey capture, that is housed internally in a distinctive body cavity known as a rhynchocoel. The phylum has over 1150 species, but relatively few are symbionts on crustaceans (Gibson 1995). Three genera, *Carcinonemertes* (16 spp.), *Ovicides* (5 spp.), and the monotypic *Pseudocarcinonemertes homari*, are relatively well-known obligate egg predators on their crab and lobster hosts (see Jensen and Sadeghian 2005; Shields et al. 2015). These worms live on the egg masses, gills, and axillae of their hosts and have a reduced proboscis with a single stylet that is used to puncture host embryos (Figure 17.5d–f). Eggs are laid in mucous sheaths attached to the eggs and setae on their host's pleopods. The juveniles of some species live in the limb axillae of their hosts. Some species migrate to the gills after host embryos hatch, where they encyst to wait for the next reproductive period. Worms can transfer to the new host instar during moulting and to new hosts during copulation (Wickham et al. 1984; Shields and Wood 1993). These

worms are highly adapted to their hosts reproductive periods and have evolved different life history strategies that reflect the different developmental periods in their host embryogenesis (Shields and Kuris 1990). They also exhibit varying degrees of host specificity presumably due to their hosts' embryogenesis.

Epizootics of these egg predators have and have contributed to the decline and non-recovery of red king crab and Dungeness crab stocks (see review by Shields 2012). Prevalence levels can reach 100% and have high intensities (> 800 worms per clutch), effectively leading to a complete loss of fecundity. The worms do not tolerate low salinities that occur in estuaries (McCabe et al. 1987; Dunn and Young 2013). For hosts such as lobsters or king crabs that have long periods of embryogenesis, the relationship between egg predation and egg mortality must be examined early in host embryogenesis when worm density and feeding activity is highest because worm density can decline rapidly in late host embryogenesis as they leave the depleted clutch (Campbell and Brattey 1986; Kuris et al. 1991).

17.2.2.5 Kingdom Animalia: Phylum Nematomorpha

Nematomorpha is a small phylum of pseudocoelomate worms known as horsehair worms Their larvae are endoparasitic in arthropod hosts. When they rupture from and kill their host, they become short-lived, nonfeeding adults; hence, they are protelean parasitoids. They are almost entirely parasites of terrestrial and aquatic insects, with the exception of the genus *Nectonema* (five spp.), which infects several marine decapods (see Poinar and Brockerhoff 2001). There are very few records of these parasites but they leave an indelible image of infection with their body mass tightly coiled within the body of their hosts. These worms cause significant pathology in their hosts, including parasitic castration and compression necrosis (Nouvel and Nouvel 1934; Born 1967; Nielsen 1969). Prevalence levels approached 50% in a population of *Cancer irroratus* (Leslie et al. 1981).

17.2.2.6 Kingdom Animalia: Phylum Nematoda

Nematoda, or roundworms, comprise a highly diverse phylum of free-living and parasitic taxa,

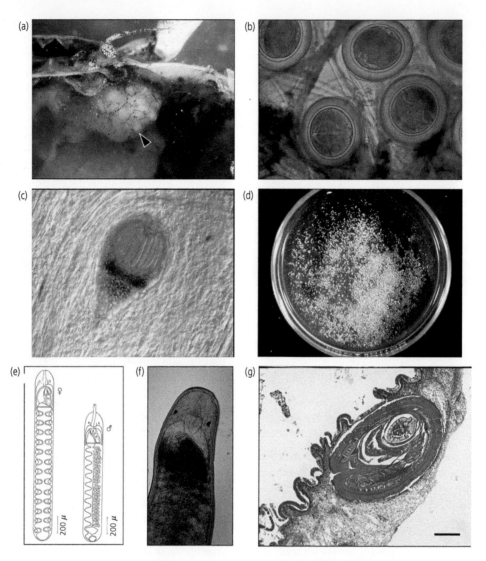

Figure 17.5 (a) Antennal gland of *Uca pugnax* packed with microphallid metacercariae (arrowhead). Dissection. (b) Thick-walled metacercariae in the connective tissues of *Leptodius sanguineus*. Wet mount. (c) *Polypocephalus moretonensis* (lecanicephalid cestode) in a ventral nerve of *Portunus armatus*. Wet smear. (d) High intensity infection of *Carcinonemertes regicides* washed free from an egg-bearing pleopod of *Paralithodes camtschaticus*. (e) Male and female of *C. regicides*. (f) *Carcinonemertes epialti* from *Cancer anthonyi*. (g) Cystacanth of *Profilicollis botulus* in connective tissue of *Homarus americanus*. H&E. Bar = 100 μm. Source: author's collection; (e) from Shields et al. 1989 and author's collection, with permission.

with estimates of at least 40,000 species in two classes and at least 14 orders (Anderson 2000; WoRMS 2020). Nematodes are constrained in their morphology and size by a flexible cuticle and a pseudocoelom, that together comprise a hydrostatic skeleton. An excellent systematic account of the phylum is given by Anderson (2000), with reviews of their life histories in invertebrates given by Adamson (1986) and Blaxter et al. (2000). Nematodes have to moult to grow, so each successive larval stage (designated L1 to L4 stages) moults, until reaching the adult instar. The transmissive stage is specific to each taxon and can be the egg, L1, L2, or L3 larva. Crustaceans serve mainly as intermediate hosts. Although many nematodes infect copepods, amphipods, and decapods,

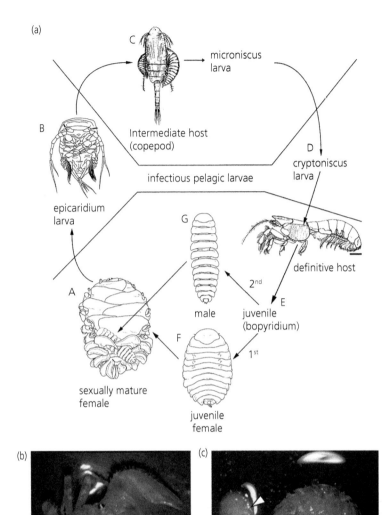

Figure 17.6 (a) Life cycle of the bopyrid *Athelges takanoshimensis*. Adult female (A) releases epicaridia (B) that attach to copepods and become microniscus larvae (C), which moult, leave the copepod and become an infectious cryptoniscus (D). The cryptoniscus moults to become a bopyridium (E) that will either become a juvenile female (F) or a dwarf male (G). Eggs develop in the marsupium of the female. (b) *Petrolisthes* sp. with a right branchial swelling. (c) Dissection of (b) showing a female *Aporobopyrina sp.* (arrowhead) as the cause of the swelling. Source: (a) from Cericola and Williams 2015, kindly provided by Jason D. Williams; (b and c) from author's collection.

surprisingly few have been reported in crayfish (see Edgerton et al. 2002 for review). The reasons for this are unclear, particularly given the plethora of studies on crayfish. Because nematodes mainly use crustaceans as intermediate hosts and they are difficult to identify as juveniles, there are many reports of nematode larvae in the literature, with few adequately identified to genus or species. The various host-parasite associations have had several reviews (Overstreet 1983; Busch et al. 2012; Shields et al. 2015). A few representatives are detailed below.

17.2.2.7 Order Ascarididomorpha

One of the best-known nematodes in crustaceans is *Anisakis simplex*, the 'sushi worm'. It and a closely related ascaridoid, *Pseutoterranova* spp. are broad host generalists and use euphausiids, amphipods, mysids, and decapods, which ingest the L3 larva to become infected. These parasites can use a wide range of invertebrates and fishes as paratenic, or transport hosts, until they reach a suitable definitive host, a marine mammal (for review, see Smith and Wootten 1978). Humans are accidental hosts and obtain infections from eating raw or undercooked squid or fish. Surveys of these parasites show that they can be very rare in crustacean populations but quite abundant in fish populations, indicating their capacity for bioaccumulation in fish hosts as they move up trophic levels (e.g. Marcogliese 1996). *Hysterothylacium reliquens* and *H. aduncum* are also broad host generalists, with similar life cycles and propensities for accumulation, but they use fishes as definitive hosts (Margolis and Arthur 1979; Deardorff and Overstreet 1981). The latter has been reported in amphipods, unidentified crab larvae, and a hermit crab (Marcogliese 1996; Jackson et al. 1997). These parasites are unusual in their broad range of intermediate, transport, and definitive hosts and this highlights their successful adaptation to a broad range of host microhabitats.

17.2.2.8 Order Spiruromorpha

The spiruroids include several worms that use copepods, amphipods, mysids, euphausiids, decapods, and several others as intermediate hosts. A parasite of humans, *Dracunculus medinensis*, the Guinea worm, and related species, use freshwater copepods as intermediate hosts. The L1 stage is ingested by a copepod, moults twice to become an L3, and the host is then imbibed in untreated, unfiltered drinking water. Male and female worms migrate through the human body to eventually mature and reproduce in subcutaneous tissue near the extremities. Female worms are ovoviparous and, when gravid, cause a painful blister to form that fills with L1 larvae, which are released into the water when the person attempts to relieve the inflammation and pain by soaking their infected extremity in cool water (Tayeh et al. 2017). Control efforts have significantly reduced this parasite to the point where it is now nearly extirpated and this was achieved through educational campaigns to filter drinking water, among other control methods. Although *D. medinensis* has a central African distribution, the related *D. insignis* uses mustelids as definitive hosts in North America.

Larvae of *Ancyracanthopsis winegardi* and *Skrjabinoclava inornatae* use fiddler crabs as intermediate hosts and shore birds as definitive hosts (Wong et al. 1989; Wong and Anderson 1990). Similar species have also been reported in grapsid and ocypodid crabs from the Southern Hemisphere, along the flight path of many migratory shore birds (Cremonte et al., 2007; La Sala et al. 2009; Alda et al. 2011). A species of *Skrjabinoclava* is shown to manipulate host behaviour in amphipods to increase transmission to its bird definitive host (McCurdy et al. 1999).

Cystidicolid nematodes use amphipods and decapods as intermediate hosts and shorebirds and teleost fishes as definitive hosts (Anderson, 2000). One genus in particular, *Ascarophis*, uses many freshwater and marine crustaceans and is broadly distributed globally (Uspenskaya 1953; Uzmann 1967; Petter 1970; Poinar and Thomas 1976; Brattey and Campbell 1986; Moravec et al. 2003; Cremonte et al. 2007; La Sala et al. 2009; Alda et al. 2011). They can be present at high prevalence levels in crustacean populations. Although crustaceans are primarily intermediate hosts, adult worms have been reported from amphipod hosts, which is highly unusual as the genus uses fishes as definitive hosts (Fagerholm and Butterworth 1988; Appy and Butterworth 2011). One other spiruroid, *Heptochona praecox*, develops as adults in the hepatopancreas of the freshwater crab *Ceylonthelphusa rugosa* (Poinar and Kannangara 1971).

A larval physalopterid, *Proleptus obtusus*, uses *Carcinus maenas* as an intermediate host and elasmobranchs as definitive hosts (Lloyd 1920, 1928; Perez-Calderon 1986). The parasite is not widely reported in its green crab host, which is surprising given the extensive studies on the parasites of this crab (e.g. Zetlmeisl et al. 2011). Nonetheless, a larval *Proleptus* sp. had a prevalence of 17% in *Cancer plebejus* off Chile, which is an indication that this crab is likely an important intermediate host for this parasite (George-Nascimento et al. 1994).

17.2.2.9 Order Rhabditida

Rhabditids are typically free-living, soil nematodes but several have symbiotic relations with amphipods and decapods. Their life cycle includes an L3 capable of arrested development that attaches to hosts for dispersal (Anderson, 2000). Those on crustaceans reside in protected areas, such as underneath the pereonites on amphipods (Rigby 1996) or on the mouthparts, egg masses, and in the branchial chambers of crabs (Riemann 1970; Sudhaus 1974, 1986; Nicholas 2004).

17.2.2.10 Order Mermithida

Mermithids are important parasites of insects. They have been reported in a few crustaceans, mainly terrestrial isopods and freshwater amphipods, but one species, *Thaumamermis zealandica* has been described from a marine amphipod (see Poinar et al. 2002, Williams et al. 2004) and another was recently reported from the shrimp, *Palaemon paludosus* (Warren et al. 2019). Mermithids are protelean parasitoids, killing their host when they leave it to mature as an adult. *Thaumamermis zealandica* can reach relatively high prevalence levels (30%) in populations of its host amphipod and can influence the burrowing behaviour of their amphipod hosts making them dig deeper burrows than uninfected hosts, thereby protecting them from bird predation (Poulin and Latham 2002). Although, likely related to mermithids, the systematic position of the benthimermithid, *Trophomera marionensis*, from amphipods at abyssal depths remains unresolved (Leduc and Wilson 2016; Leduc and Zhao 2019).

17.2.2.11 Order Monhysterida

Monhysterids are mainly free-living nematodes that live in terrestrial, marine, and aquatic systems. As with most free-living species, they are small (< 1.0 mm) and live as meiofauna in sediments. Several species are symbionts in crustaceans, typically living in the gill chambers of their semi-terrestrial decapod hosts (e.g. Riemann 1970), others are parasites in amphipod hosts (e.g. Poinar et al. 2010) and those in crayfish have an unclear host-association (e.g. Edgerton et al. 2002). They can reach very high intensities in the gills and gill cavities of their gecarcinid crab hosts (Shields, personal observation).

17.2.2.12 Kingdom Animalia: Phylum Acanthocephala

Acanthocephala represent a small phylum of pseudocoelomate, obligate 'thorny-headed worms' that form a clade with rotifers (Garey et al. 1996). They are named after the distinctive, thorny proboscis that is used as an attachment organ. The phylum is comprised of four classes with well over 1,400 species. They have relatively simple, indirect life cycles, using arthropods as intermediate hosts, and vertebrates as definitive hosts. The eggs are ingested by an arthropod host, they hatch as acanthellae which penetrate the gut wall, encyst and develop into cystacanths. When the arthropod is eaten by a suitable host, the cystacanth excysts in the small intestine and inserts its proboscis into the intestinal wall. They are known to use paratenic, or transport, hosts before reaching their definitive host.

Acanthocephalans have been reported from a variety of crustaceans, including isopods, amphipods, mysids, euphausiids, and decapods. They are relatively common in species that use filter feeding, suspension feeding, or coprophagy to obtain nutrition (e.g. Goulding and Cohen 2014). Acanthocephalans have been found in crustaceans from diverse ecosystems ranging from abyssal marine habitats to terrestrial systems. One example, *Profilicollis botulus*, is relatively common and can be found in a number of decapods over a broad geographic distribution (Figure 17.5g) (Nickol et al. 1999). Prevalence can be quite high in certain populations (Ching 1989; Nickol et al. 1999) and is no doubt reflective of the presence of the definitive host, diving ducks, and their extensive migrations (Thompson 1985; Ching 1989; Goulding and Cohen 2014). Sea otters are accidental hosts for this parasite and often suffer severe pathologies and mortality when infected by it (Mayer et al. 2003).

Acanthocephalans are known to induce behavioural manipulation of their hosts and many studies have examined their effects on isopods, amphipods, and a few decapods (Bethel and Holmes 1974; Helluy and Holmes 1990; Latham and Poulin 2001; Moore 1983, 2002). In addition to the typical pathologies of castration and altered pigmentation, these parasites cause alterations in behaviour that increase their risk of predation, thus enhancing their parasite-induced trophic transfer to the definitive host (e.g. Bethel and Holmes 1977; Moore and Gotelli 1990; Kolluru et al. 2011). Behavioural alterations include reversal of photo-taxis, changes to evasion tactics (Bethel and Holmes 1973), altered geotaxis, clinging behaviour (Bauer et al. 2005), and increases in salinity tolerance (Piscart et al. 2007). These changes have a neurological basis with parasitism inducing changes in serotonin and dopamine levels (Helluy and Holmes 1990; Rojas and Ojeda 2005; Poulin et al. 2003; Pérez-Campos et al. 2012; Kopp et al. 2016). Ultimately this leads to increased susceptibility to predation as infected hosts no longer respond to avoid predators (Moore 1983; Moore and Gotelli 1990). Infections also lead to altered pigmentation that can enhance the visibility of the infected hosts to visual predators (Moore 1983; Moore and Gotelli 1990; Fayard et al. 2020). Prevalence can be high enough to result in altered ecological responses between infected and uninfected host species (Macneil et al. 2003). Indeed, given the ability of the parasite to modify transmission rates to their bird hosts, they can affect food web properties in larger ecosystems (Thompson et al. 2005).

17.2.2.13 Kingdom Animalia: Phylum Annelida

Annelida is a large clade (*ca.* 22,000 species) of mainly free-living, segmented, coelomate worms They are not common symbionts on crustaceans, albeit there are several well-known symbiotic relationships, particularly with polychaetes commensal on hermit crabs (see Williams and McDermott 2004) and branchiobdellids on crayfishes (Gelder 2010). The paucity of relationships may be a result of predation pressure as many decapods can preen themselves, thus feeding on potential symbionts (Shields et al. 2015). Nonetheless, crustacean host–annelid relationships range from commensal to semi-parasitic. Indeed, branchiobdellids exhibit both a beneficial relationship at low densities and a negative, parasitic relationship at high densities (Brown et al. 2012).

17.2.2.14 Subclass Hirudinea

Leeches can be free-living predators of invertebrates or parasitic blood suckers on fishes and other vertebrates. Although they can eat microcrustaceans such as copepods and amphipods, several have phoretic relationships with decapods hosts, using them as a hard substrate for transportation, dispersal, and as a site cocoon deposition. This type of phoretic relationship is known for leeches on crabs and lobsters from many different habitats (for review see Shields et al. 2015). In these relationships, the leeches typically deposit their cocoons on the carapaces of their host, which in turn gain protection from predation. There is scant evidence that leeches feed on their decapod hosts (Overstreet 1983) but in one system, leeches had host hemolymph in their guts (Zara et al. 2009).

17.2.2.15 Subclass Oligochaeta: Family Branchiobdellida

Branchiobdellids are a clade of highly modified oligochaetes that have a close obligate symbiotic relationship with crayfish. They are ectosymbionts and appear almost leech-like in their morphology. The family is comprised of 22 genera containing around 150 species (Gelder et al. 2002; Gelder 2010). Although specialists on crayfish, a few are also known from freshwater isopods, crabs (Holt 1973), and shrimp (Ohtaka et al. 2012). They can even infest blue crabs at low salinities (< 3 psu) (Holt 1968; Gelder et al. 2002; Gelder and Messick 2006). Branchiobdellids are commensals and feed on detritus and fouling organisms but not host eggs (Jennings and Gelder 1979; Brown et al. 2012); however, at high densities, they may cause detriment to their hosts (Lee et al. 2009; Brown et al. 2012; Farrell et al. 2014). As with leeches, they deposit cocoons onto the carapace of their host and move to the new instar with host moulting (Gelder 2010).

17.2.2.16 Class Polychaeta

Polychaetes from several families live in symbiosis with decapods, particularly on anomurans, such as hermit crabs, and gebiids, such as *Upogebia* (for reviews, see Williams and McDermott 2004; Martin and Britayev 1998, 2018). One species, *Histriobdella homari*, is common in the branchial chambers and egg masses of American and European lobsters but is not correlated with egg mortality or pathology (for review see Shields et al. 2006).

17.2.3 Kingdom Animalia: Phylum Arthropoda: Subphylum Crustacea

Crustaceans serve as both hosts and symbionts with other crustaceans, but notably they frequently use crustaceans as definitive hosts, a relationship rarely seen in most protozoan and helminth parasites that use these hosts. Members of the sub-phylum Crustacea range from free-living to symbiotic; and the symbiotic associations range from loosely commensal, to obligate mutual, to highly derived parasites, see Figures 17.6–7. The parasitic crustaceans as a whole were reviewed recently (see Smit et al. 2019). The major parasitic groups include Amphipoda, Ascothoracida, Branchiura, Cirripedia (with Rhizocephala), Copepoda, Isopoda, Ostracoda, Pentastomida and Tantulocarida (Boxshall and Hayes 2019). Many of these include representatives that use other crustaceans as hosts.

17.2.3.1 Subclass Copepoda

The siphonostomatoid copepods are a large group of obligate parasites, mostly parasitic on fishes and echinoderms One family, Nicothoidae, consists of 137 known species in 22 genera, all of which infect crustaceans (Boxshall and Hayes 2019). Nicothoids are small, ovoid, cryptic egg mimics that are found in the egg masses or gills of ostracods, mysids, isopods, and a few decapods (Lemercier 1963; Bowman and Kornicker 1967; Ohtsuka et al. 2005, 2011). They have sucking mouthparts, with a stylet-like adaptation to their mandible, and feed on host eggs or suck host haemolymph depending on their habitus. On decapods, they settle on the gills as juveniles and then migrate to the egg clutch as adults. Several species are known from brachyuran crabs (see Shields et al. 2015 for review) and lobsters (see Shields et al. 2007 for review). These parasites can reach high prevalence levels on ovigerous crabs which can result in significant egg mortality (Shields and Wood 1993). On *Homarus gammarus* they can cause localised host responses in the gills from their feeding on haemolymph (Wootton et al. 2011).

The harpacticoid copepods are a very large, diverse group of mostly free-living, benthic species. Many harpacticoids can be found in the egg masses of benthic crustaceans but, with a few exceptions,

these are simply transiting over the host. However, at least 40 species in eight families are symbionts of crustaceans (Hendrickx and Fiers 2010). The Cancrincolidae have a specialised relationship as symbionts in the gill chambers of semi-terrestrial crabs (Boxshall and Hayes 2019). Although their egg production is synchronised with that of their host and they have sucking mouthparts for feeding on the gills, their host-symbiont relationship is unclear (Huys et al. 2009).

17.2.3.2 Subclass Eumalacostraca: Order Amphipoda

Amphipods are surprisingly common on the egg masses of benthic decapods but their host-symbiont relationships have received little attention. Although they are likely facultative egg predators, a few species appear to live in close association with their hosts. Ischyrocerids have been identified on several decapod hosts (for review see Vader and Tandberg 2015). In some studies, they are voracious egg predators (Kuris et al. 1991), whereas in others they are not (Dvoretsky and Dvoretsky 2010). Several also have been reported as commensals on lobsters. Given their motility, amphipods likely leave their host upon capture or only occur on ovigerous hosts, which can make gauging their relationships more difficult.

17.2.3.2 Subclass Eumalacostraca: Order Isopoda: Infraorder Epicaridea

Epicaridea is a clade of obligate endo- and ectoparasitic isopods that specialise on other crustaceans. They are a large taxon comprised of two superfamilies, 13 families, and well over 800 species, collectively more than 8% of the described isopods (Williams and Boyko 2012; Boyko et al. 2013; Boxshall and Hayes 2019). One superfamily, Bopyroidea, contains taxa that cause pathognomonic swellings in the branchial chambers of their hosts. These swollen gill chambers have been observed in fossils from the Late Jurassic (163.5–157.3 Mya), indicating a long evolutionary association (Klompmaker et al. 2014). The superfamily contains three families, all of which infect decapods. The other superfamily, Cryptoniscoidea, contains ten families with each family specialising on different crustaceans, including barnacles, isopods, amphipods, and decapods (Trilles 1999; Boxshall and Hayes 2019).

Epicarideans use crustaceans as both intermediate and definitive hosts. The parasites have modified mouthparts that form a suctorial cone for feeding on haemolymph. In most cases they have three larval stages and a two-host life cycle involving copepods as intermediate hosts and then their preferred definitive hosts (Figures 17.6a–c). The characteristic epicaridium larvae, for which the clade is named, hatch from eggs within the female's marsupium and swim off to find and attach to the copepod intermediate host. The epicaridium metamorphoses into a microniscus larva that feeds on haemolymph, until it eventually leaves the copepod and metamorphoses into a cryptoniscus larva that infests a definitive host. Once attached, it becomes the juvenile bopyridium which matures on the definitive host, usually in the branchial chamber, but the site of attachment is specific to each host-parasite association. They use epigametic sex determination, hence, the first bopyridium to settle invariably becomes a female, with subsequent isopods becoming dwarf males.

Bopyridae is the largest family with 600+ described species (Williams and Boyko 2012). They parasitise shrimp and anomurans, but two subfamilies parasitise other decapods, including brachyurans, axiioid and gebiidean shrimp, and palinuran lobsters. They are found in all of the world oceans, albeit mostly restricted to shallow waters (Markham 1986; Williams and Boyko 2012).

Entonisicidae is a small family of 36 species that infect crabs and caridean shrimp. These parasites are unusual in that they live as adults in the haemocoel of their definitive host, communicating with the external environment through a pore in the exoskeleton of their host. The females are highly modified endoparasites, often unrecognisable as isopods, with a tumour—or worm-like appearance (Williams and Boyko 2012). Dwarf males and cryptoniscus larvae are usually found on the body of the female, often within her marsupium, where they develop and fertilise eggs. Also unusual is the fact that entoniscids grow without molting, an unusual feature for arthropods.

Cryptoniscoidea is a modest-sized taxon with nine families comprised of 146 species. They are primarily parasites of ostracods, other isopods, euphausiids, and barnacles, albeit one species infects a decapod, the coral crab *Hapalocarcinus marsupialis*. This taxon has some unusual hyperparasites, particularly in rhizocephalans that infect decapods, but they are cryptic and often resemble a small swelling on their host, much like that of the barnacles they infect (Trilles 1999).

Dajidae is a small family of 56 species in Cryptoniscoidea that infect isopods, mysids, euphausiids, and several shrimp. The female is usually found as a 'rider' on the body of its host, typically on the dorsal aspect of the carapace or the juncture of the cephalon and thorax and the dwarf males are often attached to the female or found free in her marsupium. Females attach to their hosts using hook-like appendages. These parasites can get knocked of their host, leaving a melanised scar at their former site of attachment. One species has highly modified antennae that it uses to clasp onto the eyestalks of its host euphausiid (Shields and Gómez-Gutiérrez 1996).

Epicardeans are often complete or partial parasitic castrators, particularly on female hosts (O'Brien and Van Wyk 1985; Calado et al. 2005). Unlike the rhizocephalans, epicarideans rarely inhibit their host's ability to moult (O'Brien and Van Wyk 1985). Prevalence levels are often low (< 1%) but some studies have reported levels over 10% or more (Bourdon 1968; McDermott 1991; Cericola and Williams 2015). They can sometimes occur in bilateral infections in both branchial chambers of their host (Veillet 1945, Trilles 1999). *Probopyrus pandalicola* is a well-known bopyrid that infects grass shrimp, *Palaemonetes* spp. It presents a useful model due to its relatively high prevalence, ease in identification of the swollen branchial chamber, and ease of host maintenance; thus, several studies have examined the effect of the parasite on the survival of the shrimp, parasite and host seasonality, as well as the host-parasite relationship (e.g. Beck 1980a, b; Anderson 1990; Sherman and Curran 2015).

17.2.3.3 Subclass Tantulocarida

Tantulocarida is a small subclass of obligate ectoparasites that specialise on copepods, isopods, tanaids, amphipods, and cumaceans as hosts (Mohrbeck et al. 2010). The taxon is comprised of five families, with 22 genera and 38 species (Mohrbeck et al. 2010; Boxshall and Hayes 2019). They exhibit high host specificity, typically at the family level. These parasites have an unusual biphasic life cycle comprised of an asexual cycle and a sexual cycle (Boxshall and Lincoln 1987; Huys et al. 1993). The adults are mesoparasitic and feed through an absorptive rootlet system that penetrates the host through the oral disc of the tantalus stage (Figure 17.7a,b) (Petrunina et al. 2014).

17.2.3.4 Subclass Thecostraca: Infraclass Cirripedia: Superorder Rhizocephala

Rhizocephala is a clade of highly derived barnacles that are obligate parasites of crustaceans. They are parasites of decapods, particularly crabs, but they also infect shrimp, axiids, stomatopods, isopods, cumaceans, and even other barnacles. At first glance they resemble a bizarre sac that hangs off the host abdomen, which to the novice resembles the host's egg mass. This sac is the external body, or externa, of the barnacle. It is attached to the internal body, or interna, a mass of rootlets, via a narrow stalk. Morphologically, they can be identified as barnacles by their nauplius and cyprid stages. The clade is monophyletic and comprised of two orders, Kentrogonida and Akentrogonida, consisting of 10 families and 288 species (Pérez-Losada et al. 2009; Glenner et al. 2010; Boxshall and Hayes 2019; Høeg et al. 2019). The Sacculinidae is the largest family with over 175 described species that are mostly parasitic on brachyuran crabs. Although most have been reported from shallow waters, they do infect deep-water hosts (Lützen 1985). Rhizocephalans also have representatives from the Miocene, visible as evidence of feminised abdominal segments on their crab hosts (Feldmann, 1998). Excellent reviews of this clade are available (Høeg 1995; Høeg et al. 2005; Høeg et al. 2019).

Rhizocephalans are mesoparasites as the internal network of roots provide nutrition to the externa that protrudes through the host's body wall. The parasites are dioecious and have complex, direct life cycles (Figure 17.7c–e). Eggs held in the externa hatch as free-living nauplius that metamorphose

to an infectious cyprid larva, which may or may not have an internal kentrogon stage, that injects a vermigon (or a mass of female cells) into the haemocoel of the host (Glenner 2001). The vermigon, or parasitic mass, entwines around the midgut and then grows absorptive rootlets into the body and tissues of the host as the sacculina interna. The externa forms internally as the parasite matures and eventually ruptures through and protrudes out of the ventral thorax or abdomen of the host, typically causing the host to cease moulting. The externa is female, and ovaries develop within it. Kentrogonids have a complicated mating system—a superb example of cryptogonorchism—as the male cyprid metamorphoses into a trichogon that penetrates the externa and crawls into the oviduct and essentially plants itself there as a functional testis (Høeg 1990, 1995). Moreover, the terminal spines on the trichogon block the oviduct, further hindering the migration of other males into the ovaries. The trichogons of the akentrogonids inject male generative cells into the mantle cavity of the externa where they become functional testes.

Rhizocephalans have piqued the interest of many crustacean biologists because they castrate their hosts and feminise infected males. The resulting alterations in host physiology, reproduction, and behaviour are enacted through disruption of and competition for host hormones (for review see Høeg et al. 2005). Infected hosts exhibit a range of physical and behavioural pathologies including castration, anecdysis, altered secondary sexual characteristics, feminisation, stunting, altered grooming behaviour, and increased mortality. The internal rootlets penetrate organs and nerves, altering their function. Infected hosts are often more docile (Bishop and Cannon 1979; Shields and Wood 1993; Horacio et al. 2020). Feminisation results from direct damage to the androgenic gland, the sex determining organ in crustaceans (Zerbib et al. 1975; Rubiliani and Payen 1979), which, when damaged prior to maturation, leads to modifications of the secondary sexual characters during moulting (Kristensen et al., 2012). Castration arises from interruption of reproductive hormones or through the metabolic cost of infection (Andrieux 1974; Rubiliani 1983). Their transmission cues have received extensive study (for review see Høeg et al. 2005).

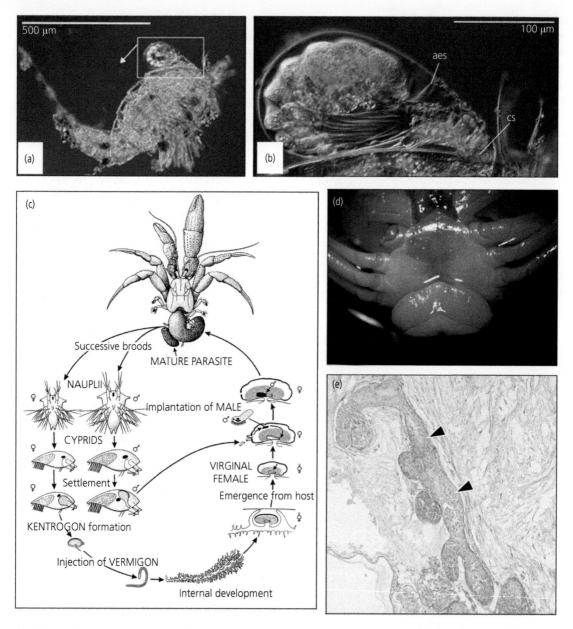

Figure 17.7 (a) Male tantulocarid *Arcticotantuls kristenseni* developing in larval trunk sac attached to copepod host. (b) Detail of *A. kristenseni* showing aesthetascs (aes) and cephalic stylet (cs). (c) Life cycle of *Peltogaster paguri* on a hermit crab, with female and male life history stages shown. (d) *Sacculina* externa on a species of *Thalamita*. (e) Rootlet of *Sacculina granifera* along a ventral nerve tract of *Portunus armatus*. H&E stain. Source: (a) and (b) from Knudsen et al. 2009, with permission; (c) from Høeg and Lützen 1995, kindly provided by Jens Høeg; (e) from author's collection.

Rhizocephalans can reach high prevalence levels in and damage host populations. For example, a prevalence level of 100% was reported for *Sacculina carcini* in local populations of *Carcinus aestuarii* (Øksnebjerg 2000; Werner 2001). Several commercial crab fisheries have reported high prevalence levels, as well as evidence of stunting and sterile matings (Sloan 1984; Hawkes et al. 1986; Walker et al. 1992; Hochberg et al. 1992; Shields and Wood 1993; Alvarez and Calderon 1996). In lithodid fisheries, infected hosts are often stunted, viewed as illegal 'shorts', and returned to the water, where they accumulate to even higher prevalence levels (Meyers 1990). Abiotic factors such as entrained water masses, constrained physiography, and salinity or temperature strata can modulate or enhance transmission (Sloan et al. 1984; Reisser and Forward 1991; Walker and Lester 1998; Boone et al. 2003; Pardal et al. 2013). Biotic factors, such as sex, size, moulting frequency, moult stage, and grooming behaviour also have been associated with increases in transmission (see Høeg 1995; Høeg et al. 2005 for review, Jensen et al. 2019). Given their high prevalence levels, their initial internal infection, and ability to castrate their hosts, rhizocephalans may have an impact on nascent aquaculture ventures (Waiho et al. 2021).

17.2.3.5 Subclass Thecostraca: Infraclass Cirripedia: Superorder Thoracica

Thoracica, the more classical representation of barnacles, have several symbiotic species that use decapods as hard, mobile surfaces. Most of these are facultative commensals, e.g. the balanomorph *Chelonibia patula,* but a few lepadomorphs are obligate symbionts: *Octolasmis* (30 spp.), *Poecilasma* (7 spp.), *Trislasmis* (1 sp.) and *Koleolepas* (3 spp). Those symbiotic on hermit crabs are reviewed in Williams and McDermott (2004). The obligate symbionts often occur on the gills, in the branchial chamber, or attached to the carapace or other external surfaces of their decapod or giant isopod hosts (Walker 1974; Jeffries and Voris 1983; Williams and Moyse 1988). Infestations can have very high densities, often with several hundred barnacles clustered in

their preferred sites (Jeffries and Voris 1983; Gannon 1990; Key et al. 1997; Voris and Jeffries 2001). At high densities, they can impinge upon their host's physiology causing elevated heart and ventilation rates (Gannon and Wheatly (1992, 1995). The presence of large numbers of barnacles may indicate the presence of other pathogens (Overstreet 1983).

17.2.3.6 Phylum Arthropoda: Subphylum Chelicerata: Class Arachnida: Subclass Acari

Mites are tiny, spider-like arachnids that are found in terrestrial, freshwater, and semi-terrestrial ecosystems. Several taxa are known from freshwater crustaceans, including amphipods, isopods, crayfish, and decapods (Bartsch 1989). The commensal mites on crayfish and hermit crabs have been reviewed by Alderman and Polglase (1988) and Williams and McDermott (2004), respectively. Parasitic mites on amphipods feed on host haemolymph and can occur in relatively high intensity infections that result in host mortality (e.g. Kitron 1980; Rigby 1996; Poulin and Rate 2001). Those on isopods may vector *Wohlbachia* sp., an unusual symbiotic bacterium, to their hosts (Rigaud et al. 2001; Bouchon et al. 2008; Cordaux et al. 2012). Others on isopods may have a phoretic relationship (Colloff and Hopkin 1986). Those on decapods are considered facultative commensals, mainly transiting over their host's surfaces. For example, in Poland the Chinese mitten crab, *Eriocheir sinensis,* was host to 22 species of oribatid and halicarid mites, with most considered facultative commensals, albeit a few may have had a closer relationship (Normant et al. 2013). Mites have also been found on marine decapods but their relationship is unknown (see Normant et al. 2013). Several facultative species have been reported from the branchial chamber of crayfish; however, one species, *Limnohalacarus wackeri,* appeared to be an obligate symbiont, only occurring in the gill chamber of its host (Zawal 1998).

17.3 Summary

- The parasites of Crustacea are extraordinarily diverse and thus require a variety of techniques for assessment and identification (Shields 2017).

- Although several parasites have counterparts in well-studied taxa, such as Platyhelminthes, many are specialised on Crustacea, particularly the protists, and thus may be more difficult for the non-specialist to identify.
- For the significant pathogens of commercially important species, additional studies are needed on life cycles, transmission, host pathology, disease modelling, and disease ecology.
- Although historically there has been a strong emphasis on the pathogens of commercially important species (e.g. Stentiford et al. 2012), several other host–parasite associations have led to significant contributions to broader ecological themes. For example, the parasites of amphipods and branchipods have been used to explore ecological relationships between competing species, to investigate the nature of parasitic castration and its effects on host populations, and to examine ecological aspects of disease.
- By incorporating parasites into further studies, additional insights can be gained into many aspects of host biology and ecology. For example, there are significant differences in the parasite fauna between marine and freshwater crustaceans. Comparable niches on freshwater and marine hosts have been exploited in different ways by their parasites and symbionts. This likely indicates different selection pressure in these environments, particularly with respect to host susceptibility, host adaptations, parasite transmission pathways, and host-parasite adaptations (Shields et al. 2015).
- Although often overlooked by ecologists, parasites are integral members of the food web, often representing a significant fraction of energy flow in aquatic communities (Lafferty et al. 2006) or imposing major costs to important members of different trophic levels. They should be included in field studies where possible.
- Given the importance of many crustacean hosts in marine ecosystems, the influence and ecological roles of their parasites cannot be overstated.

Recommended further reading

Bojko, J. and Ovcharenko, M. 2019. Pathogens and other symbionts of the Amphipoda: taxonomic diversity and pathological significance. *Diseases of Aquatic Organisms* 136: 3–36

Ho, J.S. and Perkins, P.S. 1985. Symbionts of marine Copepoda: An overview. *Bulletin of Marine Science* 37: 586–598

McDermott, J.J., Williams, J.D., and Boyko, C.B. 2010. The unwanted guests of hermits: A global review of the diversity and natural history of hermit crab parasites. *Journal of Experimental Marine Biology and Ecology* 394: 2–44

Shields, J.D., Stephens, F.J., and Jones, J.B. 2006. Pathogens, parasites and other symbionts: In *Lobsters: Biology, Management, Aquaculture and Fisheries*, Phillips, B.F. (ed.) Chapter 5, pp. 146–204. UK: Blackwell Scientific

Shields, J.D., Williams, J.D., and Boyko, C.B. 2015. Parasites and diseases of Brachyura. In *The Crustacea. Treatise on Zoology Treatise on Zoology-Anatomy, Taxonomy, Biology*. Vol. 9, Part C (2 vols) Castro, P., Davie, P.J.F., Guinot, D., Schram, F.R., and von Vaupel Klein, J.C. (eds). pp. 639–774. The Netherlands, Brill https://brill.com/view/title/19136?contents=toc-44457

Williams, J.D. and McDermott, J.J. 2004. Hermit crab biocoenoses: A worldwide review of the diversity and natural history of hermit crab associates. *Journal of Experimental Marine Biology and Ecology* 305: 1–128

Acknowledgements

I thank Prof Andrew Rowley for the invitation to contribute to this volume. This is contribution #4032 of the Virginia Institute of Marine Science, William & Mary.

References

Adamson, M.L. 1986. Modes of transmission and evolution of life histories in zooparasitic nematodes. *Canadian Journal of Zoology* 64: 1375–1384

Ahyong, S.T., Lowry, J.K., Alonso, M., Bamber, R.N., Boxshall, G.A., Castro, P., Gerken, S., Karaman, G.S., Goy, J.W., Jones, D.S., and Meland, K. 2011. Subphylum Crustacea Brünnich, 1772. In *Animal Biodiversity: An Outline Of Higher-Level Classification and Survey of Taxonomic Richness. Zootaxa*, Zhang, Z.-Q. (ed.) 3148: pp. 165–191

Aiken, D.E., Sochasky, J.B., and Wells, P.G. 1973. Ciliate infestation of the blood of the lobster *Homarus americanus*. *International Council for the Exploration of the Sea, Shellfish Commission* K 46: 1–2

Aiken, D.E., Waddy, S.L., Uhazy, L.S., and Campbell, A. 1983. A nemertean destructive to the eggs of the lobster *Homarus americanus*. *Rapports Procès-verbaux Réunion Conseil Permanent International pour l'Exploration de la Mer* 182: 120–133

Alda, P., La Sala, L., Marcotegui, P., and Martorelli, S.R. 2011. Parasites and epibionts of grapsid crabs in Bahía Blanca estuary, Argentina. *Crustaceana* 84: 5–6

Alderman D.J. and Polglase J. 1988. Pathogens, parasites and commensals. In *Freshwater Crayfish: Biology, Management and Exploitation*, Holditch, D.M. and Lowery, R.S. (eds.) pp. 167–212. UK: Croom Helm Ltd.

Alvarez, F. and Calderon, J. 1996. Distribution of *Loxothylacus texanus* (Cirripedia: Rhizocephala) parasitizing crabs of the genus *Callinectes* in the Southwestern Gulf of Mexico. *Gulf Research Reports* 9: 205–210

Alves-de-souza, C., Cornet, C., Nowaczyk, A., Gasparini, S., Skovgaard, A., and Guillou, L. 2011. *Blastodinium* spp. infect copepods in the ultra-oligotrophic marine waters of the Mediterranean Sea. *Biogeosciences* 8: 2125–2136

Amat, F., Gozalbo, A., Navarro, J.C., Hontoria, F., and Varó, I. 1991. Some aspects of *Artemia* biology affected by cestode parasitism., In *Studies on Large Branchiopod Biology and Aquaculture*, Belk, D., Dumont H.J., and Munuswamy, N (eds.), pp. 39–44. Dordrecht: Springer

Anderson, G. 1990. Postinfection mortality of *Palaemonetes* spp. (Decapoda: Palaemonidae) following experimental exposure to the bopyrid isopod *Probopyrus pandalicola*

(Packard) (Isopoda: Epicaridea). *Journal of Crustacean Biology* 10: 284–292

Anderson, R.C. 2000. *Nematode Parasites of Vertebrates: Their Development and Transmission*, 2nd ed, pp. 650. Wallingford: CABI Books

Andrieux, N. 1974. Action de l'ecdystérone sur les phénomènes de mue des crabes *Carcinus mediterraneus* sains et parasités par *Sacculina carcini*. *Comptes rendus de l'Académie des Sciences, Séries D* 279: 807–810

Appleton, P.L. and Vickerman, K. 1998. *In vitro* cultivation and development cycle in culture of a parasitic dinoflagellate (*Hematodinium* sp.) associated with mortality of the Norway lobster (*Nephrops norvegicus*) in British waters. *Parasitology* 116: 115–130

Appy, R.G. and Butterworth, E.W. 2011. Development of *Ascarophis* sp. (Nematoda: Cystidicolidae) to maturity in *Gammarus deubeni* (Amphipoda). *Journal of Parasitology* 97: 1035–1048

Armstrong, D.A., Burreson, E.M., and Sparks, A.K. 1981. A ciliate infection (*Paranophrys* sp.) in laboratory-held Dungeness crabs, *Cancer magister*. *Journal of Invertebrate Pathology* 37: 201–209

Arundell, K.L., Dubuffet, A., Wedell, N., Bojko, J., Rogers, M.S., and Dunn, A.M. 2019. *Podocotyle atomon* (Trematoda: Digenea) impacts reproductive behaviour, survival and physiology in *Gammarus zaddachi* (Amphipoda). *Diseases of Aquatic Organisms* 136: 51–62

Arzul, I., Chollet, B., Boyer, S., Bonnet, D., Gaillard, J., Baldi, Y., Robert, M., Joly, J.P., Garcia, C., and Bouchoucha, M. 2014. Contribution to the understanding of the cycle of the protozoan parasite *Marteilia refringens*. *Parasitology* 141: 227–240

Athanassopoulou, F., Speare, D., Cawthorn, R.J., MacMillan, R., and Despres, B. 2004. Pathology of *Anophryoides haemophila* (Scuticociliatida: Orchitophryidae), parasite of American lobster *Homarus americanus* kept under experimental conditions. *Aquaculture* 236: 103–117

Audemard, C., Sajus, M.C., Barnaud, A., Sautour, B., Sauriau, P.G., and Berthe, F.J. 2004. Infection dynamics of *Marteilia refringens* in flat oyster *Ostrea edulis* and copepod *Paracartia grani* in a claire pond of Marennes-Oleron Bay. *Diseases of Aquatic Organisms* 61: 103–111

Bartsch, I. 1989. Marine mites (Halacaroidea: Acari): A geographical and ecological survey. *Hydrobiologia* 178: 21–42

Basile, R., Tirelli, T., and Pessani, D. 2004. *Pebrilla paguri* (Ciliophora, Folliculinidae) on four Mediterranean hermit crab species. *Italian Journal of Zoology* 71: 329–335

Bateman, K.S., Hicks, R.J., and Stentiford, G.D. 2011. Disease profiles differ between non-fished and fished populations of edible crab (*Cancer pagurus*) from a major commercial fishery. *ICES Journal of Marine Science* 68: 2044–2052

Bauer, A., Haine, E.R., Perrot-Minnot, M.J., and Rigaud, T. 2005. The acanthocephalan parasite *Polymorphus minutus* alters the geotactic and clinging behaviours of two sympatric amphipod hosts: The native *Gammarus pulex* and the invasive *Gammarus roeseli*. *Journal of Zoology* 267: 39–43

Beck, J.T. 1980a. The effects of an isopod castrator, *Probopyrus pandalicola*, on the sex characters of one of its caridean shrimp hosts, *Palaemonetes paludosus*. *The Biological Bulletin* 158: 1–15

Beck, J.T. 1980b. Life history relationships between the bopyrid isopod *Probopyrus pandalicola* and of its freshwater shrimp hosts. *Palaemonetes paludosus*. *American Midland Naturalist* 104: 135–154

Bellon-Humbert, C. 1983. *Fecampia erythrocephala* Giard (Turbellaria, Neorhabdocoela), a parasite of the prawn *Palaemon serratus* Pennant: The adult phase. *Aquaculture* 31: 117–140

Benesh, D.P. 2019. Tapeworm manipulation of copepod behaviour: Parasite genotype has a larger effect than host genotype. *Biology Letters* 15: 20190495

Bethel, W.M. and Holmes, J.C. 1973. Altered evasive behavior and responses to light in amphipods harboring acanthocephalan cystacanths. *Journal of Parasitology* 59: 945–956

Bethel, W.M. and Holmes, J.C. 1974. Correlation of development of altered evasive behavior in *Gammarus lacustris* (Amphipoda) harboring cystacanths of *Polymorphus paradoxus* (Acanthocephala) with the infectivity to the definitive host. *Journal of Parasitology* 60: 272–274

Bethel, W.M. and Holmes, J.C. 1977. Increased vulnerability of amphipods to predation owing to altered behavior induced by larval acanthocephalans. *Canadian Journal of Zoology* 55: 110–115

Bishop, R.K. and Cannon, L.R.G. 1979. Morbid behaviour of the commercial sand crab, *Portunus pelagicus* (L.), parasitized by *Sacculina granifera* Boschma, 1973 (Cirripedia: Rhizocephala). *Journal of Fish Diseases* 2: 131–144

Bittner, K., Rothhaupt, K.O., and Ebert, D. 2002. Ecological interactions of the microparasite *Caullerya mesnili* and its host *Daphnia galeata*. *Limnology and Oceanography* 47: 300–305

Blair, D., Xu, Z.B., and Agatsuma, T. 1999. Paragonimiasis and the genus *Paragonimus*. *Advances in Parasitology* 42: 113–222

Blakeslee, A.M., Keogh, C.L., Byers, J.E., Kuris, A.M., Lafferty, K.D., and Torchin, M.E. 2009. Differential escape from parasites by two competing introduced crabs. *Marine Ecology Progress Series* 393: 83–96

Blaxter, M., Dorris, M., and Ley, P.D. 2000. Patterns and processes in the evolution of animal parasitic nematodes. *Nematology* 2: 43–55

Bojko, J. and Ovcharenko, M. 2019. Pathogens and other symbionts of the Amphipoda: Taxonomic diversity and pathological significance. *Diseases of Aquatic Organisms* 136: 3–36

Bojko, J., Stebbing, P.D., Dunn, A.M., et al. 2018. Green crab *Carcinus maenas* symbiont profiles along a North Atlantic invasion route. *Diseases of Aquatic Organisms* 128: 147–168

Boone, E., Boettcher, A.A., Sherman, T.D., and O'Brien, J.J. 2003. Characterization of settlement cues used by the rhizocephalan barnacle *Loxothylacus texanus*. *Marine Ecology Progress Series* 252: 187–197

Born, J. W. 1967. *Palaemonetes vulgaris* (Crustacea, Decapoda) as host for the juvenile stage of *Nectonema agile* (Nematomorpha). *Journal of Parasitology* 53: 793–794

Boschma, H. 1949. Ellobiopsidae. *Discovery Reports* 25: 283–314

Boschma, H. 1959. Ellobiopsidae from tropical West Africa. *Atlantide Reports* 5: 145–175

Bouchon, D., Cordaux, R., and Grève, P. 2008. Feminizing *Wolbachia* and the evolution of sex determination in isopods. *Insect Symbiosis* 3: 273–294

Bourdon, R. 1968. Les Bopyridae des mers européennes. *Mémoires du Muséum National d'histoire Naturelle. Série A* 50: 77–424

Bower, S.M. and Meyer, G.R. 2002. Morphology and ultrastructure of a protistan pathogen in the haemolymph of shrimp (*Pandalus* spp.) in the northeastern Pacific Ocean. *Canadian Journal of Zoology* 80: 1055–1068

Bowman, T.E. and Kornicker, L.S. 1967. Two new crustaceans: The parasitic copepod *Sphaeronellopsis monothrix* (Choniostomatidae), and its myodocopid ostracod host *Parasterope pollex* (Cylindroleberidae) from the Southern New England Coast. *Proceedings of the U.S. National Museum, Smithsonian Institution* 123: no. 3613

Boxshall, G. and Hayes, P. 2019. Biodiversity and taxonomy of the parasitic Crustacea. In *Parasitic Crustacea*, Smit, N., Bruce, N.L., and Hadfield, K.A. (eds.) pp. 73–134. Switzerland: Springer

Boxshall, G.A. and Lincoln, R.J. 1987. The life cycle of the Tantulocarida (Crustacea). *Philosophical Transactions of the Royal Society of London. B, Biological Sciences* 315: 267–303

Boyer, S., Chollet, B., Bonnet, D., and Arzul, I. 2013. New evidence for the involvement of *Paracartia grani* (Copepoda, Calanoida) in the life cycle of *Marteilia refringens* (Paramyxea). *International Journal for Parasitology* 43: 1089–1099

Boyko, C.B., Moss, J., Williams, J.D., and Shields, J.D. 2013. A molecular phylogeny of Bopyroidea and Cryptoniscoidea (Crustacea: Isopoda). *Systematics and Biodiversity* 11: 495–506

Bradbury, P.C. 1994. Parasitic protozoa of molluscs and Crustacea. In *Parasitic Protozoa*, Vol. 8. J.P. Kreier and J.R. Baker (eds.) pp. 139–264. San Diego: Academic Press

Bradbury, P.C. and Trager, W. 1967. Excystation of apostome ciliates in relation to molting of their crustacean hosts. II. Effect of glycogen. *Biological Bulletin, Woods Hole* 133: 310–316

Brattey, J. and Campbell, A. 1986. A survey of parasites of the American lobster, *Homarus americanus* (Crustacea: Decapoda), from the Canadian maritimes. *Canadian Journal of Zoology* 64: 1998–2003

Briones-Fourzán, P., De Cote-hernández, R.M., and Lozano-Álvarez, E. 2016. Variability in prevalence of *Cymatocarpus solearis* (Trematoda, Brachycoeliidae) in Caribbean spiny lobsters *Panulirus argus* (Decapoda: Palinuridae) from Bahía de la Ascensión (Mexico). *Journal of Invertebrate Pathology* 137: 62–70

Brown, B.L., Creed, R.P., Skelton, J., Rollins, M.A., and Farrell, K.J. 2012. The fine line between mutualism and parasitism: Complex effects in a cleaning symbiosis demonstrated by multiple field experiments. *Oecologia* 170: 199–207

Burreson, E.M. and Ford, S.E. 2004. A review of recent information on the Haplosporidia, with special reference to *Haplosporidium nelsoni* (MSX disease). *Aquatic Living Resources* 17: 499–517

Busch, M.W., Kuhn, T., Münster, J., and Klimpel, S. 2012. Marine crustaceans as potential hosts and vectors for metazoan parasites. In *Arthropods as Vectors of Emerging Diseases*, Mehlhorn, H. (ed.) pp. 329–360. Berlin, Heidelberg: Springer

Butler, M.J., III, Tiggelaar, J.M., Shields, J.D., and Butler, M.J., IV. 2014. Effects of the parasitic dinoflagellate *Hematodinium perezi* on blue crab (*Callinectes sapidus*) behavior and predation. *Journal of Experimental Marine Biology and Ecology* 461: 381–388

Butler, S.A. 1987. Taxonomy of some tetraphyllid cestodes from elasmobranch fishes. *Australian Journal of Zoology* 35: 343–371

Caballero, G. 1959. Tremátodes de las tortugas de México. VII. Descripción de un trematodo digéneo que parasita a tortugas marinas comestibles del Puerto de Acapulco, Guerrero. *Anales del Instituto de Biología, Universidad Nacional Autónama de México* 30: 159–166

Cachon, J. and Cachon, M. 1987. Parasitic dinoflagellates. In *The Biology of Dinoflagellates*, Taylor, F.J.R. (ed.) pp. 571–610. Oxford: Blackwell Scientific Publications

Cafaro, M.J. 2005. Eccrinales (Trichomycetes) are not fungi, but a clade of protists at the early divergence of animals and fungi. *Molecular Phylogenetics and Evolution* 35: 21–34

Cain, T.A. and Morado, J.F. 2001. Changes in total hemocyte and differential counts in Dungeness crabs infected with *Mesanophrys pugettensis*, a marine facultative parasitic ciliate. *Journal of Aquatic Animal Health* 13: 310–319

Calado, R., Bartilotti, C., and Narciso, L. 2005. Short report on the effect of a parasitic isopod on the reproductive performance of a shrimp. *Journal of Experimental Marine Biology and Ecology* 321: 13–18

Campbell, A. and Brattey, J. 1986. Egg loss from the American lobster, *Homarus americanus*, in relation to nemertean, *Pseudocarcinonemertes homari*, infestation. *Canadian Journal of Fisheries and Aquatic Sciences* 43: 772–780

Cannon L.R.G. and Joffe, B.I. 2001. The Temnocephalida. In *Interrelationships of the Platyhelminthes*, D.T.J. Littlewood and R.A. Bray (eds.) pp. 83–91. London: Taylor and Francis

Capriulo, G.M., and Small, E.B. 1986. Discovery of an apostome ciliate (*Collinia beringensis* n. sp.) endoparasitic in the Bering Sea euphausiid *Thysanoessa inermis*. *Diseases of Aquatic Organisms* 1: 141–146

Carnegie, R.B., Burreson, E.M., Hine, M.P. et al. 2006. *Bonamia perspora* n. sp. (Haplosporidia), a parasite of the oyster *Ostreola equestris*, is the first *Bonamia* species known to produce spores. *Journal of Eukaryotic Microbiology* 53: 232–245

Carrasco, N., López-Flores, I., Alcaraz, M., Furones, M.D., Berthe, F.C., and Arzul, I. 2007. Dynamics of the parasite *Marteilia refringens* (Paramyxea) in *Mytilus galloprovincialis* and zooplankton populations in Alfacs Bay (Catalonia, Spain). *Parasitology* 134: 1541–1550

Castellanos-Martínez, S., Pérez-Losada, M., and Gestal, C. 2013. Molecular phylogenetic analysis of the coccidian cephalopod parasites *Aggregata octopiana* and *Aggregata eberthi* (Apicomplexa: Aggregatidae) from the NE Atlantic coast using 18S rRNA sequences. *European Journal of Protistology* 49: 373–380

Cavalier-Smith, T. 2018. Kingdom Chromista and its eight phyla: A new synthesis emphasising periplastid protein targeting, cytoskeletal and periplastid evolution, and ancient divergences. *Protoplasma* 255: 297–357

Cave, C.S. and Strychar, K. 2014. Decline of *Diporeia* in Lake Michigan: Was disease associated with invasive species the primary factor? *International Journal of Biology* 7: 93–99

Cawthorn, R.J. 1997. Overview of 'bumper car' disease-impact on the North American lobster fishery. *International Journal of Parasitology* 27: 167–172

Cawthorn, R.J., Lynn, D.H., Despres, B., et al. 1996. Description of *Anophryoides haemophila* n. sp. (Scuticociliatida: Orchitophryidae), a pathogen of the American lobster *Homarus americanus*. *Diseases of Aquatic Organisms* 24: 143–148

Cericola, M.J. and Williams, J.D. 2015. Prevalence, reproduction and morphology of the parasitic isopod *Athelges takanoshimensis* Ishii, 1914 (Isopoda: Bopyridae) from Hong Kong hermit crabs. *Marine Biology Research* 11: 236–252

Chantangsi, C., Lynn, D.H., Rueckert, S., Prokopowicz, A.J., Panha, S., and Leander, B.S. 2013. *Fusiforma themisticola* n. gen., n. sp., a new genus and species of apostome ciliate infecting the hyperiid amphipod *Themisto libellula* in the Canadian Beaufort Sea (Arctic Ocean), and establishment of the Pseudocolliniidae (Ciliophora, Apostomatia). *Protist* 164: 793–810

Chatton, É. 1920. Les péridiniens parasites; morphologie, reproduction, éthologie. *Archives de Zoologie Expérimentale et Générale* 59: 1–475

Chatton, É. and Lwoff, A. 1935. Les ciliés apostomes. I. Aperçu historique et général: étude monographique des generes et des espèces. *Archives de Zoologie Expérimentale et Générale* 77: 1–453

Ching, H.L. 1989. *Profilicollis botulus* (Van Cleave, 1916) from diving ducks and shore crabs of British Columbia. *Journal of Parasitology* 75: 33–37

Cho, S.Y., Kang, S.Y., Kong, Y., and Yang, H.J. 1991. Metacercarial infections of *Paragonimus westermani* in freshwater crabs sold in markets in Seoul. *Korean Journal of Parasitology* 29: 189–191

Christensen, A.M. and Hurley, A.C. 1977. *Fecampia balanicola* sp. nov. (Turbellaria Rhabdocoela), a parasite of Californian barnacles. *Acta Zoologica Fennica* 154: 119–128

Cleary, A.C. and Durbin, E.G. 2016. Unexpected prevalence of parasite 18S rDNA sequences in winter among Antarctic marine protists. *Journal of Plankton Research* 38: 401–417

Clopton, R.E. 2002. Phylum Apicomplexa Levine, 1970: Order Eugregarinorida Léger, 1900. In *Illustrated guide to the Protozoa*, 2nd edn, Lee, J.J., Leedale, G.F., and Bradbury P. (eds.) pp. 205–288. Kansas: Society of Protozoologists

Coats, D.W. 1999. Parasitic lifestyles of marine dinoflagellates. *Journal of Eukaryotic Microbiology* 46: 402–409

Coffey, A.H., Li, C., and Shields, J.D. 2012. The effect of salinity on experimental infections of a *Hematodinium* sp. in blue crabs, *Callinectes sapidus*. *Journal of Parasitology* 98: 536–542

Colloff, M.J. and Hopkin, S.P. 1986. The ecology, morphology and behaviour of *Bakerdania elliptica* (Acari: Prostigmata: Pygmephoridae), a mite associated with terrestrial isopods. *Journal of Zoology* 208: 109–123

Cordaux, R., Pichon, S., Afia Hatira, H. et al. 2012. Widespread *Wolbachia* infection in terrestrial isopods and other crustaceans. *Zookeys*, 176: 123–131

Couch, J.A. 1983. Diseases caused by Protozoa. In *The Biology of the Crustacea*, Vol. 6, Pathobiology, Provenzano, A.J. Jr. (ed.) pp. 79–111. New York: Academic Press

Cremonte, F., Etchegoin, J., Diaz, J.I., and Navone, G.T. 2007. Larval Spirurida (Nematoda) parasitizing two crab species (*Uca uruguayensis* and *Chasmagnathus granulatus*) from the southwest Atlantic coast of Argentina. *Comparative Parasitology* 74: 88–95

Cribb, T.H., Bray, R.A., and Littlewood, D.T.J. 2001. The nature and evolution of the association among digeneans, molluscs and fishes. *International Journal of Parasitology* 31: 997–1011

Davies, C.E., Bass, D., Ward, G.M. et al. 2020. Diagnosis and prevalence of two new species of haplosporidians infecting shore crabs *Carcinus maenas*: *Haplosporidium carcini* n. sp., and *H. cranc* n. sp. *Parasitology* 147: 1229–1237

Davies, C.E., Batista, F.M., Malkin, S.H., et al. 2019. Spatial and temporal disease dynamics of the parasite *Hematodinium* sp. in shore crabs, *Carcinus maenas*. *Parasites & Vectors* 12: 472

Davies, C.E. and Rowley, A.F. 2015. Are European lobsters (*Homarus gammarus*) susceptible to infection by a temperate *Hematodinium* sp.? *Journal of Invertebrate Pathology* 127: 6–10

Deardorff, T.L. and Overstreet, R.M. 1981. Larval *Hysterothylacium* (= *Thynnascaris*) (Nematoda: Anisakidae) from fishes and invertebrates in the Gulf of Mexico. *Proceedings of the Helminthological Society of Washington* 48: 113–126

Deblock, S., Williams, A., and Evans, L.H. 1991. Contribution a l'etude des Microphallidae Travassos 1920 (Trematoda). Description de *Thulakiotrema genitale* n. gen., n. sp., metacercaire parasite de langoustes australiennes. *Bulletin du Museum Nationale D'Histoire Naturelle, Paris* 12: 563–576

Delgery, C.C., Cragg, S.M., Busch, S., and Morgan, E.A. 2006. Effects of the epibiotic heterotrich ciliate *Mirofolliculina limnoriae* and of moulting on faecal pellet production by the wood-boring isopods, *Limnoria tripunctata* and *Limnoria quadripunctata*. *Journal of Experimental Marine Biology and Ecology* 334: 165–173

Doanh, P.N., Van Hien, H., Nonaka, N., Horii, Y., and Nawa, Y. 2012. Co-existence of *Paragonimus harinasutai* and *Paragonimus bangkokensis* metacercariae in fresh water crab hosts in central Viet Nam with special emphasis on their close phylogenetic relationship. *Parasitology International* 61: 399–404

Dogiel, V. 1906. Beiträge zur Kenntnis der Peridineen. *Mittheilung aus der Zoologischen Station zu Neapel* 18: 1–45

Dunn, P.H. and Young, C.M. 2013. Finding refuge: The estuarine distribution of the nemertean egg predator *Carcinonemertes errans* on the Dungeness crab, *Cancer magister*. *Estuarine and Coastal Shelf Science* 135: 201–208

Dvoretsky, A.G. and Dvoretsky, V.G. 2010. The amphipod *Ischyrocerus commensalis* on the eggs of the red king crab *Paralithodes camtschaticus*: Egg predator or scavenger? *Aquaculture* 298: 185–189

Dyková, I., Fiala, I., Lom, J., and Lukeš, J. 2003. *Perkinsiella amoebae*-like endosymbionts of *Neoparamoeba* spp., relatives of the kinetoplastid Ichthyobodo. *European Journal of Protistology* 39: 37–52

Dyková, I., Lom, J., and Fajer, E. 1988. A new haplosporean infecting the hepatopancreas in the penaeid shrimp, *Penaeus vannamei*. *Journal of Fish Diseases* 11: 15–22

Dyková, I., Nowak, B., Pecková, H., Fiala, I., Crosbie, P., and Dvoráková, H. 2007. Phylogeny of *Neoparamoeba* strains isolated from marine fish and invertebrates as inferred from SSU rDNA sequences. *Diseases of Aquatic Organisms* 74: 57–65

Edgerton, B. F., Evans, L.H., Stephens, F.J., and Overstreet, R.M. 2002. Synopsis of freshwater crayfish diseases and commensal organisms. *Aquaculture* 206: 57–135

Edwards, M., Coates, C.J., and Rowley, A.F. 2019. Host range of the mikrocytid parasite *Paramikrocytos canceri* in decapod crustaceans. *Pathogens* 8: 252

Fagerholm, H.P. and Butterworth, E. 1988. *Ascarophis* sp. (Nematoda: Spirurida) attaining sexual maturity in *Gammarus* spp. (Crustacea). *Systematic Parasitology* 12: 123–139

Farrell, K.J., Creed, R.P., and Brown, B.L. 2014. Preventing overexploitation in a mutualism: Partner regulation in the crayfish–branchiobdellid symbiosis. *Oecologia* 174: 501–510

Fayard, M., Dechaume-Moncharmont, F.X., Wattier, R., and Perrot-Minnot, M.J. 2020. Magnitude and direction of parasite-induced phenotypic alterations: A meta-analysis in acanthocephalans. *Biological Reviews* 95: 1233–1251

Feist, S. W., Hine, P.M., Bateman, K.S., Stentiford, G.D., and Longshaw, M. 2009. *Paramarteilia canceri* sp. n. (Cercozoa) in the European edible crab (*Cancer pagurus*) with a proposal for the revision of the order Paramyxida Chatton, 1911. *Folia Parasitologica* 56: 73–85

Feldmann, R.M. 1998. Parasitic castration of the crab, *Tumidocarcinus giganteus* Glaessner, from the Miocene of New Zealand: coevolution within the Crustacea. *Journal of Paleontology* 72: 493–498

Fernández-Leborans, G. 2010. Epibiosis in Crustacea: An overview. *Crustaceana* 83: 549–640

Fernández-Leborans, G. and Tato-Porto, M.L. 2000a. A review of the species of protozoan epibionts on crustaceans. I. Peritrich ciliates. *Crustaceana* 73: 643–684

Fernández-Leborans, G. and Tato-Porto, M.L. 2000b. A review of the species of protozoan epibionts on crustaceans. II. Suctorian ciliates. *Crustaceana* 73: 1205–1237

Fields, D.M., Runge, J.A., Thompson, C. et al. 2015. Infection of the planktonic copepod *Calanus finmarchicus* by the parasitic dinoflagellate, *Blastodinium* spp: Effects on grazing, respiration, fecundity and fecal pellet production. *Journal of Plankton Research* 37: 211–220

Frischer, M.E., Fowler, A.E., Brunson, J.F. et al. 2018. Pathology, effects, and transmission of black gill in commercial penaeid shrimp from the South Atlantic Bight. *Journal of Shellfish Research* 37: 149–158

Frischer, M.E., Lee, R.F., Price, A.R., et al. 2017. Causes, diagnostics, and distribution of an ongoing penaeid shrimp black gill epidemic in the US South Atlantic Bight. *Journal of Shellfish Research* 36: 487–500

Frischer, M. E., Lee, R.F., Sheppard, M.A. et al. 2006. Evidence for a free-living life stage of the blue crab parasitic dinoflagellate, *Hematodinium* sp. *Harmful Algae* 5: 548–557

Gannon, A.T. 1990. Distribution of *Octolasmis muelleri*, an ectocommensal gill barnacle, on the blue crab. *Bulletin of Marine Science* 46: 55–61

Gannon, A.T. and Wheatly, M.G. 1992. Physiological effects of an ectocommensal gill barnacle, *Octolasmis muelleri*, on gas exchange in the blue crab, *Callinectes sapidus*. *Journal of Crustacean Biology* 12: 11–18

Gannon, A.T. and Wheatly, M.G. 1995. Physiological effects of a gill barnacle on host blue crabs during short-term exercise and recovery. *Marine Behavior and Physiology* 24: 215–225

Garey, J.R., Near, T.J., Nonnemacher, M.R., and Nadler, S.A. 1996. Molecular evidence for Acanthocephala as a subtaxon of Rotifera. *Journal of Molecular Evolution* 43: 287–292

Gates, A.R., Sheader, M., Williams, J.A., and Hawkins, L.E. 2017. Infection with cerebral metacercariae of microphallid trematode parasites reduces reproductive output in the gammarid amphipod *Gammarus insensibilis* (Stock 1966) in UK saline lagoons. *Journal of the Marine Biological Association of the United Kingdom* 98: 1–10

Gelder, S.R. 1999. Zoogeography of branchiobdellidans (Annelida) and temnocephalidans (Platyhelminthes) ectosymbiotic on freshwater crustaceans, and their reactions to one another *in vitro*. *Hydrobiologia* 406: 21–31

Gelder, S.R. 2010. Branchiobdellida, section III. In *Ecology and classification of North American freshwater invertebrates*, J.H. Thorp and A.P. Covich (eds.) pp. 403–409. London: Academic Press

Gelder, S.R., Gagnon, M.L., and Nelson, K. 2002. Taxonomic considerations and distribution of the Branchiobdellida (Annelida: Clitellata) on the North American continent. *Northeastern Naturalist* 9: 451–468

Gelder, S.R. and Messick, G. 2006. First report of the aberrant association of branchiobdellidans (Annelida: Clitellata) on blue crabs (Crustacea: Decapoda) in Chesapeake Bay, Maryland, USA. *Invertebrate Biology* 125: 51–55

George-Nascimento, M., Carmona, R., and Riffo, R. 1994. Occurrence of larval nematodes *Proleptus* sp. (Spirurida: Physalopteridae) and *Anisakis* sp. (Ascaridida: Anisakidae) in the crab *Cancer plebejus* Poeppig, in Chile. *Scientia Marina* 58: 355–358

Georgiev, B.B., Sánchez, M.I., Vasileva, G.P., Nikolov, P.N., and Green, A.J. 2007. Cestode parasitism in invasive and native brine shrimps (*Artemia* spp.) as a possible factor promoting the rapid invasion of *A. franciscana* in the Mediterranean region. *Parasitology* Research 101: 1647–1655

Gestal, C., Guerra, A., Pascual, S., and Azevedo, C. 2002. On the life cycle of *Aggregata eberthi* and observation on *Aggregata octopiana* (Apicomplexa, Aggregatidae) from Galicia (NE Atlantic). *European Journal of Protistology* 37: 427–435

Gestal, C., Pascual, S., and Hochberg, F.G. 2010. *Aggregata bathytherma* sp. nov. (Apicomplexa: Aggregatidae), a new coccidian parasite associated with a deep-sea hydrothermal vent octopus. *Diseases of Aquatic Organisms* 91: 237–242

Gibson, R. 1995. Nemertean genera and species of the world: An annotated checklist of original names and description citations, synonyms, current taxonomic status, habitats and recorded zoogeographic distribution. *Journal of Natural History* 29: 271–561

Ginsburger-Vogel, T. 1991. Intersexuality in *Orchestia mediterranea* Costa, 1853, and *Orchestia aestuarensis* Wildish, 1987 (Amphipoda): A consequence of hybridization or parasitic infestation? *Journal of Crustacean Biology* 11: 530–539

Ginsburger-Vogel, T. and Desportes, I. 1979. Etude ultrastructurale de la sporulation de *Paramarteilia orchestiae* gen. n., sp. n., parasite de l'amphipode *Orchestia gammarellus* (Pallas). *Journal of Protozoology* 26: 390–403

Glenner, H. 2001. Cypris metamorphosis, injection and earliest internal development of the Rhizocephalan *Loxothylacus panopaei* (Gissler). Crustacea: Cirripedia: Rhizocephala: Sacculinidae. *Journal of Morphology* 249: 43–75

Glenner, H., Høeg, J.T., Stenderup, J., and Rybakov, A.V. 2010. The monophyletic origin of a remarkable sexual system in akentrogonid rhizocephalan parasites: A

molecular and larval structural study. *Experimental Parasitology* 125: 3–12

Glockling, S.L., Marshall, W.L., and Gleason, F.H. 2013. Phylogenetic interpretations and ecological potentials of the Mesomycetozoea (Ichthyosporea). *Fungal Ecology* 6: 237–247

Gómez Del Prado-Rosas, M.C., Álvarez-Cadena, J.N., Lamothe-Argumedo, R., and Grano-Maldonado, M.I. 2003. *Cymatocarpus solearis* a brachycoeliid metacercaria parasitizing *Panulirus argus* (Crustacea: Decapoda) from the Mexican Caribbean Sea. *Anales del Instituto de Biologia, Universidad Nacional Autónoma de Mexico, Serie Zoologia* 74: 1–10

Gómez, F., Artigas, L.F., and Gast, R.J. 2019. Molecular phylogeny of the parasitic dinoflagellate *Syltodinium listii* (Gymnodiniales, Dinophyceae) and generic transfer of *Syltodinium undulans* comb. nov. (= *Gyrodinium undulans*). *European Journal of Protistology* 71: 125636

Gómez, F., Lopez-Garcia, P., Nowaczyk, A., and Moreira, D. 2009b. The crustacean parasites *Ellobiopsis* Caullery, 1910 and *Thalassomyces* Niezabitowski, 1913 form a monophyletic divergent clade within the Alveolata. *Systematic Parasitology* 74: 65–74

Gómez, F., Moreira, D., and López-García, P. 2009a. Life cycle and molecular phylogeny of the dinoflagellates *Chytriodinium* and *Dissodinium*, ectoparasites of copepod eggs. *European Journal of Protistology* 45: 260–270

Gómez-Gutiérrez, J., López-Cortés, A., Aguilar-Méndez, M.J. et al. 2015. Histophagous ciliate *Pseudocollinia brintoni* and bacterial assemblage interaction with krill *Nyctiphanes simplex*. I. Transmission process. *Diseases of Aquatic Organisms* 116: 213–225

Gómez-Gutiérrez, J. and Morales-Ávila, J.R. 2016. Parasites and diseases. In *Biology and Ecology of Antarctic Krill*, Siegel, V. (ed.) pp. 351–386. Switzerland: Springer

Gómez-Gutiérrez, J., Peterson, W.T., De Robertis, A., and Brodeur, R.D. 2003. Mass mortality of krill caused by parasitoid ciliates. *Science* 301: 339

Gómez-Gutiérrez, J., Peterson, W.T., and Morado, J.F. 2006. Discovery of a ciliate parasitoid of euphausiids off Oregon, USA: *Collinia oregonensis* n. sp. (Apostomatida: Colliniidae). *Diseases of Aquatic Organisms* 71: 33–49

Gómez-Gutiérrez, J., Strüder-Kypke, M.C., Lynn, D.H., et al. 2012. *Pseudocollinia brintoni* gen. nov., sp. nov. (Apostomatida: Colliniidae), a parasitoid ciliate infecting the euphausiid *Nyctiphanes simplex*. *Diseases of Aquatic Organisms* 99: 57–78

Gonzalez, S.T. 2016. Influence of a trematode parasite (*Microphallus turgidus*) on grass shrimp (*Palaemonetes pugio*) response to refuge and predator presence. *Journal of Parasitology* 102: 646–649

Gooding, E.L., M.R. Kendrick, J.F. Brunson, P.R. Kingsley-Smith, A.E. Fowler, M.E. Frischer, and J.E. Byers. 2020. Black gill increases susceptibility of white shrimp, *Litopenaeus setiferus* (Linnaeus, 1767), to common estuarine predators. *Journal of Experimental Marine Biology and Ecology* 524: 151284

Gornik, S.G., Cassin, A.M., MacRae, J.I. et al. 2015. Endosymbiosis undone by stepwise elimination of the plastid in a parasitic dinoflagellate. *Proceedings of the National Academy of Sciences USA* 112: 5767–5772

Goulding, T.C. and Cohen, C.S. 2014. Phylogeography of a marine acanthocephalan: Lack of cryptic diversity in a cosmopolitan parasite of mole crabs. *Journal of Biogeography* 41: 965–976

Green, J. 1974. Parasites and epibionts of Cladocera. *Transactions of the Zoological Society of London* 32: 417–515

Grimes, B.H., and Bradbury, P.C. 1992. The biology of *Vampyrophrya pelagica* (Chatton and Lwoff, 1930), a histophagous apostome ciliate associated with marine calanoid copepods. *Journal of Protozoology* 39(1): 65–79

Grolière, C. A. and Leglise, M. 1977. *Paranophrys carcini* n. sp., ciliate Philasterina found in the haemolymph of *Cancer pagurus*. *Protistologica* 13: 503–507

Gurney, R.H., Johnston, D.J., and Nowak, B.F. 2006. The effect of parasitism by trypanorhynch plerocercoids (Cestoda, Trypanorhyncha) on the digestive enzyme activity of *Carcinus maenas* (Linnaeus, 1758) (Decapoda, Portunidae). *Crustaceana* 79: 663–675

Gurney, R.H., Nowak, B.F., Dyková, I., and Kuris, A.M. 2004. Histopathological effects of trypanorhynch metacestodes in the digestive gland of a novel host, *Carcinus maenas* (Decapoda). *Diseases of Aquatic Organisms* 58: 63–69

Haefner, P.A. Jr. and Spacher, P.J. 1985. Gill meristics and branchial infestation of *Ovalipes stephansoni* (Crustacea: Brachyura) by *Synophrya hypertrophica* (Ciliata, Apostomida). *Journal of Crustacean Biology* 5: 273–280

Hamilton, K.M., Morritt, D., and Shaw, P.W. 2010. Genetic diversity of the crustacean parasite *Hematodinium* (Alveolata, Syndinea). *European Journal of Protistology* 46: 17–28

Hanif, A.W., Dyson, W.D., Bowers, H.A. et al. 2013. Variation in spatial and temporal incidence of the crustacean pathogen *Hematodinium perezi* in environmental samples from Atlantic Coastal Bays. *Aquatic Biosystems* 9: 11

Hartikainen, H., Stentiford, G.D., Bateman, K.S. et al. 2014. Mikrocytids are a broadly distributed and divergent radiation of parasites in aquatic invertebrates. *Current Biology* 24: 807–812

Hawkes, C.R., Meyers, T.R., Shirley, T.C., and Koeneman, T.M. 1986. Prevalence of the parasitic barnacle *Briarosaccus callosus* on king crabs of southeastern Alaska. *Transactions of the American Fisheries Society* 115: 252–257

Helluy, S. and Holmes, J.C. 1990. Serotonin, octopamine, and the clinging behavior induced by the parasite *Polymorphus paradoxus* (Acanthocephala) in *Gammarus lacustris* (Crustacea). *Canadian Journal of Zoology* 68: 1214–1220

Helluy, S. and Thomas, F. 2010. Parasitic manipulation and neuroinflammation: Evidence from the system *Microphallus papillorobustus* (Trematoda)—*Gammarus* (Crustacea). *Parasites & Vectors* 3: 38

Hendrickx, M. and Fiers, F. 2010. Copepodos Harpacticoda asociados con crustaceos decapodos. *Ciencia Marina* 14: 3–30

Hendrickx, M.E. 2013. Prevalence and distribution of the dinoflagellate *Thalassomyces californiensis* Collard, 1966 (Ellobiopsidae) on *Pasiphaea emarginata* Rathbun, 1902 (Decapoda, Caridea, Pasiphaeidae), off western Mexico. *Crustaceana* 86: 693–703

Henebry, M.S. and Ridgeway, B.T. 1979. Epizoic ciliated protozoa of planktonic copepods and cladocerans and their possible use as indicators of organic water pollution. *Transactions of the American Microscopical Society* 98: 495–508

Ho, J.S. and Perkins, P.S. 1985. Symbionts of marine Copepoda: An overview. *Bulletin of Marine Science* 37: 586–598

Hochberg, R.J., Bert, T.M., Steele, P., and Brown, S.D. 1992. Parasitization of *Loxothylacus texanus* on *Callinectes sapidus*: Aspects of population biology and effects on host morphology. *Bulletin of Marine Science* 50: 117–132

Høeg, J.T. 1990. 'Akentrogonid' host invasion and an entirely new type of life cycle in the rhizocephalan parasite *Clistosaccus paguri* (Thecostraca: Cirripedia). *Journal of Crustacean Biology* 10: 37–52

Høeg, J.T. 1995. The biology and life cycle of the Rhizocephala (Cirripedia). *Journal of the Marine Biological Association of the United Kingdom* 75: 517–550

Høeg, J.T., Glenner, H., and Shields, J.D. 2005. Cirripedia Thoracica and Rhizocephala (barnacles). In *Marine Parasitology*, Rohde, K. (ed.) pp. 154–165. Victoria, Australia: CSIRO Publishing

Høeg, J.T. and Lützen, J. 1995. Life cycle and reproduction in the Cirripedia Rhizocephala. *Oceanography and Marine Biology, An Annual Review* 33: 427–485

Høeg, J.T., Noever, C., Rees, D.A., Crandall, K.A., and Glenner, H. 2019. A new molecular phylogeny-based taxonomy of parasitic barnacles (Crustacea: Cirripedia: Rhizocephala). In *Parasitic Crustacea*, Smit, N. Bruce,

N.L and Hadfield K.A. (eds.) pp. 387–419. Zoological Journal of the Linnean Society, Switzerland, Springer

Holt, P.C. 1968. The Branchiobdellida: Epizootic annelids. *Biologist* 50: 79–94

Holt, P.C. 1973. A summary of the branchiobdellid (Annelida: Clitellata) fauna of Mesoamerica. *Smithsonian Contributions to Zoology* 142: 1–40

Horacio, V.L., Ruth, E.M., Genaro, D.P. et al. 2020. Effects of the parasite *Loxothylacus texanus* on the agonistic behavior of the crab *Callinectes rathbunae*. *International Journal of Zoological Investigations* 6: 122–134

Huchin-Mian, J.P., Small, H.J., and Shields, J.D. 2017. Patterns in the natural transmission of the parasitic dinoflagellate *Hematodinium perezi* in American blue crabs, *Callinectes sapidus* from a highly endemic area. *Marine Biology* 164: 153

Huchin-Mian, J.P., Small, H.J., and Shields, J.D. 2018. The influence of temperature and salinity on mortality of recently recruited blue crabs, *Callinectes sapidus*, naturally infected with *Hematodinium perezi* (Dinoflagellata). *Journal of Invertebrate Pathology* 152: 8–16

Huys, R., Boxshall, G.A., and Lincoln, R.J. 1993. The tantulocaridan life cycle: The circle closed? *Journal of Crustacean Biology* 13: 432–442

Huys, R., Mackenzie-Dodds, J., and Llewellyn-Hughes, J. 2009. Cancrincolidae (Copepoda, Harpacticoida) associated with land crabs: A semiterrestrial leaf of the ameirid tree. *Molecular Phylogenetics and Evolution* 51: 143–156

Ianora, A., Mazzocchi, M.G., and Scotto Di Carlo, B. 1987. Impact of parasitism and intersexuality on Mediterranean populations of *Paracalanus parvus* (Copepoda: Calanoida). *Diseases of Aquatic Organisms* 3: 29–36

Jackson, C.J., Marcogliese, D.J., and Burt, M.D. 1997. Role of hyperbenthic crustaceans in the transmission of marine helminth parasites. *Canadian Journal of Fisheries and Aquatic Sciences* 54: 815–820

Jakobsen, P.J. and Wedekind, C. 1998. Copepod reaction to odor stimuli influenced by cestode infection. *Behavioral Ecology* 9: 414–418

Jankowski, A. 1973. Infusoria subclass Chonotricha. Faune URSS, 2(1): 1–355. (Akad. Nauka SSSR, St. Petersburg [formerly Leningrad].) (Not seen in the original.)

Jeffries, W.B. and Voris, H.K. 1983. The distribution, size, and reproduction of the pedunculate barnacle, *Octolasmis mulleri* (Coker, 1902), on the blue crab, *Callinectes sapidus* (Rathbun, 1896). *Fieldiana Zoology, New Series*, 16: 1–10

Jennings, J.B. and Gelder, S.R. 1979. Gut structure, feeding and digestion in the branchiobdellid oligochaete *Cambarincola macrodonta* Ellis 1912, an ectosymbiote of the freshwater crayfish *Procambarus clarkii*. *Biological Bulletin Woods Hole* 156: 300–314

Jensen, A.R., Schneider, M.R., Høeg, J.T., Glenner, H., and Lützen, J. 2019. Variation in juvenile stages and success of male acquisition in Danish and French populations of the parasitic barnacle *Sacculina carcini* (Cirripedia: Rhizocephala) parasitizing the shore crab *Carcinus maenas*. *Marine Biology Research* 15: 191–203

Jensen, K. and Bullard, S.A. 2010. Characterization of a diversity of tetraphyllidean and rhinebothriidean cestode larval types, with comments on host associations and life-cycles. *International Journal of Parasitology* 40: 889–910

Jensen, K. and Sadeghian, P.S. 2005. Nemertea (ribbon worms). In *Marine Parasitology*, Rohde, K. (ed.) pp. 387–419. Collingwood, Australia: CSIRO Publishing

Jepps, M.W. 1937. Memoirs: On the protozoan parasites of *Calanus finmarchicus* in the Clyde Sea area. *Journal of Cell Science* 2: 589–658

John, A.W.G. and Reid, P.C. 1983. Possible resting cysts of *Dissodinium pseudolunula* Swift ex Elbrächter et Drebes in the northeast Atlantic and the North Sea. *British Phycological Journal* 18: 61–67

Johnson, C.A. III, and Bradbury, P.C. 1976. Observations on the occurrence of the parasitic ciliate *Synophrya* in decapods in coastal waters off the Southeastern United States. *Journal of Protozoology* 23: 252–256

Johnson, D.S. and Heard, R. 2017. Bottom-up control of parasites. *Ecosphere* 8: e01885

Johnson, D.S., Shields, J.D., Doucette, D., and Heard, R. 2020. A climate migrant escapes its parasites. *Marine Ecology Progress Series* 641: 111–121

Johnson, P.T. 1977. Paramoebiasis in the blue crab, *Callinectes sapidus*. *Journal of Invertebrate Pathology* 29: 308–320

Johnson, P.T. 1983. Diseases caused by viruses, rickettsiae, bacteria, and fungi. In *The Biology of Crustacea*, Vol. 6, Pathobiology, Provenzano, A.J. (ed.) pp. 1–78. New York: Academic Press

Jones, S., Carrasco, N.K., Perissinotto, R., and Vosloo, A. 2016. Association of the epibiont *Epistylis* sp. with a calanoid copepod in the St Lucia Estuary, South Africa. *Journal of Plankton Research* 38: 1404–1411

Keeling, P.J., Burger, G., Durnford, D.G., Lang, B.F., Lee, R.W., Pearlman, R.E., Roger, A.J. and Gray, M.W. 2005. The tree of eukaryotes. *Trends in Ecology & Evolution* 20: 670–676

Key, M.M., Jr., Volpe, J.W., Jeffries, W.B., and Voris, H.K. 1997. Barnacle fouling of the blue crab *Callinectes sapidus* at Beaufort, North Carolina. *Journal of Crustacean Biology* 17: 424–439

Khan, R.A. and Thulin, J. 1991. Influence of pollution on parasites of aquatic animals. *Advances in Parasitology* 30: 201–238

Kim, E.M., Kim, J.L., Choi, S.I., Lee, S.H., and Hong, S.T. 2009. Infection status of freshwater crabs and crayfish with metacercariae of *Paragonimus westermani* in Korea. *Korean Journal of Parasitology* 47: 425–426

Kimmerer, W.J. and McKinnon, A.D. 1990. High mortality in a copepod population caused by a parasitic dinoflagellate. *Marine Biology* 107: 449–452

Kitron, U.D. 1980. The pattern of infestation of the beach-hopper amphipod *Orchestoidea corniculata*, by a parasitic mite. *Parasitology* 81: 235–249

Klompmaker, A.A., Artal, P., Van Bakel, B.W., Fraaije, R.H., and Jagt, J.W. 2014. Parasites in the fossil record: A Cretaceous fauna with isopod-infested decapod crustaceans, infestation patterns through time, and a new ichnotaxon. *PLoS One* 9: p.e92551

Knudsen, S.W., Kirkegaard, M., and Olesen, J. 2009. The tantulocarid genus *Arcticotantalus* removed from Basipodellidae into Deoterthridae (Crustacea: Maxillopoda) after the description of a new species from Greenland, with first live photographs and an overview of the class. *Zootaxa* 2035: 41–68

Køie, M. 1979. On the morphology and life-history of *Derogenes varicus* (Müller, 1784) Looss, 1901 (Trematoda, Hemiuridae). *Zeitschrift für Parasitenkunde* 59: 67–78

Kolluru, G.R., Green, Z.S., Vredevoe, L.K., Kuzma, M.R., Ramadan, S.N., and Zosky, M.R. 2011. Parasite infection and sand coarseness increase sand crab (*Emerita analoga*) burrowing time. *Behavioural Processes* 88: 184–191

Kopp, D.A., Bierbower, S.M., Murphy, A.D., Mormann, K., and Sparkes, T.C. 2016. Parasite-related modification of mating behaviour and refuge use in the aquatic isopod *Caecidotea intermedius*: Neurological correlates. *Behaviour* 153: 947–961

Koprivnikar, J., Ellis, D., Shim, K.C., and Forbes, M.R. 2014. Effects of temperature and salinity on emergence of *Gynaecotyla adunca* cercariae from the intertidal gastropod *Ilyanassa obsoleta*. *Journal of Parasitology* 100: 242–245

Kristensen, T., Nielsen, A.I., Jørgensen, A.I. et al. 2012. The selective advantage of host feminization: A case study of the green crab *Carcinus maenas* and the parasitic barnacle *Sacculina carcini*. *Marine Biology* 159: 2015–2023

Kuris, A.M. 1991. A review of patterns and causes of crustacean brood mortality. In *Crustacean Egg Production, Crustacean Issues*, Wenner A. and Kuris A.M (eds.) Vol. 7. pp. 117–141. Rotterdam: A.A. Balkema

Kuris, A.M., Blau, S.F., Paul, A.J., Shields, J.D., and Wickham, D.E. 1991. Infestation by brood symbionts and their impact on egg mortality in the red king crab, *Paralithodes camtschatica*, in Alaska: Geographic and temporal variation. *Canadian Journal of Fisheries and Aquatic Sciences* 48: 559–568

Kuris, A.M., Torchin, M.E., and Lafferty, K.D. 2002. *Fecampia erythrocephala* rediscovered: Prevalence and distribution of a parasitoid of the European shore crab, *Carcinus maenas*. *Journal of the Marine Biological Association of the United Kingdom* 82: 955–960

Kurtz, J., Van Der Veen, I.T., and Christen, M. 2002. Fluorescent vital labeling to track cestodes in a copepod intermediate host. *Experimental Parasitology* 100: 36–43

Kvingedal, R., Owens, L., and Jerry, D.R. 2006. A new parasite that infects eggs of the mud crab, *Scylla serrata*, in Australia. *Journal of Invertebrate Pathology* 93: 54–59

La Sala, L.F., Diaz, J.I., Martorelli, S.R., and Alda, P. 2009. Some nematodes from Olrog's gull, *Larus atlanticus* Olrog, 1958 (Aves: Laridae), and prey crabs from the Bahía Blanca Estuary, Argentina. *Comparative Parasitology* 76: 293–296

Lafferty, K.D., Dobson, A.P., and Kuris, A.M. 2006. Parasites dominate food web links. *Proceedings of the National Academy of Sciences* 103: 11211–11216

Landers, S.C. 2004. Exuviotrophic apostome ciliates from crustaceans of St. Andrew Bay, Florida, and a description of *Gymnodinioides kozloffi* n. sp. *Journal of Eukaryotic Microbiology* 51: 644–650

Landers, S.C., Gómez-Gutiérrez, J., and Peterson, W.T. 2006. *Gymnodinioides pacifica*, n. sp., an exuviotrophic ciliated protozoan (Ciliophora, Apostomatida) from euphausiids of the Northeastern Pacific. *European Journal of Protistology* 42: 97–106

Landers, S.C., Lee, R.F., Walters, T.L. et al. 2020. *Hyalophysa lynni* n. sp. (Ciliophora, Apostomatida), a new pathogenic ciliate and causative agent of shrimp black gill in penaeid shrimp. *European Journal of Protistology* 73: 125673

Larsson, J.R. 1987. On *Haplosporidium gammari*, a parasite of the amphipod *Rivulogammarus pulex*, and its relationships with the phylum Ascetospora. *Journal of Invertebrate Pathology* 49: 159–169

Latham, A. and Poulin, R. 2001. Effect of acanthocephalan parasites on the behaviour and coloration of the mud crab *Macrophthalmus hirtipes* (Brachyura: Ocypodidae). *Marine Biology* 139: 1147–1154

Lavallée, J., Hammell, K.L., Spangler, E.S., and Cawthorn, R.J. 2001. Estimated prevalence of *Aerococcus viridans* and *Anophryoides haemophila* in American lobsters *Homarus americanus* freshly captured in the waters of Prince Edward Island, Canada. *Diseases of Aquatic Organisms* 46: 231–236

Leduc, D. and Wilson, J. 2016. Benthimermithid nematode parasites of the amphipod *Hirondellea dubia* in the Kermadec Trench. *Parasitology Research* 115: 1675–1682

Leduc, D. and Zhao, Z.Q. 2019. Phylogenetic position of the parasitic nematode *Trophomera* (Nematoda, Benthimermithidae): A molecular analysis. *Molecular Phylogenetics and Evolution* 132: 177–182

Lee, J.H., Kim, T.W. and Choe, J.C., 2009. Commensalism or mutualism: Conditional outcomes in a branchiobdellid–crayfish symbiosis. *Oecologia* 159: 217–224

Lee, S.C., Corradi, N., Byrnes, E.J. III et al. 2008. Microsporidia evolved from ancestral sexual fungi. *Current Biology* 18: 1675–1679

Lemercier, A. 1963. Comparison du complexe buccal de trois Copepodes parasites, *Sphaeronella* sp., *Choniosphaera maenadis* (Bloch et Gallien) (Choniostomatidae) et *Nichothoe astaci* Audouin et Milne Edwards. *Bulletin de la Société Linnéenne de Normandie* 4: 119–139

Leslie, H.A., Campbell, A., and Daborn, G.R. 1981. *Nectonema* (Nematomorpha: Nectonematoidea) a parasite of decapod Crustacea in the Bay of Fundy. *Canadian Journal of Zoology* 59: 1193–1196

Li, C., Miller, T.L., Small, H.J., and Shields, J.D. 2011a. *In vitro* culture and developmental cycle of the parasitic dinoflagellate *Hematodinium* sp. from the blue crab *Callinectes sapidus*. *Parasitology* 138: 1924–1934

Li., C., Wheeler, K.N., and Shields, J.D. 2011b. Lack of transmission of *Hematodinium* sp. in the blue crab *Callinectes sapidus* through cannibalism. *Diseases of Aquatic Organisms* 96: 249–258

Li, C., Song, S., Liu, Y., and Chen, T. 2013. *Hematodinium* infections in cultured Chinese swimming crab, *Portunus trituberculatus*, in northern China. *Aquaculture* 396: 59–65

Li, M., Wang, J., Huang, Q., and Li, C. 2019. Proteomic analysis highlights the immune responses of the hepatopancreas against *Hematodinium* infection in *Portunus trituberculatus*. *Journal of Proteomics* 197: 92–105

Li, M., Wang, J., Song, S., and Li, C. 2016. Molecular characterization of a novel nitric oxide synthase gene from *Portunus trituberculatus* and the roles of NO/O$_2$-generating and antioxidant systems in host immune responses to *Hematodinium*. *Fish & Shellfish Immunology* 52: 263–277

Lichtwardt, R.W. 1986. *The Trichomycetes. Fungal Associates of Arthropods*, p. 343. Springer-Verlag

Lightner, D.V. 1981. Fungal diseases of marine Crustacea. In *Pathogenesis of Invertebrate Microbial Diseases*, Davidson, E.W. (ed.) pp. 451–484. Totowa, New Jersey: Allanheld, Osmun

Linton, E. 1940. Trematodes from fishes, mainly from the Woods Hole region, Massachusetts. *Proceedings of the United States National Museum* 88: 1–172

Lloyd, J.H. 1920. Some observations on the structure and life-history of the common nematode of the dogfish (*Scyllium canieula*). *Proceedings of the Zoological Society of London* 90: 449–456

Lloyd, J.H. 1928. LXXXVI.—On the life-history of the common nematode of the dogfish (*Scyllium canicular*). *Journal of Natural History, London* 1: 712–714

Lohr, J.N., Laforsch, C., Koerner, H., and Wolinska, J. 2010. A *Daphnia* parasite (*Caullerya mesnili*) constitutes a new member of the Ichthyosporea, a group of protists near the animal–fungi divergence. *Journal of Eukaryotic Microbiology* 57: 328–336

Longshaw, M. 2016. Parasites, commensals, pathogens and diseases of crayfish. In *Biology and Ecology of Crayfish*, M. Longshaw and P. Stebbing (eds.) pp. 171–250. Boca Raton, Florida: CRC Press

Loughlin, M.B., Bayer, R.C., and Prince, D.L. 1998. Lobster, *Homarus americanus*, gastric fluid is a barrier to the ciliate, *Anophryoides haemophila*, in an *in vitro* study. *Journal of Applied Aquaculture* 8: 67–72

Lützen, J. 1985. Rhizocephala (Crustacea: Cirripedia) from the deep sea. *Galathea Reports* 16: 99–112

Lycett, K.A. and Pitula, J.S. 2017. Disease ecology of *Hematodinium perezi* in a high salinity estuary: Investigating seasonal trends in environmental detection. *Diseases of Aquatic Organisms* 124: 169–179

Lynn, D.H., Gómez-Gutiérrez, J., Strüder-Kypke, M., and Shaw, C.T. 2014. Ciliate species diversity and host-parasitoid codiversification in the apostome genus *Pseudocollinia* (Ciliophora, Apostomatia, Pseudocollinidae) that infect krill, with description of *Pseudocollinia similis* n. sp., a parasitoid of the krill *Thysanoessa spinifera*. *Diseases of Aquatic Organisms* 112: 89–102

Lynn, D.H. and Strüder-Kypke, M.C. 2019. The sanguicolous apostome *Metacollinia luciensis* Jankowski 1980 (Colliniidae, Apostomatia, Ciliophora) is not closely related to other sanguicolous apostomes. *Journal of Eukaryotic Microbiology* 66: 140–146

MacKenzie, K. 1963. *Stichocotyle nephropis* Cunningham, 1887 (Trematoda) in Scottish waters. *Annals of the Magazine of Natural History* 6(3): 505–506

MacNeil, C., Fielding, N.J., Dick, J.T. et al. 2003. An acanthocephalan parasite mediates intraguild predation between invasive and native freshwater amphipods (Crustacea). *Freshwater Biology* 48: 2085–2093

Marcogliese, D.J. 1996. Larval parasitic nematodes infecting marine crustaceans in eastern Canada. 3. *Hysterothylacium aduncum*. *Journal of the Helminthological Society of Washington* 63: 12–18

Margolis, L. and Arthur, J.R. 1979. Synopsis of the parasites of fishes of Canada. *Bulletin of the Fisheries Research Board of Canada*, 199: p. 269

Markham, J.C. 1986. Evolution and zoogeography of the Isopoda Bopyridae, parasites of Crustacea Decapoda. *Crustaceana* 4: 143–164

Martin, D. and Britayev, T.A. 1998. Symbiotic polychaetes: Review of known species. *Oceanography and Marine Biology: An Annual Review* 36: 217–340

Martin, D. and Britayev, T.A. 2018. Symbiotic polychaetes revisited: An update of the known species and relationships (1998–2017). *Oceanography and Marine Biology: An Annual Review* 56: 371–448

Mayer, K.A., Dailey, M.D., and Miller, M.A. 2003. Helminth parasites of the southern sea otter *Enhydra lutris nereis* in central California: Abundance, distribution and pathology. *Diseases of Aquatic Organisms* 53: 77–88

McCabe, G.T., Emmett, R.L., Coley, T.C., and McConnell, R.J. 1987. Effect of a river-dominated estuary on the prevalence of *Carcinonemertes errans*, an egg predator of the Dungeness crab, *Cancer magister*. *Fishery Bulletin, NOAA* 85: 140–142

McCurdy, D.G., Forbes, M.R., and Boates, J.S. 1999. Evidence that the parasitic nematode *Skrjabinoclava* manipulates host *Corophium* behavior to increase transmission to the sandpiper, *Calidris pusilla*. *Behavioral Ecology* 10: 351–357

McDermott, J.J. 1991. Incidence and host-parasite relationship of *Leidya bimini* (Crustacea, Isopoda, Bopyridae) in the brachyuran crab Pachygrapsus transversus from Bermuda. *Ophelia* 33: 71–95

McDermott, J.J. 2011. Parasites of shore crabs in the genus *Hemigrapsus* (Decapoda: Brachyura: Varunidae) and their status in crabs geographically displaced: A review. *Journal of Natural History, London* 45: 2419–2441

McDermott, J.J., Williams, J.D., and Boyko, C.B. 2010. The unwanted guests of hermits: A global review of the diversity and natural history of hermit crab parasites. *Journal of Experimental Marine Biology and Ecology* 394: 2–44

Messick, G.A. and Shields, J.D. 2000. The epizootiology of the parasitic dinoflagellate *Hematodinium* sp. in the American blue crab *Callinectes sapidus*. *Diseases of Aquatic Organisms* 43: 139–152

Messick, G.A. and Small, E.B. 1996. *Mesanophrys chesapeakensis* n. sp., a histophagous ciliate in the blue crab, *Callinectes sapidus*, and associated histopathology. *Invertebrate Biology* 115: 1–12

Meyers, T.R. 1990. Diseases caused by protistans. In *Diseases of Marine Animals*, Vol. III, Diseases of Crustacea, Kinne, O. (ed.) pp. 350–368. Biologische Anstalt Helgoland, Hamburg, Germany

Meyers, T.R., Koeneman, T.M., Bothelho, C., and Short, S. 1987. Bitter crab disease: A fatal dinoflagellate infection and marketing problem for Alaskan Tanner crabs *Chionoecetes bairdii*. *Diseases of Aquatic Organisms* 3: 195–216

Miller, T.L., Small, H.J., Peemoeller, B.-J., Gibbs, D.A., and Shields, J.D. 2013. Experimental infections of *Orchitophrya stellarum* (Scuticociliata) in American blue crabs (*Callinectes sapidus*) and fiddler crabs (*Uca minax*). *Journal of Invertebrate Pathology* 114: 346–355

Misra, J.K. 1998. Trichomycetes—Fungi associated with arthropods: Review and world literature. *Symbiosis* 24: 179–220

Mohrbeck, I., Arbizu, P.M., and Glatzel, T. 2010. Tantulocarida (Crustacea) from the Southern Ocean deep sea, and the description of three new species of *Tantulacus* Huys, Andersen and Kristensen, 1992. *Systematic Parasitology* 77: 131–151

Montreuil, P. 1954. Parasitological investigations. *Rapport Annual, Station Biologique Marine Department Peches* 1953, Quebec, Contribution No. 50, Appendix 5, 69–73

Moore, J. 1983. Responses of an avian predator and its isopod prey to an acanthocephalan parasite. *Ecology* 64: 1000–1015

Moore, J. 2002. *Parasites and the Behavior of Animals*, Longshaw M. and Stebbing P. (eds.) Oxford Series in Ecology and Evolution, p. 338. Oxford: Oxford University Press

Moore, J. and Goteli N.J. 1990. A phylogenetic perspective on the evolution of altered host behaviours: A critical look at the manipulation hypothesis. In *Parasitism and Host Behaviour*, Barnard C.J. and J.M Behnke (eds.) pp. 193–233. London: Taylor and Francis

Morado, J.F. 2011. Protistan diseases of commercially important crabs: a review. *Journal of Invertebrate Pathology* 106: 27–53

Morado, J.F. and Small, E.B. 1995. Ciliate parasites and related diseases of Crustacea: A review. *Reviews in Fisheries Science* 3: 275–354

Morales-Ávila, J.R., Gómez-Gutiérrez, J., Hernandez-Saavedra, N.Y., Robinson, C.J., and Palm, H.W. 2019. Phylogenetic placement and microthrix pattern of *Paranybelinia otobothrioides* Dollfus, 1966 (Trypanorhyncha) from krill *Nyctiphanes simplex* Hansen, 1911. *International Journal for Parasitology: Parasites and Wildlife* 10: 138–148

Moravec, F., Fredensborg, B.L., Latham, A.D.M., and Poulin, R. 2003. Larval Spirurida (Nematoda) from the crab *Macrophthalmus hirtipes* in New Zealand. *Folia Parasitologia* 50: 109–114

Mouritsen, K.N., Sørensen, M.M., Poulin, R., and Fredensborg, B.L. 2018. Coastal ecosystems on a tipping point:

Global warming and parasitism combine to alter community structure and function. *Global Change Biology* 24: 4340–4356

Mullen, T.E., Nevis, K.R., O'Kelly, C.J., Gast, R.J., and Frasca, S. 2005. Nuclear small-subunit ribosomal RNA gene-based characterization, molecular phylogeny and PCR detection of the *Neoparamoeba* from western Long Island Sound lobster. *Journal of Shellfish Research* 24: 719–731

Mullen, T. E., Russell, S., Tucker, M.T. et al. 2004. Paramoebiasis associated with mass mortality of American lobster *Homarus americanus* in Long Island Sound, USA. *Journal of Aquatic Animal Health* 16: 29–38

Newman, M.W. and Ward, G.E. Jr. 1973. An epizootic of blue crabs, *Callinectes sapidus*, caused by *Paramoeba perniciosa*. *Journal of Invertebrate Pathology* 22: 329–334

Nicholas, W.L. 2004. *Crustorhabditis chitwoodi* sp. nov. (Nematoda: Rhabditidae): An intertidal species from the coast of New South Wales, Australia, with observations on its ecology and life history. *New Zealand Journal of Marine and Freshwater Research* 38: 803–808

Nickol, B.B., Crompton, D.W.T., and Searle, D.W. 1999. Reintroduction of *Profilicollis* Meyer, 1931, as a genus in Acanthocephala: Significance of the intermediate host. *Journal of Parasitology* 85: 716–718

Nielsen, S.-O. 1969. *Nectonema munida* Brinkmann (Nematomorpha) parasitizing *Munida tenuimana* G.O. Sars (Crust. Dec.) with notes on host parasite relations and new host species. *Sarsia* 38: 91–110

Normant, M., Zawal, A.J., Chatterjee, T., and Wojcik, D.A. 2013. Epibiotic mites associated with the invasive Chinese mitten crab *Eriocheir sinensis*–new records of Halacaridae from Poland. *Oceanologia* 55: 901–915

Nouvel, H. and Nouvel, L. 1934. Sur deux Crevettes parasites par un *Nectonema*. Evolution des parasites. Influence sur l'hôte. *Bulletin de la Société Zoologique de France* 59: 516–521

Nunan, L.M., Lightner, D.V., Pantoja, C.R., Stokes, N.A., and Reece, K.S. 2007. Characterization of a rediscovered haplosporidian parasite from cultured *Penaeus vannamei*. *Diseases of Aquatic Organisms* 74: 67–75

O'Brien, J. and Van Wyk, P. 1985. Effects of crustacean parasitic castrators (epicaridean isopods and rhizocephalan barnacles) on growth of crustacean hosts, pp. 191–218. In *Factors in Adult Growth*, Crustacean Issues, Vol. 3. Wenner, A.M. (ed.) Rotterdam: A. A. Balkema

O'Leary, P.A. and Shields, J.D. 2017. Fiddler crabs (*Uca* spp.) as model hosts for laboratory infections of *Hematodinium perezi*. *Journal of Invertebrate Pathology* 143: 11–17

Odermatt, P., Habe, S., Manichanh, S. et al. 2007. Paragonimiasis and its intermediate hosts in a transmission focus in Lao People's Democratic Republic. *Acta Tropica* 103: 108–115

Odhner, T. 1898. Über die geschlechtsreife Form von *Stichocotyle nephropis* Cunningham. *Zoologische Anzeiger* 21: 509–513

Ohtaka, A., Gelder, S.R., Nishino, M. et al. 2012. Distributions of two ectosymbionts, branchiobdellidans (Annelida: Clitellata) and scutariellids (Platyhelminthes: 'Turbellaria': Temnocephalida), on atyid shrimp (Arthropoda: Crustacea) in southeast China. *Journal of Natural History* 46: 1547–1556

Ohtsuka, S., Boxshall, G.A., and Harada, S. 2005. A new genus and species of nicothoid copepod (Crustacea: Copepoda: Siphonostomatoida) parasitic on the mysid *Siriella okadai* Ii from off Japan. *Systematic Parasitology* 62: 65–81

Ohtsuka, S., Hora, M., Suzaki, T., Arikawa, M., Omura, G., and Yamada, K. 2004. Morphology and host-specificity of the apostome ciliate *Vampyrophrya pelagica* infecting peagic copepods in the Seto Inland Sea, Japan. *Marine Ecology Progress Series* 282: 129–142

Ohtsuka, S., Horiguchi, T., Hanamura, Y. et al. 2011. Symbiosis of planktonic copepods and mysids with epibionts and parasites in the North Pacific: Diversity and interactions. In *New Frontiers in Crustacean Biology*, Asakura, A. (ed.) pp. 1–14. Leiden, Netherlands: Brill

Ohtsuka, S., Suzaki, T., Kanazawa, A., and Ando, M. 2015. Biology of symbiotic apostome ciliates: Their diversity and importance in the aquatic ecosystems. In *Marine Protists*, Ohtsuka, S. et al. (eds.) pp. 441–463. Tokyo: Springer

Øksnebjerg, B. 2000. The Rhizocephala (Crustacea: Cirripedia) of the Mediterranean and Black Seas: Taxonomy, biogeography, and ecology. *Israel Journal of Zoology* 46: 1–102

Overstreet, R.M. 1983. Metazoan symbionts of crustaceans. In *The Biology of the Crustacea*, Vol 6, *Pathobiology*, Provenzano, A.J. Jr. (ed.) pp. 156–250. New York: Academic Press

Pagenkopp Lohan, K.M., McDowell, J.R., Shields, J.D., and Reece, K.S. 2014. Genotypic variation in the parasitic dinoflagellate *Hematodinium perezi* along the Delmarva Peninsula, Virginia, using microsatellite markers. *Marine Biology* 161: 261–273

Pagenkopp Lohan, K.M., Reece, K.S., Miller, T.L., Wheeler, K.N., Small, H.J., and Shields, J.D. 2012. The role of alternate hosts in the ecology and life history of *Hematodinium* sp., a parasitic dinoflagellate of the blue crab (*Callinectes sapidus*). *Journal of Parasitology* 98: 73–84

Pardal, M.A., Bessa, F., and Costa, S. 2013. The parasite *Sacculina carcini* Thompson, 1836 (Cirripedia, Rhizocephala) in the crab *Carcinus maenas* (Linnaeus, 1758) (Decapoda, Portunidae): Influence of environmental conditions, colour morphotype and sex. *Crustaceana* 86: 34–47

Pasternak, A.F., Arashkevich, Y.G., and Sorokin, Y.S. 1984. The role of the parasitic algal genus *Blastodinium* in the ecology of planktic copepods. *Oceanology* 24: 748–751

Pauley, G.B., Newman, M.W., and Gould, E. 1975. Serum changes in the blue crab, *Callinectes sapidus*, associated with *Paramoeba perniciosa*, the causative agent of gray crab disease. *Marine Fisheries Review* 34–38. Washington, DC: National Marine Fisheries Service, NOAA

Peralta, A.S.L., Nonato, B.S., Matos, P., Santos, M.N.S., and Matos, E. 2005. Structural aspects of *Temnocephala lutzi* Monticelli, 1913 (Turbellaria, Temnocephalida) in red crabs (*Callinectes bocourti*). *Brazilian Journal of Morphological Science* 22: 232–233

Perez-Calderon, J.A. 1986. Occurrence of nematode parasites in *Calocaris macandreae* (Crustacea: Decapoda) from an Irish Sea population. *Journal of the Marine Biological Association of the United Kingdom* 66: 293–301

Pérez-Campos, R.A., Rodríguez-Canul, R., Pérez-Vega, J.A., González-Salas, C., and Guillén-Hernández, S. 2012. High serotonin levels due to the presence of the acanthocephalan *Hexaglandula corynosoma* could promote changes in behavior of the fiddler crab *Uca spinicarpa*. *Diseases of Aquatic Organisms* 99: 49–55

Pérez-Losada, M., Høeg, J.T., and Crandall, K.A. 2009. Remarkable convergent evolution in specialized parasitic Thecostraca (Crustacea). *British Medical Council Biology* 7: 15

Perkins, F.O. 1975. Fine structure of *Minchinia* sp. (Haplosporida). Sporulation in the mud crab, *Panopeus herbstii*. *Marine Fisheries Review* 37: 5–6

Petrunina, A.S., Neretina, T.V., Mugue, N.S., and Kolbasov, G.A. 2014. Tantulocarida versus Thecostraca: Inside or outside? First attempts to resolve phylogenetic position of Tantulocarida using gene sequences. *Journal of Zoological Systematics and Evolutionary Research* 52: 100–108

Petter, A.J. 1970. Quelques spirurides de poissons de la région nantaise. *Annales de Parasitologie Humaine et Comparee* 45: 31–46

Pickup, J. and Ironside, J.E. 2018. Multiple origins of parasitic feminization: Thelygeny and intersexuality in beach-hoppers are caused by paramyxid parasites, not Microsporidia. *Parasitology* 145: 408–415

Piscart, C., Webb, D., and Beisel, J.N. 2007. An acanthocephalan parasite increases the salinity tolerance of the freshwater amphipod *Gammarus roeseli* (Crustacea: Gammaridae). *Naturwissenschaften* 94: 741–747

Pitula, J.S., Dyson, W.D., Bakht, H.B., Njoku, I., and Chen, F. 2012. Temporal distribution of genetically homogenous 'free-living' *Hematodinium* sp. in a Delmarva coastal ecosystem. *Aquatic Biosystems* 8: p. 16

Poinar, G.O. Jr., and Brockerhoff, A.M. 2001. *Nectonema zealandica* n. sp. (Nematomorpha: Nectonematoidea) parasitizing the purple rock crab *Hemigrapsus edwardsi* (Brachyura: Decapoda) in New Zealand, with notes on the prevalence of infection and host defence reactions. *Systematic Parasitology* 50: 149–157

Poinar, G.O. Jr., Duarte, D., and Santos, M.J. 2010. *Halomonhystera parasilica* n. sp. (Nematoda: Monhysteridae), a parasite of *Talorchestia brito* (Crustacea: Talitridae) in Portugal. *Systematic Parasitology* 75: 53–58

Poinar G.O. Jr. and Kannangara, D.W. 1971. *Rhabdochona praecox* sp. n. and *Proleptus* sp. (Spiruroidea: Nematoda) from fresh water crabs in Ceylon. *Annales de Parasitologie Humaine et Comparee* 47: 121–129

Poinar, G.O. Jr., Latham, A.D.M., and Poulin, R. 2002. *Thaumamermis zealandica* n. sp. (Mermithidae: Nematoda) parasitising the intertidal marine amphipod *Talorchestia quoyana* (Talitridae: Amphipoda) in New Zealand, with a summary of mermithids infecting amphipods. *Systematic Parasitology* 53: 227–233

Poinar, G.O. Jr. and Thomas, G.M. 1976. Occurrence of *Ascarophis* (Nematoda: Spiruridea) in *Callianassa californiensis* Dana and other decapod crustaceans. *Proceedings of the Helminthological Society of Washington* 43: 28–33

Porchet-Hennere, E. and Richard, A. 1971. La sporogenèse chez la coccidie *Aggregata eberthi*. Étude en microscopic électronique. *Journal of Protozoology* 18: 614–628

Poulin, R., Nichol, K., and Latham, A.D.M. 2003. Host sharing and host manipulation by larval helminths in shore crabs: Cooperation or conflict? *International Journal of Parasitology* 33: 425–433

Poulin, R. and Rate, S.R. 2001. Small-scale spatial heterogeneity in infection levels by symbionts of the amphipod *Talorchestia quoyana* (Talitridae). *Marine Ecology Progress Series* 212: 211–216

Procop, G.W. 2009. North American paragonimiasis (caused by *Paragonimus kellicotti*) in the context of global paragonimiasis. *Clinical Microbiological Reviews* 22: 415–446

Prytherch, H.F. 1940. The life cycle and morphology of *Nematopsis ostrearum*, sp. nov. a gregarine parasite of the mud crab and oyster. *Journal of Morphology* 66: 39–65

Reisser, C.E. and Forward, R.B. 1991. Effect of salinity on osmoregulation and survival of a rhizocephalan parasite, *Loxothylacus panopaei*, and its crab host, *Rhithropanopeus harrisii*. *Estuaries* 14: 102–106

Riemann, F. 1970. Das Kiemenlückensystem von Krebsen als Lebensraum der Meiofauna, mit Beschreibung freilebender Nematoden aus karibischen amphibisch lebenden Decapoden. *Veröffentlich. Institut für Meeresforschung Bremerhaven* 12: 224–233

Rigaud, T., Pennings, P.S., and Juchault, P. 2001. *Wolbachia* bacteria effects after experimental interspecific transfers in terrestrial isopods. *Journal of Invertebrate Pathology* 77: 251–257

Rigby, M.C. 1996. The epibionts of beach hoppers (Crustacea: Talitridae) of the North American Pacific coast. *Journal of Natural History, London* 30: 1329–1336

Robaldo, R.B., Monserrat, J., Cousin, J.C.B., and Blanchini, A. 1999. Effects of metacercariae (Digenea: Microphallidae) on the hepatopancreas of *Chasmagnathus granulata* (Decapoda: Grapsidae). *Diseases of Aquatic Organisms* 37: 153–157

Rogers-Talbert, R. 1948. The fungus *Lagenidium callinectes* Couch (1942) on eggs of the blue crab in Chesapeake Bay. *Biological Bulletin Woods Hole* 95: 214–228

Rojas, J.M. and Ojeda, F.P. 2005. Altered dopamine levels induced by the parasite *Profilicollis antarcticus* on its intermediate host, the crab *Hemigrapsus crenulatus*. *Biological Research* 38: 259–266

Rowley, A.F., Smith, A.L., and Davies, C.E. 2015. How does the dinoflagellate parasite *Hematodinium* outsmart the immune system of its crustacean hosts? *PLoS Pathogens* 11: e1004724.

Rubiliani, C. 1983. Action of a rhizocephalan on the genital activity of host male crabs: characterization of a parasitic secretion inhibiting spermatogenesis. *International Journal of Invertebrate Reproduction* 6: 137–147

Rubiliani, C. and Payen, G.G. 1979. Modalités de la destruction des régions neurosécrétrices des crabes *Carcinus maenas* (L.) et *C. mediterraneus* Czerniavsky infestés par la Sacculine. *General and Comparative Endocrinology* 38: 215–228

Ruggiero, M.A., Gordon, D.P., Orrell, T.M., et al. 2015. A higher-level classification of all living organisms. *PloS One* 10: p.e0119248

Ruppert, E.E., Fox, R.S., and Barnes, R.D. 2004. *Invertebrate Zoology: A Functional Approach*, 7th edn, p. 963. Belmont, CA: Brooks/Cole

Ryazanova, T.V., Eliseikina, M.G., Kukhlevsky, A.D., and Kharlamenko, V.I. 2010. *Hematodinium* sp. infection of red *Paralithodes camtschaticus* and blue *Paralithodes platypus* king crabs from the Sea of Okhotsk, Russia. *Journal of Invertebrate Pathology* 105: 329–334

Sánchez, M.I., Georgiev, B.B., and Green, A.J. 2007. Avian cestodes affect the behaviour of their intermediate host *Artemia* parthenogenetica: An experimental study. *Behavioural Processes* 74: 293–299

Sánchez, M.I., Thomas, F., Perrot-Minnot, M.J., et al. 2009. Neurological and physiological disorders in *Artemia* harboring manipulative cestodes. *Journal of Parasitology* 95: 20–24

Sardella, N.H. and Martorelli, S.R. 1997. Occurrence of merogony of *Aggregata* Frenzel, 1885 (Apicomplexa) in *Pleoticus muelleri* and *Artemesia longinaris* (Crustacea: Natantia) from Patagonian Waters (Argentina). *Journal of Invertebrate Pathology* 70: 198–202

Saville, D.H. and Irwin, S.W.B. 2005. A study of the mechanisms by which the cercariae of *Microphallus primas* (Jag, 1909) Stunkard, 1957 penetrate the shore crab, *Carcinus maenas* (L.). *Parasitology* 131: 521–529

Sewell, R.B.S. 1951. The epibionts and parasites of the planktonic Copepoda of the Arabian Sea. *Scientific Reports of the Murray Expedition* 9: 255–394

Sherburne, S.W. and Bean, L. 1991. Mortalities of impounded and feral marine lobsters, *Homarus americanus* H. Milne-Edwards, 1837, caused by the protozoan ciliate Mugardia (formerly *Anophrys-Paranophrys*), with initial prevalence data from ten locations along the Maine coast and one offshore area. *Journal of Shellfish Research* 10: 315–326

Sherman, M.B. and Curran, M.C. 2015. Sexual sterilization of the daggerblade grass shrimp *Palaemonetes pugio* (Decapoda: Palaemonidae) by the bopyrid isopod *Probopyrus pandalicola* (Isopoda: Bopyridae). *Journal of Parasitology* 101: 1–5

Shibahara, T. 1991. The route of infection of *Paragonimus westermani* (diploid type) cercariae in the freshwater crab, *Geothelphusa dehaani*. *Journal of Helminthology* 65: 38–42

Shields, J.D. 1990. *Rhizophydium littoreum* on the eggs of *Cancer anthonyi*: Parasite or saprobe? *Biological Bulletin Woods Hole* 179: 201–206

Shields, J.D. 1992. The parasites and symbionts of the crab *Portunus pelagicus* from Moreton Bay, eastern Australia. *Journal of Crustacean Biology* 12: 94–100

Shields, J.D. 1994. The parasitic dinoflagellates of marine Crustacea. *Annual Review of Fish Diseases* 4: 241–271

Shields, J.D. 2012. The impact of pathogens on exploited populations of decapod crustaceans. *Journal of Invertebrate Pathology* 110: 211–224

Shields, J.D. 2013. Complex etiologies of emerging diseases in lobsters (*Homarus americanus*) from Long Island Sound. *Canadian Journal of Fisheries and Aquatic Sciences* 70: 1576–1587

Shields, J.D. 2017. Collection techniques for the analyses of pathogens in crustaceans. *Journal of Crustacean Biology* 37: 753–763

Shields, J.D. and Gómez-Gutiérrez, J. 1996. *Oculophryxus bicaulis*, a new genus and species of dajid isopod parasitic on the euphausiid *Stylocheiron affine* Hansen. *International Journal for Parasitology* 26: 261–268

Shields, J.D., Huchin-Mian, J.P., Leary, P.A., and Small, H.J. 2017. New insight into the transmission dynamics of the crustacean pathogen *Hematodinium perezi* (Dinoflagellata) using a novel sentinel methodology. *Marine Ecology Progress Series* 573: 73–84

Shields, J.D. and Kuris, A.M. 1990. *Carcinonemertes wickhami* n. sp. (Nemertea), an egg predator on the California lobster, *Panulirus interruptus. Fishery Bulletin, NOAA* 88: 279–287

Shields, J.D. and Overstreet, R.M. 2007. Parasites, symbionts, and diseases. In *The Blue Crab Callinectes sapidus*, Kennedy V. and Cronin L.E (eds.) pp. 299–417. Maryland: University of Maryland Sea Grant College, College Park

Shields, J.D., Stephens, F.J., and Jones, J.B. 2006. Pathogens, parasites and other symbionts. In *Lobsters: Biology, Management, Aquaculture and Fisheries*, Phillips, B.F. (ed.) Chapter 5, pp. 146–204. Oxford: Blackwell Scientific

Shields, J.D., Taylor, D.M., O'Keefe, P.G., Colbourne, E., and Hynick, E. 2007. Epidemiological determinants in outbreaks of bitter crab disease (*Hematodinium* sp.) in snow crabs, *Chionoecetes opilio* from Newfoundland, Canada. *Diseases of Aquatic Organisms* 77: 61–72

Shields, J.D., Taylor, D.M., Sutton, S.G., O'Keefe, P.G., Ings, D., and Party, A. 2005. Epizootiology of bitter crab disease (*Hematodinium* sp.) in snow crabs, *Chionoecetes opilio* from Newfoundland, Canada. *Diseases of Aquatic Organisms* 64: 253–264

Shields, J.D., Wickham, D.E., and Kuris, A.M. 1989. *Carcinonemertes regicides* n. sp. (Nemertea), a symbiotic egg predator from the red king crab, *Paralithodes camtschatica* (Decapoda: Anomura), in Alaska. *Canadian Journal of Zoology* 67: 923–930

Shields, J.D., Williams, J.D., and Boyko, C.B. 2015. Parasites and diseases of Brachyura. In *The Crustacea. Treatise on Zoology Treatise on Zoology-Anatomy, Taxonomy, Biology.* Vol 9, Part C, 2 vols. Castro P. et al. (eds.) pp. 639–774. Netherlands: Brill

Shields, J.D. and Wood, F.E.I. 1993. Impact of parasites on the reproduction and fecundity of the blue sand crab *Portunus pelagicus* from Moreton Bay, Australia. *Marine Ecology Progress Series* 92: 159–170

Short, S., Guler, Y., Yang, G., Kille, P., and Ford, A.T. 2012. Paramyxean–microsporidian co-infection in amphipods: Is the consensus that Microsporidia can feminise their hosts presumptive? *International Journal of Parasitology* 42: 683–691

Siddeek, M.S.M., Zheng, J., Morado, J.F., Kruse, G.H., and Bechtol, W.R. 2010. Effect of bitter crab disease on rebuilding in Alaska Tanner crab stocks. *ICES Journal of Marine Science* 67: 2027–2032

Silberman, J.D., Collins, A.G., Gershwin, L.A., Johnson, P.J., and Roger, A.J. 2004. Ellobiopsids of the genus *Thalassomyces* are alveolates. *Journal of Eukaryotic Microbiology* 51: 246–252

Skovgaard, A. 2004. Tumour-like anomalies on copepods may be wounds from parasites. *Journal of Plankton Research* 26: 1129–1131

Skovgaard, A., Karpov, S.A., and Guillou, L. 2012. The parasitic dinoflagellates *Blastodinium* spp. inhabiting the gut of marine, planktonic copepods: Morphology, ecology, and unrecognized species diversity. *Frontiers in Microbiology* 3: 305

Skovgaard, A. and Saiz, E. 2006. Seasonal occurrence and role of protistan parasites in coastal marine zooplankton. *Marine Ecology Progress Series* 327: 37–49

Sloan, N.A. 1984. Incidence and effects of parasitism by the rhizocephalan barnacle, *Briarosaccus callosus* Boschma, in the golden king crab, *Lithodes aequispina* Benedict, from deep fjords in northern British Columbia, Canada. *Journal of Experimental Marine Biology and Ecology* 84: 111–131

Sloan, N.A., Bower, S.M., and Robinson, S.M.C. 1984. Cocoon deposition on three crab species and fish parasitism by the leech *Notostomum cyclostoma* from deep fjords in northern British Columbia. *Marine Ecology Progress Series* 20: 51–58

Sluys, R. and Van Ginkel, W. 1989. New records of *Fecampia abyssicola* (Platyhelminthes: Rhabdocoela: Fecampiidae). *Cahiers de Biologie Marine* 30: 235–241

Small, H.J. 2012. Advances in our understanding of the global diversity and distribution of *Hematodinium* spp.—Significant pathogens of commercially exploited crustaceans. *Journal of Invertebrate Pathology* 110: 234–246

Small, H.J., Huchin-Mian, J.P., Reece, K.S., et al. 2019. Parasitic dinoflagellate *Hematodinium perezi* prevalence in larval and juvenile blue crabs *Callinectes sapidus* from coastal bays of Virginia. *Diseases of Aquatic Organisms* 134: 215–222

Small, H.J., Miller, T.L., Coffey, A., Delany, K.L., Schott, E., and Shields, J.D. 2013. Discovery of an opportunistic starfish pathogen, *Orchitophrya stellarum*, in captive blue crabs, *Callinectes sapidus. Journal of Invertebrate Pathology* 114: 178–185

Small, H.J., Neil, D.M., Taylor, A.C., and Coombs, G.H. 2005. Identification and partial characterisation of metalloproteases secreted by a Mesanophrys-like ciliate parasite of the Norway lobster *Nephrops norvegicus. Diseases of Aquatic Organisms* 67: 225–231

Small, H.J., Shields, J.D., Reece, K.S., Bateman, K., and Stentiford, G.D. 2012. Morphological and molecular characterization of *Hematodinium perezi* (Dinophyceae: Syndiniales), a dinoflagellate parasite of the harbour crab, *Liocarcinus depurator. Journal of Eukaryotic Microbiology* 59: 54–66

Smit, N.J., Bruce, N.L., and Hadfield, K.A. (eds.) 2019. *Parasitic Crustacea: State of Knowledge and Future Trends*, Vol. 3. Switzerland, Springer

Smith, A.L., Hirschle, L., Vogan, C.L., and Rowley, A.F. 2015. Parasitization of juvenile edible crabs (*Cancer pagurus*) by the dinoflagellate, *Hematodinium* sp.: Pathobiology, seasonality and its potential effects on commercial fisheries. *Parasitology* 142: 428–438

Smith, J.W. and Wootten, R. 1978. Anisakis and Anisakiasis. *Advances in Parasitology* 16: 93–163

Southward, A.J. 1950. On the occurrence in the Isle of Man of *Fecampia erythrocephala* Giard, a platyhelminth parasite of crabs. *Marine Biology Station of Port Erin, Isle of Man Annual Report*, pp. 27–10

Sparks, A.K. and Hibbits, J. 1981. A trematode metacercaria encysted in the nerves of the Dungeness crab, *Cancer magister*. *Journal of Invertebrate Pathology* 38: 88–93

Sprague, V. 1963. *Minchinia louisiana* n. sp. (Haplosporidia, Haplosporidiidae), a parasite of *Panopeus herbstii*. *Journal of Eukaryotic Microbiology* 10: 267–274

Sprague, V. and Beckett, R.L. 1966. A disease of blue crabs (*Callinectes sapidus*) in Maryland and Virginia. *Journal of Invertebrate Pathology* 8: 287–289

Sprague, V., Beckett, R.L., and Sawyer, T.K. 1969. A new species of *Paramoeba* (Amoebida, Paramoebidae) parasitic in the crab *Callinectes sapidus*. *Journal of Invertebrate Pathology* 14: 167–174

Sprague, V. and Couch, J. 1971. An annotated list of protozoan parasites, hyperparasites, and commensals of decapod Crustacea. *Journal of Protozoology* 18: 526–537

Stentiford, G.D., Bateman, K.S., Stokes, N.A., and Carnegie, R.B. 2013. *Haplosporidium littoralis* sp. nov.: A crustacean pathogen within the Haplosporida (Cercozoa, Ascetosporea). *Diseases of Aquatic Organisms* 105: 243–252

Stentiford, G.D. and Feist, S.W. 2005. A histopathological survey of shore crab (*Carcinus maenas*) and brown shrimp (*Crangon crangon*) from six estuaries in the United Kingdom. *Journal of Invertebrate Pathology* 88: 136–146

Stentiford, G.D., Feist, S.W., Stone, D.M., Bateman, K.S., and Dunn, A.M. 2013. Microsporidia: Diverse, dynamic, and emergent pathogens in aquatic systems. *Trends in Parasitology* 29: 567–578

Stentiford, G.D., Neil, D.M., Albalat, A., Milligan, R.J., and Bailey, N. 2015. The effect of parasitic infection by *Hematodinium* sp. on escape swimming and subsequent recovery in the Norway lobster, *Nephrops norvegicus* (L.). *Journal of Crustacean Biology* 35: 1–10

Stentiford, G.D., Neil, D.M., and Atkinson, R.J.A. 2001. The relationship of *Hematodinium* infection prevalence in a Scottish *Nephrops norvegicus* population to seasonality, moulting and sex. *ICES Journal of Marine Science* 58: 814–823

Stentiford, G.D., Neil, D.M., Peeler, E.J. et al. 2012. Disease will limit future food supply from the global crustacean fishery and aquaculture sectors. *Journal of Invertebrate Pathology* 110: 141–157

Stentiford, G.D. and Shields, J.D. 2005. A review of the parasitic dinoflagellates *Hematodinium* species and *Hematodinium*-like infections in marine crustaceans. *Diseases of Aquatic Organisms* 66: 47–70

Strassert, J.F., Karnkowska, A., Hehenberger, E., et al. 2018. Single cell genomics of uncultured marine alveolates shows paraphyly of basal dinoflagellates. *The ISME Journal* 12: 304–308

Stunkard, H.W. 1957. The morphology and life-history of the digenetic trematode, *Microphallus similis* (Jägerskiöld, 1900) Baer, 1943. *Biological Bulletin Woods Hole* 112: 254–266

Sudhaus, W. 1974. Zur Systematik, Verbreitung, Okologie und Biologie neuer und wenig bekannter Rhabditiden (Nematoda) 2. Teil. *Zoologische Jahrbücher. Abteilung für Systematik, Geographie und Biologie der Tiere* 101: 417–465

Sudhaus, W. 1986. Matthesonema eremitum n. sp. (Nematoda, Rhabditida) associated with hermit crabs (Coenobita) from the Philippines and its phylogenetic implications. *Nematologica* 32: 247–255

Symonds, D.J. 1972. Infestation of *Nephrops norvegicus* (L.) by *Stichocotyle nephropis* Cunningham in British waters. *Journal of Natural History* 6: 423–426

Tayeh, A., Cairncross, S. and Cox, F.E., 2017. Guinea worm: From Robert Leiper to eradication. *Parasitology* 144: 1643–1648

Taylor, D.M. and Khan, R.A. 1995. Observations on the occurrence of *Hematodinium* sp. (Dinoflagellata: Syndinidae): The causative agent of Bitter Crab Disease in the Newfoundland snow crab (*Chionoecetes opilio*). *Journal of Invertebrate Pathology* 65: 283–288

Thompson, A.B. 1985. Analysis of *Profilicollis botulus* (Acanthocephala: Echinorhynchidae) burdens in the shore crab, *Carcinus maenas*. *Journal of Animal Ecology* 54: 595–604

Thompson, R.M., Mouritsen, K.N., and Poulin, R. 2005. Importance of parasites and their life cycle characteristics in determining the structure of a large marine food web. *Journal of Animal Ecology* 74: 77–85

Threlfall, W. 1968. Note on metacercariae of *Spelotrema excellens* Nicoll in *Carcinus maenas* (L.). *Journal of Experimental Marine Biology and Ecology* 2: 154–155

Thrupp, T.J., Lynch, S.A., Wootton, E.C. et al. 2013. Infection of juvenile edible crabs, *Cancer pagurus* by a haplosporidian-like parasite. *Journal of Invertebrate Pathology* 114: 92–99

Trilles, J.-P. 1999. Ordre des isopodes sous-ordre des épicarides (Epicaridea Latreille, 1825). In *Traité de Zoologie. Anatomie, Systématique, Biologie (Pierre-P. Grassé). Tome VII, Fascicule III A, Crustacés Péracarides*, Forest, J. (ed.) 19: 279–352. Monaco: *Bulletin Institute Océanographie*

Troedsson, C., Lee, R.F., Walters, T., et al. 2008. Detection and discovery of crustacean parasites in blue crabs

(*Callinectes sapidus*) by using 18S rRNA gene-targeted denaturing high-performance liquid chromatography. *Applied and Environmental Microbiology* 74: 4346–4353

Tuckey, T.D., Swinford, J., Fabrizio, M.C., Small, H.J., and Shields, J.D. 2021 Penaeid shrimp in Chesapeake Bay: Population growth and black gill disease syndrome. *Marine and Coastal Fisheries* 13: 159-173.

Urrutia, A., Bass, D., Ward, G., et al. 2019. Ultrastructure, phylogeny and histopathology of two novel haplosporidians parasitising amphipods, and importance of crustaceans as hosts. *Diseases of Aquatic Organisms* 136: 87–103

Uspenskaya, A.V. 1953. Life cycle of nematodes of the genus *Ascarophis* Van Beneden (Nematodes–Spirurata). *Zoologicheskii Zhurnal* 32: 828–832 [not seen by author]

Utari, H.B., Senapin, S., Jaengsanong, C., Flegel, T.W., and Kruatrachue, M. 2012. A haplosporidian parasite associated with high mortality and slow growth in *Penaeus* (Litopenaeus) *vannamei* cultured in Indonesia. *Aquaculture* 366: 85–89

Uzmann, J. R. 1967. Juvenile *Ascarophis* (Nematoda: Spiruroidea) in the American lobster, *Homarus americanus*. *Journal of Parasitology* 53: 218

Vader, W. and Tandberg, A.H.S. 2015. Amphipods as associates of other Crustacea: A survey. *Journal of Crustacean Biology* 35: 522–532

Vasileva, G., Redón, S., Amat, F., et al. 2009. Records of cysticercoids of *Fimbriarioides tadornae* Maksimova, 1976 and *Branchiopodataenia gvozdevi* (Maksimova, 1988) (Cyclophyllidea, Hymenolepididae) from brine shrimps at the Mediterranean coasts of Spain and France, with a key to cestodes from *Artemia* spp. from the Western Mediterranean. *Acta Parasitologica* 54: 143–150

Veillet, A. 1945. Recherches sur le parasitisme des crabes et des galathées par les rhizocéphales et les épicarides. *Annales de l'Institut Océanographique, (Paris)* 22: 193–341

Vieira, J.C., Blankespoor, H.D., Cooper, P.J., and Guderian, R.H. 1992. Paragonimiasis in Ecuador: Prevalence and geographical distribution of parasitisation of second intermediate hosts with *Paragonimus mexicanus* in Esmeraldas province. *Tropical Medicine and Parasitology* 43: 249–252

Violante-González, J., Monks, S., Quiterio-Rendon, G., García-Ibáñez, S., Larumbe-Morán, E. and Rojas-Herrera, A.A. 2016. Life on the beach for a sand crab (*Emerita rathbunae*) (Decapoda, Hippidae): Parasite-Induced mortality of females in populations of the Pacific sand crab caused by *Microphallus nicolli* (Microphallidae). *Zoosystematics and Evolution* 92: 153

Visse, M. 2007. Detrimental effect of peritrich ciliates (*Epistylis* sp.) as epibionts on the survival of the copepod

Acartia bifilosa. *Proceedings of the Estonian Academy of Sciences. Biology Ecology* 56: 173–178

Voris, H.K. and Jeffries, W.B. 2001. Distribution and size of a stalked barnacle (*Octolasmis muelleri*) on the blue crab, *Callinectes sapidus*. *Bulletin of Marine Science* 68: 181–190

Waiho, K., Glenner, H., Miroliubov, A. et al. 2021. Rhizocephalans and their potential impact on crustacean aquaculture. *Aquaculture* 531: 735876

Walker, G. 1974. The occurrence, distribution and attachment of the pedunculate barnacle *Octolasmis mülleri* (coker) on the gills of crabs, particularly the blue crab *Callinectes sapidus* Rathbun. *Biological Bulletin Woods Hole* 147: 678–689

Walker, G., Clare, A.S., Rittschof, D., and Mensching, D. 1992. Aspects of the life cycle of *Loxothylacus panopaei* (Gissler), a sacculinid parasite of the mud crab *Rhithropanopeus harrisii* (Gould): A laboratory study. *Journal of Experimental Marine Biology and Ecology* 157: 181–193

Walker, G. and Lester, R.J.G. 1998. Effect of salinity on development of larvae of *Heterosaccus lunatus* (Cirripedia: Rhizocephala). *Journal of Crustacean Biology* 18: 650–655

Ward, G.M., Bennett, M., Bateman, K. et al. 2016. A new phylogeny and environmental DNA insight into paramyxids: An increasingly important but enigmatic clade of protistan parasites of marine invertebrates. *International Journal for Parasitology* 46: 605–619

Ward, G.M., Neuhauser, S., Groben, R. et al. 2018. Environmental sequencing fills the gap between parasitic haplosporidians and free-living giant amoebae. *Journal of Eukaryotic Microbiology* 65: 574–586

Warren, M.B., Dutton, H.R., Whelan, N.V., Yanong, R.P., and Bullard, S.A. 2019. First record of a species of Mermithidae Braun, 1883 infecting a decapod, *Palaemon paludosus* (Palaemonidae). *Journal of Parasitology* 105: 237–247

Werner, M. 2001. Prevalence of the parasite *Sacculina carcini* Thompson 1836 (Crustacea, Rhizocephala) on its host crab *Carcinus maenas* (L.) on the west coast of Sweden. *Ophelia* 55: 101–110

Wiackowski, K., Hryniewiecka-Szyfter, Z. and Babula, A. 1999. How many species are in the genus *Mesanophrys* (Protista, Ciliophora, facultative parasites of marine crustaceans)? *European Journal of Protistology* 35: 379–389

Wickham, D.E., Roe, P., and Kuris, A.M. 1984. Transfer of nemertean egg predators during host molting and copulation. *Biological Bulletin Woods Hole* 167: 331–338

Williams, C.M., Poulin, R., and Sinclair, B.J. 2004. Increased haemolymph osmolality suggests a new route for behavioural manipulation of *Talorchestia quoyana* (Amphipoda: Talitridae) by its mermithid parasite. *Functional Ecology* 18: 685–691

Williams, J. D. and Boyko, C.B. 2012. The global diversity of parasitic isopods associated with crustacean hosts (Isopoda: Bopyroidea and Cryptoniscoidea). *PLoS One* 7: e35350

Williams, J.D. and McDermott, J.J. 2004. Hermit crab biocoenoses: A worldwide review of the diversity and natural history of hermit crab associates. *Journal of Experimental Marine Biology and Ecology* 305: 1–128

Williams, R. and Moyse, J. 1988. Occurrence, distribution, and orientation of *Poecilasma kaempferi* Darwin (Cirripedia: Pedunculata) epizoic on *Neolithodes grimaldi* Milne-Edwards and Bouvier (Decapoda: Anomura) in the northeast Atlantic. *Journal of Crustacean Biology* 8: 177–186

Wing, B.L. 1975. New records of Ellobiopsidae (Protista (Incertae sedis)) from the North Pacific with a description of *Thalassomyces albatrossi* n. sp., a parasite of the mysid *Stilomysis major*. *Fishery Bulletin* 73: 169–185

Winters, A.D. and Faisal, M. 2014. Molecular and ultrastructural characterization of *Haplosporidium diporeiae* n. sp., a parasite of *Diporeia* sp. (Amphipoda, Gammaridea) in the Laurentian Great Lakes (USA). *Parasites & Vectors* 7: 343

Wong, P.L. and Anderson, R.C. 1990. *Ancyracanthopsis winegardi* n. sp. (Nematoda: Acuarioidea) from *Pluvialis squatarola* (Aves: Charadriidae) and *Ancyracanthus heardi* n. sp. from *Rallus longirostris* (Aves: Rallidae), and a review of the genus. *Canadian Journal of Zoology* 68: 1297–1306

Wong, P.L., Anderson, R.C., and Bartlett, C.M. 1989. Development of *Skrjabinoclava inornatae* (Nematoda: Acuarioidea) in fiddler crabs (*Uca* spp.) (Crustacea) and western willets (*Catoptrophorus semipalmatus inornatus*) (Aves: Scolopacidae). *Canadian Journal of Zoology* 67: 2893–2901

Wootton, E.C., Pope, E.C., Vogan, C.L., et al. 2011. Morphology and pathology of the ectoparasitic copepod, *Nicothoe astaci* ('lobster louse') in the European lobster, *Homarus gammarus*. *Parasitology* 138: 1285

WoRMS Editorial Board 2020. World Register of Marine Species. Available from http://www.marinespecies.or gat VLIZ. [Accessed several times in 2020.] doi:10. 14284/170

Young, N.D., Dyková, I., Crosbie, P.B., et al. 2014. Support for the coevolution of *Neoparamoeba* and their endosymbionts, Perkinsela amoebae-like organisms. *European Journal of Protistology* 50: 509–523

Zara, F.J., Diogo Reigada, A.L., Domingues Passero, L.F., and Toyama, M.H. 2009. *Myzobdella platensis* (Hirundinida: Piscicolidae) is true parasite of blue crabs (Crustacea: Portunidae). *Journal of Parasitology* 95: 124–128

Zawal, A. 1998. Water mites (Hydracarina) in the branchial cavity of crayfish *Orconectes limosus* (Raf. 1817). *Acta Hydrobiologica-Polish Academy of Sciences* 40: 49–54

Zerbib, C., Andrieux, N., and Berreur-Bonnenfant, J. 1975. Données préliminaires sur l'ultrastructure de la glande de mue (organe Y) chez le crabe *Carcinus mediterraneus* sain et parasité par *Sacculina carcini*. *Comptes rendus de l'Académie des Sciences* 281: 1167–1169

Zetlmeisl, C., Hermann, J., Petney, T., Glenner, H., Griffiths, C., and Taraschewski, H. 2011. Parasites of the shore crab *Carcinus maenas* (L.): Implications for reproductive potential and invasion success. *Parasitology* 138: 394–401

Diseases of Deuterostomes

CHAPTER 18

Echinoderm diseases and pathologies

L. Courtney Smith, S. Anne Boettger, Maria Byrne, Andreas Heyland, Diana
L. Lipscomb, Audrey J. Majeske, Jonathan P. Rast, Nicholas W. Schuh,
Linsheng Song, Ghada Tafesh-Edwards, Lingling Wang, Zhuang Xue, and
Zichao Yu.

18.1 Introduction

18.1.1 Anatomy and characteristics of echinoderms

The appearance of animals in the fossil record that
are recognisable as echinoderms indicates that the
phylum is at least 570 million years old (Pisani et al.
2015) and is currently comprised of more than 7,300
recognised extant species (and about 15,000 fossil
species) that inhabit all ranges of marine systems
(Appletans et al. 2012; Mooi 2016; Giribet and Edgecombe 2020). The echinoderms and the hemichordates together form the ambulacrarians, which positions them as the major basal clade of the deuterostome lineage and sister to the chordates (Blair
and Hedges 2005; Schlegel et al. 2014). The extant
classes of echinoderms include the Echinoidea (sea
urchins and sand dollars), Ophiuroidea (brittle
stars), Holothuroidea (sea cucumbers), Asteroidea
(sea stars and sea daisies), and Crinoidea (sea lilies
and feather stars) as the basal group, plus several extinct classes (Hyman 1955). Some echinoderms
are keystone predators (many sea stars), some are
omnivores or herbivores (many sea urchins), some
are planktivores (many crinoids and brittle stars),
and others feed on or burrow through the sediment
(many sea cucumbers, heart urchins). Adult echinoderms are typically benthic with radially pen-
tameric symmetry, although multi-armed species

can deviate from this form, whereas most larvae
are bilateral and planktonic. The endoskeleton is
composed of either fitted plates or dermal ossicles
depending on the class of echinoderm. The water
vascular system is a character that is specific to
echinoderms. It is a set of fluid filled tubes and
reservoirs connected to the madreporite on the body
surface by the stone canal, or to one or more internal madreporites in the case of holothurians. The
system functions by hydraulics to extend and control the movement of the papillae and tube feet that
typically terminate with a suction cup and function in locomotion and attachment for many species
of adults. Larval motility is typically passive and
based on water motion and currents but larvae have
surface cilia including coordinated ciliary bands
that facilitate feeding and swimming. The internal coelomic spaces are filled with coelomic fluid
in adults, or blastocoelar fluid in larvae, and are
open systems without a heart or pumping structure. Most echinoderms are broadcast spawners
where species form large aggregations of adults
with external fertilisation of gametes in the water
column. Embryonic development for most echinoderms proceeds from a fertilised egg and cleavage through gastrulation as in all deuterostomes
and culminates in a planktotrophic larval stage that
serves in dispersal (Smith 1997; McEdward and
Miner 2001; McClay 2011). Some species produce

L.C. Smith et al., *Echinoderm diseases and pathologies*. In: *Invertebrate Pathology*. Edited by Andrew F. Rowley, Christopher J. Coates and Miranda M.A. Whitten,
Oxford University Press. © Oxford University Press (2022). DOI: 10.1093/oso/9780198853756.003.0018

large eggs that are highly provisioned with nutrients to support direct lecithotrophic development to a juvenile, skipping the larval phase (McEdward and Miner 2001; Byrne 2006; Raff and Byrne 2006). When mature planktonic larvae are competent to metamorphose they attach to a benthic surface with one or more tube feet or attachement complex, and metamorphose into a juvenile, although some asteroids and ophiuroids metamorphose in the plankton (Pearse and Cameron 1991; Smiley et al. 1991; Young and Sewell 2002; Sonnenholzner-Varas et al. 2018).

18.1.2 Allorejection established the echinoderm immune system as innate

The question of whether invertebrates have immune systems has been addressed through the application of allograft rejection assays. This approach was moderately straight forward, although technically challenging, and demonstrated that both sea urchins (Coffaro and Hinegardner 1977; Coffaro 1980) and sea stars (Karp and Hildemann 1976; Hildemann and Dix 1972) could reject allografts and accept autografts but that the response had no specific immunological memory (Smith and Davidson 1992). Although echinoderms are very efficient at clearing inert particles and foreign cells that are injected into the coelomic cavity or by encapsulating large particles demonstrated in short term cultures (Johnson 1969; Reinisch and Bang 1971; Coffaro 1978; Bertheussen 1981a,b; Yui and Bayne 1983; Dybas and Fankboner 1986; Plytycz and Seljelid 1993), they do not clear particles *in vivo* more efficiently or quickly upon second challenge (Smith and Davidson 1994). In general, immune responses in echinoderms are strictly innate for which these phenomenological data were confirmed by the annotation of the immune genes in the sequenced genome of the purple sea urchin, *Strongylocentrotus purpuratus*, and other echinoderms (Hibino et al. 2006; Sodergren et al. 2006; Kinjo et al. 2018; Cary et al. 2019; Nesbit et al. 2019; Jo et al. 2017; Long et al. 2016; Hall et al. 2017, see Table S13 in Zhang, Sun et al. 2017, and see http://www.echinobase.org and http://www.genedatabase.cn/aja_genome_20161129.html for genomic data on a range of echinoderm species).

18.1.3 The echinoderm immune system

18.1.3.1 The sea urchin immune system

The echinoderm innate immune systems are mediated by a variety of coelomocytes (see Section 18.1.4; Box 1.2 and Figure 1.6 in Chapter 1) that function through pathogen recognition receptors (PRRs) and associated cell signalling pathways, anti-microbial peptides and proteins (AMPs) in the coelomic fluid, and a melanisation pathway (Smith and Söderhall 1991; Smith et al. 2010; Franco et al. 2011; Smith et al. 2018). Many of the sea urchin PRRs are conserved among many animals (Bulgakov et al. 2013; Gowda et al. 2013) and include Toll-like receptors (TLRs), NOD-like receptors (NLRs), scavenger receptors that are cysteine rich (SRCRs), peptidoglycan recognition proteins (PGRPs), Gram-negative binding proteins (GNBPs), cytokine signalling molecules, plus proteins that are involved in the signalling pathways for PRRs (Pancer 2000; Hibino et al. 2006; Buckley and Rast 2012; Furukawa et al. 2016; Buckley et al. 2017). Anti-pathogen effector mechanisms include multiple AMPs, the diverse SpTransformer proteins, a wide variety of lectins, a complement pathway, a prophenoloxidase cascade, and other molecules (Smith et al. 2001; Hibino et al. 2006; Roch et al. 1992; Li et al. 2015; Smith et al. 2010; Buckley and Rast 2012; Dheilly et al. 2013; Smith and Lun 2017). Many of the genes encoding the PRRs in the purple sea urchin, *Strongylocentrotus purpuratus*, are greatly expanded including 253 *TLR* genes and 203 *NLR* genes (Hibino et al. 2006; Buckley and Rast 2012) in keeping with the expanded numbers of potential pathogens and opportunists in the marine system with which the host species are in constant contact. Other genes suggest overlap with vertebrate adaptive immunity, including homologues of *RAG1* and *RAG2* genes, which in vertebrates are involved in diversification of recognition capabilities of lymphocytes and response to pathogens and diseases (Fugmann et al. 2006; Carmona et al. 2016). Homologous genes encoding DNA repair enzymes including a terminal deoxynucleotidal transferase and an AID/APOBEC-like cytidine deaminase are also present in the genome of the purple sea urchin *S. purpuratus* (Hibino et al. 2006; Liu et al. 2018). Although *RAG* and *AID* gene homologues are present in echinoderm genomes, these animals do not have the equivalent of vertebrate lymphocytes, and their genomes do not encode split genes that might undergo somatic recombination as in immunoglobulin family genes in higher vertebrates or assembly of genes as in the variable lymphocyte receptor genes in agnathans. On the other hand, the expanded complexity of the innate immune system in echinoderms (Smith et al. 2018) stands in striking contrast to the innate system in vertebrates that has less complexity based on, for example, fewer numbers of *TLRs* and other genes encoding PRRs. As in all animals, the key functions of the immune cells in echinoderms are phagocytosis or encapsulation, as first reported by Ilya Metchnikoff (1883) for larval sea stars, destruction of phagocytosed non-self-pathogens and particles, and secretion of anti-pathogen molecules (Li et al. 2015) including homologues of the complement cascade that have a key function of opsonisation (Smith et al. 2001). The sea urchin immune system is complex and robust, and when the animal is not impacted by environmental stressors, it is very effective in protecting the host against infection.

18.1.3.2 The sea cucumber immune system

Sea cucumbers also rely on an innate immune system for defence against invading pathogens, which is similar in many ways to the sea urchin system. PRRs activate conserved host defence signalling pathways that control the expression of a variety of immune response genes encoding anti-pathogen proteins. Typical PRRs that have been documented in sea cucumbers include lectins, lipopolysaccharide-binding protein/bactericidal permeability-increasing protein (LBP/BPI), SRCRs, fibrinogen-like proteins, and TLRs (Hatakeyama et al. 1994; Bulgakov et al. 2000; Pancer 2000; Mojica and Merca 2004; Tomomitsu et al. 2004; Mojica and Merca 2005; Takuya et al. 2005; Gowda et al. 2008; Yang et al. 2009; Jiang et al. 2010; Sun et al. 2013; Dong et al. 2014; Shao et al. 2015; Xue, Li et al. 2015; Che et al. 2019). Components of the TLR signalling pathway in the edible sea cucumber, *Apostichopus japonicus*, include two TLR homologues, *Aj*TLR3 and *Aj*Toll (Sun et al. 2013), homologues of MyD88 and TRAF6 (Lu et al. 2013), and homologues of NFκB, *Aj*-rel, and *Aj*-p105 (Wang, Sun et al. 2013). Immune challenge of sea

cucumbers with LPS or *Vibrio splendidus* results in significant up-regulation of genes encoding proteins involved in TLR signalling suggesting, as in other animals (Akira and Takeda 2004), that this pathway in *A. japonicus* is involved in the innate immune defence response to invading bacteria (Dong et al. 2014). The innate immune system in the sea cucumber is similar to that in the sea urchin, and both are effective in host protection.

18.1.4 Coelomocytes and blastocoelar cells mediate echinoderm immune functions

Coelomocytes are cells in the coelomic fluid of adult echinoderms and blastocoelar cells are located in the blastocoel of larval stages; both set of cells mediate the immune response (Endean 1996; Chia and Xing 1996; Smith et al. 2010; Smith et al. 2018, see Figure 1.6 in Chapter 1). There are a variety of coelomocytes that have been described in adult sea urchins, sea stars, and sea cucumbers (Hyman 1955; Smith 1981; Ramírez-Gómez et al. 2010; Smith et al. 2018), and several blastocoelar cells described in larvae including some that localise to the ectoderm (Ho et al. 2016) (see Section 1.5.1, Box 1.2 and Figure 1.6 in Chapter 1). Phagocytes are the most abundant cells in coelomic fluid, with a variable range in sea urchins that is dependent on the species, the individual animal, and its immunological status. Polygonal and discoidal phagocytes in sea urchins are named for differences in their cytoskeletal morphology (Edds 1993; Henson et al. 1999), whereas the medium (Golconda et al. 2019) and small phagocytes (Gross et al. 2000; Brockton et al. 2008) are smaller and vary in number depending on the immunological status of the sea urchin. Red spherule cells in sea urchins contain echinochrome A and other naphthoquinones that fill the large vesicles or spherules (Coates et al. 2018; Hira et al. 2020). Colourless spherule cells also have large vesicles but do not contain echinochrome A, and these cells are a minor population of the coelomocytes in sea urchins. Vibratile cells are spherical, contain large vesicles, and are motile based on a single flagellum. Hemocytes are large cells in some species of sea cucumbers and sea stars (see Figure 1.6 in Chapter 1). Progenitor cells are small with a large nucleus, have morphology similar to naïve vertebrate lymphocytes, and are speculated to be stem cells (Ramírez-Gómez et al. 2010). Crystal cells are rare in sea cucumbers and are distinctive because of a large central vacuole that may contain a small rhomboid crystal (Ramírez-Gómez et al. 2010). The pigment cells of larval sea urchins are located in the ectoderm and are similar to the adult red spherule cells because both have large granules that contain echinochrome A and both cell types migrate to sites of infection or injury (Coffaro and Hinegardner 1977; Ho et al. 2016; Buckley et al. 2017; Schuh et al. 2020; See Section 1.5.1, Box 1.2 and Figure 1.6 in Chapter 1). Additional larval blastocoelar cells include globular cells, filopodial cells, and ovoid cells (see Figure 1.6 in Chapter 1).

Coelomocytes in adult sea urchins have a broad range of immune functions. These activities include phagocytosis of foreign particles (Smith and Davidson 1994), walling off foreign cells (Johnson 1969), encapsulation and syncytia formation (Majeske et al. 2013), secretion of clotting factors to activate clotting of the coelomic fluid (Hillier and Vacquier 2003; D'Andrea-Winslow et al. 2012), and changes in large phagocyte morphology from lamellipodial to filopodial that promote cellular clot formation (Edds 1977; Edds 1980, 1985; Henson and Schatten 1983). Coelomocytes secrete or exocytose a range of antimicrobial molecules including echinochrome A (Service and Wardlaw 1984; Coates et al. 2018; Hira et al. 2020), AMPs (Li et al. 2015), anti-bacterial proteins (Gerardi et al. 1990; Smith and Lun 2017) and produce lysozyme, reactive oxygen species, prophenyloxidase (Roch et al. 1992) and hydrogen peroxide (Ito et al. 1992). When acting together, the colourless spherule cells and large phagocytes show cytotoxic activity (Arizza et al. 2007). Immune cells in larval sea urchins act similarly to those in adults by responding quickly to pathogen invasion. They secrete echinochrome A from the pigment cells and secrete a number of immune proteins from the blastocoelar cells including, a thioester-containing protein in addition to complement proteins, scavenger receptors, Macpf that is a perforin-like homologue (Ho et al. 2016), SpTransformer proteins that show anti-bacterial activity (Smith and Lun, 2017), among others (see Smith et al. 2018). The combination of coelomocyte activity and the molecules they secrete into the coelomic fluid or blastocoelar

fluid results in a very effective immune system in both phases of the life history that protects echinoderms from a myriad of pathogens that are always present under normal marine conditions (Smith et al. 2018).

Immune function in sea cucumbers is similar to that in sea urchins and also relies on AMPs (Tan et al. 2012; Schillaci et al. 2013; Cusimano et al. 2019) and several lysozymes (Yang, Cong et al. 2007; Cong et al. 2009; Wang, Xu et al. 2011) with activity against various pathogens, including Gram-positive and Gram-negative bacteria. The complement system is also an important aspect of innate immunity and participates in recognition and defence against pathogens (Ramírez-Gómez and García-Arrarás 2010; Chen et al. 2015; Zhong et al. 2015). Phenoloxidase ($AjPO$) in the coelomic fluid is another crucial component of immune responses in sea cucumbers, and is a member of the copper containing metalloenzyme family of proteins (Jiang, Zhou et al. 2014a; 2014b). $AjPO$ mRNA increases and the activity of $AjPO$ in coelomocytes is altered significantly in response to challenge with LPS, peptidoglycan, zymosan A, and polyI:C, which implies that $AjPO$ is involved in the defence against the infection from bacteria and viruses. Complement components in *Apostichopus japonicus* include homologues of C3, $AjC3$, $AjC3–2$, and $AjC4$ that show significant increases in gene expression in response to challenge with LPS (Zhang et al. 2007; Zhou et al. 2011), and are associated with haemolytic activity *in vivo* and opsonisation and augmented phagocytosis, similar to reports in sea urchins (Clow et al. 2004). Phagocytosis of pathogens is the predominant activity of the cellular immune response of sea cucumbers (Eliseikina and Magarlamov 2002; Ramírez-Gómez et al. 2010, Zhang, Li et al. 2014). β integrin on coelomocytes from *A. japonicus* binds LPS and promotes phagocytosis through increased activities of GTPases that are required for the process of phagocytosis (Wang, Lv et al. 2018). Phagocytosis of *Vibrio splendidus* by coelomocytes involves the TLR signalling pathway based on the interleukin-1 receptor-associated kinase ($AjIRAK$-1) and its regulation by miRNA-133 (Lu et al. 2015).

Echinoderms in general have complex innate immune systems with multiple layers of functions for recognising and responding to pathogens. However, microbial success in invading and establishing infections can be based on interference with the immune response. For example, the phagocytic capacity of coelomocytes and the associated production of reactive oxygen species (ROS) are significantly reduced in sea cucumbers afflicted with some diseases (Song 2005) (see Section 18.5.1). The immune system of echinoderms is effective, multifaceted, and functions through both cellular and humoral components. The coelomocytes are key components of host protection and are capable of recognising, phagocytosing, and neutralising pathogens. Humoral immunity is mediated by a wide variety of secreted molecules in the coelomic fluid of adults or blastocoelar fluid of larvae and have overlapping activities with essential functions in defence against infection, including opsonisation and aggregation of invading pathogens. Continued investigations of echinoderm innate immunology will improve our understanding of the circumstances under which echinoderms become infected, whether the immune system and coelomocyte function may be overwhelmed leading to death, or whether they mount an effective immune response and survive and return to health.

18.2 Echinoderm pathology

Echinoderms are host to a range of diseases (Jangoux 1987a, 1987b, 1987c), although most are sub-lethal and are evident only in a subset of animals within populations. While the health of individual diseased animals can be compromised, they may recover, and the disease may not spread to conspecifics. Alternatively, in cases of highly infectious pathogens, disease can spread over vast distances via water currents resulting in local extirpation, population declines, or mass die-offs of species throughout their range (Dungan et al. 1982; Lessios 2005; Hewson et al. 2014; Lessios 2016; Harvell et al. 2019) (see also Chapter 19). A large number of microorganisms are associated with echinoids that pose either as potentially pathogenic or as opportunistic symbionts (Turton and Wardlaw 1987; Jangoux 1987a; 1987b, 1987c, 1990; Stien and

Halvorsen 1998; Gudenkauf et al. 2014). Although only a small portion of potential disease-causing microorganisms has been linked to a specific disease phenotype, this chapter is an overview of the disease symptoms of echinoderms and the breadth of pathogens that infect them. Little is known about fungal infections of echinoderms and this topic is not addressed here. The ecological outcomes of echinoderm mass die-offs, which can be significant, are addressed in Chapter 19.

18.3 Bald sea urchin disease

Bald sea urchin disease (BSUD) is named for its gross and histological disease aetiology (Figure 18.1; Johnson 1971; Maes and Jangoux 1984a, 1984b). A major characteristic of this disease is conspicuous focal lesions that appear as blackish red/purple spots on the animal surface followed by the loss of spines and other appendages from the infected regions leading to the loss of epidermal tissue in more advanced stages of the disease (Figure 18.1a); Maes and Jangoux 1984a; Jangoux 1987a; Roberts-Regan et al. 1988; Shimizu et al. 1995; Bauer and Young 2000; Becker et al. 2008). Muscle fibers in the tube feet, ampullae, and at the base of the spines become infected and disintegrate (Shimizu et al. 1995). The location of lesions occur on the oral and lateral surfaces and generally have equal distribution across ambulacra and interambulacral regions, although in some cases can be limited to one or the other area (Johnson 1971; Roberts-Regan et al. 1988; personal observation M.A Barela Hudgell). In advanced cases of the disease and in cases involving particularly virulent pathogens, the lesions may lead to ulcers on the test plate (endoskeleton or stereom) or expand and create holes in the test that result in the death of the affected animal within a few days of disease onset (Maes and Jangoux 1984a, 1984b, Jangoux 1987a; Jangoux 1990; Shimizu et al. 1995). In comparison, spotting disease (also called red spotting disease) in echinoids has different initial symptoms with discrete surface lesions that are surrounded by blackened tissue and loss of all appendages and the epidermis within the lesion (Figure 18.1b; Shimizu et al. 1995; Tajima et al 1997a; Wang, Li et al. 2006; Roberts-Regan et al. 1988, Zhang, Lv et al. 2019;

Li et al. 2020). However, in some cases spotting disease infection can spread to some of the animal surface or the entire animal and therefore has the same appearance as BSUD (Figure 18.1c; Li et al. 2020). BSUD and spotting disease may be two entirely different diseases that can infect an animal at same time (personal observation M.A. Barela Hudgell), which causes confusion in descriptions of these diseases that remains impossible to untangle in the literature. In general, BSUD is believed to be a communicable bacterial disease reported to affect 26 species of echinoids that have a worldwide distribution in both temperate and tropical marine waters and can lead to mass die-offs (see Chapter 19). It may actually be a number of diseases that all show the same symptoms.

18.3.1 Gross pathology of bald sea urchin disease

Disease development and progression of BSUD occurs in distinct steps (Johnson 1971; Maes and Jangoux 1984a; Maes et al. 1986; Jangoux 1987a). First, a discolouration of the epidermis appears at the base of the spines, along the basal regions of the tube feet, and with the tubercles that surround the mouth. The discolourations range from pinkish or green (Maes and Jangoux 1984b; Maes et al. 1986; Jangoux 1987a; Roberts-Regan et al. 1988, unpublished observation, A. Boettger) to bluish-green/olive (Mortensen and Rosenvinge 1934) or a dark reddish-purple (Shimizu et al. 1995). When the disease is induced experimentally, the development of lesions often occurs more slowly (Maes and Jangoux 1984a) or requires mechanical abrasion of the epidermis and test for infection to become established (Roberts-Regan et al. 1988). Differences in epidermal discolouration have been associated with differences in natural sea urchin pigmentation, microbial populations associated with the animal surface, and varying concentrations of red spherule cells among species (Roberts-Regan et al. 1988). However, it is feasible that these initial indications of infection may represent a combination of all factors. During the second phase of the disease, the spines and other appendages are lost (Shimizu et al. 1995; Bauer and Young 2000). Tube feet and pedicellariae atrophy become non-functional, and eventually separate from the body surface. Tube feet also display disease development by loss of their apical suction cup or the appearance of holes in those disks and the ampullae that are located inside the test at the base of the tube feet that create openings in the water-vascular system (Shimizu et al. 1995). This not only results in leakage and failure of the hydraulic functions of the water vascular system but may also be an avenue to transport the causative agent(s) throughout the internal regions of the animal. The loss of surface appendages initiates the development of small round, oval, or irregularly shaped lesions of one to a few millimeters in diameter that occasionally cover a quarter to a third of the entire body surface. This stage of disease progression also includes necrotic changes to the underlying epidermal tissue and intensified lesion discolouration that appear darker and may develop black central areas that are surrounded by a lighter peripheral ring (Roberts-Regan et al. 1988; Becker at al. 2008). In the third phase, the epidermis and superficial dermal tissue are lost, which results in entirely denuded areas of the test, with extreme cases showing the exposed upper layer of the endoskeleton that becomes either partially or fully destroyed (Jangoux 1987a; Becker et al. 2007). Denuding the endoskeleton enables secondary infections of the exposed tissues by protozoans and/or opportunistic bacteria (Bauer and Young 2000). It is noteworthy, that this description of BSUD is very similar to the description of the disease that caused the mass die-off of the long-spined black sea urchin, *Diadema antillarum*, in the Caribbean Sea in the early 1980s (Lessios et al. 1984, reviewed in Coppard and Alvarado 2013, see Section 19.3 in Chapter 19). Although echinoids may often recover from BSUD and regenerate tissues and appendages (Roberts-Regan et al. 1988), once the test is perforated the disease often results in death of the animal (Maes and Jangoux 1984a).

Most studies are in agreement with the general description of BSUD progression, and the infection in the green or variegated sea urchin, *Lytechinus variegatus*, shows similar lesions, however, in one case, the epidermal dissolution from the underlying dermis occurred first rather than last (Boettger and McClintock 2009). The epidermis displayed an initial white discolouration, which was followed by separation from the dermis and elevation towards

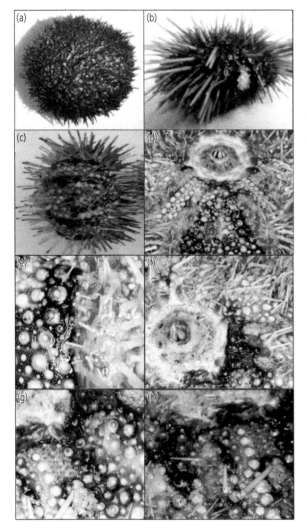

Figure 18.1 Surface infections of sea urchins. (a) Bald sea urchin disease (BSUD) in the purple sea urchin, *Strongylocentrotus purpuratus*, displays the key characteristic of the loss of all primary spines and the appearance of baldness. (b) Spotting disease in *S. purpuratus* shows a discrete lesion on the lateral surface of the animal that is surrounded by a black band of tissue. Spines and other surface appendages are only lost in the region of the lesion. (c) Spotting disease in *S. intermedius* initiates as discrete lesions (arrows) that can spread to about half of the animal with loss of spines and other appendages, showing similarities to BSUD. (d–h) More advanced infection of BSUD in *Lytechinus variegatus*. (d) The infection shows evidence of epidermal dissolution that leads to spine loss and green discoloration, which is most likely a secondary green algal invasion on the interambulacra and the teeth in Aristotle's lantern. (e) The interambulacral green lesions with exposed spine tubercles are present next to intact tube feet and spines on the ambulacra (right). (f) Brown lesions are present on ambulacra and interambulacra on the oral side of the sea urchin. (g) The brown lesions on the oral surface within the ambulacra are entirely devoid of spines and tube feet in a region close to the peristomial membrane that is located at the upper left corner of the image. (h) Black lesions on the exposed test of *Lytechinus variegatus* only occur in animals that have a heavy infection of an unknown *Vibrio* sp. that is detected in the coelomic fluid and based on a match to 16S rRNA gene sequence. The bacterium could only be identified to the genus level but is most closely related to *V. campbellii* and *V. carchariae*. Source: the image in (a) was provided by M.A. Barela Hudgell. The image in (b) was provided by L.C. Smith. The image in (c) is reprinted from Aquaculture Reports 16, 100244, Li, R. et al. (2020). *Vibrio coralliilyticus* as an agent of red spotting disease in the sea urchin *Strongylocentrotus intermedius*, with permission from Elsevier. The images in (d–h) are republished from Boettger (2001) with permission from the author.

the apical portion of the spines, before it ruptured 24 to 72 hours after the initial observation. When the epidermis ruptured the underlying tissue slowly developed green or brown-black discolouration followed by disease progression as described previously (Figure 18.1d–h). The differences in test discolouration in *Lytechinus variegatus* have been attributed to secondary invasion and colonisation of green algae on the exposed test lesions, whereas brown discolourations are likely associated with bacterial colonisation. In addition to the discolourations described for lesions on the test, the infection of *Lytechinus variegatus* also display discolouration of the teeth (Figure 18.1d, f). Differences in disease

aetiology in echinoids support the hypothesis that although BSUD results in lesions and in some cases mass mortalities, this disease displays variable disease progression and aetiologies that is likely the outcome of different pathogens.

18.3.2 Histopathology of bald sea urchin disease

In echinoids affected by BSUD, the epidermal tissues surrounding lesions including tissues at the base of spines become swollen and infiltrated with coelomocytes, particularly by red spherule cells (or eleocytes) indicating an inflammatory-like reaction (Maes and Jangoux 1984a; Shimizu et al. 1995; Mangiaterra and Silva 2001; de Faria and da Silva 2008; Wang, Chang, et al. 2013). The red spherule cells likely release echinochrome at the sites of lesions and its anti-bactericidal activity may act against the invading pathogen(s) (Wardlaw and Unkles 1978; Messer and Wardlaw 1980; Service and Wardlaw 1984; Roberts-Regan et al. 1988) as has been demonstrated *in vitro* (Coates et al. 2018). In heavily infected epidermis and interstitial tissue of the test, the necrotic surface appendages are infiltrated with rod shaped and filamentous bacilliform bacteria (Shimizu et al. 1995; Becker et al. 2008). Large accumulations of bacteria also associate with surface lesions and the extensive breakdown of the test leads to skeletal holes. In moribund animals that display gross lesions, the coelomic fluid also contains bacteria (Bauer and Young 2000; Boettger and McClintock 2009; Clemente et al. 2014), which is in contrast to sterile coelomic fluid in healthy animals that show natural anti-bacterial activity against many marine bacteria (Wardlaw and Unkles 1978; Wardlaw 1985; Stabili et al. 1996). Overall, the incidence, symptoms, and the host immune response against BSUD appears dynamic based on differences among host species, differences in the infecting and opportunistic microbes, and perhaps in part on the variety of descriptions of the disease.

18.3.3 Putative pathogens of bald sea urchin disease

Bacterial pathogens are the most common aetiological agents associated with the set of similar symptoms that appear as BSUD. Molecular analyses of diseased sea urchins have identified several bacterial genera or species that likely cause BSUD, including *Vibrio alginolyticus* and *V. anguillarum*, *Aeromonas salmonicida*, *Cytophagales*, *Flavobacteriaceae*, and *Bacteriodetes* (Gilles and Pearse 1986; Tajima et al. 1997a, 1997b; Becker et al. 2008). Although bacterial pathogens have not been identified from lesions in nine species of sea urchins (Table 18.1), the majority of echinoid species that display disease development show an association with *Vibrio* sp. (*Vibrio alginolyticus*, *V. anguillarum*,

Vibrio spp., and some of the unnamed vibrios previously included in Alpha and Gammaproteobacteria). It is noteworthy that all *Vibrio* species plus *A. salmonicida* and *Pseudoalteromonas* sp. produce either tetrodotoxin or anhydrotetrodotoxin, which are common in marine systems and act to block sodium channels of excitable membranes (Yasumoto et al. 1986). Metagenomics of 16S rRNA gene sequences to identify pathogenic groups from body surface samples of infected collector sea urchins, *Tripneutes gratilla*, in South Africa shows that many common marine bacteria are present and belong to the families of Vibrionaceae, Saprospiraceae, Flavobacteriaceae, and Sphingomonadaceae (Brink et al. 2019). These bacteria show varying concentrations in healthy compared to diseased animals, yet no single or specific combination of taxa could be identified as the key pathogen(s) for the disease phenotype. Overall, the aetiology of the BSUD is likely complex and the infection, as manifested by a similar set of symptoms, is likely an outcome of a complex set of different bacterial species.

18.3.4 Conclusion: bald sea urchin disease

BSUD appears to be a set of diseases associated with different causative agents, and display similar symptoms of appendage loss including spines, tube feet, and pedicellariae. This results in the appearance of bare epithelium, exposed test in some areas of the animal surface (Figure 18.2), and test perforation leading to sea urchin death. However, the bacterial pathogens associated with the diseases are either unknown or confusingly variable. There is no clear distinction among microbial pathogens in temperate compared to tropical locations because bacteria associated with these diseases are present in sea water over a range of temperatures. It is therefore difficult to determine whether the diseases categorised under the names of bald sea urchin or spotting disease are the same disease or different diseases that happen to display similar pathologies. The actual progression of the diseases is still moderately unclear, because the literature describing this set of diseases is inconsistent including whether surface abrasion is required to initiate disease onset. If physical abrasions are the dominant initiator, then the bacterial pathogens invading the tissues may

Table 18.1 Microbial pathogens that cause bald sea urchin disease or spotting disease in echinoid species

Infectious microbial agent	Species impacted
Unknown	*Allocentrotus fragilis, Arbacia lixula, Cidaris, Diadema mexicanum, Hemicentrotus pulcherrimus, Psammechinus miliaris, Pseudocentrotus depressus, Sphaerechinus granularis, Strongylocentrotus franciscanus*
Vibrio alginolyticus[*]	*Archaeopneustes hystrix, Diadema africana, Palaeopneustes cristatus*
Vibrio anguillarum[*]	*Echinus esculentus, Heliocidaris erythrogramma, Holopneustes purpuraescence, Mellitta quinquiesperforata, Strongylocentrotus purpuratus*
other *Vibrio* spp.[*#]	*Lytechinus variegatus, Paracentrotus lividus, Tripneustes gratilla* *Strongylocentrotus intermedius*
Aeromonas salmonicida	*Mellitta quinquiesperforata, Strongylocentrotus purpuratus*
Alcaligenes sp.	*Strongylocentrotus droebachiensis*
Acinetobacter sp.	*Strongylocentrotus droebachiensis*
Bacillus sp.	*Diadema antillarum*
Exignobacterium sp.	*Tripneustes gratilla*
Flavobacterium sp.	*Strongylocentrotus purpuratus*
Pseudoalteromonas sp.	*Meoma ventricosa*
Dactylococcopsis sp.[§]	*Echinus acutus*

[*] There are twelve different species of echinoids that are infected by a *Vibrio* species.
[#] *Alpha* and *Gammaproteobacteria* are generally cited as including *Vibrio* sp. and are therefore only listed under *Vibrio* sp.
[§] A cyanobacterium was found only on necrotic tissue and was not present in histological tissue sections of diseased *Echinus acutus* (Mortensen and Rosenvinge 1934).

simply be opportunistic invaders of injuries, and the species that invade may depend on the variable bacterial communities on body surfaces of different echinoid species in any given geographical area (Brink et al. 2019). Further studies are therefore required to elucidate whether body wall lesions are a response to abrasion, whether disease onset occurs due to an underlying pathogenic condition, whether impaired immune responsiveness from host stress is involved, or whether all unknowns act together. Experimental replication of BSUD in healthy animals should be combined with fulfilling Koch's postulates to achieve a reliable identification of the range of microbial pathogens that may all show the same infection symptoms in adult sea urchins. A few studies have addressed Koch's postulates in which *Vibrio* sp. or *Vibrio coralliilyticus* isolated from diseased animals and injected into

healthy sea urchins resulted in symptoms of the disease (Li, Dang et al. 2020). However, it is unclear how many different microbes might show the same outcome. Transcriptomes of diseased vs. healthy sea urchins indicate immune non-responsiveness in diseased animals, however, it is not known whether the pathogen or an environmental stressor compromises the immune system (Zhang, Lv et al. 2019). The combination of decreased immune responsiveness and increased pathogen virulence is likely to work in combination to drive the onset of this disease.

18.4 Sea star wasting disease

18.4.1 Symptoms of sea star wasting disease

Sea star wasting disease (SSWD) is a lethal syndrome that began in 2013 along the west coast of

North America that has caused mass mortalities of all species of sea stars. The manner in which sea stars succumb to SSWD has generated considerable scientific and public concern over the fate of these iconic marine species (Menge et al. 2016). However, local outbreaks of SSWD have been reported for years. Initially reported as an 'unknown skin disease' (Mead 1898; Christensen 1970), SSWD was described as necrotic areas on *Asterias* spp., which progressed to erosions through the body wall (Menge 1979). The wasting signs in the predatory sun star, *Heliaster kubiniji*, show the appearance of white lesions on the aboral surface of infected animals, which spread rapidly to the whole animal and result in limb autotomy. The progression of the disease is the same as that for the sunflower sea star, *Pycnopodia helianthoides* (Figure 18.2) (Dungan et al. 1982). Infected crown-of-thorns sea star, *Acanthaster planci*, shows lesions of the dermal tissue, collapsed spines, and a debilitated water-vascular system (Pratchett 1999), and symptoms in the sand star, *Astropecten jonstoni*, are similar (Figure 18.3.a,b) (Staehli et al. 2009). All species with SSWD described before and since 2013 show the same initial symptoms of white skin lesions (Figure 18.3.a), followed by loss of body turgor pressure and an overall appearance of body 'deflation', twisting arms and limb autotomy, dermal swelling, eversion of stomach, spread of lesions over the entire body, erosions in body wall tissues, perforations in the body wall resulting in the exposure of internal organs, tissue decay, and finally, animal death and disintegration (Figures 18.2, 18.3) (Staehli et al. 2009). The entire process to death takes a few days to a week after the appearance of the first lesions (Bates et al. 2009, Miner et al. 2018; Hewson et al. 2019).

It is noteworthy that the rapid body breakdown in SSWD and the appearance of white degenerating lesions is similar to the extreme 'self-destruct' phenomenon of mutable connective tissue (MCT) breakdown that has been documented in sea cucumbers and sea stars (Wilkie 1984, 2005, Motokawa and Tsuchi 2003, 2005; Motokawa 2011, Byrne 2001, 2017). The mechanical properties of the body wall of echinoderms are unusual because they are controlled by MCT that is under nervous control with a range of states that can change from hard to soft over seconds to minutes involving both viscosity and elasticity of the tissue (Wilkie 1984; Motokawa and Tsuchi 2003; Motokawa 2011). The characteristics of the rapid body degradation during SSWD in sea stars is very similar to and may involve irreversible MCT softening that contributes to the complete body breakdown (Wilkie 2005). Although the mechanisms of how MCT softening, body deflation, and behavioural changes associate with SSWD are not understood, nervous control of these changes is consistent with disease associated disruption of gene expression in the nervous tissue (Fuess et al. 2015, and see Section 18.4.3).

18.4.2 Biological causes of SSWD

Contagion among sea stars for SSWD has been demonstrated by housing a healthy sea star such as *Linckia* with a diseased *Acanthaster* sp sea star in which *Linckia* becomes infected showing symptoms of SSWD (Rivera-Posada et al. 2011a, 2011b). Identification of the pathogen from the earlier reported and local, short term episodes of SSWD was not possible because of the limited occurrence (Mead, 1892; Christensen 1970, Dungan et al. 1982; Pratchett 1999; Staehli et al. 2009; Bates et al. 2009), and because technologies to identify a putative pathogen were not available at the time, although a *Vibrio* was initially implicated in the disease for the sand star, *Astropectin jonstoni* (Staehli et al. 2009). Massive DNA sequencing approaches to compare putative pathogen sequences in diseased versus healthy sea stars identified a sea star associated densovirus (SSaDV) as a likely pathogen in the sunflower sea star, *Pycnopodia helianthoides* (Hewson et al. 2014; Hewson et al. 2018; Hewson et al. 2019; Høj et al. 2018; Harvell et al. 2019). However, SSaDV is present at varying levels in all sea stars tested, including asymptomatic animals, and is also present in the surrounding sea water (Hewson et al. 2014). The prediction of a virus as the initiating pathogen was verified by the inoculation of a viral-sized fraction of tissue lysates into healthy sea stars that replicates SSWD symptoms. These results suggest a strong link between SSaDV as a waterbourne pathogen that causes SSWD.

A parallel mass die-off of sea stars from SSWD also occurred along the Atlantic coast of North America in addition to Antarctica, Australia, and

Figure 18.2 Sea Star Wasting Disease (SSWD) in the sunflower sea star, *Pycnopodia helianthoides*. A. SSWD infects and kills sunflower sea stars within a few days. (a) A healthy sunflower sea star. (b) A sea star with mild SSWD symptoms showing a twisted arm (arrow). (c) A sea star with severely twisted arms. (d) A sea star with the beginnings of arm tip autotomy (arrows). (e) A separated arm with active tube feet. (f) Severe SSWD symptoms showing multiple lesions at the base of the arms and the beginning of tissue deterioration. (g) A dead sea star in the process of disintegration or 'melting'. B. The sunflower sea star is present in large numbers on a particular rock near Vancouver, British Columbia, Canada in October 2013 (left), but is completely absent three weeks later during the SSWD outbreak (right). Source: all images in A are from Jackie Hildering, www.TheMarineDetective. com, while on scuba at the northeast end of Vancouver Island, British Columbia, Canada. Images A (b-g) were acquired between December 2013 and November 2014. Image A (a) shows a healthy adult sea star that was imaged in April 2018 at Vancouver Island. Image B is credited to Neil McDaniel, www.neilmcdaniel.com and is reprinted with permission from Montecino-Latorre et al. © 2016 Devastating transboundary impacts of sea star wasting disease on subtidal asteroids. PLoS One 11(10): e0163190.

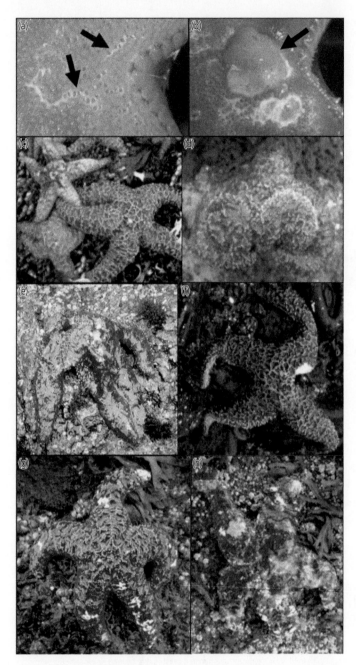

Figure 18.3 Progression of Sea Star Wasting Disease SSWD. (a, b) Early stages of SSWD in the sand star, *Astropecten jonstoni*. (a) Initial symptoms appear as small white lesions (arrows) on the aboral surface of the sea star. (b) Symptoms proceed to more and larger lesions leading to perforations of the body wall that exposes internal organs (arrow). (c-h). SSWD progression in the ochre sea star, *Pisaster ochraceous*. (c) Healthy sea stars. (d) Severe arm twisting. (e) Loss of turgor pressure or body deflation. (f) A lesion remains after autotomy of one arm. (g) Multiple lesions on two or more arms (h) Severe tissue deterioration with body wall perforations and exposed internal organs. Source: the images in (a) and (b) are reproduced with permission from Staehli et al. (2009), Cambridge University Press. Images (c, d, and f are reproduced with permission from Menge et al. © 2016. Sea star wasting disease in the keystone predator *Pisaster ochraceus* in Oregon: insights into differential population impacts, recovery, predation rate, and temperature effects from long-term research. PloS One 11(5): e0153994. Image (e) is reproduced with permission from Moritsch and Raimondi © 2018. Reduction and recovery of keystone predation pressure after disease-related mass mortality. Ecology and Evolution 8: 3952-3964. Published by John Wiley & Sons Ltd. Images (g) and (h) were provided by Melissa Miner (http://www.pacificrockyintertidal.org) and are diseased sea stars on the Pacific coast of Washington and Sitka Alaska.

New Zealand (Bucci et al. 2017; Nunez-Pons et al. 2018; Hewson et al. 2019). Similar to the sequencing analysis of pathogens for the sunflower sea star, a metaviromic analysis of three sea star species from the North American Atlantic coast, *Asterias forbesi*, *A. rubens*, and *Henricia* spp, also identified a denso-virus as the possible pathogen (Jackson et al. 2020). However, in this case the sequenced genome of the virus was only 78% identical to SSaDV indi-cating that it is a similar but different virus and has been called *Asterias forbesi*-associated denso-virus (AfaDV) (Jackson et al. 2020). From these various results it is becoming clear that with greater understanding of the sea star wasting event in North America and in other regions of the world, the more complex it appears to be, with the possibility that the disease is actually a pandemic rather than a series of isolated regional mass die-offs.

The identification of SSaDV as a pathogen that causes SSWD suggested that this association may only apply to the sunflower star, *Pycnopodia helianthoides*, thus leaving most of the other species of sea stars without an identified pathogen for the disease (Hewson et al. 2018). Other analyses of SSWD suggest the involvement of an enigmatic group of fungus-like saprobiotic protists, which could also be isolated from sea stars with wasting disease (Fiorito et al. 2016). The animals infected with this protist show complex changes in the microbial flora with reduced microbiome species richness (Lloyd and Pespeni 2018). An evaluation over time of the microbiomes associated with healthy, diseased, and dead ochre sea stars, *Pisaster ochraceus*, from SSWD show changes in microbial community composition through disease progression with a decrease in species richness of the microbiome in late stages of SSWD. This indicates a remodelling of the microbial community associated with disease progression. At the time of initial symptom onset there is a decrease in abundance of *Pseudoalteromonas* spp, a known beneficial taxa, which is followed by an increase in abundance of putatively opportunistic bacteria *Phaeobacter* spp., late in the disease (Lloyd and Pespeni 2018). Therefore, changes in microbial composition demonstrate that SSDW may disrupt key microbial assemblages and their collective functions, contributing to a microenvironment that enables pathogenic communities to develop. These changes may be based on alterations that SSWD imposes on host interactions with its microbiome, which predisposes the host for the formation of polymicrobial communities that factor into the SSWD onset and progression. Hence, SSWD may be the result of a complex interaction among multiple microbial taxa rather than a single pathogenic organism (Sweet 2020). Despite intense research on SSWD, identifying the causative agent versus opportunistic organisms is very difficult to untangle. The problem of pathogen identification is complicated by the normal versus stressed status of the host immune system because these agents may not cause disease unless the host becomes immuno-compromised.

The ochre sea star, *Pisaster ochraceus*, also suffered major mortality events from SSWD between 2013 and 2014 along the west coast of North America from Alaska to Southern California (Figure 18.3.c-h) (Eisenlord et al. 2016; Miner et al. 2018). This iconic asteroid is common in the intertidal zone and is present in a range of colours from orange to dark purple (Figure 18.2.c). It is unknown as to why the orange colour morph had significantly decreased disease and mortality compared to the purple, blue, and brown morphs (Menge et al. 2016). Consistent with the outcomes in which major die-offs from pathogens that change the makeup of populations, the genomics of adult and newly recruited juvenile *P. ochraceus* before the SSWD outbreak compared to the survivors revealed significant shifts in the allelic frequencies of multiple loci and in haplotype frequency in the populations (Schiebelhut et al. 2018). Thus, these measurable signals of selection for survival that is driven by pathogen pressure highlight the potential for the persistence of genetic changes in populations of subsequent generations that may influence the resilience and trajectory of *P. ochraceous* in future outbreaks. Since the die-off of adult *P. ochraceus*, the population recovery has shown a 100-fold increase in the settlement of juvenile sea stars compared to pre-die-off levels over large spatial ranges, which bodes well for recovery in the short-term future (Menge et al. 2016; Eisenlord et al. 2016; Moritsch and Raimondi 2018).

18.4.3 The sea star immune response to SSWD

The rise in marine epizootics underscores the need for a thorough understanding of invertebrate immune systems and their status with respect to both pathogen pressure and environmental conditions. The immune response in the sunflower sea star, *Pycnopodia helianthoides*, to SSDaV is an example of how this asteroid responds to infection that may apply to all echinoderms. The greatest increase in expression of genes in the sunflower sea star in response to inoculation with virus size fractions of tissue lysates isolated from sea stars with SSWD, encodes proteins that are broadly categorised as functioning in immunity, cytokine biosynthesis, and cell adhesion (Fuess et al. 2015). Elevated expression of specific genes that encode cytokines, proteins that function in immune signalling pathways, an IL-17 receptor, enzymes that produce echinochrome, clotting factors, complement proteins, EGF related proteins, TLRs and associated signalling proteins, among others. Genes involved in the nervous system and others involved in tissue remodeling are also differentially expressed, which is consistent with the behavioural alterations and tissue changes associated with SSWD. The meta-transcriptomics of holobiont gene expression in body wall tissues in symptomatic versus asymptomatic sun stars affected by SSWD are highly variable among replicate samples, and most of the differentially expressed genes are represented either by transcripts of associated microorganisms (particularly *Pseudomonas* and *Vibrio* relatives) or by low-level echinoderm transcripts of unknown function (Gudenkauf and Hewson 2015). However, the pattern of annotated host genes reflects increased immune responsiveness, apoptosis, tissue degradation processes, and decreased energy metabolism that may be consistent with viral infection and replication in cells.

The evaluation of SSWD in the ochre sea star, *Pisaster ochraceus*, has focused on changes in the chemical physiology of the coelomic fluid and the coelomocyte concentrations rather than changes in gene expression (Wahltinez et al. 2020). Compared to healthy ochre sea stars, animals with symptoms of SSWD have significantly elevated chloride and protein concentrations, decreased calcium concentrations, and are hyper-osmotic relative to the surrounding sea water. Many diseased ochre sea stars have significantly more coelomocytes in late stages of disease although for echinoderms with advanced diseases, including those with putative secondary bacterial invasions, no coelomocytes may be detectible in the coelomic fluid (L.C. Smith, personal observation). Cellular clot formation and reactions to bacteria in the coelomic fluid of many diseased ochre sea stars may result in decreased coelomocyte concentrations. These results indicate that SSWD induces osmodysregulation by unknown means that may include changes in energy dependent membrane molecular pumping systems, ion channels, plus increased protein from cellular necrosis and apoptosis. It is noteworthy that decreased calcium is consistent with body wall softening, as observed in sea stars with SSWD including *Pisaster ochraceus* (Figure 18.2.e); (Wahltinez et al. 2020) and in MTC disease (see Section 18.4.1) (Wilkie 1984; Byrne 2001; Motokawa and Tsuchi 2003; Wilkie 2005; Motokawa 2011; Byrne 2017). In general, the molecular and physiological changes in sea stars with SSWD are consistent with an overwhelmed immune system and progression to the gross aetiology of the disease.

18.4.4 Conclusion: SSWD

The mass mortality events from SSWD altered the populations of many sea star species. Mass infections presumably lead to pathogen replication and release into the environment with the putative outcome of significant increases pathogen concentrations in sea water, increased pathogen contact with sea stars, and exacerbated spread of the disease. Two different densoviruses appear to be likely infective agents for sea stars in multiple oceans, which is consistent with SSWD as a viral pandemic. However, identifying the initiating pathogen for all sea star species and disentangling the initiating agent from subsequent opportunistic invaders is challenging. Furthermore, a single pathogen causing a single disease may be oversimplified in many cases of echinoderm pathology as exemplified for BSUD (see Section 18.3). The immune response of diseased compared to healthy sunflower sea stars and changes in gene expression illustrates that animals mount an immune response to fight the virus. Similar changes in gene expression in the nervous tissue pose the possibility that the virus invades nerve cells, and body surface lesions suggest viral infection in skin, muscle, and body wall tissues. This deadly pathogen is specific for asteroids given that no other classes of echinoderms are susceptible either because of putative differences in the immune systems among the classes such as a missing PRR in the asteroids, or because the asteroids have a receptor that is recognised by the pathogen that enables infection and is missing or different in the other echinoderm classes. This is consistent with a virus or group of similar viruses that infect all asteroids.

Mass die-offs change the populations of the susceptible species, which changes the gene pool of those animals that survive the infection and are able to reproduce successfully. This is a standard outcome for the genetics of all species that survive pandemics or other infectious causes of population declines or bottlenecks. The SSWD pandemic has changed the populations of many sea stars with correlated changes to the genetics of the species that can be detected at the level of the population. These changes have been observed for populations of the ochre sea star, *Pisaster ochraceus*, which appear to be recovering from the SSWD pandemic. However, the sunflower sea star, *Pycnopodia helianthoides*, remains absent from many places where they were once common and are now listed as endangered. A note of optimism remains for the hope that a protected refuge exists from which the sunflower sea star may reappear (see Figure 18.2.Aa).

18.5 Diseases of the sea cucumber, *Apostichopus japonicus*, in aquaculture facilities

The sea cucumber, *Apostichopus japonicus*, is one of the most economically valuable aquaculture species in China (Chen 2004), which is raised for human consumption and used in traditional medicines in China, Korea, Japan, Vietnam, and Malaysia (Liao 1980; Toral-Granda et al. 2008; Jo et al. 2017). The annual production of *A. japonicus* exceeded 200,000 tons since 2014 (Chinese Fisheries Yearbook 2017, 2018, 2019) and the annual production value exceeded 60 billion Chinese Yuan since 2016. With the expanded production of *A. japonicus* in aquaculture, mass mortalities of these animals from disease have resulted in huge economic losses to the industry (Chinese Fisheries Yearbook 2019). This problem has led to the necessity for understanding the impacts of aquaculture on the health of this species and for developing strategies and technologies to control disease. The diseases of *A. japonicus* are diverse and highly infectious with mortalities of 80–100% for the most virulent diseases (Zhang, Xing

et al. 2018). The possible pathogens that are associated with *A. japonicus* are highly diverse and include bacteria, viruses, and parasites (Wang, Zhang et al. 2004). Sea cucumbers have an integrated and highly complex innate immune system that recognises and eliminates various invaders (Ramírez-Gómez and García-Arrarás 2010; Xue et al. 2015, see Section 18.1.3.2). For invertebrates that survive on innate immunity with no clear demonstration of specific immune memory in echinoderms, disease prevention and control through vaccines or other pharmaceutical preparations are unlikely to provide long-term protection against infection. Therefore, it is essential to understand the pathogenesis of diseases in *A. japonicus* and the host immune defence mechanisms that may be manipulated to prevent and control diseases.

18.5.1 Skin ulcer syndrome

Skin ulcer syndrome (SUS) is the most severe of the diseases that impact *A. japonicus* and there has been considerable effort to understand the pathogenesis of this infection (Zhang, Xing et al. 2018). SUS occurs in both the juvenile and adult animals with symptoms of anorexia, anterior end shaking, swollen mouth, evisceration, general atrophy, and skin ulcers that expand and lead to death (Figure 18.4.a; Table 18.2). SUS was first identified in cultured juveniles of the sandfish sea cucumber, *Holothuria scabra*, which may have been caused by a bacterial pathogen (Becker et al. 2004), and more recently in adults of *H. scabra* in Madagascar (Delroisse et al. 2020). SUS also affects cultured *A. japonicus,* and to date, a variety of pathogens have been identified from infected animals housed in different mariculture farms in different regions of China. Possible bacterial pathogens include *Vibrio splendidus*, various other *Vibrio* species, *Pseudoalteromonas nigrifaciens*, *Pseudoalteromonas* sp, *Aeromonas*, *Photobacterium*, and *Shewanella* (Zhang, Wang et al. 2006; Wang, Fang et al. 2006; Yang, Zhou et al. 2007a, 2007b; Wang, Yuan et al. 2007; Deng et al. 2009; Liu, Zheng et al. 2010; Li, Qiao et al. 2010b; Delroisse et al. 2020) although *Shewanella* sp., *V. splendidus*, and *P. tetraodonis* are the most virulent (Li, Qiao et al. 2010a). Furthermore, in wild populations of *Holothuria arguinensis*, *Vibrio gigantis*,

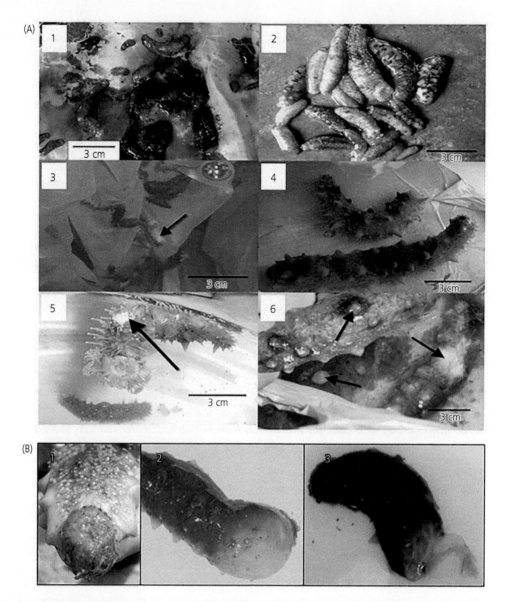

Figure 18.4 Diseases in *Apostichopus japonicus* in aquaculture facilities. (a) Skin ulcer syndrome (SUS) in *A. japonicus*. 1. Healthy juvenile sea cucumbers. 2. Juvenile with symptoms of SUS. 3. A juvenile sea cucumber in an aquaculture tank with symptoms of SUS (arrow). 4. Healthy adult. 5, 6. Adults with symptoms of SUS (arrows). (b) Acute peristome oedema disease (APED) in *A. japonicus*. 1. APED affects the oral end showing swelling and absence of feeding tentacles. 2, 3. Skin ulcers are present at the oral end, which is oriented toward the lower right. Source: the images in (a) are reprinted from Deng et al. (2009). Isolation and pathogenicity of pathogens from skin ulceration disease and viscera ejection syndrome of the sea cucumber *Apostichopus japonicus*. Aquaculture 287: 18–27, with permission from Elsevier. The images in (b) were provided by Wei Zhou.

and *V. crassostreae* are likely associated with SUS (Cánovas et al. 2019). Microbiome analysis shows that *Lactococcus garvieae* is significantly enriched in the gut of infected *A. japonicus* compared to non-infected sea cucumbers suggesting that this microbe may be a potential pathogen, and that it might be employed as an indicator of potential SUS in cultured sea cucumbers (Zhang, Xing et al. 2018). In addition to bacterial pathogens, parasitic ciliates are also possible causative agents for SUS in *A. japonicus* (Wang, Rong et al. 2005). Infecting ciliates are enriched in the respiratory trees of diseased *A. japonicus* where they appear to absorb nutrients and cause tissue damage (Liu, Fan et al. 2005; Long et al. 2006) that proceeds to secondary infections (Wang, Rong et al. 2005). Overall, bacteria, viruses, and parasites are associated with SUS but bacteria, namely *Vibrio*, *Shewanella*, and *Pseudoaltermonas* spp., are speculated to be the key pathogens for this disease.

18.5.2 Acute peristome edema disease

Acute peristome edema disease (APED) in both juvenile and adult *A. japonicus* (Wang, Chang et al. 2005) is associated with bacteria and viruses. Typical symptoms of the disease are swelling around the mouth, skin ulcers, decreased adhesion by the tube feet to surfaces, and mortality of 30–60% (Figure 18.4.b; Table 18.2). APED usually occurs in the spring when water temperatures range from 5–14°C in China (Ma et al. 2006). The dominant bacteria associated with the peristome, body surface, and body wall of affected sea cucumbers are *Marinomonas dokdonensis*, *Vibrio splendidus*, *V. tapetis*, *Vibrio* sp., and *Pseudoaltermonas* sp. (Ma et al. 2006; Liu, Zheng et al. 2010), however, only *M. dokdonensis*, *V. splendidus*, and *V. tapetis* may be the initiating pathogens. Coronavirus-like virus particles are also associated with *A. japonicas* with symptoms of APED (Wang, Chang et al. 2007), although verification of this association by inoculation of the virus into healthy sea cucumbers has not been carried out. Because bacteria have fulfilled Koch's postulates, the current belief is that they are the initiating and pathogenic agents of APED.

18.5.3 Off-plate syndrome

Off-plate syndrome occurs most often in juvenile *A. japonicus* and is highly contagious with rapid onset and high mortality among sea cucumbers housed in aquaculture facilities. The typical symptoms are a reduction in body size and a gradual loss of tube foot function and the inability to remain attached to vertical surfaces (Figure 18.5.a;

Table 18.2). Infected animals show a loss of vitality, fall off the plate or surface on which they are normally attached (hence the name of the disease), and accumulate at the bottom of the aquaculture tank or pond (Wang, Zhang et al. 2004). *Vibrio lentus* and *Vibrio* sp. have been suggested as the most likely pathogens in this disease (Wang, Zhang et al. 2004; Sun 2006), although a spherical, enveloped virus is also associated with the disease based on an ultrastructural analysis of the pathology (Song et al. 2007). However, the actual pathogen or set of pathogens for off-plate syndrome are unclear.

18.5.4 Viscera ejection syndrome

Viscera ejection syndrome (VES) in adult *A. japonicus* is highly contagious and characterised by the evisceration of the intestinal tract, which is followed by gradual body swelling and rapid skin ulceration leading to significant mortality (Figure 18.5.b.; Table 18.2) (Deng et al. 2005). Two bacteria, *Arthrobacter protophormiae* and *Staphylococcus equorum*, as well as an unknown spherical virus are associated with VES. Although both bacteria and the virus are isolated from diseased sea cucumbers, the virus can replicate the symptoms when inoculated into healthy sea cucumbers, suggesting that the bacteria may be opportunists (Deng et al. 2008, 2009). Yet, any of these potential pathogenic agents may contribute independently or collectively to disease induction and progression.

The phenomenon of evisceration in sea cucumbers has been described as a normal behaviour to distract a predator, however, it is not equivalent to evisceration in VES. Some have proposed that because evisceration is observed for both SUS and VES, that they are the same disease (Wang, Zhang et al. 2004), while others argue that these two diseases are different (Deng et al. 2009). The differences between these diseases are based on (i) the distinct differences in the overall symptoms of VES compared to SUS, (ii) VES is very contagious and has a wide transmission among sea cucumbers in aquaculture whereas SUS spreads comparatively slowly, and (iii) the pathogens isolated from *A. japonicus* with SUS are bacterial, whereas the VES symptoms can be replicated by a spherical virus (Deng et al. 2008, 2009). Although SUS and VES

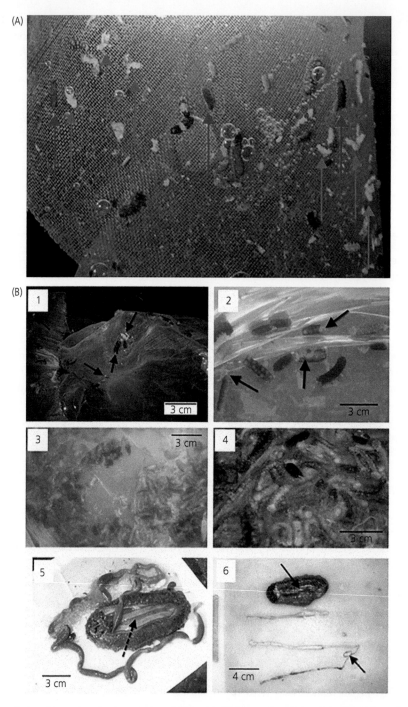

Figure 18.5 Diseases in *Apostichopus japonicus* in aquaculture facilities. (a) Off-plate syndrome in juvenile *A. japonicus*. Both diseased (red arrows) and healthy sea cucumbers (blue arrows) are shown in an aquaculture facility. (b) Viscera ejection syndrome (VES) in *A. japonicus*. 1–4. Juvenile sea cucumbers with VES show symptoms of mouth tumescence and ejection of viscera (arrows). 5. Dissection of a healthy adult sea cucumber shows normal gut with ingested contents and healthy body wall (arrow). 6. Dissection of a sea cucumber with VES shows intestines without contents (arrow) and body wall ulceration (arrow). Source: image (a) was provided by Anguo Zhang. The images in (b) are reprinted from Deng et al. (2009) Isolation and pathogenicity of pathogens from skin ulceration disease and viscera ejection syndrome of the sea cucumber *Apostichopus japonicus*. Aquaculture 287: 18–27, with permission from Elsevier.

appear to be two different diseases, additional work will be needed to understand the diseases and the initiating pathogens.

18.5.5 Diseases of larval sea cucumbers

The pathogenesis of diseases in sea cucumbers is complex and may be a consequence of interactions among microbes including the pathogens, the environment in which animals are housed, and the status of the host defence system (Hedrick 1998). Stomach ulceration is a disease of larval *A. japonicus* that usually appears at the auricularia stage of larval development when they initiate feeding (Figure 18.6.A). Stomach ulceration is more likely to occur when larvae are stressed with high temperatures (> 23°C) (Han et al. 2016) and cultures of high larval density (> 6 or 7 larvae/10 mL) (Li, Xing et al. 2006). The major symptoms include thickening and roughening of the larval stomach in which the stomach boundaries become vague plus a gradual decrease in stomach size from the normal pear-shape to one with thickened walls and decreased lumen based on epithelial cell proliferation (Figure 18.6.Ab,c; Table 18.2) (Sun 2006; Wang, Sun et al. 2006). In severe cases, the stomach disintegrates, and the entire larval body wall ulcerates leading to death (Figure 18.6Ad). Infected larvae have diverse bacterial species associated with the disease including *V. splendidus*, *Alteromonas* sp., and *Pseudoalteromonas* sp., which are considered to be the potential pathogens (Sun 2006; Wang, Sun et al. 2006).

Rotting edge disease also occurs at the auricularia stage of larval development with symptoms of darkening of the body edges or ectoderm followed by gradual autolysis until the entire larva disintegrates (Figure 18.6b; Table 18.2) (Zhang 2007). There is no indication of a viral pathogen associated with this type of infection and the histopathology of rotting edge disease suggests a bacterial pathogen (Deng and Sui 2004) identified as *Vibrio lentus* (Zhang 2007). When *V. lentus* is used to replicate the disease in larvae, symptoms of both rotting edge disease and stomach ulceration appear simultaneously suggesting that these two diseases are likely different descriptions of the same infection (Sun 2006).

18.5.6 Environmental factors in aquaculture facilities and disease in *A. japonicus*

The onset, development, and spread of diseases in *A. japonicus* are closely related to the characteristics of, or drastic changes in, the environment in which the animals live, which applies to both aquaculture settings and ecosystems in nature (Chen 2004; Li, Jiang et al. 2014). Environmental factors have significant influence on the immune function of host animals that correlates with pathogen virulence and the frequency of disease outbreaks. A metalloprotease and a hemolysin that are produced by *V. splendidus* are considered virulence factors with cytotoxic activities that alter coelomocyte interactions with *V. splendidus* by blocking phagocytosis and thereby promoting pathogen survival (Liang et al. 2016; Zhang, Liang et al. 2016). Two additional virulence factors, dihydrolipoamide dehydrogenases, are associated with the bacterial membrane and likely enhance adhesion between the microbe and *A. japonicus* cells (Dai et al. 2019). Critical environmental factors include optimal temperature, salinity, and nitrogenous compound concentrations for successful aquaculture of healthy *A. japonicus*. At the increased temperature of 28°C, *V. splendidus* has maximal expression of a metalloprotease (Zhang, Zhang et al. 2019) and maximal protein activity at 40°C (Zhang, Liang et al. 2016) that correlates with increased virulence. The environmental stress of high temperature and enriched nitrite and ammonium concentrations in aquaculture seawater may result in significantly decreased diversity of the normal gut microbiome in *A. japonicus*. The associated changes in the interactions among the microbial populations may promote infection by *V. splendidus* (Zhang, Zhang et al. 2019) that is considered the key pathogen in SUS and APED. Low seasonal water temperatures in Madagascar are also associated with increased SUS in *H. scabra* when they are raised in shallow water sea pens (Delroisse et al. 2020).

The variation, degradation, and overall nonoptimal environmental conditions in aquaculture facilities can also have negative impacts on the immune responsiveness of *A. japonicus*. Decreased water temperature of aquaculture

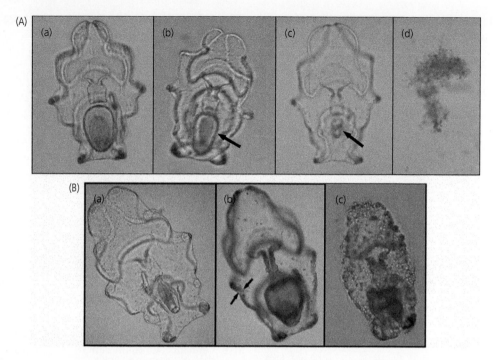

Figure 18.6 Diseases of larval *Apostichopus japonicus*. A. Stomach ulceration in *A. japonicus*. (a) A healthy auricularia larva. (b) The initial symptom shows a thickening of the stomach wall (arrow) and a narrowing of the lumen. (c) The stomach shows an overall reduction in size. (d) The terminal stage is death and cell lysis. B. Rotting edge disease in *A. japonicus*. (a) A healthy auricularia larva. (b) The body edges or walls thicken and become dark (arrows). (c) The terminal stage is death and cell lysis. Source: the images in A are reproduced from Sun (2006) with permission. The images in B are reproduced with permission from Wang, Zhang et al. (2004). Food and Agriculture Organization of the United Nations, Diseases of cultured sea cucumber, *Apostichopus japonicus*, in China (http://www.fao.org/3/y5501e/y5501e10.htm#bm36).

ponds in winter and early spring may cause severe immunosuppression (Jiang, Zhou et al. 2017). Alternatively, at high temperature conditions of 25°C, the transcriptome of muscle tissue from *A. japonicus* shows reduced expression of genes encoding immune-related proteins including some with immunoglobulin domains, C-type lectins, cell adhesion molecules, and enzymes that produce ROS. Enzymes associated with immune function such as superoxide dismutase, catalase, myeloperoxidase, and lysozyme show decreased activities under acute changes in temperature or salinity (Wang, Yang et al. 2008). Hence, *A. japonicus* appears to be susceptible to pathogens under conditions in which water temperatures are either too cold or too warm relative to the optimal temperature for the species (Li, Fang et al. 2019). The stress of elevated ammonia also reduces the immune function of *A. japonicus* based on the functional inhibition of enzymes such as superoxide dismutase, glutathione peroxidase, alkaline phosphatase, lysozyme, and catalase in the coelomic fluid (Liu, Zheng et al. 2012; Xu, Zhao et al. 2017). Suboptimal conditions of both temperature and increased ammonia correlate with higher susceptibility to pathogens and cumulative mortality of *A. japonicus* in aquaculture facilities (Liu, Zheng et al. 2012; Xu et al. 2017). In general, increased environmental stresses are likely to be important inducers of diseases of *A. japonicus* in aquaculture through a combination of negative impacts on immune function in sea cucumbers plus increases in pathogen virulence.

18.5.7 Strategies for disease prevention and control in *A. japonicus* in aquaculture facilities

Environmentally friendly strategies to reduce or control diseases of *A. japonicas* in aquaculture settings will need to focus on enhancing host immune capabilities. Immunostimulants, or immunopotentiators, are substances that activate the host immune response to enhance overall disease resistance (Shrestha et al. 2015; Wang, Sun et al. 2017). Currently, polysaccharides, nucleic acids, probiotics, and extracts from herbs are used as immunostimulants because they may replicate microbial associated molecular patterns (MAMPs). Polysaccharides are broad-spectrum immunostimulants and have been added to feed in aquatic animal breeding facilities. For example, β glucan results in significantly augmented immune responses in *A. japonicus* and correlates with improved disease resistance (Sun et al. 2008; Gu et al. 2011). Peptidoglycans, yeast polysaccharide, polysaccharides extracted from the Chinese milkvetch, *Astragalus membranaceus*, seaweed sulfated polysaccharides, and chitosan have all been used to enhance the immune response in sea cucumbers (Liu, Kong et al. 2008; Sun et al. 2008; Li, Sun et al. 2009; Wang et al. 2009; Zhao et al. 2010; Gu et al. 2011). These treatments promote phagocytosis by coelomocytes in juvenile sea cucumbers with the outcome of significant improvement in survival. Unmethylated oligodeoxynucleotides enhance the immune response in *A. japonicus* including improved phagocytic activity, respiratory burst, superoxide dismutase level, and total nitric oxide synthase activity (Gu et al. 2010). In general, MAMPs and other macromolecules from eukaryotes may all be detected by PRRs in *A. japonicus* that activate the immune system, which may overcome the immunosuppressive stresses of aquaculture facilities and enhance survival of the sea cucumbers.

Probiotics are an alternative approach to augment disease resistance in the host (Verschuere et al. 2000) through the expanded use of microbial species in the feed such as *Lactobacillus, Lactococcus, Leuconostoc, Enterococcus, Carnobacterium, Shewanella, Bacillus, Aeromonas, Vibrio, Enterobacter, Pseudomonas, Clostridium,* and *Saccharomyces* spp. (Mohapatra et al. 2013; Pandiyan et al. 2013). The probiotics used with *A. japonicus* are derived from microbes isolated from the intestinal tract of sea cucumbers (Tian et al. 2012; Song et al. 2014). Results show that when a *Bacillus* probiotic is used, *A. japonicus* shows greater resistance and survival against a variety of pathogenic bacteria including *Bacillus* and *Vibrio* (Zhang, Ma et al. 2010; Zhao et al. 2011, 2012). The C14 strain of the yeast, *Metschnikowia bicuspidata*, isolated from the intestine of *A. japonicus* improves the phagocytic activity of coelomocytes, elevates the activities of immuno-enzymes, and extracellular products isolated from the yeast inhibit the growth of pathogenic *Shewanella marisflavi* and *Vibrio splendidus* (Li, Ma et al. 2012; Liu, Ma et al. 2012). Furthermore, probiotics can also improve the quality of water and sediment in aquaculture facilities by reducing pH, levels of ammonia, nitrite, and soluble reactive phosphorus, as well as modulating the microbial community in the ambient environment (Tian et al. 2012; Chi et al. 2014; Song et al. 2014; Wang, Ran et al. 2019), all of which can improve disease prevention and control. Overall, the application of safe and environmentally friendly probiotics is an effective alternative strategy to antibiotics for disease prevention in aquacultured *A. japonicus*.

Many Chinese herbs produce abundant secondary metabolites such as organic acids, terpenes, and alkaloids, of which most are substances that activate immune responsiveness (Citarasu 2010). Hence, there is growing interest in using plant metabolites for disease control in aquaculture. The effective fractions isolated from the Mongolian milkvetch, *Astragalus propinquus*, display direct inhibition or killing of pathogenic microorganisms with the benefit that selection for microbial resistance does not occur (Ding et al. 2011; Hou et al. 2013; Wang, Bai et al. 2013; Radhakrishnan et al. 2014; Shrestha et al. 2015). In general, the activities of the herbal extracts increase total coelomocyte count, phagocytic activity, and the respiratory burst in *A. japonicus* indicating that the extracts contain promising immunostimulants for improved disease resistance (Fan et al. 2014). Overall, the use of specific Chinese herbs to control or reduce diseases in sea cucumbers avoids accumulations of residual antibiotics making this approach optimal for both the producers and the consumers of *A. japonicus*.

18.5.8 Approaches to block infection in A. japonicus in aquaculture facilities

Blocking the invasion or improving resistance to infection of *A. japonicus* by pathogenic bacteria through augmenting the immune response is another approach to control infection and disease. Lectins in *A. japonicus* function as PRRs but also exhibit some anti-bacterial activities making them possible inhibitors of infection (Hatakeyama et al. 1993; Hisamatsu et al. 2013; de Melo et al. 2014; Wei et al. 2015; Xue et al. 2015; Ono et al. 2018; Wang, Xue et al. 2018; Xiong et al. 2005). Lysozyme and antimicrobial peptides from sea cucumbers also have activity against many microbes and therefore constitute attractive candidates for protection against infection (Yang, Cong et al. 2007; Cong et al. 2009; Wang, Xu et al. 2011; Tan et al. 2012; Schillaci et al. 2013; Cusimano et al. 2019). In addition to anti-pathogen proteins from *A. japonicus*, vertebrate antibodies against pathogens have also been employed to augment the sea cucumber immune functions. For example, chicken IgY raised against *Shewanella marisflavi* induces bacterial agglutination and damages the bacterial cell membranes *in vitro*. Furthermore, when it is injected into *A. japonicus* the survival rate from infections by *S. marisflavi* is significantly improved (Xu et al. 2019). Antibodies to membrane bound virulence factors of *V. splendidus* block adhesion of the microbe to host cells and inhibit pathogen invasion by more than 50%. Finally, bacteriophages specific to *Vibrios* may be a specific and effective preventative approach for *A. japonicus* infection (Zhang, Cao et al. 2015; Li, Li et al. 2016; Li, Ren et al. 2020). Although there has been much focus on preventing disease by blocking or otherwise interfering with pathogen invasion, a variety of approaches are under consideration that have not been scaled up to the level of aquaculture.

18.5.9 Conclusion: Diseases in A. japonicus

Apostichopus japonicus is one of the most economically valuable aquaculture species in China. Improved and expanded technologies to control pathogens and reduce sea cucumber disease and pathology are urgently needed to ensure the sustainable development of the industry for food production. At present, one major problem is the confused nomenclature of the diseases that is based on symptom descriptions of the infected animals. This can lead to incorrect assumptions about whether a particular disease is caused by a specific pathogen and should be treated with a single focused approach, or whether similar symptoms are actually caused by different pathogens or combinations of pathogens. The opposite assumption may be true in which diseases that are labelled with different names because they show different symptoms may actually be caused by the same pathogen that are different versions or stages of the same disease. Future work must elucidate comprehensive epidemiological and etiological descriptions to standardise the classification, nomenclature, and complete description of the diseases. This will require a clear understanding of the pathogens and their pathogenic mechanisms that underlie the diseases of *A. japonicus* in addition to the responses by the host to fight the disease. Research that focuses on understanding the interactions between the pathogens and their associated virulence factors with the host and its immune defense will enable real progress in solving aquaculture diseases. Outcomes are expected to provide a beginning for establishing healthy and sustainable sea cucumbers raised in aquaculture systems.

18.6 Protist pathogens of echinoderms

Eukaryotes that carry out all life functions within a single cell are commonly called protists. Once combined into a single kingdom called the Protista, it has long been understood that they do not form a cohesive, monophyletic group. Some protists are closely related to multicellular animals, others to multicellular plants, and others to fungi, while others form distinct eukaryote clades (Ald et al. 2019). Although our understanding of the phylogenetic relationships among protist groups and their evolutionary history has improved in recent years, the exact structure of the protist phylogenetic tree is still unknown. Nevertheless, many protist lineages can be assigned to one of 12 major groupings or 'clades'

Table 18.2 Disease characteristics and pathogens of *Apostichopus japonicus* in aquaculture

Disease	Characteristics and Mortality	Pathogens	References
Skin ulcer syndrome (SUS)[1,2]	Weak and anorexic; skin ulcers; mortality of 50–100%.	*Vibrio*, *Aeromonas*, *Pseudoalteromonas*, *Photobacterium*, *Shewanella*	Wang, Fang et al. 2006; Zhang, Wang et al. 2006; Wang, Yuan et al. 2007; Yang, Zhou et al. 2007a, 2007b; Deng, He et al. 2009; Li, Qiao et al. 2010a; Liu, Zheng et al. 2010
Acute peristome oedema disease (APED)[1,2]	Peristomial oedema; skin ulcers; eviscera-tion; decreased tube foot adhesion; mortality of 30–60%.	*Vibrio*, *Marinomonas*, corona-like virus	Ma, Xu et al. 2006; Wang, Chang et al. 2007
Off-plate syndrome[2]	Decreased body size; lost attachment to vertical surfaces; mortality of up to 100%.	*Vibrio*, spherical virus	Wang, Zhang et al. 2004; Sun 2006; Song, Wang et al. 2007
Viscera ejection syndrome[1,2]	Evisceration followed by autolysis; mortality of more than 90%	*Vibrio*, spherical virus	Deng, Sui et al. 2005; Deng, Zhou et al. 2008; Deng, He et al. 2009
Stomach ulceration[3]	Thickening followed by dissolution of the stomach walls; mortality of 70–90%	*Vibrio*, *Aeromonas*, *Pseudoalteromonas*	Wang, Zhang et al. 2004
Rotting edge disease[3]	Darkening and disinte-gration of the ectoderm; autolysis; mortality of up to 90%	*Vibrio*	Deng and Sui 2004, Zhang 2007

[1] Disease of juvenile sea cucumbers.
[2] Disese of adult sea cucumbers.
[3] Disease of larval sea cucumbers.

that include multiple phyla, although most have not yet been formally recognised as kingdoms.

18.6.1 The alveolate clade

The alveolates form a major clade of eukaryotes that all display flattened membrane-bound vesi-cles (called alveoli) underneath the cell membrane. The alveoli together with the outer cell membrane form a tough but usually flexible pellicle that gives these cells distinctive, well defined shapes. The clade is very diverse with regard to nutritional sources with forms that are predators, photoau-totrophs, and endo- and ecto-parasites. The major phyla within the alveolate clade include the ciliates (Ciliophora), Apicomplexa, and the dinoflagellates (Dinophyceae). The Apicomplexa are exclusively parasitic, and both the dinoflagellates and ciliates have parasitic as well as free-living members. Both ciliates and apicomplexans are known to infect echinoderms, and although parasitic dinoflagel-lates cause diseases in a variety of invertebrates and fish, none are known to infect echinoderms (reviewed in Coats 1999). However, secreted toxins and/or metabolites from blooms of dinoflagellates can impact larval echinoderms (see Section 18.7.1).

18.6.1.1 Phylum Ciliophora

Members of the phylum Ciliophora are active and conspicuous members of microscopic aquatic ecosystems The ciliates, which include well-known genera such as *Paramecium* and *Stentor*, characteristically have rows of cilia, known as kineties, on the cell surface and nuclear dualism with a large, physiologically active polyploid macronucleus and a smaller diploid micronucleus whose meiotic products are exchanged during conjugation. Most also have a cytostome or 'cell mouth' surrounded by specialised feeding cilia.

Although many ciliates are found living commensally in or on echinoderms, only one species, *Orchitophyra stellarum*, is a known parasite (Figure 18.7.A). This ciliate belongs to the scuticociliate family Orchitophryidae and is closely related to other taxa that are histophagous or parasites of various invertebrates (Bouland et al. 1986; Harikrishnan et al. 2010; Pan et al. 2016; Zhang, Fan et al. 2019). *Orchitophyra stellarum* was first identified infecting the sea star, *Asterias rubens*, in France (Cépède 1907; Vevers 1951). Since then, it has been reported in many species of sea stars from the Atlantic, Mediterranean, and Pacific Oceans (Table 18.3), although it also infects crustaceans (Small 2004, 2005a; Miller et al. 2013). *Orchitophyra stellarum* can be maintained *in vitro* on infusions of yeast and animal tissue and it also exists free-living in sea water where it feeds on bacteria (Stickle et al. 2007a, 2007b). These observations plus samples collected from different hosts from different parts of the world are almost genetically identical (Goggin and Murphy 2000) suggesting that *O. stellarum* might be a fairly ubiquitous facultative parasite with widespread occurrence in habitats where decomposing organic material is abundant (Stickle and Kozloff 2008). However, more research is needed to confirm that this is a single, pan-global species by using more refined genetic markers to distinguish cryptic species in this genus. It is important to note that the lack of host species specificity for *O. stellarum* and its ability to live in an ecosystem as a free-living organism cautions against its use in biocontrol of nuisance or invasive sea stars (Byrne et al. 1997).

In echinoderms, *Orchitophyra stellarum* is present on the external surface of both sexes of sea stars but it is only a parasitic castrator of males, which is accomplished in a variety of ways. Initially the ciliate colonises the aboral surface epithelium of the sea star and then invades through the gonopores

travelling into the testes where it feeds actively on sperm. Ciliates in the testes are filled with food vacuoles, many of which contain sperm in various stages of digestion (Figure 18.7.B); (Sunday et al. 2008) that causes partial or complete castration (Figure 18.7.C). Ciliates multiply rapidly and produce dense populations within the testes, which can disrupt the germinal epithelium of the testis when ciliates burrow between spermatocyte columns and cause them to separate and release sperm into the coelomic cavity (Bouland and Jangoux 1988; Byrne et al. 1997). This activity not only damages the germinal epithelium of the testis but also confuses and thereby avoids the host immune system (Bang 1982; Taylor and Bang 1978; Bouland and Jangoux 1988; Leighton et al. 1991; Byrne et al. 1997; Goggin and Bouland 1997). In response to damage to the germinal layer, the host increases the number of coelomocytes, however this has little effect on the ciliate (Coteur et al. 2004). This is because the ciliates alter the sea star immune response by blocking the ability of the coelomocytes to aggregate, which also blocks encapsulation of ciliates, possibly due to ammonia waste secretion from *O. stellarum* (Taylor and Bang 1978). Hence, rather than recognising the ciliates as foreign and targeting them for removal, the coelomocytes initiate a destructive auto-immune response to attack, phagocytose, and clear their own sperm that are present in the wrong anatomical space (Byrne et al. 1997; Sunday et al. 2008). The outcome of this reaction is that the sea star immune response is often responsible for the marked decrease and eventual absence of sperm. When *O. stellarum* is placed in culture to which lobster haemolymph is added, the ciliates secrete metalloproteases, which degrade muscular structural proteins such as myosin heavy chains (Small et al. 2005b). Although those enzymes have been reported for infected lobsters, it is feasible that *O. stellarum* uses a similar biochemical attack mechanism to break down tissues in sea star hosts. As the disease progresses, the testes shrivel, harden, and are discoloured brown (healthy testes are rounded, soft, and creamy white). In some cases, not only are male sea stars castrated by the ciliate, but mortalities of the male sea stars also occur. Male ochre stars, *Pisaster ochraceus*, infected with *O. stellarum* also show body surface discolouration

(A)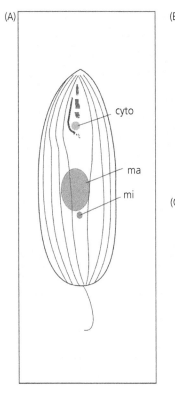

(B)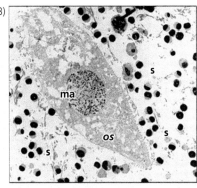

(C)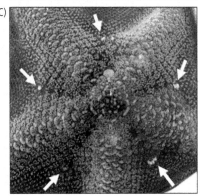

Figure 18.7 The protist parasite ciliate, *Orchitophyra stellarum*, infects the testes of male asteroids. (A) An illustration of *O. stellarum* shows the cytophage (cyto) or cellular mouth, the macronucleus (ma), and the micronucleus (mi). (B) *O. stellarum* (os) in the testes of the sea star, *Asterias amurensis*, is surrounded by sperm (s). The image from transmission electron microscopy shows the macronucleus (ma) of the ciliate. (C) A male bat star, *Patiria miniata*, from Vancouver Island that is severely infected with *O. stellarum* shows portions of testis lobes that contain ciliates extruding from the gonopores (arrows). The ciliates are also detected on the surface of infected males (see Sunday et al. 2008). Source: image (A) is modified from Bouland et al. (1986). Images (B) and (C) were provided by M. Byrne.

and decreased overall health (Leighton et al. 1991), whereas infected males of the Northern Pacific sea star, *Asterias amurensis* are more likely to autotomise arms and die, as has been observed for animals collected from the wild (Byrne et al. 1997).

Orchitophyra stellarum typically infects asteroids in the family Asteriidae and was first described in the sea stars *Asterias rubens* and *A. forbesi*, in the North Atlantic (Jangoux 1987a; Goggin and Bouland 1997). The ciliate was later reported in *Asterias amurensis* in Tokyo Bay (Byrne et al. 1997, 1998) and in several species along the Pacific coast of North America (Leighton et al. 1991; Bates et al. 2010). Although for nearly 100 years this parasite was only known from *Asterias* species, it has been identified as a parasite for least ten other genera, including species outside the Asteriidae (Sunday et al. 2008; Byrne 2017) and hosts in other phyla (Small 2004; Small et al. 2005a; Miller et al. 2013). Population surveys for some species of sea stars where the parasite occurs reveal a prevalence of females suggesting that *O. stellarum* may kill some of the males (Leighton et al. 1991; Byrne et al. 1997). Although there is a great deal of geographic variation in abundance (Zilz 2018), infection by *O. stellarum* and the intensity of its pathogenicity is generally associated with water temperature. In laboratory experiments, mean cell division time for *O. stellarum*, which is equivalent to the reproduction rate, is faster in warmer waters, which correlates with sea stars kept in warmer waters that have more heavily infected testes (Bates et al. 2010). In the wild, *O. stellarum* occurs along on the West Coast of North America but is not present in Alaska, where temperatures are presumably too cold for the parasite to survive (Bates et al. 2010; Stickle and Kozloff 2008). On the other hand, *O. stellarum* is present in northern Hokkaido Japan, where the waters are cold (Table 18.3) (Byrne et al. 1997; Goggin and Bouland 1997; Byrne et al. 1998). Global warming and increases in ocean temperatures may have profound effects on the distribution of *O. stellarum*, which in turn, may alter the populations of sea stars and the ecological structures in regions where they are found.

18.6.1.2 Phylum Apicomplexa

All members of the phylum Apicomplexa are parasites with some species responsible for notable diseases in humans (e.g., malaria and toxoplasmosis), livestock (e.g., babesiosis and coccidiosis), other vertebrates, and invertebrates (Votýpka et al. 2017). The name of the phylum refers to the unique complex of organelles at the anterior apical end of the infective or sporozoite stage of the life cycle and includes spirally arranged microtubules, polar ring(s), and secretory bodies called as rhoptries and micronemes. This apical complex structure facilitates entry of the parasite into host cells or between cells. The phylum is the sister taxon to a group of free-living photosynthetic flagellates (*Chromera*, *Vitrella*, and *Colpodella*) and many apicomplexans possess a unique organelle called the apicoplast, which is a highly reduced non-photosynthetic plastid. The apicoplast retains a few functions essential for parasite survival such as lipid metabolism, isoprenoid synthesis, heme synthesis, and iron-sulfur cluster synthesis.

The Apicomplexan phylum includes morphologically and ecologically diverse protists, such as the gregarines (Class Gregarinida), coccidia (Class Coccidia), haemosporidia (Class Haemosporidia), and piroplasms (Class Piroplasmida) (Votýpka et al. 2017). Many species that are members of these classes, including 26 gregarines and one coccidian, are parasites of 27 species of holothuroids and spatangoid echinoids (Table 18.4). There are undoubtedly more types of echinoderms infected with apicomplexans and more species of apicomplexans associated with echinoderms to be discovered, but at present, knowledge is limited. Even with the species that have been identified, information on their life cycle, biology, and host-microbe interactions are relatively unknown.

18.6.1.3 Class Gregarinida

Gregarines are a large and abundant group of apicomplexans that live exclusively in the intestines, coelom, and other extracellular spaces of invertebrate hosts, including annelids, molluscs, nemerteans, phoronids, echinoderms, sipunculids, crustaceans, hemichordates, appendicularians, and insects. The gregarine life cycle in echinoderms is monoxenous, meaning that each parasite species is restricted to a single host species (Figure 18.8), although a general scheme of interactions between the parasite and the host has numerous genus- and species-specific modifications. The host is infected when it ingests a gregarine sporocyst (Figure 18.8.A). Once in the host digestive tract, the cyst breaks open and sporozoites are released, which employ the functions of the apical complex to penetrate the epithelial cells lining the wall of digestive tract (Figure 18.8.B). For many species, the sporocysts migrate through the epithelial cells and into the hemal system or the coelom where they grow and transform into much larger trophozoites (Figure 18.8.C). After an enormous size increase, each trophozoite binds to a partner trophozoite in a species-specific orientation (head-to-head, tail-to-tail, side-to-side, or head-to-tail) in a process called syzygy (Figure 18.8.D). The joined cells develop into gamonts and become enclosed in a gametocyst (Figure 18.8.E). Several cell divisions inside the gametocyst give rise to hundreds of gametes from each of the original cells (Figure 18.8.F). Gametes from one of the original gamonts fuse with one of the gametes from the other to produce numerous zygotes (Figure 18.8.G). A sporocyst wall develops around each zygote (Figure 18.8.H), which divides to produce sporozoites (eight in many species that infect echinoderms), each becoming a sporocyst (Figure 18.8.A). Mature sporocysts are released into the environment either with feces or evisceration (Jespersen and Lützen 1971). However, they can also be encapsulated by aggregates of coelomocytes that include the pigmented red spherule cells (see Section 18.1.4; and see Box 1.2 and Figure 1.6 in Chapter 1) forming brown bodies (DeRidder and Jangoux 1984) that are expelled through the coelomo-cloacal ducts during coelomic purging in holothuroids (Doignon et al. 2003), or when the host dies and decomposes. Once in the sediment, the sporocysts are available to be taken up by a new host, and the cycle repeats. Gregarines are common in holothurians and spatangoid echinoids (Jangoux 1987a; Eeckhaut et al. 2004) because these groups are deposit feeders and may become infected simply by ingesting sediment that contains infective sporocysts. A few gregarines parasitise suspension-feeding sea cucumbers (e.g., *Pawsonia*

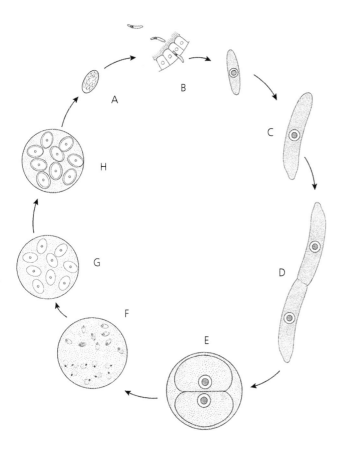

Figure 18.8 The generalised gregarine life cycle in echinoderms. A. The sporocyst stage in the environment begins the cycle. B. Sporozoites are ingested and released from the cyst in the digestive tract of the host and penetrate the gut wall. C. Sporocysts grow and transform into trophozoites. D. Two trophozoites bind in syzygy. E. The two trophozoites in syzygy develop into gamonts within a gametocyst. F. Cell division within the gametocyst produces hundreds of gametes. G. Gametes from each original gamont fuse to form zygotes. H. Each zygote secretes a wall and cell division results in eight sporozoites. The cycle repeats when the sporozoite is eaten by an echinoderm. See text for additional details. Source: this figure is based on descriptions by Coulon and Jangoux (1987); Dogiel (1906); and Theodorides and Laird (1970).

saxicola and *Cucumaria frondosa*) when sporocysts in the sediment are taken up with the water current into the cloaca during respiration and thus are deposited into and infect via the respiratory trees (Woodcock 1906; Djakonov 1923; Pixell-Goodrich 1929).

Details about the ability of gregarines to cause disease are limited and no lethality from gregarines has been described in the literature to date. There is some evidence that the growth and feeding of the trophozoites and accumulation of cysts may cause mechanical damage to the tissues of the host, such as mass infestations that cause the hemal lacunae to become so distended by cysts that the hemal fluid no longer circulates (Pixell-Goodrich 1925). One species of gregarine, *Gonospora gonadipertha*, has been described as infecting and partially destroying the gonads of the sea cucumber, *Cucumaria frondosa* (Djakonov 1923). Trophozoites of some species can grow so large that they induce the formation

of pouches in the host mesothelium and connective tissue, which rupture and release the trophozoites into the coelomic cavity (Changeux 1961; Lützen 1968; Jangoux, 1987a). Despite these reports, it is not known whether the presence of gregarines normally produce permanent damage to the host tissues or result in any mortality. What is known is that infected organisms have an aggressive coelomocyte reaction to gregarines (Pomory and Lares 1998). Coelomocytes coat the outer surface of the intestine that faces the coelomic cavity and form an unbroken layer covering the gregarines that are attached at this location during maturation. However, coelomocytes do not appear to bind directly to the trophozoites and therefore the immune response in holothuroids does not appear to stop the infection. In contrast to holothuroid infections, echinoids infected with gregarines do elicit a strong and effective immune response (Léger 1897; Pixell-Goodrich 1915; Brownell and McCauley 1971;

DeRidder and Jangoux 1984 ;Coulon and Jangoux 1987) in which coelomocytes surround and attack trophozoites, killing the parasites. Furthermore, when parasite are in the paired gamont stage, they are more sensitive to coelomocyte attack compared to single trophozoites, suggesting that the echinoid immune response prevents the successful completion of the life cycle (Coulon and Jangoux 1987).

18.6.1.4 Class Coccidia

A single described species of Coccidia, *Ixoreis psychropotae*, that parasitises echinoderms is known from the deep-sea holothuroid, *Psychropotes longicauda* (Massin et al. 1978). Like most of the other sporozoans that infect sea cucumbers, *I. psychropotae* is found in the hemal system that is associated with the gut where it may occur in very high numbers. The species is relatively unique because the gamont stage is absent from the life cycle, although the other stages are present (Levine 1984). A second coccidian been identified using DNA sequencing from the ovaries of the holothuroid *Apostichopus japonicus*, but this species has not yet been described completely (Unuma et al. 2020).

18.6.2 The Amoebozoa clade

As the name suggests, the Amoebozoa clade consists mostly, although not entirely, of organisms that are amoebae for much or all of their life cycles. Although the group lacks a clear morphological character in all members, it was identified based on phylogenetic analyses of DNA sequence data (Cavalier-Smith et al. 2016; Kang et al. 2017; Ald et al. 2019). This clade includes many of the naked (without shells, tests, or scales) amoebae with lobose pseudopods (e.g., *Amoeba proteus*), testate amoebae (e.g., *Arcella*), pelobionts (e.g., *Pelomyxa*, *Entamoeba histolytica*), as well as cellular and acellular slime molds. Amoebozoans in the genus *Paramoeba* are small, naked amoebae with dactylopodiate (more or less finger-like) pseudopodia. The genus is found in temperate marine environments in the northern and southern hemispheres. Some species are free-living, feeding on bacteria and protists, while others persist as free-living amoebae but are also facultative parasites in sea urchins, crustacea, and fish (Table 18.5).

Paramoeba cells have a characteristic structure positioned next to the nucleus known as the 'parasome', which is actually a eukaryotic endosymbiont (Kudryavtsev et al. 2011; Sibbald et al. 2017). Using the DNA sequence of the parasome endosymbiont in a phylogenetic analysis, it is identified as a member of the Kinetoplastida of the genus, *Perkinsela* (Young et al. 2014). *Perkinsela* lacks a flagellum and is surrounded by a single membrane that is not host-derived. Like other kinetoplastids, it contains a single large, DNA-rich mitochondrion and has mitochondrial RNA editing capabilities including spliced-leader trans-splicing (Tanifuji et al. 2017). The symbiotic relationship between *Paramoeba* and *Perkinsela* appears to be stable and obligatory, but whether it benefits to one or both members is unknown. Although most kinetoplastids are parasitic, it is not known whether the presence of *Perkinsela* within *Paramoeba* contributes to the parasitic life history of the amoeba (Tanifuji et al. 2017). Molecular investigations show that the phylogenetic relationships of *Perkinsela* species mirrors that of its host *Paramoeba* species, which is a strong indicator of co-evolution between the two genera (Young et al. 2014; Sibbald et al. 2017). When sea urchins are infected with *Paramoeba* they become immobile due to muscle damage, stop feeding, lose attachment to the substrate, and die (Jones and Scheibling 1985; Jones 1985; Jones et al. 1985, Jellett et al. 1988; Jellett and Scheibling 1988a, 1988b). How the amoebae get into the body of the sea urchin is not known, but the disease can be reproduced by either injection of animals or their emersion in water with cultured amoebae (Jones and Scheibling 1985).

18.6.3 The Rhizaria clade

The Rhizaria superclade includes many shelled amoebae that have threadlike (filose) pseudopodia (such as the phyla Foraminifera and Radiolaria), the phylum Cercozoa (amoebae and flagellates that feed using filose pseudopods), in addition to a group of plant parasites, Phytomyxea, and animal parasites, Ascetosporea (Ald et al. 2019). The order Haplosporida (within the Ascetosporea) are parasites of many invertebrates but only one species, *Haplosporidium comatulae*, parasitises an echinoderm

(La Haye et al. 1984). It has been identified in the tissues surrounding the digestive tract of the feather star, *Oligometre serripinna*. Infected crinoids, which contain both the intracellular and spore stages of the of the haplosporidian life cycle, have damage to the walls of the digestive tract that appears greatly reduced in thickness.

18.6.4 Plantae (or Archaeplastida)

The plant clade consists of eukaryotes whose plastids were acquired directly through a symbiosis with a cyanobacterium (Price et al. 2012). This clade consists of the Rhodophyta (red algae), the Glaucophyta (a small group of freshwater unicellular algae), and the Chloroplastida or Viridiplantae (green algae and the land plants). While most are free-living, a few unicellular members of the green algae (Chlorophyceae) are parasitic.

18.6.4.1 Class Chlorophyceae

The Chlorophycaca include the green algae, and the genus *Coccomyxa* includes both free-living species and species that live as symbionts with other organisms *Coccomyxa* species have a worldwide distribution, with representatives found in many moist terrestrial and aquatic ecosystems. There are a few marine species that are parasitic to echinoderms (Guiry and Guiry 2020) including *Coccomyxa ophiurae* that infects three species of brittle stars, *Ophiura texturata*, *Ophiura albida*, and *Ophiura sarsii* from the North Atlantic and North Sea (Mortensen 1897; Mortensen and Rosenvinge 1910, 1933), and *Coccomyxa astericola* that infects the sea stars, *Hippasteria phrygiana* and *Solaster endeca* (Mortensen and Rosenvinge 1910, 1933). In all infections, the algal cells first invade the aboral surface epidermis where they grow and divide to form small green lesions. As the algal masses grow, these lesions coalesce, and the epidermis disintegrates. Eventually the arms of the sea stars undergo autotomy, the body cavity becomes perforated, and the echinoderm dies. In sea urchins, the algae dissolve the skeletal plates of the test forming holes in which the algae can continue to grow. Unfortunately, because *C. ophiurae* and *C. astericola* have not been studied since the last work of Mortensen

and Rosenvinge (1933), no additional details are available.

18.6.5 Miscellaneous

There are descriptions of protists parasitizing echinoderms that are not sufficiently detailed to know what type of protist is the basis of the infection. For example, opaque, spherical 'amoebae' or 'flagellates' of about 10 microns in diameter have been described that attack the digestive system of larval sea cucumbers (*Isostichopus fuscus*) in aquaculture (Becker et al. 2009). In this disease, the protist enters through the body wall and appears to feed on the intestinal content or the larval tissues, followed by intestinal shrinkage and larval death. The protist might be a member of the amoebozoan or rhizarian clades but cannot be defined based on the description. Similarly, unidentified 'flagellates' have also been noted in the body cavity of the heart sea urchin, *Meoma ventricosa*, but based on the report, it is not possible to even place this pathogen into a phylum (Chesher 1969). There are other descriptions of flagellated protists in the body cavity of some echinoderms that may not be protists but instead may be vibratile cells, which are coelomocytes that have a single long flagellum and swim through the body fluids (see Section 18.1.4; and see Box 1.2 and Figure 1.6 in Chapter 1). Xing et al. (2008) suggested that the unidentified flagellates described by Hetzel (1963) and a flagellate identified by Lecall (1980) as *Cryptobia* in the coelomic fluids of the rosy feather star, *Antedon bifida*, could in both cases be vibratile cells. The flagellate *Oikomonas echinorum*, noted in the general body cavity of most European echinoids (Cuénot 1912), is now also thought to be a miss-identification of vibratile cells (Barel and Kramers 1977; Jangoux 1987a).

18.6.6 Conclusion: Protist parasites of echinoderms

The current knowledge of protist parasites in echinoderms is extremely patchy and undoubtedly incomplete. This is, in part, because many parasitic protists are small (most less than 1 mm), cryptic (becoming visible only when the host is dissected and its tissues examined with light or

Table 18.3 Distribution of *Orchitophyra stellarum* in echinoderms

Host	Location	Reference
Asterias amurensis	Japan (Tokyo Bay; Ise Bay; Otsuchi Bay; Usujiri Bay; Mutsu Bay)	Byrne et al. 1997, 1998; Goggin and Bouland 1997
Asterias forbesi	United States (Long Island Sound, New York; Isle of Shoals, New Hampshire)	Piatt 1935; Galtsoff and Loosanoff 1939; Burrows 1936, Stickle and Kozloff 2008
Asterias rubens	France (Boulogne-sur-Mer; Wilmereux); Netherlands (Wadden Sea); United States (Isle of Shoals, New Hampshire; Gulf of Maine); England (Plymouth); Belgium (North Sea); Canada (Prince Edward Island)	Cépède 1907; Schoenmakers et al. 1984; Stickle and Kozloff 2008; Vevers 1951; Jangoux and Vloebergh 1973; Smith 1936, Lowe 1978
Asterina miniata	Canada (Barkley Sound, British Columbia)	Bates et al. 2010
Evasterias troschelii	United States (Puget Sound and Manchester, Washington; Little Port Walter, Alaska)	Stickle et al. 2001b; Stickle and Kozloff 2008
Leptasterias spp.	United States (San Juan Island, Washington)	Stickle et al. 2001a
Patiria miniata	Canada (Dixon Island and Grappler Inlet, British Columbia)	Sunday et al. 2008
Pisaster ochraceus	Canada (Barkley Sound, British Columbia), United States (Post Point, Washington)	Leighton et al. 1991; Bates et al. 2010; Zilz 2018
Sclerasterias richardi	Corsica (Calvi)	Febvre et al. 1981

electron microscopy), and often present in low densities. For example, protists in the phylum Microsporidia (Fungi clade) are obligate intracellular, spore-forming parasites that are microscopic, usually just 1–4 microns in size. They are parasites of many eukaryote phyla including other protists and aquatic invertebrates such as crustacea, annelids, molluscs, cnidarians, sponges, bryozoans, and platyhelminths (Leung et al. 2015). Although none have been described in echinoderms, it may be likely that they have been overlooked rather than being truly absent. Similarly, lack of discovery of parasitic dinoflagellates (Coats 1999), myxozoans (Lom and Dyková 2013), or parasitic excavates (which includes parasites such as *Trichomonas, Naegleria*, and *Trypanosoma* that have an 'excavated' feeding groove on one side of the cell) in echinoderms may simply be a sampling problem and not true nonexistence.

Molecular approaches in which environmental samples or host tissue samples are processed for DNA sequencing and exons of genes may help detect elusive protist parasites (for review see Bass et al. 2015; Okamura et al. 2018). For example, a surprising diversity of protist parasites have

been identified based on sequencing environmental DNA of plankton collected *en masse* from the ocean photic zone (de Vargas et al. 2015) or from neotropical rainforests (Mahé et al. 2017). As approaches improve for acquiring environmental DNA sequence data sets, it may be possible to launch targeted searches for parasites of invertebrates with the outcome of a more complete accounting of echinoderm protist parasite diversity.

18.7 Microbial pathology of sea urchin larvae

The literature on the infectious disease pathology of larval echinoderms is scarce and focuses primarily on sea urchins (although see Section 18.5.5). This limits our understanding of echinoderm pathology because larval studies expand and integrate information on adults into the contexts of development, ecology, and life history. In general, larval echinoderms have an immune system that is distinct from that in the adult, yet both are the outcome of a single genome and protects the host as a planktonic larva and as a benthic adult, plus

Table 18.4 Apicomplexans that infect echinoderms

Phylum Apicomplexa	Host	Location in host	Geographic location	References
Class Gregarinida				
Order Archigregarinida				
Veloxidium leptosynaptae	Leptosynapta clarki (H[1])	Intestine	Western Pacific (Canada)	Wakeman and Leander 2012
Selenidium synapta	Synapta sp.(H)	Intestine	Mediterranean Sea (Italy)	Mingazzini 1893
Order Eugregarinida				
Gonospora gonadipertha (syn.[3], Diplodina gonadipertha)	Cucumaria frondosa (H)	Gonads	Arctic Ocean (Murmansk, Russia)	Djakonov 1923; Levine 1977
Gonospora holoflora	Holothuria floridana (H)	Intestine	Gulf of Mexico (Florida Keys)	Pomory and Lares 1998
Gonospora irregularis (syns., Cystobia irregularis, Gregarina irregularis, Diplodina irregularis)	Holothuria forskali (H); Holothuria nigra (H)	Hemal sinus	English Channel	Minchin 1393; Levine 1977; Trégouboff 1918; Woodcock 1904, 1906
Gonospora mercieri (syn., Goniospora mercieri)	Oestergrenia digitate (H) (syns, Labidoplax digitate, Synapta digitate)	Coelom	Atlantic Ocean (France)	Cuénot 1912
Gonospora minchini (syns, Cystobia minchinii, Diplodina minchinii, Lithocystis minchini)	Pawsonia saxicola (H) (syn., Cucumaria saxicola)	Hemal sinus	English Channel	Woodcock 1906; Pixell-Goodrich 1929
Gonospora stichopi (syns. Cystobia stichopi)	Parastichopus tremulus (H) (syn., Stichopus tremulus)	Along muscle bands	North Sea (Norway)	Levine 1977; Lützen 1963
Lithocystis brachycercus	Chiridota laevis (H)	Intestinal wall, coelom	North Atlantic (England, Canada)	Goodrich 1930
Lithocystis cucumariae	Pawsonia saxicola (H) (syn., Cucumaria saxicola)	Respiratory tree	English Channel	Goodrich 1930
Lithocystis foliacea	Echinocardium cordatum (E)	Coelom	Mediterranean Sea (Italy); English Channel	Pixell-Goodrich 1915; Coulon and Jangoux 1987
Lithocystis latifronsi	Briaster latifronsi (E[2])	Coelom	North Pacific (United States)	Brownell and McCauley 1971
Lithocystis microspora	Spatangus purpureus (E)	Coelom	English Channel	Pixell-Goodrich 1915
Lithocystis oregonensis	Briaster latifrons (E)	Coelom	North Pacific (United States)	Brownell and McCauley 1971
Lithocystis schneideri	Echinocardium cordatum (E)	Coelom	North Atlantic (France), Mediterranean Sea (Italy), English Channel	Giard 1876; Léger 1896, 1897, De Ridder and Jangoux 1984; Cuénot 1891; 1892, 1912, Pixell-Goodrich 19˙5; Coulon and Jangoux 1987

continued

Table 18.4 *Continued*

Phylum Apicomplexa	Host	Location in host	Geographic location	References
Urospora chiridotae (syn., *Chiridota chiridotae*)	*Chiridota laevis* (H) (syn., *Chiridota pellucida*)	Blood vessels	Arctic Ocean (Murmansk, Russia)	Dogiel 1906; Pixel-Goodrich 1925; Theodorides and Laird 1970; Dyakin and Paskerova 2004
Urospora echinocardi	*Echinocardium cordatum* (E), *Spatangus purpureus* (E)	Coelom	English Channel	Pixell-Goodrich 1915
Urospora grassei (syn., *Cystobia grassei*)	*Holothuria tubulosa* (H), *Holothuria stellati* (H)	Hemal sinus	Mediterranean Sea (France)	Changeux 1961; Levine 1977
Urospora holothuriae (syns., *Gregarina holothuriae*, *Monocystis holothuriae*, *Syncystis holothuriae*, *Cystobia holothuriae*, *Lithocystis chiajii*)	*Holothuria tubulosa* (H), *Holothuria stellati* (H)	Hemal sinus	Mediterranean Sea (France, Italy); North Atlantic (England)	Changeux 1961; Cuénot 1891; Schneider 1858; Trégouboff 1918
Urospora intestinalis	*Cucumaria japonica* (H)	Intestine	NW Pacific (Peter the Great Bay, Sea of Japan)	Bogolopeva 1953
Urospora muelleri (syn., *Syncystis muelleri*)	*Oestergrenia digitate* (H) (syn., *Synapta digitate*)	Coelom	Mediterranean Sea (France, Italy)	Cuénot 1891, 1892
Urospora neapolitana	*Echinocardium cordatum* (E)	Coelom	Mediterranean Sea (Italy); English Channel	Pixell-Goodrich 1915; Coulon and Jangoux 1991
Urospora ovalis	*Echinocardium cordatum* (E)	Coelom	Mediterranean Sea (Italy)	Pixell-Goodrich 1915; Dogiel 1910
Urospora pulmonalis	*Cucumaria frondosa* (H) (syn., *Cucumaria japonica*)	Respiratory trees	NW Pacific (Peter the Great Bay, Sea of Japan)	Bogolopeva 1953
Urospora schneideri (syns., *Cystobia schneideri*)	*Holothuria polii* (H); *Holothuria impatiens* (H), *Holothuria tubulosa* (H), *Holothuria stellate* (H)	Hemal sinus and coelom	Mediterranean Sea (France, Italy)	Changeux 1961; Cuénot 1891; Levine 1977; Minchin 1893; Mingazzini 1893
Urospora synaptae (syns., *Syncystis synaptae*, *Esarabdina synaptae*, *Lithocystis muelleri*)	*Leptosynapta galliennei* (H), *Leptosynapta inhaerens* (H)	Coelom	North Atlantic (France)	Cuénot 1891, 1892, 1912; Barel and Kramers 1970
Class Coccidia				
Ixoreis psychropotae	*Psychropotes longicauda* (H)	Intestine	North Atlantic (deep sea west of France)	Massin et al. 1978
Undescribed species	*Apostichopus japonicus* (H)	Ovary	Sea of Japan	Unuma et al. 2020

[1] H, Holothuriidae
[2] E, Echinoidea
[3] syn(s), taxonomic synonym(s)

Table 18.5 Distribution of *Paramoeba* that infect echinoderms[1]

Species	Host	Location	Reference
Paramoeba invadens	*Strongylocentrotus droebachiensis*	Canada (Nova Scotia)	Jones 1985; Jones and Scheibling 1985; Jones et al. 1985; Jellett and Scheibling 1988a, 1988b; Feehan et al. 2013; Buchwald et al. 2018
Paramoeba branchiphila (syn., *Neoparamoeba branchiphila*)	*Diadema africanum*	Spain (Canary Islands)	Dyková et al. 2011
	Heliocidaris erythrogramma[2]	Australia (Tasmania)	Dyková et al. 2007
	Paracentrotus lividus[2]	Greece (Cretan Sea)	Dyková et al. 2007
Paramoeba pemaquidensis (syn., *Neoparamoeba pemaquidensis*)	*Strongylocentrotus droebachiensis*	United States (Gulf of Maine)	Caraguel et al. 2007

[1] The *Paramoeba* genus has some nomenclature confusion. Page (1987) divided *Paramoeba* into two genera (*Paramoeba* and *Neoparamoeba*) based on the presence or absence of surface scales. Recent molecular phylogenetic analyses however, show that scales have been secondarily lost multiple times and that *Neoparamoeba* must be synonymised with *Paramoeba* to maintain monophyly (Kudryavtsev et al. 2011; Feehan et al. 2013). Thus, citations of *Neoparamoeba* in the literature and in sequence submissions to GenBank refer to the genus *Paramoeba*.

[2] *Paramoeba* has been noted as present but is not associated with disease.

during metamorphosis as it navigates the transition between these two life phases in two environments (Ho et al. 2016). Although little is known of diseases in developing echinoderms, embryonic and larval cultures are employed to evaluate environmental factors based on their impacts on embryonic and larval development in assays that evaluate sea urchin embryo toxicity (SET). SET assays are used widely to screen toxic abiotic factors including environmental marine pollutants and physical factors such as ultraviolet (UV) light spectra, CO_2 levels, ocean warming, acidification, heavy metals, organic and inorganic nanoparticles, drugs and microplastics, in addition to other pollutants in the marine system (Isely et al. 2009; Lister et al. 2010; Byrne et al. 2013; Mesarič et al. 2015; Reinardy and Bodnar 2015; Blewett et al. 2016; Gambardella et al. 2016, Tyunin et al. 2016; Martínez-Gómez et al. 2017; Magesky et al. 2018; Messinetti et al. 2018; Chiarelli et al. 2019; Lee et al. 2019; Beiras et al. 2019; Lenz et al. 2019; Ragusa et al. 2019; Stumpp et al. 2019; Murano et al. 2020). The implications of these assays, which are conducted in laboratories, is that these factors can alter larval growth, development, and recruitment to ecosystems, which are necessary to maintain healthy populations in nature. Although a summary of this extensive literature is outside the scope of this review, this section will focus on what

is known about microbial pathogens of echinoderm larvae with an emphasis on echinoids.

The first studies of echinoderm larval pathology were conducted by Ilya Metchnikoff in the 19th century (Metchnikoff 1893). After impaling larval sea urchins and sea stars with rose prickles, he observed phagocytes migrating to the site of injury and encapsulating the foreign material. This was a seminal result that initiated the field of cellular immunology. Only recently have researchers returned to larval echinoderms using modern approaches to understand their immune systems. Although little work has been performed on viral, parasitic, or prion pathogens of echinoderm larvae, the bacterial microbiomes of larval echinoderm species have been characterised by sequencing the 16S ribosomal RNA (rRNA) genes to assess the bacterial species and to provide evidence as to whether they are foreign pathogens, endogenous and primary commensals, or opportunistic commensals that might be capable of invasion. Larval sea urchins that have been evaluated include *Lytechinus variegatus*, *Strongylocentrotus purpuratus*, *S. droebachiensis*, and *Mesocentrotus franciscanus* (Ho et al. 2016; Carrier and Reitzel 2018; Carrier and Reitzel 2019, Carrier et al. 2019, Schuh et al. 2020), and larval sea stars include *Oreasteridae* sp., *Mithrodia clavigera*, and *Acanthaster planci*

(Galac et al. 2016; Carrier et al. 2018), in addition to the larval sea cucumber, *Apostichopus japonicus* (see Section 18.5.5). These microbiome data in conjunction with the elucidation of larval immune cell types (see Section 18.1.4; and see Box 1.2 and Figure 1.6 in Chapter 1) and their functions (Solek et al. 2013; Schrankel et al. 2016, Ho et al. 2016; Buckley et al. 2017; Heyland et al. 2018; Schuh et al. 2020; reviewed in Smith et al. 2018) and the increased use of SET assays (Manzo, 2004; Morroni et al. 2016) have established echinoderm larvae as simple but informative models for investigations of echinoderm pathologies. Furthermore, several studies have addressed the toxicity of microbially larvae are likely to encounter in their planktonic environment (Lopes et al. 2010; Romano et al. 2010; Varrella et al. 2014, 2016; Costa et al. 2015; Beleneva et al. 2015; Liu et al. 2016; Ruocco et al. 2016; Sartori and Gaion 2016; Leung et al. 2017; Neves et al. 2018; Albarano et al. 2019).

The evaluation of echinoderm embryos and larvae using SET assays begins with fertilised sea urchin eggs following standard procedures (Hodin et al. 2019) followed by exposure to a toxin or a pathogen of interest for 24–72 hours (Byrne et al. 2003; Manzo 2004; Byrne et al. 2008; Morroni et al. 2016). Zygotes are evaluated for normal development from cleavage to pluteus larval stage including whether they fail to develop, show a developmental arrest and die and at what stage, or reach pluteus stage with or without morphological abnormalities. This approach has largely been used to screen the level of toxic activity of potential abiotic marine pollutants, however, several groups have used it to test pathogenesis of live microbes or their associated toxins. When zygotes of the sea urchin, *Strongylocentrotus intermedius*, are exposed to bacteria (genera *Aliivibrio, Bizionia, Colwellia, Olleya, Paenibacillus, Photobacterium, Pseudoalteromonas, Shewanella, Vibrio*) isolated from the coelomic fluid from an adult sea urchin, *S. pallidus*, several species (*Aliivibrio, Vibrio, Colwellia, Shewanella,* and *Photobacterium* spp.) inhibit growth of the larvae or cause developmental arrest by 72 hours post-fertilisation (hpf) (Kiselev et al. 2013). Conversely, zygote contact with other bacterial strains increases both larval viability and pigment cell numbers, perhaps suggesting either a degree of symbiosis or the ability

of the developing sea urchin to activate its immune system and resist infection.

18.7.1 Evaluation of toxic compounds in the environment using larval sea urchins

In addition to live pathogens, several groups have addressed the toxicity of compounds secreted into the larval environment by bacteria, diatoms, and dinoflagellates (Lopes et al. 2010; Romano et al. 2010; Varrella et al. 2014, 2016; Costa et al. 2015; Beleneva et al. 2015; Liu et al. 2016; Ruocco et al. 2016; Sartori and Gaion 2016; Leung et al. 2017; Neves et al. 2018; Albarano et al. 2019). When extracts from cyanobacteria (*Cyanobium, Leptolyngbya, Microcoleus, Phormidium, Nodularia, Nostoc,* and *Synechocystis*) collected from estuaries on the Iberian Peninsula are added to cultures of the sea urchin, *Paracentrotus lividus*, larvae show numerous defects in a 48-hour SET assay including developmental arrest, decreased cell aggregation, decreased growth, and arm malformations in plutei (Lopes et al. 2010). Similarly, the lipophilic fraction of crude extracts from *Cyanobium* sp. from the Portuguese coast and tested on *P. lividus* embryos in a 48-hour SET assay shows significant decreases in embryogenesis and development to the larval stage (Costa et al. 2015). Although these effects may be widespread among picocyanobacterial taxa based on genetic and phenotypic similarities, it is unclear whether the effects of these extracts are relevant in the wild. For example, it is not known whether these compounds are secreted at effective concentrations into the natural environment of larvae or are released upon ingestion and lysis of *Cyanobium* cells in the larval gut. However, two strains of the bacterial species, *Pseudomonas aeruginosa*, release exotoxins (notably phenazine pigments), which inhibit normal larval development in *Strongylocentrotus nudus* by interfering with cytoskeleton formation (Beleneva et al. 2015). These examples illustrate that in addition to microbial colonisation and infection of echinoderm larvae, the molecules released from microbes can also have deleterious effects on development and survival.

Diatoms are a dominant group of microalgae in oceanic ecosystems that represent a major food source for zooplankton (Romano et al. 2010; Liu et al. 2016), including echinoderm larvae (Strathmann 1971; Wray et al. 2004; Hodin et al. 2019). Three polyunsaturated aldehydes (2,4-decadienal, 2,4-octadienal, and 2,4-heptadienal) commonly produced and released by diatoms (e.g., *Skeletonema costatum*, *Skeletonema marinoi*, and *Thalassiosira rotula*) and tested on larvae of the rock-boring sea urchin, *Echinometra mathaei*, in a 48-hour SET assay, all inhibit normal development to the pluteus stage in a dose dependent manner (Sartori and Gaion 2016), and higher concentrations cause developmental arrest at the morula stage. Diatoms also produce hydroxy acids (collectively termed oxylipins) as secondary metabolites that function as diatom signalling molecules (Romano et al. 2010; Varrella et al. 2014, 2016; Ruocco et al. 2016; Sartori and Gaion 2016; Albarano et al. 2019) and result in similar changes in larval development (Romano et al. 2010; Varrella et al. 2014, 2016; Albarano et al. 2019). In general, these compounds have dose dependent effects that block cleavage, cause developmental abnormalities, and inhibit growth. At higher concentrations, they are lethal through the induction of caspase-3/7 and caspase-8 that result in cell death by apoptosis (Romano et al. 2010; Ruocco et al. 2016; Albarano et al. 2019). Moreover, larval exposure to mixtures of hydroxyacids appear to function synergistically and show much more severe effects in development at lower concentrations compared to exposures to individual hydroxyacids (Varrella et al. 2014; Albarano et al. 2019). These effects are coincident with mostly down-regulated expression of ~ 50 genes that encode proteins involved in cell stress, skeletogenesis, development, and detoxification. Collectively, these findings may have meaningful ecological outcomes because diatom blooms likely increase environmental oxylipin concentrations to the toxic levels used in these studies (Ruocco et al. 2016; Albarano et al. 2019). These results may also be of concern for larval studies conducted in the laboratory because cultures of single celled algae that are used for feeding larvae can be contaminated with diverse bacterial and protist species including diatoms. For example, the oxylipin, 2–4-decadienal, has a relatively long environmental half-life of up to 14 days (Romano et al. 2010) in Guillard's f/2 enriched sea water medium that is used commonly for culturing feeder algae (Guillard and Ryther 1962; Guillard 1975). Unnoticed contamination of algal cultures with diatoms and other organisms that are fed to larvae may be an uncontrolled source of toxins and could confound studies on larval health and development and on analyses of larval infection and pathology.

Multiple strains of four benthic dinoflagellate species, *Coolia malayensis*, *C. canariensis*, *C. tropicalis*, and *C. palmyrensis*, from the coast of Hong Kong, and evaluated on larvae of the sea urchin, *Heliocidaris crassispina*, for 24 to 72 hpf all inhibit development and induce arm abnormalities and death in a dose-dependent manner (Leung et al. 2017). Similarly, embryos and larvae of the variegated sea urchin, *Lytechinus variegatus*, exposed to the dinoflagellate, *Ostreopsis c f. ovata*, that produces ovatoxin, in a SET assay of five days results in reduced fertilisation success of eggs, developmental arrest, embryonic and larval malformations, skeletal disruption, and lethality that all correlate with increasing concentrations of dinoflagellate cells (Neves et al. 2018). Some of these effects appear as soon as 1–3 hours and larvae only develop successfully at the lower concentrations *O. cf. ovata* cells. Interactions between *O. cf. ovata* and sea urchin gametes and developing embryos may occur naturally in coastal areas with the outcome of significant impacts on larval fitness. This is likely exacerbated because the reproductive period of *L. variegatus* coincides with *O. cf. ovata* blooms, and the lower concentrations of *O. cf. ovata* used in the SET assay are observed year round in tropical environments.

18.7.2 Microbial pathology of sea urchin larvae

Besides the SET assay approach for evaluating environmental toxins and pathogens, similar exposure assays have been used to characterise larval pathology from microbes and the anti-microbial immune responses of larvae (Ho et al. 2016; Buckley et al. 2017; Buckley et al. 2019; Schuh et al. 2020). Most of the work is focused on the model marine pathogen, *Vibrio diazotrophicus*, that was originally isolated from a diseased adult green sea urchin, *Strongylocentrotus droebachiensis* (Guerinot et al. 1982). Exposure of larval *S. purpuratus* to *V. diazotrophicus* induces predictable, reversible changes in cellular behaviour and morphology, including midgut epithelial wall thickening and lumen occlusion, a change in pigment cell morphology from dendritic to round, a migration of pigment cells through the blastocoel to the surface of the gut that

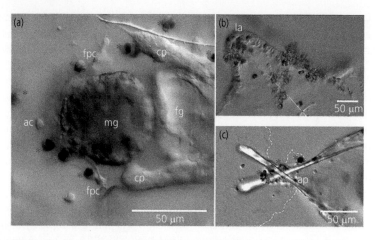

Figure 18.9 Interactions between bacteria and immune cells and the subsequent pathology in 10–15 day old larvae of the purple sea urchin, *Strongylocentrotus purpuratus*. (a) Gut inflammation mediated by *Vibrio diazotrophicus*. After 12 hours of exposure to 5 × 10⁷ *V. diazotrophicus* cells/mL in artificial sea water, the larval gut epithelium thickens, and multiple activated pigment cells (red asterisks) migrate to the midgut (mg) epithelium and a coelomic pouch (cp). Filopodial cells (fpc) extend multiple projections to the gut epithelium and interact with pigment cells. A migratory amoeboid cell (ac) is also present. fg, foregut. (b,c) *Vibrio lentus* attacks the larval ectoderm. (b) Activated pigment cells (red) interact with aggregated *V. lentus* (tan/orange) at the ectoderm of a dorsal arm of a 4-arm larva after 6 hours of exposure. la, larval arm. (c) The body rods at the body apex (ap) penetrate the ectoderm in a lysed larva after 24 hours of exposure to *V. lentus* (10⁷ cells/mL). Several activated pigment cells (red asterisks) are present. White dashed lines indicate the margins of lysed larval contents. Source: the figures were kindly provided by N.W. Schuh and J.P. Rast.

is in contact with the blastocoel, and an increase in interactions among immune cells in the blastocoel (i.e., between pigment cells, filopodial cells, globular cells, ovoid cells, and amoeboid cells) (Figure 18.9a, see Section 18.1.4; and see Box 1.2 and Figure 1.6 in Chapter 1) (Ho et al. 2016; Buckley and Rast 2019). Similar effects are observed upon exposure to other Gram-negative marine bacteria such as *Marinomonas*, *Pseudoalteromonas*, *Psychrosphaera*, and other *Vibrio* species. Eventually, *V. diazotrophicus* invades the blastocoel from the gut of larval *S. purpuratus* and is phagocytosed by filopodial cells (Ho et al. 2016).

Under normal larval development, a large number of isolated and cultured bacterial strains are associated with the 4-arm larval stage based on matches to 16S rRNA gene sequences (Ho et al. 2016; Buckley et al. 2017; Schuh et al. 2020). Multiple reports have identified *Vibrio* and their kin as members of the larval microbiota of sea urchins and sea stars (Galac et al. 2016; Ho et al. 2016; Carrier et al. 2018; Schuh et al. 2020). The cultured microbes associated with *S. purpuratus* include seven *Vibrio* species in the Splendidus clade, of which several have been tested individually in exposure experiments with larvae. Although *Vibrios* are

repeatedly isolated from seemingly healthy larvae from multiple environments, two strains of *V. cyclitrophicus* induce immune responses in *S. purpuratus* larvae, whereas a *V. splendidus* strain is lethal (Ho et al. 2016; unpublished observations, N.W. Schuh). Similar to *V. diazotrophicus*, strains of *V. lentus* (Splendidus clade) induce rapid (< 30 min) immune responses by multiple cell types upon injection into the blastocoel (Heyland et al. 2018; Schuh et al. 2020). However, unlike *V. diazotrophicus*, *V. lentus* is lethal to larvae by 24 hours whether injected into the larval blastocoel or added to the culture. The threshold at which *V. lentus* exposure becomes toxic or lethal is variable (5x10⁴ to 10⁷ cells/mL) among genetically different larval cultures. This variability in responses among different larval cultures is likely typical for exposure to other Gram-negative bacteria (unpublished observations, N.W. Schuh). Under circumstances when *V. lentus* does not kill larvae, this interaction corresponds with observed inflammatory responses in larvae including midgut epithelial hypertrophy and pigment cell activation, rounding, and migration. When *V. lentus* infects larvae, it progresses through several stages over 3 to 24 hours beginning with large aggregations of bacteria adhering to the ectoderm (Figure 18.9.b), pigment cell activation and morphological rounding at the ectoderm (Figure 18.9.b.c), protrusion of the post-oral and anterolateral skeletal rods through the ectoderm (skeleto-ectodermal penetration), and larval lysis (Figure 18.9.c), resulting in complete skeletonisation or death (Schuh et al. 2020).

It is noteworthy that not all microbial contact is detrimental. Microbial exposure during early larval development has consequences on subsequent immunity to bacterial pathogens. When larvae are raised from fertilised eggs in sterile filtered artificial sea water to which is added penicillin and streptomycin, they grow significantly larger by the 4-arm stage compared to larvae grown in the presence of marine microbes (Schuh et al. 2020; unpublished observations, N.W. Schuh). On the other hand, larvae raised in sterile culture are much less resistant to lethal infections with *Vibrio lentus* compared to larvae raised in non-sterile, artificial, or natural sea water in which adult sea urchins are housed. The development of small larvae with a high level of immune activity in association with a high microbial load compared to large larvae that develop in the presence of antibiotics and absence of microbes may be an energetic trade-off between growth and immune system activation.

18.7.3 Conclusion: pathologies of larval sea urchins

The results reviewed here illustrate several major themes in larval sea urchin microbial pathology. (i) Microbial exposure during larval development influences subsequent infection resistance suggesting that early bacterial colonisation and the detection and response by the immune system may be necessary to establish an optimal and proper relationship between the host immune system and the associated microbiota. (ii) Several endogenous *Vibrio* species, mostly from the Splendidus clade, are potentially lethal to larvae upon experimental exposure, suggesting that they may be either secondary or opportunistic pathogens. (iii) Many microbial components of planktonic and benthic environments (e.g., cyanobacteria, Gram-negative bacteria, diatoms, and dinoflagellates) produce toxins that interfere with larval development and cause disease. This is a concern for both laboratory algal feeder cultures and during algal blooms in nature, which may limit larval fitness. (iv) Under conditions where echinoderm embryos and larvae are established model organisms, the literature on larval pathology remains sparse, although information is available for abiotic factors and there is some literature that describes infections in larval sea cucumbers (see Section 18.5.5). Collectively, these findings suggest clear goals for future research. More study is needed, particularly with regard to fungal, viral, parasitic (see Section 18.6), and prion pathogens, which remain poorly or not addressed. Furthermore, future work is expected to elucidate the mechanisms that mediate bacterial colonisation and the development of resistance to pathogenic microbes by host larvae. Because echinoderm larvae and adults are essentially treated as separate organisms in the existing literature, future investigations of the relationship between larval and adult immune systems may elucidate correlations, if they exist, between pathogen exposure in larvae and infection resistance and health in juveniles and adults. The possibility that pathogens may be transmitted from adults to offspring through the gametes with associated deleterious effects has not been addressed and may be a concern.

18.8 Summary and conclusions

18.8.1 The pathogens

- Bacterial microbiomes are similar for larval and adult echinoderms (Hakim et al. 2015, 2016; Lloyd and Pespeni 2018; Zhang, Lv et al. 2019)
- A small set of pathogens infect echinoderms and can cause diseases of similar pathologies, and include:

○ bacteria
- ■ Gammaproteobacteria (particularly *Vibrio* spp.)
- ■ *Vibrio alginolyticus, V. anguillarum, V. coralliilyticus, other Vibrio* spp.
- ■ *Aeromonas*
- ■ *Pseudoaltermonas*
- ■ *Flexibacter*

○ densoviruses and other unidentified viruses
○ a few identified protist parasites
- Environmental toxins and virulence factors in sea water that are secreted from non-pathogens can affect larval echinoderm health and development
- It is difficult to differentiate among echinoderm pathologies caused by:
  ○ a single initiating pathogen
  - ■ closely related set of pathogens
  ○ multiple pathogens acting in concert
  ○ secondary opportunists that arise from uncontrolled commensals
  ○ an altered microbiome or pathobiome (Sweet 2020)
- Diseases in adult and larval *Apostichopus japonicus* tend to be caused by a similar set of bacteria
- Some diseases of larval sea cucumbers and sea urchins appear similar or identical (Ho et al. 2016; Schuh et al. 2020) and may be the same diseases (see Sections 18.5.5 and 18.7)
- The Splendidus clade of *Vibrio* spp. cause diseases with similar symptoms in other marine organisms (Nicolas et al. 1996; Gatesoupe et al. 1999; Farto et al. 2003, 2006; Gómez-León et al. 2005; Domeneghetti et al. 2014; Lasa et al. 2015, Rojas et al. 2015)

18.9 Future directions

- The echinoderm immune system is innate, complex, and robust in both adults and larvae (Smith et al. 2018; Schuh et al. 2020)
- Of the protist parasites, ciliate infections in asteroids can change the gender ratios in populations
- Mass die-offs of sea urchins and sea stars from a range of pathogens can change the ecosystems from which they disappear (see Chapter 19)
- DNA sequencing:
  ○ identified the pathogen of the sunflower sea star, *Pycnopodia helianthoides* (Hewson et al. 2014)
  ○ is expected to provide new data on echinoderm parasites
  ○ is expected to identify the pathogens in aquaculture facilities to:
    - ■ address the problem of disease outbreaks
    - ■ evaluate the prevalence of infection prior to disease symptoms

- ■ devise prophylactic treatments or respond to active infections
- ■ improve aquaculture practices and conditions
- Important factors for understanding the drivers of disease appear to be host stress that is often an outcome of environmental changes (see Chapter 19)
- The key for responding to echinoderm pathology will be to understand the details of the immune systems in multiple species of echinoderms, including:
  ○ the immune genes and their regulation
  ○ the encoded immune proteins and their functions
  ○ how to optimise or manipulate the immune system to improve echinoderm survival from debilitating and lethal diseases

Acknowledgements

The authors were supported during the writing of this chapter by funding from the Australian Research Council (DP150102771) to MB, the Natural Sciences and Engineering Research Council of Canada (400230 and 401393) to AH and (RGPIN-2017-06427) to NWS and JPR, the United States National Science Foundation (IOS-1855747) to LCS, and the National Science Foundation of China (31530069) to LS, ZX, and ZY, and the National Key Research and Development Program (2018YFD0900606) to LW. The authors are grateful to Jackie Hildering www.TheMarineDetective.com), Melisa Miner www.pacificrockyintertidal.org), Alejandro Mercado Capote (University of Puerto Rico Mayagüez, Puerto Rico), and Lee C. Heiman for providing images of echinoderms in their natural habitats, to Megan A. Barela Hudgell (George Washington University, USA) for the image of BSUD in *Strongylocentrotus purpuratus*, to Anguo Zhang (National Marine Environmental Monitorsing Center, China) for the image of off-plate syndrome in *Apostichopus japonicus*, and to Wei Zhou (Dalian Ocean University, China) for images of APED in *Apostichopus japonicus*. We thank Chenghua Li (Ningbo University, China) for constructive feedback on an early draft of the section on diseases of *A. japonicus* in aquaculture settings. We acknowledge the valuable assistance from Deborah Bezanson (Senior Librarian for the Sciences, Engineering & Assessment, George Washington University Libraries) for help with correcting references and requesting permissions for republishing some images.

References

Akira, S. and Takeda, K. 2004. Toll-like receptor signalling. *Nature Reviews Immunology* 4: 499–511

Albarano, L., Ruocco, N., Ianora, A., Libralato, G., Manfra, L., and Costantini, M. 2019. Molecular and morphological toxicity of diatom-derived hydroxyacid mixtures to sea urchin *Paracentrotus lividus* embryos. *Marine Drugs* 17(3): 144

Ald, S.M., Bass, D., Lane, C.E., Lukeš, J. et al. 2019. Revisions to the classification, nomenclature, and diversity of eukaryotes. *Journal of Eukaryotic Microbiology* 66: 4–119

Appeltans, W., Ahyong, S.T., Anderson, G., Angel, M.V. et al. 2012. The magnitude of global marine species diversity. *Current Biology* 22(23): 2189–2202

Arizza, V., Giaramita, F.T., Parrinello, D., Cammarata, M., and Parrinello, N. 2007. Cell cooperation in coelomocyte cytotoxic activity of *Paracentrotus lividus* coelomocytes. *Comparative Biochemistry and Physiology A* 147(2): 389–394

Bang, F.B. 1982. Disease processes in sea stars: A Metchnikovian challenge. *Biological Bulletin* 162: 135–148

Barel, C.D. and Kramers, P.G. 1970. Notes on associates of echinoderms from Plymouth and the coast of Brittany. *Proceedings of the Koninklijke Nederlandse Akademie van Wetenschappen (C)* 73: 159–170

Barel, C.D. and Kramers, P.G. 1977. A survey of the echinoderm associates of the north-east Atlantic area. *Zoologische Verhandelingen* 156: 1–159

Bass, D., Stentiford, G.D., Littlewood, D.T.J., and Hartikainen, H. 2015. Diverse applications of environmental DNA methods in parasitology. *Trends in Parasitology* 31(10): 499–513

Bates, A.E., Hilton, B.J., and Harley, C.D.G. 2009. Effects of temperature, season and locality on wasting disease in the keystone predatory sea star *Pisaster ochraceus*. *Diseases of Aquatic Organisms* 86: 245–251

Bates, A.E., Stickle, W.B., and Harley, C.D. 2010. Impact of temperature on an emerging parasitic association between a sperm-feeding scuticociliate and northeast Pacific sea stars. *Journal of Experimental Marine Biology and Ecology* 384: 44–50

Bauer, J.C. and Young, C.M. 2000. Epidermal lesions and mortality caused by vibrosis in deep-sea Bahamian echinoids: A laboratory study. *Diseases of Aquatic Organisms* 39: 193–199

Becker, P., Eeckhaut, I., Ycaza, R.H., Mercier, A., and Hamel, J.-F. 2009. Protozoan disease in larval culture of the edible sea cucumber *Isostichopus fuscus*. In *Echinoderms*, Harris L.G., Boettger, S.A. Walker, C.W., and Lesser, M.P. (eds.) pp. 571–573. London: CRC Press

Becker, P., Gillan, D., Lanterbecq, D., Jangoux, M., Rasolofonirina, R., Rakotovao, J., and Eeckhaut, I. 2004. The skin ulceration disease in cultivated juveniles of *Holothuria scabra* (Holothuroidea, Echinodermata). *Aquaculture* 242: 13–30

Becker, P., Gillan, D.C., and Eeckhaut, I. 2007. Microbiological study of the body wall lesions of the echinoid *Tripneustes gratilla*. *Diseases of Aquatic Organisms* 77(1): 73–82

Becker P.T., Egea, E., and Eeckhaut, I. 2008. Characterization of the bacterial communities associated with the bald sea urchin disease of the echinoid *Paracentrotus lividus*. *Journal of Invertebrate Pathology* 98(2): 136–147

Beiras, R., Tato, T., and López-Ibáñez, S. 2019. A 2-Tier standard method to test the toxicity of microplastics in marine water using *Paracentrotus lividus* and *Acartia clausi* larvae. *Environmental Toxicology Chemistry* 38(3): 630–637

Beleneva, I.A., Shamshurina, E.V., and Eliseikina, M.G. 2015. Assessment of the toxic effect exerted by fluorescent pseudomonads on embryos and larvae of the sea urchin *Strongylocentrotus nudus*. *Ecotoxicolgical and Environmental Safety* 115: 263–271

Bertheussen, K. 1981a. Endocytosis by echinoid phagocytosis in vitro. I. Recognition of foreign matter. *Developmental and Comparative Immunology* 5(2): 241–250

Bertheussen, K. 1981b. Endocytosis by echinoid phagocytes in vitro. II. Mechanisms of endocytosis. *Developmental and Comparative Immunology* 5(4): 557–564

Blair, J.E. and Hedges, S.B. 2005. Molecular phylogeny and divergence times of deuterostome animals. *Molecular Biology and Evolution* 22(11): 2275–2284

Blewett, T.A., Smith, D.S., Wood, C.M., and Glover, C.N. 2016. Mechanisms of nickel toxicity in the highly sensitive embryos of the sea urchin *Evechinus chloroticus*, and the modifying effects of natural organic matter. *Environmental Science and Technology* 50(3): 1595–1603

Boettger, S.A. 2001. The effects of inorganic and organic phosphates on aspects of physiology and reproduction in the sea urchin *Lytechinus variegatus*. PhD dissertation. Birmingham: The University of Alabama at Birmingham

Boettger, S.A. and McClintock, J.B. 2009. The effects of chronic inorganic and organic phosphate exposure on bactericidal activity of the coelomic fluid of the sea urchin *Lytechinus variegatus* (Lamarck) (Echinodermata: Echinoidea). *Comparative Biochemistry and Physiology C* 150(1): 39–44

Boettger, S.A. and McClintock, J.B. 2009. The effects of chronic inorganic and organic phosphate exposure on bactericidal activity of the coelomic fluid of the sea urchin sea urchin *Lytechinus variegatus* (Echinodermata: Echinoidea). *Comparative Biochemistry and Physiology C* 150: 39–44

Bogolopeva, I.I. 1953. Gregarines of Peter the Great Bay. *Trav Inst Acad Sci USSR* 13: 38-56 (in Russian)

Bouland, C., De Puytorac, P., and Bricourt, E. 1986. *Orchitophrya stellarum* cilié pretend astome, est un scuticocilié. *Annales des Sciences Naturelles. Zoologie et Biologie Animale* 8: 249–257

Bouland, C. and Jangoux, M. 1988. Infestation of male *Asterias rubens* L. (Echnidermata) by the ciliate *Orchitophrya stellarum* Cépède: Effect on gonads and host's reaction. *Diseases of Aquatic Organisms* 5: 239–242.

Brink M., Rhode C., Macey, B.M., Christison, K.W., and Roodt-Wilding, R. 2019. Metagenomic assessment of body surface bacterial communities of the sea urchin, *Tripneustes gratilla*. *Marine Genomics* 47(100675): 1–11

Brockton, V., Henson, J.H., Raftos, D.A., Majeske, A.J., Ki, Y,O., and Smith, LC. 2008. Localization and diversity of 185/333 proteins from the purple sea urchin—unexpected protein-size range and protein expression in a new coelomocyte type. *Journal of Cell Science* 121(3): 339–348

Brownell, C.L. and McCauley, J.E. 1971. Two new parasites (Protozoa: Telosporea) from the spatangoid *Brisaster latifrons*. *Zoologischer Anzeiger* 186: 141–147

Bucci, C., Francoeur, M., McGreal, J., Smolowitz, R., Zazueta-Novoa, V., Wessel, G.M., and Gomwz-Chiarri, M. 2017. Sea star wasting disease in *Asterias forbesi* along the Atlantic coast of North America. *PLoS One* 12: e0188523

Buchwald, R., Scheibling, R.E., and Simpson, A.G.B. 2018. Detection and quantification of a keystone pathogen in a coastal marine ecosystem. *Marine Ecology Progress Series* 606: 79–90

Buckley, K.M., Ho, E.C.H., Hibino, T., Schrankel, C.S., Schuh, N.W., Wang, G, and Rast, J.P. 2017. IL17 factors are early regulators in the gut epithelium during inflammatory response to *Vibrio* in the sea urchin larva. *eLife* 6: e23481

Buckley, K.M. and Rast, J.P. 2012. Dynamic evolution of Toll-like recetor multigene families in echinoderms. *Frontiers in Immunology* 3: 136

Buckley, K.M. and Rast, J.P. 2019. Immune acidity at the gut epithelium in the larval sea urchin. *Cell and Tissue Research* 377(3): 469–474

Buckley, K.M., Schuh, N.W., Heyland, A., and Rast, J.P. 2019. Analysis of immune response in the sea urchin larva. In *Echinoderms*, Foltz K.R. and Hamdoun A. (eds.) *Methods in Cell Biology* 150: Chapter 19, pp. 333–355. New York: Elsevier

Bulgakov, A.A., Eliseikina, M.G., Kovalchuk, S.N., Petrova, I.Y., Likhatskaya, G,N, Shamshurina, E.V., and Rasskazov, V.A. 2013. Mannan-binding lectin of the sea urchin *Strongylocentrotus nudus*. *Marine Biotechnology* 15(1): 73–86

Bulgakov, A.A., Nazarenko, E.L., Petrova, I.Y., Eliseikina, M.G., and Zubkov, V.A. 2000. Isolation and properties of a mannan-binding lectin from the coelomic fluid of the holothurian *Cucumaria japonica*. Biochemistry Biokhimiia 65: 933–939

Burrows, R.B. 1936. Further observations on parasitism in the starfish. *Science* 84: 329

Byrne, M. 2001. The morphology of autotomy structures in the sea cucumber *Eupentacta quinquesemita* before and during evisceration. *Journal of Experimental Biology* 204: 849–863

Byrne, M. 2006. Life history evolution in the Asterinidae. *Integrative and Comparative Biology* 46: 243–254

Byrne, M. 2017. Introduction to Echinoderms. In *Australian Echinoderms: Biology Ecology and Evolution*, M. Byrne and T.D. O'Hara (eds.) pp. 1–36. Clayton Australia: CSIRO Publishing

Byrne, M., Cerra, A., Nishigaki, R., and Hoshi, M. 1997. Infestation of the testes of the Japanese sea star *Asterias amurensis* by the ciliate *Orchitophyra stellarum*: A caution against the use of this ciliate for biological control. *Diseases of Aquatic Organisms* 28: 235–239

Byrne, M., Cerra, A., Nishigaki, R., and Hoshi, M. 1998. Male infertility: A new phenomenon affecting Japanese populations of the sea star *Asterias amurensis* (Asteroidea) due to introduction of the parasitic ciliate *Orchitophyra stellarum* to Japan. In *Echinoderms*, Mooi, R. and Telford, M. (eds.) San Francisco: Balkema, Lisse

Byrne, M., Lamare, M., Winter, D., Dworjanyn, S.A., and Uthicke, S. 2013. The stunting effect of a high CO_2 ocean on calcification and development in sea urchin larvae, a synthesis from the tropics to the poles. *Philosophical Transactions of the Royal Society of London B* 368(1627): 20120439

Byrne, M., Oakes, D.J., Pollak, J.K., and Laginestra, E. 2008. Toxicity of landfill leachate to sea urchin development with a focus on ammonia. *Cell Biology and Toxicology* 24: 503–512

Byrne, M., Pollak, J., Oakes, D., and Laginestra, E. 2003. Comparison of the submitochondrial particle test, Microtox and sea urchin fertilisation and development tests: Parallel assays with leachates. *Australian Journal of Ecotoxicology* 9: 19–28

Cánovas, F., Domínguez-Godino, J., and González-Wangüemert, M. 2019. Epidemiology of skin ulceration disease in wild sea cucumber *Holothuria arguinensis*, a new aquaculture target species. *Diseases of Aquatic Organisms* 135(1): 77–88

Caraguel, C.G., O'Kelly, C.J., Legendre, P., Frasca, S., Gast, R.J., Despres, B.M., Cawthorn, R.J., and Greenwood, S.J. 2007. Microheterogeneity and coevolution: An examination of rDNA sequence characteristics in *Neoparamoeba pemaquidensis* and its prokinetoplastid endosymbiont. *Journal of Eukaryotic Microbiology* 54: 418–426

Carmona, L.M., Fugmann, S.D., and Schatz, D.G. 2016. Collaboration of RAG2 with RAG1-like proteins during the evolution of V(D)J recombination. *Genes and Development* 30: 909–917

Carrier, T.J., Dupont, S., and Reitzel, A.M. 2019. Geographic location and food availability offer differing levels of influence on the bacterial communities associated with

larval sea urchins. *FEMS Microbiology and Ecology* 95(8): fiz103

Carrier, T.J. and Reitzel, A.M. 2018. Convergent shifts in host-associated microbial communities across environmentally elicited phenotypes. *Nature Communications* 9(1): 952

Carrier, T.J. and Reitzel, A.M. 2019. Bacterial community dynamics during embryonic and larval development of three confamilial echinoids. *Marine Ecology Progress Series* 166: 179–188

Carrier, T.J., Wolfe, K., Lopez, K., Gall, M., Janies, D.A., Byrne, M., and Reitzel, A.M. 2018. Diet-induced shifts in the crown-of-thorns (*Acanthaster* sp.) larval microbiome. *Marine Biology* 165:157.

Cary, G.A., Cameron, R.A., and Hinman, V.F. 2019. Genomic resources for the study of echinoderm development and evolution. In *Echinoderms*, Foltz, K.R. and Hamdoun, A. (eds.) *Methods in Cell Biology* 151: 65–88., Cambridge MA, USA: Academic Press

Cavalier-Smith, T., Chao, E.E., and Lewis, R. 2016. 187-gene phylogeny of protozoan phylum Amoebozoa reveals a new class (Cutosea) of deep-branching, ultrastructurally unique, enveloped marine Lobosa and clarifies amoeba evolution. *Molecular Phylogenetics and Evolution* 99: 275–296

Cépède, C. 1907. La castration parasitaire des étoiles de mer mâles par un nouvel infusoire astome: *Orchitophrya stellarum* n.g., n. sp. *Comptes Rendus Hebdomadaires des Séances de l'Académie des Sciences* 145: 1305–1306

Changeux, J.P. 1961. Contribution à l'étude des animaux associés aux Holothuries. *Vie et Milieu, Supplément* 10: 1–124

Che, Z., Shao, Y., Zhang, W., Zhao, X., Guo, M., and Li, C. 2019. Cloning and functional analysis of scavenger receptor B gene from the sea cucumber *Apostichopus japonicus*. *Developmental and Comparative Immunology* 99: 103404

Chen, J. 2004. Present status and prospects of sea cucumber industry in China. In *Advances in Sea Cucumber Aquaculture and Management*, A. Lovatelli, C. Conand, S. Purcell, S. Uthicke, J.F. Hamel, and A. Mercier (eds.) pp. 25–38. Food and Agriculture Organization of the United Nations, Fisheries Technical Paper No. 463. Rome, Italy

Chen, Z., Zhou, Z.C., Yang, A.F., Dong, Y., Guan, X.Y., Jiang, B., and Wang, B. 2015. Characterization and expression analysis of a complement component gene in sea cucumber (*Apostichopus japonicus*). *Journal of Ocean University of China* 14(6): 1096–1104

Chesher, R.H. 1969. Contributions to the biology of *Meoma ventrjcosa* (Echinoldea: Spatangoida). *Bulletin of Marine Science* 19: 72–110

Chi, C., Liu, J.Y., Fei, S.Z., Zhang, C., Chang, Y.Q., Liu, X.L., and Wang, G.X. 2014. Effect of intestinal autochthonous probiotics isolated from the gut of sea cucumber (*Apostichopus japonicus*) on immune response and growth of *A. japonicus*. *Fish and Shellfish Immunology* 38(2): 367–373

Chia, F., and Xing, J. 1996. Echinoderm coelomocytes. *Zoological Studies* 35: 231–254

Chiarelli, R., Martino, C., and Roccheri, M.C. 2019. Cadmium stress effects indicating marine pollution in different species of sea urchin employed as environmental bioindicators. *Cell Stress and Chaperones* 24(4): 675–687

Christensen, A.M. 1970. Feeding biology of the sea-star *Astropecten irregularis* Pennant. *Ophelia* 8: 1–134

Chinese Fisheries Yearbook. 2017. China Bureau of Fisheries. Beijing, China: Agriculture Press

Chinese Fisheries Yearbook. 2018. China Bureau of Fisheries. Beijing, China: Agriculture Press

Chinese Fisheries Yearbook. 2019. China Bureau of Fisheries. Beijing, China: Agriculture Press

Citarasu, T. 2010. Herbal biomedicines: A new opportunity for aquaculture industry. *Aquaculture International* 18(3): 403–414

Clemente, S., Lorenzo-Morales, J., Mendoza, J.C., Lopez, C., Sangil, C., Alves, F., Kaufmann, M., and Hernandez, J.C. 2014. Sea urchin *Diadema africanum* mass mortality in the subtropical eastern Atlantic: Role of waterbourne bacteria in a warming ocean. *Marine Ecology Progress Series* 506: 1–14

Clow, L.A., Raftos, D.A., Gross, P.S., and Smith, L.C. 2004. The sea urchin complement homologue, SpC3, functions as an opsonin. *Journal of Experimental Biology* 207: 2147–2155

Coates C.J., McCulloch C., Betts J., and Whalley T. 2018. Echnochrome A release by red spherule cells is an iron-withholding strategy of sea urchin innate immunity. *Journal of Innate Immunity* 10(2): 119–130

Coats, D.W. 1999. Parasitic lifestyles of marine dinoflagellates 1. *Journal of Eukaryotic Microbiology* 46: 402–409

Coffaro, K.A. 1978. Clearance of bacteriophage T4 in the sea urchin *Lytechinus pictus*. *Journal of Invertebrate Pathology* 32: 384–385

Coffaro, K.A. 1980. Memory and specificity in the sea urchin *Lytechinus pictus*. In *Phylogeny of Immunological Memory*, Manning M.J., (ed.) Developments in Immunology, vol. 10, pp. 77–80. New York, NY: Elsevier/North Holland Biomedical Press

Coffaro, K.A. and Hinegardner, R.T. 1977. Immune response in the sea urchin *Lytechinus pictus*. *Science* 197(4311): 1389–1390

Cong, L., Yang, X., Wang, X, Tada, M., Lu, M., Liu, H., and Zhu, B. 2009. Characterization of an i-type lysozyme gene from the sea cucumber *Stichopus japonicus*, and enzymatic and nonenzymatic antimicrobial activities of its recombinant protein. *Journal of Bioscience and Bioengineering* 107(6): 83–588

Coppard, S.E. and Alvarado, J.J. 2013. Echinoderm diversity in Panama: 144 years of research across the isthmus. In *Echinoderm Research and Diversity in Latin America*. Alvarado, J.J., and Solis-Marin, F.A, (eds.) pp. 107–144. Springer-Verlag, Berlin and Heidelberg, Germany

Costa, M.S., Costa, M., Ramos, V., Leão, P.N., Barreiro, A., Vasconcelos, V., and Martins, R. 2015. Picocyanobacteria from a clade of marine Cyanobium revealed bioactive potential against microalgae, bacteria, and marine invertebrates. *Journal of Toxicology and Environmental Health A* 78(7): 432–442

Coteur, G., Corriere, N., and Dubois, P. 2004. Environmental factors influencing the immune responses of the common European starfish (*Asterias rubens*). *Fish and Shellfish Immunology* 16: 51–63

Coulon, P.M. and Jangoux, M. 1987. Gregarine species (Apicomplexa) parasitic in the burrowing echinoid *Echinocardium cordatum*: Occurrence and host reaction. *Diseases of Aquatic Organisms* 2: 135–145

Coulon, P. and Jangoux, M. 1991. Cyclic occurrence of gregarine trophozoites (Apicomplexa) in the burrowing echinoid *Echinocardium cordatum* (Echinodermata, Spatangoidea). *Diseases of Aquatic Organisms* 12: 71–73

Cuénot, L. 1891. Protozoaires commensaux et parasites des échinoderms. Note préliminaire. *Revue Biologique du Nord de la France* 3: 283–300

Cuénot, L. 1892. Commensaux et parasites des échinoderms (deuxieme note). *Revue Biologique du Nord de la France* 5: 1–22

Cuénot, L. 1912. Contributions à la faune du bassin d'Arcachon. V. Echinoderms. *Arcachon Bulletin du Station Biologique* 14: 17–127

Cusimano, M.G., Spinello, A., Barone, G., Schillaci, D., Cascioferro, S., Magistrato, A., Parrino B., Arizza, V., and Vitale, M. 2019. A synthetic derivative of antimicrobial peptide holothuroidin 2 from Mediterranean sea cucumber (*Holothuria tubulosa*) in the control of *Listeria monocytogenes*. *Marine Drugs* 17(3): 159

D'Andrea-Winslow, L., Radke, D.W., Utecht, T., Kaneko, T., and Akasaka, D. 2012. Sea urchin coelomocyte arylsulfatase: A modulator of the echinoderm clotting pathway. *Integrative Zoology* 7(1): 61–73

Dai, F., Zhang, W., Zhuang, Q., Shao, Y., Zhao, X., Lv, Z., and Li, C. 2019. Dihydrolipoamide dehydrogenase of *Vibrio splendidus* is involved in adhesion to *Apostichopus japonicus*. *Virulence* 10(1): 839–848

De Faria, M.T. and Da Silva, J.R. 2008. Innate immune response in the sea urchin *Echinometra lucunter* (Echinodermata). *Journal of Invertebrate Pathology* 98(1): 58–62

De Melo, A.A., Carneiro, R.F., De Melo Silva, W., Moura Rda, M., Silva, G.C., De Sousa, O.V., De Sousa Saboya, J.P., Nascimento, K.S., Saker-Sampaio, S., Nagano, C.S., Cavada, B.S.A., and Sampaio, H. 2014. HGA-2, a novel galactoside-binding lectin from the sea cucumber *Holothuria grisea* binds to bacterial cells. *International Journal of Biological Macromolecules* 64: 435–442

de Vargas, C., Audic S., Henry, N., Decelle, J., Mahé F,, Logares R., Lara E, Berney C, Le Bescot, N., Probert I., Carmichael, M, Poulain, J., Romac S, Colin S, Aury JM, Bittner L, Chaffron, S, Dunthorn, M, Engelen S, Flegontova O, Guidi L, Horák A, Jaillon, O., Lima-Mendez, G., Lukeš, J., Malviya, S., Morard, R., Mulot, M., Scalco, E., Siano R., Vincent F., Zingone, A., Dimier, C., Picheral, M., Searson, S., Kandels-Lewis, S., Tara Oceans, Coordinators: Acinas, S.G., Bork, P., Bowler, C., Gorsky, G, Grimsley, N., Hingamp, P., Iudicone, D., Not, F., Ogata H., Pesant, S., Raes, J., Sieracki, M.E., Speich, S., Stemmann, L., Sunagawa, S., Weissenbach, J., Wincker, P., Karsenti, E. 2015. Eukaryotic plankton diversity in the sunlit ocean. *Science* 348(6237): 1261605

Delroisse, J., Van Wayneberghe, K., Flammang, P., Gillan, D., Gerbaux, P.P., Opina, N, Todinanahary, G.G.B., and Eeckhaut, I. 2020. Epidemiology of a skin ulceration disease (SKUD) in the sea cucumber *Holothuria scabra* with a review on the SKUDs in Holothuroidea (Echinodermata). *Scientific Reports* 10: 22150

Deng, H., He, C., Zhou, Z., Liu, C., Tan, K., Wang, N., Jiang, B., Gao, X., and Liu, W. 2009. Isolation and pathogenicity of pathogens from skin ulceration disease and viscera ejection syndrome of the sea cucumber *Apostichopus japonicus*. *Aquaculture* 287(1/2): 18–27

Deng, H., Sui, X., Chen, Q., Wang, Z., and Dong, Y. 2005. A new disease found in *Apostichopus japonicus* in Dalian area during wintering. *Fisheries Science* 24:40–42.

Deng, H. and Sui, X.-L. 2004. Common epidemic of *Apostichopus japonicus* in incubation period. *Fisheries Science* 3: 4

Deng, H., Zhou, Z.-C., Wang, N.-B., and Liu, C. 2008. The syndrome of sea cucumber (*Apostichopus japonicus*) infected by virus and bacteria. *Virologica Sinica* 23(1): 63–67

DeRidder, C. and Jangoux, M. 1984. Intracoelomic parasitic Sporozoa in the burrowing spatangoid, *Echinocardium cordatum* (Pennant) (Echinodermata, Echinoidea): Coelomocyte reaction and formation of brown bodies. *Helgoländer Meeresuntersuchungen* 37: 225–23

Dheilly, N.M., Raftos, D.A., Haynes, P.A., Smith, L.C., and Nair, S.V. 2013. Shotgun proteomics of coelomic fluid from the purple sea urchin, *Strongylocentrotus purpuratus*. *Developmental and Comparative Immunology* 40(1): 35–50

Ding, J., Bao, P., and Liu, X. 2011. Effect of compound Chinese herbal medicinal feed additive on the immunity of juvenile sea cucumber (*Stichopus japonicus*). *Feed Industry* 32: 14–17

Djakonov, M.D. 1923. *Diplodina gonadipertha*, n. sp. a new neogamus gregarine, parasite of the gonads of *Cucumaria frondosa* (Gunn.). *Russkiĭ arkhiv protistologii* 2: 127–147

Dogiel, V. 1906. Beitrage zur kenntnis der gregarinen. I. *Cystobia chirodotae* nov. sp. *Archiv für Protestenkunde* 7: 106–130

Dogiel, V. 1910. Bietrage zur kenntnis der gregarinen. 4. *Caillynthrochlamys phronimae* Frenz u. a. m. *Archiv für Protestenkunde* 20: 60–78

Doignon, G., Jangoux, M., Féral J.P., and Eeckhaut I. 2003. Seasonal release of the egg capsules of *Anoplodium parasita* Schneider, 1858, intracoelomic turbellarian (Platyhelminthes, Rhabdocoela) symbiotic of the sea cucumber *Holothuria tubulosa* melin, 1788 (Echinodermata, Holothuroida). *Echinoderm Research 2001*: 261–264

Domeneghetti, S., Varotto, L., Civettini, M., Rosani, U., Stauder, M., Pretto, T., Pezzati, E., Arcangeli, G., Turolla, E., Pallavicini, A., and Venier, P. 2014. Mortality occurrence and pathogen detection in *Crassostrea gigas* and *Mytilus galloprovincialis* close-growing in shallow waters (Goro lagoon, Italy). *Fish and Shellfish Immunology* 41: 37–44

Dong, Y., Sun, H., Zou, Z., Yang, A., Chen, Z., Guan, X., Gao, S., Wang, B., Jiang, B., and Jiang, J. 2014. Expression analysis of immune related genes identified from the coelomocytes of sea cucumber (*Apostichopus japonicus*) in response to LPS challenge. *International Journal of Molecular Sciences* 15(11): 19472–19486

Dungan, M., Miller, T.E., and Thomson, D.A. 1982. Catastrophic decline of a top carnivore in the Gulf of California rocky intertidal zone. *Science* 216: 989–991

Dyakin, A.Y. and Paskerova, G.G. 2004. Morphology of *Urospora chiridotae* (Sporozoa: Gregarinomorpha: Eugregarinida) from sea cucumber *Chiridota laevis* (Echinodermata: Holothuroidea: Apoda). *Parazitologiia* 38: 225–238

Dybas, L. and Fankboner, P.V. 1986. Holothurian survival strategies: Mechanisms for the maintenance of a bacteriostatic environment in the coelomic cavity of the sea cucumber, *Parastichopus californicus*. *Developmental and Comparative Immunology* 10(3): 311–330

Dyková, I., Lorenzo-Morales, J., Kostka, M., Valladares, B., and Pecková, H. 2011. *Neoparamoeba branchiphila* infections in moribund sea urchins *Diadema aff. antillarum* in Tenerife, Canary Islands, Spain. *Diseases of Aquatic Organisms* 95: 225–231

Dyková, I., Nowak, B., Pecková, H., Fiala, I., Crosbie, P., and Dvořáková, H. 2007. Phylogeny of *Neoparamoeba* strains isolated from marine fish and invertebrates as inferred from SSU rDNA sequences. *Diseases of Aquatic Organisms* 74(1): 57–65

Edds, K.T. 1977. Dynamic aspects of filopodial formation by reorganization of microfilaments. *Journal of Cell Biology* 73(2): 479–491

Edds, K.T. 1980. The formation and elongation of filopodia during transformation of sea urchin coelomocytes. *Cell Motility* 1(1): 131–140

Edds, K.T. 1985. Morphological and cytoskeletal transformation in sea urchin coelomocytes. In *Blood Cells of Marine Invertebrates: Experimental Systems in Cell Biology and Comparative Physiology*, Cohen, W.D. (ed.) pp. 53–74. Alan R. Liss, Inc.

Edds, K.T. 1993. Cell biology of echinoid coelomocytes. I. Diversity and characterization of cell types. *Journal of Invertebrate Pathology* 61: 173–178

Eeckhaut, I., Parmentier, E., Becker, P., Gomez Da Silva, S., and Jangoux, M. 2004. Parasites and biotic diseases in field and cultivated sea cucumbers. In *Advances in Sea Cucumber Aquaculture and Management*, Lovatelli, A., Conand, C., Purcell, S., Uthicke, S., Hamel, J.-F. and Mercier, A. (eds.), pp. 311–325. Rome: Food and Agriculture Organization of the United Nations Fisheries Technical Paper No. 463

Eisenlord, M.E., Groner, M.L., Yoshioka, R.M., Elliot, J., Maynard, J., Fradkin, S, Turner, M., Pyne, D., Rivlin, N., Van Hooidonk, R., and Harvell, C.D. 2016. Ochre star mortality during the 2014 wasting disease epizootic: Role of population size structure and temperature. *Philosophical Transactions of the Royal Society of London B* 371: 20150212

Eliseikina, M.A. and Magarlamov, T.Y. 2002. Coelomocyte morphology in the holothurians *Apostichopus japonicus* (Aspidochirota: Stichopodidae) and *Cucumaria japonica* (Dendrochirota: Cucumariidae). *Russian Journal of Marine Biology* 28(3): 197–202

Endean, R. 1996. The coelomocytes and coelomic fluids. In *Physiology of Echinodermata*, Boolootian, R.A. (ed.) pp. 302–328. New York: Intersciences

Fan, Y., Li, L., Yu, X.Q., Li, TB, Wang, S. X., Dio, J., and Wang, Y.Q. 2014. Application of new immunostimulants on immunity of sea cucumber *Apostichopus japonicus*. *Journal of Fujian Agriculture and Forestry University* 43(4): 408–413

Farto, R., Armada, S.P., Montes, M., Guisande, J.A, Pére,z M.J., and Nieto, T.P. 2003. *Vibrio lentus* associated with diseased wild octopus (*Octopus vulgaris*). *Journal of Invertebrate Pathology* 83: 149–151.

Farto, R., Armada, S.P., Montes, M., Perez, M.J., and Nieto, T.P. 2006. Presence of a lethal protease in the extracellular products of *Vibrio splendidus*. *Journal of Fish Diseases* 29: 701–707

Febvre, M., Fredj-Reygrobellet, D., and Fredj, G. 1981. Reproduction sexuee d'une Asterie fissipare, *Clerestories richardi* (Perrier, 1982). *International Journal of Invertebrate Reproduction* 3: 193–208

Feehan, C.J., Johnson-Mackinnin, J., Scheibling R.E, Lauson-Guay J.S., and Simpson AG. 2013. Validating the identity of *Paramoeba invadens*, the causative agent of recurrent mass mortality of sea urchins in Nova Scotia, Canada. *Diseases of Aquatic Organisms* 103: 209–227

Fiorito, R., Leander, C., and Leander, B. 2016. Characterization of three novel species of *Labyrinthulomycota* isolated from ochre sea stars (*Pisaster ochraceus*). *Marine Biology* 163: 170

Franco, C.F., Santos, R., and Coelho, A.V. 2011. Proteome characterization of sea star coelomocytes—the innate immune effector cells of echinoderms. *Proteomics* 11(17): 3587–3592

Fuess, L.E., Eisenlord, M.E., Closek, C.L., Tracy, A.M., Mauntz, R., Gignoux-Wofsohn, S., Moritsch, M.M., Yoshioka, R., Burge, C.A., Harvell, C.D., Friedman, C.S., Hewson, I., Hershberger, P.K., and Roberts, S.B. 2015. Up in arms: Immune and nervous system response to sea star wasting disease. *PLoS One* 10: e0133053

Fugmann, S.D., Messier, C., Novack, L.A., Cameron, R.A., and Rast, J.P. 2006. An ancient evolutionary origin of the *Rag1/2* gene locus. *Proceedings of the National Academy of Sciences* 103: 3728–3733

Furukawa, R., Tamaki, K., and Kaneko, H. 2016. Two macrophage migration inhibitory factors regulate starfish larval immune cell chemotaxis. *Immunology and Cell Biology* 94: 315–321

Galac, M.R., Bosch, I., and Janies, D.A. 2016. Bacterial communities of oceanic sea star (Asteroidea: Echinodermata) larvae. *Marine Biology* 163: 162

Galtsoff, P.S. and Loosanoff, V.L. 1939. Natural history and method of controlling the starfish (*Asterias forbesi*, Desor). *Bulletin of the Bureau of Fisheries*, 31: pp. 75–132. Washington: US Government

Gambardella, C., Ferrando, S., Gatti, A.M., Cataldi, E., Ramoino, P., Aluigi, M.G., Faimali, M., Diaspro, A., and Falugi, C. 2016. Review: Morphofunctional and biochemical markers of stress in sea urchin life stages exposed to engineered nanoparticles. *Environmental Toxicology* 31(11): 1552–1562

Gatesoupe F.J., Lambert C., and Nicolas, J.L. 1999. Pathogenicity of *Vibrio splendidus* strains associated with turbot larvae, *Scophthalmus maximus*. *Journal of Applied Microbiology* 87: 757–763

Gerardi, P., Lassegues, M., and Canicatti, C. 1990. Cellular distribution of sea urchin antibacterial activity. *Biology of the Cell* 70(3): 153–157

Giard, A. 1876. Sur une nouvelle espèce de sporospermie (*Lithocystis schneideri*), parasite de l'*Echinocardium cordatum*. *Comptes Rendus Hebdomadaires des Séances de l'Académie des Sciences Paris* 82: 1208–1210

Gilles, K.W. and Pearse, J.S. 1986. Disease in sea urchins *Strongylocentrotus purpuratus*: Experimental infection and bacterial virulence. *Diseases of Aquatic Organisms* 1: 105–114

Giribet, G. and G. D. Edgecombe. 2020. *The Invertebrate Tree of Life*. Princeton, New Jersey: Princeton University Press

Goggin, C.L. and Bouland, C. 1997. The ciliate *Orchitophrya* cf. *stellarum* and other parasites and commensals of the northern Pacific sea star *Asterias amurensis* from Japan. *International Journal for Parasitology* 27: 1415–1418

Goggin, C.L. and Murphy. N.E. 2000. Conservation of sequence in the internal transcribed spacers and 5.8S ribosomal RNA among geographically separated isolates of parasitic scuticociliates (Ciliophora, Orchitophryidae). *Diseases of Aquatic Organisms* 40: 79–83

Golconda, P., Buckley, K.M., Reynolds, C.R., Romanello, J.P., and Smith, L.C. 2019. Coelomocytes proliferate in the axial organ and pharynx of the purple sea urchin. *Frontiers in Immunology* 10: 870

Gómez-León, J., Villamil, L., Lemos, M.L., Novoa, B., and Figueras, A. 2005. Isolation of *Vibrio alginolyticus* and *Vibrio splendidus* from aquacultured carpet shell clam (*Ruditapes decussatus*) larvae associated with mass mortalities. *Applied Environmental Microbiology* 71(1): 98–104

Goodrich, H.P. 1930. The gregarines of *Cucumaria*: *Lithocystis minchinii* Woodc. and *Lithocystis cucumariae* n. sp. *Quarterly Journal of Microscopical Science* 73: 275–287

Gowda, N.M., Gaikwad, S.M., and Khan, M.I. 2013. Kinetics and thermodynamics of glycans and glycoproteins binding to *Holothuria scabra* lectin: A fluorescence and surface plasmon resonance spectroscopic study. *Journal of Fluorescence* 23(6): 1147–1155

Gowda, N.M., Goswami, U., and Khan, M.I. 2008. Purification and characterization of a T-antigen specific lectin from the coelomic fluid of a marine invertebrate, sea cucumber (*Holothuria scabra*). *Fish and Shellfish Immunology* 24: 450–458

Gross. P.S., Clow, L.A., and Smith, L.C. 2000. SpC3, the complement homologue from the purple sea urchin, *Strongylocentrotus purpuratus*, is expressed in two subpopulations of the phagocytic coelomocytes. *Immunogenetics* 51(12): 1034–1044

Gu, M., Ma, H., Mai, K., Zhang, W., Ai, Q., Wang, X., and Bai, N. 2010. Immune response of sea cucumber *Apostichopus japonicus* coelomocytes to several immunostimulants *in vitro*. *Aquaculture* 306(1/4): 49–56

Gu, M., Ma, H., Mai, K., Zhang, W., Bai, N., and Wang, X. 2011. Effects of dietary β-glucan, mannan oligosaccharide and their combinations on growth performance, immunity and resistance against *Vibrio splendidus* of sea cucumber, *Apostichopus japonicus*. *Fish and Shellfish Immunology* 31(2): 303–309

Gudenkauf, B.M., Eaglesham, J.B., Aragundi, W.M., and Hewson, I. 2014. Discovery of urchin-associated densoviruses (family Parvoviridae) in coastal waters of the Big Island, Hawaii. *Journal of General Virology* 95(3): 652–658

Gudenkauf, B.M. and Hewson, I. 2015. Metatranscriptomic analysis of *Pycnopodia helianthoides* (Asteroidea) affected by sea star wasting disease. *PLoS One* 10: e0128150

Guerinot, M.L., West, P.A., Lee, J.V., and Colwell, R.R. 1982. *Vibrio diazotrophicus* sp. nov, a marine nitrogen-fixing bacterium. *International Journal of Systematic Bacteriology* 32: 350–357

Guillard, R.R.L. 1975. Culture of phytoplankton for feeding marine invertebrates. In *Culture of Marine Invertebrate Animals*, W.L. Smith and M.H. Chanley (eds.) pp. 26–60. New York: Plenum Press

Guillard, R.R.L. and Ryther, J.H. 1962. Studies of marine planktonic diatoms. I. *Cyclotella nana* Hustedt and *Detonula confervacea* Cleve. *Canadian Journal of Microbiology* 8: 229–239

Guiry, M.D. and Guiry, G.M. 2020. AlgaeBase. National University of Ireland, Galway: World-wide electronic publication https://www.algaebase.org, [viewed 24 January 2020]

Hakim, J.A., Koo, H., Dennis, L.N., Kumar, R., Ptacek, T., Morrow, C.D., Lefkowitz, E.J., Powell, M.L., Bej, A.K., and Watts, S.A. 2015. An abundance of Epsilonproteobacteria revealed in the gut microbiome of the laboratory cultured sea urchin, *Lytechinus variegatus*. *Frontiers in Microbiology* 6: 1047

Hakim, J.A., Koo, H., Kumar, R., Lefkowitz, E.J., Morrow, C.D., Powell, M.L., Watts, S.A., and Bej, A.K. 2016. The gut microbiome of the sea urchin, *Lytechinus variegatus*, from its natural habitat demonstrates selective attributes of microbial taxa and predictive metabolic profiles. FEMS *Microbial Ecology* 92 (9): fiw146

Hall, M.R., Kocot, K.M., Baughman, K.W., Fernandez-Valverde, S.L., Gauthier, M.E.A., Hatleberg, W.L., Krishnan, A., McDougall, C., Motti, C.A., Shoguchi, E., Wang, T., Xiang, X., Zhao, M., Bose, U., Shinzato, C., Hisata, K., Fujie, M., Kanda, M., Cummins, S.F., Satoh, N., Degnan, S.M., and Degnan, B.M. 2017. The crown-of-thorns starfish genome as a guide for biocontrol of this coral reef pest. *Nature* 544(7649): 231–234

Han, Q., Keesing, J.K., and Liu, D. 2016. A review of sea cucumber aquaculture, ranching, and stock enhancement in China. *Reviews in Fisheries Science and Aquaculture* 24: 326–341

Harikrishnan, R., Balasundaram, C., and Heo, M.S. 2010. Scuticociliatosis and its recent prophylactic measures in aquaculture with special reference to South Korea: Taxonomy, diversity and diagnosis of scuticociliatosis:

Part I. Control strategies of scuticociliatosis: Part II. *Fish and Shellfish Immunology* 29: 15–31

Harvell, C.D., Montecino-Latorre, D., Caldwell, J.M., Burt, J.M., Bosley, K., Keller, A., Heron, S.F., Salomon, A.K., Lee L., Pontier, O., Pattengill-Semmens, C., and Gaydos, J.K. 2019. Disease epidemic and a marine heat wave are associated with the continental-scale collapse of a pivotal predator (*Pycnopodia helianthoides*). *Scientific Advances* 5: eaau7042

Hatakeyama, T., Himeshima, T., Komatsu, A., and Yamasaki, N. 1993. Purification and characterization of two lectins from the sea cucumber *Stichopus japonicus*. *Bioscience, Biotechnology, and Biochemistry* 57(10): 1736–1739

Hatakeyama, T., Kohzaki, H., Nagatomo, H., and Yamasaki, N. 1994. Purification and characterization of four Ca^{2+}-dependent lectins from the marine invertebrate, *Cucumaria echinata*. *Journal of Biochemistry* 116: 209–214

Hedrick, R. 1998. Relationships of the host, pathogen, and environment: Implications for diseases of cultured and wild fish populations. *Journal of Aquatic Animal Health* 10(2): 107–111

Henson, J.H. and Schatten, G. 1983. Calcium regulation of the actin-mediated cytoskeletal transformation of sea urchin coelomocytes. *Cell Motility* 3(5–6): 525–534

Henson, J.H., Svitkina, T.M., Burns, A.R., Hughes, H.E., MacPartland, K.J., Nazarian, R., and Borisy, G.G. 1999. Two components of actin-based retrograde flow in sea urchin coelomocytes. *Molecular Biology of the Cell* 10(12): 4075–4090

Hetzel, H.R. 1963. Studies on holothurian coelomocytes. I. A survey of coelomocyte types. *The Biological Bulletin* 125: 289–301

Hewson, I., Bistolas, K.S.I., Quijano Cardé, E.M., Buttom, J.B., Foster, P.J., Flazenbaum, J.M., Kocian, J., and Lewis, C.K. 2018. Investigating the complex association between viral ecology, environment, and Northeast Pacific sea star wasting. *Frontiers in Marine Science* 5: 77

Hewson, I., Button, J.B., Gudenkauf, B.M., Miner, B., Newton, A.L., Gaydos, J.K., Wynne, J., Groves, C.L., Hendler, G., Murray, M., Fradkin, S., Breitbart, M., Fahsbender, E., Lafferty, K.D., Kilpatrick, A.M., Miner, C.M., Raimondi, P., Lahner, L., Friedman, C.S., and Harvell, C.D. 2014. Densovirus associated with sea-star wasting disease and mass mortality. *Proceedings of the National Academy of Sciences* 111(48): 17278–17283

Hewson, I., Sullivan, B., Jackson, E.W., Zu, Q., Long, H., Lin, C., Quijano Carde, E.M., Seymour, J., Siboni, N., Jones, M.R.L., and Sewell, M.A. 2019. Perspective: Something old, something new? Review of wasting and

other mortality in Asteroidea (Echinodermata). *Frontiers in Marine Science* 6: 406

Heyland, A., Schuh, N., and Rast, J. 2018. Sea urchin larvae as a model for postembryonic development. *Results and Problems in Cell Differentiation* 65: 137–161

Hibino, T., Loza-Coll, M., Messier, C., Majeske, A.J., Cohen, A.H., Terwilliger, D.P., Buckley, K.M., Brockton, V., Nair, S.V., Berney, K., Fugmann, S.D., Anderson, M.K., Pancer, Z., Cameron, R.A., Smith, L.C., and Rast, J.P. 2006. The immune gene repertoire encoded in the purple sea urchin genome. *Developmental Biology* 300(1): 349–365

Hildemann, W.H, and Dix, T.G. 1972. Transplantation reactions of tropical Australian echinoderms. *Transplantation* 14(5): 624–633

Hillier, B.J. and Vacquier, V.D. 2003. Amassin, an olfactomedin protein, mediates the massive intercellular adhesion of sea urchin coelomocytes. *Journal of Cell Biology* 160(4): 597–604

Hira, J., Wolfson, K., Andersen, A.J.C., Haug, T., and Stensvåg, K. 2020. Autofluorescence mediated red spherulocyte sorting provides insights into the source of spinochromes in sea urchins. *Scientific Reports* 10: 1149.

Hisamatsu, K., Nagao, T., Unno, H., Goda S., and Hatakeyama, T. 2013. Identification of the amino acid residues involved in the hemolytic activity of the *Cucumaria echinata* lectin CEL-III. *Biochimica et Biophysica Acta* 1830(8): 4211–4217

Ho, E.C., Buckley, K.M., Schrankel, C.S., Schuh, N.W., Hibino, T., Solek, C.M., Bae, K., Wang, G., and Rast, J.P. 2016. Perturbation of gut bacteria induces a coordinated cellular immune response in the purple sea urchin larva. *Immunology and Cell Biology* 94: 861–874

Hodin, J., Heyland, A., Mercier, A., Pernet, B., Cohen, D.L., Hamel, J.-F., Allen, J.D., McAlister, J.S., Byrne, M., Cisternas, P., and George, S.B. 2019. Culturing echinoderm larvae through metamorphosis. In *Echinoderms*, K.R. Foltz. and A. Hamdoun (eds.) *Methods in Cell Biology* 150: 125–169. New York: Elsevier

Høj, L., Levy, N., Baillie, B.K., Clode, P.L., Strohmaier, R.C., Siboni, N., Webster, N.S., Uthicke, S., and Bourne D.G. 2018. Crown-of-thorns sea star *Acanthaster cf. solaris* has tissue-characteristic microbiomes with potential roles in health and reproduction. *Applied and Environmental Microbiology* 84: e00181–18

Hou, W., Yang, F., and Wang, W. 2013. In vitro bacteriostatic effect of Chinese herbs against causative pathogens of skin ulcer syndrome in sea cucumber *Apostichopus japonicus*. *Chinese Agricultural Science Bulletin* 5: 76–80

Hyman, LH. 1955. *The Invertebrates: Echinodermata, the Coelomate Bilateria*. Vol. IV. New York NY: McGraw-Hill Book Co. Inc.

Isely, N., Lamare, M., Marshall, C., and Barker, M. 2009. Expression of the DNA repair enzyme, photolyase, in developmental tissues and larvae, and in response to ambient UV-R in the Antarctic sea urchin *Sterechinus neumayeri*. *Photochemistry and Photobiology* 85(5): 1168–1176

Ito, T., Matsutani, T., Mori, K., and Nomura, T. 1992. Phagocytosis and hydrogen peroxide production by phagocytes of the sea urchin *Strongylocentrotus nudus*. *Developmental and Comparative Immunology* 16: 287–294

Jackson, E.W., Pepe-Ranney, C., Rohnson, M.R., Distel, D.L., and Hewson, I. 2020. A highly prevalent and pervasive densovirus discovered among sea stars from the North American Atlantic coast. *Applied Environmental Microbiology* 86(6): e2723-19

Jangoux, M. 1987a. Diseases of echinodermata. I. Agents microorganisms and protistans. *Diseases of Aquatic Organisms* 2: 147–162

Jangoux, M. 1987b. Diseases of echinodermata. II. Agents metazoans (Mesozoa to Bryozoa). *Diseases of Aquatic Organisms* 2: 205–234

Jangoux, M. 1987c. Diseases of echinodermata. III. Agents metazoans (Annelida to Pisces). *Diseases of Aquatic Organisms* 3: 39–83

Jangoux, M. 1990. Diseases of echinodermata. In *Diseases of Marine Animals*, vol. 3, Kinne, O. (ed.) pp. 439–567. Hamburg, Germany: Biologische Anstalt Helgoland

Jangoux, M. and Vloebergh, M. 1973. Contribution à l'étude du cycle annuel de reproduction d'une population d'*Asterias rubens* (Echinodermata, Asteroidea) du littoral belge. *Netherlands Journal of Sea Research* 6: 389–408

Jellett, J., Wardlaw, A., and Scheibling, R. 1988. Experimental infection of the echinoid *Strongylocentrotus droebachiensis* with *Paramoeba invadens*: Quantitative changes in the coelomic fluid. *Diseases of Aquatic Organisms* 4: 149–157

Jellett, J.F. and Scheibling, R.E. 1988a. Effect of temperature and prey availability on growth of *Paramoeba invadens* in monoxenic culture. *Applied and Environmental Microbiology* 54(7): 1848–1854

Jellett, J.F. and Scheibling, R.E. 1988b. Virulence of *Paramoeba invadens* Jones (Amoebida, Paramoebidae) from monoxenic and polyxenic culture. *The Journal of Protozoology* 35: 422–424

Jespersen, A. and Lützen, J. 1971. On the ecology of the aspidochirote sea cucumber *Stichopus tremulus* (Gunnerus). *Norwegian Journal of Zoology* 19: 117–132

Jiang, J., Zhou, Z., Dong, Y., Guan, X., Wang, B., Jiang, B., Yang, A., Chen, Z., Gao, S., and Sun, H. 2014b. Characterization of phenoloxidase from the sea cucumber *Apostichopus japonicus*. *Immunobiology* 219(6): 450–456

Jiang, J., Zhou, Z., Dong, Y., Jiang, B., Chen, Z., Gao, S., Guan, X., and Han, L. 2017. Seasonal variations of immune parameters in the coelomic fluid of sea cucumber *Apostichopus japonicus* cultured in pond. *Aquaculture Research* 48(4): 1677–1687

Jiang, J., Zhou, Z., Dong, Y., Sun, H., Chen, Z., Yang, A., Gao, S., Wang, B., Jiang, B., and Guan, X. 2014a. Phenoloxidase from the sea cucumber *Apostichopus japonicus*: cDNA cloning, expression and substrate specificity analysis. *Fish and Shellfish Immunology* 36(2): 344–351

Jiang, Z., Kim, D., Yamasaki, Y., Yamanishi, T., and Oda, T. 2010. Mitogenic activity of CEL-I, an N-acetylgalactosamine (GalNAc)-specific C-type lectin, isolated from the marine invertebrate *Cucumaria echinata* (Holothuroidea). *Bioscience, Biotechnology, and Biochemistry* 74(8): 1613–1616

Jo, J., Oh, J., Lee, H.G., Hong, H.H., Lee, S.G., Cheon, S., Kern, E., Jin, S., Cho, S.J., and Park, J.K. 2017. Draft genome of the sea cucumber *Apostichopus japonicus* and genetic polymorphism among color variants. *Gigascience* 6(1): 1–6

Johnson, P.T. 1969. The coelomoic elements of sea urchins (*Strongylocentrotus*) I. The normal coelomocytes, their morphology and dynamics in hanging drops. *Journal of Invertebrate Pathology* 13(2): 5–41

Johnson, P.T. 1971. Studies on diseased urchins from Point Loma. In *Annual Report: Kelp Habitat Improvement Project*, pp. 82–90. Pasadena CA: California Institute of Technology

Jones, G.M. 1985. *Paramoeba invadens* n. sp. (Amoebida, Paramoebidae), a pathogenic amoeba from the sea urchin, *Strongylocentrotus droebachiensis*, in Eastern Canada. *The Journal of Protozoology* 32(4): 564–569

Jones, G.M., Hebda, A.J., Scheibling, R.E., and Miller R.J. 1985. Histopathology of the disease causing mass mortalities of sea urchins (*Strongylocentrotus droebachiensis*) in Nova Scotia. *Journal of Invertebrate Pathology* 45: 260–271

Jones, G.M. and Scheibling, R.E. 1985. *Paramoeba sp.* (Amoebida, Paramoebidae) as the possible causative agent of sea urchin mass mortality in Nova Scotia. *The Journal of Parasitology* 71: 559–565

Kang, S., Tice, A.K., Spiegel, F.W., Silberman, J.D., Pánek, T., Čepička, I., Kostka, M., Kosakyan, A., Alcântara, D.M., Roger, A.J., and Shadwick, L.L. 2017. Between a pod and a hard test: The deep evolution of amoebae. *Molecular Biology and Evolution* 34: 2258–2270

Karp, R.D. and Hildemann W.H. 1976. Specific allograft reactivity in the sea star *Dermasterias imbricata*. *Transplantation* 22(5): 434–439

Kinjo, S., Kiyomoto, M., Yamamoto, T., Ikeo, K., and Yaguchi, S. 2018. HpBase: A genome database of a sea urchin, *Hemicentrotus pulcherrimus*. *Development Growth and Differentiation* 60(3): 174–182

Kiselev, K.V., Ageenko, N.V., and Kurilenko, V.V. 2013. Involvement of the cell-specific pigment genes *pks* and *sult* in bacterial defense response of sea urchins *Strongylocentrotus intermedius*. *Diseases of Aquatic Organisms* 103(2): 121–132

Kudryavtsev, A., Pawlowski, J., and Hausmann, K. 2011. Description of *Paramoeba atlantica* n. sp. (Amoebozoa, Dactylopodida)—a marine amoeba from the eastern Atlantic, with emendation of the dactylopodid families. *Acta Protozoologica* 50: 239–253

La Haye, C.A., Holland, N.D., and McLean, N. 1984. Electron microscopic study of *Haplosporidium comatulae* n. sp. (phylum Ascetospora: Class Stellatosporea), a haplosporidian endoparasite of an Australian crinoid, *Oligometra serripinna* (phylum Echinodermata). *Protistologica* 20: 507–515

Lasa, A., Avendano-Herrera, R., Estrada, J.M., and Romalde, J.L. 2015. Isolation and identification of *Vibrio toranzoniae* associated with diseased red conger eel (*Genypterus chilensis*) farmed in Chile. *Veterinary Microbiology* 179: 327–331

Lecall, L. 1980. Etude des coelomocytes d´un crinoïde, description de *Cryptobia antedona*, n.sp., zooflagellé bodonidé du coelome général d'antedon bifida (Pennant). In *Echinoderms: Present and Past*, Jangoux, M.J. (ed.) pp. 271–275. Brussels, Balkema, Rotterdam: Proceedings of the European Colloquium on Echinoderms

Lee, H.G., Stumpp, M., Yan, J.J., Tseng, Y.C., Heinzel, S., and Hu, M.Y. 2019. Tipping points of gastric pH regulation and energetics in the sea urchin larva exposed to CO_2-induced seawater acidification. *Comparative Biochemistry and Physiology A* 234: 87–97

Léger, L. 1896. L'evolution du *Lithocystis schneideri*, parasite de l'*Echinocardium cordatum*. *Comptes Rendus Hebdomadaires des Séances de l'Académie des Sciences* 122: 702–705

Léger, L. 1897. Contribution a la connaissance des sporozoaires parasites des echinodermes. Étude sur le *Lithocystis schneideri*. *Bulletin Biologique de la France et de la Belgique* 30: 240–264

Leighton, B.J., Boom, J.D.G., Bouland, C., Hartwick, E.B., and Smith, M.J. 1991. Castration and mortality in *Pisaster ochraceus* parasitized by *Orchitophrya stellarum* (Ciliophora). *Diseases of Aquatic Organisms* 10: 71–73

Lenz, B., Fogarty, N.D., and Figueiredo, J. 2019. Effects of ocean warming and acidification on fertilization success and early larval development in the green sea urchin *Lytechinus variegatus*. *Marine Pollution Bulletin* 141: 70–78

Lessios, H.A. 2005. *Diadema antillarum* populations in Panama twenty years following mass mortality. *Coral Reefs* 24: 125–127

Lessios, H.A. 2016. The great *Diadema antillarum* die-off: 30 years later. *Annual Review of Marine Science* 8: 267–283

Lessios, H.A., Robertson, D.R., and Cubit, J.D. 1984. Spread of *Diadema* mass mortality through the Caribbean. *Science* 226: 335–337

Leung, P.T.Y., Yana, M., Yiu, S.K.F., Lama, V.T.T., Ipa, J.C.H., Aua, M.W.Y., Chen, C.Y., Wai, T.C., and Lama, P.K.S. 2017. Molecular phylogeny and toxicity of harmful benthic dinoflagellates *Coolia* (Ostreopsidaceae, Dinophyceae) in a sub-tropical marine ecosystem: The first record from Hong Kong. *Marine Pollution Bulletin* 124(2): 878–889

Leung, T.L., Mora, C., and Rohde, K. 2015. Patterns of diversity and distribution of aquatic invertebrates and their parasites. In *Parasite Diversity and Diversification, Evolutionary Ecology meets Phylogenetics*, S. Morand, B.B. Krasnov, and D.T.J. Littlewood (eds.) pp. 39–57., Cambridge UK: Cambridge University Press

Levine, N.D. 1977. Checklist of the species of the aseptate gregarine family Urosporidae. *International Journal for Parasitology* 7: 101–108

Levine, N.D. 1984. Nomenclatural corrections and new taxa in the Apicomplexan protozoa. *Transactions of the American Microscopical Society* 103: 195–204

Li, C., Blencke, H.M., Haug, T., and Stensvag, K. 2015. Antimicrobial peptides in echinoderm host defense. *Developmental and Comparative Immunology* 49(1): 190–197

Li, C., Fang, H. and Xu, D. 2019. Effect of seasonal high temperature on the immune response in *Apostichopus japonicus* by transcriptome analysis. *Fish and Shellfish Immunology* 92: 765–771

Li, H., Qiao, G., Gu, J.Q., Zhou, W., Li, Q., Woo, S.H., Xu, D.H., and Park, S.I. 2010a. Phenotypic and genetic characterization of bacteria isolated from diseased cultured sea cucumber *Apostichopus japonicus* in northeastern China. *Diseases of Aquatic Organisms* 91(3): 223–235

Li, H., Qiao, G., Li Q., Zhou, W., Won, K., Xu, D.H., and Park, S.I. 2010b. Biological characteristics and pathogenicity of a highly pathogenic *Shewanella marisflavi* infecting sea cucumber, *Apostichopus japonicus*. *Journal of Fish Diseases* 33(11): 865–877

Li, J., Sun, X., Zheng, F., and Hao, L. 2009. Screen and effect analysis of immunostimulants for sea cucumber, *Apostichopus japonicus*. *Chinese Journal of Oceanology and Limnology* 27(1): 80–84

Li, J., Xing, Z., and Pan, Y. 2006. Problems and countermeasures of seedling breeding of sea cucumber. *Shandong Fisheries* 23(4): 19–20

Li, M., Ma, Y.X., Liu, Z.M., Yang, Z.P., Bao, P.Y., and Song, J. 2012. The yeast composition and antibacterial activities in sea cucumber *Apostichopus japonicus*. *Journal of Dalian Ocean University* 27(5): 436–440

Li, Q., Jiang, W., Liu, H., Xiaolong, L.I., Song, X., Wang, X., and Qiu, Z. 2014. Dynamic and correlation study on culturable bacteria and physicochemical indexes in the environment of *Apostichopus japonicus* culture ponds. *Chinese Journal of Applied and Environmental Biology* 20(3): 544–550

Li, R., Dang, H., Huang, Y., Quan, Z., Jiang, H., Zhang, W., and Ding, J. 2020. *Vibrio coralliilyticus* as an agent of red spotting disease in the sea urchin *Strongylocentrotus intermedius*. *Aquaculture Reports* 16: 100244

Li, Z., Li, X., Zhang, J., Wang, X., Wang, L., Cao, Z., and Xu, Y. 2016. Use of phages to control *Vibrio splendidus* infection in the juvenile sea cucumber *Apostichopus japonicus*. *Fish and Shellfish Immunology* 54: 302–311

Li, Z., Ren, H., Li, Q., Murtaza, B., Li, X., Zhang, J., and Xu Y. 2020. Exploring the effects of phage cocktails in preventing *Vibrio* infections in juvenile sea cucumber (*Apostichopus japonicus*) farming. *Aquaculture* 515: 734599

Liang, W., Zhang, C., Liu, N., Zhang, W., Han, Q., and Li, C. 2016. Cloning and characterization of *Vshppd*, a gene inducing haemolysis and immune response of *Apostichopus japonicus*. *Aquaculture* 464: 246–252

Liao, Y. 1980. The aspidochirote holothurians of China with erection of a new genus. In *Proceedings of European Colloquium on Echinoderm*. Jangoux, M. (ed.) pp. 115–120. Rotterdam, Netherlands: AA Balkema

Lister, K.N., Lamare, M.D., and Burritt, D.J. 2010. Sea ice protects the embryos of the Antarctic sea urchin *Sterechinus neumayeri* from oxidative damage due to naturally enhanced levels of UV-B radiation. *Journal of Experimental Biology* 213(11): 1967–1975

Liu, H., Chen, M., Zhu, F., and Harrison, P.J. 2016. Effect of diatom silica content on copepod grazing, growth and reproduction. *Frontiers in Marine Science* 3: 00089

Liu, H., Zheng, F., Sun, X., Hong, X., Dong, S., Wang, B., Tang, X., and Wang, Y. 2010. Identification of the pathogens associated with skin ulceration and peristome tumescence in cultured sea cucumbers *Apostichopus japonicus* (Selenka). *Journal of Invertebrate Pathology* 105(3): 236–242

Liu, H, Zheng, F.R., Sun, X.Q., Tang, X.X., and Dong, S.L. 2012. Effect of exposure to ammonia nitrogen stress on immune enzyme of holothurian *Apostichopus japonicus*. *Marine Sciences* 36: 47–52

Liu, M.C., Liao, W.Y., Buckley, K.M., Yang, S.Y., Rast, J.P., and Fugmann, S.D. 2018. AID/APOBEC-like cytidine deaminases are ancient innate immune mediators in invertebrates. *Nature Communications* 9(1): 1948

Liu, X., Fan, R., Tan, J., and Gao, L. 2005. Ciliates and skin ulcer syndrome of sea cucumber *Apostichopus japonicus*. *Journal of the Chinese Electron Microscopy Society* 24: 428

Liu, Y., Kong, W., Jiang, G., and Wu, Z.Q. 2008. Effects of two kinds of immunopolysaccharide on the activities of immunoenzymes in sea cucumber, *Apostichopus japonicus*. *Journal of Fishery Sciences of China* 15(5): 787–793

Liu, Z.M., Ma, Y.X., Yang, Z.P., Li, M., Liu, J., and Bao, P.Y. 2012. Immune responses and disease resistance of the juvenile sea cucumber *Apostichopus japonicus* induced by *Metschnikowia* sp. C14. *Aquaculture* 368: 10–18

Lloyd, M.M. and Pespeni, M.H. 2018. Microbiome shifts with onset and progression of sea star wasting disease revealed through time course sampling. *Scientific Reports* 8: 16476

Lom, J. and Dyková, I. 2013. Myxozoan genera: Definition and notes on taxonomy, life-cycle terminology and pathogenic species. *Folia Parasitologica* 53(1): 1–36

Long, H., Song, W., Chen, J., Gong, J., Ji, D., Hu, X., Ma, H., Zhu, M., and Wang, M. 2006. Studies on an endoparasitic ciliate *Boveria labialis* (Protozoa: Ciliophora) from the sea cucumber, *Apostichopus japonicus*. *Journal of the Marine Biological Association of the United Kingdom* 86(4): 823–828

Long, K.A., Nossa, C.W., Sewell, M.A., Putnam, N.H., and Ryan, J.F. 2016. Low coverage sequencing of three echinoderm genomes: The brittle star *Ophionereis fasciata*, the sea star *Patiriella regularis*, and the sea cucumber *Australostichopus mollis*. *Gigascience* 5(1): 20

Lopes, V.R., Fernández, N., Martins, R.F., and Vasconcelos, V. 2010. Primary screening of the bioactivity of brackishwater cyanobacteria: Toxicity of crude extracts to *Artemia salina* larvae and *Paracentrotus lividus* embryos. *Marine Drugs* 8(3): 471–482

Lowe, G.F. 1978. Relationships between biochemical and caloric composition and reproductive cycle in *Asterias vulgaris* (Echinodermata: Asteroidea) from the Gulf of Maine. PhD Dissertation, Orono: University of Maine

Lu, M., Zhang, P.J., Li, C.H., Lv, Z.M., Zhang, W.W., and Jin, C.H. 2015. miRNA-133 augments coelomocyte phagocytosis in bacteria-challenged *Apostichopus japonicus* via targeting the TLR component of IRAK-1 in vitro and in vivo. *Scientific Reports* 5: 12608

Lu, Y., Li, C., Zhang, P., Shao, Y, Su, X., Li, Y., and Li, T. 2013. Two adaptor molecules of MyD88 and TRAF6 in *Apostichopus japonicus* Toll signaling cascade: Molecular cloning and expression analysis. *Developmental and Comparative Immunology* 41(4): 498–504

Lützen, J. 1968. Biology and structure of *Cystobia stichopi*, n. sp. Eugregarinida, fanlily Urosporidae, a parasite of the holothurian *Stichopus tremulus* Gmel. *Nytt Magasin for Zoologi* 16: 14–19

Ma, Y., Xu, G., Zhang, E., Wang, P., and Chang, Y. 2006. The etiology of acute peristome edema disease in cultured juveniles of *Apostichopus japonicus*. *Journal of Fisheries of China* 30(3): 377–382

Maes, P. and Jangoux, M. 1984a. The bald-sea-urchin disease: A biopathological approach. *Helgoländer Meeresuntersuchungen* 37(1–4): 217–224

Maes, P. and Jangoux, M. 1984b. The bald sea urchin disease: A bacterial infection. In Proceedings of the Fifth International Echinoderm Conference, Keegan, B.F., O'Connor, B.D.S. (eds), pp. 313–314. Galway, Ireland. Rotterdam: A.A. Balkema

Maes, P., Jangoux, M., and Fenaux, L. 1986. The bald-sea-urchin disease ultrastructure of the lesions and nature of their pigmentation. *Annales de l'Institue Ocèanographique* (France) 62: 37–46

Magesky, A. and Pelletier, É. 2018. Cytotoxicity and physiological effects of silver nanoparticles on marine invertebrates. *Advances in Experimental Medicine and Biology* 1048: 285–309

Mahé, F., De Vargas, C., Bass, D., Czech, L., Stamatakis, A., Lara, E., Singer, D., Mayor J., Bunge, J., Sernaker, S., and Siemensmeyer T. 2017. Parasites dominate hyperdiverse soil protist communities in Neotropical rainforests. *Nature Ecology and Evolution* 1: 0091

Majeske, J., Bayne, C.J., and Smith, L.C. 2013. Aggregation of sea urchin phagocytes is augmented in vitro by lipopolysaccharide. *PLoS ONE* 8(4): e61419

Mangiaterra, M.B. and Silva, J.R. 2001. Induced inflammatory process in the sea urchin *Lytechinus variegatus*. *Invertebrate Biology* 120(2): 178–184

Manzo, S. 2004. Sea urchin embryotoxicity test: Proposal for a simplified bioassay. *Ecotoxicology and Environmental Safety* 57: 123–128

Martínez-Gómez, C., León, V.M., Calles, S., Gomáriz-Olcina, M., and Vethaak, A.D. 2017. The adverse effects of virgin microplastics on the fertilization and larval development of sea urchins. *Marine Environmental Research* 130: 69–76

Massin, C., Jangoux, M., and Sibuet, M. 1978. Description d'*Ixoreis psychropotae*, nov. gen., nov. sp., coccidie parasite du tube digestif de l'holothurie abyssale *Psychropotes longicauda* Théel. *Protistologica* 14: 253–259

McClay, D. 2011. Evolutionary crossroads in developmental biology: Sea urchins. *Development* 138: 2639–2648

McEdward, L.R. and Miner, B.G. 2001. Larval and life-cycle patterns in echinoderms. *Canadian Journal of Zoology* 79(2001): 1125–1170

Mead, A.D. 1892. Twenty-eighth annual report of the commissioners of inland fisheries, made to the general assembly at its January session. Southwick, J.M.K., Root, H.T., Willard, C.W., Morton, W.M.P., Roberts, A.D., and Bumpus, H.C. (eds.) p. 112. Freeman and Sons, Printers to the State of Rhode Island

Menge, B.A. 1979. Coexistence between the sea stars *Asterias vulgaris* and *Asterias forbesi* in a heterogeneous environment—nonequilibrium explanation. *Oecologia* 41: 245–272

Menge, B.A., Cerny-Chipman, E.B., Johnson, A., Sullivan, J., Gravem, S., and Chan, F. 2016. Sea star wasting disease in the keystone predator *Pisaster ochraceus* in Oregon: Insights into differential population impacts, recovery, predation rate, and temperature effects from long-term research. *PLoS One* 11: e0153994

Mesarič, T., Sepčić, K., Drobne, D., Makovec, D., Faimali, M., Morgana, S., Falugi, C., and Gambardella, C. 2015. Sperm exposure to carbon-based nanomaterials causes abnormalities in early development of purple sea urchin (*Paracentrotus lividus*). *Aquatic Toxicology* 163: 158–166

Messer, L.I. and Wardlaw, A.C. 1980. Separation of coelomocytes of *Echinus esculentus* by density gradient centrifugation. In *Proceedings of the European Colloquium on Echinoderms: Past and Present*, Jagoux, M. (ed.) pp. 319–323. Rotterdam: AA Balkema

Messinetti, S., Mercurio, S., Parolini, M., Sugni, M., and Pennati, R. 2018. Effects of polystyrene microplastics on early stages of two marine invertebrates with different feeding strategies. *Environmental Pollution* 237: 1080–1087

Metchnikoff, I. 1893. Lectures on the Comparative Pathology of Inflammation: Delivered at the Pasteur Institute in 1891. London: Kegan Paul, Trench, Trübner and Co., Ltd.

Miller, T.L., Small, H.J., Peemoeller, B.J., Gibbs, D.A., and Shields, J.D. 2013. Experimental infections of *Orchitophrya stellarum* (Scuticociliata) in American blue crabs (*Callinectes sapidus*) and fiddler crabs (*Uca minax*). *Journal of Invertebrate Pathology* 114: 346–355

Minchin E.A. 1893. Observations on the gregarines of holothurians. *Quarterly Journal of Microscopical Science* 34: 279–310

Miner, C.M., Burnaford, J.L., Ambrosew, R.F., Antrim, L., Bohlmann, H., Blanchette, C.A., Engle, J.M., Fradkin, S.C., Gaddam, R., Harley, C.D.G., Miner, B.G., Murrau S.N., Smith, J.R., Whitaker, S.G., and Raimondi, P.T. 2018. Large-scale impacts of sea star wasting disease (SSWD) on intertidal sea stars and implications for recovery. *PLoS One* 13: e0192870

Mingazzini, P. 1893. Contribute alla conoscenza degli sporozoi. *Ricerche della Laboratorio di Anatomi Normale dei Reale Universita de Roma* 3: 31–85

Mohapatra, S., Chakraborty, T., Kumar, V., DeBoeck, G., and Mohanta, K. 2013. Aquaculture and stress management: A review of probiotic intervention.

Journal of Animal Physiology and Animal Nutrition 97(3): 405–430

Mojica, E.-R.E. and Merca, F.E. 2004. Lectin from the body walls of black sea cucumber (*Holothuria atra* Jaeger). *Philippine Journal of Science* 133(2): 77–85

Mojica, E.-R.E. and Merca, F.E. 2005. Biological properties of lectin from sea cucumber (*Holothuria scabra* Jaeger). *Journal of Biological Sciences* 5(4): 472–477

Montecino-Latorre, D., Eisenlord, E., Turner, M., Yoshioka, R., Harvell, C.D., Pattengill-Semmens, C.V., Nichols, J.D., and Gaydos, J.K. 2016. Devastating transboundary impacts of sea star wasting disease on subtidal asteroids. *PLoS One* 11: e0163190

Mooi R. 2016. Phylum Echinodermata. In Invertebrates, 3rd ed, Brusca, R.C., Moore, W., and Shuster, S.M. (eds). pp. 968-1003. Sunderland, MA: Sinauer

Moritsch, M.M. and Raimondi, P.T. 2018. Reduction and recovery of keystone predation pressure after disease-related mass mortality. *Ecology and Evolution* 8: 3952–3964

Morroni, L., Pinsino, A., Pellegrini, D., Regoli, F., and Matranga, V. 2016. Development of a new integrative toxicity index based on an improvement of the sea urchin embryo toxicity test. *Ecotoxicology and Environmental Safety* 123: 2–7

Mortensen, T. 1897. Smaa faunistike og biologiske Meddelelser. *Videnskabelige Meddelelser fra Dansk Naturhistorisk Forening* 1897: 311–331

Mortensen, T. and Rosenvinge, L.K. 1910. Sur quelques plantes parasites dans des echinodermes. *Oversigt Kongelige Danske Videnskabernes Selskab Forhandlinger* 1910: 339–354

Mortensen, T. and Rosenvinge, L.K. 1933. Sur une nouvelle algue: *Coccomyxa astericola*, parasite dans une astérie. *Biologiske Meddelelser* 10: 1–8

Mortensen, T. and Rosenvinge, L.K. 1934. Sur une algue: Cyanophycée *Dactylococcopsis echini* N. sp. parasite dans un oursin. *Biologiske Meddelelser* 11: 1–7

Motokawa, T. 2011. Mechanical mutability in connective tissue of starfish body wall. *Biological Bulletin* 221: 280–289

Motokawa, T. and Tsuchi, A. 2003. Dynamic mechanical properties of body-wall dermis in various mechanical states and their implications for the behavior of sea cucumbers. *Biological Bulletin* 205: 261–275

Murano, C., Agnisola, C., Caramiello, D., Castellano, I., Casotti R., Corsi, I., and Palumbo, A. 2020. How sea urchins face microplastics: Uptake, tissue distribution and immune system response. *Environmental Pollution* 264: 114685

Nesbit, K.T., Fleming, T., Batzel, G., Pouv, A., Rosenblatt, H.D., Pace, D.A., Hamdoun, A., and Lyons, D.C. 2019. The painted sea urchin, *Lytechinus pictus*, as a genetically-enabled developmental model. In Echinoderms. Foltz K.R. and Hamdoun A. (eds.) *Methods in Cell Biology* 150: 105–123. Cambridge: Academic Press

Neves, R.A.F., Contins, M., and Nascimento, S.M. 2018. Effects of the toxic benthic dinoflagellate *Ostreopsis* cf. *ovata* on fertilization and early development of the sea urchin *Lytechinus variegatus*. *Marine Environmental Research* 135: 11–17

Nicolas, J.L., Corre, S., Gauthier, G., Robert, R., and Ansquer, D. 1996. Bacterial problems associated with scallop *Pecten maximus* larval culture. *Diseases of Aquatic Organisms* 27: 67–76

Nunez-Pons, L., Work, T.M., Angulo-Preckler, C., Moles, J., and Avila, C. 2018. Exploring the pathology of an epidermal disease affecting a circum-Antarctic sea star. *Scientific Reports* 8: 11353

Okamura, B., Hartigan. A., and Naldoni, J. 2018. Extensive uncharted biodiversity: The parasite dimension. *Integrative and Comparative Biology* 58(6): 1132–1145

Ono, K., Suzuki, T.A., Toyoshima, Y., Suzuki, T., Tsutsui, S., Odaka, T., Miyadai, T., and Nakamura O. 2018. SJL-1, a C-type lectin, acts as a surface defense molecule in Japanese sea cucumber, *Apostichopus japonicus*. *Molecular Immunology* 97: 63–70

Page, F.C. 1987. The classification of 'naked' amoebae of phylum Rhizopodia. *Arch Protistenkd* 133: 199–217

Pan, X., Fan, X., Al-Farraj, S.A., Gao, S., and Chen, Y. 2016. Taxonomy and morphology of four ophrys-related scuticociliates (Protista, Ciliophora, Scuticociliata), with the description of a new genus, *Paramesanophrys* gen. nov. *European Journal of Taxonomy* 191: 1–18

Pancer, Z. 2000. Dynamic expression of multiple scavenger receptor cysteine-rich genes in colomocytes of the purple sea urchin. *Proceedings of the National Academy of Sciences* 97: 13156–13161

Pandiyan, P., Balaraman, D., Thirunavukkarasu, R., George, E.G.J., Subaramaniyan, K., Manikkam, S., and Sadayappan, B. 2013. Probiotics in aquaculture. *Drug Invention Today* 5(1): 55–59

Pearse, J.S. and Cameron, R.A. 1991. Echinodermata: Echinoidea. In *Reproduction of Marine Invertebrates*. Vol. VI, *Echinoderms and Lophophorates*. Giese A.C., Pearse J.S., and Pearse V,B. (eds.) pp. 514–662. Pacific Grove CA: The Boxwood Press

Piatt, J. 1935. An important parasite of starfish. Commercial Fisheries Service Bulletin, U.S. Department of Commerce, Washington 247: 3-4

Pisani, D., Feuda, R., Peterson, K.J., and Smith, A.B. 2015. Resolving phylogenetic signal from noise when divergence is rapid: A new look at the old problem of echinoderm class relationships. *Molecular Phylogenetics and Evolution* 62(1): 27–34

Pixell-Goodrich, H.L.M. 1915. On the life-history of the Sporozoa of spatangoids, with observations on some allied forms. *Quarterly Journal of Microscopical Science* 29: 81–104

Pixell-Goodrich, H.L.M. 1925. Observations on the gregarines of *Chirodota*. *Quarterly Journal of Microscopy Science* 69: 620–628

Pixell-Goodrich, H.L.M. 1929. The gregarine of Cucumaria: *Lithocystis minchinii* Woodc. and *Lithocystis cucumariae* n. sp. *Quarterly Journal of Microscopical Science* 73: 275–287

Plytycz, B. and Seljelid, R. 1993. Bacterial clearance by the sea urchin, *Strongylocentrotus droebachiensis*. *Developmental and Comparative Immunology* 17(3): 283–289

Pomory, C.M. and Lares, M.T. 1998. *Gonospora holoflora*: A new species of gregarine protozoan parasite (Apicomplexa) in *Holothuria floridana* (Echinodermata: Holothuroidea) from the Florida Keys. *Bulletin of Marine Science* 62: 213–218

Pratchett, M.S. 1999. An infectious disease in crown-of-thorns starfish on the Great Barrier Reef. *Coral Reefs* 18: 272–272

Price, D.C., Chan, C.X., Yoon, H.S., Yang, E.C., Qiu, H., Weber, A.P., Schwacke, R., Gross J., Blouin, N.A., Lane, C., and Reyes-Prieto, A. 2012. *Cyanophora paradoxa* genome elucidates origin of photosynthesis in algae and plants. *Science* 335: 843–847

Radhakrishnan, S., Bhavan, P.S., Seenivasan, C., Shanthi, R., and Poongodi, R. 2014. Influence of medicinal herbs (*Alteranthera sessilis*, *Eclipta alba* and *Cissus quadrangularis*) on growth and biochemical parameters of the freshwater prawn *Macrobrachium rosenbergii*. *Aquaculture International* 22(2): 551–572

Raff, R.A. and Byrne, M. 2006. The active evolutionary lives of echinoderm larvae. *Heredity* 97: 244–252

Ragusa, M.A., Nicosia, A., Costa, S., Casano, C., and Gianguzza, F. 2019. A survey on tubulin and arginine methyltransferase families sheds light on *P. lividus* embryo as model system for antiproliferative drug development. *International Journal of Molecular Science* 20(9): 2136

Ramírez-Gómez, F., Aponte-Rivera, F., Méndez-Castaner, L., and García-Arrarás J.E. 2010. Changes in holothurian coelomocyte populations following immune stimulation with different molecular patterns. *Fish and Shellfish Immunology* 29(2): 175–185

Ramírez-Gómez R. and García-Arrarás J.E. 2010. Echinoderm immunity. *Invertebrate Survival Journal* 7: 211–220

Reinardy, H.C. and Bodnar, A.G. 2015. Profiling DNA damage and repair capacity in sea urchin larvae and coelomocytes exposed to genotoxicants. *Mutagenesis* 30(6): 829–839

Reinisch, C.L. and Bang, F.B. 1971. Cell recognition: Reactions of the sea star (*Asterias vulgaras*) to the injection of amebocytes of sea urchin (*Arbacia punctulata*). *Cellular Immunology* 2(5): 496–503

Rivera-Posada. J., Pratchett, M., Cano-Gomex, A., Arango-Gomex, J., and Owens, L. 2011a. Refined identification of *Vibrio* bacterial flora from *Acanthaster planci* based on biochemical profiling and analysis of housekeeping genes. *Diseases of Aquatic Organisms* 96: 113–123

Rivera-Posada, J., Pratchett, M., Cano-Gomex, A., Arango-Gomex, J., and Owens. L. 2011b. Injection of *Acanthaster planci* with thiosulfate-citrate-bile-sucrose agar (TCBS). I. Disease induction. *Diseases of Aquatic Organisms* 97: 85–94

Roberts-Regan, D.L., Scheibling, R.E., and Jellett, J.F. 1988. Natural and experimentally induced lesions of the body wall of the sea urchin *Strongylocentrotus droebachiensis*. *Diseases of Aquatic Organisms* 5: 51–62

Roch, P., Canicatti, C., and Sammarco, S. 1992. Tetrameric structure of the active phenoloxidase evidenced in the coelomocytes of the echinoderm *Holothuria tubulosa*. *Comparative Biochemistry and Physiology B* 102: 349–355

Rojas, R., Miranda, C.D., Opazo, R., and Romero, J. 2015. Characterization and pathogenicity of *Vibrio splendidus* strains associated with massive mortalities of commercial hatchery-reared larvae of scallop *Argopectin purpuratus* (Lamark, 1819). *Journal of Invertebrate Pathology* 124: 61–69

Romano, G., Miralto, A., and Ianora, A. 2010. Teratogenic effects of diatom metabolites on sea urchin *Paracentrotus lividus* embryos. *Marine Drugs* 8(4): 950–967

Ruocco, N., Varrella, S., Romano, G., Ianora, A., Bentley, M.G., Somma, D., Leonardi, A., Mellone, S., Zuppa, A., and Costantini M. 2016. Diatom-derived oxylipins induce cell death in sea urchin embryos activating caspase-8 and caspase 3/7. *Aquatic Toxicology* 176: 128–140

Sartori, D. and Gaion, A. 2016. Toxicity of polyunsaturated aldehydes of diatoms to Indo-Pacific bioindicator organism *Echinometra mathaei*. *Drug and Chemical Toxicology* 39(2): 1–5

Schiebelhut, L.M., Puritz, J.B., and Dawson, M.N. 2018. Decimation by sea star wasting disease and rapid genetic change in a keystone species, *Pisaster ochraceus*. *Proceedings of the National Academy of Science* 115: 7069–7074

Schillaci, D., Cusimano, M.G., Cunsolo, V., Saletti, R., Russo, D., Vazzana, M., Vitale, M., and Arizza, V. 2013. Immune mediators of sea-cucumber *Holothuria tubulosa* (Echinodermata) as source of novel antimicrobial and anti-staphylococcal biofilm agents. *AMB Express* 3(1): 35

Schlegel, M, Weidhase, K., and Stadler, P.F. 2014. Deuterostome phylogeny—a molecular perspective. In *Deep Metazoan Phylogeny: The Backbone of the Tree of Life: New Insights from Analyses of Molecules, Morphology, and Theory of Data Analysis*. Wägele, J.W. and Bartolomaeus, T. (eds.) pp. 413–424. Berlin/Boston: Walter de Gruyter

Schneider, A.F. 1858. Ueber einige parasiten de *Holothuria tubulosa*. *Archiv der Anatomic, Physiologie und Wissenschaftliche Medizin* 1858: 323–329

Schoenmakers, H.J.N., Goedhart, M.J., and Voogt, P.A. 1984. Biometrical and histological aspects of the reproductive cycle of the ovaries of *Asterias rubens* (Echinodermata). *The Biological Bulletin* 166: 328–348

Schrankel, C.S., Solek, C.M., Buckley, K.M., Anderson, M.K., and Rast, J.P. 2016. A conserved alternative form of the purple sea urchin HEB/E2-2/E2A transcription factor mediates a switch in E-protein regulatory state in differentiating immune cells. *Developmental Biology* 416(1): 149–161

Schuh, N.W., Carrier, T.J., Schrankel, C.S., Reitzel, A.M., Heyland, A., and Rast, J.P. 2020. Bacterial exposure mediates developmental plasticity and resistance to lethal *Vibrio lentus* infection in purple sea urchin (*Strongylocentrotus purpuratus*) larvae. *Frontiers in Immunology* 10: 3014

Service, M. and Wardlaw, A.C. 1984. Echinochrome-A as a bactericidal substance in the coelomic fluid of *Echinus esculentus* (L.). *Comparative Biochemistry and Physiology B* 79(2): 161–165

Shao, Y., Li, C., Che, Z., Zhang, P., Zhang, W., Duan, X., and Li, Y. 2015. Cloning and characterization of two lipopolysaccharide-binding protein/bactericidal permeability–increasing protein (LBP/BPI) genes from the sea cucumber *Apostichopus japonicus* with diversified function in modulating ROS production. *Developmental and Comparative Immunology* 52(1): 88–97

Shimizu, M., Takaya Y., Ohsaki, S., and Kawamata, K. 1995. Gross and histopathological signs of the spotting disease in the sea urchin *Strongylocentrotus intermedius*. *Fisheries Science* 61(4): 608–613

Shrestha, G., St. Clair, L.L., and O'Neill, K.L. 2015. The immunostimulating role of lichen polysaccharides: A review. *Phytotherapy Research* 29(3): 317–322

Sibbald, S.J., Cenci, U., Colp, M., Eglit, Y., O'Kelly, C.J., and Archibald, J.M. 2017. Diversity and evolution of *Paramoeba spp.* and their kinetoplastid endosymbionts. *Journal of Eukaryotic Microbiology* 64: 598–607

Small, H.J. 2004. Infections of the Norway lobster, *Nephrops norvegicus* (L.) by dinoflagellate and ciliate parasites. Doctoral dissertation, University of Glasgow.

Small, H.J., Neil, D.M., Taylor, A.C., Bateman, K., and Coombs G.H. 2005a. A parasitic scuticociliate infection in the Norway lobster (*Nephrops norvegicus*). *Journal of Invertebrate Pathology* 90: 108–117

Small, H.J., Neil, D.M., Taylor, A.C., and Coombs, G.H. 2005b. Identification and partial characterisation of metalloproteases secreted by a *Mesanophrys*-like ciliate parasite of the Norway lobster *Nephrops norvegicus*. *Diseases of Aquatic Organisms* 67: 225–231

Smiley, S., McEuen, F.S., Chafee, C., and Kreshnan, S. 1991. Echinodermata: Holothuroidea. In *Reproduction of Marine Invertebrates*. Vol. VI, *Echinoderms and Lophophorates*. Giese, A.C., Pearse, J.S., and Pearse, V.B. (eds.) pp. 664–750. The Boxwood Press, Pacific Grove CA.

Smith, A.B. 1997. Echinoderm larvae and phylogeny. *Annual Review of Ecology and Systematics* 28: 219–241

Smith, G.F.M. 1936. A gonad parasite of the starfish. *Science* 84: 157–157

Smith. L.C., Arizza, V., Barela Hudgell, M.A., Barone, G., Bodnar, A.G., Buckley, K.M., Cunsolo, V., Dheilly, N., Franchi, N., Fugmann, S.D., Furukawa, R., Garcia-Arraras, J., Henson, J.H., Hibino, T., Irons, Z.H., Li, C., Lun, C.M., Majeske, A.J., Oren M., Pagliara, P., Pinsino, A., Raftos D.A., Rast, J.P., Samasa, B., Schillaci, D., Schrankel, C.S., Stabili, L., Stensväg, K., and Sutton, E. 2018. Echinodermata: The complex immune system in echinoderms. In *Advances in Comparative Immunology*, Cooper, E.L. (ed.) pp. 409–501. Cham Switzerland: Springer Publisher

Smith, L.C., Clow, L.A., and Terwilliger, D.P. 2001. The ancestral complement system in sea urchins. *Immunological Review* 180: 16–34

Smith, L.C. and Davidson, E.H. 1992. The echinoid immune system and the phylogenetic occurrence of immune mechanisms in deuterostomes. *Immunology Today* 13(9): 356–362

Smith, L.C. and Davidson, E.H. 1994. The echinoderm immune system. Characters shared with vertebrate immune systems and characters arising later in deuterostome phylogeny. *Annals of the New York Academy of Science* 712: 13–226

Smith, L.C., Ghosh, J., Buckley, K.M., Clow, L.A., Dheilly, N.M., Haug, T., Henson, J.H., Li, C., Lun, C.M., Majeske, A.J., Matranga, V., Nair, S.V., Rast, J.P., Raftos, D.A., Roth, M., Sacchi, S., Schrankel, C.S., and Stensvåg, K. 2010. Echinoderm immunity. In *Invertebrate Immunity*, Söderhall, K. (ed.) *Advances in Experimental Medicine and Biology*, 708: 260–301. Austin TX: Madame Curie Bioscience Database, Landes Biosciences and Spring Science+Business Media

Smith, L.C. and Lun, C.M. 2017. The *SpTransformer* gene family (formerly *Sp185/333*) in the purple sea urchin and the functional diversity of the anti-pathogen rSpTransformer-E1 protein. *Frontiers in Immunology* 8: 725

Smith, V.J. 1981. The echinoderms. In *Invertebrate Blood Cells*, Ratcliffe, N.A. and Rowley, A.F. (eds.) pp. 513–562. New York, NY: Academic Press

Smith, V.J. and Soderhall, K. 1991. A comparison of phenoloxidase activity in the blood of marine invertebrates. *Developmental and Comparative Immunology* 15(4): 251–261

Sodergren, E., Weinstock, G.M., Davidson, E.H., Cameron, R.A. et al. The genome of the sea urchin, *Strongylocentrotus purpuratus*. *Science* 314(5801): 941–952

Solek, C.M., Oliveri, P., Loza-Coll, M., Schrankel, C.S., Ho, E.C., Wang, G, and Rast, J.P. 2013. An ancient role for Gata-1/2/3 and Scl transcription factor homologs in the development of immunocytes. *Developmental Biology* 382(1): 280–292

Song, J., Wang, P.-H., Li, C.-Y., Yang, E.-X., and Chang, Y.-Q. 2007. Ultrastructural pathology of adhesion dysfunction disease in cultured juvenile sea cucumber *Apostichopus japonicus*. *Journal of Dalian Fisheries University* 22(3): 221–225

Song, S.K., Beck, B.R., Kim, D., Park, J., Kim, J., Kim, H.D., and Ringø, E. 2014. Prebiotics as immunostimulants in aquaculture: A review. *Fish and Shellfish Immunology* 40(1): 40–48

Song, Z. 2005. A preliminary study on the phagocytosis function of coelomocytes and the antibacterial activity of coelomic fluid of sea cucumber *Apostichopus japonicus*. Master's Thesis, Dalian Ocean University

Sonnenholzner-Varas, J.I., Touron, N., and Panchana Orrala, M.M. 2018. Breeding, laral development, and growth of juveniles of the edible sea urchin, *Tripneustes depressus*: A new target species for aquaculture in Ecuador. *Aquaculture* 496: 134–145

Stabili. L., Pagliara, P., and Roch, P. 1996. Antibacterial activity in the coelomocytes of the sea urchin *Paracentrotus lividus*. *Comparative Biochemistry and Physiology B* 113(3): 639–644

Staehli, A., Schaerer, R., Hoelzle, K., and Ribi, G. 2009. Temperature induced disease in the starfish *Astropecten jonstoni*. *Marine Biodiversity Records* 2: e78

Stickle, W.B. and Kozloff, E.N. 2008. Association and distribution of the ciliate *Orchitophrya stellarum* with asteriid sea stars on the west coast of North America. *Diseases of Aquatic Organisms* 80: 37–43

Stickle W.B., Kozloff E.N., and Henk M.C. 2007a. The ciliate *Orchitophrya stellarum* viewed as a facultative parasite of asteriid sea stars. *Cahiers de Biologie Marine* 48: 9–16

Stickle, W.B., Kozloff, E.N, and Story, S. 2007b. Physiology of the ciliate *Orchitophrya stellarum* and its experimental infection of *Leptasterias* spp. *Canadian Journal of Zoology* 85: 201–206

Stickle W.B., Rathbone E.N., and Story, S. 2001b. Parasitism of sea stars from Puget Sound, Washington, by *Orchitophrya stellarum*. In *Echinoderms 2000*, Barker M. (ed.) pp. 221–226. Lisse: Swets and Zeitinger

Stickle, W.B., Weidner, E.H., and Kozloff, E.N. 2001a. Parasitism of *Leptasterias* spp. by the ciliated protozoan *Orchitophrya stellarum*. *Invertebrate Biology* 120: 88–95

Stien, A. and Halvorsen, O. 1998. Experimental transmission of the nematode *Echinomermella matsi* to the sea urchin *Strongylocentrotus droebachiensis* in the laboratory. *The Journal of Parasitology* 84(3): 658–660

Strathmann, R.R. 1971. The feeding behavior of planktotrophic echinoderm larvae: Mechanisms, regulation, and rates of suspension feeding. *Journal of Experimental Marine Biology and Ecology* 6(2): 109–160

Stumpp, M., Dupont, S., and Hu, M.Y. 2019. Measurement of feeding rates, respiration, and pH regulatory processes in the light of ocean acidification research. In *Echinoderms*. Foltz, K.R. and Hamdoun, A. (eds.) *Methods in Cell Biology* 150: 391–409. Cambridge MA: Academic Press

Sun, H., Zhou, Z., Dong, Y., Yang, A., Jiang, B., Gao, S., Chen, Z., Guan, X., Wang, B., and Wang, X. 2013. Identification and expression analysis of two Toll-like receptor genes from sea cucumber (*Apostichopus japonicus*). *Fish and Shellfish Immunology* 34(1): 147–158

Sun, S. 2006. Study on the major bacterial diseases of cultured sea cucumber (*Apostichopus japonicus*) during larval stages. Master's Thesis, Dalian Ocean University of China.

Sun, Y., Jin, L., Wang, T., Xue, J., Liu, G., Li, X., You, J., Li, S., and Xu, Y. 2008. Polysaccharides from *Astragalus membranaceus* promote phagocytosis and superoxide anion (O_2-) production by coelomocytes from sea cucumber *Apostichopus japonicus* in vitro. *Comparative Biochemistry and Physiology part C: Toxicology and Pharmacology* 147(3): 293–298

Sunday, J., Raeburn, L., and Hart, M.W. 2008. Emerging infectious disease in sea stars: Castrating ciliate parasites in *Patiria miniata*. *Diseases of Aquatic Organisms* 81: 173–176

Sweet, M. 2020. Sea urchin diseases: Effects from individuals to ecosystems. In *Sea Urchins: Biology and Ecology*, 4th edn, Lawrence J.M. (ed.) pp. 219–226. Amsterdam: Elsevier

Tajima, K., Hirano, T., Nakano, K., and Ezura Y. 1997a. Taxonomical study on the causative bacterium of spotting disease of sea urchin *Strongylocentrotus intermedius*. *Fisheries Science* 63(6): 897–900

Tajima, K., Hirano, T., Shimizu, M., and Ezura, Y. 1997b. Isolation and pathogenicity of the causative bacterium of spotting disease of sea urchin *Strongylocentrotus inter-medius*. *Fisheries Science* 63(2): 249–252

Takuya, K., Hitomi, U., Tomomitsu, H., Tadashi, T., Takuji, N., Kenichi Y., and Tatsuya O. 2005. Cytotoxicity of a GalNAc-specific C-type lectin CEL-I toward various cell lines. *Journal of Biochemistry* 137(1): 41–50

Tan, J., Liu, Z., Perfetto, M., Han, L., Li, Q., Zhang, Q., Wu, R., Yuan, Z., Zou, X., and Hou L. 2012. Isolation and purification of the peptides from *Apostichopus japonicus* and evaluation of its antibacterial and antitumor activities. *African Journal of Microbiology Research* 6(44): 7139–7146

Tanifuji, G., Cenci, U., Moog, D., Dean, S., Nakayama, T., David, V., Fiala, I., Curtis, B.A., Sibbald, S.J., Onodera, N.T., and Colp, M. 2017. Genome sequencing reveals metabolic and cellular interdependence in an amoeba-kinetoplastid symbiosis. *Scientific Reports* 7: 1–13

Taylor, C.E. and Bang, F.B. 1978. Alteration of blood-clotting in *Asterias forbesi* associated with a ciliate infection. *Biological Bulletin* 155: 468–469

Theodorides, J. and M. Laird. 1970. Quelques eugre-garines parasites d'invertebres marins de St. Andrews (Nouveau-Brunswick). *Canadian Journal of Zoology* 48: 1013–1016

Tian, G., Liu, F., Duan, D., Zhang, S., and Zhang, J. 2012. Effect of EM on purification of major water pollutants in farming sea cucumber. *Journal of Shandong Agricultural University (Natural Sciences)* 43: 381–386

Tomomitsu, H., Kouhei, S., Noriaki, M., Tokik,o F., Tatsuya, O., Hajime S., and Haruhiko, A. 2004. Characterization of recombinant CEL-I, a GalNAc-specific C-type lectin, expressed in *Escherichia coli* using an artificial synthetic gene. *Journal of Biochemistry* 135: 101–107

Toral-Granda, V., Lovatelli, A., and Vasconcellos, M. 2008. Sea cucumbers: A global review of fisheries and trade. *Fisheries and Aquaculture Technical Paper*, no. 516. Rome: Food and Agriculture Organization of the United Nations

Trégouboff, G. 1918. Étude monographique de *Gonospora testiculi* Treg. *Archives de Zoologie Expérimentale et Générale* 57: 471–509

Turton, G.C. and Wardlaw, A.C. 1987. Pathogenicity of the marine yeasts *Metschnikowia zobelli* and *Rhodotorula rubra* for the sea urchin *Echinus esculentus*. *Aquaculture* 67(1–2): 199–202

Tyunin, A.P., Ageenko, N.V., and Kiselev, K.V. 2016. Effects of 5-azacytidine-induced DNA demethylation on polyketide synthase gene expression in larvae of sea urchin *Strongylocentrotus intermedius*. *Biotechnology Letters* 38(12): 2035–2041

Unuma, T., Tsuda, N., Sakai, Y., Kamaishi, T., Sawaguchi, S., Itoh, N., and Yamano, K. 2020. Coccidian parasite in sea cucumber (*Apostichopus japonicus*) ovaries. *The Biological Bulletin* 238: 64–71

Varrella, S., Romano, G., Ianora, A., Bentley, M.G., Ruocco, N., and Costantini, M. 2014. Molecular response to toxic diatom-derived aldehydes in the sea urchin *Paracentrotus lividus*. *Marine Drugs* 12(4): 2089–2113

Varrella, S., Romano, G., Ruocco, N., Ianora, A., Bentley, M.G., and Costantini, M. 2016. First morphological and molecular evidence of the negative impact of diatom-derived hydroxyacids on the sea urchin *Paracentrotus lividus*. *Toxicological Sciences* 151(2): 419–433

Verschuere, L., Rombaut, G., Sorgeloos, P., and Verstraete, W. 2000. Probiotic bacteria as biological control agents in aquaculture. *Microbiology and Molecular Biology Reviews* 64(4): 655–671

Vevers, H.G. 1951. The biology of *Asterias rubens* L. II. Par-asitization of the gonads by the ciliate *Orchitophrya stel-larum* Cépède. *Journal of the Marine Biological Association of the United Kingdom* 29: 619–624

Votýpka, J., Modrý, D., Oborník, M., Šlapeta, J., and Lukeš, J. 2017. Apicomplexa. In *Handbook of the Protists*. Archibald, J., Simpson, A., and Slamovits, C. (eds.) pp. 1–58. Cham: Springer

Wahltinez, S.J., Newton, A.L., Harms, C.A., Lahner, L.L., and Stacy, N.I. 2020. Coelomic fluid evaluation in *Pisaster ochraceus* affected by sea star wasting syndrome: evidence of osmodysregulation, calcium homeostasis derangement, and coelomocyte responses. *Frontiers in Veterinary Science* 7: 131

Wakeman, K.C. and Leander, B.S. 2012. Molecular phylogeny of pacific archigregarines (Apicomplexa), including descriptions of *Veloxidium leptosynaptae* n. gen., n. sp., from the sea cucumber *Leptosynapta clar-ki* (Echinodermata), and two new species of *Selenidium*. *Journal of Eukaryotic Microbiology* 59: 232–245

Wang, A., Ran, C., Wang, Y., Zhang, Z., Ding, Q., Yang, Y., Olsen, R.E., Ringø, E., Bindelle, J., and Zhou, Z. 2019. Use of probiotics in aquaculture of China—a review of the past decade. *Fish and Shellfish Immunology* 86: 734–755

Wang, B., Li, Y., Li, X., Chen, H.X., Liu, M.Q., and Kong, Y.T. 2006. Biological characteristic and pathogenicity of the pathogenic vibrio on the 'red spot disease' of *Strongylocentrotus intermedius*. *Journal of Fishery Sciences of China* 30: 371–376 [in Chinese]

Wang, F., Yang, H., Gao, F., and Liu, G. 2008. Effects of acute temperature or salinity stress on the immune response in sea cucumber, *Apostichopus japonicus*. *Comparative Biochemistry and Physiology A* 151(4): 491–498

Wang, G., Yuan, J., Zhao, Y., and Yuan, M. 2007. Isola-tion, identification and drug sensitivity of the pathogens of the skin ulceration disease in *Apostichopus japonicus*. *Journal of Northwest A&F University (Natural Science Edition)* 35: 87–90

Wang, H., Xue, Z., Liu, Z., Wang, W., Wang, F., Wang, Y., Wang, L., and Song, L. 2018. A novel C-type lectin from the sea cucumber *Apostichopus japonicus* (*Aj*CTL-2) with preferential binding of d-galactose. *Fish and Shellfish Immunology* 79: 218–227

Wang, P., Chang, Y., Xu, G., and Song, L. 2005. Isolation and ultrastructure of an enveloped virus in cultured sea cucumber *Apostichopus japonicus* (Selenka). *Journal of Fishery Science of China* 12(6): 766–770

Wang, P., Chang, Y., Yu, J., Li, C., and Xu, G. 2007. Acute peristome edema disease in juvenile and adult sea cucumbers *Apostichopus japonicus* (Selenka) reared in North China. *Journal of Invertebrate Pathology* 96(1): 11–17

Wang, T., Sun, Y., Jin, L., Thacker, P., Li, S., and Xu, Y. 2013. *Aj*-rel and *Aj*-p105, two evolutionary conserved NF-κB homologues in sea cucumber (*Apostichopus japonicus*) and their involvement in LPS induced immunity. *Fish and Shellfish Immunology* 34(1): 17–22.

Wang, T., Sun, Y., Jin, L., X,u Y., Wang, L., Re,n T., and Wang, K. 2009. Enhancement of non-specific immune response in sea cucumber (*Apostichopus japonicus*) by *Astragalus membranaceus* and its polysaccharides. *Fish and Shellfish Immunology* 27(6): 757–762

Wang, T., Xu, Y., Liu, W., Sun, Y., and Jin, L. 2011. Expression of *Apostichopus japonicus* lysozyme in the methylotrophic yeast *Pichia pastoris*. *Protein Expression and Purification* 77(1): 20–25

Wang, W., Bai, Y., Wang, H., Su, C., and Yang, F. 2013. Effects of compound Chinese herbal medicine on immune enzyme in sea cucumber (*Apostichopus japonicus*). *China Feed* 16: 37–39

Wang, W., Sun, J., Liu, C., and Xue, Z. 2017. Application of immunostimulants in aquaculture: Current knowledge and future perspectives. *Aquaculture Research* 48(1): 1–23

Wang, Y., Fang, B., Zhang, C., and Rong, X. 2006. Etiology of skin ulcer syndrome in cultured juveniles of *Apostichopus japonicus* and analysis of reservoir of the pathogens. *Journal of Fishery Sciences of China* 13(4): 610–616

Wang, Y., Ron,g X., Zhang, C., and Sun, S. 2005. Main diseases of cultured *Apostichopus japonicus*: Prevention and treatment. *Marine Science* 29: 1–7

Wang, Y., Sun, S., and Rong, X. 2006. Stomach ulcer disease in auricularia of sea cucumber (*Apostichopus japonicus*) and its etiological identification. *Journal of Fishery Sciences of China* 13(6): 908–916

Wang, Y., Zhang, C., Rong, X., Chen, J., Shi, C., Sun, H., and Yan, J. 2004. Diseases of cultured sea cucumber, *Apostichopus japonicus*, in China. In *Advances in Sea Cucumber Aquaculture and Management*. Lovatelli, A., Conand, C., Purcell, S., Uthicke, S., Hamel, J.-F., and Mercier, A. (eds.), pp. 297–310. Rome: Food and Agriculture Organization of the United Nations, Fisheries Technical Paper No 463

Wang, Y.N., Chang, Y.Q., Lawrence, J.M. 2013. Disease in sea urchins. In *Sea Urchins: Biology and Ecology*, Lawrence J.M. (ed.) *Developments in Aquaculture and Fisheries Science* 38: 179–186. Amsterdam; Elsevier, Academic Press

Wang, Z., Lv, Z., Li, C., Shao, Y., Zhang, W., and Zhao, X. 2018. An invertebrate β-integrin mediates coelomocyte phagocytosis via activation of septin2 and 7 but not septin10. *International Journal of Biological Macromolecules* 113: 1167–1181

Wardlaw, A.C. 1985. Bactericidal activity of coelomic fluid of the sea urchin, *Echinus esculentus*, on different marine bacteria. *Journal of the Marine Biological Association of the United Kingdom* 65(1): 133–139

Wardlaw, A.C. and Unkles, S.E. 1978. Bactericidal activity of coelomic fluid from the sea urchin *Echinus esculentus*. *Journal of Invertebrate Pathology* 32(1): 25–34

Wei, X., Liu, X., Yang, J., Wang, S., Sun, G., and Yang, J. 2015. Critical roles of sea cucumber C-type lectin in nonself recognition and bacterial clearance. *Fish and Shellfish Immunology* 45(2): 791–799

Wilkie, I.C. 1984. Variable tensility in echinoderm collagenous tissues—a review. *Marine Behavior and Physiology* 11: 1–34

Wilkie, I.C. 2005. Mutable collagenous tissue: Overview and biotechnological perspective. In *Marine Molecular Biotechnology Echinodermata*, Matranga V. (ed.) pp. 221–250. Berlin: Springer-Verlag

Woodcock, H.M. 1904. On *Cystobia irregularis* (Minch.) and allied neogamous gregarines. (Preliminary note.). *Archives de Zoologie Expérimentale et Générale* 2: 125–128

Woodcock, H.M. 1906. The life-cycle of *Cystobia irregularis* (Minch.) together with observations on other neogamous gregarines. *Quarterly Journal of Microscopical Sciences* 50: 1–100

Wray, G.A., Kitazawa, C., and Miner, B. 2004. Culture of echinoderm larvae through metamorphosis. *Methods in Cell Biology* 74: 75–86

Xing, K., Yang, H.S., and Chen, M.Y. 2008. Morphological and ultrastructural characterization of the coelomocytes in *Apostichopus japonicus*. *Aquatic Biology* 2: 85–92

Xiong, C., Li, W., Bai, X., and Du, Y. 2005. Application of lectins as immunostimulants in mariculture of sea cucumber *Stichopus japonicus*. *Feed Industry* 26: 30–32

Xu, L., Xu, Y., He, L., Zhang, M., Wang, L., Li, Z., and Li, X. 2019. Immunomodulatory effects of chicken egg yolk antibodies (IgY) against experimental *Shewanella marisflavi* AP629 infections in sea cucumbers (*Apostichopus japonicus*). *Fish and Shellfish Immunology* 84: 108–119

Xu, S., Zhao, B., Li, C., Hu, W., Han, S., and Li, Q. 2017. Effects of ammonia nitrogen stress on the survival and activities of non-specific immune enzymes of different-sized sea cucumber (*Apostichopus japonicus*). *Progress in Fishery Sciences* 38: 173–179

Xue, Z., Li, H., Wang, X., Li, X., Liu, Y., Sun, J., and Liu C. 2015. A review of the immune molecules in the sea cucumber. *Fish and Shellfish Immunology* 44(1): 1–11

Yang, A.F., Zhou, Z.C., He, C.B., Hu, J.J., Chen, Z., Gao, X.G., Dong, Y., Jiang, B., Liu, W.D., and Guan, X.Y. 2009. Analysis of expressed sequence tags from body wall, intestine and respiratory tree of sea cucumber (*Apostichopus japonicus*). *Aquaculture* 296(3–4): 193–199

Yang, J.L., Zhou, L., Sheng, X.Z., Xing, J., and Zhan, W.B. 2007b. Identification and biological characteristics of pathogen RH2 associated with skin ulceration of cultured *Apostichopus japonicus*. *Journal of Fisheries of China* 4: 504–511

Yang, J.L., Zhou, L, Xing, J., Sheng, X.Z., and Zhan, W.B. 2007a. Identification of *Aeromonas salmonicida* associated with skin ulceration of cultured sea cucumber *Apostichopus japonicus* and characterization of the extracellular products. *Journal of Fishery Sciences of China* 6: 981–989

Yang, X., Cong, L., Lu, M., Liu, H., and Zhu, B. 2007. Characterization and structure analysis of a gene encoding i-type lysozyme from sea cucumber *Apostichopus japonicus*. *Journal of Bioscience and Bioengineering* 107: 583–588

Yasumoto, T., Yasumura, D., Yotsu, M., Michishita, T., Endo, A., and Kotaki, Y. 1986. Bacterial production of tetrodotoxin and anhydrotetrodotoxin. *Agricultural and Biological Chemistry* 50(3): 793–795

Young, C.M. and Sewell, M.E. 2002. Marine invertebrate larvae. London: Academic Press

Young, N.D., Dyková, I., Crosbie, P.B., Wolf, M., Morrison, R.N., Bridle, A.R., and Nowak, B.F. 2014. Support for the coevolution of *Neoparamoeba* and their endosymbionts, *Perkinsela* amoebae-like organisms. *European Journal of Protistology* 50: 509–523

Yui, M. and Bayne, C. 1983. Echinoderm immunity: Bacterial clearance by the sea urchin *Strongylocentrotus purpuratus*. *The Biological Bulletin* 165: 473–485

Zhang C. 2007. Aetiological study on bacterial diseases of cultured sea cucumber *Apostichopus japonicus* in China. Master's Thesis, Ocean University of China.

Zhang, C., Liang, W., Zhang, W., and Li, C. 2016. Characterization of a metalloprotease involved in *Vibrio splendidus* infection in the sea cucumber, *Apostichopus japonicus*. *Microbial Pathogenesis* 101: 96–103

Zhang, C., Wang, Y., and Rong, X. 2006. Isolation and identification of causative pathogen for skin ulcerative syndrome in *Apostichopus japonicus*. *Journal of Fishery Science of China* 30(1): 118–123

Zhang, C., Zhang, W., Liang, W., Shao, Y., Zhao, X., and Li, C. 2019. A sigma factor RpoD negatively regulates temperature-dependent metalloprotease expression in a pathogenic *Vibrio splendidus*. *Microbial Pathogenesis* 128: 311–316

Zhang, F., Gong, J., and Wang, H. 2007. Determination of activity of complement-like in sea cucumber, *Apostichopus japonicus*. *Journal of Dalian Ocean University* 22(4): 246–248

Zhang, J., Cao, Z., Li, Z., Wang, L., Li, H., Wu, F., Jin, L., Li, X., Li, S., and Xu, Y. 2015. Effect of bacteriophages on *Vibrio alginolyticus* infection in the sea cucumber, *Apostichopus japonicus* (Selenka). *Journal of the World Aquaculture Society* 46(2): 149–158

Zhang, P., Li, C., Li, Y., Zhang, P., Shao, Y., Jin, C., and Li, T. 2014. Proteomic identification of differentially expressed proteins in sea cucumber *Apostichopus japonicus* coelomocytes after *Vibrio splendidus* infection. *Developmental and Comparative Immunology* 44(2): 370–377

Zhang, Q., Ma, H., Mai, K., Zhang, W., Liufu, Z., and Xu, W. 2010. Interaction of dietary *Bacillus subtilis* and fructooligosaccharide on the growth performance, non-specific immunity of sea cucumber, *Apostichopus japonicus*. *Fish and Shellfish Immunology* 29(2): 204–211

Zhang, T., Fan, X., Gao, F., Al-Farraj, S.A., El-Serehy, H.A., and Song, W. 2019. Further analyses on the phylogeny of the subclass Scuticociliatia (Protozoa, Ciliophora) based on both nuclear and mitochondrial data. *Molecular Phylogenetics and Evolution* 139: 106565

Zhang, W., Lv, Z., Li, C., Sun, Y., Jiang, H., Zhao, M., Zhao, X., Shao, Y., and Chang, Y. 2019. Transcriptome profiling reveals key roles of phagosome and NOD-like receptor pathway in spotting diseased *Strongylocentrotus intermedius*. *Fish and Shellfish Immunology* 84: 521–531

Zhang, X., Sun, L., Yuan, J., Sun, Y, Gao, Y., Zhang, L., Li, S., Dai, H., Hamel, J.F., Liu, C., Yu, Y., Liu, S., Lin, W., Guo, K., Jin, S., Xu, P., Storey, K.B., Huan, P., Zhang, T., Zhou, Y., Zhang, J., Lin, C., Li, X., Xing, L., Huo, D., Sun, M., Wang, L., Mercier, A., Li, F., Yang, H., and Xiang,

J. 2017. The sea cucumber genome provides insights into morphological evolution and visceral regeneration. *PLoS Biology* 15(10): e2003790

Zhang, Z., Xing, R., Lv, Z., Shao, Y., Zhang, W., Zhao, X., and Li, C. 2018. Analysis of gut microbiota revealed *Lactococcus garviaeae* could be an indicative of skin ulceration syndrome in farmed sea cucumber *Apostichopus japonicus*. *Fish and Shellfish Immunology* 80: 148–154

Zhang, Z., Zhang, W., Hu, Z., Li, C., Shao, Y., Zhao, X., and Guo, M. 2019. Environmental factors promote pathogen-induced skin ulceration syndrome outbreak by readjusting the hindgut microbiome of *Apostichopus japonicus*. *Aquaculture* 507: 155–163

Zhao, W., Liang, M., and Zhang, P. 2010. Effect of yeast polysaccharide on the immune function of juvenile sea cucumber, *Apostichopus japonicus* Selenka under pH stress. *Aquaculture International* 18(5): 777–786

Zhao, Y., Mai, K., Xu, W., Zhang, W., Ai, Q., Zhang, Y., Wang, X., and Liufu, Z. 2011. Influence of dietary probiotic *Bacillus* TC22 and prebiotic fructooligosaccharide on growth, immune responses and disease resistance against *Vibrio splendidus* infection in sea cucumber *Apostichopus japonicus*. *Journal of Ocean University of China* 10(3): 293–300

Zhao, Y., Zhang, W., Xu, W., Mai, K., Zhang, Y., and Liufu, Z. 2012. Effects of potential probiotic *Bacillus subtilis* T13 on growth, immunity and disease resistance against *Vibrio splendidus* infection in juvenile sea cucumber *Apostichopus japonicus*. *Fish and Shellfish Immunology* 32(5): 750–755

Zhong, L., Zhang, F., Zhai, Y., Cao, Y.H., Zhang, S., and Chang, Y.Q. 2015. Identification and comparative analysis of complement C3-associated microRNAs in immune response of *Apostichopus japonicus* by high-throughput sequencing. *Scientific Reports* 5(17763): 1–11

Zhou, Z., Sun, D., Yang, A., Dong, Y., Chen, Z., Wang, X., Guan, X., Jiang, B., and Wang, B. 2011. Molecular characterization and expression analysis of a complement component 3 in the sea cucumber (*Apostichopus japonicus*). *Fish and Shellfish Immunology* 31(4): 540–547

Zilz, Z. 2018. Wasted and castrated: Two diseases affecting the ochre star, *Pisaster ochraceus*, in North America. Master's Thesis, Department of Biology, Western Washington University Graduate School Collection 758. https://cedar.wwu.edu/wwuet/758

CHAPTER 19

Ecological outcomes of echinoderm disease, mass die-offs, and pandemics

L. Courtney Smith, Maria Byrne, Keryn B. Gedan, Diana L. Lipscomb, Audrey J. Majeske, and Ghada Tafesh-Edwards

19.1 Introduction

Infectious disease outbreaks in marine organisms can have profound effects on ecological community structure with significant cascading effects on species diversity and ecosystem function. Mass die-offs from communicable diseases not only result in large declines in the infected species that can endanger its viability as a population but can also trigger dramatic phase shifts in the ecosystems from which they disappear. This is particularly true for many echinoderms that are keystone species or that impact foundational species in marine habitats. Although echinoderm diseases can affect individual animals, as has been noted in the past for sea star wasting disease (SSWD) (Bang 1982; Menge 1979) or bald sea urchin disease (Johnson 1971; Becker et al. 2008), little is known about how a low level of echinoderm disease may control populations in what are deemed to be healthy ecosystems. When whole populations are impacted by disease, this is often the outcome of a multitude of factors including infectiousness of the pathogen in conjunction with host stresses, of which all are often induced by changes in the environment. While most microorganisms may be harmless when present in low numbers, sudden and dramatic shifts in ocean chemistry and temperatures may result in significant host stresses that appear to correlate with increases in host disease that may lead to increased sizes of pathogen populations (Sabine et al. 2004;

Burge et al. 2013; Brink et al. 2019). These changes may not only include immune suppression of the host but also shifts in the normal commensal microbial communities (dysbiosis) associated with host animals, which can cause or exacerbate episodic disease outbreaks in host animals (Sweet 2020). Repeated environmental changes may be associated with serial repeats of diseases that occur with increasing frequency and provide little time for recovery by the affected animals. Here, we discuss examples of mass mortality events for sea urchin and sea star species that have reduced or eliminated the affected animals at vast spatial scales. We also describe the outcomes of these changes relative to the structure of the marine ecosystems that they inhabit.

19.2 Mass die-offs from bald sea urchin disease

Bald sea urchin disease (BSUD) is generally described as the loss of spines, tube feet, and pedicellariae from the surface of echinoids so that the afflicted animal appears bald (Figure 19.1); (see Chapter 18). It is believed that BSUD is a communicable bacterial disease and has been reported to affect as many as 26 species of echinoids that have a world-wide distribution in both temperate and tropical marine waters (Table 19.1); (Johnson 1971; Pearse et al. 1977; Maes and Jangoux 1984; Becker et al. 2007, 2008; Brink et al. 2019). In the

L.C. Smith et al., *Ecological outcomes of echinoderm disease, mass die-offs, and pandemics.* In: *Invertebrate Pathology.* Edited by Andrew F. Rowley, Christopher J. Coates and Miranda M.A. Whitten, Oxford University Press. © Oxford University Press (2022). DOI: 10.1093/oso/9780198853756.003.0019

Figure 19.1 Bald sea urchin disease (BSUD) of the purple sea urchin. The purple sea urchin, *Strongylocentrotus purpuratus*, on the left shows symptoms of spine loss in the interambulacral regions of the test that is consistent with BSUD (see Section 18.3 in Chapter 18). For comparison, the animal on the right does not show symptoms. Source: this image was taken at Point Loma California in 1989 and obtained from the collection of the Ron McPeak Library (https://alexandria.ucsb.edu/collections/f3 f76c59).

early 1970s, BSUD was documented for both the purple sea urchin, *Strongylocentrotus purpuratus*, and the red sea urchin, *Mesocentrotus franciscanus*, in southern California (Johnson 1971) and in the mid-1970s a localised mass mortality of *M. franciscanus* was noted in the kelp forests near Santa Cruz California (Pearse et al. 1977). Later that decade, BSUD was described for *Paracentrotus lividus* off the coast of France (Azzolina 1983a; Azzolina 1983b; Höbaus et al. 1981; Maes and Jangoux 1984) that caused mass mortalities of 27–83% of the population (Boudouresque et al. 1980). Another incidence of BSUD in *P. lividus* was noted in the Mediterranean Sea off the coast of France in 2006, although this did not result in significant mass mortality (Becker et al. 2008). In 2003 and 2004, there were additional mass mortality events for *P. lividus* in the Canary Islands (~ 10–95% die-off) (Girard et al. 2012). From 2009 to 2010, a widespread die-off of *Diadema africanum* associated with BSUD affected populations across more than 400 km of the Atlantic from Portugal to the Canary Islands (Hernández et al. 2013). During the winter of 2016–2017 there was a second recorded incidence of BSUD in *S. purpuratus* near Santa Barbara, California (unpublished observation, L.C. Smith). Recent outbreaks of BSUD in the sea urchin, *Heliocidaris erythrogramma*, have also been noted in eastern Australia (Sweet et al. 2016).

It is noteworthy that in any given location in which BSUD has been observed it has been limited to one or two species of echinoids. Furthermore, echinoids of the same species located in different geographical regions do not necessarily show signs of the same disease. For example, the keyhole sand dollar, *Mellitta quinquiesperforata*, has shown symptoms of BSUD in Venezuela but not along the Eastern coasts of the United States and Brazil, the Gulf of Mexico, Bermuda, Jamaica, or Puerto Rico where this species is common. Although the disease is most typically observed in regular echinoids, there are also examples of the disease documented for irregular echinoids including the keyhole sand dollar mentioned previously (Loffler 2018) and several species of heart sea urchins (Table 19.1); (Schwammer 1989; Nagelkerken et al. 1999; Bauer and Young 2000). Variations in the disease presence at different locations may be based on species specificity for the disease in addition to the presence of specific causative agents in disease development and putative initiating environmental stresses.

There are reports of diseases in echinoids with symptoms similar to BSUD that have either not been identified by specific names (Johnson 1971; Gilles and Pearse 1986; Roberts-Regan et al. 1988; Nagelkerken et al. 1999) or have been described as a general vibriosis speculating on the likely genus of the potential pathogen (Bauer and Young 2000). For example, a surface infection described as spotting

Table 19.1 Echinoid species that display bald sea urchin disease, their geographic locations, and putative pathogens

Echinoid species	Geographical Location	Potential pathogen	Reference
Allocentrotus fragilis	California	Unknown	Boolootian et al. 1959; Giese 1961
Arbacia lixula	Western Mediterranean Sea	Unknown	Höbaus et al. 1981; Maes and Jangoux 1984
Archaeopneustes hystrix	Bahamas	*Vibrio alginolyticus*	Bauer and Young 2000
Cidaris cidaris	France	Unknown	Jangoux 1990
Diadema africanum	Canary Islands	*Vibrio alginolyticus*	Clemente et al. 2014
Diadema antillarum	Caribbean, Bermuda, Canary Islands	*Bacillus* sp., *Clostridium* sp.	Bak et al. 1984; Lessios et al. 1984; Dyková et al. 2011
Diadema mexicanum	Western Mexico	Unknown	Benítez-Villalobos et al. 2009
Echinus esculentus	Brittany, Scotland	*Vibrio anguillarum*, *Aeromonas salmonicida*	Maes and Jangoux 1984; Tyler-Walters 2008; S.A. Boettger, unpub. obs.
Echinus acutus	Norway	*Dactylococcopsis echini* (likely secondary infection)	Mortensen and Rosevinge 1934; Jangoux 1987a
Heliocidaris erythrogramma	Eastern Australia	*Vibrio anguillarum*	Sweet et al. 2016
Hemicentrotus pulcherrimus	Japan	Unknown	Kanai 1993
Holopneustes purpurascens, Heliocidaris erythrogramma	Eastern Australia	*Vibrio anguillarum*	Sweet et al. 2016
Lytechinus variegatus	Florida panhandle	Currently unnamed *Vibrio* sp., closest RNA similarity to *V. carchariae*	Boettger and McClintock 2009
Mellita quinquies-perforata	Venezuela	*Vibrio anguillarum*, *Aeromonas salmonicida*	Loffler 2018
Meoma ventricosa	Curacao	Tetrodotoxin producing strain of *Pseudoalteromonas* sp.	Nagelkerken et al. 1999
Paleopneustes cristatus	Bahamas	*Vibrio alginolyticus*	Bauer and Young 2000
Paracentrotus lividus	Brittany, Mediterranean (France, Sardinia, Sicily, Spain, Yugoslavia)	*Alpha* and *Gammaproteobacteria* including *Vibrio* sp. and *Colivellia*	Azzolina 1983a, 1983b; Maes and Jangoux 1985; Bower 1996; Becker et al. 2008; Girard et al. 2012
Psammechinus miliaris	Normandy	Unknown	Maes and Jangoux 1984
Pseudocentrotus depressus	Japan	Unknown	Kanai 1993; Hamaguti et al. 1993; Shimizu et al. 1995
Sphaerechinus granularis	Brittany, France	Unknown	Höbaus et al. 1981; Maes and Jangoux 1984
Spatangus purpureus	Croatia	Unknown	Schwammer 1989

continued

Table 19.1 *Continued*

Echinoid species	Geographical Location	Potential pathogen	Reference
Strongylocentrotus droebachiensis	Nova Scotia, Canada, Europe	*Acinetobacter* sp., *Alcaligenes* sp.	Scheibling and Stephenson 1984; Miller 1985; Roberts-Regan et al. 1988
Strongylocentrotus franciscanus	California	Unknown	Johnson 1971; Pearse et al. 1977
Strongylocentrotus intermedius	Japan	*Flexibacter* sp., *Vibrio* sp.	Tajima et al. 1997a, 1997b; Takeuchi et al. 1999; Wang et al. 2006
Strongylocentrotus purpuratus	California	*Aeromonas salmonicida*, *Flavobacterium* sp., *Pseudomonas* sp., *Vibrio anguillarum*	Johnson 1971; Pearse et al. 1977; Gilles and Pearse 1986; Yui and Bayne 1983
Tripneustes gratilla	Madagascar	*Exiguobacterium* sp., *Alpha* and *Gammaproteobacteria*	Becker et al. 2007, 2009

This table is modified and updated from Jangoux (1987a) and Becker et al. (2008) by adding references and additional species that show similar disease etiologies.

disease in the edible sea urchin, *Strongylocentrotus intermedius*, was introduced into China from Japan in the late 1980s and has since become a common problem for sea urchin mariculture (Chang 2008). However in some cases the symptoms for spotting disease are similar to BSUD and in other cases it appears quite different (see Chapter 18). Furthermore, infection of the green sea urchin, *Strongylocentrotus droebachiensis*, with the parasite, *Paramoeba invadens*, (see BOX 19.1); (Feehan et al. 2013; Buchwald 2018, see Section 18.6 in Chapter 18) also shows spine loss and therefore can appear similar to BSUD. The confusion about BSUD and its underlying pathogens is rooted in the visible symptoms of primary spine loss and detachment from the substrate that is likely a general process of most late stage diseases in echinoids that are caused by a wide variety of pathogens.

19.2.1 Environmental stressors may drive BSUD

Changing climate conditions, particularly increased water temperatures, and frequency of severe storms may result in BSUD and may also exacerbate pathogen virulence. Tracking seawater temperatures and BSUD from 1982 to 2001 for the Channel Islands of California shows a disease increase of 40–70% between 1992 and 2001 that appeared to correlate with an increased mean summer water temperature (Lafferty et al. 2004). This association is consistent for sea urchin disease and warm water from El Niño events in which water moves

northwards along the west coast of North America including the El Niño event of 2016–2017 and BSUD in the purple sea urchin *Strongylocentrotus purpuratus* (unpublished observation, L.C. Smith). Increased seawater temperature along eastern Australia has also been attributed to BSUD in the sea urchin, *Heliocidaris erythrogramma* (Sweet et al. 2016). The incidence of BSUD in the southern end of the range for the purple and red sea urchins, *Strongylocentrotus purpuratus* and *Mesocentrotus franciscanus*, is also associated with warm water (Lester et al. 2007). Unusually warm waters at the Canary Islands in the Eastern Atlantic have been suggested as the initiator of the mass die-off of the sea urchin, *Diadema africanum*, that showed similarities to BSUD and reduced the population by 65% (Clemente et al. 2014). A similar outcome for *Paracentrotus lividus* in the Canary Islands correlated with increased sea surface temperature and decreased wave height and turbulence in the intertidal zone (Girard et al. 2012). The combination of warmer water and disease prevalence and progression was verified under controlled conditions in the lab (Lester et al 2007). In addition to climate change and warm water temperatures, anthropogenic stressors are also likely involved and may include coastal development, changes in populations due to fishing or harvesting sea urchins, or other species that interact with sea urchins, pollution and microplastics, terrestrial or agricultural runoff that can carry fertilisers and other nutrients associated with increased eutrophication and algal

---

Box 19.1 Ecological outcomes of paramoebiosis of the green sea urchin

The protist parasite, *Paramoeba invadens*, has repeatedly caused mass mortalities of the green sea urchin, *Strongylocentrotus droebachiensis*, in the shallow subtidal zone in Nova Scotia, Canada (Feehan et al. 2013; Buchwald 2018, and see Section 18.6.2 in Chapter 18). Outbreaks of the disease in sea urchins along the north Atlantic coast are associated with hurricanes, tropical storms, and warm core rings or eddies from the meanderings of the Gulf Stream (Scheibling and Lauzon-Guay 2010; Scheibling et al. 2010; Feehan et al. 2013, 2016). Short exposures to warm sea temperatures have resulted in mass die-offs of up to 100% of the population of the green sea urchin over 130 Km of coastline (Scheibling and Hennigar 1997). Given this temperature stressor associated with warm storms and the possibility of negative impacts on echinoderm immune function, continued ocean warming may have important implications for the resilience of susceptible host populations to *Paramoeba*.

In addition to warm water stress, high densities of the green sea urchin can also drive infections with *Paramoeba*. Very high densities of sea urchins appear as moving 'fronts' of animals that run parallel to the shore and progress from deeper to shallower water (Scheibling and Hennigar 1997). In places where the leading edge of the front comes in contact with the deep edges of kelp beds, sea urchin densities can reach up to 400 animals per m². The grazing activities of the sea urchins shift the kelp beds initially to gaps in the beds that coalesce to rocky barrens over extended regions of the Nova Scotia coastline (Breen and Mann 1976). Overpopulation of *S. droebachiensis* at feeding fronts has been speculated to promote infection with *Paramoeba invadens*, which can lead to mass mortalities. The subsequent repopulation of the green sea urchin through migration of adult animals from deeper regions ($\geq$ 24 m) and recruitment of larvae occurs rapidly, in as little as two years in one recorded instance (Brady and Scheibling 2005). Repeated die-offs of the green sea urchin from paramoebiosis have been recorded (Miller 1985; Scheibling 1985) and the impact of this changing sea urchin population alters the trophic interaction between herbivore and the foundational kelp, which affects the species that are associated with the kelp bed ecosystem, and shifts the near shore marine ecosystems of Nova Scotia from kelp beds to rocky barrens. Sea urchin disease and reductions in the echinoid population shifts the ecosystem back to kelp beds. These two stable states of the ecosystem structure for this rocky reef system are controlled by sea urchin populations and infection, in this case by *Paramoeba invadens*.

---

blooms and reduced water clarity, pesticides, herbicides, toxicants, and antibiotics (reviewed in Lafferty et al. 2004; Murano et al. 2020). These changes can lead to increased sea urchin stress and reduced immune function but may also increase the numbers and virulence of pathogens.

Suboptimal conditions at aquaculture breeding facilities can also result in disease outbreaks, as has been reported for *Strongylocentrotus intermedius*, a sea urchin that is widely cultured in China (Tajima et al. 1997a, 1997b; Wang, Li et al. 2006). These outbreaks and animal losses have led to research efforts to understand the causes of the disease and to find remedies to avoid future catastrophic losses (Chang 2008) that are coincident with water temperature above 20° C (Tajima et al. 1997a, 1998; Wang, Li et al. 2006). Although transcriptomes of diseased versus healthy sea urchins indicate decreased immune responsiveness, it is not known whether a pathogen or an altered environment is the initial stressor on

sea urchins that compromises the immune system (Zhang et al. 2019). The combination of decreased immune responsiveness and increased pathogen virulence are likely factors in more mass disease outbreaks and animal mortalities in farmed and wild populations.

19.2.2 Ecological outcomes of mass die-offs from BSUD

Algal grazing by sea urchins is considered a key function in marine trophic cascades that has direct impacts on the structure and stability of the entire ecosystem. The loss of, or significant reductions in, sea urchin populations from diseases with symptoms consistent with BSUD and the associated impacts on the marine ecosystems have been reported for many regions, of which all show similar outcomes. The assumed association between sea urchin disease and water temperatures can be confounded by fishing, such as harvesting spiney lobsters, *Panulirus interruptus*, which are sea urchin predators (Behrens and Lafferty 2004). The removal of spiney lobsters from the rocky reefs of the Channel Islands of California releases the population regulation on the purple sea urchin, *Strongylocentrotus purpuratus*, which leads to two major outcomes. The sea urchins overgraze the kelp forests and shift the ecosystem to rocky reef barrens that is dominated by crustose coralline algae (Figure 19.2). However, the increased sea urchin population also results in increased disease with symptoms consistent with BSUD (Figure 19.1); (Lafferty 2004) that reduces the population and may shift the ecosystem back toward kelp forest. Although the long-term state of the temperate rocky reef marine ecosystem over the last 50 years is generally that of kelp forest (Krumhansl et al. 2016), the population size of sea urchins is the primary driver for the repeated shifts between kelp forests and barrens (Behrens and Lafferty 2004; Ling et al. 2015; Byrne and Andrew 2020).

Similar outcomes of sea urchin populations and ecosystem structure are noted in 13 rocky reef systems in both the Northern and Southern hemispheres spanning 11 different regions (Ling et al. 2015). The loss of kelp forests from overgrazing by sea urchins does not necessarily drive a population crash in the echinoids but changes their food source from kelp to encrusting microalgae and small invertebrates and decreases growth rather than death from starvation. Hence, barrens may persist in the absence of disease or other means to control sea urchin population. The return to a kelp forest requires a tenfold decrease in sea urchin mass per area of reef. Similarly, the 65% reduction in the density of the sea urchin, *Diadema africanum*, to less than two animals per m^2 on the reefs of the Canary Islands resulted in a phase shift to macroalgal growth (Clemente et al. 2014). The experimental removal of the rock boring sea urchin, *Echinometra mathaei* and the larger *Echinothrix diadema* to 15% of their populations on coral reef lagoons of Kenya resulted in an increase in fleshy algae a year after sea urchin removal (McClanahan et al. 1996). However, changes in the Kenyan reef lagoons also depended on the populations of herbivorous fish, which are also important species in the lagoon ecosystem. In general, the changes to the ecological structure in these marine systems are the same outcome and are based on varying population sizes of sea urchins.

19.2.3 Conclusions: bald sea urchin disease

The symptoms of BSUD that show loss of spines and other surface appendages (Figure 19.1), loss of attachment to the substrate, and either death or recovery are described repeatedly in the literature and are likely due to infections from a variety of pathogens or mixes of pathogens. This includes, for example, the description of paramoebiosis in the green sea urchin, *Strongylocentrotus droebachiensis* (Jones et al. 1985) that appears very similar to diseases from a variety of bacterial infections (see Box 19.1) (Lester et al. 2007). The description of red spot or spotting disease may appear differently in the initial phases of the disease, but can also proceed to general spine loss and detachment from the substrate (see Section 18.3 in Chapter 18) (Li et al. 2020). The outcomes are the same for this set of BSUDs and can lead to mass die-offs that change the ecological structure because sea urchins are keystone members of marine communities based on their herbivorous grazing activity. Regardless of the agent that induces changes to the populations of sea urchins, whether due to disease, release from predation, or experimental removal, they all trigger similar state changes in a diversity of marine systems, with lower sea urchin abundance favoring growth of foundational kelp species and other fleshy macroalgae, and higher sea urchin abundance favoring a shift to crustose coralline algae and rocky barrens (Figure 19.2).

19.3 The massive die-off of the long-spined black sea urchin, *Diadema antillarum*, in the Caribbean Sea

Mass mortalities of sea urchins have occurred in the North Atlantic (Miller and Colodey 1983; Miller 1985; Feehan and Scheibling 2014), Eastern Pacific (Jurgens et al. 2015), the Eastern Atlantic in the Canary and Portuguese Islands (Girard et al. 2012; Clemente et al. 2014), and the Western South Pacific

Figure 19.2 The phase shift from a mature kelp forest to a barren habitat at the Channel Islands California. (a,b) The foundational species in a mature kelp forest in the eastern Pacific is giant kelp, *Macrocystis pyrifera*, that maintains significant species diversity in the ecosystem. (c-f) The purple sea urchin, *Strongylocentrotus purpuratus*, is a major herbivore of the rocky subtidal ecosystem. Increased populations of sea urchins devour the giant kelp plants including leaves, stipes, and holdfasts. Often, only the holdfast is devoured (c-e), which releases the entire kelp plant from the substrate and removes it from the forest. (f) The kelp forest undergoes a phase shift to barrens in which the remaining algal species that can resist overgrazing by sea urchins often includes crustose coralline algae (*Lithophyllum* sp., pink coating on the rock). The inset shows crustose coralline algae in the intertidal region, in addition to a sea anemone, *Anthopleura elegantissima*, a small chiton, *Cyanoplax dentiens*, and a juvenile six-armed sea star, *Leptasterias* sp. Source: images (a-d) were obtained from the Ron McPeak library (https://alexandria.ucsb.edu/collections/f3f76c59). Images (e) and f) were provided by Katie Davis Koehn. The inset in (f) was provided Lee C. Heiman.

along Eastern Australia (Sweet et al. 2016). However, the largest and most profound die-off event of echinoids was the widespread mass mortality of the long-spined, black sea urchin, *Diadema antillarum*, that disappeared from the coral reefs of the Caribbean Sea and the Western Atlantic between 1983 and 1984 due to an unidentified pathogen (Bak et al. 1984; Hughes et al. 1985; Gilles and Pearse 1986; Lessios et al. 1984, 1988a, 1988b, 2005, 2016; Carpenter 1988, 1990; Forcucci 1994; Gardner

et al. 1994; Scheibling and Hennigar 1997; Moses and Bonem 2001; Clemente et al. 2014;). The initial die-off in 1983–1984 was followed by repeat die-offs in 1985 and 1991–1992, which reduced the population of *D. antillarum* in the Caribbean by 85% to 100%, impacting a spatial scale of about 3.5 million km^2 (Lessios 1995). No other echinoid or members of other echinoderm classes in the region showed population crashes during these same years, suggesting that only *D. antillarum* was susceptible to this particular pathogen. The path of the infection began near the Panama Canal and moved east across the entire Caribbean basin to Jamaica and Florida, and eventually to Puerto Rico, Hispaniola, the Virgin Islands, and Bermuda (Lessios et al. 1984). Similar die-offs also occurred in Venezuela, Barbados, and the Lesser Antilles. Because this progression follows the path of surface water currents, it suggests a water-borne pathogen as a likely candidate (Lessios 1988b). The pathogen was never identified mostly due to the lack of appropriate technology during the 1980s and the decade following the initial event to address the question of the cause of the die-off.

Before the initial mortality event, the population density of *D. antillarum* on coral reefs was about 15 to 25 sea urchins per m^2 and it was the keystone herbivore that maintained the reef structure, species diversity, and coral dominance (reviewed in Steneck 2020). In the ~10 years following the initial mortality event, the population levels of *D. antillarum* remained low across the Caribbean compared to the pre-mortality population numbers, albeit with a few exceptions (Lessios 1995). About 16 years after the initial event, *D. antillarum* populations began to recover across most of the greater Caribbean (Figure 19.3) including some regions of Barbados, Jamaica, the Dry Tortugas, St. Croix, Costa Rica, and Curacao (Hunte and Younglao 1988; Chiappone et al. 2001; Edmunds and Carpenter 2001; Haley and Solandt 2001; Moses and Bonem 2001; Miller et al. 2003; Alvarado et al. 2004; Lessios 2005; Debrot and Nagelkerken 2006). However, in other areas including other regions of Barbados, Florida, St. Croix, Panama, and Puerto Rico, the post mortality populations have remained below the pre mortality level (Hunte and Younglao 1988; Chiappone et al. 2002; Miller et al. 2003; Lessios 2005; Ruiz-Ramos et al. 2011). In the 30 years since the event, *D. antillarum* populations recovered only in patches and are still only about 12% of the densities in the Caribbean prior to 1983–1984 (Lessios 2016). Overall, the recovery of *D. antillarum* across the Caribbean has been moderate and variable among locations. One hypothesis for the poor recovery and the low recruitment of juveniles is an allee effect, or consequence of low adult density and gamete dilution effects on broadcast spawning that

results in fertilisation failure (Levitan 1991). This hypothesis may well apply to *D. antillarum* because recruitment of juveniles and reappearance of populations have been either very slow or continues to fail likely because the population remains below the spawning threshold in many places. In theory, however, if there are refugia of sea urchins that survived the die-off or were not infected by the pathogen and have sufficient density for successful fertilisation, this may be the source of larvae that will eventually redistribute and repopulate the species on the Caribbean reefs. These refugia are the only likely explanation for the repopulation that has been noted in some regions of the Caribbean.

19.3.1 Ecological outcomes of the *Diadema* die-off

Before the mass die-off of the long-spined black sea urchin, *Diadema antillarum*, it was a major herbivorous grazer on the Caribbean coral reefs that regulated the algal abundance and maintained the overall diversity, complexity, and productivity of reef communities (Atkinson et al. 1973; Sammarco et al. 1974; Ogden 1976; Ogden and Lobel 1978; Sammarco 1980; Carpenter 1981; Sammarco 1982a, 1982b). Coral maintenance was accomplished by preventing macroalgal overgrowth, maintaining the reduced algal turf on reefs that promoted coral domination through recruitment of coral larvae and survival of juveniles (Moore et al. 1963; McPherson 1965; McPherson 1968; Atkinson et al. 1973; Foster 1987; Edmunds and Carpenter 2001). Historically, large populations of manatees, sea turtles, and herbivorous fishes accounted for most of the grazer function on reefs and maintained the coral dominated ecosystem in the Caribbean prior to human arrival, even though *D. antillarum* was also very abundant (Jackson 1997). After the human impact of overharvesting and disappearance of the large vertebrates, *D. antillarum* and small fishes functioned as the primary grazers (Jackson, 2001; Pandolfi et al. 2003; Steneck 2020). Hence, the disappearance of this keystone echinoid herbivore from disease resulted in immediate algal growth with broad repercussions on reef ecology that disrupted the trophic cascade with a swift and sustained phase

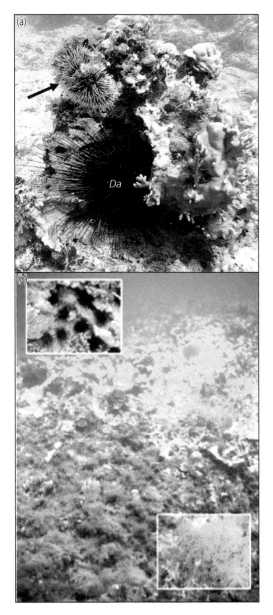

Figure 19.3 The long-spined black sea urchin, *Diadema antillarum*, is a keystone herbivore on Caribbean reefs. *D. antillarum* disappeared from the Caribbean Sea due to repeated mass die-offs during the early 1980s and the early 1990s. (a) A healthy *D. antillarum* (Da) is under coral cover and a white sea urchin, *Tripneustes ventricosus* (arrow) is present above near Culebra Island, east of Puerto Rico in the Spanish Virgin Islands, December 2018. (b) The recovery of small populations of *D. antillarum* on the shallow reefs of the north coast of Jamaica result in grazed (top) adjacent to non-grazed (bottom) regions. The top inset shows a group of *D. antillarum* in the grazed region and the bottom inset shows growth of algae in regions beyond which the sea urchins move from their daytime refuge. Source: image (a) was provided by Alejandro Mercado Capote. Source: the image in (b) is reprinted with permission from Idjadi et al. (2010) Recovery of the sea urchin *Diadema antillarum* promotes scleractinian coral growth and survivorship on shallow Jamaican reefs. Marine Ecology Progress Series 403: 91–100. © Inter-Research 2010.

shift that altered the reef biome dynamics from coral dominance to cover by macroalgae with associated reduction in species diversity (Liddell and Ohlhorst 1986; Hughes et al. 1987; Carpenter 1988, 1990; Hughes 1994; Ferrari Legorreta 2012; Vega Thurber et al. 2012; Burkepile et al. 2013; Menge et al. 2016; Schultz et al. 2016; Steneck 2020). This exact phase shift was predicted prior to the die-off event by exclusion experiments in which *D. antillarum* was removed from patches of reefs in Jamaica and demonstrated significant growth of soft algae in those patches that could out compete many coral species for space on the substrate (Sammarco 1980). Similar ecological phase shifts were also observed in Japan and Kenya when other sea urchin species were removed experimentally, were over collected for human consumption, or the populations were reduced by fish predation (McClanahan and Shafir 1990; Azuma et al. 1997). Following decreased sea urchin herbivory, algal growth killed corals of all sizes by overgrowth, blocked access to light, and abraided and smothered corals due to the accumulation of sediment at the base of algal attachment to the substrate (Edmunds and Carpenter 2001). Similarly, juvenile corals at Ningaloo Reef in Western Australia grew more slowly with lower survival under conditions of increased algal abundance (Webster et al. 2015). The recovery of the reef system in the Caribbean to coral domination is essentially prevented because coral recruitment from the plankton is markedly reduced by 75% of pre-1982 levels (Bak 1985; Arnold et al. 2010). Although other sea urchin species, *Tripneustes ventricosus* (Figure 19.3.a, arrow) and *Echinometra viridis*, increased their populations on the reefs in the region, they do not exert the same keystone herbivore function as *D. antillarum* and therefore have had little effect on the composition and structure of the ecosystem (Steneck 2020). The combination of the range of factors described here is likely to result in continued slow recovery of the Caribbean reef systems

19.3.2 The future of the Caribbean reefs

Evaluations of, and predictions for the recovery of the *D. antillarum* population along the north coast of Jamaica began in 1995 and were based on animal age

estimates according to sea urchin sizes and the sizes of the corals recruited to the regions in which sea urchin populations were low compared to adjacent regions where populations were higher (Edmunds and Carpenter 2001). In some local regions of northern Jamaican reefs, *D. antillarum* was present in populations of up to a 10-fold increase in animals compared to post die-off (Carpenter and Edmonds 2006). The appearance of discrete sea urchin grazing zones have become apparent where the macroalgae has disappeared and replaced with algal turf and crustose coralline algae (Idjadi et al. 2010). These areas are due to groups of *D. antillarum* that graze at night and form distinct circles with clearly demarcated edges next to neighboring non-grazed algal zones, which are the limits to which the sea urchins would move from their day time refuge from predators (Figure 19.3.b). In a broader analysis, sea urchins have been noted as being present in a 4100 km arc across six islands of the Caribbean showing kilometer sized regions of significant repopulations with an associated reduction in algal mass and increased coral recruitment. However, the population recovery of *D. antillarum* is not uniform throughout the Caribbean and they remain minimal at San Blas Archipelago of Panama as of 2010 (Coppard and Alvarado 2013). Although the recovery of *D. antillarum* in some areas of the Caribbean is not promising, other areas show encouraging signs of population increases and provide hope that the reefs will return to coral domination.

19.3.3 Conclusion: the mass die-off of *Diadema antillarum*

The long-spined black sea urchin disappeared from the Caribbean reefs after several mass mortality events. Although the pathogen was not identified, it was assumed to be transmitted via water because the path of disease followed the surface water currents throughout the Caribbean basin. The pathogen did not impact other species of echinoids, other echinoderms, or other invertebrates. The mass die-off reduced the population of *D. antillarum* population to below the effective reproduction level because fertilisation fails for broadcast spawners from excessive dilution of gametes when adults are distant from each other. The outcome is a very slow recovery to pre-die-off populations. The disappearance of this keystone herbivore shifted the reefs from coral dominance to macroalgal cover, which was exacerbated by the historic loss of large vertebrate herbivores from overfishing. Macroalgal overgrowth killed corals through abrasion, smothering from masses of plant tissue, sediment accumulation, and blocking light required for the symbiotic zooxanthellae. Recovery has been followed for the last 40 years and is not uniform throughout the Caribbean Sea. After 16 years post die-off, the *Didema* population is not uniform and recoveries are noted only in some places. After 30 years, the population recovery continues to be incomplete and effective recovery is variable among regions. In regions of partial recovery, discrete sea urchin grazing zones are visible in which the algal cover is reduced. Overall, the *Diadema* recovery and return to coral cover is slow for reefs dominated by macroalgae because of decreased juvenile recruitement for both *Diadema*, and for corals.

19.4 Sea star wasting disease and the mass die-off of sea stars

Sea star wasting disease (SSWD) is a lethal syndrome of asteroids that has been known for over 100 years from local die-offs. The first report was for the sand sea star, *Astropecten irregularis*, from Danish waters (Mead 1898), followed by many subsequent reports of the disease in individual sea stars or for small isolated populations in many regions of the world (Sieling 1960; Christensen 1970; Tiffany 1978; Menge 1979; Bang 1982; Dungan et al. 1982; Thorpe and Spencer 2000; Suzuki et al 2012; Bucci et al. 2017; Miner et al. 2018; Hewson et al. 2019). However, since 2013, SSWD resulted in an unprecedented mass mortality of millions of asteroids along a geographically extensive region along the west coast of North America, from Alaska to Baja Mexico (Staehli et al. 2008, Bates et al. 2009; Hewson et al. 2014; Stokstad 2014; Jurgens et al. 2015; Menge et al. 2016; Eisenlord et al. 2016; MARINe 2016). This event is notably different from previous endemic events, which were limited to local impacts on ecosystems, and is amongst the most extensive

event ever observed based on persistence, involvement of as many as 20 asteroid species, and extremely rapid progression to death (Montecino-Latorre et al. 2016; Miner et al. 2018; see Section 18.4 in Chapter 18). Symptoms of SSWD are a suite of typical gross morphological changes that first appears as necrotic regions of the skin, often white in colour, and progresses to perforated body wall and exposure of internal organs, arm twisting and autotomy, and ultimately leading to death and disintegration (Eckert et al. 2000). The sunflower sea star, *Pycnopodia helianthoides*, has been particularly impacted with an 80–100% decrease in abundance (Hewson et al. 2014, 2018, 2019; Harvell et al. 2019). It is noteworthy that other sympatric echinoderms of different classes, such as sea urchins, brittle stars, and sea cucumbers, have not been affected by the disease, indicating that the pathogen and the phenomenon are specific to sea stars.

19.4.1 Environmentally induced stresses as underlying factors in the SSWD pandemic

The repeated documentation of local sea star wasting events of short term indicate that the pathogen is not likely to be a new introduction but rather the scale of the pandemic has suggested that something else changed in the environment that acted as the disease initiator (Miner et al. 2018). Stress induced by increased water temperature has been suggested as a mechanism for compromising the immune system in asteroids leading to increased susceptibility or reduced resistance to this lethal disease (Eisenlord et al. 2016; Harvell et al. 2019). A possible stressor may have been a marine heat wave of warm water (colloquially called 'the Blob') that moved from the Bering Sea and Alaska in 2013 and was apparent in California by 2014 (Bond et al. 2015; Di Lorenzo and Mantua 2016). The effects of this marine heat wave may have been compounded by an overlapping El Niño of 2015–2016, which moved warm water from south to north (Cavole et al. 2016; Sanford et al. 2016). This large scale environmental change generally correlates with models of the die-off for the ochre sea star, *Pisaster ochraceus*, that was first identified in 2013 (Aalto et al. 2020), in addition to the die-off of the sunflower star, *Pycnopodia helianthoides* (Harvell et al. 2019). It is tempting to attribute this heat stress to the inability of sea stars to defend against and survive the presence of a pathogen that is normally in the environment at moderately low levels. Experimental analyses of increased water temperature of 8–13°C above ambient for the predatory sunstar, *Heliaster kubiniji*, show significantly accelerated progress of wasting symptoms (Dungan et al. 1982). However, it is noteworthy that SSWD also occurred in asteroids along the northeast coast of North America, which did not undergo any reported warm water events similar to the marine heat wave and the El Niño on the west coast (Menge et al. 2016; Bucci et al. 2017). Thus, while increased temperature (Dungan et al. 1982; Eisenlord et al. 2016), decreased temperature (Bates et al. 2009; Menge et al. 2016), shifts in temperature, and changes in precipitation (Hewson et al. 2018) correspond with SSWD at various locations, there remains no consistent explanation for wasting occurrence across the geographically worldwide distribution of this pandemic event. In general, evaluations of the immune responses of *P. helianthoides* to SSWD and the approaches to identify the pathogen have resulted in confusing answers. This is because the association between SSWD and the sea star associated densovirus (SSaDV) has only been verified for infected *P. helianthoides* and because the virus is also present on healthy or asymptomatic sea stars (Hewson et al. 2018). Furthermore, a different densovirus is associated with SSWD along the western Atlantic (Jackson et al. 2020) and SSaDV is only one of ten different densoviruses associated with sea stars along the eastern Pacific (Jackson et al. 2021). Hence, the 'pathogen' may actually be a set of densoviruses, however complicating factors for understanding the initiation of the SSWD pandemic likely include changes in the health status of sea stars due to an environmental stressor that compromised the immune system to make the sea stars vulnerable to the 'pathogen' that is normally present. SSWD may have been amplified to a pandemic by sea star populations with increased infection, pathogen proliferation, and release of virions into the seawater that led to increased environmental concentrations compounding the mass mortalities through contagious spread by water currents.

High densities of sea stars have also been suggested as a possible factor underlying the SSWD die-off (Fuess et al. 2015; Eisenlord et al. 2016). Conversely, the elevated incidence of SSWD in the ochre star, *Pisaster ochraceus*, has been noted at all locations that have been investigated despite widely varying densities of the species (Eisenlord et al. 2016; Menge et al. 2016). This lack of correlation between population density and disease lends support that the pathogen is waterborne and does not require direct contact among animals. Host size has also been considered as a possible factor for SSWD because larger animals in populations observed in the field are more likely to show symptoms of the disease and subsequently have greater reductions in abundance after the outbreak (Eisenlord et al. 2016). Smaller juveniles tend to show signs of the disease later than adults but once the symptoms appear, they succumb more quickly. This is consistent with the observation of a bias toward the smallest size distributions of sea stars with larger juvenile size classes notably missing, suggesting significant mortality after settlement in locations where SSWD is present (Eisenlord et al. 2016). It is likely that the ultimate explanation for the onset of SSWD pandemic will be complex, involving unknown combinations of environmental and biological factors (Sweet 2020).

19.4.2 Ecological phase shifts resulting from SSWD in the ochre sea star, *Pisaster ochraceus*

The mass mortality of multiple sea star species, many of which function as keystone predators, resulted in significant shifts in the population sizes of prey species. Upon release from predation, prey species in the intertidal and near subtidal habitats expanded, shifted their vertical locations, and altered the availability of refuge for other species (Gravem and Morgan 2017). This reduction in sea star populations from disease reinforced the findings from experimental species removal studies (Paine 1966, 1974), demonstrating that predatory sea stars exert top-down control over prey population sizes (Paine and Vadas 1969). Along the Pacific coast of North America, the ochre sea star, *Pisaster ochraceus*, is a keystone predator on the California mussel, *Mytilus californianus* (Figure 19.4), which is a foundational species that establishes and maintains the ecological structure and diversity of the rocky intertidal zone (Paine 1966, 1974). When SSWD decimated the population of *P. ochraceus* in 2014 along the Washington and Oregon coastlines, this resulted in predation release on *M. californianus* (Menge et al. 2016; Eisenlord et al. 2016). The reduction in predation was exacerbated by the recruitment of juvenile *P. ochraceus* sea stars, which projected population recovery in many sites in the years following the peak of the mass mortality event, however small sea stars have less predation impact on full size *Mytilus* prey compared to adults (Moritsch and Raimondi 2018). Consequently, the mussel beds expanded into the low intertidal zone, became the dominant species and outcompeted macroalgae, anemones, gastropods, sea urchins, and other species that are normally present and thereby changed the structure of the ecosystem (Menge et al. 2016). However, the increase in mussel bed size also favored increased populations of epibionts that settle on the expanded substrate of mussel shells, in addition to the assemblage of species that inhabit the substrate among the byssal threads at the base of the mussel beds in a zone that is protected from predation. While populations of *P. ochraceus* recovered in 2015–2017 to about 40% of pre-die-off biomass at many places north of Point Conception California, regions in southern California have not shown recovery (Eisenlord et al. 2016; Moritsch and Raimondi 2018). This north vs. south difference is likely based on ocean currents that transport larvae from the north to the south and is consistent with few juvenile recruits recorded south of Point Conception. The possibility of a rapid recovery of *P. ochraceus* for the northern populations and a return of predation pressure on mussels is projected to result in a reversal of the ecological state in the intertidal zone to the pre-2013 structure (Moritsch and Raimondi 2018).

19.4.3 The ecological phase shift resulting from SSWD and disappearance of the sunflower sea star, *Pycnopodia helianthoides*

Severe reductions to complete depletion of the predatory sunflower sea star, *Pycnopodia helianthoides*, because of SSWD released population control leading to significant increases in the purple and red sea urchins, *Strongylocentrotus purpuratus*

and *Mesocentrotus franciscanus* (Moitoza and Phillips 1979; Schultz et al. 2016; Bonaviri et al. 2017). The rapid expansion of sea urchin populations from 0–1.7 to as many as 24 animals per m^2 in the subtidal rocky substrate along the California coast had the cascade effect of sea urchins overgrazing the kelp forests (Figure 19.2.a, b) and the associated decrease in ecosystem diversity. Sea urchin feeding at the base of giant kelp, *Macrocystis pyrifera*, releases entire plants from the substrate and decimates the kelp forest habitat (Figure 19.2.c–e) (Rogers-Bennett and Catton 2019). This activity reduced kelp forest area by 90% from more than 50 km^2 of subtidal area and more than 300 km of linear coastline to less than 2 km^2. The loss of kelp, a foundational species that normally supports a great diversity of associated species, in addition to on the loss of many other algal species resulted in a massive ecosystem phase shift to sea urchin barrens dominated in some places by crustose coralline algae (Figure 19.2.f). As a consequence, the red abalone, *Haliotis rufescens*, that is also an herbivore typical of kelp forests and an inferior competitor with sea urchins for algal resources, subsequently starved and underwent a mass mortality event in 2017 with an overall population crash of up to 96% by 2018 (Figure 19.5). Currently, *P. helianthoides* is locally extinct, which unless refuge populations exist, may reduce or eliminate the production of larvae and juvenile recruits to replace the population and therefore may have long-lasting consequences. Hence, the SSWD pandemic does not only threaten the long-term persistence of this predatory sea star but its control on sea urchin populations clearly has wide ranging trophic effects on maintaining the kelp forest ecosystem.

19.4.4 Conclusion: SSWD

SSWD altered the populations of many sea star species in many regions of the world including opposite shores of North America plus coastal Antarctica of the Southern Ocean. Hence, SSWD should be considered a viral pandemic of asteroids. Environmental impacts such as incursions or movement of warm ocean water likely compromises the sea star immune system reducing the immunological protection mechanisms of the host against

Figure 19.4 A healthy ochre sea star, *Pisaster ochraceus*, feeds on California mussels, *Mytilus californianus*, at the base of a mussel bed in the intertldal zone. Source: This image was provided by Lee C. Heiman and was taken south of Pelican State Beach at Smith River California during a low tide in August 2017.

pathogens that are normally present in the environment. However, marine heat waves are not always associated with the disease in all localities, as has been noted for the two coasts of North America. Other environmental stressors such as increased population density are also inconsistently associated with asteroid die-offs. Reductions in keystone asteroid populations have signficant impacts on marine ecosystems, including loss of kelp forests and associated decreased species diversity, and significant population increases in prey species. These disruptions in trophic cascades lead to massive phase shifts in the the ecosystem structure. In some cases, recoveries of some asteroid species have been noted such as that for the ochre sea star, *Pisaster ochraceus*. However, the sunflower sea star, *Pycnopodia helianthoides*, remains absent from many places where they were once common. Yet, a note of optimism remains for the possible existence of a protected refuge from which the sunflower sea star will reappear (see Figure 18.2.Aa in Chapter 18).

19.5 Ecological outcomes of echinoderm mass mortality events

While population changes of many sea star species from SSWD are ongoing, the range of species that are affected and their population recovery rates remain variable and uncertain, particularly in the southern regions of western North America (Miner et al. 2018; Hewson et al. 2019; Harvell et al. 2019). SSWD is a significant threat to some of the most ecologically important predatory species (*P. ochraceus* and *P. helianthoides*) and there are encouraging signs for a population recovery based on the appearance of new recruits of *P. ochraceus* (Moritsch and Raimondi 2018); The SSWD epidemic and loss of keystone sea stars may result in continental-scale changes in coastal habitats depending on the timing of population recovery or its failure (Moritsch and Raimondi 2018; Harvell et al. 2019). Similarly, the loss of the long-spined black sea urchin, *Diadema antillarum*, and the ecological outcomes that have been followed since 1983,

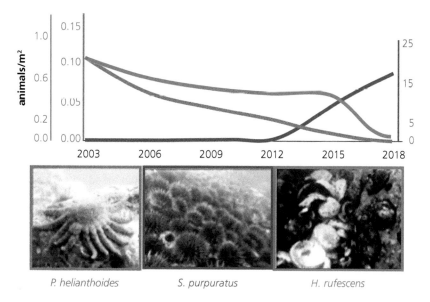

P. helianthoides   S. purpuratus   H. rufescens

Figure 19.5 The population reduction in the sunflower sea star is indirectly related to the population crash of the red abalone along the California coastline. The population reduction of the sunflower sea star, *Pycnopodia helianthoides* (green), from SSWD released predation pressure from the purple sea urchin, *Strongylocentrotus purpuratus* (purple), which increased dramatically in population. The grazing competition between the purple sea urchin and the red abalone, *Haliotis rufescens* (red), resulted in abalone starvation and population crash. Source: this figure is modified from Rogers-Bennett and Catton (2019) and shows the general changes in populations based on results for 12 sites along the California coastline.

have illustrated how the long-term disappearances of keystone echinoderms and the rate of their recovery have direct and dramatic effects on massive shifts in ecosystem structures and species diversity.

19.5.1 Abiotic impacts to echinoderm populations

In addition to pathogens that infect echinoderms and reduce their viability (see Chapter 18), abiotic agents also alter the health and survival of echinoderms. These effects include both long-term and acute changes in seawater temperature and pH, as well as environmental pollutants such as heavy metals, drugs, nanoparticles, microplastics, and other compounds (reviewed in Lafferty et al. 2004; Murano et al. 2020). A key and repeated concept is that echinoderm infection is often the outcome of stressful impacts on the animals that reduce immune function leading to increased disease and in some cases resulting in mass die-offs. A key to understanding the environmental impacts on population health is to evaluate the echinoderm immune system including the sea urchin immune

cells that function as sentinels of immunological impacts. These cells can be used *in vitro* to evaluate whether chemicals, pollutants, microplastics, changes in water chemistry or temperature alter their activities or functions, providing inference of changes to the defense systems of animals living in polluted water (Jiang et al. 2016; Matranga et al. 2000, 2005; Coteur et al. 2003a, 2003b; Pinsino et al. 2008; Stabili and Pagliara 2009, 2012; Pinsino and Matranga 2015; Brothers et al. 2016; Murano et al. 2020). Sea urchin embryos and larvae have also been used to evaluate environmental contaminants and toxins because they can be reared easily in large numbers (Hodin et al. 2019) and are in direct contact with abiotic factors in the environment (see Section 18.7 in Chapter 18). Developmental failure or abnormalities in embryos and larvae are used to assess responses to abiotic agents, changes in water chemistry, in addition to the pathogenicity of specific microbes and/or their secreted toxins (Byrne 2012; Kiselev et al. 2013; Ho et al. 2016; Buckley et al. 2017; Schuh et al. 2020). Large numbers of coelomocytes, embryos, or larvae provide systems that are repeatable and statistically robust to identify environmental stressors that that can lead to disease, changes in population sizes, and perhaps to significant shifts in ecosystem structure.

19.5.2 Ecological phase shifts

It can be difficult to pinpoint whether disease, environmental change, or both are the root cause of phase shifts in marine ecosystems. This is because of the complex correlations between the environmental change, organismal disease, and resulting changes in the trophic network of interactions among organisms in a community. Changes in the ecosystem communities that appear as drastic phase shifts may proceed as (i) environmental changes that vary temporally and spatially and may include multiple, interacting environmental stressors, (ii) which may depress the immune system of echinoderms because of physiological impacts, (iii) which may alter the microbiome of the host and its natural defenses against pathogen colonisation (Sweet 2020), which together (iv) may enable pathogens to gain an advantage because of a reduction in the multiple levels of host defenses, (v) which may result in disease. The removal of keystone species (either grazers or predators) from an ecosystem changes the populations of interacting organisms in the community that alters the trophic cascades and drives massive phase shifts in ecosystem structure. It is noteworthy that the scientific literature on marine ecology tends to ignore the significant impact of pathogens (Belkin and Colwell 2005) as to whether an ecosystem structure will be stable and remain unchanged over time, whether it will become unstable and undergo a shift to a different stable state, return to the previous state, or the rare possibility that, in association with extinctions, it will transition to a novel state (Pandolfi et al. 2020).

The unidentified pathogen that drove the unprecedented massive die-off and almost complete disappearance of the herbivorous long-spined black sea urchin, *Diadema antillarum*, from Caribbean coral reefs resulted in a swift ecological phase shift from coral reefs to macroalgal dominated reefs (Steneck 2020). The same ecological outcome from disease, in this case with symptoms consistent with BSUD, induced population reductions in the purple, red, and green sea urchins of the subtidal temperate rocky ecosystems of the Channel Islands of the

eastern Pacific and Nova Scotia in the western Atlantic. The reductions in these echinoids that are major herbivores shifted the ecosystems from barrens to kelp forest (Behrens and Lafferty 2004). In both temperate and tropical systems, the reduction in echinoid herbivory shifts the ecological structure to macroalgae. Time scale differences in which shifts and recoveries occur in the tropical versus temperate reef ecosystems is striking. The temperate rocky reefs may shift in as little as a few years (Brady and Scheibling 2005) based on varying population sizes from migrations of adult sea urchins plus larval recruitment vs. fast growth rate of kelp (Mann 1973). To the contrary, the tropical coral reefs have not generally recovered from algal dominance and shifted back to coral cover following the sea urchin die-off that was first evident in 1983. It is noteworthy that the shift toward or maintenance of macroalgae is the expected norm for ecosystem structure in many areas of temperate rocky reefs but this is the opposite of the expected norm for coral reef structure in the Caribbean in which fleshy algae is expected to be controlled by herbivores. But what is the historical or long-term ecological structure or expected norm for these two regions? The temperate rocky reefs are generally believed to be dominated by kelp forests rather than barrens based on surveys from the last 50 years (Pandolfi et al. 2003). The Caribbean reef ecosystem, on the other hand, has likely been dominated by coral assemblages for hundreds of thousands of years based on community structures of fossil reefs from the Pleistocene (Jackson 1992). Clearly, the changes in the trophic cascades in these two regions, whether they are short or long term, are maintained today by different population densities of sea urchins: kelp forests by low densities, and tropical coral reefs by high densities.

An alternative example of changes in ecosystem structure is defined by disease in predatory echinoderms in which SSWD altered the populations of asteroids worldwide that initiated a variety of complex changes to intertidal and subtidal ecosystems. The types of changes in ecosystem structure depend on the sea star species that is affected by SSWD, how the sea star functions in the ecological community, whether the sea star fully or partially succumbed to SSWD with associated decreases in the population, how quickly the sea star has or has not recovered, whether the change in the sea star population alters the populations of preferred prey, whether the preferred prey are keystone or foundational species, whether populations of foundational species are altered, either directly or indirectly. Examples of altered ecological trophic cascades from loss of asteroids along the Pacific coast of North America include changes in the structure of the rocky intertidal ecosystem (Menge et al. 2016) and the disappearance of the subtidal kelp forests (Rogers-Bennett and Catton 2019) with a correlated increase or decrease in species diversity. The length of time for which ecosystems may be maintained in one stable state or another depends on the time required for echinoderm populations to return to pre-disease levels. Although changes in populations of non-keystone asteroids from the SSWD pandemic are difficult to discern, the loss of those asteroids and the effects on complex marine ecosystems are unknown. Perhaps regions where

SSWD has not shown observable changes to trophic cascades and observable phase shifts, changes in populations of non-keystone asteroids from SSWD may have had impacts on complex marine ecosystems, such as a reduction of functional redundancy in the asteroid predator species in those regions. The example of a loss of trophic redundancy as a template for future ecosystem changes is the case of the Caribbean phase shift, which occurred due to a depauperate set of vertebrate grazers leaving only *Diadema antillarum* as the ecologically effective herbivore. This left the Caribbean basin primed for a massive phase shift upon loss of the remaining keystone herbivore. Understanding the impacts of stress on echinoderms, either from global or local changes in the environment, is key to identifying the drivers of disease and are expected to inform strategies to improve conditions for echinoderms and to promote an understanding of the marine environment for the future.

19.6 Summary and conclusions

- Mass die-offs of echinoderms
  - ○ sea urchins and sand dollars from BSUD in many oceans
    - ■ likely caused by a variety of pathogens including a protist parasite
    - ■ symptoms are consistent with the unknown pathogen(s) that caused the die-off of the long-spined black sea urchin
  - ○ sea stars of all species from SSWD in many oceans
    - ■ caused by densoviruses known for at least two species of sea stars
- Environmental stressors may reduce immunological function in hosts that precede disease leading to changes in populations
  - ○ warm water
  - ○ pollution
  - ○ eutrophication
  - ○ biological toxins
  - ○ crowding or excess populations of echinoderms
- Changes to the trophic cascades in ecological communities drive phase shifts to a different structure
  - ○ population decrease or loss of keystone echinoids from disease
    - ■ reduction in herbivory
      - • shift to macroalgal cover from coral cover
      - • expansion of kelp beds from rocky barrens

○ population decrease or loss of echinoderm predators from disease or over fishing in temperate rocky reefs
■ asteroid or spiny lobster predators
■ predation release of prey species
  • increased population of keystone prey echinoids
○ increased herbivory
■ shift from kelp forests to rocky barrens
■ increased competition with other herbivore species
  • increased population of foundational mussel prey species
○ changes in the temperate intertidal ecosystem structure
■ spacial expansion of mussel species to the subtidal
■ changes in associated species diversity
• The timeframe of recovery from an ecological phase shift depends on the recovery of the keystone or foundational species
○ The shift to barrens from kelp forest in temperate rocky reefs
■ generally short term and cyclic
■ expansion and reduction in kelp forest cover
■ correlation with moderately rapid recovery of echinoid populations following population reductions from disease
○ The shift to macroalgal domination from coral cover in tropical reefs is long term
■ slow recovery of the dominant long-spined black sea urchin
  • the shift is short term when viewed from historical and paleontological data
• A complete understanding of the marine environment will require detailed study and integration of a wide range of topics
○ The pathogens
○ The hosts
■ Their diseases
○ The impacts of ecological stressors
○ The changes in the complexity of the ecosystems
■ Altered trophic cascades
■ Ecological phase shifts
○ A detailed understanding of the host immune systems

Acknowledgements

The authors were supported during the writing of this chapter by funding from United States National Science Foundation (IOS-1855747) to LCS, and the Australian Research Council (DP150102771) to MB. The authors are grateful to Alejandro Mercado Capote (University of Puerto Rico Mayagüez), Katie Davis Koehn (University of California at Santa Barbara), and Lee C. Heiman for providing images of marine organisms. We acknowledge the valuable assistance from Roger Clarke for help with identifying the chiton in Figure 19.2, and from Deborah Bezanson, Senior Librarian for the Sciences, Engineering & Assessment, George Washington University Libraries for assistance with references.

References

Aalto, E.A., Lafferty, K.D., Sokolow, S.H., Grewelle, R.E., Ben-Horin, T., Boch, C.A., Raimondi, P.T., Bograd, S.J., Hazen, E.L., Jacox, M.G., Micheli, F., and De Leo, G.A. 2020. Models with environmental drivers offer a plausible mechanism for the rapid spread of infectious disease outbreaks in marine organisms. *Scientific Reports* 10: 5975

Alvarado, J.J., Cortés, J., and Salas, E. 2004. Population densities of *Diadema antillarum* Philippi (Echinodermata: Echinoidea) at Cahuita National Park (1977–2003), Costa Rica. *Caribbean Journal of Sciences* 40: 257–259

Arnold, S., Mumby, P., and Steneck, R.S. 2010. Running the gauntlet to coral recruitment through a sequence of local multiscale processes. *Marine Ecology Progress Series* 414: 91–105

Atkinson, C., Hopley, S., Mendelsohn, L., Yacowitz, S. 1973. Food studies on *Diadema antillarum* on a patch reef, St. Croix, U.S. Virgin Islands. In *Studies on the Activity and Food of the Echinoid* Diadema antillarum Philippi *on a West Indian Patch Reef*, Ogden, J.C., Abbott, D.P., and Abbott, L. (eds.) pp. 65–80. Special Publication Number 2., U.S. *Virgin Islands*: West Indies Laboratory, Fairleigh Dickinson University, St. Croix

Azuma, Y., Matsuyama, K., Nakata, A., Kawai, T., and Nishikawa, N. 1997. Transition of seaweed communities after removal of sea urchins in the coral plains along the coast of the Sea of Japan in Hokkaido. *Journal of the Fisheries Society of Japan* 63(5): 672–680

Azzolina, J.F. 1983a. Evolution de la maladie de l'oursin comestible *Paracentrotus lividus* dans la baie de Port-Cros (Var, France). *Rapport Commission International Exploration Science Mer Mediterrane Monaco* 28: 263–264

Azzolina, J.F. 1983b. Evolution à long terme des populations de l'oursin comestible *Paracentrotus lividus* dans la baie de port-Cros (var, France). In *Colloque international sur* Paracentrotus lividus *et les oursins comestibles*, Boudouresque C.F. (ed.) pp. 257–269 G.I.S Posidonie, Marseille France

Bak, R.P.M. 1985. Recruitment patterns and mass mortalities in the sea urchin *Diadema antillarum. Procedings of the Fifth International Coral Reef Congress* 5: 267–272. Paris: Museum National d'Histoire Naturellle Publons

Bak, R.P.M., Carpay, M., De Ruyter Van, and Steveninck, E. 1984. Densities of the sea urchin *Diadema antillarum* before and after mass mortalities on the coral reefs of Curagao. *Marine Ecology Progress Series* 17: 105–108

Bang, FB. 1982. Disease processes in sea stars: A Metchnikovian challenge. *Biological Bulletin* 162: 135–148

Bates A.E., Hilton, B.J., and Harley, C.D.G. 2009. Effects of temperature, season and locality on wasting disease in the keystone predatory sea star *Pisaster ochraceus. Diseases of Aquatic Organisms* 86: 245–251

Bauer, J.C. and Young, C.M. 2000. Epidermal lesions and mortality caused by vibrosis in deep-sea Bahamian echinoids: A laboratory study. *Diseases of Aquatic Organisms* 39: 193–199

Becker, P., Eeckhaut, I., Ycaza, R.H., Mercier, A., and Hamel, J.-F. 2009. Protozoan disease in larval culture

of the edible sea cucumber *Isostichopus fuscus*. In *Echinoderms*, Harris, L.G., Boettger, S.A., Walker, C.W., and Lesser, M.P. (eds.) pp. 571–573. London: CRC Press

Becker, P., Gillan, D.C., and Eeckhaut, I. 2007. Microbiological study of the body wall lesions of the echinoid *Tripneustes gratilla. Diseases of Aquatic Organisms* 77(1): 73–82

Becker, P.T., Egea, E., and Eeckhaut, I. 2008. Characterization of the bacterial communities associated with the bald sea urchin disease of the echinoid *Paracentrotus lividus. Journal of Invertebrate Pathology* 98(2): 136–147

Behrens, M.D. and Lafferty, K.D. 2004. Effects of marine reserves and urchin disease on southern California rocky reef communities. *Marine Ecology Progress Series* 279: 129–139

Belkin, S. and Colwell, R.R. (eds.) 2005. *Oceans and Health: Pathogens in the Marine Environment.* Vol. 233. New York, NY: Springer

Benítez-Villalobos, F., Díaz Martínez, J.P., and Martínez-García, M. 2009. Mass mortality of the sea urchin *Diadema mexicanum* in La Entrega at Bahias de Huatulco, Western Mexico. *Coral Reefs* 28: 1017

Boettger, S.A. and McClintock, J.B. 2009. The effects of chronic inorganic and organic phosphate exposure on bactericidal activity of the coelomic fluid of the sea urchin *Lytechinus varieg*atus (Lamarck) (Echinodermata: Echinoidea). *Comparative Biochemistry and Physiology C* 150(1): 39–44

Bonaviri, C., Graham, M., Gianguzza, P., and Shears, N.T. 2017. Warmer temperatures reduce the influence of an important keystone predator. *Journal of Animal Ecology* 86: 490–500

Bond, N.A., Cronin, M.F., Freeland, H., and Mantua, N.J. 2015. Causes and impacts of the 2014 warm anomaly in the NW Pacific. *Geophysical Research Letters* 42: 3414–3420

Boolootian, R.A., Giese, A.C., Tucker, J.S., and Farmanfarmaian, A. 1959. A contribution to the biology of a deep sea echinoid, *Allocentrotus fragilis* (Jackson). *Biological Bulletin* 116: 362–372

Boudouresque, C.F., Nedelec, H., and Shepherd, S.A. 1980. The decline of a population of the sea urchin *Paracentrotus lividus* in the Bay of Port-Cros (Var, France). *Travaux Scientifique Du Parc National de Port-Cros* 6: 243–251

Bower, S.M. 1996. Synopsis of infectious diseases and parasites of commercially exploited shellfish: Bald-sea-urchin disease. Fisheries and Oceans Canada. Available from: http://www.dfo-mpo.gc.ca/science/aah-saa/diseases-maladies/bsudsu-eng.html

Brady, S.M. and Scheibling, R.E. 2005. Repopulation of the shallow subtidal zone by green sea urchins (*Strongylocentrotus droebachiensis*) following mass mortality in

Nova Scotia, Canada. *Journal of the Marine Biology Association of the UK* 85: 1511–1517

Breen, P.A. and Mann, K.H. 1976. Desctructive grazing of kelp by green sea urchins in eastern Canada. *Journal of the Fisheries Research Board of Canada* 33: 1278–1283

Brink, M., Rhode, C., Macey, B.M., Christison, K.W., and Roodt-Wilding, R. 2019. Metagenomic assessment of body surface bacterial communities of the sea urchin, *Tripneustes gratilla*. *Marine Genomics* 47(100675): 1–11

Brothers, C.J., Harianto, J., McClintock, J.B., and Byrne, M. 2016. Sea urchins in a high-CO_2 world: The influence of acclimation on the immune response to ocean warming and acidification. *Proceedings of the Royal Society B* 283: 230161501

Bucci, C., Francoeur, M., Mcgreal, J., Smolowitz, R., Zazueta-Novoa, V., Wessel, G.M., and Gomwz-Chiarri, M. 2017. Sea star wasting disease in *Asterias forbesi* along the Atlantic coast of North America. *PLoS One* 12: e0188523

Buchwald, R., Scheibling, R.E., and Simpson, A.G.B. 2018. Detection and quantification of a keystone pathogen in a coastal marine ecosystem. *Marine Ecology Progress Series* 606: 79–90

Buckley, K.M., Ho, E.C.H., Hibino, T., Schrankel, C.S., Schuh, N.W., Wang, G., and Rast, J.P. 2017. IL17 factors are early regulators in the gut epithelium during inflammatory response to *Vibrio* in the sea urchin larva. *eLife* 6: e23481

Burge, C., Kim, C., Lyles, J., and Harvell, C. 2013. Special issue, Oceans and Humans Health: The ecology of marine opportunists. *Microbial Ecology* 4: 869–879

Burkepile, D.E., Allgeier, J.E., Shantz, A.A., Pritchard, C.E., Lemoine, N.P., Bhatti, L.H., and Layman, C.A. 2013. Nutrient supply from fishes facilitates macroalgae and suppresses corals in a Caribbean coral reef ecosystem. *Scientific Reports* 3(1493): 1–9

Byrne, M. 2012. Global change ecotoxicology: Identification of early life history bottlenecks in marine invertebrates, variable species responses and variable experimental approaches. *Marine Environmental Research* 76: 3–15

Byrne, M. and Andrew, N. 2020. *Centrostephanus rodgersii* and *C. tenuispina*. In *Sea Urchins: Biology and Ecology*, 4th edn, Lawrence, J.M. (ed.) pp. 281–297. Amsterdam: Elsevier

Carpenter, R.C. 1981. Grazing by *Diadema antillarum* and its effects on the benthic algal community. *Journal of Marine Research* 39(4): 749–765

Carpenter, R.C. 1988. Mass mortality of a Caribbean sea urchin: Immediate effects on community metabolism and other herbivores. *Proceedings of the National Academy of Sciences* 85: 511–514

Carpenter, R.C. 1990. Mass mortality of *Diadema antillarum*—I. Long-term effects on sea urchin population-dynamics and coral reef algal communities. *Marine Biology* 104: 67–77

Carpenter, R.C. and Edmunds, P.J. 2006. Local and regional scale recovery of *Diadema* promotes recruitment of scleracinian corals. *Ecology Letters* 9: 271–280

Cavole, L.M., Demko, A.M., Diner, R.E., Giddings, A., Koester, I., Pagniello, C.M.L.S., Paulsen, M.L., Ramirez-Valdez, A., Schwenck, S.M., Yen, N.K., Zill, M.E., and Franks, P.J.S. 2016. Biological impacts of the 2013–2015 warm-water anomaly in the northeast Pacific: Winners, losers, and the future. *Oceanography* 29: 273–285

Chang, Y. 2008. Aquaculture of edible echinoderms in China. *Global Aquaculture Advocate* 11: 38–39

Chiappone, M., Miller, S.L., Swanson, D.W., Ault, J., and Smith, S. 2001. Comparatively high densities of the long-spined sea urchin in the Dry Tortugas, Florida. *Coral Reefs* 20(2): 137–138

Chiappone, M., Swanson, D.W., Miller, S.L., and Smith, S.G. 2002. Large-scale surveys on the Florida Reef Tract indicate poor recovery of the long-spined sea urchin *Diadema antillarum*. *Coral Reefs* 21(2): 155–159

Christensen, A.M. 1970. Feeding biology of the sea-star *Astropecten irregularis* Pennant. *Ophelia* 8: 1–134

Clemente, S., Lorenzo-Morales, J., Mendoza, J.C., Lopez, C., Sangil, C., Alves, F., Kaufmann, M., and Hernandez, J.C. 2014. Sea urchin *Diadema africanum* mass mortality in the subtropical eastern Atlantic: Role of waterboune bacteria in a warming ocean. *Marine Ecology Progress Series* 506: 1–14

Coppard, S.E. and Alvarado, J.J. 2013. Echinoderm diversity in Panama: 144 years of research across the isthmus. In *Echinoderm Research and Diversity in Latin America*, Alvarado, J.J. and Solis-Marin, F.A. (eds.) pp. 107–144. Berlin and Heidelberg, Germany: Springer-Verlag

Coteur, G., Gillan, D., Joh, G., Pernet, P., and Dubois, P. 2003b. Field contamination of the starfish *Asterias rubens* by metals. Part 2: Effectson cellular immunity. *Environmental Toxicology Chemistry* 22(9): 2145–2151

Coteur, G., Gosselin, P., Wantier, P., Chambost-Manciet, Y., Danis, B., Pernet, P., Warnau, M., and Dubois, P. 2003a. Echinoderms as bioindicators, bioassays, and impact assessment tools for sediment-associated metals and PCBs in the North Sea. *Archives of Environmental Contamination Toxicology* 45(2): 190–202

Debrot, A.O. and Nagelkerken, I. 2006. Recovery of the long-spined sea urchin *Diadema antillarum* in Curaçao (Netherlands Antilles) linked to lagoonal and wave sheltered shallow rocky habitats. *Bulletin of Marine Science* 79(2): 415–424

Di Lorenzo, E. and Mantua, N. 2016. Multi-year persistence of the 2014/15 North Pacific marine heatwave. *Nature Climate Change* 6: 1042–1047

Dungan, M., Miller, T.E., and Thomson, D.A. 1982. Catastrophic decline of a top carnivore in the Gulf of California rocky intertidal zone. *Science* 216: 989–991

Dyková, I., Lorenzo-Morales, J., Kostka, M., Valladares, B., and Pecková, H. 2011. *Neoparamoeba branchiphila* infections in moribund sea urchins *Diadema aff. antillarum* in Tenerife, Canary Islands, Spain. *Diseases of Aquatic Organisms* 95: 225–231

Eckert, G.L., Engle, J.M., and Kushner, D.J. 2000. *Sea Star Disease and Population Declines at the Channel Islands*. Proceedings of the 5th California Island Symposium, pp. 390–393. US Department of the Interior, Minerals Management Service

Edmunds, P.J. and Carpenter, R.C. 2001. Recovery of *Diadema antillarum* reduces macroalgal cover and increases abundance of juvenile corals on a Caribbean reef. *Proceedings of the National Academy of Sciences* 98(9): 5067–5071

Eisenlord, M.E., Groner, M.L., Yoshioka, R.M., Elliot, J., Maynard, J., Fradkin, S., Turner, M., Pyne, D., Rivlin, N., Van Hooidonk, R., and Harvell, C.D. 2016. Ochre star mortality during the 2014 wasting disease epizootic: Role of population size structure and temperature. *Philosophical Transactions of the Royal Society of London B* 371: 20150212

Feehan, C.J., Johnson-Mackinnin, J., Scheibling, R.E., Lauson-Guay, J.S., and Simpson, A.G. 2013. Validating the identity of *Paramoeba invadens*, the causative agent of recurrent mass mortality of sea urchins in Nova Scotia, Canada. *Diseases of Aquatic Organisms* 103: 209–227

Feehan, C.J. and Scheibling, R.E. 2014. Effects of sea urchin disease on coastal marine ecosystems. *Marine Biology* 161: 1467–1485

Feehan, C.J., Scheibling, R.E., Brown, M.S., and Thompson, K.R. 2016. Marine epizootics linked to storms: Mechanisms of pathogen introduction and persistence inferred from coupled physical and biological time-series. *Limnology and Oceanography* 61: 316–329

Ferrari Legorreta, R. 2012. Building resilience of Caribbean coral reefs to macroalgal phase shifts: Identifying key habitat feature. Master's Thesis, University of Queensland, Australia

Forcucci, D. 1994. Population density, recruitment and 1991 mortality event of *Diadema antillarum* in the Florida Keys. *Bulletin of Marine Science* 54: 917–928

Foster S.A. 1987. The relative impacts of grazing by caribbean coral reef fishes and Diadema: Effects of habitat and surge. *Journal of Experimental Marine Biology and Ecology* 105(1): 1–20

Fuess, L.E, Eisenlord, M.E., Closek, C.L., Tracy, A.M., Mauntz, R., Gignoux-Wofsohn, S., Moritsch, M.M.,

Yoshioka, R., Burge, C.A., Harvell, C.D., Friedman, C.S., Hewson, I., Hershberger, P.K., and Roberts, S.B. 2015. Up in arms: Immune and nervous system response to sea star wasting disease. *PLoS One* 10: e0133053

Gardner, T.A., Hughes, T.P., Côté, I.M., Gill, J.A., Grant, A., and Watkinson, A.R. 1994. Long-term region-wide declines in Caribbean corals. *Science* 265(5178): 1547–1551

Giese, A.C. 1961. Further studies on *Allocentrotus fragilis*, a deep-sea echinoid. *Biological Bulletin* 121: 141–150

Gilles, K.W. and Pearse, J.S. 1986. Disease in sea urchins *Strongylocentrotus purpuratus*: Experimental infection and bacterial virulence. *Diseases of Aquatic Organisms* 1: 105–114

Girard, D., Clemente, S., Toledo-Guedes, K., Brito, A., Hernandez, and J.C. 2012. A mass mortality of subtropical intertidal populations of the sea urchin *Paracentrotus lividus*: Analysis of potential links with environmental conditions. *Marine Ecology* 33(3): 377–385

Gravem, S.A. and Morgan, S.G. 2017. Shifts in intertidal zonation and refuge use by prey after mass mortalities of two predators. *Ecology* 98(4): 1006–1015

Haley, M.P. and Solandt, J.L. 2001. Population fluctuations of the sea urchins *Diadema antillarum* and *Tripneustes ventricosus* at Discovery Bay, Jamaica: A case of biological succession? *Caribbean Journal of Science* 37(3–4): 239–245

Hamaguti, M., Kawahara, I., and Usuki, H. 1993. Mass mortality of *Pseudocentrotus depressus* caused by a bacterial infection in summer. *Suisanzoshoku* 41: 189–193 [in Japanese]

Harvell, C.D., Montecino-Latorre, D., Caldwell, J.M., Burt, J.M., Bosley, K., Keller, A., Heron, S.F., Salomon, A.K., Lee, L., Pontier, O., Pattengill-Semmens, C., and Gaydos, J.K. 2019. Disease epidemic and a marine heat wave are associated with the continental-scale collapse of a pivotal predator (*Pycnopodia helianthoides*). *Scientific Advances* 5: eaau7042

Hernández, J.C., Sangil, C., and Clemente, S. 2013. Sea urchins, natural events and benthic ecosystem functioning in the Canary Islands. In *Climate Change: Perspectives from the Atlantic: Past, Present and Future*, Fernández-Palacios J.M., Nascimento L.D., Hernández J.C., Clemente S., González A., and Diaz-González J.P. (eds.) pp. 487–512. Canary Islands: Servicio de publicaciones de la Universidad de La Laguna

Hewson, I., Bistolas, K.S.I., Quijano Cardé, E.M., Buttom, J.B., Foster, P.J., Flazenbaum, J.M., Kocian, J., and Lewis, C.K. 2018. Investigating the complex association between viral ecology, environment, and northeast Pacific sea star wasting. *Frontiers in Marine Science* 5: 77

Hewson, I., Button, J.B., Gudenkauf, B.M., Miner, B., Newton, A.L., Gaydos, J.K., Wynne, J., Groves, C.L.,

Hendler, G., Murray, M., Fradkin, S., Breitbart, M., Fahsbender, E., Lafferty, K.D., Kilpatrick, A.M., Miner, C.M., Raimondi, P., Lahner, L., Friedman, C.S., and Harvell, C.D. 2014. Densovirus associated with sea-star wasting disease and mass mortality. *Proceedings of the National Academy of Sciences* 111(48): 17278–17283

Hewson, I., Sullivan, B., Jackson, E.W., Zu, Q., Long, H., Lin, C., Quijano Carde, E.M., Seymour, J., Siboni, N., Jones, M.R.L., and Sewell, M.A. 2019. Perspective: Something old, something new? Review of wasting and other mortality in Asteroidea (Echinodermata). *Frontiers in Marine Science* 6: 406

Ho, E.C., Buckley, K.M., Schrankel, C.S., Schuh, N.W., Hibino, T., Solek, C.M., Bae, K., Wang, G., and Rast, J.P. 2016. Perturbation of gut bacteria induces a coordinated cellular immune response in the purple sea urchin larva. *Immunology and Cell Biology* 94: 861–887

Höbaus, E., Fenaux, L., and Hignette, M. 1981. premières observations sur les lèsions provoquèes par une maladie affectant le test des oursins en Mèditerrannèe occidentale. *Rapports et Procès-verbaux des Réunions Commission Internationale pour l'Exploration Scientifique de la Mer Méditerranée* 27: 221–222

Hodin, J., Heyland, A., Mercier, A., Pernet, B., Cohen, D.L., Hamel, J.-F., Allen, J.D., McAlister, J.S., Byrne, M., Cisternas, P., and George S.B. 2019. Culturing echinoderm larvae through metamorphosis. In *Echinoderms*. Foltz K.R. and Hamdoun A. (eds.) *Methods in Cell Biology* 150: 125–169. New York: Elsevier

Hughes, T.P. 1994. Catastrophes, phase shifts, and large-scale degradation of a Caribbean coral reef. *Science* 265: 1547–1551

Hughes, T.P., Keller, B.D., Jackson, J.B.C., and Boyle, M.G. 1985. Mass mortality of the echinoid *Diadema antillarum* Philippi in Jamaica. *Bulletin of Marine Science* 36: 377–384

Hughes, T.P., Reed, D.C., and Boyle, M.-J. 1987. Herbivory on coral reefs: Community structure following mass mortalities of sea urchins. *Journal of Experimental Marine Biology and Ecology* 113: 39–59

Hunte, W. and Younglao, D. 1988. Recruitment and population recovery of *Diadema antillarum* (Echinodermata, Echinoidea) in Barbados. *Marine Ecology Progress Series* 45: 109–119

Idjadi, J.A., Haring, R.N., Precht, W.F. 2010. Recovery of the sea urchin *Diadema antillarum* promotes scleractinian coral growth and survivorship on shallow Jamaican reefs. *Marine Ecology Progress Series* 403: 91–100

Jackson, E.W., Pepe-Ranney, C., Rohnson, M.R., Distel, D.L., and Hewson, I. 2020. A highly prevlaent and pervasive densovirus discovered among sea stars from the North American Atlantic coast. *Applied Environmental Microbiology* 86(6): e2723–19

Jackson, E.W., Wilhelm, R.C., Johnson, W.R., Lutz, H.L., Danforth, I., Gaydos, J.K., Hart, M.W., and Hewson, I. 2021. Diversity of sea star-associated densoviruses and transcribed endogenized viral elements of densovirus origin. *Journal of Virology* 95(1): e01594.

Jackson, J.B.C. 1992. Pleistocene perspectives on coral reef community structure. *American Zoologist* 32: 719–731

Jackson, J.B.C. 1997. Reefs since Columbus. *Coral reefs* 16(1): S23–S32

Jackson, J.B.C. 2001. What was natural in the coastal oceans? *Proceedings of the National Academy of Sciences* 98(10): 5411–5418

Jangoux, M. 1987a. Diseases of echinodermata. I. Agents microorganisms and protistans. *Diseases of Aquatic Organisms* 2: 147–162

Jangoux, M. 1990. Diseases of echinodermata. In *Diseases of Marine Animals*, Vol. 3. Kinne O. (ed.) pp. 439–567. Hamburg, Germany: Biologische Anstalt Helgoland

Jiang, J., Zhou, Z., Dong, Y., Jiang, B., Chen, Z., Yang, A., Wang, B., Guan, X., Gao, S., and Sun, H. 2016. The *in vitro* effects of divalent metal ions on the activities of immune-related enzymes from the sea cucumber *Apostichopus japonicus*. *Aquaculture Research* 47: 1269–1276

Johnson, P.T. 1971. Studies on diseased urchins from Point Loma. In Annual Report Kelp Habitat Improvement Project, North, W.J. (ed.) pp. 82–90. California Institute of Technology, Pasadena CA.

Jones, G.M., Hebda, A.J., Scheibling, R.E., Miller, R.J. 1985. Histopathology of the disease causing mass mortalities of sea urchins (*Strongylocentrotus* droebachiensis) in Nova Scotia. *Journal of Invertebrate Pathology* 45: 260–271

Jurgens, L.J., Rogers-Bennett, L., Raimondi, P.T., Schiebelhut, L.M., Dawson, M.N., Grosberg, R.K., Gaylord, B. 2015. Patterns of mass mortality among rocky shore invertebrates across 100 km of Northeastern Pacific coastline. *PLoS One* 10(6): e01280

Kanai, K. 1993. 'Togenukesho' of sea urchins. In Proceedings of the Symposium on Diseases in Fish and Shellfish in Kyushu and Okinawa, p. 7. Tokyo: The Japanese Society of Fish Pathology

Kiselev, K.V., Ageenko, N.V., and Kurilenko, V.V. 2013. Involvement of the cell-specific pigment genes pks and sult in bacterial defense response of sea urchins *Strongylocentrotus intermedius*. *Diseases of Aquatic Organisms* 103(2): 121–132

Krumhansl, K.A., Okamoto, D.K., Rassweiler, A., Novak, M., Bolton, J.J., Cavanaugh, K.C., Connell, S.D., Johnson, C.R., Konar, B., Ling, S.D., Micheli, F., Norderhaug, K.M., Pérez-Matus, A., Sousa-Pinto, I., Reed, D.C., Salomon, A.K., Shears, N.T., Wernberg, T., Anderson, R.J., Barrett N.S., Buschmann, A.H., Carr, M.H., Caselle, J.E., Derrien-Courtel, S., Edgar, G.J., Edwards, M., Estes, J.A., Goodwin, C., Kenner, M.C., Kushner, D.J., Moy, F.E., Nunn, J., Steneck, R.S., Vásquez, J.,

Watson, J., Witman, J.D., Byrnes, J.E. 2016. Global patterns of kelp forest change over the past half century. *Proceedings of the National Academy of Sciences* 113(48): 13785–13790

Lafferty, K.C. 2004. Fishing for lobsters indirectly increases epidemics in sea urchins. *Ecological Applications* 14(5): 1566–1573

Lafferty, K.D., Porter, J.W., and Ford, S.U. 2004. Are diseaes increaseing in the ocean. *Annual Review of Ecology, Evolution, and Systematics* 35: 31–54

Lessios, H.A. 1988a. Mass mortality of *Diadema antillarum* in the Caribbean: What have we learned? *Annual Review of Ecology and Systematics* 19: 371–393

Lessios, H.A. 1988b. Population dynamics of *Diadema antillarum* (Echinodermata: Echinoidea) following mass mortality in Panamá. *Marine Biology* 99: 515–526

Lessios, H.A. 1995. *Diadema antillarum* 10 years after mass mortality: Still rare, despite help from a competitor. *Proceedings of the Royal Society of London* 259: 331–337

Lessios, H.A. 2005. *Diadema antillarum* populations in Panama twenty years following mass mortality. *Coral Reefs* 24: 125–127

Lessios, H.A. 2016. The great *Diadema antillarum* die-off: 30 years later. *Annual Review of Marine Science* 8: 267–283

Lessios, H.A, Robertson D.R., and Cubit J.D. 1984. Spread of *Diadema* mass mortality through the Caribbean. *Science* 226: 335–337

Lester, S.E., Tobin, E.D., and Behrens, M.D. 2007. Disease dynamics and the potential role of thermal stress in the sea urchin, *Srongylocentrotus purpuratus*. *Canadian Journal for Aquatic Science* 64: 314–323

Levitan, D.R. 1991. Influence of body size and population density on fertilization success and reproductive output in a free-spawning invertebrate. *Biological Bulletin* 181: 261–268

Li, R., Dang, H., Huang, Y., Quan, Z., Jiang, H., Zhang, W., and Ding, J. 2020. *Vibrio coralliilyticus* as an agent of red spotting disease in the sea urchin *Strongylocentrotus intermedius*. *Aquaculture Reports* 16: 100244

Liddell, W.D. and Ohlhorst, S.L. 1986. Changes in benthic community composition following the mass mortality of *Diadeam* at Jamaica. *Journal of Experimental Marine Biology and Ecology* 95 (3): 271–278

Ling, S.D., Scheibling, R.E., Rassweiler, A., Johnson, C.R., Shear,s N., Connell, S.D., Salomon, A.K., Norderhaug, K.M., Pérez-Matus, A., Hernández, J.C., Clemente, S., Blamey, L.K., Hereu, B., Ballesteros, E., Sala, E., Garrabou, J., Cebrian, E., Zabala, M., Fujita, D., and Johnson, L.E. 2015. Global regime shift dymanics of catastrophic sea urchin overgrazing. *Philosophical Transactions of the Royal Soceity B* 370: 20130269

Loffler S.G. 2018. Sea urchin disease: Case study on Venezuelan shores. *Journal of Marine Microbiology* 2(1): 1–2.

Maes, P. and Jangoux, M. 1984. The bald-sea-urchin disease: A biopathological approach. *Helgoländer Meeresuntersuchungen* 37(1–4): 217–224

Maes, P. and Jangoux, M. 1985. The bald sea urchin disease: A bacterial infection. In Proceedings of the Fifth International Echinoderm Conference, Keegan B. F. and O'Connor, B. D. S. (eds.) pp. 313–314. Galway, Ireland Rotterdam: A.A. Balkema Publishers

Mann, K. H. 1973. Seaweeds: Their productivity and strategy for growth. The role of large marine algae in coastal productivity is far more imporaant thgan has been suspected. *Nature* 182(4116); 975-981

MARINe 2016. Multi-Agency Rocky Intertidal Network, Sea Star Wasting Syndrome. www.eeb.ucsc.edu/pacifi crockyintertidal/data-products/sea-star-wasting/index. html. www.seastarwastingsyndrome.org.

Matranga V., Pinsino A., Celi M., Natoli, A, Bonaventura, R, Schröder, H.C., and Müller, W.E.G. 2005. Monitoring chemical and physical stress using sea urchin immune cells. In *Echinodermata*, Matranga V. (ed.) pp. 85–110. *Progress in Molecular and Subcellular Biology. Subseries Marine Molecular Biotechnology.* Berlin/Heidelberg: Springer

Matranga V., Toia G., Bonaventura R., and Müller W.E.G. 2000. Cellular and biochemical responses to environmental and experimentally induced stress in sea urchin coelomocytes. *Cell Stress and Chaperones* 5(2): 113–120

McClanahan, T.R., Kamukuru, A.T., Muthiga, N.A., Yebio, M.G., and Obura, D. 1996. Effect of sea urchin reductions on algae, coral, and fish poulations. *Conservation Biology* 10: 136–154

McClanahan, T.R., Shafir, S.H. 1990. Causes and consequences of sea urchin abundance and diversity in Kenyan coral reef lagoons. *Oecologia* 83(3): 362–370

McPherson, B.F. 1965. Contributions to the biology of the sea urchin *Tripneustes*. *Bulletin of Marine Science* 15(1): 228–244

McPherson, B.F. 1968. Contributions to the biology of the sea urchin *Eucidaris tribuloides* (Lamarck). *Bulletin of Marine Science* 18(2): 400–443

Mead, A.D. 1898. Twenty-eighth annual report of the commissioners of inland fisheries, made to the general assembly at its January session. Southwick, J.M.K., Root, H.T., Willard, C.W., Morton, W.M.P., Roberts, A.D., and Bumpus, H.C. (eds.) pp. 112. Freeman and Sons, Printers to the State of Rhode Island

Menge, B.A. 1979. Coexistence between the seastars *Asterias vulgaris* and *Asterias forbesi* in a heterogeneous environment—nonequilibrium explanation. *Oecologia* 41: 245–272

Menge, B.A., Cerny-Chipman, E.B., Johnson, A., Sullivan, J., Gravem, S., and Chan, F. 2016. Sea star wasting disease in the keystone predator *Pisaster ochraceus* in Oregon: Insights into differential population impacts, recovery, predation rate, and temperature effects from long-term research. *PLoS One* 11: e0153994

Miller, R.J. 1985. Succession in sea urchin and seaweed abundance in Nova Scotia, Canada. *Marine Biology* 84: 275–286

Miller, R.J., Adams, A.J., Ogden, N.B., Ogden, J.C., and Ebersole, J.P. 2003. *Diadema antillarum* 17 years after mass mortality: Is recovery beginning on St. Croix? *Coral Reefs* 22(2): 181–187

Miller, R.J and Colodey, A.G. 1983. Widespread mass mortalities of the green sea urchin in Nova Scotia, Canada. *Marine Biology* 73: 263–267

Miner, C.M., Burnaford, J.L., Ambrosew, R.F., Antrim, L., Bohlmann, H., Blanchette, C.A., Engle, J.M., Fradkin, S.C., Gaddam, R., Harley, C.D.G., Miner, B.G., Murrau, S.N., Smith, J.R., Whitaker, S.G., and Raimondi, P.T. 2018. Large-scale impacts of sea star wasting disease (SSWD) on intertidal sea stars and implications for recovery. *PLoS One* 13: e0192870

Moitoza, D.J. and Phillips, D.W. 1979. Prey defence, predator preference, and nonrandom diet: The interactions between *Pycnopodia helianthoides* and two species of sea urchins. *Marine Biology* 53: 299–304

Montecino-Latorre, D., Eisenlord, M.E., Turner, M., Yoshioka, R., Harvell, C.D., Pattengill-Semmens, C.V., Nichols, J.D., and Gaydos, J.K. 2016. Devastating transboundary impacts of sea star wasting disease on subtidal asteroids. *PLoS One* 11: e0163190

Moore, H.B., Jutare, T., Bauer, J.C., and Jones, J.A. 1963. The biology of *Lytechinus variegatus*. *Bulletin of Marine Science* 13(1): 23–53

Moritsch, M.M. and Raimondi, P.T. 2018. Reduction and recovery of keystone predation pressure after disease-related mass mortality. *Ecology and Evolution* 8: 3952–3964

Mortensen, T. and Rosenvinge, L.K. 1934. Sur une algue: Cyanophycée *Dactylococcopsis echini* N. sp. parasite dans un oursin. *Biologiske Meddelelser* 11: 1–7

Moses, C.S. and Bonem, R.M. 2001. Recent population dynamics of *Diadema antillarum* and *Tripneustes ventricosus* along the North coast of Jamaica, W.I. *Bulletin of Marine Science* 68(2): 327–336

Murano, C., Agnisola, C., Caramiello, D., Castellano, I., Casotti, R., Corsi, I., Palumbo, A. 2020. How sea urchins face microplastics: Uptake, tissue distribution and immune system response. *Environmental Pollution* 264: 114685

Nagelkerken, I., Smith, G.W., Snelders, E., Karel, M., and James, S. 1999. Sea urchin *Meoma ventricosa* die-off in Curaçao (Netherlands Antilles) associated with a pathogenic bacterium. *Diseases of Aquatic Organisms* 38: 71–74

Ogden, J.C. 1976. Some aspects of herbivore-plant relationships on Caribbean reefs and seagrass beds. *Aquatic Botany* 2(C): 103–116

Ogden, J.C. and Lobel, P.S. 1978. The role of herbivorous fishes and urchins in coral reef communities. *Environmental Biology of Fishes* 3(1): 49–63

Pagliara, P. and Stabili, L. 2012. Zinc effect on the sea urchin *Paracentrotus lividus* immunological competence. *Chemosphere* 89(2012): 563–568

Paine, R.T. 1966. Food web complexity and species diversity. *American Naturalist* 100: 65–75

Paine, R.T. 1974. Intertidal community structure. Experimental studies on the relationship between a dominant competitor and its principal predator. *Oecologia* 15: 93–120

Paine, R.T. and Vadas, R.L. 1969. The effects of grazing by sea urchins, *Strongylocentrotus* spp., on benthic algal populations. *Limnology and Oceanography* 14(5): 710–719

Pandolfi, J.M., Bradbury, R.H, Sala, E., Hughes, T.P., Bjorndal, K.A., Cooke, R.G., McArdle, D., McClenachan, L., Newman, M.J., Paredes, G., and Warner, R.R. 2003. Global trajectories of the long-term decline of coral reef ecosystems. *Science* 301(5635): 955–958

Pandolfi, J.M., Staples, T.L., and Kiessling, W. 2020. Increased extinction in the emergence of novel ecological communities. *Science* 370(6513): 220–222

Pearse, J., Costa, D., Yellin, M., and Agagian, C. 1977. Localized mass mortality of red sea urchin, *Strongylocentrotus franciscanus*, near Santa Cruz, California. *Bulletin of the US Fish Commission* 75: 645–648

Pinsino, A., Della Torre, C., Sammarini, V., Bonaventura, R., Amato, E., and Matranga, V. 2008. Sea urchin coelomocytes as a novel cellular biosensor of environmental stress: A field study in the Tremiti Island Marine Protected Area, Southern Adriatic Sea, Italy. *Cell Biology Toxicology* 24(6): 541–552

Pinsino, A. and Matranga, V. 2015. Sea urchin immune cells as sentinels of environmental stress. *Developmental and Comparative Immunology* 49: 198–205

Roberts-Regan, D.L., Scheibling, R.E., and Jellett, J.F. 1988. Natural and experimentally induced lesions of the body wall of the sea urchin *Strongylocentrotus droebachiensis*. *Diseases of Aquatic Organisms* 5: 51–62

Rogers-Bennett, L. and Catton, C.A. 2019. Marine heat wave and ultiple stressors tip bull kelp forest to sea urchin barrens. *Scientific Reports* 9: 15050

Ruiz-Ramos, D.V., Hernández-Delgado, E.A., and Schizas, N.V. 2011. Population status of the long-spined urchin *Diadema antillarum* in Puerto Rico 20 years after a mass mortality event. *Bulletin of Marine Science* 87(1): 113–127

Sabine, C.L., Feely, R.A., Gruber, N., Key, R.M., Lee, K., Bullister, J.L., Wanninkhof, R., Wong, C.S., Wallace, D.W., Tilbrook, B., Millero, F.J., Peng, T.H., Kozyr, A., Ono, T., and Rios, A.F. 2004. The oceanic sink for anthropogenic CO_2. *Science* 305(5682): 367–371

Sammarco, P.W. 1980. *Diadema* and its relationship to coral spat mortality: Grazing, competition, and biological disturbance. *Journal of Experimental Marine Biology and Ecology* 45(2): 245–272

Sammarco, P.W. 1982a. Echinoid grazing as a structuring force in coral communities: Whole reef manipulations. *Journal of Experimental Marine Biology and Ecology* 61(1): 31–55

Sammarco, P.W. 1982b. Effects of grazing by *Diadema antillarum* Philippi (Echinodermata: Echinoidea) on algal diversity and community structure. *Journal of Experimental Marine Biology and Ecology* 65(1): 83–105

Sammarco, P.W., Levinton, J., and Ogden, J. 1974. Grazing and control of coral reef community structure by *Diadema antillarum* Philippi (Echinodermata: Echinoidea): a preliminary study. *Journal of Marine Research* 32: 47–53

Sanford, E., Sones, J.L, Garcia-Reyes, M., and Goddard, J.R.H., and Largier, J.L. 2016. Widespread shifts in the coastal biota of northern California during the 2014–2015 marine heatwaves. *Scientific Reports* 9: 4216

Scheibling, R.E. 1985. Increased macroalgal abundance following mass mortalities of sea urchins (*Strongylocentrotus droebachiensis*) along the Atlantic coast of Nova Scotia. *Oecologia* 68(2): 186–198

Scheibling, R.E., Feehan, C., and Lauzon-Guay, J.S. 2010. Disease outbreaks associated with recent hurricanes cause mass mortality of sea urchins in Nova Scotia. *Marine Ecology Progress Series* 408: 109–116

Scheibling, R.E. and Hennigar, A.W. 1997. Recurrent outbreaks of disease in sea urchins *Strongylocentrotus droebachiensis* in Nova Scotia: Evidence for a link with large-scale meteorologic and oceanographic events. *Marine Ecology Progress Series* 152: 155–165

Scheibling, R.E. and Lauzon-Guay, J.S. 2010. Killer storms: North Atlantic hurricanes and disease outbreaks in sea urchins. *Limnology and Oceanography* 55: 2331–2338

Scheibling R.E. and Stephenson R.L. 1984. Mass mortality of *Strongylocentrotus droebachiensis* (Echinodermata: Echinoidea) off Nova Scotia, Canada. *Marine Biology* 78: 153–164

Schuh, N.W., Carrier, T.J., Schrankel, C.S., Reitzel, A.M., Heyland, A., and Rast, J.P. 2020. Bacterial exposure mediates developmental plasticity and resistance to lethal *Vibrio lentus* infection in purple sea urchin (*Strongylocentrotus purpuratus*) larvae. *Frontiers in Immunology* 10: 3014

Schultz, J.A., Cloutier, R.N., and Cote, A.M. 2016. Evidence for a trophic cascade on rocky reefs following sea star mass mortality in British Columbia. *PeerJ* 4: e1980

Schwammer, H.M. 1989. Bald-sea-urchin disease: Record of incidence in irregular echinoids—*Spatangus purpureus*, from the SW-coast of Krk (Croatia—Jugoslavia). *Zoologischer Anzeiger* 223: 100–106

Shimizu, M., Takaya, Y., Ohsaki, S., and Kawamata, K. 1995. Gross and histopathological signs of the spotting disease in the sea urchin *Strongylocentrotus intermedius*. *Fisheries Science* 61(4): 608–613

Sieling, F.W. 1960. Mass mortality of the starfish, *Asterias forbesi*, on the Atlantic coast of Maryland. *Chesapeake Science* 1: 73–74

Stabili, L. and Pagliara, P. 2009. Effect of zinc on lysozyme-like activity of the seastar *Marthasterias glacialis* (Echinodermata, Asteroidea) mucus. *Journal of Invertebrate Pathology* 100: 189–192

Stabili, L. and Pagliara, P. 2012. Zinc effect on the sea urchin *Paracentrotus lividus* immunological competence. *Chemosphere* 89(5): 563–568

Staehli, A., Schaerer, R., Hoelzle, K., and Ribi, G. 2008. Temperature induced disease in the starfish *Astropecten jonstoni*. *Marine Biodiversity Records* 2: e78

Steneck, R.S. 2020. Sea urchins as drivers of shallow water benthic community structure. In *Sea Urchins: Biology and Ecology*, Lawrence, J.M. (ed.) pp. 195–212. Academic Press, San Diego, California

Stokstad, E. 2014. Death of the stars. *Science* 344(6183): 464–467

Suzuki, G., Kai, S., and Yamashita, H. 2012. Mass stranding of crown-of-thorns starfish. *Coral Reef* 31: 821

Sweet, M. 2020. Sea urchin diseases: Effects from individuals to ecosystems. In *Sea Urchins: Biology and Ecology*, 4th ed. Lawrence J.M. (ed.) Vol 43: 219–226. London: Academic Press, Elsevier

Sweet, M., Bulling, M., and Williamson, J. 2016. New disease outbreak affects two dominant sea urchin species associated with Australian temperate reefs. *Marine Ecology Progress Series* 551: 171–183

Tajima, K., Hirano, T., Nakano, K., and Ezura, Y. 1997a. Taxonomical study on the causative bacterium of spotting disease of sea urchin *Strongylocentrotus intermedius*. *Fisheries Science* 63(6): 897–900

Tajima, K., Hirano, T., Shimizu, M., and Ezura, Y. 1997b. Isolation and pathogenicity of the causative bacterium of spotting disease of sea urchin *Strongylocentrotus intermedius*. *Fisheries Science* 63(2): 249–252

Tajima, K., Shimizu, M., Miura, K., Ohsaki, S., Nishihara, Y., and Ezura, Y. 1998. Seasonal fluctuations of *Flexibacter* sp. the causative bacterium of spotting disease of sea urchin *Strongylocentrotus intermedius* in the culturing facilities and coastal area. *Fisheries Science* 64(1): 6–9

Thorpe, J.P. and Spencer, E.L. 2000. A mass stranding of the asteroid *Asterias rubens* on the Isle of Man. *Journal of the Marine Biological Association of the United Kingdom* 80: 749–750

Tiffany, W.J. 1978. Mass mortality of *Luidia senegalensis* (Lamarck, 1816) on Captiva Island, Florida, with a note on its occurrence in Florida Gulf coastal waters. *Florida Science* 41: 63–64

Tyler-Walters, H. 2008. Edible sea urchin (*Echinus esculentus*). Availble from: Marine Life Information Network (MarLIN): Biology and Sensitivity Key Information Reviews. see https://dx.doi.org/10.17031/marlinsp.1311.1

Vega Thurber, R., Burkepile, D.E., Correa, A.M.S., Thurber, A.R, Shantz, A.A., Welsh, R., Pritchard, C., and Rosales, S. 2012. Macroalgae decrease growth and alter microbial community structure of the reef-building coral, *Porites astreoides*. *PLoS One* 7(9): e44246

Wang, B., Li, Y., Li, X., Chen, H.X., Liu, M.Q., and Kong, Y.T. 2006. Biological characteristic and pathogenicity of the pathogenic vibrio on the 'red spot disease' of *Strongylocentrotus intermedius*. *Journal of Fishery Sciences of China* 30: 371–376 [in Chinese]

Webster, F.J., Babcock, R.C., Van Keulen, M., and Loneragan, N.R. 2015. Macroalgae inhibits larval settelement and increases recruit mortality at Ningaloo Reef, Western Australia. *PLoS ONE* 10(4): e0124162

Yui, M. and Bayne, C. 1983. Echinoderm immunity: Bacterial clearance by the sea urchin *Strongylocentrotus purpuratus*. *Biological Bulletin* 165: 473–485

Zhang, W., Lv, Z., Li, C., Sun, Y., Jiang, H., Zhao, M., Zhao, X., Shao, Y., and Chang, Y. 2019. Transcriptome profiling reveals key roles of phagosome and NOD-like receptor pathway in spotting diseased *Strongylocentrotus intermedius*. *Fish and Shellfish Immunology* 84: 521–531

CHAPTER 20

Diseases of tunicates and cephalochordates

Andrew F. Rowley and Shin-Ichi Kitamura

20.1 Introduction

The phylum Chordata consists of two inverte-brate groups: the cephalochordates and tunicates (also known as urochordates), together with the vertebrates. Cephalochordates belong to the sub-phylum Cephalochordata and compromise *ca.* 25 extant species commonly called lancelets or 'the amphioxus'. The urochordates are classified as belonging to the sub-phylum Tunicata (= sub-phylum Urochordata) and are a larger group of over 2,000 species including solitary (e.g. sea squirts) and colonial ascidians (class Ascidiacea), appendic-ularians (class Larvacea) and pelagic salps (class Thaliacea). Because both cephalochordates and uro-chordates are related to the phylogeny of the Verte-brata, they have been studied extensively in terms of their phylogeny (e.g. Lowe et al. 2015; Putnam et al. 2008; Satoh et al. 2014), developmental biology (e.g. Alié et al. 2020; Cao et al. 2019) and immunity (Franchi and Ballarin 2017; see also Chapter 1).

Tunicates, such as the sea pineapple, *Halocynthia roretzi* are farmed in Korea and Japan and consumed fresh or dried (Lambert et al. 2016). Production has been severely affected by mass mortality events since the 1980s. Ascidians, in general, also have oth-er economic significance as invasive species fouling ships and harbours (Aldred and Clare 2014) and as potential sources of novel therapeutic substances including the anti-viral didemnins (Chen et al. 2018; Reinhart et al. 1981). Cephalochordates including 'the amphioxus', *Branchiostoma* spp., are fished in some parts of the globe including the China Seas but to date have not been subject to culture production.

Despite the extensive research into both tunicates and cephalochordates, little is known of disease conditions in these animals in the wild but there are a small number of reports of diseases of those either held under aquarium conditions or subject to aquaculture production. This brief chapter reviews our current knowledge of these disease states in both groups with an emphasis on the detection and pathology of such conditions.

20.2 Principal diseases

There is a paucity of information on diseases caused by bacteria, fungi, oomycetes and viruses in tuni-cates and cephalochordates (Monniot 1990). For example, although a marine birnavirus was detect-ed in the solitary tunicate, *Halocynthia roretzi* under aquaculture conditions in Korea (Jung et al. 2001) later studies reported that there were no differences in detection rates of the virus in apparently healthy and diseased specimens (Azumi et al. 2007) ruling out its role as a primary cause of disease. Similarly, bacteria found in the tunic of ascidians appear to be part of a normal flora that may have a host defen-sive role against microbial invasion (Blasiak et al. 2014; Chapter 1). The main diseases of tunicates and cephalochordates are summarised in Tables 20.1 and 20.2 respectively.

Andrew F. Rowley and Shin-Ichi Kitamura, *Diseases of tunicates and cephalochordates*. In: *Invertebrate Pathology*. Edited by Andrew F. Rowley, Christopher J. Coates and Miranda M.A. Whitten, Oxford University Press. © Oxford University Press (2022). DOI: 10.1093/oso/9780198853756.003.0020

Table 20.1 An overview of the diseases of tunicates

Disease	Disease causing agent(s)	Host range	Geographical range	Pathology	Key references
Soft tunic syndrome	The kinetoplastid, *Azumiobodo hoyamushi*	*Halocynthia roretzi*	Korea, Japan	*Gross pathology:* Weakened and thin tunic, tunic splits leading to death of host *Histopathology:* Numerous flagellated parasites in tunic (Figure 20.1e). No evidence of interaction with host's immune cells (haemocytes)	Hirose et al. 2012; Kumagai et al. 2011l
Cup cell disease	Haplosporidian?	*Botryllus schlosseri*	Colonies held under aquarium conditions with water from the Mediterranean changed twice a week (no evidence of any recirculating technology used)	*Gross pathology:* Dark spots on zooids, breakdown of colonies, ampulla retraction, swollen and soft tunic, mortality *Histopathology:* Parasites found in haemocytes, aggregated parasites found in blood vessels, plasmodia-like stages seen in tunic	Moiseeva et al. 2004
	Haplosporidium ascidiarum	*Ciona intestinalis*	Bay of Naples, Italy	Parasites found in gut lumen; no pathobiology noted	Ciancio et al. 1999
	Copepods, *Botryllophilus ruber* and *Mychophilus roseus*	*B. schlosseri*	Roscoff, France	Copepods found in canal systems of host. No reference to any pathology	Ooishi 1999
	Copepod, *Botryllophilus conicus*	*Aplidium conicum*	Straits of Gibraltar	No reference to any pathology	Conradi et al. 1994
	Copepod, *Enteropsis fusiformis*	Compound ascidian, *Polyclinum insulsum*	Madagascar	No reference to any pathology	Ooishi 2009
	Sculpid fish	Solitary ascidians including *Halocynthia roretzi* and *H. ritteri*	Japan	Fish deposit eggs in atrium. May cause reduced growth of host	Awata et al. 2019

20.2.1 Soft tunic syndrome

The tunic is the outer covering of tunicates and is a layer containing a variety of cells together with extracellular matrix consisting of cellulose and proteinaceous fibres (Hirose 2009; see Figure 1.3 in Chapter 1). This condition was first reported in 1995 in sea pineapples (*Halocynthia roretzi*) in ascidian farms in Korea (Jung et al. 2001). By 2007, it was found in Japan perhaps as a result of importation of spat from Korea (Kumagai et al. 2011). The characteristic symptoms of the disease are the progressive softening and thinning of the outer protective tunic (Figures 20.1 a-c). Tunic hardness can be measured using a force gauge as an indicator for softening (Hirose et al 2018) and once the tunic splits, the host then dies. This condition has caused dramatic declines in sea pineapple production in both Korea and Japan. There are also reports that in wild populations of *H. roretzi* close to farms can become infected by this condition (Kumagai et al. 2013). Other species of ascidians, including *Styela clava*, *S. plicata* and *Pyura vittata* can act as carriers or hosts to the infectious agent that causes this condition (Kumagai et al. 2014; Nam et al. 2015). This condition is temperature dependent with most outbreaks of disease occurring in the spring when water temperatures reach *ca.* 10 °C (Nawata et al. 2015; Nam et al. 2017).

The characteristic features of the disease can be seen in histopathology where the tunic fibre bundles that compose the tunic matrix fail to form thick bundles seen in apparently healthy specimens. Instead, low fibre density areas are found in diseased tunics (Hirose et al. 2009; Figure 20.1c). There is no difference in the composition and distribution pattern of tunic cells between healthy and diseased individuals (Hirose et al. 2009).

Following the first reports of this disease, several attempts were made to determine the causative agent of the condition. Candidates from these included viruses (e.g. Jung et al. 2001) and possibly an intrahaemocytic protist (Choi et al. 2006). More recent definitive studies have established that the causative agent of soft tunic syndrome is a bi-flagellate kinetoplastid, *Azumiobodo hoyamushi* (Kumagai et al. 2010, 2011, 2013; Hirose et al. 2012; Figures 20.1d,e). This flagellate is culturable in cell culture medium (Kumagai et al. 2011). Presence of the disease-causing agent has been confirmed using PCR (Kumagai and Kamaishi 2013), real-time PCR (Shin et al. 2014) and loop mediated isothermal amplification (Song et al. 2014) based on the nucleotide sequences of 18S rRNA or β-tubulin genes in *A. hoyamushi*.

The route of invasion of these parasites is thought to be via the damaged tunic cuticle lining the oral siphon (Hirose et al. 2014). Kinetoplastid parasites are only found in the tunic where they cause an inflammatory influx of haemocytes (Kumagai et al. 2011). The mechanism of how the kinetoplastid damages the tunic is uncertain. In the initial reports of the nature of the causative agent, it was suggested that the parasites secrete cellulases and proteases (Hirose et al. 2012) but a later study found that the presence of parasites does not cause any degradation of cellulose fibres (Kimura et al. 2015). Host death occurs as a result of the rupturing of the tunic causing the kinetoplastids to be released into the surrounding environment where they probably undergo encystment. These cyst-like forms are able to persist in seawater for several months (Nawata et al. 2015). New infections may be aided by the production of chemotactic factors from intact tunic that attract the parasite (Nawata et al. 2018).

Potential approaches to the control of soft tunic syndrome have been explored (Park et al. 2014; Kumagai et al. 2016). Both disinfection of *H. roretzi* eggs or various stages of the parasite with sodium hypochlorite or povidone-iodine, have proven to be an effective control (Kumagai et al. 2016). Because the infection is inhibited by low water temperature, site selection for *H. roretzi* cultivation based on environmental temperature, could be a further option to control this disease condition (Nam et al. 2017).

20.2.2 Gregarine 'infections' of tunicates

There are several reports of the presence of gregarines in tunicates worldwide including temperate ascidians such as *Ciona intestinalis* (Mita et al. 2012) through to salps, *Salpa thompsoni* in the Southern Ocean (Wallis et al. 2017). In most cases, the presence of these organisms is of little detriment to the host (Monniot 1990; Rueckert et al. 2019) but in

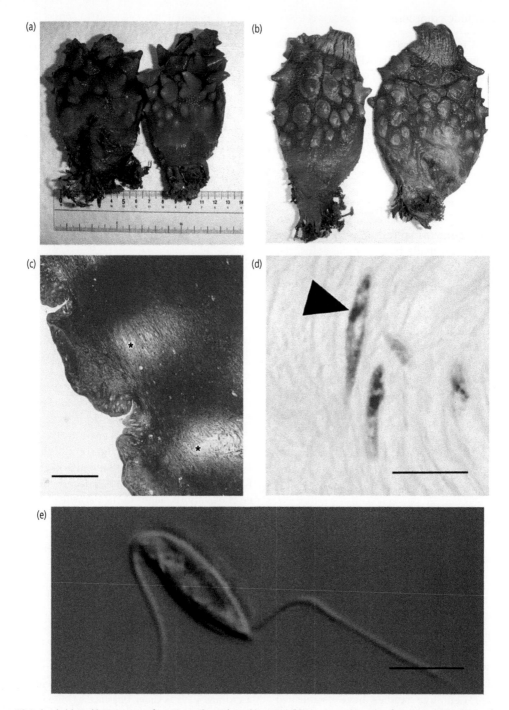

Figure 20.1 (a–e). (a) Healthy specimens of sea pineapples, *Halocynthia roretzi*. (b) Sea pineapples with soft tunic syndrome. Note soft appearance of tunic especially close to the base of these specimens. (c) Histological image of the tunic of *H. roretzi* infected with soft tunic syndrome. Note regions showing loss of normal structure in tunic fibres (*). Scale bar = 100 μm. (d) Flagellated cells (arrowhead) of *Azumiobodo hoyamushi* in the tunic of sea pineapple with soft tunic syndrome. Scale bar = 10 μm. (e) Differential interference contrast micrograph of biflagellated *A. hoyamushi* parasite. Scale bar = 5 μm. Source: micrographs courtesy of Dr A. Kumagai (c and d) and Professor E. Hirose (e).

Table 20.2 An overview of the diseases of cephalochordates

Disease	Disease causing agent(s)	Host range	Geographical range	Pathology	Key references
Vibriosis	*V. alginolyticus*	*Branchiostoma belcheri tsingtauense*	Aquarium-based infection	Lesions in tail region, no histopathology. Causes mortality. Opportunistic infection?	Zou et al. 2016

extreme cases, such as long faeces syndrome in *C. intestinalis*, they obstruct the stomach or intestine and can cause death (Mita et al. 2012). Similar observations have been made in *C. intestinalis* infected with the gregarine parasite, *Lankesteria ascidiae* taken from the wild, where the trophozoite stages attach to the epithelial cells that line the stomach resulting in their hypertrophy (Ciancio et al. 2001). As argued by Rueckert et al. (2019), gregarines have a range of associations with their hosts ranging from symbiotic through to parasitic and may be a normal component of the microbiome, so their presence is not necessarily indicative of disease.

20.2.3 Copepods in compound ascidians

Various copepods have been described in association with compound and solitary ascidians (Conradi et al. 1994; Ooishi 1999, 2009, 2014; Table 20.1) but their effect on the host is unknown and whether these are in a parasitic or commensal relationship is unclear. In a photosymbiotic ascidian *Diplosoma virens*, the parasitic copepods, *Loboixys ryukyuensis*, appeared to feed on the host tunic (Hirose 2000) and the parasite suppressed the fecundity of the host colonies (Hirose et al. 2005).

20.2.4 Haplosporidian-like parasites of tunicates

Two haplosporidian-like agents have been reported in the colonial ascidian, *Botryllus schlosseri* (Moiseeva et al. 2004) and the sea squirt, *Ciona intestinalis* (Ciancio et al. 1999). In both cases the disease-causing agents were identified on morphological

grounds alone and so the identification needs confirmation using molecular approaches.

20.2.5 Egg deposition in tunicates

Animals that deposit their eggs in live animals are called 'ostracophils'. Some species of sculpin fishes deposit eggs in invertebrates including sponges and tunicates (Awarta et al. 2019). This process occurs via specialised ovipositors and in some cases may cause reduction in seawater passage in the target host resulting in reduced growth and survival (Awarta et al. 2019).

20.2.6 Vibriosis in cephalochordates

Amphioxus held under culture conditions have been reported to be infected by *Vibrio alginolyticus* in tail lesions (Table 20.2; Zou et al. 2016). Animals challenged with this pathogen by intramuscular injection show progressive mortality. No reports exist of this infection in such animals collected from the wild suggesting that it may be an opportunistic infection instigated by holding animals in aquarium conditions.

20.3 Future directions

The tunicate microbiome consists of a wide assemblage of bacteria that has promise as sources of novel antimicrobial and antitumour compounds (Ayuningrum et al. 2019; Chen et al. 2018). They have a potential role in controlling other microbes that colonise and invade across the tunic (Blasiak et al. 2014; Chapter 1) as well as helping hosts adjust to changing environmental conditions (Dror et al. 2019). Its study is hence of importance to both tunicate diseases but also to the treatment of infectious and non-infectious diseases relevant to humans.

Tunicates may act as reservoirs of diseases of other species of aquatic invertebrates. For example, invasive tunicates including *Styela* *clava* appear to harbour the parasites of some bivalve molluscs, *Bonamia ostreae* and *Minchinia mercenariae* and may therefore be responsible for transporting these diseases to sites previously free from infection (Costello et al. 2021). Further study is needed to determine whether this route of disease transfer is important.

20.4 Summary

- There is a paucity of reports demonstrating infections of both tunicates and cephalochordates originating from the wild.
- Many infections reviewed in this chapter are from animals under culture conditions either in aquaria or in culture out at sea (e.g. soft tunic syndrome).
- The causative agents are not always fully established.
- The microbiome found in the tunic may act as sources of antimicrobial factors important in keeping tunicates disease-free.

Acknowledgements

We are grateful to the following for provision of the micrographs in Figure 20.1, Professor E. Hirose in University of the Ryukyus and Dr A. Kumagai in Miyagi Prefecture Fisheries Technology Institute, Japan. S-I K was partly supported by KAKENHI (no. 15H05251 and 20H03074) from the Japan Society for the Promotion of Science and Matsuoka Research Institute for Science.

References

Aldred, N. and Clare, A.S. 2014. Impact and dynamics of surface fouling by solitary and compound ascidians. *Biofouling* 30: 259–270

Alié, A., Hiebert, L.S., Scelzo, M., and Tiozzo, S. 2020. The eventful history of nonembryonic development in tunicates. *Journal of Experimental Zoology B (Molecular and Developmental Evolution)* 334: 1–17

Awata, S., Sasaki, H., Goto, T. et al. 2019. Host selection and ovipositor length in eight sympatric species of sculpins that deposit their eggs into tunicates or sponges. *Marine Biology* 166: 59

Ayuningrum, D., Liu, Y., Riyanti, M.T.S., Kristiana, R. et al. 2019. Tunicate-associated bacteria show a great potential for the discovery of antimicrobial compounds. *PLoS ONE* 14: e0213797

Azumi, K., Nakamura, S., Kitamura, S., Jung, S.J. et al. 2007. Accumulation of organotin compounds and marine birnavirus detection in Korean ascidians. *Fisheries Science* 73: 263–269

Blasiak, L.C., Zinder, S.H, Buckley, D.H., and Hill, R.T. 2014. Bacterial diversity associated with the tunic of the model chordate *Ciona intestinalis*. *The ISME Journal* 8: 309–320

Cao, C., Lemaire, L.A., Wang, W. et al. 2019. Comprehensive single-cell transcriptome lineages of a protovertebrate. *Nature* 571: 349–353

Chen, L., Hu, J.S., Xu, J.L., Shao, C.L., and Wang, G.Y. 2018. Biological and chemical diversity of ascidian-associated microorganisms. *Marine Drugs* 16: E362

Ciancio, A., Scippa, S. and Cammarano, M. 2001. Ultrastructure of trophozoites of the gregarine *Lankesteria ascidiae* (Apicomplexa: Eugregarinida) parasitic in the ascidian *Ciona intestinalis* (Protochordata). *European Journal of Protistology* 37: 327–336

Ciancio, A., Scippa, S., and Izzo, C. 1999. Ultrastructure of vegetative and sporulation stages of *Haplosporidium ascidiarum* from the ascidian *Ciona intestinalis* L. *European Journal of Protistology* 35: 175–182

Conradi, M., López-González, P.J., and García-Gómez, J.C. 1994. *Botryllophilus conicus* n. sp. (Copepoda: Cyclopoida: Ascidicolidae) associated with a compound ascidian from the Strait of Gibraltar. *Systematic Parasitology* 29: 97–104

Costello, K.E., Lynch, S.A., McAllen, R., O'Riordan, R.M., and Culloty, S.C. 2021. The role of invasive tunicates as reservoirs of molluscan pathogens. *Biological Invasions* 23: 641–655

Dror, H., Novak, L., Evans, J.S., López-legentil, S., and Shenkar, N. 2019. Core and dynamic microbial communities of two invasive ascidians: Can host–symbiont dynamics plasticity affect invasion capacity? *Microbial Ecology* 78: 170–184

Franchi, N. and Ballarin, L. 2017. Immunity in protochordates: The tunicate perspective. *Frontiers in Immunology* 8, e674

Hirose, E. 2000. Diet of a notodelphyid copepod inhabiting in an algal-bearing didemnid ascidian *Diplosoma virens*. *Zoological Science* 17: 833–838

Hirose, E. 2009. Ascidian tunic cells: Morphology and functional diversity of free cells outside the epidermis. *Invertebrate Biology* 128: 83–96

Hirose, E., Kumagai, A., Nawata, A., and Kitamura S.I. 2014. *Azumiobodo hoyamushi*, the kinetoplastid causing soft tunic syndrome in ascidians, may invade through the siphon wall. *Diseases of Aquatic Organisms* 109: 251–256

Hirose, E., Nakayama, K., Yanagida, T., Nawata, A., and Kitamura, S.I. 2018. Measurement of tunic hardness in an edible ascidian, *Halocynthia roretzi*, with remarks on soft tunic syndrome. *Zoological Science* 35: 548–552

Hirose, E., Nozawa, A., Kumagai, A., and Kitamura, S.I. 2012. *Azumiobodo hoyamushi* gen. nov. et sp. nov. (Euglenozoa, Kinetoplastea, Neobodonida): A pathogenic kinetoplastid causing the soft tunic syndrome in ascidian aquaculture. *Diseases of Aquatic Organisms* 97: 227–235

Hirose, E., Ohtake, S.I., and Azumi, K. 2009. Morphological characterization of the tunic in the edible ascidian, *Halocynthia roretzi* (Drasche), with remarks on 'soft tunic syndrome' in aquaculture. *Journal of Fish Diseases* 32: 433–445

Hirose, E., Oka, A.T., and Akahori, M. 2005. Sexual reproduction of the photosymbiotic ascidian *Diplosoma virens* in the Ryukyu Archipelago, Japan: Vertical transmission, seasonal change, and possible impact of parasitic copepods. *Marine Biology* 146: 677–682

Jung, S.J., Oh, M.J., Date, T., and Suzuki, S. 2001. Isolation of marine birnavirus from sea squirts *Halocynthia roretzi*. In *The Biology of Ascidians*, Sawada H., Yokosawa H., and Lambert C.C. (eds.), pp. 436–441. Tokyo: Springer-Verlag

Kimura, S., Nakayama, K., Wada, M. et al. 2015. Cellulose is not degraded in the tunic of the edible ascidian *Halocynthia roretzi* contracting soft tunic syndrome. *Diseases of Aquatic Organisms* 116: 143–148

Kumagai, A., Ito, H., and Sasaki, R. 2013. Detection of the kinetoplastid *Azumiobodo hoyamushi*, the causative agent of soft tunic syndrome, in wild ascidians *Halocynthia roretzi*. *Diseases of Aquatic Organisms* 106: 267–271

Kumagai, A. and Kamaishi, T. 2010. Development of polymerase chain reaction assays for detection of the kinetoplastid *Azumiobodo hoyamushi*, the causative agent for soft tunic syndrome in the ascidian *Halocynthia roretzi*. *Fish Pathology* 48: 42–47

Kumagai, A., Sakai, K., and Miwa, S. 2014. The sea squirt *Styela clava* is a potential carrier of the kinetoplastid *Azumiobodo hoyamushi*, the causative agent of soft tunic syndrome in the edible ascidian *Halocynthia roretzi*. *Fish Pathology* 49: 206–209

Kumagai, A., Suto, A., Ito, H., and Tanabe, T. et al. 2011. Soft tunic syndrome in the edible ascidian *Halocynthia roretzi* is caused by a kinetoplastid protist. *Diseases of Aquatic Organisms* 95: 153–161

Kumagai, A., Suto, A., Ito, H., Tanabe, T., Takahashi, K., Kamaishi, T., and Miwa, S. 2010. Mass mortality of cultured ascidians *Halocynthia roretzi* associated with softening of the tunic and flagellate-like cells. *Diseases of Aquatic Organisms* 90: 223–234

Kumagai, A., Tanabe, T., Nawata, A., and Suto, A. 2016. Disinfection of fertilized eggs of the edible ascidian *Halocynthia roretzi* for prevention of soft tunic syndrome. *Diseases of Aquatic Organisms* 118: 153–158

Kitamura, S.I., Ohtake, S.I., Song, J.Y., Jung, S.J., Oh, M.J. Choi, B.D., Azumi, K., and Hirose, E. 2010. Tunic morphology and viral surveillance in diseased Korean ascidians: Soft tunic syndrome in the edible ascidian, *Halocynthia roretzi* (Drasche), in aquaculture. *Journal of Fish Diseases* 33: 153–160

Lambert, G., Karney, R.C., Rhee, W.Y., and Carman, M.R. 2016. Wild and cultured edible tunicates: A review. *Management of Biological Invasions* 7: 59–66

Lowe, C.J., Clarke, D.N., Medeiros, D.M., Rokhsar, D.S., and Gerhart, J. 2015. The deuterostome context of chordate origins. *Nature* 520: 456–465

Mita, K., Kawai, N., Rueckert, S., and Sasakura, Y. 2012. Large-scale infection of the ascidian *Ciona intestinalis* by the gregarine *Lankesteria ascidiae* in an inland culture system. *Diseases of Aquatic Organisms* 101: 185–195

Moiseeva, E., Rabinovitz, C., Yankelevich, I., and Rinkevich, B. 2004. 'Cup cell disease' in the colonial tunicate *Botryllus schlosseri*. *Diseases of Aquatic Organisms* 60: 77–84

Monniot, C. 1990. Diseases of Urochordata. In *Disease of Marine Animals* Vol. III, O. Kinne (ed.) pp. 569–636. Hamburg: Biologische Anstalt Helgoland

Nam, K.W., Shin, Y.-K.. and Park, K.-I. 2015. Seasonal variation in *Azumiobodo hoyamushi* infection among benthic organisms in the southern coast of Korea. *Parasites & Vectors* 8: 569

Nam, K.W., Shin, Y.-K., and Park, K.-I. 2017. Cold temperatures inhibit infection (soft tunic syndrome) of sea pineapple *Halocynthia roretzi* by the kinetoplastid parasite *Azumiobodo hoyamushi*. *Bulletin of the European Association of Fish Pathologists* 37: 218

Nawata, A., Hirose, E., and Kitamura, S.I. 2018. Tunic extract of the host ascidian attracts the causal agent of soft tunic syndrome, *Azumiobodo hoyamushi* (Kinetoplastea: Neobodonida). *Diseases of Aquatic Organisms* 129: 207–214

Nawata, A., Hirose, E., Kitamura, S.I., and Kumagai, A. 2015. Encystment and excystment of kinetoplastid *Azumiobodo hoyamushi*, causal agent of soft tunic syndrome in ascidian aquaculture. *Diseases of Aquatic Organisms* 115: 253–262

Ooishi, S. 1999. Female and male *Botryllophilus ruber* (Copepoda: Cyclopoida) associated with the compound ascidian *Botryllus schlosseri*. *Journal of Crustacean Biology* 19: 556–577

Ooishi, S. 2009. *Enteropsis fusiformis*, new species (Copepoda: Cyclopoida: Ascidicolidae), living in a compound ascidian from Madagascar. *Proceedings of the Biological Society of Washington* 122: 333–341

Ooishi, S. 2014. *Botryllophilus millari*, new species (Copepoda: Cyclopoida: Ascidicolidae), living in the compound ascidian *Eudistoma caeruleum* (Sluiter) from Madagascar. *Proceedings of the Biological Society of Washington* 127: 496–509

Park, K.H., Zeon, S.-R., Lee, J.-G., Choi, S.-H., Shin, Y.K., and Park, K.-I. 2014. *In vitro* and *in vivo* efficacy of drugs against the protozoan parasite *Azumiobodo hoyamushi* that causes soft tunic syndrome in the edible ascidian *Halocynthia roretzi* (Drasche). *Journal of Fish Diseases* 37: 309–317

Putnam, N., Butts, T., Ferrier, D. et al. 2008. The amphioxus genome and the evolution of the chordate karyotype. *Nature* 453: 1064–1071

Rinehert, K.L., Gloer, J.B., Hughes, R.J. et al.1981. Didemnins: Antiviral and antitumor depsipeptides from a Caribbean tunicate. *Science* 212: 933–935

Rueckert, S., Betts, E.L., and Tsaousis, A.D. 2019. The symbiotic spectrum: Where do the gregarines fit? *Trends in Parasitology* 35: 687–694

Rueckert, S., Wakeman, K.C., Jenke-Kodama, H., and Leander, B.S. 2015. Molecular systematics of marine gregarine apicomplexans from Pacific tunicates, with descriptions of five novel species of *Lankesteria*. *International Journal of Systematic and Evolutionary Microbiology* 65: 2598–2614

Satoh, N., Rokhsar, D., and Nishikawa, T. 2014 Chordate evolution and the three-phylum system. *Proceedings of the Royal Society B* 281: 20141729

Shin, Y.K., Nam, K.W., Yoon, J.M., and Park, K.I. 2014. Quantitative assessment of *Azumiobodo hoyamushi* distribution in the tunic of soft tunic syndrome-affected ascidian *Halocynthia roretzi* using real-time polymerase chain reaction. *Parasites & Vectors* 7: 539

Song, S.M., Sylvatrie-Danne, D.B., Joo, S.Y., Shin, Y.K., Yu, H.S., Lee, Y.S., Jung, J.E., Inoue, N., Lee, W.K., Goo, Y.K., Chung, D.I., and Hong Y. 2014. Development of loop-mediated isothermal amplification targeting 18S ribosomal DNA for rapid detection of *Azumiobodo hoyamushi* (Kinetoplastea). *The Korean Journal of Parasitology* 52: 305–310

Wallis, J.R., Smith, A.J., and Kawaguchi, S. 2017. Discovery of gregarine parasitism in some Southern Ocean krill (Euphausiacea) and the salp *Salpa thompsoni*. *Polar Biology* 40: 1913–1917

Zou, Y., Ma, C., Zhang, Y. et al. 2016. Isolation and characterization of *Vibrio alginolyticus* from cultured amphioxus *Branchiostoma belcheri tsingtauense*. *Biologia* 71: 757–762

Index

Notes: Tables, figures and boxes are indicated by an italic *t, f* and *b* following the page number.